Innovative Production Machines and Systems

Fourth I*PROMS Virtual Conference

Innovative Production Machines and Systems

Fourth I*PROMS Virtual Conference
1st–14th July 2008

Edited by

D. T. Pham
E. E. Eldukhri and
A. J. Soroka

Organised by: FP6 I*PROMS Network of Excellence
Sponsored by: The European Commission

Whittles Publishing
CRC Press
Taylor & Francis Group

Published by
Whittles Publishing,
Dunbeath,
Caithness KW6 6EY,
Scotland, UK
www.whittlespublishing.com

Distributed in North America by
CRC Press LLC,
Taylor and Francis Group,
6000 Broken Sound Parkway NW, Suite 300,
Boca Raton, FL 33487, USA

ISBN 978-1904445-81-4
USA ISBN 978-14398-0117-8

Preliminary pages and index typeset by
Thomson Digital

Printed in Poland

Contents

Innovative Design

Intelligent Decision Support

Intelligent Optimisation and Planning

Special Session: Bees Algorithm

Special Session: Gentelligent

Sponsors

Schneider Electric

The DTI Manufacturing Advisory Service, MAS

Springer

Institution of Mechanical Engineers, ImechE

Innovative Manufacturing Research Centres, IMRC

British Computer Society, BCS

EU Sixth Framework Programme, FP6

The Institution of Engineering and Technology, IET

Cardiff University

The Manufacturing Engineering Centre, MEC

Preface

The 2008 Virtual International Conference on Innovative Production Machines and Systems (IPROMS) took place on the Internet between 1 and 14 July 2008. IPROMS 2008 was an outstanding success. During the Conference, over 5000 registered delegates and guests from over 90 countries participated in the Conference, making it a truly global phenomenon.

This book contains the Proceedings of IPROMS 2008. The 100 peer-reviewed technical papers presented at the Conference have been grouped into sixteen sections:

- Advanced Production Machines
- Digital Manufacturing
- Fit and Sustainable Manufacturing
- HMI & HCI
- Innovative Design
- Intelligent Decision Support
- Intelligent Optimisation and Planning
- Mechatronics and Robotics
- Micro and Nano Manufacturing
- Production Automation and Control
- Production Organisation and Management
- Reconfigurable Manufacturing
- Special Session – Bees Algorithm
- Special Session – Gentelligent
- Special Session – ICME (Intelligent and Competitive Manufacturing Engineering)
- Special Session – IWARD (Intelligent Robot Swarm for Attendance, Recognition, Cleaning and Delivery)

Many of the IPROMS 2008 papers were written by partners and associate partners in the I*PROMS EU-funded FP6 Network of Excellence for Innovative Production Machines and Systems, but there were also large numbers of authors external to the Network. In total, IPROMS 2008 authors were from 27 countries across five continents.

Numerous people and organisations have helped make IPROMS 2008 a reality. We are most grateful to the IPROMS 2008 sponsors, I*PROMS partners, Conference Programme and Organising Committees, Session Organisers, Chairmen, Authors, Referees, and the I*PROMS Central Coordination Team.

The names of contributors to the success of IPROMS 2008 can be found elsewhere in the Proceedings. Here, we would highlight the much appreciated efforts of the Special Session Organisers: E. Koc, L-E. Lorenzen, S. Otri and R. Teti.

Finally, thanks must go to our colleague Vladimir Zlatanov, the technical co-ordinator of IPROMS 2008, who implemented the entire IT infrastructure for our Virtual Conference. Without his expertise and dedication, IPROMS 2008 would have forever remained virtual.

D.T. Pham, E.E. Eldukhri and A.J. Soroka
MEC, Cardiff University

Introduction by Mr A Gentili, European Commission

IPROMS 2008 is the 4th online web-based conference on Innovative Production Machines and Systems organised by the EU-funded FP6 I*PROMS Network of Excellence. It built on the outstanding success of its predecessors, IPROMS 2005, 2006 and 2007. IPROMS 2008 attracted over 5000 registered delegates and guests from over 90 countries. During the conference, interested researchers and industrial practitioners worldwide took part free of charge. They had the opportunity to view presentations, view/download full papers and contribute to the online discussions.

As a sponsor of I*PROMS Network of Excellence, the European Commission strongly supports the organisation of this annual event. This will enable the Network to disseminate the results of its work globally for the benefit of the wider community. Moreover, it will contribute to the integration of research resources in Europe for an efficient spending of R&D budget, avoiding overlaps in European research activities and exploiting synergies.

Furthermore, the international aspect of manufacturing is becoming more and more important in an era marked by global competition and cooperation. This online web-based conference, being a pioneer in the sector, helps European stakeholders to establish links and synergies at international level, providing added value for Europe.

Finally, this event, in line with the European Manu*future* Platform, will help to create an effective and cooperative research manufacturing "society" which is a necessary condition for the establishment of a European Manufacturing and Innovation Research Area. This will be a pre-requisite for any of the European initiatives for public-private and public-public partnerships under preparation within the European Manu*future* Platform, within which Research Centres and Academia will have to play a central and crucial role between Industry and their markets.

I wish all participants to this conference success in establishing the necessary high-level international networks which are the foundation of any successful collaboration.

Andrea Gentili
Manufuture Platform
European Commission

Programme and Organising Committees

D. Pham (Chair), MEC, Cardiff University, UK
E. Eldukhri (Organising Committee Chair), MEC, *ibid*
A. Soroka (Programme Committee Chair), MEC, *ibid*
V. Zlatanov (Technical Co-ordinator), MEC, *ibid*
S. Dimov, MEC, *ibid*
A. Fahmy, MEC, *ibid*
M. Packianather, MEC, *ibid*
B. Peat, MEC, *ibid*
P. Pham, MEC, *ibid*
A. Thomas, Cardiff University, UK
P. Hines, Cardiff University, UK
M. Naim, Cardiff University, UK
N. Rich, Cardiff University, UK
R. Setchi, Cardiff University, UK
N. Aspragathos, University of Patras, Greece
K. Cheng, Leeds Metropolitan University, UK
A. Colombo, Schneider Electric, Germany
B. Denkena, IFW, University of Hannover, Germany
P. Dietz, Clausthal University of Technology, Germany
J. Efstathiou, University of Oxford, UK
F. Feenstra, TNO, The Netherlands
B. Grabot, ENIT, France
R. Harrison, Loughborough University, UK
S. Hinduja, University of Manchester, UK
M. Hoepf, Fraunhofer IPA, Germany
E. Hohwieler, Fraunhofer IPK, Germany
A. Kusiak, The University of Iowa, USA
R. La Brooy, RMIT University, Australia
A. Labib, University of Portsmouth, UK
V. Marik, Czech Technical Univ., Czech Republic
F. Meo, Fidia S.p.A, Italy
A. Nearchou, University of Patras, Greece
G. Putnik, University of Minho, Portugal
E. Oztemel, Sakarya University, Turkey
V. Raja, University of Warwick, UK
Z. Salem, University of Aleppo, Syria
T. Schlegel, Fraunhofer IAO, Germany
R. Teti, University of Naples Federico II, Italy
A. Thatcher, University of the Witwatersrand, S. Africa
M. Wu, University of Derby, UK
X. Xie, INRIA, France
X. Yu, RMIT University, Australia

Special Session Organisers

E. Koc and S. Otri, (Bees Algorithm)
R. Teti (Intelligent and Competitive Manufacturing Engineering – ICME)
L-E. Lorenzen (Gentelligent)

Session Chairs and Co-Chairs

F. Abdul Aziz, MEC, Cardiff University, UK
A. Afify, MEC, Cardiff University, UK
M. Al-Kutubi, MEC, Cardiff University, UK
N. Aspragathos, University of Patras, Greece
E. Brousseau, MEC, Cardiff University, UK
E. Charles, MEC, Cardiff University, UK
A. Colombo, Schneider Electric, Germany
D. D'Addona, University of Naples Federico II, Italy
A.A. Fahmy, MEC, Cardiff University, UK
C. Ferri, MEC, Cardiff University, UK
T. Fouweather, University of Newcastle-upon-Tyne, UK
A. Ghanbarzadeh, MEC, Cardiff University, UK
B. Grabot, ENIT, France
R. van Heek, TNO, The Netherlands
E. Hohwieler, Fraunhofer IPK, Germany
Z. Ji, MEC, Cardiff University, UK
E. Koc, MEC, Cardiff University, UK
N. Lagos, MEC, Cardiff University, UK
C.H. Le, MEC, Cardiff University, UK
J.Y. Lee, MEC, Cardiff University, UK
H. Liu, MEC, Cardiff University, UK
F. Meo, Fidia S.p.A, Italy
K.W. Ng, MEC, Cardiff University, UK
S. Otri, MEC, Cardiff University, UK
M. Packianather, MEC, Cardiff University, UK
G. Putnik, University of Minho, Portugal
V. Raja, University of Warwick, UK
G. Ruz, MEC, Cardiff University, UK
S. Sahran MEC, Cardiff University, UK
T. Shamsuddin, MEC, Cardiff University, UK
A. Soroka, MEC, Cardiff University, UK
R. Teti, University of Naples Federico II, Italy
A. Thomas, MEC, Cardiff University, UK
O. Williams, MEC, Cardiff University, UK
Y. Wu, MEC, Cardiff University, UK
V. Zlatanov, MEC, Cardiff University, UK

Referees

A. Afify, MEC, Cardiff University, UK
A. Albut, University of Bacau, Romania
M. Awadalla, Helwan University, Egypt
P. Bach, Czech Technical University in Prague, Czech Republic
S. Badiyani, University of Warwick, UK
R. Barton, Cardiff University, UK
H. Bochnig, Technical University of Berlin, Germany
D. Bormann, University of Hannover, Germany
E. Brosseau, MEC, Cardiff University, UK

P. Charbonnaud, ENIT, France
K. Cheng, Brunel University, UK
S. Coleman, University of Newcastle-upon-Tyne, UK
D. D'addona, University of Naples Federico II, Italy
M. Dumke, University of Hannover, Germany
M. Ebrahimi, University of Bradford, UK
E. Eldukhri, MEC, Cardiff University, UK
A. Fahmy, MEC, Cardiff University, UK
F. Fisser, University of Hannover, Germany
T. Fouweather, University of Newcastle-upon-Tyne, UK
S. Franke, University of Hannover, Germany
M. Gerosa, Politecnico di Milano, Italy
T. Goszczynski, PIAP, Poland
B. Grabot, ENIT, France
T. Griebach, University of Hannover, German
C. Griffin, Dublin City University, Ireland
M. Hasan, Dublin City University, Ireland
A. Hassan, University of Technology Malaysia, Malaysia
H. Henning, University of Hannover, German
M. Hoepf, Fraunhofer IPA, Germany
A. Hoque, Dublin City University, Ireland
P. Hu, Wuhan University of Technology, China
T. Huehns, Technical University of Berlin, Germany
A. Islam, Dublin City University, Ireland
Z. Ji, MEC, Cardiff University, UK
S. Kazi, University of Technology Malaysia, Malaysia
P. Kidd, Cheshire Henbury, UK
M.Kraft, Fraunhofer IPK, Germany
F. Lange, University of Hannover, Germany
H. Lee, Korea Atomic Energy Research Institute, Korea
L. Lorenzen, University of Hannover, Germany
K. Mamun, Dublin City University, Ireland
I. Mariolis, University of Patras, Greece
H. Marzi, St. Francis Xavier University, Canada
T. Masood, Loughborough University, UK
S. Mekid, University of Manchester, UK
J. Marco Mendes, University of Porto, Portugal
F. Meo, Fidia S.p.A, Italy
G. Mroz, University of Hannover, Germany
E. Oztemel, Marmara University, Turkey
M. Packianather, MEC, Cardiff University, UK
B. Peat, MEC, Cardiff University, UK
M. Perez, Fundacion Tekniker, Spain
D.T. Pham, MEC, Cardiff University, UK
A. Pithon, Federal Centre of Technological Education, Brazil
G. Putnik, University of Minho, Portugal
D. Ralf, University of Hannover, Germany
A. Razali, University of Strathclyde, UK
M. Rodmann, University of Hannover, Germany
T. Schlegel, Fraunhofer IAO, Germany

A. Soroka, MEC, Cardiff University, UK
T. Strasser, Profactor, Austria
T. Szecsi, Dublin City University, Ireland
A. Thomas, Cardiff University, UK
D. Trifonov, Bulgarian Academy of Sciences, Bulgaria
R. van Heek, TNO, The Netherlands
K. Weilandt, University of Hannover, Germany
K. Wu, University of Hannover, Germany
E. Xidias, University of Patras, Greece
P. Zacharia, University of Patras, Greece
V. Zlatanov, MEC, Cardiff University, UK

I*PROMS Central Coordination and MEC Teams
M. Haines
F.D. Marsh
C. Rees
R. Williams

Innovative Production Machines and Systems
D.T. Pham, E.E. Eldukhri and A.J. Soroka (Eds.)

An investigation of influence of machining conditions on machining error

T. Ogedengbe, S. Mekid, S.Hinduja

School of Mechanical, Aerospace and Civil engineering, The University of Manchester, Manchester M60 1QD, UK

Abstract

Machining parameters have been investigated in this paper to study their influences on the machining error on machined part. While past studies and theoretical applications were used to analyse effect of some machining parameters, experimental investigation was applied to the others and the responses were analysed using classical Analysis of Variance (ANOVA). The paper attempt to analyse the behaviour of machining error on parts machined on a three axis vertical machining center under the influence of various machining parameters. The machining parameters that influence the error on machined part was established and proposed for used within error prediction methodology.

Keywords: Machining parameters, machined part, Machining error

1. Introduction

It is generally known that the accuracy of a machined part is largely influenced by machine errors that have various sources. The three main sources could be associated with geometric, thermally induced and load induced error while other sources like machine vibration, part clamping and part deformation exists. The result is reflected in the inaccuracy of the machined component which is the relative error between the cutting tool and the machined part. Hence, Error compensation has been widely applied to improve machine tool accuracy cost-effectively.

Several researchers have studied the various sources of machine tool errors towards improving accuracy of machined part through error compensation[**1, 2, 3, 4**]. The major setback for error compensation in machine tools is that it has been difficult to predict machining error on the machined part online due to lack of dedicated probes. The machining error on a machined part could be measured using post-process measurement whereby the machined part is removed from the process and transported by a suitable device to an independent inspection machine such as a Coordinate Measurement Machine (CMM) where the part is inspected and machining error determined. The major advantage of this method is that the predicted error includes the effects of all error sources that affect the part in a single setup. However, this is usually time-consuming and might results in a risk of producing a large number of defective parts before the inspection results are known. To avoid the production of defective components, using post-process measurement, a technique for repetitive machining was proposed[**5**]. The technique involves profile measurement of the first machined part by a CMM. Based on the measured data, the corresponding contour error is calculated. The tool path is then modified with respect to the contour error and fed into the CNC machine tool for the machining of subsequent parts. Although this approach works well for the first few components it lacks adaptability. As more components are being machined the machine structure would change as a result of higher heat generation. Also, fluctuation of cutting force may affect the judgment of the technique thereby making the approach ineffective in correcting the errors when machining a large number of components. Therefore there is a need for a method that will predict the machining error effectively by adapting to changes during the machining process. Machining parameter and error on machined part could be mapped within an appropriate methodology to predict machining error on machined parts. However, since there are many machining parameters during a machining process it is important to identify, analyse and investigate them to determine those that actually contribute to the error on machined parts. Hence, the research work undertaken herein presents a critical step towards the development of a comprehensive and effective method for in-process prediction of machined part machining error. The study involves the identification and investigation of the machining parameters that influences the machining error generated on machined part.

2. Machining parameters and their influence on machining error

Fig. 1 shows the identified machining parameters and their associated error sources that could contribute to the machining error on machined part. It could be established from past works and theoretical analysis the fact that some of these parameters would influence the machining error on machined part. For instance, the part coordinate which is the location of the part to the machined on the machine tool is a function of the machine axis. Positioning accuracy of machine tool has been shown to vary along machine axes[**6, 7**], this is evidence enough to say that the part coordinate influences machining error. Also, temperature rise at critical location of machine tool directly result in thermal error. This machining parameter has been widely modelled and utilised to compensate for thermal error in machine tool [**3, 8**]. Most of the study on thermal error revealed that machine tool error changes with temperature of critical machine elements[**9**]. Hence this parameter would influence machining error from knowledge of past research.

The influence of tool type and materials of the part to be machined on the error could be examined theoretically. An appealing model for the study of tool deflection is that in which the tool is regarded as a cylindrical cantilever beam. The deflection of such beam then conforms to eqn. (1)

$$\delta = \frac{64F}{3\pi E}\frac{L^3}{D^4} \tag{1}$$

Eqn. (1) shows mathematically that for a given cutting force F, the deflection of the tool increases as the young's modulus E for the material of the tool decreases. Also the deflection increases with the slenderness parameter L^3/D^4 of the tool. Where L is the overhang length of the tool and D is the equivalent tool diameter. The tool deflection contributes to the overall machining error. Since all the factors within the equation as explained above can vary with tool type, the implication is that the tool type would influence the machining error on machined part.

The machined part is also expected to deform by a combination of deflection under the cutting force and expansion due to heat rise at tool-part interface. From finite element analysis method, the amount of deformation depends on the Young's modulus and density of the material of the part. These properties vary in materials. Hence, the part material will also influence the machining error on machined parts.

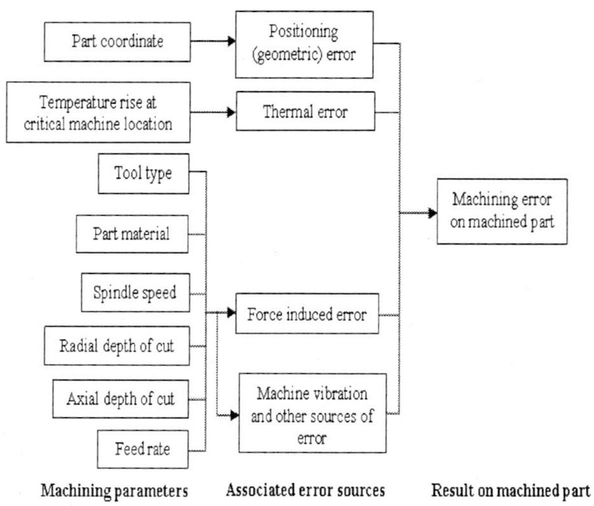

Fig 1: Relationship between machining parameter, error sources and machining error on machined part

The feed rate, spindle speed and the depth of cuts in Fig.1 is associated with force-induced error and could influence the force induced error on a machined part. However it has not been experimentally proven and there is no theoretical equation that could be used to establish the extent to which the spindle speed, feed rate, axial and radial depths of cut could influence the machining error on machined part. Based on this, the effect of varying feed rate, spindle speed and depth of cut on machining error has been experimentally investigated.

3. Experimental investigation

To carry out the investigation cutting experiments have been conducted on a 3-axis Takisawa Mac3 vertical machining center. The experiment was set up with mild steel material. In the experiment, a 22mm slot was machined on the material with 8mm diameter milling cutter along the X-axis of the machine using varying machining parameters. New SwisTech plain shank 4 flute end mill Q-Coat Coated cutter was used to reduce effect of tool wear. The cutting experiment was designed as a completely randomized full factorial experiment. The experiment has three machining parameters (spindle speed (n), axial depth of cut (d_a) and feed (f)) as factor effects examined at either low or high level. For a vertical milling center, it is assumed that the effect of axial depth of cut would be similar to that of the radial depth of cut as both are actually representing the same concept in different directions. The various values of a factor examined in an experiment are known as levels. The three factors (machining parameters) were combined in the test at two levels to give 8 test having different combinations of cutting conditions. This implies 8 diverse cutting conditions applied to the machining of the slots. The factor –level combination for the experiment and the corresponding cutting conditions are given in Table 1. Under this diverse machining conditions slot of same length were machined along the X-axis of the machine. The maximum spindle speed of the machine tool is 10,000 rpm. The machine was thermally stable and the cutting tests were performed at fairly constant environmental temperature of 20°C. The depth of cut, feed and spindle speed were selected within the operating range of the cutting tool to prevent the tool from excessive wear and breakage. The machined part was located close to the machine's home position where the error due to the part coordinate is small.

Factorial designs have been reported to be most effective for experiments involving the study of the effects of two or more factors [**10**]. Also it has been shown that the presence of interaction requires that factors be evaluated jointly rather than individually[**11**]. Hence, it is important that an experiment be designed to measure interactions. This is because failure to do so can lead to misleading and incorrect conclusion. Full factorial experiments enable all individual and joint factors effect to be estimated hence it was employed. The full factorial experimental approach was used in the design of the experiment as in Table 1. A full factorial experiment is one with all possible combination of the factors levels. All experiments and measurements are subjected to random error to some degree, therefore it is frequently necessary to combine the result of a number of repeated runs of an experiment and observations in other to obtain the required information with some precision. Since the present study is concerned mainly with the effect of varying the machining parameters it is enough to examine the parameter at two levels to reduce the number of experiments which would allow us to conduct reasonable numbers of repeat experiments thereby enabling proper consideration of noise/uncontrollable uncertainty in the experiments. Hence four repeated runs of each experiment on Table 1 were planned.

Table 1: Machining parameter levels and their combinations

Exp.	Machining condition		
No.	n (rpm)	d_a (mm)	f (mm/tooth)
1	1000	2	0.02
2	1000	4	0.02
3	2000	2	0.02
4	1000	2	0.04
5	1000	4	0.04
6	2000	2	0.04
7	2000	4	0.02
8	2000	4	0.04

Completely randomized design approach was used with the full factorial design to assign order in which the cutting experiment are conducted in random manner with respect to the factor-level combinations and their repeated run. Using a table of random numbers from [**11**], we obtained and assigned the randomized order in which the factor-level combinations and their repeat test were run. Four repeated runs of each experiment were conducted to give a total of 32 runs. This is important to prevent extraneous factors from causing bias on the result of the experiment. The Randomization of the experiments affords protection from bias by tending to average the bias effect over all levels and combinations of the factors in the experiments. When comparisons are made among levels of factors, the bias effect will tend to cancel and the true factor effect will remain.

The repeated tests which constitute experimental observation in which the factor-level combination are identical are used to estimate the uncontrolled experimental error variability so that a satisfactory estimate of experimental error can be obtained. Hence, four machining runs were conducted for each machining parameter-level combination test. After machining is completed, the parts were removed from the machine and inspected on a LK Coordinate Measurement Machine (CMM) available in Royce Laboratory.

The CMM has been calibrated to an accuracy of 1μm. The inspection processes on the CMM was automated through writing part program thereby removing human error to arrive at a reliable inspection result. The deviation of the slot length from the specified length is recorded and presented in Table 3 as the machining error/observed response for analysis of variance (ANOVA).

4. An overview of ANOVA

Analysis-of-variance (ANOVA) procedures separate or partition the variation observable in a response variable into two basic components: (a) variation due to assignable causes and (b) uncontrolled or random variation. Assignable causes refer to controlled experimental factors in this case the machining parameters while random variation includes the effects of all other sources not controlled during the experiment.

As in Table 1, the three machining parameters investigated in the experiment was symbolized as n, f and d_a representing spindle speed, feed rate and axial depth of cut respectively. A statistical model for this type of experiment could be written as

$$\varepsilon_{ijkl} = \mu_{ijk} + e_{ijkl} \qquad (2)$$

It can be said that subscript i refers to one of the levels of spindle speed, the subscript j one of the levels of feed rate, the subscript k one of the levels of axial dept of cut and the subscript l one of the four repeat tests. Therefore, ε_{ijkl} represents the experimental response which is the measured machining error, μ_{ijk} is the effects of the machining parameters (assignable causes) and e_{ijkl} is the random error effects. If I, J and K is the levels of n, f and d_a respectively and L total number of repeat tests in the experiments, then

$i = 1,\text{--},I \;\; j = 1,\text{--}, J \;\; k = 1,\text{--},K$ and $l = 1,\text{--},L$
Where $I = J = K = 2$ and L = 4

If eqn. (2) is generalised to the three factors investigated in this study, the contribution of all the assignable designated as the model sum of square (MSS) could be partitioned as

$$\begin{aligned} MSS &= SS_n + SS_f + SS_{d_a} + SS_{nf} \\ &+ SS_{nd_a} + SS_{fd_a} + SS_{nfd_a} \end{aligned} \qquad (3)$$

In eqn. (3) SS_n, SS_f and SS_{d_a} is a measure of the individual contribution of spindle speed, feed and depth of cut to the experimental response respectively, SS_{nf}, SS_{nd_a} and SS_{fd_a} measures the contribution of the respective two conditions joint effect on machining error, and SS_{nfd_a} measures the joint effects of all the three machining conditions. If all the main factor effects and interactions of the three machining conditions are accounted for, a symbolic ANOVA table presented in Table. 2 would be the output.

Table 2: Symbolic analysis of variance table for the experiment

SOV	DOF	SS	MS	F
n	$I-1$	SS_n	MS_n	F_n
f	$J-1$	SS_f	MS_f	F_f
d_a	$K-1$	SS_{d_a}	MS_{d_a}	F_{d_a}
nf	$(I-1)(J-1)$	SS_{nf}	MS_{nf}	F_{nf}
nd_a	$(I-1)(K-1)$	SS_{nd_a}	MS_{nd_a}	F_{nd_a}
fd_a	$(J-1)(K-1)$	SS_{fd_a}	MS_{fd_a}	F_{fd_a}
nfd_a	$(I-1)(J-1)(K-1)$	SS_{nfd_a}	MS_{nfd_a}	F_{nfd_a}
Error	$IJK(L-1)$	SS_e	MS_e	
Total	$IJKL-1$	TSS		

The sum of squares for the sources of variation could be calculated as in eqn. (4) to (13).

$$SS_m = \left(\sum_{i=1}^{I} \sum_{j=1}^{J} \sum_{k=1}^{K} \sum_{l=1}^{L} \varepsilon_{ijkl} \right)^2 \Big/ IJKL \qquad (4)$$

$$TSS = \sum_{i=1}^{I} \sum_{j=1}^{J} \sum_{k=1}^{K} \sum_{l=1}^{L} \varepsilon_{ijkl}^2 - SS_m \qquad (5)$$

$$SS_n = \left(\sum_{i=1}^{I} \varepsilon_{i...}^2 \Big/ JKL \right) - SS_m \quad (6)$$

$$SS_f = \left(\sum_{j=1}^{J} \varepsilon_{.j..}^2 \Big/ IKL \right) - SS_m \quad (7)$$

$$SS_{d_a} = \left(\sum_{k=1}^{K} \varepsilon_{..k.}^2 \Big/ IJL \right) - SS_m \quad (8)$$

$$SS_{nf} = \left(\sum_{i=1}^{I} \sum_{j=1}^{J} \varepsilon_{ij..}^2 \Big/ KL \right) - SS_m - SS_n - SS_f \quad (9)$$

$$SS_{nd_a} = \left(\sum_{i=1}^{I} \sum_{k=1}^{K} \varepsilon_{i.k.}^2 \Big/ JL \right) - SS_m - SS_n - SS_{d_a} \quad (10)$$

$$SS_{fd_a} = \left(\sum_{j=1}^{J} \sum_{k=1}^{K} \varepsilon_{.jk.}^2 \Big/ IL \right) - SS_m - SS_f - SS_{d_a} \quad (11)$$

$$SS_{nfd_a} = \left(\sum_{i=1}^{I} \sum_{j=1}^{J} \sum_{k=1}^{K} \varepsilon_{ijk.}^2 \Big/ L \right) - SS_m - SS_n - SS_f - SS_{d_a} - SS_{nf} - SS_{nd_a} - SS_{fd_a} \quad (12)$$

$$SS_e = TSS - SS_n - SS_f - SS_{d_a} - SS_{nf} - SS_{nd_a} - SS_{fd_a} - SS_{nfd_a} \quad (13)$$

On the symbolic ANOVA Table (Table 2), the first column SOV (source of variation) designates the partitioning of the response variability into the various components representing individual effects, interaction between factor effects and experimental error. The degree of freedom (DOF) separates the sample size into similar components that relates the amount of information obtained on each factor, the interaction and the error term of the model. The mean square (*MS*) of the sources of variation is obtained by dividing the sum of square (*SS*) by the respective degree of freedom. The F-value is then obtained by weighting the mean square of each source of variation against that of the experimental error to check the effect of varying the sources of variation in the experiment.

5. Effect of investigated machining parameter

The analysis of variance approach discussed in section 4 was adopted to analyze the influence of varying the investigated machining conditions on the machining error in this study. Table 3 gives the result of inspection on CMM for the diverse combination of machining parameters including the repeat tests.

Using Table 2 and the computational formula for the sum of squares, the complete ANOVA table for the result of machining error inspection in Table 3 is presented in Table 4. The hypothesis being tested in the ANOVA table is whether the influence of particular factors (machining parameters in this case) varied in the experiments and their interactions is zero or otherwise. Under this hypothesis, if the corresponding F-ratio value (column 5) of factors or their interaction is less than or equal to 1, the null hypothesis that varying these factors or their interaction has no effects on the response is true.

Table 3: Machining error from cutting tests

Machining Condition			Machining error (μm)			
n	f	d_a	Test 1	Test 2	Test 3	Test 4
1000	0.02	2	454	393	451	442
		4	229	180	212	206
	0.04	2	463	439	456	396
		4	296	286	247	204
2000	0.02	2	438	385	420	416
		4	273	310	248	210
	0.04	2	475	487	408	427
		4	253	240	246	211

Table 4: Analysis of variance table for the result of the experiment

SOV	DOF	Sum of squares	Mean square	F-value
n	1	288	288	0.0038
f	1	946.125	946.125	0.0123
d_a	1	168200	168200	2.1873
nf	1	406.125	406.125	0.0053
nd_a	1	840.5	840.5	0.0109
fd_a	1	45.125	45.125	0.0006
nfd_a	1	2926.125	2926.125	0.0381
Error	24	1845569	76898.71	
Total	31	2019221		

This is because the mean square of experimental error variance is either equal to or greater than that of the sources of variations. This is the case for $n, f, nf, nd_a, fd_a, nfd_a$ in Table 4. These F-value less than unity for these sources of variation implies that varying the spindle speed, feed rate and interaction effect at low and high level of the three machining conditions investigated did not influence the error of the machined part.

However, the F-value of 2.187293 for the axial depth of cut shows that the effect of changing this factor in the experiment is greater than that of the experimental error variance. The null hypothesis of no effect can therefore be rejected for this factor. Hence, it can be said that changing the axial depth of cut influences the machining error on the machined parts. A comparison of the F-value with F critical values in statistical table reveals that the F value for d_a is higher for $p=0.25$. Therefore, it is enough to say that the machining error variability attributable to changing the axial depth of cut is greater than that attributable to uncontrollable experimental error with 75% confidence ($P<0.25$). This is considered enough within the aim of the experiment to conclude that varying the axial depth of cut actually affects the error on the machined parts.

6. Conclusion

From the analysis, while the spindle speed and the feed do not influence machining error, there is sufficient evidence to justify the fact that part coordinate, temperature of critical machine element, part materials, tool type and depths of cut would influence the machining error on machined part. Therefore, the machining parameters that influence the machining error could be lumped together within an appropriate error prediction methodology and mapped with machining error (using adaptive neural network model trained with experimental data) for subsequent and effective prediction of the error on machined part while the part is on the machine.

References

1. J. Yuan and J. Ni, The real-time error compensation technique for CNC machining systems, Mechatronics 8 (1998), 359-380.
2. C. Raksiri and M. Parnichkun, Geometric and force errors compensation in a 3-axis CNC milling machine, International Journal of Machine Tools & Manufacture 44 (2004), 1283–1291.
3. H. Yang and J. Ni, Dynamic neural network modeling for nonlinear, nonstationary machine tool thermally induced error, International Journal of Machine Tools & Manufacture 45 (2005), 455–465.
4. A. C. Okafor and Y. M. Ertekin, Derivation of machine tool error models and error compensation procedure for three axes vertical machining center using rigid body kinematics, International Journal of Machine Tools & Manufacture 40 (2000), 1199–1213.
5. C. C. Lo and C. Y. Hsiao, A method of tool path compensation for repeated machining process, International Journal of Machine Tools & Manufacture Vol.38 (1998), No. No. 3, 205-213.
6. A. C. Okafor and Y. M. Ertekin, Vertical machining center accuracy characterization using laser interferometer Part 1. Linear positional errors, Journal of Materials Processing Technology 105 (2000), 394-406.
7. A. C. Okafor and Y. M. Ertekin,, vertical machining center accuracy characterization using laser interferometer Part 2. Angular errors, Journal of Materials Processing Technology 105 (2000), 407-420.
8. Z. C. Du, J. G. Yang, Z. Q. Yao and B. Y. Xue, Modeling approach of regression orthogonal experiment design for the thermal error compensation of a CNC turning center, Journal of Materials Processing Technology 129 (2002), 619-623.
9. R. Ramesh, M. A. Mannan and A. N. Poo, Error compensation in machine tools - a review Part II: thermal errors, International Journal of Machine Tools & Manufacture 40 (2000), 1257–1284.
10. S. Ghosh, Statistical design and analysis of industrial experiments, New York : M. Dekker (1990).
11. R. L. Mason, R. F. Gunst and J. L. Hess, Statistical design and analysis of experiments with application to engineering and science, John Wiley and Sons, Inc. New York (1989).

Innovative Production Machines and Systems
D.T. Pham, E.E. Eldukhri and A.J. Soroka (Eds.)

Conductive metallic tracks by Pyrolytic printing

G. Oosterhuis[a], F. Feenstra[a], R. Houben[a], Bart van der Vorst[a], L. Brouwers[a], F.Verhoeven[a]

[a] *TNO Science and Industry, Eindhoven, Netherlands*

Abstract

Printing highly conductive metal tracks in a one-step process called pyrolytic metal printing (patent pending) provides a powerful alternative to existing printing techniques for metal track production. Pyrolytic printing starts with the generation of droplets of a solution of some precursor salt, guided by a nitrogen flow through a furnace that evaporation the solvent, followed by pyrolysis of the precursor material, generating liquid metal particles. A the outlet of the furnace, aerodynamic focusing assures accurate deposition of the molten metal droplets, which yields 20 μm wide conductive metal tracks directly onto the substrate.

Keywords: Pyrolysis, Printing, Conductive tracks

1. Introduction

TNO is currently facing the challenge to print highly conductive metal tracks in a one-step process, with minimal heat load onto the substrate. A process has been developed called pyrolytic metal printing (patent pending), which has strong potential to meet these specifications. When these specifications can be met, the method will be a powerful alternative to existing techniques to deposit conducting metal tracks.

2. 2 Overview of metal printing processes

2.1. Existing metal printing techniques

There are many ways to produce conductive metal tracks of arbitrary geometry. Hence, new techniques should provide a distinct advantage with respect to the existing ones. Therefore, to facilitate comparison, a summary of the main existing metal printing techniques is given in Table 1. All direct deposition - or jetting – techniques deposit the metal (or its precursor) directly at the desired location, thus affecting the surface at the location of the printing track. Whereas deposition techniques like etching, screen printing and gravure affect the whole of the substrate body. This imposes boundary conditions on the whole of the surface, including the locations where no metal should be deposited. Such boundary condition may be: smooth geometry, chemical resistance and temperature resistance. Another distinction between direct deposition methods and other techniques is that in direct deposition, little material is wasted. Whereas in techniques that are based on material removal instead of deposition, material waste is inherent to the process.

Jetting, or direct-deposition has different application areas than techniques like etching, screen printing and gravure. Therefore, the scope of the comparison is limited to direct deposition techniques.

As can be seen in Table 1, each of the existing techniques implies the drawback of multiple process steps, material restrictions or thermal loads on the substrate. Spray pyrolysis eliminates such drawbacks but does not provide a means for accurate deposition. Pyrolytic metal printing combines the advantages of spray pyrolysis with a mono-disperse droplet generator (print head) and aerodynamic focusing.
This yields accurate and controlled deposition of the metal droplets. Because of the inherent advantages of such an approach, TNO is developing a method to realize such a printing process as is described in more detail the following sections.

Table 1 : Existing metal printing techniques

Technique	Description
Jetting silver nano-particles suspension	Jetting of silver nano-particles suspension, followed by sintering (Craigh and Laruso 2007 [1]).
Spray deposition of precursor solution or nano-particle suspension	Spray of $AgNO_3$-solution, or silver nano-particle suspension is deposited. Laser spot curing (Optomec 2007 [2])
Jetting a precursor followed by metallization	A precursor is jetted, metallization in next process step, immersion in a solution that contains metal ions, which causes a gal vanic reaction. (CIT 2007 [3])
Direct jetting of molten metal	Inside the print head, solid metal is melted and jetted on the surfa ce. Max temperature 250 °C, only low melting point metals or alloys to be used. (Microfab 2007)
Spray pyrolysis	A polydisperse spray of precursor solution is generated and subsequently heated, causing solvent evaporation and pyrolysis. Used to generate particles, no conductive tracks. Not a jetting technique. (Pingali et al 2005)
Pyrolytic metal printing (TNO)	Mono disperse droplets of a precursor solution by a print head and subsequently heated, evaporaion and pyrolysis. Molten metal droplets are focused into narrow tracks using aerodynamic focusing (Kempen 2006).

3. General process description

The general process of pyrolytic metal printing can be described according to Figure 1. Consecutively, the following process steps take place.

Droplet generation The droplet is generated using a mono-disperse droplet generator. For the printing of 2D or 3D structures, a drop-on-demand print head is required. For high speed production of straight lines, a continuous head may be used.

Droplet heat-up Once the droplet is generated, the nitrogen flow carries it into the furnace where it is heated by infrared radiation. First, the absorbed heat causes a temperature increase.

Solvent evaporation Once the droplet temperature approaches the boiling point of the solvent, the solvent will quickly evaporate, until only solid salt particles remain.

Pyrolysis Further heating will cause the salt to melt after which it reaches the pyrolysis temperature of AgNO3. Then, the reaction AgNO3 (l) $\rightarrow$ Ag(s) + NO(g) + O2(g) takes place, which yields solid silver.
Melting silver Following the pyrolysis step, the solid silver particles are heated further, up to the melting point of silver. Once molten, the droplet may be heated further until it leaves the furnace. In this phase, the heat input needs to be controlled in order to control the droplet temperature on impact, which highly influences the final track size and quality.
Focusing Near the outlet of the furnace, the droplets may deviate from their initial course as a result of the preceding process steps. Therefore, aerodynamic focusing is required to
(re-)concentrate the droplets in the center of the printed track. Focusing is achieved by accelerating the flow through a contraction at the end of the tube. During this acceleration of the gas flow, the droplet velocity also increases as a result of drag forces. The flow pattern is designed in such a way that there exists a transitional point from where the droplet motion is dominated by inertia instead of drag. CFD calculations show that the droplet trajectories exhibit a focal point, if the droplet velocity increases at least one order of magnitude in the converging part of the tube. This way of focusing yields a deposition accuracy better than 5 μm.

4. Experimental set-up

The printing set-up comprises a drop-on-demand jetting device (http://www.microfab.com /equipment/devices.html) that creates 30-100 μm droplets of a 10-20%w/w (mass percentage) aqueous AgNO3 solution. The generated droplets pass through an inspection chamber followed by a ceramic tube furnace with a temperature up to 1500°C. A nitrogen flow guides the droplets through the furnace. The substrate located several millimeters below the orifice is placed upon an x-y stage to be able to generate 2D patterns.

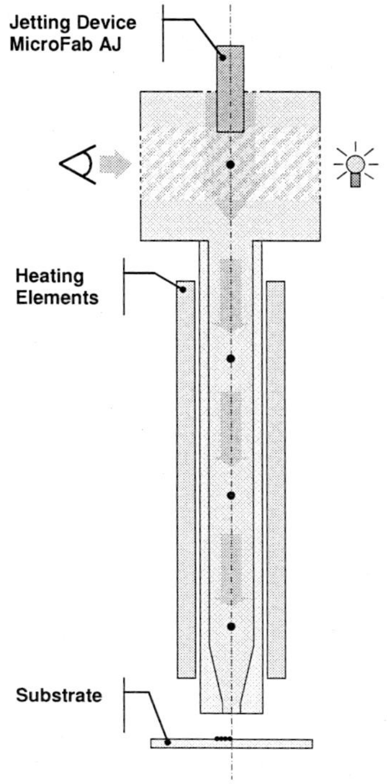

Fig. 1. Schematic overview of the pyrolitic printing process.

5. Results

Currently, printing with this set-up was successful up to the pyrolysis temperature of $AgNO_3$. The test set-up was optimized on the basis of CFD calculations on the aerodynamic focusing. As a result, the aerodynamic focusing yielded a deposition accuracy of 5 µm. This is illustrated by a printed free-standing pillar of 20µm diameter and 2mm height. (Figure 2a). The focusing results have been obtained at reduced furnace temperature. At this temperature, the salt was already molten but pyrolysis did not start yet. At higher furnace temperatures solid and molten silver droplets could be deposited in straight tracks that show good electrical conductivity (Figure 2b) as well as high purity of the deposited silver (Figure 3).

The deposition of highly conductive silver tracks as well as the aerodynamic focusing was successful, however both aspects could not yet be achieved simultaneously. At temperatures necessary to achieve molten metal droplets, the focusing accuracy was reduced, resulting in track width of approximately 300 µm, instead of the expected 20-50 µm. For yet unkown reasons, droplets break up into a poly-disperse spray with diameters ranging from 0.5-10 µm. As a result of both the smaller droplet size as well as the wider size distribution, the aerodynamic focusing accuracy decreases. Current efforts aim at understanding what phenomenon causes the break up the droplets. Particularly challenging is the fact that the whole process takes place within a single-zone 1500°C furnace which excludes most regular inspection and measurement methods.

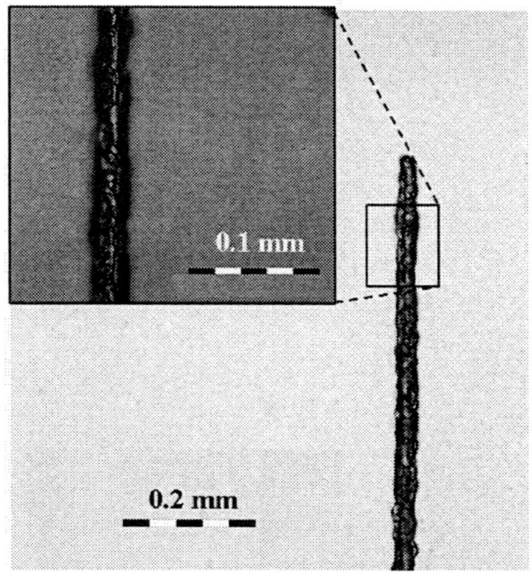

Figure 2a : Experimental printing results
Aerodynamic focusing; A free standing pillar could be printed of 2mm high and 20 µm wide.

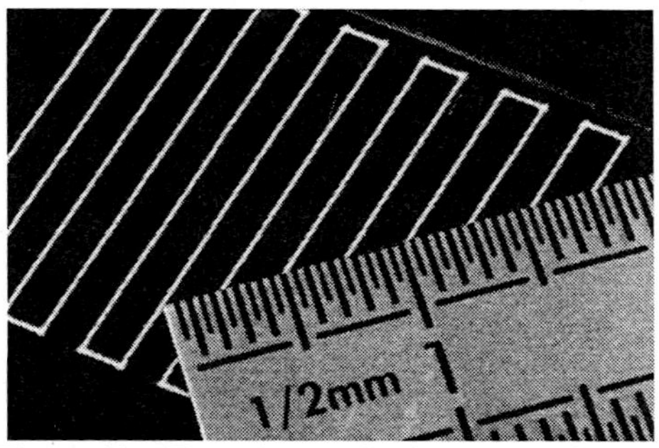

Figure 2b : Experimental printing results
Silver tracks with full metal-conductivity, printed on glass. The track width is 0.3 ± 0.1mm.

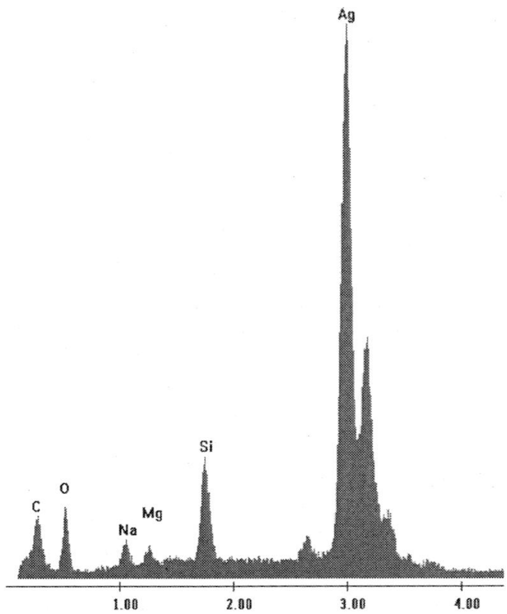

Figure 3: X-ray spectroscopy result of silver printed on glass. The deposited silver is very pure, the other peaks are attributed to the glass substrate

5 Conclusions

Evaluation of several aspects , reveals that pyrolytic metal printing provides a distinct advantage over existing techniques. Printing highly conductive metal tracks in a one-step process, with minimal heat load onto the substrate will become reality in the near future. The major challenge in the near future is to maintain the accurate focusing shown in Figure 2a at elevated temperatures that yield molten silver droplets. Further, optimizing the process towards industrial applications is key to future success. Hence major future research topics will scaling the process up towards printing speeds required for industrial implementation.

Acknowledgements

The authors are members of the EU-funded FP6 Network of Excellence for Innovative Production Machines and Sysems (I*PROMS).

References

[1] Creagh L.T. and Caruso J., 2007, Industrial Inkjet Printing of Single Layer Electronics, SGIA Membrane Switch Symposium CA, USA, available at http://www.dimatix.com/files/SGIAMembraneSwitchSymposiumJune2005.pdf
[2] Optomec 2007, http://www.optomec.com/site/m3d_home
[3] CIT 2007, http://www.conductiveinkjet.com/
[4] Microfab2007,http://www.microfab.com/equipment/pdf/ph-05_mf1.pdf
[5] Pingali, K. C., Rockstraw, David A. and Deng, Shuguang, 2005 Silver Nanoparticles from Ultrasonic Spray Pyrolysis of Aqueous Silver Nitrate, Aerosol Science and Technology, 39:10, 1010 - 1014
[6] Kempen, A. T.W., 2006, Method for applying material onto a substrate using a droplet printing technique, European patent application number WO2006004406

Innovative Production Machines and Systems
D.T. Pham, E.E. Eldukhri and A.J. Soroka (Eds.)

Hydraulic Axis Actuation Control for Precision Motion

S. El-Shalabi [a], R. Whalley [b], M. Ebrahimi [a], and B. Mayson [a]

[a] *School of Engineering, Design and Technology, University of Bradford*
[b] *Institute of Engineering, British University, Dubai, UAE*

Abstract

A new design for machine tool axis traverse, by hydraulic cylinder with spool and needle valve actuators, is presented. Regulation procedures enabling the formulation of a least effort control strategy are proposed. System integrity in the event of actuator failure is examined. Correspondingly, the maintenance of closed-loop stability, following feedback transducer loss, is also secured, demonstrating fail-safe, multivariable, feedback control. The suppression of the effects of disturbances and the response of the system to deterministic inputs is presented. Comments on the practicality of the design are incorporated and the mandatory specification of control system integrity is advocated.

Keywords: optimum, hydraulic, machine, axis, drive

1. Introduction

The use of electrical drives and ball-screw mechanisms, for the axis drives of machine tools, promotes geometrical machine tool variations, stiffness and vibration problems. These changes and disturbances affect machining accuracy, surface finish and repeatability with damaging consequences for manufacturing efficiency.

Essentially, with the incorporation of ball-screw mechanisms system performance becomes a function of table and/or cutting tool position. Endemic effects such as ball-screw wind-up, arising from combined torsion and tension loading together with coulomb friction and hysteresis degrade machining performance and are difficult to counter.

Alternatively, a traverse drive employing hydraulic cylinder actuation, with spool valve control, for example, would avoid the inherent difficulties arising from effective ball-screw length changes. However, there would now be challenging regulation issues with the deployment of high pressure, and highly oscillatory hydraulic systems.

This proposition is not new. In earlier work, see for example de Pennington et al [1], high pressure hydraulic actuation was investigated for machine tool, traverse applications. Single input-output and transient acceleration feedback was incorporated in this study, enabling the imposition of velocity and position control.

As in this exposition, a cylinder drive and spool valve control was the preferred method of actuation. Equally, the common aim of achieving a simple, cost effective, constant stiffness, rapid response table movement mechanism, is acknowledged.

Although offering these advantages, high performance hydraulic drives exhibit lightly damped transients and careful analysis, design and the guarantee of fail-safe operation are mandatory requirements. To achieve these, and demanding machine tool specification requirements, energy dissipation, via leakage control, will be proposed in the formulation, outlined herein. Hydraulic cylinder actuation is controlled by the servo-valve whilst leakage flow is regulated via the needle valve, reservoir chamber and flexible diaphragm, as shown in figure 1.

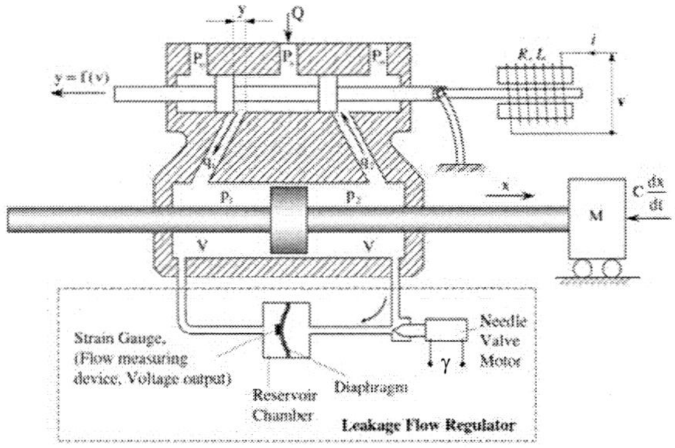

Fig. 1: Hydraulic actuator, spool valve and leakage flow regulator

The additional design freedom offered by the inclusion of transient damping may be exploited thereafter by the incorporation of optimum, multivariable control with the minimization of the regulation energy required, for disturbance suppression.

This strategy would also ensure that actuator activity was minimized and with it noise, maintenance and refit costs would be reduced. Machine tool availability would also be maximized promoting thereby cost-effective manufacturing.

The configuration of a machine tool, x axis traverse drive, using a hydraulic cylinder, is shown in figure 2.

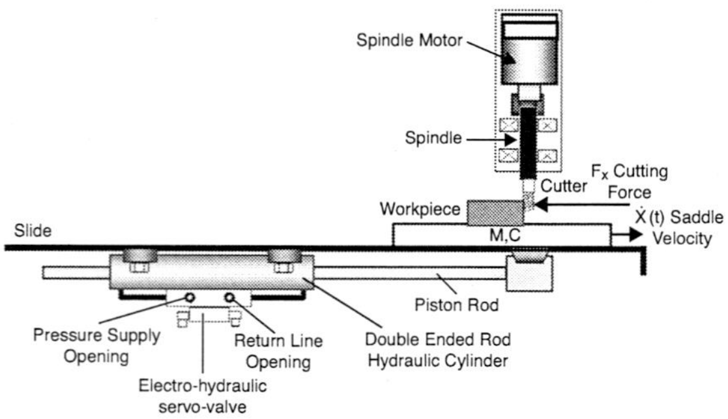

Fig. 2: Machine tool x axis traverse, hydraulic cylinder drive

The block diagram representation of the drive system is shown in figure 3.

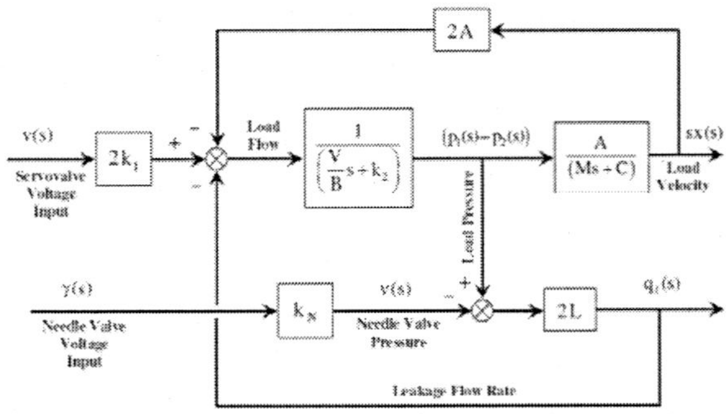

Fig. 3: Block diagram representation of the system with the leakage flow regulator

2. System Model and Open-Loop Response

For the system shown in figure 1 the hydraulic fluid flow rates through the spool valve, due to small changes in valve voltage v(t) are:

$$q_1(t) = k_1 v(t) - k_2\ p_1(t) \tag{1}$$

$$q_2(t) = k_1 v(t) + k_2\ p_2(t) \tag{2}$$

For small changes in the load velocity $\dot{x}(t)$, the flow rates given by equations (1) and (2) can also be related to the change in the volume of the hydraulic fluid in the cylinder. If the compressibility and leakage effects are included then:

$$q_1(t) = A\dot{x}(t) + \frac{V}{B} Dp_1(t) + L\left(p_1(t) - p_2(t) - v(t)\right) \tag{3}$$

$$q_2(t) = A\dot{x}(t) - \frac{V}{B} Dp_2(t) + L\left(p_1(t) - p_2(t) - v(t)\right) \tag{4}$$

The force balance for this system is given by:

$$MD\dot{x}(t) + C\dot{x}(t) = A(p_1(t) - p_2(t)) \tag{5}$$

Using the manipulation given in [4]:

$$\begin{bmatrix} sx(s) \\ q_\ell(s) \end{bmatrix} = \frac{1}{H(s)} \begin{bmatrix} a_{11}(s) & a_{12}(s) \\ a_{21}(s) & a_{22}(s) \end{bmatrix} \begin{bmatrix} v(s) \\ \gamma(s) \end{bmatrix}$$

$$a_{11} = 2k_1 A,\ a_{12} = 2ALk_N,\ a_{21} = 2L\left(Ms + C\right)2k_1, \tag{6}$$

$$a_{22} = -2Lk_N\left[\left(Ms + C\right)\left(\frac{V}{B}s + k_2\right) + 2A^2\right]$$

$$H(s) = \left(Ms + C\right)\left(\frac{V}{B}s + k_2 + 2L\right) + 2A^2$$

The model responses following 100% step input changes on the %v(s) and then on %γ(s), are presented in [4]. The uncontrolled system is highly oscillatory and has pronounced overshot problems.

Hence, multivariable control, improving system response, disturbance suppression whilst diminishing load offset errors, elevating machining quality thereby, is necessary. This is achieved by designing a multivariable controller and regenerative output feedback regulation based on the optimum, minimum control energy strategy, described by Whalley and Ebrahimi [5], [6].

3. Closed-Loop Regulation Strategy

The following procedure employs an inner and outer loop structure for design purposes. Once this structure has been determined an established, multivariable feedback realization will be evaluated. Figures 4, 5 and 6 show the construction of the inner, outer and conventional loops respectively, for 2 input-output system.

The outer loop controller will primarily be employed to secure specified steady state, reference set point and disturbance containment characteristics.

The inner loop will be utilized to attain initial, targeted, dynamic, and load disturbance recovery rates, where the system equation including, disturbance inputs is:

$$\mathbf{y}(s) = \mathbf{G}(s)\mathbf{u}(s) + \boldsymbol{\delta}(s) \qquad (7)$$

The control law for the proposed configuration, for analysis purposes, is:

$$\mathbf{u}(s) = \mathbf{k}(s)\left(\bar{r}(s) - \mathbf{h}(s)\mathbf{y}(s)\right) + \mathbf{P}\left(\mathbf{r}(s) - \mathbf{F}\mathbf{y}(s)\right) \qquad (8)$$

where in equations (7) and (8) there are m independent inputs, disturbances and outputs are described by El-Shalabi [4]

The Laplace transformed open loop system, given by equation (7), where the system model $\mathbf{G}(s)$ is assumed to be an m×m linear, proper or strictly proper realization, admits a factorization:

$$\mathbf{G}(s) = \mathbf{\Gamma}_L(s)\,\mathbf{L}(s)\,\frac{\mathbf{A}(s)}{d(s)}\,\mathbf{R}(s)\,\mathbf{\Gamma}_R(s) \qquad (9)$$

where the elements of:

$$\mathbf{L}(s),\ \mathbf{R}(s),\ \mathbf{\Gamma}_L(s)\ \mathbf{\Gamma}_R(s)\ \text{and}\ \frac{\mathbf{A}(s)}{d(s)} \in H_\infty,\ s \in \mathbb{C}$$

In equation (9), $\mathbf{L}(s)$ contain the left (row) factors, $\mathbf{R}(s)$ contains the right (column) factors and $\mathbf{\Gamma}_L(s)$ and $\mathbf{\Gamma}_R(s)$ contain the transformed, system and actuator, finite time delays, respectively, such that the m×m matrices comprising 8, as in Whalley and Ebrahimi [3]

4. Disturbance Rejection Analysis

In the following, bounds on the disturbance response, arising from changes in the sensitivity matrix, are determined. This enables the formulation of a functional, J, from which variations in the system's closed-loop disturbance attenuation can be established.

Gain ratios for which there are rapid changes in the performance index J and hence in the disturbance containment capacity of the system, also become apparent. These values indicate a lack of robustness to modelling variations and should be avoided.

The possibility of optimizing this process can now be considered. An indication that the freedom exists to do this arises from the arbitrary choice of n1, n2 etc., for the gain ratios, as detailed in [6].

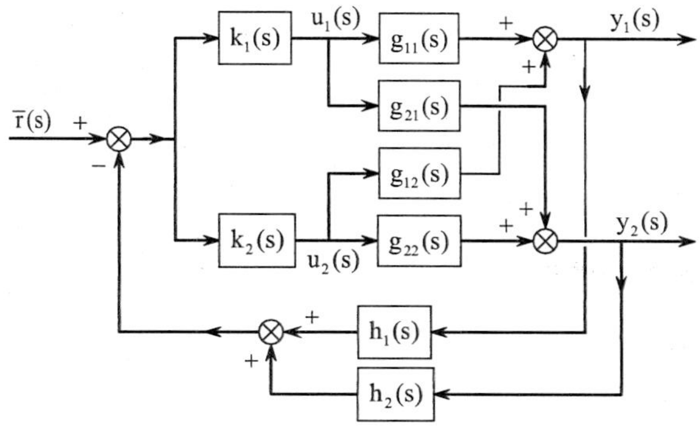

Fig. 4: Inner-loop structure for 2 input-output system

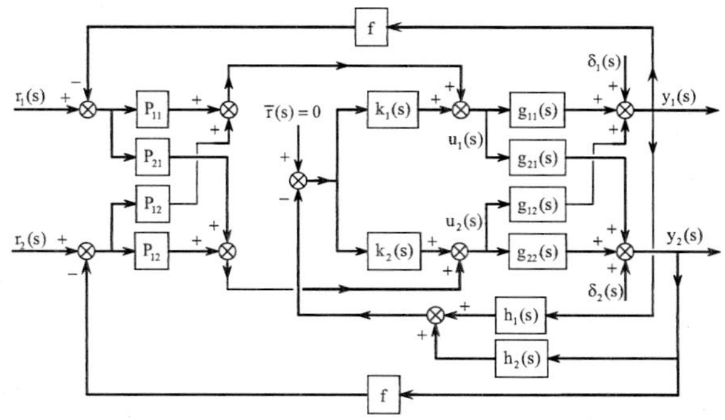

Fig. 5: Outer-loop structure for 2 input-output system

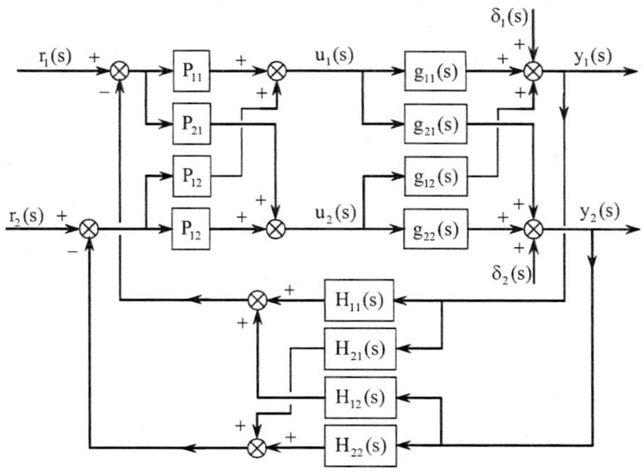

Fig. 6: Conventional-loop structure for 2 input-output system

5. System Stability and Implementation

The input-output relationship for the complete closed loop system is given by equation (7). Consequently, the input-output stability condition is dependent on the denominator of this equation.

However, controller design for the inner, outer and conventional loops are fully detailed in El-Shalabi [7]

From the simulation of this configuration, the output responses following input step changes of unity, on $r_1(t)$ and $r_2(t)$, respectively, can be computed, as shown in figure 7A and 7B.

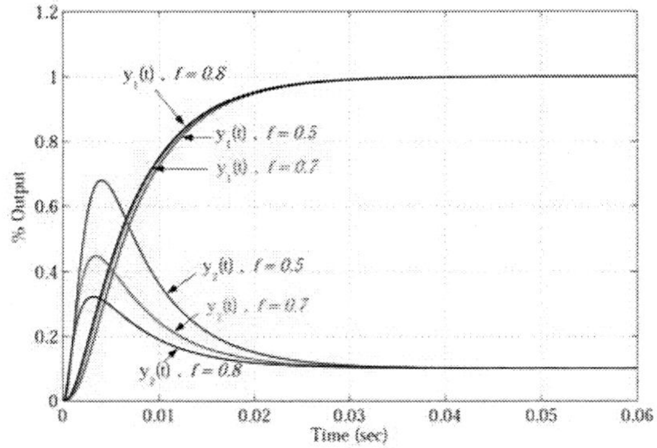

Fig. 7A: Closed-loop output response following a unit input step change on $r_1(t)$

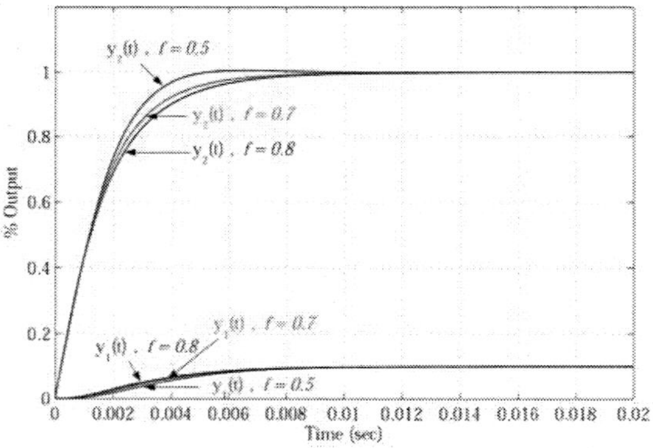

Fig. 7B: Closed-loop output response following a unit input step change on $r_2(t)$

These traces, depict acceptable response variations with no overshoot problems on the table speed , with the steady state interaction restricted to the limits specified. Thereafter, the disturbance offset graphs, for inputs steps on and then, are presented in figures 8A and 8B. The important table velocity, disturbance rejection characteristics of figure 8A, provides excellent recovery transients limiting the loading problem to less than ±20% once f is increased to accommodate this requirement.

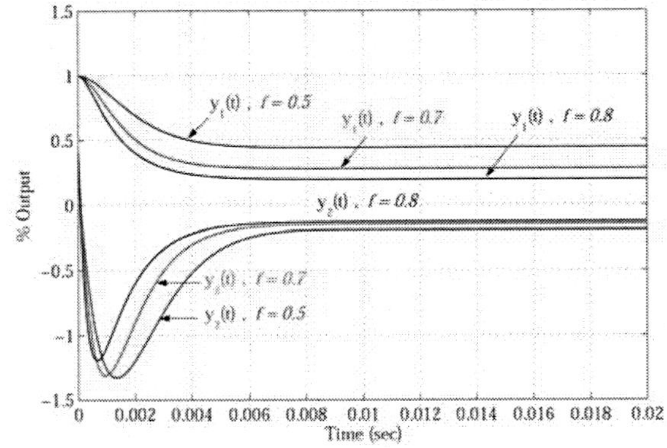

Fig. 8A: Closed-loop output response following a unit disturbance step change on $\delta_1(t)$

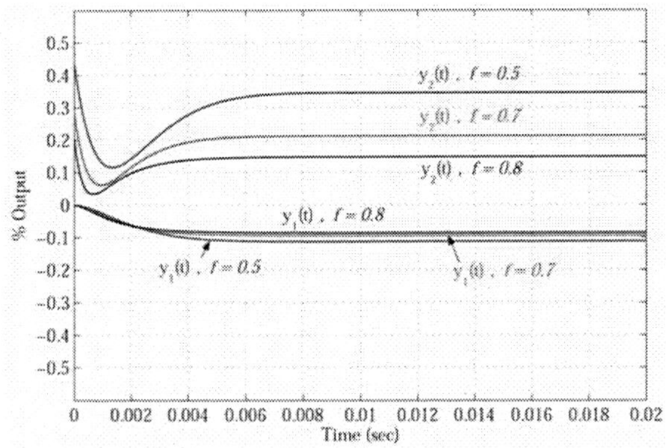

Fig. 8B: Closed-loop output response following a unit disturbance step change on $\delta_2(t)$

6. System Integrity

The specification of system integrity is normally mandatory, for safety reasons, and this requires that asymptotic stability must be maintained, in the event of transducer or actuator failure. This is an important requirement which is often conveniently ignored when debating the merits of multivariable system design methods.

To investigate integrity the gain space, as defined by MacFarlane [8], must be divided into stable and unstable regions. From this analysis, if transducers failure results in instability then the system is conditionally stable, see [2].

For the conventional loop:

$$y(s) = \left(\mathbf{I}_m + \mathbf{G}(s)\mathbf{K}(s)\mathbf{H}(s)\right)^{-1}\left(\mathbf{G}(s)\mathbf{K}(s)\mathbf{r}(s) + \boldsymbol{\delta}(s)\right) \quad (10)$$

If feedback the transducer from $y_m(s)$ fails then stability depends on

$$\det\left(\mathbf{I}_m + \mathbf{G}(s)\mathbf{K}(s)\mathbf{H}(s)\right) = 0 \quad (11)$$

and this polynomial for

$$\mathbf{H}(s) = \begin{bmatrix} h_{11}(s) & h_{12}(s) & \cdots & 0 \\ h_{21}(s) & h_{22}(s) & \cdots & 0 \\ \vdots & \vdots & \ddots & 0 \\ h_{m1}(s) & h_{m2}(s) & \cdots & 0 \end{bmatrix}$$

must yield a Hurwitz polynomial. Consequently, for complete integrity equation (11) must results in a Hurwitz polynomial for each **H**(s) with any j column, $1 \leq j \leq m$, of **H**(s) zero, as shown in El-Shalabi [7].

The system responses in the event of transducer failure are shown in figure 9.

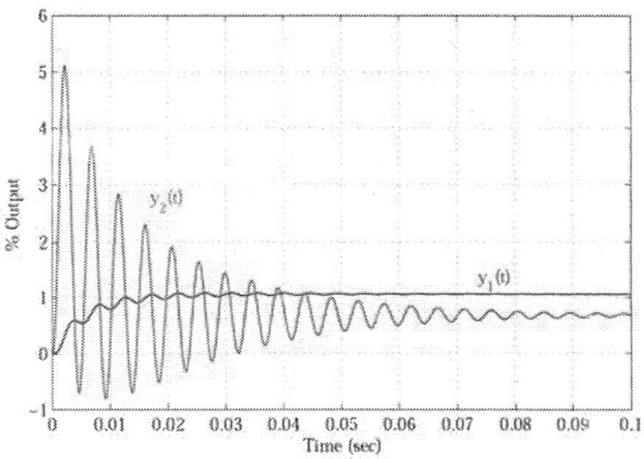

Fig. 14A: System response following a unit input step change on, following the failure of measurement transducer

Conclusion

This contribution outlines the analysis, design and simulation procedures for a closed-loop, hydraulic, machine tool axis drive system. The process incorporates a comprehensive specification including the output response requirements following input step and disturbance changes. Limitations on the output, steady state coupling, disturbance suppression and recovery were also specified and achieved.

The design was based on the formulation of a minimum energy control strategy, guaranteeing a robust, low gain feedback configuration. Owing to this the control system implementation advocated was simple, and in the case considered the feedback system could be realized by proportional gain terms alone.

In addition to the specification requirements normally considered, the integrity of the control system, in the event of either feedback transducer or actuator failure, was examined. This condition is critically important for industrial operations where "fail-safe" performance is a mandatory requirement.

Once again the minimum energy controller demonstrated effectiveness in terms of absolute integrity. The failure mode responses computed for the fault conditions specified, confirmed analytical predictions with declining performance to steady state being achieved in all of the failure mode, situations.

Finally, the relative stability assessment, using the closed-loop frequency response plots illustrated the effect of the outer-loop gain. As this characteristic shows the steady state gain of the closed-loop system increases dramatically for increasing values of the outer loop feedback, f. To maintain stability and transient performance, the frequency response curve moves to the right of the origin, as f is progressed, so that relatively small changes, in the high frequency sectors of the curve, are initiated.

Consequently, advancing the value of the outer loop gain f produces modest changes in relative stability and in the system's performance. The significant adjustment, as show by the analysis, is accorded to the improvements in disturbance suppression, with load offset being inversely proportional to (1-f), as predicted by the analysis.

References

[1] de Pennington, A., Marsland, D. W., and Bell, R., 1971, "The Improvement of the Accuracy of Electrohydraulic Cylinder Drives for NC Machine Tools by the Use of Active Feedback Compensation," 12th International Conference on Machine Tool Design and Research, Manchester, pp. 199-205.

[2] Rosenbrock, H. H., 1974, "Computer Aided Control System Design," Academic Press, London.

[3] Whalley, R., Ebrahimi, M., 2005, "Control of a Lamination Process," Proc. IMechE, Part E: J. of Process Mechanical Engineering, 219(1), pp. 69-81.

[4] El-Shalabi, S, Hydraulic Axis Actuation with High Integrity Control for Precision Motion, Internal Report 12.04,05, EDT, University of Bradford, UK.

[5] Whalley. R., and Ebrahimi, M., 2000, "Process System Regulation with Minimum Control Effort," Proc. IMechE, Part E: J. of Process Mechanical Engineering, 214(4), pp. 245-256.

[6] Whalley, R., and Ebrahimi, M., 2003, "Closed-Loop System Disturbance Recovery", Proc IMechE, Part C: J. of Mechanical Engineering Science, 217(6), pp 631-649

[7] El-Shalabi, S., 2008, "Hydraulically Actuated Axis Drive Systems for Machine Tools", Ph.D. thesis, EDT, University of Bradford, UK.

[8] MacFarlane, A. G. J, 1970, "Multivariable Control System Design Techniques", Proc IEE, Vol. 117, pp 1039-1047.

Innovative Production Machines and Systems
D.T. Pham, E.E. Eldukhri and A.J. Soroka (Eds.)

Implementation and dissemination of RM in the industry

W.P.M. Kruf[a]

[a] TNO Science and Industry, Department Rapid Manufacturing, De Rondom 1, 5612 AP Eindhoven, The Netherlands

---

Abstract

The mission of TNO (the Netherlands Organisation for Applied Scientific Research) is to apply scientific knowledge with the aim of strengthening the innovative power of industry and government. By Developing, integrating and applying knowledge we generate creative and practicable innovations: new products, services and processes, fully customised for business and government.

The Rapid Manufacturing department of TNO consist of two sections; Research and the Demo Centre [RMDC]. While the research section focuses on the realization and improvement of equipment, materials and processes, the RMDC bridges the gap regarding the market demands with redesigning parts making them suitable for RM, providing training, workshops and seminars.

The Workshop Rapid Manufacturing has been developed in order to enable companies to get acquainted to the various Rapid Manufacturing processes, materials and design rules. To explore the possibilities themselves, the companies are invited to contribute with a case of their own. Due to the success, TNO has decided to offer the workshop on an European scale. The paper describes, amongst other types of technology transfer, the format and the content of the workshop.

Keywords: Rapid Manufacturing, training, workshop

---

1. Barriers

More and more, companies need parts that are hard to make with conventional production technologies or are simply impossible to produce. The parts have a high degree of complexity, the time pressure is high and the requested amount of parts is relatively low. In these cases Rapid Manufacturing can be a solution, but some hurdles must be taken for the implementation of these layer technologies. Several barriers have been identified.

1.1 Unknown technology

It is surprising that the use, or even the existence of Rapid Manufacturing or Rapid Prototyping is relatively unknown amongst engineers, designers or even students. Most universities and higher educational schools simply do not have lectures and courses on these techniques in their regular curriculum in the first place. For many of our customers, who typically have several years of engineering experience, our course is their first acquaintance with RM.

1.2 Rapid Prototyping can be used as RM

And if one is familiar with Rapid Prototyping, the next step towards RM is seldom taken because one is unfamiliar with the new materials and processes that have been developed recently. For example carbon filled materials and high performance metal alloys allow application in a heavy duty environment.

1.3 Design strategy

Fundamental for RM is its freedom of design. RM enables the design of high added value/integrated products with maximum customer involvement. Beside the basic design rules, for example regarding the anisotropy of many materials, RM allows for a highly automated design process, in which geometry can follow the ideal topology on basis of load/stress conditions.

1.4 TQM

Total quality management, norms and predictable production are issues that still have to be addressed more explicitly. Much research effort will be undertaken to evolve RM into a mature production technology.

2. Current solutions

To raise awareness and train students and companies about the new possibilities of Rapid Manufacturing, the RM Demo Centre developed several methods for technology transfer.

2.1 Student assignments

To train students mechanical engineering of the Eindhoven University of Technology and give the right set of skills for designing in Rapid Manufacturing, a special format have been developed by Kruf [1]. Students work in groups on a 6 week assignment to design and construct an object that performs a specific function. The object has of course to be made with RM. Every week, a new design is made, printed with RM techniques and evaluated with a tutor. This way,

Fig 1.Object made by students in “Scrambled Eggs”

the students learn swiftly how to cope with the characteristic design challenges. Typically, there are 6 groups of 4 students, each student requiring 20 hours per week. Also a competition element is added; at the end of the course the teams battle for the first place. For example, one of the assignments was called “Scrambled Eggs” (see Fig. 1). Students were supposed to design a construction that would protect an egg, while it was thrown from the second floor. The winning team of the end-match was that one with the egg not broken. The first restriction was that the students were allowed to use the SLS technique only and secondly, they had a building volume assigned of only 100x100x100mm3. The challenges lied in the fact of the varying geometry, every egg has a different shape and how to enclose it. Secondly, one would like to have a large construction to absorb as much as energy as possible, while the building volume is fairly limited. It is our experience that during these challenges every group reaches the moment that there is a ‘click’ and the next level in design comprehension is achieved. Up to now, over 200 students have done the course.

2.2 Technology transfer in small clusters

An excellent example of demand driven knowledge transfer, focused on the needs of a small group of companies, can be found in the so-called Technology clusters. In these tailor made courses, companies are accompanied by RM experts in order to work on cases over a period of time, mostly 4-6 months. Some examples that have been performed are; High end equipment, Jewellery, functional RM parts, Designers.

Fig. 2. Participants of the RM workshop

2.3 *RM workshop*

The growing demand for practical Rapid Manufacturing solutions on a larger scale was the motivation to offer companies a workshop (see Fig. 2). The format was developed by Buining [2]. In this workshop, participants can acquire knowledge and design skills in a very practical manner by introducing their own case. The cases are discussed in small subgroups and result in samples made by the appropriate RM technologies. The workshop consists of three afternoons with an interval of 3 weeks. This practical and interactive approach leads to a very steep learning curve. Also because of the fact that people not only learn from the possibilities with RM, but moreover about the downsides and how to cope with them. Many times, a hybrid solution with conventional technologies can be the result. Our records show that the course is highly appreciated by the participants and is awarded with an 8/10.

After having completed the RM workshop, companies are aware of new possibilities with RM. They also can judge which parts are possible with RM. Concerning design rules, engineers will have the skill to redesign construction, optimized for producing with RM and how to handle the downsides of the technology.

2.3 *Example*

The case to discuss is daone in co-operation with a Dutch Aerospace company called DAC, Dutch Aerospace Company, which produces small aircrafts in low quantities. In June 2000 a small team of dedicated people joined together to manufacture the tooling and

Fig. 3. Aircraft DAC Ranger

components for a new all composite diesel powered aircraft called the Ranger. The subject of discussion is this two-seater with a 90 HP diesel engine with a turbo and intercooler (see Fig. 3.).

After a first analysis, in which we spotted feasible RM opportunities varying from air-intakes to paddles, we chose to design an electronics compartment to enclose the PCB responsible for the wings-flaps management (see Fig 4.).

Fig 4. The PCB in directly mounted to the body inside the tail of the aircraft.

The main issues that had to be addressed were:
- Integration of functions with microswitches;
- better protection of the electronics and cables;
- reduction of assembly time.

After several iterations in conjunction with the customer, in which prototypes have been made and tested, a final geometry has been established.
The results of the exercise were an enhanced level of safety, because of the encasement of the electronics and an improved professional look (see Fig. 5.).

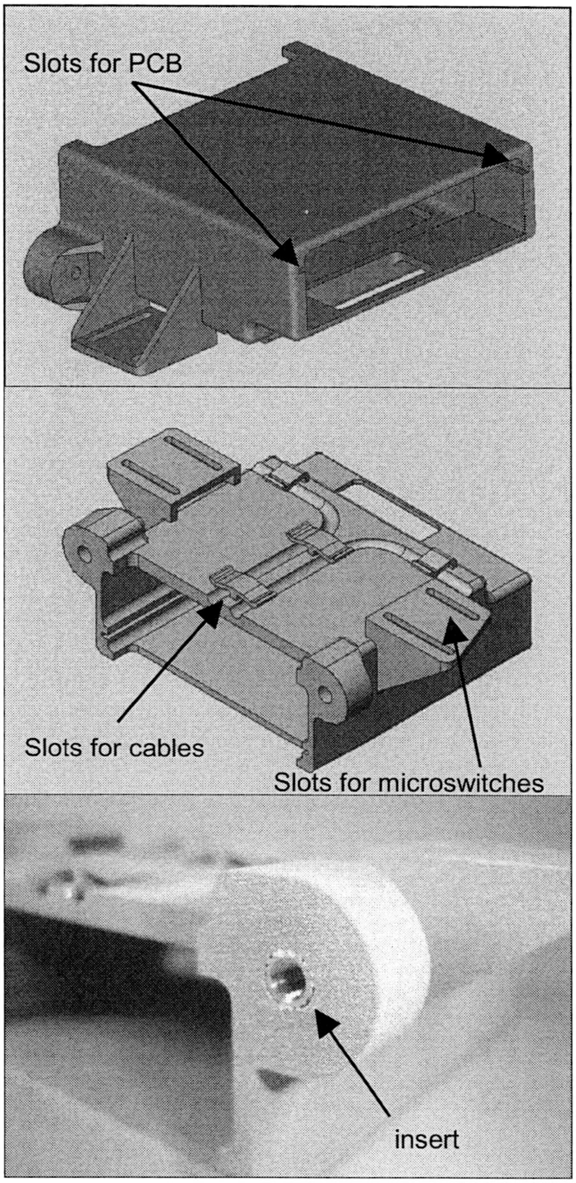

Fig 5. Final result

2.4 EU RM Workshop

Because of the increased interest for Rapid Manufacturing, TNO has decided to develop a workshop Rapid Manufacturing on a European level. The workshop allows participants to become familiar with RM in a practical way. Participants can even bring in cases from their own company and samples of products are manufactured using various RM techniques. Participants have an opportunity to experience the possibilities of the techniques themselves.

Cases are explored in small sub-groups. The workshop examined cases of more or less 'normal' components that tend to be manufactured using injection moulding. However, there are also cases of products, such as an integrated honeycomb structure, that could not be manufactured using any conventional technique.

The workshop is available in the English language and can be followed either in Eindhoven in the Netherlands, or on location in the home country. To release the workshop further, we also make this workshop available for European Research institutes, so they can provide an innovative training for their local market. TNO can provide the training material and can teach the new trainers how to carry out the training. The cooperation with European research institutes consist of the following steps:

- First training is carried out by TNO allowing the new trainers to prepare themselves.
- Next all training material is provided by TNO and even updates are provided.
- The European Research institute is then allowed to consult TNO for further questions or for the production of RM-samples.

2.5 Conclusion

Our conclusion is that technology transfer to the industry is very successful by providing workshops in this format. Companies get both theoretical and practical experience and discover the potential of Rapid Manufacturing.

References

[1] Kruf WPM, Assignments TU/e 2003-2008.
[2] Buining HA, Workshop Rapid Manufacturing 2007-2008.

Innovative Production Machines and Systems
D.T. Pham, E.E. Eldukhri and A.J. Soroka (Eds.)

Predictive Maintenance and Diagnostics of Machine Tools

F. Meo[a], M. Foursa[b], S. Kopàcsi[c], T. Schlegel[d]

[a] *Fidia S.p.A.,Corso Lombardia 11, 10099 San Mauro Torinese, Italy*
[b] *Fraunhofer Institute for Intelligent Analysis and Information System, Schloss Birlinghoven, D-53754 Sankt Augustin, Germany*
[c] *Computer and Automation Research Institute, Kende u.13-17., 1111 Budapest, Hungary*
[d] *Fraunhofer Institute for Industrial Engineering, Nobelstr. 12, 70569 Stuttgart, Germany*

Abstract

The objective of condition-based predictive maintenance applied to machine tools is the "measurement" of the slow degradation taking place in mechanical components in advance of the consequent failures and the evaluation of the residual life in order to better manage the total Life Cycle Cost. An automatic self-tuning of the system should accompany this procedure, in order to always optimize the machine tool performance as an optimal trade-off between surface quality and machining speed. These activities may be performed by intelligent software modules, communicating among them, with the machines, the sensors and the humans. The implementation of some of these modules is taking place in the frame of the project "Intelligent Networked Manufacturing System" (INT-MANUS).

Keywords: Predictive Maintenance, Tuning, Machine Tools

1. Introduction

The development of new approaches for the maintenance of production machines is a very challenging and essential topic. In this paper we present our ideas on possible improvements of current approaches. The implementation of some of the ideas is being carried out in the frame of the INT-MANUS project (NMP2-CT-2005-016550).

The main objective of the INT-MANUS project is to develop a Smart-Connected-Control Platform for manufacturing enterprises. The platform should integrate the information about the production process and manufacturing machines and provide possibilities to access the information for program agents and human personnel. The purpose of the platform is to decrease maintenance costs and increase the efficiency of production process.

With the help of the Smart-Connected-Control Platform, pro-active maintenance will be improved. Machines and personnel will enter machine states, necessary maintenance operations and their schedule into the Smart-Connected-Control platform. The semantics enabled operating system will schedule maintenance operations and inform the production plant supervisor accordingly.

In this article we will speak mostly about the applications relative to predictive maintenance and diagnostics of machine tools.

2. Maintenance approaches

The current approaches for the maintenance of machine tools are:

- **Breakdown Maintenance or Corrective Maintenance** keeps focus on time-efficient and

lower-cost repairing/replacing equipment, only when a failure occurs. This strategy assumes that failure will take place and cannot guarantee zero failure situations. The biggest disadvantages of this method are the high value of failure costs, unpredictable time when failure takes place, undetermined time for maintenance activity (supported product downtime).
- **Preventive Maintenance** is a perspective strategy, based on the knowledge of the mean-time-between failure figures. To minimize breakdowns and to avoid losses due to unexpected stops, Condition-Monitoring Maintenance (CMM) is a basic reference. This strategy focuses on keeping the equipment in good condition to minimize failures, and on repairing components before they fail.
- The objective of **Condition-based Predictive Maintenance** is the "measurement" of the slow degradation taking place in mechanical components in advance of the consequent failures and the evaluation of the residual life in order to better manage the total Life Cycle Cost. An automatical self-tuning of the system should accompany this procedure, in order to always optimize the machine tool performance as an optimal trade-off between surface quality and machining speed.

Several activities are needed to achieve this goal, all of them complex, but even more when put together. These activities may be performed by intelligent software modules, communicating among them, with the machines, the sensors and the humans. The main modules should be:

- Data Mining: it should collect data coming from sensors, perform filtering of the huge amount of data and their interpretation in order to extract a set of significant quality indicators that synthesize the machine status.
- Decision Support: from the comparison between the current values of quality indicators and the values they had when the machine was new, it should evaluate the machine status, the residual life of mechanical components and take a decision about actions to be taken:
  - No action;
  - Self-tune;
  - Schedule maintenance for substitution of mechanical parts.
- Self-tuning: it should tune the main parameters in the NC and servo drives (gains of control loops, speed, acceleration and jerk, etc.) in order to keep a good quality of machining even though by decreasing the working speed.
- Rescheduling: in case maintenance should be scheduled, it should select the optimal time for service intervention and generate an updated process plan for the interested cell, shifting the operations scheduled on that machine to the other machines in the cell.
- Communication: it should notify to human personnel in the plant evaluated data (residual life, machine status, decisions about maintenance, etc.) and exchange information with other machines.

Moreover the collection in a central, remote repository of data relative to failures coming from similar machines working in different environments under different working conditions may help in the diagnosis of failures and in the improvement to the design of the machine tool itself. It would provide a wider information database to be used by learning modules, compare data coming from similar machines, map defects to climate, working conditions, and so on, and thus help in the improvement of machine design.

3. Architecture of the system for Maintenance

Figures 1 and 2 show the logical architecture of the system. Figure 1 describes the Data Mining part in detail: sensors collect data from the Machine Tool; the Data Mining software module is organized in form of a data processing pipeline and can be split in three main phases:

1. Collect data from sensors with the suitable frequency;
2. Since these data are a huge amount, they need to be filtered;
3. Some indicators identifying the state of "health" of the mechanical parts (meaning the status of wearing out of components) can be extracted and stored. Some examples of these indicators can be: vibration critical frequency, damping factor, backlash (error at the inversion of the direction of movement), maximum and average value of current, ripple of current. The selection of parameters, to be used as input for diagnostic and monitoring algorithms (the so called feature extraction), is one of the most critical phases in the whole process.

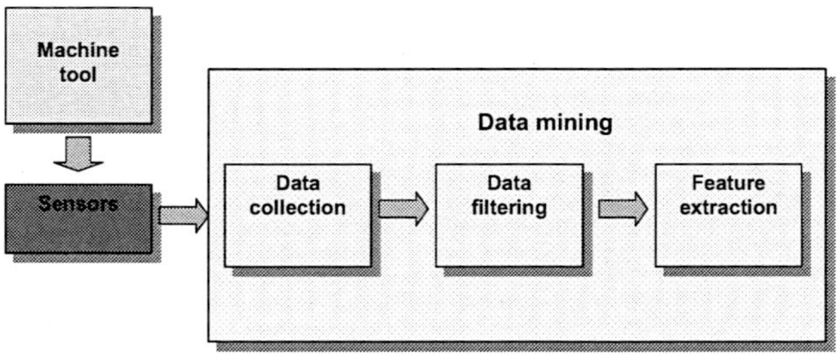

Figure 1: Logical architecture of system for maintenance: data mining

All collected data can be classified using meta-data, e.g. an RDF-based ontology to describe their type, source, dependencies and relations to other data acquired. This can be used for high-level filtering and interpretations applied later on.

Figure 2 shows a complete view of the system architecture: the Data Mining software provides indicators to a Decision Support module that selects the possible action, choosing among the following:

- Do nothing because the system is working very close to an optimal situation;
- Self-tune the system, whenever possible, in order to achieve a different trade-off between accuracy and time; in fact, for instance in the production of moulds and dies, the tuning of a machine tool is usually based on the selection of a suitable compromise: if the working operation is faster, it is normally less accurate, and vice versa. The optimal compromise depends on the kind of operation; for instance during a roughing operation, accuracy is not at all required, therefore the tuning should optimize the execution time (and cost, as a consequence). Usually the wearing out of mechanical components can be balanced through a different tuning that sacrifices the execution time but leaves the quality and accuracy of the result unchanged; typical machine parameters to be tuned for this goal are: speed, acceleration and jerk, gains of control loops.
- After a certain threshold, it is not efficient anymore to slow down the machine, and an intervention for the substitution of a mechanical part is needed. The resulting reorganisation of the plant schedule should be automatically performed.

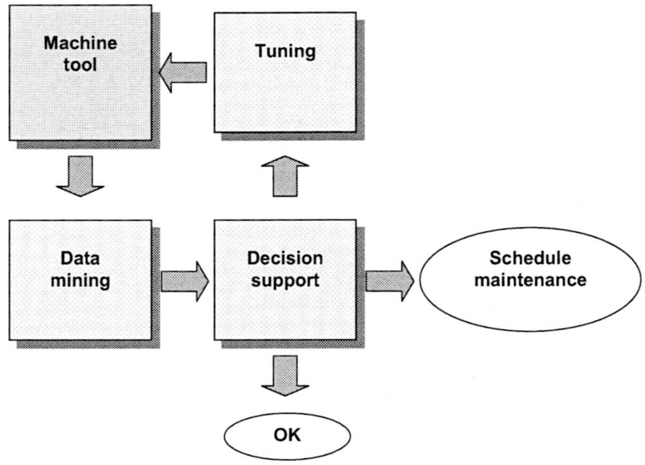

Figure 2: Logical architecture of system for maintenance: complete view

The production plant supervisor should be notified about this decision and the possible reason and mechanical parts to be substituted.

4. Architecture of the system for Diagnosis

A Remote Knowledge Repository should be added. The repository collects information coming from several machine tools (some of them of the same kind, some of them of different kind). An Evaluation Component exploits these information through algorithms able to learn from correlation of data. The result is a much more sophisticated Decision Support module.

The Knowledge Repository has to be able to transform information into knowledge, and learn by the correlation of data. For instance, applying this philosophy to diagnosis, the system will be able to correlate machine failures with factors like the climate or with the use of the machine.

Figure 3 shows the logical architecture of the system for diagnosis.

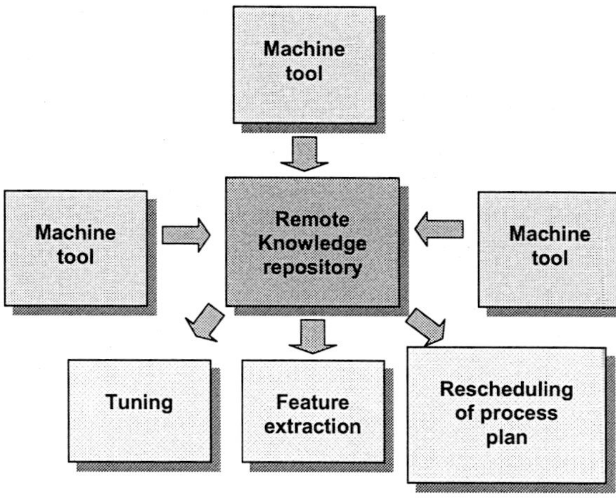

Figure 3: Logical architecture of system for diagnosis

The Knowledge Repository and associated tools for new knowledge discovering will be able to work in four main areas:

4.1 Failure detection and prevention

The goals of this area are:

- To find out the most likely reasons for the fault, and guide the service engineer to easily identify them. For instance, a defect found on a machine can help in the diagnosis of a similar defect on a similar machine.
- To identify the presence of defective stocks.
- To increase the comprehension of malfunctions and breaks.
- To identify the critical parts in the design of products and to improve it by re-designing them, this way increasing reliability and technical quality. For instance, if the comparison of data coming from similar machines shows that the same sub-component often breaks down, the conclusion can be drawn that this component should be substituted by a different one, this way improving the product design.

4.2 Evaluation of the Residual Life of Mechanical Components

The main objective of this component is the evaluation of the residual life of mechanical components, calculated as a percentage. This is a very complex task for several reasons:

- It is affected by different kinds of factors (climate, typology, rate of use etc.).
- Historical data are usually missing or cannot easily be interpreted.
- A mechanical axis is composed of a certain number of sub-components. Sometimes the influence of the degradation of one of them will be preponderant, in other times it will be negligible, and the influence of another component will be more significant.

The easiest way to perform the evaluation of the residual life is through a set of thresholds on the health indicators calculated through the Data Mining procedure. But obviously the setting of values for the thresholds is an extremely complex task even for very experienced engineers. Thus it is necessary to implement an "intelligent" way of performing the evaluation. The idea is to use fuzzy logic to dynamically modify these thresholds as a consequence of the previous evaluations made on similar machines (or on machines where the same mechanical components are used). As a consequence, this system will be provided of a learning ability.

The result will be an evaluation of the residual life and as a consequence a decision recommended to the operator.

4.3 Tuning of machine parameters

Tuning of machine parameters requires to hold a deep "understanding" in the form of models and specification values that describe abilities, borders and specialties of machine types and even of machines themselves. Some of them can be derived from specifications available – even in written form – from the manufacturer of machine tools. Others can only be acquired while the machine is running or be derived from statistical data acquired during the runtime of the machine. Other parameters may not be specific to machine but to the application field of the product, like the automotive or aerospace industry, defining different requirements towards exactness etc. All these classifications have to be integrated in one descriptive model to build relations and draw implications from these about the influence of the tuning parameters and to automatically keep the system learning and models growing while the machines are used.

4.4 Rescheduling of the process plan

Processes can be modelled using different representations like EPC (Event-driven Process Chains) for high-level processes down to NC data on machine level. For integration, all the different models, representations and different states need a common process-driven, i.e. flow-centred, meta-model. This common meta-model is needed to specify and recognize connections and interdependencies between process-steps, machines, tools, etc. Rescheduling of process plans is based on time definitions and restrictions that allow for calculating the processing time for a task on a specific machine. A very complex and not fully describable and accomplishable problem is the fulfilment of a special type of task with a different machine type not planned or even over-qualified for this task (maybe even not initially built for this task) under economic restrictions. Like using a full-featured production centre for the task of a milling machine, which will do the job but increase cost tremendously.

Additionally, these modules should evolve by learning from the results of previous operations, which will require a wider information database for better and faster learning capabilities of the software. Even more important is the definition and implementation of a common model that requires the definition of a common interface, exchange mechanisms and an interpreting software.

Static and dynamic learning and decisions making mechanisms like rule-based reasoning, case-based reasoning, fuzzy logic, neuronal networks etc. have to be evaluated in order to find a feasible solution to the operative problems in the described production environment.

5. Conclusion

Figure 4 shows the resulting architecture for the whole system. After data collection from the sensors the data will be partially processed locally and after filtering sent to the SCC platform for storing and high-level analysis. The system then will apply its decision-support techniques and if necessary send notifications to control systems. Using the control systems it will always be possible to visualize the whole industrial plant as a colour-coded map, see the production machines connected to the SCC platform, check their status and if necessary modify it. Visualization of

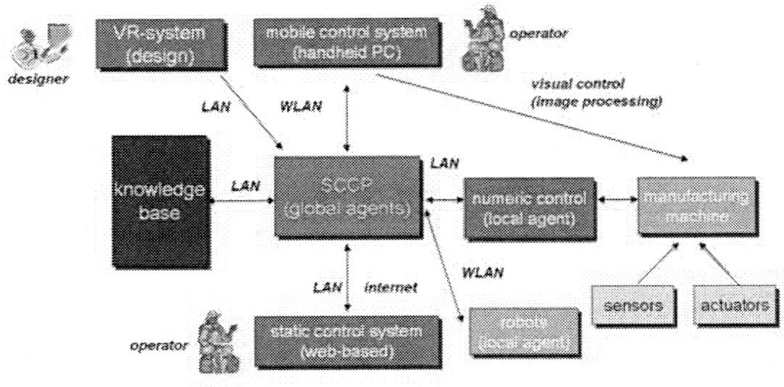

Figure 4: Complete architecture of system

robots and their routes with possibility to change them is also planned. The main modules are:

5.1 Mobile control system

This system works on a small tracked wearable PC or PDA with an additional camera, which can be used by operating personnel directly at the industrial plant. It has global view over the whole industrial plant and local view for machines or parts analysis. The system works in real- or near real-time and provides additional information support for the personnel, like escape routes in case of emergency etc.

5.2 Static control system.

There is a web-server integrated with the SCC platform, which allows connections from both local computers and other computers connected to Internet. Different security policies can be applied.

5.3 Knowledge base repository

This is basically a remote DataBase where several data are stored and that is able to transform information into knowledge, and learn by the correlation of data.

5.4 Smart-Connected-Control Platform

This communication platform allows the exchange of information among all system components: control systems, local agents, global agents, knowledge repository and visualization tools.

In this article we presented our ideas about new approaches for predictive maintenance and diagnosis of production tools, which play an important role in the

INT-MANUS project.

A success of this philosophy will represent a significant breakthrough in the maintenance and diagnosis of machine tools, but will also pave the way for different kinds of applications, all of them sharing the target of increasing the efficiency of production process.

Acknowledgements

"Intelligent Networked Manufacturing System" (INT-MANUS) is a project funded by the European Commission under the joint priority IST-NMP (NMP2-CT-2005-016550).

The authors wish to thank the other partners of the consortium: CIM-EXP (HU), Fundación Fatronik (E), ROBOSOFT (F), CRF (I), Vertigo Systems GmbH (D).

Fidia S.p.A., Fraunhofer IAO, Fundacion Fatronik, ROBOSOFT and CRF are partners of the EU-funded FP6 Innovative Production Machines and Systems (I*PROMS) Network of Excellence. http://www.iproms.org

References

[1] Emmanouilidis C., Jantunen E., MacIntyre J., 2004, Flexible Software for Condition Monitoring, Incorporating Novelty Detection and Neural Network Diagnostics, IMS Forum 2004

[2] Audisio E., Borello C., Borgarello L., Gambera M., 2004, A statistical approach to "Prognostic", IMS Forum 2004

[3] Arnaiz A., Arana R., Maurtua I., Susperregi L., 2004, Maintenance: future technologies, IMS Forum 2004

[4] Sauer O., 2004, Agent technology used for monitoring of automotive production, IMS Forum 2004

[5] Nacsa J., Haidegger G., 1997, Built-in intelligent control applications of open CNCs, IMPS 1997

[6] Carnero M.C., 2005, Selection of diagnostic techniques and instrumentation in a predictive maintenance program. A case study, Decision Support System, 38, 539-555

[7] Mobley R. K., 2002. An introduction to predictive maintenance – Second Edition. Elsevier Science, USA

[8] Wang H., 2002, A survey of maintenance policies of deteriorating systems. European Journal of Operational Research 139, 469-489

[9] Kopácsi S., Kovács G., Anufriev A., Michelini R.' 2006, Ambient intelligence as enabling technology for modern business paradigms. Science Direct

[10] Meo F., Foursa M., Schlegel T., Kopàcsi S., et al., 2006, Report on end-user requirements – Draft. Deliverable 7.1.1 of the INT-MANUS project

[11] Meo F., Foursa M., Schlegel T., et al., 2007, Report and software suite for the first research prototype. Deliverable 7.5.1 of the INT-MANUS project

Innovative Production Machines and Systems
D.T. Pham, E.E. Eldukhri and A.J. Soroka (Eds.)

The sheet thickness effect on springback amplitude of a U-shaped part made from tailor welded stripes

A. Albut

Faculty of Engineering, University of Bacau, Calea Marasesti 157, Romania

Abstract

This paper presents the results obtained by numerical simulation regarding springback phenomenon of a U-shaped part manufactured from tailor welded stripes. The final shape of the formed part is seriously affected by springback phenomenon. This paper work is trying to prove the important role the metal sheet thickness on the springback effect. The U-shaped part has different springback values for each material from the welded assembly structure. The influence of the sheet thickness on the tailor welded stripes springback is examined by finite element method using Abaqus Standard for forming process and Abaqus Explicit for springback of the obtained part.

Keywords: Tailor welded stripes, metal sheet thickness, springback.

1. Introduction

Recently, the automobile industries have been trying to develop various types of model and high-quality low-cost cars to meet the customer's requirements and to find new ways of establishing this goal effectively. For the purpose of achieving the above presented objectives, different methods using various welding processes (such as laser-welding, mash seam-welding processes, etc.) were developed. A tailor welded blank consists of two or more sheets that have been welded together in a single plane prior to forming. The sheets can be identical, or can have different thickness, mechanical properties or surface coatings [1]. Thickness of tailor welded sheets plays an important role in sheet metal forming since fracture, wrinkling and weak spots are strongly influenced by material behaviour.

The techniques of numerical analysis applicable for sheet metal forming have been considerably developed for the last several years. However, accurate prediction of the springback remains elusive [2]. In finite element method (FEM) models of metal forming, the roughness has usually been assumed to be constant; even though it is commonly observed that sheet drawn under tension over a tool radius results in the surface becoming shiny, indicating a major change in surface morphology.

Many studies presents a wide range of information about the formability and failure patterns of welded stripes. A wide range of information about the formability and failure patterns of tailor-welded stripes and the springback of non-welded sheet metal parts has been presented. However, the springback characteristics of tailor-welded stripes have hardly been found [3]. Published results on springback prediction of tailor welded stripes are minimal. The welding line was insignificant influence when is placed perpendicular to the direction of the deformation force [4].

Since the springback is also affected by the material properties, such as Young's modulus and initial yield stress, the process design for tailor-welded stripes is more complicated than a homogenous stripe.

Though novel approaches relating to the formality of tailor-welded stripes are available, the change of springback due to the characteristic of each process should be verified by finite element method.

In this study, the tailor welded stripes with two types of material having the same thickness, are used to investigate springback characteristics in U-shape bending.

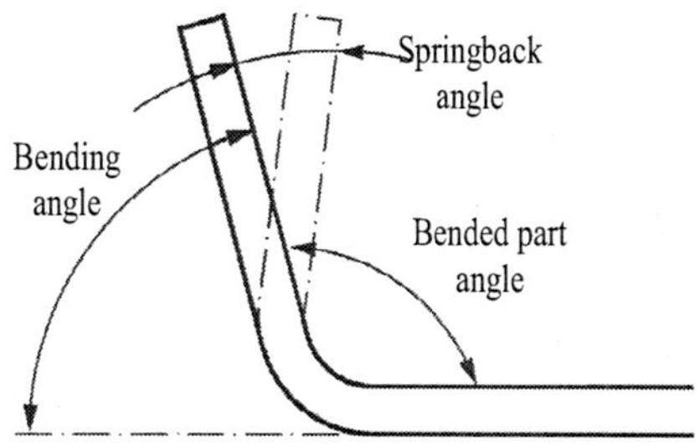

Fig. 1. Scheme of the bending process and springback

Springback (Fig. 1) is mainly influenced by the sheet thickness, the punch and die profile radii, initial clearance between punch and die, friction conditions, rolling direction of the materials, blankholder force, material properties (elastic modulus, Poisson's coefficient, constitutive behaviour in plastic field) etc. The purpose of this study was to investigate the sheet thickness influence on the springback effect of the tailor-welded stripes. To achieve this goal, simulation tests were carried with different sheet thickness of the laser welded assembly.

2. Material properties

Simulation of the forming processes based on finite element method using ABAQUS software requires as input data the mechanical properties of the used materials. Determination accuracy of the mechanical properties has an important influence on simulation results.

To obtain the mechanical properties of the base materials and of the welding line, tensile tests were performed on a universal mechanical testing machine, equipped with Hottinger force cells of 5 tf and a Hottinger – Baldwin electronic measurement system for PCs – type Spider 8. The data acquisition, processing and visualisation were performed using Catman Express software. The measurement of specific strains for determination of stress - strain curves was performed using an Epsilon extensometer for a strain rate of 0.1 s^{-1}.

The specimens were cut as a function of the rolling direction being achieved sets of specimens corresponding to the direction of 0° and were worked by milling and grinding in order to obtain the prescribed dimensions. The reference length of the specimen was equal to 50 mm. To obtain a good accuracy of the results, 3 specimens were tested for each determination.

To determine the mechanical properties of the welding line, from the original TWB, a 4 mm wide stripe which includes the welding line, has been removed using EDM wire cutting (Fig. 2).

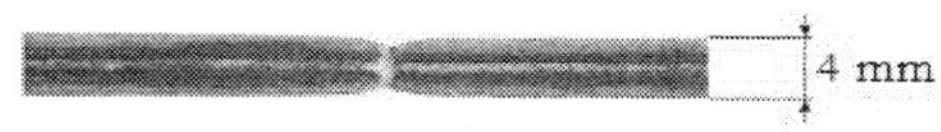

Fig. 2. Sample used to determine the mechanical properties of the welding line

The flow stress and true total strain were calculated for each recorded couple of the force and displacement. The total strain was decomposed on elastic strain and plastic strain using determined Young module.

To obtain a better accuracy of FEM modelling, especially in the range of small deformation, there was resigned the functional stress – strain curves in favour of the stress – strain curves in the numerical form.

The mechanical properties of FEPO steel and E220 steel determined for 0° material rolling direction are presented in Table 1. In the table below are presented also, the mechanical properties of the welding line. In Figure 3 are presented the stress – strain curves for FEPO ans E220 steels and for the welding line material.

Table 1.
Mechanical properties

Property	FEPO	E220	weld
Yield strength $\mathbf{R_{p0,2}}$ [MPa]	203	268	252
Tensile strength **Rm** [MPa]	381	458	417
Percentage elongation after fracture $\mathbf{A_{80}}$ [%]	–	35,3	–
Elongation for max. load $\mathbf{A_{gt}}$ [%]	31,8	20,4	17,3
Strain-hardening coefficient **n**	0,222	0,190	-
Plastic strain ratio **r**	1,860	1,420	1,29
Poisson's ratio **ν**	0,286	0,297	0,278
Young modulus **E** [MPa]	200825	204271	203253

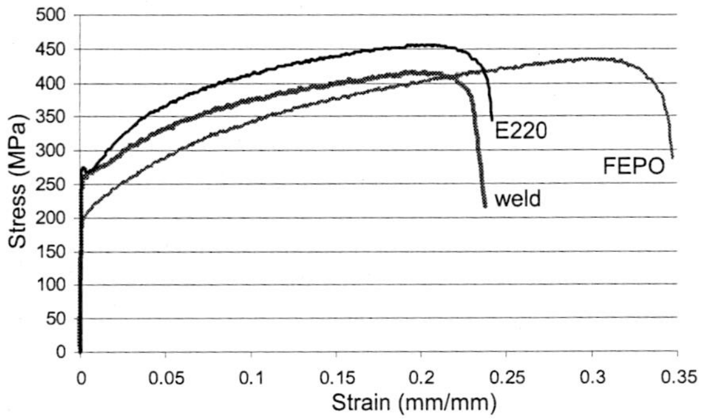

Fig. 3. Stress – strain curves for 0°

The materials microstructure has been analysed using a metallographic microscope with a magnification of 100X for base materials and 500X for welding line. The materials have uniform, typical microstructure with fine grain. The microstructure is shown in Figure 4.

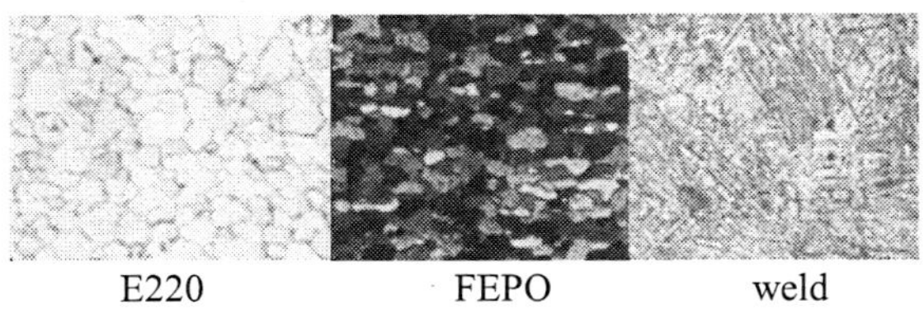

Fig. 4. The material microstructure

3. Simulation of the forming process

The simulation of U-shape part forming was made using finite element method. The objective is to create a model that allows an accurate prediction of springback intensity, stress and strain state at the end of the forming process. The analyzed geometrical parameters are sidewall radius ρ and springback angles θ_1 and θ_2.

The tailor welded stripes used in the simulation tests were made from FEPO and E220 steel. Stripes of 350 mm length and 30 mm wide, with thickness variation from 0,8 till 1,3 mm (Fig. 5).

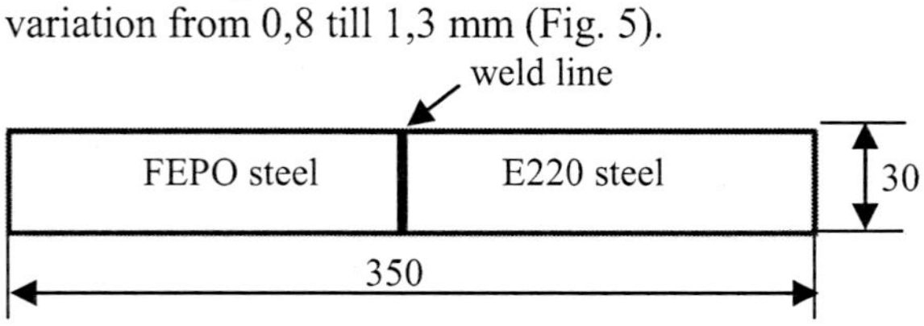

Fig. 5. Tailor welded strips with transversal weld line

The material was modelled as elastic-plastic, where elasticity is considered isotropic and plasticity is modelled as anisotropic using Hill quadratic anisotropic yield criterion.

In this simulation, fine mesh technology was applied to interested regions such as the punch profile radius and the tailor-welded strip model to improve the accuracy of the analysis for the springback.

The geometrical model is presented in Figure 6. The sheet was modelled as deformable body with 400 shell elements (S4R) on one row with 5 integration points through the thickness. The tools (punch, die and blankholder) were modelled as analytical rigid because they have the advantage of reduced calculus efforts and a good contact behaviour. Rigid body movements are controlled by reference points.

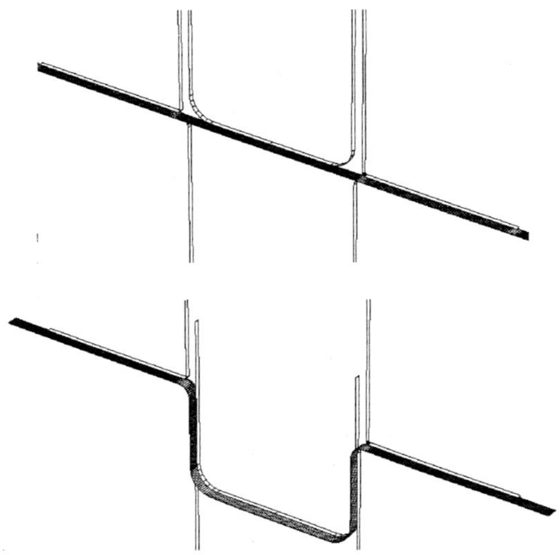

Fig. 6. Geometrical model

The boundary conditions imposed to the tools were intended to describe the experimental conditions as accurate as possible. For contact conditions a modified Coulomb friction law combined with penalty method was used.

For determination of the sheet thickness influence on springback phenomenon, the simulations have been done under the following conditions: the sample model is loaded with the material characteristics corresponding to 0° rolling direction; blank holder force F = 15 kN; the friction coefficient between the specimen and tools surfaces is 0.1, the thickness of the sample model are 0.8, 0.9, 1.0, 1.1, 1.2 and 1.3 mm.

Springback parameters that were observed during the simulation tests are presented in Figure 7:

- θ_1 – sidewall angle between real profile and theoretical profile;
- θ_2 – flange angle between real profile and theoretical profile;
- ρ – curvature radius of the sidewall.

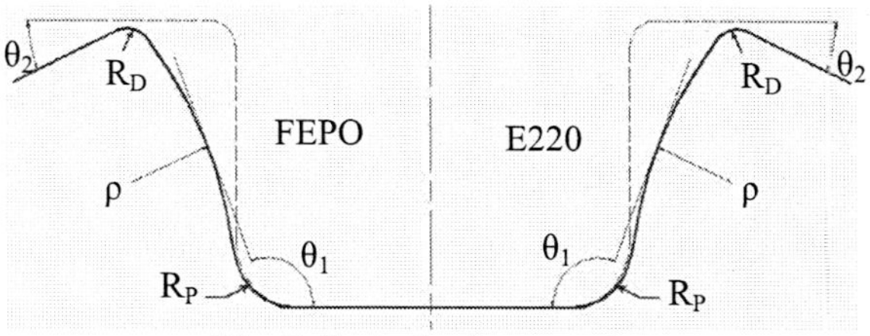

Fig. 7. Springback parameters

4. Simulation results

In Figure 8 are presented the tailor weld blanks with different thickness, after the forming tools have been removed. Parts having different thickness are affected in a dissimilar mode by the springback effect.

The values of springback parameters are recorded in Table 2.

5. Conclusion

From the analysis of Figure 8 and Table 2, the following observations can be presented concerning the influence of the sheet thickness on springback parameters:

- the modification of sheet thickness leads to important variations of springback parameters;
- increasing of the sheet thickness results in reduction of the springback effect, the final geometry of the formed part is closer to the ideal part shape;
- variation of the springback phenomenon proportional with sheet thickness is observed for both areas of the part;
- for sheet thickness over 1.0 mm the sidewall radius increases so much, that it can be considered a straight line;
- the springback parameters from the part zone made from FEPO steel record smaller values for both angles θ_1 and θ_2 with respect to the E220 part area;
- the part area made by E220 present a springback intensity higher that the part area made from FEPO steel.

References

[1] G.Y. Li, M.J. Tan, K.M. Liew – Springback analysis for sheet forming processes by explicit finite element method in conjunction with the orthogonal regression analysis, Int. Journ. of Solids and Structures, 36, 1999

[2] L. Papeleux, J.P. Ponthot – Finite element simulation of springback in sheet metal forming, J. of Mat. Proc. Tech., 125-126, 2002

[3] N. Yamamura, a.o. – Springback simulation by the static explicit FEM code, using a new algorithm for canceling the non-equilibrated forces, Simulations of Materials Proc., Swets & Zeitlinger, Lisse, 2001

[4] F. Pourboghrat, E. Chu – Prediction of spring-back and side-wall curl in 2-D draw bending, J. of Mat. Proc. Tech., 50, 1995

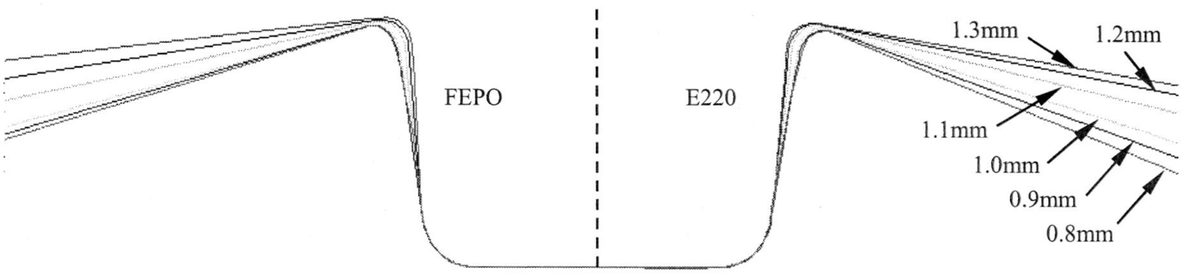

Fig. 8. Shape of the TWB deformed by springback phenomenon

Table 2. Springback parameters

TWB thickness [mm]	FEPO steel Angle θ_1 [degrees] Nominal value	Obtained value	Angle θ_2 [degrees] Nominal value	Obtained value	Sidewall radius [mm] Nominal value	Obtained value	E220 steel Angle θ_1 [degrees] Nominal value	Obtained value	Angle θ_2 [degrees] Nominal value	Obtained value	Sidewall radius [mm] Nominal value	Obtained value
0.8	90	100.6	0	14.9	∞	205.0	90	102.8	0	19.3	∞	188.4
0.9		99.7		14.1		307.6		101.0		17.5		265.7
1.0		99.0		13.4		385.8		100.1		15.6		322.9
1.1		96.8		10.3		1524.0		98.2		11.8		818.5
1.2		95.1		7.6		2875.5		96.7		9.5		2106.6
1.3		93.6		5.5		3347.3		95.6		8.3		3071.5

Innovative Production Machines and Systems
D.T. Pham, E.E. Eldukhri and A.J. Soroka (Eds.)

Use of HSS tools for milling of Titanium alloys

P. Bach

Czech technical University, Faculty of mechanical Engineering, Research Centre for Manufacturing Technology, Prague, CZ

Abstract

There is a possibility to increase MRR significantly using low cutting speeds and HSS tools for milling of Titanium alloy Ti6Al4V. The HSS tools with irregular teeth pitch gain the MRR all more. The wear of the tools is satisfactory low.

Keywords: Milling, Titanium alloy, HSS tools, Stability limit, MRR

1. Introduction

Titanium alloys are obviously machined using carbide tools. Cutting conditions are for example v_c=40-120 m/min, f_t=0, 05-0,1mm/z, a_p= 2-8 mm, a_e=0, 1- 0, 5 mm. Some researchers try to achieve a higher Metal Removal Rate under very high cutting speeds above 120 m/min. They believe the cutting forces drop thanks high temperature in cut. In all those trials a tool wear of carbide inserts used is within tens of minutes maximally. Obviously, the flank wear is not higher then 30 minutes in the best case. It is known that the low thermal conductivity of titanium alloys generates high temperatures when cutting under the speed range mentioned above, [14], [15] and many others.

In reality, stability against chatter is another problem on the way to higher MRR. Considering the speed range used for titanium, which is relatively low, a stability limit depends on the tool stiffness first of all. Naturally, the spindle and tool-holder stiffness play also an important role.

But, there is a possibility to use also HSS tools for titanium machining. This kind of tools can cut titanium till speed of 80 m/min, but the better cutting conditions are v_c=15 – 35 m/min, f_t=0,1mm/z, a_p=10 – 80 mm, a_e=0,3 – 1 mm. Under this condition the performance of the HSS tools is better than the performance of the carbide tools. The reason for this is higher stability limit. Simultaneously, flank wear is lower. Further more, modern High Speed Steels presents very promising properties regarding tool wear.

2. Theory and dynamic properties of tools

Performance of any machining operation is limited not only by machine tool power and tool geometry but also by dynamic properties of the entire machine tool structure. When milling the spindle-holder-tool system (SHT) stiffness is of great importance in this respect, [4]. Process of machining can be stable or instable. Instability means chattering of the machine tool structure as showed in [1], [3] and in many other references from [9] to [13.]. For milling, special stability diagram can be calculated to find stable depth of cut (DOC) and stable ranges of spindle speed. DOC level depends on dynamic stiffness of the tool – spindle mechanical system, [5], [6]. Titanium milling needs a stiff SHT structure powered by a strong motor. The dynamic flexibility less then 0, 55 μm/N is proper value for milling of titanium, [7]. Effect of the wave length and chatter frequency rate is considered as a cause of higher stability limit within low cutting speed. In this case a tool flank can be in touch with the wave surface generating a friction force, which gains damping of the system and consequently means rise of stability limit.

3. Testing the CPM tool performance under a low

cutting speed

The titanium alloy Ti6Al4V has been machined using HSS cutters made of various materials presented in the **Tab. 1**

Tab. 1 Cutters

Tool material	Provider	Clamping
HSS Co8,	Co. Bohler	Weldon
ASP 2030,	Co. Erasteel	Weldon
CPM REX T15,	Co. Zapp	Weldon
CPM REX 76,	Co. Zapp	Weldon

The High Speed Steel Co8 is the steel with 8% of cobalt as the main alloy element. This steel is of common use for titanium machining and serves here as a reference material. For production of the steels with the trade marks ASP and CPM REX the technology of powder metallurgy is used. The steels ASP 2030, CPM REX T15 and 76 contain cobalt too but except of this also higher percentage of other alloy elements.

All tools had the same body and edge geometry, 4 teeth, 32mm diameter and 106 mm length of cutting part. One half of tools had a regular pitch of teeth while the second half had an irregular pitch. The irregularity was within range of two grades. Helix angle was uniform. All tools used had quite obvious quality of balance. One example of HSS tool used for the presented research is in **Fig. 1**.

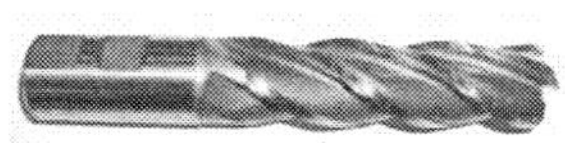

Fig. 1 The HSS cutter used for testing

4. Stability tests

Firstly, stability of the cutting process, using HSS tools has been measured for varying cutting speed, chip load f_t=0, 1 mm per tooth and a_e=1 mm.

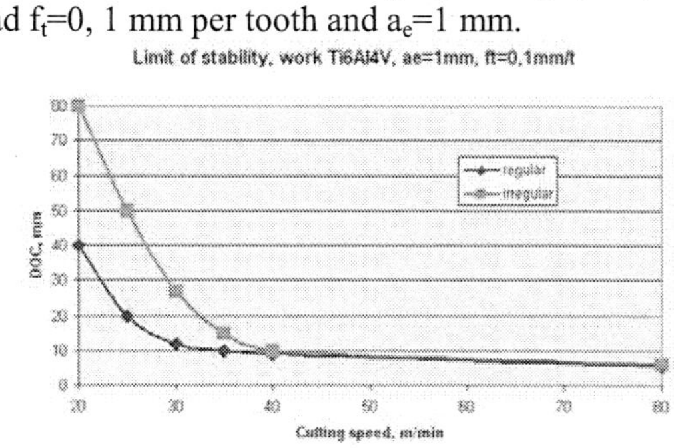

Fig. 2 The stability limit of the tools with regular and irregular pitch. Tool clamping: ISO50. Tool material: CPM.

The stability limits, measured by axial depth of cut (DOC) for low cutting speeds below the value of 40 m/min are significantly higher than those for the speeds above 40 m/min, as shown in the **Fig. 2**. This is caused by effect of so called process friction. An idea of milling titanium under low cutting speeds is based exactly on this fact, whereby performance of machining is increased.

Performance can be measured by Metal Removal Rate (MRR). It has been evaluated for stability limit using the formula

$$MRR = a_p.a_e.f_t.n.10^{-3} \qquad (1)$$

where a_p is DOC close to stability limit, a_e is radial depth of cut, n is tool speed, f_t is chip load and z is number of cutter's teeth. The technology of down-milling was used in all cases with chip load 0, 1 mm per a tooth, a_e=1 mm and speed varying according the charts in the **Fig. 2**.

Under the stability limit, we understand maximal DOC which is possible to use without chatter rising during milling. If the cutting speed is significantly low, within the range 10 – 30 m/min, the chatter doesn't occur till relatively high values of the DOC as shown in the same figures. This fact increase MRR to values despite of the cutting speed drop. The MRR values are higher than the values achieved by speeds above 40 m/min as depicted in the **Fig. 3**.

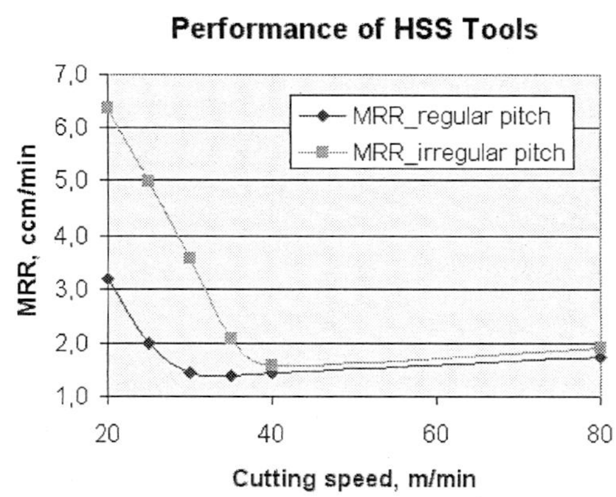

Fig. 3 MRR of a tool made of CPM REX T15

Thus, the HSS tools can achieve much higher performance than carbide tools that are used under cutting speeds above values of 40 m/min.

4. Testing HSS tool life

Naturally, tool life is a decisive parameter for practice. The investigated flank-wear VBB (average flank wear) and VB_{max} of HSS tool seems to be very promising. The results are shown in the following figures **Fig. 4**, **Fig. 5**, **Fig. 6** All the tools have been working under the same condition: cutting speed v_c=20 m/min, chip load f_t=0, 1 mm/tooth, a_e=1 mm, down milling.

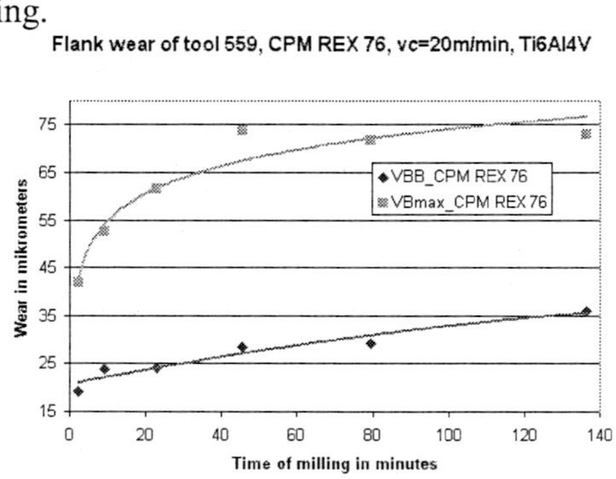

Fig. 4 Flank wear VBB and VB_{max}

It is evident that the PM materials (ASP2030, REX T15 and T76) have different characteristics of the flank wear than the cobalt steel Co8. The top wear values of PM materials are significantly lower and the curves are flat. After approximately 40 – 50 minutes, the wear has stabilizing tendency. The results of VB_{max} have not such a difference in terms of the highest values. **Fig. 6**.

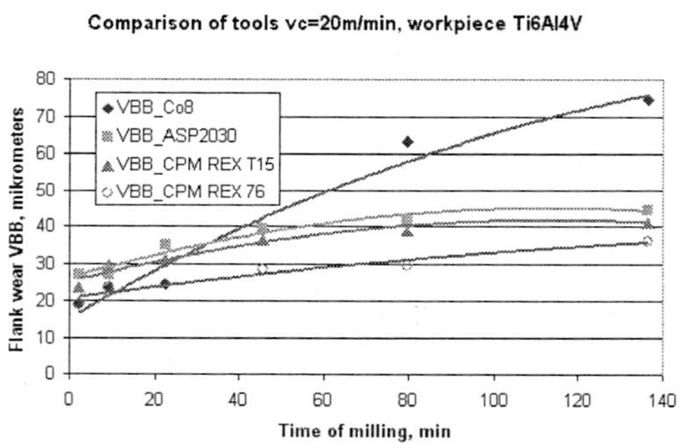

Fig. 5 Flank wear of the tested tool materials

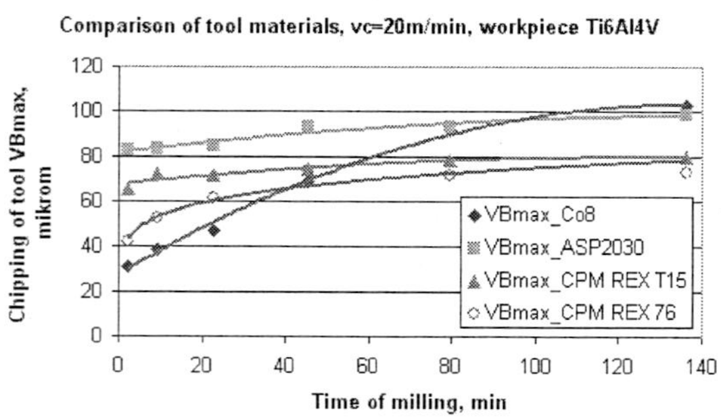

Fig. 6 VB_{max} of tested tool materials

The time of milling shown in the **Fig. 5** and **Fig. 6** has been prolonged up to 240 minutes with result in **Fig. 7**. It has turned out, that using cutting speed 20m/min, the flank wear is stable till 205 minutes. After the limit the flank-wear is climbing strongly.

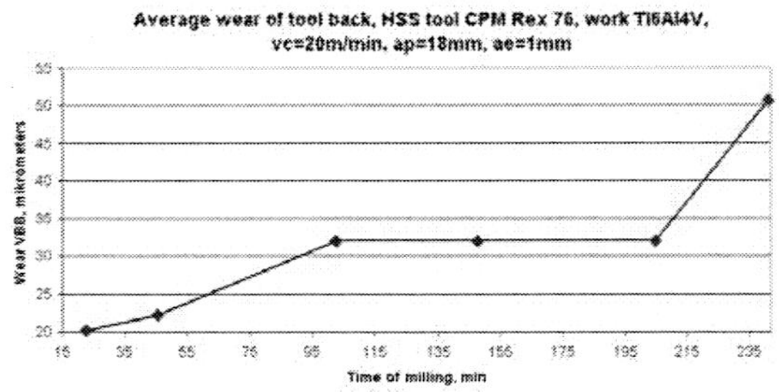

Fig. 7 Averaged flank wear

5. Regular & irregular pitch

The difference between wear of tools with regular and irregular pitch is significant as shown in **Fig. 8**. The irregular pitch of teeth makes the work of tool smoother, with lower level of vibration. Consequently, the tool life is increased.

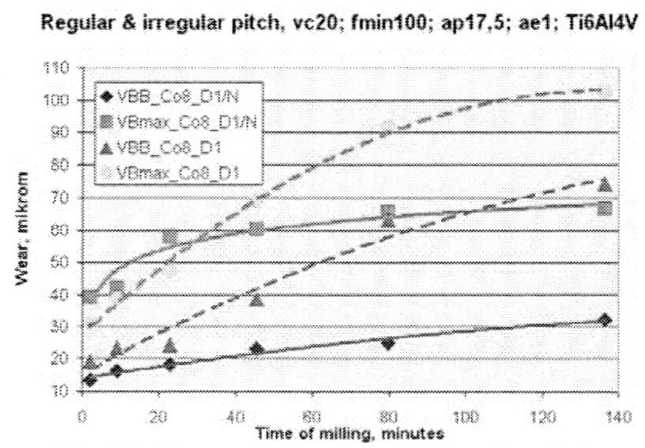

Fig. 8 Influence of regular and irregular pitch

6. Summary

The presented HSS cutters can not fully substitute carbide tools for titanium milling, but they have the following features.

Using this kind of tools, higher MRR can be achieved. Especially this geometry feature, the irregularity of pitch is very useful in this respect.

When cutting speed is very low, temperature in the cut is significantly lower comparing it to temperature, which occurs when cutting speed is above 40 m/min. Consequently, there is no hardening of the work surface. This fact allows us to use any cutting strategy in terms of cut number or chip thickness.

The HSS tools can be designed with long cutting edge, which is useful for higher MRR. The long edge allows us to use high axial depth of cut and thus specific load of the edge is low. Importance of this emphasizes [16]. Usage of small radial dept of cut leads to better cooling of cutting edge.

The flank wear is lower comparing it with the wear of carbide tools. There is very low cutting edge chipping as well. This fact is a feature of the fine-grained powder metallurgy materials.

Even better results of wear are expected from the current research of coated HSS tools.

There is one important dynamic condition to achieve good results of MRR. This is high dynamic stiffness of SHT system. This fact results from any of experiments dealing with titanium milling. Stiffness of spindles for machine tools is basic property of such a structure as showed for example in [17], [18], [19], [20], [21].

Acknowledgements

The company ROBERT ZAPP WERKSTOFFTECHNIK GMBH from Germany provided the materials CPM REX T15 and CPM REX 76. The company ZPS-Milling Tools provided materials Co8 and ASP 2030 and made all the cutters. For tool performance testing the company Kovosvit-MAS, Sezimovo Usti, CZ has provided the powerful CNC milling centre MCV 1270R equipped with spindle POWER using ISO50 for tool clamping. This research has been founded within the frame of the grand 1M6840770003 of the Ministry of Education and Youth of the Czech Republic.

References

[1] Tlusty J., Polacek M. 1963. The Stability of the Machine Tool Against Self Excited Vibration in Machining, Prod. Eng. Res. Conf. ASME, Pittsburgh, pp. 16 – 28.

[2] Slavíček J., 1965. The Effect of Irregular Tooth Pitch on Stability of Milling, Proceedings of the 6th MTDR Conference, Pergamon Press, London, pp. 15-22.

[3] Tlusty J. 1999. Manufacturing Processes and Equipment, Prentice Hall, New Jersey.

[4]Agapiou J. S., 2003. Selection of Tool Holding System for a Machine Tool Spindle, Proceedings of the 3rd SME Conference, 28 pages, USA.

[5] Bach P., 2003. Milling Machine Tool Stability and Performance, Proceedings of International Conference LAMDAMAP, pp. 25-35, Huddersfield, UK.

[6] Bach P., 2004. Modelling of Milling Spindles for Optimizing the Spindle Cutting Performance, Proceedings of 34th International Conference MATADOR, Manchester, p. 25-35, UK.

[7] Bach P., 2007. Milling of Titanium Using HSS Tools, Proceedings of the conference Milling, Brno, CR, pp. 87 -100.

[9] Daněk O., Poláček M., Špaček L., Tlustý J., 1962. Selbsterregte Schwingungen am Werzeugmaschinen. VEB Verlag Technik, Berlin.

[10] Tobias S.A., 1964. Vibration of Machine Tools. Product Engineering, Vol. 43, No.12, pp. 599-608.

[11] Tlusty J., Zaton W., Ismail F., 1983. Stability Lobs in Milling. Annals of CIRP Vol.32/1, pp. 309 – 313.

[12] Koenigsberger F., Tlusty J., 1970. Structure of Machine Tools. Pergamon Press, Oxford.

[13] Tlusty J., 1980. Criteria and Methods in Structural Analysis. Machine Tool Task Force, Lawrece Livemore National Laboratory, SME.

[14] Schulz H., 1996. Hochgeschwindigkeitsbearbeitung. C.Hanser Verlag, München-Wien.

[15] Schulz H., 2001. Scientific Fundamentals of HSC. C.Hanser Verlag, München-Wien.

[16] Trmal G. J., 2000. High Performance Machining of Aerospace Materials. Proceedings of conference MATAR, str. 127 – 134. Praha.

[17] Popoli B., 2003. Selecting the Proper Bearing System for Your High Speed Spindle Application. Proceedings of the 3rd SME Conference, (DSP), 32 pages, USA.

[18] Smith S., 2003. Spindles for Airframe Machining. Proceedings of the 3rd SME Conference, (DSP), 26 pages, USA.

[19] Brecher Ch., 2004. Grundlagen der Spindelkonstruktion. Seminar „Gestaltung von Spindle-Lager-Systeme“, No 4, 29 Seiten, FIP, Aachen.

[20] Kolar P., 2004. Mathematical Modelling of Machine Tools Spindles. Proceedings of Conference MATAR, pp. 33 – 39, Praha.

[21] Ch. Brecher, 2004. Main Spindle Systems for HSC. Proceedings of Conference MATAR, pp. 19 – 24, Praha.

Innovative Production Machines and Systems
D.T. Pham, E.E. Eldukhri and A.J. Soroka (Eds.)

Behaviour and integration of service-oriented automation and production devices at the shop-floor

J.M. Mendes[a], A.W. Colombo[b], P. Leitão[c], F. Restivo[d]

[a] *Department of Electrical and Computer Engineering, Faculty of Engineering - University of Porto, Rua Dr. Roberto Frias s/n, 4200-465 Porto, Portugal*
[b] *Schneider Electric GmbH, Steinheimer Str. 117, D-63500 Seligenstadt, Germany*
[c] *Department of Electrical Engineering, Polytechnic Institute of Bragança, Quinta Sta Apolónia, Apartado 134, 5301-857 Bragança, Portugal*
[d] *Department of Informatics Engineering, Faculty of Engineering - University of Porto, Rua Dr. Roberto Frias s/n, 4200-465 Porto, Portugal*

Abstract

Automation and manufacturing systems are changing in the direction of cooperative ecosystems with heterogeneous entities. An important feature that should be considered is the vertical integration from the business needs down to the shop-floor where the real action takes place. This paper analyses the integration of shop-floor devices into the IT-enterprise but maintaining also a certain degree of independence in terms of behaviour. Service-oriented paradigm is used as the main backbone due its proven merits in the business levels and recently also in automation and production systems. In the provided example, High-Level Petri Nets demonstrate a set of useful features, namely the partial behaviour description and analysis and some parameters in the integration. The resulting application leads to an easy integration of autonomous devices in the IT-enterprise, taking especially in account the requirements of the shop-floor level.

Keywords: Industrial Automation, Service-oriented Architectures, IT-integration, High-level Petri Nets.

1. Introduction

New Information Technologies (IT) have begun to have a strong presence in the new generation of manufacturing systems. After decades of parallel development, the paths of Information Systems Tools and Manufacturing Systems are converging to provide the impetus that will allow the integration of the total business enterprise [1]. A special attention is given where the main action relies (the production itself) and its incorporation in the IT-enterprise.

Several proposals have been made to enhance the flexibility, usability and integration, from Holonic and Multi-Agent Systems (see [2], [3] and [4]), High-Level Petri Nets (see [5]), to the more novel Service-oriented production systems (see [6] and [7]). One promising approach that has the potential to surmount the technical, organizational and financial limitations inherent in most current approaches, is to consider the set of production entities as a conglomerate of distributed, autonomous, intelligent and reusable units, which operate as a set of collaborating entities [8]. Each of these entities is typically constituted by hardware - mechatronics, control software and embedded intelligence, and is able to communicate with others. From a functional point of view, each collaborative unit can, at each time, initiate actions and dynamically interact with other ones in order to achieve both local and global objectives, considering that they are within a cross-layer infrastructure like a manufacturing enterprise (see [9] and the references therein).

Contributions have been made about the integration of service-oriented components from the factory's shop-floor into the IT-enterprise, in order to provide vertical information and control sharing. One possible solution is that these components provide the required services to the upper levels of the enterprise and thus are controlled by them. This fits with the Service-oriented paradigm, but the traditional master-slave hierarchy from the top-down perspective is not favourable when speaking about collaborative and proactive devices. A partial solution is a mix of autonomous devices from the shop-floor level that may provide services, but also may request services provided by the other levels, such as Decision Making Systems (DMS), Manufacturing Execution Systems (MES, see [10]) and Enterprise Resource Planning (ERP, see [11]).

The aim of this paper is to present a methodology for analysis of the operational behaviour of autonomous service-oriented automation and production devices at the shop-floor and their integration into a flexible IT-enterprise. The goal is to provide a more proactive control to the shop-floor devices, contrasting with the more common service-oriented factory automation systems where devices are only providers of services to the upper levels (such as DMS/MES/ERP).

The paper is structured in the following sections: first, the integration middleware is presented in section 2 to provide an easy and feasible consolidation of DMS/MES/ERP and other upper level systems with the shop-floor. Section 3 goes deeper into the shop-floor, where the automation devices are running and explains the autonomous behaviour of these service-oriented devices. An application example of these concepts using High-Level Petri Nets is given in section 4 and finally section 5 terminates the paper with the conclusions.

2. Integration middleware for service-oriented shop floor automation systems

The proposed approach describes a middleware to integrate service-oriented shop floor automation components into a flexible IT-enterprise architecture. In terms of context, it focuses on automation and production systems based on distributed and reconfigurable devices and their integration from the shop floor level into the higher levels (such as the DMS/MES/ERP systems). The resulting middleware infrastructure manages and interfaces the shop floor to the higher level by using a function vectorial and layout based process control of devices. All inter- and intra-level interactions are service-oriented.

The exposed services represent diverse and sufficient functionalities of the shop floor to be managed and integrated into the higher levels, providing a high order of control and information feedback. Some of the exposed features include, but not limited to: topology information, maintenance, several control operations, conflict resolution, and process analysis and monitoring. The approach is based on a bottom-up perspective (from the devices) to interface the top-down approach from the upper levels. There is a loose-coupled inheritance, so the higher levels are not limit to one specific business activity, but to any depending on the provided features by the middleware. The methodology applies and extends the concept of service-orientation by the adoption and integration from the device level up to the enterprise levels. Since services are available in both sides, the integration middleware is also capable of using the services of the upper-levels when required by the middleware itself or forwarded by some device from the shop-floor.

One main outcome of the approach is that the middleware works as a system integrator for service-oriented shop-floor with other service-oriented components of IT-enterprise architecture (see Fig. 1):

- Integrate service-oriented production devices (Control vendor and Machine vendor).
- Integrate shop-floor with upper-levels, such as production management systems (Control vendor and DMS/DES/ERP).

Each production device (integrating mechatronics, communication and control aspects) is a process controller that offers services that have been defined and developed by the device manufacturer. The sequence and link of services offered by each device follows properties and restrictions that are generated by the hardware (mechatronics), communication and possible control aspects of the device. The set of services offered by a device can be mapped into a set of functions (vectors). Composing the function vectors allows knowing all possible combinations (aggregation

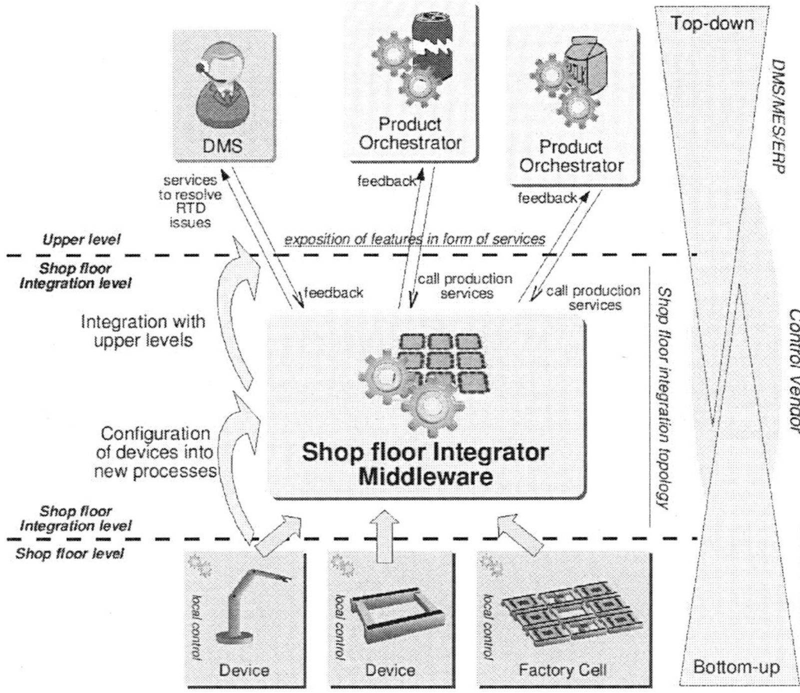

Fig. 1. Shop floor integration middleware in the IT-enterprise architecture.

of services), thus the vector space of the device contains all possible (allowed) individual and aggregated services.

Configuring a layout implies composing devices and of course, formally spoken, composing their vectorial spaces, generating new restrictions and a new set of aggregated services. Again, the analysis of the vectorial space gives all necessary acceptable specification about the complete set of services offered by a given configuration. Vectorial spaces may also represent executable processes bounded to the services that are accessible and manageable by any level that has access to them. Thus, the integrated behaviour is driven by processes that access available services. Processes can be distinguished from the automation processes (at the shop floor), integration processes, production processes to the business and enterprise processes.

After fixing a given layout, a very well defined set of services has been defined. It is possible to speak here about process definition of the shop floor topology. Putting a fixed layout running implies that the sequence of called services will have to respect mechatronics, communication and control restrictions like shared-resources, capacity of processing, concurrency, etc. that are explicitly included in the topology of the aggregated services.

The integration topology for a given automation/production system is now able to be interfaced with external service-oriented components, such as product orchestrator and Decision-making systems (DMS). For example, a defined workflow associated to a product will be able to call services (individual or aggregated), offered by the integrator topology of the shop floor only if the right interface exists and the controller associated to a product recognize which services are to be found in the shop floor integration topology.

3. Autonomous behaviour of service-oriented devices at the shop-floor

From the previously explained integration, nothing was pointed about the devices itself, i.e. their behaviour and how they contribute to the overall system. The addressed automation and production units at the shop-floor are sometimes recognized differently, for instance “modular intelligent automation unit”, “physical agent”, “Holon”, “collaborative automation units”, “Smart Mechatronic Component”, etc. (see [12], [13] and [8]).

The autonomous behaviour methodology given in this section focuses on the setup and running phases in the life cycle of collaborative service-oriented production systems. These systems are composed of distributed, reconfigurable smart production automation devices that expose their functionality as services or aggregation of them. It permits, after the initial setup of the automatic operation of the device and the exposition of their services, the collaborative interaction/cooperation in order to follow flexible and customized workflows associated to the products to be manufactured.

Under the umbrella of the service-oriented architecture paradigm, the services exposed by the production devices at the shop floor have (see Fig. 2):

- Lateral/horizontal (device - device) relationship. Calling services or responding with services as e.g., move-pallet-in, pick-gripper, transport-pallet-up, etc.
- Vertical (device – upper-level) relationship. Calling services offered by upper-level components for dynamic re-scheduling of production/automation tasks, responding with services as monitoring, diagnosis/information indexes, etc.

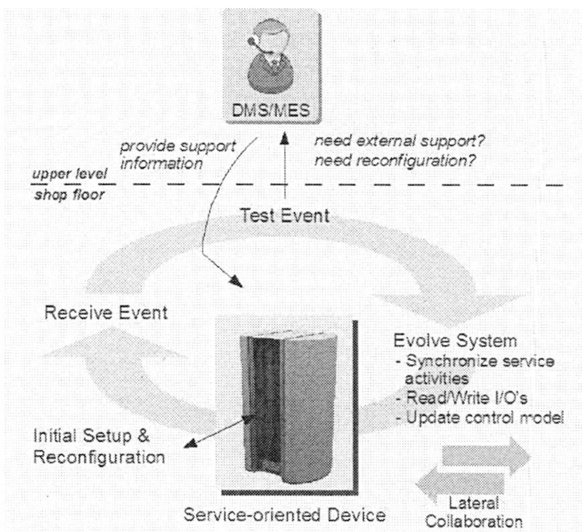

Fig. 2. Concept of the independent behaviour cf autonomous components.

The devices have a degree of autonomy in sense of auto-sustainable control and exposition of necessary services to permit lateral/horizontal collaboration with other devices (aggregating services), requesting and providing services for decision-making information from DMS/MES and integration. All interactions and resource sharing are via service-orientation. There is a loused-coupling inheritance in form of bottom-up perspective (from the devices/shop floor level), enhancing the autonomy and consequent reconfiguration capabilities.

Having the shop-floor mapped into a service-oriented system, the behaviour of each smart production automation device is part of a middleware shell formally specified by e.g. High Level Petri Net (HLPN) models and supported by routines to handle undocumented events and decide over present conflicts in the behaviour of the device. The operational behaviour of these devices follows the token-game of the HLPN (i.e. the behaviour control of the HLPN), it is then self-controlled and guided by internal/external events that link the shell with other components of the system. These events may also correspond to service calls.

Several procedures are defined for the behaviour life-cycle of these components (see Fig. 2):

- Initial setup of the device, including configuration, exposition of services, establishment of connections to other devices/components and putting it on waiting initial state.
- Events are received via service operations, internal device interface to I/O and generated directly by the control (e.g. from decisions/conflicts).
- The received event must be tested. If it corresponds to some description of the control model's actual state, then the system is evolved by updating the control model, synchronizing service activities, accessing to I/O and report diagnosis/monitoring indexes. In case of being an exception, an undocumented event or an internal conflict, some decision is required. If the device has the necessary information to resolve it, special procedures are taken to interfere in the normal system's control and new events are generated. In the case of not having sufficient control over the event, it may ask for services provided by external components of the system (e.g. DMS) to provide more useful information for a concrete decision over the problem.
- After processing the event and evolving according to the token-game of the model-based middleware shell, the system reaches the next state and it is again able to receive other events.

The application of this method results on autonomous devices that are self-controlled and have less dependency to other components, especially from the upper levels such as the Decision Making Systems. In short terms, the features of these devices are:

- Service-orientation, event-based evolution according to the rules of the token-game of a HLPN;
- Autonomous control and consequent behaviour;
- Event-based life-cycle following the rules of e.g. the token-game of a HLPN;

- Handling of documented events and exceptions to the normal control.

4. Application example using High-Level Petri Nets

The methodology is applied to a mechatronics device corresponding to a lifter with two levels and four different ports where pallets can be inputted and outputted. These ports should be used to connect to other devices, such as conveyors, but can be triggered manually by placing a pallet in the lifter (since a sensor detects it). Fig. 3 shows a representation of the lifter with all 12 possible predicted operation modes (that corresponds to the paths that a pallet may take) and that are represented in a service-oriented specification as 12 aggregated services.

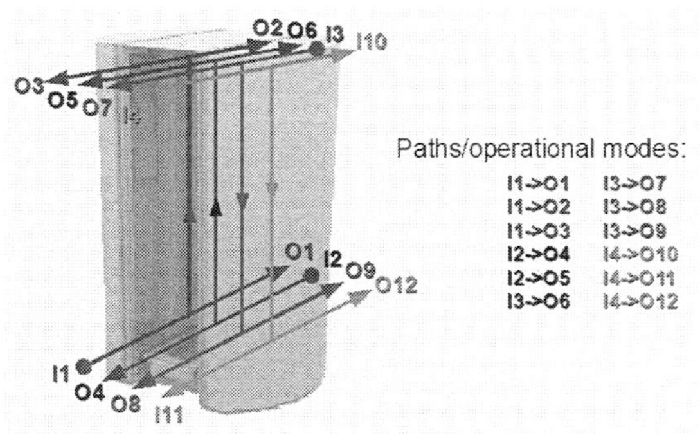

Fig. 3. Lifter with possible paths/modes.

The behavioural control of the lifter is formally represented by a High-Level Petri Net model that shows the global operation in the different modes. The atomic service *"top-left-transfer-in"* (see Fig. 4) implies production automation operations associated to reading/writing to corresponding I/O and synchronizing the service activity (e.g. a conveyor requests the service). The activation of this transition is done when: 1) the model presents a logic state true (lifter is ready to offer the service, i.e. the lifter is available, and 2) when a conveyor asks for the service or a sensor did detect a pallet in that port. Other situations that are not documented can also be handled and needs special procedures, as referenced previously.

Transfer out operations (such as the *"top-right-transfer-out"* transition) should be done synchronously with connected transfer devices (e.g. conveyor) to be able to provide a smooth transitional movement of the pallet from one device to another one. This requires that the lifter requests a transfer in service of the connected conveyor.

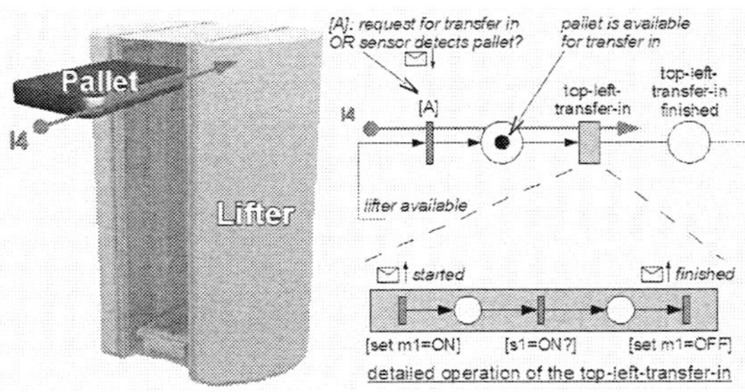

Fig. 4. Operation of the top-left-transfer-in service.

After the initial setup and configuration of the device with the control model and additional routines, the device is available offering/exposing services and waiting for events and call services.

For example, a connected conveyor is requesting the *"bottom-left-transfer-in"* service (I1). In this case, and since it is a documented event in the control model, the device proceeds to evolve the system by running the HLPN model and taking the related actions. After the *"bottom-left-transfer-in"* is successfully concluded, the system has to confront an exceptional event that is thrown by a conflict in the model (see Fig. 5).

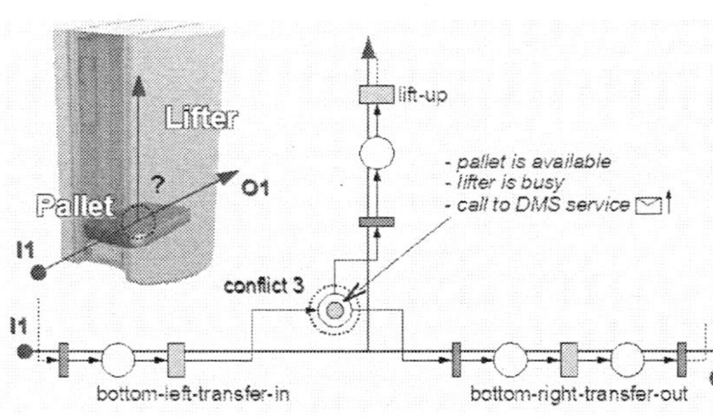

Fig. 5. Example situation in which a conflict is the actual state of the model.

If it does not have the necessary information to decide, it must call specialized components to help in the procedure. As in Fig. 2, DMS are used for this purpose. The lifter sends a request (call service) for support to the DMS, including supporting information (the ID of the pallet and possible outputs (service to be called): *"lift-up"* and *"bottom-right-transfer-out"*). Based on the workflow of the pallet, a decision is returned in form of an event to the lifter and now it should be able to resolve the conflict and evolve the system. For example, the lifter receives the suggestion to do a *"lift-up"* from the DMS, so it may proceed to do a *"lift-up"*. In any case, the final decision is up to the lifter that considers the received suggestion, but may operate differently in case of internal situations (e.g.

occupation of the lifter and/or connected conveyors).

The application example shows some of the features and benefits of the used methodologies such as independent control definition for devices, requirements in collaboration, service-oriented lateral and vertical integration and decision support (specially when the device does not have the necessary parameters to decide and should ask others for information). The control of the device was defined using HLPN due its valuable properties, but it can be anything else including the additional support for intelligent routines to enhance the normal control situations.

5. Conclusions

The goal of this paper is to provide an approach to configure flexible shop floors as an integration method and to specify the elements to develop the middleware for a transparent integration of shop floor components (e.g. devices) into a service-oriented IT-enterprise. In account is taken the formalization of the operational behaviour of autonomous and collaborative automation devices at the shop floor.

Several advantages can be distinguished from the methodology. On one hand, the inherited features of service-oriented architectures merged with traditional configuration approaches for shop floor, and on the other hand, a new and innovative integration approach is formally grounded on vectorial analysis theory. Each device has an autonomous service-oriented control behaviour that is local to the device but that is linked to the other device based on the layout configuration of the shop floor. The middleware shell makes these possible behavioural links based on offering/calling services. The decision mechanisms associated to the behaviour of aggregated services are local to the neighbouring devices but can also be influenced by information associated to the whole system, including shop floor (lateral collaboration) and upper level components (vertical collaboration) of the IT-enterprise system.

Acknowledgements

The authors would like to thank the partners of the Innovative Production Machines and Systems (I*PROMS) Network of Excellence (http://www.iproms.org) and the SOCRADES project (http://www.socrades.eu), for their support.

References

[1] Fitzgerald A. Enterprise resource planning (ERP)-breakthrough or buzzword? Proc. Third International Conference on Competitive Performance Through Advanced Technology. 1992, pp. 291-297.

[2] Bussmann S and Schild K. An agent-based approach to the control of flexible production systems. 8th IEEE International Conference on Emerging Technologies and Factory Automation. 2001, vol. 2, pp. 481-488.

[3] Deen S. Agent-based manufacturing: advances in the holonic approach. Springer Verlag, 2003.

[4] Leitao P, Colombo A and Restivo F. ADACOR: a collaborative production automation and control architecture. Intelligent Systems, IEEE, Jan-Feb 2005, vol. 20, n. 1, pp. 58-66.

[5] Colombo AW. Development and implementation of hierarchical control structures of flexible production systems using high-level Petri nets. Fertigungstechnik, Erlangen, Nuernberg, 1998.

[6] Jammes F and Smit H. Service-oriented architectures for devices - the SIRENA view. 3rd IEEE International Conference on Industrial Informatics, INDIN '05, 10-12 Aug. 2005, pp. 140-147.

[7] Lastra J and Delamer I. Semantic web services in factory automation: fundamental insights and research roadmap. IEEE Transactions on Industrial Informatics, Feb. 2006, vol. 2, n. 1, pp. 1-11.

[8] Bepperling A, Mendes J, Colombo A, Schoop R and Aspragathos A. A Framework for Development and Implementation of Web service-Based Intelligent Autonomous Mechatronics Components. IEEE International Conference on Industrial Informatics, Aug. 2006, pp. 341-347.

[9] Colombo A and Harrison R. Modular and Collaborative Automation: Achieving Manufacturing Flexibility and Reconfigurability. Int. Journal of Manufacturing Technology and Management, 2006.

[10] Qiu R and Zhou M. Mighty MESs; State-of-the-art and future manufacturing execution systems. IEEE Robotics & Automation Magazine, 2004, vol. 11, n. 1, pp. 19-25, 40.

[11] Gibson N, Holland C and Light B. Enterprise resource planning: a business approach to systems development. Proc. 32nd Annual Hawaii International Conference on HICSS-32 System Sciences, Track7, 1999, pp. 9.

[12] Leitão P. An Agile and Adaptive Holonic Architecture for Manufacturing Control. Faculty of Engineering, University of Porto, Portugal, 2004.

[13] Colombo AW. Industrial Agents: Towards Collaborative Production-Automation, -Management and -Organization. IEEE Industrial Electronics Society Newsletter, 2005, v. 52, pp. 17-18.

Innovative Production Machines and Systems
D.T. Pham, E.E. Eldukhri and A.J. Soroka (Eds.)

A theoretical model in measuring global manufacturing complexity and its potential applications

M. Flowers and K. Cheng*

Department of Advanced Manufacturing and Enterprise Engineering (AMEE), Brunel University, Uxbridge, Middlesex, UB8 3PH, UK

Abstract

The concept of global manufacturing has been studied by both academics and the manufacturing sector. However, lack of a universal understanding and measurement of the underlying complexities continues to remain problematic for many multinational manufacturing corporations. In this paper we present a theoretical model as a measure of complexity as applied to manufacturing operations. We conclude that the model leads to the requirements for measuring complexity applicable to real case global manufacturing scenarios.

Keywords: Global manufacturing, complexity, entropy, theoretical model

1. Introduction

Global manufacturing may be defined as the transfer of domestic manufacturing management resources by multinational enterprises or international corporations to low wage rate foreign countries or to international corporations striving to gain a competitive edge in international competition via economies of network [1]. Practically, the same manufacturing operations—Agile Manufacturing, just-in-time (JIT), Total Quality Management (TQM), Rapid Manufacturing [2]—and industrial economics apply in the global domain, irrespective of the manufacturing location. However, the dynamics of the global manufacturing challenges tend to threaten existing conventional manufacturing processes, making the latter inadequate for adaptation to the challenges. The global manufacturing environment [3] may be too technology-intensive and innovation-driven for the existing myriad of manufacturing techniques and philosophies to be adequate enough to revolutionize conventional manufacturing processes in the global domain.

The key to growth and survival of multinational manufacturing corporations (MNMCs) is a function of how to successfully translate the prevailing global manufacturing challenges into new opportunities, and subsequently, into the emerging global marketplace.

The growing globalization of economic and social enterprises, facilitated by the reduction of tariffs and removal of trade barriers within regions, has been the major driver for global manufacturing [4]. While the

* Correspondence to: Professor Kai Cheng, Department of Advanced Manufacturing and Enterprise Engineering, School of Engineering and Design, Brunel University, Uxbridge, Middlesex, UB8 3PH, UK
E-mail: kai.cheng@brunel.ac.uk

move toward a global economy may depend on a MNMC's particular situation and objectives—saturated domestic markets, uncertainty in the exchange rates, inflation, diminishing returns, labour shortages, etc.,—the urge to lower manufacturing resource costs and to gain a competitive advantage remain one of the major motivational factors.

The conception of global manufacturing promises an incentive of lowering cost for centralized training and personnel functions; maximum return on investment (ROI); more trade; flexible manufacturing; modest capital investment; creation of new jobs; etc.

In order to survive fierce competition posed by manufacturers from industrialized and emerging industrialized nations [5,6,7], many US-based manufacturers have resorted to establishing production-sharing operations abroad to help improve their cost competitiveness. This approach explores international division of labour strategies in developing countries, promising a two-fold benefit: new market launch strategy or competitiveness for MNMCs and unemployment reduction in the host country.

The dictates of today's demanding customers in the context of expecting to pay less for high quality products, for example, have forced manufacturers to adapt by matching customers to production capacity and capabilities. China and India, for example, provide MNMCs with low wage labour for the production of desired quality goods reasonably priced for the customer.

The first two sections of this paper present an insight into the conception of global manufacturing and its complexity. The third section derives and explores the applicability of a theoretical model as an entropic measure of complexity in manufacturing operations. In concluding, we present application cases in an attempt to demonstrate the applicability of the model in measuring global manufacturing complexity.

2. Concept of global manufacturing complexity

A complex system may be viewed as any system featuring a large number of interacting components whose aggregate activity is nonlinear—not derivable from the summations of the activity of individual entities—and typically exhibits hierarchical self-organization under selective pressures [8]. Understanding complexity, therefore, can help MNMCs to formulate theories that can predict complex global manufacturing phenomena, and help derive the tools that can deal with global manufacturing complexity. Deriving and exploring theoretical models for measuring complexity is fundamental to understanding manufacturing complexity, a motivation for this paper.

Complexity is subjective in that it is dependent on what a MNMC considers to be its main functional requirements that must be achieved in order to gain a competitive edge, an agreement with Suh [9] and Murray [10]. Complexity may be seen as heterogeneous in terms of changeability and variability or disorder in the sense of uncertainty or entropy [11,12].

The perspectives of complexity may be contested in the context of a property of global manufacturing systems, referring to that condition of a global manufacturing process which must be understood and measured in order to formulate models and simulations of real complex process-based manufacturing systems for the improvement of conventional manufacturing operations or systems in the global domain. For example, although extreme favourable manufacturing conditions exist for any multinational automotive manufacturing corporation to settle down in Mexico, there are unavoidable social and cultural challenges that are likely to impact global manufacturing operations [5,13]. Such challenges or uncertainties can increase the number of states in manufacturing operations, thereby, increasing manufacturing complexity and consequently reducing the probability of manufacturing success at a given level of investment. Thus, an initial approach to such problem solution lies in predicting these kind of uncertainties.

In an attempt to answer a question such as *what is the complexity in predicting the uncertainty in achieving a set of global manufacturing functional requirements?*, we adapt Suh's [9] presentation of complexity and define the subject-matter as *a measure of uncertainty or the probability of success in achieving a set of global manufacturing functional requirements.*

Manufacturing challenges in the global environment expect manufacturing systems to translate the challenges into opportunities and present themselves as complex adaptive systems (CAS). In this context, we may define CAS as manufacturing systems that emerge over time into a coherent form, and adapt and organize themselves without any singular entity or component deliberately managing or controlling it.

Given that the global manufacturing phenomena take into account several interrelated manufacturing and management activities or dependent components, it is imperative that MNMCs measure complexity at the individual value chain or micro level rather than in a broader sense or macro level. The results achieved at the microstate can be extrapolated for macro state scenarios.

The impact of the growing complexities on the global manufacturing phenomena could force many researchers to rethink their models. For example, some authorities [3,4,14] have successfully applied static and dynamic complexity measurements (explained further in the following sections) to manufacturing operations at the micro level. However, they have failed to explore the potential of the model in the global manufacturing domain. This paper seeks to explore this missing material.

3. A complexity measurement model

Global challenges expect manufacturing systems to adapt or become CAS. This re-emphasizes the need for MNMCs to deal with manufacturing complexity, primarily, at the micro level in order to evaluate manufacturing strategies that will lead to the evolution of tangible, result-oriented and cost-effective global entry or sourcing decisions, for instance.

The complexity of adaptability and integration of systems to changing requirements in the global environment [15] requires the management of information. The information needs of a manufacturing organization [7,16], therefore, constitute one of the external resources on which the organization has to rely in order to perform its functions.

Information theory [17,18] provides a unique, unambiguous criterion for the measure of the amount of uncertainty (entropy) in a system. The entropy [17] of a random variable is defined in terms of its probability distribution and can be shown as a good measure of randomness or uncertainty. Adapting Suh's definition [12], we define manufacturing information as a logarithmic function of the probability of achieving a set of specified global manufacturing functional requirements.

In this paper we adopt the ideas from information theory to derive a theoretical model as a measure of complexity.

3.1 Derivation of a theoretical model

Suppose we wish to explore a model that will provide a measure of complexity of a manufacturing operation at the microstate. Let us consider a discrete manufacturing system X consisting of a finite number of n independent possible states, $x_i = x_1, x_2, \ldots, x_n$ or $x_i \in X$, with an associated probability distribution

$$P = p_1, p_2, \ldots, p_n \ : \ p(x_i) \geq 0, \ \sum_{i=1}^{n} p(x_i) = 1 \qquad (1)$$

where $p(x_i)$ is the probability mass function of state x_i and the normalization condition is given by

$$\sum_{i=1}^{n} p(x_i) = 1.$$

Suppose we want to define a quantity $H(p_1, p_2, \ldots, p_n)$ which will measure the degree of uncertainty with respect to the probability distribution in a system. This presents the condition:

H is a continuous function of p_i.

Let us assume that the n independent possible states of the system have equal probability, $p(x_i) = 1/n$. This follows the condition that:

If all p_i are equal, the quantity $H(1/n, \ldots, 1/n)$ is a monotonic increasing function of n.

Based on (1), we introduce a function h defined on the interval (0, 1]. Thus, for a given n independent possible states in a given manufacturing process, the uncertainty $h(p)$ associated with the system may be expressed as

$$h(p) = \ln n \qquad (2)$$

where the logarithmic term provides the additive property required for a measure of independent uncertainty. Suppose we consider two independent manufacturing events X $(x_i \in X)$ and Y $(y_i \in Y)$ having m and n possible outcomes respectively. The probability distribution function associated with X is

$$P = p_1, p_2, ..., p_n \ : \ p_i \geq 0, \ \sum_{i=1}^{m} p_i = 1$$

and that with Y is

$$Q = q_1, q_2, ..., q_m \ : \ q_j \geq 0, \ \sum_{j=1}^{n} q_j = 1$$

Thus the uncertainty $H_{mn}(PQ)$ associated with a set of mn possible outcomes is given by

$$H_{mn}(PQ) = h_m(P) + h_n(Q) \tag{3}$$

for all

$$PQ = p_1q_1, ..., p_1q_n, p_2q_1, ..., p_2q_n, ..., p_mq_1, ..., p_mq_n$$

satisfying the Composition Law. Recalling (2), we express (3) in its logarithmic form as

$$H_{mn}(PQ) = \ln(mn) \qquad \text{or}$$

$$H_{mn}(PQ) = \ln m + \ln n \tag{4}$$

Now reconsidering our discrete system X where the probability of each state is $p(x_i) = 1/n$, implying,

$$n = 1/p(x_i) \tag{5}$$

Putting (5) into (2) yields

$$h(p) = \ln\left(\frac{1}{p(x_i)}\right) = \ln[p(x_i)]^{-1} = -\ln p(x_i) \tag{6}$$

We define a function H as the uncertainty or entropy associated with the event in the system X whose probability mass function is expressed in (1) such that

$$\mathrm{H} = H(p_1, p_2, ..., p_n) = \sum_{i=1}^{n} p(x_i)h(p) \tag{7}$$

Putting (6) into (7) yields

$$\mathrm{H} = -\sum_{i=1}^{n} p(x_i)\ln p(x_i) \tag{8}$$

For a discrete system, we rewrite (8) as

$$\mathrm{H} = -\sum_{i=1}^{n} p_i \ln p_i \tag{9}$$

It can be seen from (9) that the more variability a system exhibits, the higher the uncertainty. Quantities of the form described in (9) play a central role in information theory as measures of complexity [17,19].

3.2. Application cases

Suppose we want to measure the uncertainty or complexity within a multi-workstation manufacturing system. Applying (9) leads to

$$\mathrm{H} = -\sum_{i=1}^{N}\sum_{j=1}^{s} p_{ij} \ln p_{ij} \tag{10}$$

where H is the entropy of a manufacturing system; p_{ij}, the probability distribution of manufacturing resource i being in state j; N, the number of workstations or resources, and, s is the number of possible states at resource j.

3.2.1. Entropic measurement of complexity

Calinescu and colleagues [14] adapted and applied (10) to evolve the requirements that a complexity measurement must meet in order to address all the manufacturing complexity issues identified in a particular manufacturing scenario. They suggested that, where the scheduled states are independent of each other, manufacturing complexity can be reduced by reducing the number of processes and scheduled states. This implies that the complexity of the structure of the manufacturing operation arises from the impact the product structure or state has on the manufacturing resources that will produce it, suggesting that (10) may be seen as structural or static complexity.

3.2.2 Measurement of complex adaptive systems

In order to study the dynamics of queues and to cover a spectrum of manufacturing processes, Frizelle and Woodcock [4] considered manufacturing systems (S) as CAS and carried out three production trials at a distinct manufacturing facility. To demonstrate the relevance of queuing behaviour and causes, they extended the model in (10) to yield a second type of complexity called dynamic complexity (11) that permits comparisons to be made across the manufacturing process or operation and presents the average amount of information an observer needs to capture to understand the system:

$$H(S) = -\left\{ \begin{array}{l} P \ln P + (1-P)\ln(1-P) + \\ (1-P)\left[\begin{array}{l} \sum_{M^b}\sum_{N_j^b} p_{ij}^q \ln p_{ij}^q + \sum_{M^q}\sum_{N_j^q} p_{ij}^m \ln p_{ij}^m \\ + \sum_{M^m}\sum_{N_j^m} p_{ij}^b \ln p_{ij}^b \end{array} \right] \end{array} \right\} \quad (11)$$

where P is the probability of the system; p^q are the probabilities of queues of varying length greater than 1; p^m, the probabilities of having a queue length of 1 (Machine Running) otherwise 0; and p^b are the probabilities of the non-programmable states (*Reject*, *Rework*, *Absenteeism* and *Resource Breakdown*). Programmable states refer to *Run*, *Set-up* and *Idle*; M is the number of resources. The independence of the queues allows each resource to be measured independently for M resources. The number of states at resource j is $N_j = N_j^q + N_j^m + N_j^b$.

3.3 Adaptability of the model to the global domain

Both measures of structural or static and dynamic complexities can be extended to the global manufacturing environment in order to present analysis of the complexity in global manufacturing operations. For example, considering two manufacturing locations in the US and Germany, Kogut and Kulatilaka [20] developed a stochastic dynamic programming model that derives flexible operability for a MNMC, and, in the face of production complexity, considered the uncertainties in the exchange rate as an essential determinant for production switching. The entropic measurement of complexity (10), for example, can be used to predict the manufacturing systems behaviour at the two manufacturing facilities in order to determine when to consider production facility switching. Here, H, may be considered as the entropy of a manufacturing system operated by the US-German MNMC; p_{ij}, the probability distribution of manufacturing resource (management) in i being in country j; N, the number of workstations or manufacturing facilities or locations, and s is the number of possible states at either the US- or German-based manufacturing facilities. The terms in the governing equations can also be adapted to country-specific needs in order to evolve decisions for global scenarios.

Similarly, the dynamic complexity model in (11) can be explored for a global manufacturing scenario by redefining the terms in the governing equation described in 3.2.2 to adapt to the determinant of interest. For instance, in recalling the Kogut and Kulatilaka [20] report, adaptation of (11) can permit comparisons to be made across the manufacturing operations at both the US- and German-based manufacturing plants, providing the observer the average information (such as, design for manufacturing and assembly guidelines) required to understand the entire system in order to progress decision to switch manufacturing operations from one plant to the other.

4. Conclusions

The application of the theoretical model (11) led to a number of predicted results [3,4,12] that seem to support intuitive ideas of how a complex manufacturing system might behave: the more complex a manufacturing operation becomes, the longer it will take parts to get through the system; as a system becomes more complex, it becomes harder to make reliable promises; most complex operation within the system is liable to be the bottleneck; simplification and or control are the only options for tackling unwanted complexity, indicated by high static complexity; etc.

In exploring the lessons learned and adaptation of the theoretical model to real case global scenarios, we can infer from the above that the results from both the static complexity and dynamic complexity models, for example, demonstrate that complexity index measures variety without regards to its cause, and the presence of *Idle* state corresponds to a reduced complexity, leading to a need to perform a hierarchical analysis of the total static complexity; labour shortages may contribute to high *Idle* component of operation; and that programmable states or queue elements may be major factors for high dynamic complexity, etc. Such results, for example, can yield the requirements that the measurement of manufacturing complexity at a microstate must meet in order to address manufacturing complexity issues identified in a macrostate scenario.

The nature of the theoretical model demonstrates that static complexity can be planned and reduced by simplifying products and processes, and dynamic complexity must be controlled in order to improve process stability and schedule reliance.

We propose a future work that will focus on applying the model to real case global manufacturing scenarios in order to present the indices that stimulate global manufacturing decisions for improvement.

References

[1] Hitomi, K. (1996), *Manufacturing Systems Engineering: a Unified Approach to Manufacturing Technology, Production Management, and Industrial Economies*, Taylor & Francis, pp. 464 – 467.

[2] Bateman, J.R. and Cheng, K. (2007), *Rapid Manufacturing as a Tool for Agile Manufacturing: Application and Implementation Perspectives*, The International Society of Agile Manufacturing, Vol. 9, No. 1, pp. 39-52.

[3] McCarthy, I. P, Rakotobe-Joel, T. and Frizelle, G. (2000), *Complex Systems Theory: implications and promises for manufacturing organizations*, International Journal of Manufacturing Technology and Management, Vol. 2, Nos. 1-7. pp. 559-579.

[4] Frizelle, G. and Woodcock, E. (1995), *Measuring Complexity as an Aid to Developing Operational Strategy*, International Journal of Operations & Production Management, Vol. 15, No. 5, pp. 26-39.

[5] Fawcett, S. E., Taylor, J. C. and Smith, S. R. (1995) *The Realities of Operating in Mexico: an Exploration of Manufacturing and Logistics Issues*, International Journal of Physical Distribution & Logistics Management, Vol. 25, No. 3, pp. 49-68.

[6] Pontrandolfo, P. and Okogbaa, O. G. (1999), *Global Manufacturing: a review and a framework for planning in a global corporation*, International Journal of Production Research, Vol. 37, No. 1, pp. 1-19.

[7]Fawcett, S. E., Calantone, R. J. and Roath, A. (2000), *Meeting Quality and Cost Imperatives in a Global Market*, Int. J. of Physical Distribution & Logistics Management, Vol. 30, No. 6, pp. 472-499.

[8] Swarz, R. S. and DeRosa, J. K. (2006), *A Framework for Enterprise Systems Engineering Processes*, 19th ICSSEA.

[9]Suh, N. P. (1999), *A Theory of Complexity, Periodicity and the Design Axioms*, Research in Engineering Design, Springer-Verlag London Limited, Vol. 11, pp. 116-131.

[10]Murray, G-M. (1995), *What is complexity*, Complexity, John Wiley and Sons, Inc.

[11]Godfrey-Smith, P. (1997), *Complexity and the Function of Mind in Nature*, The British Journal for the Philosophy of Science, Vol. 48, No. 4, pp. 613-617.

[12]Suh, N. P. (2005), *Complexity: Theory and Applications*, Oxford University Press.

[13]Acosta, C. and López, A. (2003), *Global Manufacturing in the Mexican Culture Perspective: a Case Study*, AI and Society, Vol. 17, No. 3-4, pp. 364-374.

[14]Calinescu, A., Efstathiou, J., Schirn, J. and Bermejo, J. (1998), *Applying and Assessing Two Methods for Measuring Complexity in Manufacturing*, Journal of the Operational Research Society, No. 49, pp. 723-733.

[15]Hsu, C. and Rubenstein, A. (1994), *Enterprise Information for Global Manufacturers*, Proceedings 4th IEEE International Conference on Computer Integrated Manufacturing and Automated Technology, IEEE Computer Society Press, pp. 364-371.

[16]Lu, B. H., Bateman, R. J. and Cheng, K. (2006), *RFID Enabled Manufacturing: Fundamentals, Methodology and Applications*, International Journal of Agile Systems and Management, Vol. 1, No. 1, pp. 73-92.

[17]Shannon, C. E. and Weaver, W. (1949), *The Mathematical Theory of Communications*, The University of Illinois Press.

[18]Jaynes, E. T. (1957), *Information Theory and Statistical Mechanics*, Physical Review, Vol. 106, No. 1, pp. 620-630.

[19]Taneja, I. J. (2001), *Generalized Information Measures and Their Applications*, http://www.mtm.ufsc.br/~taneja/bhtml/bhtml.html. Accessed 14.10.2007.

[20]Kogut, B. and Kulatilaka, N. (1994), *Operating Flexibility, Global Manufacturing, and the Option Value of a Multinational Network*, Management Science, Vol. 40, No. 1, pp. 123-139.

Innovative Production Machines and Systems
D.T. Pham, E.E. Eldukhri and A.J. Soroka (Eds.)

An Auto ID-enabled framework for manufacturing information sharing systems

Z.D. Zhou[ab], P. Hu[a*]

[a] *School of Information Engineering, Wuhan University of Technology, Wuhan, China*
[b] *Hubei Digital Manufacturing Key Laboratory, Wuhan 430070, China*
[*] Corresponding author, E-mail: penghu@whut.edu.cn

Abstract

The current research of Auto ID (Automated Identification) boosts the applications in a broad spectrum including manufacturing and supply chain. This paper proposes an information sharing system framework dwelling on the smooth and integral information access of manufacturing information (e.g. product information, design information and sales information etc.), and the whole product lifecycle management of products. Also, Auto ID system provides a method for a product to rapidly retrieve its needed information when it is transmitted through shop floors and supply chains, during which the generated information by Auto ID readers in return can facilitate manufacturing scenarios in stages of the whole product lifecycle such as its ordering, design and production. Taking advantage of such information enhancement, a corresponding framework AIDMISS (Auto ID-enabled Manufacturing Information Sharing Systems) enabling the manufacturing information sharing process in manufacturing processes across enterprises is suggested. A case study concerning clutch manufacturing scenario is given to show the effectiveness of the proposed framework on the manufacturing enterprise.

Keywords: manufacturing information sharing, automated identification, framework

1. Introduction

With the great competition of global manufacturing, more and more enterprises start to realize the importance of improving their competing ability by taking advantage the manufacturing information (MI). And MI also plays a important role in both manufacturing technology and digital manufacturing science [1, 2].

The modern manufacturing activities are information driven [1]. In retrospect, the use of information systems in manufacturing industries has many years. The ERP, BOM, CRM are all the examples. However, the segregation of information flow exists between software providers (with different forms of data structure), external enterprises (probably with distinct operation management and information logs afterwards), and even inner departments in an enterprise. Fortunately, the diversion of different data formats in the information sharing seems to have a graceful solution with the XML technology [3]. Also, with the help of service-oriented architecture, all the heterogeneous software systems can be integrated into fresh procedure obeying a redefined work flow.

The Auto ID technology makes it possible the tracking, identify, logging of a product in its production control and lifecycle information.

Therefore, the amount of information pertaining to the product's retail and logistics (or distribution) process, using process, or even disposal process is more available and observable than ever. In such sense, the corresponding information can be integrated into the MI sharing system, which can be in return used for the better design, production management, ordering management, execution, etc., facilitating the nucleus of agile manufacturing and flexible manufacturing.

There have been many researches for MI systems [4-7]. However, with the development of Internet and Intranet that are functioning as the means of exchanging product development information [8], the rise of Grid technology can greatly enhanced those means. The Manufacturing Grid (MGrid) [9], as an advanced technology of known Networked Manufacturing, is the promising application of Grid technology in solving the effective MI sharing problems, by which MI can be seamlessly integrated, being flexible and loosely coupled. Therefore, the discussion of our proposed framework will be based on the MGrid.

This papers mainly intent to analyze and propose the Auto ID-enabled manufacturing information system (AIDMISS). Taking the advantage of Auto ID, the more observatibility and availability of product whole lifecycle information can be integrally and smoothly shared by distributed cooperative and collaborative manufacturing scenarios. Main contribution of the paper embodies: (1) the Auto ID technology for MI sharing is analyzed; (2) the information sharing framework enabled by Auto ID and inherited by Grid technology is proposed; (3) the EPCIS (Electronic Product Code Information Services) is enhanced under proposed AIDMMISS; (4) the proposed framework is extensible and can leverage the heterogeneous resources available in manufacturing industry such as Grid computing, sensor networks, etc. which can be used to evaluate the performance in an integrated view.

This paper is organized as follows: the related work is discussed in Section 2; Section 3 demonstrates the proposed system framework; Section 4 gives the case study; and Section 5 presents the conclusion.

2. Related work

The characteristics of information are complex, including mathematical forms, qualitative ones, etc. Rather, we discuss the essential qualitative characteristics of MI and the differences from the general information.

2.1. Manufacturing information sharing systems

The MI includes a large amount of information [4], specifically including: (1) the information that reflects the direct manufacturing process such as real-time status of machine tools, operation commands, and machining controlling information; (2) the recordable data in manufacturing, such as design knowledge, design document, engineering database, and multi-media; (3) the manufacturing knowledge, such as expert knowledge, product knowledge and marketing knowledge. Note that the MI categorization may exist in other ways, which is out of the scope in this paper. However, with the assumption that all the MI will be stored in the digitized form, we will focus on the research of effective MI sharing method among distributed departments in an enterprise or federation.

There are some researches dwelling on the research of MI sharing. The information model for a factory is proposed [4] using different views of a database to analyze the model. C.Y. Lee [5] has presented the TMIS (Total Manufacturing Information System), with the data bases of marketing, engineering, manufacturing, business, in the integration of manufacturing technologies and supporting functions. A WWW-based information management system has been demonstrated in [6]. J. Zhang *et al.* [7] have proposed a model for multi-agent manufacturing resource integration, and mentioned that the key step is to store all the information in the database, through which to complete the task decomposing, resource researching, resource editing, etc.

Since the collaboration and information sharing between federation enterprises or various departments in one enterprise distributed worldwide is increasing, the traditional information models are not suitable for its implementation and extension, which otherwise will handicap the efficient and integrated information sharing in MI sharing systems. Fortunately, the current Grid technology makes a graceful solution for MI sharing possible. In the Grid technology, the information architecture is virtualized, autonomic, and open [10]. Also, in the information architecture, a collection of distributed information will be managed in an integrated way, and perform autonomic management and interoperable operations.

2.2. *Auto ID technology in manufacturing*

Auto ID technology is developed by the Auto ID Center. The mostly used application of Auto ID is in the supply chain with the positive points, such as inventory management, product tacking, replenishment, counterfeit protection, retail checkout, etc [11]. Currently, the RFID is a broadly-used and promising Auto ID technology, and the typical Auto ID system aims at its networked capabilities. For example, when a product comes to the RFID reader, the reader will transmit the related information encoded in the product's tag to networked actuators. Also, the information can be transmitted to the networked database for future use or online analysis. Similarly, the other sensors used for monitoring and control can be also integrated to enhance the manufacturing process.

Apart from the application in supply chain, the RFID can be used in manufacturing control systems. In [11], the RFID tag is attached to the parts and hence the Auto ID data can improve the handling of simultaneous orders, trackablility, quality error management, outperforming the conventional assembly cell control. The intelligent product is discussed in [12]. And the further analysis of the impact of how Auto ID systems bring to distributed intelligent control is conducted in [11]. The general impact of closed loop control can be consolidated by Auto ID by adding the product sensing ability to the traditional process sensing [13]. In [14], the RFID technology is used to enhance the information availability during the communication process between agents.

Moreover, the incorporation of networked Auto ID systems will further enhance the observability of information. EPC Network [15] is proposed under such system, in which EPCIS, the Information Services in the EPC Network, mainly takes charge of the individual companies' databases that contain the details related to a product.

In order to make the Auto ID systems into the realistic application in manufacturing, its networking problem should be firstly solved. One of the successful solutions is EPC Network, the specification of networked RFID systems, suggested by Auto ID Centre, which generally consists of six parts [16]. From the EPC Network, the distributed information generated by tags on products can be integrally utilized. Moreover, EPCIS mainly takes charge of the individual companies' databases that contain the details related to a product, maintains a record of each EPC status change for that site, and communicates that status change to the EPC Discovery. Inherited by the Web Services technology, the EPCIS specification [17] has been improved to be one with well-defined interfaces and data formats, more full-fledged than it used to be. However, the current EPCIS has two negative facts: (1) it is hardly extensible for the many sorts of other data sources (e.g. binary files, semi-structure database, XML database, or metadata); (2) it rarely dwells on the smooth integration with other information platforms. The proposed system framework will improve the EPCIS due to the MGrid technology.

3. System framework

3.1. *Auto ID-enabled MI sharing systems*

In the section, we further discuss how the Auto ID systems in manufacturing field can enhance the information sharing process, although the Auto ID holds many potentials being used in a broad spectrum of applications, such as aerospace [18], supply chain [12], pharmaceutical business [19], and retails [20].

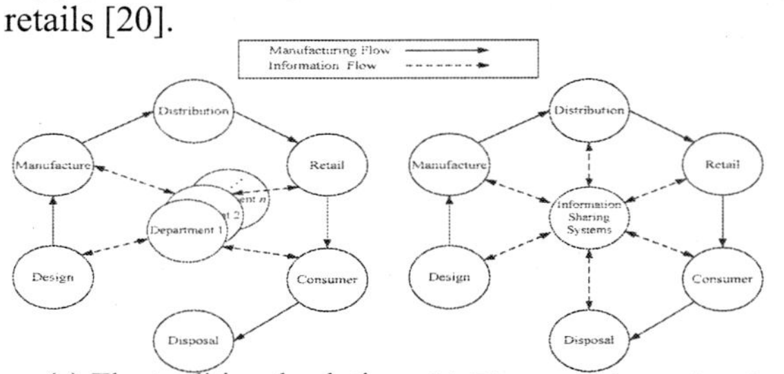

(a) The traditional solution without the MI sharing system and Auto ID systems

(b) The condition using the MI sharing systems and Auto ID systems.

Fig. 1. The enhancement of information sharing system in cost and efficiency by using Auto ID systems. Each circle represents the different stage in the product whole lifecycle.

It is possible, in the review of whole lifecycle of a product, to see how Auto ID can boost the observability of manufacturing processes. From Fig. 1, we can see the difference that the conventional information systems and the one with the facility of Auto ID systems. Taking advantage of the information availability in item-level of products, we can track and log much more information (e.g. the assembly status, transportation, usage condition, or disposal status during and after a product's production) that is blind to us before. In other words, the Auto ID system enhances information in both the upstream and downstream of manufacturing. In the

Fig. 1(a), On the one hand, the traditional solution of MI sharing system, which mostly merely implement and cover the management issues in one or several stages of the whole product lifecycle, will inevitably increase the expense in the aggregated setting of department or staff that should handle the conformity and accessibility of MI. On the other, in Fig. 1(b), it is apparent that the information system can cover the whole lifecycle of a product, and can provide the integrated way for accessing all the information at all different stages.

3.2. AIDMISS Framework

As shown in Fig. 2, the proposed model architecture is mainly conceptually composed of three parts, the *Application layer*, *Grid layer*, and *MI layer*. For the clarity, the assumption is made that MI is exposed or included in each geographically distributed department in an enterprise. Nevertheless, to a large extent, it is obvious that the department can also be seemed as a sub branch or federation enterprise, which can follow the same idea with our discussion.

The Application layer includes the manufacturing tasks, often assigned by an enterprise on the users' demands. The *MI layer* chiefly performs the information exposing; while the *Grid layer* is based on the Grid technology to provide the integrated information sharing system. In detail, the *Grid layer* mainly consists of modules as follows:

(i) ***QoS Manager:*** it aims to provide the SLA (Service Level Agreement) management, map and translate the requirements in SLAs into the performance specifications. Also, the SLAs can be input by the MGrid portal.

(ii) ***WS-BPEL Manager:*** this module is in charge of executing the workflow in business procedure and in automatic information maintenance or execution. Also, it is essential in the information sharing system, because manufacturing process is always defined by a series of procedures and because the GDS (Grid Data Service) in each department can be combined and managed by the use of WS-BPEL documents, by which will enhance the collaborative performance in efficiency. In particular for the more complex business or manufacturing process, including many concurrent or sequential sub tasks with constraints, the workflow model can ensure the accomplishment of tasks as it expected. For instance, if a department wants to retrieve or remove intelligently the similar or redundant information, a predefined WS-BPEL document can be used in such progress, resulting in restoring the information from distributed information sources or mitigating the unnecessary information stored across departments. Moreover, from the interface that workflow provides, an execution state can be visualized in MGrid Portal to administrators or decision makers.

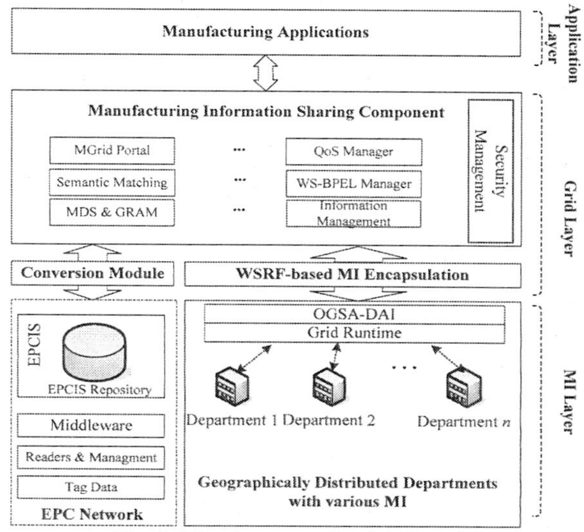

Fig. 2. The framework of AIDMISS

(iii) ***Information Management:*** it provides the information mechanism, including the component to handle with the data transmission and GDS instance in OGSA-DAI (Open Grid Services Architecture Data Access and Integration), the archive of the information, and the information retrieval and maintenance. Also, it takes in charge of the data flow transmission when a user performs querying or retrieval commands through *Manufacturing Applications* interface.

(iv) ***MDS & GRAM module:*** the GRAM (Globus Resource Allocation Manager) and MDS (Metacomputing Directory Service) in this module are used to provide the essential manufacturing job management and MI registration, searching, notification.

(v) ***OGSA-DAI and Grid Runtime:*** as an essential part of implementation of MI sharing system, Grid runtime (e.g. GT4) and the OGSA-DAI module are resided in departments that have the information to share. Specifically, according to OGSA-DAI and Grid runtime module, the execution with other Grid components (e.g. GRAM, MDS, etc.) and information sharing among various departments with different data forms (e.g. MySQL database, XML database, files and metadata, etc.) can be gracefully integrated and accessed. In fact, most of MI, such as the information in departments with different information as staff information, product information, and design knowledge information, can be stored in the data sources, e.g. databases. Accordingly, most of the MI can be integrated as GDSs through this module. Moreover, for other MI that cannot be store in databases, the XML description for the normal file is also supported in OGSA-DAI.

(vi) ***MGrid portal:*** users or administrators can perform the information searching, information editing, and information dynamic updating through the user interface. The MGrid portal can be implemented by the APIs in other modules. Following the friendly wizard interface provided in the portal, user can also perform the workflow definition and task assignment.

(vii) ***Security Management:*** it uses the security functionality provided by the GSI (Grid Security Infrastructure) in GT4 to provide the security management. Although the existing role mapping to authorized access to the data resources in OGSA-DAI can control the execution from the administrator and normal user, the Grid Layer needs the overall security management. To this end, the security management contains three parts: (1) access-level security, (2) the message-level security, and (3) transport-level security. Part (1) utilizes role mapping authorization access to database in OGSA-DAI, files or metadata. Part (2) can encode the information, such as query results, response messages, SQL statement. Part (3) targets at encoding all the contents in session in SOAP protocol.

(viii) ***WSRF-based MI Encapsulation:*** in order to transparently access the MI in an integrated way under Grid technology, their encapsulation conforming to WSRF specifications under Grid technology is necessary.

(ix) ***Conversion Module:*** it takes in charge of converting the data format conforming to the EPCIS specifications to the WSRF compatible one. This is necessary when users want to access the manufacturing applications in an integrated way, where the tags data (e.g. RFID tag data) is accessible from EPCIS.

In the Fig. 2, the product information exposed by Auto ID tags (such as RFID tags) will be processed by the readers connected and managed by the RFID Readers and Reader Management, and then processed by the *Filtering & Collection* component in the Middleware. The detailed discussion of Middleware, Readers & Management, and Tag Data in the EPC Network can be found in [15].

Furthermore, the MI layer will expose most of the MI in departments by using the OGSA-DAI, and, on the other hand, it will adopt the availability of EPC Network, which can enhance the MI by the Auto ID data from EPCIS. Specifically, the OGSA-DAI deployment involves three steps: (a) deploy new data services, (b) deploy data service resources and (c) expose data service resource. After step (b), the data services could be exposed by data services. After the step (c), the data service resource can be plugged into the information sharing system. As an explicit point, the data service can be invoked to access, query or update a data resource. To expose the file system or metadata, the additional data source configuration file should be added, including the contents of file description, file path, and file keywords. Further, for improving performance of MI sharing, the more efficient expose, such as less redundant tags in XML, the more need of an efficient format conversion mechanism to be implemented.

4. Case analysis

In this section we take the case study of AIDMISS with the facility of Auto ID data from clutch disposal process (the RFID tag is assumed to be attached to each clutch), and explore how it can enhance the ordering, design, production processes. Also, the number of clutch usages (in counts), total clutch usage (in hours), and number of vehicle starts (in counts) are stored in each clutch [21] as its tag data.

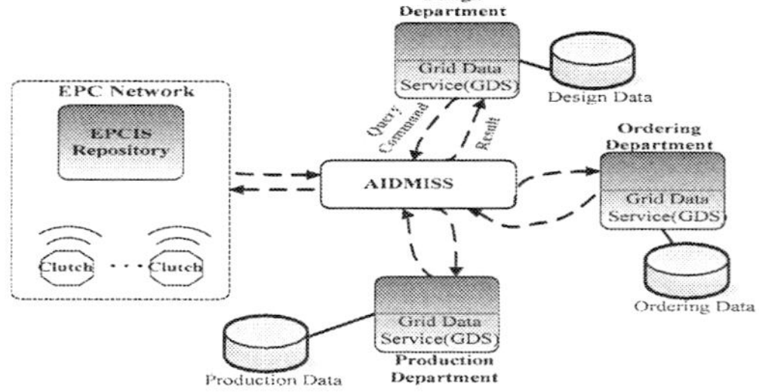

Fig. 3. An AIDMISS application in a manufacturing enterprise with geographically distributed departments connected with individual local data.

In Fig. 3, the ordering department, design department, and production department are connected to their own local databases as Ordering Data, Design Data, and Production Data. The departments are also installed with Grid environment for running AIDMISS, e.g. GDS. Also note that, in the left side of the figure, the *EPCIS Repository* takes in charge of storing the related tag data in clutches for disposal.

When the clutches arrive at the disposal centre, their tag data will be sensed and transmitted to the *EPCIS Repository* of the EPC Network. When a designer tents to get the feedback as the auxiliary information for their design evaluation and analysis, it will send query commands through AIDMISS *MGrid Portal* interface. Then the *Information Management* in AIDMISS will process the EPC Network, and the different data format between the two systems is accomplished by *Conversion Module* in AIDMISS. After the related data about the clutch, the result is transmitted back to the designer. In the same time, he/she would consider the data for future design improvement. Further, such process can be substituted by an intelligent application that can automatically make decisions from the obtained data and display them with human language, or even store them into historical design database.

Similarly, the corresponding staff in ordering department and production department can also initiate the MI sharing process by performing the manual or automated query and retrieval process. Moreover, the *WS-BPEL Manger* will prepare the BPEL document for the complex query, according to which the *Information Management* module will prepare the perform document to initiate and handle with the GDS instance in each department.

5. Conclusion

The MI sharing plays a key role in the manufacturing activities. The enhanced MI facilitated by Auto ID technology can improve the ordering, design, and production management and process. The proposed framework AIDMISS provides a promising solution to the MI sharing problem, where available MI during the product lifecycle can be accessed and shared in an integrated way under the MGrid environment; also, it is easy to be extended, implemented, and interoperable with other information systems. In the case analysis, we show AIDMISS is an efficacious framework for integrated manufacturing information sharing systems.

Acknowledgements

This work is in part supported by NSFC projects under Grant No. 50620130441 and No. 50675166.

References

[1] B. Zhang. Information Drived Digitized Manufacturing. *China Mechanical Engineering,* vol. 10, pp. 211-216, 1999.
[2] Z. Zhou, *Digital Manufacturing.* Beijing: Science Press, 2004.
[3] K. Muto, "Advanced Technology for Manufacturing Engineering Development: XML Technology on a System that Enables User to View Required Information from the Work Shop through Web Browser," *JSAE Review,* vol. 24, pp. 303-312, 2003.
[4] J. A. Harding, B. Yu, and K. Popplewell. Information modeling: an integration of views of a manufacturing enterprise. *Intl J. Production Research,* vol. 37, pp. 2777-2792, 1999.
[5] C. Y. Lee, "Total manufacturing information system: a conceptual model of a strategic tool for competitive advantage," *Integrated Manufacturing Systems,* vol. 14, 2003.
[6] S.Q. Xie, H. Huang, and Y. L. Tu. A WWW-based information management system for rapid and integrated mould product development. *Intl. J. Advanced Manufacturing Technology,* vol. 20, pp. 50-57, 2002.
[7] J. Zhang, L. Gao, and P. Li, *The Application of Multi-agent Technology in Advanced Manufactuirng.* Beijing: Science Press, 2004.
[8] S. Szykman, R. D. Sriram, and W. C. Regli. The Role of Knowledge in next-Generation Product Development Systems. *ASME Journal of SME Journal of Computation and Information Science in Engineering,* vol. 1, pp. 3-11, 2000.
[9] L. Chen, H. Deng, Q. Deng, and Z. Wu. Framework for Grid Manufacturing. *Tsinghua Science and Technology,* vol. 9, pp. 327-330, 2004.
[10] S. Bourbonnais, V. M. Gogate, L. M. Haas, R. W. Horman, S. Malaika, I. Narang, and V. Raman, Towards an information infrastructure for the grid, *IBM System Journal,* vol. 43, pp. 665-688, 2004.
[11] D. McFarlane, S. Sarma, J. L. Chirna, C. Y. Wong, and K. Ashtonb, Auto ID systems and intelligent manufacturing control, *Engineering Applications of Artificial Intelligence* vol. 16, pp. 365-376, 2003.
[12] C. Y. Wong, D. McFarlane, A. A. Zaharudin, and V. Agarwal, The intelligent product driven supply chain. In *IEEE Intl. Conference on Systems, Man and Cybernetics*, 2002.
[13] D. McFarlane. Product Identity and Its Impact on Discrete Event Observability. In *ECC*, Cambridge, UK, 2003.
[14] A. G. Higuera and A. C. Montalvo. RFID-enhanced multi-agent based control for a machining system. *Intl. J. Flexible Manufacturing Systems,* vol. 19, pp. 41-61, 2007.
[15] K. S. Leong, M. L. Ng, and D. W. Engels, "EPC Network Architecture," in *AUTOIDLABS-WP-SWNET-012*, 2003.
[16] D. McFarlane. Auto Id in Industrial Control: Future. *IFIP International Federation for Information Processing,* vol. 159/2005, 2005.
[17] EPCglobal. EPC Information Services (EPCIS) Version 1.0.1 Specification. 2007.
[18] M. Lampe, M. Strassner, and E. Fleisch. A Ubiquitous Computing Environment for Aircraft Maintenance. In *AUTOIDLABS-WP-BIZAPPS-013*, 2005.
[19] J. McQuivey and M. Feehan. RFID Could Enhance Pharmaceutical Marketing. In *http://www.rfidjournal.com/article/view/1444/1/82*, 2005.
[20] G. Roussos and B. College. Enabling RFID in Retail. *Computer,* vol. 39, pp. 25-30, 2006.
[21] A. K. N. Parlikad. The impact of product information on the effectiveness of product recovery decisions. PhD Thesis. Cambridge University, 2006.

Innovative Production Machines and Systems
D.T. Pham, E.E. Eldukhri and A.J. Soroka (Eds.)

Forecasting Options

H. Marzi, E. Marzi, and M. Turnbull

Department of Information Systems, St. Francis Xavier University
Antigonish NS B2G 2W5 Canada

Abstract

This paper reviews current methods of forecasting market trends in option prices and present a Neural Network based method to accurately predict future options. The study brings twenty years of data from S&P 500 index call options and employs non-parametric and parametric solutions and compares the outcomes of each forecasting approach. Initially a formulaic model was employed using Black-Sholes, and Artificial Neural Networks was applied for future options forecasting. The outcome of Simple Neural Networks method proved better than the Black-Scholes approach. An integration of the two approaches was developed which demonstrated better outcomes than each method did independently. The integrated model surpassed the other models and was able to better forecast the market trends.

Keywords: Option Pricing, Forecasting

1. Introduction

Forecasting option prices and the ability to predict future market trends are very important for investors in options market and traders. For this reason, a great deal of research has been done in this area to try to find the best predictor to utilize. Past studies have looked at basically two approaches. These are parametric and non-parametric. The most popular conventional parametric approach is the Black-Scholes model. This approach uses financial mathematics in an attempt to predict the option prices. The non-parametric approach is a realitively new approach to option pricing as it has been studied in just a little over a decade. This approach utilizes self-learning artificial intelligence tools as their forecasters. Neural networks are one of the most popular tools used in the area. The purpose of this study is to compare the forecasting abilities of the traditional Black-Scholes model against both a simple neural network and a hybrid neural network which utilizes aspects of the Black-Scholes model into its neural network.

Options are financial contracts between two parties that convey the right, not the obligation, to transfer the ownership of some security. There are two forms of options, call options, and put opstions. A call option is an option that gives the holder (beneficiary or buyer) the right to purchase from grantor (also called writer or seller) an underlying asset on or before a specified time (the expiration date) in the future at a specified price (the strike price). The opposite of this is a put option which gives the holder (beneficiary or buyer) the right to sell to grantor (also called writer or seller) an underlying asset at a specified price (the strike price) in the future before a specified time (the expiration date). Conventional methods for forecasting option prices use Black-Sholes formula. This formula for put options is different than for call options and the neural networks can only train for one type of option at a time.

There are three different types of options. These are European, American, and Bermudian. A European style option may only be exercised on its expiration date. An American style option may be exercised at any point up to its expiration date. The Bermudian style option may be exercised at any one of pre-selected points during the life of the option. European call options will be examined in this study as the Black-Scholes formula used is specifically for these type of options.

The Black-Scholes model utilizes the current stock price of the option, the strike price, the risk-free interest rate, the number of years to maturity, and the implied volatility on the share to try and forecast what the current call price will be. It is required to follow several assumptions to achieve relatively accurate results. These are that the stock pays no dividends, European exercise terms are used, markets are efficient, no commissions are charged, interest rates remain constant and are known, and returns are log normally distributed. Neural networks however are not bound by these assumptions. They are very adaptive and can handle out-of-sample data better than the Black-Scholes formula.

2. Options and futures

An option's value is determined by three separate factors. These are intrinsic value, time value, and the volatility of the underlying share. Intrinsic value is the value to an investor who exercises immediately and basically looks at the difference between the share price and the exercise price. Hassett and Wallison [1] specify that call options are classified as being either "in-the-money" or "out-of-the-money". "In-the-money" refers to the call options whose share price is greater than the exercise price. The investor can buy the share at the contracted exercise price and sell it at the spot price to make a profit [2]. "Out-of-the-money" on the other hand occurs when the share price is less than the exercise price and no benefits will come from immediately exercising the option.

Time value is the second factor that affects an option's value. An option that is currently "out-of-the-money" may possibly be "in-the-money" at the time of expiration. Investors will look at the probability of the option becoming "in-the-money" and also the degree to which it will be "in-the-money" when they are selecting an option to buy/sell [2].

The volatility of the underlying share is the third value factor. Basically this shows the degree of uncertainty with respect to the future price of the share. As Brooke and Fraser [2] state, the higher the volatility of the option, the wider the range of potential future prices that can occur. Therefore the value of the option increases as the volatility increases because the total risk/benefit has greatly risen. There is also a connection between volatility and time value. Brooke and Fraser (2002) [2] explain: "The shorter the time to expiration the less the uncertainty with respect to future share prices and therefore the lower the value of the volatility and vice versa."

Although these are the main factors when looking at the valuation of options, there are other varibles that could be looked at too. These are interest rate and dividend payouts. First to be examined is interest rate. An increase in the interest rate will decrease the present value of the exercise price of the option. This results in an increase of the value of a call option (or decrease of the value if it is a put option) [3]. The opposite is true if the interest rate were to decrease. The present value of the option's exercise price would increase which would result in the value of a call option to decrease (or increase if it is a put option). Faseruk and Blynski [3] describe that when dividend payouts are examined, it is shown that high dividend payouts limit the potential for growth of the underlying stock, consequently reducing the chance for the call option to trade in-the-money (while the opposite is true for put options). Dividend payouts will not be looked at in this study however as they require a different variation to the original Black-Scholes formula.

3. Parametric approach

The Black-Scholes model is used to calculate the theoretical value of European put and call options by use of a mathematical formula. It was derived by Fisher Black and Myron Scholes and published in 1973 [4]. The formula for the Black-Scholes model is as follows:

$$C = SN(d_1) - Xe^{-r_f T}N(d_2) \qquad (1)$$

Where:

$$d_1 = \frac{\ln(S/X) + r_f T + \frac{1}{2}\sigma^2 T}{\sigma\sqrt{T}}$$

$$d_2 = d_1 - \sigma\sqrt{T}$$

C is current call price, S is current share price (stock price), N(d) is the probability that a random draw from the standard normal distribution will yield a value less than d, X is the exercise price (strike price), E is the base of the natural log function, r_f the risk-free interest rate, T the number of years to maturity (time to expiration), σ the standard deviation of the return on the share (implied volatility).

Although the Black-Scholes model is still used today, recent studies, such as Amilon's [5], Board's et al. [6], and Hassett and Wallison's [1], have shown that the Black-Scholes method systematically overpriced options and that it had a poor performance in predicting post realized costs. In Hutchinson's [7] study though, he found that the Black-Scholes method did not overprice the options but actually underpriced them by a noticeable amount. This leads to the conclusion that the Black-Scholes model is not a realible method of forecasting options. This led to derivations from the original formula to try and create a better predictor by relaxing various assumptions. An example of this can be seen in Merton's derivation of the generalized Black-Scholes model where the dividend payments were allowed.

4. The Neural network paradiams

There is a wide variety of different neural networks but researchers Faseruk and Blynski [3] explain that one of the most popular models used by researchers in the financial area is the simple supervised feedforward multilayer perceptron (MLP) neural network with only one hidden layer. The reason this particular neural network is the most popular in the financial area is because it has been proven that the MLP neural network with only one hidden layer and a linear output unit is able to approximate an almost infinite array of unknown functions [3].

In 2000, Yao et al. [8] conducted their study by comparing the forecasting results of the Black-

Scholes formula and a backpropagation neural network that they created against actual market options. The option prices that they were forecasting were Nikkei 225 index futures which are traded on the Singapore International Monetary Exchange (SIMEX). Their data set included 17790 option prices for the year 1995. The inputs used were S/X and T while the output was C. They also utilized the hyperbolic tangent sigmoid as their transfer function. The results of the study suggested that for volatile markets a neural network option pricing model would outperform the traditional Black--Scholes model. However, the Black-Scholes model was still good for pricing at-the-money options. As a conclusion Yao et al. stated that those who prefer less risk and less returns may use the traditional Black-Scholes model results while those who prefer high risk and high return may choose to use the neural network model results.

Amilon [5] conducted a study in 2003 which examined whether a neural network (particularly a Multilayer Perceptron neural network) could be used to find a call option pricing formula better corresponding to market prices and the properties of the underlying asset than the Black-Scholes formula. He used daily Swedish OMX index and daily closing bid and ask prices of European OMX index call options, from the time periods June 1997-March 1998, and June 1998-March 1999. The data was spilt up to include a training set, a validation set, and a test set. Malliaris and Salchenberger [9] and Hutchinson et al. [10] also used this cross validation technique. Although in their studies, the data was only split into a training set and a validation set. Two Black-Scholes models were used as benchmarks in the study. One used historical volatility as an input while the other used implied volatility. The neural network model used was a Multilayer Perceptron neural network with one hidden layer of neurons. The performance measure that Amilon used to compare the models was with the Root Mean Squared Error (RMSE). The conclusion that Amilon arrived at was that the neural networks made better predictions than both of the Black-Scholes benchmarks.

In 2004, Bennell and Sutcliffe [11] analyzed European-style FTSE 100 call options traded on the London International Financial Futures Exchange (LIFFE) over the period of January 1st, 1998 to March 31st, 1999. Their data set consisted of 9556 option prices. The inputs used were stock price, exercise price, time to maturity, risk-free interest rate, implied volatility, and dividends. They included dividends in their computations by using Merton's derivation of the Black-Scholes formula. This was because it had been shown in previous studies that Merton's addition of dividends increased the accuracy of a neural network's predictions. The results of the study were that the artificial neural network was superior to the Black-Scholes model in both in-the-money and out-of-the-money predictions.

5. PROCEDURAL APPROACH

In this study, 5047 data points were used. They were S&P 500 European index call options from January 1986 to December 2005. For the Black-Scholes model (or formula BSF), five input variables stock price (S), exercise price (X), time to maturity (T), implied volatility (σ), and risk-free interest rate (r_f) had to be known. The variables S, X, and σ were all available on the Chicago Board Options Exchange's (CBOE) website [12]. Implied volatility is used in this study instead of historical volatility. This was because past studies had shown that the Black-Scholes with historical volatility was not as good of a predictor as when the formula used implied volatility. Implied volatility is considered a better estimation of future market volatility as it is based on traders' opinions on real market risk while historical volatility is described as the annualized standard deviation of the daily returns for individual options [3]. The daily return is $u_i = \ln(s_i/s_{i-1})$, where i is the time interval in days, s_i is the stock price at the end of the ith day, and s_{i-1} is the stock price on the previous day. For the time to maturity, it was calculated based on the expiration month of the option. It was also calculated using calender days instead of trading days. Other studies, such as Faseruk and Blynski [3] have used both and it has shown that using 365 days instead of 252 days decreases the errors for both the Black-Scholes and neural network models. For the interest rate, three month British Bankers' Association (BBA) LIBOR rates were used. These were available on the BBA's website [10]. Hull [13] explains that the LIBOR (which stands for London Interbank Offer Rate) are determined in trading between banks and change as economic conditions change. If more banks want to borrow funds than lend funds, LIBOR increases. If the opposite is true and the banks wish to lend funds more than they want to borrow funds, LIBOR decreases.

This study used two neural networks to compare against the Black-Scholes model. One was a simple neural network which received four input variables, stock price (S), exercise price (X), risk-free interest rate (r_f), and implied volatility (σ). Also the output of the neural network was the call price (C). Another approach was to use what is known as a hybrid neural network which employed the same inputs as in the simple neural network, but tried to predict the difference between the actual market price and the Black-Scholes' predicted price. The architecture of the neural network included five input neurons, one hidden layer of four neurons and one output layer containing one neuron. The activation function used for both networks is the Logistic Sigmoid function.

In order to compare the Black-Scholes, the simple neural network, and the hybrid neural network model, the mean absolute percentage error (MAPE) performance measure was used.

6. Neural Networks at training

Once the network is designed and the network weights and biases have been initialized, the network is ready for training. The training process requires a set of examples of proper network behavior to call from. These examples come in the form of network inputs p and target outputs t. During training the weights and biases of the network are iteratively adjusted to minimize the network performance measure.

The training function used for training in this study was the Levenberg-Marquardt backpropagation function. After training has occurred, it is time to simulate the network to obtain the outputs which will be used to compare against the actual market values and the Black-Scholes model. Once all of the simulations were done for both the Simple Neural Network and the Hybrid Neural Network, the predictions were then compared to the actual market prices. The errors were then compared to the errors of the Black-Scholes to see which model made the best predictor of option prices.

7. Discussion and Results

The three forecasting models utilized and compared in this study include Black-Scholes, Simple Neural Networks, and Hybrid Neural Networks. Each of these models utilized the 20 subsections of data from January 1986 to December 2005, and their forecasting results were compared against the actual market prices by employing the Mean Absolute Percentage Error performance measure MAPE. MAPE is calculated by subtracting the predicted price given by the model from the actual target price and dividing it by the predicted price. The absolute value of this number is then used in a summation function with the same calculations done to the other data points used in that subsection. Then that total is divided by the number of calculations used in the summation function to obtain the MAPE for that subsection according to (2).

$$\text{MAPE} = \frac{\sum_{i=1}^{n} \left| (y_i - x_i)/x_i \right|}{n} \qquad (2)$$

Where: y_i is the actual target price, x_i is the predicted price, and i is index of data points 0, 1, 2, … n

The MAPE has been used as performance measure in the forecasting studies of other researchers such as Faseruk and Blynski [14], and Lajbcygier et al [15], and makes it possible to compare the outcome of the current study with work or others. Each model was examined for the value of MAPE. The first that was examined was the conventional Black-Scholes model. The conventional Black-Scholes model (1) was employed as a benchmark to compare how well different approaches based on neural networks could predict option prices. BSF was used with five input variables. These were stock price (S), exercise price (X), risk-free interest rate (r_f), time to maturity (T), and implied volatility (σ). Implied volatility was used instead of historical volatility as stated earlier. Implied volatility is based on traders' opinions about real market risk while historical volatility is described as the annualized standard deviation of the daily returns for individual options. Also, it was reported by researches such as Amilon's [2], Bennell and Suteliffe's [6], and Faseruk and Blynski's [14] that using implied volatility in lieu of historical volatility will produce better prediction results. Dividends were not looked at in this study for simplicity of the Black-Scholes formula but it is a good area to look at in later studies as past research done utilizing dividends have shown excellent results in forecasting option prices.

After the Black-Scholes' outputs were calculated, they were compared to the real market prices. It was found that the Black-Scholes tended to overprice options 64% of the time. The MAPE for the Black-Scholes ranged from 0.6620% to 19.5936%. The MAPE for the complete data set was calculated to be 5.3964%. **Fig. 1** shows the comparison between the actual option prices taken from the market and the outputs derived from the Black-Scholes model over the twenty year period when divided into subsections. The entire data set of actual option prices is shown on the chart while the Black-Scholes' predictions are shown in 4-month intervals.

The next model that was created was a simple supervised feedforward multilayer perceptron (MLP) neural network that was coined Simple Neural Network or NNSimp. This neural network took in the input variables stock price (S), exercise price (X), risk-free interest rate (r_f), and implied volatility (σ). Also the output of the neural network was the call price (C). There was one hidden layer which contained four neurons and one output layer which contained one neuron. The activation function for the hidden layer was the Logistic Sigmoid transfer function and the outer layer used the linear transfer function. After training was finished, simulations were run to extract the predicted outputs. They were then compared against the actual market options.

An important conclusion that was arrived at from this is that, much like the Black-Scholes model, the NNSimp tended to overprice the options around 60% of the time. Although not a very high percentage, this result shows that there is a tendency for both the Black-Scholes and NNSimp models to overprice options. This would be valuable information in the trading world and will be compared with the hybrid neural network to observe if there is a trend for all of the option pricing models in this study. The MAPE performance measure was utilized again to obtain a standard error so the NNSimp model could be measured against the Black-Scholes model.

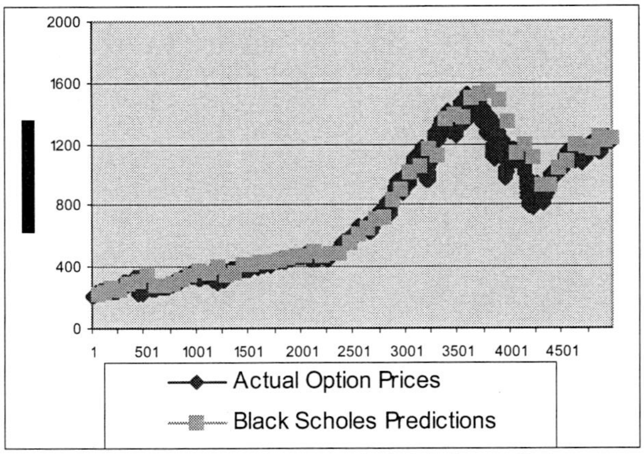

Fig. 1: Comparison between Black-Scholes prediction result and the Actual Option Prices

An important conclusion that was arrived at from this is that, much like the Black-Scholes model, the NNSimp tended to overprice the options around 60% of the time. Although not a very high percentage, this result shows that there is a tendency for both the Black-Scholes and NNSimp models to overprice options. This would be valuable information in the trading world and will be compared with the hybrid neural network to observe if there is a trend for all of the option pricing models in this study. The MAPE performance measure was utilized again to obtain a standard error so the NNSimp model could be measured against the Black-Scholes model.

The MAPE for the NNSimp model ranged from 0.2404% to 7.6821% and gave an error percentage of 2.8304% for the complete data set. This range was much lower in the NNSimp model and the overall error for the data set was almost half that of the MAPE error given by the Black-Scholes model. But the NNSimp model did not provide the lowest MAPE for all subsection of the data set. The Black-Scholes model gave a lower MAPE for seven out of the twenty subsections. Although the percentage for the overall data set is in favor of the NNSimp model, the subsections show that it was only better 65% of the time. Although the above result means Simple Neural Networks is a better predictor, it is not by enough to warrant it being called superior over the BSF.

The hybrid neural network will now be examined to see if it fared as a better predictor than both the Black-Scholes model and the NNSimp model. This hybrid neural network was coined NNHybrid. This neural network took the same inputs as the NNSimp model but instead of C, the output was the difference between the actual market price and the Black-Scholes' predicted price (C – BSF).

As with the NNSimp model, there was one hidden layer which contained four neurons and used the Logistic Sigmoid transfer function as its activation function. There was one output layer also which contained one neuron and utilized the linear transfer function. Training was done on by using the neural network toolbox functionality in Matlab 7.1 with the Levenberg-Marquardt backpropagation algorithm as the training function. After training, simulations were run in each subsection to extract the predicted outputs and they were then compared against the actual market options. The first conclusion that was observed was that the trend of overpricing options in the previous models continued within the NNHybrid model. This model tended to overprice options more than 70% of the time. The Mean Absolute Percentage Error was also taken from the outputs of the NNHybrid model compared to the actual market prices. These MAPE measures ranged from 0.0003% to 0.4988%. The MAPE for the entire data set of 5047 options ended up being 0.1539%. As predicted, the MAPE for the NNHybrid model gave a lower percentage than the MAPE from both the Black-Scholes model and the NNSimp model. It was more than eighteen times lower than the NNSimp model and thirty-five times lower than the Black-Scholes model.

Also when comparing the performance measure of each subsection, the evidence shows that the NNHybrid model gave better predictions than the NNSimp model in 95% of the cases while it better predicted option prices 100% of the cases when compared to the Black-Scholes model. The chart shown in Fig. 2 depicts the MAPE performance measure for the Black-Scholes model, NNSimp model, and NNHybrid model as to compare how low each one was. This chart clearly displays that the NNHybrid had a lower MAPE than both models almost 100% of the time.

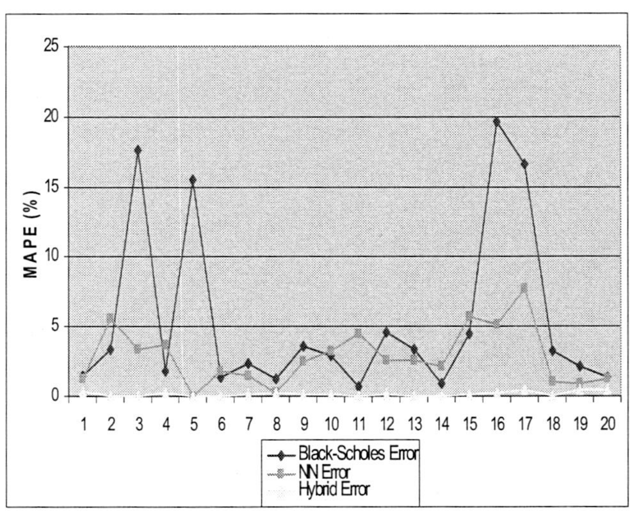

Fig. 2: MAPE Performance Measures for Each Model

Table 1 shows the MAPE differences between the three models on the entire data set. Again this clearly displays that the NNHybrid model had the lowest MAPE and was superior to the Black-Scholes and NNSimp models.

Table 1. Comparing MAPE Performance Measures in for the entire 5097 data set

	BSF	NNSimp	NNHybrid
All Data	5.3964%	2.8304%	0.1539%

With all of the facts that are taken from this study, the main conclusion is that although the neural network in this study tended to be a better predictor than the conventional Black-Scholes model, it is not by a significant enough amount to warrant it completely succeeding the Black-Scholes model. When the two methods were used together in the hybrid neural network, the resulting model was the superior forecaster. Another fact that came from this study in which traders could utilize was that each model tended to overprice the options.

This is useful knowledge because if the neural network only forecasted a slight increase in the option, the actual price may turn out to be lower than this predicted price which could actually be a decrease in the option's value. The next chapter will summarize this study and its conclusions and describes the limitations of the study and what could be examined for future research in this area.

8. Conclusion

The neural network was trained using the Levenberg-Marquardt backpropagation algorithm. When the NNHybrid model was compared against both the Black-Scholes model and the NNSimp model, a number of observations were made. The first outcome was that the NNHybrid's MAPE for the entire data set was better than both the Black-Scholes and NNSimp models. It produced a forecasting that was more accurate than NNSimp, indicated by a lower MAPE value by a factor of 18, and was 35 times lower than the Black-Scholes model. When looking at each subsections, it was also found that the NNHybrid model gave better predictions than the NNSimp model 95% of the cases while it resulted in better predictions than the Black-Scholes model 100% of the time. This concluded that the combination of the conventional Black-Scholes model and the neural network model into one hybrid neural network resulted in a superior option forecaster. Another important observation that was made during this study was that each of the models tended to overprice the options.

9. Acknowledgement

Authors wish to acknowledge financial Support of University of St. Francis Xavier, UCR-1497 for this research.

9. References

[1] Hassett, Kevin A, Wallison, Peter J. 2004. "A troubling requirement", Regulation. Washington: Vol. 27, Iss. 1; pp. 52 – 59.

[2] Brooke, GTF, Fraser, ET. 2002. "Options pricing using the Black-Scholes Model", The Investment Analysts Journal. South Africa: Number 52, Part 5.

[3] Faseruk, Alex, Blynski, Lev. 2006. "Option Price Forecasting: Black-Scholes vs. Simple and Hybrid Neural Networks", The 42nd annual MBAA International Conference. The Palmer House Hilton, Chicago, Illinois, March 15-17.

[4] Black, Fisher, Scholes, Myron. 1973. "The Pricing of Options and Corporate Liabilities", J. of Political Economy. Vol. 81; pp. 637 – 654.

[5] Amilon, Hendrik. 2003. "A neural network versus Black-Scholes: A comparison of pricing and hedging performances", Journal of Forecasting. Chichester: Vol. 22, Iss. 4; p. 317.

[6] Board, John, Sutcliffe, Charles, Ziemba, William T. 2003. "Applying operations research techniques to financial markets", Interfaces. Linthicum: Vol. 33, Iss. 2; pp. 12 – 26.

[7] Hutchinson, Martin O. 1998. "The use of fuzzy logic in business decision making", Derivatives Quarterly. New York: Vol. 4, Iss. 4; pp. 53 – 66.

[8] Yao, J., Li, Y. 2000. "Option Price Forecasting Using Neural Networks", Omega (Oxford): Vol. 28, Iss. 4; pp. 455 – 467.

[9] De Winne, R., Francois-Heude, A., Meurisse, B. 2001. "Market Microstructure and Option Pricing: A Neural Network Approach".

[10] On-line resource http://www.bba.org.uk/bba/jsp/polopoly.jsp?d=141&a=627 , last accessed April 20th, 2006.

[11] Bennell, J. and Sutcliffe, C. 2004. "Black-Scholes Versus Artificial Neural Networks in Pricing FTSE 100 options", International Journal of Intelligent Systems in Accounting, Finance and Management: Vol. 12, Iss. 4.

[12] On-line resource http://www.stockcharts.com/education/IndicatorAnalysis/, last accessed April 23rd, 2006.

[13] Hull, John C. 2003. "Options, Futures, & Other Derivatives", Prentice Hall, US.

[14] Gately, Edward. 1996. "Neural Networks For Financial Forecasting".

[15] Hanke, M. 1999a. "Neural Networks Vs. Black and Scholes: An Empirical Comparison of the Pricing Accuracy of Two Fundamentally Different Option Pricing Methods", Journal of Computational Intelligence in Finance. Vol. 7; pp. 26 – 34.

Innovative Production Machines and Systems
D.T. Pham, E.E. Eldukhri and A.J. Soroka (Eds.)

The value of PLM for the smaller companies

A.E Noyvirt

Manufacturing Engineering Centre, Cardiff University, Cardiff, CF24 3AA, UK

Abstract

Product Lifecycle Management (PLM) has become an ubiquitous strategy for improving the efficiency in big enterprises. However it has been difficult for small companies to benefit from its potential due to high entry barriers like the cost of the initial investment. This paper discusses the driving forces behind the shift in this divide facilitated by a recent wave of easier to implement PLM systems. The potential benefits of applying PLM in collaborative environments consisting of small companies are also considered.

Keywords: PLM, business process, extended enterprise, virtual enterprise

1. Product Lifecycle Management in the current business environment.

Product Lifecycle Management (PLM) as a business strategy that applies a consistent set of business solutions in support of collaborative creation, management, and use of product definition information is becoming a must not only for the companies that consider product development as their core competency but for a wide range of companies. While the benefits of PLM to big multinational corporations are already well proven, there is a growing consensus among suppliers, analysts and users, as to the potential benefits of PLM to medium and even small companies.

According to CIMdata's research [1] companies around the world are investing in PLM in record numbers. Investments in commercially available PLM-related software and services reached $20.1 billion in 2006 and are forecast to exceed $30 billion by 2011. Significant improvements in terms of shortened product development cycles, fewer errors and improved quality have been reported as resulting from the application of PLM in the big corporations.

Since a PLM solution deeply impacts the business processes and requires deep analysis, and if necessary, re-engineering of the processes itself, it is clear that good planning and extensive human recourses are required to implement and maintain PLM in a sustainable manner. This is one of the main reasons why such change thus far has only been within the reach of big players on corporate arena. Corporate reporting as a result of successful PLM implementations has highlighted consistent and significant improvements to their efficiency.

Historically, product data management (PDM) and more recently PLM solutions were only practical for large, distributed enterprises that had the extensive resources required to deploy and maintain them. The high entry barrier associated with the impact of PLM implementations on the organisation has meant that the small and medium companies have been missing on important opportunities to deliver continuously product innovations. However recently, much better solutions have began to emerge that promise to offer lower entry barrier for every company regardless of their size and financial power. Together with the pre-packaged functionality, such offerings make these new applications well positioned to satisfy existing business needs even for smaller companies. At the same time, many small organisations are increasingly recognising the need to improve the management of their intellectual property during all phases of the product lifecycle. This is due to their main focus is no longer being purely on the revenue growth, but it is shifting gradually to a combination of growth with cost reduction. Taking into account all of the above factors many managing directors have started to realise the potential for improving their bottom line and are subsequently considering the introduction of PLM systems to their organisation.

On one hand, fundamentally, the same driving forces that are behind the success of PLM on the big corporate arena are equally valid in SME environment, for example:

a) SMEs face a similar complex product innovation environment as large manufacturers, e.g. globalisation and demand for more and more complex products.

b) in a similar way to big companies the product development processes in the smaller companies are frequently inadequate to cope with the dynamically changing demands, e.g. like those for innovation and product development trends.

c) there is an urgent need, both in small and big organisations, to bridge the separate islands of automation that tend to form over time in any business and to improve the product information integrity across the company.

On the other hand, since the employees in small and medium companies are mostly co-located on a single site, the communication channels between them are as direct as they are physically possible. Such companies not realising the full potential benefits have not positioned PLM very high in their priority list. At the same time, this has also meant that they have been missing on the opportunities to save time and cost while improving quality and facilitating greater product and process innovation across their whole supply chain.

2. Recent trend towards wider adoption of PLM in smaller companies.

The combination of increased availability of out-of-the box PLM features at the right price level, and better customer awareness of the potential benefits can explain the recent trend of the wider adoption of PLM among small and medium sized companies. It is essential for every SME that it considers themselves a possible candidate for deploying PLM to build a good understanding and a clear view of the best PLM practises in their particular industrial sector. Based on this understanding and other publicly available market research in the PLM area, the company can then define the right objectives of PLM implementation that have to be clearly communicated to every level of the organisation. Since in most cases the investment decision comes from a board level, the middle-management, engineers, administrators and sales have to be part in the end-to-end process of introduction of the selected PLM system. They had to have understood clearly the benefits and accepted that there will be difficulties arising from the fact that the introduction of a PLM system redefines many long established practises and communication channels within the organisation.

In early 1990's investing in Enterprise Resource Planning (ERP) has been the top IT priority for the manufacturing sector. Today many company managers can accept that priorities have started to shift and investing in PLM is equally, if not more important than investing in ERP. According to some industry analysts ERP, Supply Chain Management (SCM), and Customer Relationship Management (CRM) can provide a short-term operational advantage since it is only matter of time when the competitors will implement similar systems and catch up. At the same time PLM can bring longer-term strategic competitive advantage since it gives a better foundation for continuous product innovation, and it is not sufficient enough for the competitors to only buy the right software in order to catch up.

It is essential in order to avoid data duplication and data re-entry to establish a good level of system integration between PLM and ERP systems. For example, a bill of materials (BOM) should reside in only one system and all other systems should be able to interrogate the master source of information when needed. In a similar sense, the integration of information will become much easier with the introduction of the new generation of modular ERP and PLM systems that allow data to be exchanged through the means of web services.

3. Important points to consider in the process of adopting PLM.

From the information integration perspective, the most difficult decision for any CEO, after having invested in ERP is whether or not it is worthwhile to opt for a PLM solution. First, in order for a company to be able to make the right decision, it needs to build a solid business case for PLM. This includes defining a roadmap of its implementation that is aligned with its overall business objectives. Second, it is important to identify which product lifecycle phases will be included in the PLM scope, and to set realistic implementation budgets based on the decision. Finally, the company needs to define a methodology for evaluation and measurement of the short and long term benefits resulting from the implementation of PLM.

Important factors to consider that will determine the overall success of a PLM system are:

a) What business requirements will be addressed?

b) Is there executive support?

c) What is the expected ROI?

d) How the data will be federated across different applications

Re-engineering inter-departmental, inter-partner and inter supply chain business processes is the most crucial phase for the successful implementation of a PLM system. Having product data distributed across legacy applications, e.g. existing ERP and SCM systems, makes introduction of a new PLM a challenging task from an integration perspective. Therefore careful planning is essential for the success of the whole process. That is why some analysts suggest that it is an easier option to invest in PLM prior to ERP. However, since many companies that have already started with an non PLM system, this advice is not always very practical and easy to implement. The right approach depends on the individual circumstances of each individual company, and it is normally the responsibility of the board of the directors to make a strategic decision based on cost benefits analysis.

Since it introduces dramatic change to the long-standing methods in place for handling data within the company, an investment decision for the introduction of PLM is not always welcomed by the professionals such as engineers and designers as well as the middle management. Therefore, in order for the PLM initiative to be a success it is important that every single employee who will be involved with the system to realise the cumulative positive effect for the company as a whole. This includes educating the potential in the time savings in the business processes, from the elimination of the mistakes in using incorrect versions of documents and from removing the bottlenecks in sharing data with others. Also, since PLM offers proven productivity benefits for individual users, e.g. quickly finding the right drawing, they can save a significant amount of their time after they have learned how to use the system. The steps to overcome the initial inertia experienced by the professionals within the company would promptly dissipate once they realise that PLM is much more convenient than just using customised spreadsheets and e-mail systems. This learning process is most likely to turn them from opponents into supporters of the new system.

There are different possible approaches in front of a small company on the decision to adopt a PLM solution. In addition to the standard all-in-one approach some vendors, e.g. PTC [2] offer an add–on approach with modules suited for different needs. It is also possible with the availability of hosted PLM services for a company to do a series of incremental projects that aim at a proper evaluation of the software and later to commit to an investment. Such an approach is time consuming but it allows any mistakes to be corrected early in the implementation process and certainly before any big investments have been made. When the most optimal configuration for PLM is reached and an accurate economic metrics for return of investment (ROI) are calculated the company can decide with confidence what is the right PLM solution for them and what extend they are prepared to invest.

4. A case study - PLM in the supply chain and collaborative environments.

In times when companies need to quickly adapt to changes within supply chains utilising a PLM system spanning across the whole extended enterprise, the approaches as described by [3] are considered an important factor for the dynamic and short-term integration between the individual participants. The implementation of such a system is a complex and non-trivial process. The large number of participating companies and the dynamic ad-hoc nature of their involvement adds an extra dimension of complexity to the implementation of an PLM system in a single organisation as discussed above. Integration of different legacy systems, e.g. ERP and CRM, highlight that the different information flows and different practices from each company participant in a virtual enterprise pose a major obstacle for the planning and the implementation of PLM. Having prior knowledge about an implementation of a similar set-up in a group of companies can significantly help the planning of PLM. The Venturing Wales project, a pilot project involving more than 30 SMEs companies in Wales working in an extended enterprise environment as described by [4]. The project has demonstrated that even smaller size companies utilising Windchill Project Link® a PLM system from PTC can benefit from the accelerated product development and much improved product data communication in their collaborative work. Although the project hasn't aimed at full data integration between the companies it has shown that the business process spanning many organisations can be successfully automated and controlled though a PLM system.

In any environment it is evident that as a product enters the design stage many organisations strive to ensure that the processes are run to schedule and to budget. As demonstrated through the utilisation of a PLM system the number of the necessary product development iterations can be significantly reduced while the duration of each iteration can been shortened leading to overall time and cost savings. In addition, because of the features offered by the PLM technology is tested by the project, significant simplification of the communication channels both within the company (see Fig.1 and Fig. 2) and across the whole virtual enterprise can be achieved. The improved project management has lead to shortened product development times and reduced total cost.

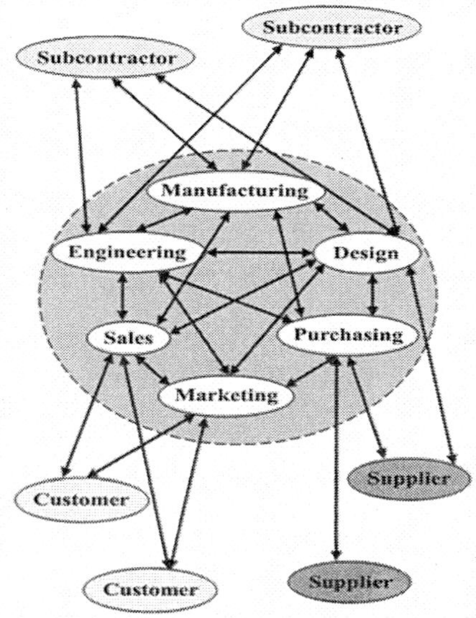

Fig. 1 Typical communication channels in a company before PLM

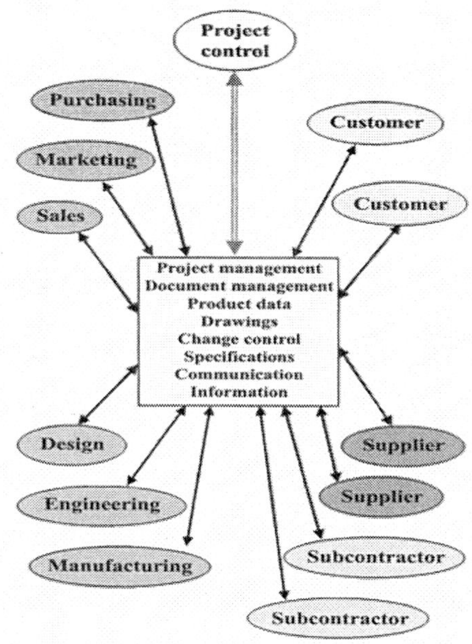

Fig.2 Typical communication channels in a company after PLM

5. Conclusions

PLM technology has evolved significantly to offer both user friendly flexible set of features and affordable solutions able to solve many evolution challenges in a company associated with complexity and high volumes of digital data. The technology has been well proven in practice to be capable of delivering tangible benefits on the corporate arena. As it is becoming rapidly within the reach of small and medium companies it should be considered by management as a possible IT initiative. Harnessing the information efficiently brought by PLM allows them to improve their performance and competitive situation. In addition, PLM has proven to be an invaluable co-ordination and communication tool along the supply chains and in collaborative virtual enterprise environments. Essential factors determining the overall success rate of a PLM deployment in any environment are careful market analysis, thorough planning and good customisation in accordance with the established business practises in the particular organisation.

References

[1] CIMdata, Position Paper – Enterprises of all sizes can benefit from PLM, www.cimdata.com

[2] Parametric Technology Corporation, PTC, www.ptc.com

[3] B. Andersen, O. Szegheo, Modeling the Extended Enterprise: A Comparison of Different Modeling Approaches, Norwegian University of Science and Technology

[4] Jeanne Ross et al., Enterprise Architecture As Strategy: Creating a Foundation for Business Execution, Cambridge, Harvard Business School Press. ISBN 1-591398-39-8, (2006)

Innovative Production Machines and Systems
D.T. Pham, E.E. Eldukhri and A.J. Soroka (Eds.)

Fit Manufacturing: A Strategy for Achieving Economic Sustainability

D.T. Pham[a], O. Adebayo-Williams[a], A.J Thomas[b], R. Barton[a], Z. Ebrahim[a], T. Shamsuddin[a]

[a] *Manufacturing Engineering Centre, Cardiff University, Wales*
[b] *Logistics Systems Dynamics Group, Cardiff Business School, Cardiff University, Wales*

Abstract

Achieving economic sustainability is a major focus for managers in manufacturing firms. The implementation of lean and agile initiatives has not really facilitated an organisation's long-term economic sustainability. The industrial landscape is littered with the carcasses of manufacturing companies who have tried to improve their productivity, competitiveness and hence economic sustainability using these initiatives. This failure calls for a shift in focus from achieving long-term competitive advantage through operational cost reduction, product flexibility and responsiveness to a "total" manufacturing initiative that is pro-active to market changes and has the ability to integrate the manufacturing efficiencies achieved through lean and agile with timeless manufacturing fundamentals required for economic sustainability. Fit manufacturing offers this advantage and provides a powerful approach for practitioners to achieve economic sustainability with limited disruption to their current manufacturing initiative.

Keywords: Fit Manufacturing, Agile, Lean, Economic Sustainability, Renewal

1. Introduction

Lean and agile are two leading manufacturing paradigms being applied in many industrial sectors today. Lean is a strategy for achieving continuous improvement in business performance through identifying a company's value stream and then systematically removing all waste. Agility on the other hand aims to achieve manufacturing flexibility and responsiveness to changing business needs in order to achieve competitive performance [1, 2].

However, the take-up of these two manufacturing initiatives by Western manufacturing firms has not really negated the economic argument to move jobs abroad especially from Western economies to low wage economies such as China and India nor has it halted the daily reported demise of manufacturing firms who have tried to improve their productivity and competitiveness using these initiatives. This is due to the fact that the effectiveness of using lean and agility to improve product cost, quality, product variety, flexibility, responsiveness and continuous improvement as the basis for achieving long term economic sustainability is fundamentally defective and restrictive in scope. The competitive advantage gained from the isolated implementation of these paradigms is significantly reduced when the market dynamics change and the struggle for survival takes over. Although lean and agile initiatives have been known to deliver short-term benefits, implementers of such initiatives have foundered once new challenges emerge. Management within companies are puzzled that of the many firms that have tried to implement lean only few succeed [3]. For instance, despite the implementation of lean manufacturing, European and U.S. automobile manufacturers continue to witness downsizing because their manufacturing operations have become less capable of meeting the challenges of today's turbulent market.

The reality remains that the competitive advantage created by lean and agile manufacturing are not timeless characteristics that are fundamental to long-term survival but are deployed to contribute towards the achievement of the company's goals and objectives of quality and reduced time to market. Therefore, the application and attainment of lean and agility for any company can be considered as having shifted from being once an 'order winner' to that of a 'market qualifier'. Lean and agility therefore are considered basic criteria that every company must operate just to remain 'players'.

These capabilities are becoming routine advantages in many industries and therefore are not sufficient to ensure long-term economic sustainability unless a manufacturing firm is capable of renewing itself by re-inventing its existing competencies while creating new ones. Hence, manufacturing companies should look at becoming "fit" through the integration of short term operational goals with long term economic survivability. Figure 1 illustrates a contemporary view of manufacturing strategies building blocks with fit manufacturing being the latest initiative.

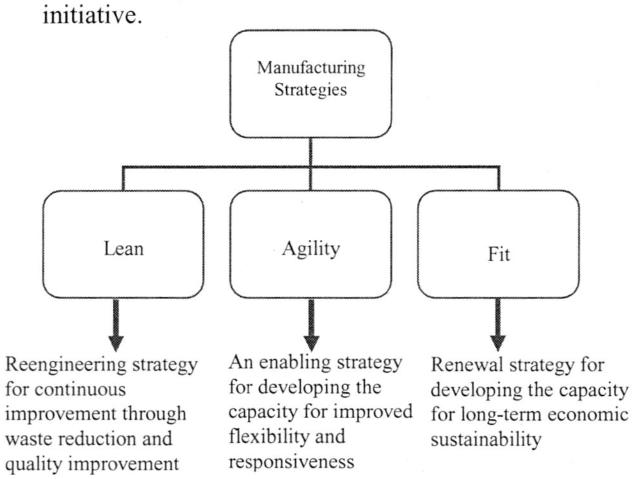

Figure 1: Manufacturing strategies building blocks

2. What is fit manufacturing?

Efforts to develop fit manufacturing were pioneered by Pham D. T. and Thomas A. J. [4, 5]. The manufacturing concept is described as a total manufacturing philosophy for an integrated approach to the use of Lean, Agility and Sustainability to achieve a level of fitness that is unique to a company. Fit was described as a manufacturing strategy which not only develops a company's latent potential to meet new market requirements but also encourages manufacturing firms to actively seek new markets through the use of an Advanced Warning Mechanism (AWM) that allows companies to identify new markets and align its manufacturing operations to take advantage of the new the opportunities. This requires "marketing and sales people to become technologically aware of a company's capabilities and act as a products gatekeeper when out in the field". Figure 2 shows the foundations of fit manufacturing [5].

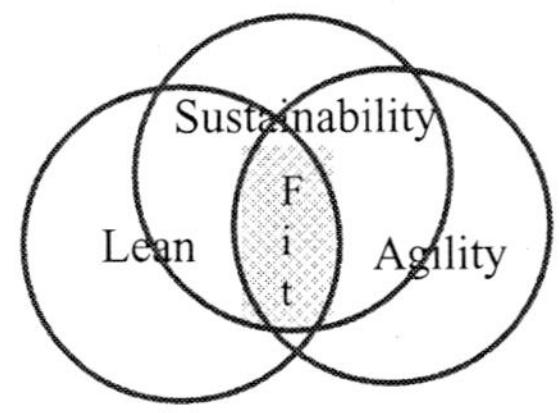

Figure 2: Fit manufacturing an integration strategy

However, fit manufacturing remains a gradually evolving paradigm. In practice, characteristics of fit manufacturing are seen in industry, but it is rarely seen that any company integrates several components of the philosophy into a world-beating strategy. The potential benefit of developing such an increased fitness level can be demonstrated by the evolution of increasingly complex organisms in nature.

Evolutionary fitness is a concept that has been widely studied in biological science. The concept of self-perpetuation in living organisms is closely associated with this fitness theory. Self-perpetuation is a phenomenon that can be described as a process to maintain and expand an identity, resisting the pressure directed toward static stability. Self-perpetuation is possible through "two principal dynamic phenomena of self-renewal, which is the ability of living things to continuously renew and recycle their components while maintaining the integrity of their overall structure, and self-transcendence, which is the ability to reach out creatively beyond physical and mental boundaries in the processes of learning, development and evolution" [6].

The need for self-perpetuation or long term economic sustainability aptly captures the essence of fit manufacturing which is to deliver sustainable benefits. How to achieve longevity through economic sustainability is an important business goal of all companies regardless of the industrial sector in which they may operate. Western manufacturers are especially vulnerable because of the continued rise of low labour cost economics such as China; and also the increased pressure in which combined forces of global competition, technology, interconnectivity and economic liberalisation make life for companies tougher than ever before. Companies need to develop new capabilities that match the new level of turbulence.

Fit manufacturing represents a strategy that imparts the values of both rapid adaptation to growing demands in terms of flexibility and speed, and simultaneously competing on the basis of production cycle time, price, quality, customisation and an effective supply chain. Fit manufacturing also encourages companies to not only identify their critical success factors, or core competitive competencies, but also stresses the need to make renewal a business wide strategy through developing distinct or unique capabilities that will set them apart from their competitors. Figure 3 shows fit manufacturing as a strategy for achieving business perpetuation through the integration of lean, agile principles and renewal strategy.

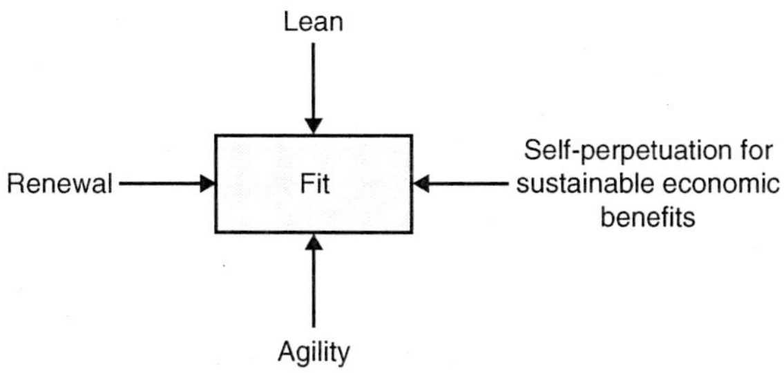

Figure 3: Components of fit manufacturing

Fit confers on manufacturing firms the *ability to remain in the long-term suitable for purpose by integrating manufacturing initiatives to fit the whole, sustainable future*. Fit represents a holistic approach that combines the strengths of lean and agile concepts with a creative capacity for sustainable long term profitable growth. An organisation that implements fit programmes is not only lean and agile but also sustainable. Fit manufacturing attempts to reduce the complexity and structural modifications requirements associated with finding sustainable optimised solutions to constant changing market conditions. Fit does this through building sustainable manufacturing systems that are creative, flexible, capable of innovation and persistent improvements through self-renewals.

A key differentiating criterion between fit and the older manufacturing initiatives of lean and agile, is the integration of the two with renewal strategy in order to achieve economic sustainability. Table 1 show the sustainable benefits delivered by each of the three inputs identified in figure 3 above.

Manufacturing Initiative	Sustainable Benefits
Lean manufacturing	Waste reduction Cost management Value maximisation
Agile manufacturing	Time to market Manufacturing flexibility Virtual partnership
Renewal strategy	Build industry leadership Business perpetuation

Table 1: Components of fit manufacturing strategy

Proponents of lean and agile manufacturing initiatives admit there are similarities and overlaps between the two concepts, fit helps to remove this controversy by specifically targeting these overlaps for areas of development. These overlaps and similarities can be explained as a consequence of the way the concepts have evolved over time (figure 4). Previous attempts at integrating lean and agiliy have led to the birth of new initiatives such as the 'Leagile' and 'Agilean' Paradigms [7]. However, there exist no previous attempt to integrate lean, agility and renewal together in order to gain a better synergy of the benefits that each of these manufacturing initiatives uniquely present.

	Craft	Mass	Lean	Agile	Fit
Integration					*
Reconfigurability				*	
Flexibility			*		
Fixed		*			
Comprehensive	*				

Figure 4: The evolution of fit. (Diagram adapted from [8])

3. Renewal strategy for fit manufacturing

Perhaps, the least developed of the three components for fit manufacturing initiative is the renewal strategy. There is no clear cut argument in literature on the nature of the concept or how it can be used to achieve long term economic sustainability. Some earlier work on the issue of renewal however, identified it as a key characteristic of long term economic sustainability [9, 10]. Renewal is seen as an organization's ability to reconfigure existing competencies and build new ones in order to outpace the industry competitors and take advantage of future opportunities. The concept of renewal obliges the creation of new ideas and dedication to the risk and promise of unexplored ideas. The drive for renewal pushes fit companies to continually experiment with new ideas and solve problems that other companies are yet to recognise, and build structures that will help them survive. The character of specific competencies created by fit companies who have managed to survive long term depends on the nature of the anticipated future competition and disruptions, some of the commonly anticipated being:

i. Change in underlying manufacturing technology including the supporting IT technologies

ii. Entrance of a new competitor to the industrial landscape or the revival of an existing one

iii. Decline in market share

iv. Changes in consumer taste

A fit manufacturing firm implementing renewal approach to economic sustainability is able to recognise: opportunities, time for change, time for new strategies and methods, time for investment in new product line, time to create new possibilities without being demanded by the outside environment, and most importantly time to create new competencies and re-invent old ones.

It is important to distinguish here that renewal strategies are different from strategies such as refocusing, downsizing, right-sizing, restructuring and business process re-engineering. These other strategies result from competitiveness problems such as falling market share, declining margins, stagnant growth and are often adopted to increase productivity, reduce costs and improve the return on investment [9]. In addition, unlike renewal which focuses on building new competencies and re-inventing old ones, the process of re-engineering focuses on the transformation of the traditional company structure in order to increase productivity and profitability. Re-engineering is closely associated with job loss and streamlining of administrative structures. Management frequently result to re-engineering process because of continued declined in productivity and economic losses. Renewal on the other hand creates capacities for long term economic growth and is a never ending process.

Renewal strategy for fit firms is recursive in nature and is never ending. It involves both innovative and re-definition actions to drive towards sustained prosperity and economic survival. Figure 5 illustrates the recursive nature of renewal and also shows the use of AWM. The Advance Warning Mechanism is used to constantly seek new unexplored opportunities in the market place and inherent potentials, and as soon as such opportunities emerge, the organisation has to determine whether to build new competencies or align current competencies to take advantage of the emerging opportunities. The AWM is used by the marketing and sales team to identify new markets which the manufacturing company can align its manufacturing and technological capabilities [5].

3.1 Apple Inc. as an organisation implementing renewal strategy

The ability to invent new industries and reinvent old ones can be seen as demonstrated by Apple in personal computers and now in digital devices – digital media players, camcorders and mobile phones including online music download. In 2001 Apple Inc. launched a new line of portable media player called iPod. This product was created

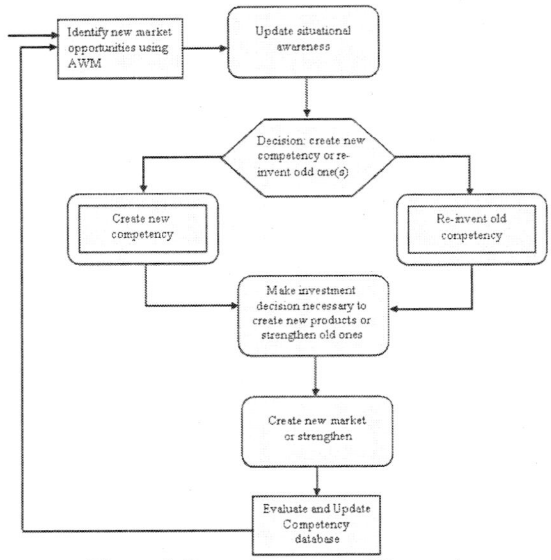

Figure 5: Recursive nature of Renewal

by utilising a number of competencies that Apple has developed overtime, such as competency in hardware engineering gained from the development of Mac computers, the iPod design, software and colour touch screen user interface competencies gained from working with partners.

In November 2007 Apple launched its revolutionary first mobile phone, tagged iPhone, in the UK. The iPhone introduces a new user interface based on a large multi-touch display, which offers users the ability to control the device with touch screen control. The competence (expertise) in screen input through the use of virtual keyboard was used by Apple in it failed Newton Message Pad launched several decades ago. The Message Pad was regarded as a failure in terms of Apple's sales predictions [10] but Apple had since grown its competence in user interface helping Apple to create iPhone virtual keyboard with touch screen capability. Apple's competency in miniaturisation was re-invested again when the company unveiled in January 2008 an ultra-thin laptop called MacBook Air. The computer is 0.76inches at it thickest point, and Apple worked with chip maker, Intel, to produce a smaller version of its Core2Duo processor for the laptop. The company also announced online film rentals for iTunes users in the US from almost every major film studio [11, 12]. As the company continues its renewal strategy of staying ahead of the competition through innovative product offerings, perhaps it is possible to project here that at some point in the future Apple will announced it entrance into the multi-billion industry of Video Games and perhaps explore opportunities in animation movies industry for 'iPoders' and other users. This will ensure that Apple keeps meeting and exceeding the expectations of the iPod generation while increasing the chances of the company's long term economic viability and sustainability. Apple did not build the new product alone through in-house development but utilised key strategic partnerships. Table 2 presents some of the companies Apple had to work with before the music iPod was brought to customer reality.

	Company Name	Some of the competencies contributed
1	Apple Inc	• Product conceptualisation • Product design and marketing • Software for digital device
2	PortalPlayers	• iPod integrated circuits
3	Pixo Inc.	• User interface software
4	Foxconn	• iPod manufacturer (OME)

Table 2: The making of Apple iPod

What is evident is that manufacturing firms aiming to remain fit or achieve manufacturing fitness through the implementation renewal strategy will possess the ability to form strategic partnerships required for new product development from conceptualisation to reality. Also essential is the capacity to make informed choices on all fronts of people, product, process, place and profit [13].

4. Building new competency through strategic partnership

Traditionally, most manufacturing firms tend to have a vertically integrated structure with all competencies required to bring a product to reality residing in-house. However, in the age of market uncertainties characterised by market turbulence, rapid advancement in information technologies, globalisation, trade liberalisation and outsourcing, with the knock-on effects of declining product life cycle and job movement from Western to low cost economies, the time is ripe for Western manufacturers to acknowledge the importance of strategic partnership in their journey toward economic longevity. Partnership is central to fit manufacturing, many manufacturing firms aspiring to be fit will require the integration of competency and skills residing in a wide varieties of partners companies. Strategic partnership in the forms of contract manufacturing, outsourcing and collaborative manufacturing can be used as effective tools for building a new competency required to explore and bring new products or services to life, helping to capture a bigger market share, sustaining the long term economic viability of the organisation and meeting the ever changing challenges of a globalised world. Under the fit manufacturing paradigm, partnership is seen as a network of competency from contributing firms; the ability to bring together such a network will be crucial in achieving manufacturing fitness and long term economic sustainability. A good grasp of agile manufacturing is particularly important in forming virtual partnerships towards competency contribution. Critical to forming effective partnership also is the ability to identify which new competency the organisation wants to grow or reinvent.

An example of a fit company implementing network of competency is BAE, the British Aerospace and Defence Company. BAE does not manufacture in-house the military hardware including the warships and fighter jets it sells to its customers in 100 countries including the US military. BAE outsource the design, prototype and manufacturing aspects of its contracts to its partners while BAE retains the competence to control, co-ordinate the supplier network, and integrate the different supplier components together to function as required. Figure 6 is an illustration of a network of competencies required to bring a new product to market. Such network offers manufacturing firms the advantage to reduce the time lag between when a new product opportunity is recognised and when the product is actually brought to reality. The competency network also helps contributing firms to be able to manage effectively the demand of shorter product life cycle and diminishing returns on new product investment. However, a disadvantage of building a network of competency is that the output of the network is dependent on the type of relationship that exists among the competency contributing firms and how this relationship is managed.

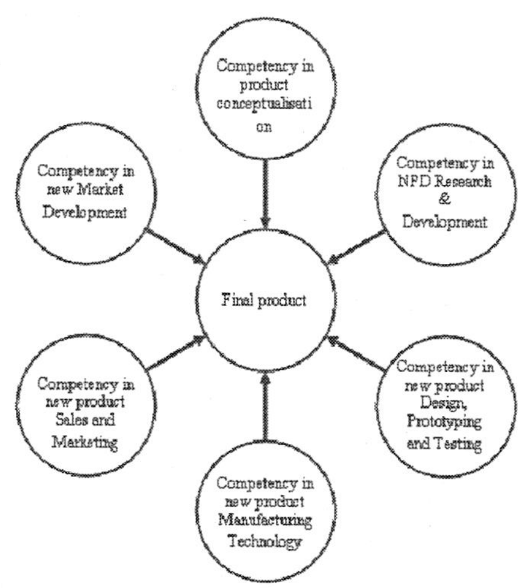

Figure 6: A form of a competency network required to bring a new product to realisation and market maturity

5. Conclusion

Building industry leadership is not realised by focusing on achieving incremental improvements offered by lean and agile manufacturing. Incremental advantages such as cost squeezing, time to market reduction, increasing customer responsiveness (flexibility) and increasing the market share by additional point are capable of sustaining current business but do not create new ones. Sustainable profitable growth is achieved through evolution of competitive that ensures the organisation stays ahead of the competition and continuously excites customers [9]. By combining the strengths of lean and agility with renewal strategy manufacturing firms can achieve fitness and business perpetuation for economic gain.

For example, the big three US auto manufacturers can be said to possess competency in fossil oil engines, but recent environmental consumers concerns are forcing the car makers to develop new competency in greener engines powered by alternative fuel such as ethanol made from plants, or powered by electric battery. The pressure to produce more environmental friendly and efficient cars are being brought about by a number of interwoven factors including soaring petrol price, unreliable supply from the Middle East, legal requirements and reducing profit from the present petrol hungry auto range. In the battle for the development of competency in greener engines, some car manufacturers such as Toyota and BMW, have already gained a head start through their latest eco-friendly cars, Toyota Prius and BMW 7-series hydrogen powered vehicle. This development demonstrates that simply focusing on lean and agile principles to meet all the requirements of today's market conditions is insufficient to ensure long term economic survivability. More relevant is how a manufacturing firm is able to re-invent itself time and time again; and build new competencies in order to remain fit for purpose and competitive. Fit has five underlying components, these are:

i. Ensuring long term beneficial self-perpetuation and fitness of the firm
ii. Being ready to go beyond core competency
iii. Capacity to function efficiently and effectively through the integration of the strength of lean and agile principles
iv. Continuous adaptation to increase fitness
v. Learning

Acknowledgements

This paper and Fit Manufacturing research is supported by the European Superman project, the I*PROMS Framework Six network of excellence and the Cardiff University IMRC.

References

[1] Womack, J.P., Jones, D.T. and Roos, D. (1990), The machine that changed the world, New York: Rawson/MacMillan

[2] Paul, T. K. (1994), Agile manufacturing: Forging new frontiers, New York: Addison-Wesley

[3] James, T. (2006), "It's all in the mind", IET Manufacturing Engineer Magazine, April/May 2006, pp.15-17

[4] Pham, D. T. and Thomas A. (2005), Fighting fit factories: making industry lean, agile and sustainable. IEE Manufacturing Engineer Magazine, April 2005, pp. 24-29

[5] Pham, D. T. and Thomas A. (2006), Fit manufacturing – a total manufacturing paradigm. Technical report, Manufacturing Engineering Centre, Cardiff University, UK.

[6] Capra, F. (1982), The turning point. New York: Banton Books

[7] Mason-Jones, R., Naylor, B. and Towill. D. R.(2000), Lean, agile or leagile? Matching your supply chain to the market. International Journal of Production Research, Vol. 38, No. 17, November 2000, pp. 4061-4070

[8] Gould, P. (1997), What is agility? Manufacturing Engineering Magazine February1997, pp. 28 - 31

[9] Prahalad, C. K. and Hamel, G. (1990), The core competence of corporation. Harvard Business Review, Vol. 68 No. 3, May-June1990, pp. 79-91

[10] Collins, J. C. and Porrass J. I. (2000), Build to last: successful habits of visionary companies London: Random House Business Books

[11] BBC News website: Apple iPhone debuts in UK stores. Website accessed on 10 Nov 2007 http://news.bbc.co.uk/1/hi/technology/7085643.stm

[12] BBC News website: Apple announces ultra-thin laptop. Website accessed on 15 Jan 2008 http://news.bbc.co.uk/1/hi/technology/7188849.stm

[13] Pham, D. T., Barton, R., Thomas, A., Williams-Adebayo, O., Shamsuddin, T. and Ebrahim, Z. (2007), FIT Manufacturing. 3rd Intelligent Production Machines and Systems (IPROMS) Virtual International Conference, 2-13 July 2007

Innovative Production Machines and Systems
D.T. Pham, E.E. Eldukhri and A.J. Soroka (Eds.)

Relationship Between Lean Manufacturing, Agile Manufacturing and Sustainability

D.T. Pham, Z. Ebrahim, T. Shamsuddin, R. Barton and O. Williams

Manufacturing Engineering Centre, Cardiff University

Abstract
Unpredictable changes in market and customer demands have required existing manufacturing concepts need to be continually updated and enhanced to enable the sustainability of the manufacturing companies. Lean and Agile manufacturing are concepts that have received much attention among industrialists and academics. This paper discusses the connection between lean manufacturing, agile manufacturing and sustainability. The criteria for Lean Manufacturing and Agile Manufacturing will be presented and arranged into four categories referred to as the basic aspects of a manufacturing system, namely: (i)Company organisation, (ii)Manufacturing processes, (iii)Hardware/Software technologies and (iv)Customer/Supplier relationships. The paper shows that there are connection between lean manufacturing, agile manufacturing and sustainability where satisfying the criteria for lean manufacturing, agile manufacturing should lead to sustainability.

Keywords : Lean Manufacturing, Agile Manufacturing, Sustainability, Fit Manufacturing.

1.0 Introduction

Rapid changes in product variety, volume and technology are the main issues in today's global manufacturing era. Lean Manufacturing (LM) and Agile Manufacturing (AM) are the most significant manufacturing concepts to provide what customers want within an the acceptable timeframe. However, it is necessary to distinguish between the LM and AM concept. McCullen and Towill [1] distinguished between the two concepts by viewing AM as the precursor to LM. In addition, Narasimhan, Swink and Kim [2] take similar view when they concluded that 'LM does not imply AM, yet AM does imply that many of principles and techniques of LM'.

The literature on LM and AM has amplified these differences into more complex characteristics. Yusuf and Adeleye [3] carried out a comparative study of LM and AM and determined twelve essential differences between LM and AM. Later, Ramesh and Devadasan [4] came out with a list of specific AM criteria as a result of their literature study. However, the differences between LM and AM have never been specifically discussed in terms of Company organization-Manufacturing processes- Hardware/Software technologies-Customer/Supplier relationships in a manufacturing system. This paper describes these four aspects as the basic elements of a manufacturing system (refer to Figure 1) where the input of a system is received from customers and produces the system's output. Liyanage[5] has clustered the manufacturing operational level of a plant into four: Human factors, organisational aspects, application technologies and work processes. However, in this paper, the human factors are considered as part of an organisation. In addition, customer/supplier relationships become necessary in the manufacturing system because customers and suppliers are the important resources in his operation. Sustainability concepts were then developed as required by society which must find acceptable such manufacturing activities.

This paper has three objectives; to provides the LM and AM criteria according to the basic elements of a manufacturing system; to determine the connections between LM, AM and sustainability concepts; to introduce the Fit manufacturing concept. There are eight sections which cover the methodology for identifying the criteria of LM and AM, the identified criteria of LM; those for AM; the sustainability concepts; the connections between the three concepts; the Fit manufacturing concept and the conclusions.

2.0 Methodology

This paper presents the results of a literature study on LM, AM and sustainability concepts. Basic statistical analysis was used to identify the criteria for LM and AM,. However, this study does not include the work on LM indicators by Sanchez and Perez [6] and on the AM criteria written by Ramesh et al. [4].

The identified criteria were then categorized into two; the Key criteria with higher mode and the Value Added criteria with lower mode. This paper describes the Key criteria as the 'most important', criteria that have highlighted the LM or AM concepts in a manufacturing system. The Value Added are the 'substantive criteria' that have added value to the manufacturing system. Both types of criteria were put into the four basic elements of a manufacturing system . Finally, the sustainability concepts were sorted out and put into one of the four elements.

3.0 The Lean Manufacturing Criteria

Sanchez and Perez [6] identified six indicators of LM which can be simulated as LM criteria. However, there are similar criteria that have also been identified by other authors [6-14]; elimination of zero-value activities; continuous improvement practised; multi-functional team; JIT production and delivery; integration of suppliers; flexible information system. Apart from the LM indicators, there are another seven criteria that have been identified by the other authors [7-9, 11, 13-14], [15] and [16]; single piece production flow; zero defect; customer involvement; flexible for changing (tools/parts/processes); small batches; flatter organization; autonomy system in operation.

Among the four elements; Company organization-Manufacturing processes, Hardware/Software technologies and Customer/Supplier relationships, the statistical results have shown the LM Key criteria that are focusing on 'Manufacturing processes' whilst the LM Value Added criteria are on 'Hardware/Software technologies' (see Tables 1a and 1b). These criteria reflect the LM definition that emphasizes wastes elimination in manufacturing operations. However, a combination of the Key criteria and the Value added criteria is vital because total waste elimination can only be achieved through the application of the technologies.

4.0 The Agile Manufacturing Criteria

Ramesh and Devadasan [4] identified twenty criteria for AM based on their literature study. However, these criteria did not apply to specific aspects of a manufacturing system in terms of Company organization-Manufacturing processes-Hardware/Software technologies and Customer/Supplier relationship. This paper identifies the similar criteria that have also been quoted by other authors [10, 14-16], [17-30]; flattened organizational structure; devolution of authority; responsiveness; adaptability and flexibility; enriching customers; rapid increase in productivity; knowledge-driven employees; fully empowered employees; open management; robust response to customers; short and effective product life cycle; short and flexible product service life; continuous design improvement; shorter manufacturing planning time; cost management; automation; information technology integration; change in business and technical processes; very efficient time management; supply chain management application.

Apart from the twenty AM criteria identified by Ramesh and Devadasan, there are another ten criteria that have been used by other authors [13, 15, 18-23, 25, 27-30], [31, 32]; adaptable; virtual corporation; reconfiguration; long-term gains; responsiveness; deployment of technology; continuous improvement practised; strategic viewed; innovative culture; customer integrated process.

The statistical result has shown that the AM Key criteria relate to all the four basic elements (refer to Table 2a and 2b). By contrast, the AM Value Added criteria focus on 'Hardware/Software technology'. The combination of the Key criteria and the Value Added criteria identify the AM concept with the technology aspect, which reflects two key points in the AM definition itself, namely flexibility and quick response.

5.0 The Sustainability Concepts

The impact of these industrial activities on the global environment requires the development of sustainability concepts. The World Commission On Environment and Development is the institute which had shed light on the idea of sustainability concepts through its book, 'Our Common Future' in 1987 [31]. However, sustainability has not been specifically discussed in the literature with respect to the LM and AM concepts. For example, De Ron [33] had defined sustainable production and its characteristics without involving LM and AM concepts. In its basic meaning, sustainability is defined as "the ability to keep (something) going over time or continuously" [34]. But, in the context of sustainability concepts, the most widely used definition is "development that meets the needs of the present without compromising the ability of future generations to meet their own needs [35]".

The growth of literature on sustainability concepts became one of the most significant issues among industrialists and academics. Furthermore, advanced thought on the sustainability concepts by industrialists and academics have been initiated through a collaborative workshop called 'The Future Manufacturing in Europe (FuTMaN), 2015-2020 [36] which took place in March 2003. In summary, the sustainability concepts for environmental impact can be linked to the manufacturing systems (refer to Figure 2). Environmental sustainability means the use of less energy resource, fewer hazards and emissions and is less toxic [5, 35] and [37-47]. The economic advantages of sustainability can be gained through manufacturing operations that take less time, have high value added and are at minimum cost [5,35, 39-40, 42, 44, 46], and [48]. Sustainability in society is seen in higher education levels, healthier life-styles and higher employment [33], [5, 37, 46], and [49, 50].

6.0 The Connections Between The Three Concepts

Pragmatically, the sustainability concepts mainly are the result of manufacturing activities. Liyanage[5] pointed out that the concept of sustainability in business is largely concentrated on broader economic, societal and environmental criteria as the bases to define the quality and standard of business activity. In addition, Labuschagne et al. [51] offered the similar view that business sustainability entails the incorporation of the objectives of sustainable development, namely social equity, economic efficiency and environmental performance. The LM concept, the AM concept and the sustainability concepts should be connected to each other. Figure 3 is the extension of Figure 2. Here, the sustainability concepts are identified as the output of the LM and AM criteria according to the four basic elements of a manufacturing system.

In terms of manufacturing organization, both the LM and AM concept are driven by team-managed criteria. But in AM, strategic view and cost management are the Value Added criteria. These criteria should be able to perform the following sustainability criteria [43]; economic stability, employment retention, quality and qualitative growth, zero inflation, long term company growth and continuous improvement.

Processes in a manufacturing system with the LM and/or AM concept are mainly free non-value activities, JIT production and delivery, continuous improvement, responsiveness, adaptability, flexibility and cost management. To achieve the following sustainability concepts, the criteria should be able to produce [39, 43, 47]; high quality flexible product at minimum cost with less emission, less toxic materials, easy disassembly afterwards and being a long-life product.

Information technology (IT) is very important for a manufacturing system. It should be used in the LM and AM concepts. In general, LM criteria emphasize flexibility in IT whilst AM criteria are about automation and reconfigurability that feature the technology. However, the criteria should be able to fulfill the sustainability concepts in the form of [33, 45, 52]; technology flexibility, high productivity, high quality, efficiency, economical, user and environmentally friendly.

The statistical result has shown the 'partnerships' with customers and suppliers is the Key criteria for both LM and AM. 'Partnerships' with customers in the LM concept means focusing on customer satisfaction, whilst the AM concept is more on delighting the customers through wider product variety and quick response. The LM concept works with long-term 'partnerships' with suppliers for mutual benefit. In contrast, the AM concept applies short-term 'partnerships' that are based on the competitiveness evaluation method. Apart from customers and suppliers, the sustainability concepts aim for 'partnerships' with other industries [52] through R&D collaboration.

7.0 The Fit Manufacturing Concept

The Fit Manufacturing concept was proposed by Pham and Thomas [53] as an integrating strategy that combines the LM and AM concepts to ensure that the manufacturing company operates in a sustainable manner and remains competitive as markets change in the future. The Fit Manufacturing system aims to develop a company's latent potential to meet new market requirements and to seek new market areas where the technological, human and financial aspects are robust enough to achieve market breakthrough [54].

Though there are connections between the LM, AM and the sustainability concepts, in reality the integration between LM and AM would not be able to achieve the sustainability concepts. For instance, there are certain problems arising from the 'de-coupling point' system as it neither considers short-term demand variability nor long-term trends or patterns such as seasonality [55]. Therefore, this paper proposes the idea of integrating LM, AM criteria and energetic/fitness system for achieving the sustainability concepts (refer to Figure 4).

8.0 Conclusions

- Team management and partnership are similar criteria in the LM and AM concepts.
- The sustainability concepts in respect of the environment, of economics and of society can be connected to the four basic elements of manufacturing systems. These connections are revealed in the LM and AM criteria.
- Apart from the LM and AM criteria, the energetic/fitness system is needed as a catalyst to a manufacturing company in achieving the sustainability concept.
- The integration between LM, AM and the energetic system would be able to 'Fit' the manufacturing company in unpredictable market conditions. This system is called as 'Fit Manufacturing'.

References

1 **McCullen, P. and Towill, D.** Achieving Lean Supply Through Agile Manufacturing. *J Integrated Manufacturing Systems*, 2001, **12**(6-7), 524-533.

2 **Narasimhan, R., Swink, M. and Kim, S.W.** Disentangling Leanness And Agility : An Empirical Investigation. *J Operations Management*, 2006, **24**(5), 440-457.

3 **Yusuf, Y.Y. and Adeleye, E.O.** A Comparative Study Of Lean And Agile Manufacturing With A Related Survey Of Current Practices In The UK. *Int J Production Research*, 2002, **40**(17), 4545-4562.

4 **Ramesh, G. and Devadasan, S.R.** Literature Review On The Agile Manufacturing Criteria. *J Manufacturing Technology Management*, 2007, **18**(2), 182-201.

5 **Liyanage, J.P.** Operations And Maintenance Performance In Production And Manufacturing Assets : The Sustainability Perspective. *J Manufacturing Technology Management*, 2007, **18**(3), 304-314.

6 **Sanchez, A.M. and Perez, M.P.** Lean Indicators And Manufacturing Strategies. *Int J Operations and Production Management*, 2001, **21**(11), 1433-1451.

7 **Holweg, M.** The Genealogy Of Lean Production. *J Operations Management*, 2007, **25**(2), 420-437.

8 **Lee-Mortimer, A.** A Lean Route To Manufacturing Survival. *J Assembly Automation*, 2006, **26**(4), 265-272.

9 **Shah, R. and Ward, P.T.** Defining And Developing Measures Of Lean Production. *J Operations Management*, 2007, **25**(4), 785-805.

10 **Bhasin, S. and Burcher, P.** Lean View As A Philosophy. *J Manufacturing Technology Management*, 2005, **17**(1), 56-72.

11 **Moore, R. and Scheinkopf, L.J.** Theory Of Constrains And Lean Manufacturing : Friends Of Foes? (Chesapeak Consulting Inc., 1998).

12 **McCurry, L. and McIvor, R.T.** Agile Manufacturing : 21st Century Strategy For Manufacturing On The Periphery? *4th Irish Academy of Management Conference,* (University of Ulster, Derry, Ireland, 2001).

13 **Philips, E.** The Pros And Cons Of Lean Manufacturing For The Small To Medium Size Fabrication Shop. *Forming and Fabricating Magazine*, pp. 1-5 (SME Inc., Lancaster, UK, 2002).

14 **Michael, A.L.** Lean Production And Sustainable Competitive Advantage. *Int J Operations and Production Management*, 2000, **20**(8), 959-978.

15 **Achanga, P., Shehab, E., Roy, R. and Nelder, G.** Critical Success Factors For Lean Implementation Within SMEs. *J Manufacturing Technology Management*, 2006, **17**(4), 460-471.

16 **Plonka, F.E.** Developing A Lean And Agile Work Force. *Int J Human Factors In Manufacturing*, 1997, **7**(1), 11-20.

17 **Richard, D. and John, M.** Agile Manufacturing. *American Society Of Mechanical Engineers*, 1995, **2-2**, 977.

18 **Abair, R.A.** Agile Manufacturing : Successful Implementation Strategies. *Annual Intl Conference on American Production And Inventory Control Society* pp. 218-219 (APICS, Washington, USA, 1997).

19 **Jain, S.** Virtual Factory Framework : A Key Enabler For Agile Manufacturing. *INRIA/IEEE Symposium on Emerging Technologies and Factory Automation, ETFA '95*, pp. 247-258 (IEEE Inc., Paris, France, 1995).

20 **Ian, C., Hossam, I., Jim, M., Simon, S., Martin, T. and David, Z.** Agile Manufacturing Transitional Strategies. In Hvolby, H.-H., ed. *4th Intl Conference on Stimulating Manufacturing Excellence in Small and Medium Enterprises, SMESME*, pp. 69-77 (Department of Production, Aalborg Univ., Denmark, 2001).

21 **Kidd, P.T.** A Strategy For The 21st Century. *IEE Colloquium on Agile Manufacturing*, pp. 1/1-1/6, (IEEE Inc., Cranfield, UK, 1996).

22 **Bunce, P. and Gould, P.** From Lean To Agile Manufacturing. *IEE Colloquium on Next Generation Manufacturing : Future Trends in*

Manufacturing and Supply Chain Management, pp. 3/1-3/5, (IEEE Inc., London, UK, 1996).

23 **Gupta, U.G. and Mittal, R.O.** Quality, Time And Innovation Based Performance Measurement System For Agile Manufacturing. *Proceedings Annual Meeting of Decision Sciences Institute*, pp. 1511-1513 (Decision Sciences Institute, Florida, USA, 1996).

24 **Sharifi, H. and Dann, Z.** Agile Manufacturing : A Structured Perspective. *Responsiveness in Manufacturing*, pp. 5/1-5/4 (IEEE Inc., 1998).

25 **Dove, R.** Tool For Analysing And Constructing Agility. *3rd Annual Agility Forum Conference/Worksyop,* (Emerald, Austin, USA, 1994).

26 **Sharifi, H., Colquhoun, G., Barclay, I. and Dann, Z.** Agile Manufacturing: A Management And Operational Framework. *J Engineering Manufacture*, 2001, **215**(6), 857-869.

27 **Esmail, K.K. and Saggu, J.** Agile Manufacturing : A Changing Paradigm. *IET Manufacturing Engineer*, 1996, **75**(6), 285-288.

28 **Gunasekaran, A.** Design And Implementation Of Agile Manufacturing Systems. *Int J Production Economics*, 1999, **62**(1-2), 1-6.

29 **Gunasekaran, A. and Yusuf, Y.Y.** Agile Manufacturing : A Taxonomy of Strategic And Technological Imperatives. *Int J Production Research*, 2002, **40**(6), 1357-1385.

30 **Sharifi, H. and Zhang, Z.** Agile Manufacturing In Practice : Application of Methodology. *Int J Operation and Production Management*, 2001, **21**(5/6), 772-794.

31 **Sikdar, S.K.** Sustainable Development And Sustainability Metrics. *J American Institute of Chemical Engineers (AIChE)*, 2003, **49**(8), 1928-1932.

32 **Tu, Y.** Production Planning And Control In A Virtual One-of-a-Kind Production Company. *J Computers In Industry*, 1997, **34**(3), 271-283.

33 **De Ron, A.J.** Sustainable Production : The Ultimate Result Of A Continuous Improvement. *Int J Production Economics*, 1998, **56-57**, 99-110.

34 Compact Oxford English Dictionary p. AskOxford.com (Oxford University Press, 2008).

35 **Shepherd, D.** Sustainability And Cement Manufacturing. *2005 IEEE-IAS/PCA Cement Industry Technical Conference Record* pp. 168-176 (IEEE Inc., PCA, IEEE IAS, Missouri, USA., 2005).

36 **Geyer, A., Scapolo, F., Boden, M., Dory, T. and Ducatel, K.** The Future Manufacturing in Europe 2015-2020 : The Challenge for Sustainability. pp. 1-19 (Institute for Prospective Technological Studies, 2003).

37 **Albino, V. and Kuhtz, S.** Enterprise Input-Output Model For Local Sustainable Development : The Case Of A Tiles Manufacturer In Italy. *J Resources, Conservation and Recycling*, 2004, **41**(3), 165-176.

38 **Baldwin, J.S., Allen, P.M., Winder, B. and Ridgway, K.** Modelling Manufacturing Evolution : Thoughts On Sustainable Industrial Development. *J Cleaner Production*, 2005, **13**(9), 887-902.

39 **Griese, H., Stobbe, L., Reichl, H. and Stevels, A.** Eco-Design And Beyond- : Key Requirements For A Global Sustainable Development. *Asian Green Electronics - Design for Manufacturability and Reliability, 2005AGEC*, pp. 37-41 (IEEE Inc., Shanghai, China, 2005).

40 **Sonntag, V.** Sustainability : In Light Of Competitiveness. *J Ecological Economics*, 2000, **34**(1), 101-113.

41 **Al-Yousfi, A.B.** Cleaner Production For Sustainable Industrial Development: Concept And Applications. *J Practice Periodical of Hazardous, Toxic, and Radioactive Waste Management*, 2004, **8**(4), 265-273.

42 **Kaebernick, H., Kara, S. and Sun, M.** Sustainable Product Development And Manufacturing By Considering Environmental Requirements. *J Robotics and Computer-Integrated Manufacturing*, 2003, **19**(6), 461-468.

43 **Seuring, S.A., Koplin, J., Behrens, T. and Schneidewind, U.** Sustainability Assessment In The German Detergent Industry : From Stakeholder Involvement To Sustainability Indicators. *J Sustainable Development*, 2003, **11**(4), 199-212.

44 The Future Of Manufacturing In Europe 2015-2020 : The Challenge For Sustainability. *FuTMaN Final Report*, pp. 1-50 (Research Industrial Technology, EU FP6 Funded Programme, 2003).

45 **Eckart, U. and Jens, K.** Sustainable Manufacturing As Factor Of Innovative Production Processes. *Intl Conference on Innovative Production and Machines Systems, I*PROMS'07* (MEC, Cardiff Univ., UK, 2007).

46 **Barlow, C.** Towards Industrial Sustainability. *IET Manufacturing Engineer*, 2003, **82**(6), 36-37.

47 **O'Brien, C.** Global Manufacturing And The Sustainable Economy. *Int J Production Research*, 2002, **40**(15 SPEC.), 3867-3877.

48 **Jayachandran, R.** The Design For Sustainable Manufacturing System : A Case Study Of It Importance To Product Variety Manufacturing. In Singh, S., Goodyer, J. and Popplewell, K., eds. *Int Conference on Innovative Production and Machines Systems, I*PROMS'06* (MEC, Cardiff Univ., UK, 2006).

49 **James, R.M., John, C.C., Mitchell, J.S., David, R.S., David, R.H., Qiong, Z., Sheryl, A.S., Valentine, U.J., John, W.S. and Jerald, L.S.** Sustainability Science And Engineering : The Emergence Of A New Metadiscipline. *J Environmental Science and Technology*, 2003, **37**, 5314-5324.

50 **Mulholland, K.L., Sylvester, R.W. and Dyer, J.A.** Sustainability : Waste Minimization, Green Chemistry And Inherently Safer Processing. *J Environmental Progress*, 2000, **19**(4), 260-268.

51 **Labuschagne, C., Brent, A.C. and Van Erck, R.P.G.** Assessing The Sustainability Performances Of Industries. *J Cleaner Production*, 2005, **13**(4), 373-385.

52 **Herron, C. and Braiden, P.M.** A Methodology For Developing Sustainable Quantifiable Productivity Improvement In Manufacturing Companies. *Int J Production Economics*, 2006, **104**(1), 143-153.

53 **Pham, D.T. and Thomas, A.J.** Fighting Fit Factories : Making Industry Lean, Agile and Sustainable. *IET Manufacturing Engineer*, 2005, **84**(2), 24-29.

54 **Pham, D.T., Pham, P.T.N. and Thomas, A.J.** Keynote Address - Integrated Production Machines and Systems : Beyond Lean Manufacturing. *Intl Conference on Manufacturing Automation, ICMA'07* (NUS, National University of Singapore, 2007).

55 **Holweg, M.** The Three Dimensions of Responsiveness. *Int J Operations and Production Management*, 2005, **25**(7), 603-622.

Table 1a : The Key Criteria of Lean Manufacturing

No.	Mode	Criteria	Description	Category
1.	10	Free Non-Value activities	- Low inventory, less set-up time, less distance part transfer, low Work-In-Progress (WIP), high percentage of common parts, short lead time.	Manufacturing processes
2.	8	JIT production and delivery	- Lower lead time, high percentage of part delivered JIT by supplier, integration between supplier & company's information system, small batches.	Manufacturing processes
3.	7	Team managed	- High no. of teams, high rotating task (multi-skilled), elected team leader by their own team, flexible organisation structure, cross functional character.	Company organisation
4.	6	Continuous improvement practised	- High benefit from suggestions, low scrap or rework parts, low machine idle time, high autonomous defect control, Total Preventive Maintenance (TPM).	Manufacturing processes
5.	6	Partnerships practised	- Customer involvement, focus on a firm's customers and their needs, parts co-design with suppliers, long term contract, good communication level.	Customer/Supplier relationships

Table 1b : The Value Added Criteria for Lean Manufacturing

No.	Mode	Criteria	Description	Category
1.	5	Single Piece Production Flow	- continuous flow.	Manufacturing processes
2.	3	Flexible Information System	- good communication in both horizontal & vertical, computer integrated equipments, autonomy system for auto response to problems.	Hardware/Software technologies
3.	2	Flexible for changing (tools/parts/processes etc.)	- assembly system.	Hardware/Software technologies

Table 2a : The Key Criteria of Agile Manufacturing

No.	Mode	Criteria	Description	Category
1.	31	Team managed	- flatted, virtual organisation, flexible & nimble, learning employees, multi-skilled, multi-functional, self committed, knowledge driven, motivated, fully empowered employees, fully utilizing the ideas & knowledge of employees, innovative culture in product development, self autonomous, open management, participation based management & susceptible to change & improvements.	Company organisation
2.	22	Responsiveness, Adaptability and Flexibility	- flexible, quick response to change, easily collapsible, control information/operating procedure, customizing production, entrepreneurial organisation, market driven, the capability to change direction with ease, social changes & competitor movements, change over/setup system, production process.	Manufacturing processes
3.	17	Partnerships practised	- customer's delight, robust & 100% response envisaged, customers integrate in designing, manufacturing, marketing & support for all product services, supply chain management principle.	Customer/Supplier relationships
4.	14	Integration of Information technology	- reengineered IT integration, information automation, distribution system, virtual corporation (one-kind-product).	Hardware/Software technologies

Table 2b : The Value Added Criteria of Agile Manufacturing

No.	Mode	Criteria	Description	Category
1.	8	Strategic viewed	- global market competition, new markets development, use & improve core competencies.	Company organisation
2.	8	Cost managed	- value focused on activity, strategy, quality & productivity based costing system, long term gains in financial performance.	Company organisation
3.	7	Automation type	- flexible, smart and adaptable automation, production technologies, modern manufacturing system (FMS and CIM), soft technologies, new, appropriate and necessary technology of differentiating with high-tech where necessary.	Hardware/ Software technologies
4.	5	New product technology	- short & effective product life cycle, easily configured /upgraded.	Hardware/ Software technologies
5.	4	Time managed	- very efficient, material movement, production schedule.	Manufacturing processes
6.	3	Quick Reconfiguration	- ability to quick reconfigure corporate structures, facilities, people, Company organisation & technology to meet unexpected & short lived market opportunities, product mix and designs.	Hardware/ Software technologies
7.	2	Short term manufacturing planned	- Just-In-Time (JIT) purchase & least dead investment.	Manufacturing processes

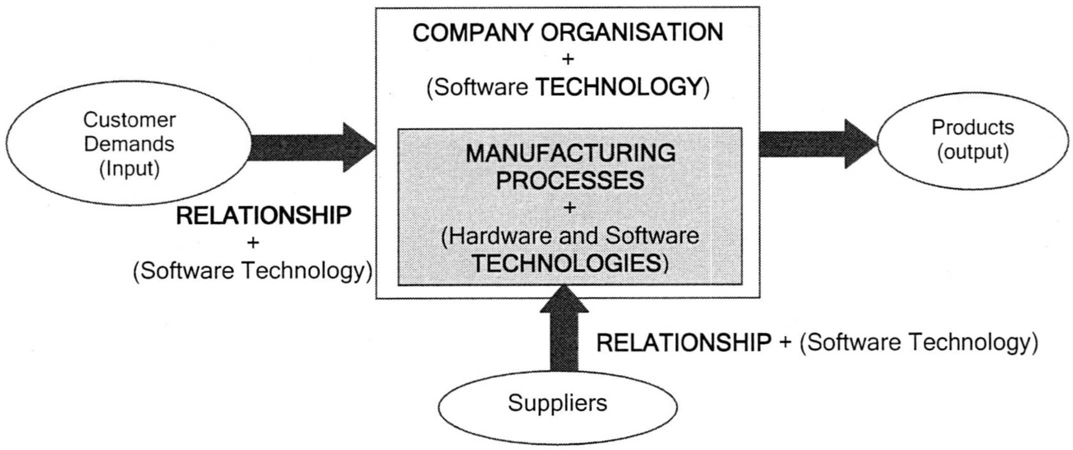

Figure 1 : The Basic Elements of Manufacturing System

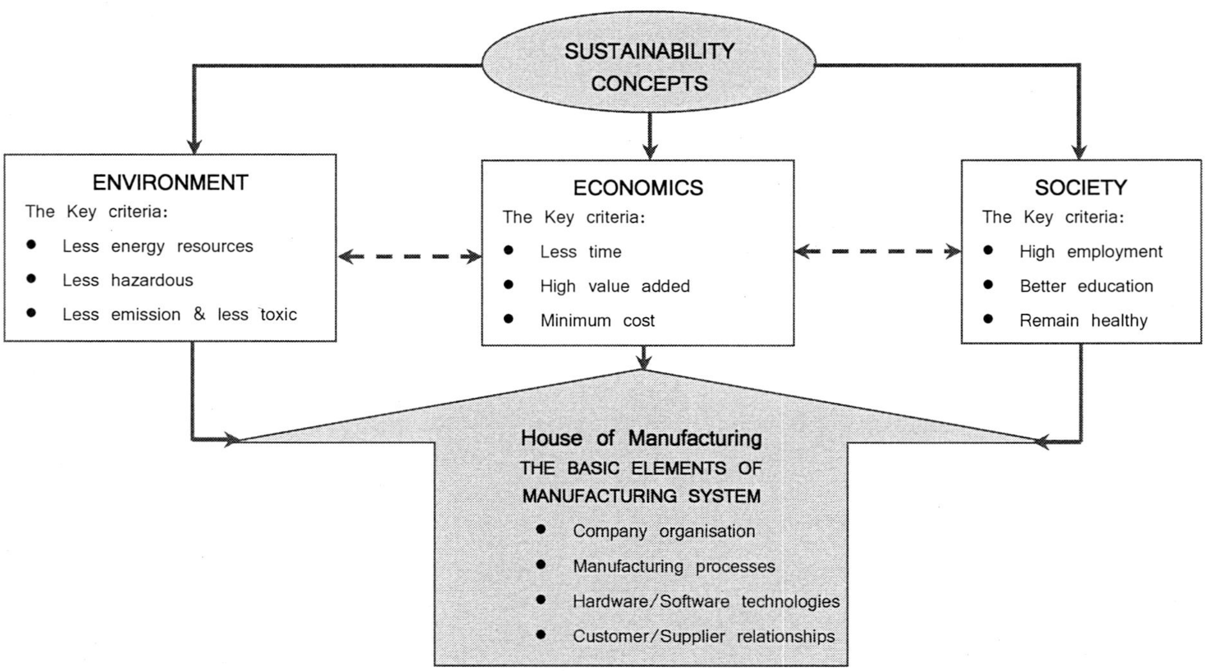

Figure 2 : The Connections of Sustainability Concepts To Basic Elements of Manufacturing System

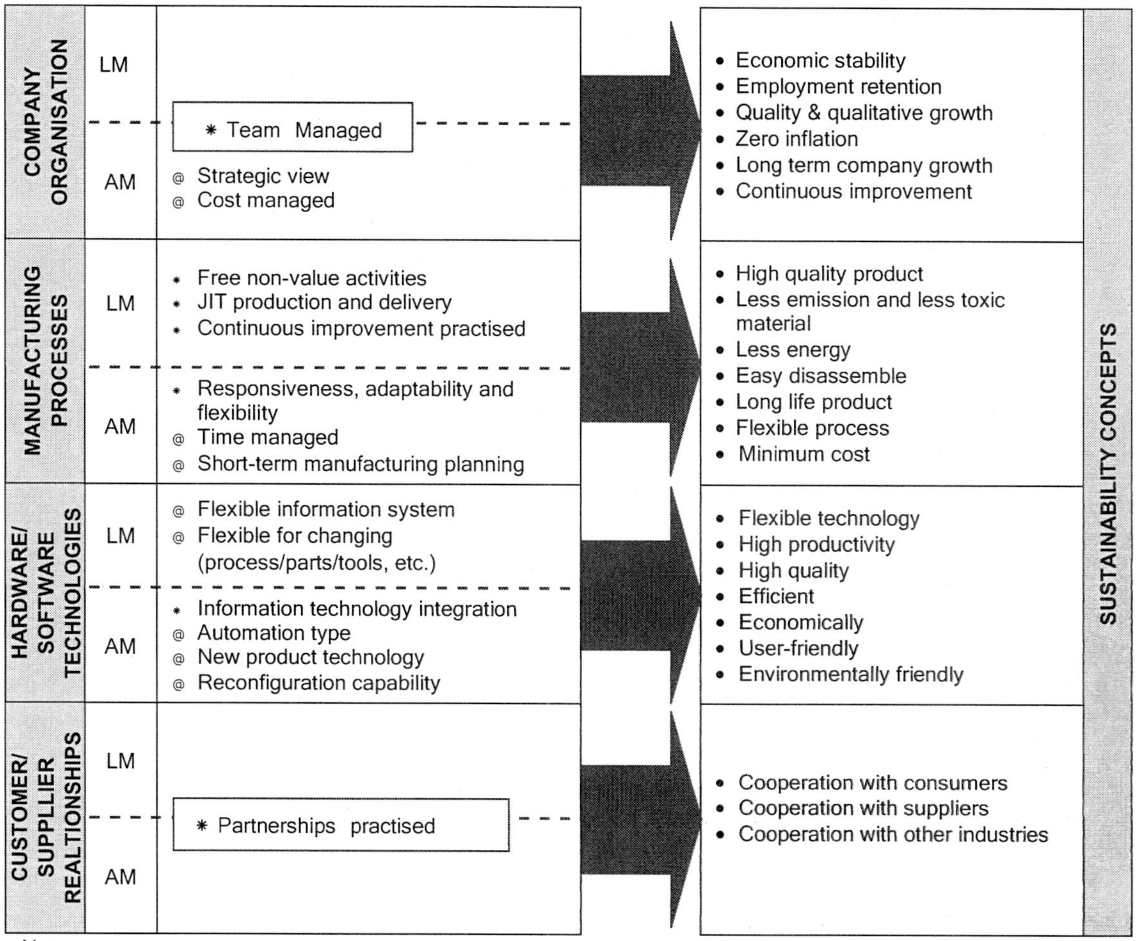

Notes;

* Key criteria

@ Value Added criteria

Figure 3 : The Connections Between LM criteria, AM criteria and the Sustainability Concepts

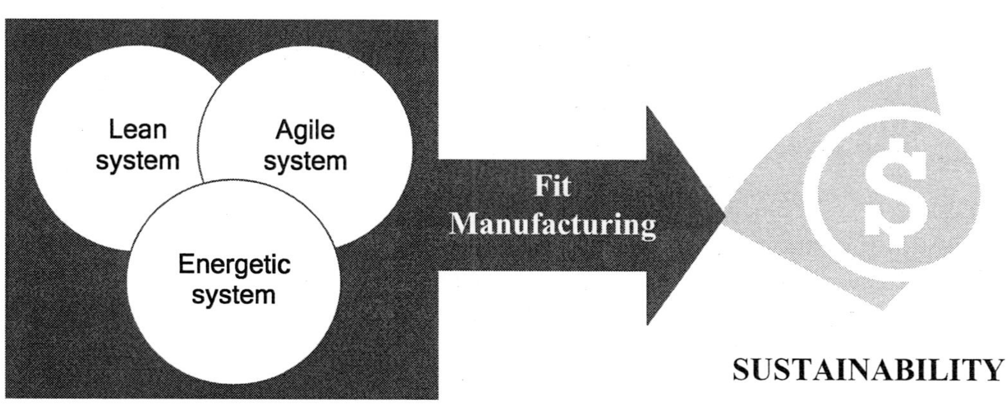

Figure 4 : The Fit Manufacturing Concept

Innovative Production Machines and Systems
D.T. Pham, E.E. Eldukhri and A.J. Soroka (Eds.)

Six Sigma Green Belt training for Researchers

S.Y. Coleman[a] , T. Fouweather[a] , H. Henning[b] , O. Uygun[c]

[a] *Industrial Statistics Research Unit (ISRU), Herschel Building, Newcastle University, Newcastle upon Tyne, UK NE1 7RU*
[b] *IFW Hannover, Leibniz Universität Hannover, An der Universität 2, Garbsen 30823, Germany*
[c] *Department of Industrial Engineering, Faculty of Engineering, Sakarya University, Esentepe Campus 54187 Sakarya, Turkey*

Abstract

Six Sigma is a management methodology for process improvement. It has been studied and adopted by many companies and institutions since it was introduced by Motorola in the late 1980s. The benefits of Six Sigma have been widely recognised and increasingly the methodology has been applied in different contexts. The application of Six Sigma in the research process, however, has not been explored. The I*PROMS network of excellence has many PhD and other researchers and is an excellent forum in which to investigate the potential for a process improvement approach to research. ISRU have been training Six Sigma black, green and yellow belt personnel for many years and have adapted their material for a pilot run of this new application area. In conjunction with partners from IFW and Cardiff, a programme has been developed and piloted on a group of I*PROMS researchers. It was decided to focus on green belt training as that encompasses the philosophy and main concepts of Six Sigma and can be taught in 10 days over two week-long sessions; this time commitment being more appropriate for research staff than black belt training which is typically five week-long sessions spread over six months. The green belt programme is suitable for enhancing and extending into a full black belt programme. This paper describes the approach and how it differs from other applications. It includes observations from the pilot study and outlines the future plans.

Keywords: Six Sigma, green belt, statistics, research, process improvement

1. Introduction

Six Sigma is one of the most widely used management philosophies for improving efficiency and profitability.[1] Six Sigma projects are run by personnel specially trained in the basics of statistics, problem solving and project management. As a test of the widespread coverage of Six Sigma, one can carry out the "web search test". Keying in the words "six sigma" into yahoo.com in March 2008 returns 29,200,000 hits. Details of six sigma, therefore, can easily be gleaned from the internet. Six Sigma is an effective vehicle to reinforce the scientific method so fundamental to research.

The original Six Sigma methodology emerged in Motorola in the US in the late 1980s. Bob Galvin, then chairman said: "Over the past decade we have reduced in-process defects by over 300 fold, which has resulted in cumulative manufacturing cost savings of over 11 billion dollars" . Such large cost savings alerted companies and institutions of all sizes to Six Sigma and interest is still growing. A search for "Six Sigma" in Google in September 2005 yielded 4,460,000 entries; a search in September 2007 yielded 20,400,000 entries.

The essence of mainstream Six Sigma is to identify areas for improvement that are of strategic importance and can be tackled in one or more projects of a reasonable size. [2] These projects are carried out by cross-functional teams of people usually led by "black belts" and typically including "green belts". Black Belts go through an intensive training combining theory and practice. Lighter training programs are provided to green belts and yellow belts. Everyone in a company has some training and participates directly or indirectly in company wide improvement initiatives. [3]

Managers, engineers and other professionals identified as candidate black belts are taught in one-week blocks with roughly a month in between for them to put the ideas into practice. Black belt training usually comprises five one-week courses focussing on the five phases of a Six Sigma improvement project: Define, Measure, Analyse, Improve and Control. Black belt training contains the most statistics, particularly in the MAIC phases but it is statistics put, within the context of the company and its improvement strategy.

There are many providers of Six Sigma training and many companies provide it for themselves. Although there is good agreement between the various curricula, there is no universally recognised British or European standard. ENBIS, the European Network of Business and Industrial Statistics has agreed an outline curriculum and the requirements for achieving black belt status through its Six Sigma Practitioners' Circle. Here is the typical content for the different phases:

- Define includes BS 7850 quality improvement tools, such as cause and effect diagrams, tree diagrams, flow charts and run charts, and a wide range of problem analysis tools, such as Kepner-Tregoe analysis, the Six Thinking Hats of Edward de Bono and TRIZ (the theory of inventive problem solving).
- Measure includes issues of data quality and access, summary statistics, probability distributions, measurement system analysis (MSA)
- Analyse includes comparative tests, such as t-tests and non parametric tests, correlation, regression, analysis of variance
- Improve includes design of experiments, risk management
- Control includes design and implementation of statistical process control (SPC)

Throughout the training there is emphasis on measuring the gains, good presentation skills and looking for opportunities to transfer the ideas to other departments.

Green belts have a less detailed training, typically two days training per phase; yellow belts typically spend one day per phase. Green belts and yellow belts may be people above or below the black belts in the hierarchy of the company. Projects should have a champion who ensures that facilities are available when required and that a project charter is established. In some organisations, black belts focus exclusively on project work after they have been trained and after about 2 years change job or go back to their previous role. Some black belts, are given further training and take on the role of Master Black Belt mentoring other black belts and designing or expanding Six Sigma initiatives.

Such an intensive period of training makes black belts highly employable and there is a good demand for them. A search on the internet shows a healthy range of job possibilities in the UK, US and elsewhere with very reasonable salaries and conditions.

The job market shows that the Six Sigma method has evolved and many companies now ask for Lean Six Sigma experts. Lean Manufacturing [4] has gained much popularity as a method to enhance the added value of production and service operations and Lean Six Sigma attempts to combine it with Six Sigma. In essence lean focuses on removing complexity, Six Sigma focuses on process improvement. Lean, however, involves less statistics and is very popular in some applications, such as healthcare.

2. Adapting Six Sigma for Researchers

The research scenario is different in many ways all of which require careful consideration.

2.1 Research institution compared to company

Research institutions may have a wide range of activities taking place. It is helpful if researchers can speak to people from different specialties. The wide application of Six Sigma reveals cases where the same scientific methods have different nomenclature in different specialties. A good example is conjoint analysis which uses factorial experiments in questionnaires. Another is product reliability which has similarities to medical survival analysis.

2.2 Researcher compared to employee

Although most researchers are part of a department or school and want to support and promote their affiliated association, they nevertheless are mostly concerned with completing their research successfully and/or achieving their PhD. Researchers are characterised by being oriented to working towards achieving something for themselves. They are usually judged by their personal success. This motivation underpins the way Six Sigma is taught to researchers. Their projects are not necessarily in line with the strategic goals of their department but are aimed at improving their research skills and ensuring progress in their longer term project.

Six Sigma emphasises the importance of a team approach. Researchers are working with a team but are also competing with team members. A multi-disciplinary team helps to smooth the way for progress and researchers should be able to draw on the support of people to guide and clear the way.

2.3 Overview of Green Belt appeal

Six Sigma green belts fulfil the role of Process Improvement or Quality Improvement team members on Six Sigma projects. A green belt team member can come from any level within an organisation and provide subject matter expertise for a project. They may be considered as leaders on certain projects that don't require the black belt level of analytical rigour. Six Sigma green belts are taught the Six Sigma DMAIC methodology which helps them to cascade Six Sigma tools and techniques throughout an organisation. Green belt competence prepares individuals within an organisation to use simple Six Sigma tools and techniques to define, measure, analyse, improve and control processes.

Although the green belt curriculum is a subset of the black belt, the content is substantial and covers many tools and techniques. On completion of the course, participants will have learned:

- A structured problem-solving methodology for addressing business improvement projects
- Basic (and some advanced) quality/ process improvement tools and techniques
- How to generate bottom-line financial results
- To understand the "Do's" and "Don'ts" of Six Sigma
- To recognise the key attributes of a successful Six Sigma program
- To understand the role of a Green Belt in the Six Sigma implementation
- To appreciate project selection process, prioritisation of projects, etc.

All of the above skills are of vital importance to the researcher.

2.4 Background to training

The trainers at ISRU developed their Six Sigma training materials in 2003 as part of a European programme with colleagues in Wroclaw University of Technology in Poland. With EU funding, over a period of six months, ISRU personnel trained Polish and Canadian employees of the US Sauer-Danfoss company alongside academic colleagues from the Centre for Advanced Manufacturing Technologies (CAMT), Wroclaw University of Technology, Institute of Production Engineering and Automation. The training material was then adapted for use by CAMT to train Polish companies in Six Sigma. This exchange has worked very well and CAMT have continued to develop their business very successfully since 2003. In summary, ISRU's Six Sigma credentials are as follows:

- training in statistical thinking since 1984
- ISRU Six Sigma programme developed with reference to General Electric's Roger Hoerl's curriculum and various authoritative sources including Breyfogle [2,3] and web sites such as www.onesixsigma.com
- further development of Six Sigma material sponsored by European Union grant in collaboration with Polish partners at Wroclaw University of Technology
- first successful ISRU Six Sigma programme delivered to US company in 2003
- £750,000 funding from European Regional Development Fund for ISRU to spread the use of Six Sigma tools to North East SMEs from 2003 to 2005
- £525,000 funding from European Social Fund for ISRU to train staff of North East SMEs in Six Sigma skills from 2003 to 2005
- Six Sigma training completed in healthcare and finance sectors
- ongoing Six Sigma programmes in manufacturing sector
- lead figures in ENBIS (European Network for Business and Industrial Statistics) Six Sigma Practitioners' Circle and RSS (Royal Statistical Society) Six Sigma study group [5]

The advantage of ISRU's extensive experience of statistics for business improvement is the large number of case studies that can be included in the training material and the broad personal knowledge that informs the teaching.

3. The I*PROMS green belt programme for researchers

3.1 Candidates

A wide range of people can benefit from applying a logical, statistically aware approach to their research. Many more non-statisticians are familiar with statistical concepts and terms because of Six Sigma and this increase in statistical awareness leads to broader interest. Statistical thinking can lead to major breakthroughs in many fields. Sometimes it is because of a quite simple application of rigorous thinking or careful tabulation. Other projects are more sophisticated and include detection of faulty measurement systems, design of experiments and advanced statistical modelling.

The Lean Six Sigma Green Belt Training and Certification is a practical 10 day course that will provide attendees with a solid understanding of basic Lean Six Sigma tools and techniques. Attendees are invited to bring along their own projects and data, in order that these can be discussed and developed during the training. Delegates do not need any prior experience of Six Sigma to attend this Green Belt training. The training is highly practical and interactive as we know from experience that this is the best way to encourage engagement and to consolidate the training.

3.2 Curriculum

The following areas are taught within the course:

- Overview of Six Sigma and its orientation to research
- Green Belt Expectations
- Working with and understanding people (including consideration of personality, such as in the MBTI)
- Running successful meetings
- Working with ideas (including Scoping Worksheet, Brainstorming, Affinity Diagrams)
- Metrics, including CTQs, KPIs
- Cost of Poor Quality
- Questionnaire Design
- Descriptive Statistics
- Data Displays & Variation Analysis
- Confidence Intervals
- Six Sigma case studies, looking for causes (including JMP Demo)
- Data Collection
- Capability
- Measurement Systems Analysis (MSA)
- Understanding variation
- Prioritisation Matrix
- Hypothesis testing (including One and Two Sample Tests)
- Nonparametric Methods
- 5 Whys
- Reliability Analysis
- Introduction to Design of Experiments (DoE)
- Analysis, Conclusions and Recommendations
- Further Experimentation and Optimisation
- Risk Analysis
- Control Charts
- Visual Management, SOPs and Control Plans

The course is interactive, examples of each area are worked through during the training and there are plenty of practical sessions.

3.3 Accreditation

An ISRU certificate of Green Belt course completion is awarded at the completion of both weeks of the course. Green Belt Certification is also available and involves attending a 1 day event at which delegates will sit a written test and deliver project presentations to a moderating panel. Achieving Six Sigma Certification with ISRU provides accreditation from a respected worldwide authority on Six Sigma. The Six Sigma Certification event will be held at the end of the two week-long training sessions.

4. Results from pilot run

There are many fringe benefits from the way different Six Sigma trainees think and learn. Because Six Sigma is a continuous improvement methodology, it is widely applicable and can therefore be taught to a diverse range of people. The training is interactive and hands-on with lots of opportunity for feedback and communication. The need to mix with different types of people has highlighted the importance of communication, an area well described by Roland Caulcutt [5]. Coaching skills are an important addition to the researchers' skill set. Tony Bendell, in his excellent presentation at the RSS Quality Forum on June 7th 2007, allocated MAC phases to left brain thinking and D and I phases to right brain thinking. Such a thorough call upon the researcher's brain is a good addition to the researcher's training.

Statistics is intrinsically difficult; people don't like the "ifs and buts" that go with it. For example, a researcher wants to improve the output of his experiment. Before he can embark on a designed experiment in the Improve phase, he needs to review how the output is measured. If the Measure phase has highlighted problems then these need to be addressed before further work can be entertained. It may be that the extra effort required to achieve a reliable measurement makes the project seem to be not worth the effort.

Common problems with statistics techniques include issues like these:

• SPC on a process shows that the process is completely out of control – much preparatory work is needed before SPC can be useful

• DoE for an important problem, but each trial costs a lot of time or resource and the outcome is either fail or not fail, rather than being a measurable continuous variable

• Process flow mapping and measurement show that measurement errors arise at the first stage of the process, but improving this stage requires involvement with another team who are not very supportive

• The problem of biased samples, where the researcher is only to ask opinions from his friends and fellow researchers, so that it is unrealistic to generalise the results to the wider population

• Aversion to statistical terms with negative connotations, such as deviation, skew, regression and mean; broken down by age and sex.

The pilot Green Belt course part 1 held 11-15 February 2008 was, however, very successful. Here are testimonials from some of the attendees:

"Two of us from Sakarya University attended the I*PROMS Six Sigma Green Belt for Researchers Training programme hosted by ISRU at Newcastle University 11-15th February 2008. We enjoyed the course and found the training very informative. We hope to apply the six sigma techniques in our research. The trainers attached great importance to applying the techniques and we can definitely see where this can be done. We look forward to attending the second part of the training which takes place in Newcastle from 19-23rd May 2008."

"We, two research assistants from IFW Hannover, Leibniz Universität Hannover, attended the I*PROMS Six Sigma Green Belt for Researchers Training programme hosted by ISRU at Newcastle University 11-15th February 2008. We found the course stimulating and very interesting. "

5. Discussion

5.1 Challenges with Six Sigma

In some institutions, Six Sigma is not well received; it is seen as another nuisance fad. Another negative aspect is that some budding statisticians may well be put off statistics because of the basic level at which it is taught and applied in Six Sigma. The subtleties of statistics are enormous and this adds to its appeal for some people but detracts from it for others. People may give up on SPC if wrongly positioned control limits result in too few or too many action signals. Similarly, DoE may not be used again if ill chosen settings for factor levels lead to one factor swamping the experiment so that the effects of other factors cannot be detected, or to the differences in results at different factor levels being too small to be detected in an experiment with a limited number of trials. DoE is fraught with potential problems; the meticulous nature of pre-experiment preparation may prove to be too time-consuming; one tiny mistake in an orthogonal array and experimental results may all be confounded so that the effects of one factor cannot be distinguished from the effects of another. Bias with the effects of time is a common potential problem.

5.2 Examples of projects from other courses

Six Sigma is good for scientists because it applies statistical thinking in unusual places. For example, a designed experiment measured speed and accuracy of operators pricing prescriptions. The experiment was designed with two three-level factors:

• the level of experience of staff: experienced, semi-experienced or novices

• the instructions given: go fast, go fast and be accurate, be accurate

Nine trials were planned with all nine combinations of instructions and experience. The measured outputs were a) speed measured as number of inputs per hour averaged over a day and b) accuracy measured as proportion of inputs without an error. The trial lasted for five days. It was found that

• speed and accuracy increase with experience level

• accuracy rates remained constant over the trials and

• there was no relationship between speed and accuracy.

Accuracy was not affected by the instructions given but speed was. If a person is asked to 'go fast', they will go faster and it will not tend to affect their accuracy level. However, if they are asked to be accurate, speed will reduce, without any noticeable effect on accuracy. It is therefore best to ask for faster speeds and accuracy levels will hold their 'natural' level for the person.

Many successful Six Sigma projects are due to their cross department focus. For example, a pump-manufacturing company was losing money. The factory was very efficient and high tech but the lead time given out to customers by the telephonist was far too long and the collection of payments so unreliable that there were continual cash flow problems. An experiment in which the company varied the time and day of telephoning to follow-up a bill showed that a phone call ten days after sending a letter led to a reduction in the wait for payment of 30 days.

Many Six Sigma techniques are applicable in areas such as web design. Kansei Engineering which attempts to capture the emotional response of customers to a product, involves questionnaire design and data analysis. Questionnaires may usefully include conjoint analysis type questions where respondents are asked to CONsider JOINTly a number a factors which may affect their decision to engage with a web site and ultimately make a commitment to a service or purchase. Post design, SPC may be employed to monitor the use of the web site and help with its continuous improvement.

5.3 Six Sigma slug example

Statistics is not always considered a natural way for people to think. However, we had a nice counter-example when teaching numerical skills to unemployed people; we discovered that one of the delegates had unknowingly run a designed experiment on how to get rid of slugs from his garden. The factors were: how many salt pellets; where to place them; was it wet weather or dry; how long to wait. He now has an optimum slug killer strategy, and he got a job!

This problem benefits from being studied within the context of Six Sigma methodology. Below is a description of the various phases.

Define: problem is too many slugs, horrible to look at and stand on and they eat plants. Enlist help of everyone in house and ensure unanimity of purpose. Consider ethical aspects and environmental implications of mass slug slaughter. Include snails.

Measure: count the number in garden at night time using torch, count slugs and snails regardless of size. Obtain baseline measurement of population.

Analyse: past experience suggests they like damp weather and travel around the whole garden. Salt pellets kill them but not sure how many to put down and how long to leave them before removing and replacing them. No other known predator, eg hedgehogs in garden.

Improve: use designed experiment with set quantity of slug pellets spread evenly or in a single pile, put out on wet days or dry days and leave for one or two days

Control: continue to put out pellets as indicated by experiment and monitor situation using SPC, apply technique to problem of greenfly in greenhouse

The Six Sigma methodology can usefully be combined with tools and techniques from "lean" methodology. It also reinforces the power of the Deming PDCA cycle, which notes the importance of review in the Plan, Check, Do and Act phases of a project.

In some companies there is pressure to dumb down the statistical content of Six Sigma training. The more committed companies will maintain content level but other companies will adopt the Define phase, process mapping and charting but not risk using DoE, regression and SPC beyond their initial training. Although this is undesirable, statisticians will be no worse off than before and hopefully the net change in the level of statistical interest will be positive.

6. Conclusions and future plans

The use of Six Sigma in the research environment is producing thought-provoking scenarios in which to develop process improvement. Researchers have specific needs and differences compared to employees in other roles, for example breaking new ground is more important than embedding new skills. The abundance of real applications is providing lots of interesting research topics, for example the proper handling of short span start up data for SPC, the role of collaboration with other colleagues and the use of teams in research.

Depending on the outcome of the complete I*PROMS Green Belt Six Sigma programme, it may be considered desirable to extend the programme to Black Belt training and to develop master classes in specific areas, such as design of experiments, reliability or optimisation to support Six Sigma Master Black Belts. Although Six Sigma has been around for some time and its popularity and penetration into new business areas and sectors is still increasing, it is important to be ready for changes in direction. The fundamentals of problem solving and quantitative analysis, however, are likely to remain fundamental features of any new programme. Training, such as this course may be a suitable activity for the I**PROMS enterprise to follow the completion of the I*PROMS network of excellence.

Acknowledgement
This work is supported by a UK Collaborative Innovation Partnership with Business and Decision, www.businessdecision.co.uk.

University of Newcastle is a partner of the EU-funded FP6 Innovative Production Machines and Systems (I*PROMS) Network of Excellence.
http://www.iproms.org

References

[1] C.Gygi, N.DeCarlo, B.Williams (2005) *Six Sigma for Dummies*. Wiley
[2] F.W.Breyfogle (2001) *Managing Six Sigma*. Wiley.
[3] F.W.Breyfogle (2003) III *Implementing Six Sigma*. Wiley.
[4] J.Bicheno (2004) The New Lean Toolbox, Picsie.
[5] S.Coleman, T.Greenfield, D.Stewardson and D.Montgomery Eds (2008) *Statistical Practice in Business and Industry*, Wiley.

Innovative Production Machines and Systems
D.T. Pham, E.E. Eldukhri and A.J. Soroka (Eds.)

A Multi-faceted Strategy to Ensure Economic, Environmental and Social Sustainability in Manufacturing SMEs

R. Barton[a], C Araújo Sá[b], J Larreina[c], B Grabot[d], D Hughes[e],

[a] *MEC, Cardiff University, UK*
[b] *IDIT – Instituto de Desenvolvimento e Inovação Tecnológica, Portugal*
[c] *Fundacion Tekniker, Spain*
[d] *Ecole Nationale d'Ingenieurs de Tarbes, France*
[e] *Department of Mechanical Engineering, UCD, Ireland*

Abstract

The pressure to increase all aspects of sustainability is increasing. Especially in the SME manufacturing sectors where physical waste represents environmental impact, material under-utilisation represents cost effectiveness and design and management capabilities are increasingly focussed on adding value without requiring additional resource. Over three years, the PREMI project has focussed on supporting SMEs trying to achieve the transition towards sustainability. Each company faces many individual aspects which the project partnership has tackled individually to create significant competitive advantage for the recruited companies. However, through the co-ordination of efforts between the project partners, dissemination of the accumulated expertise between and throughout the regions, and collaboration in creating a company-wide strategy for sustainability has yielded a more robust output in terms of increasing company sustainability. This paper describes some of the typical key elements and collates the findings to describe a wider approach to sustainability that can be adopted by any level of SME and developed accordingly. The approach is case-study oriented, reflective of a wide scope of practical work undertaken in industry and note should be made that due to space restriction there is little opportunity for academic verification at this stage, future work will tackle this area specifically.

Keywords: Recyclability, Miniaturisation, Waste Reduction, Sustainability

1. Introduction

In the course of working with over 120 companies (60 on detailed projects) focussed on aspects of sustainability, the PREMI partnership is well placed to amalgamate learning and experience in this area into a sustainable operational strategy.

Individual studies were elaborated to the rest of the partnership, and where permitted published as case studies but it was seen that there was rarely any coherent strategy to develop sustainability as a business improvement initiative within any of the SMEs visited. This paper aims to close that gap by demonstrating how such improvements can lead to greater gains when coupled with progress in other areas of the business controlling interactions and dependencies in the manufacturing operation.

1.1. Sustainability

Sustainability has a wide variety of meanings and often also understandings, but we can consider it in terms of manufacturing systems to be a term to encompass aspects of economy, ecology and social responsibility. A company can claim to be sustainable when the net effect of the operations management has positive outcomes on company profitability and social standing while reducing environmental impact.

The following sections describe key features which demonstrate capability improvement in the areas of recycling, waste prevention, waste control and operational management which, when accumulated, aim to deliver improvements in environmental, economic and social sustainability.

2. Design for Recyclability

In the current context of the sustainable development, products' recyclability seems to be a crucial social concern, especially for companies. This is reflected by a growing number of strict environmental norms and regulations they have to face in practice.

In order to make designers aware of the knowledge allowing the recyclability assessment of a product, a suitable solution (amongst others) consists in making this knowledge available within the tools that they already use during the design stage of the product, i.e., the CAD–CAM systems. However, providing the designers with such systems is nothing less than an obvious task, since the environmental norms and standards are usually described in natural language, which could be difficult to interpret.

We have proposed a methodology allowing to analyse the compliance of a product with an eco-label, with the support of a decision support system (DSS)..

2.1. Pre-requisite; type of company targeted

The targeted companies design and manufacture products and are interested in getting eco-labels on which this study focuses. Eco-labels are non compulsory regulations, but are consistent with the general framework of the ISO 14000 family standard. They provide details on the parts that are concerned with requirements a product has to satisfy.

Let us mention that the compliance of industrial products with environmental standards and norms will soon be an important competitive advantage for the companies, before becoming a legal obligation. This implies that the proposed approach could apply to any company.

The process is facilitated by the use of the software tool developed at ENIT for checking automatically the compliance of a product with an eco-label. The method described hereafter describes how to create the data required by the tool.

2.2. Description of the Methodological Implementation Scheme

- Step 1: definition of the bill of material

This step is concerned with the usual design process of a product.

The result is the data on the product contained in the bill of materials.

- Step 2: eco-labels analysis

A set of well known environmental standards available in the literature are selected, then analyzed in order to carry out the data required by the recyclability assessment that are not present in the bill of materials (defined in step 1). The definition of the method has been based on an analysis of the following standards:

  - the German Blue Angel eco-label,
  - the Nordic Swan eco-label,
  - the European Computer Manufacturers Association (ECMA) technical report on product-related environmental attributes,
  - a paper on DFE by the America Electronics Association (AEA),
  - the TCO'99 Swedish eco-label program,
  - the SIEMENS 'SN 36350-1, Environmentally Compatible Products' corporate standard,
  - the Japanese PC Green Label system.

The result is additional product data, i.e. those related to the environmental requirements (defined in eco-labels).

- Step 3: definition of an extended bill of material

Based on the 'recyclability data' obtained in the previous step, this step consists in adding these data to the classical bill of material (as defined in step 1). The additions may be of different types: new 'objects' to add to the bill of materials, data on the links between parts, or new data related to components already described in the bill of materials. The new model is then characterized by:

the product structure which provides the classical model of a product, allowing for instance to describe a bill of materials,

the product recyclability, allowing to deal with the criteria concerning the nature of materials contained in a product,

the product disassembly, in order to provide the data required for assessing the how the different parts of the product have to be assembled.

- Step 4: modelling the set of criteria a product has to complies with

The standards or eco-labels targeted have to be modelled through a set of 'criteria', so that these criteria can be applied on the data present in the extended bill of materials.

Since norms and standards are often described in natural language, it is necessary to use an appropriate conceptual modelling language, capable of translating these sources into an exploitable form (useful for the recyclability assessment): the main issue here is to extract critical knowledge which could be hidden or partial, from textual sources.

In the case of the ENIT software, the criteria have been structured into constraints that are propagated within the product model while processing an assessment.

2.3. Results

The final expected result is an operating DSS with the following characteristics:

- on-line access on the designer's workstation, with a possible integration to CAD–CAM tools,
- possibility of selection of a given standard or eco-label in a database,
- possibility of assessment of the ongoing product design according to the selected standard.

The use of this system is the following:

- When a standard of eco-label is selected in the database, the corresponding criteria are extracted;
- The criteria are then instantiated and propagated in the bill of materials: in some cases, questions to the designer are required when the data available are not sufficient for allowing to state whether a criterion is verified;
- Data that are inconsistent with the criteria are identified, and then submitted to the designer with an explanation (reference to the criterion which is not satisfied).

3. Waste Prevention via Miniaturisation

The current trend towards product miniaturisation is leading to a major increase in products performance and besides in waste prevention. Although in most of the cases, the motivation for such trend is to improve performance even to achieve new capabilities in the product, the consequence of such trend is a reduction in the generated waste mainly during the manufacturing process. For such purpose, micro-technologies are playing an important role.

3.1. Pre-requisite; type of company targeted

The targeted companies for such kind of applications, are small or medium companies, devoted to the production of medium technology products, which need to incorporate the latest micro-manufacturing capabilities to their product range in order to compete in a global economy.

However the access to these new micro-manufacturing technologies is a barrier for them. Firstly because of the required knowledge and expertise to use them, (even to be conscious of their capabilities), and secondly, due to the high cost of the required equipment. Besides, the required micro-manufacturing capabilities are only for small and specific operations, in order to increase the added value of the final product. So they are not for substituting the existing conventional technologies, but for complementing them. So, the conclusion is that this typology of company is not able to access to micro-manufacturing by itself; they strongly needs the support of technological agents.

3.2 Miniaturisation Process Steps

Step 1. Identification of miniaturisation needs:
First step was to analyse in our surrounding SMEs their need for miniaturisation. For such purpose a number of small companies, around 30 companies, were interviewed to identify their need.

Step 2. Dissemination of micro-manufacturing technologies:
Second step, which was made sometimes in parallel with the step one, was to show the micro-manufacturing capabilities to the SMEs of our surroundings. This change of information, and sharing SMEs needs with micro-manufacturing capabilities,

allowed to identify some pilot projects to check industrial feasibility of such technologies.

Step 3. Feasibility tests:
Some preliminary case studies were selected in order to demonstrate capabilities of micro-manufacturing for the real metal-working industry. Those preliminary studies were founded by PREMI founds. In most of the cases, the studies were technically successful.

Step 4. Industrial Application:
As has been told, in most of the cases the feasibility projects or case studies were technically successful. It means that the required performance of the product was achieved. Afterwards, the question for the involved companies is to integrate these new micro-manufacturing technologies within their production capabilities or to purchase it to a supplier.

Although micro-manufacturing technologies used in the project (micro-milling, micro-turning, micro-electro-discharge machining and micro-injection moulding) are from the operational point of view quite similar to the corresponding conventional processes, and also they are quite usual for our surrounding metal-mechanical industry; in most of the cases they are rejected to integrate these new capabilities.

The main reason for that decision is that the micro-manufacturing capabilities are required to produce some micro-features in small parts which shall be produced 90% by conventional means. So, the micro-manufacturing capabilities should be added to the existing manufacturing capabilities and not for substituting them. It implies a very large investment for a small portion of their production process.

The solution for such limitation is to create specific companies devoted to micro-manufacturing, in such manner that they supply to the existing companies with this new capability.

3.3 Example - Case study of spraying nozzles

This project has been done in collaboration between a surrounding SME and TEKNIKER. The final goal was to develop a new spraying system more accurate than actual in order to go in advance of the more restrictive regulations regarding the use of pesticides. Miniaturisation is required because of regulations for pesticides are everyday more restrictive, so spraying nozzles for lower flow rates are demanded in agricultural purpose, and the corresponding spraying parameters shall be achieved more accurately: flow rate, dispersion angle and quality of the pulverization.

The nozzle is the key component of the sprayer to define its main parameters (flow rate, spraying angle, spraying distribution and minimum spot size). So, project was focused in the design and manufacturing of a new set of nozzles.

Nozzles for agricultural purpose are plastic made with a complex shape, so in order to satisfy the market demand for manufacturing large number of parts and at a reasonable price; the use of a replication technology, like micro-injection moulding is a mandatory. On the other hand, the quality and performance of any replicated part depend mainly on the quality and performance of the corresponding mould. So, the original objective was split in the following specific objectives:

Design of a nozzle able of a more accurate spraying (done by the SME).

Definition of the manufacturing process for the nozzle. It implies the design of the nozzle mould according to the micro-manufacturing capabilities, and according to the final replication process (micro-injection moulding) capabilities.

The new nozzle design presents some difficulties which necessitate the use of technologies for precision design and micro-manufacturing. Main challenges of the new design were as follows:

Spraying section is generated by the contact between two opposite mould inserts, which must be positioned with micrometric accuracy not only in an horizontal plane (XY) but also in the axial direction of the mould (Z).

Inserts must be removable and adjustable in order to inject different references of nozzles.

Finally, the shape, roughness and dimensions of the mould inserts require micro-machining capabilities as micro-wire EDM, micro-milling and micro-injection moulding.

3.4 Results

In most cases, as the driving force for miniaturisation is the improvement of the product performance, that is achieved. From the operational point of view, the standard range of nozzles, which are used for flow rates larger than 1 l/min, has been substituted by the miniaturised nozzles, which allow for flow rates below 0,4 l/min. It implies a reduction in the quantity of pesticide used for the same area in the order of 20-25 %, with its corresponding effect in waste reduction and over the environment.

So, waste reduction comes from the manufacturing point of view, because of less material is used to produce the final parts and even the mould tool inserts. In the case study, the waste reduction in the manufacturing of the tool inserts is very small (less than 3%) and nearly negligible. Bigger waste reduction is achieved by the use of micro-injection moulding as replication technology.

With the micro-injection moulding technology, each nozzle is injected independently, one by one due to the capability of inject up to few milligrams. So the accuracy of the final part is higher, not only due to the better quality of the mould tool inserts, by also by the better control of the injection process itself. In comparison with conventional injection moulding, in which a set of 8 to 12 nozzles are injected each shot. Main consequences are the reduction of the sprues (waste material) and the cycle time significantly.

In our case study, the sprue in the micro-injection case is in the order of 50% of the injected part. It means a finale part weight around 1.3 grams and 0.6 grams for the sprue (waste). With the previous conventional injection process, each 8 parts, around 10.5 grams, the weight of the sprue (waste material) is between 20 to 22 grams, it means about 200% of the part weight.

The consequence of this solution was a waste prevention in a double sense: firstly by the reduction of the material needed in the manufacturing process, both, in the tool inserts manufacturing and in the plastic nozzle replication. The second way of waste prevention comes from the use of the new nozzle design in spraying pesticides. As dosing is more accurate and consequently the efficiency of the fumigation operation is higher, then less quantity of pesticide is required.

4. Waste Reduction

The reduction of industrial wastes has always been a matter of concern for every industrial company: it is a mean for the improvement of the overall manufacturing process.

During the last few years, as a consequence of the produced environmental impact, this subject attained a higher level of importance - new and more effective environment regulations have been issued, not only demanding a deeper measurement and analysis of the produced wastes, but also requiring a more effective (and expensive) waste treatment processes and leading to a set of interesting (but not fully used) results.

The periodic analysis and characterisation of the wastes produced by a company over the years, complemented by the assessment of the corresponding occupational health and safety conditions, enables the development of a powerful database that can be used as a concurrent tool for the evolution monitoring of such company. Furthermore, the result comparison between similar manufacturing processes may allow, in the next future, interesting benchmark studies covering similar industrial sectors.

4.1. Pre-requisite; type of company targeted

No pre-requisite seems to be necessary – every company or manufacturing process may be analysed and characterised from an environment point of view. Besides, all of them must be compliant with already existing environment regulations.

Nevertheless, due the importance the results may constitute for the SME´s sustainability - a key factor for the survival of an industrial company, namely with a reduced financial capacity - such companies were chosen.

4.2. Description of the Methodological Implementation Scheme

To optimise the desired data collection within PREMI duration, the following methodology was established and implemented:

Step 1 – Identification of target companies / industrial sectors. For this purpose, using the large experience of IDIT on both:

- production waste analysis, characterisation, management and valorisation (namely through air emission measurements and wastewater treatment)

- occupational health and safety assessment (namely through indoor air pollutants, lighting and noise measurements)

a contact list of the industrial SME's to be "recruited" was established, involving about one hundred companies from wide spread sectors.

Step 2 – Companies "recruitment"

To engage the desired companies, the objectives of each pilot project, and the corresponding expected benefits, nor only were presented on a bi-lateral basis (technical meetings and visits), but also during thematic workshops and seminars.

This "recruitment" work led to a set of a dozen industrial companies that agreed to co-operate regularly with IDIT, the remaining being only considered as companies "receiving technical assistance" from IDIT (because they were strictly interested on regulation compliance).

The "recruited" companies cover quite different industrial sectors, namely: cork transforming, food preparing, textile and clothes manufacturing, metallic goods manufacturing, automotive parts suppliers.

Step 3 – Pilot project implementation

The desired data was collected on each company, two or more times a year and during two or three years.

The collected data covers different items, such as:
- air emission content
- indoor air contaminants
- wastewater content and treatment demands
- noise level at workplaces
- indoor lighting level

The results, computed at IDIT after analysis and characterisation, were not only reported as technical certificates to the company involved, but also saved at IDIT for later processing (evolution monitoring and comparisons between similar companies / processes).

5. Operations Management – A Top Level Model for Sustainability

Each of the above aspects has elements of necessary management in order to drive any improvement, which is the key reason for undertaking this study. Low value products are being sourced in cheaper economies and the remaining higher value supply chains demand high capability and qualification. The principal difference is realising this value without incurring additional cost or resource, hence careful management is essential to meet the more demanding market requirements

In accordance with the above sections, there are elements of commonalty forming a logical chronology for company development which is interdependent with measures and improvements of sustainability;

1. Bring control within company sphere of operation – include design and customer facing capabilities to capture the voice of the customer, exceed their expectations by driving improvements in sustainable aspects.

2. Association with technology agent – work with all available partners and resources so that the companies key capabilities are focussed on but all available opportunities are taken.

3. Priorise: although they are convinced they must invest on environmental issues, SMEs do not feel it is a priority particularly when compared with other required investments (on productivity improvements, for example); as a consequence, these companies frequently only intend to be compliant with the existing environment regulations, as a rule discarding further involvement on waste reduction projects;

4. Collect, measure, analyse and improve on levels of data representing the key performance criteria; establish the correlation between the improvement on the manufacturing processes and the evolution of the corresponding environmental, economical and social procedures (amount and type of wastes produced; production waste management; waste treatment and valorisation).

5. repeat the process to realise longevity in the business plan and ensure social sustainability.

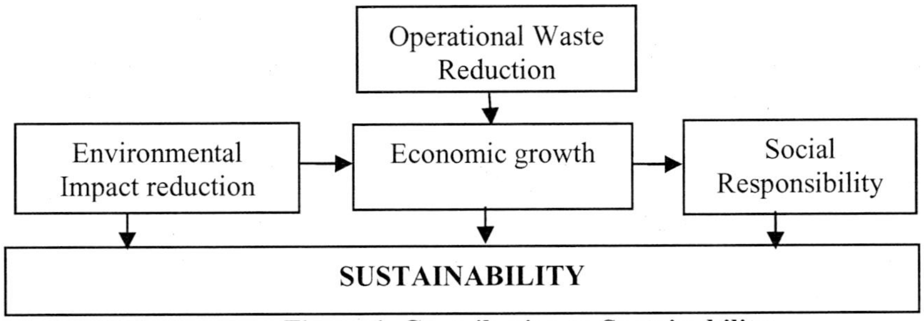

Figure 1. Contribution to Sustainability

This basic chronology is represented in the schematic figure 1 which highlights how business improvement is linked directly to ecological, economical and social parameters, rather than a purely efficiency driven initiative.

6. Conclusions

PREMI has been involved in a wide range of activities, across a range of industrial sectors and in a variety of companies. The extent of this variation has made direct correlation between cause and effect of business improvement and sustainability challenging in the context of creating a more generic implementation model, however it has become evident that the three areas comprising the operational programme for PREMI are key to realising long term sustainability, and that they can contribute either independently or as a structured programme of technology implementation that will ensure sustainability for a manufacturing company into the future.

Hence, it is proposed that an integrated waste management strategy, incorporating each of the facets of the previously described waste reduction and operational optimisation techniques, can provide an ideal for producing a balanced model for sustainability.

6.1 Future Work

The recyclability assessment leads to a multi-criteria problem. This study has been aimed at 'incorporation' of the recyclability knowledge within the designer workstation. So, a great difficulty for the future is to define a communication between the proposed system and the existing ones (i.e. CAD-CAM) in order to collect data that do not need to be redefined by the user.

Micro-manufacturing technologies requires development of special skills and expertise, improved consumer knowledge and awareness and generation of new companies which provide specialist micro-manufacturing service.

With regard to waste control, IDIT intends to establish an "environment picture" of each analysed company, when possible comparing the achieved results inside each industrial sector involved and correlating them with the manufacturing improvement evidenced by the company (particularly as a result of the adoption of innovative manufacturing technologies, new management procedures and/or internal training initiatives).

Finally, the integrated model must be embedded into test companies for continued development and evolution to build upon this practically derived basis for company growth and longevity.

Acknowledgements

The PREMI project was supported by Interreg European funding and by the ODPM.

Cardiff University, ENIT and Tekniker are members of the IPROMS FP6 NoE which works closely with many European projects such as PREMI for mutual benefit.

References

[1] Hurley E and Grant H. Probability maps. IEEE Trans. Reliab. R-124 (1995) 13 28.
[2] Singh B. In: Liu B (Ed.) Probability maps (2nd edn). Springer, Berlin, 1988, pp 154-187.
[3] Ruck MJ. High-resolution spectroscopy of late-type stars. D.Phil. Thesis, University of Oxford, UK, 1994.

Innovative Production Machines and Systems
D.T. Pham, E.E. Eldukhri and A.J. Soroka (Eds.)

Changing conditions that may lead the manufacturing companies to failure modes

K.M. Saridakis[a], A.J. Dentsoras[a], L. Chatzigoulas[b]

[a] *Dept. of Mechanical Engineering & Aeronautics, University of Patras, 26500 Rio Patras, Greece*
[b] *Papastratos SA, 18545, Piraeus, Greece*

Abstract

Modern manufacturing enterprises meet a continuous challenge to improve performance and identify opportunities for growth and long-term sustainability. Locally and internationally, leading businesses are responding to this challenge by employing manufacturing principles to enhance competitiveness in a rapidly changing and demanding global market. A significant research activity has been initiated and the manufacturing community has been motivated in order to enhance the commercial performance of industrial processes and achieve productivity gains, reduced energy and business costs, enforced brand image and reputation. The present paper reviews some of the initiatives that have been identified as pillars of successful strategies, and provides concerns related to hidden aspects that have not been included in the widely proposed holistic and integrated approaches. The positions stated in the context of the paper are justified through a research in five multinational manufacturing firms as well as business cases and examples found in literature.

Keywords: sustainability, organizational learning, lean, agile, partnership

1. Introduction

The present paper comments the current status regarding research for the operational sustainability of manufacturing enterprises and attempts to cover issues that are not taken into consideration during research or implementation activities. These issues are related with conditions that have been identified by the authors after a research in five large and medium sized multinational manufacturing enterprises located in Greece and then validated using case studies located in international bibliography. The conclusions of this research are presented as failure modes of initiatives for efficiency and long-term operational sustainability, which are positioned in order to trigger a discussion for the elements that a holistic approach should integrate, which of them should be prioritized in order to avoid false implementation or misled research activities. The following paragraphs briefly present these conditions and failure modes and introduce potential strategies or actions in order to eliminate them and enforce trends towards successful and efficient manufacturing modes.

2. Comprehension of sustainable manufacturing

The term of sustainability meets variation in its interpretation. For a large group of people it is translated to taking all these actions that will ensure the protection of the environment and will preserve the basic conditions to keep existing in the planet and thus to the business. However, in the business cycles the aforementioned interpretation of sustainability is only one dimension of their complex business models that are based also on social, economical, human and political dimensions. Pham et al. [3] refer to sustainability as "the ability to meet the needs of the present without compromising the ability of the future to meet their own needs". Extending their thinking on sustainability they propose the concept of 'fit' manufacturing as an integration of 'lean', 'agile' and sustainable manufacturing. Thomas et al. [3] propose and describe a holistic manufacturing model targeting to companies development. Their proposed holistic model integrates lean, agility and sustainability, which are identified as independent operational elements. The authors propose a different relationship among these terms. Sustainable manufacturing is a mode that can be achieved by a manufacturing corporation through providing agility to its manufacturing systems while in parallel keeping its operations effective based on lean processes. Sustainability is the main drive of each corporation's activities and agility or leanness may not be the keys to achieve it. There is another dimension, which authors call X-factor, which is related to social, economical, political and other status and which is periodically changing from time to time and from market to market. Every company should observe and investigate what is the status of this X-factor that models the external reality and significantly affects sustainability. Moreover, a fourth dimension is added, the so-called by the authors "everlasting purpose or cause". Collins and Porras studied in their book [5] common characteristics of companies that met great success for many decades and achieved tremendous grow and significant financial profits. A common characteristic of these companies has been their significant purpose that could last over the years and which could drive and inspire their entire organization. They give as an example a pharmaceutical multinational company that distributed vaccines and medicine at no cost in specific market. Later on this company dominated the market as it had gained the consumers respect and preference. This company had as a purpose/cause to take care people's health by any means, even if it comes to zero profitability. Looking at the various examples given in this book, it is very important to conclude that companies should not pursuit the financial profit, but instead to form a cause that may lead them to sustainability. This cause may related with society, working environment, the planet, the human being or other elements but in any case it should keep being an everlasting cause for the employers. An important consideration when addressing the dimensions of sustainable manufacturing is to challenge the misleading 'OR' operator in undertaken activities and decisions (e.g. one company can optimally be either lean or agile) and to chase excellence in all dimensions. Figure 1 presents the four dimensions of sustainable manufacturing as per the authors' perception.

3. Scope and success of technology implementation

In many cases the undertaken research and the advances of technology fail to be addressed in the daily business of the industrial companies. A research has been performed by Saridakis and Dentsoras [6] in order to identify if the outcome of the research activity in the field of artificial intelligence is applied in real-world engineering design problems proving their validity. In their review, they outline eight characteristics that a new element (technological tool or method) should provide in order to achieve high utilization and acceptance: 1) uncertainty management and conceptual design, 2) collaboration, communication and coordination, 3) solution extraction under lacking knowledge/uncertainty, 4) management and reuse of existing design knowledge, 5) generality and domain-independence, 6) extensibility and connectivity to other tools, 7) balance between automation and human activities, 8) simplicity. Although these characteristics are related to artificial intelligence techniques, this research provide a picture of what industries are being expected by the researchers and the technology developers.

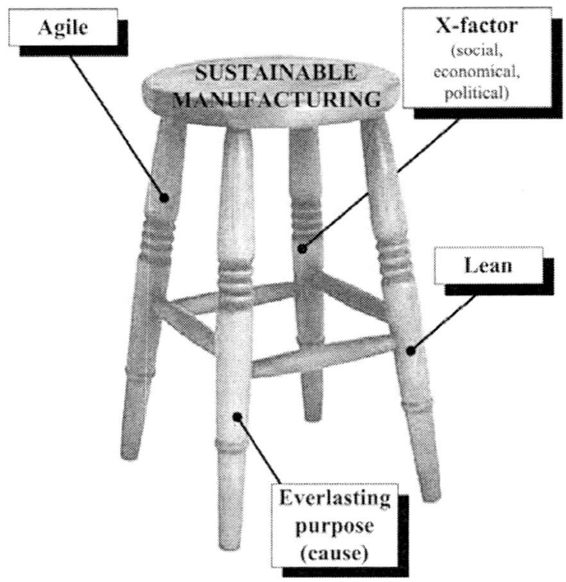

Fig. 1. Sustainable manufacturing dimensions.

Thomas et al. discuss [5] the reasons that technology implementation in manufacturing companies fails and based on their conclusions they propose a model called TIM (Technology Implementation Model). The concerns of the authors have been justified by a research on UK SMEs (small & medium sized enterprises) that have invested in advanced manufacturing technologies. In their work, they focus on two main reasons: the incapability of organizations to adopt new technologies and the lack of a formal implementation model on how the technology implementation will be road-mapped to tangible financial profits.

Even if the new technological is proven to offer a significant but theoretic beneficial effect to the company, this effect can be seriously deteriorated if human resources do not fully accept it. Therefore, it is an important implementation step to involve all related individuals that will have a role as servants of this technology element before this element is finalized in order to equally contribute to the development of its specification and characteristics. Through this step, the technological change will rave funs [9] in the company before even implementing it. Additionally, it is very important to clarify in which areas the technology may contribute to the business while at the same time satisfying individual and organizational needs. The technology should be used to do what annoys people, what requires significant time without adding value to company or the organization, and what people cannot do.

4. Escape business shrinking: Partnerships

In many business markets periodically a shrinking trend is observed. Small and medium sized companies strive for their sustainability through incorporations, collaborations and positioning to other market areas. A common reaction of large-sized manufacturing company in order to keep profitability having decreasing sales is to reduce the available manpower, a move that may have a significant effect on the company's future regarding agility and potential to create strategic advantages to outperform competition, whereas a dramatic headcount reduction that will have clear and direct financial impact, may drive the company to a loosing streak. As the large multinational companies that have numerous distributed industrial units watch markets fluctuations that directly affect their profitability, an effort is made in order to stabilize profits in specific levels that will be preserved independently of the volumes produced and the sales undertaken. Moreover, these large multinational in most cases have entered their shares in the stock markets and the investors are keen on companies with high financial turnover. This means no assets that are not being used (silent stores of obsolete equipment and false investments), no non-moving inventories (spare parts that are reserved for safety reasons and never been used) and surplus expenses (standard variable expenses and salaries) resulting to high manufacturing conversion costs and thus to low competiveness. Towards the aforementioned initiatives, the manufacturing companies can purchase outsourcing services in order to cover activities that do not coincide with their core businesses such as facilities management, and/or purchase services by original equipment manufacturers that are related to their core businesses, such as asset care of special manufacturing equipment. In this way the multinational companies may add an extra manufacturing cost by outsourcing services, but they are capable of reducing gradually their headcount and they stabilize their profit in some marginal intervals (if they sell less, they have less production volumes and they pay for less services).

The last two decades the original equipment manufacturers (OEMs) build machines that are more reliable and efficient in the past, while the manufacturing capacity is increased. Moreover, the re-configurability of these machines has been significantly improved. The latter considerations along with the shrinking effect in some markets, lead the client manufacturing companies to buy less and less machines to cover their needs. In such conditions the OEMs are forced to find other ways to keep their profitability, such as servicing on maintenance activities and on equipment special modifications. And there comes the paradox: is it reasonable for an OEM to design and sell a fully reconfigurable machine knowing that a significant profit will be lost in short terms because of not executing special modification projects requested by clients? As the OEMs identify or predict profitability losses, they proceed to a reduction of their investments related to research and development. If no advances are forced by the OEMs themselves then it is very probable to see some OEMs to evolve and some others to focus on conventional technologies and spare parts support. This ascertainment outlines situations of monopolies, which are not at all favourable in market.

A solution to this problem could be provided if OEMs and manufacturing companies work together, so as to have common performance goals and share profitability as well as sustainability. This synergy could be realized via leasing contracts. Contracts can be set-up with financial terms being related to performance outputs, while incentives and penalties should be also integrated in order to motivate high performance. In this way the manufacturing companies do not need to invest to new technology which is expensive without firstly having proved its adding value to the profitability, while the OEMs will keep possessing their equipment, and they will have the financial support to implement further development to build high technology solutions, of which they will take advantage.

5. Organizational learning: a human-centric

approach of success

Most manufacturing companies that have already identified their need to proceed to actions so as to increase their efficiency choose to invest in technology while they select one or a mix of the numerous theories (TPM, six sigma, 5s, etc.) to follow as an operation model. In many cases these steps may lead to a failure to achieve the expected results since the company's organization cannot digest the changes that occur. Moreover, it is doubtful if technology is the key to keep sustainable a manufacturing under the rapidly and constantly changing various markets environments. Technology can give a strategic advantage to a manufacturing company by increase its productivity and flexibility, but as technology evolves it is difficult to keep every resource at a technology edge it is necessary to focus on the companies resources that influence the companies overall performance and that do not change as rapidly as technology: the human resources. Ken Blanchard suggests "people are not your organization's resources-they are the organization" [8]. When a change is decided to be made in company either in terms of technology advance or the implementation of a new model, it is a common phenomenon to deploy these changes so quickly that the organization cannot adapt. It can not be a priori defined if the human resources of a manufacturing company can adapt to external environment's changes and adopt technological and other advances in order to boost their company's performance and ensure its sustainability. However, if specific conditions are met then the potential of treating efficiently such changes is growing.

All "great teams" in sports, arts and business have been basing their success on the mutual trust, the acceptance, the relationships and the synergy and they usually started off as of a group of individuals rather than a great team or organization. It takes time to develop the knowledge of working as a whole, just as it takes time to develop knowledge of walking or riding a bicycle. In other words great teams are learning organizations - groups of people who, over time enhance their capacity to create what they truly desire to create. The organizational learning theory is meeting a growing interest during the last fifteen years and is considered by many theorists a key towards successful changes that consequently may lead to sustainability. Organizational learning is based on specific disciplines [10] that outline the actions that a company should make in order to sustain profound changes. Looking more closely at the development of great teams/organizations, it becomes evident that people significantly change under deep learning cycles. These deep learning cycles constitute the essence of a learning organization- the development not just of new capacities, but of fundamental shifts of mind, individually and collectively [12]. The key of success is the empowerment of the companies' human resources. As soon as the manufacturing company sets clear and unanimously shared and understandable identity (vision, significant purpose, values that guide performance) [13] then organizational learning initiatives should be addressed before forcing any action. This is very important because in many cases the company's goals do not always coincide with the sub-functions (departments) goals, and as a rule the second are predominant. The overall company performance may be then increased by inversing the pyramid of hierarchical decision-making and give power to first-line people in order to act with freedom within specified boundaries. If each stakeholder has role in which he/she is highly competent, he/she has passion and he/she moves the economical engine then the results should be superb. Lack of hard and soft skills can be covered by applying proper in-class and on-the-job training accompanied by situational leadership [8]. Additionally, the customers should be considered as key players in a company's sustainability. A lot of theories and best practices can be demonstrated related to supply chain management, management of customer complaints, customer satisfaction, etc. However, despite the fact that the empowered first-line human workforce of the company is identified as a key force of sustainability, there are no significant actions about how the voice of the customers can be channelled directly to this driving workforce (Kansei engineering) [14]. In other words, the authors strongly believe that systems and procedures should be established at each manufacturing firm in order to link customers anticipations, complaints or reactions directly to the heart of the operating/manufacturing line (operators, first-line stakeholders) instead of queuing them to a responsible but not responsive enough respective functional department. Finally, it is important to remark that as every organizational strives to achieve its sustainability in the present, actions should be made to seed the success for the future. The future communities (industrial and social) should be raised in an environment of organizational learning in order to be able to drive prosperity and industrial non-industrial companies, and this is a major social and political issue that should be considered seriously by governments and firms.

5. Conclusions

The present paper reviews some of the latest approaches related to the sustainability of the manufacturing companies along the world and identifies some issues that should be considered ahead before implementing any proposed operational model or making significant investments. Judging from the on-going research, the sustainability initiative is driven by researchers and advisory companies whose conclusions and proposals should be passed to the manufacturing companies in order to achieve their sustainability. The authors strongly believe that the wind of change should be blown by the manufacturing companies themselves and that these manufacturing companies should deeply understand and address the issues that the current paper discusses before stepping to changes or investments. Moreover the research community should come closer to the industries in order to realize their actual needs and to clarify what should be the steps and the initiatives to build a sustainable community in small, medium and large sized manufacturing companies. Table 1 summarizes the issues that are addressed in the current paper, enlisting conditions that emerge from the reality of the current business activities. The aforementioned table as well the content of the present paper may raise questions if redirection and repositioning of strategies and initiatives for sustainability in EU and worldwide are needed for researchers and advisory companies and if implementation schemes are proper.

Acknowledgements

University of Patras is a member of the EU-funded I*PROMS Network of Excellence.

References

[1] Thomas A., Grabot B. (2006). Key technologies and strategies for creating sustainable manufacturing organisations, I*PROMS Virtual International Conference on Intelligent Production Machines and Systems.

[2] Thomas A., Weichhart G. (2006). Roadmapping as a strategic manufacturing tool, I*PROMS Virtual International Conference on Intelligent Production Machines and Systems.

[3] Pham D.T., Barton R., Thomas A., Williams O., Shamsuddin T., Ebrahim Z. (2006). Fit manufacturing, I*PROMS Virtual International Conference on Intelligent Production Machines and Systems.

[4] Thomas A., Byard P. Barton R. (2007). A holistic manufacturing model for company development, I*PROMS Virtual International Conference on Intelligent Production Machines and Systems.

[5] Collins J., Porras J. (2004).Built to Last: Successful Habits of Visionary Companies, published by Collins, ISBN: 0060566108.

[6] Saridakis K.M., Dentsoras A.J. (2006). Soft Computing in Engineering Design-A Review, accepted for publication in the Journal of Advanced Engineering Informatics.

[7] Thomas A., Byard P., Barton R. (2007). Technology implementation in manufacturing SMEs: A model for change, I*PROMS Virtual International Conference on Intelligent Production Machines and Systems.

[8] Barton R., Thomas A. (2006). Maximising the effectiveness of introducing advanced technologies, I*PROMS Virtual International Conference on Intelligent Production Machines and Systems.

[9] Blanchard K. (2006). Leading at a Higher Level: Blanchard on Leadership and Creating High Performing Organizations, published by FT Press, ISBN: 0132347725.

[10] Senge P. (2006).The fifth discipline: The art and practice of the learning organization, published by Currency, ISBN: 0385517254.

[11] Learning Organization Practitioners' Network (LOPN), http://www.lopn.net

[12] Society for organizational learning, http://www.solonline.org/

[13] Collins J.(2001). Good to great: Why some companies make the leap and others don't, published by Collins, ISBN: 0066620996.

[14] Coleman S., Pearce K. Van Lottum C. (2005). Statistics supporting the design process via Kansei Engineering, I*PROMS Virtual International Conference on Intelligent Production Machines and Systems.

Table. 1. Conditions that may lead to failure modes regarding sustainability and proposals to resolve them.

Condition	Failure modes	Recommendations
Comprehension of sustainable manufacturing	Sustainability is identified as a aspect independent of agile and lean manufacturing	Sustainability should be treated as a complex aspect that is based on well defined dimensions (e.g. agile, lean, etc) and on random and fuzzy dimensions related to external reality
	Financial profit is set as number one objective of manufacturing companies	Each company should build a significant everlasting purpose (cause) that will inspire the organization and drive its sustainability in the long run
Scope and success of technology implementation	Technology is considered that plays a key role in long term sustainability	Case by case a study is needed to be made in order to explicitly define the potential contribution and the adding value of technology elements if introduced to a manufacturing system
	High-end technology and advanced techniques cannot be digested and implemented by the majority of manufacturing enterprises	Proper technology utilization can be achieved only if end users have been convinced about the reasons of implementation and have participated in all steps for the beginning of decision making
Escape business shrinking: Partnerships	Many markets are shrinking and a common practice is to reduce headcount in order to reduce fixed manufacturing costs and retain sustainability. However, this may lead to losing strategic advantages	Proper outsourcing strategies cab be deployed in order to ensure stable profitability in fluctuating markets
	Modern manufacturing equipment is more reliable ad original equipment manufacturers sell less and thus their reduced profitability is translated into less investment in advanced machinery	Manufacturing companies should support original equipment manufacturers by building partnership that will coincide to business growth and avoidance of monopolies
Organizational learning: a human-centric approach of sustainability	Many manufacturing companies strongly resist to new operational models and changes that are decided to be forced in order to address external conditions	Companies should concentrate and practice on organizational learning before stepping to any change or implementing a new operational element
	Companies tend to identify people as simple resources that will follow any strategic decision or manufacturing model that will be communicated the management	Manufacturing organizations may turn from good to great by simply empowering their human resources and let the act with freedom within specified boundaries that are defined by clear vision, significant purpose, roles and values

Innovative Production Machines and Systems
D.T. Pham, E.E. Eldukhri and A.J. Soroka (Eds.)

FIT Sigma – An Integrated Strategy for Manufacturing Sustainability.

A.J. Thomas[a], R.A Barton[b] P.Byard[c], G Thomas[d]

[a] *Cardiff Business School, Cardiff University, CF10 3EU, UK*
[b] *Manufacturing Engineering Centre, Cardiff University, CF24 3AA, UK*
[c] *Manufacturing Advisory Service, Bridgend, Wales, UK*
[d] *British Airways Avionics Engineering, Llantrisant, Wales*

Abstract

Both Lean and Six Sigma are key business process strategies which are employed by companies to enhance their manufacturing performance. However, whilst there is significant information available on implementing Lean, Agility Six Sigma into companies these business improvement strategies are still implemented primarily in a sequential manner and little information exists regarding the integration of these approaches to provide a single and highly effective strategy for change in companies.

This paper proposes an integrated approach to Lean, Agility and Six Sigma and proposes a Fit-Sigma strategy for industry. The aim is that the effective implementation of the Fit-Sigma approach will lead to greater opportunities for companies to achieve economic sustainability through continued growth and improved manufacturing efficiency.

Keywords: Fit, Lean, Six Sigma, Sustainability

Lean and agility are widely considered as business process strategies that have facilitated high levels of sustainable growth in many manufacturing industries throughout the world. Lean is a strategy for achieving continuous improvements in business performance through identifying a company's value stream, adding value to a company's operations and then systematically removing waste – anything that does not add value to a company's operations. The application of 'Agility' was introduced into companies some years later in order to deal with the issues of volatile markets and irregular demand patterns. In this case, high levels of responsiveness and flexibility are required in order to cope with these demands.

Over the years many academics and practitioners have attempted to integrate the two approaches through proposing the concept of 'Leagility' [1] and later 'Agilean' [2] since it was hoped and proven in some cases that an integrated system of creating a Lean yet highly responsive manufacturing system was beneficial to a range of companies dealing with the increased threat of globalization and low labour cost competition.

Over the past ten years or so Six Sigma has been hailed as a key business improvement approach that is capable of achieving significant improvements in business process performance. Companies such as Motorola and GE have based their business process strategy around the Six Sigma concept.

As companies have continued to seek ways of delivering greater business performance at lower cost, the concept known at Lean Six Sigma has come to the forefront. Early developers of the Lean Six Sigma approach [3] seemed to develop a simple linear connection between Lean and Six Sigma proposing that you should Lean up the business first and then introduce Six Sigma as a mechanism to reduce variation and improve quality. Others proposed that the 'Lean' part of Lean Six Sigma could be brought in at the Improve stage of the Six Sigma DMAIC process thus effectively demoting Lean to a secondary process.

It could therefore be argued that early attempts to integrate the two concepts has failed to effectively achieve a single coherent business process system and a cohesive framework and as such, the true effect of Lean Six Sigma remains somewhat ambiguous.

The authors of this paper suggest that a simple integrated Fit-Sigma framework be developed which will allow for the effective implementation and delivery of Fit-Sigma to manufacturing companies thus assisting them to achieve a sustainable manufacturing system in the future.

The proposed Fit-Sigma framework has been developed as a result of extensive research work with twenty manufacturing companies both large and small who have undertaken either Lean, Six Sigma or Lean Six Sigma projects in the past (with varying degrees of project success). Through working with these companies the authors have been able to piece together both the best practices undertaken and the failings made through the adoption of the various business improvement strategies. This has therefore allowed for the creation of an integrated framework for future development.

Survey Data and Findings

A sample of 20 manufacturing organizations were targeted and visited for the survey. 10 companies were considered to be SMEs (meeting EU criteria for SME status) and 10 companies were large companies all forming part of multi-national organizations.

Of the 20 manufacturing companies, 15 Six Sigma Black Belt practitioners were interviewed along with a 25 Green Belt practitioners. 2 Master Black Belts were consulted along with 10 Lean implementation champions.

The aim of the interviews was to identify the problems associated with implementing, managing and operating Lean, Six Sigma or Lean Six Sigma projects.

Unstructured interviews were undertaken with each practitioner in an attempt to draw out as many key issues. The interviews were supplemented by a detailed walk through of the company operations concentrating upon the Lean and Six Sigma projects and how this work impacted upon the company's manufacturing operations. A number of key findings were identified:

1. The Lean and Six Sigma projects which were considered as poorly performing from the company's viewpoint were attributed to poor initial strategic planning where the selection of the wrong project or project area was seen as the major issue. Most companies had highlighted that if more time had been spent early on in the project through accurately planning and identifying project needs and requirements, resources would have been used more effectively and the project would have yielded greater benefits to the organizations concerned.
2. For Lean Six Sigma projects the development of a clear and well defined project framework identifying a clear set of staged points for taking the project forward including the typical tools and techniques which could be developed at each point would have been helpful and would have assisted in taking the organization forward as a whole.
3. With the cases of Lean Six Sigma implementation, the practitioners identified that projects were delivered in a piecemeal format often with teams being split between Six Sigma specialists and Lean specialists. This approach limited integration of the teams and often led to the teams trying to implement ideas and techniques which did not benefit the overall project process.
4. The 'Improve' phase was identified as a weak link with 60% of the black belt practitioners interviewed. Although the solution was deemed as being robust and effective, the implementation of the improve phase was not seen as sufficiently effective and did not deliver the fully benefits expected from this phase. The practitioners identified the lack of a distinct implementation mechanism which provided the necessary impetus to drive the change forward.
5. The lack of an effective control mechanism was also seen as a weakness in the current DMAIC

Six Sigma and Lean projects. Following the implementation phase, many of the practitioners felt that the good work undertaken during implementation of the improve phase was lost soon after as bad practices returned to the operational system. Also, the control phase was not seen as being effective enough to develop a culture towards sustainable and continuous improvement.

The Need for Fit-Sigma

UK manufacturing industry is under even greater threat from globalization and the associated problems of competing against low labour cost countries for the manufacture and supply of products. Over the years UK companies have attempted to minimise the effect of direct competition from these countries by integrating them into their own supply chain systems and to use the low cost element to their advantage. This however, provides only one small element of learning to survive and therefore a suitably robust and effective business process strategy must be at the heart of any manufacturing organisation of the future.

There should be a clear integrated strategy which combines the concepts Lean, Agility , Six Sigma and sustainability into one business approach which identifies and adopts best practice from these areas to ensure that the company remains sustainable and competitive not only now but also as markets change in the future.

A key driver for an effective Lean strategy is ensuring that the quality of product, process and service provided by a company is excellent and remains at this level throughout the product's life cycle. To achieve this, Six Sigma should be used as a key operating mechanism that sits within the Lean process in order to drive major improvements in business operations practice. Six Sigma tackles specific and highly focused business improvement projects [4] which yield significant benefits to the organization and when combined with Lean it is able to act as a mechanism for improvement that achieves not only increased productivity by doing more with less but also creates robust and highly competitive products processes and systems.

The proposal is therefore that companies should now look at employing an integratedFit-Sigma strategy in order to achieve economic sustainability through reducing costs whilst moving towards consistent perfection in their performance objectives. It is suggested that in order for companies to make the most of the integrated Fit-Sigma approach, the level or depth of application of Lean and Six Sigma will vary dependent upon the company, the market in which it operates etc. It is suggested that the first stage of any Fit-Sigma development process is that of laying down the principles of Lean whilst simultaneously applying the Six Sigma DMAIC methodology as a technique to remove the pockets of waste which can not be removed through using traditional industrial and production management techniques and systems. Fit-Sigma should therefore achieve simultaneously, improvements in the three major areas as identified in the Fit-Sigma triad shown in Figure 1.

Figure 1 The Fit-Sigma Triad

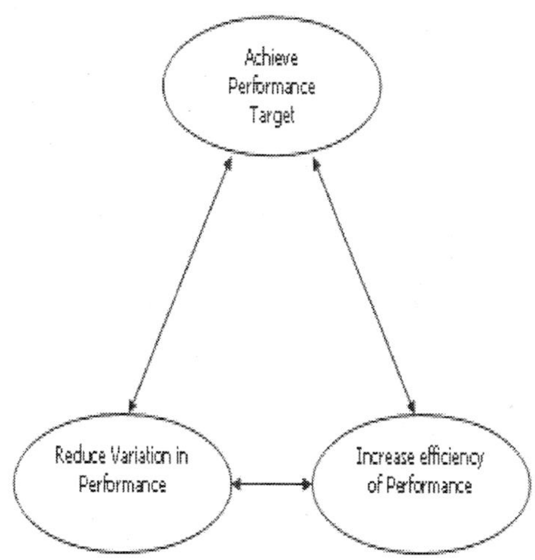

The Fit-Sigma triad identifies the need to address three business objectives namely: (i) achieve performance targets which are driven by customer requirements and demand, (ii) reduce variation in the achievement of the performance target i.e. meet the performance target every time and (iii) remove waste, improve value and drive efficient throughout the organization. Fit-Sigma therefore aims to achieve economic sustainability for an organization by meeting the three objectives within the triad in a simultaneous manner.

The Fit-Sigma Approach

It The survey identified that the current approach relating to Lean Six Sigma suggests a sequential approach to applying the concepts of Lean and Six Sigma. In many cases there seems to be two distinct operating mechanisms for applying the concepts where Six Sigma employs a five stage DMAIC process and Lean is implemented in a number of different ways each employing a different number and sequence of Lean techniques which are encased in a strategic vision based around the issue of doing more with less. Fit-Sigma will gain maximum effectiveness if a single operating strategy is used to employ Fit-Sigma simultaneously into companies. It is therefore suggested that a Six stage process called PMASEE is used to create a common implementation and operational strategy. Figure 2 shows the integration of both the Lean and Six Sigma concepts into one effective system of delivery. PMASEE is based on the traditional Six Sigma acronym of DMAIC but where DMAIC stands for Define, Measure, Analyse, Improve and Control, PMASEE both modifies and extends the DMAIC process to incorporate two important elements. PMASEE stands for:

Plan
Measure
Analyse
Solve
Execute
Embed

In Fit-Sigma the traditional Define stage is replaced by a 'Plan' stage. The reason for this is that when combining both Lean and Six Sigma there must be more time spent on planning the system in such a way that maximum benefit is achieved through the application of Lean and Six Sigma. The Planning stage subsumes the define stage but also includes the need to consider the effect on the manufacturing system if a particular process is improved (downstream bottlenecks as a result of improved system efficiency upstream, effects of improved reliability on the output of the system etc).

The second addition to the Fit-Sigma process is the introduction of an 'Execution' stage to the model. Experience through working with companies implementing such projects has shown that in many companies implementing Six Sigma that the Improve stage is not fully exploited and that the effectiveness of the solution is reduced through incorrect project execution. It is therefore suggested that the 'Execute' stage is added and is distinctly separate to the Solve stage. Therefore in this approach Improve is replaced with Solve and Execute. The 'Execute' stage allows engineers to concentrate on implementing an effective solution in order to maximize effectiveness of the system. Table 1 shows each of the stages of the Fit-Sigma approach and includes the typical tools and techniques that can be used to support the process. Note that the tools, technique and stages identified on bold lettering are suggested essential elements of the Fit Sigma process with the other elements being optional and brought in to suit the problem at hand.

Fit-Sigma must achieve significant improvements in productive performance through an effective integration policy since it is only through effectively integrating both Lean and Six Sigma and applying the concepts simultaneously that maximum leverage in performance can be achieved. With the emerging issues of low cost manufacture and responsive and flexible production systems, the strategic requirement now is to achieve quick product Turn-Round Times (TRTs) with consistency and to become leaner and more efficient in what a company does in the future.

Figure 2 The Integrated Fit-Sigma Concept

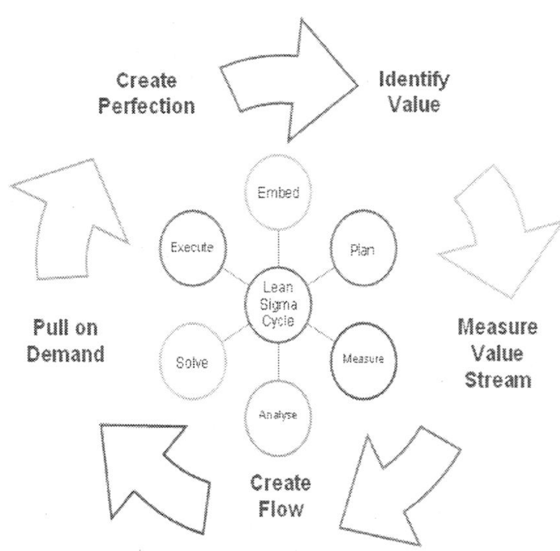

These requirements must also be achieved while ensuring high product quality and low operating costs. This is the only way that UK manufacturing industry will achieve economic sustainability in the future. In order to remain competitive, companies must achieve an appropriate level of leanness, agility achieved through the effective application of Six Sigma in order to achieve economic sustainability.

With Fit-Sigma therefore, the company will need to have the best aspects of lean, agility and product / process quality combined with the important aspect of systems re-configurability. The latter requires a company to change its strategic and technical focus quickly to respond to market trends and demands. This therefore calls on companies to have the necessary business process systems in place along with the correct technology platform and management systems.

For any particular manufacturing facility, a balance between Lean / Agility and Six Sigma is crucial. In some plants, leanness may be important, agility less so and vice versa for others. More often than not, however, companies will need to manufacture a range of products where each product line has a different demand profile and, as such, the balance between lean and agility varies each time whereas the application of Six Sigma remains the same in each case. While the level of leanness and agility will change, the issue that remains static throughout is that of sustainability and the need for a company to continually remain competitive.

Whatever the balance between Lean and Agility is for a company, what is required however is the need for the company to achieve world class levels of product quality and performance and this is where Six Sigma becomes an integral business driver [5]. What is therefore required is a suitable control system which enables a company to measure at any stage in its operating cycle, whether the Fit-Sigma process requires adjustment or improvement in order to keep operations running to target.

Without a suitable control system in place, many companies see the initial benefits gained through the adoption of Fit-Sigma lost when project teams move onto develop other areas of the company and bad practices creep back into the previously successful Fit-Sigma project. It is therefore key to develop a control system which allows managers to maintain a watching brief over these Fit-Sigma projects so that lasting effectiveness is seen and that suitable and timely intervention is triggered if a Fit-Sigma area falls below the performance targets expected. Figure 3 shows the conceptual development of a Fit-Sigma control system. The central section of the figure shows three boxes each containing an area in which Fit-Sigma should operate. Initially the Fit-Sigma process should concentrate on the current product, its design and its operational characteristics to see what waste can be removed and whether the product can be modified or redesigned in order to improve manufacturability or repair. At this stage it is quite possible that through redesign, the product may impact on the facility layout and the repair sections either positively or negatively. Therefore, feedback loops exist between each stage in order to facilitate iteration.

As stated previously, central to the idea of Fit-Sigma is that of system integration. Traditionally, companies have incrementally and systematically implemented various operational paradigms (TQM, lean, agility, etc) in a sequential manner as they become available. This often leads to an operational environment that is left fragmented as individual systems are bolted onto existing infrastructures, often causing internal conflicts within the company as the demands of one paradigm pull against those for another as system and operational complexity increases.

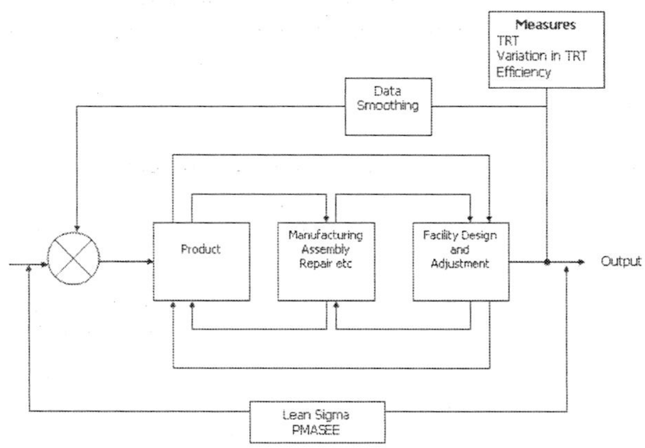

Fit-Sigma extends throughout a company, from order placement to repair and dispatch. Technology is central to the Fit-Sigma strategy concentrating on the technological interfaces between customer and company. This technology is then dovetailed into an effective strategic and operational management system that leads to the creation of an efficient, sustainable and responsive manufacturing environment. Fit-Sigma integrates key business process strategies together with a company's existing and future technology platforms and support systems (TPM, ERP, supply chain reengineering etc). This integration provides not just a single business approach but an integrated design and manufacturing system that combines the systemics of a range of business process concepts into one model that has low operational and systems complexity

Technology includes more than just the machinery and associated systems that convert the raw material into a finished product. It also covers e-commerce at the front end through to the electronic transfer of customer order requests and the complete e-manufacturing facility that takes essential customer data, design and manufacturing data and drives them forward in a simultaneous manner so that a product can be manufactured quickly and cost effectively. It is the tight integration of these various electronic platforms with the strategic and business systems that will provide the seamless cost effectiveness and rapid response required to meet customer demands. In order to provide effective backing for the technology elements in the strategy, support functions must exist that can ensure that the systems operate in the expected manner with minimum downtime and maximum efficiency [6]. These support functions include self-diagnostic systems, adaptive control systems, and so on.

Conclusions

Developing a Fit-Sigma strategy calls for consideration of three major operational areas of systems development. These are:

- the development of a company's supply chain system to ensure high quality, highly responsive and dependable supply of raw material and subcontracted products;
- the development of a lean, technologically driven and highly agile manufacturing system that is designed to convert customer requirements to finished products quickly and efficiently; and
- the development of systems that enhance sustainability by supporting and continually improving the performance of the product, the logistics and the manufacturing systems.

The requirements of Fit-Sigma can thus be split into the three main areas of lean, agility and Six Sigma. These are integrated into a holistic system structure. It is every company's basic requirement to be sustainable. Sustainability is essentially the ability of a company to stay competitive by adapting to changes in the market trends and customer requirements. Many companies consider themselves to be in a sustainable condition. However, this is sometimes not a realistic assumption to make since, whilst they continue to survive at present, they may not have the infrastructure to remain competitive in the future.

As an example of this, take a company that currently manufactures products for a niche market. The real advantage of working in this area is that there is little direct competition and therefore the company survives purely on this basis. Faced with this situation a company may not feel the need to develop a coherent business process system as they see little need for its implementation when things are going well. However, without a systematic approach to cost reduction and value creation, the company will not be able to sustain its business capabilities when competition increases. Therefore, developing and applying a coherent Fit Sigma strategy early in a company's life ensures a company remains sustainable and resistant to later competition.

References

[1] Childerhouse,P Towill,D. "Engineering supply chains to match customer requirements" Logistics Information Management; (2000) 13, 6

[2] Cox,A, Chicksand,D, Tong, Y "The proactive alignment of sourcing with marketing and branding strategies: a food service case", Supply Chain Management: An International Journal; (2007) 12, 5.

[3] George, M L,. (2002) "Lean Six Sigma – Combining Six Sigma Quality with Lean Speed", McGraw Hill, ISBN 978-0071385213

[4] Breyfogle, F.W. Implementing Six Sigma, Smarter Solutions - Using Statistical Methods, (1999).John Wiley & Sons Inc.

[5] Thomas A J and Barton R - Developing an SME based Six Sigma Strategy, Journal of Manufacturing Technology Management 18/5 2007, 417-434 ISSN 1741-038X, pp 490-512

[6] "Achieving Best Practice in Your Business – QCD Measuring Manufacturing Performance". Department of Trade and Industry Brochure, www.dti.gov.uk.(2002).

PLAN	MEASURE	ANALYSE	SOLVE	EXECUTE	EMBED
Strategic Review	**Quickscan**	**Quickscan**	**Obtain Solutions** (for Variation reduction Performance achievement Efficiency Improvement)	**Solution Implementation Design and Planning**	**Create cultural change**
Fiscal Analysis	**Value Stream Mapping**	**Identify Inhibitors to System Performance Enhancement (ToC)**	**Select most effective solution**	**Implementation Team Selection**	**Develop Procedures**
Quickscan1	**Identification of system constraints**	**Cause and Effect Analysis (CLD)**	**Check solution on upstream and downstream operations**	**Project Management**	**Train Staff**
Big Picture Mapping	Pareto Analysis	**Value Stream Mapping**	**Test solutions**	**Execute Solution**	**Develop Ongoing Discipline of Application**
Project Definition (Set Lean Sigma Objectives)	Activity Sampling	QFD (Wants and Hows)	**Check solution integrity**	**Evaluate Solution Effectiveness**	**Monitor control system to check for performance**
Project Selection	Industrial Engineering	Industrial Engineering	**Agree on Solution**	**Agree on solution performance, can it be improved?**	**Revisit area for next stage of improvement**
Resource Selection	Scrap / Throughput Analysis	Critical Path Analysis	TRIZ		**Do it again!**
Project Team Selection	OEE	Experimental Design	Computer Aided Engineering		
Problem Definition	SPC	Fault Tree Analysis	Simulation		
Project Planning	Critical Path Analysis		What-if Analysis		
Project Risk Analysis & Contingency Planning					
Cost Benefit Analysis					

Table 1 The Fit Sigma Process and Typical Tools and Techniques

1 Quickscan was developed by the Logistics Systems Dynamics Group at Cardiff Business School, Cardiff University. It is a holistic supply chain audit methodology which assists companies in identifying key areas within a business which require further development and analysis.

Innovative Production Machines and Systems
D.T. Pham, E.E. Eldukhri and A.J. Soroka (Eds.)

Key Technologies for Sustainable Production

E. Uhlmann[a,b], J. Krüger[a,b], T. Hühns[a,b], C. Hübert[a], J. König[a], R. Hollan[a], D. Oberschmidt[a,b], F. Byrne[a], C. Neumann[a], B. Duchstein[a], P. Marcks[a]

[a] *Institute for Machine Tools and Factory Management (IWF), Technical University Berlin, Germany*
[b] *Fraunhofer-Institute for Production Systems and Design Technology (IPK), Berlin, Germany*

Abstract

Innovative production engineering needs to be in a position to use human and material resources in an efficient manner in order to secure production centres. The continuously shortening life-cycles of products is in conflict with the need to increase the useful life of production systems. Flexible, adaptable, reconfigurable and self-organising equipment are the keys to solving this conflict. People, with their specific technological knowledge, must therefore be integrated into a newly defined role within the production process.

Keywords: sustainability, production, energy efficiency

1. Sustainability in Production

The sustainability concept is based on results from the work of the Brundtland-Commission, who formulated resolution methods for the linked and interdependent social, economical and ecological issues facing modern society. These resolution methods are the fundamental idea behind today's understanding of sustainability, with the key goal being the retention of the ability for current and future generations globally to satisfy their needs. The Enquete Commission, in their guideline published in 1998, summarised sustainability to be 'the concept of constantly evolving economical, social and ecological aspects. These three pillars of sustainability stand together interdependently and require long-term balanced co-ordination.' Sustainability can only then be achieved when the goals of all three pillars are equally fulfilled.

The question arises as to how these thoughts on

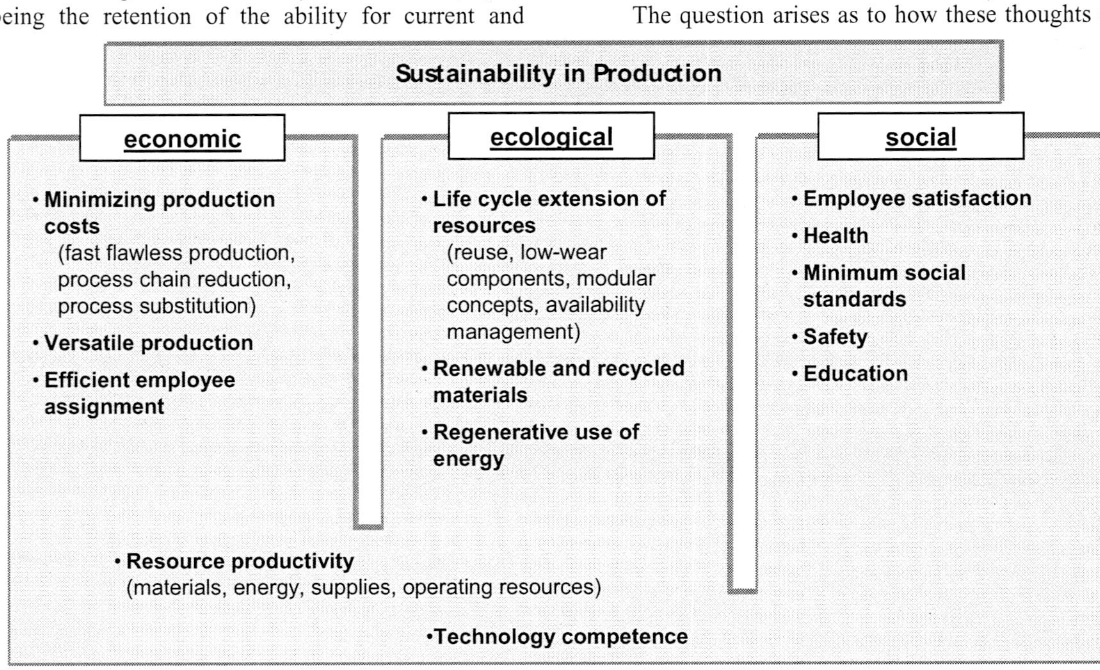

Fig. 1. Dimensions of Sustainability

sustainability can be successfully transferred into production engineering. At first glance it seems that the three dimensions offer conflicting goals. The challenge lies in the development of new technologies, which are not yet state of the art today, to resolve these seemingly conflicting goals. On the other hand resolution methods exist that equally support two or three of the sustainability dimensions simultaneously, for example the technological competence of a company and its associated resource productivity (Figure 1).

Economic production is based as before on the minimisation of production costs, and increasingly on the mutability of the production process. This mutability is necessary to allow the production of new products in a short time frame using existing equipment as far as possible. The efficient use of resources shows a positive effect on both economical and ecological aspects. Considerable drivers for the improvement of the ecological aspects of a company are the increase in the life cycles of both equipment and consumables. From the point of view of the recycling industry, the re-use of materials and the use of renewable raw materials for material production and for renewable energy sources has significant effects on ecological efficiency. As before, a substantial challenge is the avoidance, or minimisation, of harmful emissions. The benefits of sustainability can be reaped by a company through the use of the available technologies which positively affect the economic, social and ecological characteristics of the company. The factory management system must be in a position to secure and promote the social aspects personnel satisfaction, health, security and development as well to ensure a minimum social standard for staff.

2. Key Technologies in Production Engineering

The overcoming of technological barriers to achieve the said economical, ecological and social goals of sustainable production is the central motivation behind the development of hybrid production technologies. Hybrid processes are therefore distinguished by the overlay of a conventional production process with an additional thermal, chemical or mechanical mechanism, which ensures a significant performance enhancement over the performance of the conventional process alone. Through the timely superposition of these processes, increased productivity, broadening of the application spectrum of new materials as well as increased product quality through better surface properties, minimal surface damages and less shape deviation can be achieved. Both mechanical and chemical hybrid technologies increase the energy efficiency and contribute to the sustainable use of natural resources. Mechanical hybrid processes are characterised by the coupling of secondary kinematics, while chemical hybrid processes aim to alter material properties.

Examples of such hybrid processes are laser-assisted dry ice blasting and the combination of conventional deep drawing with integrated electromagnetic non-contact forming. Laser-assisted dry ice blasting has been shown to increase material removal by between 50 % and 700 % compared with conventional dry ice blasting. The process is used for the cleaning and structuring or de-coating of surfaces, which is usually only achieved with unjustifiable effort by dry ice blasting alone [1]. The conventional deep drawing process can be combined with integrated electromagnetic non-contact forming through the use of coils integrated within the forming die, which generate magnetic forces. This hybrid process reduces the number of necessary process steps and the need for individual machines, as well as decreasing the handling of steel plates between individual process steps [2].

Sustainable machining processes, which lead to a higher productivity and an increased resource and energy efficiency, can be realised through innovative tooling concepts. The deposition of thin film systems on various tool substrates and geometries is in many cases the most technologically and economically sound solution [3].

In contrast to the traditional High-Pressure-High-Temperature (HPHT) production route for the deposition of super-hard synthetic cutting materials based on diamond and cubic Boron Nitride (cBN), Chemical Vapour Deposition (CVD) or Physical Vapour Deposition (PVD) may be used. The use of such technologies opens up further applications for tool coatings and allows for the combination of high mechanical wear resistance and chemical resistance of the cutting materials with application specific design of the cutting tool geometry [4, 5]. Besides the increased attainable cutting speeds to increase the productivity and work piece quality, the use of non-environmentally friendly cooling fluids may be reduced.

The use of cooling fluids to assist in the heat dissipation and to improve the tribological procedures during chip formation is widespread in machining operations. The so-called wet machining places a substantial burden on the environment and the machine operators, and also leads to high costs. Through the employment of dry machining technologies, total machining costs can be reduced by up to 45%. The forgoing of cooling fluid entirely is however often not possible [6]. A promising alternative for efficient heat dissipation in the machining of high-temperature alloys is the development of internally cooled tools. Through the use of a closed cooling fluid loop, the release of the fluid during the machining process can be avoided. At the same time the work piece is protected from contamination, which reduces the associated environmental burden, cooling fluid removal and cooling fluid disposal costs. The associated improvement in tool life and work piece quality has significant economical and ecological impacts.

Generative machining processes have experienced a significant upsurge in recent years. In the field of metallic materials, aluminium, high speed steel and stainless steel can be machined into usable components using a direct laser sintering process. Moulds with novel geometries can, for example, be realised without the need to use conventional machining operations and small batch production can already be undertaken in a commercially viable manner. In the medium-term, generative machining processes offer an alternative to the time intensive conventional processes with both economical and ecological advantages. With respect to the protection of resources the generative technologies also offer potential benefits, such as the production of the necessary components on-site, meaning transportation and storage costs can be avoided. If it becomes possible to decrease the machining times and increase reproducibility and especially the not yet sufficient component quality, it can be expected that generative processes will provide considerable competition for conventional machining processes.

The simulation of machining operations offers the potential to fulfil the ecological, social and economic requirements of sustainability. For example, the adjustment of the suitable feed rate in the milling of complex geometries from difficult to machine materials may be optimised through simulation and thus reduce the machining time by up to 40% [7]. The resultant reduction in consumed resources allows a saving of both costs and energy.

The simulation of chip formation using the Finite-Element-Method allows the virtual testing of a range of geometries [8, 9]. This allows the pre-testing of geometries and so reduces the number of variations to be produced as prototypes, thus saving time in the subsequent machining trials. State of the art today is the individual simulation of the machining process and the machine performance. In real machining processes however, these parameters are inter-dependent and influence one another heavily. Integrated simulation, whereby the process-machine interaction is simulated, is therefore a further key technology for sustainable production in the future. This leads to the optimisation of the machine tool structure not only through the increase in the static stiffness, but also through the consideration of various process parameters, in particular the dynamic performance of the machine [10].

The wear related breakdown of the bearings in machine tools, such as the spindle bearing, are a common cause for machine down-time. Wear-free air bearings are not suitable replacements for ball bearings however due to their limited loading capacity. Bearings produced from porous materials have been developed at the Institute for Machine Tools and Factory Management (IWF). These bearings exhibit a higher loading capacity and stiffness compared with classic single nozzle bearings, and so permit the efficient implementation of air bearing technology in a broader application area. Air bearings with porous bearing material are produced today from sintered materials and the geometry of the airflow channels can therefore only be influenced to a limited extent. A post-processing compression of the bearing surface is usually undertaken. Significant narrowing and direction change of the flow channels can be observed after the compression has taken place, which may cause energy loss due to strong turbulences in the channels. The bearings developed at the IWF are produced from a fibre-reinforced ceramic composite material [11]. The production of this material takes place through a ceramic infiltration of a fibre matrix. By defining the fibre length the porous structure of the material and therefore the flow channels can be directly influenced. Energy losses can therefore be reduced, and the efficiency of the bearings increased. Such optimised bearings show better dynamic performance, which allows a higher machine and resultant component accuracy to be achieved. Material scrap and the need for reconditioning of components could be significantly reduced if the impact of thermal and dynamic interferences that act on the machine tool could be balanced by autonomous compensation. Self-compensation of thermal paths ensures in the first instance an efficient use of energy. In order to avoid the negative effects of the discrepancies between start-up and processing temperatures, machines in industry usually run around the clock, even when there are no components to be machined. Self-compensation, by using an adaptronic system, aims to resolve this issue. Experimental investigations undertaken at the IWF demonstrated the feasibility of controlled compensation for thermal distortion in the use of carbon fibre reinforced polymers, whereby carbon fibres are deemed suitable due to their mechanical and thermal properties (Figure 2). While the aluminium spindle holder expands due to heating, the carbon fibre reinforcement, through its negative expansion coefficients, forces the spindle back into its starting position. In order to realise the necessary heating power for the Carbon Fibre Reinforced Plastic (CFRP) compensation structure, the heat flow out of the spindle is not sufficient. Through externally located heating foils in combination with the heating filaments located within the composite layers, a significantly better compensation effect is achieved. An additional cooling of the spindle holder is not undertaken in this case [12]. The machine related inaccuracies related to different start-up and process temperatures are therefore minimised, meaning the machine produces only parts of comparable quality.

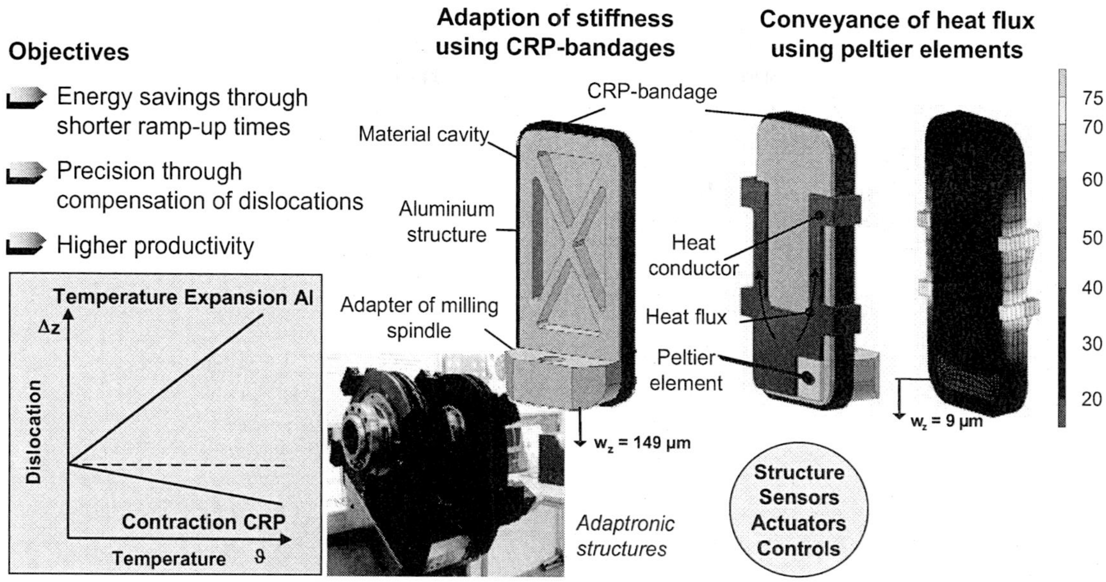

Fig. 2: Compensation of thermal dislocations

High speed machining makes light weight design solutions imperative, as the maximum limits of loading capacity are reached. In order to fulfil the high demands of a stable High Speed Cutting (HSC) process, materials must possess a high stiffness and low weight. Solutions developed at the IWF for light weight machine tools include weight optimised chuck jaws produced from CFRP and a rapidly opening machine door. A prototype of the designed machine door led to a weight saving of 25 % and reductions in the opening and closing times of 25 % and 50 % respectively [13].

The modular design of machine tools and complete machining systems offers many advantages that ensure sustainable production. The production of macro and microstructures as well as material savings will become possible when system scalability is attained through modular design (Figure 3). Through the implementation of modular design systems the re-use of components becomes a possibility. Such a modular system for micro-production is currently being developed at the IWF [14]. An example of a modular machine tool is a rope driven machine, which uses wire drivers as opposed to conventional rods for the positioning of the spindle. The modularisation of machine tools increases their reconfigurability. Through this substantial extension in their life cycle as well as reduced long term costs can be achieved. Individual solution concepts can be flexibly realised, and can be adjusted to constantly evolving demands. The realisation of modular light weight design for the reconfiguration of machine tools ensures sustainable production with the conservation of resources. An example of a reconfigurable handling system is the Dodekapod developed at the IWF. Through the expansion and reorganisation of conventional Hexapod structures, the system specific advantages can be further increased, and their principal disadvantages can be minimised. The increased size of the working chamber and the combination of the clamping and handling technologies in one system leads to flexibility, especially in disassembly systems. The application possibilities of the clamping platform, which is divided into three sections, has almost no limits with respect to the geometry of the work piece to be clamped. A multitude of work pieces can be securely clamped using only few clamping elements. The structure of the Dodekapon, as also the Hexapoden, is scalable. The simple construction as well as the large amount of repeated pieces allows a modular construction and high flexibility through simple configuration [16]. A simple machine-user interface was developed which can be implemented into a disassembly system and assist the user in the handling of complex goods.

Germany is a world-wide leader in machine and plant construction. In order to maintain this position in a sustainable manner, producers of machines and plants must find solutions to satisfy the rising customer requirements. These requirements include maximum process security, minimal machine down times, rapid fault repair, optimal use of wear components and the ability to plan maintenance procedures. Life cycle cost analysis increasingly plays a role in customer purchasing decisions, which leads to requests for guarantees of availability [17]. In order to be in a position to give an availability guarantee, the machine producer must possess an increased knowledge regarding the condition of the machines when in use by the customer. This information is not often accurately provided however, due to the fear of losing valuable 'Know-How'. Centreless grinding machines are often used in interlinked mass production systems. Due to their high productivity they are often used to avoid bottlenecks in a production cycle, and breakdowns thus occur more frequently due to the lack of redundant time allowed for such machines. In order to satisfy the increased customer demands as well as the increasing price pressures, new concepts and models to allow the producers to give an availability guarantee have become increasingly important in recent years. The Fraunhofer-Institute for Production Systems and Design Technology therefore developed an availability monitoring system, which documents the machine history and allows a condition assessment of the particularly relevant wear components related to the feed axis [18]. The data bank based approach allows a long term recording of the machine history as well as the comparison of machines from the same producer under different conditions of use. The developed software solution is today delivered as standard with machines supplied by an industry partner of the Fraunhofer Institute. This example shows that the implementation of solutions for the management of availability benefits both machine user and producer. With respect to economical advantages, this system allows the early detection of changes in dynamic behaviour of, for example feed drivers, an improvement in planning capability for maintenance procedures and the optimised use of the complete machine life-span. The optimisation of machine components for specific user profiles and the support during maintenance operations are further advantages of such a system. From an ecological point of view, resources are saved due to the customised maintenance procedures and the reduction in scrap. The simplification of maintenance through fault searches as well as the fulfilment of the demanded availability guarantees within the automotive industry are factors which positively influence the purchasing decisions of the customer, and thus influence the social dimension of sustainable production (Figure 1).

3. Resource-Efficient User-Oriented Automisation

Sustainable production systems are characterised by their optimal use of resources. The product spectrum is constantly increasing and the demanded lot size of particular products is therefore decreasing. The sustainment of competitiveness of companies in such an environment with shortening market launch times is therefore demanded. In order to rise to the challenge of such markets, highly flexible production systems are necessary. These systems must be in a position to produce a wide spectrum of product variations, and to be adjusted for the production of new products with minimal costs being incurred. Today's automisation solutions often do not fulfil these demands. A significant improvement could be reached through the use of so-called hybrid automisation. These technologies are characterised by the integration of the machine operator with the automisation system, meaning that human intelligence, flexibility and skilled workmanship are combined with the technical capabilities of the machine, such as accuracy and speed.

Despite decades of research in the field of artificial intelligence, the human is still by far and away the most adaptable and intelligent resource in the production process. This is shown particularly in complex assembly processes in which humans, with their superior motor-sensor qualities and rapid learning ability, are irreplaceable. The smaller the lot sizes of an individual production, the more the human capabilities mentioned above are needed. Not only in the field of artificial intelligence is research being undertaken to attempt to mimic these capabilities electronically. The significant increase in systems for digital image processing and pattern recognition, particularly for quality control, underlines the importance of the visual perception capability of humans in production processes. In robotics, attempts have been made for many years to replicate the sensor-motor abilities of humans and to combine these with the human flexibility in complex

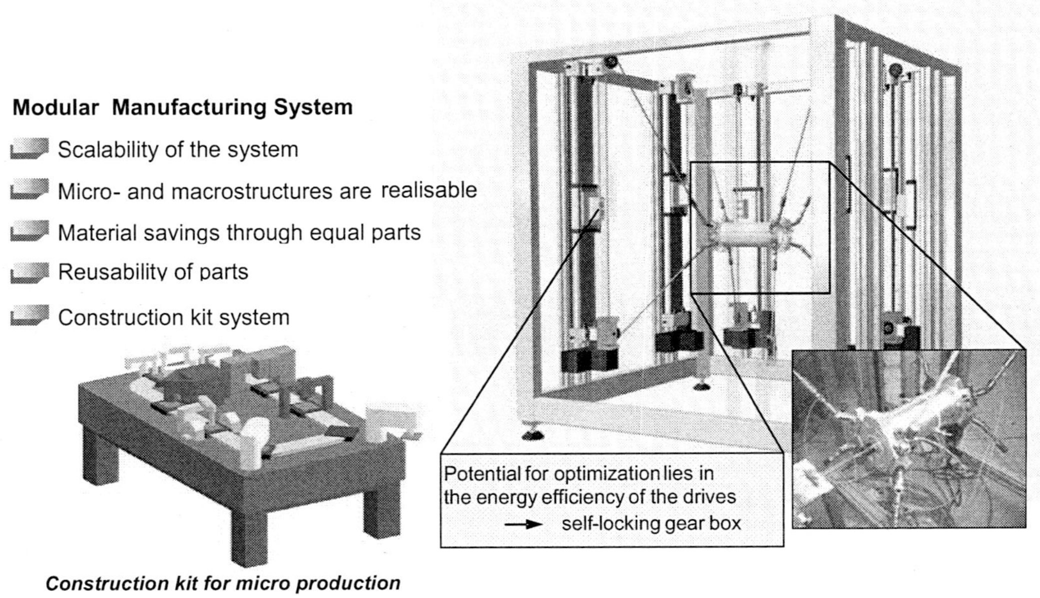

Fig. 3: Modular Manufacturing Systems

motions. Co-operational, flexible and humanoid robots are the results of such research. From a sustainability point of view, highly developed computer supported systems and humans are not necessarily in competition. When examining automisation from a sustainability point of view under the social, economic and ecological aspects, a perspective is formed whereby a closer co-operation between humans and automisation systems is necessary. This leads to a new automisation technology paradigm; human-oriented automisation.

The goal of human-oriented automisation is not the replication but the optimal support of human abilities through the automisation system. This support will be given in such a way to allow the sustainment, or improvement, of the human physical capability, for example through strength support to decrease occupational illnesses. Examples of such human-oriented automisation systems are so-called co-operative robots. These are robots which offer support during physically demanding tasks, but leave full control over the system motion to the human. The human leads the complex process, while the robot unburdens the human and at the same time learns the process chain.

The sustainability of production is supported in more than one respect through the use of human-oriented automisation. Economic advantages are gained by the improved adaptability of the partly automated systems in changeable production cycles, while ecologically the long term usability and energy efficiency of such a system are advantageous. Socially the improved integration of the human into the production process and the physical strain reduction are beneficial.

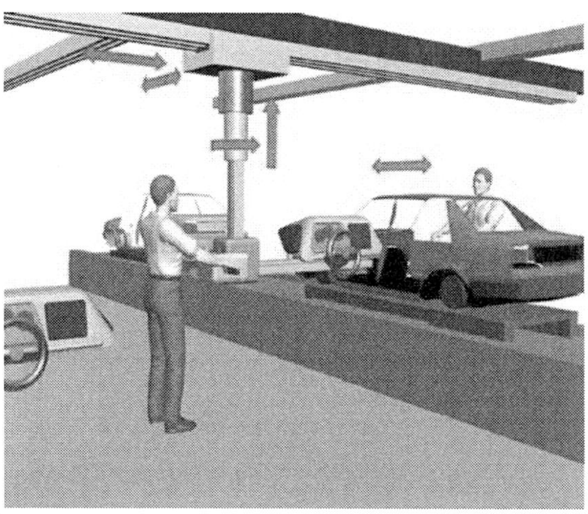

Fig. 4: KOBOT

In human-oriented automisation, the machine operator should not be replaced but rather integrated into the process. Appropriate tooling must therefore be available to the operator, for which the term 'Intelligent Assistant Systems (IAS)' has been coined. Such hybrid automisation systems integrate and support the machine operator. Through such flexibility, complex products and product spectrums can be produced with minimal cost incurrence.

IAS depict a new generation of machining and assembly systems, and can work together with the machine operator in the same working chamber. This can be either through direct physical contact (co-operative, passive robot, KOBOT, Figure 4) or without contact, with an industrial robot.

KOBOTS support the machine operator in many respects, but leave the process control in the operator's hands. Significant support functions are the active collision avoidance mechanism as well as the strength support [19]. Intelligent support for the machine operator, usually without contact with the robot, offer co-operative working environments in which the robot undertakes simple repetitive assembly tasks while the operator aids in the assembly of complex components. The role of the operator in such co-operative assembly processes is particularly in the completion of operations which require complex motor-sensor capabilities. A prerequisite for such co-operative working environments is the safety of the operator in the overlapping working area. The system ensures this through the use of an image based monitoring system. This system is based on a technology which captures three dimensional images of the human body without the need for marking of the operator's clothes [20].

The technologies discussed above are not yet available in practice, or in certain cases are only in the early stages of development. A range of different national and European research projects (e.g. the integrated EU-Project PISA) have been initiated. It should be noted that the development of the technologies is only one part of the solution. Software tools for the design, planning and optimisation of hybrid machining and assembly systems must also be developed.

4. Development of Key Technologies for Energy Efficiency

Germany is a world-leader in energy efficient production, second only to Japan. The 'Energy efficient export initiative' ('Exportinitiative Energieeffiezienz') led by the German Federal Ministry of Economics, underlines the commercial importance of German knowledge and German technologies for the efficient use of energy. Production companies have in many cases introduced measures to increase their energy efficiency, and energy cost is now a focal point. Significant savings have been made possible in recent years through the use of tools for energy management and controlling.

The identification of potential possibilities to improve the energy efficiency in production can be divided into two steps; the monitoring of energy flows in factories and the analysis of the energy use in individual production processes and systems. Through systematic use of software modules already available on the market, energy saving possibilities can be highlighted which can result in payback times of one to two years for such a system. Future steps for the optimisation of energy usage in production must go above this type of energy management if the goals set by the Federal Government of Germany, that the total commercial energy productivity be doubled by the year 2020 compared with that recorded in the year 1990, are to be reached. These steps include energy management and the controlling of energy coefficients, i.e. tailored process planning and process implementation with consideration for the energy coefficients of the processes in question.

FuE-projects with subject themes in the field of energy efficient process control, with emphasis on control systems and technology for water supply systems, have been successfully run by the Fraunhofer IPK for a number of years [21].

Authors

Prof. Dr. h. c. Dr.-Ing. Eckart Uhlmann is Head of the Fraunhofer-Institute for Production Systems and Design Technology (IPK) and Department Manager in Machine Tools and Manufacturing Technology at the Insitute for Machine Tools and Factory Management (IWF) of the Technical University Berlin.

Prof. Dr.-Ing. Jörg Krüger is Managing Director of the Institute for Machine Tools and Factory Management (IWF) of the Technical University Berlin and Department Manager of Automation Technology at the IPK.

Dipl.-Ing. T. Hühns, Dipl.-Ing. C. Hübert, Dipl.-Ing. C. Neumann, Dipl.-Ing. J. König, Dipl.-Ing. R. Hollan and Dipl.-Ing. D. Oberschmidt are department and group leaders at the Fraunhofer IPK and the IWF.

Dipl.-Ing. B. Duchstein, Dipl.-Ing. P. Marcks and Ms. F. Byrne are research engineers in the machine tools and cutting technology departments at the IWF.

References

[1] Uhlmann, E.; Hollan, R.; Veit, R.; El Mernissi, A.: A Laser Assisted Dry Ice Blasting Approach for Surface Cleaning. In: Proceedings of the 13th CIRP International Conference on Life Cycle Engineering (2006) 471-475.

[2] Uhlmann, E.; Damavandi, K.; Jurgasch, D.; Rautenstrauch, A.: Wirkung des Magnetimpulsverfahrens auf die Nachhaltigkeit in der Produktentstehungsphase. In: ZWF 22 (2007) 10.

[3] König, W.: Fertigungsverfahren Band 1: Drehen, Fräsen, Bohren. Springer-Verlag, Berlin, Heidelberg: (7th edn), 2002, pp. 59-78.

[4] Uhlmann, E.; Wiemann E.: Kubische Bornitrid-Schichten – Potenziale für die Zerspanung im Werkzeug- und Formenbau. In: Uhlmann, E. (Hrsg.), Tagungsband 3D-Erfahrungsforum Werkzeug- und Formenbau (2002) 91-96.

[5] Keunecke, M.; König, J.; Richter, V.; Wiemann, E.: Anwendung superharter Beschichtungen in der Zerspantechnik. In: Industrie Diamanten Rundschau 41/2 (2007) 38-43.

[6] Heiler, R.: Trockenbearbeitung beim Gewindebohren. VDI-Z Special Werkzeuge 147 II (2005) , 44-45.

[7] Uhlmann, E.; Mattes, A.: Optimizing the Feed Velocity of NC-Tool Paths. In: Proceedings of the 10th CIRP International Workshop on Modeling of Machining Operations (2007) 245-252.

[8] Uhlmann, E.; Mattes, A.; Jost, P.: Simulation der Zerspanung mit Hilfe der FEM. In: VDI-Z (2006) 74-77.

[9] Kammermeier, D.; Mylavaram, N.: Zerspanwerkzeuge mit CLD schneller entwickeln. In: Werkstatt und Betrieb (2005) 53-55.

[10] Uhlmann, E.; Mense, C.: New Structure Optimisation Procedure for Improving the Dynamic Behaviour of HPC-milling Machines. In: Proceedings of the 5th CIRP International Seminar on Intelligent Computation in Manufacturing Engineering (2006) 165-170.

[11] Duchstein, B.; Neumann, C.: Porous Ceramics used for Aerostatic Axial Bearings. In: Annals of WGP Production Engineering 12/1 (2005) , 105-108.

[12] Uhlmann, E.; Marcks, P.: Adaptronic Compensation of Thermal Strain at Machine Tool Spindles using CRP Bandages. In: Tagungsband Second Manufacturing Engineering Society International Conference (2007)

[13] Karrasch, I.; Wittner, M.; Duchstein, B.: Entwicklung und Konstruktion einer schnellöffnenenden Werkzeugmaschinentür – Zusammenspiel zwischen Leichtbau und Maschinenschutz. In: Uhlmann, E. (Hrsg.), Tagungsband der Berliner Runde (2007) 187-196.

[14] Uhlmann, E.; Geisert, C.; Stelzer, C.: Hybride Leistungsbündel für die Mikroproduktion. In: wt Werkstattstechnik online 97/3 (2007), 126-129.

[15] Uhlmann, E.; Tonn, N.: Integrierte Modellbildung und Simulation von Steuerung und Mechanik unter Berücksichtigung des Bearbeitungsprozesses für Parallelkinematiken mit Seilantrieben. In: Heisel, U. (Hrsg.), Buch zum SPP 1099 Fertigungsmaschinen mit 345 Parallelkinematiken – Forschung in Deutschland, Shaker Verlag, Stuttgart 2007.

[16] Friedrich, T.; Früsch, I.: Entwicklung einer 12-achsigen Parallelkinematik zur Handhabung schwerer Bauteile. In: Uhlmann, E. (Hrsg.), Tagungsband der Berliner Runde (2006) 85-94.

[17] Moser, S.: Hersteller sehen TCO überwiegend positiv. In: Problem und Lösung, 22 (2006)

[18] Uhlmann, E.; Geisert, C.: Steuerungsbasierte Erfassung zustandsrelevanter Life-Cycle Daten an CNC Werkzeugmaschinen. In: Zuverlässigkeit und Diagnose in der Produktion, VDI Fortschritt-

Berichte, 663 (2007) 1-14.
[19] Krüger, J; Bernhard, R; Surdilovic, D.: Intelligent Assist Systems for Flexible Assembly. In: Annals of the CIRP 55 (2006) 29-32.
[20] Krüger, J: Nickolay, B.; Heyer, P.: Image based 3D Surveillance for flexible Man-Robot-Cooperation. Annals of the CIRP 54 (2005) 19-22.
[21] Lisounkin, A.; Schmidt, H.-W.: Modellbasierte Prozessführung in einem Wasserwerk. Journal Wasser und Boden 51 (1999) 44-47.

Innovative Production Machines and Systems
D.T. Pham, E.E. Eldukhri and A.J. Soroka (Eds.)

Sustainable energy consumption in European manufacturing: the driver for a new research agenda addressing climate change, energy supply, and global competitiveness

P T Kidd

Cheshire Henbury, Tamworth House, PO Box 103, Macclesfield, SK11 8UW, UK
Email: paulkidd@cheshirehenbury.com

Abstract

Addressed are the topics of energy consumption and the future impacts that energy supply and use will have on European manufacturing. The scope of energy related matters encompasses three critical issues: climate change, security of supply, and depletion of world oil reserves. The first issue tends to hold centre stage, but only belatedly, but addressing this problem on its own ignores the second and third of the critical issues considered. The consequences of such an omission, for European manufacturing, are explored. Through the use of a scenario it is argued that a failure to consider all three matters might lead to a circumstance where there will be significant (further) decline of European manufacturing. This decline, it is argued, will be driven by the economics of the *comparative advantage of nations*, in which countries such as China stand poised to benefit, not because of labour cost or technological advantages, but because of their potential to create low-energy costs relative to other nations. The work presented represents the outcome of the first phase of privately funded research work that is aiming to develop a new research strategy, along with policy recommendations. This first phase has led to some foundational recommendations, presented here, upon which this new research strategy will be based. Development of this strategy will be addressed in the next phase of the work, which will be completed by early 2009.

Keywords: Sustainable manufacturing; energy supply; energy consumption;

1. Introduction

Research in European Manufacturing has, for the past 25 years and more, been driven by the concerns of globalisation and competitiveness, against a backdrop of increasing complexity. This complexity relates to a range of challenges that encompass new technologies, new business practices, new organisational concepts, more sophisticated products, more fragmented markets, increased competition, changing customer demand, etc. This is a circumstance that is well documented in the literature (e.g. see [1, 2, 3, 4, 5, 6, 7, 8, 9]).

In the early part of the 21st century a new concern for manufacturing industry has been the offshoring of manufacturing to China and other emerging Asian economies, and the growing competence of these nations, not just in the manufacturing function, but also in terms of research and development, and also new product development capability. However, one further challenge has also emerged which, ironically, given its political prominence, has yet to be given significant consideration in terms of manufacturing research. This is the challenge of sustainable energy consumption in manufacturing.

On March 8th, 2007, European Union leaders agreed to adopt a binding target on the use of renewable energy (wind, solar power, etc.) committing to the achievement of a 20% boost overall in renewable energy use by the year 2020. They also agreed to reduce energy consumption by 20%, and to cut carbon dioxide emissions by 20% from 1990 levels, by year 2020 [10].

Given these new goals for European Energy policy, it is appropriate to ask probing questions relating to Europe's manufacturing industry's capabilities to achieve an alignment with these strategic targets. Is, for example, European Manufacturing industry well positioned to contribute towards the achievement of these targets? What challenges do these targets pose for European Manufacturing? Do these targets represent the full dimensions of manufacturing related energy issues, or are there other critical elements that also need to be addressed? And, given the answers to these questions, is there a need for new research agenda?

The questions posed above are being addressed in a privately funded research project [11,12] that aims to develop proposals for a European Manufacturing Research Strategy and to inform European Manufacturing Research Policies as they relate to three key issues:

- the ongoing process of offshoring of manufacturing to Asia;
- energy supply and security; and
- the human contribution to climate change through greenhouse gas emissions.

Note that the latter two of these key issues are the main driving forces behind the EU commitment to the *20% targets*.

This paper discusses some of the results of the first phase of research; research that is considering circumstances as they relate to energy supply and consumption in the context of European Manufacturing industry, but in a global context. The paper provides an overview and an explanation of the critical energy issues, above and beyond the obvious greenhouse gas emissions that result from the burning of fossil fuels. The paper also examines a possible future scenario and its implications for European manufacturing.

2. Energy supply

That the world faces the challenge of reducing the emission of greenhouse gases, primarily Carbon Dioxide, but also other greenhouse gases such as Methane, Nitrous Oxide, Sulphur Hexafluoride, etc. is well understood, even though there is still not unanimous agreement that human activity is contributing towards climate change. That the burning of fossil fuels is a significant contributor to the increase of Carbon Dioxide in the atmosphere is also widely understood. That part of the response to the problem of climate change involves greater use of renewable energy sources and also increased energy efficiency is also understood. These can be taken as givens and the targets set by the EU in March 2007 are a response to this understanding. But the issues are not so simple, for there are other major concerns that also need to be considered, especially within the longer term perspective of manufacturing futures and foresight, which typically look ahead 15 to 20 years. From the reference point in time of 2008, such a timeframe would be the years 2020 to 2025, although, given the nature of energy related matters, addressing a longer time scale is also necessary, typically 50 years, which expands the time horizon out to 2058. However, given that manufacturing experts are not use to considering such far distant points in the future, this paper will concentrate on the more normal 15 to 20 year timeframe, with a few excursions beyond when the referenced projections exceed this time horizon.

One of the other issues of importance in this timeframe up to 2020/2025 is the depletion of fossil fuels. This consideration is primarily centred on the depletion of the world's crude oil reserves, which has significance in the timeframe of 2010 onwards. Why this is so has a very simple explanation, for 2010 corresponds to the year when, at least one (of many) world oil reserve depletion studies, predicts peak production, with reserves in decline after that year [11]. While other studies are more optimistic, and the whole topic is subject to uncertainty for several reasons (see [11]), even the most independent of studies, that is to say, the one undertaken by the US Geological Survey [13], foresees exhaustion of world oil supplies by 2055, with peak production occurring many decades before that. The US Geological Survey study however, does not take into account the growing demand for crude oil, the result of which will be a much earlier exhaustion of supplies than that predicted from the study.

A reliable source of information on world energy facts is the US Government's Energy Information Administration (EIA), which is the provider of official energy statistics from the US Government. Every year the EIA provides a detailed set of data on numerous energy-related issues. In its *2007 International Energy Outlook* [14] the EIA reports that as of January 1, 2007, proved world oil reserves were estimated at

1,317 billion barrels. As of December 2007, estimated proved reserves stood at 1,313 billion barrels [15]. In the year 2000 however, the US Geological Survey [13] estimated world oil reserves at 3,021 billion barrels. Why such a significant difference?

The answer to the above question lies in the basis for the figures. The EIA looks at data (in the public domain) for proved estimated reserves. The US Geological survey estimate however factors-in oil that has been discovered but not yet extracted, likely improvements to recovery rates, and estimates of the oil that is likely to be discovered in the future. One conclusion that might be drawn from the US Geological Survey's estimate is that actual worldwide oil reserves could up to 2.3 times greater than the estimated proven reserves.

Nevertheless, there is no room for complacency here. Already mentioned is the fact that accelerating demand for crude oil is expected to lead to depletion at a much earlier point in time. But more sobering is the context for theses oil reserve estimates when past consumption is considered. Over 900 billion barrels of oil were produced in the 40-year period between 1965 and 2005, and most of this can be accounted for by consumption in the western industrialised nations [11]. With the emergence of newly industrialising economies in Asia and South America, and the development of Eastern European economies, the expectation has to be one of rapidly increasing consumption of crude oil. Indeed, this is exactly what the EIA predicts. Its 2007 assessment of world energy consumption reports a projected increase of 57 percent from 2004 to 2030. Total energy demand in the non-OECD countries is expected to increase by 95 percent, compared with an increase of 24 percent in the OECD countries. World use of petroleum and other liquids will grow from 83 million barrels oil equivalent per day in 2004 to 97 million barrels per day in 2015 and 118 million barrels per day in 2030.

Putting these figures into context, the projected consumption of oil in the specific year 2030, is expected to be 43 billion barrels of oil. Note that this is the projected consumption for one year. If this had been the rate of consumption in the past, then the 900 billion barrels of oil that were consumed in the 40 year period between 1965 and 2005, would have been used up in a matter 20 years. In other words, by 2030 the rate of consumption will have doubled when compared with the period 1965-2005. And the point is, even putting aside concerns about the sustainability of such high levels of consumption, and the matter of the legacy that this will leave to future generations, and the potential for conflict that might arise in an oil-depleted world, there is really no guarantee that the world will be able to extract such an amount oil given the uncertainties that surround oil reserves. Moreover, what is interesting about the figures given in the EIA report [14] is that these projections take into account the fact that oil prices are expected to be significantly higher than in 2007. But it is not just oil prices that are expected to increase; the price of natural gas is also predicted to rise, and one unfortunate consequence of increasing prices for natural gas, will be to make coal a much more cost competitive fuel. This is unfortunate because burning coal for electricity generation is one of the most significant sources of Carbon Dioxide emissions.

Coal is the most carbon-intensive of the fossil fuels, and it is also the fastest growing energy source in the *International Energy Outlook 2007* projections. World energy-related Carbon Dioxide emissions are projected to grow by an average of 1.8 percent per year from 2004 to 2030. The highest growth rate will be in the non-OECD regions. Importantly from a manufacturing prospective, it is projected for China, that its energy-related Carbon Dioxide emissions will rise by 3.4 percent annually from 2004 to 2030, reflecting the country's continued heavy reliance on fossil fuels, especially coal, over the projection period. China's energy-related emissions of Carbon Dioxide are projected to exceed US emissions by about 5 percent in 2010 and by 41 percent in 2030.

Another issue of concern for Europe, is that of security of supplies, given that most of the world's supplies of crude oil and natural gas, the foundations upon which the modern industrial world is built, lie in areas that can be best described as *problematic*, for many varied reasons. In an oil depleted world, security of supply becomes an even more critical issue as oil scarcity will provide the power to unfriendly (oil-rich) governments to exert unwelcome influences over Europe and other major oil consuming nations.

So in summary, while greenhouse gas emissions are a major concern, this is not the only energy-related issue that will impact upon European Manufacturing, for in addition to this, the depletion and security of oil supplies also needs to be factored in. But in fact, the reason why all three are important lies in one common problem, and that is the energy intensity of Manufacturing Industry. Reducing the energy intensity of European Manufacturing could effectively address all three problems. This is an important strategic point, for if the focus is placed solely on reducing greenhouse gas emissions, this may lead to actions what will not address the problems of depletion and security. For example, the use of carbon capture and sequestration will prevent some Carbon Dioxide emissions, but it will do nothing to curb the increasing demand for oil, and the impact that this will have upon the depletion of global oil reserves.

3. The consequences for manufacturing of failing to address oil reserve depletion

So far the discussion has focused on the increasing demand for oil and the impending problems of oil reserve depletion. It is tempting to believe that this is an issue for future generations, and that the problem of climate change is a difficult enough challenge, without considering these other matters. This is understandable, but it is wrong to think so. To understand why, it is necessary to consider the implications of a scenario where the focus is solely on greenhouse gas emissions.

A scenario involving a single-minded attention to reducing greenhouse gas emissions is likely to involve European manufacturing industries focusing attention on *reducing its carbon footprint*. How would this be achieved? Primarily the answer lies with *Efficiency and Conservation* measures [16]. These include: improved energy efficiency (for example through more efficient motor controls and lighting); improved vehicle efficiency for road transport; more and better recycling of materials; materials substitution; greater use of by-products; waste elimination; remanufacturing; and several other means that have yet to be considered. (Another category–de-carbonisation of fuels and electricity–is also relevant to manufacturing, but poses some problems, but this is an issue that space prevents further consideration of).

The point here is not to catalogue the means of achieving reduced carbon emissions for manufacturers. The purpose is to highlight the implications of ignoring the impact that oil depletion will have on European Manufacturing if nothing is done to reduce oil dependency. So developing the scenario further, now consider the likely consequences.

There is a very high probability that the price of oil and natural gas will continue to increase as demand continues to grow, and, in the case of oil, there continues to be restrictions on production and insufficient global refining capacity. This will affect input prices, causing upward shifts in costs such as plastics, energy, fuel, heating, etc. Hence the cost base of European Manufacturing will rise as a result. This will further erode the competitiveness of European Manufacturing, as it must be remembered in this discussion, that although rising oil and natural gas prices will affect all manufacturers, wherever they may be, oil prices in Europe have been historically higher than in other parts of the world owing to the policy of European Governments to heavily tax oil imports to dampen demand (a policy which no longer seems to be working very well). A real possibility in this scenario is that the continuing rise in oil and natural gas prices will result in many smaller manufacturing firms going out of business, either a result of loss of competitiveness or as a result of the actions of larger firms. These larger manufacturing businesses will be faced with a choice that will very much depend on the prevailing economics of energy: they will either shift their manufacturing offshore to low-energy cost countries, or exit from manufacturing business completely.

Here the crux of the matter has been reached. The competitive edge of countries such as China has lain in its low labour costs, up to now. This will change however. With time, as expectations rise and skill shortages arise, this advantage will begin to erode. In addition, increasing oil prices will cause transportation costs to rise, and thus the economic case for offshoring to China will become less compelling for European Manufacturing. This argument however assumes a continuation of the underlying structure of competition. This may well be a false assumption in the scenario considered, for the following reasons.

Faced with rising labour costs, countries like China may well change the basis on which they compete in the world. This simply will be the economics of Ricardo's theory of *comparative advantage* at work: nations specialise in what they are best at producing and trading with other nations. Comparative advantage forms the basis of modern trade theory: a country has a comparative advantage in the production of a product if the country is relatively well-endowed with inputs that are used intensively in producing the product. In the opening years of the 21st century China used low-cost labour as the basis of its comparative advantage. When low cost labour ceases to be a significant factor in its comparative advantage, it will find a new one best suited to the prevailing conditions. In a high oil and natural gas-price competitive environment, its comparative advantage is most likely to lie in low energy costs. This it will achieve through its abundant coal reserves, which it is likely to deploy in two ways; burning coal in coal fired power stations and converting its coal resources to liquid fuels, using, for example, the Fischer-Tropsch process, or one of its several variants.

The economics of outsourcing will therefore turn on the cost of transportation to and from China, versus the costs of manufacturing in a China that is using its coal-based low-cost energy to attract manufacturing into China. And if Chinese coal is used to produce the liquid fuels consumed by the world's shipping industry, then there is a high probably that the case for offshoring will be overwhelming.

The knock-on effects of this for other sectors of European industry that support manufacturing–IT Vendors, consultancies, housekeeping services, and the like–could be very significant if they are unable to follow manufacturing to China. Of course the consequences for climate change through the additional greenhouse gas emissions that will result, will also be very detrimental (assuming that nothing is done to prevent the Carbon Dioxide emissions).

4. Conclusions–towards a new research agenda for European manufacturing

Given that 90 percent of the world's fossil fuel reserves are in the form of coal [17], with many nations having reserves with expected exploitation time horizons stretching into hundreds of years, a shift from an oil-based global economy back to a coal-based economy is a real possibility. In fact the International Energy Agency (IEA), in its 2007 World Energy Outlook (*WEO*) [18] have already noted that there has been a resurgence of coal as an energy source, driven primarily by booming power-sector demand in China and India. This change, the IEA notes, is a marked departure from past *WEOs*.

This shift back to coal, if it is fully realised, will give comparative advantage to nations capable of exploiting their large coal reserves to create (relatively) low-energy cost environments. China is a potential candidate for such a development, but so it the United States.

Such a significant structural change could have serious consequences for European manufacturing, but it is not apparent that this is an issue that is widely understood. It is fair to state that there has been a *failure of foresight* in relation to energy issues in manufacturing, given the belated attention that is currently being given to global warming. What is worrying about this, is that this is not the first time this has happened. The circumstance is similar to that which prevailed in the 1980s, when manufacturing research was technology driven, being preoccupied with computer integrated manufacturing and *unmanned factories*, reflecting manufacturing strategies of the past that had, even in the 1980s, already become obsolete, while what was really needed at the time was a integrated approach that addressed technology, organisation, and people [2]. It took a long time for manufacturing industry stakeholders to understand and accept this. The danger is that history might be about to repeat itself.

There is therefore some doubt whether European industry is ready to rise to the challenges that lie ahead. There is a real danger, and one that can be avoided, of pursuing the wrong strategy. Elsewhere [9] is has been argued that it is current preoccupation with competitiveness issues, which will, ironically, lead to further decline in European manufacturing. This is so because of the failure to grapple with the full dimension of the energy problem.

If Europe is to tackle the problems of climate change, energy supply and the competitiveness of its manufacturing sector, a new approach is needed. The old policies have largely failed to deliver an adequate response to the decline of European Manufacturing and the competitive challenges that it has faced over the past 25 years. A new research agenda is needed, not driven by the concerns of globalisation, but by issues that will ensure the development of European manufacturing industry within the context developing a non-fossil fuel based sector, for within such an approach lies an effective response to global competitive challenges, through novel competitive strategies, technologies, and operational practices.

What name this goes by, whether it is sustainable manufacturing, environmentally benign manufacturing, or something else, does not matter. To achieve sustainability there is a need for a strategic research framework. Such a framework should rest of a number of foundational propositions, which are:

- That the long-term goal should be to develop a non-fossil fuel based manufacturing industry;
- Key to achieving this goal is reducing both the carbon intensity and the energy intensity of manufacturing industry;
- That achieving such reductions will require a move beyond improving energy efficiency towards absolute reductions in energy consumption, and a significant de-coupling of economic growth from energy consumption;
- An alignment of manufacturing research agendas with energy policy is needed, particularly with respect to renewable energy

commitments, so as to enable achievement of a circumstance where renewable energy becomes a viable energy source for manufacturing industry;
- The scope of manufacturing research focused on achieving reduced energy intensity, needs to address, in an integrated manner, process technology, logistics and transportation, and information and communications technologies.

The above focuses on manufacturing industry as a consumer of energy, but there is also an additional dimension that needs to be addressed, since manufacturers are also suppliers of products (to businesses and consumers) that consume energy. Thus the sector has also to consider how what it produces, and the way its products are used, contribute to energy consumption. This however has, for the moment, lain outside the present scope of the study.

The ultimate step in this work is to take these foundational propositions and to transform them into a strategic research agenda.

References

[1] Piore, M.J. & Sabel, C.F. The Second Industrial Divide: Possibilities for Prosperity. New York, Basic Books, 1984.

[2] Kidd, P.T. Organisation, People and Technology in European Manufacturing, Office for Official Publications of the European Communities, Luxembourg, 1991.

[3] Pine, B.J. Mass Customization: The New Frontier in Business Competition. Harvard Business School Press, Boston, 1993.

[4] Kidd, P.T. Agile Manufacturing: Forging New Frontiers. Addison-Wesley, Wokingham, 1994.

[5] Savage, C. 5th Generation Management: Co-creating Through Virtual Enterprising, Dynamic Teaming, and Knowledge Networking. Butterworth Heinemann, Boston, 1996.

[6] Handy, C. Beyond Certainty: The Changing Worlds of Organisations. Arrow Business Books, Reading, 1996.

[7] Haeckel, S.H. Adaptive Enterprise: Creating and Leading Sense-and-Respond Organizations. Harvard Business School Press, Boston, 1999.

[8] Choi B-W, and Nagy D. (Eds) Proceedings of the IMS Vision Forum, Cheong-Moon-Gak Publishers, Seoul, 2006.

[9] Taisch, M., Thoben, K-D, & Montorio, M. Advanced Manufacturing: An ICT and Systems Perspective. Taylor & Francis, London, 2007.

[10] BBC News, EU agrees renewable energy target, http://news.bbc.co.uk/1/hi/world/europe/6433503.stm, 2007.

[11] Kidd, P.T. The Future of European Manufacturing: Driven by Globalisation or Global Warming? In: European Visions for the Knowledge Age: A Quest for New Horizons in the Information Society, Kidd P.T. (Ed). Cheshire Henbury Publications, Macclesfield, 2007.

[12] Kidd, P. T. European Manufacturing Matters: A Strategic to Response to Climate Change, Energy Scarcity, and the Chinese Manufacturing Revolution. Cheshire Henbury Publications, 2008 (forthcoming).

[13] United States Geological Survey. World Petroleum Assessment 2000: Description and Results. U.S. Geological Survey Digital Data Series - DDS-60, (http://pubs.usgs.gov/dds/dds-060/), 2000.

[14] Energy Information Administration, Annual International Energy Outlook 2007 (Energy Information Administration, Washington DC, (http://www.eia.doe.gov/oiaf/ieo/index.html), 2007.

[15] Anonymous. Worldwide Look at Reserves and Production, Oil and Gas Journal, Dec 24, 105, 48, pp 24-25, 2007.

[16] Pacala, S. & Socolow, R. Stabilization Wedges: Solving the Climate Problem for the Next 50 Years with Current Technologies. Science, 305, Aug., pp 968-972, 2004.

[17] Starr, C., Searl, M.F. & Alpert, S. Energy Sources: A Realistic Outlook, Science, 256, May, 981-987, 1992.

[18] The International Energy Agency. World Energy Outlook 2007 Executive Summary, China and India Insights, International Energy Agency, (http://www.iea.org/Textbase/subjectqueries/index.asp), 2007.

Innovative Production Machines and Systems
D.T. Pham, E.E. Eldukhri and A.J. Soroka (Eds.)

Consideration of vibration behaviour of rectangular plates

W. Rolshofen[a], P. Dietz[a], G. Schäfer[a]

[a] *Institute of Mechanical Engineering,
Technical University of Clausthal, Robert-Koch-Str. 32, 38678 Clausthal-Zellerfeld, Germany
rolshofen@imw.tu-clausthal.de*

Abstract

How physical objects can be transformed into natural, seamless, unrestricted touch interfaces has been researched at the Institute of Mechanical Engineering in the EU research project Tangible Acoustic Interfaces for Computer-Human Interaction (TAI-CHI). The objective is to design multimodal interfaces on the basis of acoustic signals that bridge the gap between the virtual and physical worlds.

An approach to localise acoustic sources in solids is investigated. Scope of work is to become objects of everyday life touch sensitive. An essential part of this is its sensor arrangement. Therefore, vibration behaviour of rectangular plates has been examined theoretically, experimentally and by numerical simulation. Consistent results exemplify observations as natural oscillation.

Keywords: vibration, natural frequency, dispersion

1. Introduction

Within the scope of research activities at the Institute of Mechanical Engineering at *Technical University of Clausthal (TUC)*, a method for acoustic source localisation in solid objects is investigated within the EU project "Tangible Acoustic Interfaces for Computer-Human Interaction (TAI-CHI)" [1, 2].

To give a reason for this activity, it is investigated how human sense of touch can be linked with machine actions. Physical objects and their complex surfaces, respectively, shall function as natural tangible interfaces, which connect virtual and real world by acoustic signals of solid objects.

The developed algorithm is by means of experimental results checked and systematically optimised subsequently. Below, analysis of vibration behaviour of rectangular plates is described what is inherent for this approach. Therefore, both theoretical calculations and numerical simulations have been done, which are compared with and experimental setup.

2. Experimental setup and data acquisition

By this approach, accelerometers are placed on a rectangular plate to detect any touch or tap on it. For the plate parameters see Tab. 1 and 2. Meanwhile, characteristic of plate vibration has been observed like it is shown in Fig. 1 exemplary. There is amplitude spectra presented which are calculated from measured acceleration. Several local maxima can be observed. These are at frequencies of 156 Hz, 272 Hz, 391 Hz, 467 Hz (not marked), 581 Hz, and 684 Hz.

In order to mark off this observation from vibration behaviour of a rectangular plate, this plate oscillation is further examined.

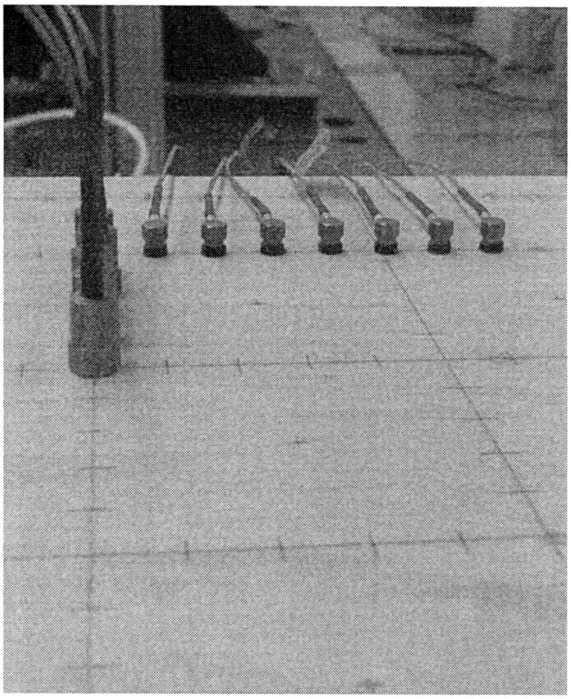

Fig. 1. An example of a sensor arrangement used in TAI-CHI project.

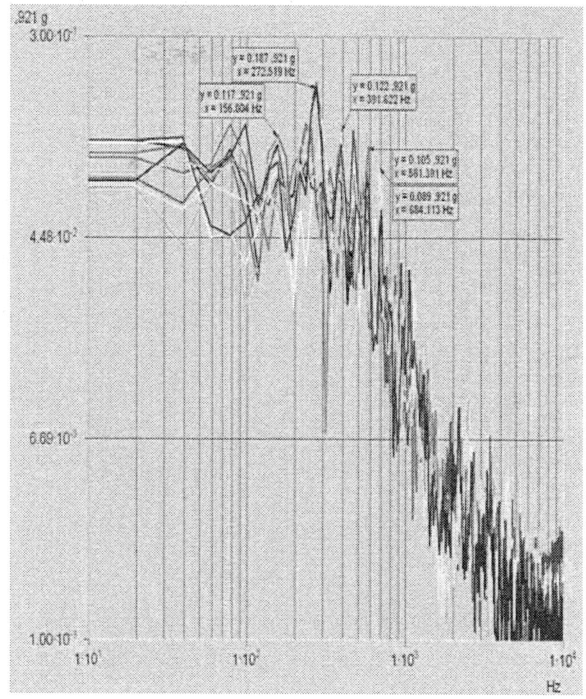

Fig. 2. Amplitude spectra calculated out of measured acceleration of sensors (s. Fig. 1) whereas average mean is used. Sticking out are several local maxima.

2.1. Bending vibration of rectangular plates

Originate in a model of a harmonic oscillator structure-borne sound of simple objects can be described.

A structure is called a plate when its wall thickness is clearly minor its extension. Thickness h is assumed to be constant across total area which is spanned by side lengths a and b. In case such an object is stimulated on its surface to vibrate, different types of acoustic waves are arising.

At this point, everybody shall direct one's attention only to *Lamb-Waves* [3]. They propagate in solid plates and they are a result of superposition of longitudinal and transversal waves. Necessary for this is that wavelength λ is very much greater-than thickness
h ($\lambda \gg h$).

Two kinds for this wave can be classified which are called extensional and flexural mode. If the upper-side of the plate oscillates symmetric to the lower-side then it is the extensional mode. Otherwise, if the upper and lower side vibrate asymmetric then it is the flexural mode. Usually, those types are denominated as dilational wave (symmetric) and bending wave (asymmetric), respectively, in literature [4,5]. Subsequently, statements are applied to asymmetric Lamb-waves. For both derivation of wave equation and boundary condition, it is referred to other publications [6,7].

With adequate suppositions of Kirchhoff's plate theory, equation of motion for an undamped plate is:

$$\underbrace{B\Delta\Delta w}_{\text{stiffness-}} + \underbrace{\rho h \frac{\partial^2 w}{\partial t^2}}_{\text{retardation-}} = \underbrace{p(x,y,t)}_{\text{external force}} \tag{1}$$

B is defined as bending stiffness of undamped plate with deflection w and density ρ as well as uniformally distributed load p. B can be calculated with modulus of elasticity E and Poisson's ratio μ as follows:

$$B = \frac{Eh^3}{12 \cdot \left(1-\mu^2\right)} \tag{2}$$

A solution for Eq. 1 is available for each boundary condition which are:

- Clamped,
- Supported,
- Free of forces.

They describe mounting of the plate and the solution has to be completed by initial conditions for *t=0.*

Free vibration of a rectangular plate results in a solution of homogeneous wave equation (s. Eq. 1). It can be shown [8,9] that for a homogeneous and supported plate (multi-directional mounted) the next equation is valid.

$$f_{m,n} = \frac{\omega_{m,n}}{2\pi} = \frac{\pi}{2} \cdot \left[\left(\frac{m}{a}\right)^2 + \left(\frac{n}{b}\right)^2\right] \cdot \sqrt{\frac{B}{\rho h}} \qquad (3)$$

Those functions which are a solution for the vibration of the plate are called eigenfunction. Related to them are natural oscillations of the plate (s. Fig. 3) and their eigen angular frequency $\omega_{m,n}$ that can be worked out by Eq. 3. Order of these eigenfunctions is labelled by m and n.

Areas can be detected where certain points vibrate with zero amplitude while others vibrate with non-zero amplitude. Those with no vibration are called nodes.

2.2. Dispersion

A significant effect during observation of vibration in solid objects is occurrence of dispersion that implies that propagation velocity is dependent on frequency. In relationship to wavelength λ and angular frequency ω, phase velocity c_B of the wave can be determined in consideration of:

$$c = \lambda \cdot f = \lambda \cdot \frac{\omega}{2\pi} = \frac{\omega}{k}$$
$$k_{m,n} = \pi \cdot \sqrt{\left(\frac{m}{a}\right)^2 + \left(\frac{n}{b}\right)^2} \qquad (4)$$

$$c_B = \sqrt{\omega} \cdot \sqrt[4]{\frac{B}{\rho h}} \qquad (5)$$

Strictly speaking this is only valid for infinite harmonic vibration. Due to the fact that those oscillations (s. Fig. 2) can be composed by different sine and cosine oscillations, particular frequency components have different velocities. In Fig. 4, both wavelength and phase velocity are plotted against frequency. These dispersion curves result from collaboration with *Laboratoire Ondes et Acoustique (LOA), Paris*, and they display air at 20 degree centigrade as well to draw comparisons with.

2.3. Used plates of steel and glass

During investigations different types of geometry and material were used. Exemplary, a plate of steel and one of glass according to their variables (s. Tab. 1 and 2) are discussed.

In Tab. 3 and 4, first eigenfrequencies are shown for each plate, which are determined with Eq. (3). Besides this calculation, two more approaches are followed up to quantify the eigenfrequency. They are presented in next paragraph.

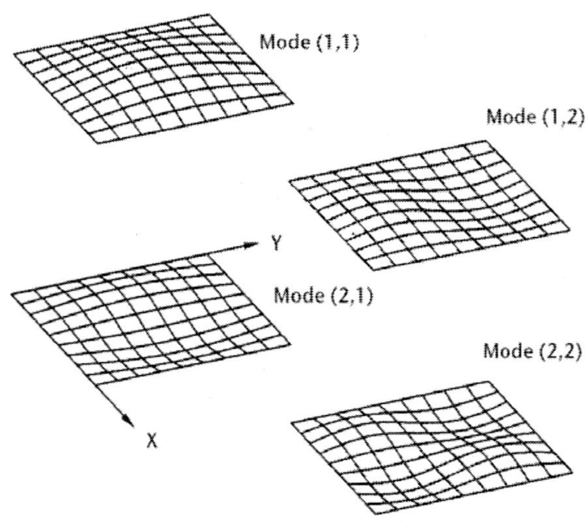

Fig. 3. Eigenfunctions of multi-directional mounted plates whereas specification declares the mode, e.g. Mode (2,1) means m=2 and n=1 [8].

Table 1
Dimensions of used rectangular plates

Material	Side length "a" [m]	Side length "b" [m]	Thickness "h" [m]
Steel	0,600	0,400	0,0014
Glass	0,436	0,332	0,0028

Table 2
Material properties for steel and glass (determined experimental)

Material	modulus of elasticity "E" [N/m²]	Poisson's ratio "μ"	density "ρ" [kg/m³]
Steel	$2,181 * 10^{11}$	0,310	8.392
Glass	$0,845 * 10^{11}$	0,201	2.569

Table 3
First eigenfrequencies $f_{m,n}$ for a plate of steel in Hz

$f_{m,n}$	n = 1	n = 2	n = 3	n = 4	n = 5
m = 1	30,7	94,6	200,9	349,9	541,4
m = 2	59,1	122,9	229,3	378,2	569,7
m = 3	106,4	170,2	276,6	425,5	617,0
m = 4	172,6	236,4	342,8	491,7	683,2
m = 5	257,7	321,5	427,9	576,8	768,3

Table 4
First eigenfrequencies $f_{m,n}$ for a plate of glass in Hz

$f_{m,n}$	n = 1	n = 2	n = 3	n = 4	n = 5
m = 1	107,4	312,1	653,4	1131,2	1745,5
m = 2	224,7	429,4	770,7	1248,5	1862,8
m = 3	420,2	625,0	966,2	1444,0	2058,4
m = 4	693,9	898,7	1240,0	1717,8	2332,1
m = 5	1045,8	1250,6	1591,9	2069,7	2684,0

2.4. Chladni Figures and numerical simulation

By means of a method introduced by Chladni, it works out to produce an optical impression of vibrating and non-vibrating areas of solid objects. To achieve this, surface of a resonating body has to be powdered with finely sand before its sounding excitation. Owing to induced vibration, sand will be swept away in moving areas and will be assembled in neutral zones [10]. In top part of Fig. 5, Chladnic experimental setup can be seen.

There is a plate of glass hanging free only with flexible ribbons and suction cups over a loudspeaker. This speaker is connected via an amplifier with a frequency synthesiser where particular stimulations can be chosen. On the plate, a squared grid is drawn for orientation. Before each test, quartz sand is dispersed which is piled in two lines lengthwise in this case.

Also, the same setting is used for plate of steel (s. Fig. 6), whereas there elastic suspension is realised through boreholes.

Moreover, numerical simulations with finite element method (FEM) have been done with properties of each plate. Thereby, free boundary conditions (no clamping) are taken as a starting point. Each numerical solution describes in particular modes of plate vibration. These can be seen at bottom of Fig. 5 and 6.

Chladni figure of glass plate is recorded at a

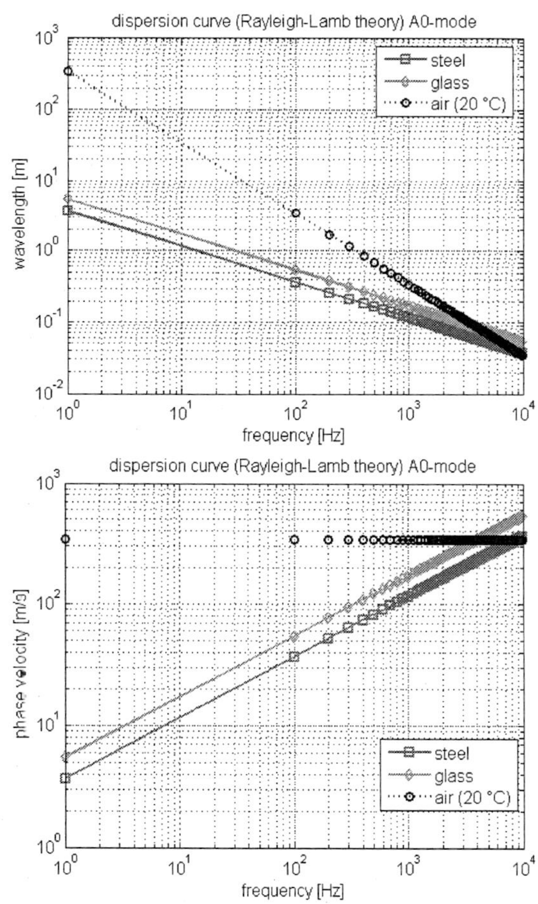

Fig. 4. Dispersion curves for rectangular plates of steel and glass (s. Tab. 1 and 2) in comparison with medium air at 20 degree centigrade what is free of dispersion.
In top part of figure wavelength is plotted against frequency and in bottom part phase velocity.

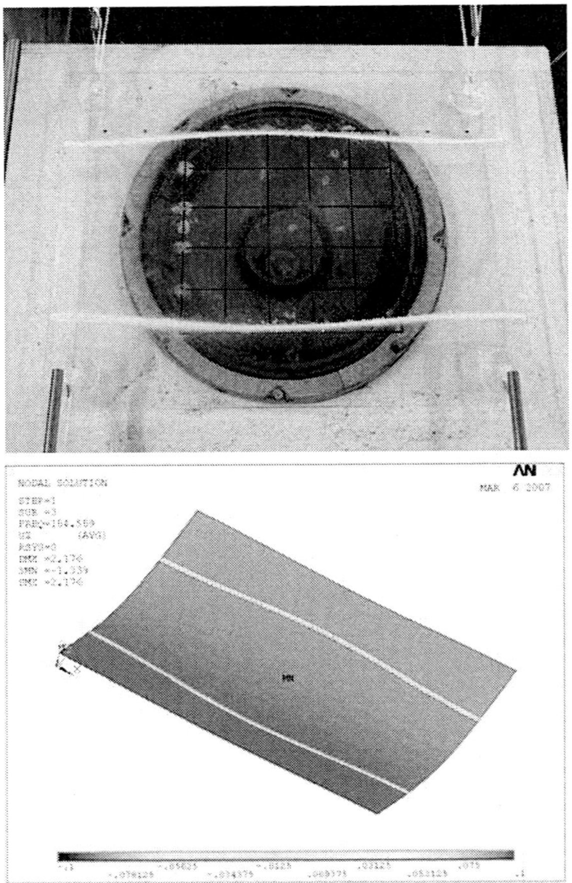

Fig. 5. Test set-up for experimental model validation of natural oscillation with plate of glass (top) as well as a result of numerical simulation (bottom).

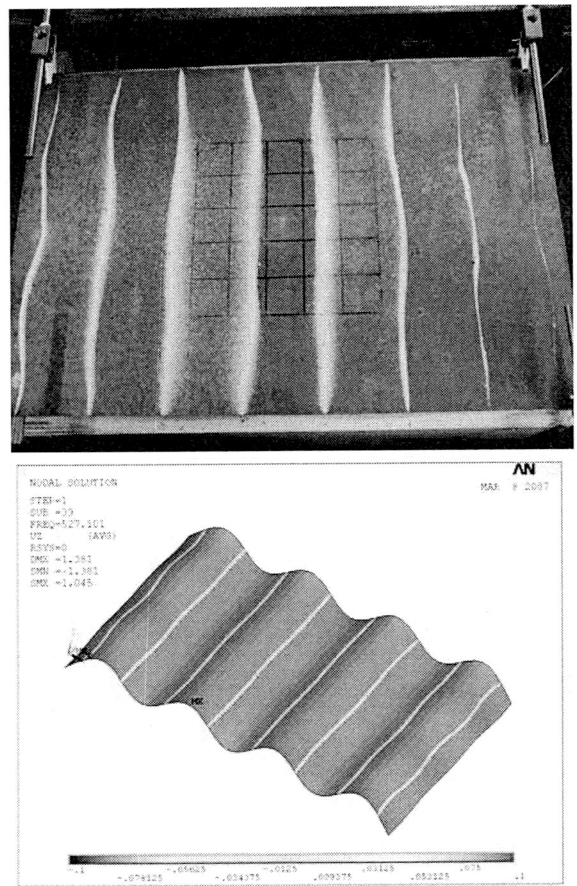

Fig. 6. Test set-up for experimental model validation of natural oscillation with plate of steel (top) as well as a result of numerical simulation (bottom).

frequency of 143 Hz and same mode is observed in simulation result at a frequency of 154 Hz. Represented results for plate of steel are at 581 Hz and 527 Hz respectively.

Differences in associated excitation frequencies can be explained with assumptions which differ from numerical model and experiment. On the one hand, it is boundary condition due to suspension and on the other these are parameters of elasticity which are deduced by experiments.

This frequency distance for test and simulation is constant for all modes what accounts for a systematic influence. It has to be remarked also, besides fixation, position of excitation and tone pitches influence on sound figures. Interestingly enough, multitude of eigenfrequencies of Tab. 3 and 4 could be determined at nearly equal frequencies with experiment and simulation. Owing to this good correlation, vibration characteristics of Fig. 2 can be explained with vibration behaviour of rectangular plates like it has been presumed.

Surely, an overlay of a vast number of modes takes place because they appear at the same or adjacent frequencies. Anyhow, only some significant ones can be detected within an amplitude spectrum. What kinds of advantages arise from this for any localisation algorithm for solid objects or which arrangement of sensors is best is still subject of investigation, currently.

In consideration of appropriate conditions, different kinds of analysis of vibration behaviour are matching together.

3. Summary

Within the scope of research about an algorithm for source localisation in solid objects, the need arose to investigate vibration behaviour of rectangular plates particularly. Therefore, starting with theoretical calculation from literature, bending waves in plates have been studied in accordance with experimental tests and numerical simulations. Furthermore, effect of dispersion in solids has been determined for material which has been is use.

On the whole, in consideration of particular suitable conditions, each implemented method reveals the vibration behaviour.

Acknowledgements

Partly, this work was financed by the European FP6 IST Project "Tangible Acoustic Interfaces for Computer-Human Interaction (TAI-CHI)" and the European Network of Excellence for "Innovative Production Machines and Systems (I*PROMS)".

The support of the European Commission and the colleagues *Guillemette Ribay (LOA), Jan Lukas Hilgermann (TUC)* and *Sergej Mamaschew (TUC)* is gratefully acknowledged.

References

[1] http://www.taichi.cf.ac.uk/

[2] Rolshofen, W.; Dietz, P.; Schäfer, G.: Application of Tangible Acoustic Interfaces in the Area of Production Control and Manufacturing. J. Jacko (Ed.): Human-Computer Interaction, Part II, HCII 2007, LNCS 4551, pp. 920–925, Springer-Verlag Berlin Heidelberg 2007, ISBN: 978-3-540-73106-1.

[3] Lamb, H.: On Waves in an Elastic Plate. Proc. of the Roy. Soc. of London, Series A, pp. 114-128, London 1917.

[4] Bergmann, L.: Der Ultraschall und seine Anwendung in Wissenschaft und Technik. S. Hirzel Verlag, Stuttgart 1949.

[5] Trendelburg, F.: Einführung in die Akustik. Springer-Verlag Berlin Göttingen Heidelberg 1961, Dritte Auflage, Erlangen 1960.

[6] Szabó, I.: Höhere Technische Mechanik. Fünfte Auflage, Springer-Verlag Berlin Heidelberg New York Tokyo 1985.

[7] Cremer, L.; Heckl, M.: Körperschall. Springer-Verlag Berlin Heidelberg 1996.

[8] Kollmann, F.; Schösser, T.; Angert, R.: Praktische Maschinenakustik. Springer-Verlag Berlin Heidelberg 2007, ISBN: 3-540-20094-0.

[9] Storm, R.: Kompendium Maschinenakustik Teil 1: Maschinenakustik – Grundlagen. Studienskript, typographics GmbH, Darmstadt 2006.

[10] Scheminzky, F.: Die Welt des Schalles, Verlag „Das Bergland-Buch", Salzburg 1943.

Innovative Production Machines and Systems
D.T. Pham, E.E. Eldukhri and A.J. Soroka (Eds.)

Design Implementation of Semantic Oriented Agent and Knowledge based approach for Intelligent Human Machine Data Manipulation

A. Zeeshan, D. Gerhard

Mechanical Engineering Informatics and Virtual Product Development Division (MIVP), Vienna University of Technology (TU Wien),
Emails: {zeeshan.ahmed, detlef.gerhard} @tuwien.ac.at
Address: Getreidemarkt 9/307 1060 Vienna, Austria

Abstract

In this research paper we discuss a semantic oriented agent and knowledge based approach towards high machine interface development, information engineering & modelling, fast data management & extraction and data representations based problems .i.e., I-SOAS. Moreover we present the conceptual and implantable architectural designs and some currently available latest rich internet application, semantic web and data management based technologies for the real time implementation of I-SOAS.

Keywords: Agent, Information Processing, Intelligent, Knowledge, Management, Semantic

1. Introduction

Now days there are four intensive existing very important aspects to be considered before and during the development of most of the business oriented product data management, product life cycle management, artificially intelligent and enterprise data base management applications .i.e.,

1. User System Communication
2. Meta Data Extraction & Processing
3. Information Management
4. Data Presentation.

These above mentioned four aspects can cause problems in the implementation of human machine interface development, Meta data extraction, information modelling, fast data extraction and management.

Targeting these above mentioned four problems we present a semantic oriented agent and knowledge based solution in the form of a conceptual iterative architecture in section 2 of this research paper. We also briefly present the implantable architectural design of proposed solution and explain its each component of in section 3 of this research paper. Moreover we also discuss some available human machine interface development, semantic web and data management based technologies for the development of proposed solution in section 4 of this research paper.

2. I-SOAS

Focusing on already mentioned four aspects in section 1 of this research paper, we have proposed a solution as the extended version of already proposed and underdevelopment solution *Semantic Oriented Agent Based Search (SOAS)* [1, 2, 3] .i.e., Intelligent Semantic Oriented Agent Based Search (I-SOAS) [4]. SOAS is a dynamic components and agents based approach, which is proposed to handle user's unstructured full text based requests by converting in to structured information models and generating semantic based queries to search particular needed information. [3].

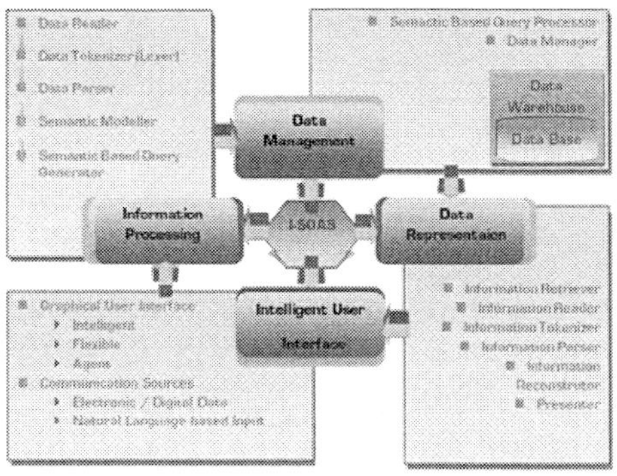

Fig 1. Intelligent Semantic Oriented Agent Based Search (I-SOAS) [4]

The proposed architecture of I-SOAS is consists of four main iterative and centrally connected components .i.e., Intelligent User Interface (IUI), Information Processing (IP), Data Management (DM) and Data Representation (DR) [4] (See Fig. 1).

2.1. Intelligent User Interface (IUI)

Intelligent User Interface is the initial component of I-SOAS, responsible for the intelligent user machine communication. Basically IUI is divided into two main categories .i.e., *Graphical User Interface* and *Communication Sources* (See Fig. 1).

In the Communication Sources, first the corresponding user is identified to enable the correct communication mode. If it is a system then electronic data communication mode is enabled and if it is human then natural language based communication mode is enabled.

The Graphical User Interface part is consists of three concepts .i.e., Intelligence, Flexibility and Agent to intelligently handle the user's requests by providing options to easily use, manipulate and redesign user interface by user itself.

2.2. Information Processing (IP)

Information Processing is the most important component of the I-SOAS because the quality of the performance of I-SOAS majorly depends on it. IP is divided into five main iterative sequential steps .i.e., *Data reading, Tokenization, Parsing, Semantic modelling and Semantic based query generation* (See Fig. 1).

IP works in a way that data is supposed to be retrieved read and organized using Data Reader, which then tokenized in to possible number of tokens using Data Tokenizer, which then parsed with respect to the used grammar using Data Parser. Tokenized data is then semantically evaluated and used in Meta data semantic model generation by Semantic Modeller, which then used by Semantic Based Query Generator to generate semantic based query to extract desired result.

2.3. Data Management (DM)

Data Management is responsible for two important functions .i.e., *Semantic based Query Processing* and *Data Management.*

In Semantic based Query Processing IP generated semantic based query is parsed to generate an SQL based query to run and extract the required relevant information from repository. The job of Data Manager is to manage the process of SQL query generation, data extraction and storage.

2.4. Data Representation (DR)

Data Representation is the last component of I-SOAS consisting of six sequential steps .i.e., *Information Retriever, Information Reader, Information Tokenizer, Information Parser, Information Reconstructor and Presenter.*

The job of this component to extract, tokenize, parse and reconstruct processed information to present the final output to user in understandable user's used natural language format.

3. I-SOAS Implementable Architectural Design

To implement I-SOAS as a real time software application, an implementable architecture is designed and presented (See Fig. 2). The designed architecture is consists of four modules .i.e.,

1. I-SOAS Graphical Interface
2. I-SOAS Data Warehouse
3. I-SOAS Knowledge Base
4. I-SOAS Processing Modelling

and three communication layers

1. Process Presentation Layer (PPL)
2. Process Database Layer (PDL)
3. Process Knowledge Layer (PKL)

These four modules and three layers are supposed to be seven independent components, which will work in an integrated form to perform the required operations and obtain user's required end results.

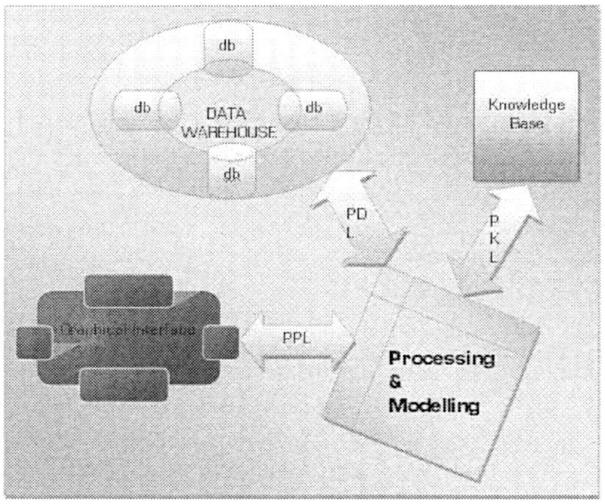

Fig 2. I-SOAS Implementable Architectural Design

3.1. I-SOAS Graphical Interface

The I-SOAS Graphical Interface is conceptually based on the concept of the first component of I-SOAS's conceptual iterative architecture Intelligent User Interface (IUI) (See Fig. 1) and practically based on the Human Machine Interface (HMI) Design concepts .i.e., Ideologies, Principles, Patterns and Guidelines [23].

The I-SOAS Graphical Interface module is the graphical user interface of I-SOAS for the user machine communication, responsible for two main jobs

1. Taking input from user and forwarding that input to Processing and Modelling Module.
2. Getting output from Processing and Modelling Module and replying back to the user.

3.2. I-SOAS Data Warehouse

The I-SOAS Data Warehouse conceptually based on the concept of third component of I-SOAS's conceptual iterative architecture Data Management (See Fig. 1). I-SOAS Data Warehouse is the repository to store, extract, transform, load and manage the organizational technical and managerial data.

The designed I-SOAS Data warehouse is supposed to work like other Data Warehouses by providing several options to produce common data models, reports, analysis, loading, security and retrieval of data without slowing down operational systems [24].

3.3. I-SOAS Knowledge Base

The I-SOAS Knowledge Base is conceptually based on the concept of third component of I-SOAS's conceptual iterative architecture Data Management (See Fig. 1).

The I-SOAS Knowledge Base is designed to capture, manage, improve and deliver knowledge based information by providing information collection, retrieval, analysis, management and organization.

3.4. I-SOAS Processing and Modelling

The I-SOAS Processing and Modelling is actually a kind of semantic based system conceptually based requirements of second component of I-SOAS's conceptual iterative architecture Information Processing (IP) (See Fig. 1).

I-SOAS Processing and Modelling module is designed in a way that it can read, organized, tokenize, parse, semantically evaluate and generate a semantic based query use and extract desired results from I-SOAS Data warehouse and I-SOAS Knowledge base using inputted set of instructions based data.

3.5. I-SOAS Process Presentation Layer (PPL)

I-SOAS Process Presentation Layer (PPL) is designed as the communication layer between Graphical Interface and Processing & Modelling Modules of the I-SOAS implementable architecture (See Fig. 2).

The job this designed layer is simply to transfer user inputted set of instructions from Graphical Interface to Processing and Modelling & Processing module and transfer back system's outputted data from Processing and Modelling to Graphical Interface.

3.6. I-SOAS: Process Database Layer (PDL)

I-SOAS Process Database Layer (PDL) is designed as the communication layer between I-SOAS Data warehouse and Processing & Modelling Modules of the I-SOAS implementable architecture (See Fig. 2).

The job this designed layer is simply to forward Modelling & Processing module's constructed query to I-SOAS Data warehouse and return retrieved results from I-SOAS Data warehouse to Modelling & Processing module.

3.7. I-SOAS: Process Knowledge Layer (PKL)

I-SOAS Process Knowledge Layer (PKL) is designed as the communication layer between I-SOAS Data Knowledge base and Processing & Modelling Modules of the I-SOAS implementable architecture (See Fig. 2).

The job this designed layer is simply to forward Modelling & Processing module's constructed semantic based query to I-SOAS Data knowledge base and return extracted results from I-SOAS Data knowledge base to Modelling & Processing module.

4. I-SOAS Technological Implementation

We are considering some available human machine interface development, semantic web and data management based technologies for the development of I-SOAS.

4.1 .Intelligent Human Machine Interface Development

At the moment for the practical implementation of Intelligent User Interface (IUI) we are considering some Rich Internet Applications (RIA) [5] based technologies .i.e., FLEX (Adobe) [6,7], AJAX [8, 9, 10, 11] OpenLaszlo [12, 13], Silverlight (Microsoft) [14]. We have chosen RIA because it provides small Installation footprint, reduces overhead for updating and distributing the application, provides automatic and transparent updates to new versions, provides usage of application from any computer with an internet connection, gives consistency to user regardless of what operating system the client and less prone to viral infection than running an actual executable.

4.2 .Information Processing and Modelling

At the moment, for the implementation of Information Processing (IP), we are considering Semantic Web based technologies for information processing and modelling. Semantic Web is an advanced version of existing web which claims to be a solution toward currently faced web problem of formatting data in machine process able format to take advantage in information retrieval by information processing and modelling. Moreover Semantic web is a mesh of data representing meanings through connectivity, expressing multiple view points and use business rules logic is making models to share information across applications. The goal of semantic web is to structure the meaningful contents of unstructured data and involve knowledge management in making more advanced knowledge modelled management systems [15].

Currently semantic web is solely depending upon a single and main building block Ontology. Ontology is the explicit and abstract modelled representation of already defined finite sets of terms and concepts, involved in knowledge engineering, knowledge management and intelligent information integration [16]. Ontology development process with respect to implementation point of view depends on some currently available ontology supported languages [20] .i.e., XML [17], RDF [18] and OWL [19].

4.3 Data Management

At the moment, for the implementation of Data Management (DM), we are considering the concepts of Data Warehousing [21] for heavy data loading, management, storage, extraction and transformation. For the practical implementation we are considering Oracle Data Warehousing Platform [22].

4.4 Data Representation

In the end, at the moment, for the implementation of *Data Representation (DR)* we are aiming to reuse the concepts of practicality of RIA and Semantic Web.

5. Conclusions

In this research paper we have briefly demonstrated I-SOAS as a potential conceptual architecture based solution towards already described four intensive currently existing problems .i.e., User System Communication, Meta Data Extraction, Management and Representation. Moreover we have also presented and explained the implantable architectural design of I-SOAS and some currently available latest rich internet application, semantic web and data management based technologies for the real time implementation of I-SOAS.

6. Future Recommendations

As this is an ongoing in process research, in future, we are aiming to focus on the in depth architectural analysis and practical implementation of the implementable architectural designs of I-SOAS using some available human machine interface development, semantic web and data management based technologies.

References

[1] A. Zeeshan and D. Gerhard, "An Agent based Approach towards Metadata Extraction, Modelling and Information Retrieval over the Web", In the proceedings of First International Workshop on Cultural Heritage on the Semantic Web in conjunction with the 6th International Semantic Web Conference and the 2nd Asian Semantic Web Conference 2007, Busan Korea, 12-15 November 2007

[2] A. Zeeshan and D. Gerhard, " Personal Assistant towards Semantic Information Retrieval", In the proceedings In the proceedings of Fifth International Workshop on Ontologies and Semantic Web for E-Learning (SWEL'07), 13th International Conference on Artificial Intelligence in Education (AIED 2007), P115, Los Angeles USA, 9–13 July 2007

[3] A. Zeeshan and D. Gerhard, "Designing a Dynamic Components and Agent based Approach for Semantic Information Retrieval", In the proceedings of 6th CIIT Workshop on Research in Computing (CWRC Fall 2007) COMSATS Institute of Information Technology, Abbottabad Pakistan, 27 October 2007

[4] A. Zeeshan, D. Gerhard: "Semantic Oriented Agent based Approach towards Engineering Data Management, Web Information Retrieval and User System Communication Problems", Accepted in 3rd International Conference for Internet Technology and Secured Transactions, Dublin Institute of Technology, Dublin Ireland, 23–28 June 2008

[5] Rich Internet application, reviewed at 06 June 2008, <http://en.wikipedia.org/wiki/Rich_Internet_application>

[6] Flex, reviewed at 06 June 2008, <http://www.adobe.com/products/flex>

[7] Adobe Flec, reviewed at 06 June 2008, <http://en.wikipedia.org/wiki/Adobe_Flex>

[8] Ajax, reviewed at 06 June 2008, <http://www.asp.net/ajax>

[9]Ajax, reviewed at 06 June 2008, <http://www.w3schools.com/Ajax/Default.Asp>

[10]Ajax, reviewed at 06 June 2008, <http://en.wikipedia.org/wiki/AJAX>

[11] Rich Internet application Ajax, reviewed at 06 June 2008, <http://en.wikipedia.org/wiki/Rich_Internet_application#JavaScript_.2F_Ajax>

[12] OpenLaszo, reviewed at 06 June 2008, <http://en.wikipedia.org/wiki/OpenLaszlo>

[13] OpenLaszo, reviewed at 06 June 2008, <http://www.donhopkins.com/home/124>

[14] Silverlight, reviewed at 06 June 2008, <http://en.wikipedia.org/wiki/Silverlight>

[15] Tim Berners-Lee, James Hendler and Ora Lassila, "The Semantic Web, A new form of Web content that is meaningful to computers will unleash a revolution of new possibilities", May 2001, Viewed February 2007, <http://www.sciam.com/article.cfm?articleID=00048144-10D2-1C70-84A9809EC588EF21>

[16] D. Fensel. Ontologies: Silver Bullet for Knowledge Management and Electronic Commerce. Springer-Verlag, Berlin, 2000

[17] The W3C Extensible Markup Language (XML), viewed February 2007, <http://www.w3.org/XML>,

[18] Klaus Tochtermann and Herman Maurer, "Semantic technologies – An Introduction", Semantic Technologies Showcase the Austrian Situation, pp. 15-20, 2006

[19] OWL Web Ontology Language, viewed February 2007, <http://www.w3.org/TR/owl-features>

[20] Zeeshan Ahmed, D. Gerhard: "Web to Semantic Web & Role of Ontology", In the proceedings of National Conference on Information and Communication Technologies (NCICT 2007), P 100, Bannu Pakistan, 9 June 2007

[21] Data warehouse, Reviewed 06 June 2008, <http://en.wikipedia.org/wiki/Data_warehouse>

[22] Architecting your Data Warehouse Platform, Reviewed 06 June 2008, <http://www.oracle.com/solutions/business_intelligence/dw_home.html>

[23]A. Zeeshan, S. K. Ganti, H. Kyhlbäck: "Design Artifact's, Design Principles, Problems, Goals and Importance ", In the proceedings of Fourth international Conference on Statistical Sciences, ISOSS 08, University of Gujrat, ID: 042, May 9-11, Gujrat Pakistan 2008

[24] Data Warehouse, Reviewed 06 August 2008, <http://en.wikipedia.org/wiki/Data_warehouse>

[25] Laura Hadley, "Developing a Data Warehouse Architecture", Reviewed 06 August 2008, <http://www.users.qwest.net/~lauramh/resume/thorn.htm>

Innovative Production Machines and Systems
D.T. Pham, E.E. Eldukhri and A.J. Soroka (Eds.)

Human-machine interaction based on multimodal interfaces for industrial environments

M. A. Pérez, I. Maurtua, C. Tubío, J. Molina

Department of Production Engineering, Fundacion Tekniker, 20600-Eibar, Spain

Abstract

Human-machine interfaces are being so far developed aiming at facilitating machine operation and minimising as much as possible errors, giving rise to quite rigid solutions. Even despite of the "friendly" qualifier, such interfaces show complex, hierarchical and inflexible structures of use.

Multimodal interfaces, where two or more ways of human-machine interaction are combined, open new chances of information exchange leading to more friendly systems. Among them, they offer more information to the user, making easy the decision taking, freeing him/her to do other tasks simultaneously and offering information complementary to the one perceived from the real world.

This paper shows developments in the field of human-machine interfaces applied in industrial environments. New ways of interaction based on speech processing and portable devices for information exchange are presented. Everything has been done over a standard Web Services (WS) system that allows using them in a multiplatform environment.

The activity has been focused on three industrial scenarios:

- Open architecture CNC for machine tool control.
- Machine tool with a proprietary control system.
- Industrial application for production control.

Keywords: Multimodal interfaces, speech processing, Web Services

1. Introduction

Most of technologies for multimodal interfaces are used out of production workshops. Here it is intended to extend the range of application of those technologies to industrial environments by focussing on speech processing and new interaction options with portable devices. Commercially available tools and devices have been used and adapted for the requirements of such environments.

Automatic speech recognition (ASR) has been successfully applied to different fields (e.g. dictation, call centres, ..). Nowadays, there is a special interest to bring it to the industrial world, for instance for machine tools operation. This implies a challenge because ASR systems are based on acoustic models trained with clean samples, i.e. free of noise. It is normal in industrial environments to have an acoustic signal contaminated with background noise. Consequently, there would be a discrepancy between the characteristics observed during the training and use phases. To overcome this situation robust tools against environmental noise have to be used.

The interaction based on portable devices is a complementary mechanism of information exchange for the worker. It allows him/her interacting with the machine and other applications receiving current and additional information. On the one hand, the worker can access remotely information shown in the machine when he/she is far from the operator's panel and, on the other hand, he/she can access relevant information usually not shown.

Most of projects related to multimodal interfaces are prototypes out of the industrial environment. For example, in ERMIS [1] the state of mind of the person is interpreted based on the way of speaking, gestures and expressions. In MIAMM [2] existing technology is used for multilingual speech recognition and haptic devices and gesture recognition for fast access to multimedia databases.

There are some examples of multimodal interfaces brought to the industrial environment. This has been noticed in the increasing number of projects related to the integration of the implied technologies that have as objective the activities at shopfloor level. Some of them are focused on guiding and training of maintenance people. STARMATE [3], SEAR [4] and a line within the global project ARVIKA [5] are examples of integration of speech processing and augmented reality in training and maintenance activities.

This kind of developments involves some non measurable benefits related to the way in which humans operate. Examples of advantages are:

- Ergonomics: use of speech as natural way of communication, less use of operators panels required and freedom to move around the machine.
- Hands free: data input and manual work can be done simultaneously.
- Flexibility of information display: the information given by the system can be seen at any place around the machine without having to be in front of the screen.
- Attention call: by means of audible massages.
- Additional information: on-line access to other relevant information related to operation, production, maintenance, etc.

2 Scenarios

Three real industrial scenarios have been selected for the identification of requirements and functionalities of the modules to be developed:

- Machine tool controlled and operated by CNC FAGOR 8070.
- Metal injection machine controlled by proprietary application DATAGRAPH running on a PC.
- Industrial software application for production control, OLANET.

The interfaces on which the developments have been focused are: speech processing (recognition and synthesis) and portable devices for information exchange such as PDA, head mounted display (HMD), tablet PC and mobile phone.

For the definition of each scenario it has been specified:

- The working environment where developments would be applied.
- Different options of human-machine interaction.
- User profiles and relationship with devices use.

2.1 Description of the current situation

For the scenario of the machine tool with CNC a milling machine with three axes has been considered. All machine devices are controlled by the CNC and its integrated PLC. Regarding human-machine interaction, the keyboard, the operator's panel. a screen and sometimes an electronic handwheel, are the only elements that the operator has to interact with the machine, what implies to stand in front of the CNC.

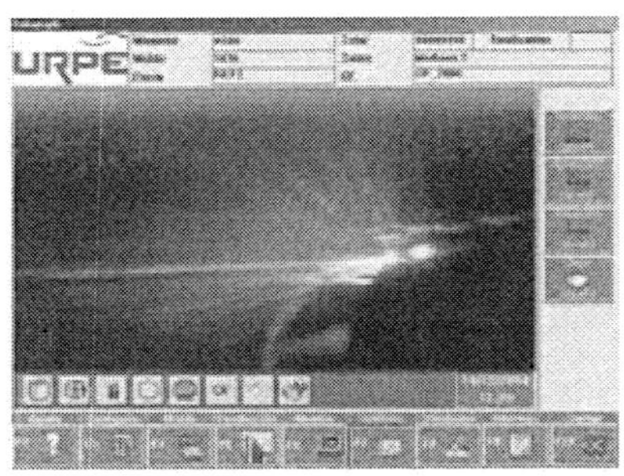

Fig. 1 DATAGRAPH control application

The control application (Fig. 1) for the metal injection machine is formed by four modules that allow, among others, parameters definition, process monitoring, production control and help for maintenance activities. Any interaction of the operator with the machine has to be done through the control application. Such interaction is limited to the use of keyboard and mouse of the PC.

The application for production control (Fig 2.) is a MES system that allows acquisition of production data and gives information to carry out production tasks. The operator interact with the application by introducing production data, consulting work orders, messages, etc., having as unique interfaces the screen and a touch panel.

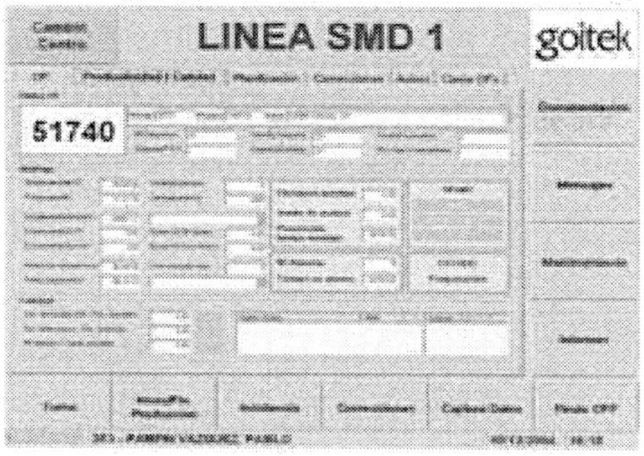

Fig. 2 OLANET application

2.2 Definition of scenarios

The definition of scenarios has consisted of identifying a number of interaction options related to the introduction of new interfaces. Different situations of human-machine interaction have been taken into account and for each one the following concepts have been specified:

- Actor or profile of the user that takes part in the human-machine interaction (operator, foreman, maintenance personnel).
- Scope or situation in which the interaction has been established.
- Context of use of the new interfaces
- Human-machine interactions for each actor using the new interfaces
- Involved technologies/devices that make possible the new interaction ways

Data have been assigned within each concept for each scenario. Table 1 shows an example of the assignments.

Table 1. Scenario: Datagraph control application

Actor	Person who sets up the process
Scope	General configuration and preparation of new work order.
Context of use	The operator configures the machine for a specific work order (WO). Process data are entered and control and tracking data are defined. During process setting up the process data are checked to be according to expectations.
Interactions	- WO preparation. (definition of moulds, configuration and modifcation of paramenters, ..) - Process monitoring (visualisation of curves and other WOs, data of pressure, temperatura, travels, …). - Access to documentation (user/operators manual, maintenance, ..).
Technologies	- Speech recognition for commands. - Voice synthesis to receive confirmations and alarms or warnings. - HMD to see information when far from the control panel (process, manuals, …) - PDA for visualisation and modification of parameters.

It is obvious that a system centred on the person and his/her interaction with the environment has to take into account usability criteria. The level of acceptance by users will depend on the easiness and stress level provided by the system. Therefore, one of the requirements is to take into account that the results are going to be used by people, thus developments have been focused on intuitive interfaces that provide for natural ways of communication.

3. Advanced interaction system

The advanced interaction system is formed by three modules developed for each scenario thus allowing information exchange among users, machines and applications. These modules are:

- Voice command module
- Modules of interaction with portable devices
- Modules integrated in machines and applications

All developments have been done based on a Web Services architecture aiming at facilitating interoperability among applications and working in different platforms.

A generic voice command module with server structure has been developed. It offers recognition and synthesis functionalities to all applications and portable interface devices. This module is dynamically configured and can be adapted to the different situations within the scenarios.

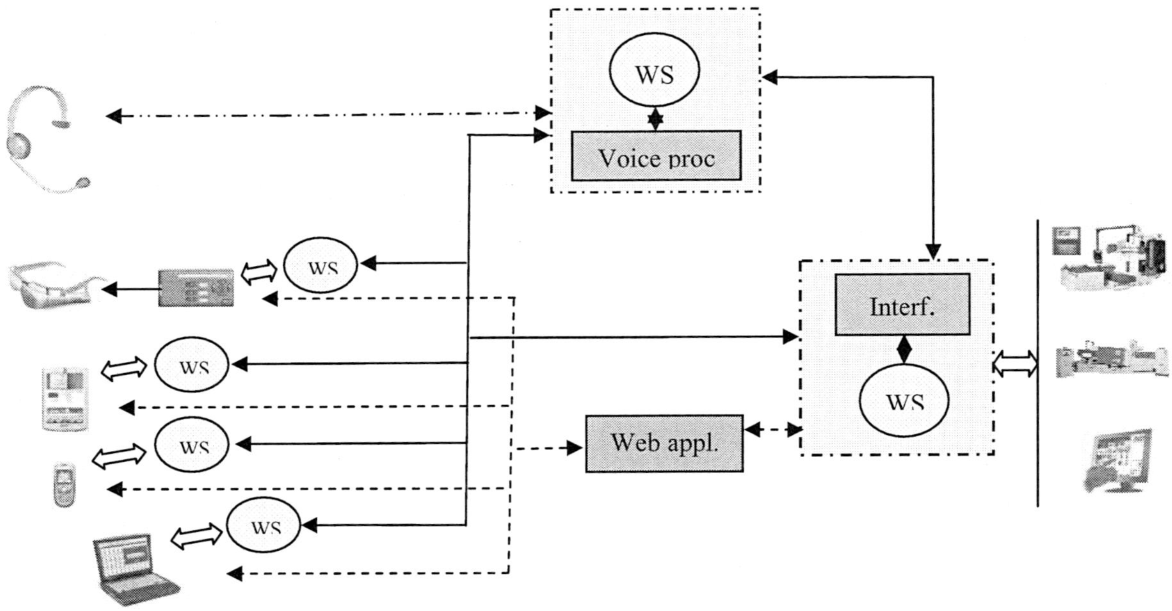

Fig. 3: Architecture of the adopted solution

Thinking of an easy reuse in other integrations the solution has been based on a client-server architecture. The voice command, interaction and specific interfaces modules work under this architecture. The communication among then is done by using standard mechanisms able to be used in different platforms. All are clients and servers among them and they are also able to exchange information with other external applications that use the appropriate standard mechanisms.

Figure 3 shows the architecture of the adopted solution. Servers are implemented as Web Services (WS) with SOAP (Service Oriented Architecture Protocol) protocol. A WS is a software component that communicates with other applications by coding messages in XML and sending them with standard Internet protocols such as Hypertext Transfer Protocol (HTTP). It is similar to a Web site without user interface, serving applications instead of people. It receives requests from an application through a XML formatted messages and sends back answer messages also in XML format.

Interaction based on Web applications has been also implemented in the portable devices. The information is given in XHTML format and mechanisms for continuous updating of data have been implemented. This allows accessing the information by using the browser already installed in the devices. The objective has been just to show different alternatives of information exchange between portable devices and machines or applications.

3.1. Voice command module

The voice command module is aimed at facilitating the interaction with the machine and applications by using speech. It allows the user sending voice commands and receiving audible information of requests and messages.

According to the required use in the scenarios, speech recognition has been focused on commands and grammars. Commands configuration is done dynamically according to the applications' requirements and possible interactions of each situation.

The speech recognition engine works with user profiles that have to be trained. Once the profile has been trained in the working environment the recognition level is very high. At the same time profiles restrict use to authorised personnel only.

The module works as a recognition and synthesis server for several applications and users. Each application/user is registered in the system and allocates a communication channel for information exchange.

3.2. Modules of interaction with portable devices

The advanced human-machine interaction system refers to several devices for display and information exchange between the user and machines or applications. The information displayed can be the same as shown in the screen or additional. The objective is to provide an information exchange system to be used at any place around the machine.

Different devices are considered for different situations and needs of information. The situations are related to the users' profiles and their activity. For instance, a production manager could be interested in receiving production data in the PDA he is carrying, when he is near a machine. In case of a maintenance worker, he could ask for the visualisation of a maintenance activity in the head mounted display he is wearing having thus the hands free to do the work.

The devices used are Head Mounted Display (HMD), PDA, tablet PC and mobile phone. All of them are commercially available devices that simply require usual communication mechanisms (Ethernet, WiFi, Bluetooth), operative systems that allow implementing Web Services and a browser for XHTML.

Two types of developments have been done: applications with Web Service (WS) structure and Web applications, just to show different options of remote interaction for information exchange.

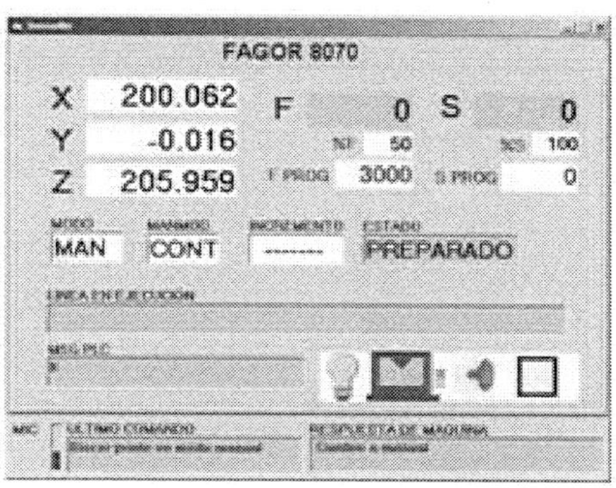

Fig. 4 CNC information in HMD

The communication and data exchange mechanism has been WS with a different implementation in each device due to the different operative systems (Windows XP, Pocket PC, Symbian). Figures 4 and 5 show examples for HMD and PDA.

An infrastructure that allows integration with other applications has been developed for each device. It includes interaction with the voice command module and with the applications of each scenario. Based on this structure particular pilot interactions among devices and machines and applications have been done.

Regarding Web applications, they are accessed with the device browsers. The information shown has the particularity of being adapted to the presentation possibilities of the devices' screen.

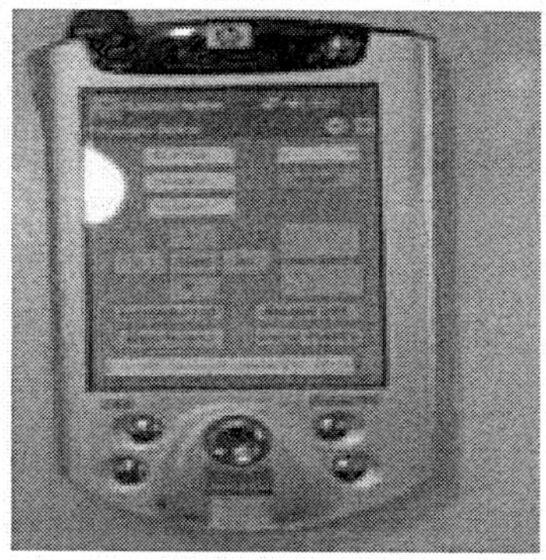

Fig. 5 Control from PDA

Web applications have been developed with AJAX technology. It allows creating interactive applications by combining three existing technologies: HTML (or XHTML) and CSS to show the information, DOM and JavaScript to dynamically interact with data and XML and XSLT to exchange data with a WS server. The information is requested to the server by using SOAP and the client uses JavaScript to process the answer.

3.3. Modules integrated in machines and applications

The implementation of the functionalities required in the three scenarios implies the development of modules in machines and applications with two purposes: to allow information exchange with the voice command module and portable devices and to act as interface between machines and applications to provide the required information with the appropriate format.

Following the same philosophy of previous cases, the solution has been also based on WS. The implemented WS architecture allows using events as mechanism to notify changes. This mechanism makes possible an asynchronous system reaction, optimising performance at the same time.

For each scenario a WS application has been developed to offer information contained in CNC, PLC and control applications. In addition to this, this WS application also implements the direct interaction by using speech. In case of CNC Fagor 8070 the two functionalities are independent because a WS based layer has been added over the already existing proprietary API to access CNC/PLC data.

4. Conclusions

New human-machine interaction options for industrial environments have been shown trying to offer alternatives to the usual ways of operation even with certain improvements. The solution presented in this paper increases the number of operation and information exchange options that people have at shop floors. This results in more freedom to move around the machine and to do other activities (hands free), more flexibility for data input, access to information independently of the user location and greater knowledge about the machine, the process and other activities related to production.

Technologies and tools so far applied in other environments have been integrated into industrial environments. To get a successful integration tools and devices have been adapted to the prototype scenarios. The solution has been developed thinking of implementation in other devices and applications different from the prototype. With this aim standard technologies that facilitate interoperability and implementation in different platforms have been used.

It has to be taken into account that the main actor is the person and, therefore, developments have to be done by considering usability criteria. The focus has been then on more intuitive interfaces intending to be closer to the natural ways of communication.

Acknowledgements

This research is possible thanks to the Basque and Spanish Government research programs. Fundacion Tekniker is partner of the EU-Funded FP6 Innovative Production Machines and Systems (I*PROMS)[1] Network of Excellence.

References

[1] Emotionally Rich Man-machine Interaction Systems – ERMIS http://manolito.image.ece.ntua.gr/ermis/

[2] Multidimensional Information Access using Multiple Modalities – MIAMM IST-2000-29487) http://www.loria.fr/projets/MIAMM/presentation.html

[3] System Using Augmented Reality for Maintenance, Assembly, Training and Education – STARMATE.

[4] Goose S., Sudarsky S., Zhang X., Navab N. *Speech-enabled Augmented Reality Supporting Mobile Industrial Maintenance – SEAR*, IEEE Pervasive Computing Jan-March 2003

[5] Augmented reality in industrial applications, http://www.arvika.de/www/index.htm

---

[1] http://www.iproms.org

Innovative Production Machines and Systems
D.T. Pham, E.E. Eldukhri and A.J. Soroka (Eds.)

Virtual Machine Tool - A Case Study on Virtual Operator Training

B. Denkena[a], C. Ammermann[a], H. Henning[a], E. Hohwieler[b], T. Schlegel[c]

[a] *IFW Hannover, Leibniz Universität Hannover, An der Universität 2, Garbsen 30823, Germany*
[b] *Fraunhofer IPK, Pascalstr. 8-9, Berlin 10587, Germany*
[c] *Fraunhofer IAO, Nobelstr. 12, Stuttgart 70569, Germany*

Abstract

In operator training for machine operations, the application of virtual supported training courses promises great benefits for industrial usage. If the Human-Machine-Interface is adequately modelled, this may lead to a decrease in ramp-up times or system failures during a running production and an immediate hand on support of the user. In the presented paper, a prototype for virtual operator training to achieve a faster and more flexible ramp-up is described. The realized prototypical system shows new possibilities of future learning and demonstrates applications of virtual machine and control models. The developed approach is compared to existing approaches in the field of computer-aided operator training. Due to its flexible structure, the system allows a global implementation and adaptation to people with different languages and cultures and different machinery. The system structure as well as the developed training concept are described in detail to illustrate the economical benefits for an implementation in industry.

Keywords: Virtual Reality, NC-Programming, Human Machine Interface

1. User training and qualification profiles

Skilled workers employed in the machining production have to cope with a multilayered activity spectrum. The combination of knowledge about production processes as well as the machines employed requires large-scale qualifications which must be permanently adapted to the available production resources. The task field of a skilled worker at numerically controlled machine tools can be represented as follows by [6]:

1. NC programming
2. Machine setting
3. Running of the NC program
4. Operating and controlling the processes
5. Quality control
6. Disturbance management
7. Maintaining and cleaning

The activities can again be grouped into the task fields of operation and controlling (1-5) as well as maintenance (additionally 6+7). For both groups, characteristic qualification profiles exist.
The training of both users and maintainers is carried out in different ways to fulfil the respective qualification profile. The training is seldom carried out by the operator of the plant itself. In most cases, it is run by the machinery supplier who offers corresponding services. Standard machines of the manufacturer are usually available in the training centres of the manufacturer. The machinery manufacturer carries out the training in the factory of the user for a specific number of employees to be instructed. The training might also be referred to special machines like transfer lines frequently used in the large-scale manufacturing. This kind of training and the imparting of knowledge for the machine user are further considered in this paper.

2. Training contents

The training carried out for future machine operators can be divided into a theoretical section (duration approx. 1 day), including a lecture given by the trainer, and a practical section at the machine introducing the participants into certain activities (2 days). The training contents comprise the following points: safety in general, safety equipment, central line operator panel plant (interlinking), operator panel single machine, tool management and measurement.
A presentation of the operating devices on the line level follows after an introduction into general safety advices and into the operation of the safety principle (Safety Integrated) and the safety equipment of the plant. The central line operator panel serves for the control of the interlinking. The remaining points concern the machine itself and contain the explanation of control functionalities. The whole training is focused on the operator panel of a single machine with the operating functions of the HMI software as well as the machine control panel with standard operating devices and those specific for the machine-tool. NC operation methods like setting (Jog)/ MDA (Manual Data Automatic), automatic as well as the machine functions like single block selection, teach-in, returned positioning and the start of a basing point are explained. The fundamental processes of the machining operation are at first theoretically explained with single operation sequences. The turning-on of the machine driving to the starting position as well as the referencing of the axes can serve as examples. The introduction of new tools into the tool magazine as well as the insertion of the proper data into the data management of the control is explained in the course of the tool management. The introduction into the measuring systems of the machine is given at the end. In the practical part, the application of these operation sequences is exercised directly on the machine.
Production equipment used as training facilities is not available for the production. Using a virtual control in combination with a virtual machine tool for the training of the operators can avoid this problem. If it works considerably to reduce the occupancy of the real machine or to make it even obsolete, the training will even be independent from actual condition, position, and utilization of the production facilities. To enable the knowledge transfer within mentioned training contents with the help of a virtual machine tool, it is a precondition to have models of both, machine and control, on an adequate level of detail.

3. Existing approaches for the computer-aided user training

More and more Virtual Machine Tools are introduced and used in control software development and for realistic NC verification [4]. In such a Virtual Machine Tool a 3D model representation for the simulation of the machine tool itself and a machining model for the simulation of the material removal at the work piece are combined. The system is controlled by a software emulation of the numerical control. With the usage of the Virtual NC Kernel (VNCK), another increase of the precision of these systems is achieved [2, 3]. Other approaches contain Internet based educational offers. Machine operator panel, HMI software and a simple, three-dimensional machine model are represented in the Internet browser [1]. Short tutorials and information domains give the user support and instructions. A further approach is the use of the hardware in- the -loop simulation for educational purposes. The hardware of the control is coupled with a machine model for the visualization and machine behaviour modelling. The approaches are distinguished either by the application of control specific hardware or by the use of insufficiently realistic operating interfaces or NC kernel. For this reason, the application of a new system for the operator training is suggested as an alternative. It combines elements of the virtual numerical control and machine visualization to a realistic training workplace which requires standard desktop PC components only for simulation and user interaction purposes.

4. System structure

The user interface contains the HMI software and its connector to the NC core as well as the virtual control panel. This virtual control panel integrates the machine instrument panel and the operator panel into one consistent user interface. The machine model is used to allow the visualization of the machine including the emulation and display of its mechatronic behaviour.

4.1 Realisation of the user interface

For the user interface of virtual operator training systems, it is important to provide as much realistic environment and system as possible for the training sessions. This applies specifically to the layout and usability of the human machine interface and machine

simulated on the virtual operator training system. Regarding individualization, the machine control panel, operator panel and the simulated human machine interface should not be changeable by the user. For example, moving and rearrangement of single windows within the user interface should be disabled. Entering user commands should not be implemented by a separate interaction device like a mouse in order to provide realistic interaction with the system.
Therefore, commands should be given directly using a

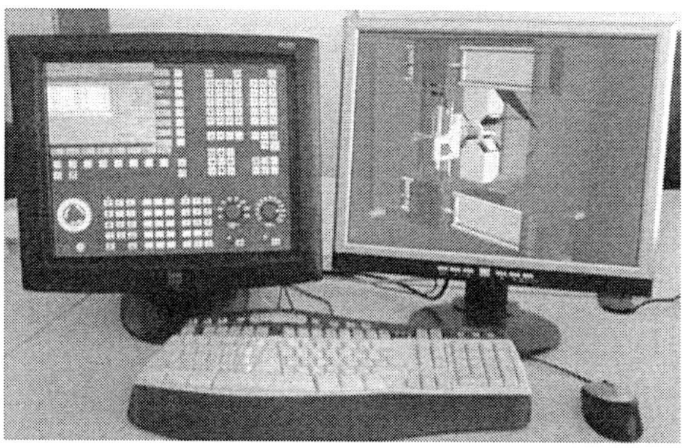

Fig. 1. View of a virtual operating training system

touch sensitive screen or a similar device for direct manipulation on which the command buttons are displayed (figure 1).To answer this recommendation, the user interface is made available on a touch screen interpreting touches by the user as mouse events.
Figure 1 shows the PC of the training workplace including the touch screen and the user interface displayed on the left side. All components of the user interface are presented as one unit. The development of the training software included the integration of components from further project partners of the IFW.

4.2 User interface conceptual evaluation

ISO 9241-10 [5] defines important aspects of usable interfaces. In the presented case of training software, some specialties are obvious. Suitability for learning is a key element for training software, which implies on one hand that the interfaces should be configured to be as easy to learn as possible. On the other hand, conformity with user expectations is even more important in the special case of NC interface emulation: The interface must be as similar as possible to the real control that is emulated to enable easy transfer of practical experience to the learning environment and from the learning experience back to practice.

For the visualization of the real behaviour of the machine and its environment, the performance of the machine should be in high conformity with user expectations, but the exact optical impression has a low impact on the learning transfer. This implies for example that photorealistic impression is not a key factor for the learning of NC operational concepts.
The main task of the virtual operator training system is learning the handling and operation of a NC machine. Suitability for the task in this case means that learning is optimally supported by covering all functions to be used in daily practice while less important factors like environmental influences are only included when they contribute to the learning goal. This also shows one of the trade-offs in a training environment: While a completely realistic model is achieving high conformity with user expectations, in some cases additional functionality for the learning environment has to be included. This can comprise guides in written form or multimedia contents to explain specific processes, functions or tasks to be solved by the trained operator.
With the criterion self-descriptiveness a trade-off is associated. While self-descriptiveness of training software eases its use and application for so far untrained users, strong deviation from the behaviour and setup of a real control will lead to issues in the transfer to practice. If the real system shows a different behaviour concept than the one learned in the virtual operator training, concepts are not anymore seamlessly transferable to the real NC interface used in practice after completion of training.
Controllability is a criterion with high influence on the NC interface, but is less considered in a training environment, because all control concepts have to be transferred from the real machine to the training environment. Only where additional functions are in place, e.g. for learning control or guidance, high controllability is an important criterion.
Often not considered enough in the development of user interfaces is the conclusion of fault and error tolerances that yield a high benefit for operational and training systems. When it is possible for users to make errors without being afraid of severe consequences, learning effects have proven to be much higher. This is a main advantage of virtual operator training environments compared to full machines. While the interface is very similar, critical side effects of the training can be eliminated. This makes the virtual operator training an important alternative to training on the machine, supporting explorative learning approaches.

Suitability for individualization is relevant and common on complex software systems with a high degree of parameterization. In the case of virtual operator training, individualization is mainly possible only where also the real environment allows for personal or group specific adaptations. In highly standardized production environments this is not recommended and commonly static user interfaces are used. Where process-driven adaptation is necessary, specific models have to be used, for example object-oriented descriptions for interactive processes. [7]

4.3 Machine model

The machine model as the main part of the virtual machine tool is a mathematical description of a machine and its components. It allows the prediction of the machine's physical properties in a very short (real) time. Basically, the machine model computes the position of the tool and its impact on the work piece in dependence on the previous state of the machine and the current target axes positions given by the numerical control. In dependency on the purpose of the simulation it contains the electrical system, the mechanical system, the actuators, the sensors and a process model. The mathematical representation distinguishes between kinematic and dynamic simulation. A kinematic model has no dynamic degree of freedom: the actual value equals the target value. The dynamic model considers all interdependencies: the actual value is a solution of a system of differential equations (including masses and velocities of all moved parts as well as their interactions).
For the purpose of a training system, a kinematic machine model and a simplified process model (the intersection of the tool and the work piece equals the actual material removal) meets all requirements. It models adequately all regular modes of operation. For the training of operators in non-regular situations (e.g. collisions) is it possible to provide specific scenarios in the virtual training system (figure 2).

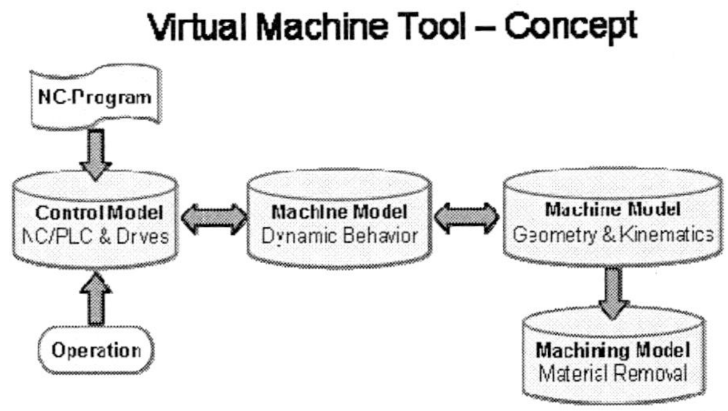

Fig. 2. Models of a virtual machine tool

The visualization module of the machine model represented at the training's workplace serves for the representation of machine geometry and machine kinematics. Set up on libraries of the NC simulation software RealNC (Siemens PLM Software), the data model to the administration of the machine model is provided. Basic CAD functions enable the representation and manipulation of the machine geometry and further components like tool or work piece. The visualization and, moreover, the possibility of navigating within the machine model easily with the help of mouse and keyboard are of special importance for the operator. By use and expansion of these available functions, a tool management, a collision control as well as a material removal calculation based on Boolean operations were implemented into the system as necessary prerequisites for user training.
The behaviour model of the machine processes the output signals of the PLC according to the deposited machine logic. Conditions of the machine hardware calculated recently are subsequently given back to the input of the PLC. The illustration of the machine and the processing of the signals can be carried out with the Siemens machine simulator. This solution is advisable, if a corresponding model from previous development steps, like construction already exists. Such a model is often more detailed than required for the training environment. If the machine model is made especially for the training, a lower level of detail can be chosen. For these purposes, a specially developed simulation solution integrated into the complete system is used. It is advantageous to avoid further independent application as well as a direct possibility to access to all functions of the behaviour model during the training. Thus, failures of certain components of the machine can for example in the future be simulated by means of the behaviour model in the context of the

virtual training.

4.4 Integration of system components

The system components are integrated into a software whose operating interface is formed by machine model and user interface. The integrated component's start-up and communication is controlled from the main application called operator panel software further on. The call of the VNCK is carried out via a COM interface (Component Object model). The HMI software is started as an independent application and automatically connects to the virtual control via internal communication protocols. The machine visualization is embedded just as the behaviour model directly in the operator panel software. A complete overview about the interaction of system components is given in [8].
The communication between the operator panel software and the VNCK takes place via COM interfaces. The visualization occurs just as the connection of the behaviour model calls via direct function. Thus, for example, the transmission of the target axis positions from the VNCK to the visualization and the calculation of material removal are carried out by the operator panel software which is also incumbent on the time management for the whole system and the synchronization of the components [9].
The connection of the machine control panel is possible via different interfaces. To enable the usage of basic MCP-functions without implementing the entire virtual PLC, the VNCK offers direct access to a number of functions which can be triggered by the machine control panel. Since the VPLC is ignored, the full MCP functionality as it is set up in the PLC program is not available. Furthermore, the connection via Ethernet is supported by the VNCK [6]. This enables the processing of all MCP-inputs as projected in the PLC and appropriate actions of the NC. Taking the PLC into consideration, a greater agreement with a real machine can be achieved. On this account, the second variant has been chosen presupposing a complete picture of the Profibus periphery in a machine behaviour model. If a lower level of detail is sufficient for the training, the PLC program and the required behaviour model can alternatively be simplified. The vital connection with the behaviour model is realized with the communication via shared memory.

5. Training concept, system functions

The function range of the virtual training system must contain the topics of the user training described above in the start-up phase in order to replace the practical part of the conventional training completely. In this case, the greatest possible saving of time is achieved. On the one hand, the training system must support the functions of all NC operation modes. On the other hand, a representation of tools is also required in the machine model for supporting the tool management in the training besides given functions of NC kernel and HMI software.
Furthermore, an intensive virtual support of the theoretical part of the training is possible which enables an improved knowledge transfer with new means. For that purpose, tutorials are used guiding the user at different operation sequences. The tutorial window is displayed directly on the user interface and describes the required steps to eliminate the disturbance. Thus, the user is able to follow these steps directly.
This feature will be used to make the user familiar with the elimination of typical problem situations arising during the running production and which have been mediated at the real machine until now. Training contents can be then removed from the practical part into the theoretical one. An example is an axis situated outside of the release area. The machine can no longer go to the starting position automatically. The operator is now informed how to change into the manual mode and moves the axis in their release areas. Used operating devices are emphasized directly on the machine operator panel so that the operator is familiarized with their location. In addition, the language of the descriptions can be adapted to the mother tongue of the user, and different languages can also be interchanged at any times. Descriptions of the individual elements of the machine operator panel represent a further operator support. These can be switched to with a key combination so that the operator is supported with occurring questions about unknown operating devices.

6. Conclusion and overview

The realized prototypical system for the virtual user training shows new possibilities of the future learning and demonstrates applications of virtual machine and control models. Low-priced standard desktop hardware has been used exclusively, so that

such a system is economically usable also in a larger number.
Primarily, advantages for the operator result from the possibility to get to know a new machine playfully. There is no danger to cause damages to the real machine. Furthermore, such a model can also serve the users during the operation of the machine for the autodidactic further education. Against the background of numerous new machines being sold abroad computer-aided tutorials provide an advantage that the language can be adapted flexibly, compared to a static lecture. For this reason, language barriers can be lowered and the knowledge transfer can be simplified. The usefulness of a virtual training decisively depends on time and effort for the modelling of the machinery manufacturer being the most frequent provider of user training. Due to its high reusability, a complex modelling pays also at standardized machines. However, the existence of models from the machine's development and the implementation of simplified models are required for economically justifiable expenses at special machines. If all these prerequisites are fulfilled, the machinery manufacturer can offer an attractive knowledge transfer to his customers by means of models and respective training services.
A complete virtual training does not increase the training quality for the usage since such a machine operation lacks the "look and feel" of the real machine. A virtual training system can complete the employee training in the previous form, though. The users can intensively be prepared for the operation of new equipment by means of the introduced concept already before the production equipment has been delivered. In this case the practical training at the machine can be carried out in a reduced form or even on-the-job beside the series production.

Acknowledgements

The investigations described in this paper were undertaken within the Network of Excellence I*PROMS (Innovative Production Machines and Systems) funded under the FP6 framework. All authors are from partner institutions within this network.
The presented work is based on the "Ramp-Up/2" project which is funded as grant number 02PS2070 by the German Office for Education and Research (Bundesministerium für Bildung und Forschung - BMBF) and supervised by Forschungszentrum Karlsruhe.

References

[1] Bastianova-Klett, F. Entwicklung einer nutzerorientierten Benutzungsschnittstelle zur Förderung und Flexibilisierung von individualisierten Lernprozessen im virtuellen Lernraum. Dissertation, Technische Universität Ilmenau, 2006

[2] Bärnklau, A. Mehr als CNC-Ausbildung. Spektrum SinuTrain, 3. Jahrgang, März 2003, pp. 10-11

[3] Blöchl, W. and Bärnklau, A. and Schröder, T. Ausbildung wird zum Wettbewerbsvorteil. NC-Fertigung 6-01, 2001, pp. 114-116

[4] Hohwieler, E. Internet-Zugang an Steuerungen für Produktionsunterstützung und Teleservice. In: Krause, F.-L. and Uhlmann, E. (Ed.) Innovative Produktionstechnik. Hanser, München, 1998, pp. 505-518

[5] ISO 9241, Ergonomic requirements for office work with visual display terminals (VDTs). International Standard, International Standards Organization, 1998

[6] Schlick, C. and Luczak, H. Facharbeiteranforderungen an Mensch-Maschine Interaktionsflächen von CNC-Werkzeugmaschinen. Jahresdokumentation 1995 der Gesellschaft für Arbeitswissenschaft. Verlag Dr. Otto Schmidt KG, Köln

[7] Schlegel, T. and Beinhauer, W. and Meo, F. Object-Orientation in Planning and Control of Decentralized Production Systems. In: VIMation Journal on Human-Machine Interaction and Models in Production. Accepted Paper, Issue 1/2008, 2008

[8] Schneider, R. Entwicklung einer virtuellen Werkzeugmaschinensteuerung. In: Denkena, B. and Brecher, C. (Ed.) Ramp-Up/2 – Anlaufoptimierung durch Einsatz virtueller Fertigungssysteme. VDMA-Verlag, Frankfurt, 2007, pp. 91-99

[9] Volkwein, G. Konzept zur effizienten Bereitstellung von Steuerungsfunktionalität für die NC-Simulation. Utz Verlag, München, 2007

Innovative Production Machines and Systems
D.T. Pham, E.E. Eldukhri and A.J. Soroka (Eds.)

Application of Design functionality rules in CAD/CAM

A. S. M. Hoque , T. Szecsi

School of Mechanical and Manufacturing Engineering, Materials Processing Research Centre, Dublin City University, Dublin 9, Ireland

Abstract

The applications of features have become standard part of assembly design, design-for-manufacturing, process planning, inspection planning, finite element mesh generation etc. In the case of manufacturing applications, features must be defined in a form that supports the planning and execution of various types of manufacturing processes on the modelled product. Manufacturing features like holes, slots, pockets, edge chamfers and keyways are widely used for machining applications.

Design functionality is one of the key factors to design a product through a feature based system. A designed product may be rejected during the manufacturing process not only due to the violation of design for manufacture rules but also due to the functionality problems. In this paper design functionality rules for feature based applications are described. Designer can select standard cutting tools from a tool library according to their process capabilities and limitations.

Keywords: Design-for-manufacture, Design Functionality.

1. Introduction

Manufacturing has always been the key to success among nations in the world economy. New technological development and market demands have major impact on manufacturing. For the last few decades, a variety of advanced product technologies, such as DNC machines, flexible manufacturing systems (FMS), robots, automated warehouses, and part transportation systems, have increased the need for complete and accurate product information [1].

Computer-aided design and manufacturing have a quite distinct origin of history from each other. Emerging computer graphics technologies influenced early CAD systems. Later, CAD systems lead to emphasis on 2D drafting and surface design issues. The origin of CAD is rooted in programming numerically controlled machine tools and the subsequent introduction of the APT language. Data transfer between CAD and CAM took place by paper drawings which were interpreted by human users. This situation still exists in many industrial companies.

Later, it was realized that a closer integration of CAD and CAM could bring about many advantages. In this regard, human reinterpretation of design information for manufacturing could reduce errors and save time. Feedback from manufacturing to design would lead to products that are easier to manufacture and thus costly rework can be avoided. Overall lead time can be shortened from design to manufacture to allow more design and manufacturing alternatives to be investigated within a given time. Thus, a superior product can be manufactured. Unfortunately, on the basis of the low-level CAD models expressed in terms of geometric elements, it is difficult to create sufficiently general and/or powerful manufacturing applications.

Much of the initial impetus of developing

feature-based systems stems from this background. Indeed, one of the primary advantages of features is that they make it possible to link design and manufacturing information in a fashion that supports the use of manufacturing information during product design, and conversely design information during manufacturing planning and actual manufacture. It is therefore not surprising that the application of features in manufacturing is more advanced than that in design. Most of the existing work in feature-based manufacturing has taken place for machining processes [2].

Nowadays for product design, process planning and machining, a great variety of feature based CAD/CAM software is available. In most feature-based systems, such as Pro/Engineer, a component is designed by starting with a simple extrusion or revolution operation (such as a cylinder or a protrusion base) and then modifying these by attaching features such as holes, ribs, slots etc. to the components base [3]. Since the designer specifies the position, size and orientation of these features and there is no limitation of size, position, orientation, dimension, the design does not contain manufacturing information like design functionality (DF) rules.

Design functionality includes:

- Parameterised geometry.
- Datum surfaces that are to be linked to existing surface in the design.
- Functionality rules.

Figure 1 shows a parameterised T-slot design feature. Upon insertion into the design, the datum surfaces of the T-slot are related to surfaces of already inserted objects, and values are assigned to the parameterised dimensions. If the parameterised dimensions do not correspond to standard cutting tools then this designed product will increase manufacturing costs during part manufacture. On the other hand it will effect the assembly operation due to the need of non-standard bolts.

The associated functionality rules ensure that the design feature is consistent with the functionality. If functionality rules are violated, the manufacturing product will be rejected. Figure 2 shows two T-slot features that do not comply with the underlying functionality rules.

The T-slot in Figure 2a is rejected because it violates a functionality rule (the width of the T-slot is larger than the width of the main body, thus a minimum wall thickness is not ensured) but for that design the DFM rules are not violated. So the designer can also check those type of rules from the rules library. Figure 2b is less straightforward: the upper section of the T-slot is too thin and it would not withstand the forces exerted by the head of the bolt when holding a devise. So the part design will be rejected during assembly [4].

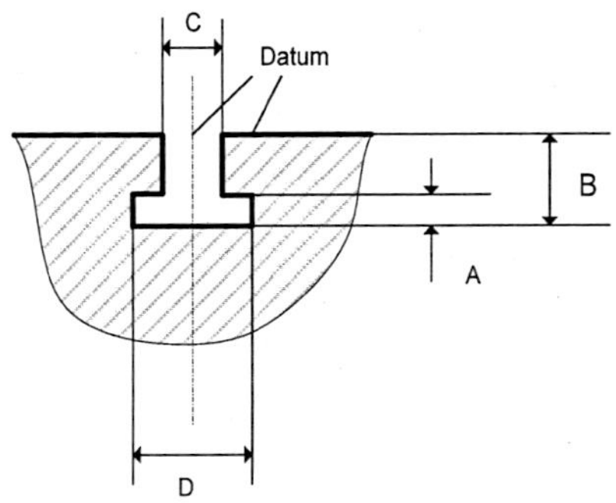

Fig.1. Parameterised T-slot design.

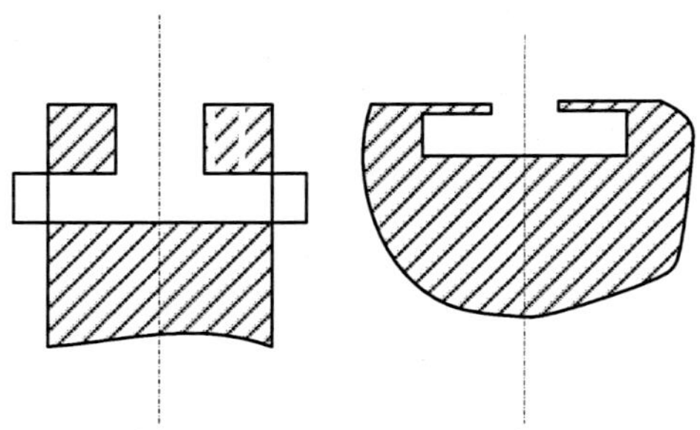

Fig.2. Rejected T-slot design.

In order to reduce the manufacturing costs, commercially available CAD/CAM systems do not violate design functionality and manufacture rules(guidelines) , a Manufacturing feature library system is developed at Dublin City University. The library enables designers to appreciate manufacturing process capabilities and limitations during the design phase. The system is linked to a commercial CAD/CAM package Pro/Engineer by the Pro/toolkit that supports parametric feature-based design [5]. In this paper only Design Functionality (DF) for Manufacturing features like holes, slots, pockets, edge chamfers, edge round keyways etc are described .

2. Manufacturing feature

A manufacturing feature is typically defined as a collection of related geometric elements which as a whole corresponds to a particular manufacturing method or processes or can be used to reason about the suitable manufacturing method or processes for creating the geometry. Manufacturing features can be classified according to the operation. For machining operations like milling, turning, drilling, features like slots, round edges, chamfered edges, pockets, keyways etc are widely used [1].

2.1. Edge round and Edge chamfer

The main purpose of edge round / edge chamfer is to avoid sharp corners. This is not only applicable for machining operation but also for castings, moulded and formed parts. Sharp edges cause more stress concentration on the part and on the tool. In the case of edge rounding for straight edges, milling operations are used. For milling operations generally two types of cutting tools are used: corner rounding cutters with Weldon Shanks and arbour-type cutters.

Table 1
Corner Rounding Cutters with Weldon Shanks [6]

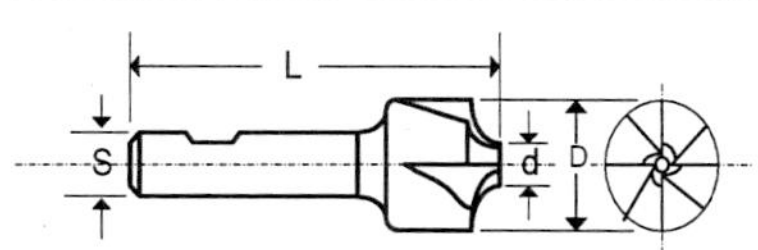

Radious R (mm)	Cutter Dia D(mm)	Diameter d(mm)	S (mm)	L(mm)
1.59	11.11	6.35	9.53	63.5
2.38	12.7	6.35	9.53	63.5
3.18	15.88	6.35	12.7	76.2
3.97	19.1	7.94	12.7	76.2
---	---	---	---	---

If the corner radius specified by the designer does not correspond to standard cutting tool parameters in Table 1 or 2 then the design will be rejected, otherwise it would lead to costly manufacture because of not using standard cutting tools.

Table 2
Corner Rounding Arbor-type Cutters [6]

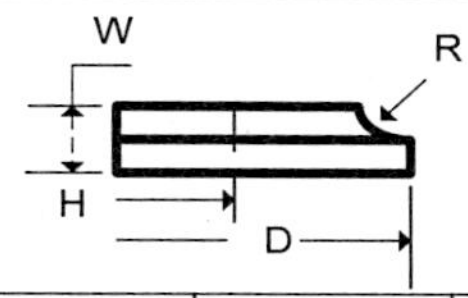

Radious R (mm)	Cutter Dia D(mm)	Width W(mm)	Diameter of Hole H(mm)
3.2	63.5	6.35	25.42
6.4	76.2	10.32	25.42
9.58	95.25	14.29	31.78
12.75	107.95	19.05	31.78
---	---	---	---

The same condition is valid for Edge Chamfer: if the designer does not use standard cutting tools it will also cause costly manufacturing.

2.2. Hole feature

There is a large variety of machining processes by which holes are produced. Generally drilling, reaming, boring etc are used for rough machined holes. The classified of different type of hole features is shown in Figure 3.

Table 3
Standard Twist drill Parameters [6]

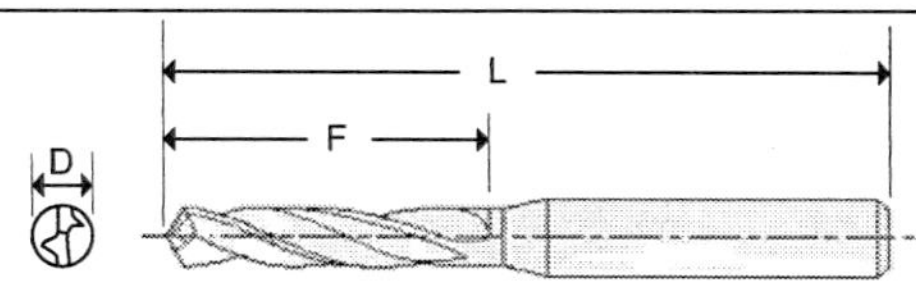

Drill Diameter D(mm)	Jobber Length	
	Flute F(mm)	Overall L(mm)
1.70	25	51
1.75	25	51
1.80	25	51
1.85	25	51
1.90	25	51
1.95	25	51
2.00	25	51
2.05	29	54
2.10	29	54
---	---	---

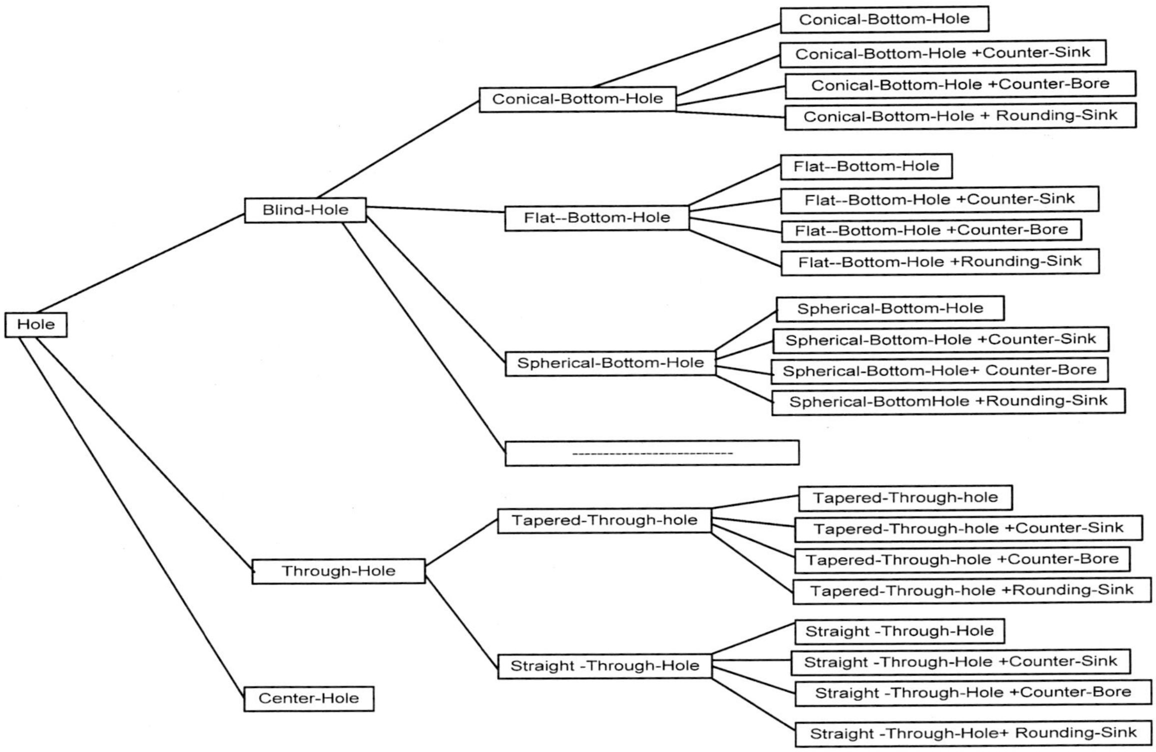

Fig.3. Classification of different types of hole feature [1].

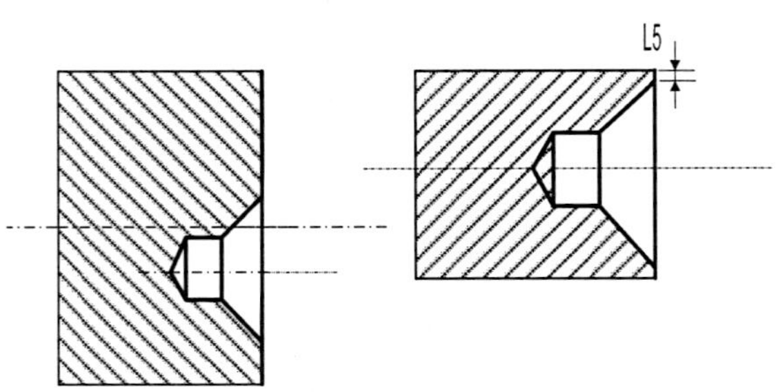

Fig.4. Rejected features due to functionality [7].

Consider the length L_1 of a drilled hole feature. Dimensions like the hole diameter of the feature are limited to a certain combination of the dimensions based on the values in Table 3. However, L_1 can not be assigned an exact value because its length depends on the drilled depth. However, due to functionality and manufacturing problems L_1 should not exceed the effective length of the cutting edge of the drill F. Improper dimensioning of L_1 will cause functionality and manufacturing problems. If $L_1 > F$ the cylindrical surface of the drill (which has no cutting edge and flute) will block the outlet of the hole during drilling, thus preventing chips to escape the hole and cutting fluid to enter the cutting zone. This is a manufacturing problem of the drilling operation.

Another example is given in Figure 4 where the design can be rejected by both the functionality and manufacturing rules. The one on the left because the axis of the centre-hole does not coincide with the axis of the rotational body, which is a functional violation of the concept of a centre-hole. The feature on the right results in a too thin wall L5, thus causing functionality problems of the centre-hole [7].

2.3. Slot feature:

Slot features are produced by milling operations. Classifications of slot features are shown in figure 5. Consider the parameterized T-slot feature from Figure 1. The dimensions of the t-slot feature depend on standard tool dimensions according to the Table 4. If the designer changes the cutter diameter D, the feature will automatically check the other dimension. During placement of the feature the system will maintain the minimum thickness of the wall.

Fig.5. Classification of slot features [1].

Table 4
T-Slot Milling Cutters with Weldon Shanks [6]

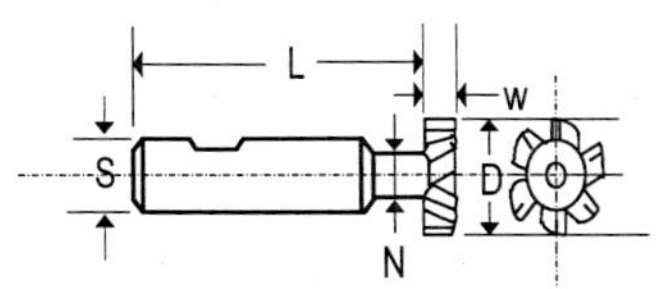

Bolt Size (mm)	Cutter Dia. D(mm)	Face Width, W(mm)	Neck Dia., N(mm)	S (mm)	L(mm)
6.35	14.29	5.95	6.75	12.7	65.88
7.94	16.67	6.75	8.33	12.7	68.27
9.53	19.84	8.33	10.32	19.05	82.55
12.7	24.61	9.92	13.5	19.05	87.31
---	---	---	---	---	---

2.4. *Keyway*

A keyway is longitudinal slot milled along the side of a cylindrical shaft, or a notch cut into the inside hole of a pulley or gear, all of which indentations are intended to accept a small rectangular or semi-circular bit of metal (known as a Woodruff key). This provides one way of locking pulleys, gears, and other parts to shafts.

There are two types of keyway: external-keyway another is internal keyway. It is basically one type of slot feature. Key size depends on the size of the shaft and hub. Table 5 shows British standard Metric Keyways for Parallel keys.

Design functionality rules for keyway depend on the standard value for keyways.

Table 5
British standard Metric Keyways for Parallel keys [6]

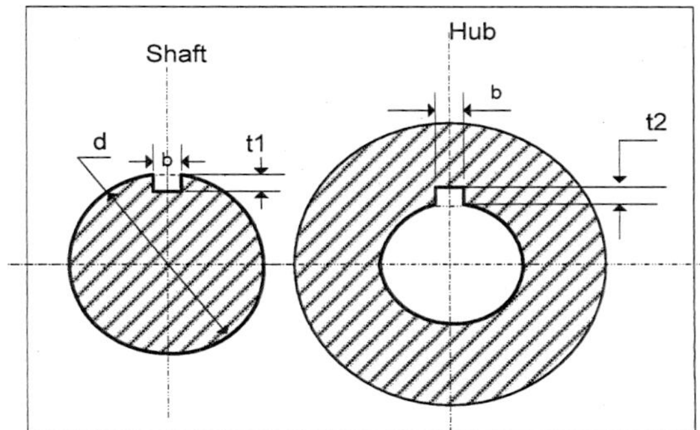

Shaft	Keyway		
Nominal Dia., d (mm)	Width b (mm)	Depth	
		t1 (mm)	t2 (mm)
6---8	2	1.2	1
8---10	3	1.8	1.4
10---12	4	2.5	1.8
---	---	---	---
22---30	8	4	3.3
30---38	10	5	3.3
---	---	---	---

3. System structure:

Pro/Engineer from PTC (Parametric Technology Corporation) provides Pro/Toolkit, a large library of C functions that allows external applications to access the Pro/Engineer database and user interface in a controlled and safe manner. The manufacturing feature library is an external application for Pro/Engineer. C language has been used for programming in the current project [8].

The system consists of several stand-alone, but interconnected modules for Hierarchical Design-for-manufacture rules, Cutting tools setup, Design functionality rules, Cutting conditions, Datum selection and Cost estimation.

Figure 6 shown the selection procedure of Edge round radius for round feature in Pro/Engineer from the manufacturing feature library

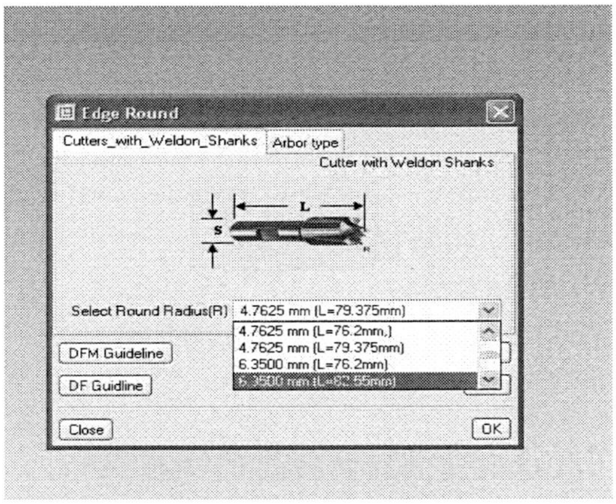

Fig.6.Selection procedure of edge round feature from manufacturing feature library.

For the placement of a centre hole the system will select the entry surface and datum axis first. If the surface is a circular planar surface then the system will calculate the diameter of the surface and compare it with the database system. If it does not mach with the database system the system will show a warning message (Figure 7). In this case the designer will select the standard cutting tools from a cutting tools dialog box.

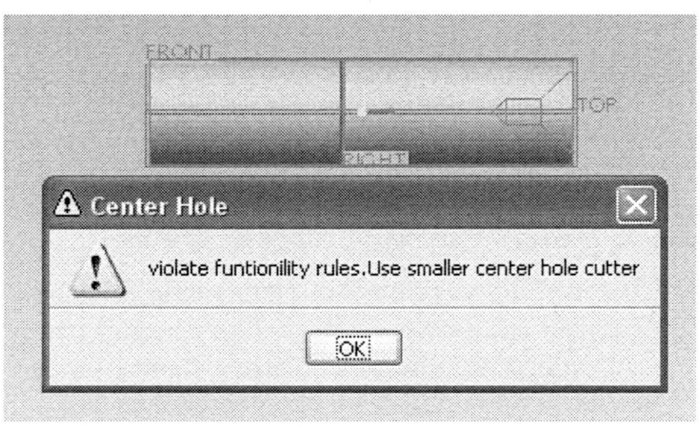

Fig.7.Displaying warning a message during the design phase.

4. Conclusions

Decreasing the cost and simplicity of a manufacturing product is essential for competitiveness and agility of a company. These desirable characteristics are mainly determined by decisions made by the product designer. Therefore early analysis and detection of possible manufacturing problems like design functionality, for example too thin wall (Figure 7) ensure that the designs are rejected or alternative solutions are suggested. So it leads to reduced production cost and time.

Acknowledgements

Dublin City University is a partner of the EU-funded FP6 Innovative production Machines and Systems (I*PROMS) Network of Excellence (www.iproms.org).

References

[1] Shah JJ and Mantyla M. Parametric and Feature-Based CAD/CAM: Concepts, Techniques and Applications. John Wiley &Sons, Inc. (1995).

[2] Hurmuzlu Y and Nwokah O. The Mechanical Systems Design Handbook: Modeling, Measurement and Control. CRC Press LLC (2002).

[3] Gao J., Zheng D. and Gindy N. Mathematical representation of feature conversion for CAD/CAM system integration. J. of Robotics and Computer-Integrated Manufacturing, Vol. 20 (2004), pp 457-467.

[4] Hoque A.S.M. and Szecsi T. Application of design - for- manufacture (DFM) rules in CAD/CAM. Proceedings of the 3-rd Virtual Conference I*PROMS, Whittles Publishing, Scotland, pp 172-177.

[5] Hoque A.S.M. and Szecsi T. Designing using manufacturing feature library. J. of Materials Processing Technology, 201(2008), pp 204-208.

[6] Oberg E., Jones F.D., Horton H.L. and Ryffel H.H. Machinery's Handbook. Industrial Press Inc (2004).

[7] Szecsi T. Implementing manufacturing feature based design in CAD/CAM. Proceedings of the 2-nd Virtual Conference I*PROMS, Elsevier, Oxford, 2006, pp 71-76.

[8] Pro/Toolkit User Guide for Wildfire 2.0, Parametric Technology Corporation, (2004).

Innovative Production Machines and Systems
D.T. Pham, E.E. Eldukhri and A.J. Soroka (Eds.)

Customer driven development of Product-Service Systems

E. Uhlmann[a], H. Meier[b], H. Bochnig[a], C. Geisert[a], K. Sadek[b], C. Stelzer[a]

[a] *Institute for Machine Tools and Factory Management, Technische Universität Berlin, Pascalstr. 8-9, 10587 Berlin, Germany*
[b] *Chair of Production Systems, Ruhr-University Bochum, Universitätsstr. 150, 44780 Bochum, Germany*

Abstract

In order to ensure competitiveness in times of globalization, manufacturers need to respond to client demands by focusing the development of industrial production machines on the customer needs. Additional value can be provided by developing and offering innovative services along the entire life-cycle of the physical product. Despite strong interdependencies between products and services there is no integrated view on their design. This paper describes the state-of-the-art regarding the fulfilment of customer requirements and illustrates an approach for the integrated design of products and services in the form of Product-Service Systems (PSS) in an industrial scope. Therefore, fundamental concepts for the customer driven development of product and service in the context of PSS are presented.

Keywords: Product-Service Systems, customer driven development, product modelling, modular service architecture

1. Introduction

The traditional view of manufacturing companies is focused on selling the physical product. Therefore, the value is based on the transfer of physical technological artefacts between a providing company and a receiving customer. As an effect this product-oriented business strategy links the company's economic growth with the consumption of natural resources. It also imposes the responsibility for the correct and efficient use of a product, its maintenance and disposal on the customer, who may not be aware of an appropriate behaviour. Also, the customer usually has limited opportunities to influence the product design in accordance with his individual needs and requirements [1,2].

Due to the ever increasing global competition as well as the demands for greater responsibility of the manufacturer for their products throughout the entire life-cycle, enterprises are forced to shift their business focus from designing and selling physical products only to offering integrated Product-Service Systems (PSS). Such combined solutions continuously provide value to the customer and therefore represent an innovative strategy to not only fulfil client demands jointly but also to compete successfully on the global market [3].

For the efficient development of PSS, the operational support by means of suitable methods and adequate computer-aided tools is essential. This article gives benefits of Product-Service Systems to fulfil customer needs and describes the differences to conventional approaches.

Furthermore, the design of products and services in a methodical and integrated manner is presented. The integration of these methods into a computer-aided assistant system shows how to support the engineer in developing innovative and customer oriented PSS.

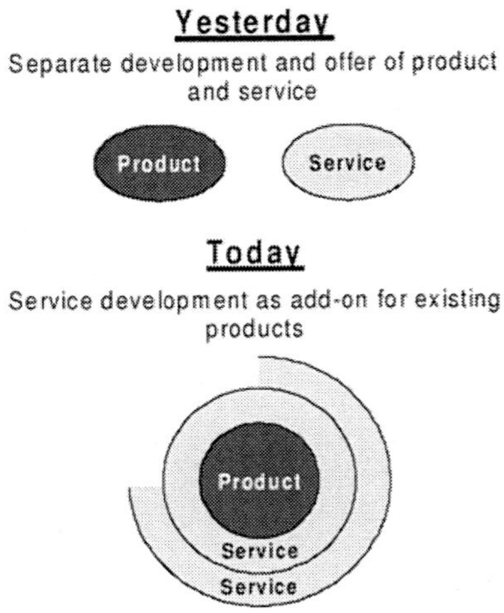

Fig. 1. States of Service Integration.

2. Comparison of PSS and state-of-the-art approaches in regard to fulfilling customer requirements

2.1. Industrial services – State-of-the-Art

The core competence of machine manufacturers is still the manufacturing of high technology machine tools. In order to establish themselves successfully in the global market and to meet the increasing requirements of their industrial customers, added value becomes more important.

One observed trend is the modularization of machines to reach higher flexibility [4,5]. Another way to meet customer needs is the development of innovative services. Many recent industrial surveys show that services have great potential in order to improve the company's market position and further to strengthen their customers' loyalty [6,7].

Nevertheless, widely offered commercial industrial services are limited to basic services like spare part service, maintenance, and repair. These services require no integration into the physical product. Therefore, their planning and design are independent from the planning and design of the product.

In this context customized services are more appropriate in order to create added value to the product. Such product accompanying services are highly process integrated and are further characterized by an increasing demand for IT-support. In this case, interfaces are needed to interlink products and services. As an example, services can be implemented via the internet. On the product side, the machine has to be equipped with a network interface. Such internet based

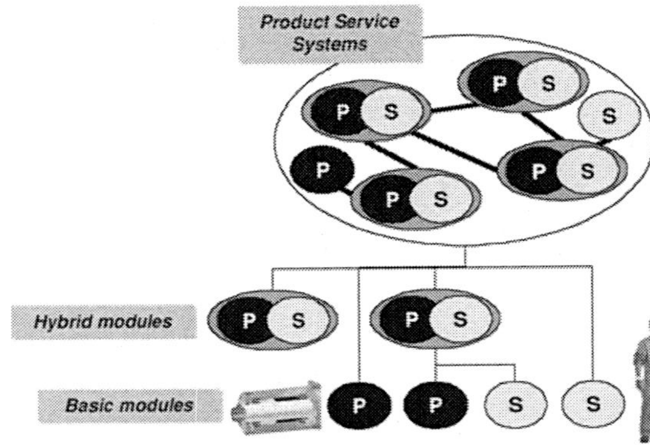

Fig. 2. Product-Service Systems.

solutions have been realized exemplarily within the research project "e-Industrial Services" [8,9].

2.2. Meeting the customer needs with PSS

The overview over the current state of industrial services shows that innovative services in the field of industrial production are needed in order to fulfil the increasing customer requirements. Since the machine manufacturer's core competences still focus on the machine technology, services are planned, designed, and realized as add-ons to the already existing machinery. Therefore, their development is driven by the customer's feedback.

The Product-Service System approach as it is proposed by the research project Transregio29: "Product Service Systems - Dynamic Interdependency of Products and Services in the Production Area" postulates that the optimal fulfilment of customer needs can only be achieved by an integrated view on products and services [10,11]. This means that products and services have to be designed jointly, due to their existing interdependencies (Fig. 1 and 2).

2.3. Dependencies between products and services

From the customer's point of view, a product has to fulfil manifold requirements besides its basic functionality. Aspects like balanced cost-benefit ratio, low life-cycle costs, high productivity, high availability, and easy handling influence the client's buying decisions. These objectives can not be achieved by selling only the physical product. Therefore, an adequate service portfolio is needed.

In order to give an example, high productivity does not only depend on electromechanical characteristics, such as fast cycle times, high speed axes and spindles. The qualification of the machine operator and the

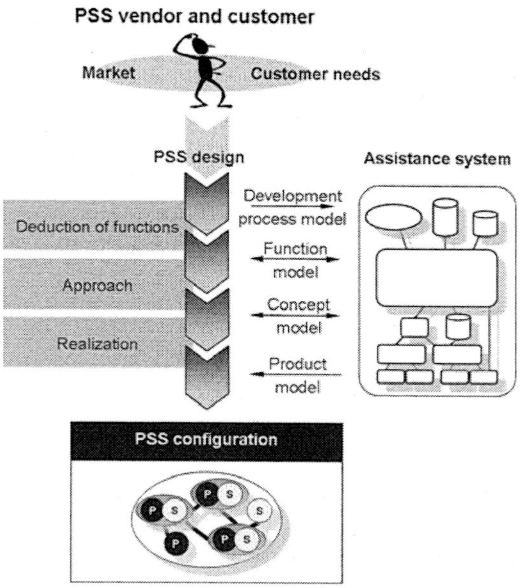

Fig. 3. Design phases of a PSS.

machine's condition also have a determining impact on the productivity. Therefore, it is obvious that both, physical products and services, such as maintenance and training, are correlated.

3. Design of PSS based on customer requirements

3.1. Product modelling of PSS

The development of Product-Service Systems requires an extensive technical knowledge and the constant consideration of the interrelationships between products and services. In order to establish the PSS approach in industrial production, innovative structures for product design are needed. Novel IT-systems have to be developed in order to support all PSS design phases. Therefore, an assistant system is being developed at the Institute for Machine Tools and Factory Management (IWF) of the Technische Universität Berlin. It will support the engineer from the planning phase to the point of the implementation of a completely developed PSS (Fig. 3). This assistant system has to map product and service modules under consideration of their dependencies and exchangeability.

A Product-Service System is characterized by its modular composition. Therefore, an object-oriented representation has been chosen as a solution approach (Fig. 4). The identification of the product components has been carried out through an exemplary analysis in the area of machine tools for micro production. The

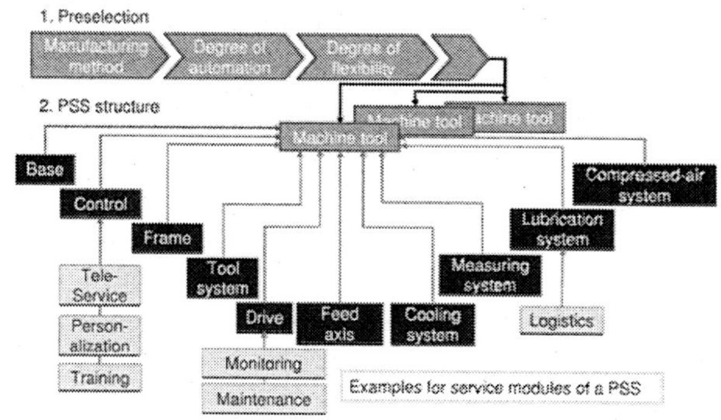

Fig. 4. Planning of product model for PSS.

identified components have been classified according to their functions, which led to the definition of product classes. Based on such product classes the interrelationships and dependencies between products and services, as well as supporting IT-systems have to be built in the PSS concept model. This is a prerequisite for the integrated product and service development.

In a top-down-approach the product classes are incrementally split and described in growing detail. With the aid of attributes all classes as well as their instances are characterized in accordance with a particular function as well as the product's specifications. Furthermore, they provide for a common interface in order to consistently connect the product and service modules. Regarding this, the solution space for a Product-Service System is given by all feasible combinations of modules that adequately fulfil the customer requirements.

The components that form the final PSS are selected in accordance with the optimal customer's benefit as well as the appropriate PSS business model: function oriented, availability oriented or result oriented. In order to achieve such a solution the entire

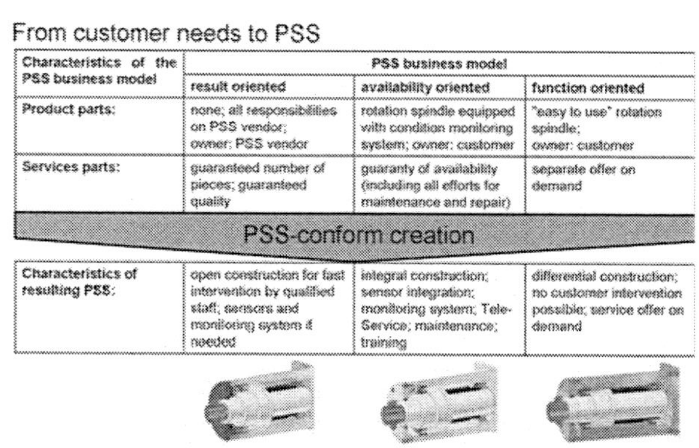

Fig. 5. PSS product models deduced from business models.

PSS development is embedded in an iterative process.

An example for the different versions of a product as part of a PSS is shown in figure 5. According to the customer requirements each version is based on a different PSS business model.

3.2. Modular service architectures for PSS

In the following a novel approach to methodically develop a Modular Service Architecture for PSS (MSA) is presented. The architecture's design is heavily influenced by the characteristics of services as well as by the principles of modularization. The development method is conducted in order to generate a variety of service modules, to construct standardized service interfaces and to establish a service platform which consistently combines these modules and interfaces.

The MSA is used in the early development phase of a PSS in order to embed different combinations of such developed service modules into a PSS concept model depending on the respective customer requirements. This modular architecture provides the opportunity of developing services which are to be integrated into a PSS problem solution, in an effective, efficient, standardized and consistent manner.

3.2.1. Modular service architecture – influencing variables

Services are defined as heterogeneous, mainly immaterial and perishable activities or processes. They are consumed simultaneously to their production [12]. Another distinguishing characteristic of services is the necessity of integrating an external factor, provided by the service recipient, into service provision. Such external factors can consist of persons, goods, rights and information, etc. [13].

Module, modularity, modularization and modular design are well known terms in industry as well as in academia. Within the scope of software engineering the term module is defined as a logical and physical unit performing a precisely defined function within an overall context. These modules are linked with each other through standardized interfaces [14].

Furthermore, modular systems are subject to the rules of decomposition, abstraction, nested hierarchy, loose coupling and limited access. The term decomposition defines the segmentation of a system into its subsystems. With the aid of nested hierarchy these subsystems can be subdivided into further part systems to the point of elemental functions or to the limits of knowledge. In regard to modularization the

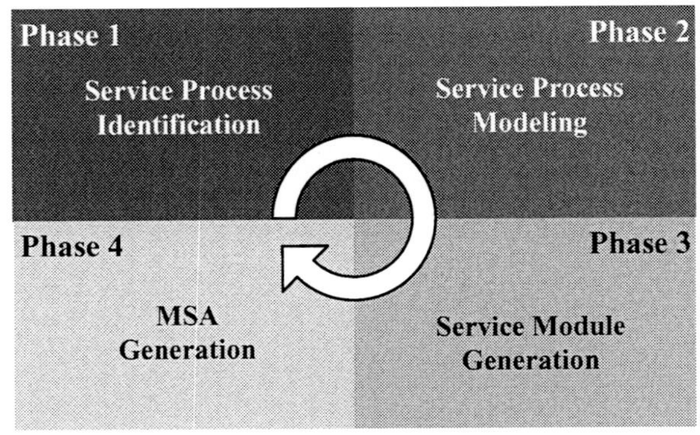

Fig. 6. The Development of a Modular Service Architecture (MSA) for PSS.

principles of loose coupling and limited access play a decisive role concerning the design of in- and output-interfaces. Loose coupling describes the request for strong dependencies between each element within one module. The principle of limited access requires for non internal relations between modules defined with the aid of public attributes. In this regard it is essential that the more the description of the module interface differs from its internals and the less those internals are revealed by in- and output information the merrier modules can be used and exchanged independent of a particular context [15].

Modular Service Architectures are therefore understood as overall service systems that are built up on the principles of decomposition, abstraction and nested hierarchy. Furthermore, they consist of unitized service modules which are linked to each other in compliance with the rules of loose coupling and limited access. Due to the process-oriented notation of the MSA, services are visualized under consideration of their intangibility as well as their integration of external factors.

Modular Service Architectures are used in order to shorten service development times and costs by decoupling, reusing, exchanging or extending standardized service modules. In this regard the ability to respond to changes at the point of sale increases. Moreover, the recombination of service modules allows for customized service configurations requiring minimum development efforts.

3.2.2. Modular PSS service architectures – development

Phase 1: The systematic development of a MSA is initialized by the analysis of a company's portfolio which provides a detailed description of products and services currently offered by that company (Fig. 6). Building on this information the service developer starts to identify existing as well as new kinds of services necessary to jointly create value. This approach constitutes a means of generating sustainable competitive advantages through services instead of merely using product-oriented business strategies.

Such identification is based on so called "7 types of services" (function creating, function keeping, optimization, logistics, planning, schooling and consulting [16]) which serve as categories containing an overall number of 86 different kinds of services.

Phase 2: Based on the information generated during phase one the developer starts to model all identified services using what is termed "event-driven process chains". Regarding these, a service is visualized with the aid of functions, so called events, technical and human resources as well as data and information. Functions describe a specific task within the overall service, while events declare that the function in question has been carried out. The data and information mentioned are dedicated to a particular function in order to render the overall service. Such visualization allows for the design of external factors as well as the integration of immaterial aspects like data and information.

After identifying and modelling all service processes the service developer has to methodically generate what is termed "service clusters". These clusters form context-related units enabling the identification of equivalent service process sections from the whole of all modelled services. Examples for such clusters are order acceptance or order invoicing which are usually carried out the same way independent of the overall function of a service. Those are called "basic clusters" Additionally, "variable clusters" contain functions which are used within several of the modelled services, e.g. a functional test repeatedly performed within different maintenance related services. As a third type, "specific clusters" are used in order to fulfil a specific function, e.g. a preventive maintenance or an installation which is only used concerning one specific service.

In order to achieve a high level of standardization, it is of particular importance that each process chain is modelled in accordance with the norms a company is or will be certified by.

Phase 3: Based on the previously established clusters so called "service modules" are to be generated. In this regard basic, variable and specific modules have to be defined. Such modules contain the previously developed clusters and are built up in compliance with the rules of loose coupling and limited access. A service module can contain an indefinite

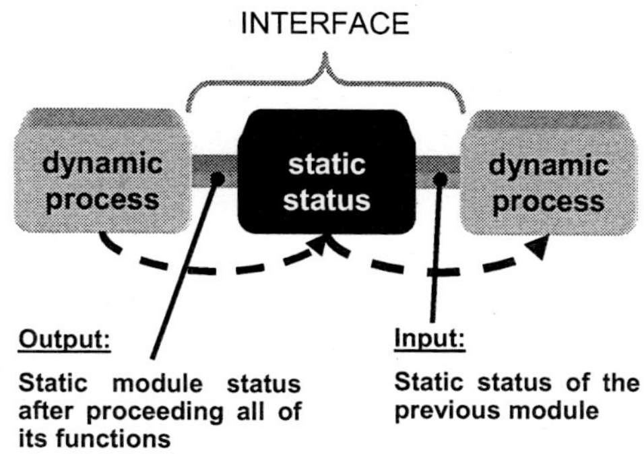

Fig. 7. Description of a Service Module Interface.

amount of service functions. Each of these functions is described by its activity, an identification number in order to determine its sequence within the overall configuration, its required human and technical resources as well as its level of automation and further a so called "linked product" which indicates the connection of the service and its related product. The behaviour of each function can be simulated by defining the time it consumes, its costs and further variables, such as the level of qualification of a human resource. An event exists as soon as all relevant variables of its function have been simulated. Similar to the event-driven process chains all functions within one module are linked to each other using standardized decision rules, e.g. AND, NOR etc..

Phase 4: In order to consistently combine the previously developed service modules, a service platform, integrating standardized interfaces and so-called "placeholders", has to be generated.

Such placeholders are linked with each other through standardized interfaces. Theses interfaces represent the static status of a module after all of its functions have been carried out. Thus, all input information necessary to perform and simulate the functions within a module are offered by using the static status of its previous service module (Fig. 7).

Each placeholder within one service platform is marked with a specific module code. Furthermore, each service module can be identified by a specific service platform ID-code. The comparison of both numbers enables the service developer to dedicate modules to its related placeholder in order to create a consistent MSA.

Hence, modular Service Architectures consist of basic, variable and specific modules which are linked to each other in compliance with the rules of loose coupling and limited access with the aid of standardized interfaces. The consistent configuration of such modules is assured by making use of a service platform.

4. Conclusion

Due to the ever increasing global competition and demands for more company responsibility of products throughout the entire life-cycle, enterprises are forced to shift their business focus from designing and selling physical products only to offering integrated Product-Service Systems (PSS). The operational support with suitable methods and adequate computer-aided tools is essential to develop PSS efficiently. Therefore, the state-of-the-art of industrial services has been analyzed and described to what extent PSS are able to satisfy customer requirements. The methodical design of products and services in the context of PSS has been introduced and the fundamentals concerning the integrated design of Product-Service Systems in the early development phase have been demonstrated. Based on this information the interdependencies between products and services can be designed in the PSS concept model, which is a prerequisite for the integrated product and service development. This development will be provided by an assistant system that maps product and service modules by considering their dependencies and exchangeability.

Acknowledgements

The presented work is part of the first, conceptual phase of the research project Transregio 29: "Product-Service Systems - Dynamic Interdependency of Products and Services in the Production Area", which is funded by the German Research Foundation (DFG).

References

[1] Tan, A.R.; McAloone, T.C., What happens to integrated product development models with product/service-system approaches?, 6th Integrated Product Development Workshop, 2006.

[2] Meier, H.; Kortmann, D., Leadership - From Technology to Use, 14thCIRP Conference on LifeCycle Engineering, 2006.

[3] Tan, A.R.; McAloone, T.C., Characteristics of strategies in product/service-system development, International Design Conference, 2006.

[4] Fleischer, J. et al., Mechanical module interfaces for reconfigurable machine tools, In: Production Engineering - Research and Development, Springer Berlin / Heidelberg, Vol. 1, No. 4 (2007), pp. 421-428.

[5] Uhlmann, E. et al., Realization of an adaptive modular control for a disassembly system, In: 2005 IEEE International Symposium on Assembly and Task Planning (ISATP), 2005, Montreal, pp. 32-35.

[6] Rainfurth, C., Dienstleistungsarbeit im produzierenden Maschinenbau - Eine Analyse am Beispiel von kleinen und mittleren Unternehmen. ISI-Schriftenreihe Innovationspotenziale, ed. Fraunhofer-Institut für Systemtechnik und Innovationsforschung ISI, Karlsruhe, Fraunhofer IRB Verlag, 2003.

[7] N.N., 2001, E-Services in der Investitionsgüterindustrie -Eine Befragung von Anbietern und Anwendern, ebc the european business consultancy, VDI Verlag, Düsseldorf.

[8] Uhlmann, E.; Hohwieler, E.; Berger, R., Providing Online Services for Machine Operation and Maintenance. In: Marco Taisch et al.: "Global Challenges in Manufacturing", Proceedings of the International IMS Forum 2004, 2004, Cernobbio, Italy, pp. 1120-1127.

[9] Hohwieler, E.; Geisert, C., Intelligent Machines Offer Condition Monitoring and Maintenance Prediction Services, in: Roberto Teti (Editor), Proceedings of the 4th CIRP International Seminar on Intelligent Computation in Manufacturing Engineering (CIRP ICME '04), 2004, Sorrento, Italy, pp. 599-604.

[10] Meier, H.; Kortmann, D.; Krug, C., Von der Technologie zur Nutzenführerschaft - Die Zukunft der Werkzeugmaschine als hybrides Leistungsbündel, ZWF - Zeitschrift für wirtschaftlichen Fabrikbetrieb, 7-8/2006, Carl Hanser Verlag, pp 431-434.

[11] Uhlmann, E.; Geisert, C.; Stelzer, C., Hybride Leistungsbündel für die Mikroproduktion, wt Werkstattstechnik online, Jahrgang 97 (3/2007), Springer-VDI-Verlag GmbH & Co. KG, Düsseldorf, pp 126-129.

[12] Matzen, D.; Tan, A. R.; Andreasen, M. M., Product/Service-Systems: Proposal for Models and Terminology, 16. Symposium "Design for X", Neukirchen, 2005.

[13] Spath, D.; van Husen, C.; Meyer, K.; Elze, R., Integrated Development of Software and Services – The Challenges of IT-Enabled Service Products, In: Spath, D.; Fähnrich, K.-P. [ed.], 2007, Advances in Service Innovations, Springer-Verlag, Berlin: pp 3-16.

[14] Kahlbrandt, B., Software-Engineering mit der Unified Modeling Language; Springer-Verlag Berlin, 2001, p 75.

[15] Böhmann, T.; Krcmar, H., Modulare Servicearchitektur, in: Bullinger, H.-J., Scheer, A.-W., Service-Engineering – Gestaltung und Entwicklung innovativer Dienstleistungen, Springer-Verlag Berlin, 2003, pp 391-415.

[16] Kortmann, D., Dienstleistungsgestaltung innerhalb hybrider Leistungsbündel, Shaker-Verlag Aachen, 2007.

Innovative Production Machines and Systems
D.T. Pham, E.E. Eldukhri and A.J. Soroka (Eds.)

Design for intuitive use: a study of mobile phones

A. Marsh, R. Setchi

School of Engineering, Cardiff University, Cardiff, CF24 3AA, UK

Abstract

This paper explores the field of design for intuitive use by researching existing ideas and conducting an experimental study to determine if mobile phones are currently viewed as intuitive to use. In the context of this study, design for intuitive use is defined as the seamless alignment of cognitive expectation with interface actuality. The experiment carried out compares mobile phones produced by two large manufacturers, to determine which areas of a mobile phone facia are commonly viewed to be used for specific tasks and to investigate which image schemata are used to perform generic interaction tasks. As indicated by the participants in the survey, the Nokia interface is currently viewed as being very intuitive to use, and their system of interface design has shaped the consumers expectations of interaction with mobile phones. The experiments conducted showed that the nearer the top of the keypad a key is positioned, the more important, or regularly used it is expected to be. The most common image schemata for interaction as indicated by the participants in the experiments are the vertical and depth schemata.

Keywords: Design for intuitive use, intuitive interactions, mobile phones, image schemata

1. Introduction

The term 'intuitive' is referred to regularly in design, technical fields and by users of products that require cognitive interaction. Intuition is defined in [1] as "the power of knowing or understanding something immediately without reasoning or being taught." Therefore, a product can be intuitive if its functions are clear and the methods needed to implement them take little or no cognitive effort. It is apparent that such products could bring a substantial competitive advantage to their manufacturers.

Design for intuitive use is used to make a process more efficient when there is more than one path that can be taken to achieve the same result. A mobile phone, for example, that responds to the owner's actions in the desired manner would not need an instruction manual, would be quick to use and would allow the customer to use the phone as a powerful tool.

This paper aims to explore existing ideas in the field of design for intuitive use and determine some of the factors that make a mobile phone intuitive to use. The experiments conducted are designed to determine preferred methods of interaction and investigate how one interacts subconsciously with the physical properties of a mobile phone.

The paper is organised as follows. Section 2 outlines current research in design for intuitive use. Section 3 introduces the experimental procedures employed and the main hypotheses tested. Section 4 contains results and a discussion of the findings. Finally, Section 5 concludes the paper.

2. Background

2.1. Intuitive interaction

There is a universal agreement that intuition is a type of cognitive processing that utilises knowledge gained through prior experience [2], and intuitive use of products involves utilising knowledge gained through other experience(s) [3]. Furthermore, an application is considered intuitively useable if the users' unconscious application of prior knowledge leads to effective interaction [4].

Intuition is a term that should not be applied to the product itself but instead to the link between interface and user. As every user is different, so too will be the level of intuition between the two. A computer mouse seems intuitive to all computer users, but the first time user may still need to be shown how to use it. However, it would be very rare for that person to need to be shown again. This would suggest that a mouse becomes intuitive very quickly and this is perhaps the ultimate aim of an intuitive interface.

It appears that intuition is learnt over time through use of technologies, so that intuition as the dictionary defines cannot exist in a designed interface. Instead, an interface that can be easily learnt has the appearance of being intuitive. Therefore, intuition is something acquired by people whilst using technology, rather than a standard of technology with which interaction is effortless.

The methodological tools developed in the area include the continuum of prior knowledge [5] and the continuum of intuitive interaction [6].

The continuum of prior knowledge developed by Hurtienne [5] shows that prior knowledge acquired before interaction with the new product comes from a variety of sources. These are classified into 5 categories: innate knowledge, embodied interaction, culture, expertise, and use of tools. The higher up the continuum, the smaller the potential number of users possessing this knowledge; lower level knowledge is used more frequently. These are the levels that are more likely to be applied unconsciously and therefore intuitively.

The continuum of intuitive interaction developed by Blackler [5, 6] outlines three principles of creating an intuitive interface:

1. Use familiar features from the same domain,
2. Transfer familiar things from other domains,
3. Increase redundancy and internal consistency of function, appearance and location within the interface.

Furthermore, each principle is linked to a set of terms. Principle 1, for example, is compared to influencing the interface with body reflectors, population stereotypes and features from existing products in the same domain. Principle 2 relates to the use of metaphors and familiar functions from other domains. Principle 3 involves using visual and audible feedback, and providing different ways of doing things so that both novices and experts, and older and younger users, can use the same interface easily and efficiently. Keeping internal consistency allows users to apply the same knowledge and metaphors throughout the interface.

Similarly, Hurtienne [4] devises a number of principles through which to design an intuitive interface. These include: suitability for the task, compatibility, consistency, gestalt laws (how the mind groups similar elements to an entity, based upon shape, colour, size and brightness of the elements), feedback, self descriptiveness, and affordances.

2.2. Cognitive ergonomic: image schema and gestalt laws

An image schema is a mental pattern that recurrently provides a structured understanding of various experiences. Image schemata can be used literally, to describe objects or actions, or metaphorically as a source domain to provide an understanding of other experiences.

Johnson [7] describes image schema as "a dynamic pattern which functions somewhat like the abstract structure of an image and thereby connects up a vast range of different experiences that manifest this same recurring structure." His list of schemata consists of twenty seven individual schemata, including centre-periphery, containment, cycle, path, force, link, part-whole, scale, and verticality schemata.

Much study of image schema stems from research in linguistic expressions. The words used to describe especially metaphorical features, act as a gateway to the structure the brain has composed to store the information. It is important to note that the term 'image' in image schema is not an image that can be drawn or be shaped in a three dimensional world. An image schema does not have the rigidity or specificity of a picture or structure, but consists of parts that can be flexed and sculpted in an infinite number of ways, sometimes interacting with other image schema, to align with perceptions, 'images' and events.

Recently, Hurtienne and Israel [8] suggested the application of image schemata and their metaphorical extensions in the design of intuitively usable interfaces. On creating an interface, its perceived structure should try to fit in as closely as possible to the mind's image schemata. It would appear that the most relevant schemata for an interface are the cycle schema, path and end of path schema and verticality schema, whilst the ergonomics of the interface (e.g. buttons, etc.) will apply to link schema, force schema and scale schema (especially with touch sensitive control).

Further studies indicate that the so called gestalt laws can be applied to image schema [9]. Gestalt is a German word meaning configuration or pattern. Theorists reject the assumption that organization is the product of learned relationships, but argue that perception itself is a basic experience. The approach focuses on the idea of the mind 'grouping' elements to perceive objects. Generally there are five laws from which others stem: proximity, similarity, closure, symmetry and common fate. The gestalt laws are often applied in product design. Buttons, for example, grouped together are perceived as having similar functions.

These theories indicate that to teach a mind how to use something, it is the information presented to it that will determine its success. Going further, the success of this information resulting in the 'correct' cognitive response lies in the similarity of the image schema formed whilst receiving information and the image schema used to implement interaction. Image schemata are flexible structures for the mental organisation of experiences and comprehension.

3. Experimental procedures and hypotheses

3.1 Questionnaire

The current style of phone sales from a high street network store typically offers a free mobile phone handset when a pay monthly tariff is purchased. These tariffs typically last for 12 or 18 months and after this time, a new tariff is sold, normally resulting in a new handset. Therefore, the vast majority of the population is expected to have owned their phone for a short period of time and to have had a number of phones in the past.

This style of selling results in the user making their way through a high number of phones in a relatively short period of time and also forces them to potentially learn a new interface every year or so. The user may, to make things easier, repeatedly choose the same phone manufacturer to replace the old handset whilst new technologies, appearances and styles are most probably the main factors that would cause the user to choose a new phone manufacturer. Users may have the ability to be in contact with the new handsets from peers, so when it is time to choose a new handset, the buyer may have already done a little market research and have made the decision for their new phone before walking into the store. The factors discussed here will become apparent, or disproved by the correlations between the handset currently used and those that have been used before.

The questionnaire published in [10] also contains a set of questions designed to find out how mobile phones are used. It is hypothesised that calls will rank the highest score for use, and the camera will score highly as well. It is not expected that internet and e-mail will have high scores as these are still relatively new technologies for the mobile phone and most people will have a computer linked to far faster internet at home. Games also, are expected to score low, simply because the youthful users, less than 17 years of age, have not been included in this survey due to ethical reasons.

The awareness and ability sections of the questionnaire will unveil how confident users feel about their current phone. Phones with a high awareness score may have an 'open' feel to their interface, where most of these features will be on display at the same time, increasing awareness, most probably in the main menu field. A high score will naturally appear if the user actually uses all the features of their phone. In theory, the ability score should not exceed the awareness score, as awareness must be present before one can have the ability to use a feature. There may however be some confusion here and perhaps pride also, where the user may rank their ability to use all the features they regularly use. The ease of 'learning' one's current phone is a direct score for specific manufacturers and will be analysed as such. If the hypothesis of most phones being owned for 1 or under years is true, then this question holds a great significance as the early learning stages may still be fresh in the user's mind.

3.2 Experiments

The experiments involve a visual test with two schematics of simplified button configurations; one based on the Nokia system, and the other, based on the Sony Ericsson system (Fig. 1), which also holds similarities to the Motorola and Samsung systems along with the iPod. The test is videoed with a digital camera. The aim is to determine what keys, or areas of the mobile phone facia are most commonly viewed to be used for specific tasks and to investigate if there are any consistencies with the methods required for existing phones and these schematics.

Participants are informed about the scene in front of them: *"You have in front of you a simple schematic of a phone facia. The visuals, key notation and screen interface are entirely that of your choice. You are currently at your 'home' screen with a locked keypad."*

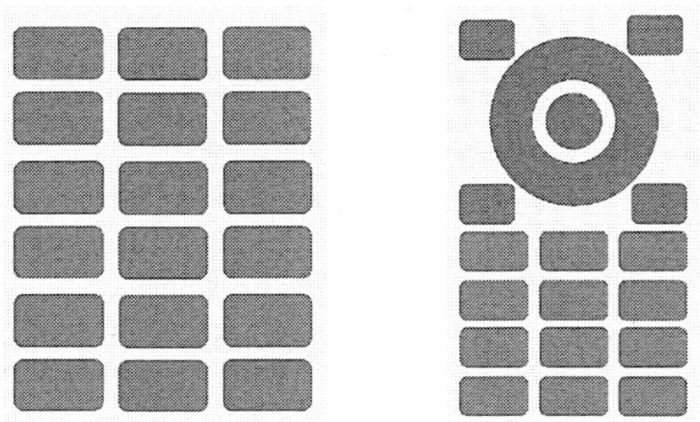

Fig. 1. Nokia and Ericsson schematics used in the test.

The participants are then asked to perform the following tasks, on each configuration one at a time, using the schematics in any way they saw fit.

1. *Unlock the key pad.*
2. *Press the 'menu' button;*
3. *Open the 'text message' folder and select 'write new message'.*
4. *Go back to the 'home' screen.*
5. *Key in the number 07856.*
6. *Delete the 6 and type a 2.*
7. *Delete all 5 numbers.*
8. *(Are you at the 'home' screen?)*
9. *Find your contacts.*
10. *Scroll down 5 names.*
11. *Scroll up 3 names.*
12. *Call this number.*
13. *Hang up.*
14. *Lock the key pad.*

The results from this task are used to determine any consistency in the areas used for locking and unlocking the keypad, menu, select, back, number keys, delete and scroll down, scroll up, call and hang up.

It is hypothesised that there should be no problems with typing of numbers due to the repeated constant standards of these keys being placed at the base of the keypad with the ordering of 1 top left and 9 bottom right with *, 0 and # along the bottom row. The Nokia schematic may cause more inconsistencies than the Sony Ericsson schematic as the only variable in the former is position, with identically sized and shaped keys, thus removing most of the gestalt laws. The results from this test may therefore indicate broad similarities in desired positions of keys only.

Having the round shape in the top central position of the Sony Ericsson schematic bares a far closer resemblance to the items from which this schematic has been derived, and may cause initial recollection of past experiences with such items. If this is the case, it would be expected that each test on this schematic will result in similar patterns of interaction.

This schematic has a clear divide of the regular shaped keys at the bottom from those above. This factor leads to a hypothesis of the bottom keys being used only for number typing whereas the top group will be used for interface manipulation, without any cross over. It will be interesting to see how many people use the circular shape as a 'wheel', in a similar manner to the iPod. None of the mobile phones widely available on the current market use this system, but the iPod has become such an iconic example of a portable electronic device that some subjects must make this link.

4. Results and discussion

A total of 46 completed questionnaires were returned and analysed. Only one of these participants did not own a mobile phone. The average was 28 years of age. There was a fair spread of age range; however the greatest number of responses was from those in the 21-24 age range. On average, each user had owned 5.5 phones in their time. Text messaging had 33% of the score for use of the mobile phone, and calls had 32%. The final third was split between all other uses displayed in the questionnaire. Nokia dominated the past market share with 49%. Sony Ericsson and Motorola shared the runner up spot, holding 15% of past sales. However, of currently owned phones, Nokia and Sony Ericsson share the top spot with 32% each and Samsung took the third spot with 16% as Motorola dropped to forth with 12%. Participants gave themselves average scores of 3.4, 3.2 and 3.6 of a possible 5, for phone awareness, ability to use all features and ease of learning current phone respectively.

74% of all questionnaires were completed by persons less than 24 years of age. Therefore, the results are primarily indicative of the perceptions of students in this age range. This generation has grown up surrounded by technology from a relatively young age, with their ownership of their first mobile phone at an average of 13 years of age. The student will have a high social use of their mobile phone, the possibility of higher peer pressures to keep up to date with styles and technology, a fair amount of 'free' time, but less free cash flow and are open and perhaps naive to special marketing techniques, which are used to target the student customer base. In contrast, results from a working business person would differ greatly from the student results with less free time, so probably more calls than texts, higher specification, the ability for more expensive handsets and business tariffs, perhaps even paid for by the employer. The results gathered from older generations can be used as a slight comparison, but no claim is made that the perceptions of these generations are properly represented.

Almost 60% of those asked said they still used instruction manuals. This question was asked before the mention of a mobile phone, but the results indicate that either the user is not willing to trust their existing ability with technology, or the majority of devices are still too complicated to use on first experience. This suggests that there is much scope for improvement in the way technology presents the way it should be interacted with, removing the need for instruction manuals.

Mobile phones have been used by those asked for an average of 6.8 years, with an average of 5.5 phones being owned in this time. This positively correlates with the result of 55% of the participants having owned their current phone for less than 1 year and a total of 87% for 2 years or under. This result follows an exponential decay profile, showing that future sales of mobile phones have the opportunity to continue to soar as they are currently doing so. If this influx of new phones to the population continues, they will have an even greater effect on shaping how one interacts with technology than this study will show.

The average score of 3.4/5 (≈70%) for phone feature awareness is a good score, but with the low use scores for the newer features, this suggests that a mobile phone is still essentially a speech and text communicator between individuals. Over half of the sample gave themselves, perhaps modestly, a score of 4/5 for their ability to use the features available. This suggests that phones are generally well mastered by the sample, however, all five participants to give a score of 1 for this question was less than 24 years of age, and all owned a Nokia mobile phone. Nokia did however score a remarkable 4.7/5 score for ease of 'learning' the current phone. Sony Ericsson had the only two 1/5 scores for ease of learning, but overall scored a second place average of 3.4/5.

Eight subjects took part in the visual experiment. On average, the experiment took a total of 10 to 15 min for the participant to complete. All of them carried out the tasks on both schematics. The Nokia schematic had a success rate of 50% for the correct interaction being carried out for the respective task. The Sony Ericsson scheme achieved 61%. Fig. 3 and Fig. 4 show the results for tasks "unlock the keypad" and "press the menu button", where the colour scheme and the numbers indicate how many participants have used the same pattern. All results are published in [10].

It has been shown that the nearer the top of the keypad a key is positioned, the more important, or regular to use it should be. These buttons are currently used to interact with the software of the phone, whilst the number pad should be exclusively used for number and text and positioned on 12, identically sized and shaped keys in a 3x4 arrangement. Future mobile phone design should keep this system to appear to be intuitive to use. Unlocking and locking a keypad should involve two keys, at opposite ends of the keypad, adapted from the Nokia system and menu should be located top left or top centre. This key should also double up as the select key when in a menu mode. There is no preferred side to locate a back button, but it should not be placed in a central position. The function of a delete key should be clearly marked as there is no preferred choice of position for this function. Scrolling should be given a top central position, with a vertical alignment. Call should always be on the left hand side and hang up on the right.

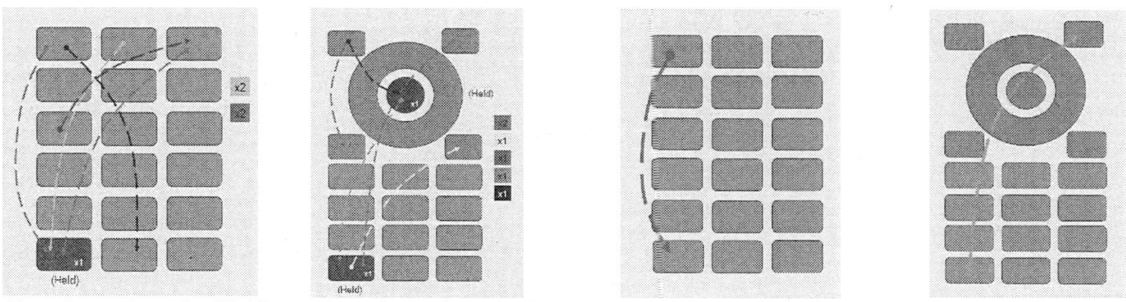

Experimental results (Nokia schematics) | Experimental results (Sony Ericsson schematics) | 'Correctly' executed task (Nokia schematics) | 'Correctly' executed task (Sony Ericsson schematics)

Fig. 3. Task "unlock the keypad".

Experimental results (Nokia schematics) | Experimental results (Sony Ericsson schematics) | 'Correctly' executed task (Nokia schematics) | 'Correctly' executed task (Sony Ericsson schematics)

Fig. 4. Task "press the menu button".

5. Conclusions

This study confirmed that intuitive use involves the seamless matching of cognitive expectation with interface actuality.

The participants in the study have shown that Nokia have set a very high standard of 'intuitive use' for the mobile phone, that other manufacturers are trying to match. Even with an unmarked, limited gestalt arrangement of keys, 50% of all interactions asked to be carried out on the schema would have been successful. Sony Ericsson, with a starkly different scheme, where gestalt laws are far more eminent had the higher score with a 61% successful interaction rate. However, Sony Ericsson were only given a 3.4/5 score for ease of learning whilst Nokia have clearly set the standard with a 4.7/5 score. This early appreciation of ease of use must have played a key factor in Nokia's early domination of the market, but now other manufacturers have learnt this fact and are far closer competitors to Nokia than just a few years ago.

The mobile phone, when used as a voice and text message communicator, is generally a very successful product in terms of being easy to use. The majority of phones are based on one of a few classic styles and there has been much pressure from the consumer for these devices to be intuitive to use. However, phone manufacturers have a delicate time ahead of them as the era of the mobile phone in a pure sense is over. New technologies are changing the function of the mobile phone and currently these new functions appear to be confusing to use on such a small device. This justifies research into design for intuitive use in the future as many different devices, along with their respective styles of interaction, are fused together into what is currently the mobile phone.

References

[1] Oxford English Dictionary, Oxford University Press, 2005.

[2] Blackler A. L., Popovic V. and Mahar, D. P. Empirical investigations into intuitive interaction: a summary. MMI-Interaktiv, 2007, vol. 13, pp. 4-24.

[3] Blackler A. L. and Popovic V. The nature of intuitive use of products: an experimental approach, Design Studies, vol. 24, 2003, pp. 491-506.

[4] Mohs C., Hurtienne J., Israel J. H., Naumann A., Kindsmüller M. C., Meyer H. A. and Pohlmeyer A. IUUI - Intuitive Use of User Interfaces. In: T. Bosenick M. Hassenzahl M. Müller-Prove and M. Peissner (Eds.), Usability Professionals 2006, Stuttgart, German Chapter der Usability Professionals' Association, 2006, pp. 130-133.

[5] Blackler A. L. and Hurtienne J. Towards a unified view of intuitive interaction: definitions, models and tools across the world, MMI-Interaktiv, 2007, vol. 13, pp. 37-55.

[6] Blackler A. L., Popovic V. and Mahar, D. P. Towards a design methodology for applying intuitive interaction, In: Friedman, K., Love T., Côrte-Real E. and Rust C. (Eds.) Proceedings WonderGround: 2006 Design Research Society Int. Conference, Lisbon.

[7] Johnson M. The body in the mind: The bodily basis of meaning, imagination, and reason. Chicago: University of Chicago Press, 1987.

[8] Hurtienne J. and Israel J. H. Image schemas and their metaphorical extensions: intuitive patterns for tangible interaction. In: 1st International Conference on Tangible and Embedded Interaction (TEI'07), Baton Rouge, ACM Press, 2007, pp. 127-134.

[9] Hurtienne J., Weber K. and Blessing L. Prior experience and intuitive use: image schemas in user centred design, In: Designing Inclusive Futures, Springer, London, 2008, pp. 107-116.

[10] Marsh A. Design for intuitive use, Final Year project MMM 130, Cardiff University, UK, 2008.

Innovative Production Machines and Systems
D.T. Pham, E.E. Eldukhri and A.J. Soroka (Eds.)

Computer aided design and analysis of the sprue system of an injection mould for optical data storage replication

D.S. Trifonov[a], Y.E. Toshev[b]

[a] *Institute of Information Technology, Bulgarian Academy of Sciences, 1113 Sofia, Bulgaria*
[b] *Institute of Mechanics and Biomechanics, Bulgarian Academy of Sciences, 1113 Sofia, Bulgaria*

Abstract

The present work offers 3D CAD models of two basic types of the sprue system of an optical disc mould that are developed. The first type of the sprue system is known as a system, in which the central hole of the polymer substrate is formed by breaking off the circle gate from the polymer substrate. In the other type of the sprue system the central hole of the polymer substrate is formed by direct punching of the gate from the polymer substrate. The gate depth and the gate position are defined as variable parameters and combined with different variants of the processing conditions in mould filling simulation. A range of improved variants of all the variable parameters is obtained with the help of iterative steps within the frame of the simulation code. A modified gate system is proposed in the process of research, which allows a gate with a lager depth parameter to be used. The modified system allows also alteration of the gate position with respect to the central hole. The best results are achieved, using the proposed modified gate system in the case, when the gate position is located symmetrically towards the central hole.

Keywords: computer aided design, simulation, optical data storage replication, nanostructures

1. Introduction

1.1 Optical data storage microscale replication and nanotechnology

Polymer injection molding is highly suitable for the mass fabrication of high precision micro- and nanostructures. Optical discs have become increasingly used for information storage in recent years. The disc information structures are both in the micro/submicro range and in the nanometer range, and in the form of pits, are moulded into the disc during the injection moulding process. The key reason for the success of Compact Discs (CD) as storage media is the fact that injection molding makes it possible to transfer Gbytes of data points (surface pits) in a few seconds onto a cheap polymer carrier. This task has become even more important for the new formats (DVD and successors) with higher storage densities, where structure sizes of below 100nm have to be replicated [1, 2]. There are different information features influencing on the CD density – linear, track and areal density [3].

1.2. Definition of the technical problem in the optical disc moulding

The production of polymer substrates for optical discs requires high dimension accuracy, exact copying of the information and low birefringence. The experience gained up to now in the field of injection moulding of optical disc substrates has shown that the mould and the sprue system occupy a central place in the production process [4, 5, 6, 7, 8].

In the process of injection moulding of optical disc substrates, a hot polymer melt is injected in the thin cooled mould cavity under high speed and pressure with the help of a circle gate located in the cavity centre. This practically means that substrate indicators such as birefringence, residual tensions and copying of the information structure are highly dependent on the behavior of the melt flow, filling the mould cavity. It becomes clear that great internal tensions are available in the final product as a result of improper location, form and dimensions of the circle gate, which result at the end in high level of the main indicator for substrate quality - the birefringence. It follows from this, that the construction solution and the scheme of the circle gate location play a significant role in the quality of the final product [9, 10]

A "statement of the problem" for optical disc moulding starts with the fundamental difficulty of filling and packing a mould cavity of much greater length than depth, without incurring high levels of moulded-in stress and orientation. For instance, a 120 mm diameter erasable media disc, which is 50 times longer melt flow path than part thickness [11, 12].

In spite of the progress in this area, the approach applied up to now by different manufacturers in solving the problem of cavity filling and formation of the central hole substrate has not declined from the basic method - filling through a thin film circle gate located at the place where the central hole is to be formed. This requires the use of a complex and precise cropping punch or another system, removing the circle gate, in order to form the central hole. On the other hand, the necessity to shape the central hole by separating the circle gate from polymer substrate, calls for minimal depth of the circle gate enabling its easy removal. This approach influences badly the level of the product internal tensions and the orientation phenomena, accompanying the process of filling [13].

1.3. Computer simulation in the optical disc moulding

Computer simulation is occupying a significant role in the most of the recent papers, which consider the optical disc moulding, but they concern mainly the advantages of the program or the creation of nonlinear viscoelastic models without an analysis of the influence of the concrete sprue system geometry on the mould filling process. Probably, the reason of this is that the sprue system is the key for useful optical disc moulding and its design represents Know-How of the companies, working in this area [14].

The purpose of the paper is to develop 3D CAD models of two basic types of the sprue system and to provide a primary investigation to determine a range of gate size parameters and correct gate location towards central hole thickness, so that an easy mould cavity filling is to be attained, using the developed 3D CAD models and computer simulation techniques.

Another purpose of the work is to give a possibility for future investigations of the influence of the sprue system design on the process of precise micro-scale information pits replication.

2. 3D CAD model of the sprue system

The 3D CAD model of an object is the first stage, when computer simulation techniques are used. An author's flow chart of the stages, necessary for the purpose of the investigations is shown (See Fig. 1).

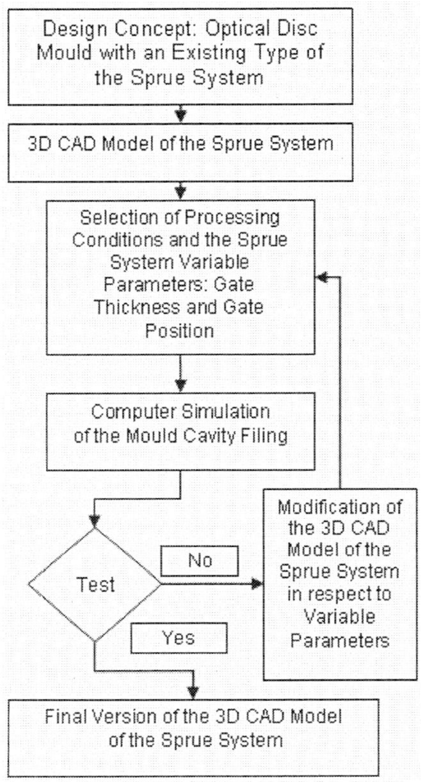

Fig. 1. Flow chart of the investigation scheme.

The present work is the first stage of the complete investigation scheme to study the optical data storage replication process in respect to investigate the effect of the sprue system design on the possibility an easy mould cavity filling to be achieved and to determine the correct sprue system design parameters to reach this goal.

Two basic types of the sprue system design of an optical disc mould are considered.

The first type of the sprue system design concept is based on a well-known type of the type of a sprue system, the central hole of the polymer substrate is formed by breaking off the circle gate from the polymer

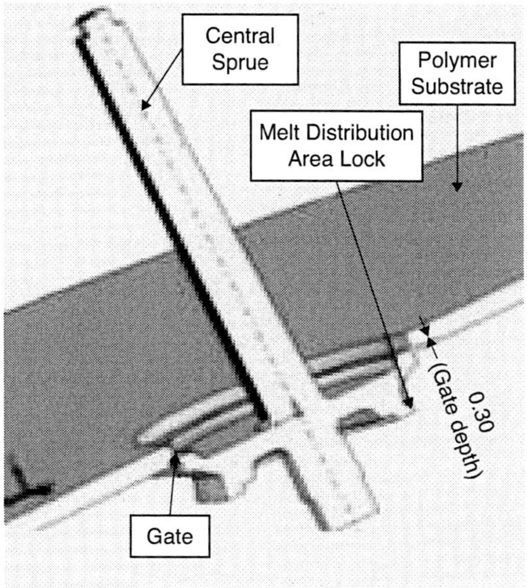

Fig. 2. 3D CAD model of the sprue system (the central hole is formed by breaking off the central gate from the polymer substrate).

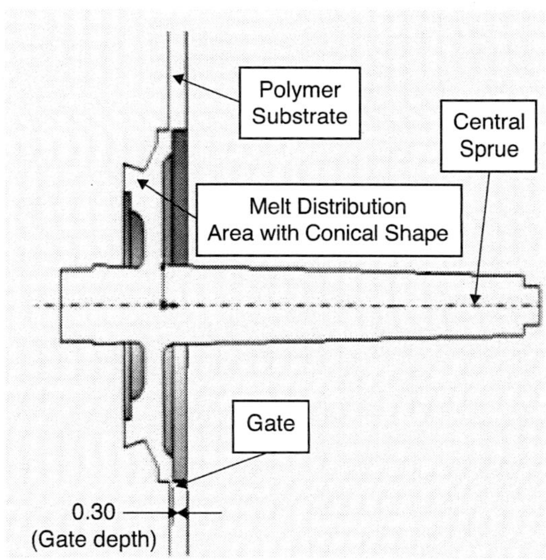

Fig. 3. Section - view of the first type of the sprue system.

substrate using a moving bush in the opposite direction of the sprue bush. (See Fig. 2 and Fig. 3).

The general disadvantage of this construction is the necessity to use a minimal depth of the circle gate for its easier removal, which contradicts to the support of the optimal conditions of the process of mould cavity filling. Another disadvantage is the impossibility to change the gate position towards the central hole. The reason for this is the necessity to locate the gate in the edge, located in the opposite direction towards the sprue bush, which is detailed described of the authors of this concept [14].

The other type of the sprue system is known as a system, in which the central hole of the polymer substrate is formed by direct punching of the central gate from the polymer substrate using a punch bush, which is moved against the air supply bush (see Fig. 4 and Fig. 5).

The disadvantages of this system are the same as the first system, considered above.

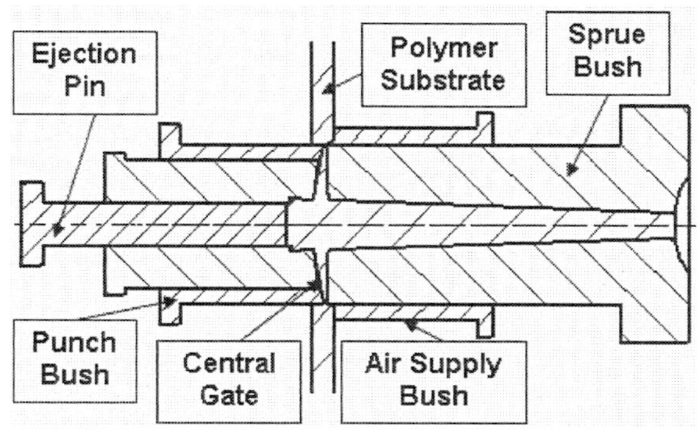

Fig.4. A Sprue system with direct punching of the central gate from the polymer substrate.

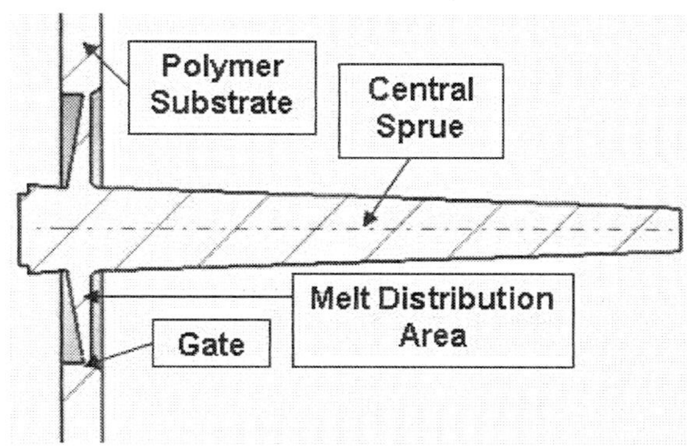

Fig. 5 3D CAD Model of the sprue system with direct punching of the central gate.

In the process of our work, on the bases of a previous author's punch system design, a modified sprue system design is realized. The central hole is formed by moving "a bending bush" and a punch bush in a direction, opposite to the sprue bush. As a result of the motion of the bending bush, the central gate is bent and at this moment the punch bush splits the gate from the polymer substrate. This allows the formation of the central hole of the polymer substrate in the case when a gate with a greater depth is used and gives a possibility to change the gate position towards the central hole (see Fig. 6).

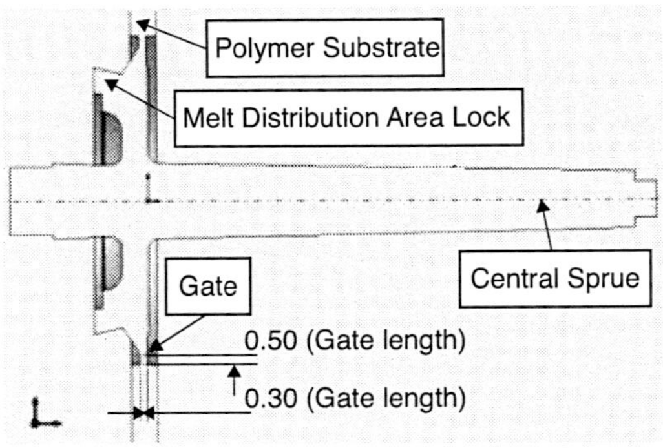

Fig. 6. 3D CAD Model of the Modified Sprue System with Gate depth=0.3 mm.

The 3D CAD models of the above considered sprue systems are built. For all of the 3D CAD models SolidWorks 2006 software is used.

The 3D CAD models of the sprue systems are built on the base of an integrated 3D CAD model of the sprue system and the polymer substrate, considered as one object. All of the considered above sprue systems consist of a central sprue, melt distribution area and a circle gate (see Fig. 5 and Fig. 6).

All the experiments are carried out in the case when the polymer substrate outer diameter is equal to 120.20 mm and the polymer substrate thickness is 1.30 mm. The gate depth and the position of the gate towards the central hole thickness are chosen as variable parameters (see Fig. 3, Fig. 5 and Fig. 6).

3. Mould filling simulation and results

A mould filling simulation program MoldflowXpress (integrated with SolidWorks) is used in the primary investigation to study the effect of the gate size and the gate position towards the central hole on the mould cavity filling process. The program can predict whether the injection moulding will be successful. The algorithm is based on the gate geometry, the plastic injection location, the type of plastic material and the processing conditions (melt and mould temperatures and injection time).

A polycarbonate plastic material Makrolon (PC) of the BAYER AG supplier is used. A PC Intel Pentium processor (2.7 GHz) with 1200 MB DDRAM is used.

A number of simulations of the two types of the gate system at values of the gate depth equal to 0.20 - 0.40 mm at intervals of 0.10 mm and at different positions of the gate towards the central hole thickness are provided. The parameters selected in three variants of the processing conditions (melt and mould temperatures) are combined: (a) T melt = 280^0 C; Tmould = 80^0 C (b) T melt = 300^0 C; T mould = 100^0 C (c) T melt = 320^0 C; T mould = 120^0 C). In many cases the results show that the mould cavity filling process is not sufficiently good at the parameters above given.

The main significant results of our investigation are shown below (see Table 1).

Table 1
Mould Filling Simulation Results

Gate depth [mm]	Processing Conditions [^{0}C]		
	Tmelt = 280; Tmould = 80	Tmelt = 300; Tmould = 100	Tmelt = 320; Tmould = 120
	Mould Filling Status		
Sprue system with breaking out the central gate			
d = 0.29 - 0.31	Difficult	Difficult	Easy
d = 0.37 - 0.40	Difficult	Difficult	Easy
Modified Sprue System			
d = 0.20 - 0.30	Difficult	Easy	Easy
d = 0.31 – 0.40	Easy	Easy	Easy
Sprue System with direct punching of the central gate			
d = 0.28 – 0.31	Difficult	Easy	Easy
d = 0.31 – 0.40	Easy	Easy	Easy

The results of the experiments with the sprue system, in which the central hole is formed breaking off the circle gate from polymer substrate show, that it is easy to fill the mould cavity, when the gate size is equal to d = 0.29 – 0.31 mm and d = 0.37 – 0.40 mm only in the case, when T melt = 320^0 C and T mould = 120^0 C.

The results of the experiments with the sprue system, in which the central hole is formed by direct punching of the circle gate from polymer substrate show, that it is easy to fill the mould cavity, when the gate depth is equal to d = 0.28 – 0.31 mm in the cases (b) T melt = 300^0 C; T mould = 100^0 C and (c) T melt = 320^0 C; T mould = 120^0 C.

The results of the experiments with the modified sprue system show, that it is easy to fill the mould cavity, when the gate depth is equal to d = 0.20 – 0.30

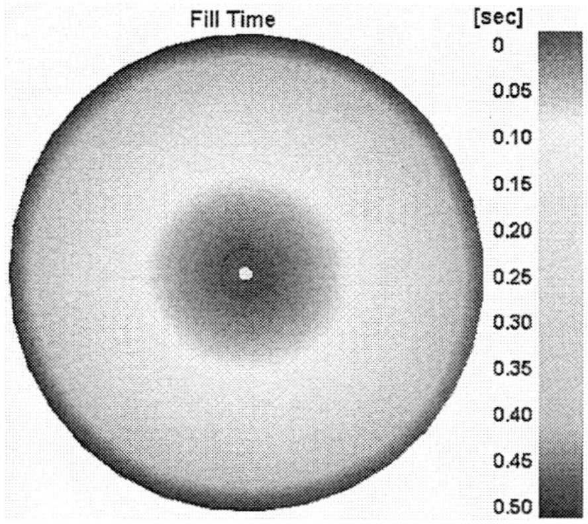

Fig. 7. Mould filling simulation: Modified sprue system: gate depth d = 0.40 mm; Tmelt = 320^0 C, Tmould = 120^0 C.

mm in the cases (b) T melt = 300^0 C; T mould = 100^0C and (c) T melt = 320^0 C; T mould = 120^0 C.

The best results are obtained with modified sprue system, when the gate depth is equal to d = 0.31 - 0.40 mm at all of the processing parameters: (a) T melt = 280^0 C; T mould = 80^0C (b) T melt = 300^0 C; T mould = 100^0C (c) T melt =3 20^0C; T mould = 120^0 C.

Only for example a plot view of the mould filing simulation is shown (see Fig. 7).

4. Conclusions

In the present work 3D CAD models of two basic types of the sprue system of an optical disc mould are developed. The first type of the sprue system is known as a system, in which the central hole of the polymer substrate is formed by breaking off the circle gate from the polymer substrate. The other type of the sprue system is known as a system, in which the central hole of the polymer substrate is formed by direct punching of the gate from the polymer substrate. In the process of our work, on the bases of a previous author's punch system design, a modified sprue system design is realized. The central hole is formed by moving "a bending bush" and a punch bush in a direction, opposite to the sprue bush. As a result of the motion of the bending bush, the central gate is bent and at this moment the punch bush splits the gate from the polymer substrate. This allows the formation of the central hole of the polymer substrate in the case when a gate with a greater depth is used and gives a possibility to change the gate position towards the central hole The gate depth and the gate location towards central hole thickness are defined as variable parameters and are combined with different variants of the processing conditions in the process of the mould filling simulation. By means of iterative steps in the frame of the simulation program MoldflowXpress (integrated with SolidWorks), a range of values of the gate size for all of the given above sprue system are obtained, in which an easy mould cavity filling is achieved. The best results are obtained, using the modified sprue system at greater depth of the central gate and when the gate is symmetrically located towards central hole thickness.

The approach presented and the 3D CAD models of the gate system can be used as a ground for future development of the mould filling simulation, oriented towards more precise determination of the resulting injection pressure, the residual stresses distribution and the temperature distribution.

Acknowledgements

We are grateful to FP6 4M Network of Excellence.

References

[1] Schift H., F. Glaus, J. Gobrecht, B. Haas, B. Ketterer , A. D'Amore, D. Simoneta, W. Kaiser, M. Gabriel, Paul Scherrer Institut, Switzerland, PSI annual report (2000).

[2] Schift H., D. Bächle, J. Gobrecht, W. Kaiser, A. D'Amore and M. Gabriel, Paul Sherrer Institut, Switzerland, PSI scientific report (2002).

[3] Tzarnorechki O., D. Borissova, IIT/WP-169B, (2003).

[4] Hatch D., D. Kazmer, The International Journal of Advanced Manufacturing Technology, **18**, No 5 (2001).

[5] Shyu G.D., A, I. Isayev and Ho-Sang Lee, Korea-Australia Rheology Journal **15**, No 4 (2003) 159-166.

[6] Trifonov D., N. Tzonev, O. Tzarnoretchki, Cybernetics and Information Technologies **2**, No 2 (2002) 113-117.

[7] Beich W.S., Photonics Spectra, March (2002) 127.

[8] Tokuhara S., Y. Onisawa, S. Iyoshi, H. Toba. Daicel Chemical Industries Ltd Research Center 1239, Aboshi - ku, Himeji, Hyogo, Japan (2002) 671-712.

[9] Keyes D.R., R. Lamonte, D. Mc-Nal1i, M. Birtritte, Photonics Spectra, October (2001) 131.

[10] Galic G., A new optical disk molding Technology. Managing Partner, Galic/Maus Ventures (1988).

[11] Dun R.M., Introduction to the molding of Optical disk substrates: Klockner Ferromatik Desma (1989).

[12] Swiss Moulds AWM targets cycle time. Optical Disc Systems, November-December (1998) 26.

[13] Zhou H., D. Li, Advances in Polymer Technology **20**, No 2 (2001) 125-131.
[14] Eichsleder M.; N. Eusemann, Kunststoffe **77** (1987) 150-153.
[15] Gábor J., K. Ovacs, T. B. Ercsey, Periodica Polytechnica Ser. Mech. Eng. **49**, No 2 (2005) 115-122.

Innovative Production Machines and Systems
D.T. Pham, E.E. Eldukhri and A.J. Soroka (Eds.)

Mapping from Product Shape Parameters into Kansei Image Using ANNs

Kesheng Wang*, Lijuan Dai

Knowledge Discovery Laboratory
Department of Production and Quality Engineering,
Norwegian University of Science and Technology,
N-7491 Trondheim, Norway

Abstract: In today's highly competitive market, the function of products has become the basic requirement of consumers. Consumers don't just buy a product; they buy value in the shape of entertainment, experience and identity. The external shape expression of products is a critical determinant of consumer response and product success. However, because of subjectivity, this particular dimension of the consumers' need is usually difficult to be expressed, quantified and assessed. In this paper, an Artificial Neural Networks (ANNs) based approach for Product Form Design (PFD) is addressed to deal with this problem. 2D draft modeling, semantic differential method and ANNs are used to set up an experiment for modeling and simulation. The experimental data are analyzed and modeled by the use of a Standard Back-Propagation (SBP) neural network. The ANN model establishes the mapping between product shape parameters and adjective image words. As a result, designers can input the product shape parameters of their new design configurations, evaluate the extent of their pleasant sensation, and then modify the shape parameters of the product until they can fit more closely to the consumers' desire. To illustrate this approach, a LCD monitor design problem is taken as a case study. Due to the generalization of the approach, it can be extended to the development of other products.

Keywords: Product shape design, industrial design, Artificial Neural Networks, Computational Intelligence.

1. Introduction

There are three components of every product: usability, aesthetics, and practicality [1]. In this technologically advanced world, usability and practicality of products are easily achieved. However, product aesthetics is attracting significant interests of the manufacturing industry. When purchasing products, there are always abundant choices with same function but different shapes in front of consumers. Most of the time, "I just feel like it" is the only reason for consumers to choose the product, rather than any other rational one. Therefore, the external shape expression of products is a critical determinant of consumer response and product success.

Because it is not easy to clearly understand how consumers perceive a product shape and how the meaning of the product shape can be effectively transmitted to them, industrial designers tend to rely on the market research and their own personal design experiences, using computer for generating 3D models or 2D drawings of physical components in the product design process. As for the psychological requirement, very few computer-aided design methods are available to support this particular dimension of product shape design. In an attempt to remedy this situation, Kansei Engineering was

* Corresponding author: E-mail: kesheng.wang@ntnu.no; Tel.: +4773597119; Fax:+4773597117

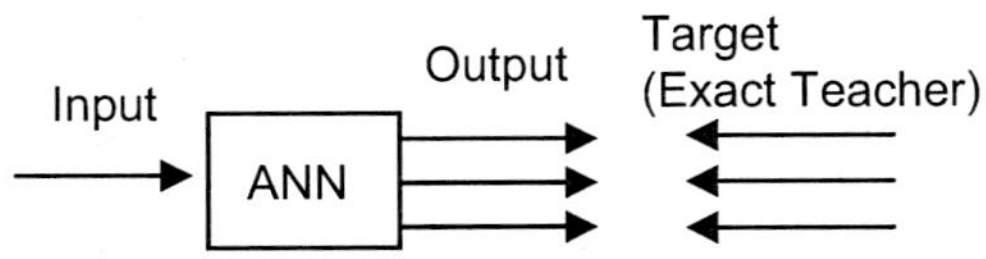

Fig. 1. Supervised learning.

invented in the 1970s by Professor Mitsuo Nagamachi. It is a method for "measuring" consumers' psychological feeling and image (kansei in Japanese) regarding new products, aiming at the implementation of the consumers' feeling and demands into product function and design [2]. Afterwards, there are many researches [3-6] have reshaped the approach under the name of Kansei Engineering, trying to quantify people's perceptions about product shapes and translate consumer's perceptions into design elements. The results have demonstrated that this approach is able to be applied to many areas, but it is not fully explored yet.

Computers are our superior in some tasks, such as rapid math calculation with high speed and accuracy, but human being's brain can overcome computers when the inputs are partially incomplete, incorrect, or even uncertain. Fortunately, there is Artificial Intelligence (AI) could be applied to simulate human being's "thinking". Artificial Neural Networks is one of the best methods simulating physiological features of the human brain and has been used for non-linear mapping by a numerical approach, due to its advantages in learning ability, storage ability, fault tolerance, inductive ability and parallel handling ability [7].

In this paper, 2D draft modeling, semantic differential method and ANNs are applied to set up an experiment in which LCD monitor is chosen as the objective product. The experimental data are analyzed by applying ANNs, which establishes the relationship between product shape parameters and adjective image words. As a result, designers can input the product shape parameters of their new design configurations, evaluate the extent of their pleasant sensation, and then modify the shape parameters of the product until they can fit more closely to the consumers' desire.

This paper is organized as follows. Section 2 describes the background of ANNs. The detail implementation procedure using NEUframe development tool and results are illustrated in Section 3. Conclusions are drawn in section 4.

2. Theoretical background

Artificial Neural Networks (ANNs) have been used extensively in design and manufacturing [8]. ANN is a model that emulates a biological neural network. It consists of Processing Elements (PEs), called neurons, and connections between PEs, called links. Associated with each link is a weight. Multiple Layer Perceptron (MLP) ANN trained by Back-Propagation (BP) algorithm is one of the most popular and versatile types of ANN. It can deal with nonlinear models with high accuracy. A three-layer supervised learning ANN consisting of input layer, hidden layer and output layer is used as an identifying model in this study.

The Process Elements (PEs) are organized in different ways to form the network structure. Each PE operates by taking the sum of its weighted input and passing the result through a nonlinear activation function, mathematically models as

$$y_i = f(a_i) = f_i(\sum_{j=1}^{n} w_{ij} - s_i) \tag{1}$$

The error is subsequently backward propagated through the network to adjust the weights of the connections and threshold, minimizing the sum of the mean squared error in output layer. The ANN computes the weighted connection, minimizing the total mean square error between the actual output of the ANN and the target output. The weights are adjusted in the presence of momentum by

$$(w_{ji})_{n+1} = (w_{ji})_n + \Delta w_{ji} \tag{2}$$

$$\Delta w_{ji}(n+1) = \eta \delta_j x_i + \alpha \Delta w_{ji}(n) \tag{3}$$

where η is learning rate, δ_j is an error value for node j and α is a momentum term. The momentum term is added to accelerate convergence.

Supervised learning mode has been the mainstream of Artificial Neural Networks (ANNs) model development and is proven highly successful in applications. It incorporates an external teacher; each output unit is told what its desired response to input signals ought to be (Figure 1).

The strength of connection between the nodes in ANN is stored as a weight-value for the specific connection. The system learns new knowledge by adjusting these connection weights. Standard Back-Propagation (BP) ANN is one of most frequently used methods for the connection of weights. There is no need to know the exact mathematical function of a system. The training data will change the weights and biases of the SBP by a learning algorithm, such as gradient descent method to build up the model of the system. Using the ANN model, the new results can be calculated in the queering process.

3. Implementation procedures

3.1. Selecting the objective product

Consumer electronic products represent a large part of electronic equipments for daily uses in the world. It is also called 3C products including computer, communication, and consumer electrics. The engineering technologies in the product category are mature and become a more standardization. The similarity among these products is very large. A Liquid Crystal Display (LCD) monitor as a component is widely used in 3C products. Since the first active-matrix LCD panel was produced in the United States in 1972, LCD monitor has been well developed worldwide until now. All the LCD manufacturers have reached pretty much the same technological level. However, the product shape design could be varied as fashion changes. In this paper, LCD monitor is chosen as the objective product.

3.2. Collecting the existing products

There are many companies involved in the design and manufacturing of LCD monitor. The most famous ones are Acer, BenQ, Epson, HP, JVC, Sony, Viewsonic, etc. As a result, it is easy to choose more than 100 existing products from the plentiful and available LCD monitors in the market.

3.3. Analyzing the relationship between shapes and images of a product

These selected 100 LCD monitors are used for collecting image perception. Semantic differential method is applied here, which consists in listing the

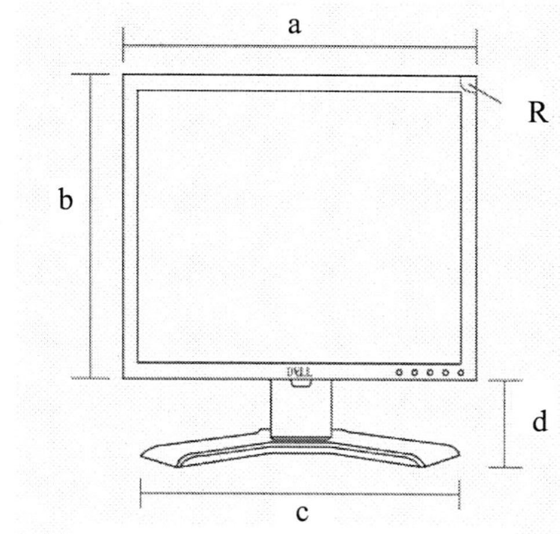

Fig. 2. The part parameters of LCD monitor.

semantic attributes of the product for analysis, and carrying out consumers' test in which the consumers assess the product according to these attributes. The attributes are often defined by pairs of anonymous adjectives that lie at either end of a qualitative scale. As a result, the adjectives consumers' use that describes their psychological feeling can be arranged in relevant pairs as the following: strong/fragile, coarse/delicate, traditional/modern, cool/warm, stable/unstable, complicated/simple, acute/mild, and masculine/feminine, etc.

3.4. Analyzing the relationship between part parameters and image evaluation of the product

The shape of LCD monitor is simple. Screen, support, and base are generally regarded as the three major components of the form of the product. To simplify the problem, 2D draft modeling is used for the image evaluation. The part parameters (b: a, d: b, c: a, and R) are used to establish the relationship with adjective image words (Figure 2).

After an analysis of the correlation between the pairs of adjectives, and a study of the meaning of the adjectives, four expressive nouns are selected as the criterion for the image score evaluation. They are:

Modernity (which may include traditional – modern, hi-tech – low-tech, etc.);

Mildness (which may include mild – acute, masculine – feminine, cool – warm, etc.);

Fragility (which may include strong – fragile, masculine – feminine, coarse – delicate, etc.);

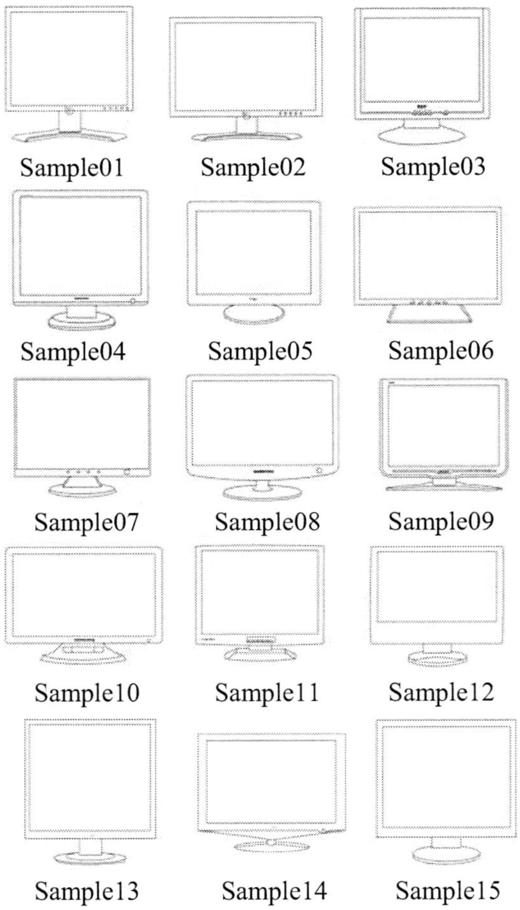

Fig. 3. 15 2D draft modeling LCD monitor samples.

Stability (which may include stable – unstable, etc.).

Then 15 samples (Figure 3) obtained by 2D draft modeling are used as the test samples for which a questionnaire (Table 1) is employed. 20 questionnaire results are taken into account to calculate the image scores for each sample respectively. For Supervised SBP networks, the network output usually lies between 0 and 1. The image scores are unified in the range.

3.5. Constructing an ANN and testing the experimental results

The obtained experiment data can be used to analyze the relationship between the part parameters and image scores of each sample product. A SBP neural network is constructed (Figure 4) by the use of a development tool, Neuframe, to analyze these data.

There are 4 input parameters and 4 output parameters in this ANN. The node number of the hidden layer is decided as follows:

$$N_h = (N_i + N_o)/2 \tag{4}$$

where *Nh*, *Ni* and *No* are the number of nodes in hidden layer, input layer, and output layer, respectively. Thus, there are 4 nodes in the hidden layer whose arrangement is shown as Figure 5. The number of neurons in hidden layer could be also obtained through heuristic or optimizing method.

Then obtained part parameters and image scores of 10 samples (from sample 06 to sample 15) are used as the training data to train the ANN network, and establish internal structures and relationships (or weights) of the network in order to learn and memorize the input/output relationship among the given training data sets.

Table 1
Example of questionnaire

Your Feeling	The Scale of Your Feeling (Please tick)				
	1	2	3	4	5
Modernity			√		
Mildness				√	
Fragility				√	
Stability		√			

-02-

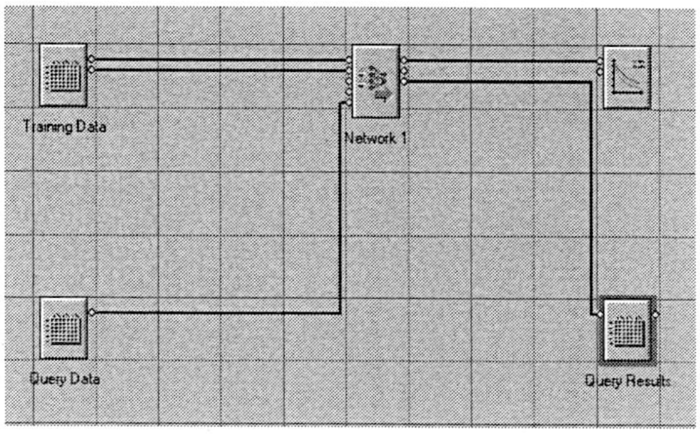

Fig. 4. A SBP ANN structure in Neuframe environment.

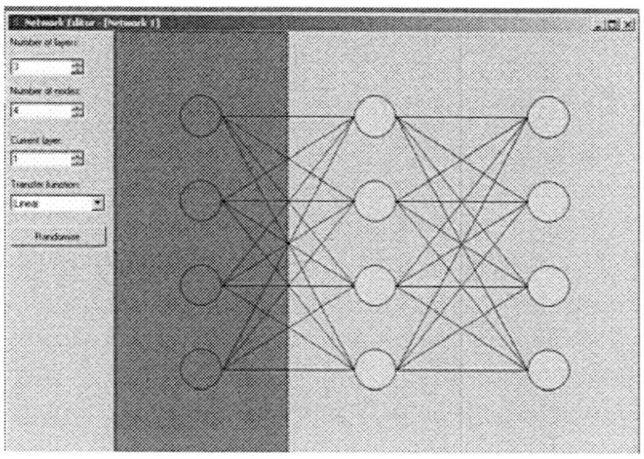

Fig. 5. The structure of the SBP ANN.

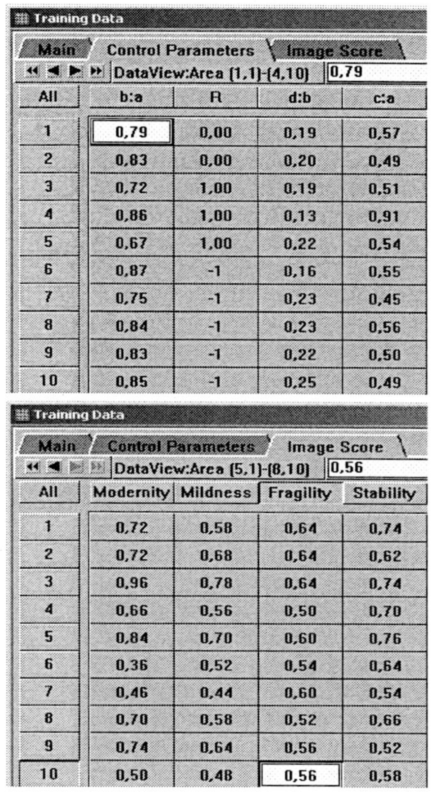

Training Data

Main | Control Parameters | Image Score

DataView:Area (1,1)-(4,10) 0,79

All	b:a	R	d:b	c:a
1	0,79	0,00	0,19	0,57
2	0,83	0,00	0,20	0,49
3	0,72	1,00	0,19	0,51
4	0,86	1,00	0,13	0,91
5	0,67	1,00	0,22	0,54
6	0,87	-1	0,16	0,55
7	0,75	-1	0,23	0,45
8	0,84	-1	0,23	0,56
9	0,83	-1	0,22	0,50
10	0,85	-1	0,25	0,49

Training Data

Main | Control Parameters | Image Score

DataView:Area (5,1)-(8,10) 0,56

All	Modernity	Mildness	Fragility	Stability
1	0,72	0,58	0,64	0,74
2	0,72	0,68	0,64	0,62
3	0,96	0,78	0,64	0,74
4	0,66	0,56	0,50	0,70
5	0,84	0,70	0,60	0,76
6	0,36	0,52	0,54	0,64
7	0,46	0,44	0,60	0,54
8	0,70	0,58	0,52	0,66
9	0,74	0,64	0,56	0,52
10	0,50	0,48	0,56	0,58

Fig. 6. The data sets in the training

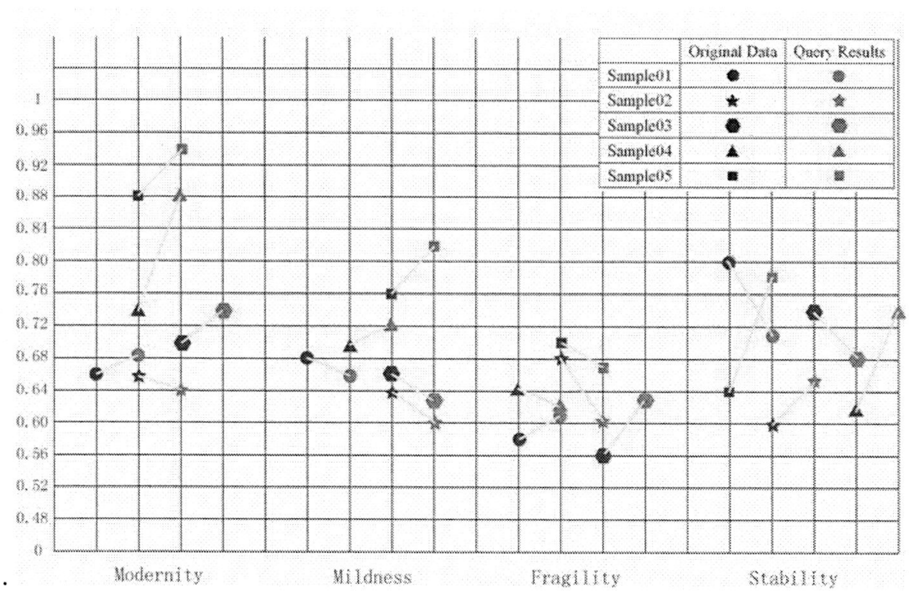

Fig. 7. Analyzing the querying results.

After the training procedure is over, we set the mode of ANN as "Query Mode" to test the other 5 samples (from sample01 to sample05), and get the testing results from the "Query Results".

Comparing the query results with the original data which are obtained from the questionnaire results, the biggest difference is sample04 in Modernity which is 0.14, sample04 and sample05 in Stability which are 0.12 and 0.14 respectively, for others, the average difference is about 0.04 (Figure 7). The accuracy of the query results increases with increase in the numbers of product shape parameters we choose. In this paper, we have four parameters focused on the form. But the kansei image of a product may be influenced by other factors such as color, material, texture, and consumers' different backgrounds. If we take all these factors into accounts, the product will be designed more perfectly. This may be a direction for future study.

4. Conclusions

In the product design stage, it is not easy for industrial designers to relate the product shape with its image sensation that is related to consumer's feeling on the product. But understanding the consumers' emotional feeling is very important and necessary in today's consumer-oriented market. To solve this problem, an experiment is performed to establish the relationship between product shape parameters and their image scores in this paper. The Supervised Standard Back-Propagation ANN is used to analyze the experiment results. There are some small errors in the results, due to the limited training data, but the results are still satisfying. This approach is an effective technique for determining product shape parameters for matching a desirable sensation image, and could be applied in industries. When selecting appropriate adjective image words and obtaining effective image score from market research, designers can use the ANN-based approach as a computer-aided design method. The designer can input the product shape parameters of their new design concepts to evaluate their sensation extent, and to modify the shape of the product until it can fit more closely to the consumers' desire.

Acknowledgement

The authors are members of the EU-fund FP6 Network of Excellence for Innovative Production Machines and Systems (I*PROMS).

References

[1] Donald A. Norman. *Emotional design: why we love (or hate) everyday things*. America. 2004.

[2] Mitsuo Nagamachi. Kansei engineering: a new ergonomic consumer-oriented technology for product development. *International Journal of Industrial Ergonomics*, 15(1995), pp 3-11.

[3] Ming C. Chuang, Chien C. Chang, Shang H. Hsu. Perceptual factors underlying user preferences toward product form of mobile phones. *International Journal of Industrial Ergonomics*, 27(2001), pp 247-258.

[4] Hongming Cai, Weiping He, Dinghua Zhang. A semantic style driving method for products' appearance design. *Journal of Materials Processing Technology,* 139(2003*)*, pp 233-236.

[5] Hsin-His Lai, Yang-Cheng Lin and Chung-Hsing Yeh. Form design of product image using grey relational analysis and neural network models. *Computers & Operations Research*, 32(2005), pp 2689-2711.

[6] Hua-Cheng Chang, Hsin-His Lai, Yu-Ming Chang. Expression modes used by consumers in conveying desire for product form: a case study of a car. *International Journal of Industrial Ergonomics*, 36(2006), pp 3-10.

[7] Shih-Wen Hsiao. Concurrent design method for developing a new product, *International Journal of Industrial Ergonomics*, 29(2002), pp 41-55.

[8] Kesheng Wang. (Second edition) *Applied computational intelligence in intelligent manufacturing systems.* Advanced Knowledge International, Australia. 2005.

Innovative Production Machines and Systems
D.T. Pham, E.E. Eldukhri and A.J. Soroka (Eds.)

An Improved Scheme for On-line Recognition of Control Chart Patterns

Adnan Hassan, Mohd. Shariff Nabi Baksh

Faculty of Mechanical Engineering, Universiti Teknologi Malaysia
81310 UTM Skudai, Johor Bahru
Malaysia

Abstract

The key feature of control charts is the provision of the method to differentiate whether a particular process is operating within a statistically stable or an unstable state. Unstable processes may produce distinct time series patterns which are useful for diagnosis and troubleshooting. However, the point when a stable process actually starts to deteriorate and reaches recognizable patterns is normally unknown. It is important that a functional on-line scheme be developed to balance between unnecessary recognition of stable processes and delayed recognition of unstable processes. This paper proposes two alternative schemes for the on-line recognition of control chart patterns, namely: (i) a scheme based on direct continuous recognition, and (ii) a scheme based on "recognition only when necessary". This study focuses on recognition of six control chart patterns plotted on the Shewhart *X-bar* chart, namely, random, shift-up, shift down, trend-up, trend-down, and cyclic. The artificial neural network (ANN) recognizer was used based on multilayer perceptrons (MLPs) architecture. The performance of the schemes was evaluated based on percentage correct recognition, average run lengths (ARL) and average recognition attempts (ARA). The findings suggest that on-line recognition should be made only when necessary. Continuous recognition is not only wasteful, but also results in poorer results. The methodology proposed in this study is a step forward in realising a truly automated and intelligent SPC chart pattern recognition system.

Keywords: Pattern recognition, Statistical process control, Control chart

1. Introduction

Most of the previous studies on control chart pattern recognition were concerned with recognition of fully developed patterns as found in Hwarng and Hubele [1], Hwarng and Chong [2], Al-Ghanim and Jordan [3], Anagun [4], Guh and Hsieh [4], Pham and Oztemel [6], Pham and Chan [7], Hassan *et al.* [8] and Hassan *et al.*[9]. Their assumption that unstable processes were represented by fully developed patterns when presented to recognisers was too simplified. Such recognisers are useful only if the starting point of the unnatural pattern is known.

Generally, process data will appear as a continuous stream of developing patterns in an observation window. The point when a stable process actually starts to deteriorate and reaches recognizable patterns is normally unknown. It is important that the recognition activity commences at the right moment. As such, unnecessary recognition of stable processes, and premature and untimely recognition attempts of unstable processes can be minimized. The recognition of any unstable processes should be made the soonest possible to enable timely diagnosis and preventive action. Any excessive delay may be catastrophic for some processes. On the other hand, stopping the deteriorating processes too early may be too costly and increase the possibility of wrong diagnosis due to insufficient information. It is important that a functional on-line scheme be developed to balance between these two extremes.

Researchers such as Tontini [10, 11] and Guh *et al.* [12, 13] have proposed on-line schemes within the context of the present discussion. Tontini [10, 11] sketchily noted that his testing system (scheme) with the focus of the study was on cumulative on-line learning to address the sensibility of the presentation order of training patterns. Guh *et al.* [12, 13] implemented a dedicated recogniser (module I) to continuously recognise the type of CCPs with recognisers in Module II to identify the pattern parameters. They noted that poor results were obtained when implemented on-line. They argued that false recognition was largely attributed to the confusion with natural variation. To overcome poor discrimination performance in the earlier study, Guh *et al.* [13] proposed a discrimination algorithm. Since the algorithm was based on the concept of cascaded recognisers, the overall performance was heavily depending on the earlier general recogniser.

There is a need for a better scheme for on-line recognition of control chart patterns where timely interpretation is critical. On-line recognition is difficult since the running patterns are normally vague and having dynamic composition of data streams prior to becoming a fully developed pattern. The objective of this paper is to present two alternative schemes for on-line recognition of developing control chart patterns (i) a scheme based on direct continuous recognition, and (ii) a scheme based on recognition only when necessary. The challenge here is to provide a balance between timely recognition of unstable patterns as they develop and minimise unnecessary recognition of stable processes.

The rest of this paper is organized as follows: Section 2 presents the scheme for continuous on-line recognition, while Section 3 discusses the scheme based on "recognition only when necessary" approach. Sections 4 and 5 present the performance of the schemes, and finally Section 6 concludes the paper.

2. Continuous on-line recognition

Figure 1 shows the generalised framework that integrates the building blocks for the scheme for direct continuous on-line recognition (Scheme A). In this design, the recognition activities are attempted continuously for all data streams as they appear in the monitoring window. Developing patterns within 75% to 90% of their development state were used for training the recognizer. Each raw data stream was represented using a minimal feature set extracted from the incoming data streams (patterns), namely autocorrelation, CUSUM, mean, mean-square-value, standard deviation and slope. The use of this minimal feature set has successfully reduced the size of input vector from 20 (raw data) to merely six. The minimal feature set was selected through extensive simulation experimental runs using fractional factorial experimental design.

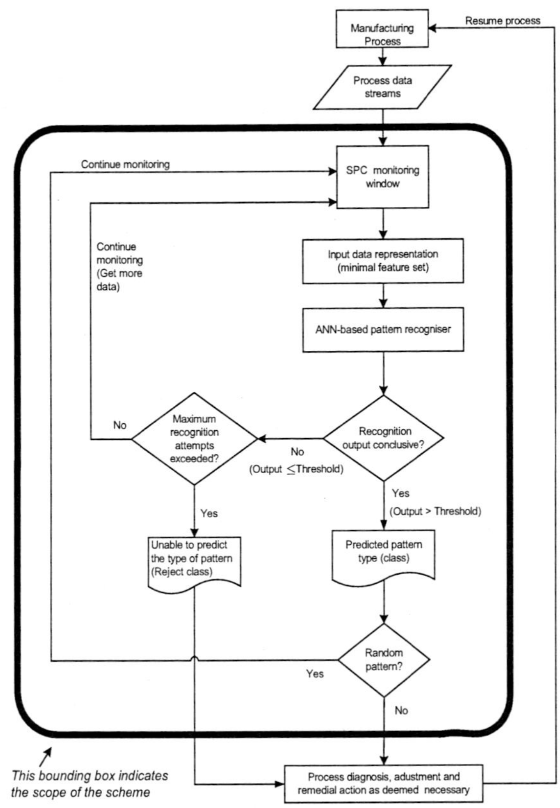

Figure 1. Scheme for direct continuous on-line recognition (Scheme A)

The extracted statistical features then presented to the ANN pattern recognizer. The ANN-based recognizer was developed based on multilayer perceptrons (MLPs) architecture. The selected ANN structure was $6 \times 10 \times 6$. The training algorithm used was BFGS quasi-Newton [14]. Details of design procedures for such MLP are widely available, for example in Patterson [15] and Haykin [16]. Detailed procedure for the recognizer training as in Hassan *et al.* [8] was adopted in this study.

The streams of process data were simulated to begin in a stable state and then they either remained stable throughout the monitoring period or progressively deteriorated into any one of the five possible unstable types, namely, shift-up, shift down, trend-up, trend-down, and cyclic.

The predicted class for the data stream was recorded whenever the maximum output exceeded the classification threshold value, $O_c = 0.8$. Generally, a low O_c leads to fewer rejections while high value of O_c increases the stringency for classification. If the maximum output is less than or equal the threshold ($\max\{O_1, O_2, \ldots O_M\} \leq O_c$), the monitoring process is allowed to proceed and more process data is obtained. This feature provides opportunity for more unstable process data to be included before making a decision. The maximum number of recognition attempts (A_c) was set to 20 corresponding to the size of the observation window. If the recognition result remained inconclusive at the 20th attempt, the pattern was classified as reject class ($M+1$). The predicted pattern is not necessarily always correct. It can be either similar to the true class or classified into some other (wrong) classes. The provision of a reject class provides flexibility in decision making, particularly to avoid being forced into making wrong classification.

3. Recognition only when necessary

The need to continuously recognise stable data stream is arguable. It is considered unnecessary in normal situations. However, the recognition of stable process becomes necessary when the system needs to recover from false alarms. Thus, this section proposes an alternative scheme by incorporating stability tests using runs rules and CUSUM as a procedure to determine the necessity to recognise prior to triggering the recognition process. This procedure was intended to avoid unnecessary recognition of stable processes and promote timely recognition of unstable processes. The detailed design for this scheme is shown in Figure 2 (Scheme B). All the components are the same as in Figure 1 except the shaded diamond which consists of Nelson's runs-rules [17] and CUSUM tests. With this configuration, all the process data streams will be tested for its stability prior to the recognition process. If a data stream is identified as coming from a stable process, then the process will continue without any recognition. As such, attempts to classify streams of process data are allowed only when they are identified as possibly coming from an unstable process. The rest of the procedure for decision-making is the same as described in Section 2.

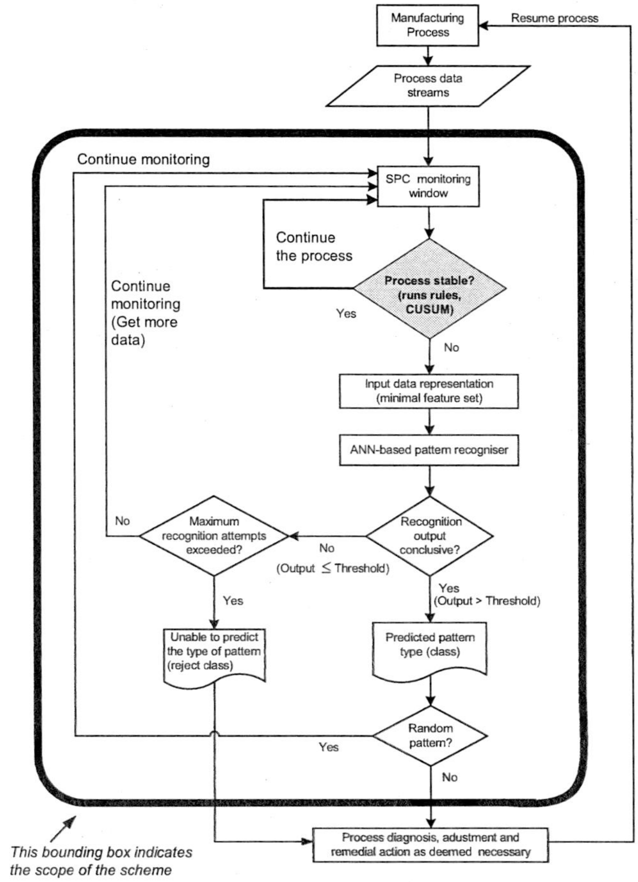

Figure 2. Scheme for recognition only when necessary (Scheme B)

4. Recognition performance

Experiments were conducted for the two alternative scheme designs using 20 different data sets, each comprising a total of 3600 developing control chart patterns (DCCPs) (600 for each type). The performance of the alternative schemes was compared based on percentage correctly recognised, ARL and ARA.

4.1. Scheme with continuous recognition

The classification performance of the scheme with direct continues recognition is summarised in a confusion matrix, Table 1. The result suggests that a significant percentage of random patterns (stable process) presented to the recognizer were either wrongly classified as unstable patterns (24.5%) or as reject class (66.8%). Only 8.7% were correctly classified as random patterns. For non-random processes (unstable data streams), the highest correct classification was for cyclic patterns (87.8%), followed by trend-up (84.8%), trend-down (80.9%), shift-down (75.6%), and shift-up (64.7%). The aggregate percentage for correct recognition of unstable data streams was 78.8%.

Table 1. Confusion matrix for the Scheme A

		Recognized Class (%)						
		RN	TU	TD	SU	SD	CY	RJ
True Class	RN	8.68	0.18	0.13	5.64	14.00	4.58	66.78
	TU	2.21	84.77	0.00	1.29	2.54	5.00	4.19
	TD	3.58	0.02	80.93	0.72	4.28	3.88	6.60
	SU	2.70	16.76	0.03	64.68	4.00	1.63	10.20
	SD	3.97	0.03	9.49	1.24	75.57	1.73	7.98
	CY	5.35	0.05	0.00	1.32	2.39	87.80	3.09

RN=Random, TU=Trend-up, TD=Trend-down, SU=Shift-up, SD=Shift-down, CY=Cyclic, RJ=Reject

The results suggest that random data streams are the most difficult to recognise compared to non-random data streams. The above findings are in agreement with Guh *et al.* [12] who reported poor performance when CCP was recognised in a real time mode.

4.2. Scheme with recognition only when necessary

Some improvements can be noticed from the results in Table 2. The percentage for correct classification of stable process has significantly improved (87.5%). This improvement was attributed to a significant reduction in unnecessary recognition and the ability of the Scheme B to recover from false alarms. For the non-random processes, the highest correct classification was maintained by cyclic patterns (85.0%), followed by trend-up (80.9%), trend-down (75.1%), shift-down (62.6%), and shift-up (51.9%). The aggregate percentage for correct classification of unstable process is 71%.

Table 2. Confusion matrix for the Scheme B

		Recognized Class (%)						
		RN	TU	TD	SU	SD	CY	RJ
True Class	RN	87.52	0.11	0.07	3.24	6.43	1.21	1.43
	TU	0.98	80.93	0.00	0.75	1.20	1.43	14.73
	TD	1.30	0.00	75.08	0.43	2.77	0.80	19.63
	SU	1.96	11.98	0.01	51.90	1.79	0.49	31.87
	SD	1.70	0.03	7.77	0.68	62.55	0.68	26.59
	CY	12.36	0.03	0.00	0.67	1.09	84.98	0.88

RN=Random, TU=Trend-up, TD=Trend-down, SU=Shift-up, SD=Shift-down, CY=Cyclic, RJ=Reject

5. ARL and ARA performance

Basically, ARL_0 measures the average run length of first occurrence of false alarm when tested for stable processes. The larger the value of ARL_0 is the better for the scheme. On the other hand, for $ARA_{correct}$ that measures the average number of recognition attempts to correctly classify the pattern, the lower $ARA_{correct}$ value is the better. Table 3 shows that Scheme B resulted in significantly better ARL_0 (212.0) compared to the Scheme A (89.3). The average number of recognition attempts ($ARA_{correct}$) was also significantly reduced from 70.3 (Scheme A) to 3.6 attempts (Scheme B). The results in Table 3 are based on the averages of 10 test data sets where each set comprised 500 streams of stable process. Each stream consisted of 600 data points (subgroup means).

Table 3. ARL_0 and ARA for stable processes

Scheme	Performance Measures*	
	ARL_0	$ARA_{correct}$
A: Direct Continuous Recognition	89.3 (3.3)	70.3 (3.3)
B: Recognition only whenNecessary	212.0 (6.1)	3.6 (0.1)

In brackets are standard deviations.

Table 4 compares the schemes performance in terms of ARL_{target} and $ARA_{correct}$ for unstable processes. ARL_{target} is the minimum number of observations required before the targeted unstable pattern is correctly recognized. The smaller the value of ARL_{target} is the better for the scheme. The results show that there was a marginal deterioration in ARL_{target} when using the Scheme B (increased from 10.76 to 11.09). However, the $ARA_{correct}$ significantly reduced from 11.76 to 4.82. Results in Table 4 were based on 20 test data sets, each consisted a total of 3000 developing unstable patterns or 600 patterns for each type.

Table 4. ARL and ARA for unstable processes

Scheme	Performance Measures*	
	ARL_{target}	$ARA_{correct}$
A: Continuous Recognition	10.76 (0.096)	11.76 (0.096)
B: Recognition Only When Necessary	11.09 (0.069)	4.82 (0.092)

* In brackets are standard deviations.

6. Conclusions

This paper investigated schemes for on-line recognition of control chart patterns (CCPs) based on continuous on-line recognition and "recognition only when necessary" approaches. The findings indicate that the unnecessary recognition of stable patterns was significantly reduced when runs rules and CUSUM were implemented prior to the recognition process. The results also reveal that these stability tests significantly improve the ARL_0. The proposed scheme has shown promising results for both stable and unstable processes. This design promotes timely recognition of unstable patterns and avoids unnecessary recognition of stable patterns. Currently, we are extending this work to further improve the overall recognition performance.

Acknowledgments

The authors thank Research Management Center, Universiti Teknologi Malaysia (UTM) and the Ministry of Science, Technology and Innovation (MOSTI), Malaysia for IRPA research grant Vot. No. 74227.

References

[1] Hwarng, H. B. and Hubele, N. F. (1993). Back-Propagation Pattern Recognizers for $\overline{X}$ Control Charts: Methodology and Performance. Computers and Industrial Engineering. Vol. 24 No. 2. pp. 219 -235.

[2] Hwarng, H. B. and Chong, C. W. (1995). Detecting Process Non-Randomness through a Fast and Cumulative Learning ART-based Pattern Recogniser. International Journal of Production Research, Vol. 33 No. 7. pp. 1817 - 1833.

[3] Al-Ghanim, A. and Jordan, J.(1996). Automated Process Monitoring Using Statistical Pattern Recognition Techniques on X-bar Control Charts. Journal of Quality in Maintenance Engineering. Vol. 2 No. 1. pp. 25 - 49.

[4] Anagun, A. S. (1998). A Neural Network Applied to Pattern Recognition in Statistical Process Control. Computers and Industrial Engineering. Vol. 35 Nos. 1 -2. pp.185 - 188.

[5] Guh, R.-S. and Hsieh, Y.-C. (1999). A Neural Network Based Model for Abnormal Pattern Recognition of Control Charts. Computers and Industrial Engineering. Vol. 36. pp. 97 - 108.

[6] Pham, D. T. and Oztemel, E. (1995). An Integrated Neural Network and Expert System Tool for Statistical Process Control. Proc. Instn. Mech. Engrs. Vol. 209 Part B. pp. 91 - 97.

[7] Pham, D. T. and Chan, A. B. (2001). Unsupervised Adaptive Resonance Theory Neural Networks for Control Chart Pattern Recognition. Proc. Instn. Mech. Engrs. Vol. 215 Part B. pp. 59 - 67.

[8] Hassan A, Baksh MSN, Shaharoum AM and Jamaluddin H. (2003). Improved SPC Chart Pattern Recognition Using Statistical Features. International journal of production research. Vol. 41, No. 7. pp. 1587 – 1603.

[9] Hassan A and Baksh MSN, Shaharoun, A.M. and Jamaluddin, H. (2006), Feature Selection for SPC Chart Pattern Recognition Using Fractional Factorial Experimental Design, 2nd I*IPROMS Virtual International Conference on Intelligent Production Machines and Systems, 3-14July2006.

[10] Tontini, G. (1996). Pattern Identification in Statistical Process Control Using Fuzzy Neural Networks. Proceedings of the Fifth IEEE International Conference on Fuzzy System. Vol. 3. pp. 2065 - 2070.

[11] Tontini, G. (1998). Robust Learning and Identification of Patterns in Statistical Process Control Charts Using a Hybrid RBF Fuzzy Artmap Neural Network. The 1998 IEEE International Joint Conference on Neural Network Proceedings (IEEE World Congress on Computational Intelligence).Vol. 3. pp. 1694 - 1699.

[12] Guh, R.-S, Tannock, J. D. T. and O'Brien, C. (1999). IntelliSPC: A Hybrid Intelligent Tool for On-line Economical Statistical Process Control. Expert Systems with Applications. Vol.17. pp. 195 - 212.

[13] Guh, R.-S, Zorriassatine, F., Tannock, J. D. T. and O'Brien, C. (1999). On-line Control Chart Pattern Detection and Discrimination- A Neural Network Approach. Artificial Intelligence in Engineering. Vol. 13. pp. 413 - 425.

[14] Demuth, H and Beale, M. (1998). Neural Network Toolbox Users's Guide. Natick: TheMathWorks

[15] Patterson, D. W. (1996). Artificial Neural Networks: Theory and Applications. Singapore: Prentice Hall.

[16] Haykin, S. (1999). Neural Networks: A Comprehensive Foundation. 2nd. ed. Upper Saddle River, N. J.: Prentice Hall.

[17] Nelson, L. S. (1984). The Shewhart Control Chart Tests for Special Causes. Journal of Quality Technology. Vol. 16. pp. 237 - 239.

Innovative Production Machines and Systems
D.T. Pham, E.E. Eldukhri and A.J. Soroka (Eds.)

Association rules discovery in manufacturing: A case study in the metal industry

A.A. Afify

Manufacturing Engineering Centre, Cardiff University, Cardiff CF24 3AA, UK

Abstract

Manufacturing activities generate large quantities of data and it is no longer practical to rely on traditional manual methods to analyse this data. Data mining offers tools for extracting knowledge from data, leading to significant improvement in the decision-making process. Association rule mining is one of the most important data mining techniques and has received considerable attention from researchers and practitioners. It aims to identify interesting relationships among a large set of data items. To date, little has been done to exploit the potentially viable association rule mining tools in manufacturing. This paper presents the basic concepts of association rule mining and describes a number of its recent applications in manufacturing. The paper also presents a case study for applying association rules in a steel bar manufacturing company.

Keywords: Data mining, association rules, manufacturing

1. Introduction

Advances in information technology, data acquisition systems, and storage technology have made it easy to collect and store large amounts of data during manufacturing. This data may be related to product design, machines, materials, processes, inventories and performance data. To utilise this abundant data resource effectively, a way of distilling information and knowledge from the data has to be found. There is a need for effective techniques to refine such data. It is no longer practical to rely on traditional manual data analysis due to the large amount of data involved. Data mining is a natural solution for transforming the data into useful knowledge. It offers tools for discovery of patterns, trends, associations, anomalies and dependencies in data. Extraction of previously unknown, meaningful information from manufacturing data provides knowledge that can improve product design or fine tune manufacturing processes [**1-5**].

Association rule mining is one of the most important techniques of data mining. It aims to extract interesting correlations, frequent patterns, associations or casual structures among a large set of data items. A typical application is market basket analysis, which studies the buying habits of customers by searching for sets of items that are frequently purchased together [**6**]. Other application areas include customer segmentation, store layout, web usage mining, software defect detection, telecommunication alarm prediction, and bioinformatics [**7-14**]. Applications to manufacturing are still under utilised and infrequently used on a large scale [**15, 16**]. Possible reasons might be the long time scales and expenses involved in introducing these new techniques, the uncertainty of the payback value, or the complexity and diversity of different manufacturing processes [**17**]. This paper reviews the basics of association rule mining and examines applications in which it has been successfully deployed.

The rest of the paper is organised as follows. Section 2 presents the basic information for association rule mining. Section 3 discusses some recent applications of association rule mining techniques in manufacturing. Section 4 presents a case study for applying association rules in the metal industry. Section 5 concludes the paper and outlines directions for future work.

2. Association rules

2.1. Problem definition

Association rule mining is a data mining method to find the interesting association or correlation among a large set of data items. A formal statement of the association rule mining problem is as follows [**18**]. Let $I = \{i_1, i_2, ..., i_m\}$ be a set of items. Let *D* be a set of transactions, where each transaction *T* is a set of items such that $T \subseteq I$. Associated with each transaction is a unique identifier, called *TID.* A transaction *T* contains *X*, a set of items in *I*, if $X \subseteq T$. An association rule is an implication of the form $X \Rightarrow Y$, where $X \subset I$, $Y \subset I$, and $X \cap Y = \varnothing$. The rule $X \Rightarrow Y$ holds in the transaction set *D* with *confidence c* if *c%* of the transactions in *D* that contain *X* also contain *Y*. The rule $X \Rightarrow Y$ has *support s* in the transaction set *D* if *s%* of the transactions in *D* contain $X \cup Y$. Confidence determines the strength of the rule and support measures the frequency of the occurring pattern.

A set of items is referred to as *itemset*. An itemset that contains *k* items is a *k*-itemset. The occurrence frequency or count of an itemset is the number of transactions that contain the itemset. If an itemset has a transaction support higher than a user-specified minimum support threshold, it is a frequent itemset.

Given a set of transactions, *D*, the problem of association mining is to find strong rules with support and confidence greater than the given minimum support and confidence thresholds, respectively. An association rule discovery algorithm can be decomposed into two successive stages [**19**]. In the first stage, all sets of frequent items are discovered. In the second stage, rules are derived from these itemsets. For instance, if set *{A, B, C}* is a frequent itemset, by considering the following rules and checking whether these rules respect the minimum confidence threshold, strong association rules can be discovered.

$$\begin{array}{ll} A \Rightarrow BC & AB \Rightarrow C \\ B \Rightarrow AC & AC \Rightarrow B \\ C \Rightarrow AB & BC \Rightarrow A \end{array}$$

It is important to generate all itemsets efficiently. An efficient association rule discovery algorithm is discussed in the next section.

2.2. The Apriori algorithm

The Apriori algorithm proposed by Agrawal et al. [**20**] is widely accepted and has been implemented in several commercial data mining packages. Apriori employs an iterative approach known as a level-wise search, where *k*-itemsets are used to explore (*k* + 1)-itemsets. The algorithm first produces frequent *1*-itemsets by simply counting the occurrence of each item in each transaction, e.g. *{A}, {B},* etc. Using the fact that for a given frequent *k*-itemset, all its *(k - 1)*-subsets must be frequent itemsets, the following two-phase strategy is adopted for generating frequent *k*-itemsets (e.g. 2-itemsets, 3-itemsets, etc). The candidate *k*-itemsets are first generated from the frequent *(k - 1)*-subsets obtained in the previous step. Then, the transaction database is scanned to delete those *k*-itemsets from the candidate itemsets that do not exceed the minimum support.

The generation of the candidate *k*-itemsets is illustrated by considering the transaction database in Table 1. The minimum support is set to 3 in this example. Figure 1 represents the frequent *1*-itemsets that are the result of counting the occurrence of each item in the database. The candidate *2*-itemsets in Fig. 2(a) are generated by 'joining' the frequent *1*-itemsets with each other. The frequent 2-itemsets in Fig. 2(b) are the result of scanning the whole database and verifying whether the supports of those candidate *2*-itemsets are above the minimum support.

The candidate *3*-itemset in Fig. 3(a) is generated from the frequent *2*-itemsets as follows. Frequent *2*-itemsets with the same first item, such as *{B C}* and *{B D},* are identified first. Then, if *{C D}*, which constitutes their second items, is a frequent *2*-itemset, *{B C D}* is a candidate *3*-itemset. No other candidate *3*-itemsets can be generated from the frequent *2*-itemsets in this example. Figure 3(b) represents the frequent *3*-itemset obtained. Apriori stops generating frequent itemsets because no candidate *4*-itemset can be constructed from the frequent *3*-itemset.

Table 1
A transaction database for association rule discovery

TID	Items
10	A B
20	B C D
30	A B C D
40	B D
50	B C D

Candidate 1-itemsets

Itemsets	Support
{ A }	2
{ B }	5
{ C }	3
{ D }	4

→

Frequent 1-itemsets

Itemsets	Support
{ B }	5
{ C }	3
{ D }	4

Fig. 1. Generation of frequent 1-itemsets.

Candidate 2-itemsets

Itemsets	Support
{ B C }	3
{ B D }	4
{ C D }	3

(a)

→

Frequent 2-itemsets

Itemsets	Support
{ B C }	3
{ B D }	4
{ C D }	3

(b)

Fig. 2. Generation of frequent 2-itemsets.

Candidate 3-itemsets

Itemsets	Support
{ B C D }	3

(a)

→

Frequent 3-itemsets

Itemsets	Support
{ B C D }	3

(b)

Fig. 3. Generation of frequent 3-itemsets.

3. Applications of association rules in manufacturing

The association rule mining has been successfully applied in a variety of manufacturing problems such as product design, quality improvement, defect detection, group technology, mass customisation, warehouse management and supply chain management. This section briefly discusses these applications.

In today's competitive environment, satisfying customer needs has become a great concern of almost every company. While there are various customer needs, the functional and affective (e.g. emotions, subjective impressions, visual perceptions, etc.) needs have been recognized to be of primary importance for customer satisfaction [**21**]. Association rule mining lends itself to the discovery of useful patterns associated with the mapping of customer needs to design features. Jiao and Zhang [**22**] developed an association rule mining system (ARMS) to improve product portfolio identification by efficient knowledge discovery from past sales and product records. The methodology aims to capture and understand customer needs effectively and accordingly to transform them into specifications of product offerings. Jiao et al. [**23**] applied association rule mining to grasp the customers' affective needs accurately and subsequently to design products that match these needs. The goodness of the resulting association rules has been evaluated in a company that produces a large variety of mobile phones.

High-quality and high-reliability products play an important role in achieving customer satisfaction. Association rule techniques have been widely used to support industrial quality improvement. Chen et al. [**24**] proposed a manufacturing defect detection method called the root-cause machine identifier (RMI) that significantly reduced the manufacturing costs and improved manufacturing performance. The method uses the technique of association rule mining for analysing correlations between combinations of machines and the defective products. Buddhakulsomsiri et al. [**25**] presented a new association rule algorithm for mining automotive warranty data. The algorithm develops useful relationships between product attributes (e.g. production date, repair date, mileage-at-repair, transmission, engine type, etc.) and causes of failure (e.g. problem related labor code). These relationships are used to effectively spot defects, identify the root causes of a particular warranty problem, and to formulate an action plan to remedy the problem and improve quality. Da Cunha et al. [**26**] adopted the technique of association rule to improve the quality of assembly operations. The focus was on the impact of assembly sequences on the quality of products. The computational results confirmed that the source of assembly sequence faults can be detected with association rules, even in the presence of noise. The rules extracted can be used to improve production quality by avoiding 'risky' sequences. Shahbaz et al. [**17**] examined the application of association rules to manufacturing databases to extract useful information about a manufacturing system's capabilities and limitations. The discovered knowledge can be used to improve the design or redesign of products and to identify manufacturing process constraints. Lau et al. [27] proposed an iterative process mining (i-PM) algorithm based on the concept of fuzzy set and association rule mining to extract interesting patterns from quantitative process data. The i-PM algorithm supports knowledge discovery and decision making by finding correlations between process parameters and finished quality. It allows quality engineers to set up different constraints so as to extract high-quality and actionable information.

Cellular manufacturing is an essential application of group technology (GT) in which families of parts are produced in manufacturing cells. Chen [28] developed a cell-formation approach based on association rule mining to find the effective configurations for cellular manufacturing systems. By applying association rules to cell-formation problems, certain sets of machines (machine groups) that frequently process some parts together can be inducted. The proposed approach shows better computational efficiency and solution quality compared with several existing algorithms. The work of Chen has been extended to solve large-scale and complex cell formation problems, where real-world production factors such as operation sequences, production volumes and alternative process routings are considered [29-31].

Mass customization is a well-established strategy that aims to produce and deliver customized products rapidly while keeping costs at the mass-production level. To advance the mass customization in the manufacturing assembly lines, Agard and Kusiak [32] presented a model for selection of the best set of subassemblies based on the analysis of prior orders received from the customers. An association rule mining algorithm was used to extract the hidden relationships among the different attributes characterising the subassembly modules of wire harness automotive parts.

Warehouse management system is a key to strengthening company logistics. Order batching in a warehouse attempts to achieve high-volume order processing operations by consolidating small orders into batches. Chen and Wu [33] developed an order batching approach based on association rule mining. With the learning of association rules, the customer demand (order) relationships can be found. The association rules often can be used to decrease the picking distance travelled in warehouses by adequately grouping orders into batches, in which they have relatively high associations with each other. Therefore, the placing of certain orders in a batch can infer that it is helpful to place certain other orders into this batch. From the results of several test problems, the proposed approach shows its ability to find quality solutions of order batching problems.

Agility deals with how responsive supply chains are to changing customer demands. Jain et al. [34] developed a new approach based on fuzzy association rule mining to support the decision makers by enhancing the flexibility in making decisions for evaluating agility in supply chains with both tangibles and intangibles attributes such as flexibility, profitability, quality, innovativeness, pro-activity, speed of response, cost and robustness.

Problem-solving is an important process that enables corporations to create competitive advantages, especially in the manufacturing industry. Liu and Ke [35] developed a mining-based knowledge support system for solving problems on a production line. Association rule mining is employed to discover decision-making knowledge rules about frequently adopted actions taken to handle specific situations. These rules are generated as knowledge support to help workers make appropriate action plans.

4. Industrial case study

This section describes an implementation of the Apriori algorithm in a steel bar manufacturing company. The company produces a wide rang of steel products which are manufactured at different sites. After a product is manufactured, it may be warehoused or shipped to the customer. The aim of this study is to find the associations between the various products delivered to the customers. Knowledge of product correlations will enable the arrangement of the warehouse storage space to keep such products closer to each other, thus reducing the time for loading the delivery truck. Moreover, the discovery of product combinations will improve the logistics efficiency when more than one production site (geographically separated) is involved in the delivery process.

Table 2 shows part of the data extracted from the original database of the company, which represents the products sold during a period of five years. The data consists of 12,442 customer orders and 50 different products. Each entry of Table 2 specifies whether a

Table 2
Part of the data used in the industrial case study

	Products manufactured at different sites															
Order No.	p1	p2	p3	p4	p5	p6	p7	...	...	...	p44	p45	p46	p47	p48	p49	p50
1	0	0	0	0	0	1	1	...	...	...	0	0	0	1	0	1	1
2	1	1	0	0	0	1	1	...	...	...	0	0	0	1	0	0	0
3	0	1	0	0	0	0	0	...	...	...	1	0	0	0	0	0	0
4	0	0	0	0	0	0	0	...	...	...	1	1	0	0	0	0	0
...	...	...	...	...	...	...	...	...	...	...	...	...	...	...	...	...	...
...	...	...	...	...	...	...	...	...	...	...	...	...	...	...	...	...	...
12439	1	0	0	0	0	0	0	...	...	...	0	0	0	0	0	1	0
12440	0	1	0	0	0	0	1	...	...	...	0	0	0	0	1	1	1
12441	1	0	0	0	0	1	1	...	...	...	1	0	0	1	0	0	0
12442	0	1	0	0	0	1	0	...	...	...	0	1	0	0	0	1	1

product has been selected (1) or not (0) by a customer.

There are many valid rules extracted from the data in Table 2. Two sample rules are presented next:

Rule 1: (p6=1) & (p7=1) => (p47=1), with support = 65.3% and confidence = 92.0%

Rule 2: (p3=0) & (p4=0) => (p5=0), with support = 82.5% and confidence = 94.0%

These sample rules are interpreted as follows:

Rule 1: When a customer ordered products "p6" and "p7", product "p47" was ordered in 8125 out of 12442 times (65.3%). This rule means that products "p6, p7 and p47" are associated in 65.3% of the orders. It provides an indication that these products should be located within close proximity in the warehouses.

Rule 2: If a customer did not order products "p3" and "p4", then product "p5" was not ordered, which occurred 10264 out of 12442 orders (82.5%). This rule implies that products "p3, p4 and p5" do not frequently appear together in the same order.

5. Conclusions

This paper has introduced the basic concepts of association rule mining and described one of its most commonly used techniques. It has also presented several studies that examine the application of association rule mining techniques in manufacturing. A case study in the metal industry has also been given. These studies have shown that association rule mining algorithms can extract information to help intelligent decision making in manufacturing industry. However, these algorithms tend to produce a large number of rules, many of which might not be interesting. Future research effort is therefore needed to find rules that are interesting and useful to decision makers. The domain knowledge in application areas can potentially provide useful criteria for identifying important rules.

Acknowledgements

This work was carried out within the ERDF (Objective One) project "Virtual Enterprise Network for Manufacturing Industry in Wales" (Venturing Wales) and within the EC-funded Network of Excellence "Innovative Production Machines and Systems" (I*PROMS).

References

[1] **Agard, B.** and **Kusiak, A.** Standardization of components, products and processes with data mining. In: *Proceedings of the International Conference on Production Research*, Santiago, Chile, 2004, 1-9.

[2] **Pham, D.T.** and **Afify, A.A.** Machine-learning techniques and their applications in manufacturing. *Proceedings of the Institution of Mechanical Engineers, Part B: Journal of Engineering Manufacture*, 2005, **219**(5), 395-412.

[3] **Harding, J.A., Shahbaz, M., Srinivas, S.** and **Kusiak, A.** Data mining in manufacturing: A review. *Journal of Manufacturing Science and Engineering*, 2006, **128**, 969-976.

[4] **Kusiak, A. and Smith, M.** Data mining in design of products and production systems. *Annual Reviews in Control*, 2007, **31**(1), 147-156.

[5] **Pham, D.T.** and **Afify, A.A.** Clustering techniques and their applications in engineering. *Proceedings of the Institution of Mechanical Engineers, Part B: Journal of Engineering Manufacture*, 2007, **221**(11), 1445-1459.

[6] **Han, J.** and **Kamber, M.** *Data Mining: Concepts and Techniques*, 2nd Edn, 2006 (Academic Press, San Diego, California, USA).

[7] **Liao, S.-H., Ho, H.-H.** and **Lin, H.-W.** Mining stock category association and cluster on Taiwan stock market. *Expert Systems with Applications*, 2008, **35**(1-2), 19-29.

[8] **Chen, M.-C.** and **Lin, C.-P.** A data mining approach to product assortment and shelf space allocation. *Expert Systems with Applications*, 2007, **32**(4), 976-986.

[9] **Song, H.S., Kim, J.K.** and **Kim, S.H.** Mining the change of customer behavior in an internet shopping mall. *Expert Systems with Applications*, 2001, **21**(3), 157-168.

[10] **Lazcorreta, E., Botella, F.** and **Fernández-Caballero, A.** Towards personalized recommendation by two-step modified Apriori data mining algorithm. *Expert Systems with Applications*, 2008, **35**(3), 1422-1429.

[11] **Song, Q., Shepperd, M., Cartwright, M.** and **Mair, C.** Software defect association mining and defect correction effort prediction. *IEEE Transactions On Software Engineering*, 2006, **32**(2), 69-82.

[12] **García, M.N.M., Román, I.R., Peñalvo, F.J.G.** and **Bonilla, M.T.** An association rule mining method for estimating the impact of project management policies on software quality, development time and effort. *Expert Systems with Applications*, 2008, **34**(1) 522–529.

[13] **Klemettinen, M., Mannila, H.** and **Toivonen, H.** Rule discovery in telecommunication alarm data. *Journal of Network and Systems Management*, 1999, **7**(4), 395-423.

[14] **Georgii, E., Richter, L., Rückert, U.** and **Kramer, S.** Analyzing microarray data using quantitative association rules. *Bioinformatics*, 2005, **21**(1), ii123-ii129.

[15] **Braha, D.** *Data Mining for Design and Manufacturing: Methods and Applications*, 2001 (Kluwer Academic Publishers, Boston, Massachusetts, USA).

[16] **Wang, K., Tong, S., Eynard, B., Roucoules, L.** and **Matta, N.** Review on application of data mining in product design and manufacturing. In: *Proceedings of the 4th IEEE International Conference on Fuzzy Systems and Knowledge Discovery*, 2007, Vol. **4**, 613-618.

[17] **Shahbaz, M., Srinivas, S., Harding, J.A.** and **Turner, M.** Product design and manufacturing process improvement using association rules. *Proceedings of the Institution of Mechanical Engineers, Part B: Journal of Engineering Manufacture*, 2006, **220**(2), 243-254.

[18] **Agrawal, R., Imielinski, T.** and **Swami, A.** Mining association rules between sets of items in large databases. In: *Proceedings of the ACM SIGMOD Conference on Management of Data*, Washington DC, USA, 1993, 207-216.

[19] **Agrawal, R.** and **Srikant, R.** Fast algorithms for mining association rules in large databases. In: *Proceedings of the 20th International Conference on Very Large Data Bases*, Santiago de Chile, Chile, 1994, 487-499.

[20] **Agrawal, R., Mannila, H., Srikant, R., Toivonen, H.** and **Verkamo, A.I.** Fast discovery of association rules. In: *Advances in Knowledge Discovery and Data Mining* (Eds U.M. Fayyad, G. Piatetsky-Shapiro, P. Smyth and R. Uthurusamy) 1996, pp. 307-328 (AAAI Press, Menlo Park, California, USA).

[21] **Khalid, H.M.** and **Helander, M.G.** A framework for affective customer needs in product design. *Theoretical Issues in Ergonomics Science*, 2004, **5**(1), 27-42.

[22] **Jiao, J.** and **Zhang, Y.** Product portfolio identification based on association rule mining. *Computer-Aided Design*, 2005, **37**(2), 149-172.

[23] **Jiao, J., Zhang, Y.** and **Helander, M.** A Kansei mining system for affective design. *Expert Systems with Applications*, 2006, **30**(4), 658-673.

[24] **Chen, W.-C., Tseng, S.-S.** and **Wang, C.-Y.** A novel manufacturing defect detection method using association rule mining techniques. *Expert Systems with Applications*, 2005, **29**(4), 807-815.

[25] **Buddhakulsomsiri, J., Siradeghyan, Y., Zakarian, A.** and **Li, X.** Association rule-generation algorithm for mining automotive warranty data. *International Journal of Production Research*, 2006, **44**(14), 2749-2770.

[26] **Da Cunha, C., Agard, B.** and **Kusiak, A.** Data mining for improvement of product quality. *International Journal of Production Research*, 2006, **44**(18), 4027-4041.

[27] **Lau, H.C.W., Ho, G.T.S., Chu, K.F., Ho, W.** and **Lee, C.K.M.** Development of an intelligent quality management system using fuzzy association rules. *Expert Systems with Applications*, 2008, In press.

[28] **Chen, M.-C.** Configuration of cellular manufacturing systems using association rule induction. *International Journal of Production Research*, 2003, **41**(2), 381-395.

[29] **Chen, M.-C., Wu, H.-P.** and **Lin, C.-P.** A data mining based clustering approach to group technology. In: *Proceedings of the 2003 IEEE International Conference on Robotics and Automation*, Taipei, Taiwan, 2003, 3554-3558.

[30] **Liu, C.** A data mining algorithm for designing the conventional cellular manufacturing systems. In: *Proceedings of the 20th Australian Joint Conference on Artificial Intelligence*, Gold Coast, Australia, 2007, 715-720.

[31] **Mahamaneerat, W.K.** and **Shyu, C.-R.** Domain-concept association rules mining for large-scale and complex cellular manufacturing tasks. *Journal of Manufacturing Technology Management*, 2007, **18**(7), 787-806.

[32] **Agard, B.** and **Kusiak, A.** Data mining for subassembly selection. *Journal of Manufacturing Science and Engineering*, 2004, **126**, 627-631.

[33] **Chen, M.-C.** and **Wu, H.-P.** An association-based clustering approach to order batching considering customer demand patterns. *Omega*, 2005, **33**, 333-343.

[34] **Jain, V., Benyoucef, L.** and **Deshmukh, S.G.** A new approach for evaluating agility in supply chains using fuzzy association rules mining. *Engineering Applications of Artificial Intelligence*, 2008, **21**(3), 367-385.

[35] **Liu, D.-R.** and **Ke, C.-K.** Knowledge support for problem-solving in a production process: A hybrid of knowledge discovery and case-based reasoning. *Expert Systems with Applications*, 2007, **33**(1), 147-161.

Innovative Production Machines and Systems
D.T. Pham, E.E. Eldukhri and A.J. Soroka (Eds.)

Design of an expert information system for decision-making in the textile industry

I. Goicoechea[a]

[a] *Department of engineer design, University of Vigo, ETSIngeneiros Industriales. Vigo. Spain*

Abstract

The overarching goal of this study is to create an expert information system for decision-making in the textile industry. Based on "Knowledge Engineering":
- Knowledge of the design of a new product,
- Knowledge of the current manufacturing process,
- Knowledge of how the machinery currently operates,
- Knowledge of the machinery's flexibility,
- Knowledge of machine times
With this expert system prototype we can provide a useful information tool for the designer who wants to introduce new products or make changes to its current configuration. It can help him make decisions in this design tasks and will provide him at all times information on the repercussions his design modifications will have on his production system.

Keywords: *Knowledge Management, expert system*

1. Introduction

The incessant evolution in design within the textile sector within an increasingly demanding market leads us to reflect on the new designs that emerge every season in the launch of new products and their consequences for the manufacturing process. What is more, it is widely known that delivery deadlines for a new campaign to launch new products in the textile sector are irrevocable, as long as a company wants to survive in the market, of course.

Even a minor change in the type of seam, colour, pocket shape, etc. entails changes in the current production process, which should be able to adapt quickly and easily in order to meet quality standards and delivery deadlines, and of course achieve success once again in each launch season. Likewise, these changes in the manufacturing process have to take place in such a way that they do not jeopardise either the quality of the end product or the factories' productivity. By examining how these new designs affect production and how these changes are addressed, which to date have been anything but flexible and have required superhuman efforts, the goal is to design an expert decision-making system in the initial stages of product design in an attempt to intertwine the planning, technical and sales functions that are currently separated. These improvements can be achieved with much greater ease, and thus with greater yield on the investment, by using knowledge-based systems than with the systems currently in use. Particularly promising is the use of expert systems to address problems that require interdisciplinary experience. *Chang Kyu (1996)*

2. Development of the study

To develop this study, exploratory research was conducted, applying bibliographic survey techniques and strategies to locate, identify and access the sources of information provided by the creation of an information tool or system for designers which is needed when taking decisions in the design phase.

The methodology used to develop this decision-making model begins with an analysis of the product design and the process, as well as the dependency between the two. It continues with the search for a tool for the designer, so that he or she can always be apprised of the repercussions that a change in design will have on production.

To achieve this, we have scanned different theories applicable to production, from which we have gleaned the following as the best alternative: a knowledge-based expert system (Gómez, 1997) that is capable of diagnosing the problems that might arise in production before making a change in design and the possible measures to adopt, even including perhaps retouching the design so that the changes in production are not so drastic.

3. Development of a product in the textile sector. Current issues.

In the textile sector, developing a product varies slightly in terms of a traditional product's life cycle, and thus in all the times in each of the stages in this life cycle.

There are two product launch campaigns each year: summer and winter. Approximately four months elapse from the start of product design (choosing fabrics, trends, etc.) until the first prototypes are manufactured (which serve as a point of departure for representatives to sell).

After that, two months tend to elapse between when the representatives sell these products to the customers (acquiring purchase commitments from customers) and when the company's Department of Process Engineering adapts the process to the new products, with all the problems this entails: adapting machinery, purchasing new machinery, redistributing the process (layout), optimising the workforce. All of this is done with the goal of achieving high productivity rates and thus lowering the cost of production as much as possible.

Next comes the automatic manufacturing stage for the first garments to be delivered to customers' stores, which tends to last four to five months. The most

Deflection data for beams under vertical loading[a]

Material	Deflection x (mm)	Deflection y (mm)
Concrete	0.123	0.524
Steel	0.145	0.246
Composite	0.133	0.415

critical period is these two months from when the prototypes leave the factory to when the best way to mass produce them is found. In some cases, there may be part of the design that is not "manufacturable" in mass quantities. That is, anything can be manufactured, but managing to "manufacture" garments productively is not always so easy.

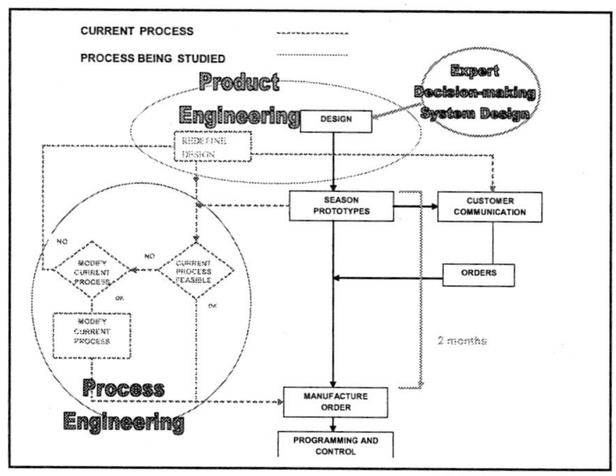

Fig. 1 schema of the current situation of a typical company in the textile sector in its new product launch stage.

4. Design

In order to design this system, we part from the information gathered from the expert opinion from the people of production. A series of production data from machines have been collected such as limitations encountered and solutions found resulting from years of experience. All this information has been converted into simple rules that are easy to understand and easy for the user to interact with them, i.e. the designer. In this way, the vast majority of data and features are simply written in Boolean form. An important aspect of a knowledge base is its capacity, expressed as a number of rules it contains. (Goicoechea, 2005)

Data Structure: Data is structure in the following manner:

Models

- The starting point are the models
- Each model can have various components (integrating parts of a model)
- Each component can have various runs (machines through which a component travels until it is ready)
- Runs comprise all the machines to be used and the corresponding adjustable parameters to be set on each machine
- Each model is dependent on other models. These dependencies are automatically generated and the user can vary the order of appearance in a screen modifying the weight of said dependency. Dependencies of lesser weight appear at the end.

Machines

- On the other hand, there are machines with their own parameters

Expert System

The expert system consists in an automatic search of dependencies between different models. Each time a run is modified, the programs find possible dependencies with other models and updates its list of cross-dependencies.

The user is able to interact with the expert system indicating which dependencies are more important (have more weight) for making future decisions. To this end, the user selects the dependency in question and modifies its weight and saves the change. Weights can go from 0 to 100.Larger weight dependencies leave the list first. (Goicoechea, 2006)

This system is thought mainly to help the designer in quick and accurate diagnostic tasks for determining most common failures from a change in design and how it is going to affect production. It is not intended to replace production technicians in the organization nor time improvement officers, but to help them in their tasks. This is how helping them know how a change in design affects production, helping them make improvements in the design, and redesign once the production problems are known and helping them make decisions regarding production costs.

There will sure be problems for witch the system has no solution, and therefore it will be important to make periodical updates on the knowledge data base for the system to be really useful. This is important because it gathers experiences from expert production technicians and because the diagnostic system is made up of a knowledge base that gathers all the information pertaining to production.

5. Summary

With this expert system prototype we can provide a useful information tool for the designer who wants to introduce new products or make changes to its current configuration. It can help him make decisions in this design tasks and will provide him at all times information on the repercussions his design modifications will have on his production system.

As a starting point, this prototype enables the designer to be aware of how important are his decisions and how they affect production. As the design is modified, the designer can see how the runs are modified too (runs are the set of machines through which the products travel) and can see how they are affected by each adjustable and critical parameters. Thus he can make modifications that may escape to the Client's eye but that are necessary for the feasibility of the new design (i.e. for manufacturing the new product with a good production yield).

6. Future developments

Since this is a prototype to check the need and feasibility, we can highlight that the program can be fully developed and extended to cater for all of the Company's products. Further research can be done in useful industrial applications:

- Relating the runs of the product along all the machines with production times of the new productive system.
- Knowing at every moment the level of occupation of the existing machine and detect when creating a new product if any machine is going to have a bottle neck and thus take action before problems occur in production.
- Also in regard with production times, it could be known in anticipation the production costs and therefore know the profit of each product and/or estimate the price and feasibility of the new design.
- The tool is also intended to be user friendly for designers to scroll through the visual screens with existing models so they can click on the objects and make modifications. The area clicked is zoomed in and the runs of the machines are shown together with critical parameters and the possibility to include new machines in an easy manner (all by the use of images).

Other plants of the Company as well as provider are also intended to be included in the tool. All we need to know it their production system (their know-how) and to reach an agreed quality system. We could do further research into this area and step into Concurrent Engineering. (Aguayo, 2003)

7. References

[1] *Chang Kyu & Dae Lee* (1996). *Knowledge-base construction of a garment manufacturing expert system.* International Journal Clothing Science and Technology, v.8 Nº. 5, 1996, p. 11-28. MCB Univer. Press, 0955-622.

[2] Gómez, A.; Juristo, N.; Montes, C. & Pazos, J.(1997): *Knowledge Engineer*. Study Ecole. Ramón Areces, S.A, Madrid, 1.997. ISBN: 8420541923

[3] Aguayo, F. & Soltero, R. (2003). Industrial Design Metodology. *By Concurrent engineer. Ed.* Rústica. ISBN: 8478975322

[4] Goicoechea, I (2005). *Design of an expert information system for making decisions in the textile sector.* Proceedings of XVII Joint International Congress Ingegraf – XV ADM. Ingegraf 2005. ISBN.: 84-96377-41-5. Congress Minute Book. 1-3 June 2005, Sevilla, Spain

[5] Goicoechea, I (2006). *Design of an expert information system for making decisions in the textile sector*). Proceedings of International Congress on Project Engineering. 13-15 September 2006, Valencia, Spain.

Innovative Production Machines and Systems
D.T. Pham, E.E. Eldukhri and A.J. Soroka (Eds.)

Enterprise Wide Knowledge Leverage

Ercan Öztemel [a] and Seher Arslankaya [b]

[a] *Marmara University, Engineering Faculty, Industrial Engineering Department.*
[b] *Sakarya University, Engineering Faculty, Industrial Engineering Department.*

Abstract

The main aim of knowledge management models is to organize the knowledge activities that are necessary to increase competitive advantage and turn the market share into a continuous and permanent superiority as well as to utilize the intellectual capital of the enterprise for the clients, products and the services. With knowledge leverage activities, knowledge sharing contributes to the transformation of individual knowledge into institutional knowledge. This paper presented an approach so called EKKM which is developed for knowledge management. Knowledge leverage is one of the key components of the proposed model. This paper presented knowledge leverage issues and possible activities for both internal and external leverages.

Keywords: Knowledge management, information systems, knowledge leverage

1. Introduction

As managing knowledge is now as much important as generating it, the knowledge management activities are becoming more attractive than ever before, especially in transferring the society from industrial to knowledge era. Note that knowledge management includes creating, storing, implementing, improving, sharing and recording the knowledge as well as increasing the knowledge capital to possible extend. One of the most important aspects of knowledge management is therefore to establish some mechanisms and methodologies to allow the right knowledge to be at the right place with the right person at the right time within the enterprises. To achieve this, it is necessary to manage the intellectual capital of the enterprises as much effectively as possible. There have been several models developed for this purpose (see for example [1-4]). While some of these models have been intensifying on using technology and producing knowledge, the others focus on the use of knowledge itself. It is now obvious that a successful knowledge management clearly requires a set of activities from defining the processes such as searching producing and updating the knowledge up to establishing a knowledge culture within the enterprise including the evaluation and assessment which will trigger improvement activities. One of the other important aspects of these models is that they are designed to be utilized mainly at operational level pointing out the importance of managerial participation. It would obviously be seen that there is a need for a model which will include strategic, tactic and operational level knowledge management activities. In order to fulfill this requirement, an extensive Enterprise Knowledge Management Model (EKKM) was developed [5]. Note that, the main motivation behind this model is to improve a strong and healthy organizational structure for the enterprises. This paper mainly focuses on knowledge leverage activities which is part of the overall EKKM model. However, for the sake of completeness, EKKM is briefly explained prior to that.

2. Enterprise Knowledge Management Model

Fig. 1 indicates basic components of the proposed model, EKKM. The model is developed on the basis of the following facts. It is now obvious that a successful knowledge management clearly requires a set of processes such as searching for knowledge, producing the knowledge and updating it etc. An effective knowledge infrastructure is not enough for just handling of all of these processes. It is also necessary to create knowledge representation schemes so that the knowledge is understood by all involved. Using and planning knowledge are also the most important parts of knowledge management model proposed. In addition to this definition, knowledge management strategies and related organizational structure are required to be defined for successful implementation. This will assure that the knowledge management activities are considered at an enterprise level.

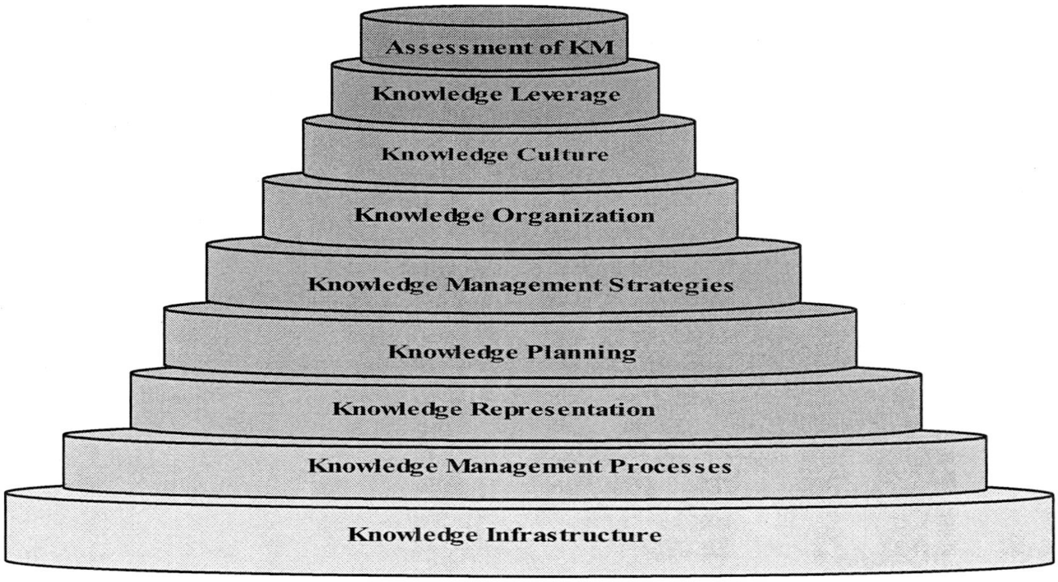

Fig. 1: Enterprise Knowledge Management Model (EKMM)

Knowledge management activities should be considered as way of life in a successful enterprise. This clearly indicates the need for a knowledge culture to be developed. This in turn increases the knowledge sharing capabilities in such a way that one increases the other's knowledge. This is called as "***knowledge leverage***". Above all of these, the model proposed to assess the knowledge management and periodically evaluate them in order to make sure that activities required are properly carried out. By this assessment, the respective problems can be identified and necessary improvements can be realized. The proposed model is also called as "*knowledge tower*" as elements of the model can be considered baseline for each other as depicted in Fig 1. As can be seen from the Figure the model consists of the following elements for which the infrastructure is the foundation of all of the other knowledge related activities to be carried out effectively.

1. Knowledge Infrastructure,
2. Knowledge Management Process,
3. Knowledge Representation,
4. Knowledge Planning,
5. Knowledge Management Strategies,
6. Knowledge Organization,
7. Knowledge Culture,
8. Knowledge Leverage,
9. Assessment of Knowledge Management.

Oztemel and Arslankaya (2007) provides detailed explanation of these components and their relationship [6]. As the main focus of this paper is on "knowledge leverage", the following section of the paper concentrated on this component.

3. Knowledge Leverage

EKKM clearly makes sure that the organization has the ability to measure the contribution of knowledge to competence. The availability of knowledge inputs needed by individuals to perform tasks is guaranteed in quantitative terms. The organizations can use performance measures to improve content management and technology infrastructure. It is therefore possible to measure knowledge Return Of Investment (ROI) through making it integral to decision-making process [7]. However, when making decision, every decision maker can only handle the problem through his/her knowledge already gained through experience. It is vital importance to make sure not only for all decision makers but also all stakeholders to possess enough knowledge in order to grasp the ideas and rational behind the decisions. This can only be provided through a set of knowledge sharing and systematic knowledge transferring activities which can be called as ***knowledge leverage***. In other words, knowledge leverage can be defined as helping others to improve his or her self knowledge. This may make sure that the level of knowledge and its utilization can be synchronized within the enterprise. Moreover, this may obviously prevent delays in decision making and increase the level of contributions off all stakeholders to decision making process. As all people will be understanding each other and using the same knowledge language this will also increase the effectiveness as well as efficiency in defining or utilizing the right knowledge.

EKMM highlights the importance of employees having the same level of knowledge for being in a cooperative environment and performing team working. In most of the case, this is not possible due to different responsibilities and different level of expertise possessed. EKMM suggests a set of activities in order to prevent lack of knowledge of the staff increasing their capability of making joint activities. It also highlights the importance of customer being knowledgeable in the areas of business as well as internal processes which are usually not open to outside organization. Increasing the knowledge of customer is therefore an essential activity within EKMM. Several knowledge leverage activities including brain storming, forming focus groups, defining proper communication channels, creating a sustainable of knowledge transferring and sharing environment, establishing problem solving teams, holding customer training programs, performing walkthrough solutions together with customers, creating knowledge banks and making the knowledge available to the staff as well as customers etc. are recommended. Knowledge leverage activities can therefore be in two folds including;

1. Internal knowledge leverage,
2. External customer knowledge leverage.

3.1. Internal knowledge leverage

Leverage for internal customers generally covers leverage activities carried out on the employees of the enterprises. EKKM recommends the following activities for this purpose. It should be noted that these activities could be carried out in any organization for problem solving. However, the main motivation of these activities should not only be to solve the problems but also to make others to be capable of solving similar problems as well. The main focus of these activities should therefore be on increasing problem solving capabilities rather then eliminating the problems at all. The activities should start by clearly indicating this objective and all should expect to deliver their knowledge to others or to accept other's knowledge to increase the experience and capability for competency.

• **Collocating on the job training activities for the purpose of leverage**: The most efficient way to enhance the knowledge of institutional employees is the intra-institutional "on-the job" training. Training of employees in this way can be provided by own staff. In other words, an employee having much knowledge and being more experienced in a specific area can share his/her knowledge with other employees. Since the employees sharing the same kind of problems can understand themselves better than anyone else, the training for this purpose may create extensively useful knowledge. It can also increase existing knowledge even for the trainees. The enterprises should carry out these training activities continuously in order to increase the level of knowledge of the employees rather then to show them how to do their job.

• **Holding brain storming sessions for the purpose of leverage:** Among the most basic elements of knowledge leverage is to establish brain storming activities among the employees and also making this an integral part of the organizational culture. The brain storming activities allows everybody to clear up their opinions without any hesitations and to evaluate different opinions which can in turn creates enterprise level knowledge. Brain storming activities also makes it possible for the employees to believe in what has to be done in order to overcome certain type of problems. This will also make sure that the employees will seek for extra knowledge by themselves in order to be more active in the sessions. It is also important to note one more time that, the brain storming activities should be carried out in the sense of establishing capability and experience rather then solving certain problems.

- **Creating leverage focus groups:** Besides the realization of daily and routine events within the enterprise, the formation of joint working groups called *focus groups* is another way of efficient leverage. This may help employees to perform suggestion of their expectations about different subjects and the transformation of their perspectives and experience into organization wide knowledge. Focus groups can be formed in various spheres such as the development of new products and processes, the organization of customer relations and the evaluation of the effects of current affairs on the institution etc. It is important to note that there should not be any restriction in formation of the focus groups. That means, it should be possible to establish focus groups in every area where knowledge leverage is essential and beneficial. Additionally, the working time of focus groups should not be restricted and they should last until the employees are fully satisfied and convinced in receiving the related knowledge. If this implementation produces a knowledge incentive organization then these focus groups should be motivated and encouraged by the top management.
- **Conducting leverage oriented workshops:** Another method of knowledge leverage can be the realization of workshops. That is to create 4 or 5 hours joint working and discussion forum for intra-institutional and extra-institutional specialists and people concerned in specific issues especially in the areas where respective knowledge is lacking. It is of great importance to share these results by the people concerned as well As other related parties within the enterprise.
- **Defining proper communication channels:** For an efficient knowledge management, communication channels should be established in such a way that all employees will have a chance to share their knowledge with others without any procedural barriers. Thanks to the developments in informatics and communication technologies as the storage and transfer of knowledge is facilitated technically with no problem. Communication can become quicker and more reliable. However, creating info channels without proper knowledge exchange schemes may not produce benefits in terms of knowledge leverage. Knowledge sources may be identified, respective knowledge should be updated and continuous maintenance should be granted for the successful leverage.
- **Organizing environment in which knowledge is going to be transferred:** Another important element of knowledge leverage is the easy transferability of knowledge. The transfer of knowledge can be performed not only unilaterally but also bilaterally. Through continuous transfer of knowledge possessed by an organization or the employees, knowledge grows continuously and new knowledge is produced. By this, employees transfer their own knowledge to others who is lacking that knowledge. Likewise, in cases when their knowledge is insufficient, they can enhance their own level of knowledge by others' knowledge provided through leverage sessions. Furthermore, knowledge can be transferred not only within the enterprise but also outside the enterprise as well. With the improvement of knowledge related technologies, the transfer of explicit and tacit knowledge is facilitated. These technologies accelerate the storage of knowledge and facilitate access to it.
- **Creating walk thorough teams:** Similar to the above methods, organizations can establish teamwork for solving problems together with the experts and focusing on the knowledge required to solve the problems and transferring that to the solvers. The systematic and continuous development of problem-solving capability can be assured through improvement teams. No need to mention that the support and encouragement of these teams by managers within the institution are essential.
- **Establishing specific knowledge leverage sessions:** Holding informative meetings for the enhancement, renewal and updating of knowledge of employees at certain intervals in the institution, finding out in which fields individuals have deficiencies and the formation of recommendations

to eliminate these deficiencies will also enhance the efficiency of knowledge leverage activities.

• **Providing back- up personnel:** For any possible extraordinary cases, there should always be alternatives of employees in the organization. In other words, when an employee is out of work for any reason, there should definitely be a back-up staff to be able to substitute him/her so that the business won't be disrupted. Note that, the staff backup in the institution contributes to institutional knowledge leverage. In order to facilitate this a rotational working scheme can be beneficial.

3.2. External knowledge leverage

Besides enhancing the knowledge level of their own employees, enterprises have to enhance the knowledge levels of their customers as well. EKKM recommends mutual understanding and cooperation with the customers. This ma lead to success in customer's knowledge leverage. The acceptance of recommendations by both parties as a result of leverage activities is important. Similar to internal leverage activities, leveraging can not be considered for only solving problems. It should definitely be focusing on increasing knowledge level of customers to be capable of grasping business related problems and defining solution procedures or understanding the way that the company is solving problems in order to get used to business related issues. Keeping this in mind, EKKM recommends the following activities.

• **Defining proper communication channels**: It is essential to establish a continuously-operating communication medium to communicate with customers in the most efficient way. Knowledge leverage is not a one time activity. Continues communication is essential.

• **Forming environment in which knowledge is going to be transferred**: The specification related to knowledge transfer to the customer who will be leveraged should define the media enabling an unproblematic exchange.

• **Collocating training activities in service**: Training of customers as well as their employees enables both to speak the same language and approach problems and events in the same manner. Therefore, customer training is one of the most important elements of external knowledge leverage.

• **Defining a way of solving the problems together with customers**: Enterprises have difficulty in making their customers to be aware of business specific problems. Forming joint problem-solving teams with customers may enable customers to realize the problems and see the intention of the organization to solve them. This surely increases customer satisfaction as well as situation awareness of the customers which could be good approach for leveraging.

• **Holding specific brain storming sessions with the customers:** In achieving institutional and sector development, holding joint brain storming activities with customers can be regarded as an important means for customers and institutional employees to share their knowledge and for mutual enhancement of their knowledge. Similarly, these brain storimng activities should directly address respective knowledge to be transferred in both ways rather than only solving the problems.

• **Establishing specific knowledge leverage sessions with the customer:** Like internal leverage, external leverage sessions can be held with customer in order to increase the existing knowledge on specific issues. The areas where the customer having lack of knowledge can be determined and related activities are carried out to eliminate the gap encountered.

• **Establishing knowledge warehouses available to customers:** The provision of the continuation of customer's knowledge is as important as its enhancement. It is important to develop a structure which will continuously provide the customer with the required knowledge. This can be satisfied with a knowledge warehouse where the customer can easily reach for any knowledge needed for him to make his decision related to products and services. This may surely prevent misunderstandings and misconception of the business rules as well as other business related issues.

• **Carrying out activities concerning the possible future needs of the customer and sharing those with them:** Carrying out activities and making estimations for the possible future needs of customers enable both the enhancement of customer's knowledge level and directing the enterprise to develop related strategies. Enterprise and customer can share their vision for the future which may produce mutual interest as well as benefits.

• **Evaluating corporate sector together with the customers and to form focus groups:** The evaluation of developments and the exchange of opinions concerning the developments in the joint sector with the customer is an important pillar of customer leverage. Enterprises may read the market in different ways and may concentrate on different aspect of the market which may help the customers to sustain competitive advantage.

4. Conclusion

Today, when transformation from industrial society to knowledge society is being experienced, knowledge and the importance of knowledge have increased more than ever before. The enterprises which reveal their intellectual accumulation of knowledge and use it properly become more successful and are able to maintain their competitive advantage. Since the efficient use of knowledge and the provision of its production of value are as important as the specification of the related knowledge, it is necessary to continuously monitor, evaluate and improve the activities of knowledge management. As it is obviously realized, the evaluation of knowledge management actually requires carrying out activities on many subjects which cover quite a wide spectrum and which vary from organizational change to technological innovation. Sharing knowledge and experience within the institution and among the shareholders of the institution is of importance so as to increase the knowledge level of individuals as much as the activities of knowledge management. With knowledge leverage activities, this sharing contributes to the transformation of individual knowledge into institutional knowledge. This paper presented an approach so called EKKM which is developed for knowledge management. Knowledge leverage is one of the key components of the proposed model. This paper concentrated on knowledge leverage issues and possible activities for both internal and external leverages are highlighted. With knowledge leverage activities, it is aimed at preventing lack of knowledge as well as related insufficiency and that are hidden within the organization and preventing the success.

References

[1] O'Leary, D. E., "A multilingual knowledge management system: A case study of FAO and WAICENT", Decision Support Systems 45 (2008) 641–661, Available online 27 July 2007 www.elsevier.com/locate/dss

[2] Malhotra, Y., "Integrating Knowledge Management Technologies in organizational Business Process: Getting Real Time Enterprises to Deliver Real Business Performance", Accepted for publication in the journal of knowledge Management (Emerald) Special Issue on 'Knowledge Management Technology', Q3, 2004.

[3] Lin, C., Yen, C. D., Tarn, D.D.C., "An industry-level knowledge management model—a study of information-related industry in Taiwan", Information & Management 44 (2007) 22–39, www.elsevier.com/locate/im, 2007

[4] Davenport, Th.H. and L. Prusak (2000): working knowledge: how organizations manage what they know. Harvard Business School Press (paperback edition), In: Proceedings of Organizational Knowledge, Learning and Capabilities, Innsbruck, April 2004. http://infolab.uvt.nl/pub/smitsm-2004-82.pdf (Available on 02.02.2006)

[5] Arslankaya S. "Enterprise Knowledge Management Model", Unpublished Phd Dissertation, Sakarya University, Institute of Natural Sciences, 2007

[6] Öztemel, E., Arslankaya, S., " Etkin Bilgi Yönetimi Kriterleri ve Etkinlikleri", 27. YAEM Yöneylem Araştırması/Endüstri Mühendisliği Kongresi, 2004

[7] V P Kochikar, PhD, *Principal Knowledge Manager* Infosys Technologies Limited Electronics City, "The Knowledge Management Maturity Model - A Staged Framework for Leveraging Knowledge", Bangalore 561229, India *Kochikvp@inf.com*, 2007

Innovative Production Machines and Systems
D.T. Pham, E.E. Eldukhri and A.J. Soroka (Eds.)

Process models for a sugar mill crystallisation stage knowledge based supervisory support system

R. Dodd[a], R. Broadfoot[b], X.Yu[a], A. Chiou[c]

[a] *School of Electrical and Computer Engineering, RMIT University
Melbourne, Victoria, Australia
Email: roland.dodd@student.rmit.edu.au, x.yu@rmit.edu.au*
[b] *Sugar Research and Innovation, Queensland University of Technology
Brisbane, Queensland, Australia
Email: r.broadfoot@qut.edu.au*
[c] *School of Computing Sciences, Central Queensland University
Rockhampton, Queensland, Australia
Email: a.chiou@cqu.edu.au*

Abstract

This paper discusses the core process models of a knowledge based supervisory control system designed to provide expert knowledge in the control and management of a sugar mill crystallization stage. Forecasting stock tank levels, within the sugar factory crystallization stage, is beneficial to ensure that there are sufficient quantities of materials in stock to allow completion of production schedules without disruption.

Integration of projected vacuum pan feed rates with syrup rate production models and pan phase detection models allows forecasting of stock tank levels for liquor and molasses to forewarn of potential problems with the current operating strategies and advise corrective procedures. These models form an integral part of an overall pan stage knowledge based system to assist in providing a better decision-making strategy for crystallization stage operations.

Keywords: Decision support, knowledge based systems, process control, crystallisation stage, process models

1. Introduction

Raw sugar production from cane is a nominally continuous operation, with 120-168 hours of processing per week, extending over 20-25 weeks of the harvest season. The crystallisation stage, often loosely referred to as the pan stage, is the most complex part of the factory process where there are several batch wise and continuous crystallisation steps taking place concurrently [1].

The overall strategic management of the pan stage is quite difficult because of the very large number of process streams of varying compositions and crystal growth rate characteristics which must be managed [2].

Often the pan stage is managed in a sub-optimal manner because an overview of operations encompassing various sections - cane receival section, juice processing stations, the pan stage and centrifugal station - is not available. Furthermore, the pressures on the Australian sugar industry to reduce the costs of sugar manufacture and increase the consistency of producing sugar of high quality require a smarter strategy for operation.

This paper reports several core process models in an industrial supervisory support system for a sugar mill crystallization stage. Currently, there is no such supervisory support system for pan stage operations neither in the Australian sugar industry nor, as best as known to the collaborators, in the world sugar industry.

This system will provide a standardised approach for pan operations by integrating key process models and combining them with the collective expertise and knowledge of pan operators [3]. The benefits include a formal structure for decision making, increased productivity, and improved use of the equipment capabilities and resources.

This paper is organized as follows. Section 2 discusses the importance of stock tank level prediction for factory operations. Section 3 presents an overview of the pan stage supervisory control system and explains how the developed process models fit into overall system framework. Section 4 discusses an overview of the models developed and how these work collectively to predict stock tank levels within the pan stage environment. Section 5 then presents a discussion and conclusions.

2. The sugar crystallisation process

The sugar crystallisation process takes place in large batch fed vessels known commercially as batch vacuums pan; in continuous vacuum pans; or in batch and continuous stirred cooling vessels referred to as crystallisers. Typical factories have between six and ten of these pans and about sixteen hours of residence time in the crystallisers. The pan stage has an incoming stream of syrup and intermediate molasses recycle streams. Output products from the pan and centrifugal stages are raw sugar and final molasses.

Intermediate storage tanks are used between various stages of the crystallisation process in both batch operations for temporary storage, and in continuous mode for flow smoothing. These storage tanks for the syrup and molasses process streams act as short-term buffer storage. It is important to realize that these tanks cannot, in the medium to long term, overcome mismatches between various parts of the factory. The levels within these tanks rise and fall to reflect the net material flow at any instant. At the extremes, the tanks become empty or will overflow. Neither of these outcomes is desirable, as they require disruptive remedial actions including the possibility of factory shutdowns until the problem is resolved.

Decisions need to be a clear cut action as a consequence of the batch operation. Decision errors often cannot be corrected and may result in severe disruptions to processing for extended periods. It is important to determine when the levels of particular stocks indicate that a modification in operating strategy is necessary to maximize process throughput.

The existing decision making process is reliant upon operators. Typically pan stage operators and supervisors are required to estimate future tank levels. They intuitively have to account for current stock tank levels and use their knowledge of pan stage processes, along with current and likely future decisions to reach a conclusion on the outcomes of future levels. This problem is further exacerbated by the forward look ahead duration. Pan stage operators will typically try to forward estimate stock tank levels not only for their current work shift but also into the next pan stage operators shift also.

Very limited research carried has been carried out into stock tank level prediction with previous approaches reliant upon stochastic prediction methods [4, 5] for level forecasting.

3. The pan stage knowledge based supervisory support system framework

The pan stage knowledge based supervisory support system provides a standardized approach for pan operations by integrating data from a variety of information sources from different sections of the sugar mill, along with dynamic process models of the pan stage and the collective knowledge and expertise of pan stage operators. The development and integration of the process models leads to a challenge in the design and development of the pan stage expert system. Previous research [3] acknowledges that no conventional software engineering methods exist to provide a solution to this problem.

The system specifications for key pan stage supervisory control support system features require that the system recommendations and expert advice provide four core control strategies [6].

The four core control strategies for the primary system output are:

1. Stock tank management;
2. Pan duty management;
3. Pan control strategy; and
4. Pan schedule management;

Additionally, these core control strategies are supported by a further two secondary system outputs:

5. Prediction of future pan stage operating conditions for system recommendations compared to current operations; and
6. Explanatory and justification capabilities.

A core feature of this system is a predictive mechanism to determine future pan stage operating conditions. The intermittent nature of stock tank level fluctuations means an advisory scheme warning of future levels and corrective strategies to current stock usage is of beneficial use to pan stage operators and supervisors to assist in management and best practices for pan stage operations. The stock tank level predictive models will aid in the stock tank management goals forming part of the primary system control strategies and offer support in the prediction of future pan stage operating conditions.

The stock tank predictive models are a major component of dynamical process models shown in Fig. 1 of the pan stage expert system framework. The pan stage process models are tightly integrated into and work in tandem with the expert system rule base and are key system technologies.

This framework is based upon a hybrid fuzzy logic based expert system design but with some innovative changes in order to merge the dynamic interrelational process models of the pan stage with the fuzzy rule base and to achieve the defined control strategies and support functions. Further details of the pan stage supervisory support system framework is available in [7] with specific software architecture in [8].

4. Pan stage process models

A major challenge in the development of the pan stage supervisory control support system is the development of the dynamic interrelational process models of the sugar mill crystallisation stage to relate the various sections together. An innovative modeling approach has been taken. By developing empirical boil-on rate models for each vacuum pan feed stream

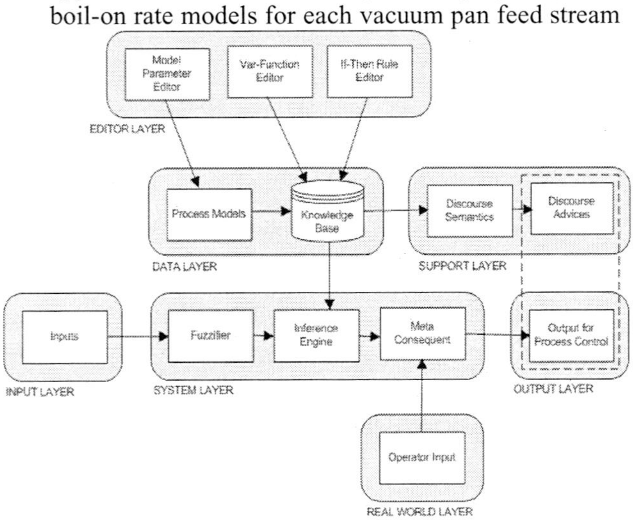

Fig. 1. Smart supervisory control system framework diagram.

through the different stages of its strike, the liquor, A molasses and B molasses feed rates for all the pans at a given point in the schedule can be summated given knowledge of the pan stage schedule. This model takes advantage of the repetitive pan stage schedule and consistency in successive pan strikes.

Once the model for each pan is established then the boil-on rates for each feed stream at the different stages of the pan stage schedule can be determined by summing the liquor, A molasses and B molasses feed rates for all the pans at that point in the schedule. Given the expected syrup production rate and C sugar remelt production rate to the liquor tank during this interval, the predicted tank levels can be determined for the liquor tank.

Similarly the predicted tank levels for the A and B molasses streams can be calculated from the production rates and the molasses at the centrifugal station and the sum of the consumption rates on the individual pans at a specific point in the pan stage schedule.

The problem of stock tank interaction on the pan stage, a mass balance [8] over a generic stock tank yields the governing differential equation as:

$$\frac{dLevel}{dT} = Supply - Demand \qquad (1)$$

where, *Level* is the stock tank level, *Supply* is the process return streams to the stock tank, *Demand* is the process feed streams from the stock tank.

Eq. 1 leads to the following liquor tank model:

$$L_{t+1} = L_t + \int_t^{t+1} S\ dt + \int_t^{t+1} R\ dt - \int_t^{t+1} F\ dt \qquad (2)$$

where, L is the liquor tank level, S is the syrup process stream input, R is the remelt sugar process stream return, F is the syrup feed rates for all vacuum pans, t is a specific point in the overall pan stage schedule.

Eq. 1 also leads to the following generic molasses tank model, with application to both A molasses and B molasses stock tanks:

$$M_{t+1} = M_t + \int_t^{t+1} C\ dt - \int_t^{t+1} V\ dt \qquad (3)$$

where, M is the molasses tank level, C is the centrifugal molasses process stream return, V is the molasses feed rates for all vacuum pans, t is a specific point in the overall pan stage schedule.

The following section describes the liquor tank model implementation of Eq. 2, however the molasses tank model presented in Eq. 3 is quite similar. The molasses tank model however needs to incorporate several other specific models in order to forward predict the centrifugal molasses process return stream. These additional models are not covered in this paper.

In order to realize the liquor tank predictive model presented in Eq. 2 a sequence of sub models are required. Each of these models provides key data in realising the overall stock tank predictive model.

4.1. Syrup prediction model

The syrup prediction model, as established in previous research [7] albeit not implemented, determines the future flow of syrup to the pan stage for bins of sugar cane entering the factory after the harvesting process. The prediction of the quantities of sucrose and impurities in syrup allows a forward forecasting of the future pan stage loading of syrup. The syrup prediction model uses cane receival data combined with juice processing station information and pan stage operator knowledge to forward predict syrup quantities and composition.

This model is of key importance as syrup comprises the basic input to the pan stage with direct feed to the pan stage liquor tank. Given the 96 minute delay from cane entering the factory and being processed to its associated syrup flows to the pan stage, this forms a fundamental boundary on the forward prediction of the overall liquor stock tank prediction model. Further forward prediction beyond this interval uses average syrup production levels with reliance on similar syrup production rates being achieved through continued sugar cane crushing rates and factory processing performance.

A 90 minute forward forecast of sucrose and impurity quantities is presented in Fig. 2 for Racecourse sugar mill (Mackay, Australia) cane data at 11:45pm on 03/09/2003. The syrup production rate is very consistent at approximately 25 tons of syrup quantity being delivered to the pan stage at each 15 minute interval. This information is further used as part of the following 8 hour forecast of liquor tank levels.

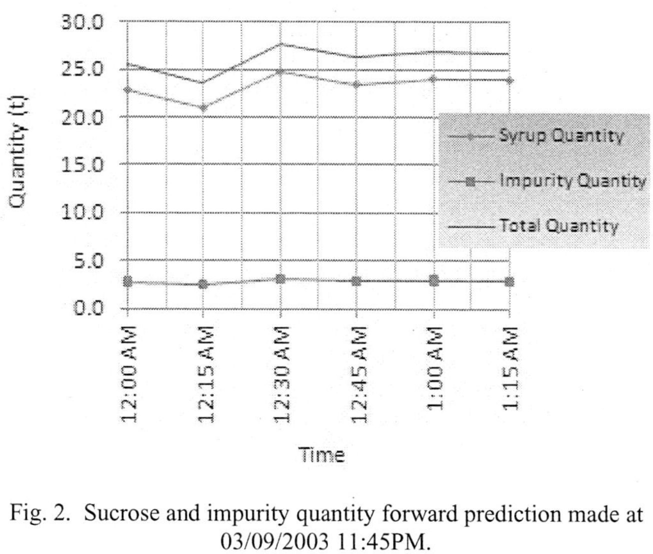

Fig. 2. Sucrose and impurity quantity forward prediction made at 03/09/2003 11:45PM.

4.2. Empirical vacuum pan models

Individual pan production rates have been modelled by constructing empirical relationships [9] for the rate at which each pan takes feed material (liquor, A molasses or B molasses) during the different phases of the pan's operation. Using this method it is possible to construct a piece-wise model of pan feed rate characteristics during each phase of the pan's operation for each of the vacuum pans. This series of models is maintained and stored as series of database records for each of the vacuum pans assigned duties.

For each phase, information on its duration and syrup, A molasses and B molasses feed rates is stored along with required time from the start of the pan strike to when the actual phase commences. An example of the phase sequence for a batch vacuum pan on A massecuite production duties is shown in Table 1. This phase sequencing information for each batch vacuum pan has been derived and implemented from prior research [9].

4.3. Pan phase determination models

This model determines the status of the vacuum pans by determining what phase of the strike each pan is currently in. Such a determination is made using the previous drop time for the pan strike and forward calculating using phase information to determine how much time has elapsed and at what point of what phase the pan is currently operating in. Current pan duties are then determined and the appropriate empirical pan model lookup for the vacuum pan is performed.

Table 1
Empirical pan model sequence of phases example[a]

Time since start of strike (minutes)	Phase duration (minutes)	Syrup feed rate (t/h)	Phase information
0	8	0	Footing
8	26	122.48	Liquor feed
34	16	0	No feed
50	23	109.80	Liquor feed
73	14	0	No feed
87	20	99.99	Liquor feed
107	16	0	No feed
123	25	107.28	Liquor feed
148	56	52.32	Liquor feed
204	18	0	No feed
222	9	0	Before pan drop
231	14	0	Pan drop and preparation

[a]Data specific to Racecourse sugar mill pan number 4 on A massecuite production duties.

With the current operating phase determined, a repeatable sequence of phases that the pan will proceed through during each strike is dynamically generated for the required forward prediction time. This information is then coupled with the previously described empirical pan models to determine the feed materials used during future time periods. This model takes advantage of the repetitive pan stage schedule and consistency in successive pan strikes.

Figure 3 depicts the repetitive nature of strikes for Racecourse sugar mill vacuum number pan 4, which was previously presented in Table 1. Once this schedule of phases is available for each pan strike, a forward forecast of future pan feeds requirements is undertaken. The forward forecast employs a unique mapping component to portion the feed rates into quantities at discrete 15 minute future intervals based upon the overall duration and projected starting time of each phase.

4.4. Stock tank prediction models

Once the model for each vacuum pan has been established through the empirical pan models, the boil-on rates for each feed stream at the different stages of the pan stage schedule can be determined by summing

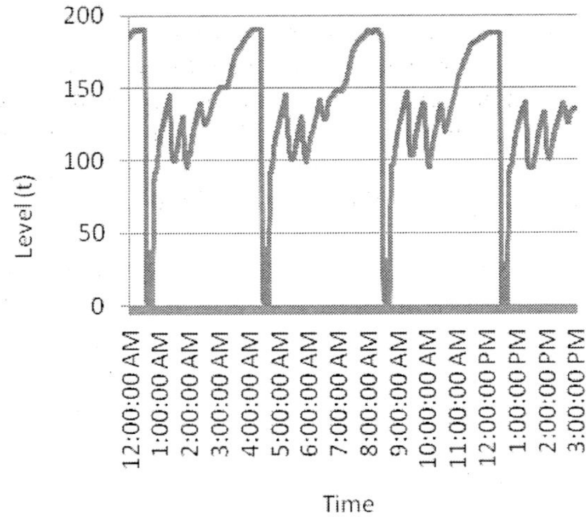

Fig. 3. Racecourse sugar mill pan number 4 performing A massecuite duty strikes on 04/09/2003.

the liquor, A molasses and B molasses feed rates for all the pans at that point in the schedule. Given the expected liquor production rate and C sugar remelt production rate to the liquor tank during this interval, the predicted tank levels can be determined for the liquor tank. An eight hour forward prediction of liquor tank levels has been made at 11:45pm on 03/09/2003 for Racecourse sugar mill using Eq. 2. The results are presented in Fig. 4 against actual liquor tank levels.

5. Discussions and conclusions

The overall strategic management of the pan stage is quite difficult because of the very large number of process streams of varying compositions and crystal growth rate characteristics which must be managed. Often the pan stage is managed in a sub-optimal manner because an overview of operations encompassing factory sections is not available.

This paper details the implementation of process models for integration with a pan stage supervisory control support system for sugar mill crystallization stage management. An innovative series of industrial models work cohesively to perform stock tank level prediction. These models have been presented along with their major features and specific results.

With syrup forming the primary liquor tank input, a syrup predictive model that forward predicts future quantities from cane receival and juice processing station information has been implemented to determine syrup quantities from crushed cane. This information is brought together with sequenced phase information on future pan strikes and a determination of the current operational point in the pan stage schedule. A forward

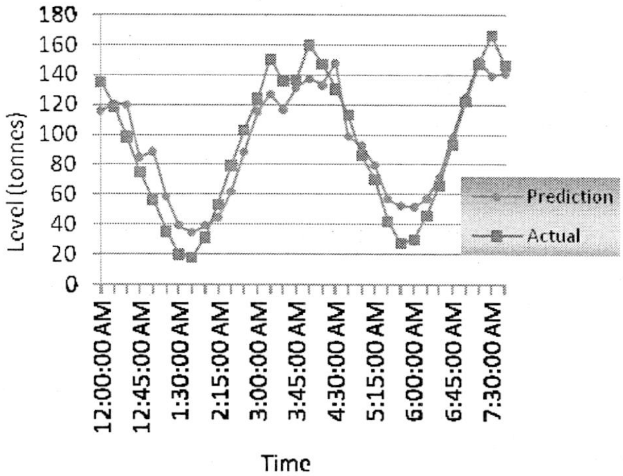

Fig. 4. A full eight hour forward prediction of liquor tank levels made at 11:45pm on 03/09/2003.

determination using boil-on feed material rates from empirical pan models is then implemented and integrated with a schedule of phases to be undertaken by each vacuum pan over the forecast period.

The liquor tank model prediction resulting from this approach generally trends well with actual stock tank levels displaying good performance of the developed models and indicate both the viability of such a method and the merits of the approach.

Overall prediction of stock tank *peaks* and *troughs*, as depicted in Fig. 4, for level prediction are of key significance since normal factory operation is prevented during tank overflow or empty conditions and serious remedial actions are required.

The developed models form a core component of the pan stage supervisory control support system, allowing a forward prediction of stock tank levels at a future point in the pan stage schedule and assisting in forewarning of problems with the current operating strategies through a forecasting of stock tank level and indication of future disturbances to pan stage operating conditions. Early decisions, such as changes to vacuum pan steam rates, or allocation of *swing* pans to different massecuite duties can then be made to avoid production rate difficulties, and maintain good operational performance with respect to pan stage production, quality and recovery objectives.

Acknowledgements

The project contributors would like to thank the staff at Racecourse Sugar Mill, Mackay, Australia for their cooperation and support. This project is supported by an Australian Research Council linkage grant. The funding assistance provided by the Sugar Research Institute, Brisbane, Australia is also acknowledged and appreciated.

References

[1] R. Broadfoot and A. Beath, (1998). "Modeling of batch sugar centrifugal performance", in Proceedings of 26th Australasian Chemical Engineering Conference, 1988, p.294-303.

[2] K. Miller and R. Broadfoot. "Crystal growth rates in high grade massecuite boilings", in Proceedings of the Australian Society of Sugar Cane Technologists Conference, (1997), p.441-447.

[3] R. Dodd, X. Yu, R. Broadfoot and A. Chiou, (2002). "Development of a pan stage expert system for pan stage operations in sugar factories", in Proceedings of Information Technology in Regional Areas, 2002, Central Queensland University, Rockhampton, Australia, p.368-375.

[4] K. Miller, (1987). "Short Term Forecast of Process Tank Levels for a Sugar Mill Crystallisation Stage", Capricorn Institute of Advanced Education, Department of Mathematics and Computing, Rockhampton, Australia. Internal project report.

[5] K. Miller, (1988). "Forecasting Pan Stage Tank Levels", in Proceedings of Australian Society of Sugar Cane Technologists, 1988, Cairns, Australia, p.203-209.

[6] X. Yu and R. Broadfoot, (2001). "Developing a Smart Supervisory Control System for Pan Stage Operations in Sugar Factories", Australian Research Council linkage project application number LP0230710.

[7] R. Dodd, R. Broadfoot, X. Yu and A. Chiou, (2005). "Development of Smart Supervisory Control System in a Sugar Mill Crystallisation Stage", in Proceedings of I*PROMS Virtual International Conference, 2005, p.527-534.

[8] R. Dodd, R. Broadfoot, X. Yu and A. Chiou, (2008). "Framework for a Smart Supervisory Control System for a Sugar Mill Crystallisation Stage", to appear in Proceedings of the 6th IEEE International Conference on Industrial Informatics, Daejean, Korea.

[9] Bureau of Sugar Experiment Stations, The Standard Laboratory Manual for Australian Sugar Mills: Principles and Practices, Vol.1, BSES, Brisbane, Australia, 1984.

[10] R. Dodd, R. Broadfoot, X. Yu and A. Chiou, (2005). "Empirical Pan Modeling of Vacuum Pans for a Sugar Mill Crystallization Stage", in Proceedings of Australian Society of Sugar Cane Technologists, 2005, Bundaberg, Australia, p.423-436.

Innovative Production Machines and Systems
D.T. Pham, E.E. Eldukhri and A.J. Soroka (Eds.)

ENERGY EFFICIENT REAL TIME SCHEDULING OF DEPENDENT TASKS

M. H. Awadalla

Department of Communication and Electronics, University of Helwan, Cairo, Egypt

Abstract

In the design of embedded systems, it is very important to reduce energy consumption, and thus prolong battery life in everywhere existed battery-powered embedded systems. Dynamic voltage scaling (DVS) processors, which support many operating voltages and speeds, can efficiently reduce energy consumption by making appropriate decisions on the processor speed/voltage during the scheduling of real time tasks. This paper addresses the problem of energy efficient real time task scheduling where the tasks are dependent due to exclusive access shared resources. Furthermore, the paper proposes enhancements over the existing multi-speed switching algorithm (MSA) where the proposed algorithm achieves more energy saving, and has the capability to function with both static resource policy (SRP) and dynamic priority ceiling protocol (DPCP).

Keywords: Energy efficient scheduling, dynamic voltage scaling, real time, embedded systems, dependent tasks.

1. Introduction

Currently, embedded systems are involved in most details of our life such as cell phones, pocket PCs, multimedia devices, PDAs (personal digital assistants), digital cameras, and medical implants, which all are battery powered. As the applications on these devices are being complicated, the energy consumption is also effectively increasing. So, minimizing energy consumption is a critical issue in the design of these embedded systems, and techniques that reduce energy consumption have been studied at different levels in details [1]. Dynamic voltage scaling (DVS) is a technique that varies the supply voltage and clock frequency (speed) based on the computation load to provide desired performance with the minimal amount of energy consumption in ubiquitous embedded systems.

The power consumption has two essential components: dynamic and static power. The dynamic power consumption, which is the main component, has a quadratic dependency on supply voltage [1] and can be represented as:

$$P_{dynamic} = C_{ef} \,.\, V^2_{dd} \,.\, F \tag{1}$$

Where C_{ef} is the switched capacitance, V_{dd} is the supply voltage, and F is the processor clock frequency (sometimes referred as speed S) which can be expressed in terms of supply voltage V_{dd} and threshold voltage V_t as following:

$$F = k \,.\, (V_{dd} - V_t)^2 / V_{dd} \tag{2}$$

The static power consumption is primarily occurred due to leakage currents (I_{leak}) [2], and the static (leakage) power (P_{leak}) can be expressed as:

$$P_{leak} = I_{leak} \,.\, V_{dd} \tag{3}$$

When the processor is idle, a major portion of the power consumption comes from the leakage. Currently leakage power rapidly become the dominant source of power consumption in circuits and persists whether a computer is active or idle [2], and much work has been done to address this problem [3]. So, lowering supply voltage is one of the most effective ways to reduce both dynamic and leakage power consumption. As a result, it reduces energy consumption where the energy consumption is the power dissipated over time. However, DVS aims at reducing energy consumption by reducing the supply-voltage/speed of the processor provided that timing constraints are guaranteed. In other words, DVS makes use of the fact that there is no benefit of finishing a real time job earlier than its deadline. DVS processors have two types [1]: *ideal* and *non-ideal*. An ideal processor can operate at any speed in the range between its minimum and maximum available speeds. A non-ideal processor has discrete speeds with negligible or non-negligible speed transition overheads. Another classification defines four different types of DVS systems: *ideal*, *feasible*, *practical*, and *multiple* [4].

2. Related Work

During the last decade much work has been done in the field of energy efficient scheduling, but Weiser et al. [5] are considered the pioneers in that field where they expected the DVS technique, then Yao et al. [6] have proposed an optimal static (offline) scheduling algorithm by considering a set of aperiodic jobs on an ideal processor. Furthermore, many dynamic and static scheduling algorithms [7] have been proposed and applied on uniprocessor systems. Also multiprocessor and distributed systems have been considered [8]. However, the problem of DVS with dependent tasks because of shared resources has been first addressed in [9]. Jejurikar and Gupta [9] have proposed two algorithms for scheduling fixed priority, Rate Monotonic (RM) scheduler, tasks using priority ceiling protocol (PCP) described in [10] as resource access protocol. They have computed static slowdown factors which guarantee that all tasks will meet their deadlines. In their first algorithm, critical section maximum speed (CSMS), they have let the critical sections to be executed at maximum processor speed and they have computed slowdown factors for executing non critical sections. The second algorithm, constant static slowdown (CSS), computes a uniform slowdown factor for all tasks and for all sections (critical and non-critical) saving speed switches occurred in the first algorithm (CSMS). The same authors [11] have then extended their previous algorithms (CSMS and CSS) to handle dynamic priority, Earliest Deadline First (EDF) scheduler, tasks using dynamic priority ceiling protocol (DPCP). The dynamic priority ceiling protocol is an extension of original priority ceiling protocol to deal with dynamic priority tasks (EDF scheduling). Jejurikar and Gupta [12] have also proposed a generic algorithm that works with both EDF and RM schedulers, and they have introduced the concept of frequency inheritance in their algorithm. Zhang and Chanson [13] have worked on the same problem (scheduling of dependent tasks) and proposed three algorithms for energy efficient scheduling of dependent tasks with shared resources over EDF scheduler, where they have used stack resource policy (SRP) as resource access protocol. The SRP can handle static and dynamic priority tasks (EDF and RM schedulers), reduces context switches over PCPs, and is easy implemented. The first algorithm is the same as CSS for EDF scheduler proposed in [10] because they all have derived the static slowdown factor directly from the EDF schedulability test with blocking time in [11]:

$$\forall i, 1 \le i \le n, \quad \frac{B_i}{D_i} + \sum_{k=1}^{i} \frac{C_k}{D_k} \le 1 \qquad (4)$$

The second algorithm is the dual speed (DS) algorithm. The main concept of this algorithm is using two speeds (L, H) and switching between them. Initially the algorithm operates with the low speed L, and switches to the high speed H as soon as a blocking occurs. The last algorithm is the dual speed dynamic (online) reclaiming algorithm which dynamically collects the residue time from early completed jobs and redistributes it to the other pending jobs to further reduce the processor speed and achieve more energy saving. Then the same authors [13] developed their previous algorithms to achieve more energy saving and also to function with RM scheduler in addition to EDF scheduler.

3. SYSTEM MODEL

The periodic task is considered as a sequence of jobs released at constant intervals (called the period). Each task τ is characterized by the following parameters: The release time (r), the period (T), the relative deadline (D), the computation time (C), and the blocking time (B). The tasks are scheduled on a single DVS processor that supports variable frequency (speed) and voltage levels continuously, i.e. DVS processors can operate at any speed/voltage in its range (ideal).

3.1 MULTI-SPEED ALGORITHM

Multi-speed (MS) algorithm is a blocking aware scheduling algorithm with non-preemptible critical sections using SRP as resource access protocol. The

MS algorithm can be considered as an extension of dual speed (DS) algorithm proposed by Zhang and Chanson [14], where the difference between the two algorithms is that MS algorithm uses many speeds (low speed S_L and multiple high speeds S_m where $1\leq m \leq n$) instead of two speeds (low speed L and one high speed H) in DS algorithm. Like DS algorithm, MS algorithm initially starts with the low speed S_L then it switches to one of high speeds as soon as a blocking occurs. The high speed to which the MS algorithm switches is determined according to the blocking task, i.e. each blocking task τ_m has its own high speed S_m, and according to the blocking task, the algorithm switches to the convenient high speed. The low speed S_L, which is exactly the same as low speed L in DS algorithm, is the optimal lowest speed with which all tasks can be scheduled without missing any deadline, and it is derived from the plain EDF schedulability test without shared resources:

$$\sum_{i=1}^{n}\frac{C_i}{D_i}\leq S_L \tag{5}$$

The high speed S_m for a blocking task τ_m is derived as in DS algorithm from the EDF schedulability test with shared resources and SRP protocol:

$$\forall k, 1\leq k < m, \sum_{i=1}^{k}\left(\frac{C_i}{D_i}\right)+\frac{B_m}{D_k}\leq S_m \tag{6}$$

Where the blocking time B_m here is the maximum time (length of critical section of τ_m) through which a low priority task τ_m can block another high priority task due to exclusive access to shared resource and unlike mentioned before, index m refers to the blocking task (low priority task). The above mentioned speeds S_L and S_m have to satisfy the condition:

$$S_L\leq S_m\leq 1 \tag{7}$$

MS algorithm ends the high speed interval when the deadline of the blocking task is reached or the processor becomes idle. An example is implemented to illustrate the MS algorithm and the contribution of this paper. A hard real time system with three tasks is considered as following: τ_1={1 ,4 ,4 }, τ_2={1.5 ,12 ,12 }, τ_3={3 ,24 ,24 }. The arrival times and critical sections of the three tasks within the least common multiple (LCM) of periods are shown in Figure 1(a). According to (6) the low speed is S_L= (1/4+1.5/12+3/24) = 0.5 which represents the processor utilization factor $U=\sum C/T$. There are two blocking tasks in this example: τ_2 that can block higher priority task τ_1 for maximum time B_2=1.5 and

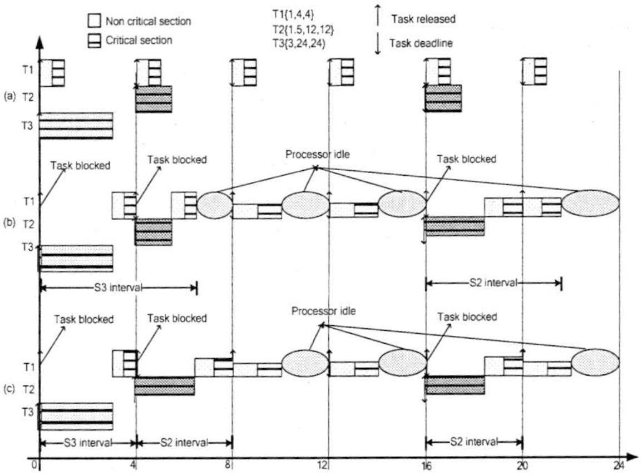

Fig.1 (a) Task set description. (b) MS algorithm. (c) IMS algorithm.

τ_3 that can block higher priority tasks τ_1 and τ_2 for maximum time B_3=3. So, there will be two high speeds (S_2, S_3) according to these two blocking tasks, and these two speeds are computed according to (6), where S_2 = (1/4+1.5/4)=0.625, and S_3 = max (1/4+3/4.1/4+1.5/12+3/12) = 1. The two speeds (S2, S3) also satisfy the condition (7), where $0.5 \leq S_2 \leq 1$ and $0.5 \leq S_3 \leq 1$. The rectangles represent the processing of tasks (jobs) by CPU where the vertical dimension represents the processor speed, and the horizontal dimension represents the execution time elapsed for processing tasks according to their WCETs and the processor speed. It is clearly noted that the area of the rectangles of the jobs of the same task is the same due to that a task always takes the same number of execution cycles which equals to processor speed multiplied by elapsed time. Back to Figure 1(b), τ_3 is released before τ_1 with enough time (ε) to lock the shred resource. When task τ_1 is released, it will be blocked by the lower priority task τ_3 due to exclusive access to shared resource (according to SRP, a task may be blocked when it is released, and as soon as it starts, it can not be blocked). So, MS algorithm switches to the high speed S_3 because the blocking task is τ_3. At time t = 4, when the second job of task τ_1, it is also blocked by the lower priority task τ_2 released before it with enough time (ε) to lock the shred resource. MS algorithm which operates with high speed S_3 switches to the maximum of the two high speeds (S_3, S_2) which is S_3, and MS algorithm ends this high speed interval and switches to the low speed S_L at time t = 6.5 when the processor becomes idle. At time t = 16, when the fifth job of task τ_1 is released, it will be blocked by the second job of task τ_2, and MS algorithm switches to the high speed S_2 and ends this

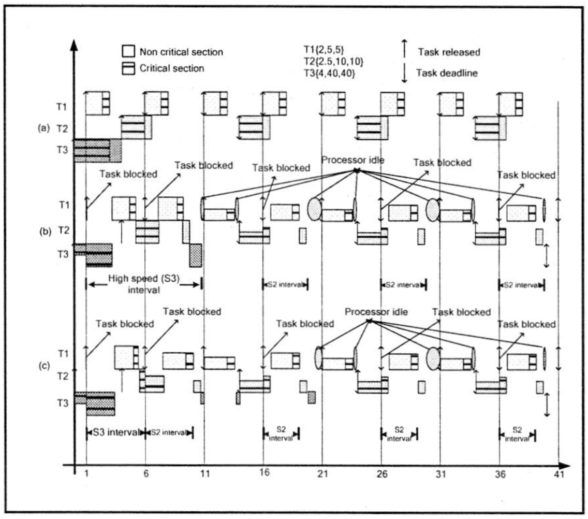

Fig.2. (a) Task set description. (b) MS algorithm. (c) IMS algorithm.

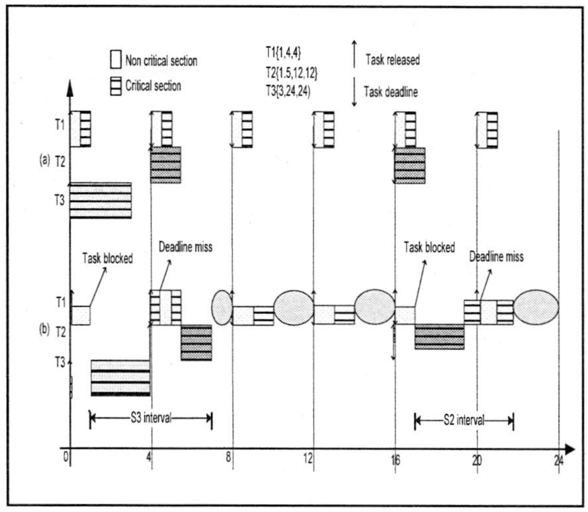

Fig.3. (a) Task set description. (b) Deadline missed

high speed interval when the processor becomes idle.

3.2 Improved Multi-Speed Algorithm

MS algorithm ends the high speed interval when the deadline of the blocking task is reached or the processor becomes idle causing more energy consumption. This highlights the key idea of the first contribution of the improved MS (IMS) algorithm proposed in this paper to end the high speed interval by the blocked (high priority) task deadline which is mostly earlier than the blocking (low priority) task deadline. Also the high speed interval has to be ended when the processor becomes idle or a lower or equal priority task (job) is selected to execute. To test the performance of IMS algorithm, the previous example has been performed using IMS algorithm, Figure 1(c), as shown in Figure1, the processor idle time is reduced from 33% in MS to 25% in IMS which reflects the improvements achieved in the system performance. To validate the effect of IMS, another hard real time system with three tasks is addressed:

$\tau_1 = \{2, 5, 5\}$, $\tau_2 = \{2.5, 10, 10\}$, $\tau_3 = \{4, 40, 40\}$

The arrival times and critical sections of the three tasks within the least common multiple (LCM) of periods are shown in Figure 2(a). According to (6) the low speed is $S_L = (2/5+2.5/10+4/40)=0.75$ which represents the processor utilization factor U. There are two blocking tasks in this example: τ_2 that can block higher priority task τ_1 for maximum time $B_2=2$ and τ_3 that can block higher priority tasks τ_1 and τ_2 for maximum time $B_3=3$. So, there will be two high speeds (S_2, S_3) according to these two blocking tasks, and these two speeds are computed according to (6),

where $S_2 = (2/5+2/5) = 0.8$, and $S_3 = \max(2/5+3/5, 2/5+2.5/10+3/10) = 1$. The two speeds (S2, S3) also satisfy the condition (7), where $0.75 \le S_2 \le 1$ and $0.75 \le S_3 \le 1$. Figure 2(b) shows that MS algorithm which ends the high speed interval when processor becomes idle (at times t=10.75, t=19.75, t=29.75, and t=39.75) or the deadline of blocking task is reached, while Figure 2(c) shows the IMS algorithm which ends the high speed interval when the blocked task deadline is reached (at time t=6), the processor becomes idle, or a lower or equal priority is selected to run (at times t=10, t=19.125, t=29.125, and t=39.125). Even though the high speeds and low speed are close to each other in this example, there is again an improvement in the system performance based on IMS compared with MS where the processor idle time is reduced from 10% to 6.5%.

4. MSA and Dynamic Priority Ceiling Protocol

The success of MS algorithm requires that the critical sections be non-preemptible, and MS algorithm may cause deadline miss when it is used with dynamic priority ceiling protocol (DPCP) because a high priority task can preempt a critical section of another low priority task if the high priority task does not need to use the shared resource. However, the blocking occurs at the moment when the high priority task needs to access the resource, and the switching to one of high speeds S_m is delayed until the moment of accessing the shared resource as shown in Figure 3. It shows the same example in Figure 1, but it uses DPCP instead of SRP as resource access protocol. Task τ_3 that began to execute its critical section is preempted by the released

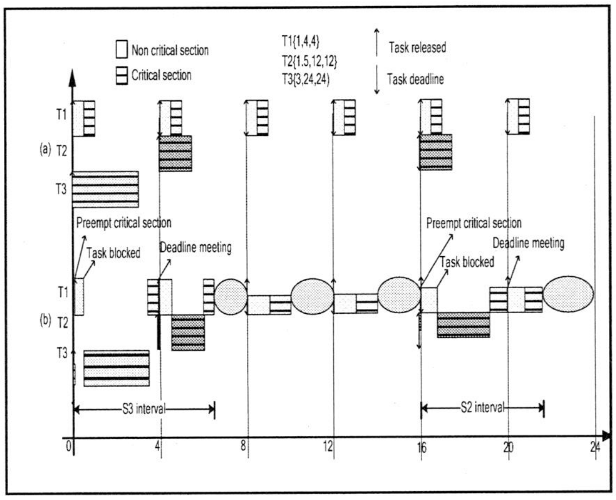

Fig. 4: (a) Task set description. (b) Deadline meeting when IMS algorithm is used with DPCP

high priority task τ_1 and task τ_1 executes its non-critical section. Task τ_1 is just blocked when it needs to access the shared resource, i.e. at time t=1, where the high speed interval S_3 starts, and τ_2 resume executing until it ends its critical section to be preempted again by τ_1.

Then, task τ_1 resume executing, but it cannot meet the deadline, where the first job (j_1) misses its deadline because of delaying the start of the high speed interval (blocking instant) due to using DPCP as resource access protocol as shown in Figure 3(b). It is clear that to avoid the occurrence of deadline miss, the switching to the high speed must occur earlier as in using SRP as resource access protocol, and this is the second contribution of IMS algorithm. To achieve that, IMS algorithm proposes that the switching to a high speed should occur when a critical section is preempted or a blocking takes place. Figure 4 shows the adaptation of IMS algorithm with DPCP as resource access protocol to avoid the deadline miss and achieve energy saving as done with SRP, where the switching to the high speed occurs not only when the high priority task is blocked, but also when a critical section is preempted. Unlike SRP, DPCP increases the context switches as it is obvious in Figures 1(c) and 4, and this is the key advantage of SRP over PCPs as mentioned earlier. Figure 5 shows the proposed algorithm IMS which achieves more energy saving and works with SRP and DPCP as resource access protocols.

// END_H is the end of S_m interval
// If the system is not in a high speed S_m interval,
End_H = -1 // J_m is the current job.
// S_{curr} is the current processor speed.
Initialization:
END_H = -1; Calculate S_L // by relation (6)
Set Speed (SL)
When a task joins or leaves the system:
Calculate S_L
for m=1 to number of tasks
Calculate S_m // by relation (7) and (8)
When job $J_{i,k}$ arrives:
if Priority($J_{i,k}$) > Priority(J_m)
if Preempt_ J_m ( ) is successful
if Preempt_Critical_Section( ) is successful
if $S_m > S_{curr}$
Set Speed(Sm)
END_H=max (END_H, deadline of $J_{i,k}$)
end if; end if; Execute $J_{i,k}$
else if $S_m > S_{curr}$ // $J_{i,k}$ is blocked
Set Speed(Sm)
END_H=max (END_H, deadline of $J_{i,k}$);
end if; end if;
When END_H is reached, the processor is idle, or a lower (equal) priority task is selected to run:
END_H = -1;
Set Speed (SL)

Fig. 5: Improved multi-speed (IMS) algorithm

As shown in Figure 5, IMS algorithm recalculates the low speed S_L and the high speeds S_m (one high speed for each blocking task) when a task joins or leave the system. The algorithm initially starts with the low speed S_L and switches to one of high speeds according to the blocking (running) task if a high priority task is blocked or a critical section is preempted. Then, it ends the high speed interval if the deadline of blocked (high priority) task is reached, processor is idle, or task, whose priority is equal or lower than the priority of the blocking task, is selected to run.

5. Results and Discussion

Referring to Figures 1 and 2, reducing the time during which the processor is idle comes from lowering the processor speed for longer time intervals. This, in turn, reduces the energy consumption dramatically due to quadratic dependency between power and processor speed. To verify that, a comparison study has been performed by computing the energy consumed in CSS, MS, and IMS using the simplified power model $P=S^2$ used in [11], where the blocking time B changes from 0 to the highest amount at which the task set is schedulable (when S_m=1). Of course, the high speeds S_m changes from S_m=S_L (when

B=0) to $S_m=1$, while the low speed S_L does not change. As it is clear from Figure 6, IMS is the most energy efficient algorithm especially with high blocking times, where the difference between the low and high speeds (S_L, S_m) increases significantly. The comparison is repeated for the second example, it is noticed that, as shown in Figure 7, IMS exhibits a slight improvement over MS with the highest blocking time due to the small difference between high and low speeds. As a result, when the blocking time is low (the high speed are almost the same as the low speed), the three algorithms exhibit the same performance. When the blocking time increases (the difference between the high and low speeds also increases), IMS behaves better than the other two algorithms (CSS and MS) especially when this difference is significant. Furthermore, IMS algorithm has the advantage over MS algorithm that it can work with SRP and DPCP as resource access protocols as shown in Figures 1 and 4.

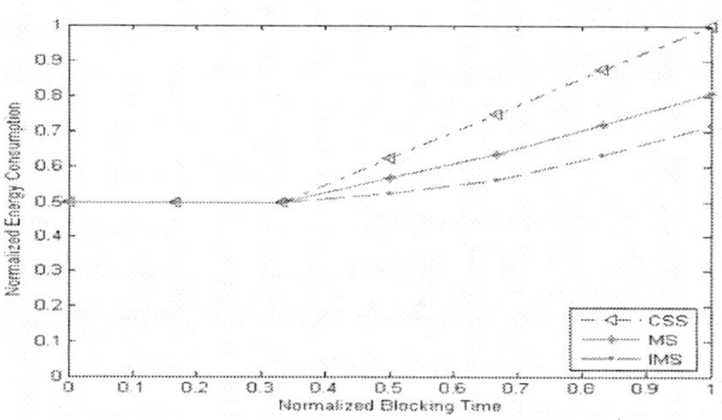

Fig.6. Energy consumption versus blocking time changes when processor utilization factor U=0.5

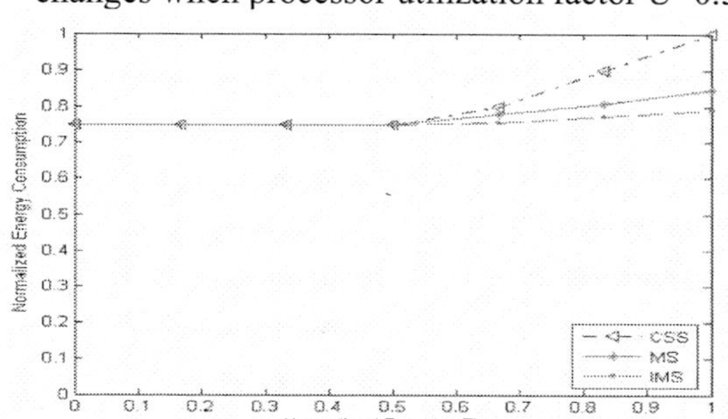

Fig.7. Energy consumption versus blocking time changes when processor utilization factor U=0.75

Conclusion

The paper has addressed the problem of real time scheduling of dependent tasks due to exclusive access shared resources taking into account the reducing of energy consumption as a main goal. The paper has proposed improvements over the existing multi-speed (MS) algorithm, where the proposed algorithm, IMS, has shown more energy saving than MS. Furthermore, it has adaptation to work not only with SRP but also with DPCP as resource access protocols.

References

[1] J. Chen, and C. Kuo. "Energy-efficient scheduling for real-time systems on dynamic voltage scaling platforms". *13th IEEE Int. Conf. on Embedded and Real-Time Computing Systems and Applications,* pp. 28-38, 2007.

[2] V. Venkatachalam, and M. Franz "Power reduction techniques for microprocessor systems" *ACM Computing Surveys (CSUR),* Volume 37, Issue 3, pp: 195-237, 2005.

[3] W. Liu. "Techniques for leakage power reduction in nanoscale circuits: A survey". IMM-Technical Report-2007-04, Informatics and Mathematical Modelling, Technical University of Denmark, DTU, 2007.

[4] G. Qu. "What is the limit of energy saving by dynamic voltage scaling? *IEEE/ACM International Conference on Computer Aided Design, (ICCAD 2001),* pp. 560–563, 2001.

[5] M. Weiser, B. Welch, A. Demers and S. Shenker. "Scheduling for reduced cpu energy". *In Proc. of Symposium on Operating system Design and Implementation (OSDI),* pp. 13–23, 1994.

[6] F. Yao, A. Demers and S. Shenker. "A scheduling model for reduced cpu energy". *In Proceedings of IEEE Annual Symposium on Foundations of Computer Science,* pp. 374–382, 1995.

[7] X. Hu, and J. Quan. "Fundamentals of power-aware scheduling". *J. Henkel and S. Parameswaran (eds.), Designing Embedded Processors – A Low Power Perspective,* pp. 219–229, 2007.

[8] A. Andrei, P. Eles, and B. M. Al-Hashimi. "Voltage Selection for Time-Constrained Multiprocessor Systems" *In Proc. Of Designing Embedded Processors,* pp. 259–284, 2007.

[9] R. Jejurikar and R. Gupta. "Energy aware task scheduling with task synchronization for embedded real time systems". *In Proc. of Int. Conf. on Compilers, Architecture and Synthesis for Embedded Systems (CASES),* pp. 164–169, 2002.

[10] L. Rajkumar, and J. Lehoczky, "Priority inheritance protocols: An approach to real-time synchronization," *IEEE Computer Transactions,* Vol. 39, No. 9, pp. 1175–1185, 1990.

[11] R. Jejurikar and R. Gupta, "Energy aware EDF scheduling with task synchronization for embedded real time operating systems," *in Workshop on Compilers and Operating System for Low Power,* 2002.

[12] R. Jejurikar and R. Gupta. "Energy aware task scheduling with task synchronization for embedded real time systems". *IEEE Transactions on Computer Aided Design of Integrated Circuits and Systems,* Vol. 25, No. 6, pp. 1024 – 1037, 2006.

[13] F. Zhang and S. T. Chanson. "Blocking-aware processor voltage scheduling for real-time tasks". *ACM Transactions in Embedded Computing Systems,* pp. 307–335, 2004.

Innovative Production Machines and Systems
D.T. Pham, E.E. Eldukhri and A.J. Soroka (Eds.)

Genetically oriented clustering using variable length chromosomes

P.Th. Zacharia and N.A. Aspragathos

Department of Mechanical & Aeronautical Engineering, University of Patras, Patras 26500, Greece

Abstract

In most cases, due to the plethora of data values, it is unrealistic for domain experts to mine useful knowledge from the database. Motivated by this, a novel approach for optimized clustering is developed in this paper. The proposed approach is genetically oriented to mine vital information incorporated in large databases avoiding entrapment in local optima and sensitivity to initialization. The aim of the proposed method is to find an optimum set of clusters that can properly classify all training data without much computational burden. The contribution is twofold: firstly, it gleans the valuable information hidden behind the database and secondly, it evolves automatically the appropriate number of cluster centers, as well as the partitioning of the data, without a priori assumptions on the cluster centers.

In this paper, the effectiveness of a Genetic Algorithm with variable length chromosomes is demonstrated for clustering data sets into an unknown number of clusters. The flexibility of the proposed variable length Genetic Algorithm to detect the optimum number of clusters and the corresponding partition is evaluated through various experimental tests. The results of the proposed algorithm are compared with those obtained by the well-known fuzzy c-means algorithm, which is applicable only for a predefined fixed number of clusters and the subtractive clustering method, where data-points are considered as candidate cluster centers.

Keywords: Clustering, Genetic Algorithm, Variable-length chromosome

1. Introduction

In real life situation, data arises from simulation, measurements and centralization procedures. Clustering is widely used in the field of intelligent manufacturing for data mining, fuzzy control and decision making. To alleviate computational complexities involved in mining valuable information hidden in large databases, clustering techniques are used due to its automatic generation capability.

Clustering [1],[2] is an unsupervised classification mechanism, where a set of objects (patterns) usually multidimensional in nature, are classified into groups (clusters), such that members of each group are similar according to a predefined criterion. Clustering forms the basis for many classification and system modeling algorithms and has been successfully applied in sensor networks [3] and robot fabric placement [4]. The proposed approach is preliminary work for training the robot sewing process through ANFIS.

The most well-known data clustering technique is the Fuzzy C-Means (FCM), where each data point belongs to a cluster to some degree that is specified by a membership grade. The technique was originally introduced by Bezdek [5] as an improvement on earlier clustering methods. The FCM algorithm is an iterative optimization algorithm that minimizes the sum of the Euclidean distances over all the clusters. However, FCM has two major limitations: it requires the a priori specification of the number of clusters and it often gets stuck at suboptimal solutions, strongly depending on the choice of the initial values.

Chiu [6] presented the Subtractive Clustering Method (SCM), which is a modified form of the Mountain method [7] for cluster estimation, which has the drawback of being computationally expensive. Chiu's algorithm considers each data point as a potential cluster center and calculates a measure of the likelihood that each data point would define the cluster center, based on the density of the surrounding data points. However, an inherent problem with the SCM is that it will give a good result only if the desired cluster center coincides with or is close to one of the data points. Moreover, for some applications, the number of data points can be prohibitively large [8]. Another drawback is that SCM demands the determination of a threshold for defining the cluster centers. This implies user intervention, which entails less flexibility.

As already mentioned, common clustering techniques often suffer from the problem of being trapped at local extrema and being sensitive to initialization. This gives rise to the use of stochastic optimization schemes, such as Genetic Algorithms, that have been successfully used in solving large, multi-modal and highly complex problems in several areas. Thus, the power of Genetic Algorithms (GAs) can be utilized to classify large multidimensional data sets with no prior knowledge about their distributions [9].

Several authors [10],[11],[12],[13],[14] have applied Genetic Algorithms for clustering using fixed-length chromosomes. They combine the concept of the Genetic Algorithm along with the FCM clustering method. The concept of clustering using a variable string length GA has been used in [15]. This algorithm uses real-coded chromosomes to encode the cluster centers and is based on the concept of FCM clustering method.

In this paper, a genetically-guided clustering approach is developed for determining the optimum number of cluster centers as well as the corresponding partition. The approach belonging to the partitioning methods is inspired from the idea presented in [6] and aims at overcoming the limitations and drawbacks of the existed methods. Unlike in [6], the selection of cluster centers is not limited to the data points in our approach; instead, the cluster centers are 'independent' points that are suggested by the proposed Genetic Algorithm.

In our work, the real challenge is to automatically evolve the appropriate number of clusters, as well as the partitioning of the data without a priori assumption on the number of cluster centers. For this purpose, variable chromosomes are used, encoding different number of cluster centers as a sequence of binary numbers. The use of variable-length chromosomes in the construction of the GA contains a vast potential for flexibility, reduction of computational complexity as well as computational time.

The remainder of the paper is outlined as follows. Section 2 describes the clustering model and Section 3 illuminates the structure of the proposed Genetic algorithm using variable length chromosomes. Section 4 demonstrates and discusses the efficiency of the proposed method through multiple experimental tests. Finally, Section 6 summarizes the contribution of the paper in the field of clustering.

2. The clustering model

Consider a data set $\boldsymbol{X} = \{\boldsymbol{x}_1, \boldsymbol{x}_2, ..., \boldsymbol{x}_i, ...\boldsymbol{x}_n\}$ implying n data points. Each data set $\boldsymbol{x}_i \in \Re^s$ is a feature vector consisting of s measurements $A_i = \{A_{i1}, A_{i2}, ..., A_{is}\}$ describing the attributes of the object represented by $\boldsymbol{x}_i$. Without loss of generality, we consider that the data points are normalized in each dimension, so that the coordinate ranges in each dimension are equal. Let $\boldsymbol{C} = \{\boldsymbol{c}_1, \boldsymbol{c}_2, ..., \boldsymbol{c}_m\}$ be the set of cluster centers that classify the data set $\boldsymbol{X}$, where m is the number of cluster centers.

The key idea is that a point is likely to be a cluster center when its potential is high. In particular, a point is a possible cluster center when located in a region with high density of data points and is distant from the other cluster centers.

Therefore, a group of data points *attracts* a point and directs it towards a position of high potential. The neighborhood, where data points have great influence, is defined by a radius r_α. The potential of the i_{th} cluster center is computed as follows:

$$p_i^{att} = \sum_{j=1}^{n} e^{-\alpha\left\|c_i - x_j\right\|^2} \quad (1)$$

where $\alpha = 4/r_\alpha^2$ and r_α is a positive constant.

Eq.(1) denotes that the influence of a neighboring data point changes exponentially with the square of the distance, implying that data points outside the radius r_α have little influence on the potential of the cluster center.

Considering all cluster centers, the attractive potential function is given by:

$$P_{att} = \sum_{i=1}^{m} p_i^{att} \tag{2}$$

On the other hand, a cluster center should *repulse* other cluster centers from being spaced too close to it. The main idea underlying the definition of the repulsive potential is to create a potential barrier around the region of each cluster center that cannot be traversed by the rest cluster centers. On the other hand, it is usually desirable that the repulsive potential should not affect the location of the cluster center when it is sufficiently far away from other cluster centers. One way to achieve these constraints is to define the repulsive potential of the i_{th} cluster center as follows:

$$p_i^{rep} = -p_i^{att} \cdot \sum_{\substack{j=1 \\ j \neq i}}^{m} e^{-\beta \left\| c_i - c_j \right\|^2} \tag{3}$$

where $\beta = 4/r_\beta^2$, r_β is a positive constant and is often selected to be $r_\beta = 1.5 r_\alpha$ to keep away the other cluster centers.

Eq.(3) denotes that a cluster center will have greatly increased repulsive potential, when other cluster centers are closely spaced. Considering all cluster centers, the repulsive potential function is given by:

$$P_{rep} = \sum_{i=1}^{m} p_i^{rep} \tag{4}$$

Finally, the total potential P of all cluster centers is computed by Eq.(5) as the sum of the two elementary potential functions for all given cluster centers.

$$P = P_{att} + P_{rep} \tag{5}$$

The optimization problem addressed in this work concerns the maximization of the potential P (expressed by Eq.(5)) to provide the optimum number of cluster centers and the corresponding optimum partition.

3. The GA-guided algorithm

Genetic Algorithms (GAs) [16] are probabilistic search methods that employ a random, yet directed, search for finding the globally optimal solution. Unlike the traditional optimization techniques that require one starting point, GAs search for a solution from a broad spectrum of possible solutions. As a result, the search is not biased towards a locally optimal solution. In contrast to random sampling algorithms, they have the ability to direct the search towards relatively promising regions in the search and converge to the near global solution.

Although GAs are not as fast as gradient methods, they are perfectly suited for problems with large and complex search spaces and are able to provide robust search. Keeping in mind the complex nature of the clustering problem, Genetic Algorithms (GAs) have been chosen as a tool for optimization. The basic components of the GA are outlined below.

Chromosome encoding scheme

In the proposed GA, variable length chromosomes are used, instead of the traditional fixed length chromosomes. The use of variable length chromosomes increases the flexibility of the approach, since the number of clusters is not defined a priori, but is yielded by the algorithm.

Concerning the string encoding, the bit string representation of solutions has dominated genetic algorithm research, since the binary alphabet offers the maximum number of schemata per bit of information of any encoding [17]. In addition, binary encoding does not demand special modifications for applying the genetic operators to tackle the concept of variable-length chromosomes. Therefore, in this paper, the chromosomes concatenate the individual cluster centres using the binary encoding.

Each chromosome can have a certain maximum length, which is equal to the maximum possible number of clusters m_{max} and is defined by the user. In a typical variable-length string, only some of the total number of positions are used to store the cluster centres. The other positions remain *don't care* (represented by the symbol '#'). A possible chromosome with m cluster centres, composed of binary numbers, is in the form of:

$$\underbrace{0\ 1\ 1...0}_{c_1}\ \underbrace{1\ 1\ 0...1}_{c_2}\ ....\ \underbrace{0\ 0\ 0...1}_{c_i}\ ....\ \underbrace{0\ 0\ 1\ ...1}_{c_m}\ \underbrace{\#\#\#...\#}_{don't\ care}$$

where $1 \le m \le m_{max}$ and c_i represents a set of s input-output data attributes $A_i = \{A_{i1}, A_{i2}, ..., A_{is}\}$.

The evaluation mechanism

The value of the fitness function for each chromosome indicates the ability of the chromosome to survive and be reproduced in the next generation. Since the addressed problem is a maximization problem, the fitness function is equal to the objective function. Therefore, the fitness function is given by:

$$fitness = P_{att} + P_{rep} \tag{6}$$

Reproduction

Reproduction is a simple copy of an individual from one generation to the next one without any modification. In the proposed GA, the proportional selection is based on the roulette wheel strategy, where the chromosomes that will be copied are selected with rates proportional to their fitness.

Crossover

Crossover combines the features of parent chromosomes in order to produce new ones for the next generation. Although the chromosomes are variable in length, they have been converted to a fixed length by using *don't care* (#) symbols. Thus, the conventional one-point crossover [18] can be applied without the need for special modifications arisen in other works, where the cluster centers are considered indivisible. For example, two parent chromosomes:

$$\underbrace{0\ 1\ 1\ 0\ ||\ 1\ 0\ 1\ 0\ 0\ \#\#\#}_{\text{Parent1}} \qquad \underbrace{1\ 1\ 0\ 0\ ||\ 1\ 1\ \#\#\#\#\#\#}_{\text{Parent2}}$$

Where (||) denotes the cut-point, produce the following offspring:

$$\underbrace{0\ 1\ 1\ 0\ ||\ 1\ 1\ \#\#\#\#\#\#}_{\text{Offspring1}} \qquad \underbrace{1\ 1\ 0\ 0\ ||\ 1\ 0\ 1\ 0\ 0\ \#\#\#}_{\text{Offspring2}}$$

It should be mentioned that the cut-point is placed between the first and the last gene of the parent chromosome with the minimum length.

Mutation

Mutation performs a random modification on some individuals with a small-predefined probability (mutation rate). The mutation operator is applied changing a random gene of digital value '0' to '1' and vice versa. This change concerns only the positions occupied by binary symbols; *don't care* (#) symbols do not undergo changes.

Control parameters

The most important control parameters are the following:

Initial population: The genetic algorithm starts the search by randomly generating a pool of a specified number of solutions. The physical length of all strings is $m_{max} \times d$, where d is the number of bits that represents each cluster center c_i depending on the required precision. However, only $m \times d$ locations are used by a string representing clusters as the rest locations remain *don't care*. The number m is different for each chromosome and is randomly generated.

Termination criterion: The most common way of stopping the algorithm is by defining the maximum number of iterations (generations). It should be mentioned that the best solution appears for several iterations before a fortuitous crossover or mutation produces a better solution.

4. Simulation results

To validate the feasibility of the proposed algorithm, a series of experiments were conducted on a number of artificial data sets to determine the cluster centers. These data sets are generated using a uniform distribution. The proposed method is general and can be applied in the *s*-dimensional space. However, all the experimental tests that are cited in this Section are referred to the 2D space for the purpose of clarity and easy conceivability of the results.

Fig. 1 shows an artificial data set (Data Set#1) comprising 245 data points, randomly distributed into five circles of equal radius, as well as the cluster centers obtained by the proposed GA. In this set, each data point consists of 2 attributes (1 input – 1 output). Next, experimental tests have been carried out to compare the performance of the proposed variable-length GA with the FCM and the SCM algorithm.

Fig. 2 shows indicatively the cluster centers after applying the SCM with radius of influence, $r_{\alpha} = 0.3$ and the FCM with a predefined number of cluster centers equal to 5. It is clear that the resulting cluster centers are close to those obtained by the proposed GA.

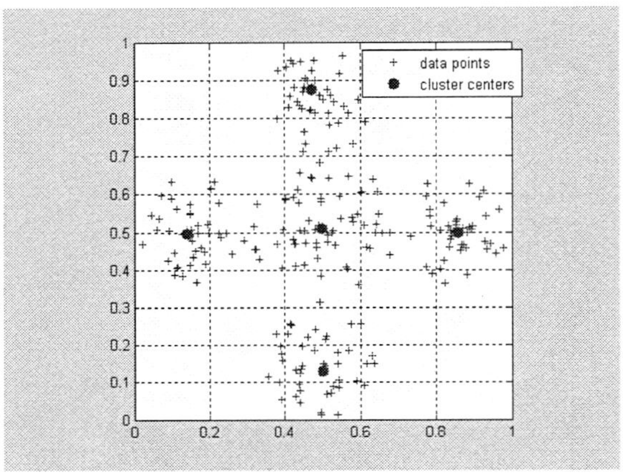

Fig. 1 Data Set#1:Simulation result using the proposed GA

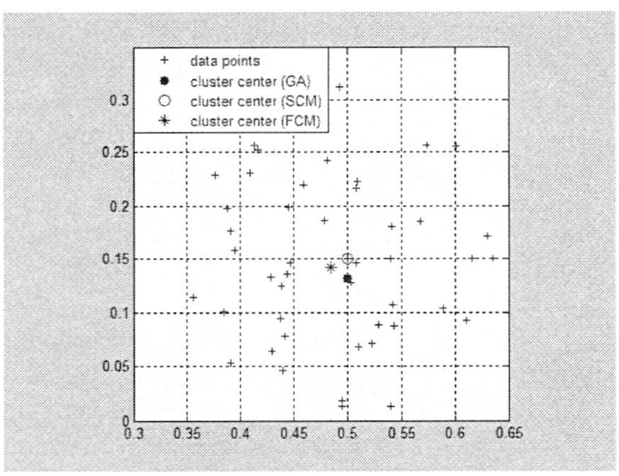

Fig. 2 A cluster of Data Set#1 containing the cluster centers yielded by the three methods

The proposed algorithm is implemented for 220 data points (Data Set#2), where the six unequal clusters are circular in nature (Fig. 3). The FCM and the SCM algorithm have also been applied for this data set, but are not presented due to the space limit.

Another data set of 225 data points (Data Set#3) is used for applying the proposed GA, where the five equal circles are distant from each other (Fig. 4). The RMS deviations obtained using the three clustering methods for each Data Set are gathered in Table 1. The deviations are small, implying that the resulting cluster centres obtained by the three methods are close to each other.

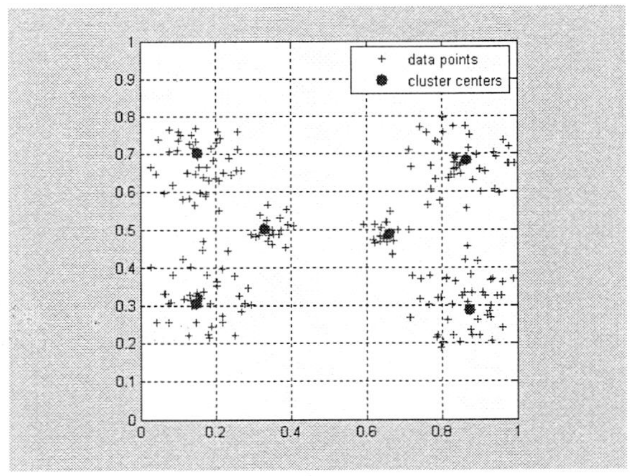

Fig. 3 Data Set#2:Simulation result using the proposed GA

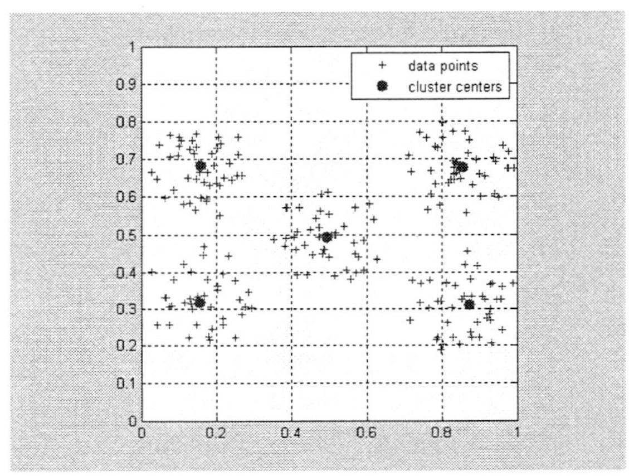

Fig. 4 Data Set#3:Simulation result using the proposed GA

Table 1
RMS deviations between the three methods

Compared methods	Deviation for Data Set #1	Deviation for Data Set #2	Deviation for Data Set #3
GA - SCM	0.0142	0.0210	0.0090
GA - FCM	0.0132	0.0116	0.0063
SCM - FCM	0.0111	0.0146	0.0064

The experimental results show that the proposed algorithm is capable of finding the optimum number of cluster centers, as well as the corresponding partitions without being trapped at local optima. Experimental tests in the 3D-space have also been carried out, but are not presented due to the space limit. The results proved that the algorithm is efficient in finding the cluster centers in multi-dimensional data sets.

5. Conclusions

In this paper, a novel method is introduced for clustering data into an unknown number of clusters. The main idea is to search for the set of cluster centers that aggregates the maximum potential value. The contribution is twofold. On the one hand, the method attains to glean vital information from large databases. On the other hand, it evolves automatically the appropriate number of cluster centers, as well as the partitioning of the data, without a priori assumption on the cluster centers.

The proposed method is based on a Genetic Algorithm applying variable-length chromosomes encoding different number of cluster centers. The chromosomes concatenate the individual cluster centers using the binary encoding, since it offers the maximum number of schemata per bit of information of any encoding.

Unlike the traditional clustering methods referred in the introduction, the proposed approach is able to escape local optima and converge to the near global solution. In contrast to SCM, where the search for cluster centers is limited to the data-points, our method searches for 'independent' points. Furthermore, the utility of variable-length chromosomes in the GA structure enhances the flexibility of the proposed approach.

The simulation results show that the proposed approach is an effective and efficient method for detecting the optimum number of cluster centers. Comparing the proposed algorithm with the conventional FCM and SCM algorithms, our approach exhibits better performance in the sense that it can automatically evolve the optimum number of cluster centers, as well as the corresponding partitions.

Acknowledgements

This work is carried out under the project "XPOMA-Handling of non-rigid materials with robots: application in robotic sewing" PENED 01 funded by the General Secretariat for Research and Technology of Greek Government. University of Patras is partner of the EU-funded FP6 Innovative Production Machines and Systems (I*PROMS) Network of Excellence.

References

[1] Anderberg M.R. Cluster Analysis for Application, Academic Press, New York, 1973.
[2] Tou J.T. and Gonzalez R.C. Pattern Recognition Principles, Addison-Wesley, Reading, MA, 1974.
[3] Saad E.M., Awadalla M.H., Darwish R.R., Distributed and adaptive clustering architecture for dynamic sensor networks, IPROMS 2007, 304-309.
[4] Zoumponos G. and Aspragathos N. Fuzzy logic path planning for the robotic placement of fabrics on a work table, Robot Cim-Inf Manuf. 24/2 (2008), 174-186.
[5] Bezdek J.C. Pattern Recognition with Fuzzy Objective Function Algorithms, Plenum Press, New York, 1981.
[6] Chiu S. Fuzzy Model Identification Based on Cluster Estimation. J. Intell. Fuzzy Syst. 2/3 (1994), 267-278.
[7] Yager R.R. and Filev D.P. Learning of fuzzy rules by mountain clustering. Proceedings of SPIE Conference on Applications of fuzzy Logic Technology, 246-254.
[8] Pal N.R. and Chakraborty D. Mountain and Subtractive Clustering Method: Improvements and Generalizations. Int.J. Intell. Syst. 15 (2000), 329-341.
[9] Pakhira M.K., Bandyopadhyay S. and Maulic U. A study of some fuzzy cluster validity indices, genetic clustering and application to pixel classification. Fuzzy Sets and Systems 155 (2005), 191-214.
[10] Scheunders P. A genetic c-means clustering algorithm applied to color image quantization. Pattern Recognition 30/6 (1997), 859-866.
[11] Hall L.O., Özyurt I.B. and Bezdek J.C. Clustering with a genetically optimized approach. IEEE Trans. on Evol. Comput. 3/2 (1999), 103-112.
[12] Di Nola A., Loia V. and Staiano A. Genetic-based spatial clustering. IEEE International Conference on Fuzzy Systems 2, (2000), 953-956.
[13] Maulik U. and Bandyopadhyay S. Genetic algorithm-based clustering technique, Pattern Recognition 33 (2000), 1455-1465.
[14] Youn J.-I., Eun H.-J. and Kim Y.-S. Fuzzy clustering for documents based on optimization of classifier using the genetic algorithm. ICCSA 2005, 10-20.
[15] Maulik U. and Bandyopadhyay S. Fuzzy partitioning using a real-coded variable length genetic algorithm for pixel classification. IEEE Trans. on Geosci. Remote Sensing 41/5 (2003), 1075-1081.
[16] Holland J.H. Adaptation in natural and artificial systems, university of Michigan Press, Ann Arbor, MI, 1975.
[17] Goldberg D. Genetic Algorithms in Search, Optimization & Machine Learning. Addison-Wesley, 1989.
[18] Michalewitz Z. Genetic Algorithms + Data Structures = Evolution Programs. 3rd edition, Springer-Verlag, 1996.

Innovative Production Machines and Systems
D.T. Pham, E.E. Eldukhri and A.J. Soroka (Eds.)

Knurling quality modelling using artificial neural network (ANN)

Dr. Khalil Aslam Awan

Institute of Space Technology, P. O. Box 2750, Islamabad, Pakistan

---

Abstract

Knurling is a machining operation typically performed on a lathe machine. Knurling is a process by which a pattern is imposed on a work piece. Knurling is applied on work piece for good holding grip, for repairing or increasing the size of a component or for aesthetic reasons so that the product will look more attractive. A good quality knurl on a part usually depends on the skill of the worker or if it is made on CNC machine on the good programming of the machine. Unfortunately there are very few studies in literature which quantify the good or bad quality of a knurl. In this research different available approaches for measuring the quality of knurling are reviewed and it is suggested that knurling quality may best be defined by measuring its depth (i.e. depth of the knurl teeth). The research also focus on different parameters that influence the quality of knurl (like work piece material and diameter, tool approach and pitch). This research uses Design of Experiment (DoE) to study the impact of these parameters on the knurling quality. A factorial design 2^4 with three replicates is selected and experiments done. Using the results obtained by experiments, an Artificial Neural Network (ANN) is designed to predict the knurling quality. The ANN model's results indicate that the ANN model predicts within one percentage error of the actual values.

Keywords: Knurling, Quality, Machining, Modeling, Design of Experiment (DoE), Artificial Neural Network (ANN).

---

1. Introduction

Knurling is widely used manufacturing operation in industry. Knurling is typically conducted on a lathe, whereby patterns like visually attractive diamond shaped (cress-cross) or straight are cut or rolled into metal.

Knurl cutting is usually accomplished on lathe by using the same automatic-feed mechanisms that are used to cut screw threads. Knurling can be thought of as simply a series of threads cut at extremely coarse pitch and in both the left-hand and right-hand directions.

Knurl rolling is more common and usually accomplished using one or more very hard rollers that contain the reverse of the pattern to be imposed. The knurling roller is push against the workpiece to produce desired pattern on the workpiece as shown in Fig. 1. A "straight" knurl (not cress-crossed) can obviously be accomplished with a single roller as shown in Fig. 2. A cress-cross pattern can be accomplished using any of:

- A single roller that contains the reverse of the complete desired pattern. These are available to form either "male" or "female" patterns,
- A left-handed straight roller followed by a right-handed straight roller (or vice-versa), or
- One or more left-handed rollers used simultaneously with one or more right-handed rollers.

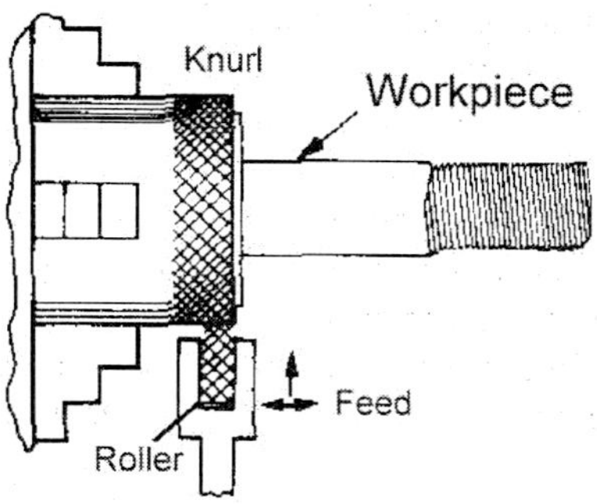

Fig. 1. Knurling operation.

Most common "Knurls' produced by knurling operations are used for decorative and grip surfaces, repair of undersized shafts and oversize bores, and driving serrations and splines. Knurling may also be used as a repair method, because a rolled-in knurled surface has raised-up areas surrounding the depressed areas, these raised areas can make up for wear on the part. This repair method is feasible on pistons of internal combustion engines, where the skirt of a worn piston was expanded back to the nominal size using a knurling process.

Knurling can also be used when a high precision component will be assembled into a low precision component, for example a metal pin into a plastic molding. The outer surface of the metal pin is knurled so that the raised detail 'bites' into the plastic irrespective of whether the size of the hole in the plastic closely matches the diameter of the pin. It is usually produced by forcing a knurling die into the surface of a rotating part, displacing material from the original diameter.

Rolled knurls are somewhat more complicated to design than cut knurls because the outer diameter of the work piece must be chosen to allow the roller to roll an integral number of patterns around the workpiece. By comparison, for cut knurls, the spacing of the cuts isn't preset and can be adjusted to allow an integral number of patterns around the workpiece no matter what the diameter of the workpiece.

Typical knurling rollers are shown in Fig. 2.

Fig. 2. Knurling Rollers.

Two methods based on the direction of the knurling roller feed onto the workpiece are used:

1. The knurling roller is feed (Bump) into the work in the radial direction 90o to the axis of the workpiece.
2. The tool is mounted on the turret or tailstock and feed into the workpiece along axis direction i.e. along the axis of the workpiece. Straddle-or synchronized roller type holder are used

The two types of tool holders are shown in Fig. 3.

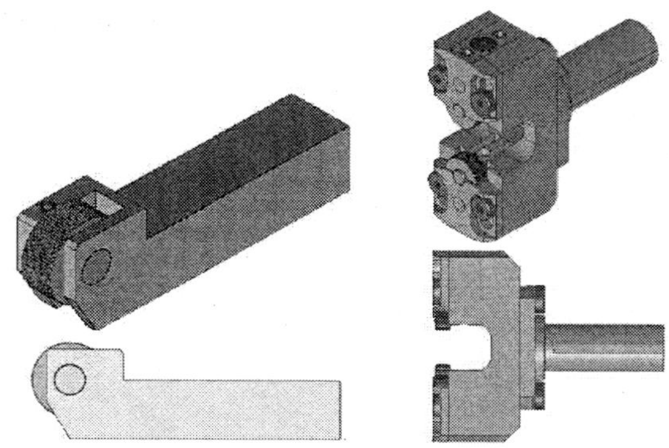

(a) Radial Feed (Bump) (b) Axial Feed

Fig. 3. Knurling tool holders

2. Design of Experiments (DoE)

There are many factors which may influence the quality of knurling but before we start investigating these factors we have to define a method of measurement of quality of knurling. In industry the common indication of good quality knurling is identified by visual inspection. The knurling tool teeth should track the knurls properly and depth of knurls should be as much as possible. In this investigation we define the quality of knurl as the difference of initial diameter of workpiece before knurling and final diameter after knurling as this will give a good indication as how much the knurling teeth are penetrating into the workpiece. Also if good tracking is happing during knurling the diameter of the final knurl will be higher.

The initial investigation indicated that the surface Speed of knurling wheel, Feed, Tool teeth per inch (TPI), workpiece initial diameter, Approach of tool (radial or axial to workpiece, see Fig. 3) and workpiece material hardness may influence the knurl quality. As the number of factor which may influence the quality are large, after initial investigation four factors were identified as the most important i.e. Workpiece hardness, Tool approach, Tool TPI and workpiece initial diameter. A survey of the machining conditions in industry indicated the ranges given in table 1 in which knurling is usually done. Level 2 factorial design [1] is used. Four factors at 2 level are considers and every experiment is replicated three times to remove noise error 2^4 x 3= 48 experiments were conducted.

Following procedure is used for conducting the experiment

- Initial factors which may influence are identified
- Knurling quality is defined as difference of initial diameter and final diameter
- Initial screening indicate four most important factors (Table 1)
- Factor levels were chosen by surveying the industrial practices
- Sequential experimental strategy is adapted [1]
- Factorial design 2^4 with 3 replicate are selected (total 48 experiments)
- Experiments are randomized to eliminate noise, operator/machine effect etc. effects (Table 2)

The experiments were conducted on EMCO Concept 155 CNC lathe. The tools used are shown in Fig. 3. Aluminium and Copper is selected as these metals are most widely used in industry as well as in education institutions.

Table 1.
Factors and Levels for Experiment.

Factors	Unit	Level High (+)	Level Low (-)
Material Hardness	HRB	44 (Al-6061T)	34 (Cu-ASME B152)
Tool Approach		Axial	Radial
Initial Diameter*	mm	25+1.0	25-1.0
Tool TPI	Teeth per in	25	20

* The diameter of the workpiece is taken as 25 mm but due to tolerance actual values (±1.0) are used instant of fixed values of higher or lower range.

Table 2.
Design of Experiment and Result

StdOrder	RunOrder	Material HRB	Approach	Ini Dia (D_i) mm	TPI	Final Dia (D_i) mm
4	1	34	Axial	25.79	25	26.12
33	2	34	Axial	25.79	25	26.11
31	3	34	Radial	25.64	20	25.74
23	4	44	Radial	24.14	20	24.24
34	5	34	Radial	24.43	25	24.66
43	6	34	Axial	24.14	20	24.53
7	7	44	Axial	24.00	25	24.41
13	8	44	Radial	25.86	25	25.93
2	9	44	Radial	24.14	20	24.12
15	10	34	Radial	24.43	25	24.44
32	11	44	Radial	25.93	25	25.93
20	12	44	Axial	25.64	20	25.98
46	13	44	Radial	25.86	25	25.92
5	14	34	Radial	25.71	20	25.90
28	15	34	Axial	24.21	20	24.54
36	16	44	Axial	25.64	20	26.05
30	17	44	Radial	24.21	20	24.28
35	18	44	Axial	25.57	20	25.94

24	19	34	Radial	25.64	20	25.87
17	20	34	Radial	24.43	25	24.69
42	21	34	Axial	24.14	20	24.59
45	22	34	Axial	25.57	25	26.09
14	23	44	Axial	24.36	25	24.75
11	24	44	Axial	24.36	25	24.61
19	25	34	Radial	24.36	20	24.46
37	26	34	Axial	24.50	25	24.98
10	27	44	Axial	26.00	25	26.34
39	28	34	Axial	25.71	20	26.26
22	29	44	Radial	25.71	20	25.74
38	30	34	Radial	26.00	25	26.19
18	31	44	Radial	24.50	25	24.51
40	32	34	Axial	24.50	25	24.84
6	33	34	Axial	25.71	20	26.16
44	34	34	Radial	25.93	25	26.04
3	35	34	Radial	24.21	20	24.52
47	36	44	Radial	24.43	25	24.59
26	37	34	Axial	24.50	25	24.73
16	38	44	Axial	24.29	20	24.82
8	39	44	Axial	25.93	25	26.20
25	40	44	Radial	24.57	25	24.57
41	41	44	Axial	24.14	20	24.55
21	42	44	Radial	25.86	20	25.87
48	43	34	Axial	25.71	20	26.10
29	44	44	Radial	25.79	20	25.93
9	45	34	Radial	26.00	25	26.14
1	46	44	Axial	24.14	20	24.47
27	47	34	Radial	24.29	20	24.46
12	48	44	Axial	26.00	25	26.26

3. Artificial Neural Network (ANN)

Artificial Neural Network (ANN) is a data mining techniques which is especially advantageous when modelling complex processes or systems where the actual working mechanics of the process is not very clear. ANN has high resistance to noisy or missing data, accommodation of multiple non linear variables for unknown interactions, good generalization capability [2,3,4]. Our literature survey of knurling processes show that there is no analytic model exist, ANN offers a good alternate approach.

Various ANN models exist. For this work we selected feed forward back propagation model as it is the best general purpose model. It is a supervised learning scheme by which a layered feed forward network is trained to become a pattern-machining engine. This research uses an open source JOONE (Java based Object Oriented Neural Engine) freely available at www.joone.com [5].

After number of trials, the best ANN model for four inputs and one output was constructed as shown in Fig. 6. Several approaches can be used to determine the architecture of a net. In general, a net with one hidden layer is recommended as the starting setup [6,7,8]. After determining the number of hidden layers, number of neurons on each of theses layers are determine. It is recommended that the number of neurons may be half of sum of input and output neurons [9]. So in our case we are using 5 hidden neurons. The hidden neurons may also be calculated by using the formula: $h = i \log_2 P$, where P is the number of training patterns, h is the number of hidden neurons and i is the number of input neurons [10]. It is also important to select the architecture of the ANN such that the residual error should be with in 10% of the actual reading.

In this research as shown in Fig. 6, one hidden layer is selected with 5 neurons. The input layer has 4 neurons corresponding to four factors given in table 1. Output layer has one neuron i.e. the knurling quality in terms of the difference between initial diameter of the workpiece and final diameter of the knurl.

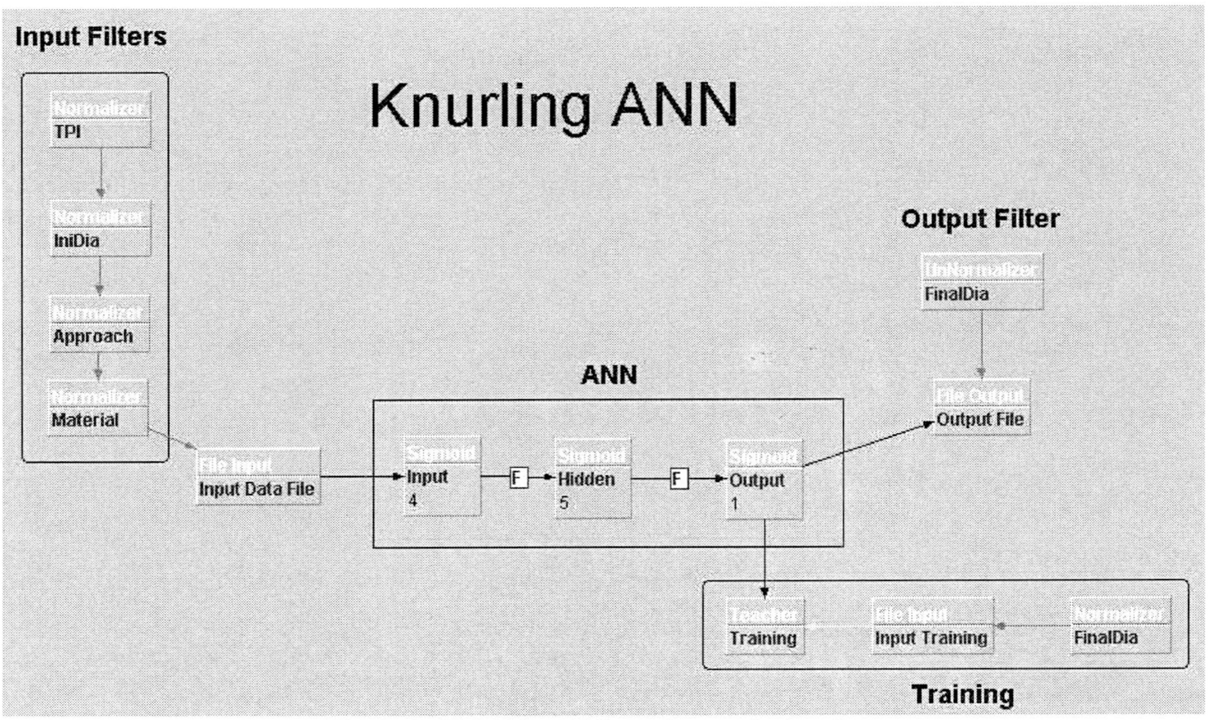

Fig. 6. Artificial Neural Network setup for knurling.

4. Results and Conclusion

There is no well defined way of quantifying the quality of knurling in industry. The most common way to check the quality of knurl is by visual inspection. In this research it is identified that up to some extent the quality of knurl can be given as the comparison of diameter of the work piece before knurling and diameter after knurling (i.e. the depth of knurl teeth). An Artificial Neural Network (ANN) model is developed to predict the final diameter of the knurl and hence predict the quality of knurling. The parameter like initial diameter of the work piece, Material, tool approach and TPI are the most important variable affecting the quality of knurl. Fig. 7 shows the experimental values of final diameter vs. ANN predicted values. It is clear that ANN gives much better prediction as the absolute percentage error (APE) is much smaller (less then 1% of actual values) as shown in Fig. 8.

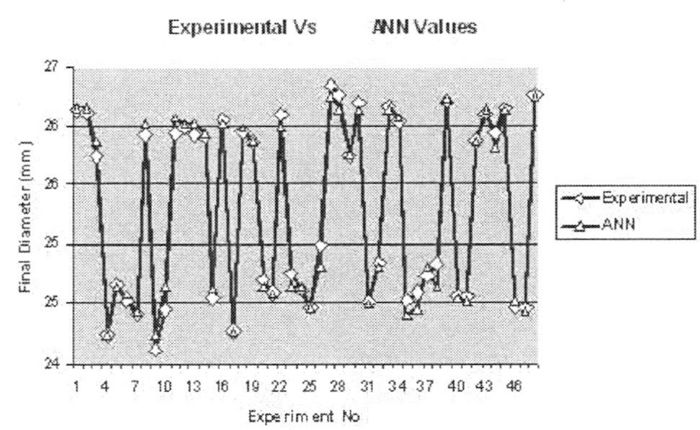

Fig. 7. Experimental vs. ANN Values.

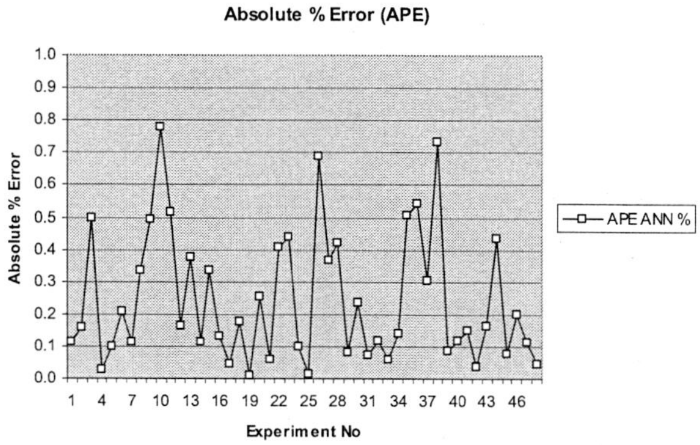

Fig. 8. Absolute Percentage Error (APE)

References

[1] Montgomery, D. Design and Analysis of Experiments, 6th Edition, Wiley, New York, 2005.

[2] Coit, D W; Jackson, B T; Smith, A E. Static neural network process models: Consideration and case studies. International Journal of Production Research, 36(11):2953-2967, 1998.

[3] Mitchell, T M. Machine Learning. McGraw-Hill, New York, 1997.

[4] Hertz, J; Krogh, K; Palmer R. Introduction to the Theory of Neural Computation. Addison-Wesley, Redwood City, CA, 1991.

[5] Joone. Java Object Oriented Neural Engine. www.joone.com, 2007

[6] Cybenko, G. Approximation by super positions of a sigmoidal function. Mathematics of Control, Signals and Systems, 2(4):3003-314, 1989.

[7] Bounds, D C; Lloyd P J; Mathew B; Waddell G. A multiplayer perceptron network for the diagnosis of low back pain. Proceedings of the 2nd IEEE Annual International Conference on Neural Network, San Diago, CA, IEEE Press. Piscataway, NJ June 21-24, pp. II.481-II.489, 1988.

[8] Lawrence, J. Introduction to Neural Networks, California Scientific Software Press, Nevada City, CA, 1994.

[9] Lawrence, M; Fredrickson J. BrainMaker User's Guide and reference manual, 7th Edition, California Scientific Software Press, Navada City, CA, 1998.

[10] Marchandani, G; Cao W. On hidden nodes for neural nets. IEEE Transactions on Circuits and Systems. 36(5): 661-664, 1989.

Innovative Production Machines and Systems
D.T. Pham, E.E. Eldukhri and A.J. Soroka (Eds.)

Discrete event system simulation approach for a nuclear facility operational analysis

H.J. Lee[a], S.H. Kim[a], H.S. Park[a], B.S. Park[a]

[a] *Korea Atomic Energy Research Institute, 1045 Daeduck Street, Daejeon 305-353, Korea*

Abstract

This paper aims to introduce a facility operation modeling and simulation mainly based on a discrete event system modeling scheme. Recently, facilities have become so complex that their operational status cannot be estimated by simple calculations. A facility can consist of a lot of individual unit processes which may be a time driven system or an event driven system and a lot of transportation between these processes. However, a transportation between processes can usually be considered as a discrete event system and not a continuous system. Therefore, on a macro-scale, a facility operation can be simulated on the basis of a discrete event system modeling if we do not consider how an individual process behaves. This study introduces the current status of the studies on an operation modeling and simulation regarding typical nuclear facilities along with some examples. Also, this paper provides insights about how a discrete event system can be applied to a nuclear facility. These kinds of modeling and simulation are very important because they can contribute to a facility design in terms of an optimal process and a sensor layout, a generic understanding for regulators, stakeholders and operators, a system diagnosis through comparing measurement data with simulation results and furthermore it can contribute to an efficient operation scheduling.

Keywords: Nuclear facility, Operation modeling and simulation, Discrete event system, Headend process

1. Introduction

Spent nuclear fuels from a pressurized water reactor (PWR) are being accumulated year by year. Accordingly, international interests have been focused on R&D for a spent fuel disposal or reprocessing. Korea started a study on what is called a pyroprocess [1] about 10 years ago and has been acknowledged as a leading group in this field along with the US. Korea also believes that pyroprocess technology as a competitive technology compared to an aqueous process will solve the accumulated spent fuel problem in Korea. The pyroprocess technology separates transuranic elements from spent nuclear fuels and then recycles them within a fast sodium reactor. Domestically as well as internationally, needs for the construction of a spent fuel reprocessing facility are increasing in order to demonstrate the feasibility of the technology. However, this facility requires a lot of construction cost so a long term schedule must precede a facility construction. Detail design and optimal layout of a process in the facility can save on money and time through a modeling and simulation, which contributes to a verification of a facility design before a construction as well as an operation scheduling, an annual throughput estimation, and a system diagnosis after a construction. Nuclear facility operation is related to a discrete event system (DES) macroscopically rather than a continuous variable dynamic system (CVDS) according to a system classification. In other words, a nuclear facility operation is very similar to systems which are driven by asynchronous events rather than a time change and which current state can be described with discrete states rather than continuous variables. However, microscopically, each unit process may need to be described by a CVDS. In this paper, main interests are focused on a modeling and simulation of a nuclear facility operation in terms of a DES approach. This paper deals with the current status of the studies on a relevant nuclear facility operation modeling and simulation and describes how to apply discrete event system modeling and simulation methodology to a relevant nuclear facility, for example, a pyro-headend process facility.

2. Brief review of discrete event systems

The definition of a discrete event system can be described according to Cassandras and Lafortune [2]. Discrete event systems imply two meanings: one is discrete state systems, and the other is event driven systems. In discrete state systems, the state variables are elements of a discrete set. In continuous state systems, the state variables are generally any real (or complex) values. In event driven systems, it is only the occurrence of asynchronously generated discrete events that forces instantaneous state transitions. In between event occurrences the state remains unaffected. In time driven systems, the state continuously changes as the time changes. For an easy understanding of a discrete event system, we compared it with a continuous variable dynamic system, which has two key properties that a system must satisfy: a continuous state and a time driven. On the other hand, discrete event systems satisfy the following properties: a discrete state and an event driven. Apparent comparison of the sample paths for a CVDS and a DES is shown in Fig. 1. For the CVDS, the state space X is a set of real numbers $\Re$, and $x(t)$ can take any value from this set. The function $x(t)$ is the solution of a differential equation of the general form $\dot{x}(t) = f(x(t), u(t), t)$, where $u(t)$ is the input. For the DES, the state space is a certain discrete set $X = \{s_1, s_2, s_3, s_4, s_5\}$. The sample path can only jump from one state to another whenever an event occurs. No mechanism is provided to specify how events might interact over time or how their occurrence time might be determined. There are various examples of a DES drawn from the real world and common engineering

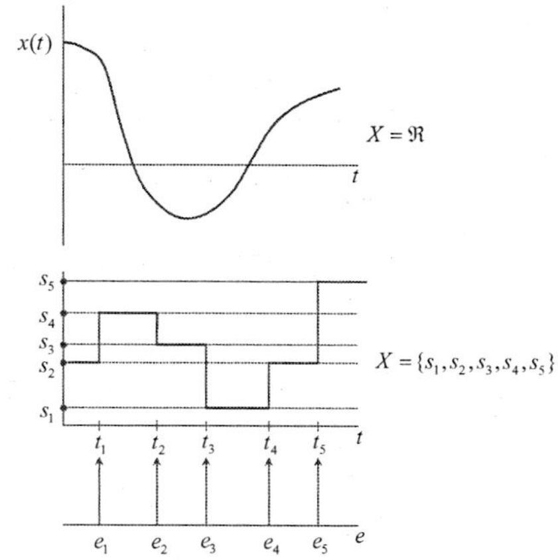

Fig. 1. Comparison of the sample paths for Continuous-Variable Dynamic Systems (CVDS) and Discrete Event Systems (DES).

experience. Representative examples are queuing systems, computer systems, communication systems, manufacturing systems, traffic systems, etc. In order to study the logical behavior of a DES, we need formal ways like differential or difference equations in a CVDS. Such modeling formalisms in a DES are automata and Petri nets. However, automata and Petri nets theories will not be dealt with in this paper. The scope of this study focuses on the application of a DES scheme using commercial software.

3. Current Status

There are a few examples regarding studies on an operational modeling and simulation in relevant nuclear facilities. Idaho national laboratory (INL) has been performing various operational modeling and simulation studies based on a discrete event system modeling and simulation about its own nuclear facilities such as its Experimental Breeder Reactor-II (EBR-II), Hot Fuel Examination Facility (HFEF), Fuel Conditioning Facility (FCF), etc. Through a DES modeling and simulation, INL analyzed an operational scheduling, annual throughput, process equipment capacity and bottleneck process, and as a result, INL suggested an efficient and enhanced the facility operation strategy [3-7]. For example, Fig. 2 shows the material flows in the FCF and Fig. 3 represents the analysis results of various what-if scenarios through a DES simulation. This study suggested that the annual throughput in the FCF can be enhanced by 8 times from 0.6 to 5 MTHM. Besides with a DES simulation, INL showed that a fault diagnosis could

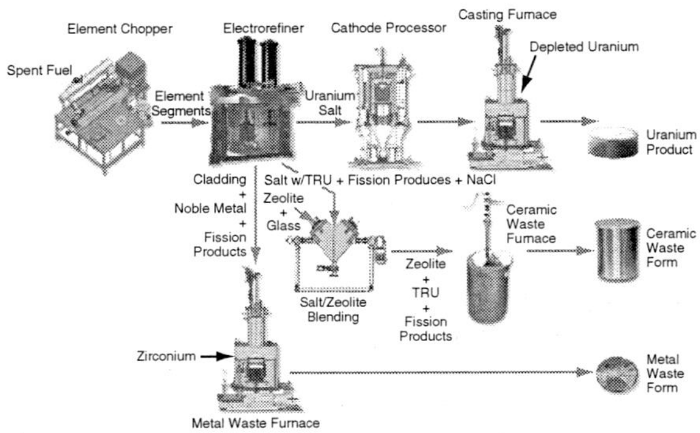

Fig. 2. Material flow between major processes at fuel cycle facility.

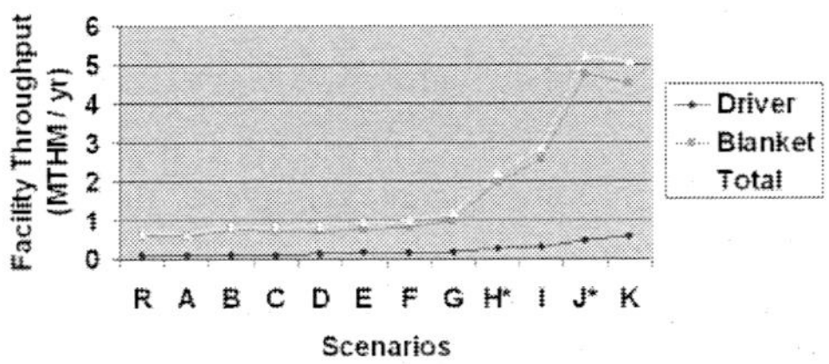

R: Reference scenario (current)
A, B, ..., K: Improved operational scenarios (not described)

Fig. 3. Annual throughput analysis via a discrete event system.

be done via a unit process modeling in a real time [8]. Sandia National Laboratory (SNL) has developed and utilized a number of simulation models to represent the processing, transportation, and disposal of a radioactive waste [9]. SNL has developed a supply chain model for the cradle to grave management of a radioactive waste and used this model to assist the Department of Energy (DOE) in developing a cost effective, regulatory compliant and efficient approach to dispose of a radioactive waste from 25 sites across the country over the next 35 years. The simulation model was developed on the basis of a DES simulation scheme. Argonne National Laboratory (ANL) also designed a commercial spent fuel reprocessing plant by a DES modeling and simulation scheme.

4. DES Application to a Nuclear Facility

In this section, application of a DES modeling and simulation to a nuclear facility will be explained. Except for the reaction dynamics of a unit process in a nuclear facility, a mechanical material transportation between process equipments can be shown as a DES. For example, if 'A' process and 'B' process occur sequentially, the end of the 'A' process triggers the 'B' process, that is, the start and end of a process plays the role of an event that makes other processes transit from one to another state. The following example is a headend process operation of which detailed modeling and simulation procedures will be presented.

4.1. Introduction to a headend process

The headend process is the first part of the pyroprocess. The headend process starts with a cask assembly connection to the headend facility and ends with separated fuel pin segments. This process does not embrace any chemical reaction process but only mechanical processes such as a withdrawal, cutting, transportation, etc. Whole process flow can be modeled as a discrete event system; therefore, the headend process is appropriate to understand the DES modeling and simulation procedures. Fig. 4 presents the flow of the headend process. A solid line loop means an assembly level process and a dotted line loop means a fuel pin segment level process. The first solid line loop accommodates an assembly transfer, inspection and storing sequentially and all the processes are performed at an assembly level. The second solid line implies a combination process at an assembly and a fuel pin level. The dotted line embedded in the second solid line implies a fuel pin level process. This loop is repeated for the number of fuel pins in an assembly. Each unit process in a headend process is as follows: 0. cask connection

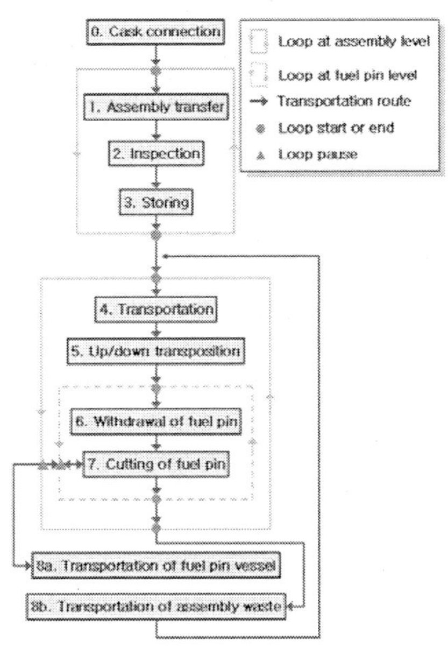

Fig. 4. Headend process flow.

process is to connect a cask for a transportation of assemblies with an interface system door in a headend process facility to transfer the assemblies. 1. assembly transfer process is to transfer the assemblies to the next process equipment, i.e., inspection equipment for an examination of an anomaly. 3. storing process is to store the transported assemblies in a temporal storage. 4. up/down transposition process is to lay down the assembly using what it is called down-ender equipment. 5. withdrawal of a fuel pin process is to withdraw a fuel pin from a horizontally fixed assembly in the down-ender equipment. 6. cutting of a fuel pin process is to cut the withdrawn fuel pin into a desired segment. 7a. transportation of a fuel pin vessel process is to transport a fully charged vessel to a designated position and to replace it with a new vessel. 7b. transportation of an assembly waste is to transport an assembly structure without fuel pins to a designated position.

4.2. DES modeling environment

There are many commercial softwares for modeling a DES like the headend process. In this paper, we utilized Stateflow [10] and SimEvents [11] developed by MathWorks Inc. The Stateflow is the Simulink [12] toolbox for a logic driven system modeling and simulation, and historically, it is based on the state chart modeling technique in ref. [13]. The SimEvents is a DES modeling and simulation toolbox in a Simulink environment which is a CVDS modeler. The SimEvents has various DES relevant blocks that enable you to undertake DES modeling tasks very easy by a drag-and-drop procedure. However, the SimEvents has a difficulty in processing various state transitions, which can be complemented by Stateflow. In Simulink, a communication across blocks is based on signals. In SimEvents, it is based on both signals and entities. The "entity" concept is motivated from the view of a DES as an environment consisting of "users" and "resources": users request resources in order to perform various tasks, occupy these resources for a certain amount of time, and then relinquish them so that other users may access them [14]. For example in a manufacturing system, users are parts and resources are machines in a factory. SimEvents consists of a number of libraries containing blocks with different system functionalities. The main libraries are the following:

1. Generators: Blocks which generate entities, or function calls (i.e., events that call Simulink blocks), or random variates.

2. Queues: Blocks where entities can be temporarily stored while waiting to access a resource.

3. Servers: Blocks that model various types of resources.

4. Routing: Blocks that control the movement of entities as they access queues and servers.

5. Gates: Blocks that control the flow of entities by enabling/disabling the access of entities to certain blocks.

6. Event Translation: Blocks that enable a communication between SimEvents and Simulink by translating events into function calls.

7. Attributes: Blocks that assign and modify data to entities. Various control actions are then made based on the values of these data, allowing the blocks to differentiate between the entities they process.

8. Subsystems: These allow for a combination of blocks to be executed upon an occurrence of specific events (not upon Simulink sample times).

9. Timers and Counters: Blocks that measure event occurrence times or a time elapse between events, and blocks that count the occurrences of particular event types. These data are supplied to a standard display or scope blocks in Simulink or specialized scopes designed specifically for SimEvents.

Above event translation blocks are necessary for a hybrid system modeling. One of the problems in accomplishing a DES or a hybrid system modeling and simulation is whether an event concurrency [15] can be processed, i.e., whether two or more events that arise concurrently can be processed. The order in which these events are executed is controlled by means of a priority scheme which is part of the underlying DES design. Such an event ordering functionality does not exist in a time-driven environment. However, SimEvents in a Simulink environment can provide this functionality for modeling hybrid systems.

4.3. DES modeling for a headend process

Unit processes in a headend process are basically modeled as server blocks in SimEvents. Transportations between unit processes are modeled as gates which have a function of blocking entity flows. To make a decision on opening or closing gates, Stateflow blocks are sometimes required. Simple decisions to control gates can be done via SimEvents blocks but complex logical decisions are difficult to implement via only SimEvents blocks.

For complex, and a logical control of gates, logics made using Stateflow are very effective. Overall flow of a headend process is shown in Fig. 4. Implementations of a loop control in Fig. 4 can be done in various ways. A use of two latch blocks which is a typical way to implement a loop control is shown in Fig. 5. Two latch blocks control a gate prior to a withdrawal server in a way that the gate is closed while a withdrawal or cutting process is being performed and the gate is open while neither a withdrawal nor cutting process is being performed. Simulink and SimEvents provide a hierarchical modeling approach which allows you to build and efficiently maintain highly complex models (consisting of thousands of objects) without losing the ability to explain the models to a non-technical audience. Fig. 6 is the top model for a headend process and Fig. 7 is a child model of the centered block in Fig. 6. Also, Fig. 7 has several child models using Stateflow which are not shown.

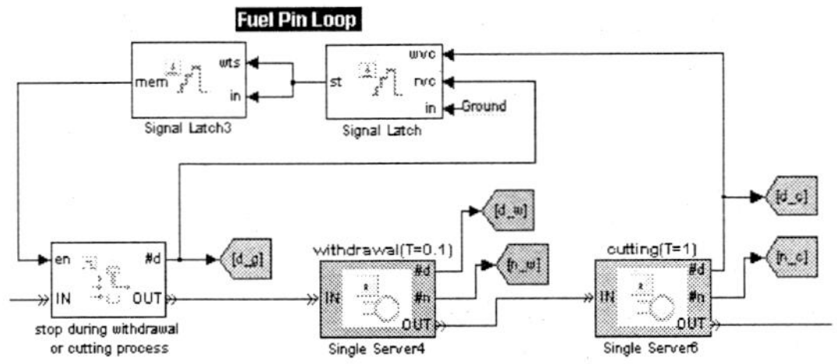

Fig. 5. Two latches for a fuel pin loop control.

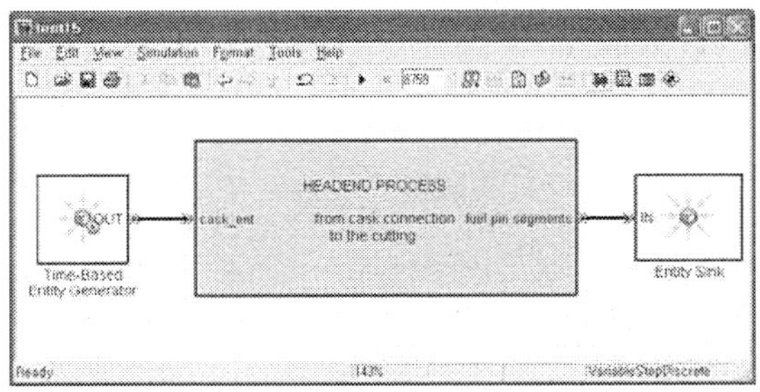

Fig. 6. Top model of the headend process.

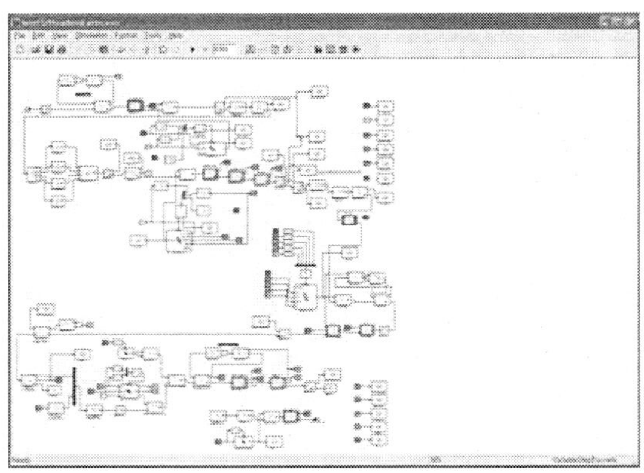

Fig. 7. A child model of the top model.

5. Simulation

There are a couple of assumptions for a headend process simulation because a headend process facility does not exist at present. First of all, an elapsed time corresponding to each unit process should be estimated. Server blocks have a dialog window for inserting basic input requirements such as the time. Additional assumptions are the number of assemblies which can be contained in a cask, the number of storage racks, the cask connection frequency, the number of remote handling tools for an assembly transportation, labor work schedule and a process equipment failure. Modeling of a labor work scheduling and a process equipment failure was not included in the following model for simplicity. Table I shows the elapsed time for each unit process and Table II shows the additional assumptions for the headend process simulation. Simulation was performed under the above assumptions. The average elapsed time from a connection to a storing process is shown in Fig. 8 on an assembly basis. In Fig. 8, dotted data represents the event loggings at each time when a storing process is finished. The x-axis indicates the time history for a year in hours. Five sets of data of which each set consists of four assembly data. The second cask contains the 5th to 8th assemblies and the corresponding average elapsed time from a connection to a storing process which can be theoretically estimated by the following equation. Detail derivation of the following equation will be omitted here.

$$T_{connection}+(T_{carry}+T_{inspection}+T_{storing})[(1+2+3+4)(nth\ Quo\ 4)+(nth\ Rem\ 4)(1+(nth\ Rem\ 4))/2]/nth \qquad (1)$$

In the above equation, '*Quo*' means quotient, for example, '*x Quo y*' means a quotient when x is divided by y. '*Rem*' means a remainder, for example, '*x Rem y*' means a remainder after x is divided by y. 'n^{th}' means the sequence of an assembly. By using (1), it takes 12.04, 11.93, 12.31 and 13 hours, respectively, for the 5th to 8th assembly to be processed from a connection to a storing process.

Table 1
Elasped Time for Each Unit Process.

Process	Elapsed Time (h)	Resource
Connection	5	Cask
Carry	0.2	Assembly
Inspection	2	Assembly
Storing	1	Assembly
Transportation	1	Assembly
Transposition	1	Assembly
Withdrawal	0.1	Fuel Pin
Cutting	1	Fuel Pin
Transportation	1	FuelPin Segment
Transportation	1	Assembly Structure

Table 2
Other Assumptions.

Assumptions	Values
The number of assembly contained in cask	4 ea
The number of storage rack	4 ea
Cask connection frequency	5 times/year
The number of assembly transportation device	1 ea
Labor work schedule	24 hours/day
Process equipment failure	No failure occurs

The results from (1) are the same as the results from the simulation shown in Fig. 8. The more complex a process is, the more difficult it is to obtain a closed form solution like (1).

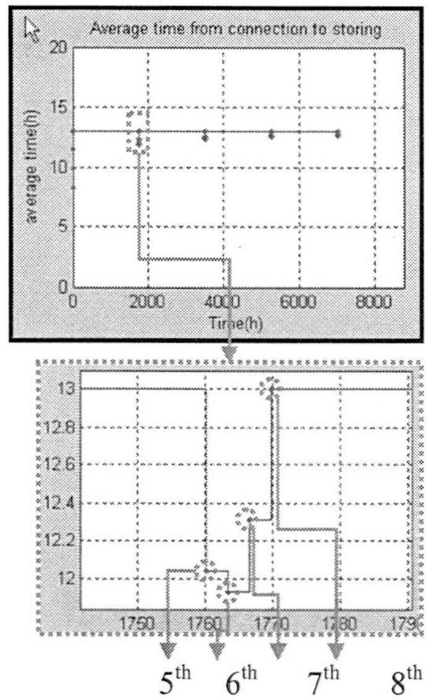

Fig. 8. Average elapsed time from a connection to a storing process.

Acknowledgements

The authors are grateful for the support provided by a grant from Atomic Energy R&D Program of Ministry of Science and Technology in Korea.

References

[1] S. W. Park and et al., Development of advanced spent fuel management process, KAERI/RR-2427/2003, 2004.

[2] C. G. Cassandras and S. Lafortune, Introduction to discrete event systems, 2nd ed., Springer, 2008.

[3] H. E. Garcia, A. Houshyar and G. R. Imel, "Planning and supervision of reactor defueling using discrete event techniques," *INRIA/IEEE Conference on Emerging Technologies and Factory Automation*, 1995, pp. 81-92.

[4] H. E. Garcia and A. Houshyar, "Discrete event simulation of fuel transfer strategies for defueling a nuclear reactor," *Simulation*, vol. 70, no. 2, 1998, pp. 104-118.

[5] H. E. Garcia, "Operation analysis and improvement of a spent nuclear fuel handling and treatment facility using discrete event simulation," *Computers & Industrial Engineering*, vol. 38, 2000, pp. 235-249.

[6] A. Houshyar, "Review of Extend Performance in Modeling a Nuclear Fuel Transfer Activity," *Simulation*, vol. 68, no. 6, 1997, pp. 403-412.

[7] A. Houshyar and G. Imel, "A Simulation model of the fuel handling system in a nuclear reactor," *Computers industrial engineering*, vol. 30, no. 1, 1996, pp. 117-135.

[8] S. Phongikaroon and T. S. Yoo, "Identification of statistical invariance for anodic signals of Mk-IV electrorefiner," *Global 2007*, 2007, pp. 1083-1089.

[9] J. Trone, A. Guerin and A. D. Clay, "Simulation of waste processing, transportation, and disposal operations," *the 2000 winter simulation conference*, 2000, pp. 1085-1089.

[10] MathWorks, *Stateflow and Stateflow Coder User's Guide*, MathWorks, Inc., 2007.

[11] MathWorks, *SimEvents User's Guide*, MathWorks, Inc., 2007.

[12] MathWorks, *Simulink User's Guide*, MathWorks, Inc. , 2007.

[13] D. Harel, "Statecharts: a visual formalism for complex systems," *Science of Computer Programming*, vol. 8, 1987, pp. 231-274.

[14] M. I. Clune, P. J. Mosterman and C. G. Cassandras, "Discrete event hybrid system simulation with SimEvents," *the 8th int. workshop on discrete event systems*, 2006, pp. 386-387.

[15] C. G. Cassandras, M. I. Clune and P. J. Monsterman, "Hybrid system simulation with simevents," *the 2nd IFAC Conf. on Analysis and Design of Hybrid Systems*, 2006, pp. 267-269.

Innovative Production Machines and Systems
D.T. Pham, E.E. Eldukhri and A.J. Soroka (Eds.)

Integrated modelling approach in support of change-capable PPC strategy realisation

T. Masood[a,b], R.H. Weston[a,b]

[a] Manufacturing System Integration (MSI) Research Institute, Loughborough University, Leics, LE11 3TU, UK
[b] Centre of Excellence in Customised Assembly (CECA), Loughborough University, Leics, LE11 3TU, UK

Abstract

The increasing demand for customization, reduced time to market and globalization are the real challenges for today's manufacturing enterprises (MEs). Therefore MEs can reduce these competitive pressures by becoming more and more change-capable. The agile and lean manufacturing philosophies must complement the application of reconfiguration techniques. However, choosing and applying the best philosophies and techniques are far from being well understood and well structured processes because most MEs deploy complex and unique configurations of processes and resource systems, and seek economies of scope and scale in respect of a number of distinctive product flows. It follows that systematic methods of achieving model driven reconfiguration and interoperation of component based manufacturing systems are required to design, engineer and change next generation MEs. This paper discusses research aimed at developing and prototyping a model-driven environment for the design, optimization and control of enterprises with an embedded capability to handle various types of change in an example of a production planning and control (PPC) scenario. The developed environment supports the engineering of common types of strategic, tactical and operational process found in many MEs. Also reported are initial findings of manufacturing case study work in which coherent multi-perspective models of a specific ME have facilitated process reengineering and associated resource system configuration and interoperation. In order to understand the system prior to realisation of any PPC strategy, multiple process segments of organisations need to be modelled. The paper considers key PPC strategies and describes a novel systematic approach to create coherent sets of unified models that facilitate the engineering of PPC strategies. Case study models are presented with capability to enable PPC decision making processes in support of complex organisation design and change (OD&C). The paper outlines key areas for future research including the need for research into unified modelling approaches and interoperation of partial models in support of complex OD&C.

Keywords: Organisation design and change (OD&C), reconfigurability, change-capable, production planning and control (PPC), enterprise modelling (EM), simulation modelling (SM).

1. Introduction

The increasing demand for customization, reduced time to market and globalization are the real challenges for today's MEs. This has induced a need for faster, better and cheaper production. Therefore MEs can reduce these competitive pressures by becoming more and more change-capable. One key common response is for MEs to have a broadened product portfolio. But to compete they must deploy an effective and change capable set of human and technical resource systems. Also because of falling product lifetimes and growing customisation requirements the deployment of these resource systems will increasingly need to give rise to economies of scope and mass customization [1]. For many companies around the world, staying in business necessitates:

- meeting specific customer requirements innovatively and effectively
- reducing the time-to-market of products
- manufacturing quality products at competitive cost.

The present day customer typically imposes constraints on MEs via their specific and changing quantity, quality, cost and delivery product and service demands. The implications of this customer focus are that; time to market is shortening, products are tailored to meet a breadth of customer needs, and demand is variable [2,3,4]. However, MEs cannot simply respond by deploying new processes and resources. Generally they need also to redeploy (i.e. re-configure and re-integrate) their use of existing processes and resources such that they can respond competitively to the ongoing business dynamic created by the specific and changing demands of their customer base.

The identification of methods by which manufacturing improvements can be achieved is ongoing and has led to a range of approaches in recent years including Lean and Agile Manufacturing. In addition, progressive improvements to information system capabilities continues to offer the belief that significantly improved support for effective decision making can be achieved [5,6].

Hence it has been observed that modern manufacturing systems must be; flexible/agile, reactive, integrated, and cost efficient [3]. It also follows that ME personnel must have (individual and collective) in-depth understandings about specific processes and resource systems and that these processes must be flexible enough to change whenever the need rises. The complexity of manufacturing systems is reaching that of many natural (e.g. economic and political) systems, thus ongoing re-design and re-engineering of such systems require the use of systematic approaches which deploy various types of system model to understand current and possible future behaviours and to inform systems engineering decision making.

This paper presents an application of the integrated modeling approach presented in [7] in an example of a change-capable PPC scenario in a manufacturing SME. The developed environment supports the engineering of common types of strategic, tactical and operational process found in many MEs in support of reconfigurable manufacturing.

2. Need for change-capable model driven PPC strategy realisation

An efficient production planning and control (PPC) system is crucial in making a manufacturing system 'live', responsive and change capable. It is also of vital importance to satisfy dynamic customer demands and expectations in highly competitive manufacturing environments [8]. In general, any given choice of PPC strategy will impact on the way in which a ME realises management of customer demand; planning and meeting material requirements; and capacity planning, scheduling and sequencing of jobs. Appropriate choice of PPC strategy should help a business to: minimize Throughput Time (TT) and Lead Time (LT); reduce Work in Progress (WIP); keep inventory costs at a minimum; improve responsiveness to change in demand (resulting in changes in product and process some times); and improve Delivery Date (DD) adherence. These are important objectives, and choosing the right PPC approach and system is hence a crucial strategic decision which should be consciously made by managers of MEs [8].

It has been observed that modern manufacturing systems must be; flexible/agile, reactive, integrated, and cost efficient [3]. It also follows that ME personnel must have (individual and collective) in-depth understandings about specific processes and resource systems and that these processes must be flexible enough to enable change whenever the need arises. Such requirements place further emphasis on the PPC system, which in general will need to be dynamically adaptable to achieve satisfactory (local and distributed) utilization of production resources and materials [9]. Thus the design and re-design of such systems requires the adoption of systematic approaches which deploy various types of system model to understand current and possible future ME behaviours and to inform ongoing decision making as environmental and organisational changes occur. Production planning and control in support of configurable manufacturing systems inherently requires organisations to have 'change capability' characteristics; suitable distribution of production resources; and some degree of autonomy of production resources [10].

There have been recent advancements to extend the coverage of public domain open systems architectures and to bridge the gap between static modelling and dynamic or simulation modelling. Computer Integrated Manufacturing Open Systems Architecture (CIMOSA) has been central to most of such developments as also mentioned in [7]. A natural focus of these modelling efforts has been to make the models 'live' and responsive to the upcoming rapid changes. Here the idea is to have modelled real elements of complex and changing sets of processes (and their underlying resource systems) interoperate in a readily integrated and change capable manner which satisfies specific dynamic requirements of the business environment in which the processes are used. The PPC system plays a vital role to depict the 'liveness', responsiveness and reconfigurability of a manufacturing system.

3. Change-capable model-driven approach in support of PPC strategy realisation

This paper describes a systematic approach to creating coherent sets of unified models that can interoperate to replicate and predict changing organisational behaviours in support of PPC. The modelling approach used to support PPC strategy realisation is based upon the methodology presented in [7] and is illustrated in Fig. 1.

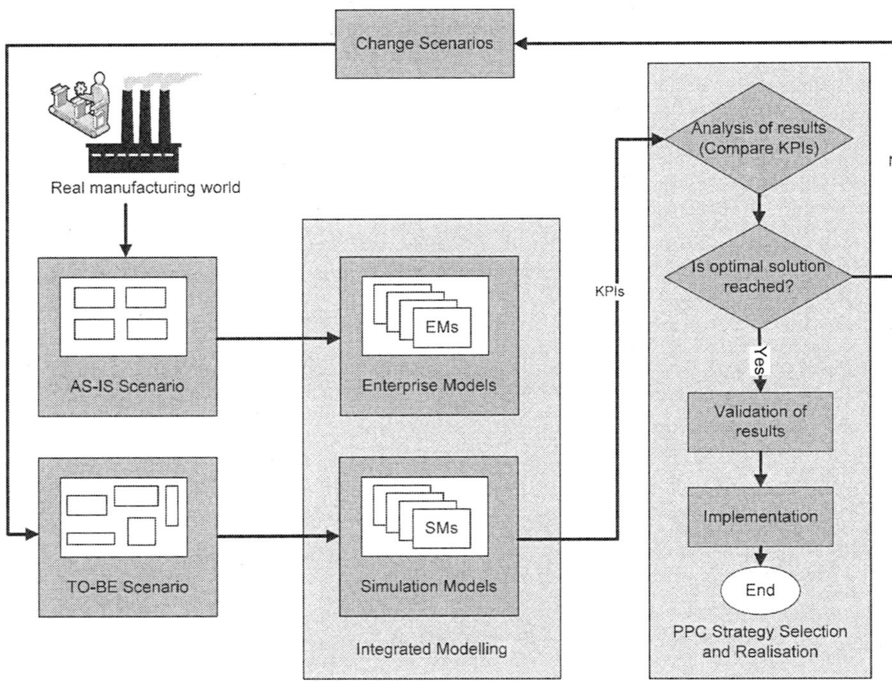

Fig. 1. Change-capable modelling approach in support of PPC strategy realization.

4. Case Study

The case study company (referred to as ABC onwards) is a furniture manufacturing SME with 50 employees. It operates primarily within the UK but has European suppliers of raw material. It manufactures over 300 different furniture products from pine wood; including a range of tables, cabinets, beds, wardrobes and other furniture items that are designed for house hold and business users. The application of a number of different PPC concepts is being investigated at ABC. A coherent set of enterprise and simulation models were created to explicitly represent and computer execute behaviours of ABC's business processes, (human and technical) resource systems and dynamic patterns of multi-product workflows.

Machining, assembly, spraying and finishing are the main production processes involved in ABC's furniture production. ABC's current production system implements a make-to-order (MTO) strategy in which production operations are triggered by groupings of customer orders based upon logistical criteria related to customers categorized by UK location. A fleet of lorries are used by ABC to deliver customer orders to specified regions of the UK. Production 'runs' are constrained by the capacity of these lorries or by a pre-set maximum order collating time of four weeks (whichever constraint is exceeded first).

According to ABC's current PPC strategy, the company compiles orders received from customers on a weekly basis. The production starts with the issuance of a production order (known as 'sales order picking list' in ABC) to the assembly shop. The production order is then reviewed by the assembly shop coordinator who generates a 'Machine Shop Production Order', that requests needed parts from ABC's machine shop. The job list of the machine shop is regularly updated and priorities are set in order to maintain smooth running of the assembly shop. The production order is then moved through the assembly shop after receiving parts to be assembled from the machine shop where assembly processes are carried out. Fig. 2 shows interactions of plan and control business process in sub interaction diagram of the business management domain of ABC.

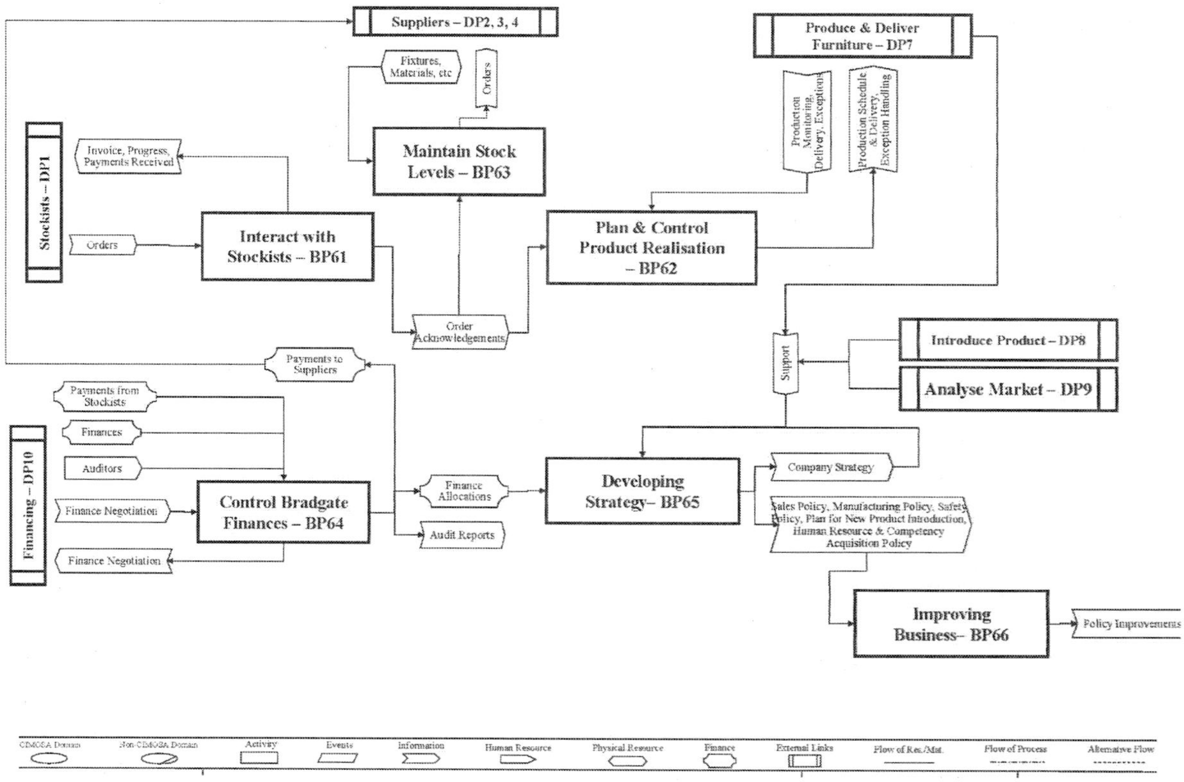

Fig. 2. An exemplary enterprise model of ABC's Plan and Control Business Process

A high variety of more than 300 products in ABC contain both less frequent and highly demanded products during certain periods which were unknown. There were no forecasting methods being adopted primarily because product demands are unpredictable. After a number of visits to the company and number of discussions with company management and operators, initial data was collected.

The assembly area works on *'real-time scheduling'* approach in which production schedules are prone to change at any time. The next job to do is decided after finishing the predecessor based on the current status of frequently changing job list, level of resources and due dates. The (human) resources are considered very flexible across different sections, sometimes even across departments.

The assembly area has three production supervisors, one being the overall assembly coordinator, the other two being the respective section sub-supervisors for table and cabinet sections. The picking list is received from customer order processing department twice a week by the assembly coordinator.

Then he prepares '*partial picking*' lists by highlighting the relevant furniture items for each assembly area and pass it to the work benches. Every time a workbench finishes a job the next job is picked from the highlighted list according to earliest due date (EDD) and minimum setup time (MST) sequencing rules. The similar items are grouped together and some items are made in batches to save the setup time if they are in the same delivery run.

In addition to the picking list, a number of times a week a machine shop request list is prepared and passed to the machine shop by the assembly coordinator. When the relevant partial picking list is received, the corresponding assembly section checks the component stock levels against the requirements due to new order and then if there is a need an order for that component is added to the machine shop request list.

It should be noted that there is a time lag between the receipt of the partial picking list and the actual start of the assembly production. Therefore this time lag, that is usually around one week, is used to manufacture the required components by the machine shop. It is also used to make the auxiliary cabinet parts such as drawers, doors, and plinths before the assembly operations starts. Each cabinet subsection calculates the required amount of these items as soon as a partial picking list is arrived and the necessary orders are passed to the auxiliary items maker. This synchronisation is achieved by the careful planning efforts by the assembly production supervisors.

In this case study conceptual AS-IS and TO-BE models were developed after number of visits to the case study company, conducting a number of useful discussions with company management and operators and collecting useful data. The products were grouped on basis of commonality of their operations and process times; for example cabinet assembly products were divided into 11 groups. An exemplary simulation model of cabinet assembly of the case study system developed in Plant Simulation ® is shown in Fig. 3.

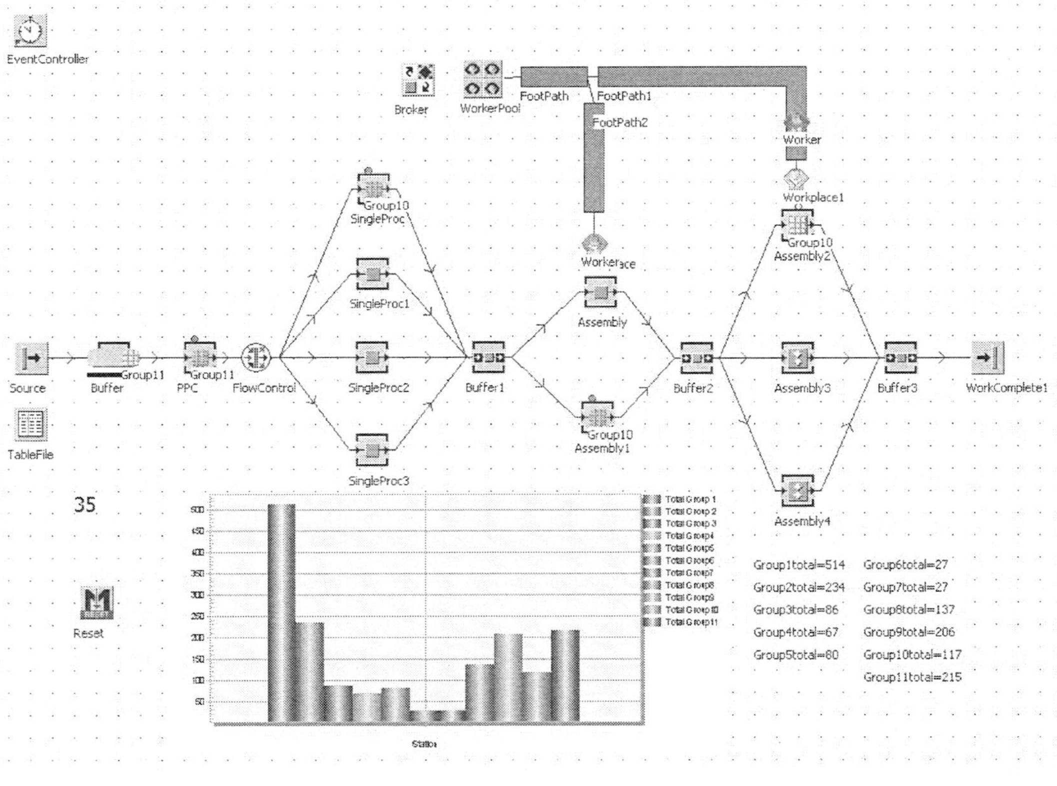

Fig. 3. An exemplary simulation model of cabinet assembly during the model run in Plant Simulation ®

Enterprise and simulation models were developed in order to enable decision making about ME possible realisation of an improved PPC strategy which could make this assembly process segment (and its underlying resource systems) become more reconfigurable and interoperable. One of the key performance indicators is lead time as also mentioned in [7]. Fig. 4 summarizes lead time KPIs of the proposed cabinet group assemblies.

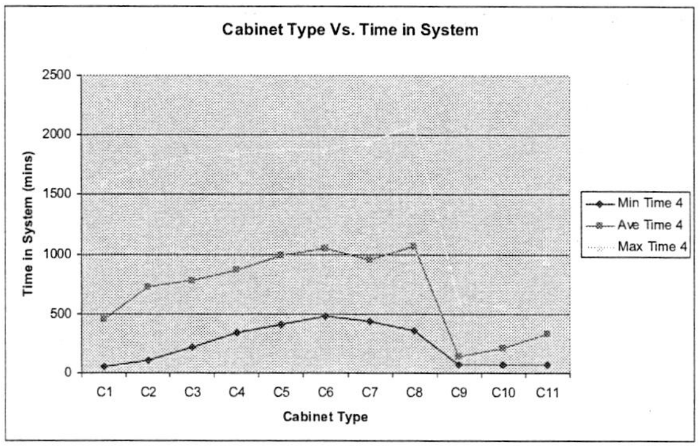

Fig. 4: Lead time results for different cabinet types

It was noted that the combined use of EM and SM techniques helped to gain in depth understanding about the ABC's PPC methods, their shortcomings and possible ways of improvements as tabulated in [7]. Such PPC decision making supports OD&C process. There is a need to develop methods and technologies for unified modelling and interoperation of partial models in support of complex OD&C. The future research also relate to the use and updating of enterprise and simulation models.

5. Conclusions

This paper presents an application of the integrated modeling approach presented in [7] in an example of a change-capable PPC scenario of cabinet assembly in a manufacturing SME. The developed environment supports the engineering of common types of strategic, tactical and operational process found in many MEs in support of reconfigurable manufacturing. The coherent multi-perspective models of a specific ME have facilitated process reengineering and associated resource system reconfiguration and interoperation. It follows that systematic methods of achieving model driven reconfiguration and interoperation of component based manufacturing systems are required to design, engineer and change next generation MEs. The basis of a systematic approach to creating coherent sets of unified models that facilitate the engineering of PPC system is described. Example enterprise and simulation models have been presented in the paper which supported model-driven reconfiguration of manufacturing systems. The paper has identified key areas for future research which relate to the use and updating of enterprise and simulation models.

References

[1] Vernadat, F.B., *Enterprise Modelling and Integration (EMI): Current Status and Research Perspectives.* Annual Reviews in Control, 2002. 26: p. 15-25.

[2] Suri, R., *QRM and POLCA: A Winning Combination for Manufacturing Enterprises in the 21st Century, Technical Report.* 2003, Centre for Quick Response Manufacturing, http://www.engr.wisc.edu/centers/cqrm/pub/qrm21st.htm.

[3] Ladet, P. and F.B.E. Vernadat, *The Dimension of Integrated Systems Engineering*, in *Integrated Manufacturing Systems Engineering.* 1995, Chapman & Hall: London. p. 3-9.

[4] Vernadat, F.B., *Enterprise Modelling and Integration: Principles and Applications.* 1st ed. 1996, London, UK: Chapman & Hall. 513.

[5] Maropoulos, P.G., *Digital enterprise technology – defining perspectives and research priorities.* International Journal of Computer Integrated Manufacturing, 2003. 16(7-8): p. 467-478.

[6] Young, R., *Informing Decision Makers in Product Design and Manufacture.* International Journal of Computer Integrated Manufacturing, 2003. 16(6): p. 428-438.

[7] Masood, T. and R.H. Weston, *An integrated model driven approach in support of next generation reconfigurable manufacturing systems*, in *The 4th International Conference on Innovative Production Machines and Systems (IPROMS 2008).* 2008. The Internet.

[8] Stevenson, M., L.C. Hendry, and B.G. Kingsman, *A review of production planning and control: the applicability of key concepts to the make-to-order industry.* International Journal of Production Research, 2005. Vol. 43(No. 5): p. 869–898.

[9] Lima, R.M., R.M. Sousa, and P.J. Martins, *Distributed production planning and control agent-based system.* International Journal of Production Research, 2006. Vol. 44(Nos. 18–19): p. 3693–3709.

[10] Rahimifard, A. and R.H. Weston, *The enhanced use of enterprise and simulation modelling techniques to support factory changeability.* International Journal of Computer Integrated Manufacturing, 2007. Vol. 20(No. 4): p. 307-328.

Innovative Production Machines and Systems
D.T. Pham, E.E. Eldukhri and A.J. Soroka (Eds.)

Simultaneously motion planning and scheduling for two autonomous vehicles

E.K. Xidias[a], A.C. Nearchou[b], N.A. Aspragathos[a]

[a] *Department of Mechanical & Aeronautics Engineering, University of Patras, Greece*
[b] *Department of Business Administration, University of Patras, Greece*

Abstract

This paper considers a planning and scheduling problem for two autonomous vehicles (AVs) operating simultaneously in 2D industrial environments cluttered by obstacles. The vehicles are requested to serve all the work centers while avoiding collisions with the obstacles during their motions. The objective is to determine the shortest possible collision free tours for the AVs. A new method is introduced to solve the problem accomplished in two successive phases: first, the AVs' environment is mapped into a 2D B-Spline surface embedded in 3D Euclidean space using the Bump-Surface concept. Then, the overall problem is expressed as a constrained optimization problem which is solved by a genetic algorithm which is applied on the generated surface to search for the optimum tours that satisfies the requirements of the vehicles' mission. The performance of the proposed method is investigated and discussed through characteristic simulated experiments.

Keywords: autonomous vehicles, motion planning, scheduling

1. Introduction

In recent years, considerable attention has been paid to industrial applications, where the autonomous vehicles (AVs) are requested to serve a set of work centers in environments cluttered with prohibited areas in minimum total travel time [1]. However, creating the appropriate advanced technology for AVs raises many critical complex combinatorial optimization problems, such as the motion planning problem and the scheduling problem. This paper examines the scenario where two AVs are demanded to serve all the work centers in the shop floor in a manner similar to the traveling salesman problem (TSP) [2]. That is, each vehicle starting from its depot, passing through a number of work centers from each one exactly once and returning back to its depot. The objective is to determine the minimum total travel time for the AVs that serves all the work centers in the 2D shop floor. It must be noticed that the AVs are not allowed to serve the same work centers.

The attainment of this objective passes through the solution of two known combinatorial optimization problems: motion planning (MP) and vehicle routing and scheduling (VRS) both of them known to be intractable. To all our knowledge, no previous work in the literature faces the two problems together. Traditionally, Vehicle Scheduling Problem (VSP) and Motion Planning Problem (MPP) are studied separately. The reason for this is that integrating scheduling and motion planning forms a very challenging NP-hard combinatorial optimization [3].

VSP is usually regarded in the literature as a variant of the time-constrained travelling salesman problem [4]. A single vehicle starts from a depot, visits a set of stations, passing through each one of them exactly once, and returns to the depot; while the overall routing schedule satisfies some pre-specified time requirements. More general versions of the problem may take into account the capacity constraints of the vehicle or may allow multiple vehicles and time windows [5]. VSP is usually presented by an undirected graph and its solution is obtained by searching this graph for an optimum route satisfying the related time constraints. All versions of VSP lead to an NP-hard COP and therefore the trend is to face these problems through robust heuristics algorithms [6].

MPP methods for multiple mobile robots can be classified as either centralized or decentralized [7]. In the centralized approaches, the robots and their degrees of freedom are considered altogether, and therefore, usually entail a high dimensional composite configuration space [8]-[10]. In the decoupled approaches [11], [12], a collision-free path for each robot is determined independently and then the paths are synchronized to resolve possible collisions among the robots. In the basic decoupled approach, planning is done in two phases. In the first phase, for each robot, a path is computed which is collision-free with respect to the obstacles in the environment excluding the other robots. Collisions between the robots are resolved in the second phase by velocity tuning, i.e. the velocities of each the robot along their paths, computed in the first phase, are selected by avoiding any collision between the robots along their respective paths. Recently, Xidias and Aspragathos [13] introduced a new method for solving MPP for multiple car-like robots moving in 2D terrains. The key-element of this method is the representation of the robot's workspace through a single mathematical entity using the Bump-Surface concept [14]. The MPP solution is searched on a 2D Bump-Surface embedded in 3D Euclidean space in such a way that its inverse image into the initial robot's environment satisfies the MPP objectives and constraints.

This paper presents a new approach for solving the combinatorial problem of VSP and MP for two AVs moving and operating in 2D industrial environments. The objective is to determine the minimum total travel time for the AVs that serves all the work centers in the 2D shop floor. The proposed approach first uses the Bump-Surface concept to map the vehicles' environment into a 2D B-Spline surface embedded in 3D Euclidean space, and then, the overall problem is expressed as a constrained optimization problem which is solved by a modified GA.

2. Problem Statement and Notations

Let two autonomous vehicles moving in a 2D industrial environment cluttered with prohibited areas and with a set of N work (see an example in Fig. 1). Let also the following assumptions:

- The prohibited areas have fixed and known geometry representing for instance, machines, walls, etc., in the vehicles' environment.
- The positions of the depots and the work centers are fixed and known.
- Depot is characterized by a location for vehicle arrivals and departures.
- Each vehicle's tour always starts from the depot, goes through a number of work centers exactly once and terminates at the depot.
- Each vehicle is represented by a point and the prohibited areas are expanded appropriately in terms of the Configuration Space [8].
- The vehicles are moving with constant velocity $|\tilde{u}|$.
- The vehicles' have to serve all the work centres.
- Since, each vehicle moves with a velocity of constant magnitude $|\tilde{u}|$ it can be assumed that a time optimal route corresponds to the minimum length route.

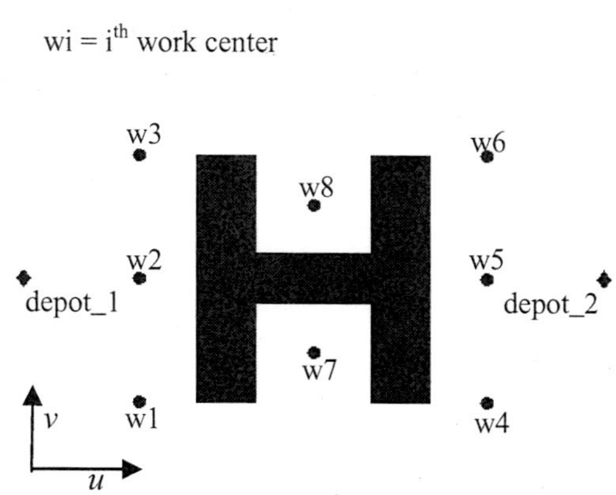

Fig. 1. A 2D industrial environment cluttered with one prohibited area, 2 depots and 8 work centers.

Under these assumptions, the problem addressed in this paper is: given the geometry of the environment, the positions of the depots and the work centers in the environment, generate a route for each vehicle satisfying the following criteria and constraints: (A) each vehicle should avoid collisions either with the other vehicle or with the prohibited areas, (B) the vehicles should serve all the work centers visiting each one exactly once in minimum total travel time and (C) the vehicles should not serve the same work centers.

3. Bump-surface: mapping the vehicles' environment

Bump-Surface [14] represents the entire 2D industrial environment of a vehicle with a 2D B-Spline surface embedded in $\Re^3$. In this way the vehicle's workspace is defined to be a 2D parametric surface in $\Re^3$ Euclidean space capturing both the free and the prohibited areas of the industrial environment. Also, we consider that the vehicle's environment is enclosed within a square of 1unit x 1 unit area.

Given a 2D industrial environment the construction of the Bump-Surface is derived after applying the Z-value algorithm [14]. This algorithm considers that the normalized environment is discretized into uniform subintervals along its u and v orthogonal directions respectively, forming a grid of points $p_{mn} = (x_{mn}, y_{mn}, z_{mn}) \in [0,1]^3$, $0 \le m,n \le M-1$ where M denotes the grid size. The z_{mn} coordinate of each grid point p_{mn} takes a value in the interval $(0,1]$ if the corresponding $(x_{mn}, y_{mn}) \in [0,1]^2$ of the grid point p_{mn} lies inside a prohibited area and a zero value otherwise.

The Bump-Surface is represented by a tensor product B-Spline surface with uniform parameterization $\tilde{S}:[0,1]^2 \to [0,1]^3$,

$$\mathbf{S}(u,v) = \sum_{m=0}^{M-1}\sum_{n=0}^{M-1} N_{m,q}(u) N_{n,f}(v) p_{mn}, 0 \le u,v \le 1 \quad (1)$$

where q, f denote the degree in the u, v directions of the Bump-Surface respectively, p_{mn} are the grid points and $N_{m,q}(u)$, $N_{n,f}(v)$ are the basis functions. It must be noticed that, the Bump-Surface consists of "flat" areas ($S_{z_{mn}}(u,v)=0$) and "bump" areas where ($S_{z_{mn}}(u,v) \in (0,1]$). Fig. 2 illustrates the corresponding Bump-Surface for the 2D industrial environment is shown in Fig. 1.

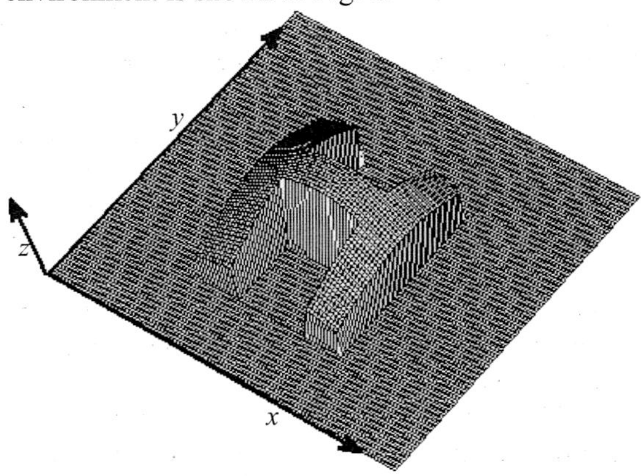

Fig. 2. The corresponding Bump-Surface.

4. Motion planning on the Bump-Surface

Let as consider that each vehicle traces a route $\mathbf{R}^j(t) = (u^j(t), v^j(t))$, $j = 1,2$ given as a first degree B-Spline curve:

$$\mathbf{R}^j(t) = \sum_{\delta=0}^{K^j-1} N_{\delta,1}(t)\mathbf{p}_\delta^j, \; 0 \le t \le 1, j = 1,2 \quad (2)$$

where $\{\tilde{\mathbf{p}}_0^j, \tilde{\mathbf{p}}_1^j, \dots, \tilde{\mathbf{p}}_{K-2}^j, \tilde{\mathbf{p}}_{K-1}^j\}$ are control points for the j-vehicle defined as in the following:

- $\tilde{\mathbf{p}}_0^j = \tilde{\mathbf{p}}_{K^j-1}^j$ denoting the depot point for the j-vehicle.
- $\{\tilde{\mathbf{p}}_1^j, \dots, \tilde{\mathbf{p}}_{K^j-2}^j\} \subseteq \left(\begin{matrix}\{Work\ Centres \equiv N_j\} \cup \\ \{intermediate\ points \equiv \mathbf{g}_l^j,\ l = 1,\dots,b_j\}\end{matrix}\right)$

  defined in the parametric space of $\tilde{S}$.
- K^j is the total number of control points $\mathbf{p}_\delta^j$, $\delta = 0,\dots,K^j-1$ in the generated $\mathbf{R}^j(t)$ route.
- $N_{\delta,1}(t)$ is the B-Spline basis function.
- $b_j = r \cdot (N_j + 1)$ denotes the total number of the intermediate points $\mathbf{g}_l^j \in [0,1]^2$.
- r is the predefined number of points between pair of work centres.

- $N_j \leq N$, where $N_1 + N_2 = N$, is the number of the work centres to be served by the *j*-vehicle in the generated route.

Therefore, a solution to the MPP consists of generating the $K^j - 2$ control points $\mathbf{p}_\delta^j$ for *j*-vehicle such that the route $\mathbf{R}^j(t)$, *j*=1, 2 satisfies the aforementioned criteria and constraints **A**, **B** and **C**.

2.1. Optimal planning and scheduling

A feasible *j*-route that avoids the prohibited areas should be searched in the "flat" areas of the Bump-Surface. A route that "climbs" the bumps of the Bump-Surface results to an invalid route in the real (initial) 2D industrial environment that penetrates the prohibited areas. By construction, the arc length of R approximates the length of its image $S(\mathbf{R}^j(t))$ on $\tilde{S}$ as long as $\mathbf{R}^j(t)$ does not penetrate the prohibited areas. Therefore, it is reasonable to search for a "flat" route on $\tilde{S}$ in order to satisfy objective **A**.

The arc length of the image of the $\mathbf{R}^j(t)$ onto $\tilde{S}$ is given by:

$$L^j = \sum_{i=0}^{N_p-2} \left\| S(\mathbf{R}_{i+1}^j) S(\mathbf{R}_i^j) \right\|, \; j = 1,2 \qquad (3)$$

where $N_p - 1$ are equal sequential chords which approximate the curve $\mathbf{R}^j(t)$. Furthermore, the two vehicles should not collide. This requirement is expressed by,

$$\left(u_1(t), v_1(t)\right) \neq \left(u_2(t), v_2(t)\right) \qquad (4)$$

where, $\left(u_j(t), v_j(t)\right), j = 1,2$ is position of the *j*-vehicle at time *t*.

In addition, each *j*-vehicle should follow a *j*-route which connect a number of work centers $N_j \leq N, j = 1,2$, where $N_1 + N_2 = N$, and its *j*-depot while they should not serve the same work centers. This requirement is expressed by,

$$\left\{\tilde{\mathbf{p}}_0^1, \tilde{\mathbf{p}}_1^1, \ldots, \tilde{\mathbf{p}}_{K^1-2}^1\right\} \cap \left\{\tilde{\mathbf{p}}_0^2, \tilde{\mathbf{p}}_1^2, \ldots, \tilde{\mathbf{p}}_{K^2-2}^2\right\} = \varnothing \qquad (5)$$

Therefore, the overall optimization problem is written as,

$$\mathbf{min} \; E = L^1 + L^2 \qquad (6)$$

$$\text{subject to,} \begin{cases} \left\{\tilde{\mathbf{p}}_0^1, \tilde{\mathbf{p}}_1^1, \ldots, \tilde{\mathbf{p}}_{K^1-2}^1\right\} \cap \left\{\tilde{\mathbf{p}}_0^2, \tilde{\mathbf{p}}_1^2, \ldots, \tilde{\mathbf{p}}_{K^2-2}^2\right\} = \varnothing \\ \left(u_1(t), v_1(t)\right) \neq \left(u_2(t), v_2(t)\right) \end{cases}$$

Note that, minimizing E with respect to the control points $\mathbf{p}_\delta^j$, leads to a collision free *j*-route for each *j*-vehicle which satisfies the criteria **A**, **B** and **C**.

5. The optimization heuristic: a modified genetic algorithm

A modified genetic algorithm (GA) has been designed and implemented in order to solve the optimization problem given by Eq. (6). The adoption of GAs in this work was motivated by the fact that these heuristics have been theoretically and empirically proven to provide robust search in large and complex search spaces which can be multimodal and non-linear. The characteristics of the proposed GA are analyzed and described in following paragraphs.

A mixed integer and floating-point representation scheme was selected for use. That is, a chromosome is a string consisting of two parts: the first (left) part contains only integer numbers, while the second part contains floating-point numbers. Each chromosome corresponds to unique vehicles' routes on the B-Spline surface. More specifically, each chromosome has variable length and consists of $N_j + b_j \cdot 2$ genes. $b_j \cdot 2$ (2 denotes the dimensions of the environment) is the overall number of the intermediate points in the generated *j*-route. The integer part of the string represents the order with which the *j*-vehicle visits the N_j work centres. The real-valued part of the string is composed by $b_j \cdot 2$ floating numbers representing the intermediate points $g_l^j \in [0,1]^2$, *l*=1,…,b_j which define the B-Spline curve (route) of the *j*-vehicle.

Reproduction is a simple copy of an individual from the population of the current generation to the population of the next generation. In this work, the proportional selection strategy is adopted. According to this strategy, the chromosomes are selected to reproduce their structures in the next generation with a rate proportional to their fitness.

Crossover joins together parts of several individuals in order to produce new ones for the next generation. The individuals are randomly selected according to a user-defined probability (crossover rate). For the first part of the chromosome, that with the integers, the Order Crossover (OX) followed by a suitable repairing mechanism was selected for use, while for the second part (with the floating-point numbers) of the chromosome, the one-point crossover was adopted.

Mutation: For the first part (with integers) the inversion operator is used, while for the second part (with the real-valued numbers) a boundary mutation was used. Inversion selects two positions along the string at random and then inverts the sub-section of the values between these two positions. Boundary mutation changes the value of a gene with a random number chosen from the permitted range of coordinates on the Bump-surface.

In this paper the following fitness function was used,

$$fitness = \begin{cases} \frac{1}{E}, & \left\{\tilde{\mathbf{p}}_0^1, \tilde{\mathbf{p}}_1^1, \ldots, \tilde{\mathbf{p}}_{K^1-1}^1\right\} \cap \\ & \left\{\tilde{\mathbf{p}}_0^2, \tilde{\mathbf{p}}_1^2, \ldots, \tilde{\mathbf{p}}_{K^2-1}^2\right\} = \varnothing, \\ & \left(u_1(t), v_1(t)\right) \neq \left(u_2(t), v_2(t)\right) \\ 0, & otherwise \end{cases} \tag{7}$$

6. Experimental results and results

Two characteristic case studies were conducted to demonstrate the performance of the proposed method. The simulations were implemented in Matlab and run on a Pentium IV 3.2 GHz PC. The grid size was set to 80x80 and $|\tilde{u}| = 1$. The suitable settings for the GA's control parameters were experimentally determined and defined as follows: *population size* =200, *maximum number of generations* =300, *crossover rate* =0.75, inversion rate =0.095, and *boundary mutation rate* =0.004. Furthermore, in the experiments a 2-degree B-Spline curves and a (2, 2)-degree B-Spline surface were used to represent the generated paths of the vehicles and the Bump-Surface, respectively.

Test case I: Consider the shop-floor layout depicted in Fig. 1. The AVs mission are to deliver products to the (8 in total) work centers. The AVs are moving in an environment cluttered with one non-convex prohibited area. We set *r* (i.e. the number of intermediate points between each pair of work centres) equal to 1. The solution route for the 1-vehicle is depot_1-w3-w2-w1-w7-depot_1 and the solution route for the 2-vehicle is depot_2-w4-w5-w6-w8-depot_2, Fig. 3.

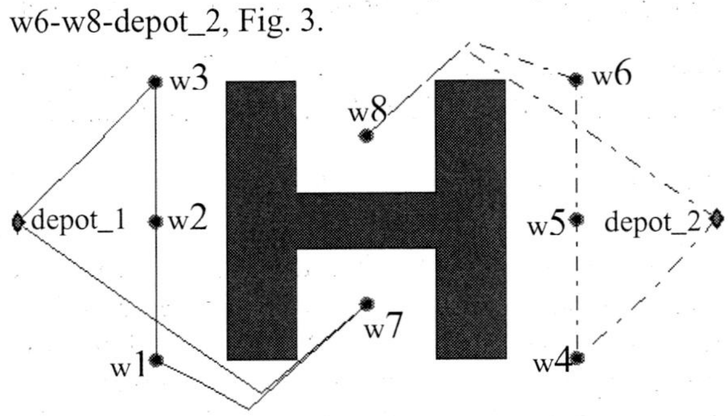

Fig. 3. The solution routes.

Test Case II: Consider two AVs which their mission are to serve 8 wotk centers in an 2D shop foor cluttered with four prohibited areas (two non-convex and two convex), Fig. 4 . *r* (i.e. the number of intermediate points between each pair of work centres) was equal to 2. The solution route for the 1-vehicle is depot_1-w1-w3-w4-w2-w8-w7-depot_1 and the solution route for the 2-vehicle is depot_2-w5-w6-depot_2, as illustrated in Fig.5.

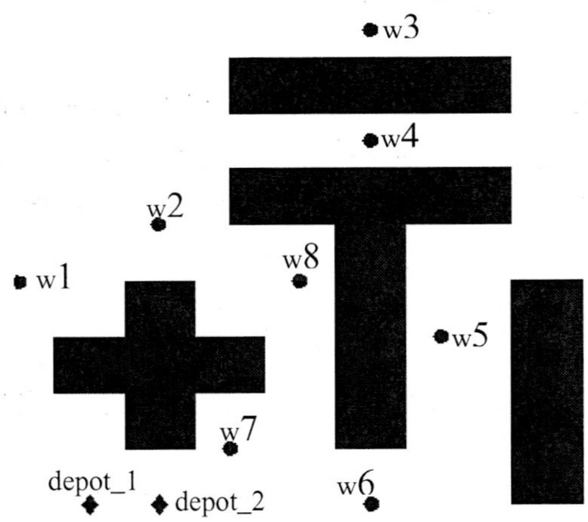

Fig. 4. The 2D environment.

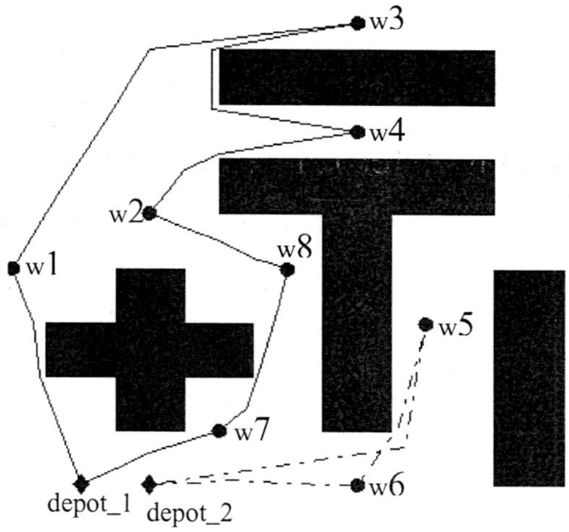

Fig. 5. The solution routes.

As can be seen from Fig. 5, the proposed method (due to the flexibility of the B-Spline curves and the robustness of the genetic search), is able to produce collision-free routes in complicated environments with narrow corridors. The generated solution routes satisfy (in a near optimum way) all the established in section 2 path planning and scheduling criteria.

7. Conclusions

In this paper, a new method for determining near-optimal, collision-free routes for two autonomous vehicles moving in 2D factory environments is presented. During their motion the vehicles are demanded to serve all the work centers in the shop floor. The proposed method consists of two phases: first, the 2D vehicle's environment is mapped into a 2D B-Spline surface embedded in 3D Euclidean space using a robust geometric model. Then, a modified genetic algorithm is applied on the generated surface to search for optimum routes that satisfy the requirements of the vehicles' mission. Simulated experiments showed a high performance for the method in terms of the quality of the solutions obtained.

Acknowledgment

University of Patras is partner of the EU-funded FP6 Innovative Production Machines and Systems (I*PROMS) Network of Excellence.

References

[1] Halme A., Koskinen K.. Intelligent Autonomous Vehicles. Elsevier Science & Technology, 1995.

[2] Lawler, E.L., Lenstra J.K., Rinnooy Kan A.H.G., and Shmoys D.B., (eds) (1985) The Traveling Salesman Problem: A Guided Tour in Combinatorial Optimization. Wiley, New York.

[3] Vis Iris F.A.. Survey of research in the design and control of automated guided vehicle systems. European Journal of Operational Research, 170, 677-709, 2006.

[4] Baker E.. An exact algorithm for the time-constrained travelling salesman problem. Operations Research, 31, 938-945, 1983.

[5] Solomon M.. Algorithms for the vehicle routing and scheduling problem with time windows constraints. Operations Research, 35, 254-265, 1987.

[6] Tsitsiklis J.N.. Special cases of travelling salesman and repairman problems with time windows. Networks, 22, 263-282, 1992.

[7] Latombe J. C., "Robot Motion Planning", Kluwer Academic Publishers, Boston, 1991.

[8] L. Perez. Spatial Planning: A Configuration Space Approach. IEEE Trans. On Computers, pp 108-120, 1983.

[9] G. Sánchez and J.C. Latombe. On delaying collision checking in PRM planning: application to multi-robot coordination. Int. J. of Robotics Research, pp 5-26, 2002.

[10] M. Egerstedt and X. Hu. Formation constrained multi-agent control. In Proceedings of the IEEE Conference on Robotics and Automation, 3961–3966, Seoul, Korea, May 2001.

[11] Y. Guo and L. E. Parker. A distributed and optimal motion planning approach for multiple mobile robots. In Proceedings of the IEEE International Conference on Robotics and Automation Washington, DC, pp 2612–2619, 2002.

[12] M. Bennewitz, W. Burgard and S. Thrun. Optimizing schedules for prioritized path planning of multi-robot systems. In Proceedings of the Int. Conf. on Robotics and Automation, pp 271- 276, 2001.

[13] Elias K. Xidias and Nikos A. Aspragathos. Motion planning for multiple nonholonomic robots using the Bump-Surface concept. Accepted for publication in *Robotica Journal, 2007.*

[14] Azariadis P., Aspragathos N.. Obstacle Representation by Bump-Surfaces for Optimal Motion-Planning. Journal of Robotics and Autonomous Systems 51(2-3): 129-150, 2005.

Innovative Production Machines and Systems
D.T. Pham, E.E. Eldukhri and A.J. Soroka (Eds.)

Architecture Investigation of Energy Harvester for Wireless Sensor Nodes

Z.Zhu, S.Mekid and R. Pietruszkiewicz

School of Mechanical, Aerospace and Civil Engineering, University of Manchester, Manchester,M60 1QD, U.K.

Abstract

A research highlight is to design an energy harvesting system with small geometric characteristics, which works indoor for wireless sensor nodes. After analyzing the power consumption of each component in a sensor node and evaluating experimentally the amount of energy harvested indoor, the paper proposes a possible architecture of energy harvesting system and provides design considerations.

Keywords: Energy harvester, Wireless sensor node, Vibration, Solar panel, Piezoelectricity buzzer.

1. Introduction

The lifespan of wireless sensor networks (WSNs) constrains greatly their applications. It is because sensor nodes are powered by batteries and hence operate on an extremely frugal energy budget. Further, they must have a lifespan on the order of months to years, since battery replacement is not an option for networks with thousands of physically embedded nodes. The network lifespan can be maximized only by incorporating energy-awareness into every stage of wireless sensor network design and operation [1].

So far there is a lot of research to extend the lifespan of WSNs on techniques such as scavenging energy from environment, wireless recharging technology [2], and reducing power consumption of wireless sensor networks. However, constraint conditions always exist in these techniques mentioned above. For example, there is a lower bound for reducing power consumption of WSNs, and wireless recharging technology depends on application environment as well as the distance between sensor nodes and power source. It seems that energy harvesting technology can make a longer lifespan of WSNs than the other two methods do as long as energy harvesting sources always exist.

Energy harvesting system normally consists of energy harvesting sources, energy converting subsystem, and energy storing components. Energy harvesting sources are mechanical energy resulting from vibration, stress and strain; thermal energy from furnaces and other heating sources even biological; solar energy from all forms of light sources, ranging from lighting, light emissions and the sun; electromagnetic energy that are captured via inductors, coils and transformers; wind and fluid energy resulting from air and liquid flow; and chemical energy from naturally recurring or biological processes [3]. Energy converting subsystem converts all kinds of external energies into electrical power. Energy storing components are divided as two: one is to use super capacitor storing electrical energy converted; the other is to use rechargeable battery storing energy.

The research highlight in this paper is to

investigate energy harvesting system architecture for indoor wireless sensor nodes. The structure of this paper is listed as follows: Section 2 introduces the related work; Section 3 analyzes the power requirement for each subsystem in a wireless sensor node; Section 4 describes the efficiency of energy captured from environment; Section 5 proposes a possible architecture of energy harvesting system; Section 6 discusses several design consideration; finally conclusion and future work are given in Section 7.

2. Related work

There are a lot of research and implementations on energy harvesting systems that are applied to wireless sensor networks, which have been demonstrated successfully. Kymissis et. al. proposed a parasitic power harvesting in shoes and developed a self-powered RFID tag with piezoelectric elements [4]. A 1.9GHz radio frequency transmit beacon was implemented based on solar and vibration energy scavenging [5]. A compact piezoelectric pushbutton and associated minimal circuitry was designed to wirelessly transmit a digital ID code to the immediate region (e.g., 50-100 foot radius) upon a single button push, without the need of batteries or other energy sources [6]. Heliomote is a solar energy harvesting system to behave as a plug-in to the Berkeley/Crossbow motes and autonomously manages energy harvesting and storage, enables near-perpetual, harvesting aware operation of the sensor node [7]. Abhiman Hande, et. al. implemented an indoor solar energy harvesting system to power Crossbow router nodes [8].

3. Power consumption of a wireless sensor node

The output of energy harvesting system should be designed to match with the power requirement of a wireless sensor node. It is necessary to analyze power consumption of each component in the sensor node before designing.

The canonical architecture of a wireless sensor node is shown in Fig.1, which is comprised of four subsystems for communicating, computing and controlling, sensing, and powering.

Communicating subsystem consists of transceiver, antenna, and peripheral components. Its power consumption is evaluated by such parameters as voltage supply, transmitting current, receiving current, and current at power down mode. Table.1 provides power parameters of radio frequency modules based on Zigbee/IEEE802.15.4 protocol specification.

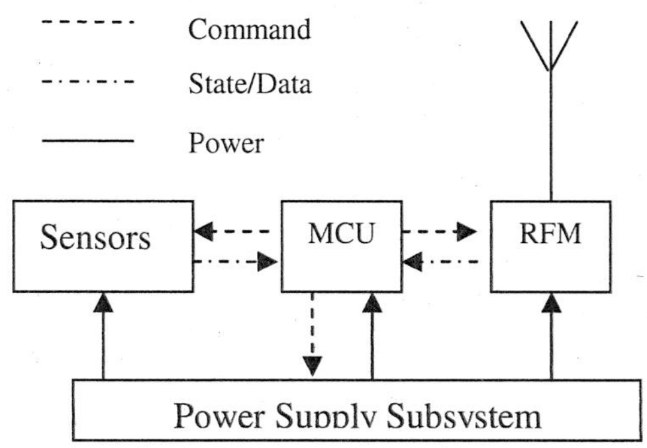

Fig.1 System architecture of a wireless sensor node

CC2420, a RF module provided by Chipcorn, is with lower power consumption than other modules according to Table 1, and is applied in many solutions such as Microchip, Silicon Labs, etc.

Table.1 Power parameters of typical RF modules

RFM	V (V)	I_{RX} (mA)	I_{TX} (mA)	I_{PD} (uA)
CC2420	2.1~3.6	18.8	17.4	0.9
MC13192	2.0~3.4	42	35	1
UZ2400	2.7~3.6	18	22	2
xBee	2.8~3.4	50	45	<10
xBee-PRO	2.8~3.4	55	270	<10
NanoPAN5360	2.4~3.6	35	78	1.5
NanoPAN5361	2.4~3.6	35	78	1.5

Note: V (supply voltage), I_{RX} (RX mode current), I_{TX} (TX mode current), I_{PD} (current at power down mode)

Computing and controlling subsystem centers a microprocessor. During the design of a wireless sensor node, a microprocessor is selected in terms of requirement of applications. Many components affecting power consumption are integrated into microprocessors, for example, ADC, DAC, RAM, timer, and so on. In fact, some internal components are not used in the design, but they still consume energy, so it is important to select a microprocessor for energy saving. Table.2 lists power parameters of part microprocessors. Clock frequency and voltage supply are important parameters to significantly affect power consumption of a microprocessor.

Table2 Power parameters of part microprocessors

Microprocessor	I (mA)	V (V)	f (MHz)	I_{PD} (uA)
C8051F930	4.25	0.9	25	0.05
PIC18F4620	16	4.2	40	0.1
MC9s08GT	6.5	3	16	2.5
ATMEGA128L	5.5	3	4	<5
MSP430CG4618	0.4	2.2	1	0.35
ML610Q431	0.65	1.1	4	0.25

Note: I (supply current), V (supply voltage), f (run frequency), I_{PD} (current at power down mode)

Sensing subsystem generally includes one or more sensors, and it is possible for these sensors to be homogeneous or heterogeneous. In general, the power of sensors is lower than that of other subsystems in a sensor node, but energy consumption on sensors still is very larger because their work time is very long during the lifespan of a sensor node.

Powering subsystem normally uses a DC-DC device to convert single voltage of power source into different voltage values to satisfy the requirements of different components. Power consumption exists in transferring efficiency and DC-DC device itself.

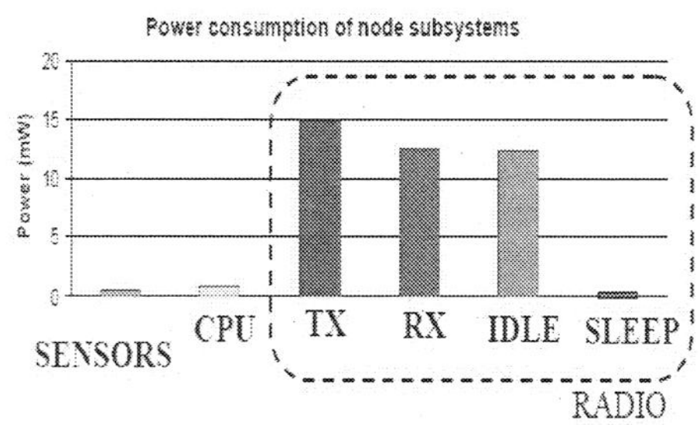

Fig.2 Power consumption distribution in a wireless sensor node

Deborah Estrin, in Conference Mobicom 2002, described the power consumption distribution for every subsystem of a wireless sensor node shown in Fig.2 [9]. It is obvious for the transceiver to consume a lot of energy, but the sensor to use small energy.

According to analysis mentioned above, the conclusion is that a wireless sensor node needs larger than 30mA current and 3volts or so voltage to support its routine work.

4. Efficiency of capturing energy from environment

It is important for analyzing efficiency of capturing energy from environment in order to confirm whether energy harvesting system designed can support a wireless sensor node to work smoothly or not. Performance of an energy harvesting source can be evaluated with such parameters as power density, current and voltage supplies. Since the experiment background is assumed as a workshop with machines similar to motor, the energy harvesting source consist of fluorescent light and vibration. So solar panels and piezoelectricity buzzers are used to scavenge ambient energy. Experimental processes and corresponding results are described as follows:

4.1. Solar panels

This experiment purpose is to evaluate performance of solar panels under different environments such as indoor and outdoor, different distance between light source and solar panel, and different light sources, and to confirm it is possible for the output of solar panel to support a wireless sensor node working normally. Two solar panels are selected for testing their voltage and current supplies under different conditions. Their sizes are 5.5cm X 5.5 cm and 15cm X 5.5cm respectively.

Table 3 Condition: Outdoor

Condition	Open Voltage (V) 5.5x5.5 panel	Short Current (mA) 5.5x5.5 panel	Open Voltag e (V) 15x5.5 panel	Short Current (mA) 15x5.5 panel
Cloudy	4.27	10.51	8.66	11.36
Sunlight	4.56	64.3	10.54	59.8

Experiment results outdoor are listed in Table3. Part features of solar panel can be observed from experiment data: first, it is obvious for light strength to impact its output current rather than to impact its output voltage. Second, ratio of output power between two solar panels almost is equivalence with cloudy and sunlight. Third, output current seems not to be impacted by area of solar panel.

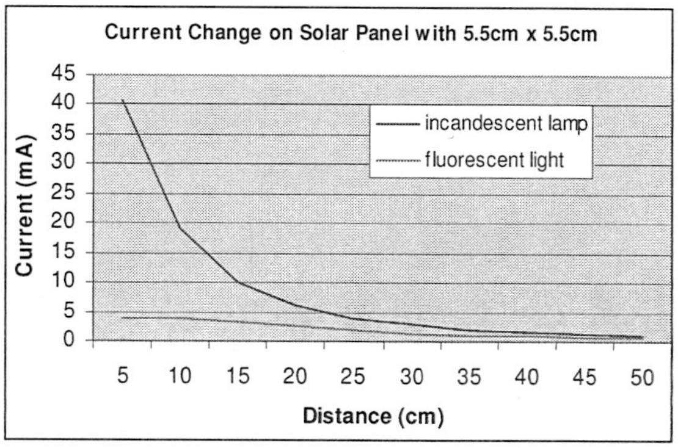

Fig.3 Current change on the solar panel with 5.5cm x 5.5cm

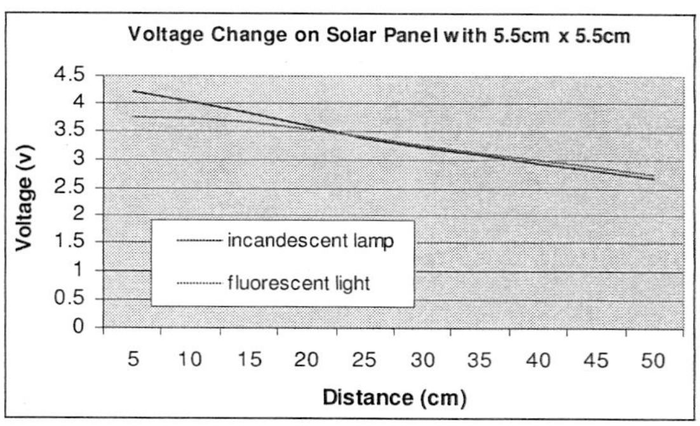

Fig.4 Current change on the solar panel with 5.5cm x 5.5cm

Indoor experiments are divided as two cases: one is to test voltage and current outputs of solar panels under incandescent lamp for different distance between light source and solar panels; the other is to repeat the same experiments under fluorescent light. For the solar panel with 5.5cm x 5.5cm, current and voltage changes based on different light sources are shown in Fig.3 and Fig.4 respectively. It is obvious the solar panel scavenges more energy from incandescent lamp than from fluorescent light. When the distance is 50cm, the open voltage and short current of the solar panel are 2.68v and1.0mA respectively under incandescent lamp, and 2.75v and 0.58mArespectively under fluorescent light.

Conclusion from experiments mentioned above can be described as follows:

- Solar panels can be modeled as a constant voltage source.
- There is a big difference of harvesting energy for different solar panels under different light sources.
- Harvesting energy from a single solar panel is too small to drive a sensor node directly.

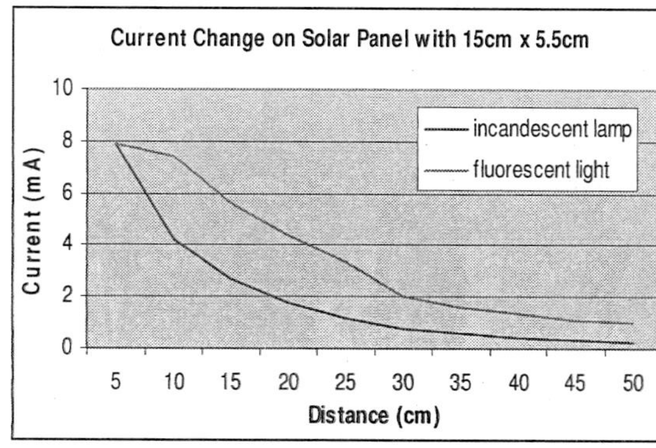

Fig.5 Current change on the solar panel with 15cm x 5.5cm

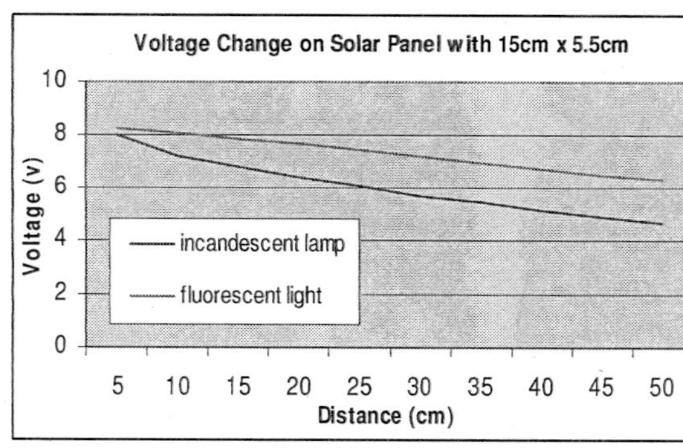

Fig.6 Voltage change on the solar panel with 15cm x 5.5cm

Piezoelectricity buzzers

In general, a piezoelectric buzzer is used to sound after it is powered. Reversely in the experiment it is put in vibration environment, and then expected to produce voltage and current. Three piezoelectric buzzers with different resonant frequencies are selected, as shown in Fig 7. Experimental results are listed in Table4.

Frequency range in experiments is selected from

20 to 2,000Hz, within which the vibration frequency produced by machine often occurs. Experiments prove that three piezoelectricity buzzers produce their maximum outputs both current and voltage at 1010Hz or so, even though with different frequency resonances. The DC outputs are 2.44mA and 11.21V respectively after the output of a piezo transducer (41.PT55L030H) via a rectifier bridge. The output of three-buzzer combiner seems to produce a current sum and a voltage mean among all buzzers. The experiment also proves that energy harvested by piezoelectric buzzers is not enough to drive a wireless sensor node.

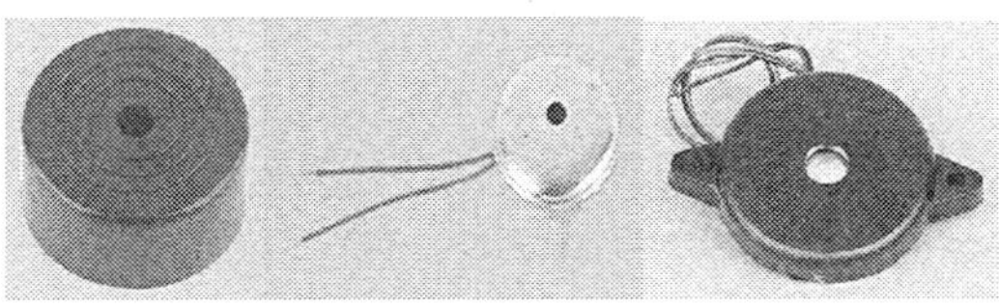

AI-4228-TWT-R TFM43 41.PT55L030H

Fig.7 Piezoelectricity buzzers with different resonant frequencies

Table3. Experimental data for piezoelectricity buzzers

Device	Resonant	Output Current @1kHz	Output Voltage @1kHz
Piezo Ceramic Buzzer AI-4228-TWT-R	2800 MHz	0.66mA	0.8V
Piezo Audio Transducer TFM43	4.7kHz	1.67mA	3.45V
Piezo Transducer 41.PT55L030H	4.4KHz	2.69mA	10.2V
Three Buzzer-Combiner		5.12mA	3.04V

5. A possible architecture of solar energy harvesting systems

According to energy consumption analysis of a wireless sensor node, power supply subsystem should provide voltage supply 3.3V or more, and more than 30mA current supply for such subsystems as calculating, communicating, and sensing. Because there are some existing unexpected factors, certain redundancy quantity should be considered for these parameters.

From the results of experiments mentioned above, single energy harvesting device only can produce several volts voltage and several milliamps current under indoor condition, so it is impossible for a single energy harvesting source to provide enough power to directly drive a sensor node to work. Also it is not a good design to increase greatly the number of energy harvesting devices for capturing enough energy to support the entire system because the volume of a node inflates sharply. But it is enough to use the energy to recharge battery. For example, the specification of Panasonic batteries indicates 3.3V rechargeable voltage and less than 4mA rechargeable current for VL serials.

A possible solution, combining rechargeable batteries and scavenging energy mechanism, is proposed in terms of reasons listed as below: The power consumption of a wireless sensor node will become smaller and smaller as the development in SoC (system on chip) technology and research reducing power consumption of IC chips. The lifespan of rechargeable batteries becomes longer and longer, and its capacity to store energy becomes larger and larger. According to the self-organization feature of wireless sensor networks, a sensor node exits from the WSN where it once worked, if its energy runs out, but it still can join into the network again when its energy is restored. During dropping out from the network, the sensor node can hoard of energy little by little, which is scavenged from outside environment.

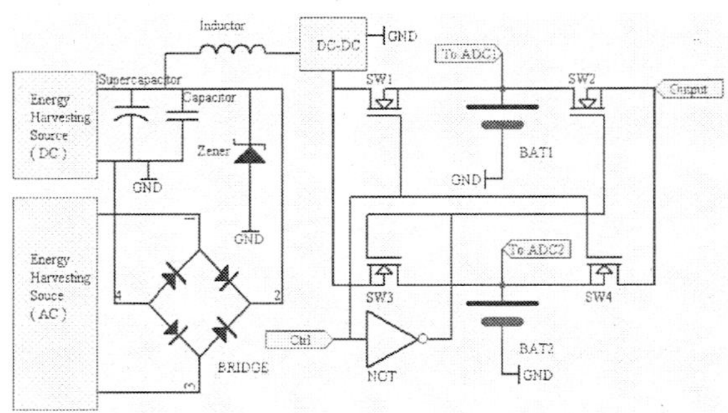

Fig. 8 A possible architecture of energy harvesting system

The framework of the solution is shown in Fig.8. Energy harvesting source can be selected from solar panel, piezoelectricity component, or thermo-generator. It is possible to use their combiner within allowing range of a sensor node volume. Energy from energy harvesting source is stored in a supper capacitor. In order to avoid shorting the lifespan of rechargeable battery, its manufacturer provides a strict specification for recharging voltage and current supplies. For getting recharging voltage with a fixed value, it is unavoidable to use a DC-DC converter in the circuit, which owns normally a larger range of input voltage, and appointed or adjustable output voltage. A zener diode is used to protect the supper capacitor and DC-DC device. A rechargeable battery with small volume and larger capacity is recommended since it is easy to form a cordwood structure for different applications. A microprocessor should monitor the voltage values of two batteries so that corresponding switch signal is sent according to their states.

6. Design consideration

There are two issues to impact the design of an energy harvesting system. One is how to capture maximum energy from environment; the other is how to extend the lifespan of rechargeable battery as long as possible.

It is feasible to design a combiner of energy harvesting sources for maximizing ambient energy captured according to application environment. For example, a combiner consisting of solar panel and piezoelectric generator is used to capture energy from fluorescent light and machine vibration in a workshop with larger motors. A combiner of multi-buzzer with different frequency resonance can capture more energy from vibration.

Reducing power consumption of an energy harvesting system is another method to maximizing environment energy captured by improving circuit design and selecting components with low power.

In the scheme the lifespan of a wireless sensor node is as long as that of recharging batteries. So it should be considered firstly to select rechargeable batteries with a long lifespan. Rechargeable batteries normally are divided as four types: Nickel Cadmium (NiCd), Nickel Metal Hydride (NiMH), Lithium based (Li+), and Sealed Lead Acid (SLA). Li+ batteries have a longer cycle lifetime, and involve a lower rate of self-discharge [7]. Secondly, effective methods for extending the lifespan of rechargeable batteries are to design suitable recharging circuit and switch scheduler reasonably, and to reduce battery aging caused by frequent charge-discharge cycles.

7. Conclusion and future work

It is a challenging problem to design an energy harvesting system applied in indoor environment for wireless sensor nodes. After analyzing the power consumption of a wireless sensor node and estimating energy harvested by solar panels and piezoelectricity buzzers, the paper proposed a possible architecture of energy harvesting system and discussed design consideration. The future work is to model and optimize the architecture of energy harvesting system, to explore how to capture ambient energy as more as possible, to extend the lifespan of rechargeable batteries as long as possible, and to finally implement an energy harvesting system with high performance for wireless sensor nodes indoor.

References

[1] Raghunathan V, Schurgers S, Park S, and Srivastava MB. Energy aware wireless microsensor networks. Signal Processing Magazine, IEEE, Volume 19, Issue 2, Mar. 2002, pp. 40-50.

[2] http://web.mit.edu/newsoffice/2007/wireless-0607.html

[3] http://www.aldinc.com/pdf/EH300Brochure.pdf

[4] Kymisis J, Kendall C, Paradiso J, and Gershenfeld N. Parasitic power harvesting in shoes. Proc. IEEE International Conference on Wearable Computing, pp. 132–139, 1998.

[5] Roundy S, Otis B, Chee YH, Rabaey J, Wright P. A 1.9GHz RF Transmit Beacon using Environmentally Scavenged Energy, ACM International Symposium on Low Power Electronics and Design, 2003.

[6] Paradiso JA and Feldmeier M. A compact, wireless, self-powered pushbutton controller. Proc. ACM International Conference on Ubiquitous Computing, pp. 299–304, 2001.

[7] Raghunathan V, Kansal A, Hsu J, Friedman J, and Srivastava MB. Design Considerations for Solar Energy Harvesting Wireless Embedded Systems. IEEE International Conference on Information Processing in Sensor Networks (IPSN), 2005.

[8] Hande A, Polk T, Walker W, and Bhatia D. Indoor Solar Energy Harvesting for Sensor Network Router Nodes. Article in press, *Journal of Microprocessors and Microsystems - Special Issue on Sensor Systems*, December 2006.

[9] Estrin D. Wireless Sensor Networks Tutorial Part IV: Sensor Network Protocols. Mobicom, Sep. 23-28, 2002. Westin Peachtree Plaza, Atlanta, Georgia, USA.

Innovative Production Machines and Systems
D.T. Pham, E.E. Eldukhri and A.J. Soroka (Eds.)

Intelligent Control of ankle-foot complex with functional electrical stimulation

I.P. Veneva[a] , R.S. Antonova[a], N.B. Zlatov[b]

[a] *Institute of Mechanics, Bulgarian Academy of Sciences, Sofia, Bulgaria*
[b] *Manufacturing Engineering Centre (MEC) School of Engineering, Cardiff University, UK*

Abstract

Functional electrical stimulation (FES) is a mobility assisting technology commonly used to help persons with weakness of ankle dorsiflexor muscles due to peripheral or central nervous system disorders. FES is a method for control of muscle contractions that produce specific movements required by the patient.

This paper presents a control device that allows the intensity of stimulation to be modulated throughout the gait cycle using pulse-width modulation (PWM). A microcontroller and signals received from two sensor arrays incorporated in the insoles of the two legs control the stimulation pulse-width applied to the user during the stimulation cycle. During each gait cycle a microcontroller estimates the forward speed and modulates flexion and extension in order to achieve quite normal walking dynamics.

The proposed intelligent control was used to control different models of personalized ankle-foot orthoses designed using reverse engineering and rapid prototyping by the joint team at Bulgarian Academy of Sciences/Cardiff University.

Keywords: Control, Biomechanics, Mechatronics, Robotics Foot drop, FES, PWM Modulation, Stimulation

1. Introduction

Electrical stimulation of muscle is based on sound physiological principles, which are reviewed here. In the intact efferent pathways, a somatic motor neuron originating in the brain or central nervous system will innervate a group of muscle fibres. This is shown in Fig.1. The neuron and the fibres that it innervates are known as the motor unit. The motor unit forms the basic unit of contraction. An electro-chemical signal is transmitted down the neuron to the fibres of the motor unit that is to contract. The signal is continued down the fibres, causing an electro-chemo-mechanical response which results in the contraction of the muscle fibre. An externally applied voltage potential to the motor unit can trigger the same electro-chemo-mechanical response. Therefore, electrical stimulation can cause a muscle to contract. Given that a contraction can be induced artificially, the muscle can then be used as an actuator to improve patient function.

Functional electrical stimulation is a methodology that uses shorts bursts of electrical pulses to generate muscle contraction (Bajd et al., 1999). These pulses generate action potentials in motor neurons attached to a muscle, which cause that muscle to contract.

There are currently three types of stimulators:

surface (transcutaneous), needle (percutaneous), or implanted electrodes (Bajd et al., 1999).

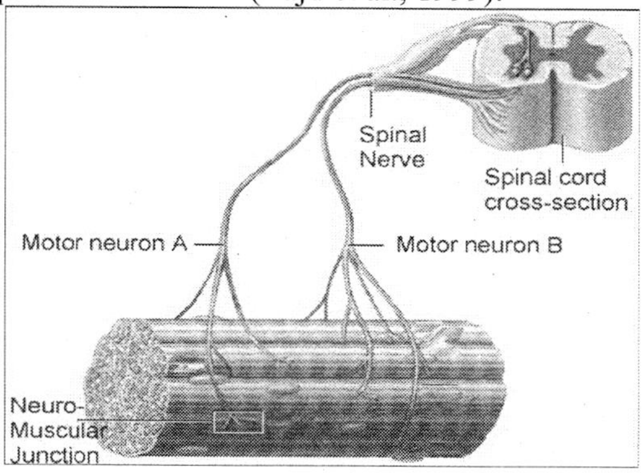

Fig. 1. Muscle with two innervating motor units

Transcutaneous stimulation consists of self-adhesive or nonadhesive electrodes placed on the subject's skin, as shown in Figure 2. Percutaneous stimulation uses wire electrodes introduced into muscle with an epidermal needle (Bukett et al., 1998). Implanted electrodes are attached to the motor neurons or the muscles close to the motor neurons (Cameron et al., 1997).

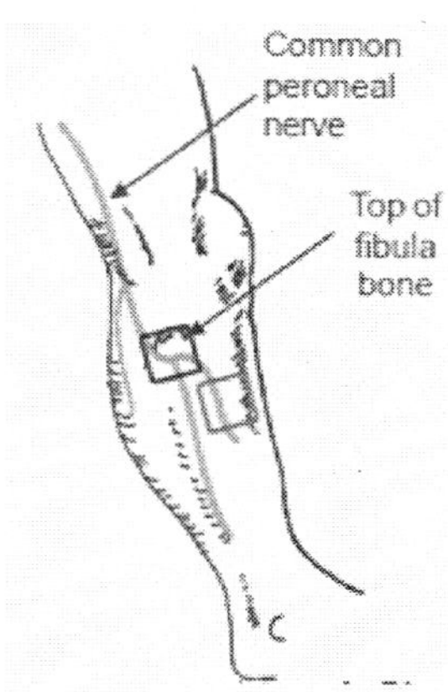

Fig. 2. Electrodes placed on the subject's skin.

The first walking FES neuroprosthesis was proposed by Liberson to help drop-foot subjects. The prosthesis stimulates the peroneal nerve (Fig.2), causing a flexion reflex of the hip, knee and ankle that allows the user to move the disabled leg (Liberson et al., 1961). Another system was developed in the Ljubljana Rehabilitation Institute that included a wireless link between the heel switch and stimulator and a small implantable radio receiver for the electrodes (Vodovnik et al., 1978). Since then, several FES systems have been developed to help drop foot subjects. Among them are the Fepa, MikroFES, WalkAid (Neuromotion, Inc. Edmonton, AB), Odstock 2 (Fig.3), and ETHZ-Paracare. All these systems use surface stimulation triggered by a push button, foot switch or gait phase recognition sensor. The push button, attached to a walker, is pressed by the subject and initiates the walking stimulation sequence. The foot switch control, located in the shoe sole under the heel of the subject, automatically triggers the sequence on heel-strike (Taylor et al., 1999). The disadvantage is false triggering from foot sliding or weight shifting (Popovic et al., 1998).

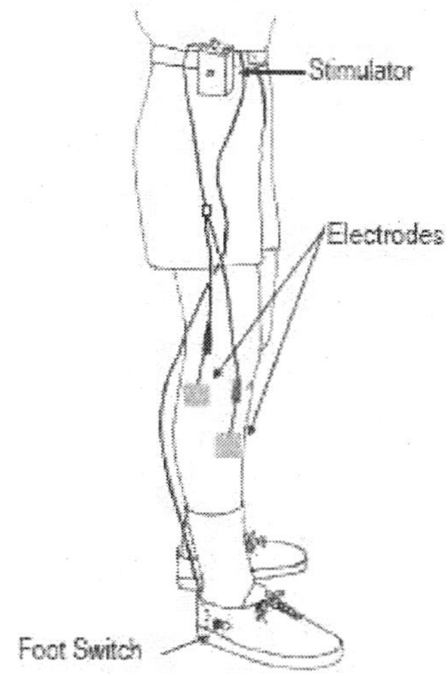

Fig.3. Odstock Dropped Foot Stimulator

The gait phase recognition sensor consists of three force-sensitive resistors (FSR), a gyroscope, and a rule-based observer. This system is able to identify four gait phases of walking: heel-off, swing phase, heel-strike, and mid stance (Popovic et al., 2001). Most of these devices are used for short-term rehabilitation in a clinical environment, though they have been used as permanent orthosis (Bajd et al., 1999). Though some of these systems have been fitted to hundreds of subjects, only the WalkAid has FDA approval.

Closed-loop control for the correction of drop foot has been examined by Chen et al. (2001). However, the intensity of the stimulation over each swing, after initial ramp-up, was constant. Mourselas and Granat (2000) investigated fuzzy control of ankle dorsiflexion during gait cycle. They varied the stimulation intensity during swing only, to provide a constant ankle angle; however, no attempt at creating an ankle angle trajectory, varying throughout gait cycle, was made.

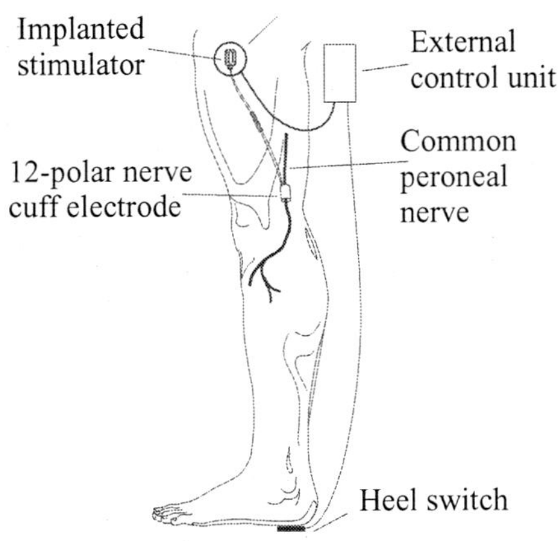

Fig.4. Implanted system with stimulation to the peroneal nerve

None of the currently developed microprocessor-based stimulators has provided a means of modulating the stimulation intensity over the gait cycle.

This paper presents a control device that allows the intensity of stimulation to be modulated throughout gait cycle using pulse-width modulation. The aim of the work is to present an approach to control the Functional electrical stimulator. A microcontroller and signals received from two sensor arrays incorporated in the insoles of the two legs control the stimulation pulse-width applied to the user during the stimulation cycle.

2 Method

During the swing phase of each gait cycle, the stimulator transmits electrical current to stimulate the ankle dorsiflexors or the common peroneal nerve to achieve toe clearance. Its function is similar to conventional Ankle-Foot Orthosis.

The purpose of the stimulator is to deliver stimulation to the peroneal nerve, which is modulated throughout the stimulation cycle, during gait cycle. The stimulation is triggered by use of a foot switch to detect heel rise, which marks the beginning of the stimulation cycle. The software generates a rectangular waveform defining the frequency and pulse-widths of the stimulation pulses at the implant. The widths and frequency of the rectangular pulses of the waveform output by the control unit determine the widths and frequency of the stimulation pulses.

The stimulator must adapt to the current walking cadence of the user. Walking is a cyclical process; different demands are placed on the musculoskeletal system at different phases of the gait cycle. The duration of the gait cycle varies with walking cadence; the delivery of stimulation must be synchronous with this cycle, and thus the stimulator must adapt to this changing walking cadence.

For the muscle to contract its motor neurons must be intact i.e., the muscle should not be "denervated". In order to achieve a continuous muscle contraction, also called tetanization, the FES system must produce pulses at a minimum of 20 Hz. Otherwise, the muscle twitches continuously and does not generate a steady output force. Both monophasic and biphasic current or voltage pulses can be used to stimulate motor neurons. The common belief, however, is that the injected charge should be removed from the body and not allowed to accumulate. Most surface electrodes use biphasic current pulses, which changes the positions of the anode and cathode during stimulation.

2.1. System Description

The system (see Fig.6) has two basic components: The external sensors array and The Control Unit. The Control Unit generates a stimulation pulse-width profile for ankle flexion/extension motions and controls the functional electrical stimulator. The system uses a position transducer mounted on the ankle and the back side of the shoe to detect flexion or extension.

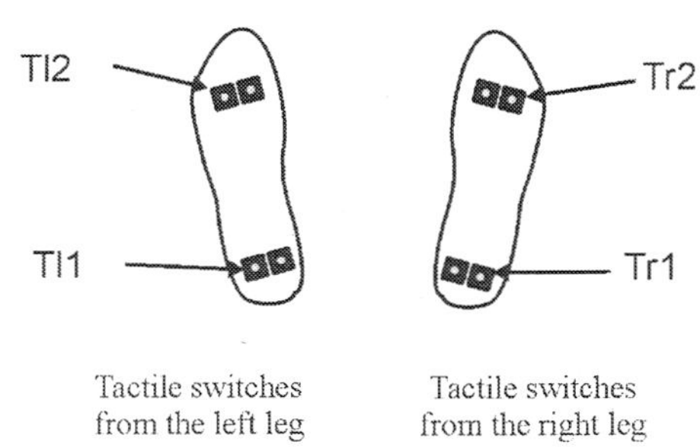

Fig. 5. Two sensor arrays incorporated in the insoles of the two legs

The flexion and extension motions are controlled by eight tactile sensors incorporated in the insoles of the two legs which trigger the foot position (Fig. 5). These sensors data are used in every step of the control algorithm in order to optimize the heel-to-forefoot transition during the stance phase or the swing phase of walking. The system is able to identify heel-strike, heel-off and four gait phases of walking: swing phase, mid stance and two double stances (Fig. 7).

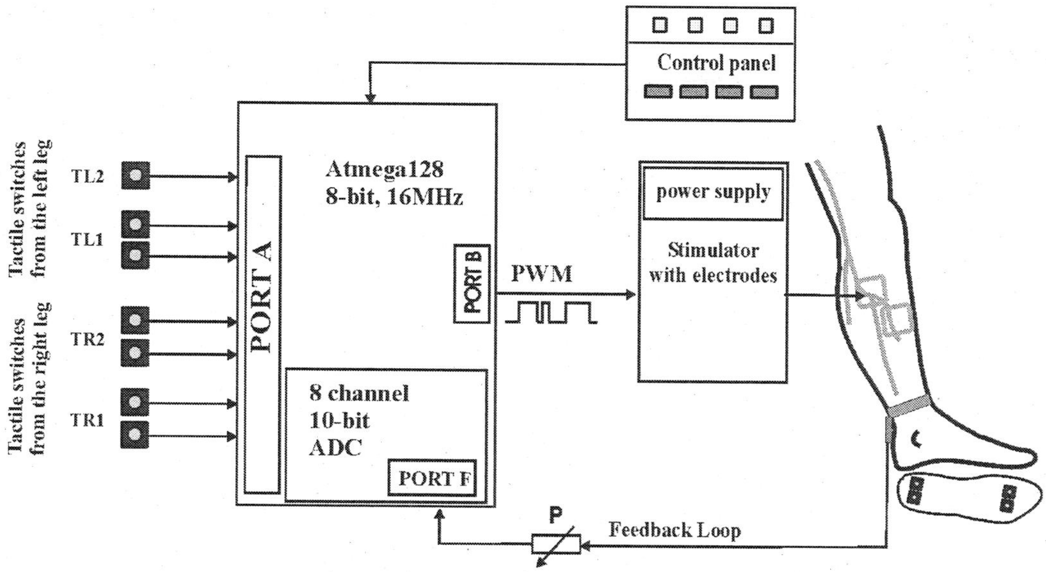

Fig. 6. Block-scheme of the system for control of the ankle-foot complex with ATmega128

The control of functional electrical stimulator with position sensors in the feedback can be implemented on sufficiently powerful microcontroller, featuring basic hardware peripherals such as Analog to Digital Converter (ADC) and a timer with PWM output (Veneva , Toshev, 2007). The Atmel ATmega128 covers the requirements well - with resources left for communication using UART and RS232 protocol.

The PWM channel is connected to the functional electrical stimulator to control the stimulation pulse-width profile for flexion-extension motions by varying the duty cycle of the PWM output.

For the stimulator the position control is handled by electronics according the outputs from a set of two sensor arrays - TR incorporated in the insole of the right leg and TL in the insole of the left leg. The sensors change their outputs throughout the stance and the swing phases of walking.

In a position control mode, additional feedback elements are added to the system to sense the position. A linear potentiometer on the ankle and the back side of the shoe is mounted to detect flexion – extension. The position sensor transforms the linear position of the joint to a voltage, which is then sampled by a corresponding analog-to-digital converter (ADC). This result is compared with the table values.

Secondary functions for the electronics in the Control Unit is to ensure that the speed and overload detection are as desired by closed loop control.

2.2 Control Algorithm

As the biomechanical demands on the dorsiflexor muscles vary throughout the gait cycle, the system must be capable of varying the stimulation pulse-width of these muscles as the stimulation cycle progresses.

During each gait cycle, by measuring the total time TL (for the left leg) and TR (for the right leg) when the foot remains in contact with the ground the microcontroller estimates the forward speed and modulates the swing phase flexion and extension in order to achieve quite normal lower limb dynamics (fig. 8).

Thus the stimulation pulse-width profile was automatically modulated to match patient specific gait requirements to permit smooth and natural heel strike to forefoot transition.

2.3. Locomotion Cycle Determination

This is the most important step of the analysis procedure since all other data are based on the beginning of the stride cycle. A locomotion cycle, T is defined as the period of time for two steps, and it is measured from the initial heel contact of one foot to the next initial heel contact of the same foot. To determine the locomotion cycle starting with the left leg the foot switch TL1 was loaded. At this point, the controller would switch from swing state to heel contact.

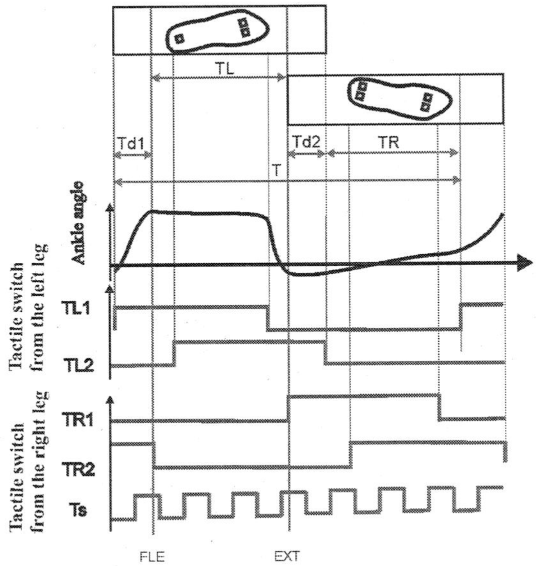

Fig. 7. Signals from the external sensors

2.4. States and Triggers

For the controller the ankle angle and the switches will be the triggers, used to change the PWM output between states. Following is the description of each of the three states implemented in the controller:

Flexion during the Heel Strike - If TR1=0 TR2=0 TL1=1. The goal is to eliminate slap foot. The ankle must be in dorsiflexion i.e. the angle must be smaller than neutral ($0<\alpha<90$). The time for generating the pulse-width profile for flexion should adapt to the gait speed of the user.

PWM switch off during the Stance phase - If TR1=1 TR2=1 TL1=0 TL2=0. The objective is to not disturb the gait of the user because drop foot does not affect the rest of stance. The pulse-width profile is set to zero i.e. there is not flexion.

Flexion during the Swing - If TR1=0 TR2=0 TL1=1 TL2=1. The goal is to lift the foot and to clear the toe. The angle must be smaller than neutral ($0<\alpha<90$).

Finally, a safe state was created which would shut off the device in any unexpected circumstance.

3. Results and Discussion

The proposed control system was designed and tested. Seven external sensors and a linear potentiometer measure ankle joint position in real-time and send signals to the microcontroller (ATMega128).

The system allows specification of the stimulation pulse-width profile and stimulation parameters, e.g. frequency, threshold pulse-width, maximum pulse-width. The parameters are specified and downloaded on the microcontroller low level program (into the control unit).

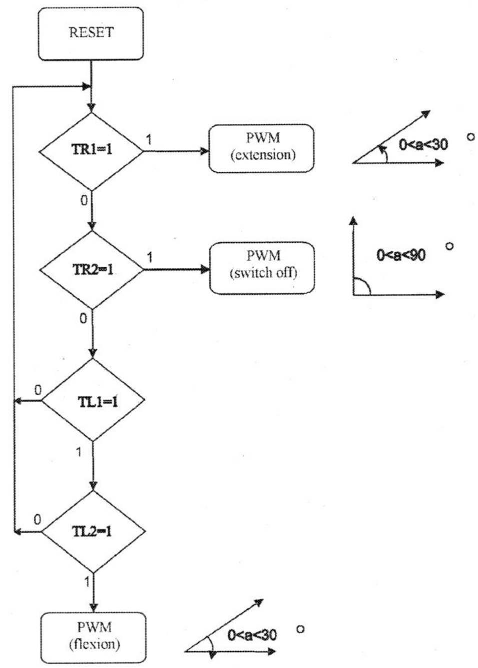

Fig. 8. Control algorithm

The Control unit controls the application of stimulation during gait cycle, using a foot switch as the triggering mechanism. The stimulation pulse-width delivered to the peroneal nerve is modulated during the stimulation cycle. The stimulation frequency and pulse-width for the stimulation waveform are determined by the rectangular pulse waveform delivered by the Control unit to the implant. PWM signals with different duty cycle, obtained experimentally are shown in Fig.9.

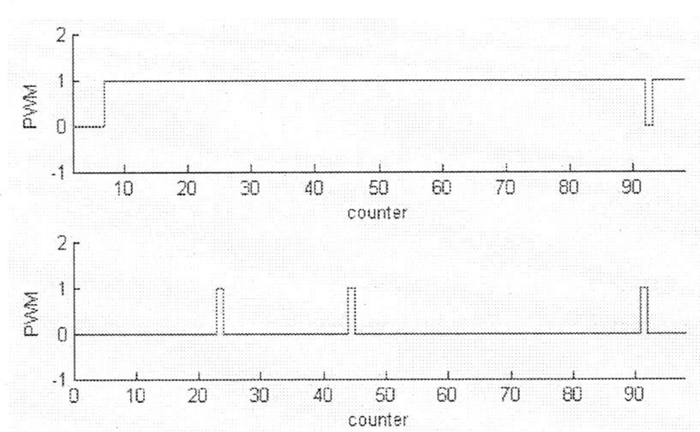

Fig.9. PWM signals with different duty cycle (90% - up, and 10% - down)

This rectangular waveform is transmitted by an inductive link to the implanted stimulator, providing energy and timing information. The implanted stimulator generates charge balanced pulses of the same frequency and pulse-width as the rectangular waveform of the external unit, and these pulses are sent to the nerve cuff electrode.

The inter-individual variations in gait pattern between individuals with are high. The ideal stimulation pattern may be different for each user of a Drop Foot Stimulator. Therefore a convenient method of varying the stimulation pulse-width profile must be provided to the clinician.

4. Conclusion

A microprocessor-controlled system with control algorithm has been developed that allows more flexible user interfaces and easier control for an external and an implanted stimulator system with Functional electrical stimulation.

The proposed device will have the possibilities to modulate the stimulation intensity throughout gait, for the correction of drop foot, to provide gait control during real-use situations, enhance the functional performance and improve patient comfort. The adaptation schemes presented here automatically modulate joint torque throughout the stance and the swing phases of walking. The described device will enable patients to walk in a safe comfortable and smooth manner using only local sensing and computation.

Along with the functional benefits of FES, there are also therapeutic benefits to electrical stimulation. Electrical stimulation has been found to increase muscle bulk, increase flexibility of the joints, and increase blood flow. Increased flexibility is often accompanied by improved locomotion. It is known that muscle atrophy due to disuse can lead to the development of pressure sores, commonly known as bed sores. Electrical stimulation, through the pumping action of contracting skeletal muscle can increase blood flow to the muscle and lead to improved tissue health as determined by tissue oxygen levels.

Acknowledgements

The study was supported by Bulgarian National Research Fund (Grants 1407/04) and French-Bulgarian project "EGIDE –Rila".

References

[1] Bajd, T., Kralj, A., Stefancic, M., and Lavrac, N. (1999). "Use of functional electrical stimulation in the lower extremities of incomplete spinal cord injured patients." Artificial Organs, 23(5), 403-409.

[2] Bukett, J. R., Peckham, P. H., Thrope, G. B., Braswell, S. D., and Keith, M. W. (1998). "A flexible, portable system for neuromuscular stimulation in the paralyzed upper extremity." IEEE Transactions on Biomedical Engineering, 45(4), 463-475.

[3] Cameron, T., Loeb, G., Peck, R., Schulman, J., Strojnik, P., and Troyk, P. (1997). "Micromodular implants to provide electrical stimulation of paralyzed muscles and limbs." IEEE Transactions on Biomedical Engineering, 44(9), 781-790.

[4] Holsheimer, J., Bultstra, G., Verloop, A. J., Van Der AA, H. E., and Hermens, H. J. (2000): 'Implantable dual-channel peroneal nerve stimulator'. Proc. Ljubljana FES Conference, pp. 42-44

[5] Liberson, W. T., Homquest, H. J., Scot, D., and Dow, M. (1961). "Functional electrotherapy: Stimulation of the peroneal nerve synchronized with the swing phase of the gait of hemiplegic patients." Archives of Physical Medicine and Rehabilitation, 42, 101-105.

[6] Popovic, M., Keller, T., Pappas, I. P. I., Vietz, V., and Morari, M. (2001). "Surface-stimulation technology for grasping and walking neuroprosthese." IEEE Engineering in Medicine and Biology, 82-93.

[7] Taylor, P. N., Burridge, J. H., Dunkerley, A. L., Wood. (1999 December). "Clinical use of the odstock dropped foot stimulator: its effect on the speed and effort of walking." Archives of Physical Medicine and Rehabilitation, 80, 1577-1583.

[8] Veneva I., Toshev Y., Control algorithm for ankle-foot orthosis and applications, Proceedings of the 9th International Robotics Conference, Varna, Bulgaria, 2007, pp. 225-230

[9] Veneva I., Toshev Y., Integrated Control of ankle-foot orthoses, Proceedings of the 1st International Conference on Computational Biomechanics and Biology, Pilzen, Czech Republic, 2007

[10] Vodovnik, L., Kralj, A., Stanic, U., Acimovic, R. and GROS, N. (1978): 'Recent Applications of functional electrical stimulation to stroke patients in Ljubljana', Clin. Orthopaed., 131, pp. 64-70

Innovative Production Machines and Systems
D.T. Pham, E.E. Eldukhri and A.J. Soroka (Eds.)

Study the Biodynamic of the Human Hand-Arm Models under Vibration

S. Kazi, A. As'arry, M. Z. Md Zain, M. Mailah, M. Hussein
Department of Mechanical Engineering, Universiti Teknologi Malaysia,
81310 Skudai, Johor, Malaysia.
E-mail: skazi75@gmail.com , zarhamdy@fkm.utm.my Fax: +6 07 2225661

Abstract

This paper describes the development of hand/arm mechanical models, to investigate the biodynamic response characteristics in a 2 and 3-DOF of the human hand and arm system. The assessment of the vibration behavior is examined by using two models. Typically, the human is modeled as a second order mass-spring-damper system. A two-degree-of-freedom (2-DOF) and three-degree-of-freedom (3-DOF) biomechanical models have been developed to derive the vibration transmission characteristics of the hand-arm. The resulting mode shapes are discussed from the viewpoint of the vibration response behaviour of the hand-arm are presented in the time and frequency domains.

Keywords: Hand-arm vibration, hand-arm models.

1. Introduction

The subject of hand biomechanics is a well-studied one. The biodynamic of human hand-arm system is a branch of biomechanics that applies laws of physics and engineering concepts to describe the motions and forces on the system, as well as their relationships. Biomechanical models of the human hand are of great importance in the biomedical and medical ergonomics fields. Even with the technological advances, the human hand is the primary source of dexterous activities. Hand motions are also extensively used in high technology areas such as virtual reality (VR) and tele-operation (Earnshaw et al., 1993).

The human hand is an amazing organ of versatility and dexterity. Not only can it be used to grasp everyday objects, but it also can perform delicate maneuvers, like surgeon's hand. The basic function of the hand can be grouped into grasping activities or non-grasping activities. Human hand-arm vibration models have been developed to characterize the vibration behavior of the hand-arm system. It is essential to understand the basic of biodynamic response of the human hand-arm before developing the hand-arm model [1-10]. Human hand-arm system is a highly damped system (Reynolds and Soedel, 1972). Human hand-arm can be model as the second and third order by using the mechanical model with the combination of mass, spring and damper system. Researchers have attempted to study the biodynamic response of the hand-arm in terms of its vibration transmissibility for the purpose of quantifying the nature of vibration transmitted to the different segments of the hand and arm. This paper describes an existing two--Degree of Freedom (2-DOF) and three-Degree of Freedom (3-DOF) hand-arm system that was taken from [1]. In the field of robotics for example in developing robot arm, this research is usually focusing more on the development of a simple model that represents the arm/hand dynamics as an input/output relationship. It is not concerned with the complex neuromuscular interactions within the human arm/hand [5]. Research on hand biomechanics has focused on the tendon and the reactive forces on finger joints quite early. Due to the limitations in dynamic motion measurement, most of the hand models are concentrated on the static hand posture. Dynamic effects of hand movements have been considered very recently. Human hands, being rigid bodies with mass and volume

(Ozkaya and Nordin, 1991), are affected by the forces invoked by motion. Dynamic equations of motion should be considered in rapid hand motions. Mussa-Ivaldi et al. [6] delivered small displacements to the hand while a test subject maintained a particular posture. The force on the hand was measured and the stiffness was calculated from the force and displacement data. They found that the hand stiffness is strongly dependent on the arm configuration. Gomi et al. [10] concluded that experimentally determined human-arm stiffness values vary greatly between subject, tasks, experimental apparatuses, and perturbation patterns.

The aim of this study is to observe the characteristic of frequency and displacement with the time and frequency domain, when force is applied. The resulting mode shapes are discussed from the viewpoint of the vibration response behavior of the hand-arm. The second and third order mass-spring-damper system can be easily simulating using MATLAB Simulink. The applications of mechanical equivalent or biodynamic models of the human hand and arm offer considerable potential to carry out assessments through both analytical and experimental analyses where the involvement of human subjects could be considerably reduced.

2. Conceptual Method

The dynamics of the human hand-arm can be represented by the mechanical models as shown in Fig: 1 & 2 respectively. It shows a two-DOF and three-DOF of hand-arm model. These models can be divided into two and three different parts which are for hand, fore arm and upper arm. The motion of the human hand-arm in 2-DOF and 3-DOF are represented by (x_{1a}, x_{2a}) and (x_{1b}, x_{2b}, x_{3b}) similarly, the masses in 2-DOF and 3-DOF are represented by (M_{1a}, M_{2a}) and (M_{1b}, M_{2b}, M_{3b}) respectively. The input signal force act at hand in 2-DOF and 3-DOF is labeled as F_{1a} and F_{1b} respectively. The stiffness of the human hand-arm in 2-DOF and 3-DOF are represented by (K_{1a}, K_{2a}) and (K_{1b}, K_{2b}, K_{3b}) respectively while the damping of the human hand-arm in 2-DOF and 3-DOF are represented by (C_{1a}, C_{2a}) and (C_{1b}, C_{2b}, C_{3b}) respectively.

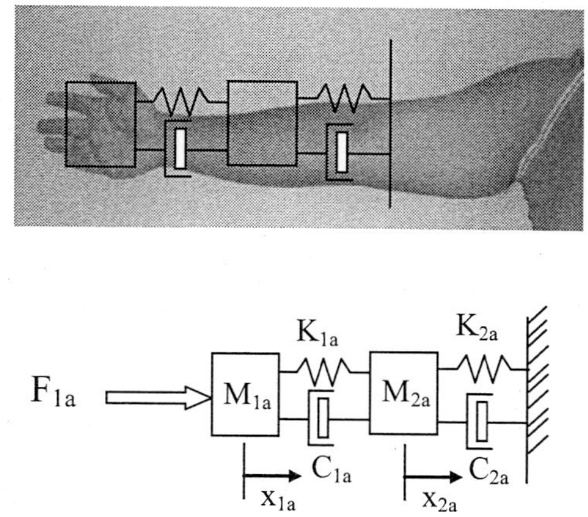

Fig. 1. 2-DOF hand-arm Model

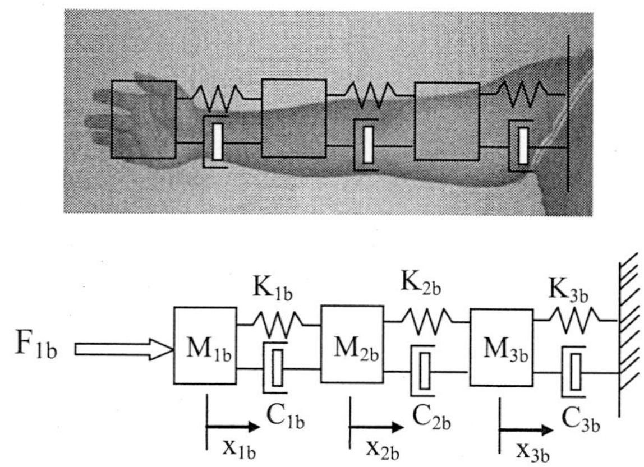

Fig. 2. 3-DOF hand-arm Model [1]

2.1. Mathematical model for 2 DOF:

The input force was act at the hand. The equation of motion for mass of the human hand, M_{1a}, can be written as:

$$M_{1a}\ddot{x}_{1a} = F_{1a} - K_{1a}(x_{1a} - x_{2a}) - C_{1a}(\dot{x}_{1a} - \dot{x}_{2a}) \quad (1)$$

For mass of the human fore arm, M_{2a}, the equation of motion can be written as:

$$M_{2a}\ddot{x}_{2a} = K_{1a}(x_{1a} - x_{2a}) + C_{1a}(\dot{x}_{1a} - \dot{x}_{2a}) - K_{2a}(x_{2a}) - C_{2a}(\dot{x}_{2a}) \quad (2)$$

2.2. Mathematical model for 3 DOF:

In this model, the input force was also applied at the hand. The equation of motion for mass of the human hand, M_{1b}, can be written as:

$$M_{1b}\ddot{x}_{1b} = F_{1b} - K_{1b}(x_{1b} - x_{2b}) - C_{1b}(\dot{x}_{1b} - \dot{x}_{2b}) \qquad (3)$$

The mass of the human fore arm, M_{2b}, the equation of motion can be written as:

$$\begin{aligned} M_{2b}\ddot{x}_{2b} = {} & K_{1b}(x_{1b} - x_{2b}) + C_{1b}(\dot{x}_{1b} - \dot{x}_{2b}) \\ & - K_{2b}(x_{2b} - x_{3b}) - C_{2b}(\dot{x}_{2b} - \dot{x}_{3b}) \end{aligned} \qquad (4)$$

For mass of the human upper arm, M_{3b}, the motion equation can be written as:

$$\begin{aligned} M_{3b}\ddot{x}_{3b} = {} & K_{2b}(x_{2b} - x_{3b}) + C_{2b}(\dot{x}_{2b} - \dot{x}_{3b}) \\ & - K_{3b}(x_{3b}) - C_{3b}(\dot{x}_{3b}) \end{aligned} \qquad (5)$$

In this study, both mechanical models of hand-arm system using the same parameter, where the value are $M_{1a} = M_{1b} = 0.65E\text{-}02$ kg, $M_{2a} = M_{2b} = 2.74E\text{-}01$ kg, $M_{3b} = 3.82E\text{-}00$ kg, $K_{1a} = K_{1b} = 8.20E+03$ N/m, $K_{2a} = K_{2b} = 1.00E+02$ N/m, $K_{3b} = 5.56E+05$ N/m, $C_{1a} = C_{1b} = 7.76E+01$, $C_{2a} = C_{2b} = 2.10E+01$ and $C_{3b} = 6.02E+01$ [1].

3. Results and Discussion

In this section, the proposed study behaviors of the hand-arm models are implemented and tested within simulation environment of the vibration, corresponding results are presented. The hand-arm response characteristics due to white noise signal as shown in Fig. 3, it is considered as a form of 'input' to the system. The results show that the vibration response between the 2-DOF and 3-DOF during white noise vibration exposure for different parts of the hand-arm system are shown in Fig. 4 and Fig. 5 respectively. Comparisons of the responses of the behaviour of vibration in hand-arm show that a same agreement between 2-DOF and 3-DOF results in the resonance frequencies were achieved. It is noted that the results for the first two modes of vibration 2-DOF of the system converged to 27.77 Hz and 41.65 Hz whereas, 13.8 Hz and 35.5 Hz for the first and second modes of 3-DOF respectively.

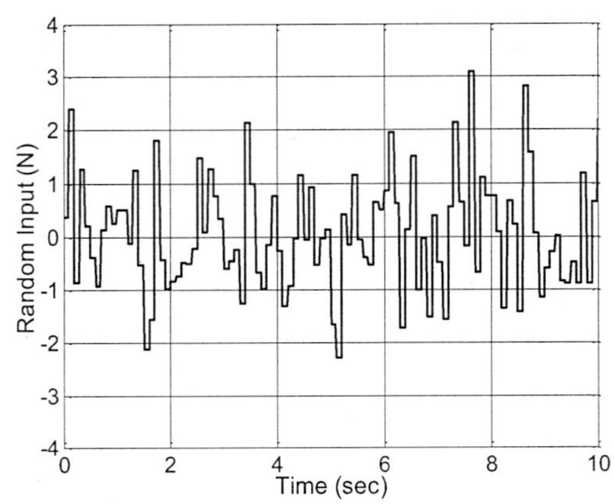

Fig. 3. White noise input

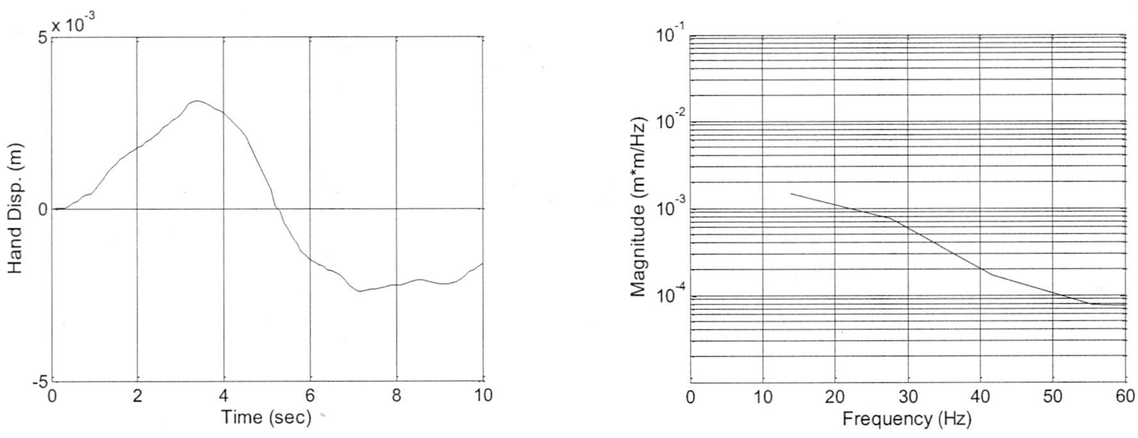

Time domain

Frequency domain

(a) Hand Displacement

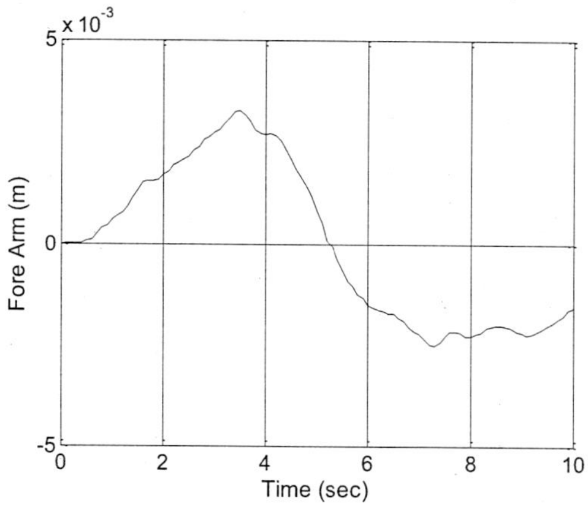

Time domain

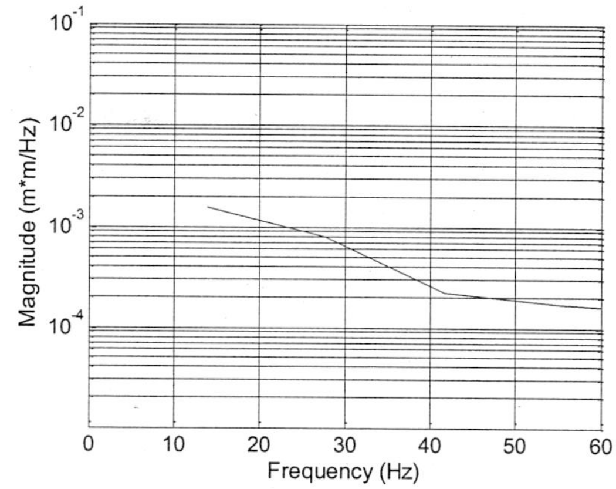

Frequency domain

(b)) Force Arm Displacement

Fig. 4. Response of the simulated behaviour of hand-arm in 2-DOF.

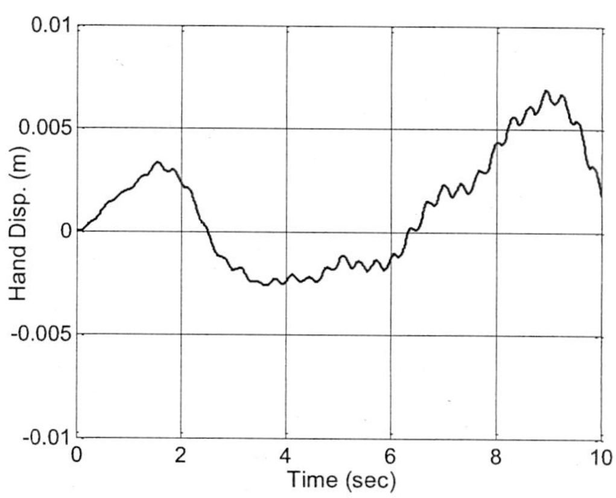

Time domain

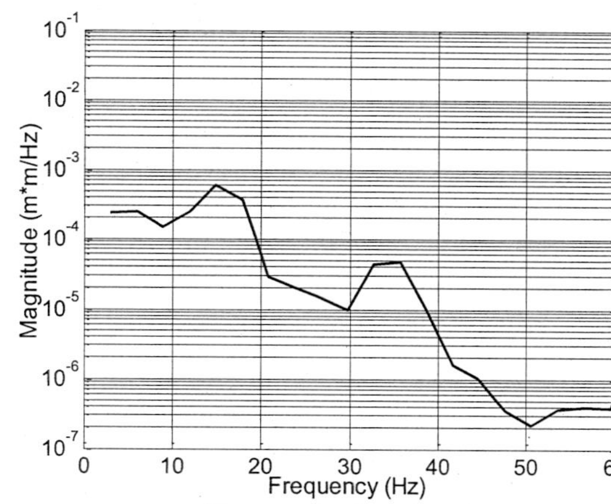

Frequency domain

(a) Hand Displacement

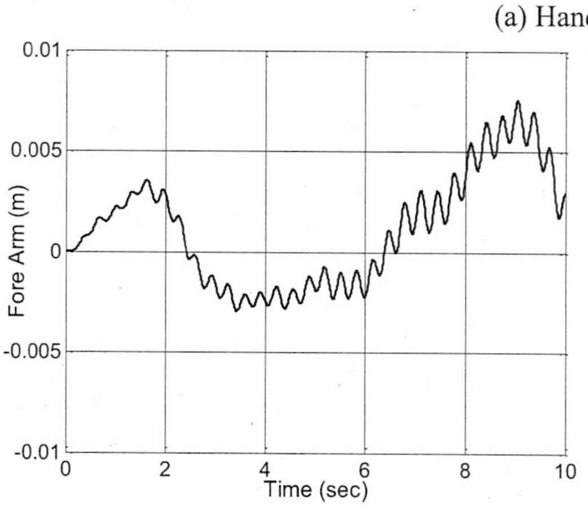

Time domain

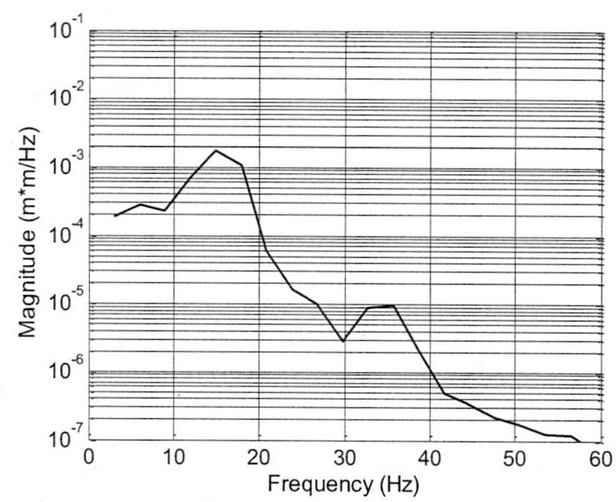

Frequency domain

(b)) Force Arm Displacement

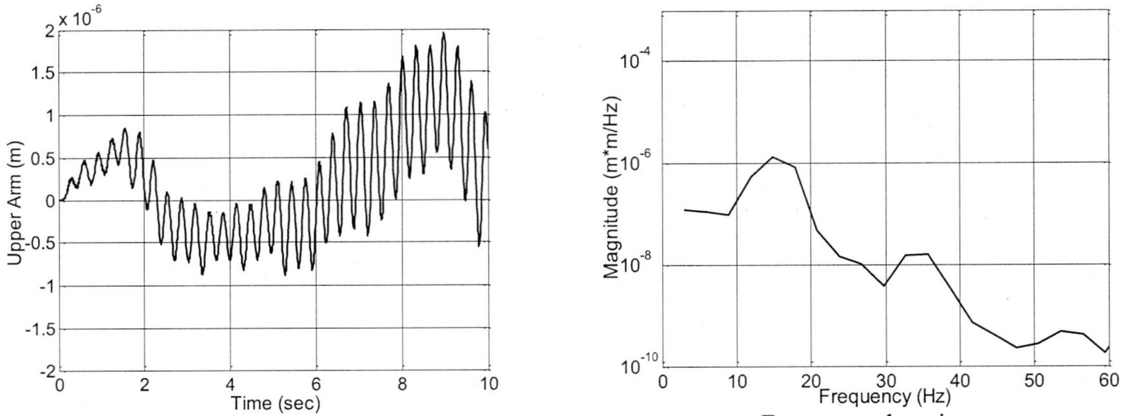

Time domain

Frequency domain

(c) Upper Arm Displacement

Fig. 5. Response of the simulated behaviour of hand-arm in 3-DOF.

The displacement and magnitude data acquired from the graphs are analyzed to evaluate the response along all the different parts when the input force applied at hand (2-DOF and 3-DOF).

The graphs in Fig: 4(a) and Fig: 5(a) displays the hand displacement corresponding to the time domain when random input force was applied within the range of 3N to -2 N. The maximum displacement of hand motion is approximately 0.0031m observed at 3.4 sec and 0.006m observed at 9 sec respectively. For the fore arm as shown in Fig: 4(b) and Fig: 5(b), the maximum displacement is approximately 0.0032m observed at 3.5 sec and 0.008m observed at 9 sec respectively. Fig: 5(c) shows that the maximum displacement of upper arm motion is too small about 2×10^{-6} m and occurred when time domain at 9 sec.

For the 3-DOF model tested, where the force was also applied at the hand, it can be conclude that the high displacement motion at the fore arm compare to the hand and upper arm. It is due to the fore arm received the force from the hand and the counter act force from the upper arm. However, the displacement value of hand and fore arm approximately same. Meanwhile, the small displacement value of the upper arm because of the location of upper arm is near to fixed.

4. Conclusion

The human hand and arm response to vibration has been investigated in terms of biodynamic responses. The vibration human hand-arm models have been presented. An analysis of two and three-DOF biomechanical models were carried out to enhance an understanding of the modal deflection patterns of the human hand-arm system. The deflection patterns of the masses due to the hand, the forearm and the upper-arm, corresponding to different modes of vibration, were analyzed in terms of the vibration behavior of different segment of the hand-arm. From the results in this study, it could conclude that there is not much different displacement when two signals was applied in different location point if fact the frequency domain was same.

Acknowledgements

Financial support by the Malaysia Esciences Fund (Vot. 79246) is gratefully acknowledged.

References

[1] R. Gurram, S. Rakheja, G.J. Gouw, "Mechanical impedance of the human hand-arm system subject to sinusoidal and stochastic excitations", Internal Journal of Industrial Ergonomics 16, 1995, pp. 135-145.

[2] C. Thomas, S. Rakheja, R. B. Bhat and I. Stitharu, "A study of the modal behavior of the human hand-arm system", Journal of Sound and Vibration, 1996, pp. 171-176

[3] John E. Speich, Liang, Shao, Michael Goldfarb, "Modeling the human hand as it interacts with a telemanipulation system", Journal Mechatronics, 2005, pp. 1127-1142.

[4] Kosuge K, Fujisawa Y, Fukudu T. "Control of mechanical system with man-machine interaction". In: Proceedings of the IEEE/RSJ international conference on robotics and automation. 2002.

[5] Kazerooni H. Human-robot interaction via the transfer of power and information signals. IEEE Trans Syst Man Cybern 1990;20(2):450-63

[6] Mussa-Ivaldi F, Hogan N, Bizzi E. Neural, mechanical, and and geometric factors subserving arm posture in humans. J Neurosci 1985;5(10):2732-43

[7] Flash T, Mussa-Ivaldi F. Human arm stiffness characteristics during the maintenance of posture. Exp Brain Res 1990;82:315-26.

[8] Tsuji T, Goto K, Moritani M, Kaneko M, Morasso P. Spatial characteristics of human hand impedence in multi-joint arm movements. In: Proceedings of the IEEE/RSJ/GI international conference on intelligent robots and systems. 1994. p. 423-30

[9] Dolan JM, Friedman MB, Nagurka ML. Dynamic and loaded impedance components in the maintenance of human arm posture. IEEE Trans Syst Man Cybern 1993;23(3):698-709.

[10] Gomi H, Koike Y, Kawato M. Human hand stiffness during discrete point-to-point multi-joint movement. In: Procedings of the IEEE international conference on engineering in medicine and biology society. 1992. p.1628-9.

[11] Earnshaw, R.A., Gigante, M.A., Jones, H., 1993.Virtual Reality Systems. Academic press, Harcourt Brace & Company.

[12] Ozkaya, N., Nordin, M., 1991. Fundamentals of Biomechanics. Van Norstrand, New York.

[13] Reynolds D.D. and Soedel, W. 1972: Dynamic response of the hand-arm system to sinusoidal input, Journal of sound and vibration, 21, pp. 339-353.

Innovative Production Machines and Systems
D.T. Pham, E.E. Eldukhri and A.J. Soroka (Eds.)

Genetic Programming based Optimisation of Soccer Agents

Christopher Lazarus and Huosheng Hu

Department of Computing and Electronic Systems
University of Essex, Wivenhoe Park, Colchester CO4 3SQ, UK
Email: clazar@essex.ac.uk, hhu@essex.ac.uk

Abstract

This paper describes our approach on evolving a robot controller to generate a set of goalkeeper behaviours in the RoboCup domain based on the genetic programming (GP) technique. The goalkeeper agent is a software construct that operates in a simulated environment provided by the Soccer Server. Our research is to determine the conditions that need to be met for an evolved goalkeeper agent to perform adequately in a defensive situation. A framework is developed to test the agent's performance in the simulator. The impact of the sensors information that operates as the input to our GP implementation is addressed. The framework by incorporating it within two established soccer agents is evaluated, and then the offline performance of each agent is tested. The experimental results show that our approach is able to produce a robust robot controller and the framework is flexible enough to be readily incorporated into other existing soccer agent frameworks.

Keywords: Evolutionary Robotics, Genetic Programming, Simulated Soccer.

1. Introduction

RoboCup competition is an international robotics event, including several different leagues [1], such as simulation, small-sized, medium-sized, Sony AIBO, humanoid and RoboCup rescue leagues. This paper reports on work done for soccer agents in the simulation league.

Programming a team of robot to perform soccer competition tasks is nontrivial task. Our efforts focus on evolving behaviours for a goalkeeper agent by adopting the Genetic Programming (GP) technique and the focus of the research is on the design of a set of specifications for the desired behaviours that are able to emerge due to evolutionary pressure set out by the designer [2]. The work presented here is based on our previous work on evolving navigation and obstacle avoidance behaviours based on the optimisation of a single objective function [3]. However, our current work involves the optimisation of multiple objectives [4].

There have been many attempts by researchers to develop and improve ways on robot evolutions. Their solutions can be similar if not better than hand-coded solution, i.e. the traditional robot navigation solution. These works include evolution of both simulated and real robots. RoboCup is a rich domain for the exploration of these techniques, and gives us a platform where we can compare the performance of their techniques in same kind of environments. Given that more than one robot is involved in a soccer match, RoboCup also enables us to explore multi-robot cooperation algorithms. Several teams have used GP in evolving their soccer teams [2], and a comparison of these works can be seen in [5].

Competition between the GP evolved teams and

hand-coded teams have been possible due to the open nature of the RoboCup soccer competition [1, 6]. Performances of GP evolved teams have usually been average. This paper shows that one of the reasons is not due to the GP implementation itself but due to other factors, such as architectural and other auxiliary functions. These functions could be the sensor interpretation routines, agent localisation and subsequently object detection routines.

The rest of this paper is organised as follows. Section 2 presents the methodology and a conceptual architecture developed. Section 3 describes the implementation of our agent architecture. The experimental results are presented and analysed in Section 4. Finally, a brief conclusion and further work are given in Section 5.

2. Methodology

The focus of our research is to find out the effects of using multiple objective fitness functions in our GP implementation. To accomplish this, we have devised a training ground situation to test the evolved behaviours of our goalkeeper. The architecture consists of a soccer server, a coach trainer agent, a hand-coded attacker agent and an evolving goalkeeper agent, as shown in Fig. 1.

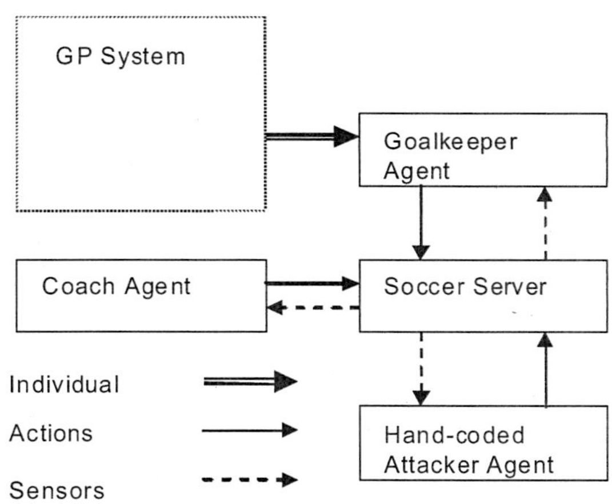

Fig. 1 System Configuration

The soccer server [6] accepts action commands every 100ms and sends back sensory information and runs the simulation. The coach agent performs the synchronisation among all the agents by relaying messages to let them know when a trial begins and ends. It also resets the positions of agents and ball. The hand-coded attacker agent is a simple agent that kicks the ball towards the centre of the goal. The goalkeeper agent is the one that has the capability to evolve based on the GP system that generates its control program.

Our control experiment uses a random goalkeeper agent that sends randomly generated commands to the soccer server instead of being generated by GP. Its fitness is evaluated during the experiment. Both random and evolving goalkeeper experiments are carried out to collect the data, and their performance is compared to measure the effectiveness of the GP implementation.

In designing our functions and terminals, we noted that if the performance of the GP algorithm is not quite what we expect then more complex functions and terminals may be added to the gene pool. One of the attractions of using GP is that detailed knowledge of the problem domain is not essential. The focus of the problem shifts from the careful design of the behaviours to the designing of specifications that encourages emergence of the desired behaviours [2]. In this case, specifications can be equated as the fitness criterion of each individual within the GP population.

Multiple fitness functions are used in our current work. A goalkeeper has a single overriding objective, which is to stop any goal from being scored into its own net. To achieve this, there is a set of other objectives that need to be met. The first objective concerns navigation, another is localisation (position must be kept near the centre of the goal), another is blocking the ball trajectory, the ability to kick the ball out of the goal area and lastly to catch the ball when possible. Therefore the aim of our optimisation is to transform a vector of objectives to a scalar version and then proceed with the assignment of fitness values to each individual.

3. Implementation

Our GP system adopts lil-gp [7] and the client networking library libsclient [8]. A coach agent is developed to provide synchronisation and monitoring during the experiment. The coach agent resets the experiment setting before the start of each trial case. This involves placing the goalkeeper, ball and attacker in their respective starting positions. Once in place, the ball is kicked by the attacker in a straight line directly towards the goal. There are 10 different ball placements for each evaluation (see Fig. 2). Each individual in the GP population representing the goalkeeper is tested on its performance in defending the goal from 10 different ball-starting positions.

The attacking area in front of the goal is divided equally into 10 sectors. At the start of each experiment, the 10 starting positions for the attacker are randomly generated within each of the 10 attacking sectors. Each attacker starting position is then tested sequentially in a clockwise order for each individual in the population. This variation in the attacker's starting position should enable a robust controller to be evolved. The evolved behaviours are not biased towards a particular attacking direction. We evaluated each individual by parsing the expression trees and sending a command every 10ms to the soccer server, instead of normal 100ms [6] to speed up the simulation, and only 1 command is allowed per 10ms.

Fig. 2 A segment of the soccer field showing a goalkeeper at location (50, 0)

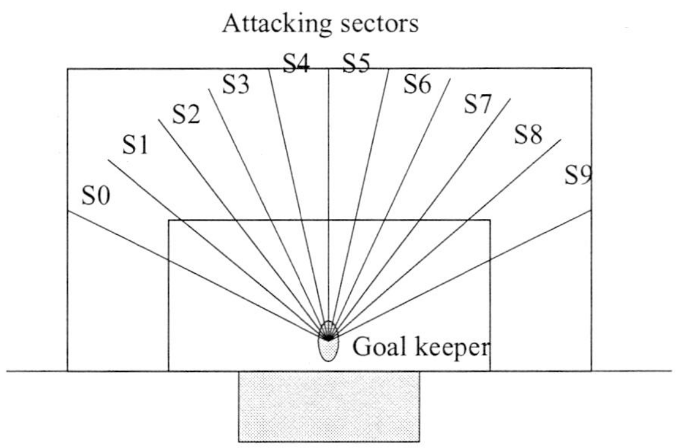

3.1 Evaluation time

The length of time steps allowed for an attacker to attack and score a goal has a direct impact on the length of time each evaluation takes. For the purpose of the experiments each evaluation takes 40 time steps. Each time step corresponds to a set period of 10ms. We allow the attacker 40 time steps per trial to give it enough time to try to score a goal.

3.2 Fitness measure

The fitness evaluation we adopted is called the Weighted Sum Approach

$$f = \sum_{i=1}^{n} w_i f_i \qquad (0.0 \leq f \leq 1.0) \qquad (1)$$

where f denotes the overall fitness of an individual, n denotes the number of fitness measures, w_i denotes the weights and f_i denotes the objectives to be optimised [4].

3.2 Fitness cases

There are 5 fitness cases that influence the evolution of desired behaviours: (i) f_1 = Ball saving; (ii) f_2 = Localisation; (iii) f_3 = Ball locate; (iv) f_4 = Movements; (v) f_5 = Positioning.

The rationales for the choice of the fitness measures in Table 1 are that we want the evolved goalkeeper to be actively moving towards the ball and intercepting it. The movements and positioning fitness cases reflect this desire. Fitness f_4 and f_5 might seem to overlap in functionality but without fitness f_4, the movement of the ball instead of the goalkeeper might be the cause for the proximity of the goalkeeper to the ball. Therefore, we encourage the selection of individuals that are active so that there are cases where the proximity of a goalkeeper to a ball is caused by the goalkeeper's own movements. We gave fitness f_4 a higher weight because of similar reasons. Nevertheless, fitness f_4 and fitness f_5 are ineffective if they don't result in a ball saving event. So we give a higher weight for behaviours that resulted in a ball saving event.

Table 1 GP Parameter Settings

Max Generation	50
Population	
Type	Generational
Size	100
Initialisation	Half and Half
Selection Type	Tournament with size 7
Genetic Operators	
Crossover rate	0.9
Mutation Rate	0.1
Function Set	iflte, add, sub, mult, div, sin, cos, catch, dash, kick, turn, put, ADF0
Terminal Set	play_dir, play_pos, dist_to_ball, dir_to_ball, dir_to_ogoal, dist_to_ogoal, dist_to_opp, dist_to_opp, rand, get0, get1, ARG0, ARG1

Fitness case 1 is given by the equation

$$f_1 = 1 - \frac{g}{t_{max}} \qquad (2)$$

where g is the goal scored by an attacker and t_{max} is the maximum trial case per evaluation

Fitness f_2 is used to measure the goalkeeper's ability to estimate its own location within the playing field. If both initial and ending position of the goalkeeper within each trial is estimated then localisation is deemed as successful and fitness f_2 is allocated.

Fitness f_3 measures the goalkeeper's ability to locate the ball by checking whether the goalkeeper is able to estimate the distance between itself and the ball at the start and end of a trial. The goalkeeper is only able to do this if the ball is in its field of vision. Therefore this fitness can also be equated with the ability of the goalkeeper in turning towards the ball.

Fitness f_4 is as follows

f_4 if p_s is equal to p_e then

no movement is detected so $f_2 = 0$

else movement detected so $f_2 = 1$

where p_s is the position of goalkeeper at the start of an evaluation, and p_e is the position of goalkeeper at the end of an evaluation

Fitness f_5 uses a logarithmic scale function of the distance between the goalkeeper and the ball at the start of an evaluation and also at the end of an evaluation, ensuring that the relationship between the distance between the goalkeeper and the ball is logarithmic. This ensures that the fitness decreases following a logarithmic curve as the distance between the ball and the goalkeeper becomes larger.

f_5 if $(1-\left(\frac{\log(d_e)}{\log(d_s)}\right)$ is less than 0 then

set f_3 to 0

else set f_3 to $1-\left(\frac{\log(d_e)}{\log(d_s)}\right)$

where d_s is the distance between goalkeeper and ball at the start of an evaluation greater than 1. d_e is the distance between goalkeeper and ball at the end of an evaluation greater than 1.

3.3 Termination-criteria

Termination of the GP run is set to the number of maximum generation predetermined at the beginning of the GP run.

4. Experimental Results and Analysis

Table 2 presents the experiments that have been carried out in this research and the extended experiments' results are shown in Table 4. A summary of noise effects on 3 agents is summarised in Fig. 4.

The first two experiments, where the goalkeeper's behaviour was generated randomly, are used as a benchmark for subsequent experiments. The starting position for the goalkeeper is set at location 50, 0 on the playing field. Three sets of weights were used during the experiments.

Table 2 Experiment runs

Experiments	Behaviour generated by	Memory	ADF	Weights
1	Random	N/A	N/A	Equal
2	Random	N/A	N/A	Lexicographic
3	GP	Yes	No	Equal
4	GP	Yes	No	Lexicographic
5	GP	No	Yes	Equal
6	GP	No	Yes	Lexicographic
7	GP	Yes	Yes	Equal
8	GP	Yes	Yes	Lexicographic
9	GP	Yes	Yes	SQRT (Lex.)

As shown in Table 3, one set of weights is designed to give equal importance to all of the 5 fitness measures. The second set of weights is designed to bias the importance of the fitness measures in a lexicographical manner in terms of fitness. The third set of weights is derived from normalised square roots of the lexicographic weights. The third set of weights has a smoother distribution compared to the second set. This is shown in Fig. 3. This allows for the fitness measures to be more uniformly distributed across the fitness range but would still be skewed according to the fitness measures ranking.

All of the experiments based on lexicographic weights have lower overall fitness when compared with runs that are based on equal weights. Lexicographic weights biased the fitness measurement too much towards fitness 1 which is ball saving. This results in GP not being able to capitalise on the contributions made by the other fitness measures. This also shows that even though ball saving is the main objective; it is not enough on its own to allow GP to explore potential solutions. By specifying that all of the fitness measures have equal weights, we see that the fitness is distinctively higher compared to using lexicographic weights. We can also see this assertion by comparing both of the random runs. The random run that uses equal weights has higher fitness compared to the random run that uses lexicographic weights.

Experiments 2, 3, 4 and 5 were run to observe whether the GP performance would be enhanced by the addition of either memory or ADF. The results show that the GP with ADF has higher fitness than the GP with memory when equal weights are used. In contrasts, the GP with memory has higher fitness than the GP with ADF when lexicographic weights are used. Looking at both the equal weights and the lexicographic weights groups, we can see that in both groups, the highest performing GP is the one that implements both memory and ADF in their architecture. The addition of either ADF or memory in isolation does improve the performance of the GP but its performance increases further if both of them are implemented together.

Table 3 Weights used during the experiments

Weights Fitness	Equal	Lexicographic	Square root (Lexicographic)
1	0.2	0.9	0.685942
2	0.2	0.09	0.216914
3	0.2	0.009	0.068594
4	0.2	0.0009	0.021691
5	0.2	0.00009	0.006859

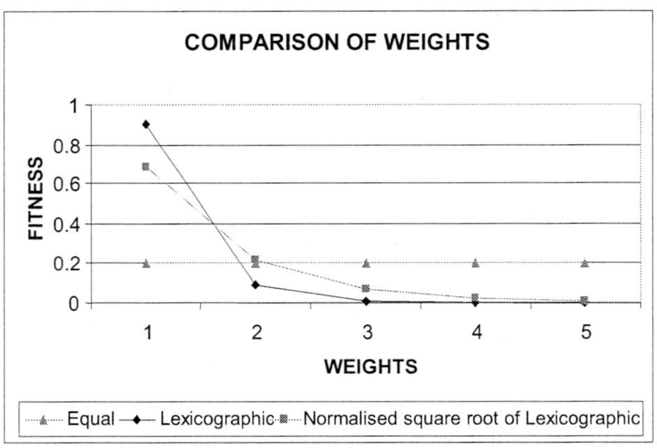

Fig. 3 Fitness distributions when lexicographic weights are normalized

When we compared performances of best of run individuals in contrasts to populations, we start to see that the distinction caused by different weights set becomes less obvious. Experiments that use the lexicographic weights, has best of run individual nearing the performances from the experiments that uses equal weights. An additional experiment (experiment 9) that uses the normalised lexicographic weights was run to observe the effects of smoothing the distribution of the fitness measures. The population fitness for this experiment shows a marked improvement. However its population fitness is still below the population fitness for experiments that uses equal weights.

The next set of experiments was conducted to show whether our GP implementation has any effect when conducted with other agent implementations that has better sensor interpretation routines. For this work, we decided to test the GP framework using our own team, Essex Wizards [9] and a competitor, UvA Trilearn [10]. Our GP framework is incorporated into these teams' agent framework without much change due to the flexibility of the lil-gp [7] architecture.

In terms of mean population fitness, we see that the fittest population results from the agent using the UvA Trilearn framework. Again the lowest mean population fitness is from the libsclient agent. This shows that other features of a soccer agent such as synchronisation, localisation and sensor interpretation algorithms has a direct impact on the fitness of the evolved behaviour tree.

Each of these evolved solutions is then tested using the same experiment scenarios. In these experiments the agent's behaviour trees are no longer being evolved. The best evolved behaviour tree at the end of the GP run for each agent is loaded at the start of the experiment. The agent then acts using this behaviour tree during the course of its run. An agent is assessed by performing 500 trials per run. Noise resulting from incorrectly kicked balls is discarded from evaluation. Fig. 4 shows a comparison of noise detected during the experiments. We see that better optimised agent architecture results in lower noises.

Our agent, who uses the libsclient library, is not as optimised as the other two agents. We can see that the most noise is produced by the libsclient agent with the randomly generated behaviour tree (see 2-1 in Fig. 4). This is reduced when the same agent uses a hand-coded behaviour tree (see 2-4 in Fig. 4). The least noise is observed in the evolved behaviour tree (see 2-7 in Fig. 4). GP has managed to optimise the behaviour tree by adapting to the level of noise in the simulation. This same task can be achieved by making sure that the synchronisation algorithm between the agent and the server are optimised.

An agent's performance is calculated based on the number of balls saved (caught and kicked by the goalkeeper agent). Table 4 and Fig. 4 show a comparison of performance between three different agent implementations. Each trial involves an attacker agent kicking a ball towards the goal. The scenario is similar to that shown in Fig. 2. We observe that the top two performances come from both the Essex Wizards and UvA Trilearn hand-coded agents. This result shows that their agents are highly optimised. Our GP evolved behaviour trees produced disappointingly low performances when compared to the two hand-coded agents. This is to be expected since the two agents have performed exceptionally during past RoboCup competition.

Another reason is that the two hand-coded agents make use of high-level behaviours which are built on top of a set of lower level behaviours. Our GP agent is only given a set of low level commands (see Table 1). Our own hand-coded agent based on the libsclient library was not designed to be in the RoboCup competition hence it is not optimised. It is designed to be a control for comparison of low level behaviours. The hand-coded agent based on libsclient uses behaviours based only on low level commands to generate its behaviour. The libsclient evolved agent's performance is slightly better than the libsclient hand-coded agent's performance (see 2-4 vs. 2-7 in Table 4).

Table 4 Experiments (Performance of Agents) using best behaviour tree evolved during previous GP run

	Libsclient	Essex Wizards1999	UvATrilearn 2001
Random	(2-1) 7.72%	(2-2) 7.65%	(2-3) 4.61%
Hand-coded	(2-4) 5.14%	(2-5) 93.93%	(2-6) 79.4%
Evolved	(2-7) 5.8%	(2-8) 4.62	(2-9) 4.21

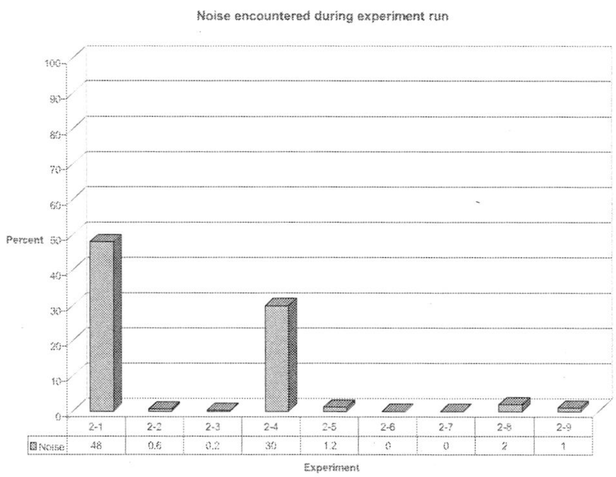

Fig. 4 Comparison of noise effects on the 3 agents

5. Conclusions and Future Work

This paper has shown that the selection of weights in the Weighted Sum Approach is important to the performance of the GP implementation although most weight selection problems are solved heuristically. Considering the simulation results, each of the fitness measure is equally important. Our attempts to bias the fitness measures by adjusting the weights shows lower overall population fitness. Nevertheless, the fittest individual within the population where the weights are not equally distributed still manages to be fitness-competitive to the fittest individual using equal weights. The addition of memory and ADF as part of an agent's memory and an individual's program tree respectively for the duration of the run increases the GP's performance. This corresponds with the increase in difficulty of the task that the goalkeeper agent is given.

The overall results show that although GP manages to evolve a good set of behaviours based on low level commands, which are still worse than behaviours optimised by human experience. Our future work is to investigate the way to improve the performance of the evolved goalkeeper further.

References

[1] H. Kitano, M. Tambe, P. Stone, M. Veloso, et al., The RoboCup Synthetic Agent Challenge, 97. Proc. of Int. Joint Conf. on Artificial Intelligence, 1997.

[2] D. Andre and A. Teller, "Evolving Team Darwin United," in *RoboCup-98: Robot Soccer World Cup II*, M. Asada, Ed.: Springer-Verlag, 1998.

[3] C. Lazarus and H. Hu, "Using Genetic Programming to Evolve Robot Behaviours," Proc. of the 3rd British workshop on Towards Intelligent Mobile Robots, Manchester, 2001.

[4] C. M. Fonseca, P. J. Fleming, "Multi-objective Optimisation," in *Evolutionary Computations 2*, T. Bäck, D. Fogel, B, and Z. Michalewicz, Eds. UK: IOP Publishing Ltd., 2000.

[5] J. Lundberg, "Survey over Genetic Programming Approaches to RoboCup", http://www.dsv.su.se/~johank/courses/int4/robocup/papers/1999/int7/lundberg.pdf, 2003.

[6] I. Noda, "Soccer Server System", http://sserver.sourceforge.net/NEW/, 2003.

[7] B. Punch and D. Zongker, "lil-gp Genetic Programming System", http://garage.cps.msu.edu/ software/lil-gp/lilgp-index.html, 1998.

[8] I. Noda, "libsclient", http://www.isi.edu/soar/galk/RoboCup/Libs/libsclient-4.01.tar.gz, 1997.

[9] H. Hu, K. Kostiadis, M. Hunter, M. Seabrook, "Essex Wizards '99 Team Description," *RoboCup-99: Robot Soccer World Cup*, 2000.

[10] R. D. Boer, J. R. Kok, and F. C. A. Groen, "UvA Trilearn 2001 Team Description," RoboCup 2001: Robot Soccer World Cup V, 2001.

Innovative Production Machines and Systems
D.T. Pham, E.E. Eldukhri and A.J. Soroka (Eds.)

TherapainiS –A socially assistive robot for the elders

A.D. Louloudi[a], V.C. Moulianitis[a,b]

[a] *Department of Product and System Design Engineering, University of the Aegean, Syros, Greece*
[b] *Mechanical Engineering and Aeronautics Dept, University of Patras, Patras, Greece*

Abstract

This paper presents a design study of a robotic system which aims at the health care of senior people. This study makes an attempt to conjugate the technological trend, which is to promote the use of personal robots, with the social demand of older person's welfare. TherapainiS, an autonomous mobile robot emerges as a suggestion to the above challenge. This concept serves three basic needs: health surveillance, social interaction and safeguarding. These needs yield the characteristics of a socially assistive robot.

Keywords: Socially Assistive Robotics, domestic healthcare

1 Introduction

According to most demographic researchers [7],[14], the population of the western countries is getting older without preserving the quality of life. The constantly growing part of senior population strongly claims day by day the right to remain an active group of the society and old age is becoming a major social issue to be dealt with. Nowadays, preserving and improving the quality of our lives and especially the lives of older people is a social need.

In today's societies senior people tend to have lower social status and less power than they had in the ancient communities. In ancient cultures, as well as in Asian ones old age were considered to be empowered by wisdom and the elders were highly respected and entrusted members of the community. They were the ones who gave advice, made all the important decisions.[19]

Today many people consider the process of aging as a disease that must be radically treated. The inevitable procedure of body decay is something that is no longer viewed as natural evolution and that is so because medical and nutritional progress have proved that the inevitable can be relented even avoided for a period of time.[19]

Still the above can't change the steady increase of health and medical needs that accompany senility. On the other hand, the number of persons who are involved in the health care professions is expected to be reduced over the coming years due to the fact that these occupations remain as difficult and demanding as ever. This prediction accompanied by the new beliefs that are being established in patients care lead to reliance on the new technologies and more specifically on the new telematic and robotic technologies. The technological field promises solutions in many problematic areas and there are already varied completed applications that solve a wide range of specialized health care issues.[11]

Current projects in the European Union, the USA and Japan indicate a trend to develop advanced robotic systems in the service of the elders and the disabled. The idea of providing physical and health assistance, companionship and even safeguarding seems to be a central feature in the researcher's field [8],[9],[10][15],[16],[18],[21]. In addition, according to the International Federation of Robotics, about two million personal robots[9] were in use around the world in 2004, and another seven million will be installed by 2008 [13].

While much attention has been given to robots that provide assistance to people through physical contact (assistive robotics) and to robots that entertain through social interaction (socially interactive robotics), a new domain is emerging, named "Socially Assistive Robotics" [5] which is defined as the intersection of Assistive Robotics and Socially Interactive Robotics. It refers to robots that provide assistance through interaction and with no physical contact.

The state of the art systems that can be characterized as socially assistive robots are presented in Table 1. These examples are notional and from a healthcare point of view. Beyond the environment where they are designed to be used (hospital, domestic), all of them are focused on telemedicine issues and on human robot interaction. Their functions are various because of the strictly determined needs they are meant to serve.

HelpMate [22] is a mobile robotic system used primarily in hospitals, laboratories and other health-care settings. It is being used today to transport pharmaceuticals and supplies throughout health-care institutions cost-effectively and allowing pharmacists and nurses to direct more energy to patient care.

RP-7 [21] is a wireless mobile Remote Presence Robot that allows professionals in the healthcare community to consult with hospital-based patients and healthcare staff in a more rapid and efficient manner, thus improving hospital through-put, physician efficiency and providing high quality care.

Care-O-bot [20] is a mobile service robot designed to provide home support and instructional help in a person's daily life as well as to promote self-initiative. The latest prototype is equipped with a manipulator arm and adjustable walking supporters, through which it can provide mobility aid. It can perform household tasks, as serving drinks, setting the table, operating the microwave. It is also able to fetch and carry objects and additionally to support in grasping, holding, and lifting objects and tools. It is responsible of the control of the technical home infrastructure too. Among its functions, there is the capability of managing senior's day time (daily routine, time for taking medicine etc.) and the supervision of his vital signs. Finally it can communicate with medical and public facilities (physician, authorities,etc.) and it has the ability of performing an emergency call.

Table 1
Socially Assistive Robots

Name/Provider	Description	Type
HelpMate HelpMate Robotics Inc.	Can cart around hospital items (food trays, lab specimens, pharmaceuticals, X-rays, bandages and blankets).	HelpMate robot (hospital environment)
RP-7 InTouch Health	Through tele-presence, a doctor can check on patients from another building or another country	Rounding robot (hospital environment)
Care-O-Bot Fraunhofer Institute for Manufacturing Engineering and Automation (IPA)	Interacts with humans and assists in typical housekeeping tasks.	Service robot (domestic environment)
Pearl or **"Nursebot"** Carnegie Mellon University	Interacts and assists the elders with mobility problems. Reminds people about routine activities, guides them through their environments.	Robotic Assistant (domestic environment)
RoboDOMO University of Nice Sophia Antipolis	Can provide video monitoring and medico-technical assistance for older or disabled people.	Robotic Assistant (domestic environment)

Pearl, the Nursebot, [23] is a mobile personal robotic assistant for the elderly that has two primary functions: reminding people about routine activities such as eating, drinking, taking medicine, using the bathroom or visit a doctor and guiding them through their environments. The robot additionally provides remote tele-presence for professional caregivers and assists with tasks that would be difficult for people with limited mobility. Specifically designed to help old people, Pearl has a humanoid aspect that enables her as a product that most closely resembles the image conjured by the term "robot."

RoboDOMO [21] is a robot assistant for elders who have reduced mobility or who live alone. This system demonstrates the successful integration of robotics technology and mobile telecommunications. The idea surrounding the system is to furnish an instrument that can provide around the clock video monitoring and medico-technical assistance for older or disabled people to improve their ability to live alone. The project is built in three phases. Phase one demonstrates the feasibility of employing the cellular network to remotely control the robot. In the next phase researchers hoped to integrate a service package called "3P" for *Peur, Paresse, Plaisir* (Fear, Idleness, Pleasure), this involved the integration of wireless sensors designed to do things like fall detection and measure temperature. In the final phase of the project, test models will be introduced in Nice.

In this paper the design of a new robotic system for assisting the elder in domestic environment is presented. TherapainiS is the concept of a socially assistive robot whose goal is to provide help to those who need it. This paper is organized as following; First, the system overview is presented, describing the system users, the environment that the system works as well as the purpose it serves. The hardware follows, that can be used to develop this system since it is not manufactured yet, and the functions of it. Finally concluding remarks is closing this paper.

2 System Overview

TherapainiS is the result of the design of a robotic system performing domestic healthcare assistance. Major system's goal is to make the senior feel independent through the machine without having the feeling of being "controlled" by a technological entity. In the present paper the first stage of this project's successful development will be introduced. In the following paragraphs the target population, the environment as well as the purpose of the development of TherapainiS will be presented. Figure 1, represents an overview of the system's architecture

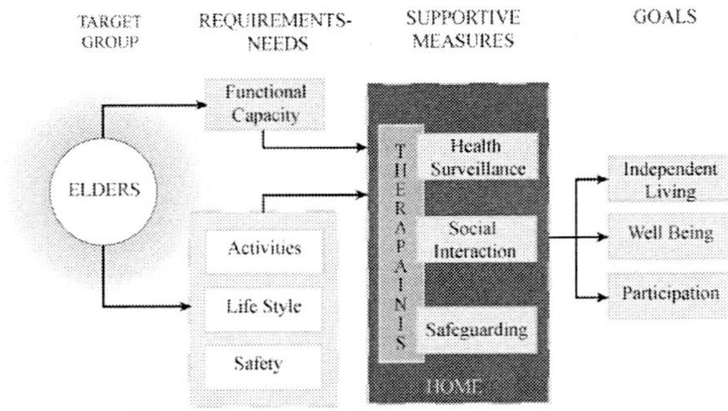

Fig. 1. System architecture overview

2.1 System Users

The present robotic system is addressed to people who are in need of domestic assistance. Such people usually live alone and require some form of supervision and help. They can be characterized as the direct users of this system and fall under the category of the elderly. Although the system was designed focusing on the elderly part of the population, this does not exclude the possibility of its use by a disabled person.

Apart from the direct user, previously mentioned, there is also the indirect user: he/she who can operate the system. The operators of the system are the persons who deal directly with the care of disabled adults. Such persons are the doctors, the nurses, the relatives and should be firmly identified by the system. They should also have accordingly restricted access.

2.2 Environment

The system is designed to be used in the domestic environment. In this case an important factor for the efficiency of the system is that the house should ideally include as a part of its structure some kind of advanced assistive technology.

2.3 Purpose

The purpose of this system is to facilitate the lives of older people and help them become as autonomous as possible. As a result of that, they do not need to lose their familiar social environment, they stay together with their families without burdening them and so they avoid institutionalization or social exclusion. In other words TherapainiS is a system that opens the way towards the insurance of a decent and healthy living for over aged people while improving their life quality.

3 Hardware Design

Figure 2 shows a 3D view of the TherapainiS concept. The robot is built on top of the Pioneer 3DX base, an agile, versatile intelligent mobile robotic platform, for general purposes [1],[2]. The robot is equipped with a SICK PLS laser rage finder [3], which is an extremely accurate distance measurement sensor, over a 180 degree area up to 80 meters away. In order to avoid the blind spots with no coverage that even laser sensors have, it is loaded with two bumpers (front, rear) so as to consolidate the system's navigation.

The robot is accoutered by one onboard PC, linked to a Wi-Fi network, via which the system can be connected to the internet and other devices.

A touch sensitive color display clamshell is mounted on top of the robot. The touch screen is equipped with a camera, microphone and speakers to enhance communication as well as with a fingerprint reader for security reasons.

Finally, TherapainiS has the ability to charge its battery automatically, by navigating and connecting itself to its docking/charging station when is necessary [2].

4 Functions

The goal of this project is the development of a personal robotic system that serves three primary needs: health surveillance, social interaction and safeguarding. According to the social assistive character [5], these needs become essential for the robot.

4.1 Health surveillance

The passage of the healthcare sector from the hospital to the home environment arises some important issues. Firstly, the household visits by healthcare professionals (e.g., caregivers, doctors)

Fig. 2. Robotic assistant TherapainiS

are extremely rare, because of their insufficient number and of the involved high cost. Additionally, there is the inability to collect and communicate data from people who live in private homes. Such data include health measurements (e.g., blood pressure, glucose levels and body temperature), statistics on health rate and medication (health history, medication and health progress). However, it is important not to forget that a large number of elders suffer from cognitive impairment and most commonly from memory lapses. They forget to take their medicine or visit the doctor and in time, which means that they might not be able to take care of themselves.

TherapainiS, can systematically collect data via onboard devices (thermometer, glucose meter and blood pressure meter). It can also organize the medication by automatic pill dispenser and work as a reminder.

Furthermore, it can work as a communicatory link between the doctor and the patient. It uses the internet technology to establish a tele-presence and interact directly with remote patients.

4.2 Social Interaction

A great number of elderly people are forced to live alone, deprived of social interaction. The ability to communicate, to be entertained and informed can act for them, as a method to delay the deterioration of health related problems [6].

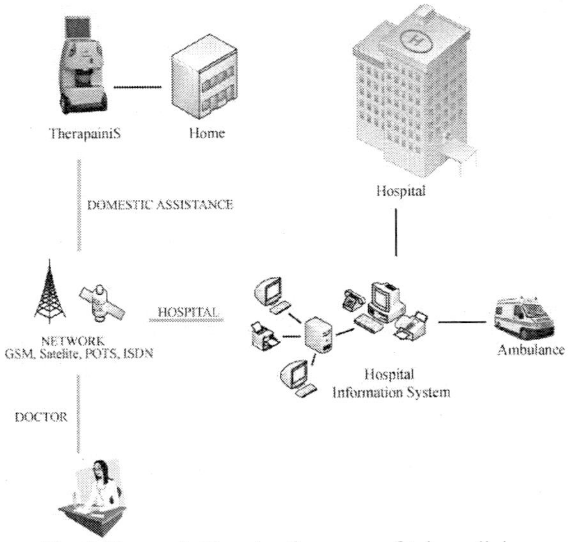

Fig. 3 TherapainiS under the scope of telemedicine

4.2.1 Communication

People grow old into communities where a social character is a fundamental feature. It is vital for every human being to interact with others so as to feel alive and active member into society.

In order to accommodate this need, TherapainiS through internet technology, has plenty applications to give (VoIP, teleconference, tele-presence). It enhances the remote contact between people and in the case of health care providence, it must be admitted that one of the most important factors that make the telemedicine obtainable is the ability to communicate from distance, as presented in figure 3.

4.2.2 Entertainment/Information

Although internet has already become part of our lives, elders hesitate to accept its role. It is important to facilitate them with applications that raise their attention. One of the most common occupations of the elders is watching television. TV is the main mean of entertainment and information. TherapainiS, carries a TV tuner, e-radio and e-Press. In addition, beyond the basic internet search engines, it offers access in medical databases where someone can be informed on health issues.

Furthermore, it offers interactive games for entertaining and educational purposes.

Table 2
Needs and functions

Health Surveillance	Systematic Data Collection	Medicine Organization and Reminder
	Thermometer Blood pressure meter Glucose meter	Pill dispenser
Social Interaction	Communication	Entertainment/Information
	VoIP Tele-conference Tele-presence	e-Radio e-Press TV-tuner Games
Safeguarding	Emergency Alert System	

4.3 Safeguarding

As seniors become physically and cognitively impaired the possibility of health emergencies and other emergency situations is growing. For example, a fall or even a stroke can suddenly disable them so that they are no longer able to call for help when they need it.

Emergency alert systems can add an important element of safety for many seniors who live by themselves. Systems like that can enhance an independent and safe life at ones home and delay the need of a day-long supervision [4].

Table 2, depicts the three main needs that the robot is aimed to serve. These needs are not the only challenges that affect the design process. There are issues and limitations related to the environment (indoor), the ergonomics (anthropometric characteristics) and the human robot interaction. This paper is an introduction to the concept of the robot TherapainiS.

5 Conclusions

A vital element that much affects, the development of this design study is the necessity to recognize the importance of providing care for the elders. Nowadays, the privilege of independent living, well being and social participation is frail and it's about time to be claimed.

Among robotic systems that operate in a domestic environment, Therapainis conjugates the social assistive character and the robotic technology with onboard appliances, already known as assistive technology products that the seniors are familiar with. This constitutes a difference with most of the resembling robots which tend to communicate and exchange information with exterior devices.

Robotics is capable of offering a great chance to accommodate all needs implicated. Still, it is meaningful to realize that a robot could never replace the human touch. What makes TherapainiS unique is that it repulses the robotic character which conjures it as an entity and challenges the opportunity to become an integral and reliable unit of a senior's daily life.

All of the facts lead to the conclusion that the clever way to take advantage of technology may be facing the challenges raised by the "modus vivendi".

The next phase of this project is to develop a prototype and test it in laboratory conditions.

Acknowledgements

University of Patras is partner and University of Aegean is associate partner of the EU-funded FP6 Innovative Production Machines and Systems (I*PROMS) Network of Excellence.

References

[1] MobileRobots Inc, Pioneer 3 Operations Manual, with MobileRobots Exclusive Advanced Robot Control & Operations Software ,2006

[2] ActivMedia Robotics, Mobile Robots, http://www.activrobots.com

[3] SICK, Sensor Intelligence, technical description, LMS200/211/221/291 Laser Measurement Systems, 2006

[4] ABLEDATA, Fact Sheet on Emergency Alert Systems, 2007

[5] Feil-Seifer D., Mataric M. J, Defining Socially Assistive Robotics, IEEE, June 28 - July 1, 2005

[6] Carnegie Mellon University, Nursebot Project, Robotic Assistants for the elders http://www.cs.cmu.edu/~nursebot/

[7] Eurostat, International Day of Older Persons, Eurostat Press Office, 29/9/2006

[8] Polack M.E., Intelligent Technology for an Aging Population: The use of AI to assist elders with cognitive impairment, AI Magazine,2005

[9] Baltus G., Fox D., Gemperle F., Goetz J., Hirsh T., Magaritis D., Montemerlo M., Pineau J., Roy N., Schulte J., and Thrun S.. Towards personal service robots for the elderly. In Proceedings of the Workshop on Interactive Robots and Entertainment, Pittsburgh, PA, April-May 2000.

[10] Mahoney R., Can der Loos H., Lum P., and Burgar C., Robotic stroke therapy assistant. Robotica, 21:33–44, 2003.

[11] Whitten P., Shypher B., Evolution of telemedicine from an applied communication perspective in the United States, telemedicine and e-health, volume 12, number 5, 2006.

[12] Kinsella A., E-Health and This Generation of Home Health Care Patient, January 19, 2000.

[13] Scientific America, A robot in every home, January, 2007 http://www.sciam.com/article.cfm?chanID=sa006&articleID=9312A198-E7F2-99DF-31DA639D6C4BA567

[14] Bayles F., Gadgets help baby boomers navigate old age, USA TODAY, 17/11/2003 http://www.usatoday.com/news/nation/2003-11-16-gadgets-cover_x.htm

[15] Cook D., Providing for Older Adults Using Smart Environment Technologies, IEEE, 05.2007 http://www.todaysengineer.org/2007/May/smart_homes.asp

[16] Tyrer H., Alwan M., Demiris G., Zhihai He, Keller J., Skubic M., Rantz M., Technology for Successful Aging, 28th IEEE EMBS Annual International Conference, 2006

[17] Labonté D., Michaud F., Boissy P., Corriveau H., Cloutier R., Roux M., A Pilot Study on Teleoperated Mobile Robots in Home Environments

[18] Haigh K., Kiff L., Myers J., Guralnik V., Krichbaum K., Phelps J., Plocher T., Toms D., The Independent LifeStyle AssistantTM (I.L.S.A.): Lessons Learned, December 1, 2003

[19] Giddens A., Sociology, Gutenberg, 2002.

[20] Fraunhofer Institute for Manufacturing Engineering and Automation (IPA), Care-O-bot, http://www.care-o-bot.de/english/

[21] AARP International: Robots Roll in an Aging Society, March 01, 2007 , http://www.aarpinternational.org/agingadvances/agingadvances_show.htm?doc_id=553904

[22] Cardinal Health, Inc., Pyxis Corporation Completes Acquisition of HelpMate Robotics, Inc., a Developer of Health-Care Robotics Systems, http://www.cardinal.com/content/news/123099_66867.asp

[23] Rotstein G., A Pearl for the elderly, April 04, 2004 http://robots.stanford.edu/press/gazette.html

Innovative Production Machines and Systems
D.T. Pham, E.E. Eldukhri and A.J. Soroka (Eds.)

Weed detection and classification for autonomous farming

S. Kodagoda[a], Z. Zhang[a], D. Ruiz[b], G. Dissanayake[a]

[a] *ARC Centre of Excellence for Autonomous Systems (CAS), University of Technology, Sydney, AUSTRALIA*
[b] *Australian Centre for Precision Agriculture (ACPA), University of Sydney, Sydney, AUSTRALIA*

Abstract

Autonomous weed control concepts have recently being extensively researched due to the advantages that they possess. One of the critical modules of such systems is the sensing and classification of weeds within crops. In this paper, we systematically chose the sensing setup and cues to be used for classification of two common weed species (Bidens pilosa L. and Lolium rigidum L.) in a wheat crop. An automatic cue selection followed by classification procedure is proposed. Some classification results are presented while discussing problems leading to future direction of research.

Keywords: Weed detection, spectral analysis, classification

1. Introduction

There are a number of field operations that can be executed by autonomous vehicles, giving more benefits than conventional man-guided machines [1]. These vehicles would be able to work unattended over long periods of time, carrying out useful tasks such as cultivation and seeding, weed control, soil scouting, application of fertilizers, irrigation, and harvesting. Automatic guidance may avoid over-application of chemicals and fertilizers, reducing environmental impact [2]. As weed populations have been found to be distributed heterogeneously in time and space within agricultural fields, weed control systems based on vision have been developed to spray specifically the weed infested areas in real-time, reducing treatment costs as well as herbicide loading to the environment [3-4].

The authors are collaborating with [5] to realize an autonomous weed control system based on CASPA weeder shown in Fig. 1. The CASPA weeder has the capabilities of remote control, joystick based control, sensing and data logging. It can be programmed for autonomous navigation. It has accurate GPS localisation system and weighs 89kg with approximate size of 800L x 550W x 400H (mm).

Fig. 1 CASPA weeder

It is planned to implement a fully insulated and isolated electrocution cradle extending out at the back of the robot to be used to destroy the weeds. It will have five independent electrodes at 20kV covering 250 mm width and a spiked castor wheel as the ground electrode. The authors' main contribution to this project is to synthesise a low cost, real time crop-weed classification algorithm.

In the literature, crop-weed classification falls into two basic categories: spectral data based classification and computer vision based classification. Spectral data based technology relies on the difference of spectral response of each plant species. Jurado et al [6] use a NIRS monochromator to measure plants' spectral reflectance in a lab environment. The spectral reflectance data was then analyzed statistically. They have shown that the spectral difference in the band of 750nm to 950nm is suitable for discriminating sunflower and wheat stubble. However, the experimental results also contained overlapping spectral responses from dissimilar plants. They proposed to use an airborne hyper spectral camera for better classification results albeit higher costs of the equipment. Borregaard et al [7] adopted a line scan spectral device to implement on site weed detection. In their work, a line scan spectrometer was mounted pointing down observing plants. The line scanning spectrometers provide additional spatial and textural information to that of normal spectrometers with a spot foot print. Classification methods such as LDA, QDA, PCA, and PLS were used to classify the spectral data. However, the accuracy of Borregaard's algorithm was between 60% and 90%. Eddy et al [8] used a hyper spectral camera with a resolution of 640x480. Each pixel from this camera can have a spectral resolution of 10nm within the spectral band of 400nm to 1000nm. A feed forward neural network was trained to classify the spectral data, and a detection rate of 88%-95% was achieved. The hyper spectral camera can capture more information than that of line scan spectrometers. However, the high cost of the hyper spectral camera prevent its feasibility of using in low cost weed detection systems.

Computer vision based techniques have also been extensively exploited for crop-weed discrimination. A color image is rich in information providing cues such as color, texture, shape,.... etc. Perez et al [9] have chosen color and shape as appropriate visual cues. The cues were then analyzed by a heuristical approach, Bayes classification and k-nearest Neighbor classification. They managed to achieve a detection rate of 71% to 89%. Aitkenhead et al [10] proposed to use shape features in a neural network to achieve a detection rate of 50%-90% for crop/weed classification. Hemming et al [11] could only achieve a detection rate of 50% to 85% using shape features. Large variation of the detection performance indicates that the accuracy of the shape parameter calculation is not robust. This can be due to occlusions introduced by proximity leaves.

Lolium rigidum L (commonly known as ryegrass) is a widespread grass weed in cereal fields, causing important yield reductions and having evolved resistance to five major groups of herbicides. In wheat fields in southern Australia yield reduction can be up to 50% at weed densities of 200-500 plants/m2 [12]. *Bidens pilosa* L (commonly known as cobbler's peg) is an annual broad leaf weed widely distributed in tropical and subtropical regions of the world and is reported to be a weed of 31 crops [13]. Therefore, in this work our particular attention is paid to classification of wheat, Bidens and Lolium species.

The rest of the paper is arranged as follows. Section 2 discusses the spectral analysis of the concerned plant species. Section 3 discribes the experimental system. Crop weed classification algorithm is given in Section 4. In Section 5, experimental results are given. Section 6 concludes the paper.

2. Spectral analysis

It is a well-known fact that various plant species have different optical and light absorption properties. That is mainly due to pigments in the UV and visible wavelengths, chemical composition in the NIR range, and numerous leaf tissue structures. Leaf pigments, which contain Chlorophyll absorb large amount of light in the UV band, blue (450nm) and red (680nm) part of the spectrum, whilst slightly lower absorption in the green (550nm) band (see Fig. 2). The leaf pigments do not have a good NIR light absorption property, hence the leaves reflect or diffuse large amount of radiation. The transmitted radiation is further affected by the leaf internal structure, which can be used for classification of diverse plant species.

Fig.2 shows the spectral response of three types of plant species under laboratory conditions. The plants were grown in trays and spectral responses were measured with artificial lighting. Oceanoptic spectrometer was used and measurements were taken on leaves of the species. Fig. 2 shows a large variation in the responses of three species within the spectral band, 750nm to 950nm (NIR). It also shows up another band around 550nm (green). The steep slope 700nm to 750nm is called the "red edge".

After analyzing several number of data sets taken from the same type of species, it was noted that the spectral responses within a particular plant species can significantly vary causing a straightforward segmentation based on spectral analysis is erroneous. Fig.3 shows the spectral variations of each type of species and the obvious overlapping regions. For example, considering the spectral band 750nm to 900nm, it can be seen that the spectral variation of wheat is large enough to contain some Bidens and rye grass responses.

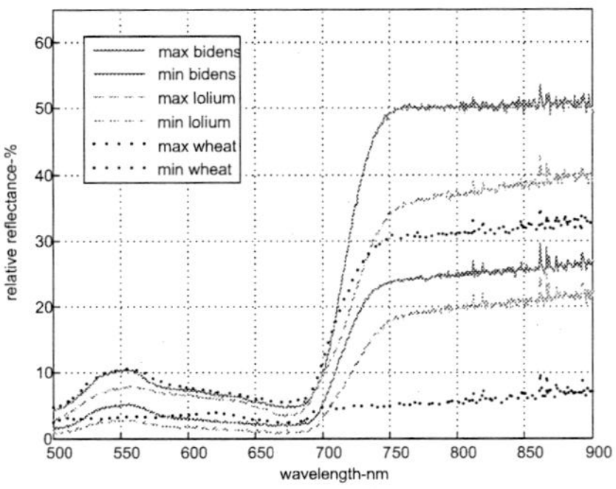

Fig. 2 Variance in spectral reflectance of different plant species

This variation in spectral response can be due to age of leaves, their orientation, spatial position of the leaves and plant health. The spectral responses of a wheat leave obtained at different orientations from the light source are shown in Fig. 3. It could be noted that the spectral responses become weak with larger angles to the light source. In some cases, relatively larger footprint of the spectrometer caused the measurements to be affected by the background, especially for thin leaves such as rye grass.

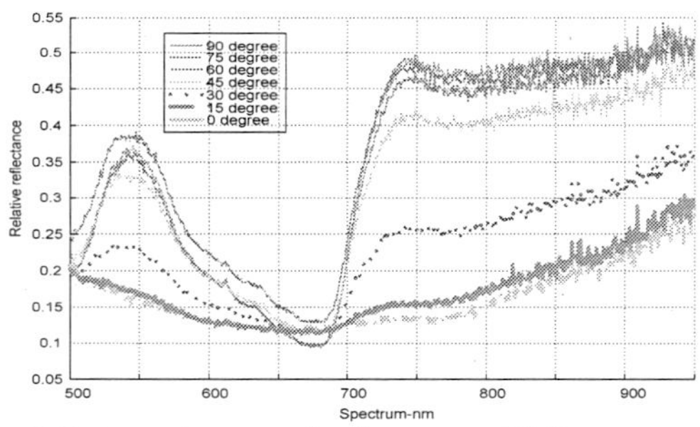

Fig. 3 Variance in spectral reflectance of different measuring angles. Angles are measured from the wheat surface. The light source is positioned at 90 degrees to wheat surface.

The spectrometers may not be a viable solution for online weed/crop classification due to either their higher costs or due to practical constraints. One such practical limitation is the requirement of a reference spectral response at the measurement point for calculating the ratio. Further, it needs a sweeping mechanism for obtaining the special information due to its limited circular footprint. The spectrometer probe also needs to be closely located to the surface before taking measurements. Therefore, it is proposed to use a color camera and a NIR (Near Infra Red) camera to capture the discriminative bands, which are around 550nm and 750nm – 950nm.

3. Experimental setup

Experiments were carried out in a laboratory consisting of a camera set up, data logging software and trays of three plant species namely wheat, Bidens and Lolium. As in Fig. 4, the experimental setup is consisted of a color camera, a NIR camera, a LED lighting source, a laptop computer and data logging software. NIR camera is a monochrome camera fitted with a visual light block filter. Lenses of both cameras were chosen to be identical. The lighting source is specially designed consisting of 5x4 LED arrays with spectral band of 780nm. The colour images were captured with fluorescent lighting and the NIR images were captured with the LED lighting. The image resolution is 640x480.

Fig. 4 Experimental setup

4. Crop/Weed classification algorithm

Vision cues have been extensively used in the literature for crop/weed classification. In this section, an unsupervised crop/weed classification algorithm based on vision information alone is presented. Fig. 5 shows a block diagram of the proposed crop/weed classification algorithm. It mainly consisted of two parts, namely model generation and classification.

4.1. Automatic model generation

In order to classify plant species, we need to choose appropriate vision cues and models defining the species. We adopt a strategy, where such a selection is carried out automatically requiring minimal input from the operator. This leads the algorithm to be used in diverse classification applications enhancing the generality. Further, it will help an unskilled worker to carry out the operations without having to consult vision experts in selecting appropriate parameters.

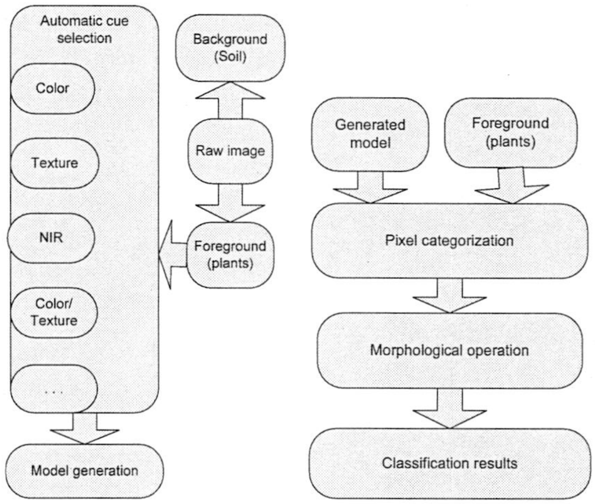

(a) Model generation (b) classification

Fig. 5 Classification algorithm

The automatic model generation algorithm is given in Fig. 5(a). Once a raw image is received, a simple color based classification is used for detection of plant/soil (see Fig. 6 (b)). Foreground consisting of pixels belonging to plant species are further processed for calculating various visual cues such as, color, texture, NIR, etc. The cues and their combinations are grouped into several number of clusters based on k-means algorithm. The estimated clusters are further analyzed by Mahalanobis distance (MD), which is a good measure for discrimination of the clusters. The cues or their combinations corresponding to larger MDs are chosen as the appropriate cues or their combinations for the classification problem. The MD is a better choice than Euclidian distance (ED) between cluster centers as latter does not consider the distribution of the clusters. For example, high ED between cluster centers does not necessarily leading to good classification results due to possible larger overlap between the cluster distributions. Given the problem, once the cues or their combinations are chosen as most appropriate, data surrounding the cluster centers are used to determine the model of a particular plant species.

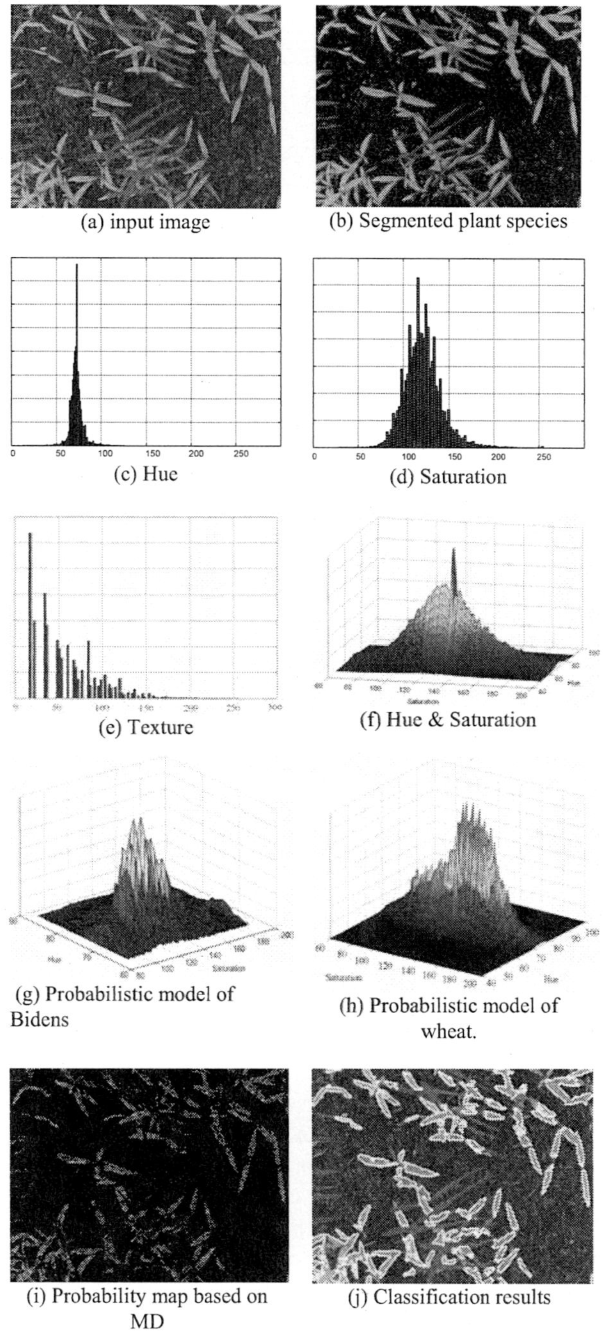

(a) input image (b) Segmented plant species

(c) Hue (d) Saturation

(e) Texture (f) Hue & Saturation

(g) Probabilistic model of Bidens (h) Probabilistic model of wheat.

(i) Probability map based on MD (j) Classification results

Fig. 6 Automatic model generation and classification: wheat/ Bidens

4.2.Classification algorithm

Once the model is established, it is possible to synthesize a classification algorithm based on [14], which is shown in Fig. 5(b). By going through each foreground pixel, the MD is calculated with the established model. Pixels that have MD values less than 1 are given a value of 1 and the other pixels are given a value of 1/MD. This is resulted in an "image" describing how confidence a pixel to be belonging to the established model. The confidence image is then thresholded and morphological operations are carried out to improve the connectivity of highly probable pixels while rejecting outliers. The final image shows classified pixels based on the established model.

5. Experimental results

In this section, some classification results are presented with an analytical discussion. The plant species considered here are wheat, Bidens and Lolium. Cues considered are, Hue, Saturation, Texture and their combinations. For all scenarios, we assumed the number of clusters to be 2 for the k-means algorithm.

5.1. Classification of Wheat and Bidens

Fig.6 (a) shows a color image of wheat and Bidens plants captured in a laboratory condition with artificial lighting. We used a classification algorithm given in Section 4. Fig.6 (b) shows the classified plant species from soil based on color. The distribution of *hue*, *saturation*, *texture*, and *hue and saturation* of the image Fig. 6 (b) are shown in Fig. 6(c,d,e,f) respectively. Texture reveals the spatial distribution of an image, typically repeated patterns. The common methods of determining texture are Gabor filter, run-length statistics and co-occurrence matrices. In this work, we have used the output of Gabor filter as a measure of texture. It could be noted that the hue is tightly distributed and the saturation has a broad distribution. Although hue and saturation as individual components do not show separable clusters, they form discriminative clusters once combined as in Fig. 6 (f). Therefore, the model generation algorithm has chosen hue and saturation combination to be the most appropriate for Bidens/wheat classification. The resulting models of Bidens and wheat are shown in Fig.6 (g) and (h) respectively. Fig. 6 (i) shows the probability map image generated using the model of the Bidens (Fig. 6 (g)) followed by classification results in Fig. 6 (j).

Further, classification results based on Bidens model is shown in Fig.7. The results are appealing with 97% detection rate and just 2% of false alarms (see Table 1). Fig. 8 shows classification results based on the derived wheat model. The probabilistic model of wheat is ambiguous as shown in Fig. 6 (h), when comparing with that of Bidens (Fig. 6 (g)). Therefore, the classification results (Fig. 8) shows degraded performances than that of using Bidens model (as in Table 1, it has a detection rate of 77% and false alarm rate of 30%).

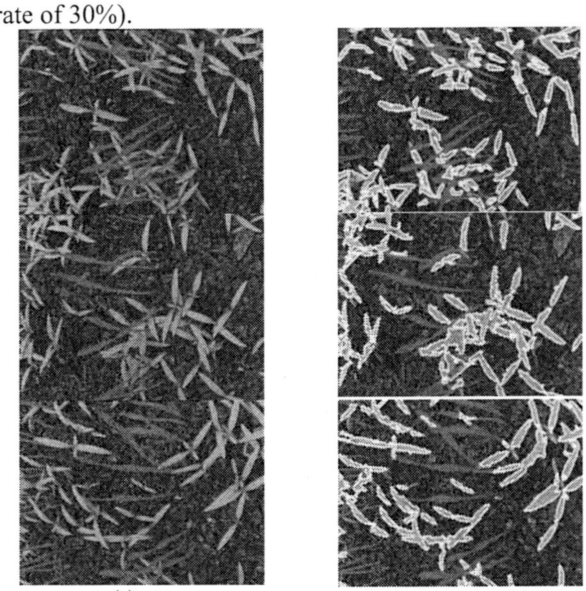

(a) raw (b) Segmented Bidens

Fig. 7 Classification based on Bidens model

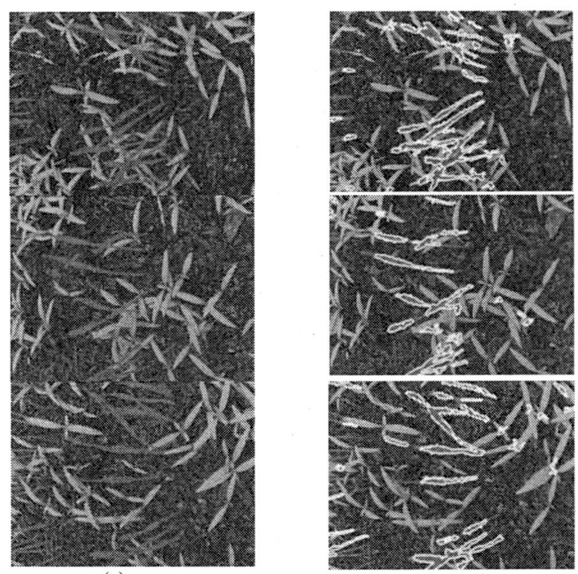

(a) raw (b) Segmented wheat

Fig. 8 Classification based on wheat model

Table 1: Classification results of wheat and Bidens

	detection rate %	false alarm rate %
wheat	77	30
Bidens	97	2

5.2. Classification of wheat and Lolium

We used the same algorithm for wheat/Lolium classification without much success (Fig. 9). As given in Table 2, while using the wheat model, it has a reasonably good detection rate of 85% accompanied by bad false alarm rate of 82% bringing overall classification results to be unacceptable. While using the Lolium model, the detection rate is too low (26%) and the false alarm rate is reasonable high (27%) causing the overall classification to be unacceptable. That is mainly due to the similarities of the two species in the distributions of *hue*, *saturation*, and *texture* cues. Therefore, we are currently working on incorporating NIR images and their cues for improved classification accuracies in such challenging situations.

(a) Raw image (b) Segmented Lolium (c) Segmented wheat

Fig. 9 Classification results: wheat/Lolium

Table 2: Classification results of wheat and Lolium

	detection rate %	false alarm rate %
wheat	85	82
Lolium	26	27

6. Conclusions and future work

Automatic crop - weed classification is a crucial module in autonomous weed control. In this paper, we have investigated a method of classifying wheat from Bidens. We have also shown that visual cues such as color and texture alone do not provide enough information to obtain high classification accuracies especially for wheat and Lolium. This made us to choose NIR image cues along with color camera cues. It leads to other challenges, related to the behavior of the lighting source and multimodal correspondence problem, which we are currently working on.

ACKNOWLEDGEMENTS

This work is supported by the ARC Centre of Excellence program, funded by the Australian Research Council (ARC) and the New South Wales State Government. The authors like to thank Dr Jay Katupitiya and Mr. Kim Son Dang for providing a picture of CASPA weeder and it's description.

References

[1] Pedersen S. M., Fountas, S., Have H. and Blackmore B. S., Agricultural robots-systems analysis and economic feasibility. Precision Agriculture (2006), 7:295-308.

[2] Blackmore S., Fountas S.,Tang L. and Have H., Design specifications for a small autonomous tractor with behavioural control, J. Int. Comm. Agric. Eng. (CIGR). Manuscript PM 04 001. vol 6. July 2004.

[3] Gerhards R. and Oebel H., Practical experiences with a system for site-specific weed control in arable crops using real-time image analysis and GPS-controlled patch system. Weed Research (2006), 46:185-193.

[4] Gee C. H., Bossu J., Jones G. and Truchetet F., Crop/weed discrimination in perspective agronomic images. Com. and Electr. in Agri. (2008), 60:49-59.

[5] Eaton R., Katupitiya J., Cole A. and Meyer C., Architecture of an automated agricultural tractor: Hardware, software and control systems, Proceedings of the 2005 IFAC World Congress. Prague: Elsevier, July 2005.

[6] Jurado-Exposito M., Lopez-Granados F., Stenciano S., Garcia-Torres L., and Gonzalez-Sndujar J. L., Discrimination of Weed Seedlings, Wheat (Tritium aestivum).

[7] Borregarrd T., Nielsen H., Norgarrd L. and Have H., Crop-weed Discrimination by Line Imaging Spectroscopy, J. agric. Eng. Res. vol. 75, pp 389-400, 2000.

[8] Eddy, P.; Smith, A.; Hill, B.; Peddle, D.; Coburn, C.; Blackshaw, R.; Proceedings of Geoscience and Remote Sensing Symposium, 2006. IEEE International Conference; p116 – 119; 2006.

[9] Pérez A. J., López F., Benlloch J. V. and Christensen S., Colour and shape analysis techniques for weed detection in cereal fields, Computers and Electronics in Agriculture, 25(3):197-212, 2000.

[10] Aitkenhead M. J., Dalgetty I. A., Mullins C. E., McDonald A. J. S. and Strachan J. J. C., Weed and crop discrimination using image analysis and artificial intelligence methods, Computers and Electronics in Agriculture 39:157-171, 2003.

[11] Hemming J. and Rath T., Computer-Vision-based Weed Identification under Field Conditions using Controlled Lighting, J. agric. Engng Res., 78(3): 233-243, 2001.

[12] Lemerle D., Verbeek B. and Orchard B., Ranking the ability of wheat varieties to compete with *Lolium rigidum*. Weed Research 41 (3): 197–209, 2001.

[13] Holm G. L., Plucknett D. L., Pancho J. V. and Herberger J. P. , The World's Worst Weeds. Distribution and Biology, FL, USA, 1991.

[14] Bradski G., Kaehler A. and Pisarevsky V., Learning-Based computer vision with Intel's open source computer vision library, Intel Technology Journal, 9(2):119-130,2005.

Innovative Production Machines and Systems
D.T. Pham, E.E. Eldukhri and A.J. Soroka (Eds.)

Effects of microstructure on surface roughness and burr formation in micromilling: A review

D.T. Pham, S.S. Dimov, K.B. Popov, A.M.A. Elkaseer

Manufacturing Engineering Centre, Cardiff University, Cardiff, UK

Abstract

Micromilling concerns the machining of parts with features that range from tens of micrometres to a few millimetres in size, with chiploads usually varying from below one micron to a few microns. Due to the fact that most commonly used materials have crystalline grains similar in size to chiploads, the cutting process sometimes takes place inside the individual grain itself. This phenomenon can be considered one of the main differences between microcutting and macrocutting. Recently it has been revealed that microstructure, especially grain size, has a dominant effect on the quality of the machined part. So far, there has been no detailed study of the influences of microstructure in the micromilling of polycrystalline materials. This paper reviews previous work on this topic. Special attention is given to surface finishes and burrs, as these are important issues for micro machined parts. The paper concludes with a summary of some of the key research problems in micromilling and suggests a systematic study of the effects of microstructure on surface generation and burr formation in the micromilling of polycrystalline materials.

Keywords: micromilling, material microstructure, grain size effect, surface finish, surface generation, burr formation

1. Introduction

Miniaturisation has become a common feature of today's technology. There is a constant requirement to make existing products smaller and produce microcomponents with tolerances in the submicron range [1]. Broad areas of industrial applications of micro components include automotive and transport systems, information technology, telecommunication, health care technology and biotechnology. Specific applications include microscale fuel cells, micromoulds, deep X-ray lithography masks, fiber optics, micronozzles for high temperature jets and microelectronic chips [1-5]. These applications require very tight tolerances and high quality surface finish. Due to the variety of micro applications, there is a need to use a wide range of engineering materials, including aluminium alloys, stainless steel, titanium, brass, plastics, ceramics, and composites. Mechanical micro-cutting, as an ultra-precision machining process, is becoming increasingly important for its capability for producing parts with three-dimensional features ranging from a few microns to a few hundred microns in a wide range of materials [6].

While macro and micro milling are kinematically similar, the mechanics (i.e., the material removal mechanisms) of micromachining and conventional machining have significant differences. These differences arise from changes in underlying physical phenomena due to the scaling effects associated with the tool geometry, tool material, and workpiece material characteristics [7].

The relationship between the cut geometry and the workpiece microstructure in microscale machining is also markedly different from that in macroscale machining. In microscale machining, where chiploads may range from submicron levels to a few microns and depths of cut may be from a few microns to perhaps 100 μm, the cut geometry and the grains of the workpiece material are now comparable in size [2]. Therefore, considering material microstructure, micromilling can be defined as a mechanical machining process with uncut chip thicknesses less than the average grain size of the smallest grain type in the material microstructure [8].

Microstructural effects in microscale cutting are not well understood and further research is required to answer the many challenging questions on the nature of those effects.

This paper reviews some of the work done on this topic. Special attention is given to the effects of material microstructure on surface roughness and burrs as they are two of the most critical factors for a micro machined part.

2. Material microstructure

When the tool dimension or a feature to be generated is of the same order as the grain size, material cannot be considered isotropic and homogeneous and the cutting mechanism differs substantially from conventional machining [9]. Most polycrystalline materials are regarded as a collection of grains with random orientation and anisotropic properties [2, 3, 9-13].

Many researchers have observed changes in various parameters, such as cutting force, chip formation, and surface roughness over multi phases or multi grains in micromachining [9]. Vogler et al. [14] conducted a series of full-slot endmilling tests with two fluted carbide endmills on both single-phase ferrite and pearlite and multiphase iron. They observed low frequency cutting forces in both pure ferrite and pure pearlite. On the other hand, high frequency forces appeared in the multi phase microstructure, which can be due to the heterogeneity of the workpiece microstructure. This in turn leads to different physical characteristics (such as variable friction coefficients) causing vibrations to increase and highly fragmented chips to occur [10].

Weule et al. [11] studied the effect of the material microstructure of steel workpieces and suggested a pre heat treatment to the workpiece before micromachining to obtain a homogeneous workpiece.

Simoneau et al. [8] observed surface defects as dimples on the machined surface of plain carbon steel, surface dimples do occur as a result of the dual phase structure of the workpiece material and always occur at a hard-to-soft grain transition. They do not occur when cutting moves across a soft-to-hard grain boundary and they explained that phenomenon by considering the transfer and evolution of the cutting energy during the chip formation process. That process itself is a result of a large amount of cutting energy being absorbed by the softer grain in the material microstructure well before the actual grain is cut. The early and large absorption of cutting energy causes a large plastic strain at the grain boundary in the softer material. On the other hand, as the cutting process moves from a soft to a hard material the strain mismatch, which occurred earlier, is not present. Also to examine the effects of microstructure during microscale cutting, Simoneau et al. [15] performed micromachining tests for normalised AISI 1045 steel and a refined AISI 1045 steel. A comparison of the experimental results demonstrated that by shifting the size of the material microstructure, the scale of cutting could be moved. For the refined 1045 steel microstructure, the shift in grain size translated to a shift in the scale of cutting, the resulting chip morphology and the transition between macro-, meso- and micro-scale cutting occurred at a smaller uncut chip thickness as compared to the cutting of normalised 1045 steel. Also they developed a new heterogeneous FE model to model the multi phase microstructure materials [15-18].

Furukawa and Moronuki [19] found different cutting mechanisms for polycrystalline, single crystal or amorphous materials and for brittle or ductile materials, and they suggested the use of about ten times larger undeformed chip thickness than the grain size for a specific material to avoid the crystallographic effects of the grains.

2.1. Material microstructure processing

Many researchers have suggested different ways to optimise the microstructure of the material to be more suitable for the microcutting process [9, 11, 15, and 20]. Refinement of the grain size structure is considered one of the accurate methods which could lead to more favourable conditions during cutter-material interaction and thus result in a better machining response, especially at the microscale [9].

Severe plastic deformation (SPD) technology has developed widely in recent years, and this technology mainly includes equal-channel angular pressing (ECAP), high-pressure torsion (HPT) and accumulative roll-bonding (ARB). Bulk sub-structured or nano-structured ultra-fined grained (UFG) materials can be fabricated by SPD methods. Compared with conventional materials, bulk UFG materials have improved hardness, strength and plasticity. Materials of this kind are promising for use in microelectronics and microelectro-mechanical systems [21].

Equal-channel angular pressing (ECAP) has attracted the attention of many researchers because of its effectiveness in producing bulk UFG materials of grain sizes in the range 200–500 nm. With this technique, a sample is pressed through a die with two intersecting channels equal in cross section and deformed via simple shear at the intersection of the channels [22]. There have been many reports on the UFG materials produced by the ECAP process, which include facecentered cubic (FCC) (Al alloys, Cu and Ni) and body centered cubic (BCC) (low carbon steel) metals.

3. Surface roughness

Surface roughness is one of the most important properties in any machining process and in micromilling it is really critical as the product needs to be of a very high surface quality. Many researchers have focussed on the surface roughness obtained in the micromilling process.

Mintegi [1] reported that Vogler et al. studied surface generation in micromilling and found it to be strongly affected by the tool-edge radius and the feed rate. The authors observed that for a 2 μm edge radius, as the feed rate was reduced to a certain value, the surface roughness started to increase, indicating that an optimal feed rate exists that will produce the smallest roughness value (see Fig. 1). The authors claimed that the existence of an optimal feed rate was due to a trade-off between the traditional effects of feed marks as the feed rate is increased and the minimum chip thickness effect resulting in tool passes without material removal as the feed rate is reduced.

Wang et al. [23] studied experimentally the surface roughness of brass machined by micro-end-milling on a miniaturised machine tool. They used full factorial design to construct experimental models, then applied the statistical methods of ANOVA and RSM to analyse the experimental data. They found that surface roughness increased linearly with increases in tool diameter and spindle speed; feed rate played an important role when the other parameters are constant.

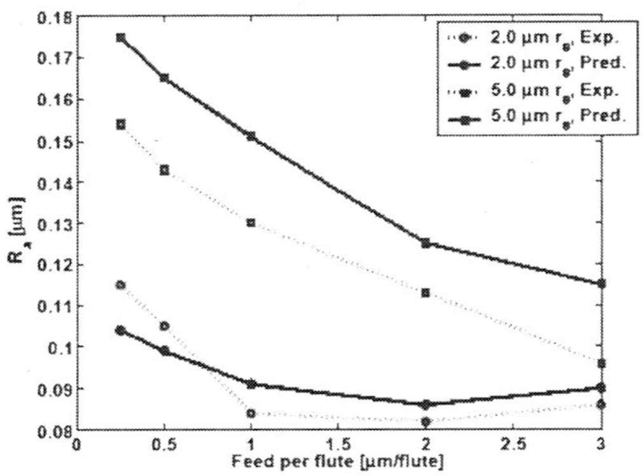

Fig. 1. Effect of edge radius on surface roughness in Pearlite [14].

Filiz et al. [7] performed an experimental investigation of micro-machinability of OFHC pure copper (101) in terms of tool wear, cutting forces, surface roughness, and burr formation. The experiments were conducted on an ultra-precision miniature machine tool using tungsten carbide cutting tools. The authors concluded that for low feed rates (up to twice the edge radius), the surface roughness of the bottom of the channels showlittle variation with feed rate and spindle speed. For higher feed rates, the surface roughness increases due to the increasing dominance of the geometric effects.

Uhlmann et al. [24] performed micromilling experiments for sintered tungsten-copper composite materials with different microstructures, and observed that a homogeneous microstructure and a homogeneous grain-size distribution have a positive effect on the attainable dimensional accuracy and surface quality (see Fig. 2).

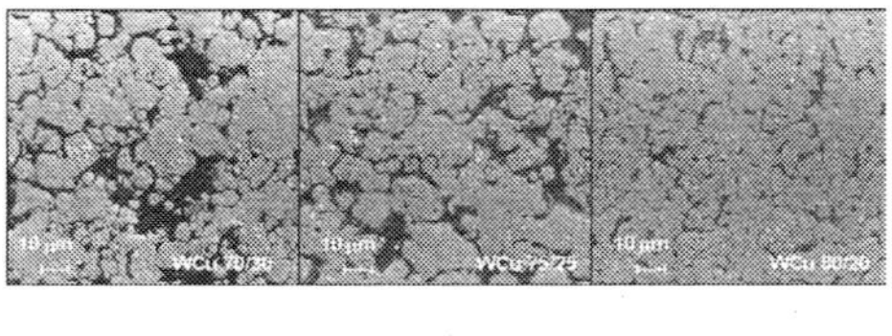

Fig. 2. Microstructure of WCu [30].

Ng et al. [25] experimentally investigated the micro- and nano-scale orthogonal cutting of aluminium 7075-T6 and studied the effects of unreformed chip thickness and cutting speed on the chip geometry, cutting forces and surface roughness. The authors found that the shear angle increased initially and then decreased as the undeformed chip thickness was increased. Also, the mean friction coefficient exhibits a size-effect that can be explained by the strengthening of the material due to a decrease in tool-chip interface temperature with decreases in undeformed chip thickness. In addition, the authors observed that the machined surface at an undeformed chip thickness smaller than the edge radius of the tool (10–60 nm) was rougher than at larger undeformed chip thicknesses for both low and high cutting speeds (10 m/min and 150 m/min).

Filiz et al. [26] performed an investigation on the viability of micromilling for fabricating micro-scale piercing elements (microbarbs) from biocompatible materials that would be compatible for medical attachment applications. They experimentally studied the effect of machining conditions on micromilling forces, surface roughness, and burr formation and performed ANOVA analyses on the results for PMMA workpieces machined using V-shaped and T-shaped micro-endmills. For the surface roughness of the PMMA V-channels, they found that only the main effect of the spindle speed was statistically significant. This indicates the possibility of increasing the material removal rate without compromising on surface roughness.

Bissacco et al. [27] performed an experimental investigation into the effects of size on surface generation by ball nose and flat end micro milling hardened steel tools. They found that increasing the ratio between cutting edge radius and chip thickness had two effects on the tool-workpiece material interaction at the micro scale. These two effects are responsible for the limited reduction in roughness; the first effect is strictly related to the geometry of the tool and consists in the accumulation of plastically deformed material in the main ridges of the machined surface. The second effect consists in the smearing of material behind the tool, forming small waves in the feed direction.

Liu [6] attributed the increase in roughness from a minimum level in the micromilling of Al5083-H116 to the increase in side flow caused by the strain-gradient induced strengthening of the material directly ahead of the tool. He developed a 2D surface roughness model that takes into account the effect of plastic side flow, tool geometry, and process parameters.

Liu [28] developed a more detailed surface generation model for prediction of the side wall and floor surface of micro endmilling. His model consists of two distinct components: a deterministic/ geometric component from the process geometry and a stochastic component from the ploughing related phenomena. However, his model did not include the effect of the workpiece material microstructure on the produced surface.

Popov et al. [20] studied the machining response of mechanically and metallurgically modified Al alloy when milling thin features in micro components. Their investigation showed that through refinement of material microstructure it was possible to improve significantly the surface integrity of the machined micro features. Popov et al. concluded that the roughness of micro features produced by micromilling was highly dependent on the material grain size. For example, the surface roughness of thin features in micro components improved more than three times as a result of the reduction in grain sizes and material anisotropy of the Al alloy used in the experiments, from 100–200 μm to 0.6 μm. The authors also found that the 'favourable' crystalline texture of the material with regard to the machining direction and the narrow grain size distribution could lead to surface roughness improvements. However, such improvements were an order of magnitude smaller than those achievable through a refinement of the material grain structure. Finally, the authors suggested further research to benefit from the grain size effects in manufacturing micro components in a range of micro engineering applications. For example, they proposed studying the use of UFG brass as a workpiece material.

4. Burr formation

Burrs in machined workpieces are real "productivity killers." They make assembly difficult and require additional finishing operations (deburring) which can damage the part [29]. Burrs can be classified as entrance, exit, top and bottom burrs (Fig. 3) [30]. Up-milling and down-milling top burrs have different characteristics. Dornfeld [29] showed that, in micromilling, up-milling burrs are small and ragged, whereas downmilling burrs are larger and wavier.

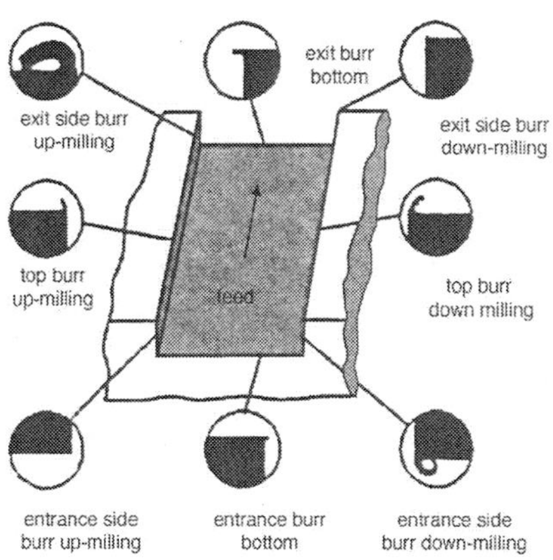

Fig. 3. Categorisation of burr types [30].

Gillespie [30] observed burr formation at the micro scale and discovered macro scale burr removal techniques could not be applied at the micro scale. Gillespie and Blotter stated that there are three generally accepted burr formation mechanisms: lateral deformation; chip bending and chip tearing.Wang et al. [10] found that in polycrystalline materials, which comprise different metal grains, the physical characteristics are obviously different. Consequently, the friction coefficient is different and according to the equation developed by the authors which relates the minimum chip thickness to the friction coefficient, the minimum chip thickness is different also. Fig. 4 depicts a case where there are four grains dispersed along the cutting edge. Because of their different minimum chip thicknesses, the chip formation statuses are different. For grains with high friction coefficients, chips are formed. However, for grains with lower friction coefficients, only little burrs are created or the material is just compressed.

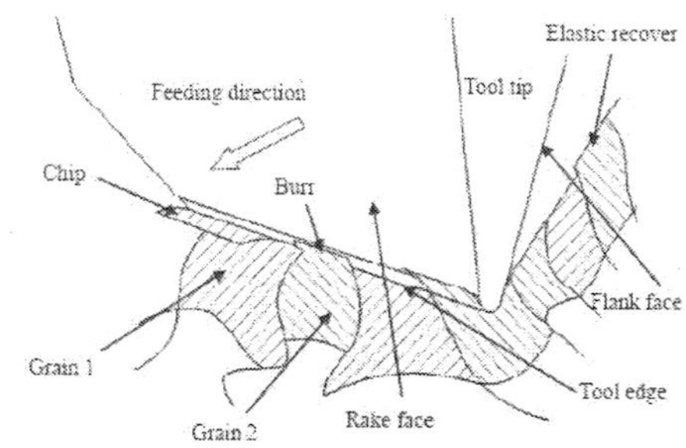

Fig. 4. Cutting zone [10].

Min et al. [31] performed experiments on single-crystal oxygen-free high conductivity (OFHC) copper workpieces. They used two-flute uncoated WC end mills, 150 microns in diameter, in a slot-milling fashion to create a series of circular slots in each of the workpieces. They found a clear difference in entrance and exit burrs at the top edges of the micromachined slots. Fig. 5 shows SEM images of the micromilling burrs varying with respect to the crystal orientation of the workpiece. Also, the influence of up and down milling can be clearly seen. A significant variation of burr height with crystallographic orientation has been found in the micromachining of single crystal copper. The authors made the following observations concerning machining in particular crystallographic orientations:

- Micromilling process parameters did not have as significant an effect on burr formation as crystallographic orientation.
- There was a distinct variation in burr height as a strong function of crystallographic orientation, particularly for the (110) and (111) cases.
- The (100) machining case did not have as clear a correlation possibly due to less anisotropy of the slip systems.

The authors suggested paying further attention to the influence of workpiece microstructure on micromachining. Also they proposed more refined testing of other crystallographic orientations to see the effect on surface and edge condition. They suggested investigating burr formation in other micromachining processes, such as microdrilling, and establishing analytical relationships between crystallographic orientation, cutting direction, and the resulting surface and edge quality.

(a) (b)

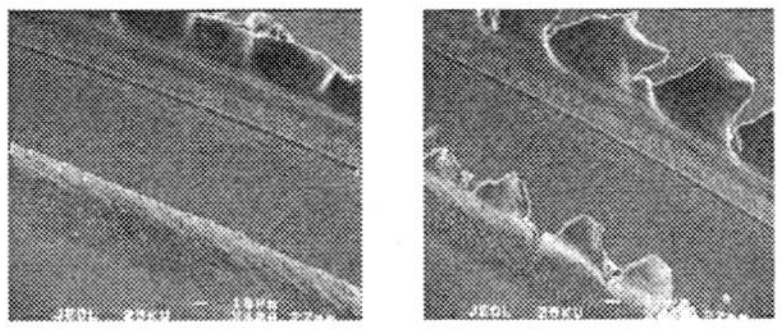

Fig. 5. SEM images of burr shape variation on (100)Plane at (a) 135° 135 and (b) 180° (ft= 1μm, Vc= 17 m/min) [31].

5. Recommendations

Although there have been several studies in the micromilling area, most of the work to date has been experimental in nature. There are still many challenging research issues that need addressing. Some of the key actions to be undertaken are:

- Developing a more accurate model for the micromilling cutting mechanism considering the effects of microstructure on the chip morphology and chipping mechanism. The suggested model should be valid for a wide range of materials considering factors like minimum chip thickness, heterogeneity of the material, ploughing and elastic recovery and grain sizes of different materials.
- Introducing ultra refining methods to optimise the material microstructure and validating them a wide range of material.
- Combining optimisation of the machining process conditions (cutting speed, feed rate, depth of cut and use of coolant), tool geometry (rake angle, lip relief angle and cutting edge radius) and material properties (material microstructure, strength, ductility, strain hardening coefficient and temperature dependence) to develop a comprehensive optimisation model.
- Developing a new analytical model for surface roughness and burr formation considering the microstructure effects. This model should consider micro burr formation and quantitatively model its effect on the total surface generated.

6. Conclusion

This paper has reviewed previous research into the effects of microstructure on the micromilling process and the surface quality obtained. Further theoretical, modelling and experimental work is needed. Special attention should be given to the effects of the microstructure of the material on surface roughness generation and burr formation, as these are critical in the case of micro machined parts. The ultimate aim of the further studies proposed in this paper will be to build a comprehensive model of the micromilling process to help researchers fully understand its governing principles.

References

[1] Mintegi E. MICROM: a revolutionary monitoring system to detect tool breakage and collisions, enhance machine cycles and introduce a new probing concept in micromilling. PhD thesis, Mondragon University, Spain, 2007.

[2] Liu X., Devor R.E., Kapoor S.G. and Ehmann K.F. The Mechanics of Machining at the Microscale: Assessment of the Current State of the Science. Journal of Manufacturing Science of Engineering, Transactions of the ASME, 2004, 126 (4), pp. 666-678.

[3] Chae J., Park S.S. and Freiheit T. Investigation of Micro - cutting operations. Int. Journal of Machine Tools and Manufacture, 2006, 46 (3-4), pp. 313-332.

[4] Gowri S., Ranjith P., Vijayaraj R., Balan A.S.S. Micromachining: technology for the future. Int. Journal of Materials and structure Integrity, 2007, 1 (1,2), pp. 161- 179(19).

[5] Ehmann F. A synopsis of U.S. micro-manufacturing research and development activities and trends. Multi-Material Micro Manufacture conference. Borovets, Bulgaria, 2007, pp. 7-13.

[6] Liu K. Process modelling of micro-cutting including strain gradient effect. PhD Thesis, George W. Woodruff School of Mechanical Engineering Georgia, Institute of Technology, USA, 2005.

[7] Filiz S., Conley C., Wasserman M. and Ozdoganlar O. An experimental investigation of micro-machinability of copper 101 using tungsten carbide micro-endmills. Int. Journal of Machine Tools & Manufacture, 2007, 47, pp. 1088–1100.

[8] Simoneau A., Ng E. and Elbestawi M.A. Surface defects during microcutting. Int. Journal of Machine Tools & Manufacture, 2006, 46, pp. 1378–1387.

[9] Dornfeld D., Min S. and Yakeuchi Y. Recent advances in mechanical micromachining. Annals of the CIRP, 55 (2), 2006, pp. 745-768.

[10] Wang J., Gong Y., Abba G., Chen K., Shi J. and Cai G. Surface generation analysis in micro end – milling considering the influences of grain. Design, Test, Integration and Packaging for MEMS/MOEMS Conference, Stresa, Italy, 2007.

[11] Weule H., Huntrup V. and Tritschler H. Micro-cutting of steel to meet new requirements in miniaturization. Annals of CIRP, 50 (1), 2001, pp. 61-64.

[12] Vogler M. On the modelling and analysis of machining performance in micro-endmilling. PhD thesis, University of Illinois at Urbana-Champaign, Urbana, Illinois, USA, 2003.

[13] Miao J., Chen G. Lai X., Li H. and Li C. Review of dynamic issues in micro-end-milling. Int.

Journal of Advanced Manufacturing Technology, 31(9-10), 2007, pp. 897-904.
[14] Vogler M., Kapoor S. and Devor R. On the modeling and analysis of machining performance in microendmiling, part 2: cutting force prediction, Journal of Manufacturing Science and Engineering, 2004, 126, pp. 695–705.
[15] Simoneau A., Ng E. and Elbestawi M. Grain size and orientation effects when microcutting AISI 1045 sreel. Annals of the CIRP, 65 (1), 2007, pp. 57-60.
[16] Simoneau A., Ng E. and Elbestawi M. Chip formation during microscale cutting of a medium carbon steel. Int. Journal of Machine Tools & Manufacture, 2006, 46, 467–481.
[17] Simoneau A., Ng E. and Elbestawi M. Modeling the effects of microstructure in metal cutting. Int. Journal of Machine Tools and Manufacture, 2007, 47, pp. 368–375.
[18] Simoneau A., Ng E. and Elbestawi M. The effect of microstructure on chip formation and surface defects in microscale, mesoscale, and macroscale cutting of steel. Annals of the CIRP, 55 (1), 2006, pp. 97-102.
[19] Furukawa Y and Moronuki N. Effects of material properties on ultra precise cutting processes. Annals of the CIRP, 1998, 37 (1), pp. 113-116.
[20] Popov K.B., Dimov S.S., Pham D.T., Minev R.M., Rosochowski A. and Olejnik L. Micromilling: material microstructure effects. Proc. IMech, 2006, 220, part B, pp. 1807-1813.
[21] Xu S., Zhao G., Ma X. and Ren G. Finite element analysis and optimization of equal channel angular pressing for producing ultra-fine grained materials. Journal of Materials Processing Technology, 2007, 184, pp. 209–216.
[22] Lee B.S., kim M.H., Hwang S.K., Kwun S.I. and Chae S.W. Grain refinement of commercially pure zirconium by ECAP and subsequent intermediate heat treatment. Materials Science and Engineering A 449–451, 2007, pp. 1087–1089.
[23] Wang W., Kweon S.H. and Yang S.H. A study on roughness of the micro-end-milled surface produced by a miniatured machine tool. Journal of Materials Processing Technology, 2005, 162 – 163, pp. 702–708.
[24] Uhlmann E., Piltz S. and Schauer K. Micro milling of sintered tungsten–copper composite materials. Journal of Materials Processing Technology, 2005, 167, pp. 402–407.
[25] Ng C., Melkote S., Rahman M. and Kumar A. Experimental study of micro- and nano-scale cutting of aluminum 7075-T6. Int. Journal of Machine Tools & Manufacture, 2006, 46, pp. 929–936.
[26] Filiz S., Xie L., Weiss L. and Ozdoganlar O.B. Micromilling of microbarbs for medical implants. Int. Journal of Machine Tools and Manufacture, 2008, 48, pp. 459–472.
[27] Bissacco G., Hansen H.N. and De Chiffre L. Size effects on surface generation in micro milling of hardened tool steel. Annals of the CIRP, 55 (1), 2006, pp. 593-596
[28] Liu X. Cutting mechanisms in micro-endmilling and their influence of surface generation. PhD thesis, University of Illinois at Urbana-Champaign, Urbana, Illinois, USA, 2003.
[29] Dornfeld D. Micromachining and Burr Formation for Precision Components. Proceedings of MTTRF, 2005, Annual Meeting.
[30] Robinson G.M. and Jackson M.J. A review of micro and nanomachining from a materials perspective. Journal of Materials Processing Technology, 2005, 167, pp. 316–337.
[31] Min S., Dornfeld D., Inasaki I., Ohmori H., Lee D., Deichmueller D., Yasuda T. and Niwa K. Variation in Machinability of Single Crystal Materials in Micromachining. Annals of the CIRP, 55 (1), 2006, pp. 103-106.

Innovative Production Machines and Systems
D.T. Pham, E.E. Eldukhri and A.J. Soroka (Eds.)

Review, issues and gap in micro-forming development

A.R. Razali

Department of Manufacture & Engineering Management, University of Strathclyde, James Weir Building, 75 Montrose Street, Glasgow, G1 1XJ. U.K

Abstract

Lots of researches have been initiated and conducted worldwide covering micro-forming field and showing that micro-forming is on promising way towards its application in industrial production [1, 2, 3]. Moreover, metal forming offers some attractive characteristics that are superior to those of other processes, for example, machining and chemical etching, considering such features as high production-rates, better material integrity, less waste, lower manufacturing costs, etc. Therefore, micro-forming could be a better option in the mass-manufacture of micro-products at reduced cost, provided if proper manufacturing facility is developed. Although at this time it becomes obvious that there were some basic knowledge on micro-forming and practically some application may seen visible in prototype stage [1, 2, 3, 4], still an extra effort and contribution should be taken and made in order to realize and bring micro-forming technology to the industries. This paper presents the issues related in micro-manufacturing and micro-forming specifically. Although the micro-forming processes have received vast development, nevertheless, in terms of the material handling itself, this has been left far behind. Therefore, in this paper, the gap is highlighted and hopes in future effort is made to fill the gap.

Keywords: Micro-forming, micro-forming handling, micro-machines, micro-manufacturing.

1. Introduction

The term micro-manufacturing concept in the context of miniature factory is understood to be micro-factory and it is a relatively new concept concerning manufacturing system [3, 5, 6]. Micro-factory can be defined as a small manufacturing system conceived as a means of achieving higher throughput with less space and reduced consumption of both resources and energy via downsizing of production processes [7]. This means that all the equipment were necessarily to be reduced in the scale of the equipment (micro-machines) which could, in-turn reduce the energy consumptions, preliminary and overhead costs, material requirements, reduce pollutions and create a more user friendly production environment. As the scales of the equipment were reduced, therefore the mass of the equipment itself could be reduced dramatically and this will lead to the increasing of tool speed and at the same time will result in improvement of the production rates by lowering down manufacturing cycle. Among the invented micro-manufacturing system is micro-forming machine which is intended to serve a production of micro-forming-products

2. Examples of developed micro forming machines

[2] has developed micro-hole punching machine which meet the requirements of process accuracy for 25μm size holes. Due to critical tool and die alignment on every punch shot with 1μm tolerance,

the machine was designed with the micro punch system mounted on the highly accurate linear motion guide system. This system enables the micro punch to move with straight motion accuracy which is advantageous compared to conventional system in terms of higher accuracy. Punching process is conducted manually and was designed based on a kinematics design principle that eliminates redundant constraints which could lead to lower achievable accuracy. In order to get extremely accurate alignment of 1μm punch-to-die clearance, machine vision system was employed to finely adjust horizontal position of a die hole under the punch tool tip. This new concept two-way imaging vision system can image planar configuration of alignment of both punch tip and die hole which are facing each other. However, this machine is unsuitable for mass-production since the entire systems were controlled manually and no automation was found on the positional axis and feeding mechanism.

[8] has developed micro-drawing process to fabricate drawn shaped microwires 100-400.MU.m in diameter by round-to-shape drawing. In particular, the above requirements are focused on and examined for the case of 7/3-Brass wires, which are frequently used in the electric machine and electronic field, and titanium wires, the market of which is expected to expand in the future. According to the author, shaped wires have been used in various applications such as splines, rails for an X-Y table, connector pins and windings wire in motors. Recently, ultrafine shaped wires have been desired, in line with the miniaturization and weight reduction of products such as cellular phones and personal computers. In some micromotors, performance increase can be expected when shaped microwires are used instead of round wires. In addition, high-quality shaped wires are also required for the development of microrobots and medical equipment.

In [9], cold drawing techniques of composite microwires is explained. In the study, the authors have developed composite microwires of external diameter 42μm, with thickness ratios of 1:4:1 and 1:10:1. The approach begins with fabricating a total of 82 dies in series varying from 3mm to 42μm in the hole diameter. The die hole is designed to be supported by synthetic diamond inserts and tungsten carbide during cold-drawing process. This ensures the hole is reinforced accordingly and not easily broken down by wear and tear from the process. The drawing process is done by rotational motion of the spindle which is provided by both manual and automated means. Automated rotation is achieved by attaching the rotating spindle to a DC motor.

[10] has developed micro-extrusion of ultra-fine grain aluminium. In the research, the authors studied ultra-fine-grain aluminium produced by severe plastic deformation during extrusion through several processes.

Another research project on micro-forming machine includes the development of micro-forming machine which was developed in EU FP6 Integrated Project so called MASMICRO under supervision of Dr. Yi Qin of University of Strathclyde [3]. The objective of the project is to develop an integrated solution for the European miniature/micro-manufacturing industry for mass production size and capability. This also includes transferring technology program, knowledge and training package for the industry in order to develop the skills in industry stage.

This effort has leads to the creation of micro-stamping machine which uses linear motor concepts to actuate the pressing tool which is similar in concept to micro-stamping machine made by [11]. This type of press converts the rotary action created by the electric motor into a translation motion of the press slide with the aid of mechanical transfer elements. As the slide is not coupled mechanically to a rotation movement, the displacement/time and force/displacement behaviour can be individually defined and programmed for every phase. Disrupting factors, such as the play of gear elements, no longer apply. The press force is available at any given point of the stroke. In this way, the new press combines the advantages of mechanical and hydraulic presses and greatly enhances both quality and accuracy. At the same time, the reduction in wear and tear results in reduced maintenance needs and raises both reliability and useful life [12].

[13] reported calculation of strain components of material of copper during micro rolling. The material thicknesses are 1.25-12.5μm. Nevertheless, no equipment of micro-rolling is explained since the authors use finite element analysis to conduct the calculation.

[14] reported micro-forming of thin metal sheets for MEMS manufacturing using short pulse laser. The micro-forming experiments reported in the research were performed on AISI 304 alloy. The test piece is fixed on a holder by means of computer controlled stage which allows the laser spot position is changed easily. A Q switched Nd:YAG laser at 1064nm laser wavelength operating at 10Hz and

providing 9.4ns FWHM, 1.05Joule pulses laser light which is then focused to the designated area by means of reflecting mirror and focusing lenses. From the research, the authors concluded that the laser micro-forming on thin strip (micro-bending using short pulses laser light) is feasible and the demonstrated experiments.

3. Issues and arguments on to-date micro-forming development

3.1. Micro-forming systems

Present micro-forming systems consists of several manufacturing cell which each cell capable of producing one or more manufacturing processes before the next manufacturing cell is initiated. Combination of these cells produces bulky in size of manufacturing system which can be said not greatly proportionally to the produced micro-parts [1, 2, 9, 13, 14, 15, 16, 17]. Existing cell still can be considered bulky if compared to desktop micro-manufacturing systems which is less or equal to 1/10 of present machinery size [5]. Nevertheless, no developed desktop machine usability can be qualified, thus, still requires further research and development in terms of its feasibility, reliability and commercial aspect. These machines are only exist in prototype stage and further effort to bring these machines to be available commercially and reliable to be used have become priority.

Integration between cells also becomes an issue to present micro-forming. Most of the micro-forming machines has separate controller for each cell which has made integration between the two or more cell becomes complicated. Higher precision and accuracy level which can be achieved through turn-key solution in one machine control, no longer available for micro forming system which has multi-controller as well as different architecture. Moreover, these various architecture controllers definitely speak different languages. Most of the developed micro-forming machine is developed for a single and open-loop process and not fully utilizing the advance capability of PIC, PLC and PC based interface. This in turn makes the system easier to be manually controlled without concern for input from sensors and other actuators.

3.2. Micro-forming processes

Lack of idea on new machining or forming concept is very significant in micro-forming processes. Most of the micro-processes were tends to mimicking conventional processes; towards miniaturization and some of them has great limitation in terms of its potential such as limited number of machining axes, limited capability on roughing process only, etc. compared to well-developed macro-scale processes. A new approach on non-contact process should be emphasized to replace physical contact process due to its significant benefits and has potential to be miniaturized [18].

There is limit on how far conventional macro-scale machining can be scaled down for miniaturization. Beyond certain dimensions, factors that can be ignored on conventional machining suddenly play a big part in micro-machining. Vibration, temperature, tool offset, rigidity and chip removal are more important because these factors have greater influence on a micron-sized part. Therefore, non-contact micro-machining is foreseen to have very good potential in eliminating these factors, hence producing reliable and consistent machining condition.

Another issue occurred on current micro-forming process technology is in terms of process automation. Standalone and manual processes on the developed prototypes [1, 2, 19] have made every aspect on the process needed manual adjustments. Most of the processes such as principal processes; pressing, milling, turning etc, handling processes; material loading and unloading, tool positioning and aligning were all manually configured and controlled by separate dedicated controller to obtain precise and accurate motion and alignment. This time consuming process has made micro-process is only qualified on low yield rate and yet to be far from conventional potential.

There is also no parameter study ever conducted on micro-forming machines development in order to study the best parameter and configuration for machining since the main stream of the development focuses only on process miniaturization. It can be concluded that, micro-forming manufacturing process still experiencing lack of novelty and ingenious new process design, and as well as lack of deeper study on its best machining parameter and configuration. Ample of design and development opportunity is foreseen to create a micro-forming machine which does not only mimic the existing conventional process, but has fully automated ultra precision capability.

3.3. Materials for micro-forming

Parallel with the fairly early year of micro-forming machines development, only ductile and soft material with low strength properties were chose and study as testing material; mostly brass, copper and aluminium [2, 6, 19]. Qualification of machining soft and ductile materials is easily achievable due to this type material exhibits low properties in mechanical strength which is tends to deformed easily under low applied load/force. Nevertheless, soft materials are very limited in its usage and some manufacturers have found that these materials are not durable enough to meet the increasing demand of reliability and long life. Therefore, switching to harder material is the only option available. To-date, in development of micro-machines, there is still lack of experiment or testing been conducted on harder alloys such as tungsten and carbon steel, composite materials such as ceramics and metal matrix composites etc. which these materials are vastly used and found commercially [20]. This situation has left ample of space for further improvement and development of micro-forming machine in terms of tooling, machining configurations and parameters.

3.4. Tools and equipment for micro-forming

Novel linear actuation concept on punching has been proposed by [2, 3, 11, 21]. [19] has qualified punching micro-holes on stainless steel using proportional solenoid. In [3, 11] pressing process is qualified by using linear motor and linear encoder as feedback mechanism which gives flexibility in terms of travel distance and pressing speed; closed-loop process for control feedback is feasible.

The key issue regarding tooling needed for micro-forming process is micro-tools. Micro-manufacturing in general, at present, 10μm end mills tool has been realized which is made from carbide (PMT). 25-50μm milling and drilling tools currently becomes typical and can be found commercially [22, 23]. Although micro tooling development has started more than a decade ago, there is still limitation exist on the tooling which limits its application [20]. Only 5-10 times of aspect ratio (tool's diameter to drilling depth) is qualified and some has even lower than 5. Deeper plunging and drilling will result to tooling breakage, hence, makes it unsuitable for aerospace and automotive industries; which exhibits very high strength material with low mass. Nevertheless, qualification on achievable precision and accuracy on the drilled hole has not yet been studied extensively. Furthermore, issue on aligning micro tools up to sub-micron precision and accuracy has not yet been explored extensively because no automated machine available at present capable of aligning the tools up to sub-micron precision [24].

In forming process specifically punching process, tool clearance plays great role in determining achievable accuracy on the punched part. Huge clearance which is relevant for thick material on conventional process is no longer relevant on thin strip/foil. According to [24, 25] the best tolerance between punch and die for punching process was at range of 5-10% of strip thickness. 1-2μm tooling tolerance has been qualified by manual adjustment using piezoelectric-actuator.

Most of the micro-forming processes developed at present only capable of machining soft materials. This is due to the usage of soft tooling such high speed steel (HSS) as it was widely used in macro-processes. No harder materials can be qualified under the usage of soft tooling. The only way out to avoid the usage of micro-tools is by approaching and exploring non-contact machining process. Non-contact machining is the only promising solution available for micro-process and the potential is seen similar to the conventional manufacturing processes [14, 15].

3.5. Handling and equipment

Precision positioning is the main problem encountered in handling of micro-parts. Issue on involved external forces in physical contact such as electrostatic, sticking effect, Van Deer Waals force have became the key issue and lots of study has been made to understand well the situation [26, 27]. Nevertheless, to overcome the problem, non-contact approach has been proposed to replace physical contact methods [14, 15]. Moreover, attention has to be paid to sensor accuracy since available sensors nowadays bulk in size and achievable precision is basically up to tenths of micron. Bulk in size makes the sensors hard to be placed on a tiny workspace, while this level of precision is not feasible for micro-parts application which requires at least sub-micron precision and accuracy. In addition, most of current machine calibration precision capability is far greater than demanded precision and accuracy for micro-handling application.

3.6. Engineering applications

Most of the machines only existed in prototype and further development and refinement has to be made in order to qualify them. Current available prototypes still experiencing lack in flexibility and robustness to meet industrial application. Some of them can only perform rough cutting/machining without considering geometry tolerance, machining precision and accuracy. Most of the machine still has no machining parameters study conducted for best machining configuration for each and different material. No pool of knowledge on machining parameters ever studied and developed for the machines, hence leaving a big hole and question marks on machine troubleshooting aspect. Moreover, limitation on machining axes has made the machines become inflexible to carry out industrial application. In addition, micro-production rate is yet to be far from macro-scale conventional potential due to many adjustments has to be made and rectified manually; which is time consuming processes.

4. Research gap

Micro-forming machines nowadays still experiencing slow in development towards commercialization compared to other micro-machines. From Fig. 1, micro-forming process consists of 7 major processes namely as extrusion, drawing, stamping, punching, rolling, electromagnetic forming and laser/heat forming. Among them, significant development has been experienced by stamping and punching processes, while an early effort has also been made to develop micro-laser forming technique [14] and micro-drawing [28]. Each process has its own handling mechanism, tooling, and the process to produce the part. Nevertheless, integration of these three factors have never been conducted and attempted and becomes one of the key issues in micro-machines world. Current development of micro-forming machines have separate controller for each process, thus, leaving space for integration or automation development.

Fig. 1 depicted the gap existed in current micro-manufacturing especially micro-forming area. From the previous reviewed effort made on micro-forming, most of the discussed prototypes usually use manual handling where human involvement is required. This significantly shows that handling development is still left far behind the others. Boxes with dotted purple coloured line show no effort ever embarked to develop the system, while shaded purple box shows area in which this research intended to embark on. The green dotted boxes represent no micro-parts ever produced by the respective processes.

One of the main conclusions can be drawn from this review is that, there is one thing in common where all the micro-forming machineries share and do not have; integrated autonomous high-precision handling mechanism which is to load and unload material to the system. As what has been reviewed by [29, 30], most of the recent micro-forming works tends to establish the manufacturing process itself by developing prototype. Thus, many improvements on the tooling and process are demonstrated. Nevertheless, an effort has to be made to balance the development of micro-forming machine in terms of its micro-tooling, micro-process and especially micro-handling. This is due to handling mechanism for micro-forming machineries still experiencing lack of development, but research in the other areas experiencing some technological development, prototyping and advancement.

Acknowledgements

The support from the European Commission for conducting research in "Integration of Manufacturing Systems for Mass-manufacture of Miniature/micro-products (MASMICRO)" (www.masmicro.net) (NMP2-CT-2004-500095) is acknowledged. Beneficial discussion and the exchanging of experience with the EU researchers and engineers in micro-manufacturing are particularly acknowledged.

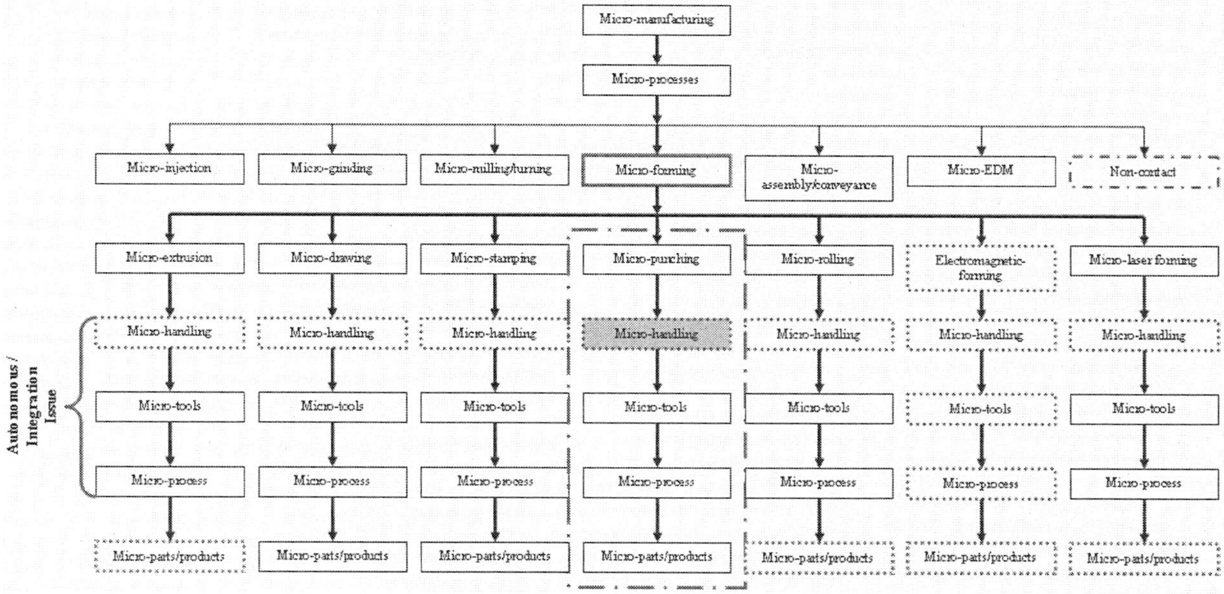

Fig. 1: Gap in micro-forming development.

References

[1] CHERN, G.-L. & RENN, J. C. (2004) "Development of a novel micro-punching machine using proportional solenoid". *Journal of Material Processing Technology*, 25, 89-93.

[2] BYUNG, -. Y. J., RHIM, S.-H. & OH, S.-L. (2005) "Micro-hole fabrication by mechanical punching process". *Journal of Materials Processing Technology*, 170, 593.

[3] QIN, Y. (2006) "MicroForming - Machine and tool development needs and perspectives". *Journal of Materials Processing Technology*, 177, 8-18.

[4] MISHIMA, N., ASHIDA, K., TANIKAWA, T. & MAEKAWA, H. (2002) "Design of a microfactory". *Proceedings of the ASME Design Engineering Technical Conference — 7th Design for Manufacturing Conference*. Montreal, Que., Canada.

[5] OKAZAKI, Y., MISHIMA, N. & ASHIDA, K. (2002) "Microfactory and micro machine tool". *The 1st Korea-Japan Conference on Positioning Technology*. Daejeon, Korea.

[6] OKAZAKI, Y., MISHIMA, N. & ASHIDA, K. (2004) "Microfactory – Concept, History, and Developments". *Journal of Manufacturing Science and Processing*, 126, 837-844.

[7] CLAESSEN, U. & CODOUREY, A. (2002) "Microfactory". Section Head CSEM CH 6055 Alpnach Switzerlan. Switzerland.

[8] MUNECHIKA, M., KAZUNARI, Y. & KEISUKE, K. (2005) "Manufacturing Technology of Brass Micro Shaped Wire by Drawing". *Journal of Japan Research Institute for Advanced Copper-Base Materials and Technologies*, 44, 123-127.

[9] LI, X. P., SEET, H. L., LEE, K. S. & LIU, L. Q. (2007) "Cold drawing of micro Ni80Fe20/Cu composite wires". *Journal of Materials Processing Technology*, 192-193, 350-354.

[10] ROSOCHOWSKI, A., PRESZ, W., OLEJNIK, L. & RICHERT, M. (2007) "Micro-extrusion of ultra-fine grained aluminium". *International Journal of Advance Manufacturing Technology*, 33, 137-146.

[11] SCHNEIDER, R. & GROCHE, P. "Method for optimization of forming presses for the manufacturing of micro parts".

[12] SCHULER (1998) *Metal Forming Handbook*, Springer-Verlag Berlin Heidelberg New York.

[13] LEE, Y. & BYON, S. M. (2007) "Calculation of strain gradient in flow formulation using strain surface function and its applications to micro rolling". *Journal of Materials Processing Technology*, 192-193, 218-224.

[14] OCAN˜A, J. L., MORALES, M., MOLPECERES, C., GARCı´A, O., PORRO, J. A. & GARCı´A-BALLESTEROS, J. J. (2007) "Short pulse laser microforming of thin metal sheets for MEMS manufacturing". *Applied Surface Science*, 254, 997-1001.

[15] ATAKA, T. (1999) "The Experimental Microfactory System in Japanese National R&D project".

[16] SUDA, M., FURATA, K., SAKUHARA, T. & ATAKA, T. (2000) "The Microfactory System Using Electrochemical Machining". Chiba, Japan.

[17] GAUGEL, T., DOBLER, H., MALTHAN, D., BENGEL, M. & WEIS, C. (2001) "Minifabrik für Laserdioden und Biochips".

[18] AUSLENDER, V. L., GORNAKOV, I. V., RADCHENKO, V. M., SHTARKLEV, E., TKACHENKO, V. O. & VORONIN, L. A. (2007) "Electron beam forming system for irradiation of large diameter cylindrical products". *Radiation Physics and Chemistry*, 76, 1792-1794.

[19] CHERN, G.-L., WU, Y.-J. E. & LIU, S.-F. (2006) "Development of a micro-punching machine and study on

the influence of vibration machining in micro-EDM". *Journal of Materials Processing Technology*, 180, 102-109.
[20] ARONSON, R. B. (2003) *The new world of micromanufacturing.*
[21] http://www.fanuc.com
[22] http://www.kyocera.com
[23] http://www.minitools.com
[24] KIBE, Y., OKADA, Y. & MITSUI, K. (2007) "Machining accuracy for shearing process of thin-sheet metals-Development of initial tool position adjustment system". *International Journal of Machine Tools & Manufacture,* 47, 1728-1737.
[25] BAEUMLER, F. (2001) "Keeping pace with today's punching requirements: Advancements in tool design, metallurgy lead to more accurate holes, improved part quality". *Punching Article.*
[26] TOMAS, J. (2007) "Adhesion of ultrafine particles—A micromechanical approach". *Chemical Engineering Science,* 62, 1997-2010.
[27] ROUGEOT, P., REGNIER, S. & CHAILLET, N. (2005) "Forces analysis for micro manipulation". *Computational Intelligence in Robotics and Automation, IEEE.*
[28] MANABE, K., SHIMIZUA, T., KOYAMAB, H., YANGA, M. & ITOC, K. (2007) "Validation of FE simulation based on surface roughness model in micro-deep drawing". *Journal of Materials Processing Technology*, xxx, xxx-xxx.
[29] ENGEL, U., GEIGER, M., KLEINER, M., ECKSTEIN, R. & TIESLER, N. (2001) Microforming. *CIRP Annals - Manufacturing Technology,* 50, 445-462.
[30] ENGEL, U. & ECKSTEIN, R. (2002) Microforming — from basic research to its realization. *Journal of Materials Processing Technology,* 125-126, 35-44.

Innovative Production Machines and Systems
D.T. Pham, E.E. Eldukhri and A.J. Soroka (Eds.)

Automatic quality control of seam puckers based on shadow detection

I.G. Mariolis and E.S. Dermatas

Department of Electrical Engineering & Computer Science, University of Patras, 26500 Patras, Greece

Abstract

A novel method performing automatic seam quality control based on visual information is introduced. More specifically, oblique illumination is applied and 2-dimensional grayscale images of seam specimens are acquired. The amount of shadowing present in each image is being estimated through an innovating technique which employs dynamic thresholding and first order statistics. The shadow estimation is used as a measure of seam quality and a simple linear model is constructed for classification by means of ordinary least squares (OLS) regression. The proposed method has been evaluated in case of 112 seam specimens classified by human experts into five discrete quality grades and a correct classification rate of 84.82% has been produced matching up to the experts' performance. Moreover, in case disagreement by one grade is acceptable, the classification results become error free.

Keywords: seam quality, machine vision, shadow detection, regression

1. Introduction

Automatic seam quality control is of major importance for the clothing manufacturing industry. The need of objective and efficient evaluation of seam quality is imperative, since this task is currently accomplished by human experts and therefore is extremely time and cost consuming. On the contrary, automatic quality control can be applied efficiently in various stages of the sewing procedure, either on-line or off-line and apart from objectively assess seam quality it can assist in adjusting sewing machines' parameters as well.

1.1. Related work

In previous work, Clapp et al [1] used beta rays to determine fabric density which served as a quality measure, while Dorrity [2] acquired the ratio between thread motion and sewing machine cycle by means of piezoelectric sensors and compared it to an optimal value. Bahlmann et al [3], employed spectral analysis and a self-organizing feature map network, in order to classify the seam specimens into five discrete grades of quality. A similar approach, based on a radial basis function network for classification was proposed by Mariolis et al [4]. Aibara et al [5] used fractal dimensions for the classification of the seam pucker on the back of suits. Bahlmann, Mariolis and Aibara, based their methods on visual information, in accordance to human experts' inspection module. Koehl et al 's approach [6] was also based on visual information but in that case, a 3D digitizing system has been used to create 3D models of samples. After converting 3D models into 2D images and standardizing them, feature vectors were extracted from test samples and standards of seams. The feature extraction was based on multi-scale wavelets analysis, spectral analysis, texture analysis and fractal analysis.

1.2. Seam Puckering

One of the main characteristics of poor quality

seams is the presence of seam puckering. This phenomenon takes place when the smoothness along the seam line is disturbed by waves and is caused by one or more of the following reasons:

- Structural jamming (Yarn Displacement)
- Excessive thread tension (Tension Puckering)
- Uneven ply feeding (Machine Puckering).

This work focuses on seam smoothness and seam puckering evaluation in particular. International Standards Organization has set a standard setup [7] for such evaluation committed by human experts. This standard involves seam specimens rather than seams in real garments and considers five grades of quality, from grade 5 (best) to grade 1 (worst).

1.3. Present work

Since this work deals with automatic seam puckering evaluation, the presented ISO standard is employed for more objective comparison of the results to that of human experts.

Seam puckering results to surface roughness and by appropriate illumination settings, self shadowing occurs. Thus this contribution proposes a method for objective seam quality evaluation based on the volume of these shadows.

A brief review of the suggested method is following. As a first step, oblique illumination is applied and 2-dimensional images are acquired. Then shadowed areas of those images are automatically extracted by means of dynamic thresholding. In the last step, a linear model is created through ordinary Least Squares (OLS) regression using shadow information to assign the corresponding quality grade to each seam specimen.

The structure of this paper is as follows. In section 2 the image acquisition module is presented, describing the geometry of the scene and the illumination setup. Section 3 focuses on the process of shadow detection and section 4 deals with OLS regression. Finally, section 5 contains the experimental results, while the paper concludes in section 6.

2. Image acquisition

2.1. Seam specimens' position

The position and orientation of the specimen can be arbitrary. It is very important, however, to ensure that each specimen has been placed in the same manner. This way the comparison between different specimens won't be biased by the forces acting upon them. A simple strategy to achieve the aforementioned condition is to let each specimen hanging in a vertical position (see Fig. 1). Thus eventually the specimens will be stretched only by their own gravitational force.

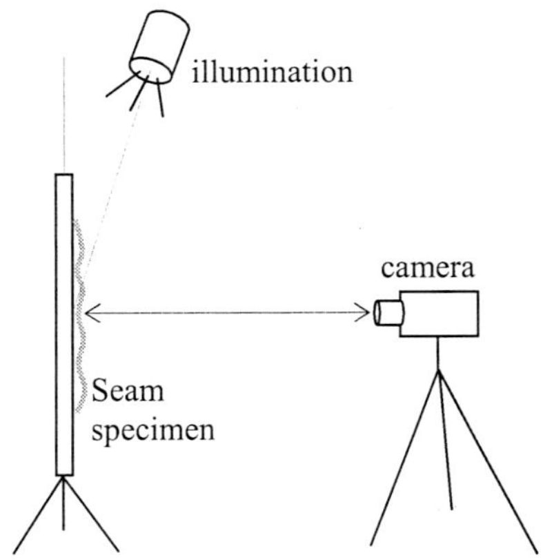

Fig. 1. Acquisition setup.

2.2 Illumination settings

Placing the illumination source is always an important part of the acquisition procedure. In our case the selection of the appropriate illumination setup (see Fig. 1) becomes even more critical, since it should be such that any rough areas are enhanced. This is accomplished by imposing oblique illumination. If the angle is shallow enough, any buckled areas cast shadows over the specimens (see Fig. 2). Finally, the illumination projection on the specimens' plane is collateral to the seam line, since buckling mainly occurs in that direction.

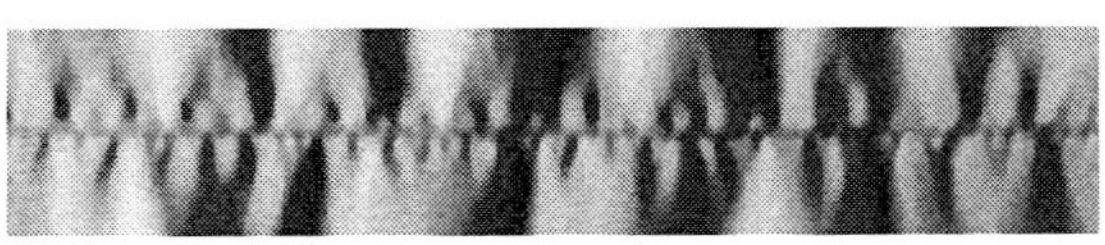

Fig. 2. Self shadowed specimen of grade 1.

3. Shadow detection

3.1. Dynamic Thresholding

Oblique illumination introduces uneven lighting conditions, thus the use of a global intensity threshold for detection of dark shadowed areas becomes inefficient. Instead of examining the grayscale image I(i,j) as a whole, a rectangular window W_{DT} is used and a local threshold is estimated based upon the intensity values of the pixels contained by the window [8]. The window is moving along the image rows and columns in such a way that every pixel of the original image has occupied its center. A threshold value is assigned to every pixel, which is actually the local threshold estimated when that pixel was at the center of the window. The size of the window should be small enough to ensure uniform illumination. On the other hand, it should also be large enough to exclude totally shadowed areas, since then there wouldn't be any threshold to estimate. These conditions, far from being mutually excluded, are easily met as it will be demonstrated in the experimental stage.

3.2. Mixture of Gaussians

In order to define the local threshold for every sub-image formed by the window's movement, the histogram of these sub-images is examined. In absence of shadowed pixels, the histogram (see Fig. 3), approximately follows Gaussian distribution (see Eq. 1),

$$\varphi(x)=\sigma^{-1}(2\pi)^{-1/2}\exp[-(x-\mu)^2/2\sigma^2] \quad (1)$$

where exp stands for the exponential function. In this equation, μ and σ denote the mean and standard deviation of the distribution.

In case of shadow, however, the histogram becomes bimodal and as the amount of shadowing increases the separate modes become more distinct (see Fig. 4).

The bimodal histogram is assumed to be a mixture of two Gaussian distributions and their means and standard deviations are usually estimated through an expectation maximization (EM) algorithm [9]. Once the distributions are estimated,the intensity value for which they cross each other is selected as the desired threshold (see Fig. 5).

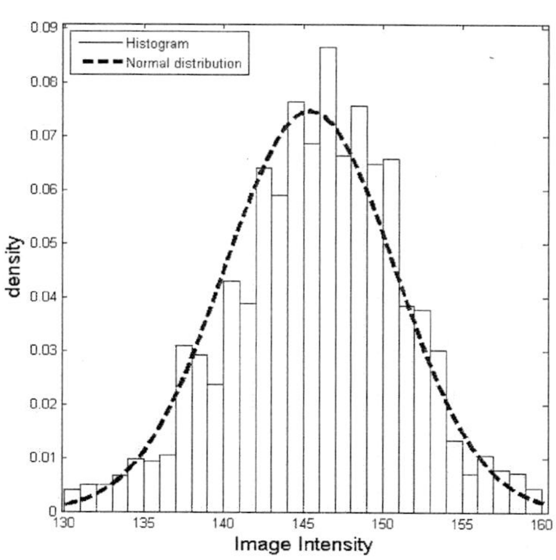

Fig. 3. Normal distribution fitted on the Histogram of a sub-image clear of shadows.

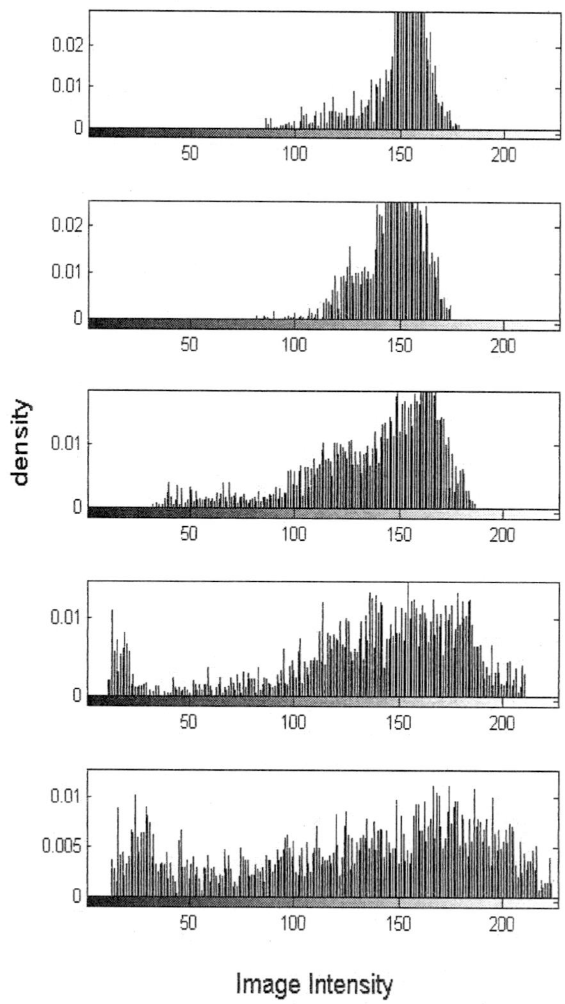

Fig. 4. Histograms of sub-images containing shadows. Each image contains more shadows than the previous one and its modes are more distinct.

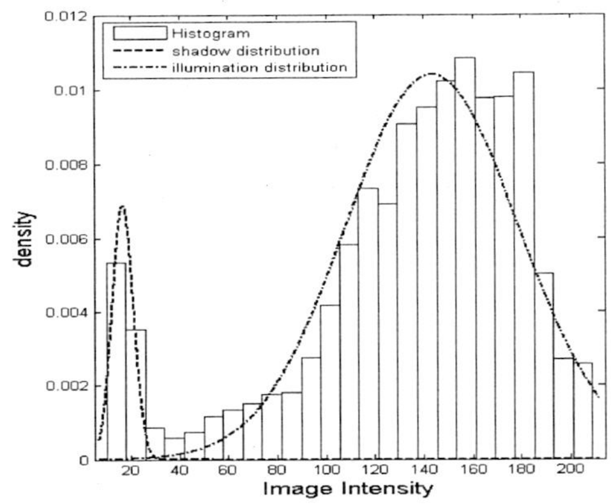

Fig. 5. Histogram of sub-image containing shadows fitted with two normal distributions using EM algorithm, crossing each other at the intensity value 30 defining T_{int}.

3.3. Proposed shadow detection technique

The presented thresholding procedure needs a great number of calculations, since the expectation maximization (EM) algorithm must be repeated for every sub-image. The amount of calculations can be significantly reduced if instead of the EM algorithm, the isodata method [10] is used in each individual thresholding (see Fig.5). The only drawback is that in case of little shadowing, the bimodality is not well-formed and the method does not perform well. In that case instead of applying the isodata method it is more efficient to treat the histogram as unimodal, estimate the mean and standard deviation by the samples and check for symmetry violation. This way the threshold is selected as the first intensity value, starting from the mean, that does not have a symmetric match.

Therefore the proposed dynamic thresholding procedure is using the isodata method in case of bimodal histograms, while in case of unimodal histograms checks for symmetry violations. However, a criterion is needed for discriminating between the two cases. In order to increase calculation efficiency, instead of employing a statistic test (i.e. chi-square statistic), a global threshold T_{std} is introduced as a criterion. It is the sample standard deviation of each histogram that is compared to that threshold. If the estimated deviation is lower than T_{std} the histogram is treated as unimodal, otherwise it is treated as bimodal. The value of T_{std} depends on the acquisition setup and the fabric's texture and derives experimentally.

Using the proposed procedure, the intensity threshold $T_{int}(i,j)$ is calculated for every pixel and a binary image is being produced. The bright areas of the binary image consist of pixels with intensities lower than their threshold, namely, $I(i,j) < T_{int}(i,j)$ and they are areas assumed to be in shadow. The amount of shadow present in every original image sums up to the number of bright pixels in its corresponding binary image (see Fig.6). Finally the ratio between this number and the total number of pixels in each image is calculated forming a feature SR_I bounded by 0 and 1.

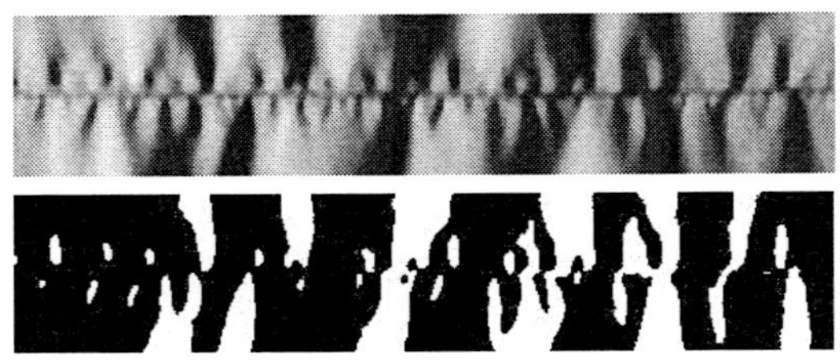

Fig. 6. Shadow detection in case of a grade 1 specimen.

4. Classification using OLS regression

In terms of classification theory the remaining process can be formulated as follows: 1-dimensional patterns SR_I representing seam specimens must be classified into five ordered categories. However, due to the fact that the number of the categories is relatively high and an ordering relation lies among them the whole process can be approximated as a linear regression problem. This is accomplished by introducing a dependent variable, which in our case should be the quality grade and assuming a linear relation between this and the 1-dimensional patterns. The main drawback of this approach is the arbitrary mapping of the quality grades since the values 1 to 5 merely denote ordering information and cannot account for information about the intervals among adjacent categories.

Nevertheless, the advantages of OLS regression such as the small number of used parameters, the low calculation cost and the existence of goodness of fit measures compensate for the incoherence introduced into the model by the arbitrary mapping.

As denoted by its name, OLS estimates the regression coefficients by minimizing the sum of the squared errors between the estimated and the desired values. When the assumptions of the Gauss-Markov theorem hold, this estimate is not only unbiased and consistent but also the one with the minimum variance producing the optimum results [11].

5. Experimental results

In order to evaluate the presented method, 112 seam specimens have been collected, composed by satinet kind of fabric. The size of each specimen has been approximately 20×4 cm. A committee consisting of three experts annotated each specimen by assigning a grade between 1(worse) and 5(best), denoting the quality of the seam. Since there have been three ratings for each specimen it was the median that defined its actual grade.

As described in section 2, grayscale images of the specimens have been acquired. The angle formed by the illumination direction and the specimens' plane has been approximately 25°, which is shallow enough and produces shading even in presence of mild puckering phenomena. The size of the 112 acquired grayscale images has been 468×90 pixels corresponding to a resolution of 22.5 pixels/cm. This relatively low resolution is adequate, since there aren't any significant puckering phenomena present in scales lower than 0.1cm, at least for the specimens in question.

The amount of shadows present in each image has been estimated according to the procedure presented in section 3. The size of the window W_{DT} has been set to 90×45, which is large enough to oversize any shadowed area and at the same time is approximately uniformly illuminated. The unimodality threshold value T_{std} has been set to 20, based on the maximum standard deviation of grade 5 specimens. The correlation coefficient between the extracted feature vector **SR** and the corresponding Quality Grades has been calculated producing a value of 0.964 which is very close to unity and justifies the use of linear regression.

Throughout the experimental procedure, the Quality Grades belong to {1,2,3,4,5}, assuming equal intervals between adjacent quality classes and also that the patterns are classified to the nearest quality grade according to their regression results. Patterns not included in the regression can always be linearly transformed using the regression coefficients and used for further evaluation of the proposed method.

At the first stage of the experiments, the entire set of 112 available patterns has been included in the regression. The produced R-square is 0.929, while the variance analysis of the regression (see Table 1) also indicates a good fit. Using the regression results and nearest neighbor classification, a correct classification rate (CCR) of 84.82% has been

Table 1
Analysis of Variance (ANOVA) table of the regression

	Sum of Squares	df [a]	Mean Square	F [b]	Sig.
regression	184.93	1	184.93	1435	0
residual	14.17	110	0.129		
total	199.11	111			

[a] df stands for degrees of freedom.

[b] The significance value (Sig.) of the F statistic is approximately zero indicating that the independent variable explains well the variation in the dependent variable.

produced. More specifically, 95 specimens were classified to their actual quality grade, while the classification of the rest 17 specimens depicted only by one grade. Table 2 contains the confusion matrix of the classification, while Table 3 presents recall, precision, F-measure separately for each Quality Grade and kappa statistic [12] for the total classification results.

Table 2
Confusion Matrix of the Actual to the Estimated by OLS Quality Grades, using 112 patterns for regression [a]

Est. Quality Grade	Actual Quality Grade 1	2	3	4	5
1	**20**	1	0	0	0
2	2	**22**	4	0	0
3	0	1	**19**	4	0
4	0	0	0	**19**	0
5	0	0	0	5	**15**

[a] The diagonal values, denoting correctly classified specimens, are dominating producing a trace of 95.

Table 3
Metrics for classification evaluation

	Actual Quality Grade 1	2	3	4	5
recall	0.91	0.92	0.83	0.68	1.00
precision	0.95	0.79	0.79	1.0	0.75
F-measure	0.93	0.85	0.81	0.81	0.86
k-statistic	0.81				

All four measures of Table 3 belong to [0,1] and the closer they are to unity the higher the performance of the classifier is considered.

At the second stage of the experiments, OLS classification has been furthered evaluated through the statistically reliable method of stratified K-fold cross validation [13]. In case of ten (disjoint) folds the number of training patterns of each instance is 101. The process is repeated for ten instances and the produced CCR is calculated as the average CCR of all instances. This way none of the patterns used in the test set was used in training and the performance evaluation is not biased. Nevertheless, the produced CCR has been equally high (84.17%). Moreover, if we repeat the process after swapping the number of the training patterns with that used for testing, namely using just 11 patterns for estimating the regressors, and the remaining 101 for testing, a CCR of 82.04% is produced. This indicates that even if very few training patterns are available the proposed method can still be efficient.

6. Conclusions

This work proposes a novel approach to automatic seam quality control. First an innovating shadow detection technique is introduced. It is designed to perform under uneven illumination conditions, since shadowing has been deliberately imposed in the scene through unbalanced illumination. The extracted shadowed area forms a feature suitable for seam quality assessment, which is highly correlated to the quality grades of the specimens. Then OLS regression is applied making use of this linear relation and resulting to correct classification rate of about 85%.

The performance of the proposed method is quite satisfactory since it is similar to that of human experts. Moreover as shown by the regression with just 11 specimens, the method can perform well even if very few training patterns (i.e. specimens already evaluated by human experts) are available. Future work could focus on the determination of the exact spacing between the categories. This way the mapping can become optimal and the regression even more efficient.

Acknowledgements

This work has been supported by the General secretariat of Research and Technology and ELKEDE, project PENED2001 No. 3049.

References

[1] Clapp TG. Titus KJ. Olson LH. Dorrity JL. The on-line Inspection of Sewn Seams. National Textile Center Annual Report (1995) Available from http://ntc.tx.ncsu.edu/html/REPORTS/YEAR-FOLDER/online.html Accessed: 2004-03-11.

[2] Dorrity JL. New developments for seam quality monitoring in sewing applications. IEEE Trans. on Ind. Appl. 31 (1995) 1371-1375.

[3] Bahlmann C. Heidemann G. Ritter H. Artificial Neural Networks for Automated Quality Control of Textile Seams. Pattern Recognition 32(6) (1999) 1049-1060.

[4] Mariolis IG. Dermatas ES. Automated quality control of textile seams based on puckering evaluation. 37th Int. Symp. on Nov. in Text., Ljubljana, Slovenia, June 2006.

[5] Aibara T. Mabuchi T. Kenji O. Automatic evaluation of the appearance of seam puckers on suits. Proc. SPIE 3652 (1999) 110.

[6] Koehl L. Miou JC. Zeng X. Selecting Relevant Features from Fabric Images for Automated Quality Control of Seam Puker Using Data Analysis and Human Experts Grading. Book Chapter in Computational Textile. Springer Berlin / Heidelberg, 2007, 39-54.

[7] Draft International Standard, ISO/DIS 7770; Textiles—Method of test for evaluating the appearance of seams in durable press products after domestic washing and drying. Draft International Standard, ISO/DIS 7770, 1981.

[8] Gonzalez RC. and Woods RE. Digital Image Processing 2nd Edition (DIP/2e). Prentice Hall, 2002.

[9] Titterington D. Smith A. Makov U. Statistical Analysis of Finite Mixture Distributions. John Wiley and Sons, Chichester, UK, 1985.

[10] Ridler TW., Calvard S: Picture thresholding using an iterative selection method. IEEE Transactions on System, Man and Cybernetics. SMC-8 630-(1978) 632.

[11] Plackett RL. Some Theorems in Least Squares. Biometrika **37** (1950) 149-157.

[12] Cohen J. A coefficient of agreement for nominal scales. Educational and Psychological Measurement 20 (1960) 37–46.

[13] Kohavi R. A study of cross-validation and bootstrap for accuracy estimation and model selection. Proc. of the 14th Int. Joint Conf. on Art. Intel. 2 (12) (1995) 1137–1143.

Innovative Production Machines and Systems
D.T. Pham, E.E. Eldukhri and A.J. Soroka (Eds.)

Calibrating force vs position control applied to a milling operation using two hybrid automata

P. Charbonnaud[a], F. J. Carrillo[a] and D. Ladevèze[b]

[a]*Laboratoire Génie de Production, Ecole Nationale d'Ingénieurs de Tarbes, BP 1629, 65016 Tarbes Cedex, France.*
[b]*CACG, Chemin de l'Alette, BP 449, 65004 Tarbes Cedex, France.*

Abstract

This paper proposes a restructuring strategy for detecting the process operating mode and for calibrating the control law according to process objectives which are pre-defined at management level. An active accommodation of the process is based on controller restructuring when the objective has to change due to a loss of process performance, whilst a passive accommodation is based on a supervised robust control method. The restructuring strategy is an active accommodation method of control law according to some operating modes. The application concerns a milling operation, the proposed model of which makes it possible the monitoring of the cutting process. The regulation depends on actions resulting from the detection-accommodation additional loop. This loop is specified by two hybrid automata. The simulation results show the interest and the feasibility of the method.

Keywords: Control accommodation, detection, machining, switching control, hybrid automaton.

1. Introduction

Today, industrial machining control is based on current, feedrate and feed cascade regulators. In cases of flexion and vibration, the numerical control does not always ensure the quality of parts, accurately. Nevertheless, these phenomena are inherent behaviours for which the existing position controller implemented in the industrial machining centre does not compensate these problems. Various studies have shown the advantages of controlling the cutting force for turning, drilling and milling [1], [2]. Adaptive control is often proposed to regulate the cutting force [3], [4]. As a consequence, the cutting process must be better controlled to achieve a high reactivity degree [5]. But, no control method was proposed to restructure the control law according to milling process operating modes. The milling process is a complex phenomenon, which depends on the material, on the cutting operation, on the machine and on the cutting tool [6]. For a static model of the cutting process, the specific pressure is a very important indicator which can be considered as a linear time-varying parameter for monitoring. Specific pressure enables the qualitative determination of operating zones. Other studies are based on the spindle current approach to diagnose defaults on the work-piece or on the tool [7].

In this paper, the proposed restructuring strategy deals with milling process operating modes. It combines passive accommodation linked to controller robustness and active accommodation which depends on restructuring actions resulting from the detection-accommodation additional loop. Section 2 describes the modelling of face milling operating modes. Section 3 presents the restructuring strategy. Section 4 deals with restructuring strategy design and simulation. Finally, the simulation results show the interest and the feasibility of the method which enables control restructuring between position and force laws.

2. Modelling of face milling operating modes

Face milling is a form of metal cutting in which a multi-edge tool performs circular cutting. For face milling, a D diameter tool, fitted with Z inserts and submitted to a spindle speed N, moves parallel to the surface machined at a feedrate V_t (see Fig. 1). The mean chip thickness h_m is given for a feed per tooth f_Z, a radial depth of cut a_r, a tool diameter

D, a engagement φ_e and lead angle χ_r by: $h_m = (2f_z a_r / D\varphi_e)\sin\chi_r$. The mean cutting force could be determined by the specific pressure for one millimetre chip thickness $k_{c_{1,1}}$, the nonlinear coefficient m_c, the mean number of engaged teeth z_{em} and the axial depth of cut a_a : $F_{cm} = a_a z_{em} k_{c_{1,1}} h_m^{1-m_c}$ where z_{em} is given by $z_{em} = \varphi_e Z / 2\pi$ and φ_e by the piece and tool geometry: $\varphi_e = a\cos(1 - 2a_r / D)$. The feed per tooth is linked to the table feedrate $V_t = f_z ZN$. Choosing the cutting speed V_C determines the spindle speed $N = 1000V_C / \pi D$.

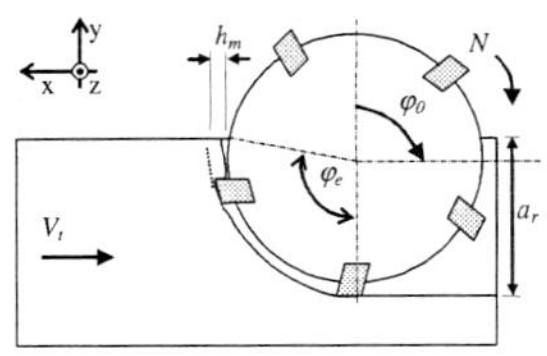

Fig. 1. Face milling operation.

The milling plant discrete transfer functions are $F_{cm}(k) = H_F(q^{-1})U_F(k)$, and $X(k) = H_X(q^{-1})U_X(k)$ where q^{-1} is the backward shift operator, U_F and U_X correspond to the same actuator command input. The actuator dynamics H_X represents the current and feedrate cascade control loop of the system consisting of the gear-box, the table, a DC Motor and its controller. The actuator discrete transfer function between the measured position X and the tension U_X corresponding to the feedrate is: $X(k) = H_X(q^{-1})U_X(k)$. A pseudo random binary sequence extra signal was applied into the actuator to identify an auto-regressive with exogenous input model (ARX model) of $H_X(q^{-1}) = B_X(q^{-1}) / A_X(q^{-1})$. Zeros and poles of the transfer function H_X are given in Tab. 1.

Table 1
Zeros and poles of H_X and H_F

Zeros H_X	Poles H_X	Zeros H_F	Poles H_F
-1.647	-0.514	-1.647	0.721±0.299i
0.561	0.721±0.299i	-0.167	-0.514
0	1	0.561	-0.0657±0.0863i

The specific pressure $k_c = k_{c_{1,1}} / h_m^{m_c}$ may be expressed as a ratio of the mean cutting force F_{cm} and of the mean chip section $a_a h_m$: $k_c = F_{cm} / a_a h_m$. Then, $F_{cm} = k_{c_{1,1}} a_a h_m^{1-m_c}$. F_{cm} is rewritten as: $F_{cm} = k_{c_{1,1}} a_a k_{nl} V_t^{1-m_c}$ with $k_{nl} = (h_m / V_t)^{1-m_c}$.

For experiments, it is convenient to limit the passband of F_{cm}. The measurement conditioning is achieved by a two poles low-pass Butterworth analog filter ($\omega_n = 200 rad.s^{-1}$). The overall model necessary to design the force controller will be approximated by a linear one. Then, if H_V is the transfer function between table feedrate and the tension U_X the force discrete-time transfer function, sampled at $T_c = 5ms$, is given by: $H_F(q^{-1}) = \dfrac{B_F(q^{-1})}{A_F(q^{-1})}$ that includes the zero order hold, the servo H_V and the conditioning transfer function $H_B(s) = \dfrac{k_{c_{1,1}} a_a k_{nl} \omega_n^2}{s^2 + 2\xi\omega_n s + \omega_n^2}$.

Zeros and poles of H_F are given in Tab 1. The sampling period T_c was chosen lower than the period of revolution (60/N) to carry out an efficient control.
The robust control design method applied to $H_F(q^{-1})$ and $H_X(q^{-1})$ transfer functions will enable the generation of the linear controller which will have the best stability properties for real plant. The material parameters have been identified. Validated parameters are $k_{c_{1,1}} = 1500$ and $m_c = 0.4$. A face milling operation with chatter which is difficult to manage and that is important to cope with was addressed in this work. The chosen parameters are given in Tab. 2.

Table 2
Parameters of a face milling operation with chatter.

Material	Operation	Machine	Tool
C35	$a_a = 3mm$	$V_c = 170m/\min$	$D = 63mm$
	$a_r = 20mm$	$N = 853rpm$	$Z = 5$
	$f_z = 0.2mm/tooth$	$P = 10KW$	$\chi_r = 44.5°$

3. Restructuring strategy with robust controller

In Fig. 2, the case of force and position control restructuring is depicted where Σ_A and Σ_M are respectively the control accommodation and monitoring set-points. When faulty behaviour without safety consequence is detected and while supervision set-points can be carried out with satisfying performance by passive accommodation, the control law remains the same; otherwise an active accommodation changes the control strategy.

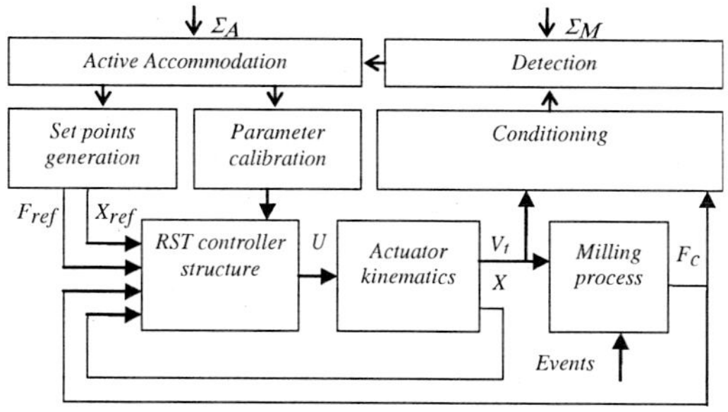

Fig. 2. Restructuring control strategy.

Generally, manufacturing operations have to be carried out respecting the work-piece position. The control law must be restructured when great vibrations, due to wear, become regenerative. The control objective is changed to finish the operation avoiding an abnormal level of vibrations. Transfer functions H_X and H_F have both one zero (zo =-1.647) outside the unit circle (|z|=1). The use of a robust pole placement controller design method is required. In this case, Landau [8] proposes an interesting robust two degrees of freedom RST controller synthesis method. In Fig. 3, the RST control structure is shown in presence of output perturbation *p* and of measurement noise *b*.

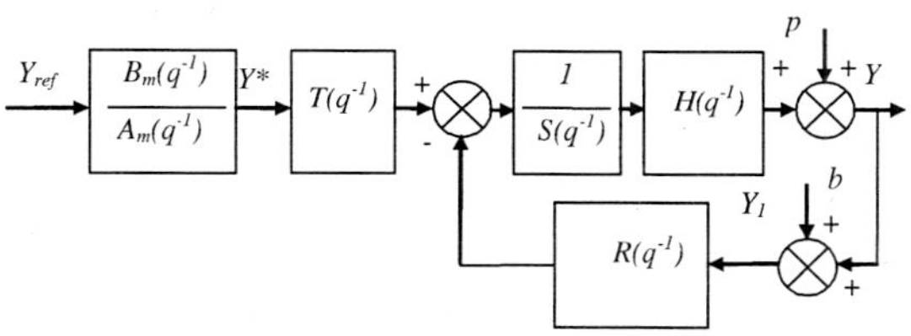

Fig. 3. RST controller structure.

The RST controller expression is given by:

$$S(q^{-1})U(k) = T(q^{-1})Y^*(k+1) - R(q^{-1})Y_1(k), \qquad (1)$$

where $Y^*(k+1) = \dfrac{B_m(q^{-1})}{A_m(q^{-1})}Y_{ref}(k)$ represents the reference model and $Y_{ref}(k)$ is a reference tracking signal, $U(k) = U_X(k) = U_F(k) =$ is the control signal and $Y(k)$ is the output signal. Tracking dynamics $B_m(q^{-1})/A_m(q^{-1})$ is obtained by a discrete-time equivalent of a continuous-time second order system with damping coefficient and natural frequency as shown in section 4. The closed loop transfer function is: $W(q^{-1}) = T(q^{-1})B(q^{-1})/P(q^{-1})$, where $P(q^{-1})$ is chosen to improve closed loop regulation dynamics: $A(q^{-1})S(q^{-1}) + B(q^{-1})R(q^{-1}) = P(q^{-1})$

$P(q^{-1})$ represents dynamics according to dominant poles P_D to which auxiliary poles P_A can be added for robustness improvement $P(q^{-1}) = P_D(q^{-1})P_A(q^{-1})$.

The method proposed by Landau et al. [9] is based on the iterative shaping of input-output sensitivity functions.

$S_{y,p} = A(q^{-1})S(q^{-1})/P(q^{-1})$

$S_{u,p} = -A(q^{-1})R(q^{-1})/P(q^{-1})$

$S_{y,b} = B(q^{-1})R(q^{-1})/P(q^{-1})$

An initial step was to solve the Bezout equation for a first choice of regulation dynamics with an integral action given by: $S(q^{-1}) = H_S(q^{-1})(1-q^{-1})S'(q^{-1})$.

$R(q^{-1}) = H_R(q^{-1})R'(q^{-1})$. H_S and H_R are first or second order polynomials which modify the frequency response of $S_{y,p}$ and $S_{u,p}$. For the initial step, $H_S(q^{-1}) = 1$ and $H_R(q^{-1}) = 1$. RST coefficients obtained with an integral action were computed giving a modulus margin and a delay margin, pointing out good robustness properties as explained in [9]. According to the methodology, the output sensibility function $S_{y,p}$ must be inside the templates which are designed to insure a modulus margin $\Delta M > 0.5$ and a delay margin $\Delta\tau > T_c$. Landau shown that the margin ΔM implies that the gain margin is at least 2 (6 dB) and the phase margin is more than 29°. The controller will be robust when the attenuation of the disturbance-input sensitivity function $|S_{u,p}| < 1$ at high frequencies is also obtained. When both conditions on $S_{y,p}$ and $S_{u,p}$ are satisfied, the robust controller design is finished. If $S_{y,p}$ is not inside templates, several possibilities of tuning are available to modify H_S or H_R or P and are more deeply described in chapter 8 of [8]

Active accommodation to some faulty behaviour with large magnitude may be carried out by restructuring the control law. In this case, operating mode and control objective are changing. The strategy is specified and implemented by two hybrid automata. In Fig. 4.a, the active accommodation automaton describes four states and their transition conditions. Automaton is sampled at the control period $T_c = 5ms$. After initialization, the first state is Position Control *PC*. When vibrations are diagnosed, d_{FC} is true, the control state is restructured toward the Force Controller *FC* for regulating the mean cutting force F_{cm}. If the feed per tooth f_z is lower than $f_{z\min}$ or if F_{cm} is higher than $F_{\max}$, the operation will be stopped *AO*. If f_z is higher than $f_{z\max}$ or if vibrations are not diagnosed, d_{PC} is true, the control is restructured toward the position controller *PC*. When the automaton state is *FO*, the end machining position X_l is reached. Associated to the control automaton, a monitoring automaton, sampled at $T_s = 1ms$, enables the state diagnosis. In Fig. 4.b, four states have been considered. The mean cutting force magnitude is a good indicator of tool engagement variations. During the state d_{PC}, there is no force regulation and the reference cutting force F_{ref} is assigned to F_{cm}. A two hypotheses decision test is used to detect if the null hypothesis H_0, $\sigma = \sigma_0$, is true or not where σ is the variance of the modelling error detailed in section 4. The alternate hypothesis H_1, $\sigma = \sigma_1$, is true if vibrations are detected. If the progressive test on variance concludes $\sigma = \sigma_1$, the alternate decision d_{H_1} is true, the wait state att_d_{FC} is activated. If the variation between the reference tracking force signal F_{ref} and the mean cutting force estimation $\hat{F}_{cm}$ becomes too great, the tool engagement is considered not constant and the automaton is re-initialized. After a number N_{H_1} of consecutive decisions d_{H_1}, denoted nd_{H_1}, the automaton state d_{FC} is activated and the presence of vibrations is diagnosed. When a null hypothesis is detected, the state $attd\_d_{PC}$ is activated. After a number N_{H_0} of consecutive decisions d_{H_0}, denoted nd_{H_0}, the automaton state d_{PC} is activated. Notice that nd_{H_0} ++ denotes $nd_{H_0} = nd_{H_0} + 1$.

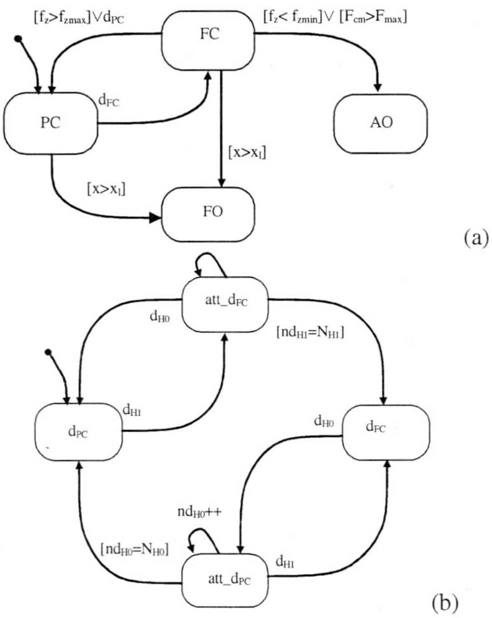

Fig. 4. (a) Active accommodation automaton at T_c, (b) Monitoring automaton at T_s.

4. Restructuring strategy design and simulation

H_X can be expressed as stable B_{XS} and unstable B_{XU} polynomials by $B_X(q^{-1}) = B_{XS}(q^{-1})B_{XU}(q^{-1})$

$B_{XS}(q^{-1}) = 0.0000182(1-0.56q^{-1})$,
$B_{XU}(q^{-1}) = q^{-1} + 1.64q^{-2} = b_1q^{-1} + b_2q^{-2}$,
$A_X(q^{-1}) = 1 - 1.927q^{-1} + 0.794q^{-2} + 0.445q^{-3} - 0.313q^{-4}$.

The RST controller is given by:

$$S_X(q^{-1})U_X(k) = T_X(q^{-1})X^*(k+1) - R_X(q^{-1})X(k)$$

where $X^*(k+1)$ represents the reference position model and $X_{ref}(k)$ is a reference tracking position signal. The closed loop transfer function denominator $P_X(q^{-1}) = P_{X0}(q^{-1})B_{XS}(q^{-1})$ determines closed loop regulation dynamics with
$P_X(q^{-1}) = A_X(q^{-1})S_X(q^{-1}) + B_X(q^{-1})R_X(q^{-1})$.

$P_X(q^{-1})$ was chosen as a discrete-time version of a continuous-time second order system with $\xi = 0.99$ and $\omega_0 = 100 rad.s^{-1}$. The first trial was to add two conjugate complex poles issued from a continuous-time second order of which the radian frequency was $\omega_0 = 260 rad.s^{-1}$ and the damping factor $\xi = 0.3$ in order to decrease the maximum of $S_{X,p}$ for a corresponding frequency (see Fig. 5.). But it was not sufficient, we have added auxiliary poles $P_{AX} = (1-0.5q^{-1})^2$. The induced shape of the position control-perturbation sensitivity function $S_{U_X,p}$ shows that the design is robust (see Fig. 5). The modulus margin $\Delta M = 0.63$ for the frequency 107 rad.s^{-1}, the gain margin $\Delta G = 2.9$ for the frequency 297 rad.s^{-1}, a phase margin $\Delta\varphi = 45°$ and a delay margin $\Delta\tau = 7ms$ (107 rad.s^{-1}) which is about $1.5\,T_c$, are pointing out good robustness performances. In Fig. 5, the disturbance-output sensitivity function is inside templates. Noise-output sensitivity function is below the upper template.

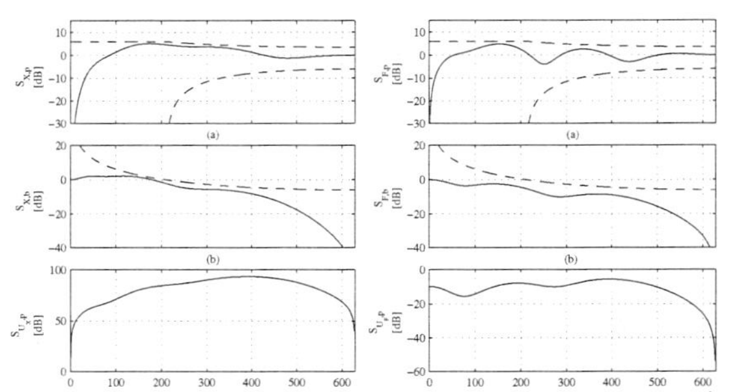

Fig. 5. Sensitivities functions.

The disturbance-input sensitivity function has a decreasing shape. Robustness conditions are satisfied. To compensate the stable zeros of the process for tracking improvement $S_X(q^{-1}) = S(q^{-1})B_{XS}(q^{-1})$

Choosing $T(q^{-1}) = P_{X_0}(q^{-1})B_{XS}(q^{-1})/B_{XU}(1)$, the closed loop transfer function with tracking dynamics is expressed by: $W_m(q^{-1}) = \frac{B_{mX}(q^{-1})B_{XU}(q^{-1})}{A_{mX}(q^{-1})B_{XU}(1)}$.

To improve tracking performances for a ramp input, $\frac{B_{mX}(q^{-1})}{A_{mX}(q^{-1})}$ is given by a pre-compensator $C_1(q^{-1})$ designed to compensate the tracking error with a structure given by: $C_1(q^{-1}) = 1 + (1 - q^{-1})C_0$, where C_0 is a constant to be determined. Notice that $C_1(1) = 1$.

The complete transfer function becomes:

$$W_m(q^{-1}) = \frac{C_1(q^{-1})B_{XU}(q^{-1})}{B_{XU}(1)},$$

with the following position tracking error:

$$\varepsilon(q^{-1}) = c_X(q^{-1}) - X(q^{-1}) = [1 - W_m(q^{-1})]c_X(q^{-1})$$

where c_X is the position input set-point. After calculation, the following condition was obtained $C_0 = \frac{b_1 + 2b_2}{B_{XU}(1)} = 1.6212$ insuring a null static error for step and ramp setpoints. $C_1(q^{-1}) = 2.6212 - 1.6212q^{-1}$ was chosen to guarantee a null static ramp error after a small transitory signal. The following RST controller values were obtained:

$$R_X(q^{-1}) = 10979 - 11370q^{-1} - 7068q^{-2} - 14425q^{-3}$$
$$-7553q^{-4} - 680q^{-5} + 4045q^{-6} - 1971q^{-7},$$
$$S_X(q^{-1}) = 1 + 0.0717q^{-1} - 0.296q^{-2} - 0.316q^{-3}$$
$$-0.166q^{-4} - 0.187q^{-5} - 0.106q^{-6},$$
$$T_X(q^{-1}) = 54366 - 154114q^{-1} + 174439q^{-2} - 98534q^{-3}$$
$$+27771q^{-4} - 3124q^{-5},$$

where $T_X(q^{-1}) = T(q^{-1})C(q^{-1})$ and

$$P_X(q^{-1}) = 1 - 2.2161q^{-1} + 1.8377q^{-2} - 0.6756q^{-3} + 0.0929q^{-4}$$

Similarly to the position controller design, the two degrees of freedom RST polynomial controller is given by relation (1) where $F^*(t+1)$ represents the reference force model and $F_{ref}(t)$ is a reference tracking force signal. H_F is expressed by:

$$B_F(q^{-1}) = 0.66q^{-2} + 0.82q^{-3} - 0.48q^{-4} - 0.10q^{-5},$$
$$A_F(q^{-1}) = 1 - 0.79q^{-1} - 0.24q^{-2} + 0.28q^{-3} + 0.04q^{-4} + 0.004q^{-5}.$$

After calculation, the following force RST controller satisfying the robustness conditions (see Fig. 5) is obtained:

$$R_F(q^{-1}) = 0.154 - 0.065q^{-1} + 0.089q^{-2} + 0.121q^{-3}$$
$$-0.247q^{-4} + 0.009q^{-5} + 0.080q^{-6} + 0.011q^{-7} + 0.001q^{-8},$$
$$S_F(q^{-1}) = 1 - 0.513q^{-1} + 0.178q^{-2} - 0.352q^{-3}$$
$$-0.060q^{-4} - 0.336q^{-5} - 0.088q^{-6} - 0.146q^{-7} + 0.025q^{-8},$$
$$T_F(q^{-1}) = 1.20 - 1.465q^{-1} + 0.498q^{-2},$$

with $P_F = 1 - 1.3084q^{-1} + 0.4449q^{-2}$ and the given reference model: $A_{mF}(q^{-1}) = 1 - 1.308q^{-1} + 0.445q^{-2}$, $B_{mF}(q^{-1}) = 0.077 + 0.059q^{-1}$. The modulus margin $\Delta M = 0.53$ (-5.5 dB) for the radian frequency 120.9 rad.s^{-1} induces a gain margin $\Delta G = 2.2$ (6.9 dB) for the radian frequency 31.1 rad.s^{-1}, a phase margin $\Delta\varphi = 47.6°$ and a delay margin $\Delta\tau = 9.5ms$ which is about $1.9T_c$ for the frequency 87.7 rad.s^{-1}, are pointing out good robustness performances. In Fig. 5, $S_{F,p}$ is inside templates, $S_{F,b}$ is below the upper template, $S_{U_F,p}$ has a decreasing shape. Robustness conditions are satisfied.

The mean cutting force F_{cm} could be estimated by conditioning the measured force components sampled at the period T_S: The mean cutting force F_{cm} may be valued by the norm:

$$\hat{F}_{cm}(kT_s) = \sum_{i=0}^{i=L-1} \left|\hat{F}_c((k-i)T_s)\right|$$

where the window size L is given by the integer part of $60/NT_s$. The restructuring condition is based on the chatter vibration detection. An analytical model of the cutting force depends on good knowledge of work-piece geometry and on the parameter estimation for each operating condition. Generally, without *a priori* knowledge, the cutting force model could be deduced from experimental data. $F_{cM}(kT_s)$ is expressed as a function of estimated cutting force interpolated according to the measured angular tool position φ_0 (*see* Figure 1) on the previous spindle revolution by:

$$F_{cM}(kT_s) = \nu(\varphi_0)\hat{F}_c((k-L)T_s) + (1 - \nu(\varphi_0))\hat{F}_c((k-L+1)T_s)$$

where $\nu(\varphi_0) = \frac{\varphi_0(kT_s) - \varphi_0((k-L)T_s)}{\varphi_0((k-L+1)T_s) - \varphi_0((k-L)T_s)}$.

At time kT_s, the modelling error e_M is estimated by:

$$e_M(kT_s) = \hat{F}_c(kT_s) - F_{cM}(kT_s).$$

When vibrations are not regenerative, $e_M(kT_s)$ is only sensitive to the vibration magnitude, increasing for chatter frequencies. The modelling error $e_M(kT_s)$ is a pertinent indicator to reveal regenerative vibration. The model $F_{cM}(kT_s)$ is satisfying only for a constant engagement of the tool. For the monitoring automaton, a Wald test on the variance σ of the modelling error e_M is carried out with m the mean of e_M and r the number of samples, as:

$$\frac{\ln(\frac{P_L}{1-P_F})+r\ln(\frac{\sigma_1}{\sigma_0})}{\sigma}<\sum_{k=1}^{r}(e_M(k)-m)^2<\frac{\ln(\frac{1-P_L}{P_F})+r\ln(\frac{\sigma_1}{\sigma_0})}{\sigma}.$$

For a milling operation without default, the null hypothesis H_0 is defined by $\sigma_0 = 125$; in presence of vibrations, the alternate hypothesis H_1 is tuned to $\sigma_1 = 150$. Lack and False alarms probabilities P_L and P_F are fixed at 1%.

The aim of this work is to manage the milling operation in presence of chatter. Without this strategy the operation may fail when important vibrations appear and even could produce the tool breakage. The parameters of the milling operation, given in Tab. 2, were chosen to test the performances of the proposed active restructuring strategy.

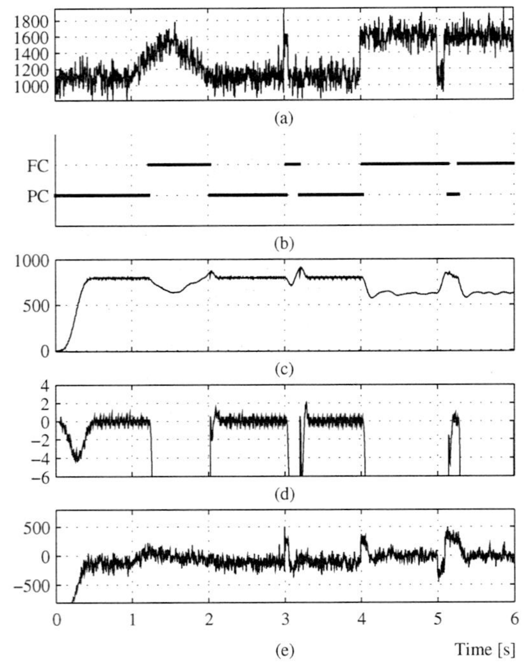

Fig. 6. Restructuring strategy simulation (a) specific pressure $k_{c_{1,1}}$ [N.mm^{-2}], (b) control automaton states, (c) feedrate control U_t [m/min], (d) table position error [μm], (e) force error [N].

The plant was monitored to switch on the robust force controller when the force error is greater than 200N. The effect of the restructuring strategy is shown in Fig. 6. Perturbations of $k_{c_{1,1}}$ considered as a time-varying parameter (Fig. 6.a) are rejected by the online RST robust controller. At the switching time, the bump on the control U_t (Fig. 6.c) is limited by the restructuring mechanism. In Fig. 6.b, switching between FC and PC can be observed and its consequence on the table position error [μm](Fig. 6.d) and on force error [N] (Fig. 6.e) shows the interest of the strategy.

5. Conclusion

The restructuring strategy proposed herein gives an answer to the problem of accommodation according to some faulty behaviour of the milling process due to chatter. A first operating mode is to regulate the table position. When the vibration magnitude becomes too important, the operating mode is changed to regulate the cutting force. Control objectives are guaranteed by the passive accommodation linked to the controller robustness. Design of the restructuring strategy based on the real time constraints was proposed. Simulation results show the interest and the feasibility.

Acknowledgements

The authors thanks the I*PROMS NoE colleagues for their advising comments.

References

[1] Landers R.G., A.G. Ulsoy and Y.-H. Mab, "A comparison of model-based machining force control approaches" International Journal of Machine Tools & Manufacture, vol. 44, pp. 733–748, 2004.

[2] Y. Tang, R. G. Landers, S.N. Balakrishnan, "Hierarchical optimal force–position–contour control of machining processes" Control Engineering Practice, vol. 14, pp. 909–922, 2006.

[3] S. J. Rober, and Y. C. Shin X, "Control of cutting for end milling processes using an extended model reference adaptive control scheme," Journal of Manufacturing Science and Engineering, vol. 118, pp. 339-347, 1996.

[4] F. J. Carrillo, and F. Rotella, "A robust adaptive controller for constant turning force regulation" Control Eng. Practice, vol. 5, n° 12, pp. 1673-1682, 1997.

[5] M. A. Elbestawi, Y. Mohamed, and L. Liu, "Application of some parameter adaptive control algorithms in machining" Journal of Dynamic Systems, Measurement and Control, vol. 112, pp. 611-616, 1990.

[6] M. Ben Lazrak and P. Charbonnaud,"Simulation and Detection of Faulty Behaviors in the Milling Process", IFAC Symposium on Information Control in Manufacturing, vol. 3, pp 427-432, 1998.

[7] Y. Altintas, "Prediction of cutting forces and tool breakage in milling from feed drive current measurements" Transactions of the ASME, vol. 104, pp. 386-392, 1992.

[8] I. D. Landau, R. Lozano, and M. M'Saad, Adaptive Control, Springer, 1998.

[9] I. D. Landau, C. Cyrot, and D. Rey, "Robust control design using the combined pole-placement/sensitivity function shaping method," in Proc. European Control Conference, Groningen, the Netherlands, vol. 3, pp. 1693-1697, 1993.

Innovative Production Machines and Systems
D.T. Pham, E.E. Eldukhri and A.J. Soroka (Eds.)

MATLAB-Based 3D Dynamic Model of a Powered Wheelchair

A. Fattouh[a], Y. Dadam[b], D. T. Pham[b]

[a] *Laboratory of Automatic Control and Automation Faculty of Electrical and Electronic Engineering University of Aleppo – Aleppo – Syria*

[b] *Manufacturing Engineering Centre, Cardiff University, Queen's Buildings, The Parade, Newport Road CARDIFF CF24 3AA, Wales, UK*

Abstract

In this paper, a detailed dynamic model of a powered wheelchair is obtained. This model not only takes in consideration the kinematics equations of a powered wheelchair but also includes all other parts of a powered wheelchair such as the power module and the DC motors. Then a 3D model of the powered wheelchair is designed using virtual reality modeling language (VRML). This 3D model can be integrated in any 3D VRML environment including a domestic environment. Finally, the dynamic model is used with the 3D model of the powered wheelchair in a simulation system. This can later be used to simulate the behavior of the powered wheelchair in different environments using different control algorithms.

Keywords: Powered Wheelchair; Dynamic Model, 3D Model, Virtual Reality.

1. Introduction

Computer simulation using dynamic models of a powered wheelchair offers tremendous advantages to the typical prototype development cycle. These include optimization of design parameters, visualization of complex motions, and simulation of dangerous conditions without any danger to test subjects or equipment [1, 2, 3]. It can also be used to improve the skills of wheelchair users [4] and to test the accessibility of the wheelchair to desired places or buildings [5, 6]. However in order to get a reliable simulation, detailed descriptions of the dynamic model with a 3D model of the wheelchair and the environment are needed.

Several control algorithms of powered wheelchairs have been developed in the literature using only kinematics equations for the wheelchair [7, 8]. This approach could be useful from an automation point of view in order to simplify the model and to get simplified control laws. However, a detailed dynamic model of the powered wheelchair is also needed for the simulation objectives

2. Analysis of a Powered Wheelchair

The powered wheelchair considered in this paper consists of a chair mounted on a chassis driven by four wheels, two free running back wheels and two front fixed drive wheels. The front wheels are connected to two separate DC motors which are controlled by a personal computer (PC) or directly by the user via a joystick as shown in Figure 1. Two odometers are attached to the motor axes which permit an estimation of the motor's angular speed. The system could also contain ultrasonic sensors and a camera in order to estimate the position of wheelchair and to discover the unknown environment.

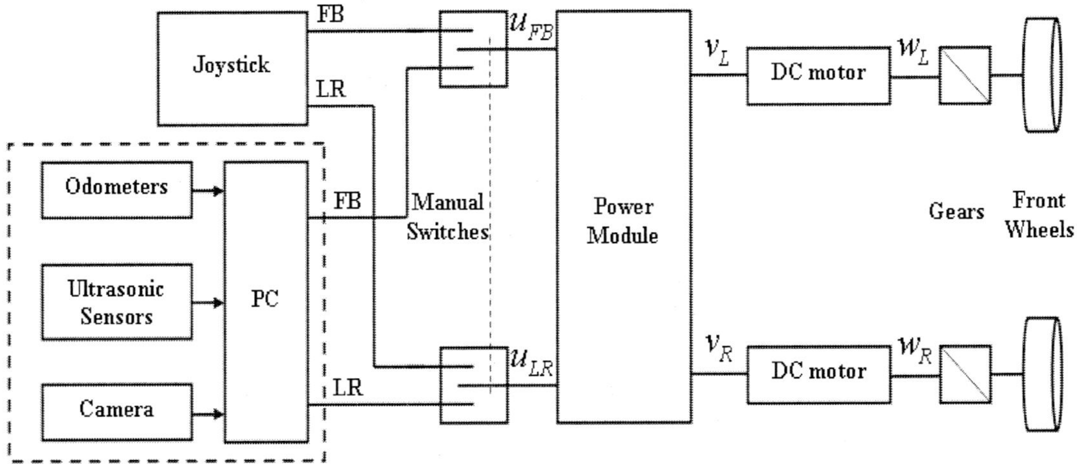

Fig. 1. Block diagram of a powered wheelchair

In Figure 1, FB stands for Forward-Backward, LR stands for Left-Right, u_{FB} and u_{LR} are desired references, v_L and v_R are left and right motor's applied voltages, w_L and w_R are left and right motor's angular velocities respectively.

2.1 The Joystick Module

The joystick module consists of a free stick which can move forward-backward and left-right. The stick movement generates two voltages, one is interpreted as a speed reference and the other one is interpreted as a steering reference as shown in Figure 2.

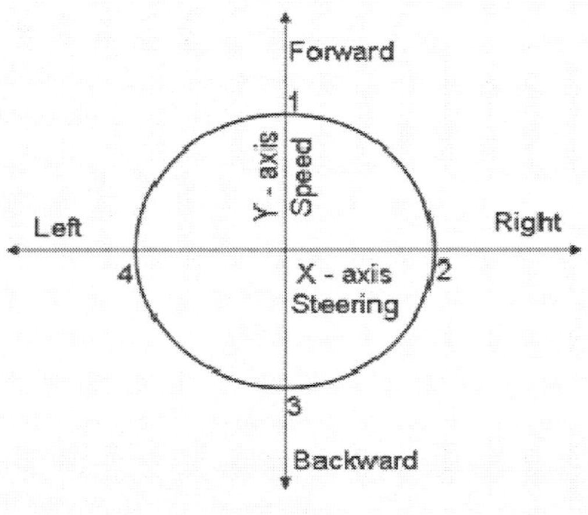

Fig. 2. Joystick's axes interpretation

At the midpoint, the joystick output is $(u_{FB}, u_{LF}) = (2.5, 2.5)$ volts, and in the other points shown in Figure 2 they are as follows: in point 1 we have $(u_{FB}, u_{LF}) = (1, 2.5)$ volts, in point 2 we have $(u_{FB}, u_{LF}) = (2.5, 1)$ volts, in point 3 we have $(u_{FB}, u_{LF}) = (4, 2.5)$ volts and in point 4 we have $(u_{FB}, u_{LF}) = (2.5, 4)$ volts. In other points, the joystick output is a linear combination of these voltages.

2.2 The Power Module

The power module is a dc-dc converter which has a ±12 volt as input and produces a regulated output voltage according to the control signal (u_{FB}, u_{LR}).

The block diagram of a dc-dc converter is shown in Figure 3 where V_g is the battery's input voltage and V is the output voltage.

The transfer function of the dc-dc converter shown in Figure 3 is given by Equation (1):

$$H_{PM}(s) = \frac{V(s)}{V_g(s)} = \frac{DR}{LCRs^2 + Ls + R} \tag{1}$$

where L, C are the converter's inductance and capacitance respectively, R is the load output and D is the duty cycle of output waveform.

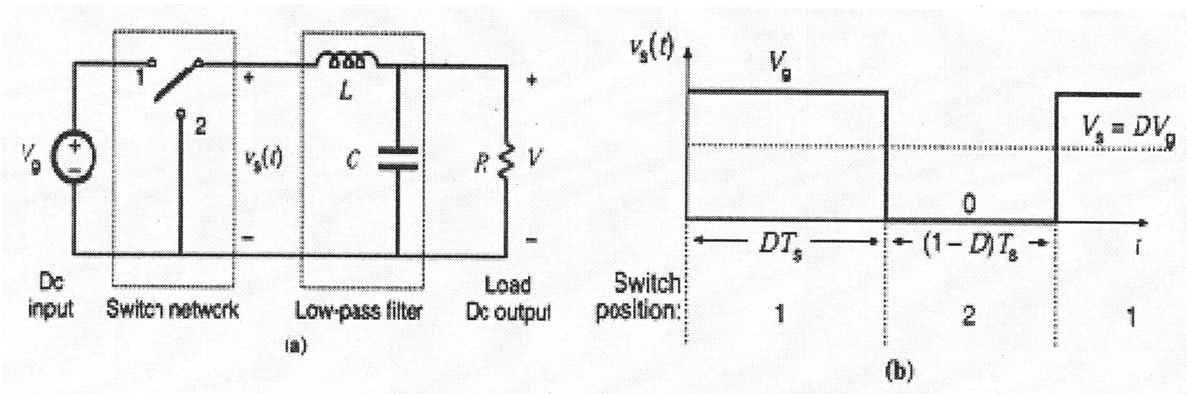

Fig. 3. DC-DC converter: (a) schematic, (b) switch voltage waveform.

The value of D must be calculated as a function of control signal (u_{FB}, u_{LR}) and it depends on different technical approaches.

2.3 The DC Motor

The Direct Current Motor (DC Motor) is a machine that converts electrical power into mechanical power. The input of a DC motor is a voltage V and its output is an angular velocity w. Figure 4 shows a basic model of a DC motor with a load via a gearbox.
The transfer function of a DC motor is given by:

$$H_m(s) = \frac{w_i(s)}{v_i(s)} = \frac{K_T \frac{N_1}{N_2}}{(L_A + R_A)\left(\left(J_M + \left(\frac{N_1}{N_2}\right)^2 J_L\right)S + \left(B_M + \left(\frac{N_1}{N_2}\right)^2 B_L\right)\right) + K_T K_V} \quad (2)$$

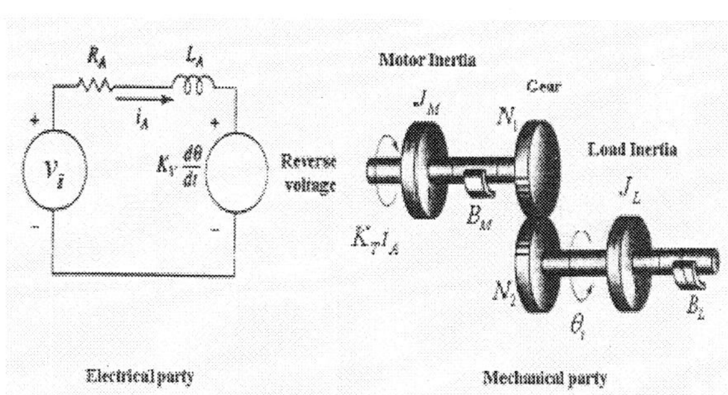

Fig. 4. Basic model of a DC motor with a load.

where i stands for L [left] or R [right], v_i [volt] is the applied voltage, w_i *[rad/s]* is the shaft angular speed, N_1/N_2 is the gear ratio $[N_2>N_1]$, K_T is the armature constant, K_v *[Nm/amp]* is the electromotive force constant, L_A [Henry] is the electric inductance, R_A [ohm] is the electric resistance, J_M, J_L [Kgm^2/s^2] are the moment of inertia of the motor and load and B_M, B_L [Nms] are the friction constant of the motor and load respectively.

It should be noted that there are some time delay between modules in the block diagram in Figure 1. This time delay should be taken into consideration in order to get an accurate dynamic model of the powered wheelchair.

3. Kinematics Equations of Wheelchair

The wheelchair can be driven to any position by the front wheels velocities. It can be coordinated by the (x, y) the position of the wheelchair center in the Cartesian space and ϕ the heading angle of the wheelchair from the x-axis as shown in Figure 5. w_{rw}, w_{lw} are the actuated angular velocities on the right and left wheel respectively, then the right and left wheel linear speeds without slipping are $v_i = r \times w_{iw}$ and the linear and angular velocity of the wheelchair are $v_t = (v_r + v_i)/2$ and $w = (v_r - v_l)/2b$ respectively where r is the wheel radius, b is the distance from the point c to each wheel, v_t is the advance speed of the wheelchair without slipping and w is the angular velocity of the wheelchair.

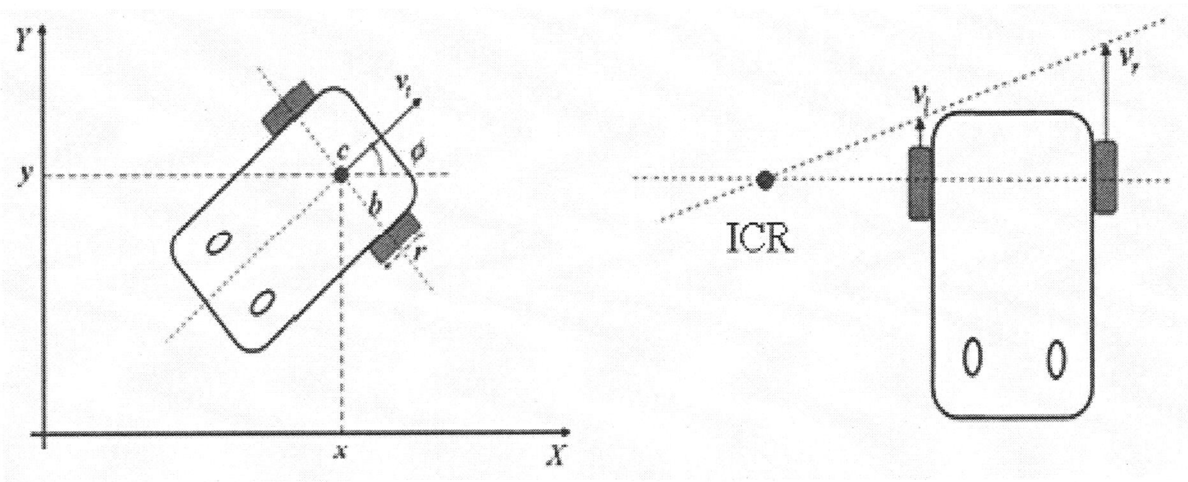

Fig. 5. The geometry and Center of Rotation of the wheelchair.

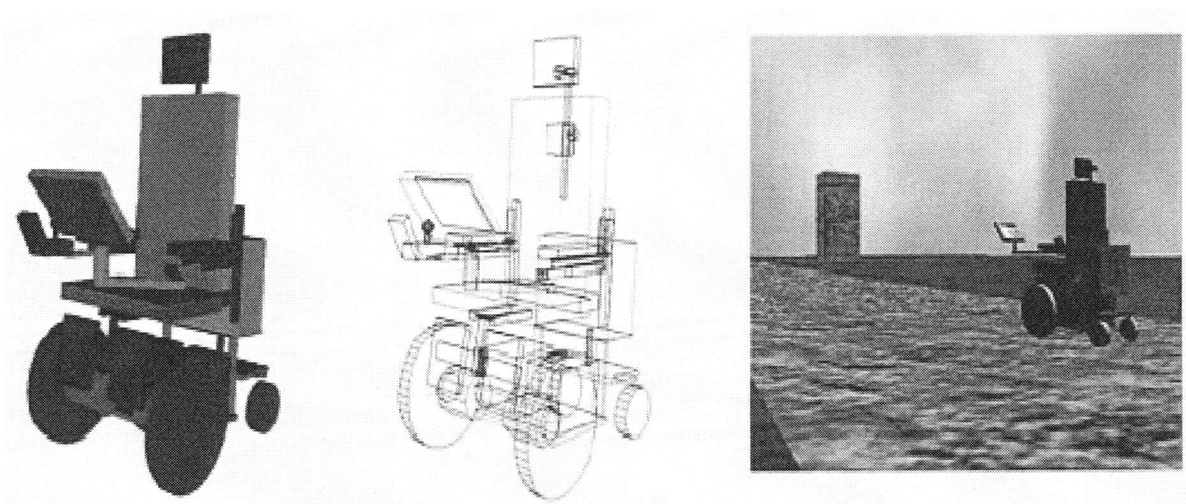

(a) **(b)**

Fig. 6. 3D model of a wheelchair and its integration in a 3D environment.

Using the above relationships the absolute velocity of point c is related to v_t and w by the following kinematics equations:

$$x^{\bullet} = v_t \cos\phi \tag{3}$$

$$y^{\bullet} = v_t \sin\phi \tag{4}$$

$$\phi^{\bullet} = w \tag{5}$$

It should be noted that the center of rotation is the intersection point of the axis between the two front wheels and the axis linking the two endpoints of wheel speed vectors as shown in Figure 5.

4. A 3D Model of the Wheelchair

In order to simulate the wheelchair behaviour we need either a 2D model or a 3D model of the wheelchair. In this work, a 3D model of the wheelchair was created using the Virtual Reality Modelling Language as shown in Figure 6a. Each part of the wheelchair is modelled in a separate node so that it can be controlled separately in the simulation system. This 3D model can be easily integrated in any 3D desired environment as shown in Figure 6-b.

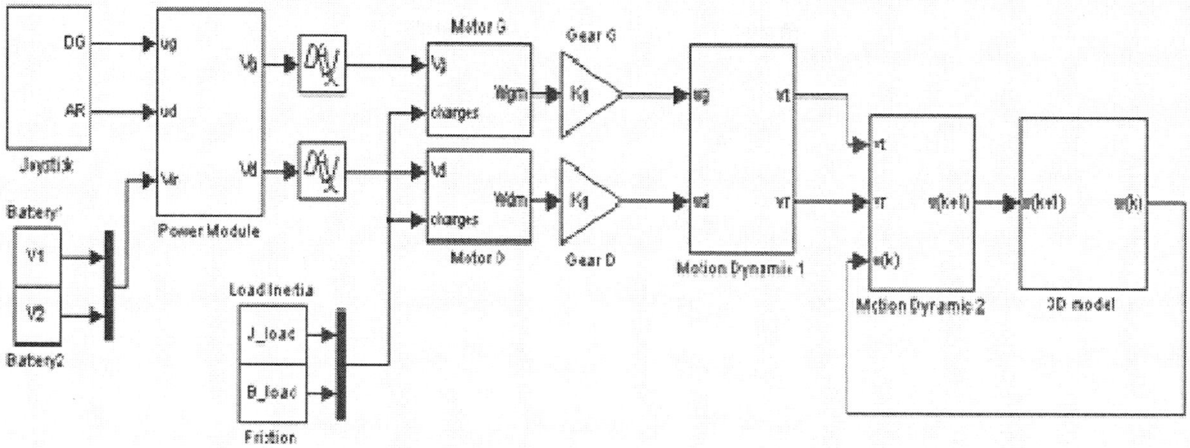

Fig. 7. The Simulation system.

5. Simulation of the Powered Wheelchair

In this section, the dynamic model of the powered wheelchair developed in previous sections as well as its 3D model will be integrated in a simulation system in order to simulate its behaviour. The simulation system is shown in Figure 7.

The simulation was carried out using Euler fixed step integration method with $1ms$ step size. The wheelchair navigated in the virtual environment with quasi-real dynamic. This simulation system can be used to test the accessibility of the wheelchair to some places or buildings. It can also be used to test different algorithms for guidance and obstacle avoidance.

6. Conclusion

In this work, a detailed dynamic model for a powered wheelchair was obtained. The powered wheelchair has been firstly analyzed from automatic point of view. Next the wheelchair motion equations have been derived. Then a 3D model of the wheelchair has been constructed using virtual reality modelling language. Finally, a simulation system has been built which included all dynamic equations previously developed. This simulation system can be used to test the accessibility of the wheelchair to selected places or buildings. It can also be used to test different algorithms for guidance and obstacle avoidance.

References

[1] T. Inoue, H. Hirose, Y. Sudoh, and K. Hase, **Development of a simulator of powered wheelchair.** *Proc. of the RESNA'98 Annual Conference*, p. 182-184, USA, 1998.

[2] H. Niniss and A. Nadif, **Simulation of the behavior of a powered wheelchair using virtual reality.** *Proc. 3rd Intl. Conf. Disability, Virtual Reality & Assoc. Tech.*, p. 9-14, Alghero, Italy, 2000.

[3] F. Lin, L. Ye, V. G. Duffy and C.-J. Su, **Developing virtual environments for industrial training.** *Information Sciences,* Vol. 140 p. 153-170, 2002.

[4] C.-H. Tsai, S.-W. Yang and Y.Y. Fan, **Using virtual reality biofeedback to improve manipulative skill of manual wheelchair users.** *Journal of Biomechanics,* Vol. 39, p.S170, 2006.

[5] D. Stredney, W. Carlson, J.E. Swan, and B. Blostein, **The determination of environmental accessibility and ADA compliance through virtual wheelchair simulation.** *Presence: Teleoperators and Virtual Environments*, 4(3):297-305, Summer 1995.

[6] C. S. Han, K. H. Law, J.-C. Latombe and J. C. Kunz, **A performance-based approach to wheelchair accessible route analysis.** *Advanced Engineering Informatics,* Vol. 16, p. 53-71, 2002.

[7] G. Campion, D. Novel, and G. Bastin, **Controllability and state feedback stabilization of non holonomic mechanical systems,** *volume 162 of Lecture Notes in Control and Information Science*. New York: Springer-Verlag, 1991.

[8] R. A. Cooper, D. M. Spaeth, D. K. Jones, M. L. Boninger, S. G. Fitzgerald and S. Guo, **Comparison of virtual and real electric powered wheelchair driving using a position sensing joystick and an isometric joystick.** *Medical Engineering & Physics,* Vol. 24, p. 703–708, 2002.

Innovative Production Machines and Systems
D.T. Pham, E.E. Eldukhri and A.J. Soroka (Eds.)

Optimization in Manufacturing Plant Layout Design

H. Marzi, and E. Marzi

Department of Information Systems, St. Francis Xavier University, Antigonish NS B2G 2W5 Canada

Abstract

The aim of this paper is to present an optimized layout for production of microelectronic components. A manufacturing group revisits their production layout and plans an improved layout to avoid existing unanticipated difficulties that were revealed through analysis of the existing layout. The final proposal is a manufacturing layout, utilizing process cells to implement "just in time" technique, highlighting optimization of WIP and throughput time. The present design provides smooth and continuous flow of production for each product type, designing production cells and linking them with particular product types and allocating overheads on a more realistic basis.

Keywords: Factory Layout, Just in time, Cell Manufacturing

1. Introduction

The creation of a layout that has an undefined path across management, engineering and finance, is a complex task composed of difference factors, some which could be enumerated as technology, process, personnel, transportation, management rules, theories, experience, etc. Each of these has subdivisions and this adds to the complexity of the matter. For example: process; whether it is automatic, manual or perhaps a mixture of both transportation- what sort of facility is going to be used for the material movement. Perhaps the most difficult task is deciding in a suitable layout design for taking advantage of special manufacturing techniques of the category of FMS, JIT etc. In each step one major factor, namely, investment limitation could be limiting conditions for any proposal.

2. Hybrid Technology

2.1. Introduction to Hybrids

A hybrid microcircuit is a set thick films, thin films, semi conductors and distinct components, collected on an insulator substrate [1]-[3]. The layer of conductors, resistors and capacitors is screened on a ceramic type substrate. Distinct compounds such as resistors, capacitors and inductors are then also attached to the substrate.

Sufrace Mount Technique, abbreviated as SMT is also referred to as printed or etched circuit board (PCB, orECB) which is made by a thin copper layer etched in chemical compounds, then the discrete components are placed and attached to the board by one of the soldering methods [4].

Integrated circuits (IC) are composed of resistors, capacitors, diodes and transistors on semi-conductor material with restrictions due to the size of components.

2.2. Hybrid Product Classes in Microelectronics

The wide variety of hybrid products in Microelectronics companies makes the production service complex [1]-[4]. Due to the variety which has been accompanied by high difference in the volume of each product type, the tasks of servicing department become more complex. This complexity in the task starts form the sales office which is also responsible for new agreements and contracts then continues to the production scheduling and stretches itself to financial difficulties; such as overheads analysis and cost of sales.

Available classes of hybrid products are; "passive", "C & D", "A & B", and "E" hybrids. Passive hybrids are the class of products with only screened conductor(s) and resistor(s) on the substrate without any active components. "C & D" hybrids are a class of special orders with low production volume this group of products are processed in the clean area and are under direct control of the development department. "A & B" hybrids are a group of active hybrid microcircuit. "E" hybrids are special high volume microcircuits. The products of this type have a longer production cycle as they pass through more operation stages.

3. Layout

3.1. Layout Considerations

The layout of a factory is the arrangement of the productive factors such as machines, materials, storage, transportation, inspection, etc. to optimize production. Layout arrangements vary with the nature of the industry. Classically three types of layouts can be distinguished, and often a combination of all three.

1. Fixed position /material layout
2. By-process/function layout
3. Line Production layout

3.2. Layout Proposals

The layout has been arrived at by considering the nature of the products, analyzing the existing layout and then employing the correct procedures. The outline process chart of the hybrid manufacturing Fig. 1 reveals that there are no subassembly production lines but the nature of the industry demands a split in production activities into two segments; passive and active.

The passive hybrid circuits undergo multiple screening, drying, and firing operation which still forms a straight flow process. Due to the repetitive nature of the processes up to this stage, layout-by-process is the most suitable type for this part of the production line, so that all screening operations with the same type of process are grouped and located in the screen room and all firing stages which consist of the curing screened substrates in furnaces are comprised in a number of furnaces grouped together in the furnace room.

A review of the flow process chart for the passive product Fig. 2 shows that exiting arrangement for material movement at this stage of production line has desirable form with minimum delay and transport. In addition, it should be noted that, both screening and firing must take place in a dust free area. The best arrangement of production equipment is the grouping of screeners and furnaces together one working area with screeners located close to the feeding end of the furnaces to eliminate transportation of substrates to the firing area after screening.

On the other hand, active circuits result from the assembly of active discrete components on screened substrates followed by lead frame mounting. The series of action taking place at this stage demand an arrangement for the required equipment in sequence of operation. This condition justified the suitability of line production type of layout for this part of the production line. Hence in Fig. 2 a number of assembly lines equal to the number of pick and place machines is provided.

Laser trimmers which are used after firing (passive trimming) or after assembly (active trimming) can be located between these two zones, However since laser trimming must be carried out in a clean area, this equipment can be located with the screeners and furnaces. Finally a series of manual operations and inspections occur on the benches which are adjacent to the assembly area.

Description	Symbols				
	A	B	C	D	E
Tiles Stored in raw material Store				*	
Transport to laser scribe machine			*		
Laser scribing	*				
Held for sending to the screening stage					*
Transport to the screen process			*		
Screen Conductor side 2	*				
Dry	*				
Check conductor print thickness		*			
Firing	*				
Screen conductor resistor side 1	*				
Dry	*				
Check resistive print thickness		*			
Check TCR		*			
Check temperature coefficient of resistance		*			
Firing	*				
Electrical Check		*			
Screen Overglaze side	*				
Dry	*				
Check Overglaze		*			
Firing overglaze	*				
Carried to burnishing			*		
Burnishing	*				
Carried to Marking			*		
Marking	*				
Carried to Laser trim room			*		
Hold for trimming					*
Laser trimming	*				
Laser wash	*				
Quality Control		*			
Carried to Storage			*	*	

Fig. 1. Flow process chart of passive stage of part "44426"

3.3.Preferences of the Proposed Layout

This layout will result in the elimination of unnecessary transport and delays, while the progress of the process and products are visible, resulting in the possibly having closer control on the production line. The advantages include: 1). Reduction in handling, 2). Reduction in W.I.P., 3). Reduction in uncontrolled delays , 4). Better control of production lines, and 5). Possibility of eliminating bottlenecks

The significant changes in the production activities can be considered from the flow process charts of part no. 44426 in the proposed layout Fig. 4 by comparing with the present situation Fig. 1 and Fig. 3. However, since changes can occur after the screening and firing process, Fig. 4 includes activities started after the firing of the over glaze.

The proposed layout had been drawn up assuming that there are no constraints already existing in the new factory and may therefore be considered as the ideal arrangement. The cost of providing clean areas for the furnace and laser room nearly at the center of the factory is significantly higher than if they are located near the outside. Therefore a management decision aimed at lower investment had dictated that, then areas for the production of C & D hybrids the laser room and the furnace room were designed and constructed before final layout approval. In addition, laser trimming systems are to be located in the same area imposing yet another constraint.

4. Proposed layout

4.1.Cell process

The location of certain groups of machines is already fixed. This implies that the following production and servicing section are fixed: Laser scribe room, screen and furnace room, Laser room, Stores, Assembly area. Any layout choices are now limited to consideration of the interaction between section and arrangements of equipment inside the sections. This focuses attention on the assembly area. The aim is to design the best arrangement for the equipment within this area to minimize or eliminate any unnecessary activities in production processes.

A cell system has been considered because it aims to reduce transport, delay and work in progress in assembly area. The idea arises from the nature of intermittent operations before and after the component and lead frame assembly. At each of these two operations there are some activities which complement each other as they go together to accomplish the required process. The first series of operations are, solder cream screening, component assembly, reflow solder cream and wash. At present, each of these operations are carried out at different stages. The equipment used for these processes are DEK, pick and place robots, Brown reflowers and Kerry washers respectively.

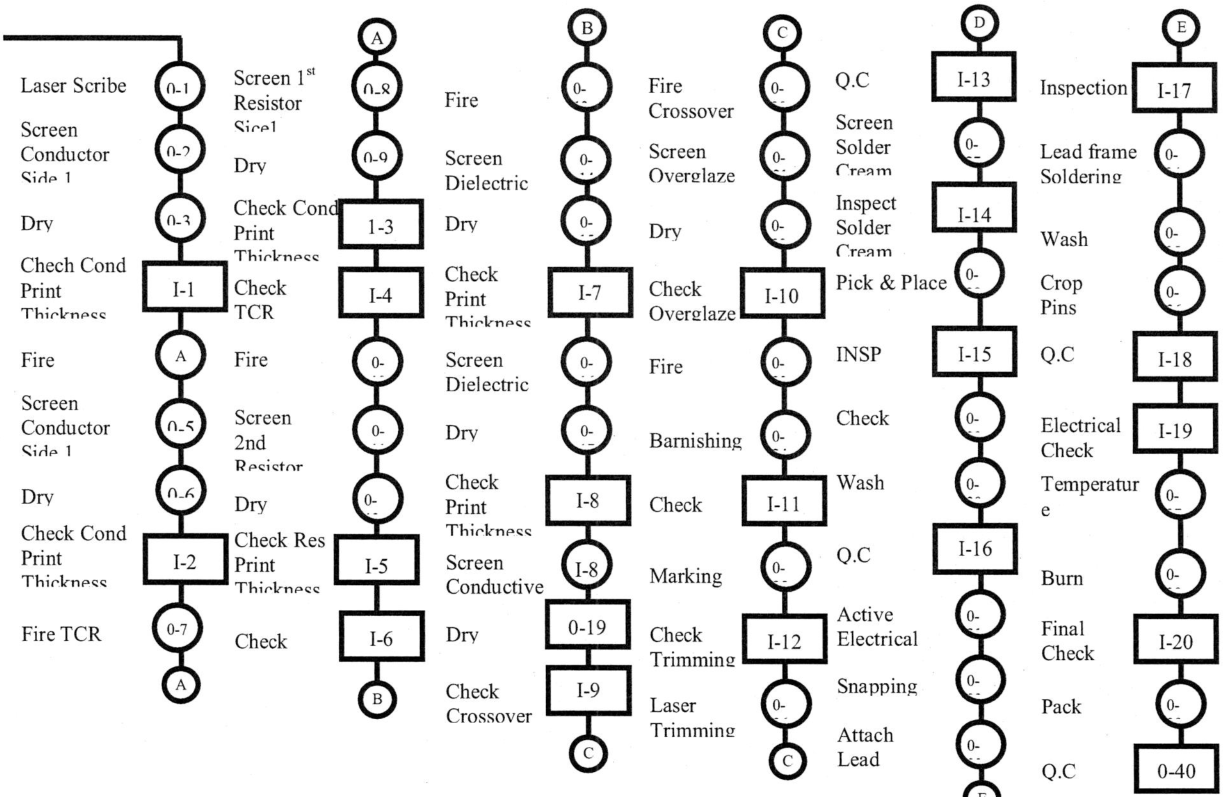

Fig. 2. Outline Process Chart for Group "E" Hybrids

Description	Symbols				
	A	B	C	D	E
Printed Substrate Stored				*	
Carried to Screener			*		
Solder Cream Screeing	*				
Check Solder Cream		*			*
Carried to Pick and Place machine			*		
Component Assembly	*				
Carried to reflow			*		
Reflow Components	*				
Carried to Wash			*		
Chlorothene Wash	*				
Visual Inspection		*			
Carried towards Lead frame assembler			*		
Snap into single circuits	*				
Lead frame Attachment	*				
Held for Soldering					*
Lead frame Soldering	*				
Chlorothene Wash	*				
Visual inspection		*			
Crop Pins to Specify Length	*				
Carried to test stations			*		
Electrical Check (general)		*			
100% Burn	*				
Final Electrical Check (wayne)		*			
Final Electrical Check (general)		*			
Final inspection		*			
Carried for Packing			*		
Packing	*				
Carried to Quality Control Room			*		
Q.C Visual inspection		*			
Q.C Electrical Inspection (wayne)		*			
Q.C Electrical inspection (general)		*			
Q.C Temperature *Cycle Sample*		*			
Carried to Store			*		
Stored				*	

Fig. 3. Flow Process chart of active part no "44426"

The proposal suggest the formation a cell in which a pick and place machine is the core, and the screener, reflow and wash equipment are located beside it so that the system can be operated by one operator.

Hence there are delay and transports between the different targets of the assembly. The idea is to bring all the different stages of lead frame assembly together and to couple them so that after each step, the product can be directly fed to the next station, and finally at the end of the processes, the lead frame test station is located by the crop machine, eliminating the issue of transport and delay.

4.2. Work Study for Cell Systems

Although the main operations taking place during the component assembly in the cell systems are the same, the time taken at each stage has been reduced significantly.

Description	Symbols				
	A	B	C	D	E
Firing Overglaze	*				
Burnishing	*				
Marking	*				
Laser trimming	*				
Laser wash	*				
Quality Control		*			
Solder cream screening	*				
Component assembly	*				
Reflow component	*				
Chlorothene wash	*				
Visual Inspection		*			
Lead frame assembly	*				
Lead frame soldering	*				
Chlorothene wash	*				
Visual inspection		*			
Crop pins to specified length	*				
Carried to test stations			*		
Electrical check (general)		*			
100% Burn	*				
Final electrical check (wayne)		*			
Final electrical check (gen.)		*			
Final inspection		*			
Carried for packing			*		
Packing	*				
Q.C. Visual inspection		*			
Q.C. Electrical inspection		*			
Q.C. Temperature Cycle sample		*			
Carried to store			*		
Store				*	

Fig. 4. Flow Process of part no "44426" in proposal- "Layout of Hybrid Products"

One factor affecting this is the fact that after each operation in the new procedure, the product is fed directly to the next machine rather than depositing it in a pallet and transporting to the next station. Therefore, the time taken for unloading a product after an operation is shared by the loading time to the next station and so on. Further still, the system gives the opportunity for the operator to do the rest of their manual jobs.

Table 1. Time study of the component assembly in the cell system (*each element in this table will be referred to by its labeled number. ie. Screen cycle would be 2)

Product Part No. 44426, 44414, 44431				
Pick and Place Machine: Philips				
No	Activity Description	Carried out by	Basic time per magazine (mins)	
			44426	444-14 &31
1.	Load Screener (DEK)	Operator	0.10	0.10
2.	Screen cycle	Machine	0.13	0.13
3.	Upload screener load pick &place	Operator	0.20	0.20
4.*	P&P initial movement	Machine	0.02	0.02
5.*	P&P cycle	Machine	1.70	0.82
6.	Unload P&P	Operator	0.10	0.10
7.	Load reflow	Operator	0.33	0.33
8.	Reflow cycle	Machine	1.59	1.59
9.	Unload reflow	Operator	0.20	0.20

It is considered that unloading the screener (DEK) and loading the pick and place device are carried out in one action. Similarly, loading and unloading the pick and place device and loading the reflow machine are performed in one action. Furthermore, it can be considered that there is no time recorded for washing, since after the reflow stage the products are placed into a suitable container. It will be shown later that the operation has enough time during machine cycles to load this container in to Kerry wash to take them out after the process.

The comparison between the time studies of different part numbers reveals that due to the similarities in the nature of the jobs the time recorded for each operation is identical to that in the other products, unless the activities are carried out by pick and place machines. Therefore, in the case of different part numbers only activity numbers 4 and 5, (marked with an asterisk) in Table 1 are different.

4.3.Flow process chart

One of the advantages of the cell system is the reduction in the number of production stages during assembly [5]-[8]. The eliminated stages are delays and transportation processes. These can be best demonstrated by comparing the flow process charts of the sample product (part no. 44426), and comparing it to the exiting conditions, Fig. 1.

4.4.Multiple Activity Chart

To demonstrate the possibility of operating within a cell system, it is required that activities of each element in the cell be studied in conjunction with the whole of the group by use of multiple activity charts. The multiple activity chart of assembly of hybrid circuits in the cell system, comprises recorded action of the four elements in the cell, the solder cream screener (DEK), the pick and place machine, the reflow equipment and the operator. The sequence of action within the cell is recorded from the start of operation and shows the build up cycle, this finally comes to the routine sequence of actions.

4.5.Build up Cycle

In the following description, the activities are referred to by the labels given in Table 1. Table 2 represents different stages in the build up cycle. As it can be considered, the last stage "J", in the table, is the same as step "I". Therefore a complete assembly cycle takes 2.33 minutes for part no. 44426 and 2.22 minutes in the case of 44414 and 44431.

4.6.Comparison between cell system assembly and present procedure

This section discusses the direct advantages of the cell system proposal [9]-[11]. The most obvious part, which is the time saving can be an appropriate starting point. To start with some conversions must be made, since the recorded cycle times in Table 1 are represented in basic time per magazine, while data in Table 3 are in terms of standard time per 100 pieces.

Table 2. Activities within build up cycle which leads to the routine sequence of component assembly

Stage	Activity	Basic Time per magazine (mins)	
		44426	44414
A	1,2,3,4	0.43	0.43
B	Repeat "A"	0.43	0.43
C	1,2,3,5	2.13	1.25
D	(1, 2) (i/w 5c)*,3,5	1.9	1.02
E	Repeat "D"	1.9	1.02
F	Repeat "D"	1.9	1.02
G	Repeat "D"	1.9	1.02
H	(1, 2) (i/w 5G)*,3,5,6,7,8,9	4.12	3.24
I	(1, 2) (i/w 5H)*,3 (i/w 8I),5 (partly i/w 8H and 9H), 6,7,8,9	2.33	2.22
J	(1, 2) (i/w 5I)*,3 (i/w 8I),5 (partly i/w 8I and 9H), 6,7,8,9	2.33	2.22

Assuming each magazine contains:

For part no. 44426: 4 tiles where each tile contains 12 single circuits giving 48 single circuits.

For part no. 44414 or 44431: 4 tiles where ach tile contains 14 single circuits giving 56 single circuits

Now the cycle time in standard assembly cycle time in the cell system is in terms of standard minute per 100 pieces.

For part 44426: (2.33 min) x (1.23) x 100/48 = 5.97 min

For part 44414 / 44431: (2.22 min) x (1.23) x 100/56 = 4.88 min

Moreover, at present, component assembly is completed by 3 operators while in the cell system, it can be carried out by one operator. This reduction in the number of operators in conjunction with the time study gives real image of the significant of the proposal. Hence Table 3 gives a comparison between the two cases in terms of the time and labor cost [12]-[16].

Table 3. Cycle time and labor cost comparison between present and cell system component assembly for same sample products

Pick and Place	Part No.	Standard Time/100 pcs (min)			Labor cost. 100 pcs ($)		
		Present	Cell System	Saving %	Present	Cell System	Saving %
Philips	4426	6.74	5.97	11	1.12	0.33	71
	44414 or 44431	5.78	4.88	16	0.96	0.28	71
Nos318	444411	48.74	27.03	45	8.09	1.50	81
Zevatec	44432	10.63	9,57	10	1.76	0.53	70

4.7. Other advantages

The cell system proposal can bring a number of significant changes which can be listed as: Achievement of just-in-time technique, reduction/elimination of WIP, reduction in the manufacture throughput time, reduction in the overheads absorbed, improvement in monitoring of quality, improvement in quality, reduction in transport, improvement in material handing, Identification of waste, reduction in material losses, and allocation of overheads on a more realistic basis.

5. Information and results

Here, two cell systems are considered. They are composed of a screener, a pick and place robot, reflow solder cream machine and a washer. The set of machinery suggested for use in production of the part number 44414, 44426, 44427 are a DEK 245 screener, a Philips pick and place, a Brown reflow solder cream machine and a Kerry washer. For part 44411 the only difference is in the pick and place machine which is a "Zevatec" rather than a Philips. It should be mentioned that as a general rule the depreciation cost is considered on a two shift basis. The depreciation cost of machinery in the cell systems under consideration are allocated using the following formula and are listed in Table 4.

$$\text{(Part no. cycle time, min)} * \frac{\text{(Annual Depreciation cost, \$)}}{\text{(Annual working time, min)}}$$

Table 4 Description Cost of Cell System Machinery

Machine	Annual Depreciation Cost
Deck 245	1537
Philips	81101
Zeratec	32440
Brown reflow	1867
Kerry Wash	578

6. Conclusion

The proposed cell system layout for the production of the hybrid microcircuits would result in improvements in throughput time, WIP quality control and production cost and also presents an opportunity to reconsider the allocation of overheads on a more realistic basis.

Although the cost reduction in the production process is one of the direct results of the proposal, the indirect savings have even greater importance. The cell system approach will result in improvements in two different activities, Manufacturing/production and Finance.

Manufacturing/production: the advantage of implementing a cell system layout effectively reduces the number of activities with within the production processes due to: reduction in throughput time, reduction in the WIP, reduction in the transportation and delay, reduction in the inspection stations, and utilization of just-in-time.

Finance: the reduction in throughput time and labor cost has been estimated to produce an annual saving of 1% to 25%. The proposed layout would facilitate overhead allocation because the process cells are linked to a particular product type and overheads can be distributed on a more realistic basis, taking into account production cycle time and equipment utilization. This results in the identification of the 'true' product cost and more effective pricing. It is noteworthy that many systems have an inherent resistance to change causing problems in adapting and implementing new layouts.

References

[1] Jones R.D. "Hybrid Circuit Design and Manufacturing", Electrical Engineering and Electronics/12, Marcel Dekker Inc. 1982.

[2] Gupta T.K. "Handbook of Thick and Thin Film Hybrid Microelectronics" John Wiley & Sons Inc. 2003

[3] Magin C.H. and MC Clelland S. "Surface Mount Technology", IFS (Publication) LTD, 1987.

[4] Mcclelland, S. Advancing Surface Mount Technology December 1988 Springer-Verlag Berlin and Heidelberg GmbH & Co. KG

[5] "Introduction to Work Study", Third (Revised) Edition, International Labour Office Geneva, 1979

[6] Muther R. "Practical Plant Layout", MC Graw – Hill Book Company, Inc.

[7] Maynard H.B. "Industrial Engineering Handbook", McGraw – Hill Book Company

[8] Schonberger R.j. "Japanese Manufacturing Techniques", The Free Press, 1982.

[9] Baggott J, "Cost and Management Accounting", Helman Professional Publishing Lt.d

[10] "British Standard " – 3138-1967

[11] Wild R, Production and Operations Management. Cassell 5th Ed 1995.

[12] Thuesen H.G., Fabrycky w.j. Thuesen G.J. "Engineering Economy" 4th Ed., Prentice Hall 1971

[13] Hartley R. V., "Cost and Managerial Accounting ", Second Edition Allyn and Bacon, Inc., 1986

[14] Coombs, Clyde F., Jr. (Ed.) (2007). Printed Circuits Handbook, Sixth Edition" ', McGraw-Hill Professioal

[15] Finch, Byron J. Operations Now: Profitability, Processes, Performance. 2nd ed. Boston: McGraw-Hill/Irwin, 2006.

[16] Stevenson, William J. *Operations Management.* 8th ed., Boston: McGraw-Hill/Irwin, 2005.

Innovative Production Machines and Systems
D.T. Pham, E.E. Eldukhri and A.J. Soroka (Eds.)

Review and Analysis of Fuel Cell System Modelling and Control

K.K.T. Thanapalan[a], G.P. Liu[a,b], J.G. Williams[a], B. Wang[a], D. Rees[a]

[a] *Faculty of Advanced Technology, University of Glamorgan, Pontypridd CF37 1DL, UK*
[b] *CSIS Lab, Institute of Automation, Chinese Academy of Sciences, Beijing, PR China*

Abstract

This paper presents a review and analysis of a fuel cell system modelling and controller design for electric fuel cell vehicle applications. The paper begins by presenting the basic principal of PEM fuel cell dynamics. The work proceeds to investigate the fuel cell models and controller design techniques that have been applied to develop control law for fuel cell systems.

Keywords: Fuel cell, modelling, control, power system.

1. Introduction

Recently there is a growing interest in fuel cell systems research, since fuel cells are clean and efficient sources of electricity, and have a wide range of vehicle applications see for example [1], [2], [3], [4], and [5]. Demands for higher fuel economy, performance, and reliability as well as reduced emissions push industry to seek improvements in fuel cell technology, before its considered to be an alternative to other energy sources. To this end, research and developments are being carried out to investigate and improve the performance, stability and reliability of the fuel cell system. The complexity of the fuel cell system requires elaborate and innovative studies for proper configuration, component sizing, and control system development to fully explore the potential of this advanced technology.

A fuel cell is an electrochemical energy conversion device. It converts the chemicals hydrogen and oxygen into water, and in the process it produces electricity. The other electrochemical device that we are all familiar with is the battery. A battery has all of its chemicals stored inside, and it converts these chemicals into electricity. This means that a battery eventually goes dead. But, with a fuel cell, chemicals constantly flow into the cell so it never goes dead-as long as there is a flow of chemicals into the cell, the electricity flows out of the cell. It is this supply of reactants which presents one of the several challenges encountered by the fuel cell system investigators. Most fuel cells in use today use hydrogen and oxygen as the chemicals. There are several different types of fuel cells, each using a different chemistry. Fuel cells are usually classified by the type of electrolyte they use. Some types of fuel cells work well for use in stationary power generation plants. Others may be useful for small portable applications or for powering cars. For example, Solid Oxide Fuel Cell (SOFC) technology is most suited to applications such as stationary power etc.; the high operating temperature of SOFCc produces heat suited well to cogeneration applications [6]. The Proton Exchange Membrane Fuel Cell (PEMFC) can be used to powering cars. PEMFC uses one of the simplest reactions of any fuel cell. Hydrogen and Oxygen (air) are supplied on each side of a cell. The cell consists of an electrolyte membrane (PEM) with a catalyst layer on each side. When hydrogen is led to the first catalyst layer, the anode (the negative post of the fuel cell), hydrogen molecules are split into their basic elements, a proton and an electron. The protons migrate through the electrolyte membrane to the second catalyst layer, the cathode (the positive post of the fuel cell). Here they react with oxygen to form water (see Fig.1). At the same time the electrons are forced to travel around the membrane to the cathode side, because they can not pass the membrane. This movement of electrons thus creates an electrical current [7]. Most of the time cells are stacked together to makes fuel cells scalable and flexible in design, thus giving a wide range of possibilities of usage.

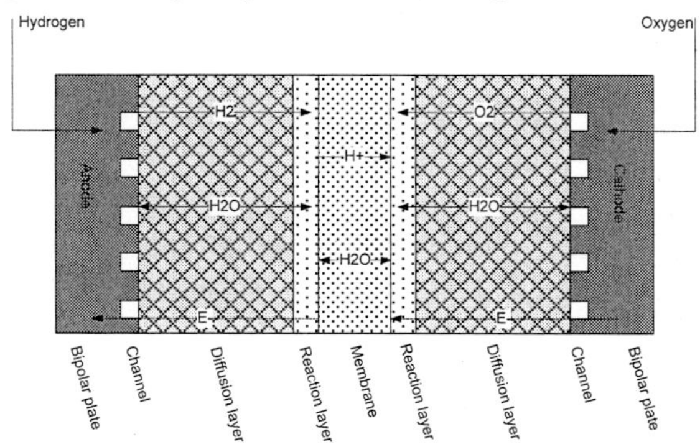

Fig.1 Basic principle of PEM fuel cell

2. Fuel cell system models

A variety of literature has been published concerning various aspects of fuel cell system modelling, dynamics and operation. Broader details of the subject can be found in various reports, which summaries the modelling aspects and problems experienced with fuel cells. These reports also suggest some technical solutions and analysis methods for some of the problems. These range from simple empirical (one-dimensional) circuit based ones [8] to more complex (three-dimensional) fluid-dynamics-based ones [9]. The former are lumped–parameter in nature, with the cell's dynamic behavior described by ordinary differential equations, while the latter are distributed-parameter in nature, with the cells dynamic behavior described by partial differential equations. In general, for the purposes of controller design, lumped-parameter system models are preferred due to the abundance of theory relevant to such systems [10]. Previous PEMFC modelling efforts aimed at transient behavior improvement [11], [12], propose a third-order lumped-parameter physical –based model that is more detailed than the first-order lumped parameter model used in [8]. Lumped parameter models of similar complexity were also used for studying the dynamic behavior and control of SOFCs and Molten Carbonate Fuel Cells (MCFCs) in [13], [14], [15]. In general, fuel cell system models in the literature are mostly steady state models, see for example [16], [17], [18], [19], which are typically used for component sizing [20], [21], [22], cumulative fuel consumption or hybridization studies and forward looking simulation models [23]. These models represent each component such as compressor, heat exchanger and fuel cell stack voltage in terms of its static performance or efficiency map. The only dynamics considered is the vehicle inertia. The authors usually assume perfect condition in the cathode, for example humidity and stack temperature. Although the dynamic fuel cell behavior is not included, these studies established a good basis for understanding the fuel cell vehicle integration [24]. Considerable numbers of research papers address the effects of transient variations in the fuel cell system performance. Excellent examples are the dominant but slow temperature effects on the stack efficiency [7], and the effects of reformed hydrogen feed rates in the stack response. The authors in [25] raise a number of interesting issues associated with the dynamic interactions of the flow, heat and water subsystems but did not provide a comprehensive set of equations for their study.

Other types of models have been proposed to investigate fuel cell technology via simulation. The basis of these models can be fluid dynamics, electrochemical reaction [26], and thermal etc. for example, [27] proposes an electrochemical model for a 5 kW fuel cell. However, to analysis the dynamic problems of controlling a fuel cell system and develop control schemes for alleviating these problems, it is necessary to have a dynamic model for fuel cell system. The dynamic model should be well suited for stability and control analysis, which may involve linearised equations around possible nominal points. In [28] Control- oriented model for an automotive PEM fuel cell system is presented. The use of a PEM fuel cell power system for vehicles has been considered by several researchers; see for example, [29], [30], [31], [32]. Mathematical models presented in most of these works are a set of complicated nonlinear equations. Despite this large number of publications on fuel cell modelling, relatively few are suitable for control studies [33], [34]. The study of a control oriented model is a critical first step for the understanding of system behavior, and the subsequent design and analysis of model based control systems [35]. To this end, the authors have made a significant contribution to the field; see for example [36], [37], [38], [39],[40],41],[42],[43]. In this paper, a brief summary of a PEM fuel cell system dynamic model that is suitable for control study is presented.

The system model contains four interacting sub-systems; cathode and anode flow, the membrane hydration, and stack voltage (see Fig.2). The electrochemical reaction at the membranes is assumed to occur instantaneously and the temperature is assumed to be constant. For the cathode and anode mass flow behavior, the principle of conservation of mass is used to obtain governing equations

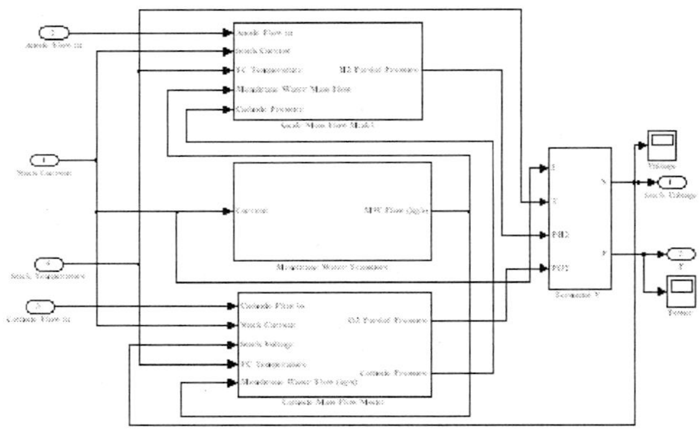

Fig.2 Structure block diagram of fuel cell system

Cathode flow:

$$\dot{m}_{O_2} = W_{O_2,in} - W_{O_2,out} - W_{O_2,reacted} \quad (1)$$

$$\dot{m}_{w,ca} = W_{v,ca,in} - W_{v,ca,out} + W_{v,gen} + W_{v,mbr} \quad (2)$$

where $W_{O_2,in}$, $W_{O_2,out}$ and $W_{O_2,reacted}$ are the oxygen mass flow in , out and reacted respectively. Each of these components is expressed as follows:

$$W_{O_2,in} = \gamma \psi^{-1} W_{a,in}$$

$$\psi = 1 + k_t T M_v M_{O_2}^{-1} (P - k_t T)^{-1}$$

$$W_{O_2,out} = \gamma \psi^{-1} W_{a,out} \quad W_{O_2,reacted} = M_{O_2} \frac{n.i_{FC}}{4F}$$

$$W_{v,ca,in} = \left(1 - \psi^{-1}\right) W_{a,in}$$

$$W_{v,ca,out} = \left(1 - \psi^{-1}\right) W_{a,out}$$

$$W_{v,gen} = M_v \frac{nI_{st}}{2F}$$

Anode flow:

$$\dot{m}_{H_2} = W_{H_2,in} - W_{H_2,out} - W_{H_2,reacted} \quad (3)$$

$$\dot{m}_{w,an} = W_{v,an,in} - W_{v,an,out} - W_{v,mbr} \quad (4)$$

where, $W_{H_2,in}$, $W_{H_2,out}$ and $W_{H_2,reacted}$ are the hydrogen mass flow in, out and reacted respectively. Each of these terms is modelled as follows:

$$W_{H_2,in} = \phi^{-1}.W_{H_2,an,in}$$

$$\phi = 1 + k_t.TM_v M_{H_2}^{-1}(P - k_t.T)^{-1}$$

$$W_{H_2,out} = \phi^{-1}.W_{H_2,an,out}$$

$$W_{H_2,reacted} = M_{H_2} \frac{ni_{FC}}{2F}$$

$$W_{v,an,in} = (1-\phi^{-1})W_{H_2,an,in}$$

$$W_{v,an,out} = (1-\phi^{-1})W_{H_2,an,out}$$

Membrane hydration:

The mass flow of vapor across the membrane $W_{v,mbr}$ is calculated using mass transport principles and membrane properties [36], [37].

$$W_{v,mbr} = M_v A_{fc} n\left(\frac{n_d I_{st}}{F} - D_w \frac{c_{v,ca} - c_{v,an}}{t_m}\right) \quad (5)$$

Voltage:

The FC voltage is calculated by subtracting the fuel cell losses or overvoltages from the fuel cell open circuit voltage, E, and is given by

$$v_{fc} = E - v_{act} - v_{ohm} - v_{conc} \quad (6)$$

Where

$$E = \frac{1}{2F}\left\{\Delta G + \Delta S(T - T_r) + RT\left(\ln(P_{H_2}) + \frac{1}{2}\ln(P_{O_2})\right)\right\}$$

$$v_{act} = v_0 + v_a\left(1 - e^{-c_1 i}\right)$$

$$v_{ohm} = i \cdot R_{ohm} = i\frac{t_m}{\sigma_m}$$

$$v_{conc} = i\left(c_2 \frac{i}{i_{max}}\right)^{c_3}$$

where membrane conductivity, σ_m, is a function of membrane humidity and fuel cell temperature T_{fc}:

$$\sigma_m = b_1 \exp\left(b_2\left(\frac{1}{T_0} - \frac{1}{T_{fc}}\right)\right)$$

parameters b_1 and b_2 heavily depends on membrane humidity, and are can be modified accordingly depends on specific applications.

3. Control strategies

In this section we described the design of controllers for fuel cell systems, in particular for fuel cell vehicle applications. There are three major control subsystem loops in the fuel cell system that regulates the air/fuel supply, the water management and the heat management [24], [44].

It is important to realize that no matter how complex the dynamic model, it never fully describes the fuel cell, and therefore system response, precisely in terms of system parameters. A fuel cell's inherent nonlinearities and time varying characteristics inevitably pose difficult problems for system identification and control, estimating either linear or more complex nonlinear models, which are only valid over an operating range. Such models are typically used to investigate system stability, sensitivity, observability and controllability to allow controller designs. However if investigated in real time, time-dependent parameter can be more readily tracked to minimize the system error. Therefore, in this case, to achieve the control objectives, it is better to use adaptive controllers which can serve as a feedback law to achieve the control objectives with the time varying nonlinear properties, as well as external disturbances which is represented by factors such as loading of an electric vehicle motor. Authors in [45] implemented an observer based feedback controller to protect the stack from the previously described oxygen starvation effects during loading. This uses a linear quadratic technique based on a linearised state-space model. Other authors using such techniques to accommodate the nonlinearities include [46] who integrated a linearised PEM fuel cell model into a power plant, and simulated a load frequency controller. An optimal controller then accommodates the control application based on the disturbances. A fuzzy controller for the humidity management subsystem which adjusts the fan according to the requirement was deployed in [47]. Meyer and Revenkar [48] highlights the requirement for further research into system models and controllers for multiple input, multiple output (MIMO) systems covering the whole operating range, a problem which is in part addressed within the reminder of this section.

The performance of fuel cells and the vehicle applications they are embedded into depends on a delicate balance of the correct temperature, humidity, reactant pressure, purity and flow rate. A Fuzzy control strategy is developed in [40] to improve the performance of the fuel cell system which is built based on the system error and its rate of variations. In [42] an online self tuning PID controller for PEM fuel cell system is described. By contrast in [43] a fuzzy set-point weighted PID controller is designed to improve the performance of the fuel cell system. For the weighted PID control, the set point weight value is continuously updated using a fuzzy inference system. The underlying idea of the controller design is to use a fuzzy based system to support the operation of a PID controller and it's claimed that the system dynamic performance with the fuzzy set-point weighted controller is better than the PID and fuzzy controller. However, from the literature it can be seen that the fuzzy controller design is widely used for performance enhancement in fuel cell systems for example, [8], [40], [41], [47], [49]. In [6] for the SOFC, a control strategy using fuzzy logic controller for the voltage control of the fuel cell at the output of the DC/DC converter is presented. A fuzzy logic based controller is designed, because the cost of the controller design and implementation is low and it has a high performance/cost ratio. The fuzzy logic controller (FLC) is used to overcome inherent disadvantages such as uncontrollable large overshoot and large current ripple. FLC does not need an accurate mathematical plant model [49]. Therefore, it is applicable to a process where the plant model is unknown or ill defined. Fuzzy control is also nonlinear and adaptive in nature and offers robust performance under parameter variations and load disturbances [50], [51].

It is clear that good performance of the fuel cell system is closely related to the kind of control is used, so a study of different control alternative is justified [52], [53]. A fuel cell system integrates many components into a power system, which includes DC/DC converters, batteries, and ultracapacitors in the system and, incase the fuel cell is not fed directly with hydrogen; a reformer must also be used. Therefore, there are many control loop schemes, the number of which depends on the configuration of the system. The lower control level takes care of the main control loops inside the fuel cell, which are basically fuel/air feed, humidity, pressure and temperature. The upper level is in-charge of the whole system, integrating the electrical conditioning, storage and reformer (if required). Many control strategies have been proposed in the literature, ranging from feed-forward control [53], Linear quartertic regulator [54], Neural Networks [55], [56] and Model Predictive Control [57], [58]. Authors in [52] explore the possibility of making use of the known inherent fault-tolerant capabilities of Model Predictive Control. Many of these papers focus on the low level control of the fuel cell to fulfill at-least one of the three main objectives such as maximum efficiency, voltage control and starvation prevention. Furthermore, in [59] a novel scheme to control fuel cell terminal voltage is presented. This control scheme prevents any substantial drop in fuel cell terminal voltage during sudden load change or disturbance. It also discusses topology and control strategy for interfacing an energy storage device like an ultra-capacitor to a fuel cell. This improves the dynamic response of the fuel cell to sudden changes in load, hence enabling a fuel cell powered automobile to accelerate suddenly and also respond in a better manner to sudden change in load conditions. It's argued that, using a bi-directional converter as an interface between fuel cell and ultra-capacitor gives better control over fuel cell voltage during transients. However, these designs are still at the theoretical stage and without real time testing the validity of these control strategy for real fuel cell system applications are still under investigation. Several analysis tools developed to investigate these type of controllers, for example in [60] are MATLAB based analysis tools for real-time control systems. A few other papers [61], [62] present the experimental study of control strategies, which include parameter estimation and diagnostics. Control and stability enhancement of fuel cells systems [14], [63] is considered by several researchers; in particular investigation of the effect of fuel cells on stability of distribution system is a major research topic in these studies. Reference in [64] describe a simplified fuel-cell model and explain the effect of the mix between fuel cell and gas turbine generation on system stability.

Fuel cells are presently being considered for use in automotive systems to replace the internal combustion engines in buses, vans, and ultimately passenger cars [65]. An auxiliary energy storage device is required for start-up and for storing the energy captured through regenerative braking in electric vehicle applications, because the present fuel cell technology lacks energy storage capability [66], [67]. A storage unit such as a battery or ultra-capacitors can be integrated into the system in order to overcome the problem. In addition a DC/DC converter is interconnected to the system to boost the fuel cell voltage to the required voltage for the motor controller. The major challenge in designing automotive fuel cell power systems is converting the electrical output from the fuel cell into a usable power and designing a suitable DC/DC converter. Many DC/DC converter topologies with their control strategies are presented in the literature see for example [66], [67]. An overall fuel cell system configuration within an automotive application is shown in Fig.3.

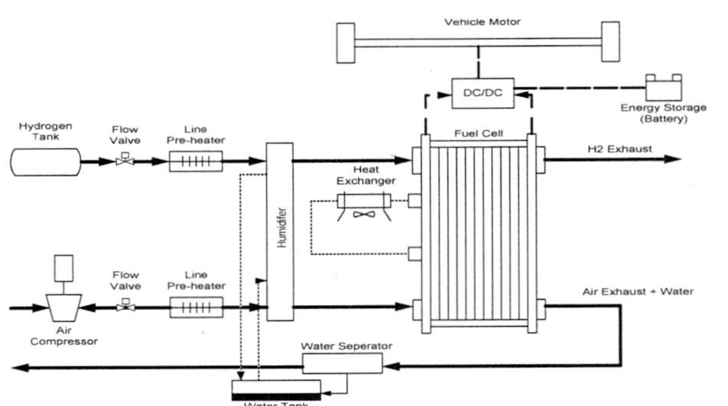

Fig.3 An overall configuration of fuel cell vehicle

The extensive studies in the controller design methods evidence that the fuel cell system control is a very active research area. The research in this area is mainly motivated by the recognition that the current control methods can not fully meet the desired design requirements on fuel cell system performance, stability, and robustness etc. Any controller designs which gives a satisfactory performance on fuel cell system behavior is worthy to be considered for implementation.

4. Conclusions

In this paper, a review and analysis of a fuel cell system modeling and controller design has been presented. In particular, the review is mainly focused on the PEM fuel cell hybrid electric vehicles system. Its configurations, topologies and control strategies are also discussed.

References

[1] M.Parten, T.Maxwell, "Development of a PEM Fuel Cell Vehicle", In the Proc. of IEEE Vehicular Technology Conference '01, Atlantic City, NJ 2001

[2] R.Lasseter, "Dynamic Models for Micro-Turbines and Fuel Cells", In the Proc of IEEE Power Eng. Soc., vol.2, pp761-766, 2001.

[3] W.Friede, S.Rael, B.Davat, "Mathematical Model and Characterization of the Transient Behaviour of a PEM Fuel Cell", IEEE Trans on Power Electronics, vol.19(5), pp1234-1241, 2004

[4] J.Garnier., *et al,* "Dynamic PEM Fuel Cell Modelling for Automotive Applications", In the Proc. of IEEE Vehicular Technology Conference '03, Orlando, Florida, 2003

[5] K.Sedghisigarchi, A.Feliachi, "Dynamic and Transient Analysis of Power Distribution Systems With Fuel Cells-Part I: Fuel-Cell Dynamic Model", IEEE Tans on Energy Conversion, Vol 19, No2, June 2004

[6] A.Sakhare, A.Davari, A.Feliachi, "Control of Stand Alone Oxide Fuel Cell using Fuzzy Logic", In the Proc of the 35th Southeastern Sym on Sys Theory' 2003, pp473-476, March 2003.

[7] W.Turner, M.Parten, D.Vines, J.Jones, T.Maxwell, "Modelling a PEMFuel Cell for use in a Hybird Electric Vehicle", In the Proc. of IEEE Vehicular Technology Conference '99, Houston, Texas, 1999.

[8] Y.Kim, S.Kim, "Electrical modelling and fuzzy logic control of a fuel cell generation system", IEEE Trans Energy Conversion, vol 14, pp239-244, June 1999

[9] W.He, "Power-generation characteristics of molten carbonate fuel-cell stacks by using computational fluid dynamics technique, Part1: Modelling equations and program implementation,", int J. Energy Res., vol 23, pp 1345-1357, 1999

[10] L.Chiu, B.Diong, R.Gemmen, "An Improved Small Signal Model of the Dynamic Behaviour of PEM Fuel Cells", IEEE Trans on Industry Applications, vol 40, No 4, July/August 2004

[11] N.Brooks, T.Baldwin, T.Brinson, J.Ordonez, C.Luongo, "Analysis of Fuel Cell Based Power Systems Using EMTDC Electrical Power Simulator", In the Proc of the 36th Southeastern Sym on Sys Theory' 2004, pp270-274, 2004.

[12] R.Gemmen, "Analysis for the effect of inverter ripple current on fuel cell operating condition", J.Fluids Eng., vol 125, No3, pp576-585, 2003

[13] R.Gemmen.,E.Liese.,J.G.Rivera.,J.Brouwer.,"Development of dynamic modelling toolts for solid oxide and molten carbonate hybrid fuel cell gas turbine systems", In the Proc of ASME/IGTI Turbo Expo, 2000

[14] K.Sedghisigarchi, A.Feliachi", "Control of grid-connected fuel cell power plant for transient stability enhancement", In the Proc of IEEE Power Eng Soc, 2002, pp383-388

[15] E.Liese, R.Gemmen, "Dynamic modelling results of a 1MW molten carbonate fuel cell/gas turbine power system", In Proc ASME/IGTI Turbo Expo, 2002, pp341-349

[16] S.Varigonda, J.Eborn, S.Bortoff, "Multivariable control design for the water gas shift reactor in a fuel processor", Proc of the ACC, Boston, June/July 2004

[17] H.Gorgun, M.Arcak, S.Varigonda, S.Bortoff, "Nonliner observer design for fuel processing reactors in fuel cell power system", Proc of the ACC, Boston, June/July 2004

[18] S.Caux, J.Lachaize, M.Fadel., L.Nicod., "PEMFC Air loop model and control", In the Proc of the IEEE Conf on Vehicle Power and Propulsion, September 2005

[19] M.Kim, H.Peng, C.C.Lin., E.Stamos., D.Tran., "Testing, Modelling and Control of a Fuel Cell Hybird Vehicle"ACC'05, Portland, USA, pp3859-3864, 2005

[20] F.Barbir, B.Balasubramanian, J.Neutzle, "Trade-off design analysis of operating pressure and temperature in PEM fuel cell systems", Proc of the ASME Adv Energy sys Division, vol 39, pp 305-315, 1999

[21] D.D.Boettner., G.Paganelli., Y.G.Guezennec., G.Rizzoni., M.J.Moran "Component power sizing and limits of operation for proton exchange membrane (PEM) fuel cell/battery hybrid automotive applications", Proc of 2001 ASME Int Mech Eng Congress and Exposition, 2001.

[22] D.Friedman.,A.Egghert.,P.Badrinarayanan., J.Cunningham., "Balancing stack, air supply and water/thermal management demands for an indirect methanol PEM fuel cell system", In the Proc of the Society of Automotive Engineers, World Congress 2001, SAE paper No 2001-01-0535.

[23] J.Pukrushpan, H.Peng, A.Stefanopouloa, "Simulation and Analysis of Transient Fuel Cell System Performance Based on a Dynamic Reactant Flow Model", Proc of the IMECE'02, Louisiana, USA, November 2002

[24] J.Pukrushpan, A.Stefanopoulou, H.Peng, "Modelling and Control for PEM Fuel Cell Stack System", Proc of the American Control Conference, Anchorage, AK May 2002

[25] S.Pischinger., W.Bornscheuer., H.Kindl., J.Ogrzewalla., A.Wiartalla., "Integrated Air Supply and Humidification Concepts for Fuel Cell Systems", In the Proc of the Society of Automotive Engineers, World Congress 2001, SAE paper No 2001-01-0233.

[26] R.F.Mann, J.C.Amphlett., M.A.I.Hooper., H.M.Jensen., B.A.Peppley., P.R.Roberge.,"Development and application of a generalised steady state electrochemical model for a PEM fuel cell", J.of Power Sources, Vol 86, pp 173-180, 2000

[27] M.El-Sharkh., A.Rahman., M.S.Alam., A.A.Sakla., P.C.Byrne., T.Thomas., "Analysis of Active and Reactive Power Control of a Stand-Alone PEM Fuel Cell Power Plant" IEEE Trans on Power Systems, Vol 19(4),Nov 2004

[28] A.Miotti, A.Domenico, Y.Guezennec, S.Rajagopalan, "Control-Oriented Model for an Automotive PEM Fuel Cell System with imbedded 1+1D membrance water Transport", In the Proc of IEEE Conf on Vehicle Power and Propulsion, September 2005

[29] Q.Yu, A.Srivastava, S.Choe, W.Gao, "Improved modelling and control of a PEM Fuel cell power system for vehicles", In the Proc of IEEE Southeast Conf 2006, pp331-336

[30] M.DeFrancesco, E.Arato, "Start-up analysis for automotive PEM Fuel cell Systems", J.of Power Sources, Vol 108, pp 41-52, 2002

[31] B.Bouneb, A.Cruden, D.Grant, J.Mcdonald, "Design and Implementation of a DSP-based Fuel cell/Battery hybrid power source", In the Proc of 2nd Conf on Power Electronics, Machines and Drives, 2004, pp408-413

[32] J.Pukrushpan,A.Stefanopoulou.,S.Varigonda, .M.Pedersen., S.Ghosh., H.Peng., "Control of Natural Gas Catalytic partial Oxidation for Hydrogen Generation in Fuel Cell Applications",IEEE Trans on Control Systems Technology, vol 13(1), January 2005

[33] V.Tsourapas, A.Stefanopoulou, J.Sun, "Dynamics, Optimization and Control of a Fuel Cell Based Combined Heat Power (CHP) System for Shipboard Applications", ACC, Portland, OR, USA, June 2005

[34] J.Pukrushpan, H.Peng, A.Stefanopoulou, "Control-oriented modelling and analysis for automotive fuel cell systems", J. of Dynamic systems, measurement and control, vol 126, no.1, pp 14-25, March 2004

[35] J.Golbert, D.Lewin, "Model-Based Control of Fuel Cells: Regulatory Control", J. of Power Sources, October 2003

[36] Thanapalan.K.,Williams.J.G.,Liu.G.P.,Rees.D., Modelling of a PEM Fuel Cell System" *In the proc. of IFAC World Congress'08, Seoul, Korea,* 2008

[37] Thanapalan.K., Wang.B., Williams.J.G., Liu.G.P., Rees.D "Modelling, Parameter Estimation and Validation of a 300W PEM fuel cell system", *In the proc. of UKACC'08, Int. conf. on Control 2008, Manchester, UK,* 2008

[38] Ding.X.R., Liu.G.P., Wang.X.C, Tan.M, "Modelling and Identification of a PEM fuel cell humidification system", *In the proc. of ICC, Int. conf. on Control 2006, UK,* 2006

[39] Xiong.Y, Liu.G.P., Wang.X.C., Tan.M "Air supply control of a small PEM fuel cell" *In the proc. of UKACC, Int. conf. on Control 2006, Glasgow, UK,* 2006.

[40] Williams.J.G., Liu.G.P., Chai.S, Rees.D., "Intelligent control for improvements in PEM fuel cell flow performance", *Int. Journal of Automation and Computing,* 2008, 05(2), pp145-151

[41] Tang.S., and Liu.G.P., "Real-time simplified variable universe fuzzy control of PEM fuel cell flow systems", *In the proc. of the European Control Conference 2007,* Kos, Greece, 2007

[42] Williams.J.G., Liu.G.P., Thanapalan.K., Rees.D., "Design and Implementation of On-line Self-Tuning Control for PEM Fuel Cells" *In the proc. of EVS-23; Sustainability-The future of Transportation, California, USA,* 2007, pp 359-375

[43] Thanapalan.K., Liu.G.P., Williams.J.G., Rees.D "Fuzzy Controller Development for A PEM Fuel Cell System", *In the proc. of UKACC'08, Manchester, UK,* 2008

[44] J.Adams, W.Yang, et al, "The development of Fords P2000 fuel cell vehicle", In the Proc of the Society of Automotive Engineers, World Congress 2000, SAE paper No 2000-01-1061.

[45] J.T.Pukrushpan., A.G.Stefannopoulou., H.Peng., "Control of Fuel Cell Breathing: Initial Results on the Oxygen Starvation Problem", IEEE Control Systems Magazine, 2004, pp30-46.

[46] A.Paradkar., A.Davari., A.Feliachi, T.Biswas, "Integration of Fuel Cell into the Power System Using An Optimal Controller Based on Disturbance Accommodation Control Theory", J.of Power Sources, 2004, vol 128(2), pp218-230.

[47] J.O.Schumacher., P.Gemmar., M.Denne., M.Zedda., M.Strueder., " Control of Miniature Proton Exchange Membrane Fuel Cell Based on Fuzzy Logic", J.of Power Sources, 2004, vol 129(2), pp143-151.

[48] R.T.Meyer, S.Revenkar, "A survey of PEM fuel cell system control models and control developments" *In the proc. of FuelCell'06., The 4th Int. conf. on fuel cell Science, Eng and Technology, Irvine, California, June, 2006.*

[49] M.Zhijun, Z.Xinjian, C.Guangyi, "Design and Simulation of fuzzy controller for PEMFCs" In the Proc of IEEE Int Conf on Industrial Technology, 2005, pp220-224

[50] F.Jurado, M.Valverde, "Fuzzy logic inverter flux control of fuel cell plants in distributed generation", 11th Int Conf on Harmonics and Quality of Power, 2004

[51] F.Jurado, M.Valverde, "Novel Inverter Flux Control ForFuel Cell Using Fuzzy Logic", In the Proc of 4th Int Conf on Power Electronic and Motion Control, 2004, vol.2, pp714-718

[52] A.Rosich.,V.Puig.,J.Quevedo.,"Fault-Tolerant Constrained MPC of PEM Fuel Cells", IAR Annual Meeting, 2006, pp1-6

[53] J.T.Pukrushpan., A.G.Stefannopoulou., H.Peng., "Control of Fuel Cell Power Systems: Principles, Modelling, and Analysis and Feedback Design", series Advances in Industrial Control, 2004, Springer.

[54] P.Rodatz., G.Paganelli., A.Sciarretta., L.Guzzella., "Optimal power management of an experimental fuel cell/supercapacitor-powered hybrid vehicle", Control Engineering Practice, 2005, vol.13, pp41-53.

[55] P.E.M. Almeida., and M.G.Simoes., "Neural optimal control of PEM fuel cells with parametric CMAC networks", IEEE Trans. on Industry Applications, 2005, vol 41(1), pp 237-245

[56] M.Y.El-Sharkh., A.Rahman., M.S.Alam., "Neural networks-based control of active and reactive power of a stand-alone PEM fuel cell plant", J.of Power Sources, 135, pp88-94, 2004

[57] C.Bordons., A.Arce., A.J. Del Real., "Constrained Predictive Control Strategies for PEM fuel cells", In the Proc. of ACC'06, ACC, Minnesota, USA, 2006.

[58] A.Vahidi., A.G.Stefannopoulou., H.Peng., " Current Management in a Hybrid Fuel Cell Power System: A Model Predictive Control Approach", IEEE Trans. on Control Systems Technology, vol.14(6), 2006.

[59] A.Drolia, P.Jose, N.Mohan "An Approach to Connect Ultracapacitor to Fuel Cell Powered Electric Vehicle and Emulating Fuel Cell Electrical Characteristics using Switched Mode Converter", In the Proc of the 29th IEEE Int Conf of the Industrial Electronics Society 2003,vol.1, pp897-901

[60] A.Cervin, D.Henriksson, B.Lincoln, K.Arzen, "Jitterbug and True time : Analysis Tools for Real-Time Control Systems", 2nd Workshop on Real-Time Tools, Copenhagen, Denmark, August 2002

[61] H.Lee, K.Jeong, B.Oh, "An Experimental study of Controlling Strategies and drive forces for hydrogen fuel cell hybrid vehicles", Int. Jof Hydrogen Energy, Vol 28, pp215-222, 2003

[62] A.Forrai, H.Funato, Y.Yanagita, Y.Kato, "Fuel-cell Parameter Estimation and Diagnostics", IEEE Trans on Energy Conversion, vol 20(3), September 2005

[63] K.Sedghisigarchi, A.Feliachi, "Dynamic and Transient Analysis of Power Distribution Systems with Fuel Cells-Part II: Control and Stability Enhancement", *IEEE Trans on Energy Conversion, Vol 19, No2, June 2004*

[64] C.Hatziadoniu, A.Lobo, et al "A simplified dynamic model of grid-connected fuel-cell generators", IEEE Trans Power Delivery, vol 17, pp467-473, April 2002

[65] U.G.Bossel, "Solid oxide fuel cells for transportation," In the Proc of the 3rd European SOFC forum, France, 1998.

[66] G.Su, F.Peng, D.Adams, "Experimental Evaluation of a Soft-Switching DC/DC Converter for Fuel Cell Vehicle Applications", In the Proc of the IEEE Int Conf on Power Electronics in Transportation, 2002, pp39-44

[67] M. Marei, S.Lambert, R.Pick, M.Salama, "DC/DC Converters for Fuel Cell Powered Hybird Electric Vehicle" In the Proc of IEEE Int Conf on Vehicle Power and Propulsion, September 2005, pp126-129

Innovative Production Machines and Systems
D.T. Pham, E.E. Eldukhri and A.J. Soroka (Eds.)

An Integrated Approach to planning for autonomous grit-blasting robot in complex bridge environments

W.K. To, G. Paul, N.M. Kwok, D.K. Liu

ARC Centre of Excellence for Autonomous Systems (CAS)
University of Technology, Sydney
NSW 2007, Australia

Abstract

This paper describes an integrated approach to robot manipulator path and motion planning in complex bridge environments. It incorporates grit-blasting specific considerations including blasting effect, coverage, path length and robot arm joint movement. A genetic algorithm is implemented for path planning with the use of environment data to increase planning efficiency. A customized gradient based method is applied in selecting collision free joint configurations for the identified path. A grit-blast coverage model is also developed for discrete non-planar 3D coverage determination to verify the performance of the planned path and motion.

Keywords: Grit-blasting Coverage, Manipulator Path/ motion Planning, Genetic Algorithm

1. Introduction

The removal of paint and rust via grit-blasting, followed by the reapplication of a protective coating is an essential part of bridge maintenance, which significantly reduces the likelihood of structurally damaging rust. Conventional bridge maintenance process involves the temporary establishment of a blast chamber with human workers manually performing the rust and paint removal operations. However, due to newly introduced Occupational Health & Safety regulations, it is no longer desirable for humans to perform this task. Hazards arise from commonly used lead based paints. With the advancement of robotic technologies, there is an increased interest in the application of autonomous robotic systems to perform a significant portion of the repetitive and dangerous bridge maintenance task.

The application of robotics in steel bridge maintenance follows from [1] where it is conducted by means of a tele-operated arm mounted on a truck atop the bridge. This tele-operated robotic system could achieve partial coverage in certain bridge environments through skilled operator control. Application of autonomous robotic systems for bridge maintenance demands an intelligent integration of (1) methods to obtain detailed bridge maintenance environment and steelwork surface maps [2][3], (2) fusion of material type classification [4][5], (3) robot path and motion planning, and (4) collision avoidance. The goal of robot arm path and motion planning is to produce a collision-free path and motion solution while aiming for maximum productivity and coverage. In complex bridge maintenance environments, the risk of robot-environment collision is high [6] and there is the need to cover specific surfaces with the predetermined surface finish. Additionally the grit-blasting apparatus attached to the robot manipulator is cumbersome and causes difficulties in dexterous movement, which results in further performance limitations. Consequently the robot manipulator joint movement must be considered and minimised.

Genetic algorithms (GA) have been applied for robot path and trajectory planning [7-9] in Cartesian and joint spaces for discretised way point movements. In bridge maintenance, tracking of blasting coverage based on a planed robot path/ trajectory is also required. Coverage tracking is demonstrated in robotic spray painting systems in car manufacturing [10-11] based on paint disposition model.

The contribution of this paper is an integrated approach to grit-blasting manipulator path/motion planning. This approach incorporates essential application specific considerations including blasting effect, coverage, path length and robot arm joint movement. It will be verified through a simulated coverage modelling and evaluation followed by experimental implementation. The breakdown of this paper is as follows: Section 2 provides grit-blast problem formulation, Section 3 presents the proposed planning approach and its architecture, and the coverage modelling; and Section 4 presents results from experiments conducted using the robotic manipulator in a to-scale constructed bridge environment model.

2. Application Specific Problem Considerations

For the purpose of the bridge maintenance grit-blasting application an autonomous system is deployed to work underneath bridges. A robot arm, specifically the 6DOF Denso VM-6083D-W is mounted on a moving platform and affixed with a grit-blasting nozzle. The laboratory configuration is shown in Figure 1a and the simulated model of the setup is depicted in Figure 1b.

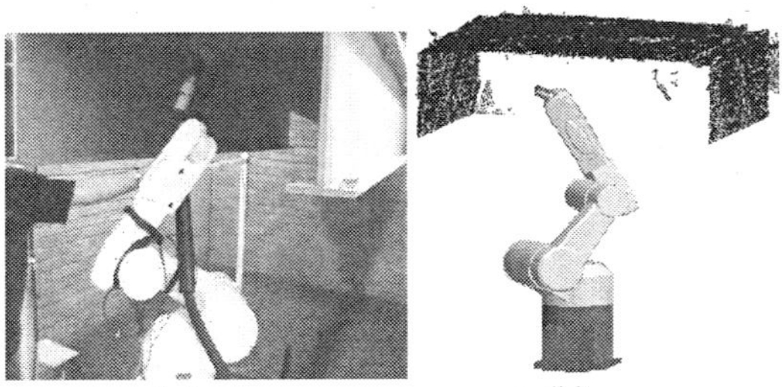

(a) (b)

Figure 1. a) To-scale model b) Simulation environment model

The autonomous robotic system operates in a 3D complex environment with non-repetitive grit-blasting tasks. In order to successfully perform the required maintenance tasks adaptively, the following grit-blasting considerations have been identified as significant for the bridge maintenance planner.

(1) Cumulative Robot Joint Movement – drastic successive changes in robot joint movement result in stress on the hose management: attributed to the rigidity of the pressurised hose resisting retraction and extension;
(2) Blasting path length – minimal blasting path length will result in maximum productivity and reduction in grit wastage;
(3) Coverage – due to the complexity of bridge maintenance environments, robot movement is significantly restricted. Coverage of blast targets must be quantified and maximised;
(4) Blast effect –blast stream length (the distance between the nozzle and the blasting surface), blast angle and blast velocity have significant effect on surface finish;
(5) Safety – the manipulator affixed with the maintenance apparatus must move safely and avoid collisions in the complex environment.

3. Efficient Grit-blast Planning and coverage determiniation

3.1. Path and Motion Planning Approach Overview

In order to plan the manipulator path and motion for the grit-blasting task, a number of successive functional procedures are performed within the path and motion planning routine. The path and motion planning routine can be described by the following procedural flow. The planner initially acquires environment data and generates appropriate blasting target points that purposefully overlap and cover a specific blasting target area. The optimal sequence of movement between target points is identified and appropriate joint configurations selected for the sequence of target points movement. The coverage achieved by the target point movement sequence is determined and modelled. As shown in Figure 2 these procedures are: 1) blast target point creation, 2) optimisation of initial path plan, 3) generation of collision free joint configurations, 4) coverage modelling of the output path solution.

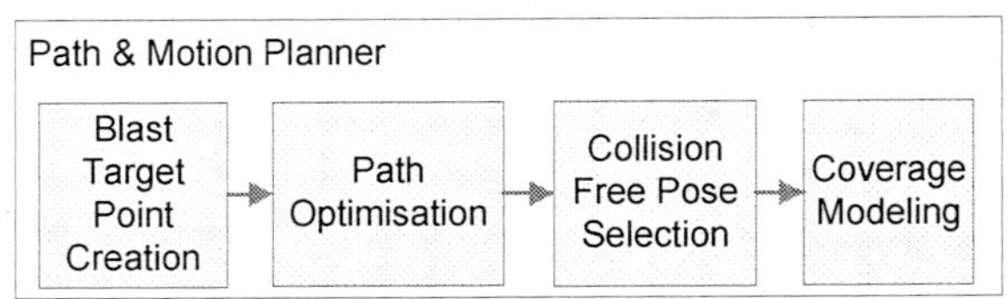

Figure 2. Functional Proceedures for Blasting

3.2. Blast Target Point Creation

The robot movement instruction is defined as a sequence of discretised movements through target points. The existing environment data generated by [3] is utilised to create the discrete blast target [2] points with fixed blast coverage radius. This method used principal component analysis (PCA) on the environment point cloud to achieve grit-blasting complexity reduction by simplifying the number of parameters required to represent a surface. The result of this algorithm is the creation of an order of target points that represents the environment. Blast sensitive area avoidance is integrated into the target point generation process; target points can only be created where the blast radius does not overlap a blast sensitive area.

Blast target points are defined by Cartesian coordinate and plane equation for the representation of position and orientation employed in joint configuration determination process. The PCA creates target points in a process where predecessor target point is adjacent to the next created point. The targets points are constructed such that the travel through each individual target centre by the blast stream will result in the effective baseline coverage of the specified target area. The initial movement sequence of target points is ordered according to the PCA target point creation order as it provides the best performance prior to further optimisation. This sequence obtained from the map building stage and the environmental knowledge can be described as a progressive sweep of the specified target surface. It will then be used in the path/motion planning stage.

3.3. *Genetic Algorithm Based Target Point Sequencing*

A simple GA is applied for the generation of target point sequence. The GA optimisation objectives include the travel distance and joint movement. A variation of the linear aggregating function [12] is chosen for the lower computational effort compared with contemporary multi-objective approaches [13-14]. Environmental knowledge obtained in the map building stage shall be utilised by the GA in the form of initial target point sequence created as an artefact of the target point generation process. Path planning steps are given in Figure 3.

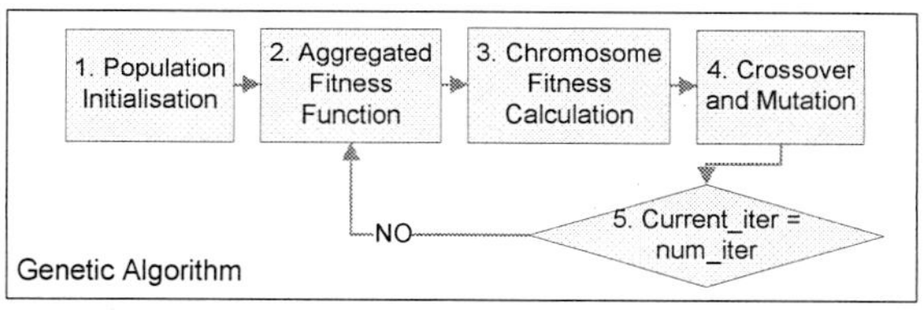

Figure 3. Optimisation by GA with Normalised Parameters

(1) Population Initialisation: The algorithm is initialised with a fixed population size of N chromosomes. In the applied context of path plan optimisation, a chromosome C is defined as a fixed size array containing target point indexes Ci = {Cj}, Cj ∈ [1,M], i = 1,...N, j=1,...M, where M is the max number of target points, and a target point is denoted as $P_j = [X_j, Y_j, Z_j]^T$ The purpose of the initial chromosome is to provide the first generation. Given the constraint on the number of generations before the algorithm terminates, it is logical to direct the search towards a local or possible global minima.

(2) Fitness Function: The fitness function g^i of the ith chromosome from point j to point j+1 is defined as the weighted sum of the travel distance/path length and the joint movement:

$$g^i_{j,j+1} = \alpha d^i_{j,j+1} + \beta m^i_{j,j+1} \quad (1)$$

where α and β are weighting coefficients to be determined, applied to normalise the travel distance and joint movement. The travel distance d is defined as the Euclidean distance travelled from target point j to point j+1:

$$d^i_{j,j+1} = \left\| P_{j+1} - P_j \right\| \quad (2)$$

The joint movement, m, is defined as the weighted sum of angles of the robot's six joints q_1 to q_6 required to move from point j to point j+1. w_1 to w_6 are the weighting coefficients assumed to have the same value.

$$m^i_{j,j+1} = \sum_{l=1}^{6} w_l \left| (q_{l,j+1} - q_{l,j}) \right| \quad (3)$$

(3) Chromosome Fitness Calculation: The fitness (f^i) value of each chromosome is the summation of the aggregated fitness values calculated from Equation (1) for the movement along the chromosome target point index sequence.

$$f^i = \sum_{j=1}^{m-1} g^i_{j,j+1} \quad (4)$$

(4) Crossover and Mutation: Successive populations are generated using genetic operators: reproduction, crossover and mutation.

(5) Termination condition checking: If the generation limit has not been reached, then proceed to step 2. Else terminate the algorithm.

Weighting Factor Value Calculation: the weighting values are calculated by evaluating the fitness distribution in population. The weighting factors α and β are defined as a ratio value calculated from the distribution means μ_T and μ_J

$$\mu_T = \frac{1}{N} \sum_{i=1}^{N} \sum_{j=1}^{M-1} d^i_{j,j+1} \quad (5)$$

$$\mu_J = \frac{1}{N} \sum_{i=1}^{N} \sum_{j=1}^{M-1} m^i_{j,j+1} \quad (6)$$

$$\alpha = \frac{\mu_T}{\mu_T + \mu_J} \quad (7)$$

$$\beta = \frac{\mu_J}{\mu_T + \mu_J} \quad (8)$$

3.4. Collision Free Pose Selection

A gradient based pose selection using least squared optimisation modified from [15], has been used to provide rapid joint configuration for grit blasting application. The objective of pose selection is to determine a valid joint configuration solution that satisfies the specified pose selection criteria including distance away from the desired coordinate, difference in orientation from current manipulator configuration and joint limitations. Additionally the grit-blast specific factors, i.e blast stream length and blast angle, are also considered in the pose selection criteria to satisfy the blast effect. Termination of the pose selection occurs when the cost function gradient is no longer improving. Implicitly the safety of the autonomous robotic system is managed by the pose selection with environment obstacle collision avoidance, using a force field method based upon [16], implemented as a validation criterion.

The current joint configuration of the robot manipulator is used as the starting guess to direct the search towards a joint configuration solution in the proximity of the current configuration. Therefore the resulting joint configuration solution is likely to be the least joint variation to complete the manipulator configuration transition. Pose selection before the GA-based path planning is performed on the initial path plan to provide joint angle values (q_1 to q_6) for joint movement. After an optimal path (the optimal sequence of target points on a blast surface) is found, pose selection is again undertaken to yield a set of joint configurations for optimised target point to target point movement.

3.5. Coverage Calculation

The modelling of coverage and wasted movements in the form of overlap is performed on the planned optimal path obtained from the GA to validate the quality of the planned path. The proposed technique provides a scalable approximation approach to coverage calculation through the representation of coverage as discrete points in 3D space. The accuracy of coverage modelling is scaleable through the variation of the proximity between discrete vertices to attain better resolution in the coverage map representation. This technique is capable of representing coverage and overlap for non-planar surfaces as apparent for a complex bridge maintenance environment. To achieve non-planar coverage calculation the technique utilises the 3D point cloud map extracted from the environment data [1, 2]. The point cloud map contains vertices and faces with millimetre resolution, arranged in triangles to represent environmental surfaces.

The process of coverage calculation involves the determination of all the individual blast line coverage created as the result of the grit-blasting spot traverses from the current target point to the next consecutive target point in the sequence specified by the path. The overall surface coverage is the accumulated coverage value from all blast lines created. The calculation of an individual blast line created from an arbitrary current target point P_j to next target point P_{j+1} is the inclusion of all 3D vertices that satisfy the two boundary conditions as shown in Figure 4.

(1) $d_{x \to line} < \sigma$ where $d_{x \to line}$ is the distance of the perpendicular line that joins the potential coverage map point P_x to the line $d_{i \to j}$ and σ the blast line width.

(2) $(d_{x \to i} + d_{x \to j}) < (d_{i \to j} + \sigma)$ is the boundary condition that describes the hemisphere area that extends past P_i and P_j

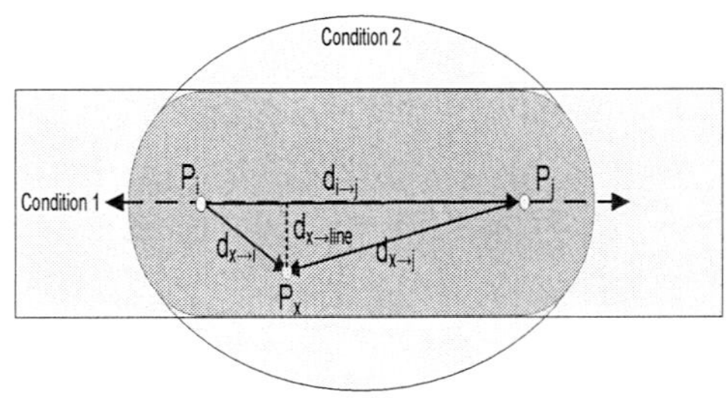

Figure 4. Blast line coverage calculation between two arbitary points

The map points that are identified as being traversed by the grit-blast stream due to the movement from P_i to P_j will be updated with new blast coverage effect values. The blast coverage effect value can be considered as the fourth dimension of each map point and is a function of the distance the map point is from the centre line of the blast stream. The blast coverage effect model is based upon the blast coverage achieved from a typical venturi grit-blasting nozzle [17]. The model proposes that the blast effect coverage values for map points within a defined distance from the centre line are assigned a fixed blast quality unit ω. For map points that fall within the blast quality roll-off range h, the blast coverage value is specified by the function r as shown below:

$$r(h) = \omega(1 - e^{h/k}) \tag{9}$$

where k is the exponential roll off factor and

$h_{min} < h < h_{max}$ describes the region of the blast roll-off subjective to factors including nozzle size/ type.

4. Results and Discussion

The results below present a comparison of the optimised path against the generated initial path solution for a selected test area from the built map of the 3D laboratory environment.

4.1. Target Point Sequencing and path planning

In this simulation, a target surface is selected manually from the built environment map. The initially obtained path, i.e. target point sequence, and the optimal path found by the proposed integrated approach are shown in Figure 5a and Figure 5b, respectively. It can be seen that the initial path is progressively scanning along an axis, and the optimised path traverses the blast surface with a less systematic sequence of movements. Although the results appear counterintuitive, the constraints applied to the system results in optimal path pattern different to that of a human operator.

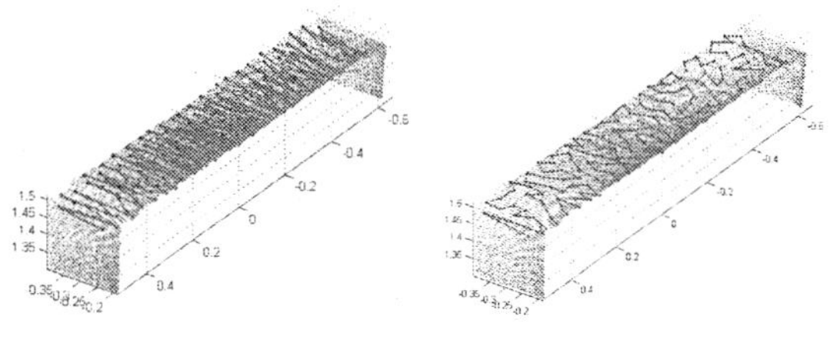

(a) (b)

Figure 5. Blasting Stream pattern on target surface a) Initial path b) optimal path obtained from the poposed approach

The path planning performance and quality are measured by the overall travel distance, overall joint movement and coverage. The initial path is used as the basis for comparison to verify the improvement instigated by the proposed path and motion optimisation approach. The comparison is performed using the same target area as shown in Figure 5

Figure 6a presents a comparison of the cumulative travel distance between the initial path and the optimal path. The cumulative travel distance is the summation of all target point to target point travel distance in the target point index sequence as specific to the paths. Numerically it can be seen from the figure that the optimal path provides less overall travel distance to traverse through the same set of target points than the initial path. The improvement in travel distance reduction was approximately 6.4 metres or a reduction of 54% of the initial path. This result demonstrates that the proposed approach has achieved overall travel distance improvement from the initial estimated path.

Figure 6b presents a comparison of the cumulative joint movement between the initial path and the optimal path. The cumulative joint movement is the summation of all point to point joint movement angles calculated from in the target point index sequence as specific to the paths. Similar to the results of the cumulative travel distance, the proposed approach has shown significant improvement over the initially estimated path. The results indicate a decrease of approximately 63 Radians or 80% of the initial path.

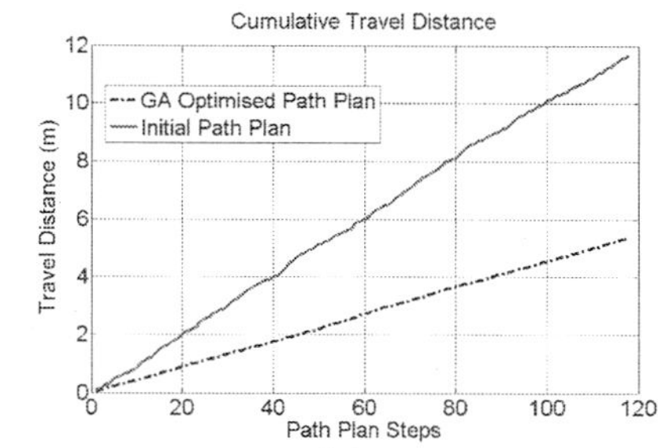

(a)

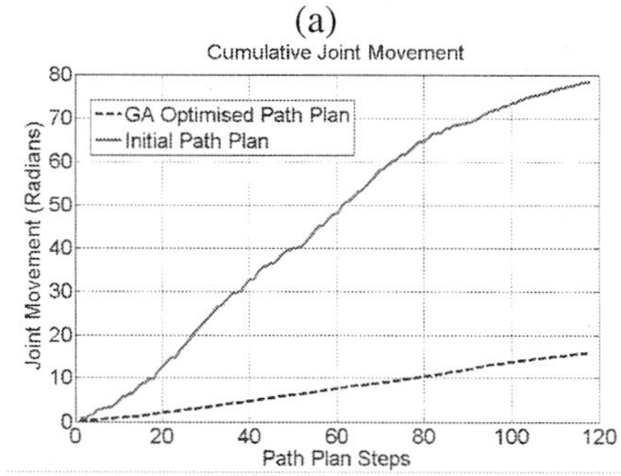

(b)

Figure 6. a) Cumulative Travel Distance b) Cumulative Joint Movement

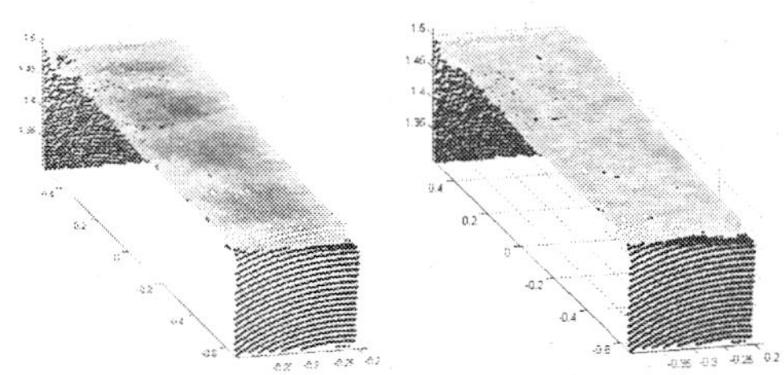

(a) (b)

Figure 7. Coverage Evaluation: a) Coverage from the initial Path b) Coverage from the Optimla Path

4.2. Coverage Evaluation

Figure 7 shows the coverage obtained from the initially estimated path and the optimal path. The coverage is visually represented in greyscale with shade intensity corresponding to repetition of coverage. Both paths are shown to have achieved 100% baseline coverage value for the target surface area due to the target point creation process producing an arrangement of target points that provide 100% coverage. However, the optimal path produces less excessive surface coverage (or over-blasting area), primarily attributed to the reduction of travel distance over the initially estimated path.

5. Discussion

Experimental results have demonstrated the improvement to path planning as a result of the application of the proposed approach. Future work will explore methodologies for effective segmentation of a blasting surface to maximise the effectiveness of the path and motion planning approach. Therefore the future work will include the implementation of Pareto framework for further refinement in the GA.

6. Conclusions

This paper has presented an integrated approach for robotic path planning. The algorithm and the coverage modelling are particularly useful in autonomous structural maintenance systems which apply environmental exploration knowledge for path planning. The proposed approach in this paper has been demonstrated that it can provide a solution that can be physically executed to perform grit-blasting task. It has the ability to generate blasting target points and search an optimal path for any target surface specified by an operator, and then select robot joint pose to implement the planned path safely.

ACKNOWLEDGMENT

This work is supported by the ARC Linkage Grant (ARC-LP0776312), by the ARC Centre of Excellence for Autonomous Systems (CAS), the Roads and Traffic Authority (RTA) and the University of Technology, Sydney.

REFERENCES

[1] Moon S., Bernold L.E., (1997) "Vision-Based Interactive Path Planning for Robotic Bridge Paint Removal" *Journal of Computing in Civil Engineering* 11(2): , pp 113-120.

[2] Paul G., Liu D.K. Kirchner N., (2007) "An Algorithm for Surface Growing from Laser Scan Generated Point Clouds", *Robotic Welding, Intelligence and Automation*, Springer-Verlag, pp 481-491

[3] Paul G., Liu D.K., Kirchner N., Webb S.,(2007) "Safe and Efficient Autonomous Exploration Technique for 3D Mapping of a Complex Bridge Maintenance Environment", 24th International Symposium on Automation and Robotics in Construction, Kochi, India, pp 99-104

[4] Kirchner N., Liu D.K., Taha T., Paul G., (2007) "Capacitive object ranging and material type classifying sensor", Int. Conf. on Intelligence Technologies: Intelligent Technology in Robotics and Automation, Sydney, pp 130-135,12-14

[5] Kirchner N., Taha T., Liu D.K., Paul G., (2007) "Simultaneous Material Type Classification and Mapping Data Aquisition Using a Laser Range Finder", Int. Conf. on Intelligence Technologies: Intelligent Technology in Robotics and Automation, Sydney, pp 124-129,12-14

[6] Xu J., Liu D.K., Fang G., (2006), "An efficient method for collision detection and distance queries in a robotic bridge maintenance system", *Robotic Welding, Intelligence and Automation*, Springer-Verlag, pp 71-82

[7] Monteiro D. C., Madrid M. K., (1999), "Planning of robot trajectories with genetic alogrithms", ", Proceedings of the First Workshop on Robot Motion and Control, 28-29 June, 1999, pp 223-228

[8] Liao X., Wang G. G., (2003), "Evolutionary path planning for robot assisted part handling in sheet metal bending", Robotics and Computer Integrated Manufacturing, pp 425-430

[9] Mahjoubi H., Bahrami F., Lucas C., (2006), "Path Planning in an Environment with Static and Dynamic Obstacles Using Genetic Algorith: A Simplified Search Space Approach", IEEE Congress on Evolutionary Computation, pp 2483-2489

[10] Chen H., Shen W., Xi N., Song M., Chen Y., (2002), "Automated Robot Trajectory Planning for Blast Painting of Free-Form Surfaces in Automotive Manufacturing", Proceedings of the 2002 IEEE International Conference on Robitcs & Automation, pp 450-455

[11] Conner D. C., Greenfield A., Atkar P. N., Rizzi A. A., Choset H., (2005), "Paint Deposition Modeling for Trajectory Planning on Automotive Surfaces", IEEE Transactions on Automation Science and Engineering, vol 2 (4), pp 381-392

[12] Carlos A., Coello Coello., (2006), "Evolutionary Multi-Objective Optimization: A Historical View of the Field" IEEE Computational Intelligence Magazine, pp 28-36

[13] Gao Y., Lei S., Yao P., (2002), "Study on Multi-Objective Genetic Algorithm", Proceeding of the 3rd World Congress on Intelligent Control and Automation, pp 646-650

[14] Carlos A., Coello Coello., (1999), "A Comprehensive Survey of Evolutionary-Based Multi-objective Optimization Techniques", Knowledge and Information Systems, Knowledge and Information Systems, pp 129-156

[15] Sujan V.A. Dubowsky S., (2005) "Efficient Information-based Visual Robotic Mapping in Unstructured Environments", *The International Journal of Robotics Research*, vol. 24, pp.275-293,

[16] Chotiprayanakul P., Liu D.K., Wang D., Dissanayake G., (2007), "A 3-Dimensional Force Field Method For Robot Collision Avoidance In Bridge Maintenance" *24th International Symposium on Automation and Robotics in Construction*, India, pp 139-145

[17] Settles G.S., Garg S., (1996) "A Scientific View on the Productivity of Abrasive Blasting Nozzles", Journal of Blast Thermal Technology, Springerlink, pp. 35-41

Innovative Production Machines and Systems
D.T. Pham, E.E. Eldukhri and A.J. Soroka (Eds.)

Multi-agent based integrated quality systems

E. OZTEMEL[a], E.K. TEKEZ[b]

[a] *Marmara University, Department of Industrial Engineering, Goztepe Campus, Kadıköy, Istanbul, Turkey*
[b] *Sakarya University, Department of Industrial Engineering, Esentepe Campus, Serdivan, Sakarya, Turkey*
ercan.oztemel@bte.mam.gov.tr, etekez@sakarya.edu.tr

Abstract:

The rapidly changing needs and opportunities of today's global market require interoperability to integrate different manufacturing functions in order to share knowledge and collaborate among organizations. Interactions are the key means for sharing knowledge among agents to coordinate towards certain manufacturing goals. In order to successfully and profitably operate in a rapidly changing market, it is necessary to integrate different manufacturing functions, evaluate manufacturing processes and improve the process performance in an integrated manner. This paper set outs a methodology for modeling agent based quality systems in manufacturing environments. The proposed methodology is supported with a case study.

Keywords: Agent, multi-agent, quality systems, integrated manufacturing, intelligent manufacturing.

1. Introduction

The rapidly changing needs and opportunities of today's global market require interoperability to integrate different manufacturing functions in order to share knowledge and collaborate among organizations. Interactions are the key means for sharing knowledge among agents to coordinate towards certain manufacturing goals. In order to successfully and profitably operate in a rapidly changing market, it is necessary to integrate different manufacturing functions, evaluate manufacturing processes and improve the process performance in an integrated manner.

By definition, multi-agent systems are suitable for distributed manufacturing environments. Agents have basic characteristics including modularity, decentralization, capability of handling ill-structured and complex problems, autonomy and independent decision making as well as timely working etc. [1]. Multi-agent systems, when their agents work together, give rise to better task sharing and problem solving mechanisms than centrally dictated schemes [2]. That is to say that multi-agent systems exhibit intelligence through the collective behavior of all manufacturing elements integrated.

There have been several studies on multi-agent manufacturing systems. However, integrating agent based quality systems and process improvements with other manufacturing agents are still research issue and discussed here in this paper. Some of the literature survey is provided below.

Chira et al. [3] proposed an agent-based architecture to support multi-disciplinary design teams that cooperate in a distributed design environment. Using ontologies and multi-agent systems, the proposed framework aims to optimize design process operation and management by supporting the dialogue among distributed design actors. Similarly, Kim et al. [4] proposed guidelines and detail procedure for developing three advisor agents for guiding the process definition, process execution and process measurement. In their study, 6-Sigma based process management system is introduced and discussed.

Recently, Lee et al. [5] proposed a procurement system across other disciplines and retrieved information with relevant parties so as to have a better co-ordination between supply and demand sides of a certain manufacturing process. Their paper demonstrated how to analyze the data with an Agent-based Procurement System (APS) to re-engineer and improve the existing procurement processes. The intelligent agents take the responsibility of searching the potential suppliers, negotiation with the short-listed suppliers and evaluating the performance of suppliers based on the assessment criteria with mathematical model.

Tatara et al. [6], on the other hand, proposed another framework which is well suited for embedding fault-tolerance capabilities in the hierarchical agent-based control structure in order to manage "control failure situations" arising in parts of the control network. Agents could have control over various resources, either in the form of process materials, energy, information, or computation.

This paper set outs a methodology for modeling agent based quality systems with possible interactions. The proposed quality system in this paper is part of an integrated reference model called REMIMS which is explained in detail elsewhere [7, 8].

As explained in the related references, REMIMS is designed in such a way that, it may promote effective management among different manufacturing functions, knowledge sharing, good communication and cooperation. To facilitate this in REMIMS, a *knowledge protocol* in a *knowledge network* is developed (see [9], for more detail). In the proposed model, the knowledge utilized by the agents are stored in so called *knowledge forms* and updated continuously in time by the respective agents.

2. Knowledge network (KN)

A ***"Knowledge Network"*** is an integral part of the proposed multi agent system which allows the

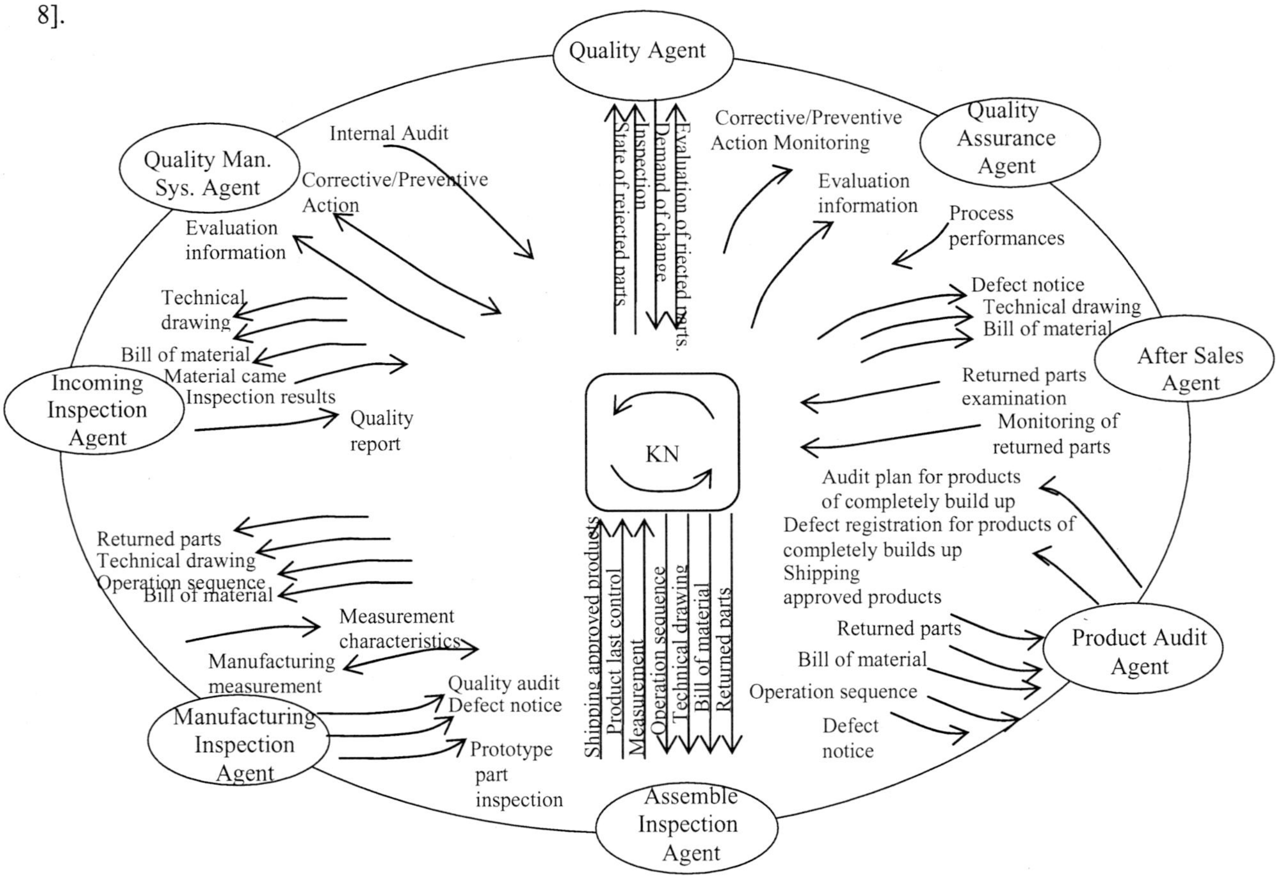

Fig. 1: Examples of several knowledge exchange between quality agents of REMIMS

manufacturing agents to communicate its outputs to others. Each agent is responsible for performing its own tasks and carries the required information to the integrated system using this network. It should be noted that an agent may produce certain knowledge which would be a strict requirement by others. Similarly it may need some inputs which could be produced by other agents within the integration framework. Fig. 1 shows an example of this. A standard knowledge exchange procedure could therefore be an extremely useful in order to handle automated exchange of information as well as knowledge.

Knowledge Protocols as encountered by REMIMS is by for the agents to communicate one another. In this schema, the agents exchange knowledge using a four-layer protocol including an identity layer, a query layer, a response layer and a remark layer (see Fig. 2). This protocol conveys the knowledge placed in so called ***Knowledge Forms*** where every type of manufacturing related information and knowledge are stored. Each form has its own architecture based on the respective manufacturing functionality. The knowledge forms can usually be stored using XML format which makes it easy to be utilized by respective agents. However any standard format could be employed. Each form is identified through a specific coding scheme which is explained in detail by Oztemel and Tekez (2008).

Note that, the content of each form can only be changed by the authorized agents identified by the Reference Model and as defined specifically in a ***knowledge library.*** Nevertheless, the forms could be available to all agents within the integrated framework and can be utilized by them. Several implications of utilizing knowledge forms are provided by Tekez (2006). Knowledge forms are stored in a ***"Form Base"*** which includes the forms, knowledge library and a ***priority relation matrix***. This matrix identifies the priorities of agents working in the integrated environment.

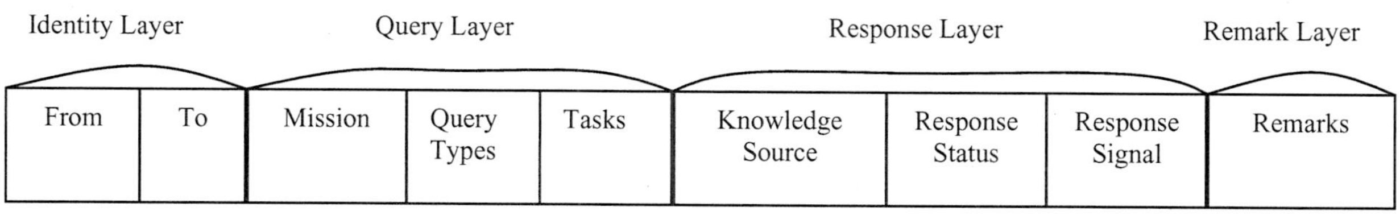

Fig. 2: Format of inter-agent knowledge exchange (knowledge protocol)

3. Modeling quality system integrated in REMIMS

Quality system of REMIMS is represented by an agent called Quality Agent (QA) which is in turn supported by a set of sub-agents including Quality Management Systems Agent (QMSA) and Quality Assurance Agent (QASA). QMSA also composed of Incoming Inspection Agent (IIA), Manufacturing Inspection Agent (MIA), Assemble Inspection Agent (AIA), Product Audit Agent (PDA) and After Sales Agent (ASA). Fig. 3 represents these agents and their relationship with other manufacturing agents.

The quality agents are responsible for the inspections of the products and all related components as well as inspection of both incoming raw materials from suppliers and components or materials produced in house. Besides they are also responsible for monitoring the process performance and quality problems that could arise after sales.

Particularly, one of the fundamental functions of Quality Agent is to examine possible statements on operations at product design, shop floor and assemble line in order to increase the effectiveness and capability of handling the demand changes whenever necessary. Another task of this agent is to evaluate whether quality problems of rejected parts are sufficiently big in terms of characteristics subject to quality control. The agent is designed to facilitate process improvement activities as well. Note that the process improvement is achieved by either measuring the product characteristics or measuring the process parameters. Brief descriptions of the quality agents utilized in the proposed model are given below.

- QASA is responsible for monitoring the performance of process against process targets. All agents in REMIMS evaluate their own activities and notify that to QMSA and QASA.
- QMSA is responsible for handling interval audits of the related to working processes of all agents generating internal *audit report forms*. The audit report form demonstrates consequences and notifies the related agents for performing corrective actions wherever necessary. Besides, if it discovers a defect, it requests direct preventive actions.
- IIA is responsible to examine materials coming from suppliers. If the inspection result is OK, an "*OK Card*" otherwise a "*Reject Card*" is attached to the material and *Quality Report* is produced.

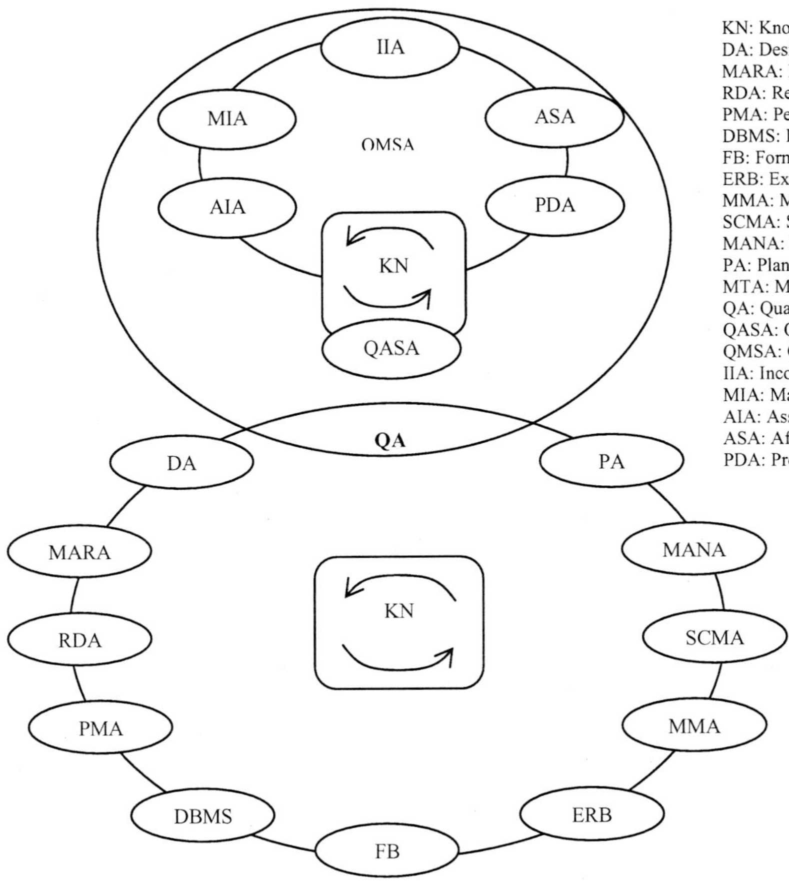

KN: Knowledge Network,
DA: Design Agent,
MARA: Marketing Agent,
RDA: Research and Development Agent,
PMA: Performance Monitoring Agent,
DBMS: Data Base Management System,
FB: Form Base,
ERB: External Resource Base,
MMA: Material Management Agent,
SCMA: Supply Chain Management Agent,
MANA: Manufacturing Agent,
PA: Planning Agent,
MTA: Material Transportation Agent,
QA: Quality Agent,
QASA: Quality Assurance Agent,
QMSA: Quality Management Systems Agent,
IIA: Incoming Inspection Agent,
MIA: Manufacturing Inspection Agent,
AIA: Assemble Inspection Agent,
ASA: After Sales Agent,
PDA: Product Audit Agent.

Fig. 3: Nested architecture of REMIMS

- MIA and AIA are responsible for handling quality audits in shop floors and assemble lines respectively. If there is any unsuitability, they notify the related agents. Another task of these agents is to determine measurement characteristics for control. Moreover, MIA does inspection for prototype parts which is produced in the factory. Besides, MIA is interested in measurement of the parts which could not be measured in the manufacturing area. AIA does final control of products and determines products for shipment.
- Similarly, PDA deals with audit plan, as well as checks audits of the products which are completely built up.
- ASA monitors quality problems after sales and examines returned parts. After performing the evaluation it decides what to do.

4. Case Study

4.1. Description of the Case

In the company where the case study was carried out, a series of quality control activities are defined, planned, and applied in order to manufacture products and components with the required quality. Problems occurred due to complaints are investigated and immediate preventive actions are employed in order to secure the quality of the products and processes.

4.2. Knowledge forms and their implementation in the case study

Note that in this particular case study, only activities for the inspection of the incoming parts are taken into account. Process improvements are explained in terms of interactions of three agents mainly, Incoming Inspection Agent (IIA), Quality

Management Systems Agent (QMSA), and Quality Assurance Agent (QASA).

QMSA handles internal audit of incoming inspection processes. According to the result of its assessment, it demands corrective action through the ***Action Demand Form (ADF)***. As seen Table 1, ADF also includes knowledge to respond to IIA. On the other hand, QMSA monitors the corrective / preventive actions for possible improvements using a specifically designed ***Action Monitoring Form (AMF)***.

The AMF includes possible improvement activities carried out by the respective agents. The results of improvement actions are also sent to QASA for its information. Example given Figure 4a indicates that QMSA requests IIA to demand change of its activities. The knowledge to be utilized is stored in the knowledge form coded F0601000002**001**000. Similarly, IIA plans and make sure that necessary actions carried out in returns to the request. When corrective actions are completed, IIA responds to QMSA through the form coded F0601000002**003**000 (note that form number is increased to indicate the change) in Figure 4b of Table 1.

Furthermore, IIA also evaluates its own activities and notifies the results to QMSA and QASA through ***evaluation form***. Utilizing the evaluation form of IIA and action monitoring form of QMSA, QASA evaluates the process performance and generated ***process performance form*** reporting this (see Table 2).

06010000 From QMSA	06010100 To IIA	0 Query	010 Demand	003 Change	F0601000002001000 Knowledge form for ADF	00 Wait	00 Wait	007 Change request

Fig. 4a. Knowledge protocol as an example of query message from QMSA to IIA

06010100 From IIA	06010000 To QMSA	1 Resp.	006 Information	008 Read	F0601000002003000 Knowledge form for ADF	01 Accepted	01 Done	14 Eval. result

Fig. 4b. Knowledge protocol as an example of answer message from IIA to QMSA

Table 1
The content of action demand form

ACTION DEMAND FORM: F0601000002003000			
Unit:	**Form No:**	**Version No:**	**Date:**
Incoming Inspection	003	000	20 January 2008
Reason:	Part 56775 does not confront with quality characteristics		
Part Number:	56775	**Success/Failure:**	Failure
Action Name which will be realized	☒ Corrective ☐ Preventive		
Action Name		**Planning Date**	**Realized Date**
%100 inspection		20 march 08	20 march 08

Table 2
The content of a process performance form

PROCESS PERFORMANCE FORM: F0602000002001000					
Form No: 001		**Version No:** 000	**Date:**20 March 2008		
Unit	**Process**	**Indicator**	**Target (T)**	**Actual (A)**	**Process Effectiveness (1+((T-A)/A))**
IIA	Incoming Inspection	Audit score for completely built up vehicle (decrease)	46	32	1,44
		The rate of rejected parts came from shop floor or assemble line in house (PPM – part per million) (decrease)	25000	27000	1,074
		Closing timely correction actions which are opened to incoming inspection (decrease)	60 gün	60 gün	1

5. Conclusions

Manufacturing systems are complex arrangements by measurable parameters that must be recognized in order to evaluate the performance of the system. This paper proposes an integrated modeling framework for quality systems in manufacturing companies. The proposed model contributes to improve quality, reduce production cycle time and delivery time, minimize the costs in a manufacturing company offering the following capabilities:

- Mechanism to exploit the information and knowledge and to diagnose the system, identifying the possibilities for process improvement and planning the continuous improvement actions.
- Integrating capability the different types of the knowledge such as quality, cost, capacity, productivity etc.
- The flexibility of the proposed method enables the use of different manufacturing systems.
- This system incorporates interactions among manufacturing agents. It takes attractions customizability attribute of the proposed model that is state of being able to be adjusted.
- This model can answer dynamic environment which changes circumstances continuously enabling feedback loop that supports improvement in performance measurement.

Acknowledgments

We would like to thank the managers of the company where the case study is performed for their invaluable support. We would also especially like to thank the Quality Manager of the company for his contribution and cooperation. Note that this study is carried out by IPROMS team in Sakarya University.

References

[1] Parunak,H. V. D. In: What can agents do in industry, and Why? An overview of industrially oriented R&D at CEC, ***Proceedings of the 1998 Second International Workshop as the series Lecture Notes in Computer Science,*** CIA'98, Vol. 1435/1998. Paris: Springer-Verlag GmbH, 1998.

[2] Lim, L. W., R., & Weidong, L., & Chin, K.. Distributed intelligent-based architecture for adaptive manufacturing systems. ***SIMTech Technical report***, AT/02/024/MET. Singapore Institute of Manufacturing Technology, 2002.

[3] Chira O., Roche C., Tormey, T. D., Brennan, A., An agent-based approach to knowledge management in distributed design, ***Journal of Intelligent Manufacturing***, 17, 2006, 737-750.

[4] Kim, J. A., Choi, S. Y., Jung, R. Process Improvement with Process Management Agent, ***International Journal of Software Engineering and Its Applications,*** Vol.1, 2007, No.1.

[5] Lee, C.K.M, Lau, H.C.W., Ho G.T.S, Ho, W. Design and development of agent-based procurement system to enhance business intelligence, ***Expert Systems with Applications***, in press, 2007 (Available online at www.sciencedirect.com)

[6] Tatara, E., Çınar, A., Teymour, F. Control of complex distributed systems with distributed intelligent agents, ***J. of Process Control***, 17, 2007, 415-427.

[7] Oztemel, E., Tekez, E. K. A Reference Model for Intelligent Integrated Manufacturing Systems: REMIMS, ***International IMS Forum 2004 on Global Challenges in Manufacturing***, Milano, Italy, 17-19 May 2004, pp.1260-1267.

[8] Tekez E., A reference model for intelligent integrated manufacturing systems (REMIMS), Unpublished PhD thesis, Sakarya University, Turkey, 2006.

[9] Oztemel, E., Tekez, E. K., Integrating manufacturing systems through knowledge exchange protocols within an agent based knowledge network, ***Robotics and Computer-Integrated Manufacturing***, in press, 2008.

Innovative Production Machines and Systems
D.T. Pham, E.E. Eldukhri and A.J. Soroka (Eds.)

A methodology for measuring product innovation

E. Öztemel[a,b], M.B. Ayhan[a]

[a] *Department of Industrial Engineering, University of Marmara,Istanbul,34740 ,TR*
[b] Also affiliated with: *MAM/BTE,Tubitak,Gebze/Kocaeli,TR*

Abstract

In this study, a general definition about innovation as well as vital importance of measuring techniques for product innovation is explained. A new approach for creating an innovation index for the improved products is established through the comparison of profit and sales values of improved and existing products. In order to achieve this index the terms "innovation earnings" and "innovation benefits" are introduced. The Degree of product innovation is formulated using these terms along with Research and Development expenditure.

Keywords: Product Innovation, Innovation, Measuring Innovation, Degree of Innovation

1. Introduction

20th century was considered to be the industrial age where machines were dominant in human's life.. Due to recent development this has changed and continues to change. The important factors of the 21st century will surely be the knowledge and the technological changes depending on the knowledge. These two dynamics is and will dominate the decision making processes of the companies. While the increasing importance of the knowledge and the technological changes are exposing the knowledge economies and knowledge societies, the knowledge itself becomes the main focus of the management in both industrial and public organizations at it brings up the changes in all areas. Hence, the companies should take into account and established a clear management schemes along this line. Utilizing knowledge can be beneficial if it turns out to be producing innovative products as well as processes.

It should be noted that the technological changes are directly affecting manufacturing systems. It is therefore extremely important to integrate them into the manufacturing facilities in a well determined and planned manner. Technological changes which trigger the innovation mainly arise from the Research and Development (R&D) activities which are problem-focused provide important achievements. Technological changes not only decrease the cost but also increase the company performance in utilizing energy, raw material, labor and machines etc.

In order to understand the relation between technology and innovation their definitions should be studied carefully. Note that, the technology is briefly defined as all the physical processes which transform the inputs to outputs and the social arrangement accompanying this transformation [1]. Similarly, the technological change is all about the improvements of this transformation process including the development of the existing and new products. Based on these definitions, the concept of innovation is described as the development and application of new ideas to create value [2]. In this paper a methodology to measure the degree of product innovation is introduced.

2. Innovation

There have been different definitions for the term *"innovation"* according to different situations, where it is implemented and analyzed. Generally speaking, the innovation is defined as turning knowledge into economic activity. It is a process of discovery, learning, and application of new technologies and techniques from many sources [3]

On the other hand, in Oslo Manual, innovation is divided into two segments, as the product innovation and the process innovation. Technological product and process (TPP) innovations comprise implemented technologically new products and processes and significant technological improvements in those[4]. According to the same manual, product innovation can take two broad forms such as technologically new product and technologically improved product. A technologically new product is a product whose technological characteristics or intended uses differ significantly from those of previously produced products. A technologically improved product is an existing product whose performance has been significantly enhanced or upgraded. [4].

In another study about the public sector in London, innovation is defined as the successful exploitation of new ideas [5]. New ideas themselves may take any number of forms, including new products and processes, new organizational techniques, new markets, and new sources of supply. The key point in this definition is the term "successful exploitation". That is, every new idea or product cannot be assumed as innovative unless they succeed in market. In the concept of software innovation for example, it is referred as the act of creating something new, normally something breakthrough in nature that did not have significant prior existence in the same form [6]. Although the definitions can change from case to case, within the knowledge-based economy; innovation definitely is play a central role.

In contrast to the different definitions of innovation, there is not a widely accepted methodology on how to measure that because of its nebulous nature. It should not be expected to occur by happenstance or accidental, but a sustainable, measurable business process. Note that however achieved, the innovation is hard to measure as it is inherently subjective and possess uncertainty as well as being imprecise, like measuring creativity or art. It requires the answer of the following questions. Who exactly does the innovation, and why do they do it? What is the value created, how is it realized, and by whom? However, exact measures of these activities are hard to identify and collect.

Productivity is one type of innovation. During the past fifty years, productivity has been driving force for manufacturing society to experiment improvements. Since *machines* are leaving their places to *knowledge* in organizations, the innovation is not only about ideas or patents. It is also about people working on new things in new ways. But existing productivity measures are not able handle these. To be able understand how innovation influences the economy, new measures which will expose if it is really making a difference is needed [7].

Some metric units are suggested to measure the innovation in a recent study [8]. These can be listed as number of ideas generated, evaluated, converted to products, and the revenue generated from new products. In another study, innovation measures are stated as; Labor Productivity, Patent Filling, New Venture Creation, and the Size of High Technology Sectors [2]. Although these studies provide promising metrics to measure the innovation, it seems that the fundamental definition of innovation is somehow ignored as it is defined as the changes adding a value to the products or processes. For example, Patent Filling figures may not be a metric sufficiently enough to measure the innovation. Because, improvement on the products product which does not add a value may be patented.

2.1. Measuring innovation

Depending on an extensive analysis, the important factors to calculate the *degree of innovation* are found out to be the earnings from the new technologies as opposed to the expenditures for the research and development activities. The simple subtraction from these two monetary units will only demonstrate the Innovation Earning and it will not be adequate for the comprehensive model. Since the innovation type – whether it is a breakthrough or a differential / incremental innovation- is also important. Therefore, the Innovation Earning should be multiplied with a factor (α), which will measure the novelty of the innovative efforts. This will be called as innovation index hereafter.

Innovation Index: If the research and development activities resulted with a product which does not exist before, it is clearly a breakthrough innovation and the Innovation Index can be assigned for that project equal to 1 which indicate 100% innovativeness. If it is a differential or an incremental innovation of a previously existing product, then we can compare the incremental product with the existing one by using their profit rate and also the sales quantity. These two factors may define if the new improvements create a value for the company. Basic assumption is that what ever does not create a value, it is not considered to be innovative.

If only the sales price of the new incremental product is taken into account disregarding profit per product, then there will be some misinterpretation of the price and cost decreases. For example, let us say that the price and cost diminishes by 10% and 50% respectively. In this case if only price is considered then the decrease may be interpreted as "no innovation". However the innovation may cause the cost reduction and profit per product increase accordingly. In order to prevent this kind of misinterpretations, profit by product is also considered as key elements of the methodology proposed in this study. This is not illogical, since we define the innovation as *the changes that add value to the product.* So, how can this value be measured? This could be done by comparing the profit of the new product after innovation with the profit of previously existing product. This information can be used for form the Innovation Index. However, that is not enough. The market price of the new product clearly, should not only set by the company for the sake of market deviations but rather established based on the properties and the novelty of the product.

Another issue is the inflation rate. If the existing product is still being manufactured at the release time of the newly developed product, then there is no need to consider the inflation rate. On the contrary, if the existing product is not still being manufactured, then the last selling price of that product is taken and a simple present value analysis should be carried out to find the present value of profit of the existing product. This is given by the following equation [9].

$$\text{Present Value} = \text{Past Value} * (1 + i)^n \qquad (1)$$

i: Yearly Interest Rate
n: Time Period (Year)

The other component of the innovation index is rate of sales quantity. That is the ratio of the sales quantity of new product over that of existing product. This component is added to prevent the misleading to "No Innovation" especially in case of in case of profit decrease but significant increase in sales. If the innovation can provide new functionalities well accepted by the market but also increased the costs to some extent, sales quantity may increase but not profit by product.

Calculating innovation index will not be complete unless the Innovativeness Factor is taken into account. As profit or sales quantity change may still occur without innovative actions due to market dynamics. In order to prevent this and innovativeness factor is introduced. This factor can take the value of 1 representing that there has been innovative efforts on the product and the value of 0 otherwise. The Following Equation can then be established based on the above discussions.

$$\text{Innovation Index } (\alpha) = (P_2 / P_1) * (S_2 / S_1) * \gamma / 4 \qquad (2)$$

Where $0 \leq \alpha \leq 1$

P_2 : New Product's Profit
P_1 : Existing Product's Profit
S_2 : New Product's Sales Quantity
S_1 : Existing Product's Sales Quantity

$$\gamma : \text{Innovativeness Factor} = \begin{cases} 1 & \text{if innovation occurs} \\ 0 & \text{if no innovation or no increase at least } \textit{profit} \text{ or } \textit{sales} \end{cases}$$

This equation is applicable only with the following assumptions.

Assumption 1: Any profit (or sales quantity) increase greater than 100% is assumed to be 100% increase.

Assumption 2: Any profit (or sales quantity) decrease greater than 100% is assumed to be 100% decrease.

The methodology proposed intends to define best possible capability of following the degree of product innovation which is reflected by 100%. The main aim of this methodology is to measure the scale of innovation where 100% indicates the best case whereas 0% indicates the worst case.

The validity of the equation can be checked by analyzing for all proper situations for the two components of Innovation Index formula as given in Table 1. For the proof of concept, initial Profit and Sales Quantities are set to 100 and 1000 respectively. Table 1 indicates possible 9 situations that could arise

Table 1: Proper Case Analysis for calculating Innovation Index

Case	% Change in Profit	New Profit (P_2)	% Change in Sales	New Sales (S_2)	Innovation Index (α)
1	80% Increase	180	80% Increase	1800	(1.8*1.8)/4=0.81
2	80% Increase	180	No Change	1000	(1.8*1)/4=0.45
3	80% Increase	180	80% Decrease	200	(1.8*0.2)/4=0.09
4	No Change	100	80% Increase	1800	(1*1.8)/4=0.45
5	No Change	100	No Change	1000	(1*1)/4=0.25 → 0
6	No Change	100	80% Decrease	200	(1*0.2)/4=0.05 → 0
7	80% Decrease	20	80% Increase	1800	(0.2*1.8)/4=0.09
8	80% Decrease	20	No Change	1000	(0.2*1)/4=0.05 → 0
9	80% Decrease	20	80% Decrease	200	(0.2*0.2)/4=0.01 → 0
Best Case	100% Increase	200	100% Increase	2000	(2*2)/4=1

in any company regarding the changes in both profit and sales quantity. In the best case both may increase 100% where as in the worst case 100% reduction could be faced (under assumption 1 and 2). In case of 100% increase innovative index have the maximum value which is 4 as calculated below. $\alpha = (P_2/P_1)*(S_2/S_1) = (200/100)*(2000/1000) = 4$.

In order to set innovation index between 0 and 1 which will be easy to interpret, the calculated value is divided by 4.

In this equation, earnings from the new technologies can be calculated as the revenue from the introduction of the new or improved technology implemented on the products excluding all related manufacturing and overhead costs. Note that, the company may perform huge amounts of R&D activities, and return of investment may not have been materialized. Hence; only the R&D Investment cost, of which return on investment is realized, will be taken into consideration. This cost may be calculated with utilized budget allocated to R&D Department. However there is still a problem of R&D investment clashing of more than one project. In other words, R&D investment can be made for more than one project. For example, one project may result with a breakthrough new product, but the rests may still continue. In this case, R&D investment cost will have to be shared on the resulted newly developed or improved products.

In order to make the above issues more clear, it would be better to present a numerical example. Assume that, a company invested $10M in R&D activities, but the return on investment from the new technologies earnings occurred only $5M for the time period analyzed. So, innovation earning becomes negative which is $-5M. Consequently the *Degree of Innovation* will be negative for the time period analyzed. However, if the time period is extended than this negative value may turn into a positive value. This argument highlights two important issues. First one is the importance of selecting time period precisely and the next one is executing this analysis repetitively for the flowing periods as well. Selecting the time period clearly depends on the type of R&D activities and their return on investment periods.

Table 1 also indicates no change situations such as rows 5, 6, and 8 as well as the worst case. In These situations, there is a clear sign of *"no innovation"*. For example if there is no change in the profit and sales quantity is also decreasing what ever done to the product can not be called innovation due to the definition provided above. The innovation index automatically set to 0 for these cases. Note that, the first two cases of Table 1 indicates clearly that changes in the product increase either profit or sales quantity or both. But the third instance is a little bit interesting. Also note that while the profit is increasing and the sales quantities are decreasing, the innovation index becomes very low. This is not surprising. Hence there could be an innovation but the customers may not respect it very much. This situation is also valid for the 7th case. At this instance, sales quantities significantly rise however profit decreases. There could still be an innovation because customers respect that in high rates however the profit rate is not so much. These two components balance the innovation index. If the increase in one of them is greater than the decrease in the other, then the innovation index will increase accordingly.

Furthermore, the innovation index calculation should take all the R&D activities into account in case of developing more than one product.

Innovation Earning: Innovation Earning indicates the value generated by innovation with respect to R&D investments. It is calculated as;

$$\text{Innovation Earning} = \text{New Technology Earning} - \text{R\&D Invest} \quad (3)$$

For the high technological R&D activities the time for the return of investment may increase up to 9-10 years, on the other hand for small technological developments return on investment time can be 6-12 months. Therefore one of the initial steps of the whole study will be defining the time period according to the manufacturing sector, and related R&D objectives.

Furthermore, similar to Innovation Index, Innovation Earnings should be completed for all the projects and hence, it is then better to take Total Innovation Earnings account if the company is performing more than one R&D activity. Summation of all innovation earnings will shed light on the general performance of the firm about the innovation process.

Innovation Benefit: Innovation Benefit is the combination of both Innovation index and Innovation earning and calculated as

Innovation Benefit = Innovation Earning * Innovation Index

Since, the firm can earn great amounts of money from new developed products but in fact this may not happen due to innovation, this should not be considered as Innovation Benefit.

If the new product includes small differences from the previous one, then it is a differential innovation and its innovation index will be less than 1. This will lead the innovation benefit to decrease near to zero. By the help of Innovation Benefit, the benefit of the innovation is assessed. Note that this is more than the earnings received from the innovation. It comprises the novelty of the innovation as well as the earnings

Degree of Innovation: Degree of Innovation indicates the relative performance of the company for being innovative. That is the ratio of Innovation Benefit over R&D Invest. That is to find out for each unit of R&D investment cost, how much amount of innovation benefit is achieved. The degree of innovation is calculated as;

Degree of Innovation= Innovation Benefit / R&D Investment (4)

In order to evaluate the model of measuring the degree of innovation and example can be elaborated as given below. Assume, the following information is provided for a specific company.

INPUTS
- R&D Investment = $10M
- New Technology Earnings = $15M
- Existing Product's Profit =$100
- New Product's Profit =$100
- Existing Product's Sales =1000
- New Product's Sales =2000

Base on this information the following information can be generated and degree of innovation can be calculated as provide below.

OUTPUTS
- Innovation Index = ($100/$100) * (2000/1000) / 4 =0.5
- Innovation Earning = $15M - $10M =$5M
- Innovation Benefit = Innovation Earning * Innovation Index = $5M * 0.5 = $2.5M
- Degree of Innovation = Innovation Benefit / R&D Invest =$2.5M / $10M = 0.25

According to the above results, the new technology earnings is high enough relative to R&D investment cost this indicates that companies' earnings are due to some innovation. This is also reflected by the innovation benefit which is 2.5 which yields the degree of innovation to become 0.25. This means for each unit ($) of investment 0.25 unit contribution comes from the innovation activities.

3. Conclusion

By the help of R&D activities, product and process improvements or innovations occur. Although there are so many definitions about innovation concept, there is a lack of well organized measurement model for the innovation. In this study only the product innovation–not the process innovation- is regarded and its measurement is settled on a mathematical formulation.

In conclusion, to calculate the degree of innovation of a manufacturing firm; profits and sales quantities of improved and existing products are compared to find the innovation index. This index is multiplied with Innovation earning which is the difference of new technology earnings and the R&D Investment costs. The result of this multiplication forms the Innovation Benefit. When innovation benefit is divided by the R&D Investment cost, degree of innovation of the firm is found. Note that the study presented in this paper is a part of an ongoing research and the methodology proposed is yet to be braced through several case studies and practical implementations.

Using the degree of innovation as a main source of change, the companies can evaluate their innovative actions and their capability to follow the changes in order to sustain competitive advantage. The higher the degree of innovation is the higher capability of following the change. This information can also provide an indication of investments adding value to the products or not. The study continues to improve meaning of the degree of innovation as well as defining data collection procedures to make the calculations easy for the management levels.

References

[1] Kırım, A.: "Türkiye imalat sanayiinde teknolojik değişim", TOBB, Ankara, TR, 1990, pp 15-45.

[2] Evalueserve.: "Innovation in Switzerland", *Thought Leadership Forums*,(2002).

[3] Tang, J.: "Competition and Innovation Behaviour", R*esearch Policy,* Vol.35, 68-82 (2006).

[4] OECD.: "The Measurement of Scientific and Technological Activities. Proposed Guidelines for Collecting and Interpreting Technological Innovation Data." Oslo Manual, 2nd Edition, DSTI, OECD, European Commission and Eurostat, Paris, France, (1995).

[5] Wilson, R.: "Innovation in London", *Working Paper19*,(2007).

[6] Misra, S. C.; Kumar, V.; Kumar, U.; Mishra, R.: "Goal-Driven Measurement Framework for Software Innovation Process", *Journal of Information Technology Management,* Vol.16, Issue.3, 30-42, (2005).

[7] Kingsland, B.: "Proposal for New Innovation Measurement", U. S. *Department of Commerce, Economics, and Statistics Administration,* Federal Register, Vol.72, No.71, p.18627, (2007).

[8] Phillips, J.; Hering, D.: "Innovate on Purpose", *NetCentrics Corporation* (2005).

[9] Sepulveda, J. A.; Souder, W. E.; Gottfried, B. S.: "Theory and Problems of Engineering Economics", *Schaums's Outline Series,* McGraw Hill Inc, New York, USA (1984).

Innovative Production Machines and Systems
D.T. Pham, E.E. Eldukhri and A.J. Soroka (Eds.)

Collaborative networked study: fundamentals, applications and discipline formation

A.J.C. Pithon

Department of Production Engineering, Federal Center of Technological Education, Rio de Janeiro, Brazil

Abstract

In the current scenery of markets opening and economical globalization, the enterprises have been great challenges to increase their efficiency levels and, like this, they maintain if competitive. In that context, the concepts of collaborative network appear. The notion of network is nowadays a central issue in many fields including social sciences, communications, computer science, physics, sociology, anthropology and biology. A collaborative network is constituted by a variety of entities (e.g., organizations and people) that are largely autonomous, geographically distributed, and heterogeneous in terms of their operating environment, culture and goal. A large variety of collaborative network have emerge during the last year as a result of the challenges faced by both the business. Nowadays, several courses in the area of collaborative Networked are already being taught or organized at different universities worldwide. In the long term, the role of education in this field may be relevant towards the establishment of the common knowledge and terminology required for proper deployment of reference scenarios. A discipline of collaborative networked shall focus on structure, behaviour, and evolving dynamics of network of autonomous entities that collaborate to better archive common goals, that is, social Network, Virtual Enterprise, Virtual Organization, etc.

Keywords: Collaborative network, social network, virtual enterprise

1. Introduction

The "term" network is nowadays a central issue in many fields like; social sciences, communications, physics, computer science, virtual organization, virtual enterprise, etc. Among the various types of network, of special relevance are collaborative networks. A collaborative network (CN) is constituted by a variety of entities (e.g., organizations and people) that are largely autonomous in terms of their: operating environment, culture, social capital and goals. Nevertheless these entities collaborate to better achieve common of compatible goals, and whose interactions are supported by computer network [1].

The implementation of collaborative process has accelerated in recent years as a consequence of both the new challenges posed to companies and organizations by the fast changing socio-economic conditions and the new developments in ICT sector.

In fact during the last three decades the information and communication technologies have been playing a growing role in organizations, namely as an instrument to support integration and flexibility. As a result of these developments, new scientific discipline emerged or where consolidated as in the case of collaborative Network [2].

Nowadays, several courses in the specific area of Collaborative Network are already being taught or organized at different universities worldwide. For instance, the University of Lisbon (Portugal) offers a 1 semester course on Virtual enterprise to the 5th year students of electrical and Computer engineering since 2002 [3]. Similarly, the Federal University of Santa Catarina (Brazil) offers course of Automation and Systems Engineering and the Costa Rica Institute of technology [4] started offering Virtual Organization coursers to their students, as well as, on February 2007, Federal Center of Technological education – CEFET/RJ in Brazil, stared a pilot course.

In the context, the objective of this paper is describe the experience with a pilot initiative of teaching a complete course on Collaborative Network (EDA 3022) at the Federal Center of technological Education – CEFET/RJ in the Master Technological program in first trimester of academic year of 2007. The proposal of this course is presented later on.

2. Collaborative Network

The word "net" is very old and it comes from the latin "retis", meaning interlacement of threads with regular openings that form a type of cloth. Starting from the interlacement notion, mesh and reticular structures, the word net went winning new meanings along the times, passing to be used in different situation.

Leaving of defined concept for [5], "nets are organizational systems capable to gather individuals and institutions, of democratic form and share, around similar causes. Flexible structures and established horizon, the work dynamics in net suppose performances collaborative and sustained by the will and likeness of their members, being characterized as a significant organizational resource for the social structuring"

In agreement with [6], enterprise network are formed initially with objective of reducing uncertainties and risks, organizing economical activities starting from the coordination and cooperation among companies.

Most of the authors study the perspective of nets and her use, as a road to study the organizations, appears to the organizations as social nets and they should be analyzed as such. A social net has there to be with a group of people, organizations, etc., linked through a social relationships group of a specific type. In this perspective, the structure of any organization should be understood and analyzed in terms of multiple nets of internal and external relationships. In that sense, the organizations are nets and the organizational form depends on the characteristics, interests and needs participant companies [7].

The organizational nets can be considered a consequence of concepts and beginnings of social nets and they can be divided in intra and inter-organization.

We started to detail the several types of formations of nets below:

- Social Network: focus on relationship among social entities, is used widely in the social and behavioural sciences, as well as in economics, marketing, and industrial engineering;
- Virtual Organization: comprising a set of (legally) independent organizations that share resources and skills to achieve its mission/goal, but that is not limited to an alliance of for profit enterprises. A Virtual Enterprise is therefore, a particular case of Virtual Organization;
- Virtual Enterprise: a temporary alliance of enterprises that come together to share skills or core competencies and resources in order to better respond to business opportunities, and whose cooperation is supported by computer networks;
- Extended Enterprise: a concept typically applied to an organization in which a dominant enterprise "extended" its boundaries to all or some of its suppliers. An Extended Enterprise can be seen as a particular case of Virtual enterprise;
- Agile Enterprise: is the ability of an organization to adapt proficiently in continuously changing, unpredictable business environment;
- Joint Venture: is an entity formed between or more parties to undertake economic activity together. The parties agree to create a new entity by both contributing equity, and they share revenues, expenses, and control of the enterprise;
- Cluster: geographical concentrations of interlinked enterprises that act is a same section of specialized suppliers, providers services and associated institutions, tends in common, besides the location, the contribution for development products of region. They are orientated by beginnings as cooperation, complementarities, community's sense and competition.

In that sense, Collaborative Network course is being implanted, trying to give an approach of concepts mentioned above.

3. Discipline Structure

Collaborative network was included in 2007 as an optical discipline in the Master Technological Program at the CEFET/RJ and it is part of curricular structure and the academic regime. These structure curricula are divided in a cast of obligatory and elective disciplines. The distribution of those disciplines in the curriculum of the course can be seen in the table 1.

Table 1 – Distribution of disciplines in Master's degree course

Obligatory Disciplines		
Code	**Denomination**	**Credit**
MAD 3007	Statistical Methods I	3
EDA 3024	Organization	3
EDA 3025	Technology	3
EDA 3300	Seminar for Master dissertation	0
EDA 3301	Research for Master dissertation	0
Elective Disciplines		
EDA 3029	Integrated Analysis of Life Cycle	3
EDA 3302	Study Activity Integrated	0
EDA 3033	Evaluation Projects	3
EDA 3030	Evaluation Technological Projects	3
EDA 3012	Systems Reliability	3
EDA 3013	Structural Reliability	3
EDA 3014	Human Reliability	3
EDA 3022	Corporate Finances	3
EDA 3021	Integration of Administration Systems	3
EDA 3008	Intelligence Computational	3
EDA 3010	Statistical Methods II	3
EDA 3011	Statistical Methods III	3
EDA 3026	Modelling of Phenomena	3
EDA 3022	Collaborative Network	3

The minimum duration for accomplishment of the course is 18 (eighteen) months and the maximum is 24 (twenty for) months. The course is divided in 3 (three) fundamentals steps, to know: obtaining the credits, development of research and defence of the dissertation.

During a period of 12 weeks, each week includes 4h theory, these lectures comprised a presentation of the main concepts, state of the art and supporting technologies, and discussion of major trends and challenges. Although an optional discipline, this attracted 10 students in summer 2007, which is a significant number considering that was the first time that discipline was offered.

Offering this discipline, the Master's degree in Technology of CEFET/RJ answers directly to the student's longings, i.e., offers possibility the same ones know and learn on the emerging organizational concepts that will face in his professional life. The next sections describe in general terms the structure of Collaborative network discipline.

4. Structure of Collaborative Network Discipline

The specific contents of the course are described bellow. For each subject the main bibliographic references that may be used.

In general, adjusted depending on the interest and skills of specific students group.

1. Network
   a) Introduction and concepts [1,2,3].
2. Network Models
   a) Social network: definitions, application and examples [1,4,5];
   b) Intra-organization Network: definition, application and examples [6];
   c) Inter-organization Network: definition, application and examples [6];
   d) Interpersonal Network: definition, application and examples [6].
3. Virtual Organization [7,8]
   a) Agile Enterprise [9,11];
   b) Virtual Enterprise [8,10,11];
   c) Extended Enterprise [12];
   d) Life cycle [13,14];
   e) Contracts in Virtual Organization (legal form) [15];
   f) Broker in Virtual Enterprise Formation: [11,16];
   g) Teams in Virtual Enterprise: [11,16].
4. Cooperative Network
   a) Joint Ventures [18];
   b) cluster [19].
5) Awareness in Network
   a) introduction e definitions [20]

5. Objectives of the Course

Collaborative network discipline has how main objectives to do the students are capable of:

• In-depth study of CN paradigm focusing on related information technologies and concrete applications;

• Identify and develop possible subjects for pos-graduation theses/projects;

• Explore the application of CN concepts within a national context;

• Develop in the students basic skills to carry out applied research and publish the corresponding scientific results;

• Aim at the establishment of a research group on CN at the Collaborative Work Nucleus.

6. Bibliography used in the discipline

[1] Yoguel, G. & Kantis, H. Reestructuración Industrial y Eslabonamientos Productivos: El rol de las pequeñas y Medianas Firmas Subcontratistas. Buenos Aires, CEPAL, 1990.

[2] Noria, N. Is a network perspective a useful way of studying organizations? In: Nohria, Nitin & Eccles, Robert G. (ed.) Network and Organizations: structure, form and action. Boston, Massachusetts: havard business School Press, pp. 1-22. 1992.

[3] Loiola, E., Moura, S. análise de redes: Uma contribuição aos Estudos organizacionais. In: Fischer, T. (org). Gestão Conteporânea: cidades estratégicas e organizações locais. Rio de Janeiro, pp. 53-68. 2007.

[4] Burt, R.S., Minor, M.J. Applied Networks Analysis. Berverly Hills, Sage. 1983.

[5] Martelo, R. M. Análise das Redes Sociais – aplicação nos estudos de transferência da informação. Revista Ciência e Informação. Vol 30, n°1, pp.71-81. 2001.

[6] Grandori, A. & Soda, G. Inter Firm Networks: Antecedents, Mechanism and Forms. 1995.

[7] Jägers, H., Jansen, W. Characteristics of Virtual organization, In: Organization Virtualness, pascal Sieber and Joachin Griese (ed), Simowa Verlag Bern. Pp. 65-76. 1998.

[8] Camarinha-Matos, L.M., Afsarmanesh, H. The Virtual Enterprise concept, In: Infrastructure for Virtual enterprise – Networking Industrial Enterprise, L. M. camarina-Maos and H. Afsarmanesh (ed), Kluwer Academic. Pp. 3-14. 1999.

[9] Lee, G.H. Designs of Components and manufacturing Systems for Agile Manufacturing. International Journal of Production research. Pp. 1023-1044. 1998.

[10] Osório, L. Barata, M., Gibon, P. communication Infrastructures Requirements in a Virtual Enterprise. In: Infrastructure for Virtual Enterprises – Networking Industrial Enterprises, L.M. Camarinha-Matos and H. Afsarmanesh (ed). Kluwer Academic. Pp. 65-76. 1999.

[11] Pithon, A.J.C. Projeto Organizacional para a Engenharia Concorrente no Âmbito das Empresas Virtuais. Ph.D. Thesis, University of Minho, Guimarães, Portugal. 2004.

[12] Browne, J. The Extendet Enterprise – Manufacturing and the Value chain. In: Balanced Automation Systems – Architectures and design Methodologies, H. Afsarmanesh (ed). 1995.

[13] Fuchs, M. Design and Implementation of value Systems: The Lifecycle Perspective. Institute for Technology Mabagement, Universty os St. Gallen, Switerland. 1999.

[14] Zimmermman, F. structural and Manageral Aspects of Virtual Enterprise. University of Bamberg, Business Information Systems. 1996.

[15] Teixeira, B. M. Proposta de um Modelo Contratual para as Empresas Virtuais no Âmbito da Legislação Brasileira. M.Sc Thesis, CEFET/RJ, Rio de Janeiro. 2007.

[16] Ávila. P., Putnik G., et al. In: Brokerage function in Agile Virtual Enterprise Integration – A Literature Review. Collaborative Business ecosystems and Virtual Enterprise. L. M. Camarinha-Matos (ed). Kluwer Academic publisher. 2002.

[17] Lipnak, J., Stamps, J. Virtual Teams: People Working Across Boundaries with Technology, John Wiley & Sons, Ins. 2000.

[18] Pimenta, E. G. Joint Ventures – Contratos de Parceria Empresarial no Direito Brasileiro, Juarez de Oliveira editora. 2005.

[19] Porter, M. Clusters and the new economics of competition. Harvard Business Review. 1998.

[20] Gutwin, C. Greenberg, S. The Importance of Awareness for team Cognition in Distributed Collaboration. In: E. Salas and S.M. Fiore (ed) team Cognition: Understanding the factors that drive Process and Performance. Pp. 177-201. 2004.

[21] Cheng, K. E-Manufacturing: Fundamentals and Applications (Ed.) WIT Press London, March 2005.

7. Discipline Practical Activities

The activity developed at classroom is based in the games as tool that makes possible business existence simulation in their several activities, including from the planning to execution. This technique is very used, mainly in the administration courses, masters degree and MBA's and it portrays, in an informal way and to entertaining, strategies to be adopted by managers in the execution of the tasks, mainly in those that it involves work team, focus of our experiment.

The accomplished experience tried to show the importance of concepts learned at classroom, concerning the structure of a collaborative Network and all their interfaces concerning obtaining of their results. It fits to detach that the exercise happened in real time and attendance, because, it is not very common once the nets operate in virtual way.

The team, composed for team students, a consultant (invited teacher) and the director of enterprise (teacher of the discipline), it was under the consultant's orientation, positioned with the hands given in circle form, halfway some of other ones, in the classroom.

The disposition in circle is due to fact the communication if to give in a lot of directions and his flow is not controlled, just administered, that is, doesn't posses a hierarchy. The consultant distributed numbers written in a piece of paper (from 1 to 10), so that it was put in the ground ahead of the feet of each team member, passing to be this identification numbering of each one. To proceed, the consultant distributed string pieces with approximately 2,5 meters in length, that crossing in diagonal, it tied for the tips each member the other, tends these to tie one of the tips of string in right pulse (Figure 1). In this dispositions and already connected, the consultant explained the game rule that consisted beforehand, in the choice manager on the part of team. Tends the manager been chosen, the consultant request to the same as does the annotations related with what will happen.

At this time, he consultant begins the change of positions of each one in the circle, and in the end of those changes, he grew up a true knot in the string, as seen in the Figure 1. That knots the events that happen the any net type acts, since for beginning, the social nets are not static. The manager's paper was driving the actions that they would be developed by group, in the sense of untying the knot. It was not allowed that in any hypothesis, the string that tied each member of the group was cut and the knot that tied him to the undone pulse. The established time so that the manager drove his task was 5 minutes.

Therefore in beginning, it was noticed that manager didn't stop the control task and nor command of the team, letting his insecurity to appear for the group. With the control loss on the part of manager, the group began showing in a disordered way, random and empiric with objective of undoing knot. Visibly amazed, the team unchained a series of disconnected manifestations that disrupt the search of solution for the problem. The time of five minutes became exhausted and the result was not reached.

Before this problem, the consultant undid the net and it substituted the manager for other member of the group. Starting from this substitution, the group through a disassemble process, got to complete the exercise in the stipulated period of five minutes.

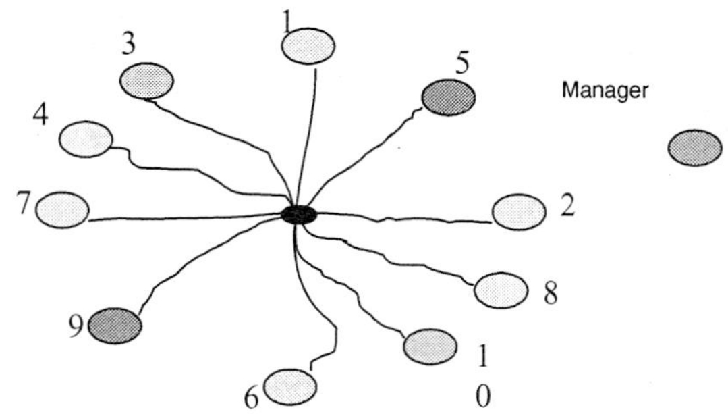

Figure 1 – Teams formation

At the end of the experiment, the teacher of discipline made an analogy with the net concepts, where the collaboration and the trust are fundamental for the development of net.

8. Conclusion

Introduction of Collaborative network discipline in master technological Program at CEFET/RJ, although an optional discipline, is being successful and contributing to preparation of a new generation of engineers and research able to play a major role in the deployment of Collaborative network concept in Brazilian university.

The assertiveness of result this experience is in the number of students that intend to include concepts of collaborative network as theoretical establish (or regarding analysis) in their theme of dissertation research.

As a result, the course is planned to be given again in coming semesters in this program.

Regarding future horizon, the possibility of offering a Virtual Course on Collaborative Network in Spanish (using e-learning techniques) aiming at integration Universities and Institutes of Technology of Latin America and Caribbean countries, can be evaluated.

Finally, it is clear more textbooks in this area are necessary in order to facilitate education tasks.

References

[1] Camarinha-Matos, L. M., Afsarmanesh, H. Journal of Intelligent Manufacturing, Vol 1, N° 16, pp.439-452, 2005.

[2] Camarinha-Matos, L. M., Cardoso, T. Education on Virtual Organizations: Na experience at UNL, In Virtual

Enterprise and Collaborative Networks, Kluwer Academic Publisher, Boston, 2004.
[3] Garita, C. A case study of VO education in Costa Rica, In Virtual enterprise and Collaborative Network, Kluwer Academic Publisher, Boston, 2004.
[4] Leon, M.E. Uma Análise de Redes de Cooperação das Pequenas e Médias Empresas do Setor de Telecomunicações. Dissertação de Mestrado, Engenharia de Produção da Escola Politécnica da Universidade de São Paulo, 1998.
[5] Cândido, G.A., Abreu, A. F. Os conceitos de redes e as relações inter-organizacionais: um estudo exploratório. In ENANPAD, Florianópolis, Brasil, 2000.

Innovative Production Machines and Systems
D.T. Pham, E.E. Eldukhri and A.J. Soroka (Eds.)

Concurrent engineering: principles and conceptual methodology for process innovation

J. Adebayo

Jade Holding Ltd., The Meridian, 4 Copthall House, Station Square, Coventry, CV1 2FL, UK

Abstract

Concurrent Engineering (CE) has become an important paradigm in product development and a widely accepted concept within the industrial and academic arena. However, the large number of different facets of tools and methods, and amount of information available in CE makes it difficult in terms of concept selection. This paper reviews CE and presents a defined methodology from contemporary literature and introduces an innovative and systematic approach to CE.

Keywords: Concurrent engineering, methodology, principles and process innovation.

1. Introduction

In today's business world, corporations must be able to react to the changing market needs rapidly, effectively, and responsively. They must be able to reduce their time to market and adapt to the changing environments. Decisions must be made quickly and they must be done right the first time out. Corporations can no longer afford to waste time repeating tasks, thereby prolonging the time it takes to bring new products to market. Therefore, concurrent engineering has emerged as a way of bringing rapid solutions to product design and development process [1].

Concurrent engineering (CE) is indisputably the wave of the future for new product development for all companies regardless of their size, sophistication, or product portfolio. In order to be competitive, corporations must alter their product and process development cycle to be able to complete diverse tasks concurrently. This new process will benefit the company, although it will require a large amount of refinement in its implementation. This is because CE is a process that must be reviewed and adjusted for continuous improvements of engineering and business operations [1].

CE, according to early pioneers, is that the engineering and design phase in product development often involves a series of activities traditionally executed by different functional departments in a phased, sequential approach. The activities should be clustered, integrated and managed concurrently rather than sequentially to save time [2-4]. Broughton [5] summarizes the aim of CE as "to achieve reduced times, improve quality and cost by the integration of design and manufacturing activities, and by maximizing parallelism in working practices". Several benefits have been associated with the adoption of CE [6,7], but there is no clearly defined universal methodology for it.

2. What is concurrent engineering?

In traditional engineering a relatively short time is spent defining the product. A relatively long time is spent designing the product and a surprisingly long time is often spent redesigning the product. The key to shortening the overall design time [8] is to better define the product and better document the design process.

Traditionally, the development of a product had been seen as a cycle of Plan-Do-Check-Act-Adjust, usually this is a slow, costly and low quality approach, leading to a lot of engineering changes, production problems, product introduction delays, and a product that is less competitive than desired. CE is a process in which appropriate disciplines are committed to work interactively to conceive, approve, develop, and implement product programs that meet pre-determined objectives. The CE concept brings together multidisciplinary teams, in which product developers from different functions work together and in parallel from the start of a project with the intention of getting things right as quickly as possible, and as early as possible. A relatively recent term, CE is applied to the engineering design philosophy of cross-functional cooperation in order to create products which are better, cheaper, and more quickly brought to market. The CE phenomenon reunites technical and non-technical disciplines such as engineering, marketing and accounting. Focusing mainly on customer satisfaction [8], the CE trends interrelate in defining the product to be manufactured.

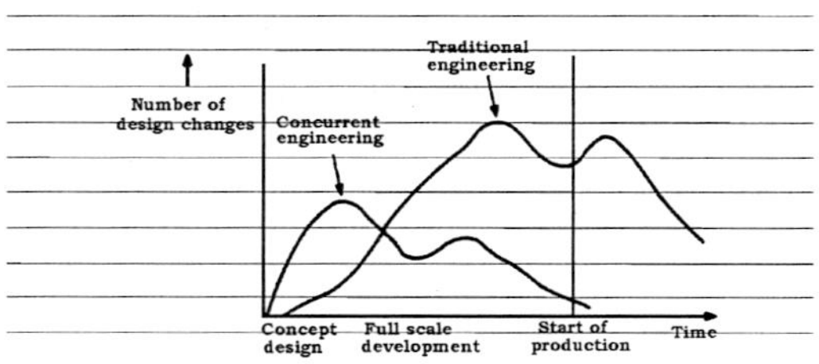

Fig. 1. Example of design changes as a function of time for an American and Japanese automobile.*

CE is a systematic approach to the integrated, simultaneous design of both products and their related processes, including production [9]. Cletus [10] defined CE in the context of a systematic approach to integrated and concurrent development of a product and its related processes, that emphasizes response to customer expectation and embodies team values of cooperation, trust sharing in such a manner that decision making proceeds with large intervals of parallel working by all life-cycle perspectives early in the process, synchronized by, comparatively, brief exchanges to produce consensus [10]. Chapman *et al.* [11] referred to CE as a philosophy of product development which integrates multiple design issues—a method of product design which integrates multidisciplinary personnel into the design team and a method to lead people where design issues are represented in the people.

An example of the use of CE can be found in General Electric Aircraft Engines Division's approach for the development of the engine for the new F/A-18E/F [16]. The approach entails several collocated, multi-functional design and development teams to merge the design and manufacturing process. The teams achieved 205 to 60% reductions in design and procurement cycle times during the full scale component tests which preceded full engine testing. Problems that surfaced earlier were dealt with more efficiently than they would have been with the traditional development process. Cycles times in the design and fabrication of some components have dropped from an estimated 22 weeks to 3 weeks.

Boeing's Ballistic System Division used CE to develop a mobile launcher for the MX missile and as a result reduced design time by 40% and cost by 10% in building the prototype. Polaroid Corporations' Captiva instant camera also implemented the CE approach, leading to increased production in working prototypes [16].

2.1 Goals of concurrent engineering

The goal of CE is to improve the interactive work of different disciplines affecting a product [11]. According to Cline [12], CE has seven goals which are based on its principles (see Table 1).

Table 1
A model to illustrate Goals of Concurrent Engineering

Project Identification
Ensure a single direction for corporate development to avoid shifting priorities, false project starts, and pre-emptied project efforts.
Project Scope
Estimate the project's effort, time, and cost so executives can make an informed decision about project's worth.
Requirement and Analysis
Build and validate a model of the business problem domain to ensure the correct problem and customer needs are defined.

* Source: Engineering Modeling and Design, Chapman, Bahill & Wymore. CRC Press Inc., Boca Raton, FL, 1992.

Table 1 *contd.*
A model to illustrate Goals of Concurrent Engineering

System Design
Provide technical solution that meets the customer needs and enhances corporate business value.
Development Planning
Define a work plan for implementing ideal technical solution.
Construction
Install or construct the solution, using a mini-release, risk-driven and priority-driven approach, to run concurrently with testing and low-level design.
Installation and Assessment
Ensure a solid and methodical way of moving the product into the customer's environment as smoothly as possible.

3. Methodology for innovation

A conceptual methodology has been developed from the seven principles above [12]. From the conceptual methodology (see Fig. 2), it can be seen that the process of undertaking CE is a systematic approach to an integrated, concurrent design of products, process and installation of plant, and processing of products, considering the related downstream aspects and elimination of non-value adding activities.

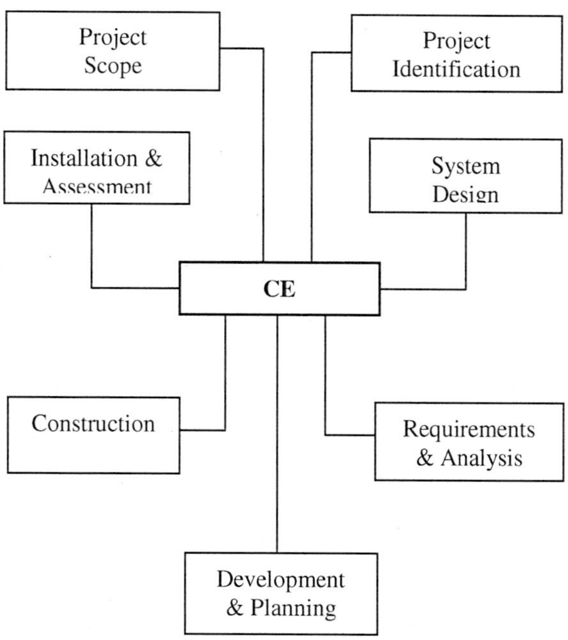

Fig. 2 Conceptual Methodology for Concurrent Engineering

Project Scope: Confirmation of customer's expectation with executive and corporate goals supports the project's success by avoiding unrealizable date or unallocated resources.

Project Identification: Executive commitment and goal for project, approval, prioritization and scheduling for all projects.

Installation and Assessment: Emulate the production environment for capacitance, stress, and performance; improve process for future predictability.

System Design: Design and validate a technical solution at the high level; define metrics to predict time of implementation and development effort, and to be used later for process improvement; build a test plan from requirement, not from the design or code.

Construction: Perform regression testing to maintain quality and avoid rework; track personnel effort, time, and defect rate to monitor project; separate developers from any testing; define a procedure by which QA and the customers approve the release.

Requirements and Analysis: Get requirements jointly with the use and write a specification that can be implemented, tested and explained; provide traceability between the customers needs, system solution, and testing to enable change management; justify project needs are best met by comparing purchase, build, or hybrid solutions—cost/benefit analysis or vendor proposals.

Development and Planning: Collect work plans for testing, customer acceptance, development, and documentation, and ratify with all involved; get written consensus on the project plan specification from all involved; define a strong QA policy to ensure process compliance and product correctness; establish configuration management for changes, defect resolution, and project control.

The CE approach can be achieved through a multidisciplinary team approach which intends to motivate process designers, plant designers and installations and participants throughout various design processes, to consider all elements of the product life cycle from the inception through to disposal of product, taking into consideration quality, cost, time-effectiveness, rapid adaptation of system to evolving business environments and end user requirements [13-15].

4. Capability of CE as process innovation

CE, in the context of a process innovation, revolves round three fundamental elements, sometimes referred to as the three Ts—tools training and time—of CE. The T elements of CE are structured as shown in Fig. 3:

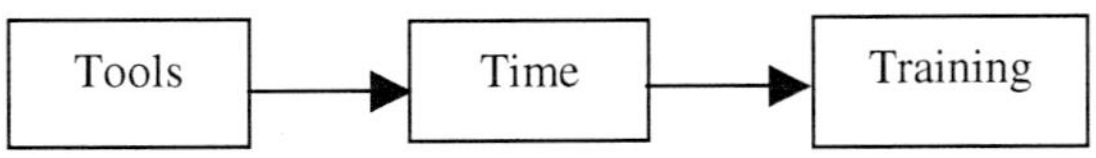

Fig. 3 Process Innovation for Concurrent Engineering

The Tool element involves the material infrastructure; Training, relates to the human aspect and includes educating personnel on the use of appropriate tools, and Time, considers realistic expectations in terms of setting targets.

The complexity and wide range of specialized disciplinary areas involved in modern manufacturing makes it an interdependent activity, often involving hundreds of personnel. In these circumstances tools are of utmost important [17,27] For centuries, a blueprint from the design office has been the agent for production planning and, subsequently, actual production. Today, CE requires that paper blueprints be replaced with an electronic three-dimensional solid model using a featured-based parametric modeling capability state-of-the-art computer-aided design (CAD) package.

Personnel training is a key factor in realizing the bottom line goals of organizations. Several companies have been innovative in their approach to the challenge for employee training. LearnShare, for example, is a consortium created cooperatively through Motorola, Owens-Illinois, Reynolds Metals, General Motors, Owens Corning, Iams, LOF (Pilkington), 3M, John Deere, and Steelcase [18]. Continuous improvement is an integral part of CE efforts.

As companies find that it is imperative to reduce the cycle time for new product development, they begin to think in terms of parallel activities. Reduced cycle time is just one of the many factors that affect the profit equation: it takes most companies approximately eight months simply to become comfortable with a new CAD/CAM system [19,28]. The idea of bringing the company's suppliers and customers within the boundary of the organization can create barriers in terms of time implementation.

5. CE and other design methodologies

Within the frame of this paper, CE is taken to represent the approach of collaborative product development with input from each functional group involved in the product development process. However, there are other methodologies which are also used in the designing of products but different in approach to CE, and they are discussed below:

QFD Methodology: Developed initially by Akao in Japan in 1966, quality function deployment (QFD) provides a formal structure for ensuring that the customer's requirements are carefully heard, and then directly translated into a company's requirements for product development [26]. The QFD process maps customer driven requirements to technical requirements, allowing competitive evaluation and weighing of factors to achieve a technical evaluation. The process was developed to this level by Bob King [20] and further used by Clausing [22].

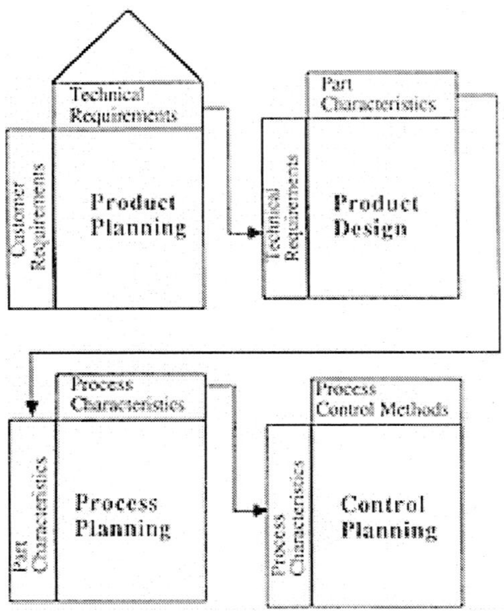

Fig. 4. The Four-Phase Approach to QFD.*

* Source: Product and Process Development Methods, by V. M. Karbhari: Japanese Technology Evaluation Centre 1994.

Total Development Process: Clausing's methodology [22] is more structured and attempts to evaluate each element on the basis of its impact on competitiveness, but it does focus on teamwork and collaboration. The methodology is organized around the following ten cash drains of product development:

- technology push, but where's the pull;
- disregard for the voice of the customer;
- eureka concept;
- pretend designs;
- pampered products;
- hardware swamps;
- here's the product -- where's the factory;
- we've always made it that way;
- inspection; and
- give me my targets -- let me do my thing.

Clausing further prescribes the use of Pugh's concept selection method [23] and the House of Quality [24] as tools to be used. Although a very useful and illuminating methodology, it lacks the tools for actual application.

Total Quality Design: Developed at the Center for Composite Materials at the University of Delaware, this methodology attempts to synthesize the best of the other methodologies with an emphasis on hearing the voice of the customer and developing design interactions early in the design process. The methodology consists of five main elements (Fig. 5.).

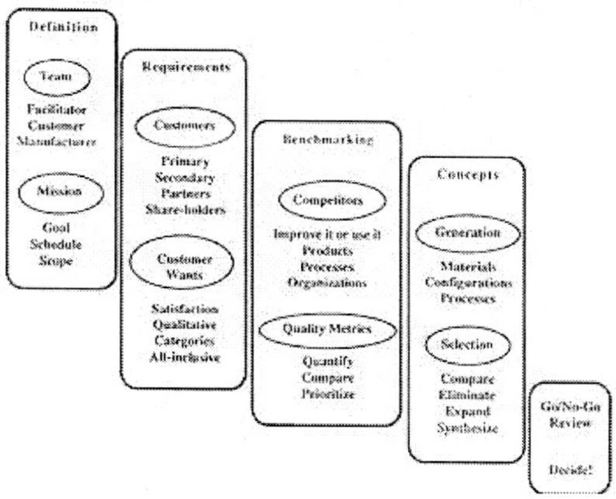

Fig. 5. The Elements of the TQD Process.*

Cross' Methodology: This methodology [25] is very adaptive and actually prescribes little except a set of three rules:

- adopt a framework;
- select design methods to flesh out the framework; and
- continually review and update the framework during the development effort.

Hubka's Methodology: Termed a general procedural model of design engineering, the Hubka Methodology [21] is a highly structured and detailed model. Although fairly detailed, it is perhaps more useful as a teaching aid than a methodology for composites product development.

6. Conclusions and future work

In this paper, an attempt has been made to present a conceptual methodology for CE. The difference in the traditional method of development of a product had been seen as a cycle of Plan-Do-Check-Act-Adjust. The implication of the methodology is to eliminate the non-value adding activities and improve flexibility in the system. The main objective of this paper has been to motivate and instigate researchers and practitioners to apply CE concept which has been ignored in the past for improving the overall performance of their organizations. A future work will be to explore the business process management in small and medium sized enterprises so as to present a model for the implementation of the methodology.

Acknowledgements

The author is grateful to Michael Flowers of KLK (Germany) GmbH for his constructive and helpful comments.

References

[1] University of California at Berkeley, Synthesis Coalition, http://best.me.berkeley.edu/~pps/pps/concurrent.html. Last accessed: June 2008.

[2] Clark, F. and Fujimoto, T. (1989), *Overlapping problem solving in product development*, Managing International Manufacturing, North-Holland Publishing Co., Amsterdam, pp.127-52.

* Source: Product and Process Development Methods, by V. M. Karbhari: Japanese Technology Evaluation Centre 1994.

[3] Kusiak, A. (1993), *Concurrent Engineering: Automation, Tools and Techniques*, John Wiley & Sons, New York, NY.

[4] Shtub, A., Bard, J.F. and Globerson, S. (1994), *Project Management, Engineering Technology, and Implementation*, Prentice-Hall, Englewood Cliffs, NJ, pp.247-92.

[5] Broughton, T. (1990), *Simultaneous engineering in aero gas turbine design and manufacture*, Proceedings of the 1st International Conference on Simultaneous Engineering, pp.25-36.

[6] Clealand, D.I. and Bursic, K.M. (1992), *Strategic Technology Management-Systems for Products and Processes*, American Management Association, New York, NY.

[7] Dowling, P.J. (1994), *Communication or isolation?* The Structural Engineer, Vol. 72 No.20, pp.329-33.

[8] Ettlie, J.E. and Stoll, H.W. (1990), Managing the Design-Manufacturing Process, McGraw-Hill,. New York, NY.

[9] Gunasekaran, A. (1998), *Concurrent Engineering, a competitive strategy for process industries*, Operational Research Society, Vol 49, pp. 758-765.

[10]Cletus, K.J. (1992), *Definitions of Concurrent Engineering,* CERT Technical Report Series, Research Note, CERC-TR-RN-92-003, Concurrent Engineering Research Centre, West Virginia University, Mortgantown, USA, pp 1-5.

[11]Chapman, W. L., Bahill, A. T., and Wymore, A.W., (1992). *Engineering Modeling and Design*, CRC Press Inc., Boca Raton, FL.

[12] Cline, A. (2000), *Concurrent Engineering, Primary Principles,* Carolla Development.

[13]Nevins, J.L. and Whitney, D.E. (1989). *Concurrent Design of Products and Processes: A strategy for the Next Generation in Manufacturing,* McGraw-Hill Publishing Company: New York.

[14]Love, P.E.D. and Gunasekaran. A. (1989). *Concurrent engineering in the construction industry,* Concurrent Engineering: Research and Application (CERA) 5(2): 155-162.

[15]Gunasekaran, A. and Adebayo, J. (2000). *GEC Alsthom Learns Many Lessons from BPR*, Production and Inventory Management Journal, Vol. 41, No. 2, pp. 9-13.

[16]Stark, J. (2007). Management Consultants in performance of product development process, John Stark Associates, www.johnstark.com/fwcce.html. Last accessed: June 2008.

[17]Kelly, D. S. (1998). *Web enabled engineering collaboration*. Journal of Industrial Technology, 14(2), 45-47.

[18]Corry, R. (1998). *The LearnShare consortium: A model for the future*. In Proceedings of the WebNet 98 world conference of the WWW, Internet and Intranet. Orlando, FL.

[19]Mark, P., & Riley, K. (1995). *Aligning technology for best business results*. Santa Cruz, CA: Design Insight.

[20] King, B (1987), *Better Designs in Half the Time – Implementing QFD Quality Function Deployment in America, 1987,* , GOAL/QPC,, Methnen, MA

[21]Hubka, V. (1982). *Principles of Engineering Design*. Springer-Verlag.

[22]Clausing, Don P. (1988). *The Total Development Process: Turning the Ten Cash Drains into a Cash Flow*. Unpublished work.

[23]Pugh, S. (1981). *Concept Selection: A Method that Works*. ICED Conference '81 (Rome).

[24]Hauser, J., and Don P. Clausing. (1988). *The House of Quality*. Harvard Business Review (May-June 1988).

[25] Cross, N. (1989). *Engineering Design Methods: Strategies for Product. Design*, John Wiley & Sons, Manchester.

[26]Liu, D.Z., Liu, M. and Zhong, P.S. (2004), 2004 Method of Product Development Process Analysis and Reengineering For Concurrent Engineering. Trans Tech Publications, Switzerland.

[27]Khalfan, M.M.A, and Anumba, C.J (2000), *A comparative review of concurrent engineering readiness assessment tools and models*, Concurrent Engineering 2000 (CE 2000) Conference, pp.578-85.

[28]Islam, M.N., (2003). Concurrent Engineering Offers the Best Solution to Tolerance Allocation Problems, *New Engineer Journal*, 6(1), pp. 14-15.

[29]Yassine, A. and Braha, D. (2003) *Four Complex Problems in Concurrent Engineering and the Design Structure Matrix Method*, Concurrent Engineering Research & Applications, Volume 11, No. 3.

Innovative Production Machines and Systems
D.T. Pham, E.E. Eldukhri and A.J. Soroka (Eds.)

Distributed Manufacturing Simulation with High Level Architecture: A Case Study

Özer Uygun[a], Cemalettin Kubat[a,]

[a] *Department of Industrial Engineering, Faculty of Engineering, Sakarya University, Esentepe Campus 54187 Sakarya, Turkey*

Abstract

Manufacturing activities, in today's global business conditions, are conducted in a physically distributed environment. Taking advantage of simulation in such a situation is not easy unless implementing a distributed simulation. High Level Architecture (HLA) is a distributed simulation architecture which is originally developed for defence simulation, however nowadays is also used for manufacturing applications. In this study a distributed manufacturing simulation case study was discussed. HLA is used in order to make it possible to interact between simulations.

Keywords: Distributed manufacturing simulation, HLA

1. Introduction

Manufacturing activities in today's global environment are conducted with collaboration of several manufacturing companies and workshops that are generally distributed geographically or physically. Simulation manufacturing activities in such an environmental condition and making benefit of simulation is not easy. Traditional simulation is difficult to use in a distributed manufacturing environment. So, distributed simulation tools and architecture is needed in order to simulate physically distributed manufacturing conditions.

High Level Architecture (HLA) is a distributed simulation architecture which was originally developed for defence simulation applications. However, it is also used for civil applications including manufacturing especially after it was accepted as a distributed simulation standard (Std 1516) by IEEE in 2000. The main aims of the HLA are interoperability and reusability which are also necessary for distributed manufacturing simulations.

In this study a distributed manufacturing simulation is implemented dealing with a case study. HLA is used as a communication tool as well as an enabler for interoperability between individual simulations.

The next section describes HLA briefly. The third part of the paper gives literature of some HLA applications in a distributed manufacturing simulation. The case study is described and detailed in the fourth part. Then, the implementation is addressed in the fifth part describing software programmes used, information representation, and the files developed for simulation execution. The results of the study are discussed in the conclusion part of the paper.

2. HLA: A brief introduction

IEEE accepted HLA as a standard architecture for distributed modelling and simulation in 2000. The framework and the rules of HLA are described in IEEE Std 1516-2000 standard document [1]. The all individual simulation in terms of HLA terminology is called federate and integration of federates with HLA constitutes the distributed simulation and is called federation.

Federates interact to each other over HLA Run

Time Infrastructure (RTI). RTI provides six services for managing overall information exchange between federates and for enabling interoperability and reusability of federates [2]. *Federation management* services provide basic functions required to create and operate a federation. *Declaration management* services support efficient management of data exchange through the information provided by federates defining the data they will provide and will require during a federation execution. *Object management* services provide creation, deletion, identification and other services at the object level. *Ownership management* services support the dynamic transfer of ownership of object and/or attributes during an execution. *Time management* services support synchronization of runtime simulation data exchange. *Data distribution management* services support the efficient routing of data among federates during the course of a federation execution [2, 3]. Using the runtime services described by HLA interface specification, it is possible to carry out the functions of the distributed system represented as a federation.

Information to be exchanged between federates in a federation should be represented object-oriented using a standard structure which is necessarily required in HLA implementations. IEEE Std 1516.2-2000 standard document describes the HLA-Object Model Template (OMT) specification [4]. Using the OMT, each federate defines, in its simulation object model (SOM), the data that it is willing to share (publish) with other federates and the data it requires from other federates (subscribe). The federation object model (FOM) combines the federate SOMs into a single object model for the federation and defines the overall data to be exchanged (published and subscribed) between federates.

3. HLA based distributed manufacturing simulation applications

Since traditional simulation applications are sometimes not capable of meeting the requirements of today's distributed manufacturing, a general distributed simulation architecture utilisation is required in order to meet the needs. HLA based distributed manufacturing simulation applications can be seen in the literature. IMS-MISSION project was one of the pioneering and leading research projects in the distributed manufacturing area. The aim of this project was to support simulation applications of globally distributed or virtual enterprises using HLA [5].

McLean and Riddick [6] present an overview of neutral reference architecture for integrating distributed manufacturing simulation systems with each other and other manufacturing software applications, as well as with manufacturing data repositories. Hibino et al. [7] developed HLA based distributed simulation system to easily evaluate a very large manufacturing system by synchronizing several different simulators.

Taylor et al. [8] report on the progress made towards the emergency of standards to support the integration of heterogeneous discrete-event simulations (DESs) created in specialist support tools called commercial-off-the-shelf (COTS) discrete-event simulation packages (CSPs). The general standard for heterogeneous integration in this area has been developed from research in distributed simulation and is the IEEE 1516.

McLean et al. [9] present an overview of neutral reference architecture and data model for integrating distributed manufacturing simulation systems with each other and other manufacturing software applications as well as with manufacturing data repositories. They established their study on the basis of HLA.

User support system and its requirements is another significant consideration in distributed manufacturing simulation. Hibino and Fukuda [10] focused on a user support system for the realization of the consistent management and support. They provide a definition of a procedure for a distributed simulation while storing and re-using parameters defined. They also designed the remote execution of a distributed simulation system with a single manufacturing system designer.

Supply chains are typical examples for distributed manufacturing. HLA is proposed as modeling and simulation middleware in order to be utilized in manufacturing supply chains [11,12,13]. Bandinelli et al. [14] gave an overview of distributed simulation standards, models and/or architectures which can be an effective tool for supply chain performance analysis, describing the technological choices related to each of them, and to propose how and when a distributed supply chain simulation framework is to be used. They explained the HLA as an enabling technology in supply chain simulation implementations. Kubat and Uygun also proposed an HLA based supply chain management system for Sakarya automotive suppliers [15].

4. Case Study

A scenario is generated and described in order to explain how HLA can be utilised in a distributed manufacturing simulation. An illustration of the related scenario is shown in Fig. 1. According to the case study, the main manufacturer needs parts A and B in order to assemble part C which is the finished product. The main manufacturer produces part A, while it procures part B from a supplier. The overall system is modelled as a federation that consists of two federates. One of the two federates is the manufacturer federate and the other one is the supplier federate. Federates interact with each other via HLA run-time infrastructure (RTI) making benefit of services provided by the RTI.

The main manufacturer opens a production order in order to produce part C. When a production order is opened, the information is sent to the supplier. While the main manufacturer begins to produce part A, the supplier also begins to produce part B based on the information it received from the main manufacturer. The process times of part A and B can obviously be different. Since the main manufacturer needs to know if the supplier finished the required amount of part B, the supplier sends the necessary information to the manufacturer. Then the main manufacturer can begin to assemble the finished product (part C). The time when to begin to produce part C, depends on the times when part A and B are finished. If part A is finished later than part B then the beginning time of part C production depends on the end time of part A production or vice versa. The transportation time of part C to the main manufacturing place is eliminated which is very easy to be included, so that it is possible to compare the finishing times of part A and B. Otherwise, the beginning time of part C would always be based on the finishing time of part B production. Going though this way, it is sometimes based on part A's finishing time and sometimes on part B's finishing time. The simulation system can be run as many times as wished.

5. Implementation

In the implementation of the system, two federates are created (manufacturer and supplier) and brought together in a HLA federation. Related federates interoperate with each other over HLA-RTI.

RTI1.3NG-V6 run time infrastructure software provided by DMSO (US Defence Modelling and Simulation Office) is used in this implementation. The software was installed on a computer with Windows XP SP2 operating system. This RTI software requires MS Visual C++ 6.0 SP5 compiler to be installed on the computer.

Before simulation algorithm is programmed using C++ and the RTI software features, it is necessary to define object models, for federates and the federation (SOM and FOM). OMDT (Object Model Development Tool) software package is used in defining information models that is going to be interacted within the federation. OMDT tool enables to identify object class structure, attributes of the objects, interaction class structure, parameters of the interactions, data types, and some other specifications. Table 1 shows the object class structure that is used in the case study addressed. As can be seen in the table, PartC is the

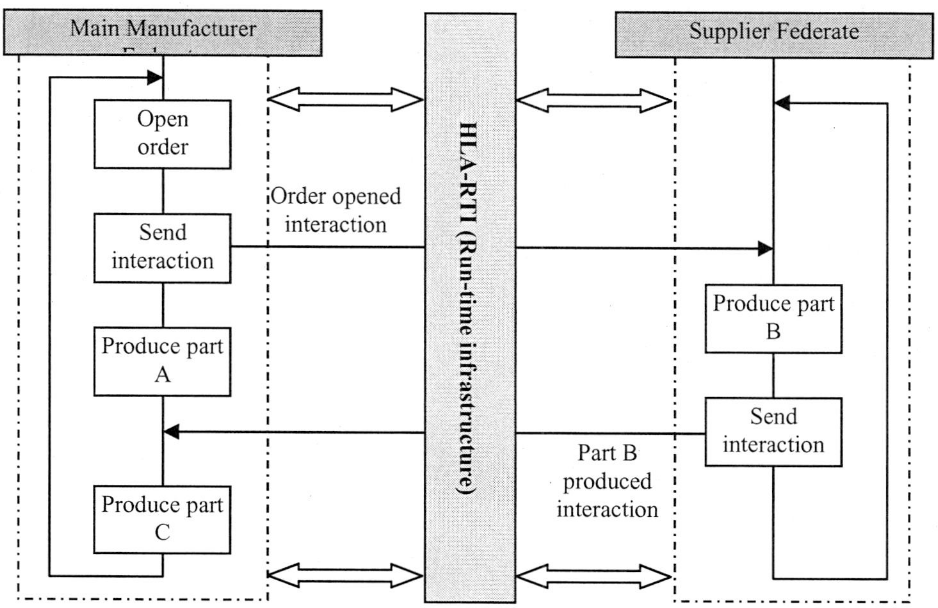

Fig. 1. The distributed simulation model of the scenario

main class while PartA and PartB are the sub-classes of PartC. Sub-classes can get the characteristic from its upper class because of the inheritance feature of the object oriented approach.

Table 1
Object class structure table

Class1	Class2
PartC	PartA
	PartB

The attributes of the objects defined are shown in Table 2. ID attribute of PartC is valid for PartA and PartB due to inheritance from PartC as well as quantity. However for each part a start time and a finish time attributes are required.

Table 2
Attributes of the objects

Object	Attribute
PartC	ID
	Quantity
	StartTimeC
	FinishTimeC
PartA	StartTimeA
	FinishTimeA
PartB	StartTimeB
	FinishTimeB

In the execution of the federation two interactions are defined and utilised. The fist one is *OrderOpened* interaction which is used to send information to the supplier and triggers it for part B production. The second one is *PartBProduced* interaction which is used to send information from the supplier to the manufacturer to let it know if the part B is finished and when.

One of the ways of interaction between federates is publishing of object attributes or subscribing to the attributes. The other way is using interactions which trigger an event. The data is kept in the attributes values. In this application information is exchanged updating and reflecting of the attributes values. However, interoperation is supported by both publishing and subscribing of the attributes, and by sending interactions.

After object model definition, federates and the federation is created and necessary algorithm coded using C++ codes. Object model is used in the application converting the model to the FED file which is the file format that the RTI software supports. HLA services are also used within C++ codes by including RTI software functions or procedures.

Two federates were developed in order to execute the federation: *manufacturer* federate and *supplier* federate. Several files were developed for each federates and several procedures within each file. Table 3 describes the files developed or modified for manufacturer federate.

Table 3
Files that is utilised within manufacturer federate

File	Description
production.cpp	This is the main file of the federate. Procedures developed in rti_interface.cpp file are used in this file as functions.
rti_interface.cpp	The procedures and functions that enable manufacturer federate to interact with RTI are developed in this file.
rti_interface.h	This is the header file where the functions used in the rti_interface.cpp file are described in terms of function parameters.
Mnf_Federate_ Ambassador.cpp	This is the ambassador file of the manufacturer federate. Each federate must have an ambassador file which is provided by the RTI software but should be modified according to the application needs. The ambassador file is the part of the federate that interact with the run time infrastructure.
Mnf_Federate_ Ambassador.h	This is the header file where the functions defined in the ambassador file are described in terms of function parameters.
part.h	The results of the interaction between federates are caught in the functions of the ambassador file. However the data caught in the ambassador file should be used in other files of the federate. One way of doing this is defining global variable classes and utilising them in other files. part.h is developed for that purpose and included to other files as a header file.
part.cpp	This is the file in which functions are developed related to the variable classes that are defined in part.h file.
production.fed	This is the FED file of the federate in which all the required information (to be published or subscribed) is defined. RTI needs SOM or FOM files in FED file format.
RTI.rid	This file is created during the RTI software installation and is necessary for federation execution. However it should be modified according to the application needs.

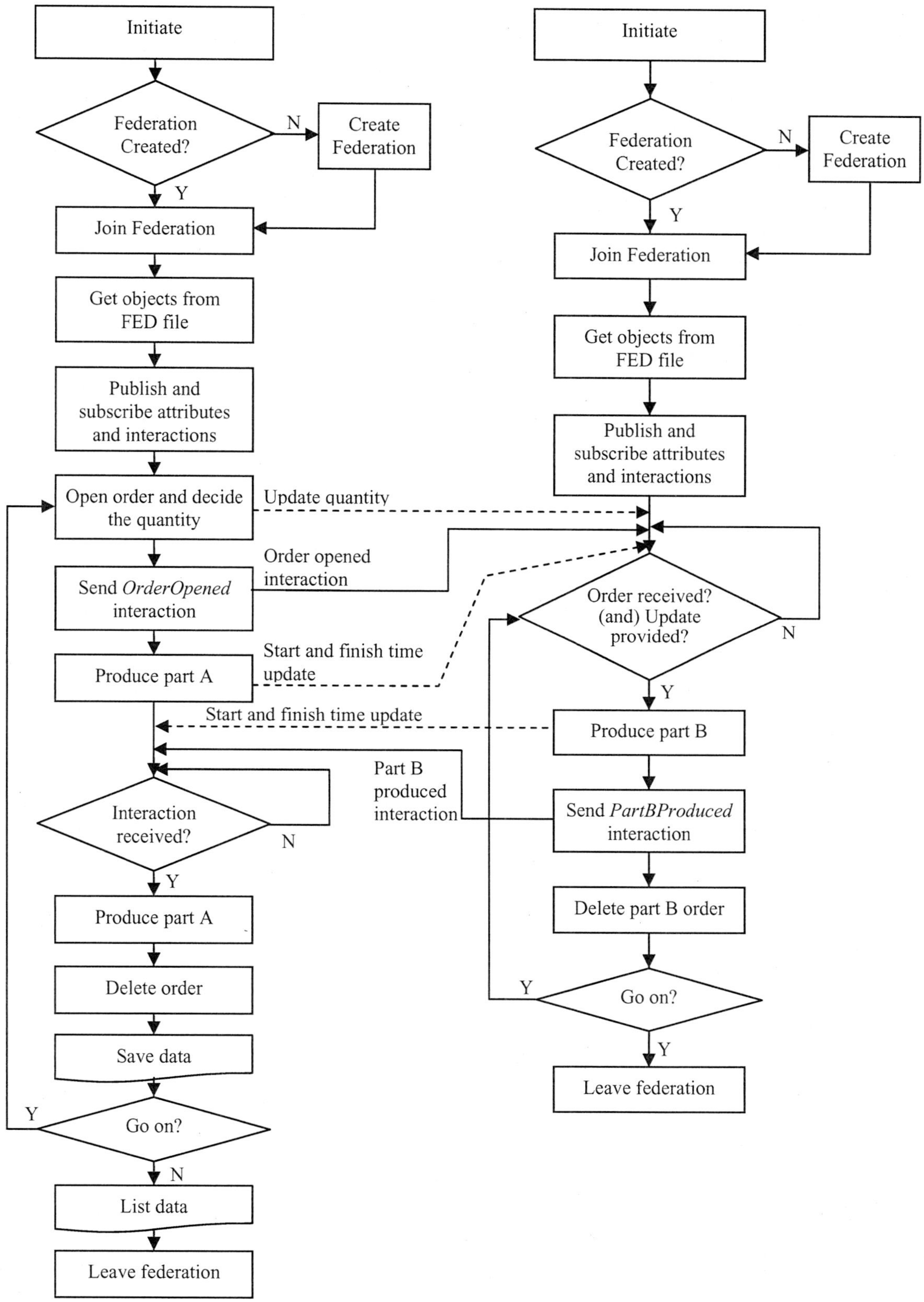

Fig. 2 . Flow chart of the federation and interoperation of federates in the federation

Similar files were also developed for the supplier federate. So, general simulation algorithm was coded in the files and RTI services are implemented using RTI functions within the files. Therefore the federation is executed and interoperation is provided between federates.

The progress of the simulation time is based on the events. Opening order, finish times of part A, B, and C are the events of the related simulation system and are used for progressing of the federation time. General flow of the simulation, and interaction between the manufacturer and the supplier federates are illustrated in Fig. 2. The information exchanged by attributes value update is represented with a dashed line while sending interactions are represented with a straight line in the figure.

6. Conclusion

In this study, a distributed manufacturing simulation case is considered. The implementation of the simulations is conducted using HLA technical methods and processes. Some details of the implementation are also described such as information representation, federate creation and files developed within federates. Description of developed files is also detailed. General flow of the simulation (the federation) and interoperation between federates is illustrated. In terms of proof of concept necessary application details are addressed.

The main aim of the paper is to describe in general how a distributed manufacturing simulation can be built using HLA. Numerical results and performance of the overall simulation is out of the scope and ignored in the study. Performance alignment is not considered so the numerical results are not given. The paper gives an outline for HLA based distributed manufacturing simulation in order to take advantage of distributed information exchange and interoperation between separate simulations.

Acknowledgements

Sakarya University is a partner of the Innovative Production Machines and Systems (I*PROMS) Network of Excellence. http://www.iproms.org

References

[1] IEEE, IEEE Standard for Modeling and Simulation (M&S) High Level Architecture (HLA)-Framework and Rules. The Institute of Electrical and Electronics Engineers Inc. Published 11 December 2000.

[2] IEEE, IEEE Standard for Modeling and Simulation (M&S) High Level Architecture (HLA)-Federate Interface Specification. The Institute of Electrical and Electronics Engineers Inc. Published 9 March 2001.

[3] Dahmann SJ. The High Level Architecture and Beyond: Technology Challenges. In Proceedings of Thirteenth Workshop on Parallel and Distributed Simulation, 1-4 May 1999, p64-70

[4] IEEE, IEEE Standard for Modeling and Simulation (M&S) High Level Architecture (HLA)-Object Model Template (OMT) Specification. The Institute of Electrical and Electronics Engineers Inc. Published 9 March 2001.

[5] MISSION, Deliverable D24 Final Report from MISSION Project, November 12, 2001

[6] McLean C and Riddick F. The IMS Mission Architecture for Distributed Manufacturing Simulation. Proceedings of the 2000 Winter Simulation Conference, 10-13 December, Orlando, FL

[7] Hibino H, Fukuda Y, Yura Y, Mitsuyuki K, Kaneda K. Manufacturing Adapter of Distributed Simulation Systems Using HLA. Proceedings of the 2002 Winter Simulation Conference, 8-11 December, San Diego, CA

[8] Taylor SJE, Wang X, Turner SJ, Low YH. Integrating Heterogeneous Distributed COTS Discrete-Event Simulation Packages: An Emerging Standards-Based Approach. IEEE Transactions on System, Man and Cybernetics-Part A: System and Humans, Vol.36, No.1, January 2006, p109-122

[9] McLean C, Riddick F, Lee YT. An Architecture and Interfaces for Distributed Manufacturing Simulation, Simulation, Vol. 81, Issue 1, January 2005, p15-32

[10] Hibino H and Fukuda YA. User Support System for Manufacturing System Design Using Distributed Simulation. Production Planning & Control, Vol. 17, No. 2, March 2006, 128–142.

[11] Mertins K, Rabe M, Jakel FW. Distributed Modelling and Simulation of Supply Chains. International Journal of Computer Integrated Manufacturing, Vol. 18, No. 5, July–August 2005, 342 – 349

[12] Terzi S, Cavaliery S. Simulation in the Supply Chain Context: A Survey. Computers in Industry 53 (2004) 3–16

[13] Lendermann P, Julka N, Gan BP, Chen D, McGinnis L F, McGinnis JP. Distributed Supply Chain Simulation as a Decision Support Tool for the Semiconductor Industry. Simulation, Vol. 79, Issue 3, March 2003, p126-138

[14] Bandinelli R, Rapaccini M, Tucci M, Visintin F. Using Simulation for Supply Chain Analysis: Reviewing and Proposing Distributed Simulation Frameworks. Production Planning & Control, Vol. 17, No. 2, March 2006, 167–175.

[15] Kubat C, Uygun Ö. HLA based supply chain management for Sakarya automotive suppliers. In the Proceedings of 3rd I*PROMS Virtual International Conference, 2-13 July 2007, pp 413-418

Innovative Production Machines and Systems
D.T. Pham, E.E. Eldukhri and A.J. Soroka (Eds.)

Robust design of workload based order release mechanisms

N.O. Fernandes[a], S. Carmo-Silva[b]

[a] *Dept. de Engenharia Industrial, Instituto Politécnico de Castelo Branco , 6000 Castelo Branco – Portugal.*
[b] *Dept. de Produção e Sistemas, Universidade do Minho, 4710-057 Braga - Portugal*

Abstract

A variety of order release mechanims have been developed for workload control. While previous research has pointed out strengths and weaknesses of each mechanim, the choice of the appropriate 'way' for the release of orders to the shop floor remains an open issue. In this paper, the effectiveness of workload based order release mechanims, in a job shop environment, is assessed from the poit of view of its robusteness when facing changes in manufacturing environmental conditions. A plan of experiments based on the Taguchi method was performed. Simulation results indicate that the order release mechanisms performance seems to be more sensitive to the operating parameters setting than to the order release strategies adopted.

Keywords: order release, production control, simulation

---

1. Introduction

For manufacturing companies to stay competitive in the global market of today, manufacturing strategies have to be focused on speed of response to customer requirements. This means short delivery times, on time deliveries and flexibility to customer requirements. Thus the make-to-order (MTO) sector of industry is of continued importance [1].

Workload Control (WLC) is a production planning and control concept suitable for job shops in MTO sector [2]. Its main principle is to keep the length of queues on the shop floor at appropriate levels in order to create short and predictable throughput times and thus to meet the promised deliver dates. Predictable throughput times are particularly important for a good timing of release and capacity adjustments and for quoting realistic due dates. It is possible to identify three hierarchical levels, related with order flow control, at which the control of these queues can be attempted [3], namely order entry, order release and dispatching. At each level, a decision must be made relatively to the orders allowed to proceed to the next phase and whether this requires capacity adjustments.

Orders release has been described as a main control element within WLC. It determines the moment and the orders (jobs) to be released into the shop floor. Before being released, orders are collected in a centrally controlled pre-shop pool. The pool acts as a buffer absorbing fluctuations of the incoming order stream. Collected orders are only released if they fit workload norms for the required capacity groups (e.g. groups of resources). The decision to release an order is usually based on its urgency and influence on the current shop floor situation [4]. It is recognised that stable direct workloads on capacity groups can only be realized by releasing an appropriate set of orders to the shop floor. This is known as load balancing.

Several mechanisms have been developed for workload based order release. Stevenson et al. [5] review literature on this matter. Two of the most known mechanisms proposed in the literature are the Load Oriented Order Release (LOOR) [6] [7] and the Lancaster University Management School (LUMS) approach [8] [9]. However, as pointed out in [10] one ` best way' for order release does not exist. Moreover, one mechanism may perform well under a particular manufacturing environment and poor when the shop operating conditions alter. This suggests that the order release mechanism robustness, when facing changes in manufacturing environmental conditions, should be considered as a main issue in assessing its potential. However, this is often overlooked by researchers.

This paper seeks to explore this issue by assessing the potential of single features of the order release mechanisms, such as the strategy used to account workload over time and the strategy to control workload. Additionally the paper aims at improving the basis for setting parameters such as the *time limit* and the *release period*, showing its impact on shop performance. The robust design concept of Taguchi was used to assess the robustness of these strategies and parameters. The results of this study should contribute to the choice of appropriate order release mechanisms in practical situations.

The remainder of the paper is organised as follows: section 2 addresses the overall research methodology; section 3 presents, analyses and discusses the results of the simulation study; and finally in section 4, concluding remarks are made and directions for future research work are presented.

2. Research Methodology

A simulation study was carried out using Arena software. During simulation runs, data were collected with reference to the system steady state. The simulation runs last for 27600 time units. For each simulation run 90 independent replications were performed. To avoid initialization bias a warm-up period of 9600 time units was used. Common random numbers were used as a variance reduction technique across all experiments.

2.1. Simulation model

A job shop without an explicit bottleneck has been considered. The shop consists of six work centres, each containing a single multi-purpose machine. In the simulation model, the job's routing is randomly chosen from a set of twenty routings patterns, each of which with an equal probability of

Table 1
Work centre job routing matrix

Routing pattern	Operation number					
	1	2	3	4	5	6
1	2	4	6	1	5	3
2	1	3	5			
3	2	3	5	4		
4	5					
5	4	2	5	6	1	
6	2	5	4	6	1	3
7	1	3	2	6		
8	2	6				
9	2	5	4			
10	3	1	5	4	6	2
11	6	2	3			
12	2	6	1	3	2	
13	2	3	6			
14	4	1	2	5	3	
15	1					
16	4	3	6	5	1	
17	4					
18	3	4	6	5		
19	4	1	6			
20	4	1				

occurrence, see Table 1. The mean number of operations per routing pattern is 3.6.

Due dates are assigned to jobs on their arrival to the system and are modelled as a random variable. They are established to ensure that, under immediate release, the number of tardy jobs falls between 5% and 10%. After the assignment of the due date, jobs are placed in the pre-shop pool, waiting for release. Job release decisions are made periodically in the beginning of each *release period.* Each job is considered for release according to the earliest planned release time. Planned release times are determined by backward scheduling from the due date, using the work centres lead times. Work centre lead times were established through some pilot simulation runs, also using immediate release.

The order release mechanism ensures that a job candidate for release is released only if it fits the established workload norms for the required capacity groups. Every job in the pool with a planned release time within a *time limit* is a candidate.

Released jobs follow a first-in-first-out (FIFO) dispatching rule on the shop floor in all machines. Setup times have been considered sequence independent and assumed as part of the operation processing time.

2.2. Experimental design

Table 2 gives an overall view of the control factors, studied, i.e. independent variables, and the assignment of the corresponding levels.

The *workload accounting* over time, defines the method of accounting the load of a released job, establishing when and how much of this load should be allocated (or removed) to (from) each capacity group. Workload accounting was tested at two levels, namely: atemporal and probabilistic. Under the atemporal approach the adjusted aggregated load method introduced by Oosterman et al. [11] was used. Under the probabilistic approach the load conversion method introduced by Bechte [6] was used. Both methods are detailed in [3].

Workload control was studied at two levels, namely: upper workload bound and workload balancing. The first strategy is a typical approach to workload control, used in mechanisms such as LOOR and LUMS, allowing the release of a job into the shop floor, only if workload in all capacity groups of the job's routing does not exceeds an upper limit or workload norm. In the second strategy, workload control is achieved by balancing workload across capacity groups. Releases are made only if the job contributes to a better load balancing between capacity groups. Workload is allowed to exceed the upper limit of each capacity group in the job's routing in 20%. This value was established after some pilot simulation runs, and leads to good results in this study. The workload balancing strategy may compensate for the fact that, when workload across capacity groups is poorly balanced, rigid upper workload limits for the heavily loaded resources may block the release of work to the under-loaded resources, introducing idle time.

The balance index (BI) employed was the follow:

$$BI = \frac{\sum_i F_{ij}}{\max_i \{F_{ij}\} m} \quad (i = 1,\dots m) \qquad (1)$$

where, F_{ij} represents the accounted workload on capacity group i resulting from the releasing of job j into the shop floor, and m the number of capacity groups. The best balancing situation is obtained for BI equal to one.

The *release period* determines the time interval between releases. This factor was tested at two levels, namely: 8 time units and 4 time units. Note that a short release period may reduce the time a job spends in the pool, and eventually in the system. However, is also likely to reduce the possibilities for load balancing, due to an expected smaller number of jobs in the pool, and hinders the release of large jobs.

Table 2
Control factors

Factors	code	Level 1	Level 2
workload accounting	A	atemporal	probabilistic
workload control	B	upper bound	balancing
release period	C	8 time units	4 time units
time limit	D	infinite	2 release periods

Table 3
Noise factors

Factors	code	Level 1	Level 2
system utilization	X	90%	80%
processing time variability	Y	2-Erlang	Exponential
time between arrivals variability	Z	Exponential	Constant

The *time limit* is used in the release procedure to restrict the set of jobs considered for release to those that are urgent. This factor was tested at two levels, namely: infinite and equal to two release periods. An infinite time limit means that all available jobs in the pool, independent of their panned release times, are considered for release, in the beginning of each release period. A finite value to the time limit means that only the jobs having a planned release time falling within the time limit are considered for release. All others must wait. However, a resulting reduction of balancing opportunities may be expected from a finite time limit. In order to avoid the delay of jobs in the pool, the time limit should always be at least as large as the release period.

Table 3 gives an overall view of the noise factors studied and the assignment of the corresponding levels.

Operations processing time variability is a common source of perturbation in manufacturing systems. Variability of processing times can be due to several reasons, e.g. to the natural influence of human work, variations in lot sizes, materials supply, etc. This factor was studied at two levels.

Table 4
Plan of the experiment and time in system results

																outer array $L_4(2^3)$				
															X	1	2	2	1	
															Y	1	2	1	2	
							inner array $L_{16}(2^{15})$								Z	1	1	2	2	
Test	A	B	AB	C	AC	BC	ABC	D	AD	BD	ABD	CD	ACD	BCD		Time in system				
i	1	2	3	4	5	6	7	8	9	10	11	12	13	14	15	y_{i1}	y_{i2}	y_{i3}	y_{i4}	η (dB)
1	1	1	1	1	1	1	1	1	1	1	1	1	1	1	1	36.6	24.0	16.9	67.1	-32,2
2	1	1	1	1	1	1	1	2	2	2	2	2	2	2	2	53.0	38.8	31.9	90.4	-35,3
3	1	1	1	2	2	2	2	1	1	1	1	2	2	2	2	30.0	20.5	14.0	43.5	-29,3
4	1	1	1	2	2	2	2	2	2	2	2	1	1	1	1	54.5	43.0	36.5	71.2	-34,5
5	1	2	2	1	1	2	2	1	1	2	2	1	1	2	2	35.4	24.2	17.0	55.8	-31,2
6	1	2	2	1	1	2	2	2	2	1	1	2	2	1	1	50.8	38.5	32.0	74.7	-34,3
7	1	2	2	2	2	1	1	1	1	2	2	2	2	1	1	29.3	20.6	14.0	39.4	-28,8
8	1	2	2	2	2	1	1	2	2	1	1	1	1	2	2	52.4	42.7	36.5	64.9	-34,0
9	2	1	2	1	2	1	2	1	2	1	2	1	2	1	2	38.0	24.6	17.0	91.5	-34,3
10	2	1	2	1	2	1	2	2	1	2	1	2	1	2	1	54.4	39.5	32.0	112.8	-36,6
11	2	1	2	2	1	2	1	1	2	1	2	2	1	2	1	32.0	21.5	14.1	60.7	-31,3
12	2	1	2	2	1	2	1	2	1	2	1	1	2	1	2	55.5	43.8	36.6	89.4	-35,5
13	2	2	1	1	2	2	1	1	2	2	1	1	2	2	1	35.5	24.2	16.9	62.3	-31,8
14	2	2	1	1	2	2	1	2	1	1	2	2	1	1	2	52.1	39.0	31.9	86.7	-35,0
15	2	2	1	2	1	1	2	1	2	2	1	2	1	1	2	31.4	21.4	14.1	52.3	-30,4
16	2	2	1	2	1	1	2	2	1	1	2	1	2	2	1	50.7	39.5	32.7	75.5	-34,4

To represent a situation of high variability in processing times an exponential distribution was used, and to represent a more stable situation the 2-Erlang was considered. According to [11] the 2-Erlang distribution approaches well the observations made in real life job shops. In both distributions, the mean processing time was the same and equal to 1 time unit.

The average planned system utilization can be measured as a percentage of the overall shop capacity used and is dependent on the amount of work released by the planning system into the pre-shop pool. According to [10] this has an important effect on the systems performance. In our study the utilization factor was tested at two levels: 90% and 80%, by setting the appropriate time between arrivals of orders.

The planning system may or may not control the orders arrivals to the pool. The former is based on smoothed schedules, while the latter is indicative of a non effective planning system, i.e. no smoothing is carried out. These two types of orders arrival are simulated in this study. Controlled orders arrivals are simulated using constant time intervals between successive arrivals, while uncontrolled orders arrivals are simulated using exponential distributed time intervals. In both situations an identical mean is used.

The approach used to plan the experiments is based on the Taguchi method. The method explicitly models the effects of manufacturing environmental variations on the four control factors, and thus provides guidance for implementing the order release mechanisms in uncertain environments. The experimental design employed an inner $L_{16}(2^{15})$ orthogonal array for control factors and an outer $L_4(2^3)$ orthogonal array for environmental (noise) factors, Table 4. The inner array has 16 rows, corresponding to the number of tests, with 15 columns at two levels. Factors and interactions are assigned to the columns. The outer array has 3 rows, corresponding to different manufacturing environment conditions, where the order release mechanisms should operate. A trial number specifies a test condition with respect to control factors, but the outer array specifies four test conditions with respect to the noise factors to that trial. This means that the experimental design is made of 64 separate tests, and an identical number of simulations runs are needed.

One key feature of the Taguchi method is the use of a robustness measure called signal-to-noise (S/N) ratio, measured in decibels (db). The S/N ratio aggregates information on the average performance and its variability. It is treated as a response of the experiment and is a measure of the variation within a trial when noise factors are present. The S/N ratio (η) for a smaller-the-better characteristic is as follows [13]:

$$\eta = -10\log\left(\frac{1}{n}\sum_{j=1}^{n} y_{ij}^{2}\right) \tag{2}$$

where, y_{ij} is the individual response value from the ith combination of control factors, and j^{th} combination of noise factors and n the total number of combinations of noise factors for each combination of control factors.

The smaller-the-better S/N ratio is used because our main performance measure under study is time in system. Time in system is defined as the time a job spends waiting in the pre-shop pool plus the shop flow time. We also recorded shop flow time, which is defined as the time that elapses between job release and job completion. However, the latter was used as an instrumental variable that allows us to develop all test conditions at the same value of shop flow time, i.e. for equally tight work load norms.

The benefits of reducing time in system are related with reducing the overall response time to customers. Reducing the shop flow time has also intrinsic benefits, which implies a smaller WIP (*work in process*) and therefore reduced tied up capital.

3. Results and Discussion

This section presents and discusses the results of the simulation study described in the preceding section. Simulation results with the calculated signal-to-noise (S/N) ratios measured in decibels are summarized in Table 4. The main factors effects are depicted in Table 5.

According to the Taguchi method, the best level for a factor, i.e. the level with the lowest variation across the combination of noise factors, is the level with the highest S/R ratio value [12]. Based on the Table 5 data it can be seen that the levels that minimize time in system are: the atemporal workload

Table 5
S/N ratios (dB)

Factors	Level 1	Level 2
workload accounting	-32,45*	-33,66
workload control	-33,63	-32,48*
release period	-33,83	-32,27*
time limit	-31,15*	-34,95

*best values

accounting over time (level 1); workload balancing strategy (level 2), a release period of 4 time units (level 2) and an infinite time limit (level 1).

Table 6 shows the results of an analysis of variance (ANOVA). In this table, pooling was done until the degrees of freedom of the error term became close to half the total of degrees of freedom. The purpose of this analysis is to determine the factors and combination of factors that significantly affect performance of the order release mechanisms. This analysis was undertaken for the level of significance of 5%, i.e. for a 95% confidence level. The last column of the table indicates the percent contribution (p) of each factor on the total variation, indicating then the degree of influence on the system performance. According to the data, the time limit and the release period are the major factors affecting the order release mechanisms performance, with p=68.84% and p=11.54%, respectively. Workload accounting and workload control strategies also affect performance, but with lower percent contributions. Interactions between main factors do not present significant contribution percentages, being smaller than the experimental error of 1.19%. An exception is the interaction CD, between release period and time limit, with a 3.49% contribution. However, part of this interaction may be due to the experimental setting, in particular, to the fact that, one of the levels of time limit factor was set in the experimentation as a function of the release period, i.e. two release periods.

In reviewing the results the following issues deserve special attention for the selection and implementation of workload based order release mechanisms in practice:

(1) The *release period* and the *time limit* are parameters of special importance, influencing in a strong manner the performance of the order release mechanisms. Therefore, when implementing the WLC concept, focus should be in establishing the levels of these parameters.

Table 6
ANOVA using the simulated results as S/N ratios

Factor or interaction	DOF [a]	SS [b]	Variance	Test F	p^c
A	1	5.87	5.87	88.32	6.95
B	1	5.28	5.28	79.45	6.24
C	1	9.70	9.70	145.92	11.54
D	1	57.57	57.57	865.79	68.84
AB	1	0.58	0.58	8.77	0.62
BC	1	0.61	0.61	9.12	0.65
AD	1	0.47	0.47	7.04	0.48
CD	1	2.98	2.98	44.78	3.49
Error	7	0.47	0.07		1.19
Total	15	83.53			100

[a] degrees of freedom; [b] sum of squares; [c] percentage contribution

(2) The workload balancing and the atemporal workload accounting over time showed to be robust strategies, when facing changes in the manufacturing environmental conditions. Therefore, good results may be expected from the use of these strategies in the order release mechanisms.

4. Conclusions

Order release mechanisms are strategically important for the economic success of companies having a great influence on systems operations behaviour. However, the performance of these mechanisms highly depend on operating parameters settings and order release strategies adopted. In this study, the effectiveness of operating parameters and order release strategies within the WLC concept, were evaluated through a simulation study.

Two important conclusions could be drawn. First, order release mechanisms performance is particularly influenced by parameters such as the 'time limit' that restricts the set of jobs considered for release and the 'release period' that determines the time interval between releases. Second, results show that 'workload balancing' and 'atemporal' workload accounting over time are robust strategies, suitable to be used in order release mechanisms for job shops in the make-to-order (MTO) sector of industry.

Although some insights point out the utility of the study for practical use, it is important to extend it to a wider spectrum of order release strategies and operating parameters in order to get an in dept understanding of the full complexity behind the order release mechanisms.

Acknowledgements

University of Minho is partner of the EU-funded FP6 Innovative Production Machines and Systems (I*PROMS) Network of Excellence. http://www.iproms.org

References

[1] Hendry L., Land M., Stevenson M. and Gaalman G. investigating implementation issues for workload control (WLC): A comparative case study analysis. International Journal of Production Economics. 112-1 (2008) 452-469.

[2] Haskose A., Kingsman B.G. and Worthington D. Performance analysis of make-to-order manufacturing systems under different workload control regimes. International Journal of Production Economics. 90-2 (2004) 169-186.

[3] Breithaupt J.W., Land J. and Nyhuis, P. The workload control concept: theory and practical extensions of Load Oriented Order Release. Production Planning and Control. 13-7 (2002) 625-638.

[4] Henrich P., Land M. and Gaalman G. Exploring applicability of the workload control concept. International Journal of Production Economics. 90-2 (2004) 187-198.

[5] Stevenson M, Hendry L and Kingsman B.G. A review of production planning and control: the applicability of key concepts to the make-to-order industry. International Journal of Production Research. 43-5 (2005) 869-898.

[6] Bechte W. Theory and practice of load-oriented manufacturing control. International Journal of Production Research. 26-3 (1988) 375-395.

[7] Wiendahl H.P. Load Oriented Manufacturing Control. Springer, New York, 1995.

[8] Hendry L. and Kingsman B.G. A decision support system for job release in make-to-order companies. International Journal of Operations and Production Management. 11 (1991) 6-16.

[9] Stevenson M, Hendry L. Aggregated Load-oriented workload control: A review and re-classification of a key approach. International Journal of Production Economics. 104-2 (2006) 676-693.

[10] Cigolini, R., Perona, M. and Portioli, A., Comparison of order review and release techniques in a dynamic and uncertain job shop environment, Int. Journal of Production Research, 36-11 (1998) 2931-2951.

[11] Oosterman B., Land M. and Gaalman G. The influence of shop characteristics on workload control, Int. Journal of Production Economics; 68 (2000) 107-119.

[12] Ross, P. Taguchi Techniques for Quality Engineering: Loss Function, Orthogonal Experiments, Parameter and Tolerance Design. McGraw-Hill, New York, 1988.

Innovative Production Machines and Systems
D.T. Pham, E.E. Eldukhri and A.J. Soroka (Eds.)

Smart Connected and Interactive Production Control in a Distributed Environment

T. Schlegel[a], M. Foursa[b], F. Meo[c], J. Agirre Ibarbia[d], G. Haidegger[e], I. Mezgar[e], A. Herrmann Praturlon[f], J. Canou[g]

[a] *Fraunhofer Institute for Industrial Engineering, Nobelstr. 12, Stuttgart, Germany*
[b] *Fraunhofer IAIS, Sankt-Augustin, Germany*
[c] *FIDIA S.p.A., Torino, Italy*
[d] *Fatronik, San Sebastian, Spain*
[e] *CIM-EXP, Budapest, Hungary*
[f] Centro Ricerche Fiat, *Torino, Italy*
[g] Robosoft, *Bidart, France*

Abstract

The European research project INT-MANUS embedded in the I*PROMS European network of excellence addresses the increasing demand for flexibility and adaptivity, which is summarized by rapid reconfigurations of complete factories as well as related aspects in Human Computer Interaction (HCI), Software, and Production Systems. The project's main goal is to develop a new technology for the production plants of the future, the Smart Connected Control Platform (SCCP). This platform allows controlling a factory with the help of an open distributed learning agent platform that integrates machines, robots, and human personnel.

Keywords: Smart-Connected-Control Platform (SCCP), Advanced Maintenance, Virtual Reality

1. Introduction

Today's production and manufacturing systems are challenged by an increasing demand for flexibility and adaptivity which is summarized by the rapid reconfiguration of a factory. While improvements are already being undertaken with respect to technical requirements, interactive processes and interaction technology in a highly dynamic environment still needs further attention.The European research project INT-MANUS [1] has been launched from the I*PROMS European network of excellence [2] in order to address these issues in addition to aspects of Human Computer Interface (HCI), Software, and Production Systems.

The Smart-Connected-Control Platform (SCCP) for manufacturing enterprises allows controlling a production plant with help of an open distributed learning agent platform that integrates machines, robots, and human personnel. The Smart Connected Control Platform will allow higher automation capabilities in production, delivering a decentralized concept for connecting machines and interaction devices. Ubiquitous augmented reality technology will provide better control abilities for factories, starting with an enhanced communication infrastructure. Integrating interfaces and support for human personnel in every aspect of the production process thereby leads to a better control of the production and manufacturing process.

Each peer (machine tool, robot or other device) is equipped with a local control system that is in charge of local control and communication with the Enterprise Service Bus (ESB). Data collected from sensors are partially processed by the local control systems and after filtering sent over ESB to SCCP services for storing and higher-level analysis. The system then applies its decision-support techniques and if necessary sends updates to control systems and notifications.

Connecting numerous devices to the SCCP communication infrastructure, a common access platform is needed. This component, the platform peer, fulfils an important task. It receives (e.g. sensor) data from the device attached and makes them accessible to other peers. Therefore the data have to be collected, classified and referenced to the semantic data description model, a part of the model ontology available in the network. Two processes for the distribution of semantic referenced/annotated data are designated. The Platform Peer can proactively send the data (including a classification reference) to a peer that is known to need them. The more recommended process is having the Platform Peer providing events for each data that can be observed by interested peers. This way the Platform Peer does not need explicit knowledge of workflow and data routing. Thus the semantic backbone will mainly use the event based data transmission.

2. SCCP Production Service Bus (PSB)

The SCCP communication protocol is defined in the Production Service Bus (PSB) as illustrated in figure 1. The PSB provides a service infrastructure to guarantee an easy and reliable transmission of semantic referenced data. A PSB consists of two layers: Peer Infrastructure Services allow implementing a decentralized peer index and contain functions that help to keep the network consistent and connected.

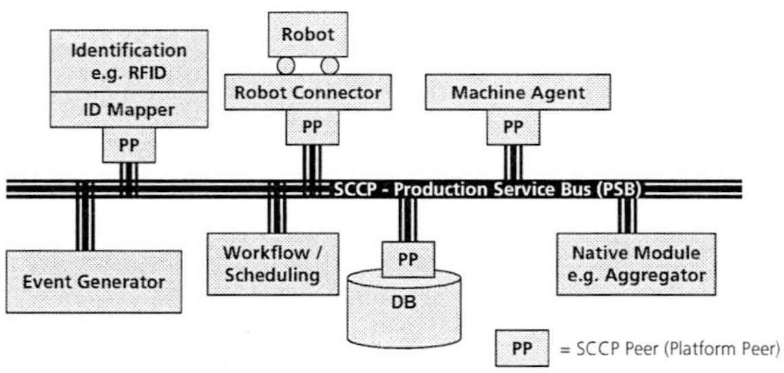

Fig. 1. PSB Bus System

The PSB describes processes for typical network situations (register, Index Node drop out, network split ...) that have to be implemented by all participating peers. The second PSB layer describes services allowing SCCP event based data transmission. On the basis of these services every peer is able to provide events to observe and can send data according to events occurred. Additionally proactive data transmission is possible as well. Object-oriented models [11] and processes provide semantic messaging and decentralized workflow control.

3. Mobile Control System (MCS)

The main task of the INT-MANUS MCS is to allow on-site operators to supervise production processes and control manufacturing machines remotely by a handheld computer. The system has three operating modes:

- The Global View Mode offers an interactive map, where the position of the machines, robots and other tracked objects can be seen (see Fig. 2). The map can be scrolled and zoomed easily. To simplify recognisability, all objects are represented by colour-coded geometrical figures: machines are represented by rectangles, while users and robots by circles. Double clicking on an object shows more detailed information of a particular object. The mode may be easily extended for hierarchical visualization of the whole factory.
- The Local View Mode offers the possibility to get a closer view of a chosen object. For manufacturing machines it shows speed, feed, name and progress of a part program and allows an operator to change some parameters, to stop the machine, and to schedule it for maintenance.
- In the Product View Mode, an operator is able to select a part/product and to see a geometrical model of it. He can send a command to the SCCP for scheduling its production. With the use of an external miniature camera, which can be connected via USB to the handheld computer, all produced objects can be photographed. The operator can visualize a geometrical model of the product and overlay it with the live image thus comparing the produced part with its model. Moreover, the operator may identify the object automatically on the basis of Scale Invariant Feature Transform (SIFT) algorithm [6], which is

able to extract repeatable characteristic feature points from images and generate respective descriptors.

The MCS is developed under Windows™ OS with Microsoft Visual Studio™ 2005 and is mainly written in C++. The Qt framework is used for the implementation of the system's graphical user interface. The MCS has two communication modules: one for monitoring and controlling manufacturing machines via the SMART tool and the other one for information exchange with supervisory semantic services of the SCCP. Where necessary for the communication, C++ data structures are serialized in XML with the help of the gSOAP toolkit. The system is presented in detail in [5].

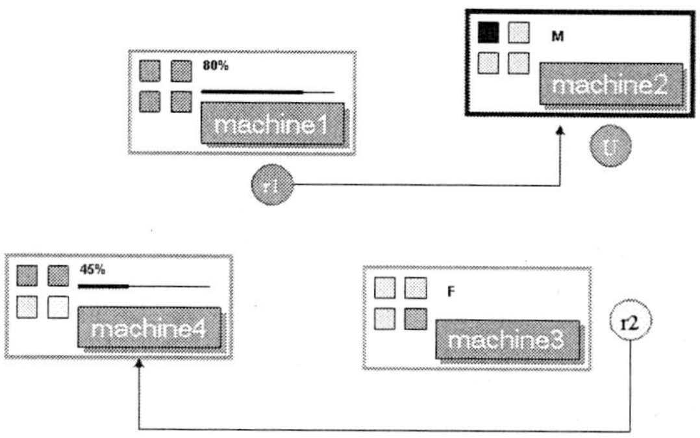

Fig. 2. Interactive visualization of the shop-floor

4. Virtual Reality-based Customization System (VRCS)

The VR-based customization system allows configuring products from parts and immediately scheduling their production through the SCCP services. The system is based on the virtual reality framework [7] and supports both large-scale Virtual Environments and standard personal computers. The input for the toolkit is an abstract grammar, which describes all possible variants of the base product. The grammar is realized in an XML-based format as a tree structure of all variants of parts, which may form the product, with additional information, such as time and costs needed for a part production, links to respective geometry files, lists of compatible parts. The output of the system is a set of parts that have to be produced for the customized product as well as information about connections of all involved parts. These results can be represented both in binary and visual forms and sent to the SCCP for rescheduling of production process.

The user-friendly interface of the VRCS is based on widgets and allows performing the following operations:

- add parts to the scene by selecting them from the menu;
- remove parts from the scene after selecting them;
- connect compatible parts, disconnect them, see current connections in a semi-transparent mode;
- change illumination and material properties;
- save results and send them to the SCCP supervisory services.

The toolkit supports multiple workspaces, thus allowing simultaneous design of multiple products and easy comparison. Furthermore, in the TwoView VR display [8] the toolkit supports simultaneous work of two users. The system renders individual perspectively correct stereoscopic views for each user. Therefore both users observe correct relative alignment of physical and virtual objects. This allows a designer and a customer to work together in a collaborative environment (see Fig. 2). The toolkit has been developed under Linux OS and is mainly written in C++ and Scheme programming languages. The system is described in detail in [9].

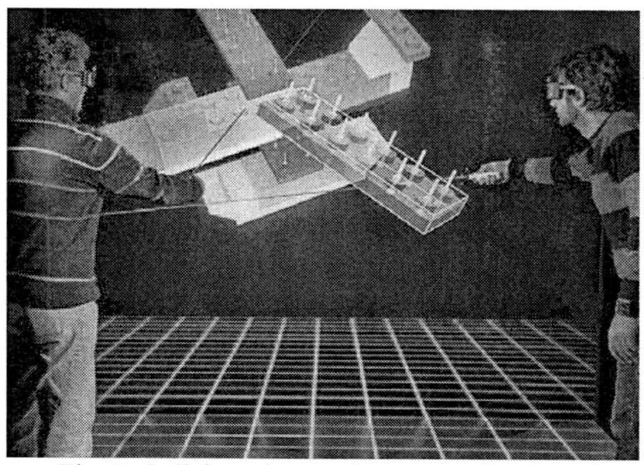

Fig. 3. Collaborative product customization

5. Integration of Machines and manufacturing equipment in the Interactive Production Control

Nowadays, the Machines and manufacturing equipment are controlled by different type of control systems. In the case of machine-tools they have Open Numerical Controls, controls that offer several interfaces allowing access to the managed internal data. Other manufacturing equipments are managed by different kind of controls but in general they also offer interfaces allowing access to the managed internal data. These interfaces make it possible the integration with higher levels of the automation pyramid.

In order to provide richer functionalities we use a framework based on Industrial Agents (SMART framework) that is especially devoted to a manufacturing environment. In SMART, the agents are considered as intelligent entities that can collaborate through communication and update mechanisms. Microsoft .NET and web-based interfaces were used as the base technology. This framework offers functionality for data collection, monitoring and data logging. The SMART distributed scenario is shown in the figure.

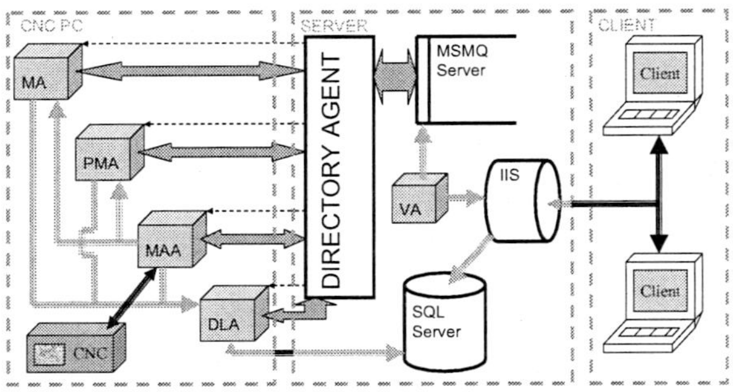

Fig. 4. SMART - Distributed scenario

As a result of the requirements of the INT-MANUS project, a new interface with the numerical control has been decided, based in the GSOAP Toolkit. This implementation of the SOAP protocol allows its use in different programming environments. Therefore using GSOAP we keep the current functionality and we extend the platform to new clients, for example GSOAP can use plain C++ programming.

In the process of keeping the agent base platform but having a lighter version it is possible to re-design the application to use a more state of the art developing paradigm such as Service Oriented Architecture (SOA). The Services could be intrinsically unassociated units of functionality, which have no calls embedded in them to other services. It is just another way of component inter-operation similar to the agent paradigm. Instead of having services embedding calls to each other in their source code, some protocols are defined to describe how one or more services can talk to each other. This architecture then relies on a high-level process to link and sequence services, in a process known as orchestration, to meet global system requirement.

Agents and SOA modules are collaborative pieces of software; they all explain how to interoperate. Agents are autonomous whereas SOA modules require an expert to make the orchestration. The integration of this new approach to the application will benefit from the fact that SMART agents relay on Web Services, so making it possible to change one agent based architecture to service based architecture.

Interactive processes are modelled using a semantic description that is directly related to the production context via instances. The identification is achieved through technologies such as RFID (**Radio Frequency Identification**) and optical recognition. Semantic Message communication is the key feature used in the decentralized communication process. Semantic annotation and content-based routing will enable loose coupling between components and flexible extensibility even at run-time. By semantic annotation, full compatibility can be maintained throughout different protocol versions. Furthermore, semantically encoded performance indicators or maintenance data can be routed along the same communication infrastructure.

6. Robotic Transport System

The Robot developed by ROBOSOFT for the INT-MANUS project is based on a RobuLAB80 indoor mobile platform which was designed to carry high payload. The RobuLab80 mobile platform has been improved for the INT-MANUS project.

The **robot software control architecture** does not only address the capabilities of the robot to assume its own displacement. The robot displacement capabilities are a part of the global control architecture which allows also communication between SCCP, supervisor (fleet manager) and robot. The communication between the robot side (robot and supervisor) and SCCP is made using the .Net technology from Microsoft. Running on the .Net environment ROBOSOFT has chosen the Microsoft Robotic Studio [10] product to develop its robotics control software called RobuBOXTM [3]. The basic principle is to divide a robotic application into communicating services [4], which can run on any CPU of the control system. The control architecture of the INT-MANUS robot allows advanced maintenance by considering the failure detection of robot components (hardware and software).

The interface between the SCCP and the robot is made through a supervisor with fleet management capabilities (**Fleet Supervisor**). It allows to send a mission to the robot but also to monitor the state of the fleet of robots working in the factory. This fleet manager allows also the optimization of the route attached to a mission.

To allow a complete flexible and reconfigurable

navigation of the robot, a set of software agents will be developed to complete this fleet manager.

These software agents will concern the detection of problematic situations and the simulation of rerouting and reassignments of the robot mission in case of failure on a production machine. The detection of problematic situations encountered by the fleet of mobile robots while carrying out their assigned tasks, such as traffic problems, unexpected obstacles or inaccessible targets will allow reconfigurable navigation.

The localization of mobile robot **(Robot Localization)** can be sometimes a problem for the navigation in flexible environment. The most classical method, wire guiding, is definitely not adapted to this type of navigation. To allow the flexibility of navigation, a localization method using reflective landmark was developped. this solution requires a minimum configuration of infrastructure for maximum of possible reconfigurability. As an experimental validation of reflective landmark free application, a SLAM algorithm will be implemented. This will be tested in order to validate the reliability of this solution for a fully flexible and reconfigurable navigation of mobile robot in the factory.

7. Evaluation in the industrial environment at CRF

For the industrial environment, the project brings high innovation in the field of customized production, allowing to achieving a complete flexibility in terms of variety of parts produced, with fast reprogramming and reconfigurability. In the case of automotive production, parts may be quickly customized, modifying the geometries and reprogramming in a smart way the whole process chain, thanks to a VR representation of the impact on the whole production flow of the plant. The process and material flow for the new product is generated, with all the details on process time and machining scheduling inside the production plant. Finally, production can start. As it is important to test the integration of the different components, the whole prototype will be setup in order to simulate the production of an automotive component, such as an engine part, which requires an optimised planning of the production. In fact, in plants where high series are produced (many thousands/year), like in the automotive environment, no delays are allowed due to failures of machines and tools, because it would mean loss of production and delivery delays, with high costs for penalties to be paid by suppliers, if they are not able to provide the products on time. It follows that quick failure detection and an accurate maintenance program are of fundamental importance for the industrial overall performance. The Advanced Machining Laboratory is used to set up the demonstrator, connecting the machine tools and having a first assessment of the customisation possibilities of the parts and of the production process. The system can rely on 3D cad models of the components, which can be used and varied by an user. If the modifications fulfil the machine capabilities, the parts will be produced accordingly. The tracking system will keep an eye on every movement in the shop floor and allow optimising the production flow.

8. Maintenance Concepts

Predictive Maintenance concepts are transferred from definition-oriented to a context-based approach supported by an adaptive system. The objective of condition-based predictive maintenance applied to machine tools is the "measurement" of the slow degradation taking place in mechanical components in advance of the consequent failures and the evaluation of the residual life in order to better manage the total Life Cycle Cost.

The architecture selected for this achievement is shown in the following figure:

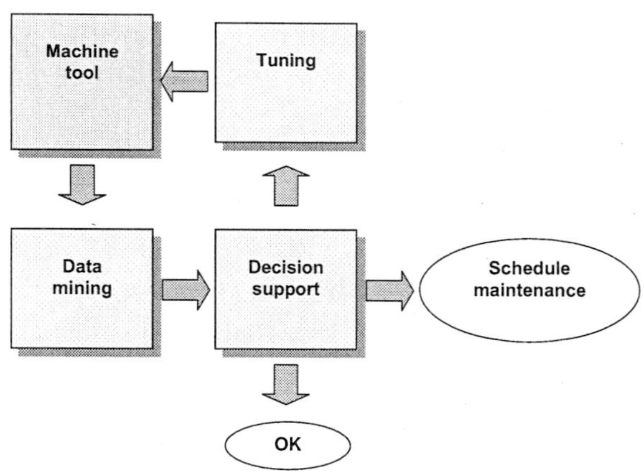

Fig. 5. Maintenance Architecture

Sensors collect data from the Machine Tool; the Data Mining software module is organized in form of a data processing pipeline and can be split in three main phases, comprising data collection from sensors with the suitable frequency, data filtering and extraction of some indices indicating the state of "health" of the mechanical parts (meaning the status of wearing out of components) like vibration critical frequency, damping factor, backlash (error at the inversion of the direction of movement), maximum and average value of current, ripple of current.

The Data Mining software provides indicators to a Decision Support module that selects the possible action, choosing among the following:

- Do nothing because the system is working very close to an optimal situation;
- Self-tune the system, whenever possible, in order to achieve a different trade-off between accuracy and time; in fact, for instance in the production of moulds and dies, the tuning of a machine tool is usually based on the selection of a suitable compromise: if the working operation is faster, it is normally less accurate, and vice versa. The optimal compromise depends on the kind of operation; for instance during a roughing operation, accuracy is not at all required, therefore the tuning should optimize the execution time (and cost, as a consequence).
- After a certain threshold, it is not efficient anymore to slow down the machine, and an intervention for the substitution of a mechanical part is needed. The resulting reorganisation of the plant schedule should be automatically performed. The production plant supervisor should be notified about this decision and the possible reason and mechanical parts to be substituted.

In a following phase, several machine tools (some of them of the same kind, some of them of different kind) are connected to a Remote Knowledge Repository where all data are stored. Stored data are relative to:

- Previous faults;
- Results of previous evaluations of residual life;
- Results of previous tuning operations;
- Results of previous rescheduling operations.

The Knowledge Repository has to be able to transform information into knowledge, and learn by the correlation of data. For instance, applying this philosophy to diagnosis, the system will be able to correlate machine failures with factors like the climate or with the use of the machine.

9. Summary and Conclusion

INT-MANUS provides a decentralized peer2peer platform for production control, integrating advanced maintenance, robot system for transport, RFID, access to numeric controls and AR / VR technology for enhanced user experience. The project will develop further two prototypes which will focus on the semantic technologies and messaging as well as advanced recognition and interaction technologies supported by a knowledge base.

Acknowledgements

INT-MANUS Project is funded by the European Commission and has emerged out of the European Network of Excellence I*PROMS, supported by its virtual Institute VIMATION.eu.

References

[1] INT-MANUS project page: www.int-manus.org
[2] I*PROMS Network of Excellence: http://www.iproms.org/
[3] White Paper robuBOXTM on Robosoft website: "http://www.robosoft.fr".
[4] DSSP specification from Microsoft website: http://download.microsoft.com/download/3/8/2/382EAA0D-576E-4F2D-803A-22255A0C7251/DSSP.pdf
[5] Foursa M., d'Angelo D., Narayanan R., Innovative Control System for Industrial Environment, accepted for publication in the proceedings of the IV International Conference on Autonomic and Autonomous Systems, ICAS 2008, March 2008, Gosier, Guadeloupe
[6] Lowe, D. G.: Distinctive Image Features from Scale-Invariant Keypoints, International Journal of Computer Vision, 60, 2, p.: 91-110, 2004
[7] Avango web-page (http://www.avango.org): object-oriented framework for Virtual Reality applications
[8] Riege K., Wesche G., Holtkaemper T., and Froehlich B.: The bent pick ray: An extended pointing technique for multi-user interaction. Proc. IEEE Symposium on 3D User Interfaces 2006, pp. 62-65, Alexandria/USA, 2006.
[9] Foursa M., Wesche G., d'Angelo D., Bogen M., Herpers R.: A Two-User Virtual Environment for Rapid Assembly of Product Models within an Integrated Process Chain, accepted for publication in the proceedings of the X Symposium on Virtual and Augmented Reality, SVR 2008, May 2008, Joao Pessoa, Brazil
[10] Microsoft Robotic Studio website: "http://msdn.microsft.com/robotics".
[11] Schlegel, T.; Beinhauer, W.; Meo, F.: Object-Orientation in Planning and Control of Decentralized Production Systems. In: VIMation Journal on Human-Machine Interaction and Models in Production, Issue 1/2008, pp. 54-64, 2008

Innovative Production Machines and Systems
D.T. Pham, E.E. Eldukhri and A.J. Soroka (Eds.)

An integrated model driven approach in support of next generation reconfigurable manufacturing systems

T. Masood[a,b], R.H. Weston[a,b]

[a] Manufacturing System Integration (MSI) Research Institute, Loughborough University, Leics, LE11 3TU, UK
[b] Centre of Excellence in Customised Assembly (CECA), Loughborough University, Leics, LE11 3TU, UK

Abstract

Shortened product life cycles and globalization have induced dynamism and uncertainty into world markets. Hence manufacturing enterprises (MEs) can gain competitive advantage from being reconfigurable. But appropriate application of agile and lean Manufacturing philosophies must complement the application of reconfiguration techniques. However, choosing and applying the best philosophies and techniques are far from being well understood and well structured processes because most MEs deploy complex and unique configurations of processes and resource systems, and seek economies of scope and scale in respect of a number of distinctive product flows. It follows that systematic methods of achieving model driven configuration of component based manufacturing systems are required to design, engineer and change next generation MEs. This paper discusses research aimed at developing and prototyping a model-driven environment for the design, optimization and control of reconfigurable manufacturing enterprises with an embedded capability to handle various types of change. The developed environment supports the engineering of common types of strategic, tactical and operational process found in many MEs. Also reported are initial findings of manufacturing case study work in which coherent multi-perspective models of a specific ME have facilitated process reengineering and associated resource system configuration. The paper outlines key areas for future research including the need for research into unified modelling approaches and interoperation of partial models in support of complex organisation design and change (OD&C).

Keywords: Organisation design and change (OD&C), reconfigurability, enterprise modelling (EM), simulation modelling (SM).

1. Introduction

Significant change is ongoing in environments in which modern MEs must operate competitively. This has induced a need for faster, better and cheaper production. Quick and timely responses of various production system types are vital for MEs to remain competitive.

One key common response is for MEs to have a broadened product portfolio. But to compete they must deploy an effective and change capable set of human and technical resource systems. Also because of falling product lifetimes and growing customisation requirements the deployment of these resource systems will increasingly need to give rise to economies of scope and mass customization [1]. For many companies around the world, staying in business necessitates:

- ❑ meeting specific customer requirements innovatively and effectively

- reducing the time-to-market of products
- manufacturing quality products at competitive cost.

The present day customer typically imposes constraints on MEs via their specific and changing quantity, quality, cost and delivery product and service demands. The implications of this customer focus are that; time to market is shortening, products are tailored to meet a breadth of customer needs, and demand is variable [2,3,4]. However, MEs cannot simply respond by deploying new processes and resources. Generally they need also to redeploy (i.e. re-configure and re-integrate) their use of existing processes and resources such that they can respond competitively to the ongoing business dynamic created by the specific and changing demands of their customer base.

The identification of methods by which manufacturing improvements can be achieved is ongoing and has led to a range of approaches in recent years including Lean and Agile Manufacturing. In addition, progressive improvements to information system capabilities continues to offer the belief that significantly improved support for effective decision making can be achieved [5,6].

Hence it has been observed that modern manufacturing systems must be; flexible/agile, reactive, integrated, and cost efficient [3]. It also follows that ME personnel must have (individual and collective) in-depth understandings about specific processes and resource systems and that these processes must be flexible enough to change whenever the need rises. The complexity of manufacturing systems is reaching that of many natural (e.g. economic and political) systems, thus ongoing re-design and re-engineering of such systems require the use of systematic approaches which deploy various types of system model to understand current and possible future behaviours and to inform systems engineering decision making.

2. Need for reconfigurable model driven manufacturing systems

Business process analyses, aided by the use of Enterprise Modelling and Integration (EM&I) techniques, has potential to reduce the risks and increase chances of realising successful business process operation and interoperation. It also has potential to enable organisations to capitalise knowledge and react to change effectively and efficiently. Systematic decomposition and analysis of complex systems is possible with the aid of supporting EM & I architectures, approaches and tools.

However realising the potential of EM & I technologies gives rise to far from trivial problems. The skill with which EM & I technologies are used in conjunction with other modelling technologies such as Simulation Modelling (SM) and IT systems engineering technologies, will determine the extent to which benefits of improved ME systems design and interoperation can in practice be realised. For example the models developed using these various technologies need themselves to be reconfigurable and interoperable in order to synchronise their development and deployment within targeted change-capable environments.

There have been recent advancements to extend the coverage of public domain open systems architectures and to bridge the gap between static modelling and dynamic or simulation modelling. Such enhancements include Rahimifard and Weston's enhanced use of EM and SM techniques [7], Chatha's E-MPM [8], and Monfared's approach [9] which has been developed at MSI Research Institute. It is noted that Computer Integrated Manufacturing Open Systems Architecture (CIMOSA) [4] has been central to most of the developments referenced here. A natural focus of these modelling efforts has been to make the models 'live' and responsive to the upcoming rapid changes. Here the idea is to have modelled and real elements of complex and changing sets of processes (and their underlying resource systems) interoperate in a readily integrated and change capable manner which satisfies specific dynamic requirements of the business environment in which the processes are used.

During recent times technological innovation has induced very significant change in industry. It has also impacted significantly on the way that manufacturing enterprises (MEs) operate and compete with each other. In general MEs are very complex entities: designed, managed and changed by people and their designed systems; to realise people and system requirements; by deploying technological resources in systematic, timely and innovative ways that generate competitive behaviours. Because typically MEs have multi purposes and stakeholders it is difficult to decide how best to develop the technological systems they deploy and it is difficult to change them rapidly and in ways that enhance overall ME competitive behaviours. Whereas comparatively it will be simpler, easier and faster to design and realise change in systems with a single well defined purpose and small set of stakeholders. Hence there is a great need to seek to deploy decomposition principles aimed at breaking down complex systems into readily understood and reusable technical building blocks; which can be used as interoperating 'components' (or modules) of wider scope and complex MEs that can be reconfigured as requirements change.

Globalisation is one outcome from technological innovation within MEs. With sufficient resources, many entrepreneurs can now physically or virtually relocate themselves and their products (knowledge, experience, ideas and artefacts) to various locations around the globe. This has enabled knowledge sharing on a worldwide basis and globalisation of technical systems. Hence many MEs now take the form of an appropriate configuration of less complex entities with a global reach as a result.

The research reported in this paper builds upon previous research in MSI Research Institute to achieve an enhanced and synergistic use of enterprise modelling, simulation modelling, causal loop modelling and workflow modelling in support of the ongoing configuration and interoperation of manufacturing systems. These techniques come from different origins and potentially can support the creation and use of different types of model. However there is little reported in the literature on (1) their complementary capabilities, and (2) their collective use so as to provide semantically rich models that can support numerous aspects of ME design and change. Hence in recent years the present authors and their research colleagues have conceived, incrementally developed and case study tested a set of unifying concepts and methods.

The new approach to creating configurable and interoperable manufacturing systems assumes a need for process oriented thinking when designing present day configurable organisations and that various modelling technologies are required in support of organisation design and enactment. Hence, an overall research aim is to develop a model-based unified environment, which lends structure to and supports multi-perspective decision making involved in organisation design and enactment. The scope of research found to be necessary to satisfy this need encompassed:

- Understanding the nature and role of processes in manufacturing organisations;
- Achieving semantically rich process modelling within the broader context of enterprise modelling decomposition principles;
- Unification of a number of state of the art modelling techniques.

On completion of this research the following main deliverables are being delivered and case study tested on a progressively comprehensive and improved basis:

(1) a methodology including modelling stages, concepts and methods for capturing relevant aspects of organisations, and techniques and tools that can be deployed to develop unified models of processes, work flows and their needed resource system;

(2) a unified modelling framework and modelling concepts that describes those aspects of an enterprise that need to be understood and captured while modelling processes so that various models of these aspects can interoperate in respect of current and possible future ME compositions.

3. A model-driven approach in support of reconfigurable manufacturing systems

The modelling approaches have significant roles to play in enabling decision making in dynamic manufacturing organisations. Analysis of business processes aided by Enterprise Modelling and Integration (EM&I) can reduce risks and increase chances of implementing successful business processes. Business process thinking can also enable organisations to capitalise knowledge so that they can react by changing operations in an effective, efficient and timely manner. Systematic decomposition and analysis of complex systems is possible with the aid of supporting architectures and by using complementary modelling techniques which can include Enterprise Modelling (EM), Causal Loop Modelling (CLM), Simulation Modelling (SM) and Work Flow Modelling (WfM) [10]. The models of ME processes, resource systems and workflows created need themselves to be reconfigurable and interoperable in order to synchronise between virtual and real elements of processes and systems that need to interoperate within dynamic (often uncertain) environments [11]. In principle, by achieving a unification of different modelling approaches new opportunities will arise to make the models live and responsive to upcoming but 'yet to be determined' rapid changes [11].

This paper describes a systematic approach to creating coherent sets of unified models that can interoperate to replicate and predict changing organisational behaviours in support of reconfigurable manufacturing systems. The modelling approach used to support reconfigurable manufacturing systems is illustrated in Fig. 1.

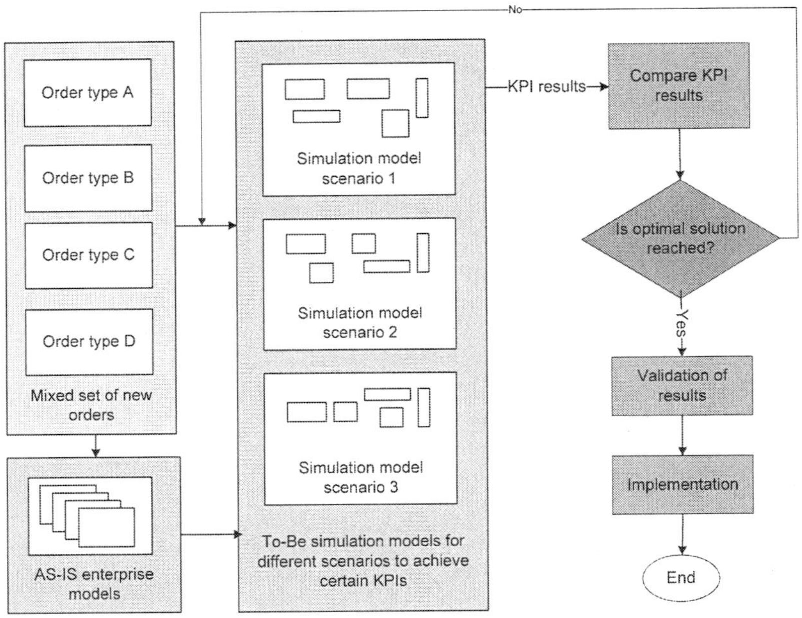

Fig. 1. A model driven approach in support of reconfigurable manufacturing systems

The purpose of this section is to describe how a simulation modelling methodology was developed to computer execute models of manufacturing systems under different configurations. This methodology is centred on simulation modelling but uses a combination of enterprise and simulation modelling techniques to create a coherent set of models that can structure and inform:

- Understanding of the problem domain;
- Development of conceptual model in relation to problem domain;
- Information and data collection;
- AS-IS model building for present configuration;
- TO-BE model building for different configurations;
- Verification and validation of models;
- Experimental work and analysis based upon key performance indicators (KPIs) leading to system selection.

A set of computer and manual simulation technologies and systems are used to simulate the problem scenarios and provide suitable interface to users. The subscript packages were used for simulation purposes:

- SIMUL8®
- Plant Simulation ®
- iThink®
- MS Visio®
- CIMOSA Diagramming Templates
- Causal Loops

The interfaces enables the user to:

- Capture requirements data;
- Input data into the model;
- Reconfigure a simulation model to meet various specified needs;
- Conduct various experiments based upon different scenarios;
- Collect results and
- Conduct analysis of results.

4. Case Study

The case study company (referred to as ABC onwards) is a furniture manufacturing SME with 50 employees. It operates primarily within the UK but has European suppliers of raw material. It manufactures over 300 different furniture products from pine wood; including a range of tables, cabinets, beds, wardrobes and other furniture items that are designed for house hold and business users. The application of a number of different reconfiguration concepts is being investigated at ABC. A coherent set of enterprise and simulation models were created to explicitly represent and computer execute behaviours of ABC's business processes, (human and technical) resource systems and dynamic patterns of multi-product workflows.

Reduction in lead time (and in effect throughput time) is one of the key performance indicators in engineering of a reconfigurable manufacturing system. The current system was analysed for future implementation of a hybrid Make-To-Stock (MTS) / Assemble-To-Order (ATO) system. When populated order behaviours are available, the predictability of orders is better achievable using forecasting methods. Such a MTS/ATO system is practicable in environments where standard parts are available to assemble using postponement approach. With product (re)design, it is achievable to make standard parts which are partially driven by MTS strategy and partly by ATO strategy. Implementation of a hybrid MTS/ATO system needs better product and process (re)design, availability of stock, availability of Bill of Material (BOM) data, availability of forecasted orders, commonality of parts (standardisation of parts or modularisation) and better supplier relationships. Provided these supports are available, reduction in lead time is achievable which leads to customer satisfaction and then may result in an increased frequency of orders. All above-mentioned factors enhance the chances of success rate of implementation of hybrid MTS/ATO system.

Enterprise and simulation models were developed in order to enable decision making about ME possible selection of an improved PPC strategy which could make this assembly process segment (and its underlying resource systems) become more configurable and interoperable. Fig. 1 shows ME modelling approach (based on using enterprise and simulation models) which was adopted in order to enable improved decision making in ABC. The case study and exemplary models have been presented in further detail in [12] for an example of PPC strategy realisation.

It was noted that the combined use of EM and SM techniques helped to gain in depth understanding about the ABC's production methods, their shortcomings and possible ways of improvements. Such reconfiguration decision making supports OD&C process. There is a need to develop methods and technologies for unified modelling and interoperation of partial models in support of complex OD&C. The future research also relate to the use and updating of enterprise and simulation models. Table 1 summarizes key existing ABC approaches, authors' observations and proposed strategies related to reconfiguration.

Table 1
Summary of observations and proposed reconfiguration strategies in ABC

Production Strategies	Existing ABC Scenario	Observations	Proposed strategies
PPC Hierarchy	❑ No aggregate planning based on forecasting	❑ No visibility of order behaviour ❑ Slow to react	❑ Needs forecasting ❑ Historical data analysis
Customer Order Decoupling	❑ Make to Order (MTO)	❑ Majority of production starts with order arrival	❑ Assemble to order ❑ Inclusion of Postponement theory
Push vs. Pull System	❑ Push system	❑ Longer lead time	❑ Pull system where possible ❑ Production based upon commonality of parts ❑ Parts standardisation
Production Scheduling	❑ Accumulated order list due to logistic constraints	❑ High variety induced in system bottleneck	❑ Break up of job list by removing the constraint
Production Sequencing	❑ First In First Out (FIFO) ❑ Minimum set up time	❑ Does not reflect global objectives ❑ Does not take into account overall lead time performance across Assembly Shop and ABC	❑ Adoption of dynamic sequencing rules o Minimum number of remaining operations o Minimum remaining process time

5. Conclusions

This paper discusses research aimed at developing and prototyping a model-driven environment for the design, optimization and control of enterprises with an embedded capability to handle various types of change in different production configurations. The developed environment supports the engineering of common types of strategic, tactical and operational process found in many MEs. Also reported are initial findings of manufacturing case study work in which coherent multi-perspective models of a specific ME have facilitated process reengineering and associated resource system configuration and interoperation. It follows that systematic methods of achieving model driven configuration and interoperation of component based manufacturing systems are required to design, engineer and change next generation MEs. The basis of a systematic approach to creating coherent sets of unified models that facilitate the engineering of reconfigurable system is described. A summary of observations and proposed configuration strategies have been presented in the paper which supported model-driven reconfiguration of manufacturing systems. The paper has identified key areas for future research which relates to the use and updating of enterprise and simulation models.

References

[1] Vernadat, F.B., *Enterprise Modelling and Integration (EMI): Current Status and Research Perspectives.* Annual Reviews in Control, 2002. 26: p. 15-25.

[2] Suri, R., *QRM and POLCA: A Winning Combination for Manufacturing Enterprises in the 21st Century, Technical Report.* 2003, Centre for Quick Response Manufacturing, http://www.engr.wisc.edu/centers/cqrm/pub/qrm21st.htm.

[3] Ladet, P. and F.B.E. Vernadat, *The Dimension of Integrated Systems Engineering*, in *Integrated Manufacturing Systems Engineering.* 1995, Chapman & Hall: London. p. 3-9.

[4] Vernadat, F.B., *Enterprise Modelling and Integration: Principles and Applications.* 1st ed. 1996, London, UK: Chapman & Hall. 513.

[5] Maropoulos, P.G., *Digital enterprise technology – defining perspectives and research priorities.* International Journal of Computer Integrated Manufacturing, 2003. 16(7-8): p. 467-478

[6] Young, R., *Informing Decision Makers in Product Design and Manufacture.* International Journal of Computer Integrated Manufacturing, 2003. 16(6): p. 428-438.

[7] Rahimifard, A. and R.H. Weston, *The enhanced use of enterprise and simulation modelling techniques to support factory changeability.* International Journal of Computer Integrated Manufacturing, 2007. Vol. 20(No. 4): p. 307-328.

[8] Chatha, K.A., *Multi-Process Modelling Approach to Complex Organisation Design. Thesis (PhD)*, in *Wolfson School of Mechanical and Manufacturing Engineering, Loughborough University.* 2004: Loughborough, UK. p. 294.

[9] Monfared, R.P., *A Component-Based Approach to Design and Construction of Change Capable Manufacturing Cell Control Systems. Thesis (PhD).* in *MSI Research Institute.* 2000, Loughborough University: Loughborough, UK. p. 266.

[10] Weston, R.H., et al. *Simulation Model Interoperability in support of complex organisation design and change.* in *European Simulation and Modelling (ESM) Conference.* 2006. Toulouse, France.

[11] Weston, R.H., et al. *Simulating Dynamic Behaviours in Manufacturing Organisations.* in *International Conference on Industrial Engineering and Systems Management (IESM 2007).* 2007. Beijing, China.

[12] Masood, T. and R.H. Weston, *Integrated modelling approach in support of change-capable PPC strategy realisation*, in *The 4th International Conference on Innovative Production Machines and Systems (IPROMS 2008).* 2008. The Internet.

Innovative Production Machines and Systems
D.T. Pham, E.E. Eldukhri and A.J. Soroka (Eds.)

A Technology Roadmap on Service-Oriented Cross-layer Infrastructure for Distributed smart Embedded devices

M. Gerosa[a], A. Cannata[a], M. Taisch[a]

[a] *Department of Management, Economics and Industrial Engineering, Politecnico di Milano, P.zza Leonardo da Vinci 32, 20133 Milano, Italy*

Abstract

SOCRADES (Service-Oriented Cross-layer Infrastructure for Distributed smart Embedded devices) is a European research and advanced development project. Its primary objective is to develop a design, execution and management platform for next-generation industrial automation systems, exploiting the Service Oriented Architecture paradigm both at the device and at the application level. As an important output of the SOCRADES project, a Technology Roadmap is being developed. The aim of the Technology Roadmap is to aid in forecasting the features of future environment and how technological research has to be used to achieve such long term objectives.

Keywords: Roadmap, SOA, Embedded devices, wireless

1. Introduction

Manufacturing plays a vital role in the European economy and society, remaining fundamental to creating stable employment. The development of competitiveness in manufacturing industry is essential for European prosperity; therefore the adoption of manufacturing innovations is needed in order to promote performance improvement of European manufacturing processes. However the present context is characterized by a wide supply of innovative technologies for manufacturing process improvement.

The SOCRADES (Service-Oriented Cross-layer Infrastructure for Distributed smart Embedded devices) [12] project is a European research and advanced development project. Its primary objective is to develop a design, execution and management platform for next-generation industrial automation systems, exploiting the Service Oriented Architecture paradigm both at the device and at the application level. The SOCRADES integrated project will create new methodologies, technologies and tools for the modelling, design, implementation and operation of networked systems made up of smart embedded devices. Achieving enhanced system intelligence by co-operation of smart embedded devices pursuing common goals is relevant in many types of perception and control system environments. In general, such devices with embedded intelligence and sensing/actuating capabilities are heterogeneous, yet they need to interact seamlessly and intensively over a (wired or wireless) network.

The middleware technologies to be developed in this project will be based on the Service-Oriented Architecture (SOA) paradigm, will encompass both wired and wireless networking technologies, and will provide open interfaces enabling interoperability at the semantic level. A SOCRADES service will be a software component encapsulating device-specific functionality. This functionality is advertised to the outside world, so as to be located and invoked by other networked devices and/or applications without the latter being aware of how the functionality is implemented. As an important output of the SOCRADES project, a Technology Roadmap is being developed. The aim of the Technology Roadmap is to aid in forecasting the features of future environment and how technological research has to be used to achieve such long term objectives.

The first release of the SOCRADES Technology

Roadmap has just been completed. In the present paper we will present the methodology, structure and results.

2. Methodology and structure

In order to identify what are the future improvements due to the implementation and developments of Service-Oriented manufacturing, the project context has been divided into four relevant Technology Areas (TA): (i) Ad-hoc networking services platform – Service-oriented Architectures, (ii) Wireless sensor/actuator networking infrastructure, (iii) Service-centric infrastructure – Enterprise integration, (iv) System Engineering & Management. First aim of the Roadmap was to identify, for each one of the above mentioned Technology Areas, some Expected Features (EFTA: Expected Features of Technology Area): these are relevant technology characteristics, that are expected to become available in the future thanks to research and developments.
Several main sources of information for the identification of EFTAs are considered in the SOCRADES Technology Roadmap:

a) State of the art: analysis of existing Roadmaps and White Papers available on the topics. A comprehensive list of such roadmaps is available in the references [1-11].

b) Workshops, carried out with experts coming from both inside and outside the SOCRADES Consortium.

c) Delphi study, still under development: the results will be included in the next releases of the roadmap.

In the following paragraphs the Expected Features identified for each one of the four Technology Areas will be presented and briefly explained

3. Ad-hoc networking services platforms – Service-oriented Architectures

A service-oriented architecture (SOA) is a set of architectural tenets for building autonomous yet interoperable systems. The proposed definition includes two keywords: autonomous and interoperable. The principal characteristics setting apart autonomous systems are that: they are created independently of each other, they provide self-contained functionality, i.e., their functionality would be useful even if it was not associated with any higher-level systems. Interoperability is favoured by: clearly abstracting the interface that a service exposes to its environment, from the implementation of that service; making this interface visible to others, together with policies and constraints for its use. The EFTAs for the Ad-hoc networking services platform – SOA Technology Area are the following:

- **Orchestration:** the practice of sequencing and synchronizing the execution of services, which encapsulate business or manufacturing processes, is denominated orchestration. An orchestration engine implements the application logic necessary to orchestrate atomic services, and provides a high-level Web Service interface for the composed process.
- **Decision Support System (DSS):** since Orchestration by itself cannot solve every problem arising during the operation of the system, a DSS is expected to increase, even more than now, its importance in the system. Some issues can be handled locally at the lower levels (through orchestration), but by implementing all functions in same level could neutralise the flexibility of the system, so external DSS implemented with MES, Expert Systems, MAS etc are expected to become necessary to address complexity and reconfigurability
- **Context-aware services:** the capacity of services being aware of devices, of factors such as service location or service state and in general of their environment and reacting by adapting their functionality. Semantic description of services and ontology for equipment etc. should be the proper tools in order to describe context.
- **Lean data generation and processing:** in order to obtain efficiency through the utilization of SOA, methods and tools are needed for data structuring and handling inside each machine (sensors, electrical signals, etc). Device developers have to decide how they want to expose their specific data as particular service attributes.
- **Standardization of basic functionalities provided by services:** .communication technologies are pretty much standardized but this is not the same for standardization of basic functionalities provided by services. A separation between the standardization of the communication technologies and the standardization of basic functionalities provided by services is needed. This specific standardization is expected to evolve in various domains and cross-domain.
- **Common Language:** a common language is expected to be widely diffused. The protocol should be at least XML-based and support OWL mark-up in order to enhance it with explicit semantics. Having a common basis for seamless operation of standard

functions such as discovery, description, addressing, invocation etc. will be required.

- **Choreography:** A complementary concept to service orchestration is choreography. The orchestration level is concerned with the workflow-oriented execution and sequencing of atomic processes, but does not take into account the different types of conversation patterns required to invoke the services associated with those atomic processes. Choreography, instead, considers the rules that define the messages and interaction sequences that must occur in order to execute a given process through a particular service interface. Choreography evolution will be related to complexity due to the increasing dimension of systems.
- **Chronology-aware Service Composition:** time is not yet represented in the description of service, while making service composition; so duration of the composed service could not be known. Methodology to infer composed times (capacity) of execution of two (or more as an expected evolution) services with time properties is expected to become relevant.
- **Knowledge processing and reasoning:** when ontologies that contradict each other are used, there could be potentially catastrophic consequences for inference results. So the issue is ontologies reconciliation for agent systems, especially at run-time. Reconciliation of ontologies is very complex to be achieved since now a suitable tool may require a considerable amount of computational resources, which are not available in embedded systems.
- **Large and complex:** increasing size involves increasing number of services and number of interactions. With a greater number of nodes and links the overall complexity of the system increases implying issues in managing the system.
- **Run-time behaviour of a SOA:** concurrency model is the core component of a real-time program. Whilst the concurrency model of high-integrity systems is now well understood and has found representation in subsets of languages, more expressive subsets are needed. A relevant subtopic of this EFTA is also the Prediction of specifications: the ability to perform an analysis in order to predict how specification will evolve, during the run-time behaviour of the system.
- **Localization of functionalities:** in a distributed system it is fundamental to have tools and methods for efficient service localization in highly distributed environment. A way to reach this EFTA should be to use Clustering technique for localization.
- **Intellectual Protection:** Another concern for systems that facilitate interoperability by openly sharing knowledge is that of Intellectual Property (IP) protection and IP interfaces. Through SOA, IP should be guaranteed since only the service is seen, not how the service is furnished, however improvement are expected.
- **Dynamic Deployment of Services:** in the present state of art this cannot be done. Following the dynamically deployment, web service should represent a key solution to reach orchestration and choreography.

4. Wireless Sensor/Actuator Networking Infrastructure (WSAN)

Wireless technologies are already widely used in industrial environments. Currently the focus is on monitoring and human machine interface applications. Further development of wireless communication technologies in order to become reliable enough for application in closed-loop control. As wireless networks are basically open systems, interoperability of the different wireless technologies deployed in the same area is becoming an increasingly important issue. The EFTAs for the Wireless Sensor/Actuator Networking Infrastructure are now described:

- **Quality of Service:** This is a macro-EFTA which includes: (i) Real-time service: discrete manufacturing requires data sampling rates in the millisecond range, motion control even in the sub-millisecond range, while process control applications are satisfied with hundreds of milliseconds. These are challenges to current WSANs as, firstly, wireless links by nature are less stable than wired ones and, secondly, such real-time requirements imply a higher degree of energy consumption than non-real-time and energy is a very limited resource. (ii) Determinism: reliable delivery of data packets with a guaranteed (mainly maximum) delay. As the wireless medium is an open one and subject to inference by machines or other wireless networks, this requirement is difficult to fulfill especially for hard real-time applications in discrete manufacturing or motion control. (iii) Reliability of sensors: in particular concerning the integration of support for dependability and real-time. High-level dependability mechanisms (including application-level ones) must complement low-level mechanisms such that distributed applications running on the sensor network as a whole exhibits reliable behaviour even if parts of the network fail. (iv) Trust in service: which in general can be considered at different levels, in this case at low level. (v) Efficient communication: in order to avoid energy wasting. Energy use is linked to frequency of communication and use for each communication. So efficient communication can improve energy saving, helping also in reaching energy autarky.

- **De-centralization:** This describes the trend to shift intelligence and processing tasks towards the field level devices. Decentralization adds robustness, flexibility and scalability to the system and it increases granularity; of course, this is not a dogma for all devices, some might still be communicating directly to the central unit.

- **Energy autarky (self-sufficient devices):** in a WSAN a way of powering the single devices in a decentralized manner has to be found. This might be (very powerful) batteries, ways of energy harvesting, energy transmission and optimization of energy usage and power management.

- **Self-X Features:** For reasons of easier engineering, easier maintenance, etc. features like self-organization, self-optimization, self-healing, self-configuration, self-stabilization, self-describing etc. are of highest importance. In their maximum extent this would mean a WSAN which does not need any engineering, maintenance, configuration, or human care at all.

- **Enhanced Service oriented Features:** due to the limited amount of resources, SOA features nowadays can only be implemented on the gateway. As soon as the energy challenge is solved, it will also be possible to apply common IT technologies.

- **Efficient Data Processing:** This is a macro-EFTA which includes: (i) Push-based information transmission: information is transmitted when ready (push logic), not when required (pull logic) (ii) Fusion/aggregation of data processed within the network: information have to be propagated within the network by using rules of aggregation/fusion to avoid to overload some nodes of the network and to provide useful information instead of raw data. (iii) Context-aware MAC and Routing: a system that extracts, interprets and uses context information and adapts its functionalities to the current context will be required in the future.

- **Sensor calibration:** calibration in WSAN is more frequent than in wired networks. It is important to understand when to do calibration. The problem is not only due to WSAN, itself but to the increase of number of sensors enabled by the WSAN.

- **Localization:** provision of localization information in distributed networks can be both a service offered by the network to the applications (object tracking, mapping, etc.) and a technique for improvement of the network itself.

- **Data storage and search:** collecting data it is crucial to minimize communication cost and delay for the application tasks. To address this, efficient storage and querying of sensor data are both critical and challenging issues in WSAN.

- **Scalability:** number of nodes should vary depending on the need of the application. Existing approaches are still far away from the optimal desired scalability.

- **Robustness:** devices in a WSAN should be resistant to the potentially harsh environmental conditions (e.g., such as high and/or low temperature) and should integrate seamlessly in the environment. Since the application trend requires more sensors and the operational environment gets more hostile, the importance of research on this EFTA increases.

- **Mobility:** in a WSAN devices can be moving relative to each other. Mobility depends on the application and can vary depending on the proprieties of the mobility scenario.

- **Security (cross-layer):** WSAN have to be secured in many ways to ensure their reliable operation. Besides the security in WSAN, the focus will be in the security in switching from a layer to another, cross-layer security.

- **Interoperability (heterogeneity):** the emergence of WSAN standards should facilitate the deployment of increasingly large industrial wireless networks at both plant and field levels.

- **WSAN technologies for closed-loop control:** in the future it is expected that WSAN technologies should be able to address closed-loop control needs through higher robustness, prioritization schemes, etc. The central components like PLC will become more and more unnecessary for controlling production processes.

- **WSAN deployment tools:** With the advent of interoperable WSAN technologies, new standard installation, deployment and maintenance tools will be available. Also, specific planning tools should be made available to cope with complex installations and provide pre-deployment simulation capabilities.

5. Service-centric infrastructure - Enterprise Integration

This TA is related to the integration. In particular the focus is first on the link between application - level functionality and device - level functionality through a common, unifying technological approach, based on the service - oriented architecture (SOA) paradigm, implemented using Web Services and associated technologies. Then the focus is on the integration of aggregated device - level services with higher - level Web Services and business processes situated at the level of business applications - in particular Enterprise Resource Planning (ERP) systems. Business application systems increasingly benefit from the adoption of service oriented architectures (SOAs) allowing to flexibly compose components of heterogeneous software systems across traditional system boundaries and swiftly adapt business processes to changing requirements. The EFTAs for the Service-centric infrastructure – Enterprise Integration are now described:

- **Flexible Production with Enterprise Support:** a characteristic of the future manufacturing plant will be its connectedness to all its vital components: workers, machines, and products. Cross-Enterprise approaches considering the end-to-end process from the supplier via the plant to the customer need to be developed. Especially the vertical dimension where automation, manufacturing execution, and enterprise systems will be seamlessly integrated into each other is of key importance. Material flow along the value chain and Information flow along the product life cycle are considered significant future features to be supported.
- **Device to Business Integration (D2B Integration):** device manufacturers have been dramatically increasing the amount of embedded software in their products, and as a result not only they are able to handle a wide range of computing and communication tasks, but are able to also cooperate and provide their functionality as a service. Therefore device can participate in real-world business applications by providing information coming from their domain and also consuming services available at enterprise level. In the future it is expected to have Open Architectures for realizing Device2Business integration.
- **Cross-layer Adaptive Modelling:** As software development is becoming more complex model-driven engineering technologies offer a promising approach to address the inability of third-generation languages to alleviate the complexity of platforms and express domain concepts effectively. Development of approaches that effectively handle service modelling and management of intelligent distributed business processes in highly populated web service enabled device infrastructures is needed. Business logic traditionally resided at high-level systems (e.g. ERP) but in the future it will be distributed in several layers (e.g. ERP, middleware, network, device level).
- **Security / Service Policy Compliance:** business applications are moving from stand-alone systems to enterprise service-oriented architectures (enterprise SOA). The openness and heterogeneity of such systems is requiring a security approach different from that of traditional systems and architectures.
- **Industrialization of software development:** it is needed to "industrialize" at a fine grained level the process of creating software that will give us the ability to rapidly configure, adapt and assemble independently developed, self describing components to produce families of similar but distinct systems, applicable in different domains that share common core functionality.

6. System engineering & management

A number of system modelling and logic design environments are now emerging with the potential to not only automatically generate control logic software but to allow the support of multiple aspects of the complete systems engineering life cycle. Such tools promise to enable higher level design, improved reuse of control logic applications and the possibility to visualise applications in a virtual environment. They also have the potential of supporting distributed control system configurations. The EFTAs for the System engineering & management are now described:

- **Efficient/effective (re-)configuration:** the ability to configure systems (machines, lines, etc.) built from SOA-enabled modules both statically and dynamically in a standardized manner, e.g., with predictable system performance. Flexible use of systems components coupled with a high-level system configuration capability (e.g. ERPs).
- **High level process definition:** the ability to describe the overall behavior of systems composed of many distributed devices in a high level process description language, which directly relates to the specific process the user is concerned with.
- **Collaborative, integrated, distributed business-driven engineering:** systems capable of being configured and managed in a global business context. Support for globally distributed engineering teams.. The added pervasiveness of SOA-enabled manufacturing devices and the effective coupling of embedded systems to business drivers is key to greater agility.
- **Open, lifecycle support:** engineering systems

capable of being used at all phases of the machine lifecycle, all provided in a consistent manner and in a vendor independent way. In future, as the lifecycle of systems continually gets shorter the provision of lifecycle support rather than a collection of separate support tools is necessary. SOA presents a standards based and neutral platform on which these lifecycle systems can be both built and linked.

- **Service-oriented engineering:** engineering support provided primarily through the provision of services rather than on-site engineering activities. SOA will aid in the standards based development of this type of remote support by linking various systems both physically and in conceptual terms.

-. **SOA Marketplace:** SOA could be used to expose services to the systems. SOA enables IP (Intellectual Protection). IP enables marketplace for tools to be developed (maintenance, diagnosis, etc.), creating new business models. This marketplace model will present the users of services and systems in the domain to new pricing models that may be more suitable to companies such as Small to Medium Enterprises.

- **SOA-enabled Digital Factory:** Planning and modelling, improving cooperation between the enterprises, and internally. This is a macro-EFTA which includes: (i) Seamless integrated digital engineering: the ability to mix the engineering of digital, (i.e., 3D dynamic models) and real system components in a seamless manner. (ii) Fully digital mock-up of machines: mechanical structure simulation. Process simulation. Prediction and validation of final pieces' optimal results in design-time. Global dynamic simulation of machines. Validation of the suitable fitting between control approach and mechanical structure. Smart configuration. (iii) PLM (Product Lifecycle Management): tools to design, analyze and manage products from the stage of initial conception to the retirement stage.

- **Synthetic environments integration:** synthetic environments integrate the last developments in the area of virtual reality which allow representing environments close the reality to the practical extension of the synthetic environments in the manufacturing domain.

7. Conclusions and Further Development

Roadmapping exercise is fundamental for technological research orientation towards long-term manufacturing objective. This is especially true when technologies integrate as in SOCRADES (e.g. embedded systems, wireless and SOA, etc). An integrated (but detailed) view is needed to catch all the opportunities emerged from the combination of these technologies.

The main results of the Roadmap first release are the identification of the Expected Features for each Technology Area. EFTAs represent the basic building blocks through which it will be possible to motivate the importance of R&D in SOCRADES-related topics for future manufacturing aims.

Further improvements will be carried out to complete SOCRADES Technology Roadmap. A first step will be the analysis of the impact of EFTAs on competitiveness of future manufacturing. Hence, general long-term objectives or challenges for manufacturing should be identified and the links between these aims and the EFTAs should be made more explicit. A second step will be the EFTA prioritization, explicating the importance of each EFTA in achieving identified objectives. Priority allocation is important in order to identify and suggest trajectories towards future competitiveness in manufacturing.

References

[1] ARTEMIS Strategic Research Agenda Working Group. ARTEMIS Strategic Research Agenda. 2006.

[2] ARTIST Advanced Real-Time Systems. Selected topics in Embedded Systems Design: Roadmaps for Research

[3] Bourgine P, Johnson J (editors). Living Roadmap for Complex Systems Science (2006). http://www.once-cs.net.

[4] De Bosschere K. High-Performance Embedded Architecture and Compilation Roadmap. On P. Stenstrom (Ed.): Transactions on HiPEAC. Springer-Verlag. 2007. pp. 5–29.

[5] FuTMaN: the Future of Manufacturing in Europe 2015-2020. The Challenge for Sustainability. 2003

[6] ITEA Technology Roadmap for Software-Intensive Systems 2nd edition, (2004)

[7] Mantys Next Generation Machine Tools Technological Roadmap 2005.

[8] Marron PJ, Minder D and WiSeNts Consortium. Embedded WiSeNts Research Roadmap. (2006).

[9] Manufuture 2020. Assuring the Future of Manufacturing in Europe. Report of the High Level Group 2004.

[10] Richardson CE, et al. Sensor Technology Roadmapping Efforts at iNEMI. IEEE Trans. on Comp. and Pack. Tech., Vol. 28, No. 2, June 2005, pp. 372-375.

[11] United States Department of Commerce. The NIST 2010 strategic plan. Version 1B. June 2004.

[12] SOCRADES website http://www.socrades.eu.

Innovative Production Machines and Systems
D.T. Pham, E.E. Eldukhri and A.J. Soroka (Eds.)

Distributed Control Concept for a 6-DOF Reconfigurable Robot Arm

T. Strasser, M.N. Rooker and G. Ebenhofer

Robotics and Adaptive Systems, PROFACTOR GmbH, Im Stadtgut A2, 4407 Steyr-Gleink, Austria

Abstract

Current trends such as shorter product lifecycles, reduced time-to-market and mass-customization require new paradigms and approaches for production lines and machines. In the long term, production and manufacturing companies will only be able to survive in the face of increasing globalisation if they can react flexibly and quickly to changing customer and market demands. New paradigms and approaches are waiting to fulfil these requirements, but their implementation requires completely new technologies.

Reconfiguration, both at machine (physical) and control technology (logical) level is a promising candidate to achieve the flexibility required by these paradigms by technical means. This makes a shift from centrally controlled, highly interlinked, and often tightly interlocked production systems to distributed, modular, collaborative components essential. This paper describes the distributed control concept for a 6-DOF reconfigurable robot arm as a demonstration platform for reconfigurable approaches at PROFACTOR.

Keywords: Production Automation and Control, Distributed Control, Reconfigurable Robots

1. Introduction

Today's markets and economies are becoming increasingly volatile, unpredictable, and change radically/abrupt. The accelerating speed of innovation with reduced model life cycles & increased product variety induces ever more frequent changes. Manufacturing companies in the 21st century will face frequent and unpredictable market changes. Economic viably production of small lot-sizes is required especially in Europe. Manufacturing technology and systems and therefore their ICT-infrastructure must undergo radical improvements to keep pace with this trend: New, more flexible technologies and embedded agile components which support the possibility to flexibly compose systems from machines to entire plants are required.

Today's manufacturing systems based on state-of-the-art technology are inadequate to meet the above requirements for more flexible manufacturing technologies. The manufacturing industry mainly uses a portfolio of Dedicated Manufacturing Systems (DMS) and Flexible Manufacturing Systems (FMS) in production. DMS are characterised by high throughput, but poor flexibility. This manufacturing system is tailored towards a specific product and adoption to new products is almost impossible. Thus a DMS represents in many cases a problem when the production life cycle of the corresponding product reaches its end. A FMS is characterised by high flexibility, but also by high costs due to expensive machinery, preventing it from extensive deployment [1]. Both manufacturing concepts are mainly controlled using centralised approaches based on Programmable Logic Controller (PLC) and Computerised Numerical Control (CNC) technology.

On the other hand an adaptable system incorporates a production capacity that is adjustable to fluctuations in product demand, is adaptable to new product functions, and is designed to be upgradeable with new process technology to accommodate evolving product specifications and government regulations. Current systems, even so-called flexible manufacturing systems, do not have these characteristics. A new manufacturing paradigm—called Reconfigurable Manufacturing Systems (RMS)—will position manufacturing companies for the 21st century and in the future. The aim of RMS is to design systems, machines, and controls that are cost-effective and can respond rapidly to changes in market and product demands. Methodologies for the systematic design and rapid ramp-up of RMS are the cornerstones of this new manufacturing paradigm. The new reconfigurable manufacturing paradigm provides exactly the needed functionality and capacity.

One of the key enablers of RMS are Intelligent Mechatronical Components (IMC) which can be considered as basic building elements/blocks [2] of RMS because they can easily be replaced. IMC are a combination of mechanics, electrics and an embedded ICT infrastructure for a certain manufacturing task. Most of the current approaches utilise a distribution set up on the granularity of PLCs and do not support the flexibility at actor/sensor levels although it is fundamentally necessary [3] for composable, reconfigurable components of automation and control systems. The major problem for the industrial usability that does not allow turning such agile systems into reality is that state-of-the-art ICT infrastructure (PLCs and CNC Controller) does not sufficiently support distribution, reconfiguration, modularization and scalability [2],[3]. More than that, up to now reconfiguration is expensive and time consuming as production lines must be stopped during reconfiguration processes and the setup of the new ICT infrastructure is very complex and not prepared for online reconfiguration [3] (i.e. reconfiguration during operation is hardly to achieve at these levels at the moment). To facilitate online reconfiguration of RMS in an economical way, an innovative ICT-infrastructure together with innovative, distributed and embedded software control modules that are dynamically reconfigurable are necessary.

The aim of this paper is to present a demonstration platform for a reconfigurable machine (i.e. reconfigurable robot arm) based on IMC in order to demonstrate the possibilities and features of such an approach. The main focus of this publication lies on the distributed control architecture. The paper is organised as follows: chapter 2 introduces the mechanical and electrical layout of the reconfigurable robot arm. The robot model and the control architecture are explained in chapter 3. The distributed control concept and its implementation are explained in chapter 4. The summary and conclusions are given in chapter 5 of this paper.

2. Reconfigurable Robot Arm

2.1. Mechanical Setup

The robot arm that will be used in this demonstration is a modular 6-DOF reconfigurable robot arm that is composed out of IMC, as depictured in Fig. 1. The robot consists out of six separate joints (i.e. IMC), which are PowerCube modules from Schunk company [6]. They are connected with each other via special connector elements (i.e. golden parts in Fig. 1).

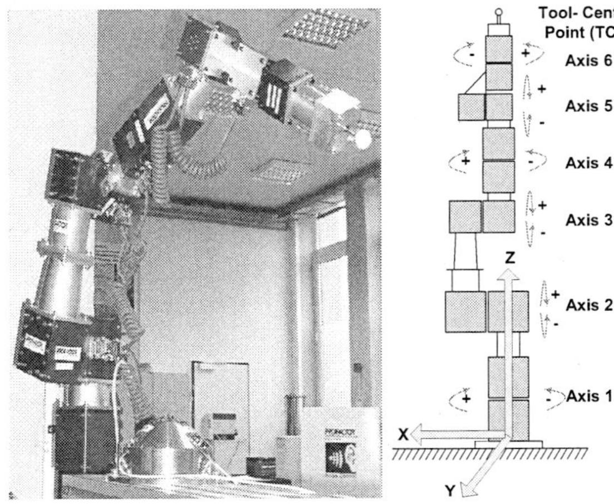

Fig. 1. Mechanical Setup of the Reconfigurable Robot Arm.

2.2. PowerCube Mechatronical Components

The PowerCubes are IMC that can be easily applied for manufacturing and assembly tasks which require fast and flexible adaptation of processes. Fig. 1 shows a PowerCube joint axis. For further information and technical documentation, for example the communication protocol, we refer to the homepage [6] of Schunk company.

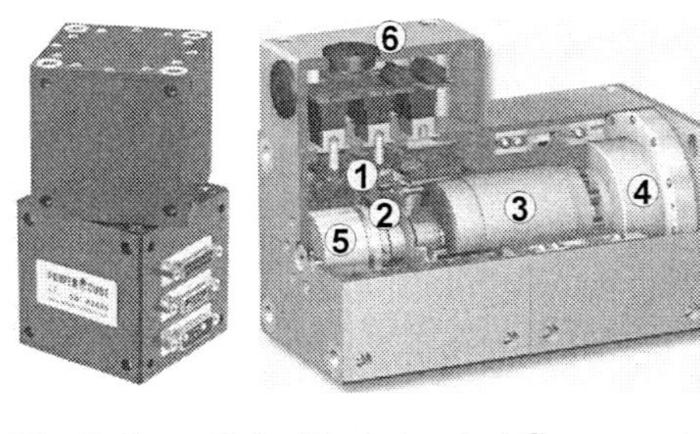

Fig. 2. PowerCube Mechatronical Components.

The following list gives a brief description of the PowerCube functionalities:

- Brushless servo motor with Harmonic Drive gear head
- allows its usage for robotic based tasks,
- Incremental encoder for positioning and velocity control,
- Limit switches-, voltage-, current- and temperature monitoring,
- Integrated reference- and limit switches,
- Magnetic brake,
- Output of internal encoder signal,
- Input of external encoder signal.

Furthermore Fig. 2 shows also the internals of the PowerCube which are:

1) Electronics with integrated control and amplifier part,
2) Encoder for measuring the actual position,
3) Electro motor for high torques,
4) Harmonic Drive gear with gear ratio of 161:1,
5) Break for holding the actual position in case of a halt or in case of losing power, and
6) Covering.

The robot used for the demonstration consists out of three different PowerCube modules. Axes 1 and 2 are of type PR-110, axes 3 and 4 of type PR-90 and axes 5 and 6 are of type PR-70. Furthermore, Table 1 gives an overview of specific parameters that are set for the different PowerCubes in the reconfigurable Robot arm.

Table 1
Specific PowerCube parameters used for the robot arm

Parameters	PR-70	PR-90	PR-110
Nominal Torque	0.181 Nm	0.558 Nm	1.1 Nm
Maximal Torque	0.568 Nm	1.6 Nm	3.3 Nm
Voltage	24 ± 1 V	24 ± 1 V	24 ± 1 V
Max. current	15 A	30 A	30 A
Nom. work torque	23 Nm	72 Nm	142 Nm
Max. work torque	73 Nm	206 Nm	425 Nm
rpm	5000 min^{-1}	4300 min^{-1}	4500 min^{-1}

The embedded micro-controller provides different functionalities in order to operate the PowerCube (e.g. halt, turn, stop etc.). Furthermore each PowerCube has its own axis controller included. The embedded axis controller is responsible for the position and velocity control of each module. In order to communicate with a PowerCube the following fieldbus interfaces are provided:

- Control Area Network (CAN)
- Profibus
- RS 232

3. Robot Model and Control Architecture

3.1. Dynamic Robot Model

In order to derive the control concept for the robot, a robot model is calculated. Therefore, the kinematic model is based on a kinematic chain as depicted in Fig. 1. The robot consists of a serial kinematic chain with 6-DOF through its modular assembly based on the PowerCube.

In order to calculate the position and orientation of the endpoint of the robot (Tool Centre Point - TCP) in the Cartesian and in the Joint Coordinate system of the PowerCubes, the forward and inverse kinematics were calculated. For a 6-DOF robot and with a central gripper on the last joint, the solution for the inverse kinematics can be calculated analytically.

Without a central gripper or with more than 6-DOF, a solution for the inverse kinematics can only be calculated using an approximation algorithm. For comparison, the Newton approximation algorithm was also used for calculating the inverse kinematics. It turned out that the analytic way can be calculated faster than the approximation one. Therefore the first was used in this demonstration. For the calculation of the equation of motion

$$M(q)\ddot{q} + G(q,\dot{q}) - Q = 0 \qquad (1)$$

with the vector of generalised module coordinates q, the robot mass matrix M, the gyroscopic matrix G and the vector of generalised forces Q, the projection equations according to [11] were used.

3.2. Motor Model

The torque produced for each robot axis is generated by the DC motor of the PowerCube. For the calculation of the motor model an equivalent circuit diagram of a permanent excited DC motor was used. Fig. 3 shows the electrical equivalent of the DC motor.

The resulting PowerCube torque including the gear ratio of the Harmonic Drive gear can be calculated according the following equation (by negligence the motor impedance L_A):

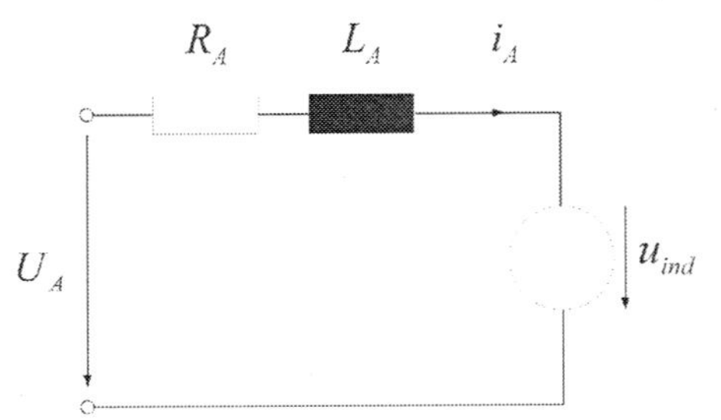

Fig. 3. Equivalent circuit diagram for the permanent excited DC motor.

$$M_{PC} = \frac{k_m}{R_A}(U_A - k_m i_G \dot{q}) \tag{2}$$

with

M_{PC} ... PowerCube torque (at motor axis incl. gear),
k_m ... Proportional factor motor torque and motor current (i.e. $M = k_m i_A$),
U_A ... Motor voltage, and
i_G ... Gear ratio.

3.2. Control Concept

The control concept of the modular, reconfigurable 6-DOF robot is depicted in Fig. 4.

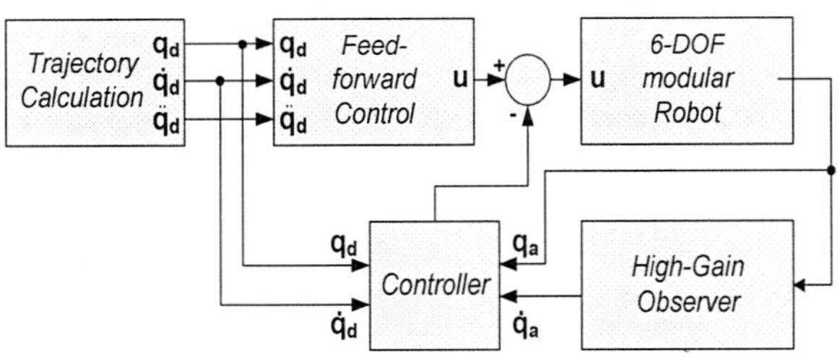

Fig. 4. Control Concept.

The goal of the control concept is to move the TCP of the robot along an interpolated path from a chosen point A to point B. The robot motion should be smooth. For the trajectory generation a *sin*²-based acceleration profile was applied (see Fig. 5). The Trajectory Calculation block in Fig. 4 calculates the desired angular position q, the desired speed $\dot{q}$ and the desired acceleration $\ddot{q}$ for each PowerCube module.

In order to get better control quality, a feedforward term (Feedforward Control) was introduced into the control concept [12]. It is calculated based on the robot equation of motion (see above) and the force term (i.e. motor torque) generated by the PowerCubes:

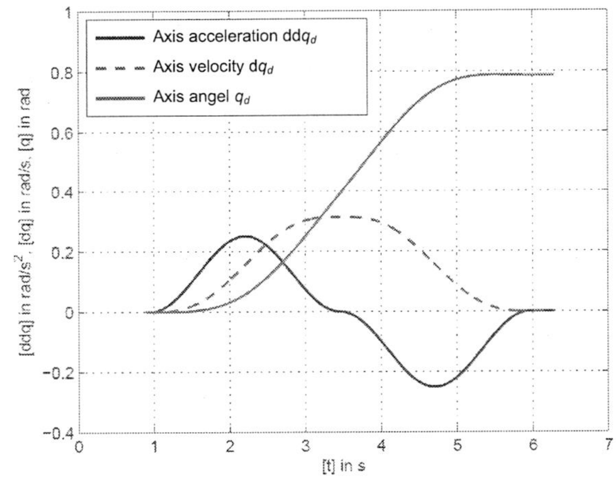

Fig. 5. Trajectory with sin²-based acceleration profile.

$$Q = Bu \tag{3}$$

This leads to the feed-forward term *u* of the control concept:

$$u = B^{-1}\left[M(q)\ddot{q} + G(q,\dot{q})\right] \tag{4}$$

For the feedback control term (i.e. robot controller) a PD control approach is used. The trajectory in the used control law includes the angular position and the angular speed of the joints (PowerCube modules). This results in the following PD controller (controller parameters K_P and K_D):

$$u = K_P(q_d - q_a) + K_D(\dot{q}_d - \dot{q}_a) \tag{5}$$

with q_d the desired position and q_a the actual position of the joints.

Since each PowerCube has only an encoder for the measurement of its angular position, a direct measurement of the angular speed is not possible. But for the used control concept it is necessary to know the angular speed of each PowerCube.

In order to calculate the angular speed, a High-Gain Observer is used. This approach is common for mechanical structure since it uses the equation of motion. Fig. 4 shows the control concept including the High-Gain Observer for calculating the angular speed of each PowerCube based on the angular position.

4. Distributed Control Concept

4.1. Embedded Hardware and Software Setup

Fig. 6 shows the hardware (embedded control) setup of the reconfigurable robotic system. Respectively two robot axes are controlled by PC/104 embedded PCs [7] equipped with a Debian Linux [8] and an OSADL real-time kernel patch [9].

The whole robot control concept explained in chapter 3 is implemented in IEC 61499 Function Blocks [15] in order to support reuse and reconfiguration on the Function Block level. The FORTE [5] IEC 61499 compliant runtime environment is installed on each of the PC/104 controllers and used for the execution of the IEC 61499 Function Block control application.

The data exchange between the PC/104 and the PowerCubes is done via CAN [10]. The engineering and visualisation PC is connected via standard Ethernet to the PC/104 controllers.

The engineering of the IEC 61499 robot control program is carried out via the 4DIAC-IDE [5]. The visualisation and HMI are also modelled with IEC 61499 Function Blocks. Therefore the Java-based FBDK [4] is used as runtime environment for the execution of the visualisation program.

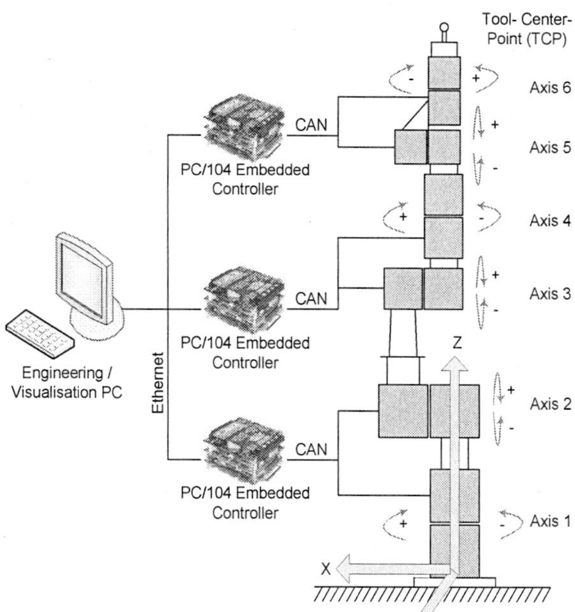

Fig. 6. Embedded Hardware Setup.

4.2. Distributed Control Implementation

The above introduced control concept (see Fig. 4) and the chosen hardware structure of the modular 6-DOF robot (as depicted in Fig. 6) leads to the IEC 61499 system configuration as presented in Fig. 7 [13].

The following two separate IEC 61499 applications have been developed:

- Robot Control Application (RCA)
- Robot HMI Application (RHA)

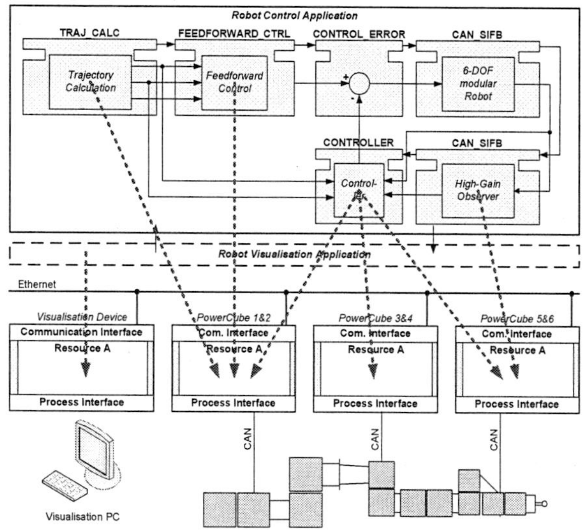

Fig. 7. IEC 61499 System Configuration for the 6-DOF reconfigurable robot arm.

The RCA contains the implemented control concept (see above) as IEC 61499 application. Each component of the control concept has its own IEC 61499 FB representative (i.e. TRAJ CALC, FEEDFORWARD CTRL etc.).

For visualisation and parameterisation issues, an own RHA was developed. The RHA is also built out of special IEC 61499 HMI Service Interface Function Blocks (SIFBs). The two applications communicate with each other via Publish/Subscribe Function Blocks. The IEC 61499 system model for the robot controller consists of four different remote devices. One remote device is executed on the Visualisation PC (equipped with the FBDK), the other three devices are executed on the PC/104 embedded hardware (equipped with the FORTE). The communication between all devices is done via Ethernet. The whole RHA is mapped to the visualisation device while the trajectory calculation and the feedforward control Composite Function Block (CFB) are mapped to the PowerCube 1&2 Device.

The High-Gain Observer CFB is mapped to the PowerCube 5&6 Device. The other control modules (CAN SIFB and CONTROLLER) are represented as sub-applications in the RCA and are distributed to the three PowerCube devices.

4.3 Motion Control Library

For the interaction of the PC/104 embedded controllers and the PowerCube modules an IEC 61499-based motion control library was developed (example see Fig. 8). The interface specification of the motion control function blocks is based on the Motion Control Specification of the PLCopen association [14].

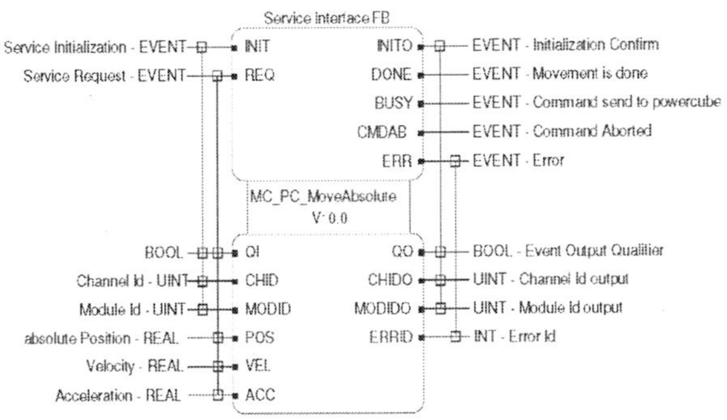

Fig. 8. IEC 61499 Move Absolute Motion Control Block.

5. Summary and Conclusions

In this paper the need for reconfigurable approaches in manufacturing was discussed. In order to demonstrate the advantages of such systems and machines a demonstration platform (i.e. reconfigurable robot arm) was introduced. The reconfigurable robot consists of PowerCube IMC.

Furthermore a robot control concept was presented which can be distributed to different embedded hardware. For the implementation of the distributed control concept the IEC 61499 standard was used. The distributed control hardware consists of PC/104 embedded controllers equipped with a real-time Linux.

Together with the modular hardware of the robot arm and the IEC 61499 based control concept it is relatively easy to make modifications to the configuration of the reconfigurable robotics system (i.e. hardware as well as software reconfiguration becomes possible).

Acknowledgements

The work leading to this invention has received funding from the European Community's Sixth Framework Program (FP6/2002-2006) under grant agreement n° 026697-2. Further information is available at: www.pisa-ip.org

PROFACTOR is partner of the EU-funded FP6 Innovative Production Machines and Systems (I*PROMS) Network of Excellence. Further information is available at: www.iproms.org

References

[1] Koren et al, "Reconfigurable Manufacturing Systems" Annals of the CIRP Vol. 48/2/1999.

[2] T. Strasser, K. Fessl, A. Hämmerle, M. Ankerl, "Rapid Reconfiguration of Machine Tools for Holonic Manufacturing Systems", 16th IFAC World Congress, Prague, Czech Republic, 2005.

[3] V. Vyatkin, J. Christensen, J. Martinez Lastra, "OOONEIDA: An Open, Object-Oriented kNowledge Economy for Intelligent Distributed Automation", IEEE Transactions on Industrial Informatics, Vol. 1, n. 1, February 2005.

[4] HOLOBLOC Inc., "HOLOBLOC.com – Function Block-Based, Holonic Systems Technology," Access Date October 2007. [Online]. Available: www.holobloc.com

[5] 4DIAC Consortium, "4DIAC - Framework for Distributed Industrial Automation and Control, Open Source Initiative," Access Date October 2007. [Online]. Available: www.fordiac.org

[6] Schunk GmbH & Co. KG, "PowerCube Mechatronic Components," Access Date October 2007. [Online]. Available: www.powercube.de

[7] PC/104 Embedded Consortium, "PC/104 Embedded PC Modules," Access Date October 2007. [Online]. Available: www.pc104.org

[8] Debian, "Debian GNU/Linux," Access Date October 2007. [Online]. Available: www.debian.org

[9] OSADL eG, "Open Source Automation Development Lab," Access Date October 2007. [Online]. Available: www.osadl.org

[10] CiA - CAN in Automation, "CAN - Controller Area Network," Access Date October 2007. [Online]. Available: www.can-cia.de

[11] H. Bremmer, "Lecture notes robotics," Johannes Kepler University Linz, Tech. Rep., 2003.

[12] M. Schnabler, "Modelling of a multi-axis kinematic for the verification of a measurement system for trajectory analysis," Master's thesis, Johannes Kepler University Linz, 2007.

[13] D. Christ, "Prototype of a distributed, reconfigurable control system," Master's thesis, University of Applied Sciences Zwickau, 2007.

[14] PLCopen Association, "Motion Control Specification", Access Date March 2008. [Online]. Available: www.plcopen.org

[15] IEC TC65/WG6, "IEC 61499: Function Blocks for Industrial Process Measurement and Control Systems". Parts 1 – 4, International Electro-technical Commission IEC Std., Rev. 1.0, 2004/2005.

Innovative Production Machines and Systems
D.T. Pham, E.E. Eldukhri and A.J. Soroka (Eds.)

Considerations on efficiency of reconfigurable systems for validation of heat meters in SME

T. Goszczynski, E. Jachczyk

Industrial Research Institute for Automation and Measurements – PIAP, Al. Jerozolimskie 202, 02-486, Warsaw, Poland

Abstract

Validation of heat meters, used for financial settlements between heat supplier and residents of apartments in buildings with central heating systems, is the most time consuming part of their production. For SME engaged in this product it is very important to have appropriate instruments and procedures for validation of different type of heat meters. In this paper procedures and instruments for reconfigurable system are proposed and analysed taking into account its efficiency and uncertainty of measurements.

Keywords: heat meters, validation tests, reconfigurable systems, modular systems, SME

1. Introduction

Heat meters are instruments intended for measuring the heat that, in a heat-exchange circuit, is absorbed or given up by a heat conveying liquid. These heat meters are used for financial settlements between heat supplier and residents of apartments in buildings with central heating systems and are submitted for control by legal metrology services. The meters shall comply with the requirements formulated in standards [1] and [2]. For manufacturing of heat meters the main limitation in productivity lies in process of their validation. Production and/or validation of heat meters are one of attractive activities for SMEs. The SMEs have to exhibit large flexibility to cope with frequent changes that may involve the reconfiguration of used system. The process of validation is time consuming and therefore optimisation and automation of it is critical for producers of these instruments. Selecting the most appropriate modular validation system can offer enormous benefits in terms of product quality, cost reduction and manufacturing productivity. However, selecting the right system for a product depends on a large number of factors, and is not a simple problem. This paper proposes a possible approach to support decision-making in the area for heat meter validation systems and a methodology of selecting instruments for modular systems considering technological efficiency and economic approach. The paper addresses two I*PROMS2008 topics: Production Organisation and Management and Reconfigurable Manufacturing Systems. The other sections are organised as follows: Sections 2, 3 and 4 give the theoretical framework of heat meter validation, Section 5 describes reconfigurable automated system for validation and in Sections 6 and 7 efficiency consideration and production organisation topics are discussed. The last Section provides some conclusions and gives suggestions for future work.

2. Theoretical framework

Heat meter sub-assemblies are: the flow sensor, the temperature sensor pair and the calculator. Flow sensor is a sub-assembly through which the heat-conveying liquid flows, at either the flow or return of a heat-exchange circuit, and which emits a signal, being a function of the volume of the flowing liquid. Temperature sensor pair is a sub-assembly, which senses the temperatures of the heat conveying liquid at the flow and return of a heat-exchange circuit. Calculator receives signals from the flow sensor and the temperature sensors and calculates the quantity of heat exchanged. In a most popular type of heat meter (examples shown in Fig. 1) flow sensor is verified on the water stand and calculator with inseparable temperature sensor pair on the special validating system.

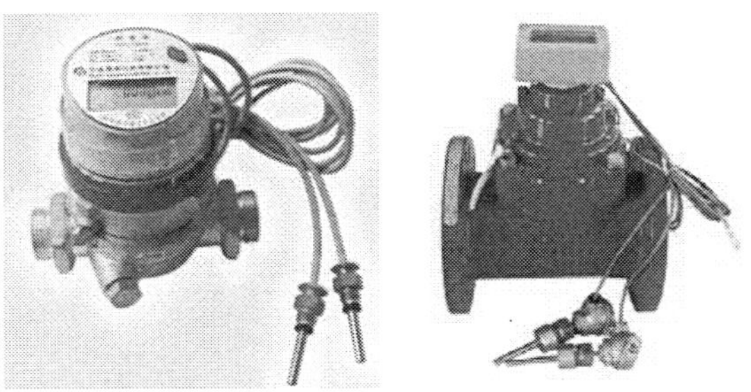

a) b)

Fig. 1. Two examples of heat meters with inseparable temperature sensor pair

When validating the calculator with inseparable temperature sensor pair, the two temperature sensors should be immersed in two thermostatic baths. Within each of the three temperature ranges given in [1] accuracy of heat calculation should be examined. Heat meter calculates actual values of heat power and counts energy in time, using algorithms given in [3]. Heat energy Q measured by heat meter is a difference of energy of heat conveying liquid on the input of the heat exchanger Q_{in} and on the output of it $Q_{out.}$ This energy measured in time from 0 to τ_1 can be presented as:

$$Q=\int_0^{\tau_1}[t_1 C_w(t_1)\rho(t_1)v_1(t_1)- \ t_2 C_w(t_2)\rho(t_2)v_2(t_2)]d\tau \quad (1)$$

where: t_1 – temperature of water on input, t_2 – temperature of water on output (return temperature), $Cw(t_1)$ – heat capacity on input, $Cw(t_2)$ – heat capacity on output, $\rho(t_1)$ – water density on input, $\rho(t_2)$ – water density on output, $v_1(t_1)$ – volume flow rate of water on input, $v_2(t_2)$ – volume flow rate of water on output, τ – time.

In process of validation a heat meter shall be tested in three points of temperature difference $\Delta t = (t_1 - t_2)$:

1. for return temperature from 40 °C to 70 °C:
$$\Delta t_{min} \le \Delta t \le 1{,}2\ \Delta t_{min} \quad (2)$$
2. for return temperature from 40 °C to 70 °C:
$$10\ ^{\circ}C \le \Delta t \le 20\ ^{\circ}C \quad (3)$$
3. $$\Delta t_{max} - 5\ ^{\circ}C \le \Delta t \le \Delta t_{max}. \quad (4)$$

3. Further analysis of the theoretical issues

It is known [4, 5] that two coefficients $k_1(t_1, t_2)$ and $k_2(t_1, t_2)$, which are functions of t_1 and t_2, can be presented as:

$$[t_1 - t_2]\ k_2(t_1, t_2)\ v_2(t_2) = \\ = t_1\ Cw(t_1)\ \rho(t_1)\ v_1(t_1) - t_2\ Cw(t_2)\ \rho(t_2)\ v_2(t_2) \quad (5)$$

$$[t_1 - t_2]\ k_1(t_1, t_2)\ v_1(t_1) = \\ = t_1\ Cw(t_1)\ \rho(t_1)\ v_1(t_1) - t_2\ Cw(t_2)\ \rho(t_2)\ v_2(t_2) \quad (6)$$

After applying it to (1) we have two equations: one, for measurement of volume on output

$$Q = \int_0^{\tau_1}[t_1 - t_2]\ k_2(t_1, t_2)\ v_2(t_2)\ d\tau \quad (7)$$

and second, for measurement of volume on input

$$Q = \int_0^{\tau_1}[t_1 - t_2]\ k_1(t_1, t_2)\ v_1(t_1)\ d\tau \quad (8)$$

The validation system is destined for measurement of heat meters accuracy according to one of these equations.

4. Evaluation of the uncertainty of measurements

Uncertainties of the system should be (with a combined standard uncertainty i.e., estimated standard deviation) less than 1/5 of the maximum permissible error of the heat meters with a level of confidence of 95% and a coverage factor k = 2. Therefore the maximum permissible relative uncertainty of the system is defined as:

$$U_{perm} = \pm(0{,}2 + 0{,}8\frac{\Delta t_{min}}{\Delta t}) \quad (9)$$

Now after rearranging the equation (7) we have:

$$\Delta Q = \Delta t\, k_2(t_1, t_2) v_2(t_2) \Delta\tau \quad (10)$$

Since uncertainty depends on systematic effects, its standard B type components are:

u_Q – uncertainty of measurement of ΔQ,
$u_{\Delta t}$ – uncertainty of measurement of Δt for sensor pair,
u_k – uncertainty of calculating values of coefficient $k_2(t_1, t_2)$,
u_v – uncertainty of measurement of pulses,
$u_{\Delta\tau}$ – uncertainty of time measurement.

Combined standard uncertainty with a 95 % level of confidence and a coverage factor k = 2 is:

$$U_Q = 2\sqrt{u_Q^2 + u_{\Delta t}^2 + u_k^2 + u_v^2 + u_{\Delta\tau}^2} \quad (11)$$

Using known limits of deviations $\pm a_i$ for consecutive parameters of measurement, the standard B type [6, 7] uncertainty of every parameter is:

$$u_Q = \frac{a_Q}{\sqrt{3}}, u_{\Delta t} = \frac{a_{\Delta t}}{\sqrt{3}}, u_k = \frac{a_k}{\sqrt{3}}, u_v = \frac{a_v}{\sqrt{3}}, u_{\Delta\tau} = \frac{a_{\Delta\tau}}{\sqrt{3}} \quad (12)$$

Components of uncertainty u_Q, u_k, u_v and $u_{\Delta\tau}$ are small and do not exceed 0,01 %.

Uncertainty of measurement of temperature difference can be calculated from:

$$u_{\Delta t} = \sqrt{\frac{a_{sen1}^2}{3} + \frac{a_{sen2}^2}{3} + \frac{a_{t1}^2}{3} + \frac{a_{t2}^2}{3} + \frac{a_{nt1}^2}{3} + \frac{a_{nt2}^2}{3}} \quad (13)$$

where:
a_{sen1}, a_{sen2} – one-year stability error of legal sensors,
a_{t1}, a_{t2} – short time error of resistance meter (transfer relative accuracy),
a_{nt1}, a_{nt2} – temperature t_1 and t_2 stability error (in thermostatic baths).

Combined standard uncertainty of system from (11) with a 95 % level of confidence and a coverage factor k = 2 is recounted in Table 1 into relative to temperature difference as U_{rel}.

Type of instruments chosen in the system will affect uncertainty as well as initial and maintenance costs of the stand. Examples are presented in Table 1. For the instruments set No. 2 temperature sensors are less expensive but the sensors must be legalized every 4-month and for instruments set No. 1 only once a year.

5. Automation of validation procedure

For validation system it is needed not only to measure with appropriate uncertainty but also to achieve maximum efficiency with reasonable cost of instruments.

Table 1. Uncertainty results for 2 different sets of instruments used in the system

	Instruments set No. 1		Instruments set No. 2	
	a (mK)	a^2 (μK)	a (mK)	a^2 (μK)
$a_{sensor1}$	2,5	6,25	10	100
$a_{sensor2}$	2,5	6,25	10	100
a_{meter}	1,8	3,24	5	25
a_{bath1}	10	100	10	100
a_{bath2}	10	100	10	100
Σ		215,74		425
$U_{\Delta t}=(\Sigma/3)^{0,5}$		8,4		11,9
$U_Q = 2\,U_{\Delta t}$		16,8		23,8
	U_{perm} %	U_{rel} %		U_{rel} %
Δt = 3 K	±1,0	±0,56		±0,79
Δt = 10 K	±0,44	±0,168		±0,238
Δt = 20 K	±0,32	±0,084		±0,119

U_{perm} – maximum permissible uncertainty

Modularity is a good solution for this purpose. The use of modular structure architecture of the validation system has many advantages. The end user and system integrator can more easily configure the system and later reconfigure it to meet the customer's future needs. Modularity is also a good solution for later upgrades or modifications of the system. Most of the modules should be standard, with known catalogue prices; this helps in calculating the cost of investment. The aims are also to minimize special product-related customisation. Therefore selection of modules is one of the first steps. It was checked that in thermostatic baths of middle range one reference sensor and up to 8 heat meter sensors could be immersed. For measurement of the two standard reference sensors resistance one precision resistance meter can be used with IEEE-488 bus for communication with computer. The 8-channel multiplexer connected to the standard reference sensors and to precision resistance meter with 4-wire technique of measurement should be used. Another multiplexer for connection heat meters pulse outputs to pulse counters should have number of channels appropriate to number of heat meters in a group.

Heat meters working in verification mode need to have a group of pulses received on their inputs in order to generate on its outputs number of pulses proportional to quantity of heat measured. Calculators could be physically connected to the system using the multiplexer. Only two are in the same phase connected to pulse generators and 2 pulse counters and in every such phase temperature in the two baths is measured using reference sensors connected one after another to the same resistance meter. In this way uncertainty due to temperature stability in the baths is minimized. The measurement can be made twice: before and after generating pulses to heat meter and average value taken into account. Time needed for acquiring output pulses from heat meter (calculator) is programmed in the stand's database, as well as other specific parameters of the heat meter under test (defined by producer). Schematic diagram of one of the proposed system [8] is shown on Fig. 2.

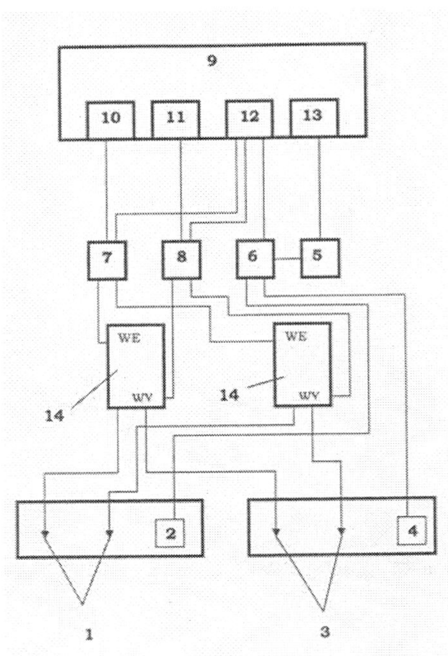

Fig. 2. Scheme of the system for validation of hybrid heat meters

Key:
1, 3 – heat meter sensors immersed in thermostatic baths; 2, 4 – standard reference sensors; 5 – digital resistance meter; 6, 7, 8 – sections of multiplexer; 9 – computer; 10, 11 – card with generators and counters; 12 – digital outputs card; 13 – interface IEEE-488 card, 14 – heat meters under test

Since every heat meter should be checked at 3 points of its temperature characteristic and since stabilizing temperature of bath takes time, it is recommended to realize measurements for few groups of heat meters after stabilizing temperature in baths. Measurement data are stored in a database and after every change of temperature in baths the procedure of measurement for 10 groups of meters under test is realized. The prototype realization of system used for confirmation of data on accuracy and time consideration is shown on Fig. 3.
It is based on scheme introduced in Fig. 2 and uses instruments listed in Table 2. Controller construction permits applying different types of multiplexers.

Fig. 3. The prototype realization of heat meter validation system

6. Validation process efficiency

One of the methods should be chosen as appropriate for any SME depending on type of heat meters. The producer of heat meters can choose from the following methods of validation:
manually one after another

$$t_{tot1} = t_1 + t_{10} + t_6 + t_{10} + t_6 + t_{10} + t_7 \quad (14)$$

automatically in group of 8 items (4 x 2)

$$t_{tot4x2} = (t_1 + 4\,t_{10} + t_6 + 4\,t_{10} + t_6 + 4\,t_{10} + 8t_7) / 8 \quad (15)$$

automatically in group of 80 items (10 x 4 x 2)

$$t_{tot10x4x2} = (10t_8 + t_6 + 10t_8 + t_6 + 10t_8) / 80 \quad (16)$$

automatically in group of 80 items in 10 x 8 manner

$$t_{tot10x8} = (10t_9 + t_6 + 10t_9 + t_6 + 10t_9) / 80 \quad (17)$$

where:
t_1 – connections and time for stabilizing temperature of immersed sensors
t_2 – bath control
t_3 – pulses generation

Table 2. Instruments list used in the prototype realization of heat meter validation system

No.	Instrument	Type	Quantity	Producer
1	digital multimeter	2010	1	KEITHLEY
2	controller	TEC-LEG-H	1	PIAP
	multiplexer card	PCLD-785B	1	ADVANTECH
	interface cards		2	ESCO
	power supply 5V/12V	PS5/12	1	ESCO
3	thermostatic baths		2	Heto Holten
4	heat meters sensors in bath		2	
5	reference sensors Pt100	935-14-16	2	ISOTECH
6	heat meter	F531		ABB
7	heat meter	compact III		TECHEM
8	PC computer	CAM	1	NCM
	digital output card	ILEG – PP – K	1	ESCO
	interface IEEE-488 card	KPCI-488	1	KEITHLEY
	generator/counter card	PCI – CTR10	1	Computer Measurement
9	bath controller		2	Heto-Holten

Table 3. Time of validation (in seconds/heat meter) for different methods of validation

Type*	t1	t2	t3	t4	t5	T6	t7	t8	T9	t10	ttot1	ttot4x2	ttot10x4x2	ttot10x8
A	60	10	20	20	10	1800	30	540	360	60	3930	577,5	247,5	180
B	60	10	20	180	10	1800	30	1180	520	220	4570	817,5	487,5	240
C	60	10	20	300	10	1800	30	1660	640	340	5050	997,5	667,5	285
D	60	10	20	20	10	3600	30	540	360	60	7530	1027,5	292,5	225
E	60	10	20	300	10	3600	30	1660	640	340	8650	1447,5	712,5	330
F	120	10	40	20	20	3600	30	1560	450	90	7710	1080	675	258,75
G	120	10	40	20	20	1800	30	1560	450	90	4110	630	630	213,75
H	120	10	20	10	10	3600	60	1640	650	50	7580	1050	705	333,75
I	180	20	60	300	20	3600	120	4000	1540	400	9100	1642,5	1590	667,5

* Types from A to I in the table represent different heat meters from different vendors (see also Table 4).

t_4 – test pulses counting
t_5 – temperature measurements in baths
t_6 – change of temperature in baths
t_7 – disconnections of heat meters

$$t_8 = t_1 + 4t_{10} + 8t_7 \quad (18)$$

$$t_9 = t_1 + t_{10} + 8t_7 \quad (19)$$

$$t_{10} = t_2 + t_3 + t_4 + t_5 \quad (20)$$

Values of time needed for validation (recounted for one meter) are given for few examples of meter types in Table 3 and instruments needed in Table 4.

7. Efficiency considerations

Above consideration give some data to overall life cycle cost analysis and efficiency of heat meter production taking the validation as a part of it. When considering efficiency of heat meters production the manufacturing cost and performance should be analysed. Appropriate methodology should be used [9] integrating cost of ownership (COO) and overall equipment efficiency (OEE). Engineers need to calculate all the costs arising during the lifetime of the equipment. The life cycle of the system in the design phase is based on scenarios. When modelling the future scenarios, it is possible to estimate the needed changes to the system and thus the cost effects. Since the basic assembly process is the same, only the product specific system parts need to be changed: sensors, meters, computer digital cards and multiplexer. COO comprises all the cost during the system life cycle divided by the number of good units produced.

Table 4. Instruments list for modular system of validation (method names as in Table 3)

Method	ttot1	ttot4x2	ttot10x4x2	ttot10x8
Thermostatic baths	2	2	2	2
Computer	1	1	1	1
Data base			1	1
Reference sensors	2	2	2	2
Resistance meter	1	1	1	1
8 channel multiplexer		1	1	1
Pulse generator	1	1	1	1
Impulse counter	1	2	4	8
16-channel multiplexer		1	1	1

Thus COO depends on the production throughput rate, cost, equipment reliability, throughput yield, and equipment utilization. The OEE is being used to assess the manufacturing efficiency of the running systems. OEE is a key performance indicator of how machines, production lines or processes are performing in terms of equipment availability, performance or speed and quality produced. Modular structure proposed in [10] and new methodology and reconfigurability of the system proposed here can be applied at an individual workstation, piece of equipment or process comparison. It can also be used for the production line level if the user enters some of the key parameters based on the bottleneck machine – validation system in this case. In the deeper analysis, both hierarchical levels should be studied, the most important workstations and also the production line level. Real factory data and accounting data must be used to develop consecutive models and obtain results in the following evaluation rounds and new system designs.

8. Conclusions

In present paper solutions that are emerging in validating heat meters and their impact on efficiency were analysed. Understanding these issues and their mathematical derivations and algorithms are essential for later works on this matter. The authors proposed methods of analysis for accurate calculation of the system parameters and methodology for building reconfigurable systems. They also have undertaken the experimental trials, which preliminarily prove the presented approach being very promising, although more trials need to be undertaken and results to be critically reviewed. Therefore in the next step analysis of automatic connections and disconnections of heat meters from the system, especially immersion of their sensors in thermostatic baths, will be taken. There is planned also cooperation with some SMEs in life cycle cost analysis and efficiency of heat meters production.

Acknowledgements

Industrial Research Institute for Automation and Measurements – PIAP is from 10 years engaged in research projects concerning validation systems for heat meters and its sub-assemblies. The results of project "Validation system for hybrid heat meters" C/06316 founded by Polish Ministry of Science are presented in [10]. The reported here work is continuation of the research in the direction of developing reconfigurable systems and their analysis. Authors would like to thank dr. J. Korytkowski for his help. Industrial Research Institute for Automation and Measurements – PIAP is a partner of the EU FP6 Innovative Production Machines and Systems (I*PROMS) Network of Excellence.

References

[1] European Standard EN 1434-5:1997/A1: 2002 (E) Heat meters – Part5: Initial verification tests.

[2] European Standard EN 60751 (IEC Publication 751) *Industrial platinum resistance thermometer sensors.*

[3] ORGANISATION INTERNATIONALE DE METROLOGIE LEGALE. International Recommendation. *Heat meters.* OIML R 75. Edition 1988 (E). Paris.

[4] Korytkowski J., *Approximation of non-linear algorithms of heat meters using platinum temperature sensors.* PAK, Nr 8, 1985 (in Polish).

[5] Korytkowski J., Goszczynski T., Jachczyk E., *Methods of computer testing system for examining the accuracy of temperature probes of heat meter.* Automation'98 Conf., March 11-12, 1998, Warsaw, Poland, pp. 335-342 (in Polish).

[6] Turzeniecka D. *Essentials of uncertainty estimates.* PAK,. Nr 9, 1998, pp. 327-330 (in Polish).

[7] Jaworski J. M., *Un-accuracy, error and uncertainty of measurement.* PAR, Nr 7-8, 1999, pp. 12-16 (in Polish).

[8] Goszczynski T., *System for compact heat meters measurements* Patent pending P361637 – Patent Office RP.

[9] Heilalaa J., Montonena J., Helinb K., Salonena T., Väätäinena O., *Life Cycle and Unit Cost Analysis for Modular Re-Configurable Flexible Light Assembly Systems* – Intelligent Production Machines and Systems, 2nd I*PROMS Virtual International Conference 3-14 July 2006.

[10] Goszczynski T., Korytkowski J., *System for validation of hybrid heat meters.* PAR, nr 9, 2006, pp 11-12, (in Polish).

Innovative Production Machines and Systems
D.T. Pham, E.E. Eldukhri and A.J. Soroka (Eds.)

Using a Hybrid PSO-Bees Algorithm to train Neural Networks for Wood Defect Classification

D.T. Pham and M. Sholedolu

Manufacturing Engineering Centre, Cardiff University, The Parade, Cardiff CF24 3AA

Abstract

This paper describes a system for automatically classifying wood boards according to whether they are defect-free or whether they have a certain type of defect. At the heart of the system is a neural classifier that can be trained to distinguish between these different categories of boards. In this work, training of the classifier was performed using the Particle Swarm Optimisation-Bees Algorithm, a very effective optimisation tool that can quickly find global optima in complex search spaces.

Keywords: Automated Visual Inspection, Neural Networks, Particle Swarm Optimisation, Bees Algorithm.

1. Introduction

Wood veneer boards are manufactured on fast production lines where boards can move at speeds exceeding 20m/s. Inspecting the boards for surface defects that can cause downstream quality problems is thus a task that is taxing for human operators. Early work aimed at automating these tasks by introducing computer-controlled visual inspection systems has involved the use of conventional signal processing and pattern recognition techniques. More recently, automated visual inspection systems with neural network classifiers have been developed [1]. This paper presents a system employing a Multi-Layer Perceptron network as a pattern classifier. Multi-Layer Perceptrons are usually trained by back propagation. However, that training technique which is based on gradient information sometimes produces classifiers with poor performances because of the existence of local optima where the gradient is null. The Multi-Layer neural classifier developed in this work was trained using a Hybrid Particle Swarm Optimisation (PSO)-Bees Algorithm, an algorithm belonging to the population-based category that is effective against local optimum problems. The hybrid algorithm combines the fast convergence property of the PSO algorithm and the inherent ability of the Bees Algorithm to avoid local optima. The remainder of the paper is organised as follows. Section 2 describes the wood surface defect classification problem. Section 3 provides an introduction to intelligent swarm-based optimisation. Section 4 outlines the Hybrid PSO-Bees Algorithm and its use to train a Multi-Layer Perceptron to classify the main types of defects found on wood veneer boards. Section 5 presents the results obtained and section 6 concludes the paper.

2. Wood defect classification problem

There are twelve common types of defects on wood veneer surfaces. These are shown in figure 1, together with a photograph of defect free (clear) wood.

As mentioned above, defect classification was performed with a trained Multi-Layer Perceptron. Features were first extracted from different wood images containing known defect types or no defects and the Multi-Layer Perceptron was taught to distinguish between the features of those images. In total, as performed in previous work [2, 3], seventeen features were extracted from the wood images and used to train the Multi-Layer neural classifier.

The wood defect classification problem thus reduces to that of mapping a given set of seventeen features extracted from an image onto one of the image categories shown in figure 1. How a Multi-Layer Perceptron is trained to perform this mapping will be discussed later in the paper.

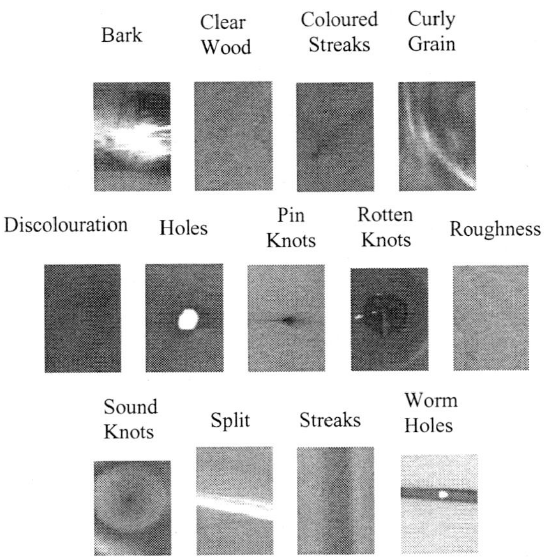

Fig. 1. Categories of veneer wood images.

3. Intelligent swarm-based optimisation

Intelligent swarm-based optimisation algorithms (SOAs) mimic nature's methods to find optimal solutions. They are inspired by the observation of the behaviour of natural systems (consisting of agents such as birds, fish, bees, wasps, ants, termites, and by swarming, flocking, herding, and shoaling phenomena in vertebrates). SOAs can handle complex optimisation problems previously not possible to solve. The abilities of these systems transcend those of the constituent individual agents due to their emergent strategy which entails local interactions between simple agents and the environment to combine and self-organise in a way to accomplish the required task.

A key advantage of SOAs over traditional techniques is robustness and flexibility, by using a population of solutions for every iteration instead of a single solution. These properties make Swarm Intelligence a successful paradigm for systems to deal with multi-modal dynamic multi-dimensional problems. If an optimisation problem has a single optimum, SOA population members can be expected to converge to that optimum solution. On the other hand, if an optimisation problem has multiple optimal solutions, an SOA can be used to capture them in its final population. SOAs include the PSO algorithm [4] and the Bees Algorithm [5].

The original PSO algorithm is a stochastic optimisation algorithm inspired by the behaviour of a flock of birds or a school of fish. It is believed that there is a form of social influence and social learning in such groups of animals. Individuals in a group emulate the success of their neighbours in attaining some reward. The collective behaviour that emerges is that of discovering high rewards (optimal solutions) in a complex search space.

With the PSO algorithm, each individual is a point (or particle) in the search space. In other words, each individual represents a candidate solution to the optimisation problem being addressed. The algorithm drives towards the optimal solution by controlling the movements of individual candidate solutions.

The Bees Algorithm is an optimisation algorithm inspired by the natural foraging behaviour of honey bees. The algorithm performs a combination of neighbourhood and global random search. Bees are recruited to exploit the most promising sites. Additionally, a proportion of the bees population is sent randomly to explore the solution space.

4. The PSO-Bees Algorithm and multi-layer perception training

4.1 The PSO-Bees Algorithm

The Hybrid PSO-Bees Algorithm was developed to solve the problem of premature convergence known to exist in the basic PSO algorithm. If a particle's current position coincides with the global best position, the particle will only move away from this point if its previous velocity and inertia weight are non zero. On the other hand, if the previous velocities are close to zero, all the particles will stop moving once they catch up with the global best particle, which leads to the premature convergence of the algorithm.

The PSO-Bees Algorithm avoids the problem of premature convergence by incorporating adaptive neighbourhood and random search into the PSO algorithm giving it the ability to escape from stagnant states in order to reach the global optimum or minimum of the objective function.

In the PSO-Bees Algorithm, three sets of particles make up the entire swarm namely:

- regular particles
- neighbourhood particles

- random particles

The regular particles search the problem space as in the basic PSO. The neighbourhood particles are for searching the neighbourhoods of promising particles including the global best particle previously found by the regular particles. The initial size of the neighbourhoods is kept unchanged provided that the neighbourhood particles are able to find better solutions in the neighbourhoods. If a neighbourhood particle has a better fitness compared to the selected particle it becomes a candidate for neighbourhood search in the next iteration. On the other hand, if the neighbourhood search does not yield any progress, the size is reduced to make the local search more exploitative, focusing the areas around the most promising particles. If there is no improvement from reducing the size of the neighbourhoods, it is assumed that the global best particle has been found. Finally, randomised particles are added to search the problem space more widely.

The Hybrid PSO-Bees Algorithm is related to the Quantum PSO algorithm [6] although the latter does not explicitly distinguish between neighbourhood and random particles. The PSO algorithm presented in [7] does involve both neighbourhood search around promising particles and random search. However, the random search particles in this case are distributed only in the vicinity of the global best particle.

Figure 2 shows the pseudo code for the PSO-Bees Algorithm in its simplest form.

```
Initialise PSO-Bees population
Repeat:
for each particle n:
   Update particle position using equations (1) & (2)
   if f(P_n) > f(Pbest_n)
      then Pbest_n = P_n
   if f(Pbest_n) > f(Gbest)
      then Gbest = Pbest_n
//Bees Algorithm Section
//Neighbourhood Search
Generate neighbourhood particles neigh
for each particle neigh:
   if f(P_neigh) > f(P_Selected Candidate)
      then P_Selected Candidate = P_neigh
   Subsume P_neigh into regular population with initial
   velocity of replaced candidate
endfor
//Random Search
Generate random particles rand
for each particle rand:
   if f(P_rand) > f(P_Selected Candidate)
      then P_Selected Candidate = P_rand
   Subsume P_rand into regular population with average
   velocity of swarm
endfor
reset each rand
endfor
Until stopping condition is true
```

Fig. 2. Pseudo code of the PSO-Bees Algorithm.

In every iteration, the position and velocity of each particle is updated according to the following best quantities:

- *Pbest*, the best position the particle has visited so far. This represents a local optimum.
- *Gbest*, the best position visited so far by any particle in the population. This represents the current global best solution.

Velocity and position updating is carried out using the following equations:

$$V_{n+1} = wV_n + c1 * rand_1 * (Pbest_n - P_n) + c2 * rand_2 * (Gbest - P_n) \quad (1)$$

$$P_{n+1} = P_n + k.V_{n+1} \quad (2)$$

where:

V_n (V_{n+1}) is the particle velocity in iteration n (n+1)

P_n (P_{n+1}) is the particle position (solution) in iteration n (n+1)

$Pbest_n$ is the "personal" best position for particle n

Gbest is the "global" best position

k is the time step

$rand_1$, $rand_2$ are random numbers between 0 and 1

c1, *c2* are weighting factors (each usually a number in the range 0 to 4)

w is the 'inertia' weight. A large value of *w* facilitates global searching while a small value encourages local searching. Usually, *w* is allowed to decrease as the optimisation progresses.

The number of particles in a PSO-Bees Algorithm population is usually in the range 20 to 40 and the weighting factors c1 and c2 typically have the value 2. These factors determine the maximum jump that a particle can make in one step. Too large a jump can result in oscillation, while too small a displacement can cause slow convergence or even trapping of particles at local minima.

Figure 3 illustrates the operation of the PSO-Bees Algorithm for a simple one-dimensional optimisation problem.

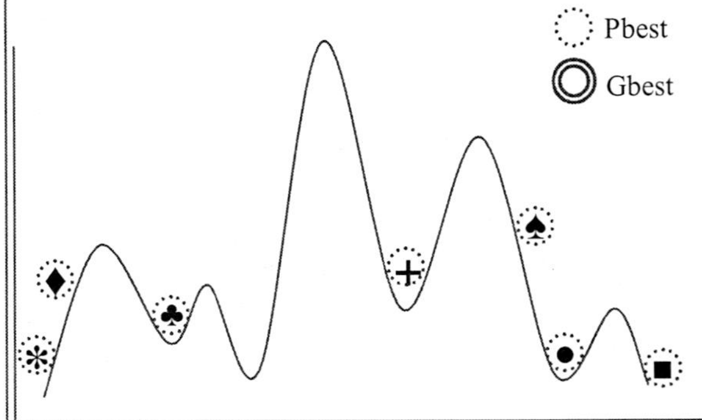

(a) Randomly initialise swarm population

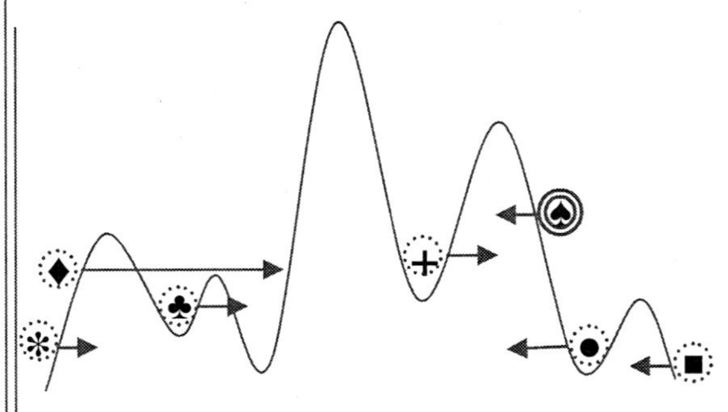

(b) Evaluate fitness

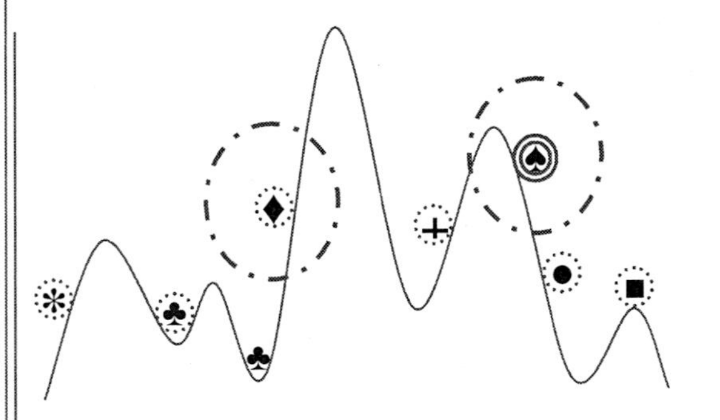

(c) Select candidates for neighbourhood search

Fig 3. Operation of the PSO-Bees Algorithm (To be continued)

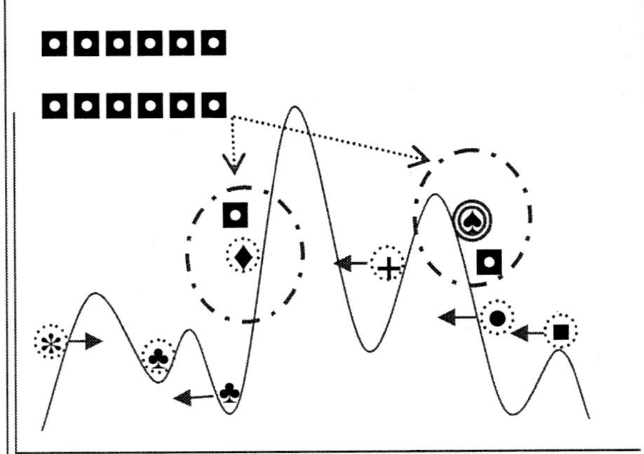

(d) Enlist neighbourhood particles to search around the selected candidates

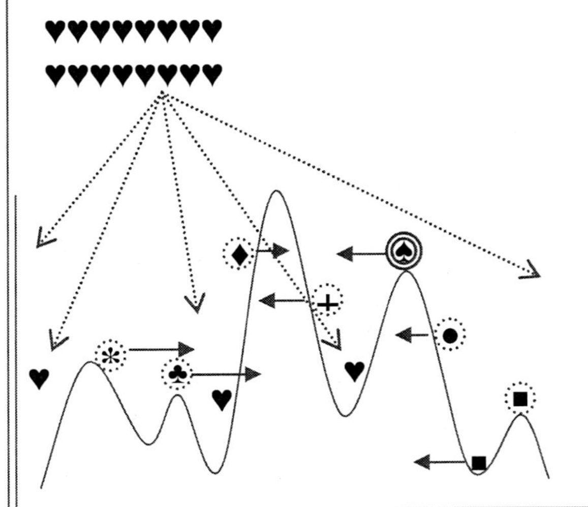

(e) Enlist random particles to search problem space

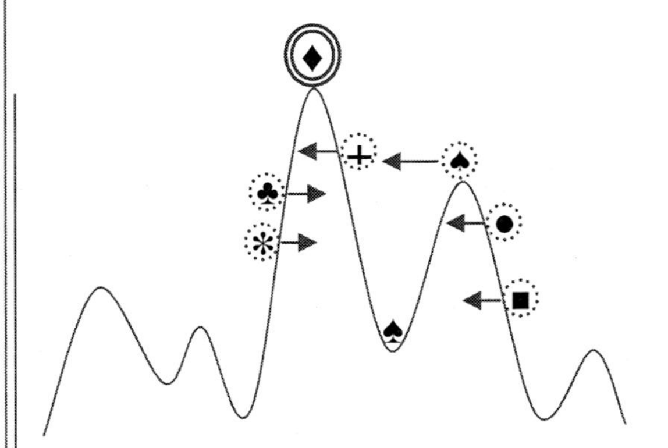

(f) Global optimum solution found

Fig. 3. Operation of the PSO-Bees Algorithm (Cont'd)

4.2 Multi-Layer Perceptron training

The Multi-Layer Perceptron network used had three layers: an input layer, a hidden layer, and an output layer of neurons. The neurons between adjacent layers are fully linked by connections, the weights of which are to be determined through training. The training of a Multi-Layer Perceptron to carry out a mapping task such as that of transforming a feature vector into an image category is essentially an optimisation problem. The aim is to select the values of the connection weights of the neural network so as to minimise the total mapping error calculated over a set of training feature vectors for which the corresponding image categories are known.

In an application of the PSO-Bees Algorithm to the training problem, each particle is a multi-dimensional weight vector P_n representing a candidate classifier. When a training feature vector is provided, the weight vector is used to calculate the response of the classifier. The difference between that response and the correct known response is the classification error corresponding to that particular training feature vector. The average of the squared error for all the feature vectors in the training set gives the overall performance (fitness) of the candidate classifier. By adjusting the position of each particle P_n according to equations (1) and (2), the PSO-Bees Algorithm changes the weight vectors and hence the performance of each candidate classifier, eventually directing the swarm of particles towards the position with the minimum classification error.

The optimum is considered found and the algorithm stops when the mean squared error has fallen below a given threshold. Alternatively, the algorithm also stops when the maximum number of iterations is reached.

5. Wood defect classification results

A classifier structure with 17 input neurons (each component to receive a component of the feature vector), 13 output neurons (each neuron corresponding to an image category) and 51 hidden neurons were adopted. The number of hidden neurons was same as that used by Packianather and Drake [1] who had employed the Taguchi Design Experiment technique to determine the most appropriate value for this parameter. The input neurons acted only as buffers and performed no processing function, transmitting directly the values of the features (regularised between -1 and +1) to the hidden layer neurons and then onward to the output neurons.

Processing was carried out by the hidden and output neurons. The output function employed was the hyperbolic tangent function. A constant bias was added to the activation of each neuron prior to the calculation of the neuron output. Thus, the classifier comprises in total 1594 connections: 17 x 51 from the input layer to the hidden layer, 51 x 13 from the hidden layer to the output layer and 51 + 13 bias connections.

The classifier was trained using a set of 185 feature vectors. The trained classifier was tested on a different set of 47 feature vectors. Table 1 gives details of the training and test vectors.

Table 1
Training and test sets

Image Class	Total	Used for Training	Used for Testing
Bark	20	16	4
Clear wood	20	16	4
Coloured streaks	20	16	4
Curly grain	16	13	3
Discolouration	20	16	4
Holes	8	6	2
Pin knots	20	16	4
Rotten knots	20	16	4
Roughness	20	16	4
Sound knots	20	16	4
Splits	20	16	4
Streaks	20	16	4
Wormholes	8	6	2
Total	232	185	47

Table 2 shows the parameter values empirically chosen for the PSO-Bees Algorithm. The positions of the particles were initialised by setting all weight values randomly within the range 0 to 1.

Table 2
PSO-Bees Algorithm Parameters

PSO-Bees Algorithm Parameters	Symbol	Value
Inertia weight	w	1(max), 0(min)
Dimension of particles	D	1594
• Stopping criteria : Mean Squared Error	MSE	6
• Maximum number of iterations	I	1,100
Maximum change a particle can make in one iteration	V_{max}	2
Weighting factors c1, c2	$c1,c2$	2
Population size	S	40

The PSO-Bees Algorithm with the parameters given in Table 2 was applied 30 times to train 13 different classifiers. The mean classification accuracy obtained with the algorithm was 92.16% while that produced by the conventional PSO algorithm was 89.79%. By comparison, the accuracy for 13 Multi-Layer Perceptron classifiers trained by back propagation was 86.52% [8]. Clearly, the Hybrid PSO-Bees Algorithm gave classifiers with a superior performance.

6. Conclusion

This paper has described the new PSO-Bees Algorithm and its application to the problem of training neural classifiers for wood board defect classification. Despite the high dimensionality of the problem – each particle represented 1594 parameters that had to be determined, the PSO-Bees Algorithm trained classifiers were able to identify defects more accurately than did classifiers trained using the original PSO algorithm and the well-established back propagation method.

Acknowledgements

This work was funded under the SUPERMAN ERDF Objective 1 Project and the EC FP6 I*PROMS Network of Excellence.

References

1. Packianather, M.S. and P.R. Drake, *Comparison of Neural and Minimum Distance Classifiers in Wood Veneer Defect Identification.* IMechE Part B, J. of Engineering Manufacture, 2005. **219**: p. 831 - 844.
2. Conners, R.W., McMillin, C. W., Lin, K., Vasquez-Espinosa, R. E., *Identifying and Locating Surface Defects in Woods.* IEEE Trans. on Pattern Analysis and Machine Intelligence, 1983. **PAMI - 5/6** (Part of an Automated Lumber Processing System): p. 573 - 583.
3. Koivo, A.J. and C.W. Kim, *Classification of Surface Defects on Wood Boards.* IEEE International Conference on Systems, Man and Cybernetics, Atlanta, GA, 1986: p. 1431 - 1436.
4. Eberhart, R.C. and J. Kennedy, *A New Optimiser using the Particle Swarm Theory.* Proceedings of the Sixth International Symposium on Micro Machine and Human Science, 1995: p. 39 - 43.
5. Pham, D.T., Ghanbarzadeh, A., Koc, E., Otri, S., Rahim, S., Zaidi, M., *The Bees Algorithm, A Novel Tool for Complex Optimisation Problems*, in *2nd Int. Virtual Conf on Intelligent Production Machines and Systems (IPROMS 2006).* 2006b, Oxford, Elsevier.
6. Sun, J., B. Feng, and W. Xu, *Particle Swarm Optimization with Particles Having Quantum Behaviour*, in *Proc. Congress on Evolutionary Computation*. 2004. p. 325 - 331.
7. Zhang, X. and T. Li, *Improved Particle Swarm Optimization Algorithm for 2D Protein Folding Prediction*, in *The 1st International Conference on Bioinformatics and Biomedical Engineering*. 2007, IEEE/CNF: Wuhan, China. p. 53 - 56.
8. Packianather, M.S. and P.R. Drake. *Post-processing Multilayered Feed-forward Neural Network Outputs for Identifying Wood Veneer Defects*. in *5th CIRP International Seminar on Intelligent Computation in Manufacturing Engineering (CIRP ICME '06)*. 2006. Ischia, Italy.

Innovative Production Machines and Systems
D.T. Pham, E.E. Eldukhri and A.J. Soroka (Eds.)

Fuzzy Selection of Local Search Sites in the Bees Algorithm

D.T Pham and A. Haj Darwish

Intelligent Systems Laboratory, Manufacturing Engineering Centre, Cardiff University
Queen's Buildings, The Parade, Newport Road, Cardiff CF24 3AA, Wales, UK.

Abstract

The Bees Algorithm is a swarm-based algorithm that mimics the natural food foraging behaviour of honey bees. The algorithm essentially involves both random exploration of the solution space and more focused exploitation of promising local search sites. A basic version of the Bees Algorithm has been applied to function optimisation and a variety of other practical problems. This paper describes an enhanced version implementing fuzzy greedy selection of local search sites. The paper presents the results obtained for a number of benchmark problems to demonstrate the robustness and self-organising ability of the new Bees Algorithm.

Keywords: Bees Algorithm, Fuzzy greedy selection, Swarm Intelligence, Optimisation

1. Introduction

The rapid development of the engineering sciences and more complex processes in industry and manufacturing mean traditional optimisation techniques are no longer adequate to solve complex multi-variable optimisation problems with large numbers of parameters. These tend to be solved by computational intelligence algorithms such as heuristic algorithms, to find local or global optimum for unconstrained and constrained optimisation problems.

Most heuristic algorithms are developed to mimic the natural behaviour of living creatures. Such natural behaviour is the result of millions of years of natural evolution.

Some of these behaviours are the food foraging mechanisms which are found in the biological swarms of such as ants, termites, bacteria, wasps, and honey-bees. This foraging behaviour offers ideas as seeds for algorithms that solve complex problems such as continuous optimisation and combinatorial problems.

Recent literature shows that the foraging behaviour of honey-bees has been modelled as an optimisation algorithm known as the Bees Algorithm [1], where a bee seeks to find the best possible flower patch. A foraging strategy involves finding a flower patch, deciding whether to enter it and search for food, and when to leave it.

The Bees Algorithm (BA) [1] has been applied in a vast number of problems, such as continuous function optimisation [1], artificial neural network training [2], for clustering problems [3], tuning fuzzy logic membership functions [4], and for some combinatorial problems such as job scheduling [5] and printed circuit board assembly planning [6].

This paper is organised as follows: section 2 reviews the basic Bees Algorithm, section 3 presents the proposed algorithm and shows a detailed description of the algorithm structure, section 4 presents the benchmark suite used to test the proposed algorithm and shows the results obtained and finally section 5 concludes the paper.

2. The basic Bees Algorithm

The Bees Algorithm [1] uses global random search for exploration and local random search for exploitation. The Bees Algorithm [1] manipulates data as floating point numbers instead of the binary-coding of the search space as used in genetic algorithms [7].

This algorithm requires a number of parameters to be set, namely: number of scout bees (n); number of patches selected for the neighbourhood search (out of n visited patches) (m); number of top-rated (elite) patches among m selected patches (e); number of bees recruited for the best e patches (nep); number of bees recruited for the other (m-e) selected patches (nsp); the initial size of each patch (ngh) (a patch is a region in the search space that includes a visited patch and its neighbourhood) and the stopping criterion. The algorithm starts with the n scout bees being placed randomly in the search space.

These parameters must be set before the Bees Algorithm begins optimising the functions. The large number of the mentioned tuneable parameters used is one of the drawbacks of the basic Bees Algorithm. Later sections show how the proposed algorithm reduces the parameters needed to run the algorithm.

Figure (1) shows the pseudo code of the basic Bees Algorithm.

1- Initialise population with random solutions.
2- Evaluate fitness of the population.
3- While (stopping criterion not met).
//Forming new population.
4- Select patches for neighbourhood search.
5- Recruit bees for selected patches (more bees for best e patches) and evaluate their fitness.
6- Select the fittest bee from each patch.
7- Assign remaining bees to search randomly and evaluate their fitness.
8- End While.

Fig. 1. Pseudo code of the basic Bees Algorithm.

3. The enhanced Bees Algorithm

The proposed algorithm is called the enhanced Bees Algorithm. This algorithm is inspired by the natural foraging behaviour of honey bees to find the optimal solution and uses the basic Bees Algorithm as a core.

The enhanced Bees Algorithm depends on fuzzy greedy selection and hierarchical abandonment.

The algorithm requires only two parameters to be set, namely:

Number of scout bees (n) and maximum number of worker bees in each patch (nw).

These two parameters can be equal to each other. The initial size of patches (ngh) is a percentage of the size of the search space. In the enhanced Bees Algorithm each explored patch is assigned with a local memory. This memory is shared among worker bees in a patch. It stores the best fitness of a patch and the recent patch size.

Figure 2 shows the pseudo code of the enhanced Bees Algorithm.

---

1. Initialise population with random solutions.
2. Evaluate fitness of the population.
3. Form fuzzy greedy system with initial values of fitness and rank.
4. While (stopping criterion not met).
//Forming new population.
5. Select patches for neighbourhood search (each patch has its own memory to store its patch size).
6. Recruit bees for selected patches (more bees for best patches) and evaluate their fitness.
7. Select the fittest bee from each patch.
8. Shrink the patch size when it is needed (dividing the patch size by 2).
9. Hierarchically abandon when trapped in a local peak.
10. Assign remaining bees to search randomly and evaluate their fitness.
11. Update fuzzy greedy system parameters.
12. End While.

Fig. 2. Pseudo code of the enhanced Bees Algorithm.

3.1. The fuzzy greedy selection

3.1.1. Fuzzy Logic

Unlike classical logic which requires a deep

understanding of a system, exact equations and precise numeric values, fuzzy logic [8, 9] incorporates an alternative way of thinking, which allows the modelling of complex systems using a higher level of abstraction originating from our knowledge and experience. Fuzzy Logic allows for the expression of this knowledge with subjective concepts such as "high" and "low" which are mapped into exact numeric ranges.

The Sugeno method [8] of fuzzy inference was introduced in 1985, similar to the Mamdani method [9] in many respects. The first two parts of the fuzzy inference process, fuzzifying the inputs and applying the fuzzy operator, are exactly the same. The main difference between the two is that the Sugeno output membership functions are either linear or constant.

3.1.2. Fuzzy greedy selection

Greedy selection of flower patches is a feature of greedy algorithms [10] which were used to solve optimisation problems and always make the choice that looks best at that moment. A greedy algorithm leads to an optimal solution because a locally optimal choice leads to a globally optimal solution.

Greedy algorithms are natural and usually simple and fast. As its name implies, a greedy optimisation builds a solution by taking the best possible choices. The natural behaviour of honey-bees [11] is also greedy, where honey-bees tend to recruit more bees for the best possible patches.

The basic Bees Algorithm uses greedy selection to choose (m) best patches and (e) elite patches out of the explored patches by scout bees. The visited patches are usually put in a list to be sorted in descending fitness value to form a sorted candidates list.

The selection from a candidate list is done in the basic Bees Algorithm depending on the rank of each visited patch. The parameters of the selection process are chosen empirically by the user to determine (e) elite patches and (m) best patches beside the rest of the parameters (nep), (nsp) and (ngh). This makes the process more complex for non-expert users.

A Fuzzy Greedy Selection system is constructed to decide the number of the selected patches and the number of recruited workers per each selected patch. It is a form of fuzzy decision-making system which tends to eliminate patches with low fitness or low rank and from now on there is no need to preset the values of (m),(e),(nep) and (nsp), since they are determined automatically by the fuzzy greedy selection system.

3.1.3. The structure of the fuzzy greedy selection system

The system consists of two fuzzy inputs, two constants for output, and the Sugeno inference system.

Each one of the input variables consists of two triangular membership functions called “low” and “high”.

The first input variable is the fitness value and the second input variable is the rank value. Figure (3) illustrates the membership function shape of the input variables.

The output consists of two constants: low with value (0) zero and high with value (nw), which is the maximum number of worker bees per patch.

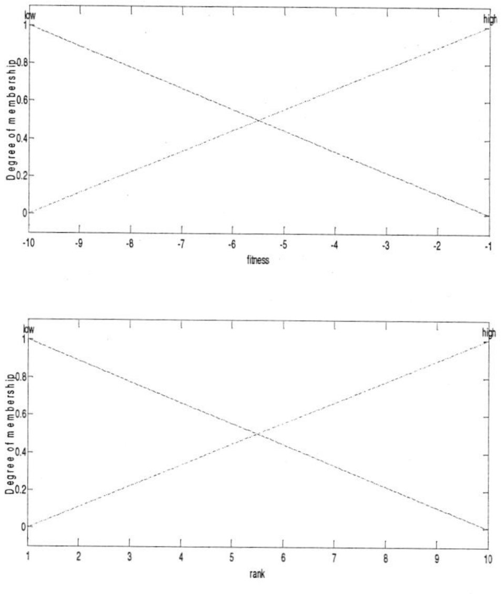

Fig. 3. The membership function shape of the input variables.

The inference system is a zero-order Sugeno fuzzy model (special case of Mamdani model) with a weighted average for aggregation.

The Sugeno model is used in the proposed algorithm since it is a more compact and computationally efficient representation than a Mamdani model.

The initial universe of discourse for the fuzzy Sugeno inference system comes from the maximum and the minimum values for fitness of the random visited patches and their ranks. The universe of discourse for the fuzzy system is updated at the end of each loop after sorting the candidate list. This repeated update makes the fuzzy greedy selection dynamic.

The initial random visited patches are put in a list and sorted to form a descending order candidate list (sorting depends on fitness). This list gives the rank for each visited patch and the value of its fitness.

The system chooses the selected patches depending on their rank and fitness and at the same time gives how many worker bees are recruited for the selected patches.

This selection is done by the rules which are shown in figure (4).

1. If (fitness is low) and (rank is low) then (rec is low)
2. If (fitness is low) and (rank is high) then (rec is low)
3. If (fitness is high) and (rank is low) then (rec is low)
4. If (fitness is high) and (rank is high) then (rec is high)

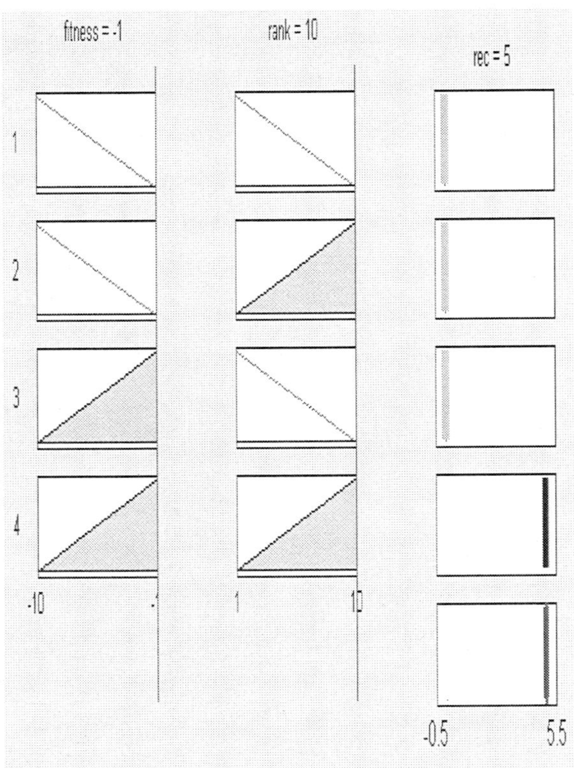

Fig. 4. Fuzzy system rules

The structure of the rules gives the system its greedy behaviour, since recruitment is only if both fitness and rank are high.

The output is rounded and gives the total number of bees in a selected patch.

Figure (5) illustrates the representative surface of the fuzzy system.

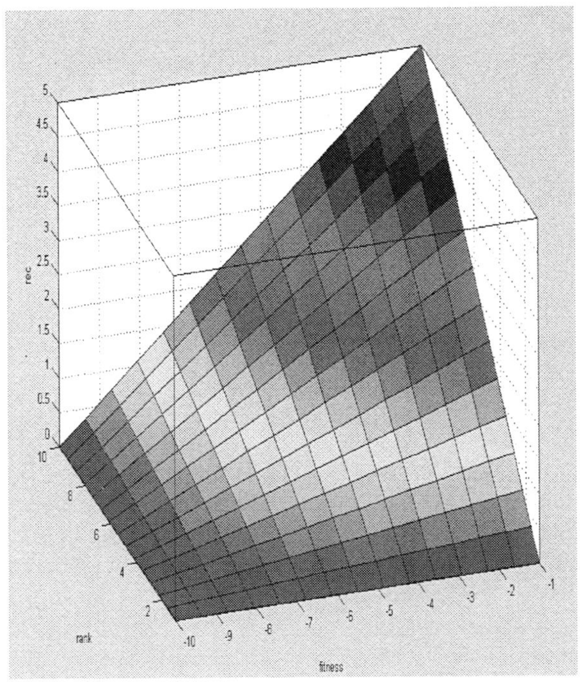

Fig. 5. The fuzzy system surface

3.2. Hierarchical abandonment

Abandon patches' in the Bees Algorithm is similar to random-restart hill climbing [10] or shotgun hill climbing. A random-restart hill climbing algorithm is supervised by an outer monitoring procedure which runs a hill climbing algorithm with a new initial state each time it stops improving. The best-found state is stored aside to be replaced with a better result from a successor run. This idea prevents the algorithm from becoming trapped in local optima.

Ghanbarzadeh [12] in his PhD thesis describes the abandonment method, which is used when the Bees Algorithm is trapped in local optima and involves eliminating only the trapped patch from the candidate list.

This hierarchical abandonment is applied not only to the trapped patch but also to all patches whose ranks are lower than the trapped ones, to be replaced by new explored patches. This proposed method increases the greedy behaviour [11] of the algorithm.

Table 1
Test Functions

No	Function Name	Interval	Function	Global Optimum
1	De Jong	[-2.048, 2.048]	$\max F = (3905.93) - 100(x_1^2 - x_2)^2 - (1 - x_1)^2$	X(1,1) F=3905.93
2	Goldstein & Price	[-2, 2]	$\min F = [1 + (x_1 + x_2 + 1)^2 (19 - 14x_1 + 3x_1^2 - 14x_2 + 6x_1x_2 + 3x_2^2)]$ $X[30 + (2x_1 - 3x_2)^2 (18 - 32x_1 + 12x_1^2 + 48x_2 - 36x_1x_2 + 27x_2^2)]$	X(0,-1) F=3
3	Branin	[-5, 10]	$\min F = a(x_2 - b x_1^2 + c x_1 - d)^2 + e(1 - f)\cos(x_1) + e$ $a = 1, b = \frac{5.1}{4}\left(\frac{7}{22}\right)^2, c = \frac{5}{22} X 7, d = 6, e = 10, f = \frac{1}{8} X \frac{7}{22}$	X(-22/7,12.275) X(22/7,2.275) X(66/7,2.475) F=0.3977272
4	Martin & Gaddy	[0, 10]	$\min F = (x_1 - x_2)^2 + ((x_1 + x_2 - 10)/3)^2$	X(5,5) F=0
5a	Rosenbrock (2D) (a)	[-1.2, 1.2]	$\min \ F = 100\,(x_1^2 - x_2)^2 + (1 - x_1)^2$	X(1,1) F=0
5b	Rosenbrock (2D) (b)	[-10, 10]	$\min \ F = 100\,(x_1^2 - x_2)^2 + (1 - x_1)^2$	X(1,1) F=0
6	Rosenbrock (4D)	[-1.2, 1.2]	$\min F = \sum_{i=1}^{3} \{100(x_i^2 - x_{i+1})^2 + (1 - x_i)^2\}$	X(1,1,1,1) F=0
7	Hypersphere	[-5.12, 5.12]	$\min \ F = \sum_{i=1}^{6} x_i^2$	X(0,0,0,0,0,0) F=0
8	Griewank	[-512, 512]	$\min \ F = \frac{1}{4000} \cdot \left(\sum_{i=1}^{10} (x_i - 100)^2 \right) - \left(\prod_{i=1}^{10} \cos\left(\frac{x_i - 100}{\sqrt{i}} \right) \right) + 1$	X($\vec{1}00$) F=0

4. Experiments

The enhanced Bees Algorithm has been applied to the eight benchmark functions given in [1] and the results compared with those obtained using other optimisation algorithms. The test functions and their optima are shown in Table 1.

Table 3 presents the results obtained by the enhanced Bees Algorithm and those by the deterministic Simplex method (SIMPSA), the stochastic simulated annealing optimisation procedure (NE SIMPSA), the Genetic Algorithm (GA) and the Ant Colony System (ANTS).

The numbers of points visited shown are averages for 100 independent runs.

The optimisation stopped when the difference between the maximum fitness obtained and the global optimum was less than 0.1% of the optimum value, or less than 0.001, whichever was smaller. In cases where the optimum was 0, the solution was accepted if it differed from the optimum by less than 0.001. If a solution is found that satisfies one of these conditions, the algorithm is said to have succeeded in finding the optimum.

Table 2
The enhanced Bees Algorithm parameters

Function no	n	nw
1	4	20
2	5	2
3	4	3
4	4	3
5a	3	20
5b	4	25
6	10	3
7	5	3
8	5	5

Table 3
Results for test functions

funct. no.	SIMPSA		NE-SIMPSA		GA		ANTS		Basic Bees Algorithm		Enhanced Bees Algorithm	
	succ %	mean no. of evaluations	succ %	mean no. of evaluations	succ %	mean no. of evaluations	succ %	mean no. of evaluations	succ %	mean no. of evaluations	succ %	mean no. of evaluations
1	****	****	****	****	100	10160	100	6000	100	868	100	830
2	****	****	****	****	100	5662	100	5330	100	999	100	212
3	****	****	****	****	100	7325	100	1936	100	1657	100	184
4	****	****	****	****	100	2844	100	1688	100	526	100	124
5a	100	10780	100	4508	100	10212	100	6842	100	631	100	689
5b	100	12500	100	5007	****	****	100	7505	100	2306	100	1448
6	99	21177	94	3053	****	****	100	8471	100	28529	100	33367
7	****	****	****	****	100	15468	100	22050	100	7113	100	526
8	****	****	****	****	100	200000	100	50000	100	20998	100	8224

Table 1 shows the benchmark suite which was used to test the proposed algorithm, optimum values, and their intervals. The benchmark suite consists of 8 functions.

Table 2 shows the used parameters of the enhanced Bees Algorithm for each one of the functions in the benchmark suite.

Table 3 shows the results of applying several optimisation algorithms to the same benchmark suite and it also shows the average number of evaluations needed to get the required optimum values using the enhanced Bees Algorithm.

From the tables it is possible to see that the enhanced Bees Algorithm needs a smaller number of evaluations to get the optimum for most of the functions than in the benchmark suite, used with the minimum number of tuneable parameters needed to run the algorithm.

5. Conclusion

The enhanced Bees algorithm has been constructed from a combination of the Bees Algorithm as a core and a fuzzy system for the greedy selection, and was applied for function optimisation as a maximisation problem. The obtained output proved that the enhanced Bees algorithm is robust and reliable for function optimisation. Another advantage of the algorithm is the small number of tuneable parameters.

Fuzzy logic lets the algorithm behave in a similar way to the natural behaviour of honey-bees in food foraging and it reduces the parameters needed to run the algorithm.

References

[1] Pham D.T., Ghanbarzadeh A., Koç E., Otri S., Rahim S., and Zaidi M. The Bees Algorithm - A Novel Tool for Complex Optimisation Problems. Proc 2nd Virtual International Conference on Intelligent Production Machines and Systems (IPROMS 2006). D.T. Pham, E.E. Eldukhri and A.J. Soroka (eds), Elsevier (Oxford) (2006), ISBN-10:08-045157-8, pp. 454-459.

[2] Pham D.T., Otri S., Ghanbarzadeh A., and Koc E. Application of the Bees Algorithm to the Training of Learning Vector Quantisation Networks for Control Chart Pattern Recognition. Proc 2nd Information and Communication Technologies. 2006. Damascus, Syria, pp. 1624 - 1629.

[3] Pham D.T., Otri S., Afify A., Mahmuddin M., and Al-Jabbouli H. Data Clustering Using the Bees Algorithm. Proc 40th CIRP International Manufacturing Systems Seminar. 2007. Liverpool, UK.

[4] Pham D.T., Haj Darwish A., Eldukhri E.E., and Otri S. Using the Bees Algorithm to tune a Fuzzy Logic Controller for a Robot Gymnast. Proc 3rd International Virtual Conference on Intelligent Production Machines and Systems (IPROMS 2007). 2007: CRC Press.

[5] Pham D.T., Koc E., J.Y.Lee, and J.Phrueksanant. Using the Bees Algorithm to Schedule Jobs for a Machine. Proc 8th international Conference on Laser Metrology, CMM and Machine Tool Performance, LAMDAMAP 2007. 2007. Cardiff: Euspen, UK, pp. 430-439.

[6] Pham D.T., Otri S., and Haj Darwish A. Application of the Bees Algorithm to PCB Assembly Optimisation. Proc 3rd International Virtual Conference on Intelligent Production Machines and Systems (IPROMS 2007). 2007: CRC Press.

[7] Pham D.T. and Karaboga D. Intelligent Optimisation Techniques: Genetic Algorithms, Tabu Search, Simulated Annealing and Neural Networks. 1999, London, Heidelberg and New York: Springer Verlag. 302.

[8] Sugeno M. Industrial Applications of Fuzzy Control 1985: Elsevier Science Ltd 269.

[9] Mamdani E.H. and Assilian S. An Experiment in Linguistic Synthesis with a Fuzzy Logic Controller. International Journal of Man-Machine Studies, 1975(7): pp. 1-13.

[10] Russell S. and Norvig P. Artificial Intelligence: A Modern Approach. 2nd Edition. 2004: Prentice-Hall.

[11] Seeley T.D., Visscher P.K., and Passino K.M. Group Decision Making in Honey Bee Swarms. American Scientist, 2006. 94(3): pp. 220-229.

[12] Ghanbarzadeh A. THE BEES ALGORITHM A Novel Optimisation Tool, in Manufacturing Engineering Centre. 2007, Cardiff University: Cardiff, United Kingdom. pp. 158.

Innovative Production Machines and Systems
D.T. Pham, E.E. Eldukhri and A.J. Soroka (Eds.)

Generating Branded Product Concepts: Comparing the Bees Algorithm and an Evolutionary Algorithm

D.T. Pham, M.C. Ang, K.W. Ng, S. Otri, A. Haj Darwish

Manufacturing Engineering Centre, Cardiff University, Cardiff CF24 3AA, UK

Abstract

The task to generate product design concepts to maintain a particular brand identity whilst meeting functional requirements is challenging to designers. Shape grammars have been shown to be able formally to describe the creation of branded product shapes using a set of shape rules. These shape rules are applied manually to generate a family of new design concepts that maintain the brand identity of the product. However, shape grammars are not meant to evaluate whether the generated product concepts can meet specified functional requirements. When a shape grammar is combined with an optimisation technique such as an evolutionary algorithm, a computational procedure for generating branded design concepts that can meet functional requirements is established. This procedure evolves the set of shape rules whilst evaluating how well the outcome of the new design concepts meet a specified functional requirement. A new optimisation technique, the Bees Algorithm, was reported to perform better than many existing techniques. This paper compares the Bees Algorithm with an evolutionary algorithm when they are combined with a shape grammar in generating concepts for a new Coca-Cola bottle design. The Bees Algorithm was found to outperform the evolutionary algorithm in this investigation.

Keywords: Bees Algorithm, branded product design, evolutionary algorithms, shape grammars

1. Introduction

A product is an artefact developed for selling to customers by an enterprise [1]. A competitive enterprise is able to identify customer's needs and customise products to meet those needs quickly at required quality. As a consumer market develops, competition arises, and technology matures. Technology is taken for granted [2]. Enterprise and its competitors will have products that have similar technology claims. The consumer market is soon filled with products that are nearly similar.. For example, all shampoo products claim to be very good for your hair. This leads consumers to look at other differentiating elements. One of the important differentiating elements is brand [3, 4]. In response, the enterprises have to brand their products distinctively and promote their product brand identity to gain market share [5, 6].

In developing a new improved product from an existing product, there is a need to gain benefits additional to those established by the existing product [7]. Such establishment benefits include the brand image depicted by the existing product shape and its design parameters. The importance of generating product concepts which consider product shape whilst meeting all functional requirements has been recognised by a number of researchers [8-10]. A branded product shape must be used to portray or reflect the identity of a brand. It is crucial that a branded product shape is considered and developed during the product development stage. This branding strategy applied to product shape is transformed into branding requirements in product specification. Thus, product specification for a branded product has both functional and branding requirements.

The task of generating product concepts depends on the experience, creativity and the knowledge of designers [8]. The ultimate aim of the product concept generating process is to derive solution concepts that meet the requirements identified in product specifications. Generating product design concepts to conform to a particular product brand identity and at the same time to fulfil functional requirements is a challenging task [9]. Therefore, there is a need to provide a support for the designer so that they may generate branded product shape/form better during product development process.

2. Shape grammars

Shape grammars were first introduced by Stiny and Gips in 1972 [11]. Shape grammars are generative design systems that allow users to produce shapes syntactically [11, 12]. It is a formal method to generate shapes through a sequence of rule applications [12].

Shape grammars are used in architecture, arts and recently in product design to generate a family of shapes that conform to certain styles and brand identities [13-16]. They have been used to design the shape of consumer products. The first example in the literature was a coffeemaker grammar [17]. The coffeemaker grammar was able to generate four existing branded models of coffeemaker. Other examples are the Dove soap bar grammar [16], Harley-Davidson Motorcycle grammar [14] and Buick automobile grammar [13], and Cola-Cola bottle grammar [18].

Shape grammars are able to produce potentially a large number of shapes to be used during the early product concept development process. However, shape grammars are not meant to assess whether the generated shapes are meeting functional requirements. In order to assist a designer to produce potentially viable solutions, a number of researchers combined adaptive and evolutionary computing, such as genetic algorithms and simulated annealing, with shape grammars to generate 2D shapes and 3D structures to fulfil structural [19, 20] and functional requirements [21]. For a branded product design problem, Ang [22, 23] has combined an evolutionary algorithm with the Coca-Cola shape grammar to generate and explore solution concepts to meet a specified bottle volume (functional requirement) and branding requirements. However, the work has not been compared with any other optimisation techniques.

3. The Bees Algorithm

Recently, an advanced adaptive computing technique, the Bees Algorithm [24] has been introduced and it shows promise. A number of successful applications using the Bees Algorithm have been reported. The Bees Algorithm has been used to solve continuous problems [24] and also on combinatorial ones [25]. In all the reported works, the quality of the results is shown to be better than the results generated by genetic algorithms. The Bees Algorithm is also used to solve a multi-objective optimisation problem [26]. These applications showed the strength of Bees Algorithm and pose an intriguing reason for further investigation. In this paper, a Coca-Cola grammar combined with the Bees Algorithm and the results are to be compared with earlier work using an evolutionary algorithm [23]. Similar volume requirements are used in this investigation.

4. Coca-Cola bottle case study

The Coca-Cola shape grammar was produced to capture the contours of the trademark Coca-Cola bottle across different timelines [18]. The Coca-Cola bottle shape grammar provides information on specific bottle sections, characteristics of the shape and contour in each section, and the relationships between sections. There are no specific rules to generate values for different diameters and heights. Each bottle section is parameterised using diameters and heights. Diameters and heights provide the start point and end point of each connecting curve. Curves in bottle sections are formed by at least three points. In the implementation, the diameters and heights of each section are set within finite ranges. These parameters were used in the calculation of the volume for the bottle shapes produced by the prototype system.

There are a total of seven rule groups in the Coca-Cola bottle grammar (Fig. 1). Starting from the rules for building the main body, there are other rules for construction of the upper part, modification of the main body, construction of the bottom, construction of the lower part, construction of the label region and construction of the cap. A rule group may contain more than one rule, for example, the construction of the upper part contains three separate rules that produce different shapes on top of the main body.

Based on the shape rules of the Coca-Cola bottle grammar, there is more than one rule in each rule group but only one rule can be selected for execution in a given computation step. In the construction of a bottle, the use of a rule from groups RG1, RG2, RG3 and RG4 is compulsory to produce a valid bottle design because every bottle must have a body, an upper part, a bottom and a cap. The rules in groups RG5, RG6 and RG7 are executed to produce variation in the bottle designs and their use is optional. Thus, the Bees Algorithm can randomly select shape rules within the same group to alter the bottle shape.

Associate parameters are used to describe the heights and diameters of each bottle section (refer Fig. 2). For example, three parameters are used to describe the upper part section: bottom diameter ($\Phi_{1,2}$), top diameter ($\Phi_{2,2}$) and height ($H_{3,2}$). These associated parameters, diameters and heights were set in the range of [*minwidth*, *maxwidth*] and [*minheight*, *maxheight*]. With the arrangement as shown in Fig.3, a total number of 16 associate parameters are used to define diameters and heights in each bottle part. To reduce the dimension of parameters, only the main body diameter and height was being generated by the Bees Algorithm. The remaining associate parameters are computed individually, based on ratio relationships between the main body and each respective bottle section.

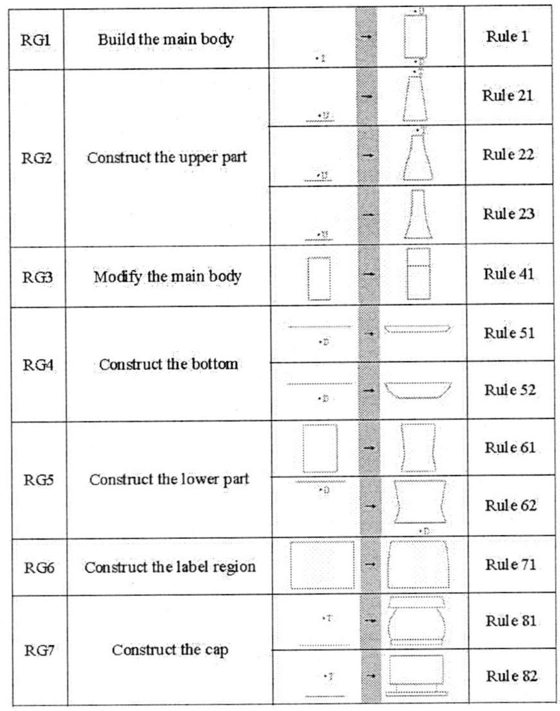

Fig. 1. Coca-Cola bottle shape grammar (adapted from [18]) .

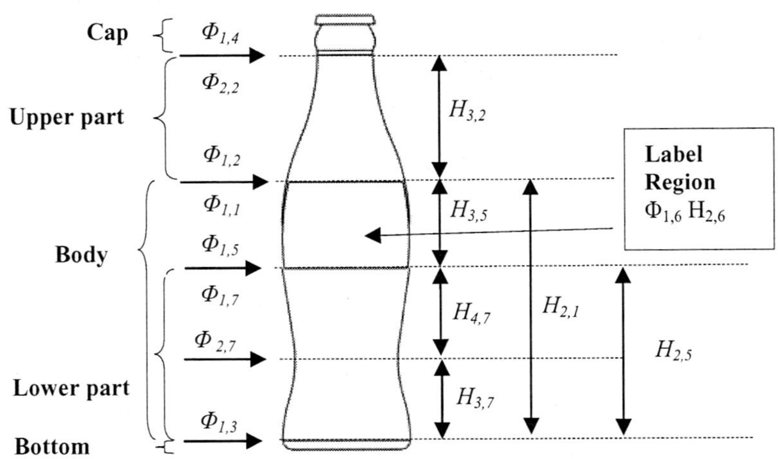

Fig. 2. Parameterisation of Coca-Cola bottle (reproduced from Ang [23])

The resulting shape of bottle sections after each execution of a shape rule is varied. They were approximated excluding the bottom and cap to facilitate the calculation of the total volume. In this particular case study, the Bees Algorithm was used to generate rule sequences and associated parameters to achieve the total volume. The objective of this study is to minimise the difference between the bottle volume and target volume of 500 ml.

RG1	*RG2*	*RG3*	*RG4*	*RG5*	*RG6*	*RG7*
$\Phi_{1,1}$	$\Phi_{1,2}$	$\Phi_{1,3}$	$\Phi_{1,4}$	$\Phi_{1,5}$	$\Phi_{1,6}$	$\Phi_{1,7}$
$H_{2,1}$	$\Phi_{2,2}$			$H_{2,5}$	$H_{2,6}$	$\Phi_{2,7}$
	$H_{3,2}$			$H_{3,5}$		$H_{3,7}$
						$H_{4,7}$

Fig. 3. Rule group and their associate parameters (adapted from [23]).

5. Results

A sample of different combinations of parameters was used by the Bees Algorithms and the results are shown in Table 1. The results demonstrate that the Bees Algorithm is able to find optimal solutions that achieve a 500 ml target volume within a 0.1 ml tolerance. The bottle shapes were generated using parameters and shape rules optimised by the Bees Algorithm. The contour of this family of bottles has a similar style to the existing trademark Coca-Cola bottle shape.

An evolutionary algorithm was also applied to generate bottle shape using Coca-Cola shape rules and associated parameters [23]. The genotype was a 5x7 matrix similar to that shown in the table Fig.3. Single point crossover was performed on a pair of selected parents and followed by an intermediate recombination operator of Breeder Genetic Algorithm to recombine the diameters and height at the crossover point. Mutation was allowed to be same performed on either, altering the shape rule within the rule or altering the parameter value in specified range in a random manner. Recombination and mutation were executed with the respective recombination probability (*Pr*) and mutation probability (*Pm*). Further details of the evolutionary algorithm implementation can be found in Ang's work [23].

Different population size, maximum generation, recombination probability, mutation probability were tested for evolutionary algorithm implementation. The population size of 200 and the maximum generation of 200 were selected and the results are shown in Table 2. These settings allowed the evolutionary algorithm to achieve the bottle volume within 0.1 ml tolerance and did not consume high computation cost. Further investigations were conducted to obtain the best parameters combination for evolutionary algorithm and used to compare the performance with the Bees algorithm. It is found that the Bees algorithm can achieve bottle volume in less number of evaluations. The diagram in Fig 4 shows a comparison between the Bees Algorithm and the evolutionary algorithm. The Bees Algorithm parameters were: swarm size (n) = 100, elites (e) = 2, selected (m) =10, number of bees around elite site, $N2$= 10, number of bees around other selected site, $N1$ = 4, and initial patch size (ngh) = 0.01.The evolutionary algorithm parameters were: population size (Sp) = 200, maximum generation (Gm) = 200, recombination probability (Pr) = 0.9, mutation probability (Pm) = 0.1.

Table 1
Generating bottle volume and shape using the Bees Algorithm

	(a)	(b)	(c)	(d)	(e)	(f)	(g)	(h)	(i)	(j)	(k)
Swarm size (n)	10	10	10	10	20	10	30	20	20	30	100
Elites (e)	1	1	1	1	1	1	1	1	1	1	2
Selected (m)	2	2	2	2	2	5	5	10	10	10	10
$N2$	3	3	3	3	3	3	3	3	3	3	10
$N1$	2	2	2	2	2	2	2	2	2	2	4
Initial patch size (Ngh)	0.5	0.1	0.025	0.01	0.01	0.1	0.05	0.1	0.05	0.1	0.01
Best volume	500.00	499.98	500.01	499.99	500.05	500.02	499.99	499.99	500.10	500.00	500.00
Bottle shape	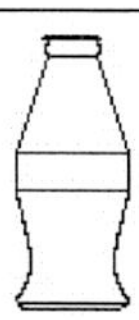	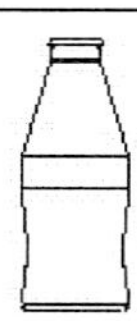	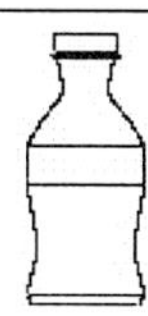								

Table 2
Generating bottle volume and shape using an evolutionary algorithm when 200 population size and 200 maximum generation

	(a)	(b)	(c)	(d)	(e)	(f)	(g)	(h)	(i)	(j)	(k)
Pr	0.05	0.05	0.05	0.1	0.1	0.1	0.5	0.5	0.5	0.9	0.9
Pm	0.05	0.1	0.5	0.01	0.1	0.05	0.01	0.05	0.1	0.1	0.05
Best volume	500.10	499.95	499.99	499.90	500.00	500.04	500.01	500.02	500.09	500.01	499.98
Bottle shape	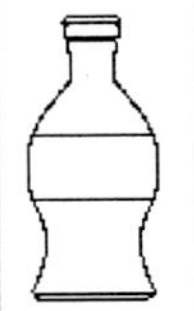	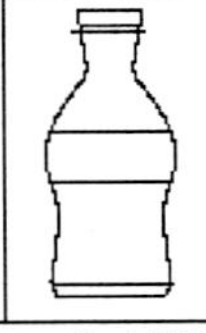	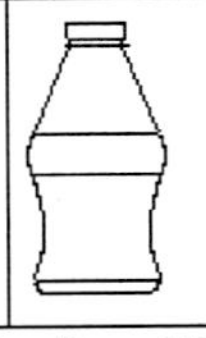	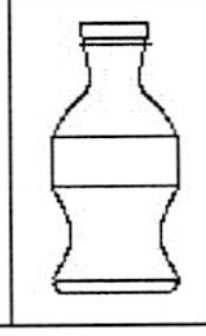	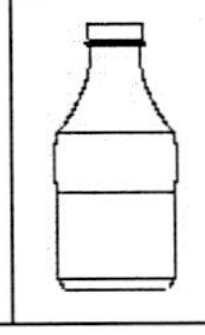	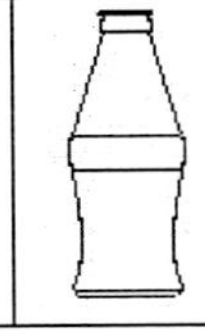	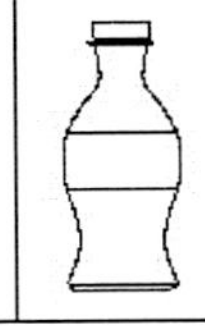	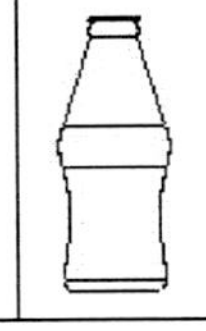	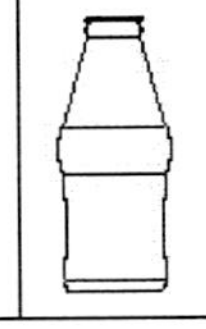	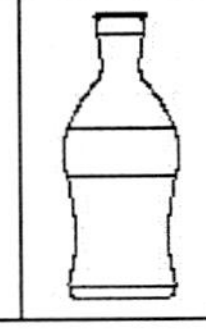	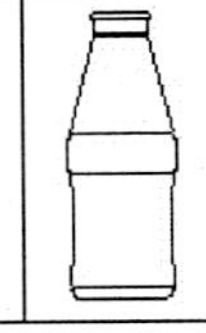

Pr = Recombination probability, *Pm* = Mutation probability

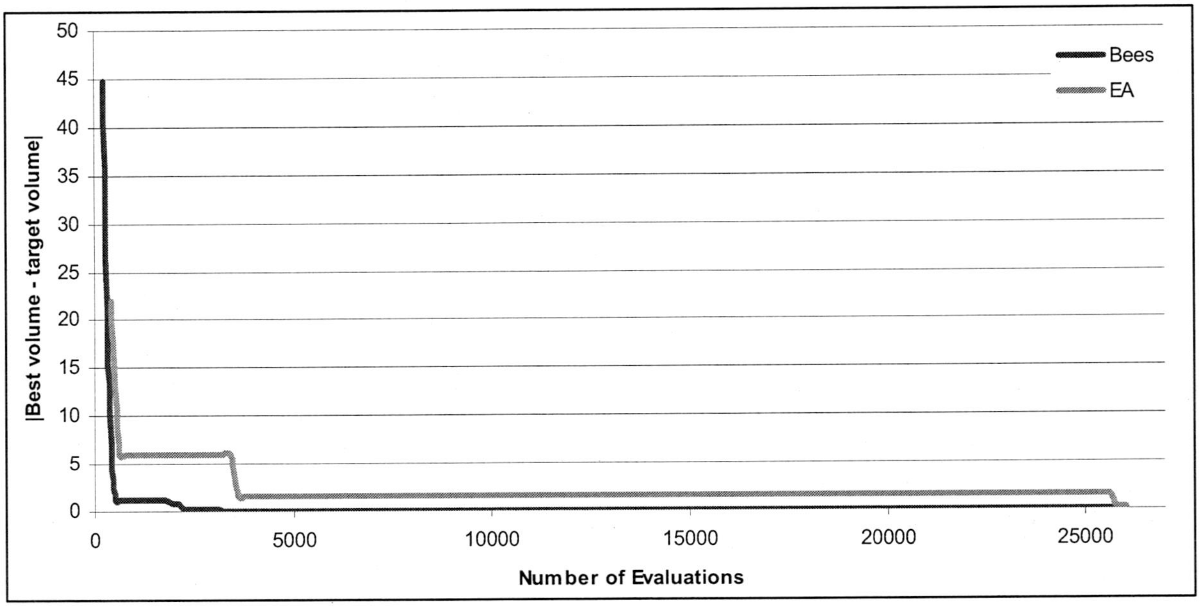

Fig. 4. The Bees algorithm vs. evolutionary algorithm

6. Conclusions

Both the Bees Algorithm and the evolutionary algorithm are found to be able to generate bottle shapes that have similar style with the existing Coca-Cola trademark bottle shape. The Bees Algorithm outperforms the evolutionary algorithm in that it was able to achieve the target volume in a smaller number of evaluations. This may due to the fact that the evolutionary algorithm is mimicking the Darwinian evolution process and that this process is a slow one. On the other hand, the Bees Algorithm is mimicking the foraging process of the bees and it is a relatively more aggressive process. Previous work [23] and this current work shows that optimisation techniques can be used with shape grammar rules to produce designs that conform to a particular brand and fulfil functional requirements.

Acknowledgements

The authors are members of the EU-funded FP6 Network of Excellence for Innovative Production Machines and System (I*PROMS).

References

[1] Ulrich KT and Eppinger SD. Product Design And Development. McGraw-Hill, Boston, 2003.

[2] Norman DA. The invisible computer: why good products can fail, the personal computer is so complex, and information appliances are the solution. MIT Press,

Cambridge, Mass., 1998.
[3] Aaker DA and Joachimsthaler E. Brand Leadership. The Free Press, New York, 2000.
[4] Aaker DA. Building Strong Brand. The Free Press, New York, 1996.
[5] Schmitt BH and Simonson A. Marketing Aesthetics: The Strategic Management Of Branding, Identity And Image. The Free Press, New York, 1997.
[6] Wheeler A. Designing Brand Identity: A Complete Guide To Creating, Building And Maintaining Strong Brands. John Wiley & Sons, United States of America, 2003.
[7] Kreuzbauer R and Malter AJ. Embodied cognition and new product design: changing product form to influence brand categorization. Journal of Product Innovation Management, 22, (2005) pp 165-176.
[8] Stompff G. The forgotten bond: brand identity and product design. Design Management Journal, 14, (2003) pp 26-33.
[9] Hekkert P. The experience of design. D3 Desire Designum Design: 4th European Academy of Design Conference Proceedings. Aveiro, Portugal, 2001. pp 396-401.
[10] Bloch PH. Seeking the ideal form: product design and consumer response. Journal of Marketing, 59, (1995) pp 16-29.
[11] Stiny G and Gips J. Shape grammars and the generative specification of painting and sculpture. In Freiman, C (Ed.) Proceedings of IFIP Congress. Amsterdam, North Holland, 1972. pp 1460-1465.
[12] Stiny G. Introduction to shape and shape grammar. Environment and Planning B: Planning and Design, 7, (1980) pp 399-408.
[13] McCormack JP, Cagan J and Vogel CM. Speaking the Buick language: capturing, understanding, and exploring brand identity with a shape grammar. Design studies, 25, (2004) pp 1-29.
[14] Pugliese MJ and Cagan J. Capturing a rebel: modeling the Harley-Davidson brand through a motorcycle shape grammar. Research in Engineering Design-Theory Applications and Concurrent Engineering, 13, (2002) pp 139-156.
[15] Chen X. Relationships between product form and brand: a shape grammatical approach. Ph.D. Thesis, University of Leeds, UK, 2005.
[16] Chau HH. Preserving brand identity in engineering design using a grammatical approach. Ph.D. thesis, University of Leeds, UK, 2002.
[17] Agarwal M and Cagan J. A blend of different tastes: the language of coffeemakers. Environment and Planning B: Planning and Design, 25, (1998) pp 205-226.
[18] Chau HH, Chen X, McKay A and de Pennington A. Evaluation of a 3D shape grammar implementation. In Gero, J (Ed.) First International Conference on Design Computing and Cognition (DCC '04). MIT, USA, Kluwer Academic Publishers, 2004. pp 357-377.
[19] Gero JS, Louis SJ and Kundu S. Evolutionary learning of novel grammars for design improvement. Artificial Intelligence for Engineering Design, Analysis and Manufacturing, 8, (1994) pp 83-94.
[20] Shea K, Cagan J and Fenves SJ. A shape annealing approach to optimal truss design with dynamic grouping of members. 119, (1997) pp 388-394.
[21] Chouchoulas O. Shape evolution: an algorithmic method for conceptual architectural design combining shape grammars and genetic algorithms. Ph.D.Thesis, University of Bath, UK, 2003.
[22] Ang MC. Branded product design: Combining evolutionary algorithms with shape grammars. Ph.D. Thesis, University of Leeds, UK, 2007.
[23] Ang MC, Chau HH, McKay A and de Pennington A. Combining evolutionary algorithm and shape grammar to generate branded produce design to meet functional requirement. Second International Conference on Design Computing and Cognition (DCC'06). Technical University of Eindhoven, Netherlands, 2006. pp 521-539.
[24] Pham DT, Ghanbarzadeh A, Koç E, Otri S, Rahim S and Zaidi M. The Bees algorithm - a novel tool for complex optimisation problems. In Pham, DT, Eldukhri, EE & Soroka, AJ (Eds.) Proceedings of the 2nd Virtual International Conference on Intelligent Production Machines and Systems (I*PROMS 2006). Elsevier, 2006. pp 454-459.
[25] Pham DT, Afify AA and Koç E. Manufacturing cell formation using the Bees Algorithm. In Pham, DT, Eldukhri, EE & Soroka, AJ (Eds.) Proceedings of the 3rd Virtual International Conference on Innovative Production Machines and Systems (I*PROMS 2007). Whittles Publishing, 2007. pp 523-528.
[26] Pham DT and Ghanbarzadeh A. Multi-Objective Optimisation using the Bees Algorithm. In Pham, DT, Eldukhri, EE & Soroka, AJ (Eds.) Proceedings of the 3rd Virtual International Conference on Innovative Production Machines and Systems (I*PROMS 2007). Whittles Publishing, 2007. pp 529-533.

Innovative Production Machines and Systems
D.T. Pham, E.E. Eldukhri and A.J. Soroka (Eds.)

Application of the Bees Algorithm to Fuzzy Clustering

D.T. Pham, H. AL-Jabbouli, M. Mahmuddin, S. Otri and A. Haj Darwish

Intelligent Systems Laboratory, Manufacturing Engineering Centre, Cardiff University, Cardiff, CF24 3AA, UK

Abstract

This paper discusses the application of the Bees Algorithm to fuzzy clustering. The Bees Algorithm is used to optimise the performance of the fuzzy C-Means (FCM) algorithm and improve its clustering results. Computational experiments show that the Bees Algorithm gives a significant improvement over the FCM algorithm on its own and better results compared to the FCM algorithm combined with a Genetic Algorithm.

Keywords: Bees Algorithm, Genetic Algorithm, Fuzzy Data Clustering, Optimisation.

1. Introduction

Data clustering is known as the problem of grouping similar data points together, such that similar data points are in the same group. Traditional data clustering, also called crisp clustering, uses hard thresholds to group data points into separate groups while fuzzy clustering uses fuzzy logic to create overlapped groups of data.

In fuzzy clustering, each data object can belong to more than one cluster at the same time with a different possibility degree or membership function value. The membership function of a cluster may vary from 1 to 0. Conversely, membership function in crisp clustering is based on either belonging to the cluster, which normally is 1 and is otherwise 0. These membership functions are depicted in Figs. 1 and 2.

A Fuzzy clustering algorithm is known as one that produces more realistic results compare to crisp clustering techniques, and has been widely used especially in pattern recognition[1] and medical applications[2]. In this paper, the Bees Algorithms, which has been used to improve crisp clustering results by integrating it with the k-means algorithm [5], is now integrated with the (FCM) algorithm, one of the most well-known fuzzy clustering algorithms, to solve the problem of trapping in a local optimum.

The rest of the paper is organised as follows: section 2 reviews the (FCM) algorithm and the Bees Algorithm; section 3 describes the proposed fuzzy clustering algorithm that combines the Bees Algorithm with (FCM) algorithm; section 4 shows the results obtained. Finally, section 5 concludes the paper.

2. FCM and the Bees Algorithm

2.1. FCM

The fuzzy C-means (FCM) algorithm is the most popular algorithm in fuzzy clustering. It was introduced and developed by Dunn [3] then improved by Bezdek [1]. The fuzzy C-means algorithm is based on minimising of an objective function (Eq. 1) that indicates the sum of distances from each cluster centre to the data points in that cluster, so that the smaller the value of J, the better the clustering.

$$J_m = \sum_{i=1}^{N} \sum_{j=1}^{C} u_{ij}^{m} \left\| x_i - c_j \right\|^2$$

(Eq. 1)

where:
m is the fuzziness degree of any real number greater than 1.
uij is the degree of membership of xi in the cluster j
cj is the centre of the cluster j
N is the number of data objects
xi is the ith d-dimensional measured data object.
||*|| is any norm expressing the similarity between any measured data and the centre

The total number of clusters is a required parameter. When performing fuzzy clustering, the FCM algorithm produces fuzzy clusters and the membership degree of each data object to each cluster by repeating the following steps:

1- Initialise $U = [u_{ij}]$

2- At k-step: calculate the centres vectors $C^{(k)} = [c_j]$ with $U^{(k)}$

$$c_j = \frac{\sum_{i=1}^{N} u_{ij}^{m} . x_i}{\sum_{i=1}^{N} u_{ij}^{m}}$$

3- Update $U^{(k)}$, $U^{(k+1)}$

$$U_{ij} = \frac{1}{\sum_{k=1}^{c} \left(\frac{\left\| x_i - c_j \right\|}{\left\| x_i - c_k \right\|} \right)^{\frac{2}{m-1}}}$$

4- If $\left\| U^{(k+1)} - U^{(k)} \right\| \langle \sigma$ then stop, otherwise go to step 2.

Fig. 3. FCM Algorithm

where:
K is the iteration step
σ is a termination criterion between 0 and 1

The main problem of this calculus-based optimisation approach is that it can become trapped in a local optimum. In this case, the Bees optimisation algorithm has been used to overcome the problem.

2.2. Honey Bees in nature

The Bees Algorithm is an optimisation algorithm proposed by Pham et al [4] based on the food foraging behaviour of honey_bees in nature. The Bees Algorithm has become one of the most successful optimisation algorithms due to its successful implementation in various applications for optimisation problems such as neural network training [5], identifying homogenous data clustering [6] and a preliminary design problem [7].

The foraging process of bees begins in a colony by scout bees being sent to search for promising flower patches. Flower patches with large amounts of nectar or pollen that can be collected with less effort tend to be visited by more bees, whereas patches with less nectar or pollen receive fewer bees.

During the harvesting season, a colony continues its exploration, keeping a percentage of the population as scout bees. When they return to the hive, those scout bees that have found a patch rated above a certain quality threshold deposit their nectar or pollen and go to the "dance floor" to perform a dance known as the "waggle dance".

This mysterious dance is essential for colony communication, and contains three pieces of information regarding a flower patch: the direction in which it will be found, its distance from the hive and its quality rating (or fitness). This information helps the colony to send its bees to flower patches precisely, without using guides or maps. Each individual's knowledge of the outside environment is gleaned solely from the waggle dance. This dance enables the colony to evaluate the relative merit of different patches according to both the quality of the food they provide and the amount of energy needed to harvest it. After waggle dancing on the dance floor, the dancer (i.e. the scout bee) goes back to the flower patch with follower bees that were waiting inside the hive. More follower bees are sent to more promising patches. This allows the colony to gather food quickly and efficiently.

While harvesting from a patch, the bees monitor its food level. This is necessary so that they can decide upon the next waggle dance when they return to the hive. If the patch is still good enough as a food source, then it will be advertised in the waggle dance and more bees will be recruited to that source.

2.3 The Bees Algorithm

The basic steps of the Bees Algorithm are explained in fig 4. The Algorithm requires a number of parameters to be set. These are; number of scout bees (n), number of sites selected for neighbourhood search (m), number of best "elite" sites out of m selected sites (e), number of bees recruited for the best e sites (nep), number of bees recruited for the other (m-e) selected sites (nsp) and a stopping criterion.

1. Initialise the solution population (each initial solution, or 'bee', being a set of randomly placed cluster centres).
2. Evaluate the fitness of the population.
3. While (stopping criterion is not met)
   // forming new population.
4. Select sites for neighbourhood search.
5. Recruit bees for selected sites (more bees for the best e sites) and evaluate fitnesses.
6. Select the fittest bee from each site.
7. Assign remaining bees to search randomly and evaluate their fitnesses.
8. End While.

Fig. 4. The Bees Algorithm

3. Proposed Fuzzy Clustering Method

The proposed clustering method exploits the search capability of the Bees Algorithm to overcome the local optimum problem of the FCM algorithm. Specifically, the task is to search for appropriate cluster centres such that the objective function (J) is minimised.

The basic steps of the proposed clustering operation are described in detail below:

1. Read data from the file.
2. Initialise the solution population.
3. Fill array *A* using formula u_{ij} and evaluate the fitness of the population using Eq. (1).
4. While (stopping criterion is not met)
5. Select sites for neighbourhood search.
6. Recruit bees for selected sites (more bees for the best e sites) and fill array *A* using formula u_{ij} and evaluate the fitness using Eq. (1).
7. Select the fittest bee from each site.
8. Assign remaining bees to search randomly then fill array *A* using formula u_{ij} and evaluate their fitnesses using Eq. (1).
9. End While.

Fig. 5. The Proposed Algorithm

The algorithm starts by reading data and providing a random population of clusters centres as in step 2. In step 3, a calculation is made using equation U in order to fill the array A with a membership function for all data objects. A repeat of generating a new population of solutions is done in steps 4-9, by selecting the best sites of former population as in step 5 then sending more bees to them. Step 6 fills the array A using equation U for each sent bee. In step 7, selection of the fittest bees is chosen from each site. Assigning the remaining bees to search randomly and fill the array A is step 8.

4. Experimental Results

This section presents the results for the Bees-Algorithm-based fuzzy clustering method compared with the FCM algorithm and GA -based fuzzy clustering method.

Table 1 shows the parameter values for each algorithm used in this test, while the value of m is set to 2 in all algorithms. The algorithms were applied to the following real data sets (Iris [8], Control Charts [9],

Wood Defects [10] and Crude Oil [11]). The main characteristics of these data sets are summarized in Table 3.

The clustering criterion J Eq. (1) is used to evaluate the performance of the algorithms: the smaller the value of this metric, the better the fuzzy clustering results. The algorithms were executed 10 times and the average, minimum and maximum values of E were collected. The results obtained for each algorithm are listed in Table 2.

The results in Table 2 show that the Bees and GA enhanced algorithms outperform the traditional FCM algorithm in most cases. The Table also shows that the Bees Algorithm produces better results than the GA in all cases. It also shows that the FCM algorithm performs better on the Control Charts data set compared to the other algorithms.

Table 1
Parameters used in the clustering experiments

Algorithm	Parameters	Value
FCM	Maximum number of iterations	1000
GA	Crossover probability, μ_c	0.8
	Mutation probability, μ_m	0.001
	Population size, P	100
Bees Algorithm	Number of scout bees	21
	Number of sites selected for neighbourhood search	8
	Number of elite bees	2
	Number of bees recruited for the elite sites	5
	Number of bees recruited for the other selected sites	2
	Number of iterations	300

5. Conclusion

Combining the Bees Algorithm with the FCM algorithm improved the fuzzy clustering results compared to the traditional C-means algorithm in most cases. It also proved that the Bees Algorithms produces better results than those of the GA combined with FCM.

One of the main concerns about the current algorithm is that it requires a long computing time. Future work should therefore concentrate on improving the computational speed of the Bees Algorithm.

Table 2
Results (clustering criterion) for the tested clustering algorithms

Data set	Algorithm	Mean	Min	Max
Iris	FCM	61.72	60.58	65.81
	GA	65.54	63.34	69.23
	Bees	**60.58**	**60.58**	**60.58**
Ctrl Charts	FCM	**530.99**	**530.99**	**530.99**
	GA	802.08	791.23	819.28
	Bees	549.08	546.87	552.15
Wood Defects	FCM	64077395.11	59195094.29	66225010.49
	GA	168035.20	157508.00	174784.00
	Bees	**165132.52**	**153866.53**	**173523.00**
Crude Oil	FCM	1330.43	1242.30	1560.57
	GA	1252.61	1238.25	1286.04
	Bees	**1239.50**	**1237.24**	**1241.56**

Table 3
Data sets used in the experiments

Data Set	# of Object	# of Feature	# of Class
Iris	150	4	3
Control Charts	1500	60	6
Wood Defects	232	17	13
Crude oil	59	5	3

References

[1] Keller B.J.M., Krishnapuram R., and Pal N.R. Fuzzy Models and Algorithms for Pattern Recognition and Image Processing. 1. 1999: Springer.

[2] Wang S., Zhou M., and Geng G. Application of fuzzy cluster analysis for medical image data mining, in Mechatronics and Automation, 2005 IEEE International Conference. 2005. pp. 631- 636.

[3] Dunn J.C. A Fuzzy Relative of the ISODATA Process and Its Use in Detecting Compact Well-Separated Clusters. Journal of Cybernetics 1973. 3: pp. 32-57.

[4] Pham D.T., Ghanbarzadeh A., Koc E., Otri S., and Zaidi M. The Bees Algorithm – A Novel Tool for Complex Optimisation Problems, in 2nd Virtual International Conference on Intelligent Production Machines and Systems (IPROMS 2006). 2006, Elsevier. Cardiff, UK. pp. 454-459.

[5] Pham D.T., Ghanbarzadeh A., Koc E., and Otri S. Application of the Bees Algorithm to the training of Radial Basis Function Networks for control chart pattern recognition, in 5th CIRP International Seminar on Intelligent Computation in Manufacturing Engineering (ICME '06). 2006, CIRP: Ischia, Italy. pp. 711-716.

[6] Pham D.T., Otri S., Afify A., Mahmuddin M., and Al-Jabbouli H. Data clustering using the Bees Algorithm, in 40th CIRP International Seminar on Manufacturing Systems. 2007: Liverpool.

[7] Pham D.T., Castellani M., and Ghanbarzadeh A. Preliminary design using the Bees Algorithm, in 8th International Conference on Laser Metrology, CMM and Machine Tool Performance, LAMDAMAP. 2007, Euspen: Cardiff, UK. pp. 420-429.

[8] Asuncion A. and Newman D.J. UCI Machine Learning Repository. 2007, University of California, School of Information and Computer Science: Irvine, CA.

[9] Pham D.T. and Oztemel E. Control chart pattern recognition using neural networks. Journal of Systems Engineering, 1992. 2, 4: pp. 256-262.

[10] Pham D.T., Soroka A.J., Ghanbarzadeh A., Koc E., Otri S., and Packianather M. Optimising Neural Networks for Identification of Wood Defects Using the Bees Algorithm, in 4th International IEEE Conference on Industrial Informatics (INDIN'06). 2006, IEEE Computer Society: Singapore. pp. 1346-1351.

[11] Johnson R.A. and Wichern D.W. Applied Multivariate Statistical Analysis 1982: Prentice Hall, Englewood Cliffs, NJ

Innovative Production Machines and Systems
D.T. Pham, E.E. Eldukhri and A.J. Soroka (Eds.)

Flowshop Sequencing Using the Bees Algorithm

D.T. Pham and Ebubekir Koç

Manufacturing Engineering Centre, Cardiff University, Cardiff CF24 3AA, UK

Abstract

This paper describes the application of the Bees Algorithm to the problem of flowshop sequencing with the aim of minimising the makespan. Following a formulation of the problem, the paper presents the results for a number of benchmark flowshop sequencing tasks. These demonstrate the efficiency and robustness of the Bees Algorithm as a tool for combinatorial optimisation.

Keywords: Bees Algorithm, Permutation Flowshop Scheduling Problem, Swarm Intelligence.

1. Introduction

Many real-life optimisation problems are so complex that they cannot be solved exactly within polynomially bounded computation times. Search algorithms are needed that are capable of finding near-optimal solutions in reasonable running times. The Bees Algorithm developed by the authors' team [1, 2] is a novel search algorithm capable of locating good solutions efficiently. Belonging to the category of 'intelligent' optimisation tools, the algorithm is inspired by the food foraging behaviour of honey bees.

The algorithm has been successfully applied to different optimisation problems such as continuous function optimisation [1, 2], artificial neural network training [3, 4], engineering design [5], multi-objective optimisation [6], data clustering [7], job scheduling [8] and manufacturing cell formation [9].

For permutation flowshop sequencing, the problem to which the Bees Algorithm was applied in this investigation, m different machines are set up in series and each job must be processed on every machine (in the order to $1 \ldots m$). The processing order of the n jobs is the same for every machine. Permutation of the n jobs gives a makespan (the time taken to complete the job sequence) and the aim is to find the permutation that gives the minimum makespan [10].

The permutation flowshop sequencing problem (PFSP) was first introduced in [11] and has attracted the attention of many researchers since then. It has been proved that permutation flowshop sequencing with makespan minimisation is an NP-hard problem [12]. Therefore, using different heuristic optimisation techniques, researchers have attempted to find high-quality solutions in a reasonable computational time instead of looking for the optimal solution [13, 14, 15, 16, 17]. To obtain high-quality solutions, many meta-heuristic optimisation techniques have been tried. They include the Genetic Algorithm [18, 19, 20], Tabu Search [10, 21, 22], Simulated Annealing [23, 24], Particle Swarm Optimisation [25] and Ant Colony Optimisation [26, 27].

This paper is organised as follows. Section 2 gives a formulation of the PFSP. Section 3 outlines the Bees Algorithm and describes its implementation for solving a PFSP. Section 4 details the benchmark problems used to test the efficacy of the algorithm and presents the results obtained. These show that the algorithm can reliably handle complex multi-modal optimisation problems without being trapped at local solutions. Section 5 concludes the paper.

2. Formulation of the permutation flowshop sequencing problem

A formulation of the PFSP is provided in [25]. Given the processing times p_{jk} for job j on machine k, and a job permutation $\pi = \{\pi_1, \pi_2, \ldots, \pi_n\}$, the n jobs $(j = 1, 2, \ldots, n)$ will be sequenced through m machines $(k = 1, 2, \ldots, m)$ using the same permutation π. Let $C(\pi_j, m)$ denote the completion time of job π_j on machine m. The completion time of an n-job-m-machine problem can be calculated as follows [25]:

$$C(\pi_1,1) = p_{\pi_1,1} \quad (1)$$

$$C(\pi_j,1) = C(\pi_{j-1},1) + p_{\pi_j,1} \qquad j = 2, \ldots, n \quad (2)$$

$$C(\pi_1,k) = C(\pi_1,k-1) + p_{\pi_1,k} \quad k = 2, \ldots, m \quad (3)$$

$$C(\pi_j,k) = \max\{C(\pi_{j-1},k), C(\pi_j,k-1)\} + p_{\pi_j,k} \qquad j = 2, \ldots, n \quad k = 2, \ldots, m \quad (4)$$

Then, among the set Π of all permutations, a permutation π^* can be found such that the makespan $C_{max}(\pi^*)$ is:

$$C_{max}(\pi^*) \leq C(\pi_n, m) \quad \forall \pi \in \Pi \quad (5)$$

3. PFSP and the Bees Algorithm

The Bees Algorithm is an optimisation algorithm inspired by the natural foraging behaviour of honey bees [1, 2]. Figure 1 shows the pseudo code for the algorithm in its simplest form. For a PFSP, the algorithm will try to find a permutation with minimum makespan C_{max}.

As explained in [1, 2], the algorithm requires the following parameters to be set: number of scout bees (n), number of sites selected out of n visited sites (m), number of the top sites among the m selected sites (e), number of bees recruited for the top e sites (nep), number of bees recruited for the other (m-e) selected sites (nsp) and stopping criterion.

---

1. Initialise population with random solutions.
2. Evaluate fitness of the population.
3. While (stopping criterion not met)
   //Form a new population.
4. Select sites for neighbourhood search.
5. Recruit bees for selected sites (more bees for the top e sites) and evaluate fitnesses.
6. Select the fittest bee from each site.
7. Assign remaining bees to search randomly and evaluate their fitnesses.
8. End While.

Fig.1 Pseudo code of the basic Bees Algorithm

The algorithm starts with an initial population of n scout bees. Each bee is a symbolic string representing a sequence of machines and jobs. Details of the completion times for the jobs on the different machines are given in a machine-job matrix (see Fig. 2). For an m x n machine-job matrix, a string with a length of $m + n$ is needed to encode each candidate solution. The first m bits of the string represent the sequence of machines that appear in the rows of the matrix and the last n bits of the string represent the sequence of jobs appearing in the columns of the matrix.

In step 2, the fitness computation process is carried out for each site visited by a bee by calculating the corresponding makespan (see Eqs. 1, 2, 3, 4 and 5).

In step 4, the m sites with the highest fitnesses (the shortest makespans) are designated as "selected sites" and chosen for neighbourhood search.

In steps 5 and 6, the algorithm conducts searches around the selected sites, assigning more bees to search in the vicinity of the best e sites. Selection of the best sites can be made directly according to the fitnesses associated with them. Alternatively, the fitness values can be used to determine the probability of the sites being selected. Searches in the neighbourhood of the best e sites which represent the most promising solutions are made more detailed by recruiting more bees for the best e sites than for the other selected sites.

m/n	1	...	n
1	$C(\pi_1,1)$	...	$C(\pi_k,1)$
⋮	⋮	...	⋮
m	$C(\pi_1,k)$	...	$C(\pi_j,k)$

Fig. 2. A machine-job matrix.

Together with scouting, this differential recruitment is a key operation of the Bees Algorithm.

In step 6, for each site, only the bee with the highest fitness will be selected to form the next bee population. In nature, there is no such a restriction. This restriction is introduced here to reduce the number of points to be explored.

In step 7, the remaining bees in the population are assigned randomly around the search space to scout for new potential solutions.

At the end of each iteration, the colony will have two parts to its new population: representatives from the selected sites, and scout bees assigned to conduct random searches. Steps 4-7 are repeated until either the best fitness value has stabilised or the specified maximum number of iterations has been reached.

4. Experimental results

The Bees Algorithm for solving PFSPs was implemented in C++ and run on an Intel P4 2.4 GHz PC with 1GB memory. For the Taillard benchmark data set [10], the Bees Algorithm, a genetic algorithm (GA) [25] and a particle swarm optimisation (PSO) algorithm were employed [25]. The parameters of the Bees Algorithm are shown in Table 1 for each data set. The population size was made equal to twice the number of jobs as was the case for the other algorithms [25]. The Bees Algorithm used the 'insert and exchange' operator to perform neighbourhood search.

For the GA, permutation representation was adopted and the crossover and mutation probabilities were taken as 1.0 and 0.05 percent respectively [25]. To perform two-cut crossover, one individual was selected randomly and the other, by tournament selection with size of 2. Tournament selection with size of 2 was also used for constructing the population for the next generation. The insert operator was used as the mutation operator [25].

Table 1. Bees Algorithm parameters for PFSP

Data Set	**Parameters**				
	n	**m**	**e**	**nep**	**nsp**
20x5	40	20	5	200	100
20x10	40	20	5	400	100
20x20	40	20	5	600	100
50x5	100	50	30	400	200
50x10	100	70	10	600	300
50x20	100	80	10	1000	500
100x5	200	120	10	500	200
100x10	200	120	10	1000	300
100x20	200	120	20	1500	400

The PSO parameters, c_1, c_2, w_0, and α, were set as 2, 2, 0.9, and 0.975, respectively [25]. The performance of the algorithm was expressed by two indices: 1) percentage deviations (Δ) and 2) standard deviation. To obtain the average performance of the algorithm, ten runs were carried out for each problem instance. Δ is the percentage deviation from the reference value reported by Taillard [10]. The average Δ_{avg} for the ten runs is computed as follows:

$$\Delta_{avg} = \sum_{i=1}^{R}\left(\frac{\left(F_{BA} - F_{ref}\right)}{F_{ref}} \times 100\right) \Big/ R \qquad (6)$$

where F_{BA} is the fitness value generated by the Bees Algorithm in each run, F_{ref} is the reference fitness value generated by Taillard [10], and $R = 10$ is the total number of runs, respectively. Table 2 summarises the results obtained. In the table, Δ_{std} denotes the standard deviation in Δ over the R runs.

Table 2. Benchmark results

	GA[25]		PSO_{spv}[25]		PSO_{vns}[25]		Bees Algorithm	
Data Sets	Δ_{avg}	Δ_{std}	Δ_{avg}	Δ_{std}	Δ_{avg}	Δ_{std}	Δ_{avg}	Δ_{std}
20x5	3.13	1.86	1.71	1.25	0.28	0.49	0.28	0.3
20x10	5.42	1.72	3.28	1.19	0.7	0.46	0.13	0.12
20x20	4.22	1.31	2.84	1.15	0.56	0.34	0.17	0.41
50x5	1.69	0.79	1.15	0.70	0.18	0.22	0.1	0.35
50x10	5.61	1.41	4.83	1.16	1.04	0.64	1.03	1.07
50x20	6.95	1.09	6.68	1.35	1.71	0.48	1.48	0.57
100x5	0.81	0.39	0.59	0.34	0.11	0.17	0.1	0.06
100x10	3.12	0.95	3.26	1.04	0.67	0.33	0.58	0.13
100x20	6.32	0.89	7.19	0.99	1.28	0.39	2.55	0.45
Mean	**4.04**	**1.156**	**3.734**	**1.018**	**0.725**	**0.391**	**0.720**	**0.384**

As can be seen in Table 2, the Bees Algorithm outperformed the GA, PSO_{spv} and PSO_{vns}. Compared to the GA and PSO_{spv}, the Bees Algorithm gave results that were more stable and closer to Taillard's. However, the algorithm produced only slightly better results compared to PSO_{vns}. This similarity in the results may be due to the closeness between the local search method adopted by the Bees Algorithm and the variable neighbourhood search (VNS) technique implemented in PSO for conducting local search. However, although VNS employs the insert-and-exchange operator, unlike the Bees Algorithm, it performs this operation only on the best solution. For the Bees Algorithm, the selection of several sites (m) for local search provides more information than VNS can. This property can help the Bees Algorithm to escape from local optima.

5. Conclusion

The Bees Algorithm has been used for a wide range of problems. This paper has presented one of the first applications of the algorithm to combinatorial optimisation. Experimental results obtained for a set of benchmark permutation flowshop sequencing problems have indicated that the algorithm generally outperforms other techniques in terms of the quality of the job sequences produced.

Several opportunities exist for further research. First, the neighbourhood search technique employed by the algorithm needs refining. A new search operator may improve the performance of the algorithm. Second, while the random search operation can play an important role in problems with smooth search surfaces, it is not very useful in combinatorial problems. This point would need to be firmly established. Third, the Bees Algorithm employs several tunable parameters which may be difficult for the user to select. An important line of research, therefore, is to find ways to help the user to choose appropriate parameters.

Acknowledgements

The research described in this paper was performed as part of the Objective 1 SUPERMAN project, the EPSRC Innovative Manufacturing Research Centre Project and the EC FP6 Innovative Production Machines and Systems (I*PROMS) Network of Excellence.

References

[1] Pham DT, Ghanbarzadeh A, Koç E, Otri S, Rahim S and Zaidi M. The Bees Algorithm. Technical Note, Manufacturing Engineering Centre, Cardiff University, UK, 2005.

[2] Pham DT, Ghanbarzadeh A, Koç E, Otri S, Rahim S and Zaidi M. The Bees Algorithm – A novel tool for complex optimisation problems. In DT Pham, EE Eldukhri & AJ Soroka (Eds.), Proceedings of the 2nd Virtual Int. Conference on Intelligent Production Machines and Systems (IPROMS 2006), Elsevier, Oxford, 2006, pp. 454-459.

[3] Pham DT, Koç E, Ghanbarzadeh A and Otri S Optimisation of the weights of multi-layered perceptrons using the Bees Algorithm. Proceedings of 5th Int. Symposium on Intelligent Manufacturing Systems, 2006, pp. 38-46.

[4] Pham DT, Soroka A, Ghanbarzadeh A, Koç E, Otri S and Packianather M. Optimising neural networks for identification of wood defects using the Bees Algorithm, In: Proc. of the IEEE Int. Conference on Industrial Informatics, Singapore, 2006, pp. 1346-1351.

[5] Pham DT, Soroka A, Koc E, Ghanbarzadeh A and Otri S. Some Applications of the Bees Algorithm in Engineering Design and Manufacture. In: Proceedings of the IEEE Int. Conference on Industrial Informatics, Singapore, 2006, pp. 62-70.

[6] Pham DT and Ghanbarzadeh A. Multi-Objective Optimisation using the Bees Algorithm. In DT Pham, EE Eldukhri & AJ Soroka (Eds.) Proceedings of the 3rd Virtual International Conference on Innovative Production Machines and Systems (IPROMS 2007). Whittles Publishing, Dunbeath, Scotland, and CRC Press, Boca Raton, USA, 2007, pp 529-533.

[7] Pham DT, Otri S, Afify A, Mahmuddin M, and Al-Jabbouli H. Data Clustering Using the Bees Algorithm. Proc 40th CIRP International Manufacturing Systems Seminar, 2007, Liverpool, UK.

[8] Pham DT, Koç E, Lee JY, Phrueksanant J. Using the Bees Algorithm to schedule jobs for a machine. Proc. of the 8th Int. Conf. and Exhibition on Laser Metrology, Machine Tool, CMM & Robotic Performance, LAMDAMAP, 2007, Cardiff, UK, pp 430-439.

[9] Pham DT, Afify AA and Koç E. Manufacturing cell formation using the Bees Algorithm. In DT Pham, EE Eldukhri & AJ Soroka (Eds.) Proceedings of the 3rd Virtual International Conference on Innovative Production Machines and Systems (IPROMS 2007). Whittles Publishing, Dunbeath, Scotland, and CRC Press, Boca Raton, USA, 2007, pp 523-528.

[10] Taillard E. Benchmarks for basic scheduling problems. European Journal of Operational Research, 64, 1993, pp 278-285.

[11] Johnson SM. Optimal two-and three-stage production schedules. Naval Research Logistics Quarterly, 1, 1954, pp 61-68.

[12] Kan R. A.H.G: Machine scheduling problems: Classification, complexity and computations. Nijhoff, The Hague, 1976.

[13] Campbell HG, Dudek RA, Smith ML. A heuristic algorithm for the *n* job, *m* machine sequencing problem. Management Science 16(10), 1970, pp B630-B637.

[14] Dannenbring DG. An evaluation of flow shop sequencing heuristics. Management Science. 23(11), 1977, pp 1174-1182.

[15] Framinan JM, Leisten R. An efficient constructive heuristic for flowtime minimisation in permutation flow shops. OMEGA, 31, 2003, pp 311-317.

[16] Liu J, Reeves CR. Constructive and composite heuristic solutions to the P|| Σ Ci scheduling problem. European Journal of Operational Research, 132, 2001, pp 439-452.

[17] Nawaz M, Enscore EE, Ham I. A heuristic algorithm for the m-machine, n-job flow shop sequencing problem. OMEGA, 11(1), 1983, pp 91-95.

[18] Reeves C. Improving the efficiency of tabu search for machine sequencing problem. Journal of Operational Research Society, 44(4), 1993, pp 375-382.

[19] Reeves C. A genetic algorithm for flowshop sequencing. Computers and Operations Research, 22(1), 1995, pp 5-13.

[20] Reeves C, Yamada T. Genetic algorithms, path relinking and the flowshop sequencing problem. Evolutionary Computation, 6, 1998, pp 45-60.

[21] Nowicki E, Smutnicki C. A fast tabu search algorithm for the permutation flowshop problem. European Journal of Operational Research, 91, 1996, pp 160-175.

[22] Watson JP, Barbulescu L, Whitley LD, Howe AE. Contrasting structured and random permutation flowshop scheduling problems: search space topology and algorithm performance. ORSA Journal of Computing, 14(2), 2002, pp 98-123.

[23] Ogbu F, Smith D. The application of the simulated annealing algorithm to the solution of the n/m/Cmax flowshop problem. International Journal of Production Research, 42(3), 2004, pp 473-491.

[24] Osman I, Potts C. Simulated annealing for permutation flow shop scheduling. OMEGA, 17(6), 1989, pp 551-557.

[25] Tasgetiren MF, Sevkli M, Liang Y-C, Gencyilmaz G. Particle swarm optimization algorithm for permutation flowshop sequencing problem. 4th International Workshop on Ant Algorithms and Swarm Intelligence (ANTS2004), Brussels, Belgium. 2004, pp. 382-389.

[26] Rajendran C, Ziegler H. Ant-colony algorithms for permutation flowshop scheduling to minimize makespan/maximum lateness of jobs. European Journal of Operational Research, 155(2), 2004, pp 426-438.

[27] Stützle T. An ant approach to the flowshop problem. In: Proceedings of the 6th European Congress on Intelligent Techniques and Soft Computing (EUFIT'98). Verlag Mainz. Germany, 1998, pp 1560-1564.

Innovative Production Machines and Systems
D.T. Pham, E.E. Eldukhri and A.J. Soroka (Eds.)

Using the Bees Algorithm to solve a stochastic optimisation problem

D.T. Pham, M.A. Negm, S. Otri

Manufacturing Engineering Centre, Cardiff University, Cardiff CF24 3AA, UK

Abstract

This paper describes the use of the Bees Algorithm to handle a stochastic optimisation problem. A new fitness evaluation method based on computing the average fitness value for each bee over a number of trials rather than the value for a single trial. This enables the algorithm to be applied to situations where the fitness value changes even when the location visited by a bee remains the same. The application studied in this paper involved setting the parameters for the Ant Colony System algorithm to facilitate the solution of a Travelling Salesman Problem.

Keywords: Bees Algorithm, Stochastic Optimisation, Ant Colony Optimisation, Travelling Salesman Problem.

1. Introduction

The Bees Algorithm is a powerful population-based algorithm able to solve complex multi-variable functional optimisation problems efficiently. So far, the algorithm has been applied only to deterministic situations where the fitness value at any given point is fixed. However, there are problems where the fitness value can change randomly at a point. An example of such a problem is that of finding the optimum parameter setting for a stochastic optimisation algorithm such as the genetic algorithm (GA) or the Ant Colony System (ACS). This paper describes the use of the Bees Algorithm for solving such a stochastic optimisation problem. To enable the algorithm to perform its task successfully, a new fitness evaluation method has been adopted. This essentially involves computing the fitness at each location visited by a bee a number of times and obtaining the average fitness value for those visits.

The paper is organised as follows. Section 2 reviews previous work on parameter setting for an optimisation system, the stochastic optimisation problem studied in this paper. Section 3 outlines the operation of the Bees Algorithm. Section 4 presents the proposed fitness evaluation method. Section 5 describes the application of the Bees Algorithm and the proposed fitness evaluation method to the setting of the parameters for the Ant Colony System to facilitate the solution of a Travelling Salesman Problem (TSP). Section 6 concludes the paper.

2. Previous work on parameter setting for an optimisation system

There have been several approaches to solving the problem of finding the best set of parameters for an optimisation system. Most of the efforts so far have been based on statistical analysis. Some researchers have been working on meta optimisation procedures in general. For instance, Adenso-Diaz and Laguna [1] have developed a statistical procedure for a meta optimisation problem. The procedure is generally applicable to different meta optimisation algorithms but is limited to problems with a maximum of five parameters and cannot be guaranteed to find the best set of parameters. The approach by Coy *et al.* [2] is based on statistical design of experiments. Their approach requires the parameters to be initialised to near the optimal point. Bartz-Beielstein and Markon [3] have proposed a method combining statistical design of experiments and tree-based regression analysis. Birattari [4] has described a procedure that empirically evaluates a set of candidate configurations by discarding bad ones as soon as statistically sufficient evidence is gathered against them.

Other researchers have focused on optimising the parameters of Ant Colony Systems. Their efforts could be divided into three groups: analytical methods for determining the best setting for some of the parameters, experimental methods of finding the proper parameter set and methods of matching parameters to problems.

As an example of work in the first group, Dorigo *et al.* [5] have presented a formula for the optimum number of ants based on the value of the pheromone decay coefficient ρ and the exploitation/exploration parameter q_0. Watanabe and Matsui [6] have proposed an adaptive mechanism for controlling the parameter candidate sets based on the pheromone concentration for improving the Ant Colony Optimisation (ACO) algorithm. Zecchin *et al.* [7] have developed parametric guidelines for the application of ACO to water distribution system optimisation. Hao *et al.* [8] have chosen to optimise three parameters, ρ, q_0 and β (β is a parameter which determines the relative importance of pheromone versus distance). They have developed a parameter study strategy based on Particle Swarm Optimisation. A hybridised algorithm between GA and ACS-TSP has been proposed by Pilat and White [9] to solve the TSP faster but their attempt failed to produce better results than the original ACS. Ling *et al.* [10] have offered a new parameter selection strategy based on Fish Swarm techniques to optimise three parameters ρ, q_0 and α (α is another parameter which determines the relative importance of pheromone versus distance). Pellegrini *et al.* [11] have used the F-Race algorithm to find the optimum values of four parameters (ρ, α, q_0 and number of ants m) for the Max-Min ACS (an ACS where an upper limit and a lower limit are set for the pheromone level on a trail).

As an example of work in the second group, Pilat and White [9] have proposed a meta-ACS using a GA as another layer wrapping the ACS to optimise three of its parameters. Socha [12] has proposed computational studies of ACS parameters. Solnon [13] has conducted computational studies of some parameters as a pre-processing step. Clark [14] has presented a design of experiments based on exhaustive search to find the optimum values of β, ρ and q_0 for a particular TSP.

As an example of work in the third group, Clark [14] has presented a new approach to finding the best ACS parameters for TSPs. He has classified TSPs based on three measurements of complexity and has proposed matching TSP classes with the appropriate sets of parameters. Figlali [15] has investigated the setting of ACS parameters for job-shop scheduling problems.

3. Bees Algorithm

Fig. 1 shows the pseudo-code of the basic Bees Algorithm which is described in more detail in [16].

```
   1. Initialise population with random solutions.
   2. Evaluate fitness of the population.
   3. While (stopping criterion not met)
// forming new population
   4. Select sites for neighbourhood search.
   5. Recruit bees for selected sites (more bees
      for the best e sites) and evaluate fitness.
   6. Select the fittest bee from each patch.
   7. Assign the remaining bees to search
      randomly and evaluate their fitness.
   8. End While.
```

Fig. 1. Pseudo-code of the Bees Algorithm

4. New fitness evaluation method for the Bees Algorithm

The proposed fitness evaluation method will allow the Bees Algorithm to solve stochastic optimisation problems. The basic Bees Algorithm involves calling the fitness function of the site where a bee is placed. In deterministic problems like those previously handled by the Bees Algorithm, the fitness is fixed if the site visited by a bee is unchanged and hence evaluating it once is enough. Stochastic problems require many evaluations at the same point and the calculation of the average value. The number of bees sent to each site is not constant and it is based on feedback information calculated from the bees already recruited to the same point. The feedback information that will determine the number of bees sent to the same point is based on the slope of the line formed by linear regression of the fitness values of the recruited bees. The algorithm will stop sending bees to the same site when the slope indicates a near horizontal line. By adopting this technique, a more accurate average fitness for bees visiting the same site can be obtained using a small number of evaluations.

The accuracy of this average value will depend on the minimum number of bees sent to the same point and the minimum acceptable slope value.

```
    1.  Initialise population with random solutions.
    2.  Evaluate fitness of the population by calling
        the Fitness subroutine.
    3.  While (stopping criterion not met)
// forming new population
    4.  Select sites for neighbourhood search.
    5.  Recruit bees for selected sites (more bees
        for the best e sites) and evaluate fitness by
        calling the Fitness subroutine.
    6.  Select the fittest bee from each patch.
    7.  Assign the remaining bees to search
        randomly and evaluate their fitness.
    8.  End While.
// Subroutine for bees fitness evaluation.
    1.  Evaluate fitness()
    2.  While (minimum slope value not reached)
    3.  Evaluate the fitness and store its value.
    4.  End While.
    5.  Calculate the mean fitness from the stored
        values.
    6.  End Subroutine.
```

Fig. 2. Pseudo-code of the modified Bees Algorithm for solving stochastic problems.

Fig. 2 shows the pseudo-code of the Bees Algorithm with the new fitness evaluation method. The algorithm is the same as the original Bees Algorithm except for the way in which the fitness of a bee is evaluated. Here, fitness is computed by an external subroutine which replaces the direct evaluation method of the original algorithm. The value of the minimum slope will determine the accuracy and speed of the algorithm. The algorithm will be slow and the accuracy, high, if the slope is small. If a higher value for the slope is chosen, the running time will decrease but the accuracy will also drop.

An analysis was performed on a random stochastic optimisation problem to demonstrate the relationship between the slope of the regression line and the accuracy of the average fitness value obtained by the algorithm. Fig. 3 shows the variation in fitness values at a point visited by five bees at different times. In this figure, the slope of the regression line is not equal to zero, which means more evaluations and fitness values are needed to obtain an accurate average value. Table 1 gives the averages obtained for 5, 22 and 90 runs as 658, 472 and 471 respectively. This means that five evaluations are not sufficient to calculate accurate average fitness values.

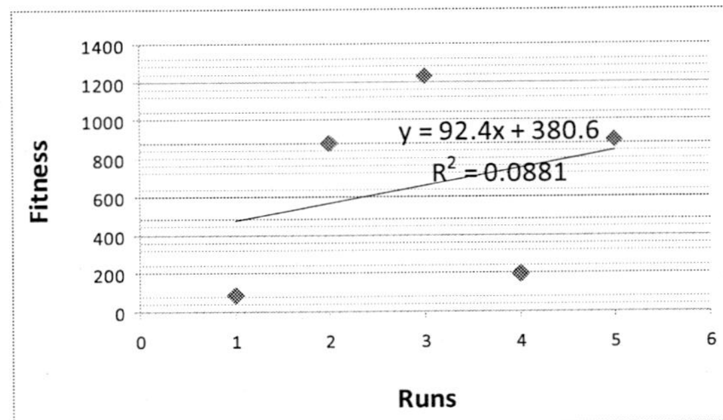

Fig. 3. Variation of fitness values of the same point for 5 independent runs.

Table 1
Results for 5, 22 and 90 independent runs.

Max (5 runs)	Min (5 runs)	Av. (5 runs)	Av. (22 runs)	Av. (90 runs)
1232	88	658	472	471

Fig. 4 shows the fitness values obtained over 22 independent runs for the same stochastic problem. The slope of the line formed is almost zero, i.e. the line is nearly horizontal. Table 2 gives the average fitness for the 22 independent runs compared to the average fitness for 90 independent runs. It can be seen that the fitness values are nearly the same in the two cases.

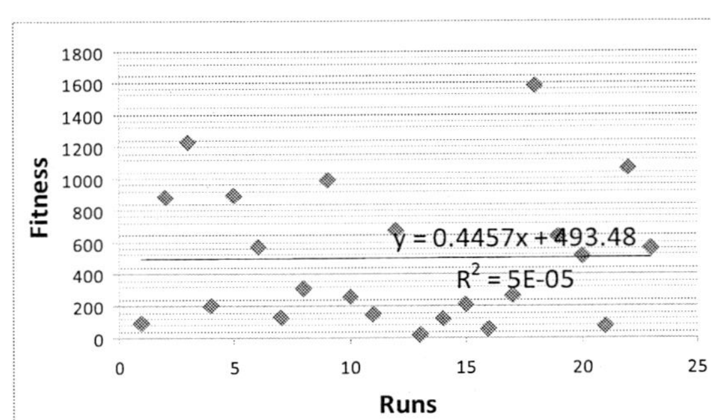

Fig. 4. Fitness values of the same point for 22 independent runs.

Table 2
Results for 22 and 90 independent runs.

Max (22 runs)	Min (22 runs)	Average (22 runs)	Average (90 runs)
1584	22	472	471

Fig. 5 shows the fitness values and Table 3, the statistics, for 90 independent runs.

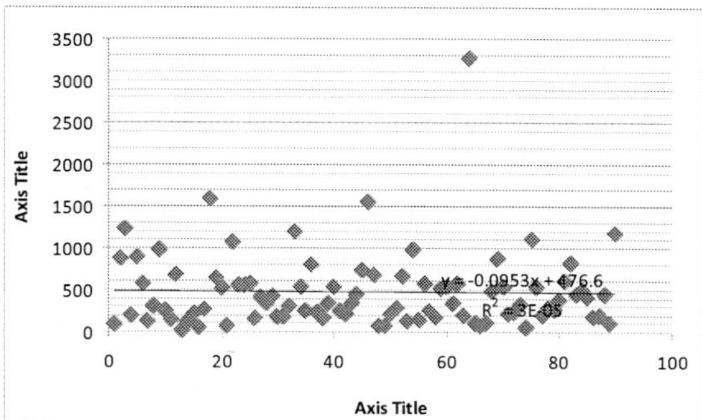

Fig. 5. Variation of fitness values of the same point for 90 independent runs.

Table 3
Results for 90 independent runs.

Max	Min	Average
3267	22	471

From the results obtained, it can be seen that the new fitness evaluation method enables the Bees Algorithm to handle fitness variations in stochastic optimisation problems. The slope of the line formed by applying linear regression to the fitness values of bees recruited to explore a particular point is an indicator of whether the number of bees sent to that point is sufficient to give a stable average fitness. Using this feedback information in the Bees Algorithm should result in accurate average fitness values in the minimum number of evaluations.

5. Optimising ACS parameters for solving TSP

5.1. ACS and TSP

Ant Colony Optimisation (ACO) imitates the foraging behaviour of real ants in their search for the shortest path between the nest and the food source. The first optimisation algorithm based on ant behaviour is the Ant System (AS) by Dorigo *et al.* [17]. Several combinatorial optimisation problems, such as the Travelling Salesman Problem (TSP) and Quadratic Assignment Problem (QAP), have been solved using the Ant System algorithm. The Ant Colony System (ACS) [5] is the most successful extension of the original Ant System and has been tested and researched the most.

The Travelling Salesman Problem (TSP) is an NP-hard problem which involves finding the shortest path connecting *n* cities. Each city must be visited once and only once. Fig. 6 shows an example of a TSP, the Oliver 30 TSP [18] in which *n* is equal to 30.

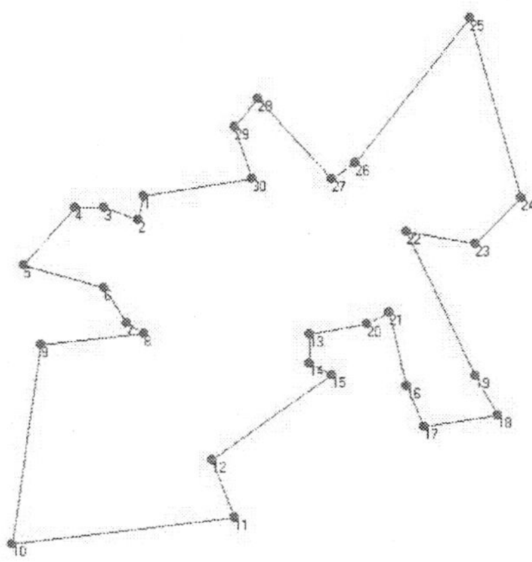

Fig. 6. The Oliver 30 Travelling Salesman Problem.

The distance between cities i and j is dij. dij is given by

$$d_{ij}=[(x_i-x_j)^2 + (y_i-y_j)^2]^{1/2} \quad (1)$$

where x_i and y_i are the co-ordinates of city i.

5.2. Meta-heuristic methods and speed of convergence

The time and number of evaluations required to find the best solution in meta-heuristic optimisation depend on two factors, one of which is controllable and the other is not. The controllable factor is the parameter setting for the meta-heuristic method. The uncontrollable factor is the stochastic nature of the optimisation method itself.

For instance, when the ACS is used to solve a TSP, it is initialised by manually setting the value of its parameters such as the number of ants and the exploitation-exploration ratio. A source of randomness is the distribution of ants over the available cities at the start of the optimisation.

The speed of convergence is illustrated in Figs. 7

and 8 for two different experiments with different parameter values. Fig. 7 presents the results for the first experiment, showing variations in the number of function evaluations made by ACS to find the optimum solution of a particular TSP in 50 independent runs. (In Fig. 7, the number of function evaluations is called 'fitness', although, strictly, the correct term would be 'cost' as the higher this number is, the poorer the particular solution will be). The set of parameters given in Table 4 was adopted for all the 50 runs. From Table 5, the average value of the fitness is 463.

Table 4
Parameters used in the first experiment.

m	B	ρ	q_0
11	4	0.2	0.7

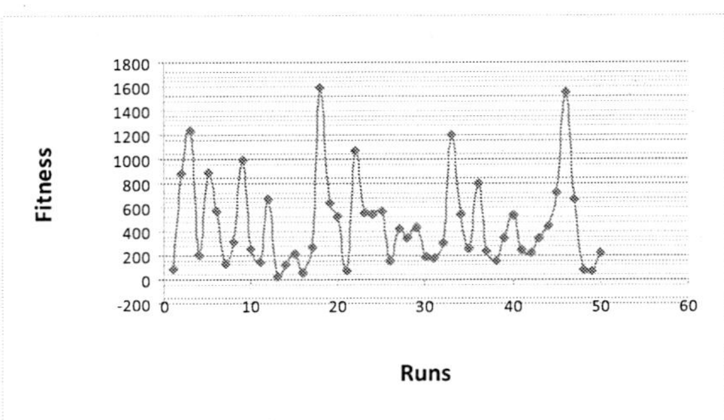

Fig. 7. Fitness values obtained in the first experiment over 50 independent runs.

Table 5
Results for the first experiment.

Runs	Min	Max	Average
50	22	1584	463

Fig. 8 presents the results for the second experiment. The set of parameters used is shown in Table 6. From Table 7, the average value over 50 independent runs is 2370.

Table 6
Parameters used in the second experiment.

m	B	ρ	q_0
25	13	0.7	0.2

These two experiments demonstrate that parameter setting for a meta-heuristic optimisation method is a stochastic problem which is affected directly by the chosen parameters values. The following section shows that this problem has been solved using the modified Bees Algorithm.

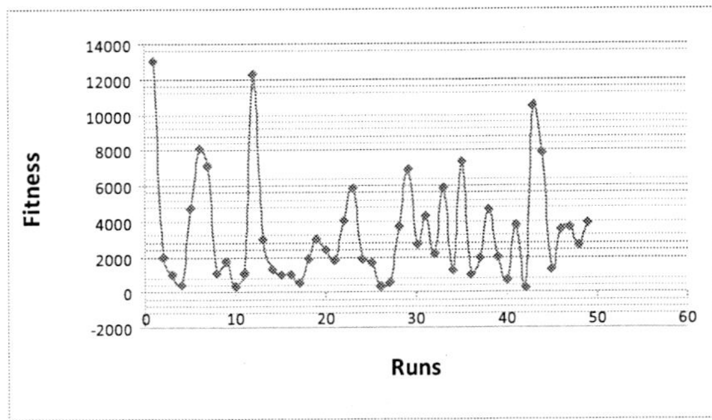

Fig. 8. Fitness values obtained in the second experiment over 50 independent runs.

Table 7
Results for the second experiment.

Runs	Min	Max	Average
50	275	13050	2370

5.3. Using the modified Bees Algorithm for Ant Colony System parameter setting

For this work, the Bees Algorithm was 'wrapped around' the ACS. Fig. 9 shows the flow of information between the algorithms. It can be seen that bees represent sets of ACS parameters and, again, fitness values are the total numbers of function evaluations required by the ACS to find the optimum solution of the TSP.

The test problem was the Oliver30 TSP (see Fig. 6). The values of parameters m, β, ρ and q_0 of the Ant Colony System were obtained using the Bees Algorithm enhanced with the proposed fitness evaluation method. The search spaces for these parameters are given in Table 8.

Table 8
Search spaces of the ACS parameters.

Value	m	β	ρ	q_0
Min Value	2	0	0	0
Max value	number of cities	14	1	1

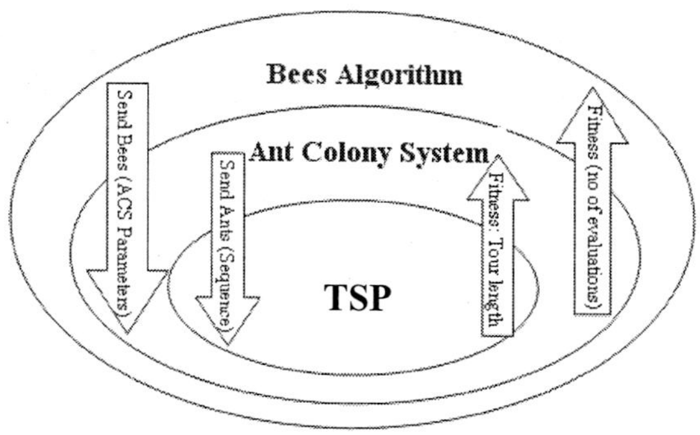

Fig. 9. Bees Algorithm wrapping ACS-TSP.

Computation experiments were performed on a PC with a Pentium Dual Core processor and 1GB Memory. C#.Net 2008 and the .net framework 3.5 were used for program development. It took three days for the Bees Algorithm to find the best parameters for the ACS when solving the Oliver 30 TSP.

Table 9 shows the best set of parameters obtained by the Bees Algorithm and the parameters adopted by Dorigo *et al.* [5] and Pilat *et al.* [9].

Table 10 provides details of experiments performed to find the optimum values of ACS parameters for different TSPs.

Table 9
Parameter sets obtained by the Bees Algorithm and the parameters suggested by others.

Author	M	β	ρ	q_0	Average evaluations
Dorigo *et al.*	10	2	0.1	0.9	7839
Pilat *et al.*	10	6	0.2	0.7	7443
Pham *et al.*	12	4	0.2	0.8	2984

6. Conclusion

This paper has described the use of the Bees Algorithm to solve a stochastic optimisation problem, the determination of the parameter setting for an Ant Colony System for finding the shortest route in a Travelling Salesman Problem. The key point for this application of the Bees Algorithm is the multiple evaluation of the fitness of each parameter setting tried. The results obtained have shown that the parameter setting produced by the Bees Algorithm outperformed the parameter sets that were suggested by other parameter settings methods.

Further work could investigate the possibility of using the variation in the values of the fitness with a particular parameter setting as an indicator of the stability of that setting and another optimisation criterion. For such an investigation, it might be useful to adopt a multi-objective optimisation technique such as that presented in [19].

Acknowledgements
The authors are members of the EU-funded FP6 Network of Excellence for Innovative Production Machines and Systems (I*PROMS).

References

[1] Adenso-Diaz B.,Laguna M. Fine-Tuning of Algorithms Using Fractional Experimental Designs and Local Search. *Operations Research,* 2006, **54**(1): pp. 99-114.

[2] Coy S.P., Golden B.L., Runger G.C., Wasil E.A. Using Experimental Design to Find Effective Parameter Settings for Heuristics. *Journal of Heuristics,* 2001,**7** (1): pp. 77-97.

[3] Bartz-Beielstein T.,Markon S. Tuning Search Algorithms for Real-World Applications: A Regression Tree Based Approach. *Proc. Evolutionary Computation,* 2004, Portland, OR, USA, pp. 1111-1118.

[4] Birattari M., Sützle T., Paquete L.,Varrentrapp K. A Racing Algorithm for Configuring Metaheuristics. *Proc. Genetic and Evolutionary Computation Conference,* 2002, New York, USA, pp. 11-18.

[5] Dorigo M.,Gambardella L.M. Ant Colony System: A Cooperative Learning Approach to The Traveling Salesman Problem. *IEEE Transactions on Evolutionary Computation,* 1997, **1**(1): pp. 53-66.

[6] Watanabe I.,Matsui S. Improving the performance of ACO algorithms by adaptive control of candidate set. *Proc. Evolutionary Computation Conference,* 2003, pp. 1355-1362.

[7] Zecchin A.C., Simpson A.R., Maier H.R., Nixon J.B. Parametric study for an ant algorithm applied to water distribution system optimization. *IEEE Transactions on Evolutionary Computation,* 2005, **9**(2): pp. 175-191.

[8] Hao Z.-F., Cai R.-C., Huang H. An Adaptive Parameter Control Strategy for ACO. *Proc. International Conference on Machine Learning and Cybernetics,* 2006, Dalian, China, pp. 203-206.

[9] Pilat M.L., White T. Using Genetic Algorithms to Optimize ACS-TSP. *Proc. of the Third*

International Workshop on Ant Algorithms, 2002, Brussels, Belgium, pp. 282-287.

[10] Ling W., Luo H. An Adaptive Parameter Control Strategy for Ant Colony Optimization, *Proc. International Conference on Computational Intelligence and Security,* 2007, Washington, DC, USA, pp. 142-146.

[11] Pellegrini P., Favaretto D., Moretti E. On MAX - MIN Ant System's Parameters. *Ant Colony Optimization and Swarm Intelligence,* 2006, Brussels, Belgium, pp. 203-214.

[12] Socha K. The Influence of Run-Time Limits on Choosing Ant System Parameters. *Genetic and Evolutionary Computation,* 2003, Chicago, IL, USA, pp. 198.

[13] Solnon C. Boosting ACO with a Preprocessing Step. *Applications of Evolutionary Computing,* 2002, Kinsale, Ireland, pp. 41-50.

[14] Gaertner D., Clark K. On optimal Parameters for Ant Colony Optimization Algorithm. *Proc. of the International Conference on Artificial Intelligence,* 2005, Las Vegas, USA, pp. 83 - 89.

[15] Figlali N., Özkale C., Engin O., Figlali A. Investigation of Ant System Parameter interactions by Using Design of Experiments for Job-Shop Scheduling Problems. *35th International Conference on Computers and Industrial Engineering,* 2005, Istanbul, Turkey, pp. 745-750.

[16] Pham D.T., Ghanbarzadeh A., Koç E., Otri S., Rahim S., Zaidi M. The Bees Algorithm - A Novel Tool for Complex Optimisation Problems. *Proc. 2nd Virtual International Conference on Intelligent Production Machines and Systems (IPROMS2006),* 2006, Cardiff, UK, pp. 454-459.

[17] Dorigo M., Maniezzo V., Colorni A. Ant System: Optimization by a Colony of Cooperating Agents. *IEEE Transactions on Systems, Man, and Cybernetics, Part B,* 1996, **26**(1): pp. 29-41.

[18] Oliver I.M., Smith D.J.,Holland J.R.C. A Study of Permutation Crossover Operators on The Traveling Salesman Problem. *Proc. of the Second Int. Conf. on Genetic Algorithms,* 1987, Cambridge, MA, USA, pp. 224-230.

[19] Ghanbarzadeh A. *The Bees Algorithm: A Novel Optimisation Tool,* PhD Thesis, 2007, Cardiff University, Cardiff, UK.

Table 10
Experiments performed to find the optimum parameter set for Ant System methods

Author	Hao *et al.*	Pilat *et al.*	Ling *et al.*	Gaertner *et al.*	Pellegrini *et al.*	Pham *et al.*
Application	TSP	TSP	TSP	TSP	TSP	TSP
Parameter Setting Strategy	PSO	GA	Artificial Fish Swarm Algorithm	Exhaustive search	F-Race algorithm	The Bees Algorithm
Datasets Studied	eil51, pr76, kroA100, lin105, ts225, pr226, lin318, pcb442, att532, rat783	eil51, eil76, pr43, kroA10 0ry48p, ft70	Oliver 30, att48, eil51, eil76, lin105, ts225, a280, rat783	Oliver 30	Unknown	Oliver 30
Number of runs	5	Problem dependent	5	10	4	Until statistical condition is met
Type of ACO	ACS	ACO	ACO	ACO	MMACS	ACS
Optimised parameters	β, ρ and q_0	β, ρ and q_0	ρ, α and Q*	β, ρ and q_0	m, β, ρ and α	m, β, ρ, q_0
Stopping criteria	1000 iterations	1000-3000 iterations depending on the problem	Iterations (not specified)	1000 iterations or the best solution found	Time boundary	Best solution found or 2000 iterations

* Q relates to the total amount of pheromone deposited on a path.

Innovative Production Machines and Systems
D.T. Pham, E.E. Eldukhri and A.J. Soroka (Eds.)

Multi-objective Environmental/Economic Power Dispatch using the Bees Algorithm with Pareto optimality

D.T. Pham, J.Y. Lee, A. Haj Darwish and A.J. Soroka

Intelligent Systems Laboratory, Manufacturing Engineering Centre, Cardiff University, Cardiff CF24 3AA, UK

Abstract

This paper presents the use of the Bees Algorithm for the Environmental/Economic (Power) Dispatch (EED) problem which is formulated as a nonlinear constrained multi-objective optimisation problem. In this problem, both fuel cost and nitrogen oxides emission are to be simultaneously minimised. Simulation results presented for the standard IEEE 30-bus system using the Bees Algorithm with Pareto optimality are compared to those obtained using other approaches. The comparison shows the superiority of the Bees Algorithm and confirms its suitability for solving the multi-objective EED problem.

Keywords: Multi-objective, Bees Algorithm, Pareto optimality, Environmental/Economic Dispatch

1. Introduction

The aim of the classical economic dispatch problem has only been to minimise the total fuel cost. However, this objective can no longer be considered singly due to rising concerns about the environment from the emissions produced by fossil-fuelled electric power plants.

Environmental/Economic Dispatch (EED) is a multi-objective problem with conflicting objectives because reducing emissions is in conflict with minimising the cost of generation. Many approaches have been proposed to solve this multi-objective problem. More recently, multi-objective evolutionary algorithms have been applied to the problem. For example, Abido used the Non-dominated Sorting Genetic Algorithm (NSGA) [1], Niched Pareto Genetic Algorithm (NPGA) [2], Strength Pareto Evolutionary Algorithm (SPEA) [3] and NSGA-II [4] on the standard IEEE 30-bus system.

This paper discusses the application of the Bees Algorithm [5] to the EED problem to minimise fuel costs and nitrogen oxides (NOx) emission simultaneously. The Bees Algorithm is a new population-based optimisation tool inspired by the foraging behaviour of honey bees.

The paper is organised as follows. Section 2 reviews previous work on multi-objective optimisation and intelligent swarm-based optimisation; section 3 describes a version of the Bees Algorithm modified to suit the EED problem; section 4 defines the Environmental/Economic Dispatch problem; section 5 presents the results obtained; section 6 concludes the paper.

2. Multi-objective optimisation and intelligent swarm-based optimisation

2.1. Multi-objective optimisation

Most realistic optimisation problems require the simultaneous optimisation of more than one objective function, and it is unlikely that the different objectives would be optimised by adopting the same set of parameters. Usually, a multi-objective problem (MOPs) is solved by combining the various objectives into one scalar objective whose solution is a Pareto optimal point for the original MOP. Most algorithms have been developed in a linear framework (i.e. with linear objectives and linear constraints), but some techniques such as Weighted Sum, Homotopy, Goal Programming, Normal-Boundary Intersection (NBI) and Multilevel Programming are also applicable to nonlinear problems. In this work, Pareto optimality was basically adopted as the solution technique and, in addition to this, the Weighted Sum technique was used to conduct one of the proposed neighbourhood searches.

2.2. Intelligent swarm-based optimisation

Swarm-based optimisation algorithms (SOAs) mimic nature's methods to drive a search towards the optimal solution. Methods which have been utilised include natural selection (in the Genetic Algorithm or GA [6]), stigmergy (in Ant Colony Optimisation or ACO [7]), animal flocking behaviour (in Particle Swarm Optimisation or PSO [8]) and food foraging behavious (in the Bees Algorithm [5]).

3. The Bees Algorithm

3.1. Proposed Bees Algorithm

The basic Bees Algorithm as described in [5] is not suitable for a MOP, especially if the aim is to achieve Pareto optimality. Moreover, although basic neighbourhood search can work satisfactorily for a MOP, new types of neighbourhood search could produce enhanced solutions. This section presents a version of the Bees Algorithm that has been adapted for multi-objective optimisation and two neighbourhood search methods implemented to promote the generation of improved solutions.

Fig. 1 shows the pseudo code for the proposed Bees Algorithm. The basic Bees Algorithm uses 6 parameters [9, 10]. The new version requires only 3 parameters to be set, namely:

1) the number of scout bees (n)
2) number of recruited bees ($n1$)
3) initial patches size (ngh)

```
1. Initialise population with random solutions
2. Evaluate fitness of the population
3. While (Stopping criterion not met)
4. Select non-dominated patches for
   neighbourhood search
5. Recruit bees for non-dominated patches and
   evaluate their fitnesses
6. Select the fittest bee from each patch
7. Assign the remaining bees to search
   randomly and evaluate their fitnesses
8. End while
9. Create the Pareto optimal set
```

Fig. 1. Pseudo code for the proposed Bees Algorithm for multi-objective problems.

The algorithm starts with n scout bees which are randomly sent to the search space. After fitness evaluation, those patches with non-dominated scout bees are designated "selected patches" for further neighbourhood search to be conducted on them. After this neighbourhood search, only the bee with the highest fitness is elected in each of the selected patches to form the next bee population. The remaining bees in the population are randomly assigned in the search space to scout for new potential solutions. These steps are repeated until a stopping criterion is met. The Pareto optimal set is created from all the non-dominated bees in each iteration. In addition to the basic neighbourhood search technique [9, 10], two other approaches were tested in this work as described below.

3.2. Neighbourhood search methods

3.2.1. Basic neighbourhood search

During a basic neighbourhood search, a recruited bee is sent to a selected patch and its fitness is evaluated. The recruited bee is then compared with the scout bee in the patch and, if the recruit is fitter than the old scout bee, the latter is replaced by the new bee. Otherwise, the old scout bee is retained. The advantage of this approach is it enables a patch and its position to be updated whenever a bee is recruited to the patch that performs better than the incumbent bee.

This provides a quicker and more efficient route to an optimal solution. Fig. 2 shows the flowchart for this search method.

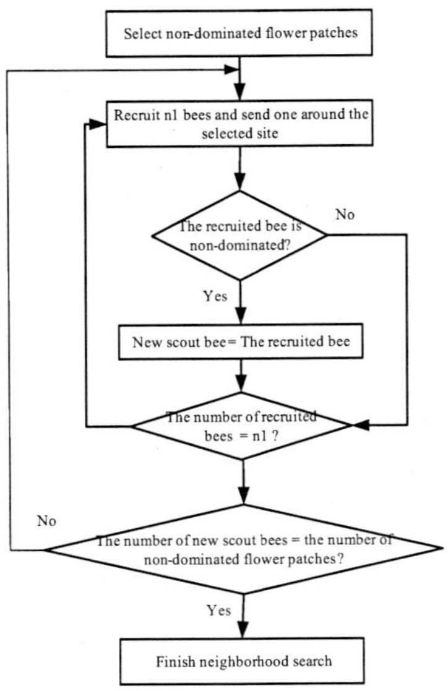

Fig. 2. Flowchart for the basic neighbourhood search method.

3.2.2. Random selection neighbourhood search

Instead of posting recruited bees to a selected patch one at a time, this approach involves sending all the *n1* recruited bees to the selected patch simultaneously and then evaluating their fitnesses. Non-dominated bees are then chosen from among the recruited bees. If there are no non-dominated bees, then the old scout bee will be retained for this patch. If there is only one non-dominated bee, it will replace the old scout bee. However, if there are two or more non-dominated bees, one of them will be chosen at random to replace the old scout bee for this patch. Fig. 3 gives the flowchart for this approach.

3.2.3. Weighted Sum neighbourhood search

This approach is similar to the Random selection neighbourhood search method. However, where there are two or more non-dominated bees in the selected patch, the Weighted Sum method involves using a linear combination of the objective functions to decide which non-dominated bee is to be selected to replace the old scout in this patch. Fig. 4 shows the flowchart for this approach.

4. Environmental/Economic Dispatch

The EED problem involves the conflicting objectives of emission reduction and fuel cost optimisation and it is formulated as described below.

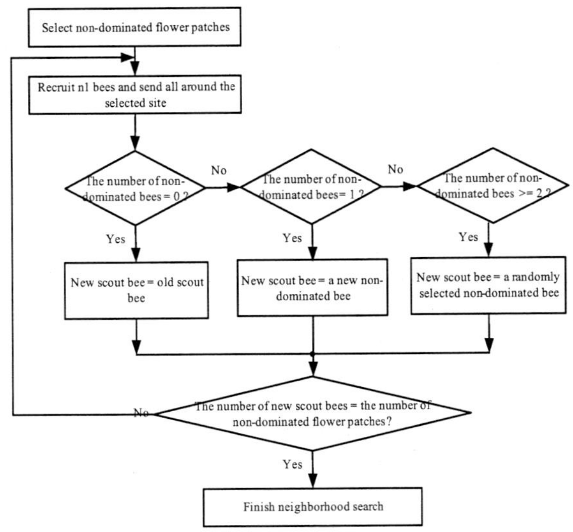

Fig. 3. Flowchart for the Random Selection neighbourhood search method.

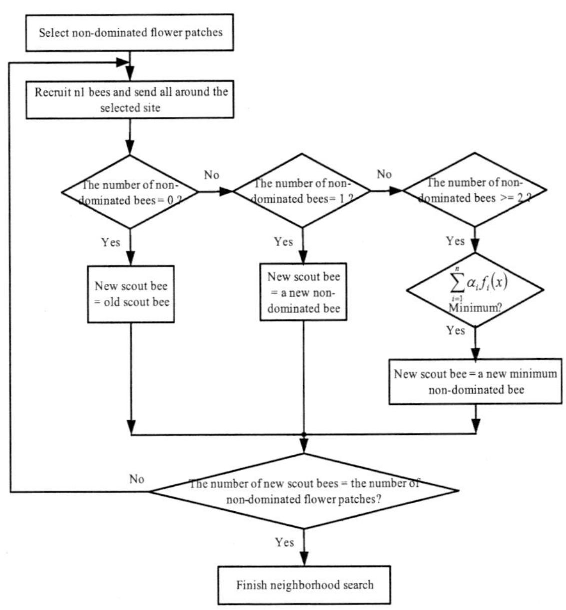

Fig. 4. Flowchart for the Weighted Sum neighbourhood search method.

4.1. Objective Functions

4.1.1. Fuel cost objective

The economic dispatch problem in power generation is to minimise the total fuel cost while satisfying the total demand. The equation used the optimal combination for this problem is [4, 11]:

$$C = \sum_{i=1}^{n} \left(a_i + b_i \times P_{Gi} + c_i \times P_{Gi}^2 \right) \quad (2)$$

where

C : total fuel cost ($/hr),

a_i, b_i, c_i : fuel cost coefficients of generator i,

P_{Gi} : power output by generator i, and

n : number of generators.

4.1.2. NOx emission objective

The total NOx emission caused by fossil fuel is expressed as:

$$E_{NOx} = \sum_{i=1}^{n} \left(a_{iN} + b_{iN} \times P_{Gi} + c_{iN} \times P_{Gi}^2 + d_{iN} \times \exp(e_{iN} \times P_{Gi}) \right) \quad \text{ton/hr} \quad (3)$$

where a_{iN}, b_{iN}, c_{iN}, d_{iN}, and e_{iN} are NOx coefficients of the ith generator emission characteristics.

4.2. Constraints

The optimisation problem is bounded by a number of constraints as detailed below.

4.2.1. Power balance constraint

The total power generated must supply the total load demand and the transmission losses expressed as:

$$\sum_{i=1}^{n} P_{Gi} - P_D - P_L = 0 \quad (4)$$

where

P_D : total load demand and

P_L : transmission losses.

In this work, P_L was taken as 0.

4.2.2. Maximum and minimum limits of power generation

The power generated P_{Gi} by each generator is constrained between its minimum and maximum limits stated as:

$$P_{Gi\min} \le P_{Gi} \le P_{Gi\max} \quad (5)$$

where

$P_{Gi\min}$: minimum power generated, and

$P_{Gi\max}$: maximum power generated.

4.3. Multi-objective formulation

The multi-objective Environmental/Economic Dispatch optimisation problem is therefore formulated as:

$$Minimise[C, E_{NOx}] \quad (6)$$

Subject to: $\sum_{i=1}^{n} P_{Gi} - P_D = 0$ (power balance), and

$P_{Gi\min} \le P_{Gi} \le P_{Gi\max}$ (generation limits)

4.4. System parameters

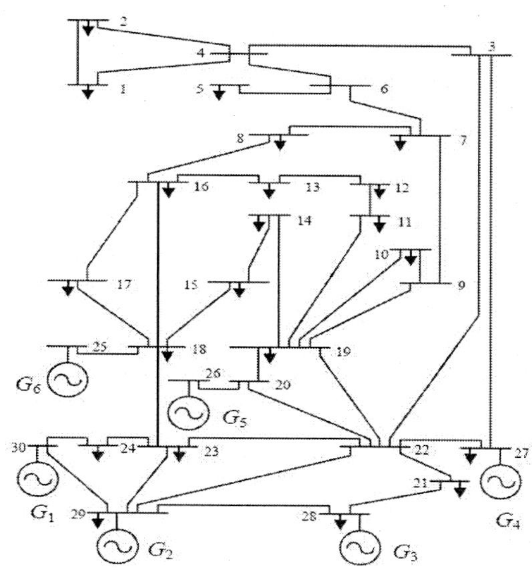

Fig. 5. Single-line diagram of IEEE 30-bus test system.

Simulations were performed on the standard IEEE 30-bus 6-generator test system using the Bees Algorithm and Fig. 6 shows it. The power system is interconnected by 41 transmission lines and the total system demand for the 21 load buses is 2.834(p.u.). Fuel cost and NOx emission coefficients for this system are given in Table 1 and 2 respectively.

Table 1
Fuel cost coefficients

Unit i	a_i	b_i	c_i	$P_{Gi\min}$	$P_{Gi\max}$
1	10	200	100	0.05	0.50
2	10	150	120	0.05	0.60
3	20	180	40	0.05	1.00
4	10	100	60	0.05	1.20
5	20	180	40	0.05	1.00
6	10	150	100	0.05	0.60

Table 2
NOx emission coefficients

Unit i	a_{iN}	b_{iN}	c_{iN}	d_{iN}	e_{iN}
1	4.091e-2	-5.554e-2	6.490e-2	2.0e-4	2.857
2	2.543e-2	-6.047e-2	5.638e-2	5.0e-4	3.333
3	4.258e-2	-5.094e-2	4.586e-2	1.0e-6	8.000
4	5.326e-2	-3.550e-2	3.380e-2	2.0e-3	2.000
5	4.258e-2	-5.094e-2	4.586e-2	1.0e-6	8.000
6	6.131e-2	-5.555e-2	5.151e-2	1.0e-5	6.667

5. Results

The performance of the algorithm was expressed by three indices:

1) percentage of relative deviations (Δ)
2) standard deviation
3) average number of solutions

To obtain the average performance of the algorithm, 10 runs were carried out for each of the neighborhood searches. This gave statistics showing how the percentage of relative deviation (Δ) differs from those for previous approaches such as Linear Programming (LP) [12], Multi-Objective Stochastic Search Technique (MOSST) [13], Non-dominated Sorting Genetic Algorithm (NSGA) [1], Niched Pareto Genetic Algorithm (NPGA) [2], Strength Pareto Evolutionary Algorithm (SPEA) [3] and Non-dominated Sorting Genetic Algorithm-II (NSGA-II) [4]. Δ_{avg} was computed as follows:

$$\Delta_{avg} = \sum_{i=1}^{R}\left(\frac{\left(F_{BA} - F_{ref}\right)}{F_{ref}} \times 100\right) \Big/ R \tag{7}$$

where F_{BA} is the fitness function values generated by the Bees Algorithm in each run, F_{ref} is the reference fitness function value generated by previous approaches, and R is the total number of runs. $\Delta_{\min}$ and $\Delta_{\max}$ denote the minimum and maximum values of the percentage of relative deviation in fitness function value over R runs, respectively. Tables 3 and 4 show the parameter values used for this experiment. The accuracy is 0.00001.

Table 3
Parameters for the Bees Algorithm

	Value
n : Number of scout bees	50
ngh : Initial patch size	0.05
n1 : Number of bees allocated to selected patches	5

Table 4
Parameters for Weighted Sum

	Value
W1: weight for fuel cost	1
W2: weight for NOx emission	1 1,000 100,000

Tables 5 to 8 present the simulation results. It should be noted the better results when compared to the previous approaches are reported in negative percentage relative deviations in all Tables given in this paper. Tables 5 and 6 show the average, minimum and maximum percentage relative deviations (Δ_{avg}, Δ_{min} and Δ_{max}) of the best Fuel cost and the best NOx emission between the Bees Algorithm and previous approaches. Tables 5 (a) – (d) and 6 (a) – (d) give the results for each search of the basic neighbourhood, Random selection and Weighted Sum. As seen in Table 5, Δ_{min} of all approaches by the Bees Algorithm are negative, which means they outperform all previous approaches such as LP, MOSST, NSGA, NPGA, SPEA and NSGA□. The average standard deviation value obtained of 0.08 is a low figure which can be considered as small enough for reliable and consistent results. In particular, Δ_{avg} of Weighed Sum neighbourhood search in Fig. 5 (d), when w2 is 1, is superior to all previous approaches. In addition, corresponding Emission at that time for most of them also outperforms previous approaches.

Emission values shown in Table 6 (c), Δ_{min}, are slightly better than or the same as for previous approaches with the exception of MOSST as the average standard deviation is virtually zero. In Tables 6 (a), (b), (c) and (d), the Δ_{avg}, and Δ_{min} of corresponding fuel costs obtained by the Bees Algorithm are much better than those for all previous approaches.

Table 5
Best Fuel cost
(a)Basic neighbourhood search

	Best Fuel Cost			Corresp. Emission		
	Δ_{avg}	Δ_{min}	Δ_{max}	Δ_{avg}	Δ_{min}	Δ_{max}
LP	-0.98	-1.02	-0.92	-2.33	-3.58	-1.03
MOSST	-0.91	-0.95	-0.85	-1.85	-3.11	-0.54
NSGA	-0.03	-0.07	0.03	-2.12	-3.37	-0.82
NPGA	0.02	-0.02	0.08	-1.38	-2.65	-0.07
SPEA	0.04	-0.00	0.10	-1.53	-2.80	-0.23
NSGAII	0.04	-0.00	0.10	-1.70	-2.97	-0.40

(b)Random selection neighbourhood search

	Best Fuel Cost			Corresp. Emission		
	Δ_{avg}	Δ_{min}	Δ_{max}	Δ_{avg}	Δ_{min}	Δ_{max}
LP	-1.00	-1.02	-0.98	-0.61	-2.19	0.90
MOSST	-0.93	-0.95	-0.91	-0.12	-1.71	1.40
NSGA	-0.06	-0.07	-0.03	-0.39	-1.98	1.11
NPGA	-0.00	-0.02	0.02	0.35	-1.25	1.87
SPEA	0.01	-0.00	0.04	0.20	-1.40	1.72
NSGAII	0.01	-0.00	0.04	0.03	-1.57	1.54

(c) Weighted Sum neighbourhood search (w2=100,000)

	Best Fuel Cost			Corresp. Emission		
	Δ_{avg}	Δ_{min}	Δ_{max}	Δ_{avg}	Δ_{min}	Δ_{max}
LP	-1.00	-1.02	-0.97	-0.64	-1.39	0.40
MOSST	-0.93	-0.95	-0.90	-0.15	-0.90	0.90
NSGA	-0.06	-0.08	-0.02	-0.43	-1.18	0.62
NPGA	-0.00	-0.02	0.03	0.32	-0.43	1.37
SPEA	0.01	-0.00	0.05	0.17	-0.59	1.22
NSGAII	0.01	-0.01	0.05	-0.00	-0.76	1.05

(d) Weighted Sum neighbourhood search (w2=1)

	Best Fuel Cost			Corresp. Emission		
	Δ_{avg}	Δ_{min}	Δ_{max}	Δ_{avg}	Δ_{min}	Δ_{max}
LP	-1.02	-1.02	-1.02	-0.51	-1.12	0.04
MOSST	-0.95	-0.95	-0.95	-0.02	-0.63	0.54
NSGA	-0.07	-0.07	-0.07	-0.30	-0.91	0.26
NPGA	-0.02	-0.02	-0.02	0.45	-0.16	1.01
SPEA	-0.00	-0.00	0.00	0.30	-0.32	0.86
NSGAII	-0.00	-0.01	0.00	0.13	-0.49	0.69

Although the gains per hour are relatively small, cost savings and emission reductions become much more significant on an annual basis.

Table 7 shows the average number of Pareto solutions. The average number of solutions over 10 runs was 21.5 for basic neighbourhood search and around 100 for the other approaches. For general situations, the basic neighbourhood search method giving a small number of solutions might be satisfactory. In other cases, the Random Selection method with a much higher number of solutions is more appropriate. The number of Pareto solutions can be determined by tailoring the neighbourhood size and the number of iterations according to the characteristics of the Bees Algorithm. It was found that using a small neighbourhood with a large number of iterations leads to an increase in the number of Pareto solutions. Moreover the standard deviation in Table 7 is virtually zero showing that the Bees Algorithm is stable and robust.

Table 6
Best Emission
(a)Basic neighbourhood search

	Best Emission			Corresp. Fuel cost		
	Δ_{avg}	Δ_{min}	Δ_{max}	Δ_{avg}	Δ_{min}	Δ_{max}
LP	0.09	-0.02	0.34	-1.01	-1.94	-0.61
MOSST	0.12	0.01	0.37	-1.71	-2.63	-1.31
NSGA	0.03	-0.08	0.28	-0.96	-1.88	-0.56
NPGA	0.04	-0.07	0.29	-0.95	-1.88	-0.55
SPEA	0.11	0.00	0.36	-0.84	-1.77	-0.44
NSGAII	0.11	0.00	0.36	-0.81	-1.74	-0.41

(b)Random selection neighbourhood search

	Best Emission			Corresp. Fuel cost		
	Δ_{avg}	Δ_{min}	Δ_{max}	Δ_{avg}	Δ_{min}	Δ_{max}
LP	0.04	0.03	0.08	-0.47	-1.13	0.10
MOSST	0.07	0.06	0.11	-1.16	-1.82	-0.61
NSGA	-0.03	-0.03	0.02	-0.41	-1.07	0.15
NPGA	-0.01	-0.02	0.04	-0.40	-1.07	0.16
SPEA	0.06	0.05	0.10	-0.30	-0.96	0.27
NSGAII	0.06	0.05	0.10	-0.26	-0.92	0.30

Table 6 (cont'd)
(c) Weighted Sum neighbourhood search (w2=100000)

	Best Emission			Corresp. Fuel cost		
	Δ_{avg}	Δ_{min}	Δ_{max}	Δ_{avg}	Δ_{min}	Δ_{max}
LP	0.02	-0.02	0.03	-0.49	-0.91	-0.02
MOSST	0.05	0.01	0.06	-1.18	-1.60	-0.72
NSGA	-0.04	-0.08	-0.03	-0.43	-0.85	0.04
NPGA	-0.03	-0.07	-0.02	-0.42	-0.84	0.05
SPEA	0.04	0.00	0.05	-0.32	-0.74	0.16
NSGAII	0.04	0.00	0.05	-0.28	-0.70	0.19

(d)Weighted Sum neighbourhood search (w2=1)

	Best Emission			Corresp. Fuel cost		
	Δ_{avg}	Δ_{min}	Δ_{max}	Δ_{avg}	Δ_{min}	Δ_{max}
LP	0.12	0.03	0.24	-0.08	-1.19	0.95
MOSST	0.15	0.06	0.27	-0.78	-1.88	0.24
NSGA	0.06	-0.03	0.17	-0.02	-1.13	1.01
NPGA	0.08	-0.02	0.19	-0.02	-1.12	1.02
SPEA	0.14	0.05	0.26	0.09	-1.02	1.12
NSGAII	0.14	0.05	0.26	0.13	-0.98	1.16

Table 7
Standard deviation and average number of solutions

	Standard deviation		Avg. no. of solutions
	In Table 5	In Table 6	
(a)	0.17	0.00	21.50
(b)	0.07	0.00	107.20
(c)	0.08	0.00	108.80
(d)	0.01	0.00	95.70
Avg. STD	0.08	0.00	

Table 8(a) gives the power output for each generator and the total minimum fuel cost with the corresponding total NOx emission. Table 8(b) shows the minimum total NOx emission and the related fuel cost. Both the total minimum fuel cost and the total minimum NOx emission are lower than those obtained by existing approaches.

Ten runs of the Bees Algorithm consisting of 200 iterations each were conducted and the results of one run are shown in Figs. 6 and 7 demonstrating a good diversity in Pareto optimality. In Fig. 6, although the basic neighbourhood search outperforms in minimising Fuel cost, the Random Selection neighbourhood search provides far greater diversity. Fig. 7 depicts three Pareto sets from two different

Table 8
Minimum cost and minimum emission by the proposed Bees Algorithm

(a)		(b)	
P_{G1}	0.2220	P_{G1}	0.3953
P_{G2}	0.1084	P_{G2}	0.4596
P_{G3}	0.3061	P_{G3}	0.5456
P_{G4}	1.0156	P_{G4}	0.4061
P_{G5}	0.5224	P_{G5}	0.5265
P_{G6}	0.3621	P_{G6}	0.5010
Min. Cost ($/hr)	600.1201	Corresp. Cost ($/hr)	635.6809
Corresp. Emission (ton/hr)	0.2220	Min. Emission (ton/hr)	0.1942

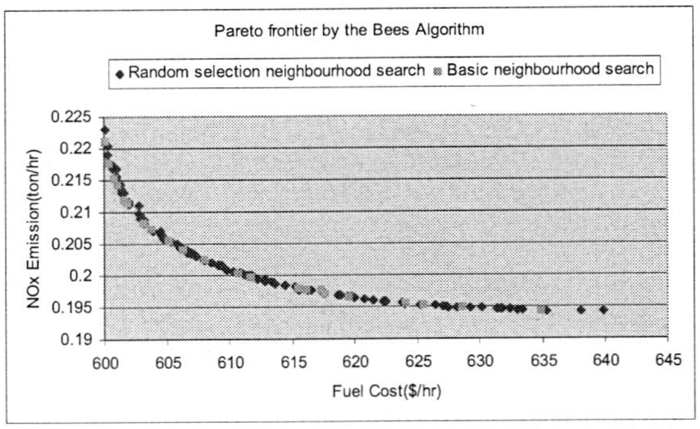

Fig. 6. Pareto frontier from Random Selection and basic neighbourhood search

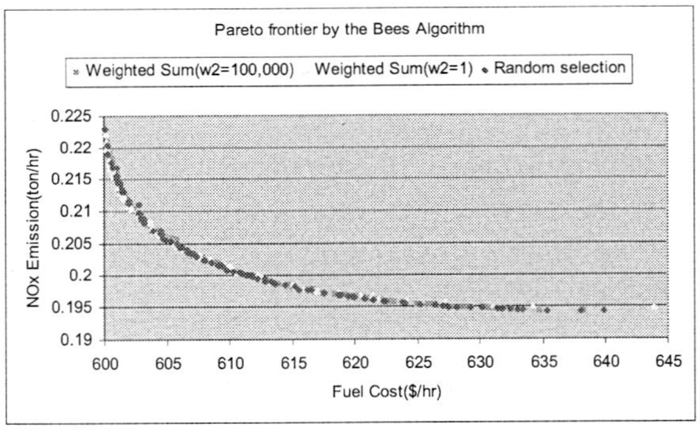

Fig. 7. Pareto frontier from Random Selection and Weighted Sum neighbourhood search.

approaches which are a Random Selection and a Weighted Sum neighbourhood search was run twice, the first weighing (w2=1) designed to minimise fuel cost and the second run (w2=100,000) aimed at minimal NOx emission. Fig. 7 proves the case for the Weighted Sum neighbourhood search approach, where w2=1 the minimum fuel cost was identified and where w2=100,000 the NOx emissions were minimised.

In summary, it has been shown that the Bees Algorithm can obtain the Pareto frontier for the EED problem to which it is ideally suited. Pareto optimality required just one single run for a satisfactory number of solutions, whereas the Weighted Sum [4] method needed numerous runs to obtain a suitable outcome.

6. Conclusion

This paper has described the application of the Bees Algorithm to the multi-objective Environmental/Economic Dispatch problem with a view to achieving Pareto optimality. The number of parameters utilised was half of that in the basic Bees Algorithm, these being the number of scout bees, the neighbourhood size and the number of recruited bees. Three different methods of conducting neighbourhood search were adopted, namely, the Basic, Random Selection and Weighted Sum methods. The algorithm was tested on the standard IEEE 30-bus system.

From the results obtained, all three neighbourhood search approaches delivered good outcomes. In particular, the minimum fuel cost and minimum emission obtained using the Weighted Sum neighbourhood search method were lower than those for all existing approaches. Moreover, the Bees Algorithm also gave a Pareto frontier with an excellent diversity for Pareto optimality whilst proving stable and robust with a virtually zero standard deviation.

This work has confirmed the suitability of the Bees Algorithm for solving the multi-objective EED problem, simultaneously achieving financial savings and reducing the emission of greenhouse gases into the atmosphere.

Acknowledgements

The authors are members of the EU-funded FP6 Network of Excellence for Innovative Production Machines and Systems (I*PROMS).

References

[1] Abido M.A. *A novel multi-objective Evolutionary Algorithm for Environmental/Economic Power Dispatch.* Electric Power Systems Research, 2003. **65**: pp. 71-81.

[2] Abido M.A. *A Niched Pareto Genetic Algorithm for multi-objective Environmental/Economic Power Dispatch.* Electrical Power and Energy System, 2003. **25**(2): pp. 97-105.

[3] Abido M.A. *Environmental/Economic Power Dispatch Using Multiobjective Evolutionary Algorithms.* IEEE Transactions on power system, 2003. **18**(4).

[4] King R.T.F.A., Rughooputh H.C.S., and Deb K. *Evolutionary Multi-objective Environmental/Economic Dispatch: Stochastic Versus Deterministic Approaches*, in *Evolutionary Multi-Criterion Optimization*. 2005, Springer-Verlag: Berlin Heidelberg. pp. 677-691.

[5] Pham D.T., Ghanbarzadeh A., Koç E., Otri S., Rahim S., and Zaidi M. *The Bees Algorithm -- A Novel Tool for Complex Optimisation Problems*, in *Intelligent Production Machines and Systems*. 2006, Elsevier Science Ltd: Oxford. pp. 454-459.

[6] Pham D.T. and Karaboga D. *Intelligent Optimisation Techniques: Genetic Algorithms, Tabu Search, Simulated Annealing and Neural Networks*. 1999, Springer Verlag: London, Heidelberg and New York.

[7] Dorigo M. and Stützle T. *Ant Colony Optimization*. 2004, MIT Press: Cambridge.

[8] Eberhart R., Shi Y., and Kennedy J. *Swarm Intelligence*. 2001, Morgan Kaufmann: San Francisco.

[9] Pham D.T. and Ghanbarzadeh A. *Multi-Objective Optimisation using the Bees Algorithm*, in *Innovative Production Machines and Systems*. 2007, Whittles: Caithness. pp. 529-533.

[10] Lee J.Y. and Haj Darwish A. *Multi-objective Environmental/Economic Dispatch Using the Bees Algorithm with Weighted Sum*, in *EKC2008 Proceedings of the EU-Korea Conference on Science and Technology*. 2008, Springer: Berlin Heidelberg. pp. 267-274.

[11] King R.T.F., Rughooputh H.C.S., and Deb K. *Stochastic Evolutionary Multiobjective Environmental/Economic Dispatch. Proc IEEE Congress on Evolutionary Computation*. 2006 pp. 946-953.

[12] Yokoyama R., Bae S.H., Morita T., and Sasaki H. *Multiobjective optimal generation dispatch based on probability security criteria.* IEEE Transactions on Power Systems, 1988. **3**(1): pp. 317-324.

[13] Das D.B. and Patvardhan C. *New Multi-objective Stochastic Serch Technique for Economic Load Dispatch.* IEEE Proceedings, 1998. **145**(6): pp. 747-752.

Innovative Production Machines and Systems
D.T. Pham, E.E. Eldukhri and A.J. Soroka (Eds.)

Acquisition of Discrete Component and Loading Information in the Component's Edge Region Using Innovative Sensor Technology

G. Mroz, W. Reimche, Fr.-W. Bach

Institute of Materials Science, Leibniz University Hanover
An der Universität 2, 30823 Garbsen, Germany

Abstract

During the life cycle of a product's components, identification problems arise owing to insufficient product information. Moreover, despite regular product maintenance, unforeseen component failures lead to operational breakdowns and, to some extent, to endangering people and the environment.

An inherent component tagging is implemented by locally changing the microstructural properties using local heat treatment. By this means, information about the component can be inherently stored in coded form in the component's edge region.

To detect and store the component's load history subjected to static and dynamic stresses, so-called sensor materials are suitable which exhibit load dependent microstructural transformations; such materials are austenitic manganese steels and chrome-nickel steels, which form stress or deformation induced martensitic microstructures under load.

Regarding component identification, the development of a high resolution sensor technology based on eddy-currents and harmonic analyses is presented for the detection and characterisation of locally introduced microstructural changes in the component's edge region.

A sensor technology was developed for the acquisition and evaluation of the component's endured loading history, subject to bending, tortional and multiaxial loads, as a function of the loading and number of load reversals. Using this technology together with a surface scan, a component's integral region and its local condition can be monitored with regard to the endured stresses and can also be evaluated as a function of the sensor material's properties.

Keywords: Component tagging, local heat-treatment, loading history, microstructure, phase change, eddy-current technology, harmonic analysis

1. Introduction

Up to the present time, insufficient possibilities of identifying components have led to numerous problems with respect to traceability, product liability as well as process and component optimisation in engineering companies. In particular, these problems arise during manufacturing, in operation and for the recovery or disposal of components. Thus, identification problems can arise during the component's operation due to the unavailability (or a lack) of product information. This can lead to cost-intensive enquires or recalls of entire product generations. For example, the quality standards according to DIN/ISO 9000 stipulate that products, particularly those which pose a risk to humans, conform to strict traceability requirements. Therefore, it should be possible at all times to access production relevant data, that is, genetic component information. Up to now, the basis of data acquisition relevant for traceability is, apart from the accompanying documents, an appropriate component designation. According to the present state of the art, this designation is essentially realised by using additional components such as labels with bar-codes or using RFID transponder technology [1].

With respect to identifying components, one currently observes a lack of procedures for evaluating the integrity of safety relevant, highly loaded components. Thus despite an increasing state of knowledge in research and technological areas as well as the implementation of designs supported by simulation and periodic tests of technical equipment, abrupt breakdowns frequently occur due to unforeseeable component failures. This leads to high follow-up costs for the plant concerned. Moreover, notwithstanding regular product maintenance, a risk of endangering people and the environment exists. Here, corrective measures are appropriate with which the component autonomously collects and provides information about its loading history. Such measures are, at present, still unavailable.

2. Objective

In order to be able to assess both the technical feasibility of the intended "genetically intelligent" (gentelligent) GI- component identification as well as the possibility of modelling the interrelationship between load sensitive material and component behaviour, initially at the laboratory scale, the objective is as follows; to integrate inherent codes, sensors appropriate to the problem and load sensitive materials into examples of GI-components. Based on this, actual GI-components for specific safety relevant applications, which are provided with gentelligent codes and the corresponding sensors, are then to be technically implemented and proven in the scheduled demonstrations of the collaborative research centre (Fig. 1).

3. Innovative Sensor Technology for Accessing Material Properties

To rapidly and non-destructively harvest the loading history from components and inserted tags, a new sensor technology is developed with regards to integrated and high resolution data acquisition. For this purpose, an innovative technology is implemented, which is based on eddy-currents and harmonic analyses, for the detection and characterisation of locally introduced microstructural changes in the component's edge region. The information content in the component

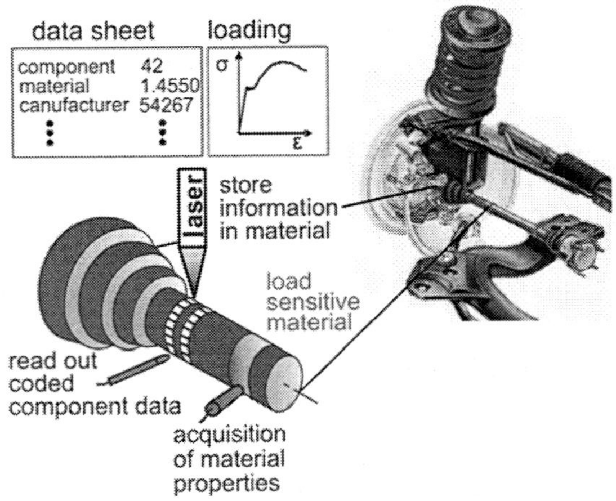

Fig. 1. Automotive drive shaft as a GI-component with gentelligent properties.

is accessed from its surface by scanning technology and the eddy-current signal information is graphically depicted from the measured level. This enables a local correlation of the eddy-current signal with the component's surface and a determination of the material properties using an analysis of the harmonic eddy-current signal units at the impedance levels (Fig. 2).

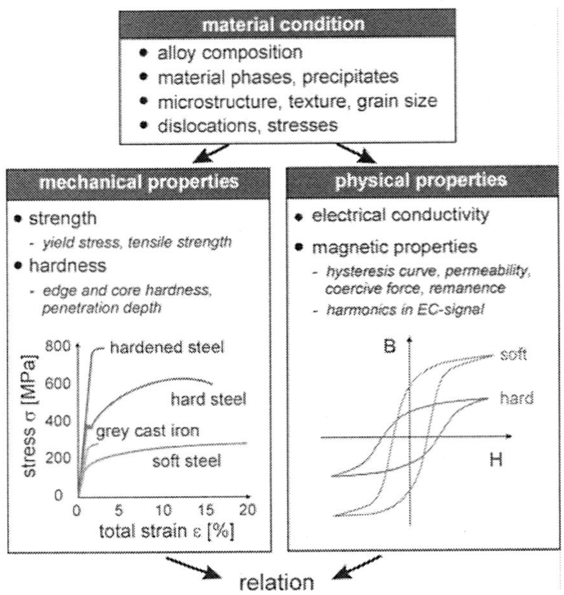

Fig. 2. Indirect determination of component properties.

To create the eddy-current signal to be harmonically analysed, an alternating magnetic field is generated by a sinusoidal alternating current in an excitation coil. This field interacts with the material. Depending on the test frequency f, the permeability μ, the conductivity σ and the test piece geometry, magnetising processes and eddy-currents develop in the material which lead to the creation of a secondary magnetic field (Fig. 3).

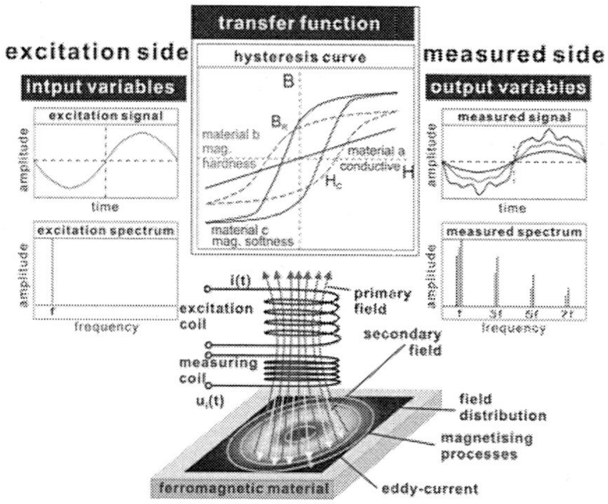

Fig. 3. Harmonic analysis of the eddy-current signal und the signal representation.

The superposition of both magnetic fields is detected using receiver coils, whereby the induced measured voltage is distorted due to the non-linear relationship between the magnetic field strength H and the magnetic flux density B. Owing to this distortion, the measured signal contains higher harmonic signal units which can be analysed in terms of their amplitudes and phase values or their real and imaginary parts [2]. By a selective acquisition and evaluation of the higher harmonics of the eddy-current signals, the change in magnetic parameters can be sensitively detected in ferromagnetic materials as well as the change in the phase transformation in austenitic materials owing to the formation of martensite [4,6,7].

4. Inherent Component Tagging

In contrast to existing information carriers and systems like barcode-labels, transponder-chips or structured surfaces, the edge region of a metallic component itself is to serve as the carrier of permanently stored, genetic component information. Besides the dispersal of particles with specifically desired properties (e.g. wear resistance, hardness, magnetic properties), components' edge regions can also be modified with respect to their hardened or tempered properties by mechanical cold-forming or, alternatively, by a local heat-treatment.

A high-energy fibre laser is employed which creates a high energy focused laser beam. This is used for locally surface heat-treating components, melt alloying the edge region or particle dispersal.

Introducing thermal energy by using, for example, a laser focused on a component's edge region which has been hardened and tempered or cold-work hardened, a tempered or recrystallised microstructure arises with locally modified electrical and magnetic material properties (Fig. 4).

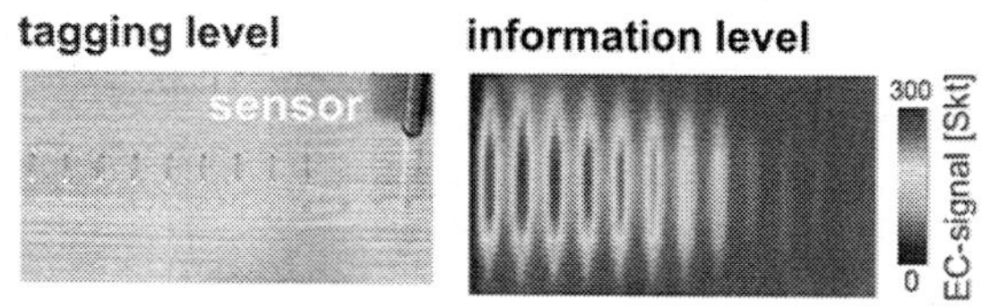

Fig. 4. Localised microstructural changes from locally introducing energy (left) and eddy-current surface scan (right).

These significantly modified physical properties of the material microstructure in the component edge region can be read out using a high resolution eddy-current sensor. Besides the level of microstructural modification, its penetration can also be specifically modified by varying the intensity and duration of the energy input (tempering temperature, depth effect).

This approach also enables a control of the extent and depth of the microstructural changes to be used as additional informational dimensions to augment the processed 2-dimensional component surface for carrying information. Tests carried out (at the Institute of Materials Science) have confirmed the practical applicability of the process which is particularly suitable for inherent component tagging. Ongoing research work (at the Institute of Materials Science) in the area of material characterisation, which uses non-destructive testing technology, concerns both the physical relationships between the changes of electrical and magnetic properties due to the manufacturing and the operational conditions, as well as the mechanical-technological material characteristics for testing the quality of components [9].

5. Requirements Concerning Material Selection

Metastable manganese and chrome-nickel steels are suitable as sensor materials for registering component overloads. In order to provide both safe and reliable confirmation of the stress which a component was subjected to and to evaluate the components condition, these materials are to directly store information about this loading history in the component. Here, the face-centred cubic lattice is transformed into a tetragonal, distorted body-centred cubic martensitic lattice caused by the shearing of the crystal due to plastic deformation. The phase transformation of the metastable austenite, with paramagnetic properties, into α' martensite, with ferromagnetic properties, can be determined using non-destructive, magneto-inductive measuring procedures [3].

6. Characterisation of Strain Induced Phase Transformations

The investigations to characterise strain induced phase transformations in austenitic steels were carried out on both quasi-statically loaded flat tensile specimens and on low cycle fatigue (LCF) specimens. In the following, the test method is described and the results ensuing from the investigations are presented.

6.1. Determination of the sensor material's behaviour subject to quasi-static loading.

To determine the phase changes from metastable austenite into martensite in austenitic materials subject to plastic deformation, tests on flat tensile specimens were carried out with a high resolution eddy-current sensor using scanning technology. For this purpose, specimens of X 6 CrNiTi 18 10, material No: 1.4541, were loaded in a test machine and stepwise plastically strained from 5 to 25 % in the gauge length region. To confirm units of martensitic phase material, the specimen surface was scanned using an eddy-current sensor guided by a handling system. The measured values of the location dependent eddy-current signals detected were prepared and depicted as scan images. Thus, it is possible to graphically depict and relate the intensity as well as the local elongation and orientation to the formation of martensite from the eddy-current indicators on the component surface.

The material X 6 CrNiTi 18 10, material No: 1.4541, shows a high sensitivity particularly at deformation levels of 15 to 25 % regarding the formation of martensite and an almost linear relationship between the plastic strain and the eddy-current signal's third harmonic (Fig. 5).

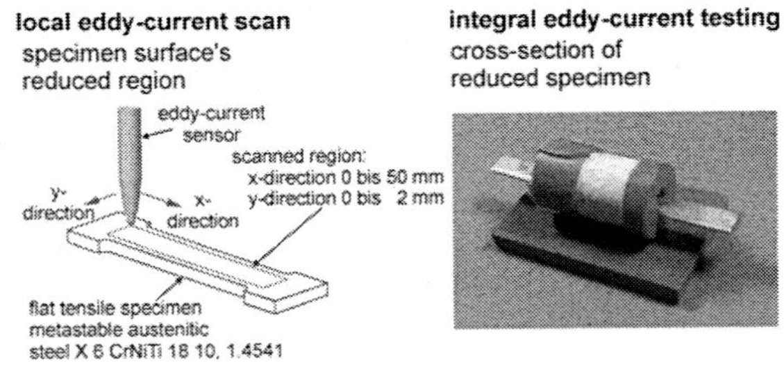

Fig. 5. Schematic test setup.

6.2. Investigation of the sensor material's behaviour subject to dynamic torsional and bending loads

For the investigation, acquisition and evaluation of martensite formation in metastable austenitic materials subject to dynamic loading with elastic and plastic deformation, reduced cylindrical specimens were subjected to bending and torsion using a suitable fatigue testing rig. The dynamic loading tests were carried out in resonance test machines in which cylindrical specimens were subjected to pure alternating loads R=-1 in bending and torsion (Fig. 6).

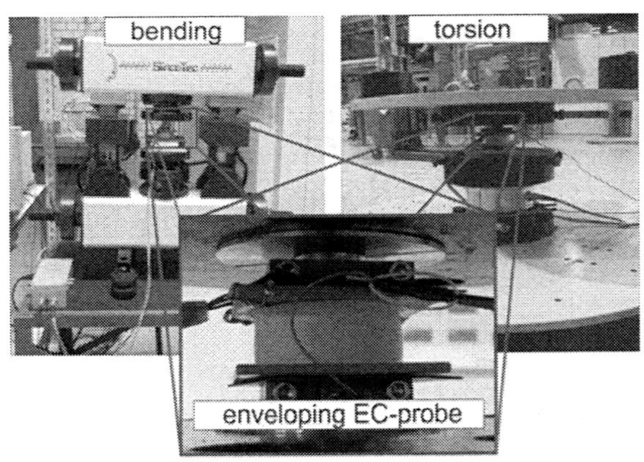

Fig. 6. Eddy-current equipment in the bending and the torsion testing rig.

In the loading test, the reduced specimen region was monitored using an enveloping eddy-current sensor with regards to the martensite formation as a function of the loading and the number of load cycles. To detect local martensite formation in the cylindrical specimens as a function of the loading condition in relation to both material inhomogeneities as well as asymmetrical stress distributions, the specimens were removed and their surface regions were surveyed in a SCAN-device using a high resolution eddy-current probe (Fig. 7).

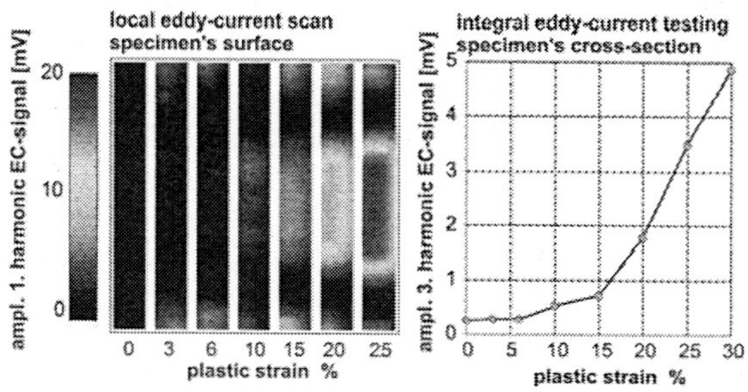

Fig. 7. Eddy-current signal changes as a function of the plastic strain in flat tensile specimens.

The specimens were subjected to cyclic bending moments in steps from ±100 Nm to ±370 Nm each for 10^5 load reversals (LR) to establish at which stress or strain induced loading, martensite formation initiates. Following the initiation of strain induced, local martensitic formation at bending loads above ±350 Nm, the cylindrical specimen was further loaded into the range from ±370 bis ±400 Nm each for $2x10^4$ LR up to fracture. An incipient crack occurred in the cylindrical specimen at a bending load of ±400 Nm. A representation of the loading history is shown in Fig. 8 by means of the signal change in the eddy-current's third harmonic as a function of the bending load. At low bending loads up to ±300 Nm, the specimen's deformations are reversible and the loading stresses lie in the elastic Hookean range. Low loading stresses in the region I lead to elastic deformations and have no influence on martensite formation. The material remains paramagnetic.

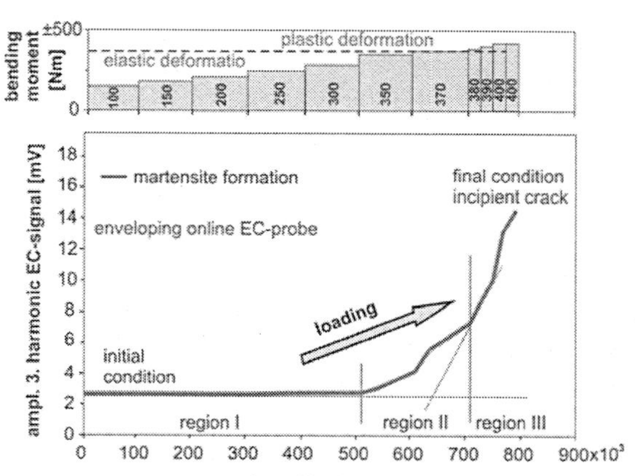

Fig. 8. Martensite formation in cylindrical specimens subject to dynamic loading in the bending fatigue testing

At bending moments of ±350 to ±370 Nm (region II), the transition from elastic to plastic deformation takes place. At the edge region of the specimen subject to dynamic loading, local plastic deformation occurs and, as a result, martensite forms. At high loads above ±380 Nm, extensive plastic deformation occurs subject to dynamic loading which results in larger rates of martensite formation. The material exhibits increasing ferromagnetic properties.

The change in the eddy-current's first harmonics as a function of the bending and torsional loading and of increasing number of loading cycles is depicted in Fig. 9. At a low test frequency of 200 Hz, the microstructural changes caused by the plastic deformation were detected from the change in the cylindrical specimen's magnetic properties. At bending loads of ±350 Nm or greater, a significant increase of the α'-martensite can be observed. In the dynamically loaded bend specimen, high tensile and compressive stresses with plastic deformation occur as a result of the bending stress distribution particularly alternating between 90° and 270°. This plastic deformation leads to extensive martensite formation.

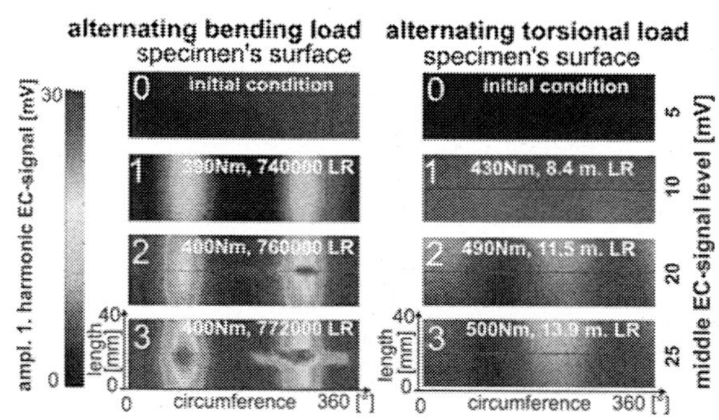

Fig. 9. EC-signal changes of the first harmonics as a function of the dynamic bending and torsional loading,

Using the eddy-current signal amplitudes in the (eddy-current) scans of specimen X 6 CrNiTi 18 10, material No: 1.4541, increasing martensite formation favouring the principal load directions 90° and 270° is detectable with increasing dynamic loading and number of load reversals. At 772 000 LR, crack initiation occurs that subsequently leads to fracture of the cylindrical specimen.

Under torsional alternating loads of ±430 Nm, a local change in the material properties can also be identified due to martensite formation with increasing load and number of load reversals. In contrast to the bending loading, the cross-section of the specimen's material is subjected to shear stresses under torsional loading. In this loading case, the load's stresses are distributed quite uniformly in the specimen's edge region and produce relatively uniform martensite formation with increasing load and number of load reversals. Here, small material inhomogeneities arising from the specimen's manufacture have an influence on the metastable material conditions and therefore merely on the martensite formation.

7. Conclusion

During the life cycle of a product's components, identification problems arise owing to insufficient product information. Moreover, despite regular product maintenance, unforeseen component failures lead to operational breakdowns and, to some extent, to endangering people and the environment.

These risks can only be limited using conventional monitoring technology by expending large amounts of time and money. Here, modern methods for inherently tagging and rapidly, non-destructively harvesting the loading history from components can offer the prospect of identifying and evaluating a component's condition and thereby estimating the risk as well as reducing the breakdown costs.

An inherent component tagging is implemented by locally changing the microstructural properties using local heat treatment. By this means, information about the component can be inherently stored in coded form at the component's edge region.

To detect and store the component's load history subjected to static and dynamic stresses, so-called sensor materials are suitable which exhibit load dependent microstructural transformations; such materials are austenitic manganese steels and chrome-nickel steels which form stress or deformation induced martensitic microstructures under loading.

Regarding component identification, the development of a high resolution sensor technology based on eddy-currents and harmonic analyses is presented for the detection and characterisation of locally introduced microstructural changes in the component's edge region.

For the acquisition and evaluation of the component's endured loading history; i.e. subject to bending, tortional and multiaxial loads, as a function of the loading and number of load reversals, a sensor technology was developed. Using this technology with a surface scan, a component's integral region and its local condition can be monitored with regard to the endured stresses and also be evaluated as a function of the sensor material's properties.

Acknowledgements

Our thanks go to the German Research Foundation for supporting this work within the framework of the collaborative research centre 653 "Gentelligent components' life cycle".

References

[1] Vincenz M. Möglichkeiten und Grenzen heutiger Transpondertechnologien in der Logistik, Fördern und Heben, Vol. 53, 2003, p. 540-542

[2] Reimche W. Bernard M., Scheer C., Bach Fr.-W. Zerstörungsfreie Bauteilprüfung in der Prozesskette Präzisionsschmieden, Sonderforschungsbereich 489, TP B6, Arbeits- und Ergebnisbericht 2000-2002.

[3] Bassler H.-J. Eifler D. Wechselverformungsverhalten und verformungsinduzierte Martensitbildung bei dem metastabilen austenitischen Stahl X6CrNiTi 18 10. Werkstoffkundliche Berichte, Lehrstuhl für Werkstoffkunde, Universität Kaiserslautern 1999.

[4] Sellger R. et al. Influence of strain rate on deformation behaviour, austenite structure and kinetics of the γ/α transformation of steel, Technical Steel Research, 2001.

[5] Reimche W. Weber W., Reichert Ch., Stegemann D. Nachweis von Ermüdung und Materialfehlern in dickwandigen Rohrleitungskomponenten mit elektromagnetischen Prüftechniken, 26. MPA-Seminar, 5. and 6. October 2000.

[6] Abrassart A. Stress-induced $\gamma \rightarrow \alpha'$ martensitic transformation in two carbon stainless steels application to trip steels, Metallurgical and MaterialsTransactions, Vol. 4, 1973, p. 2205-2216.

[7] Dietrich H., Heimann W., Strom H. Einfluß der Kaltverformung auf die mechanischen und magnetischen Eigenschaften von nichtrostenden Stählen, TEW-Technische Berichte, Vol. 2, 1976, p.61-69.

[8] Boll R. Weichmagnetische Werkstoffe - Einführung in den Magnetismus, Vacuumschmelze GmbH, Hanau, 4. Auflage, 1990.

[9] Feiste K.L. Entwicklung der Harmonischen-Analyse von Wirbelstromsignalen zur Charakterisierung mechanischer Kugelgraphitgusseigenschaften, Dissertation, Hannover, 2003

[10] Leng J. and. Asundi A. Structural health monitoring of smart composite materials by using EFPI and FBG sensors, Sensors and Actuators, Physical, Vol. 103, ISSN 0924-4247, 2003, p.330-340

[11] McMaster R.C. The present and future of eddy current testing. Materials Evaluation, Vol. 60, ISSN 0025-5327, 2002, p. 27-28,30-32,34,36-37.

Innovative Production Machines and Systems
D.T. Pham, E.E. Eldukhri and A.J. Soroka (Eds.)

Advanced production of direct applied wave guides

M. Dumke[a], T. Fahlbusch[b], L. Overmeyer[a]

[a] *Institute of Transport- and Automation Technology, Leibniz University of Hannover, An der Universität 2, 30823 Garbsen, Germany*

[b] *Institute for Measurement and Automatic Control, Leibniz University of Hannover, Nienburger Str. 17, 30167 Hannover, Germany*

Abstract

The application of optical structures in data transfer and measurement steadily increases. Polymer optical fibres (POFs) are superior to electrical conductors, due to their light weight, their resistance to damage and electromagnetic interference. Just like their glass equivalents they provide a high data transfer rate. By using direct dispensing 3D-light guiding structures on surfaces and the integration of light guiding structures into metallic components can be accomplished. The first part of the article describes the main properties of POFs and there influences of processing with dispensing technology. In the course of this article the technical implementation and structural properties are described. A short overview on printing of wave guides is given.

Keywords: optical polymer, direct writing of wave guides, dispensing technology

1. Introduction

A new production process for light guiding structures has been developed. Within this method it is possible to create light guiding structures made of polymer optical fibres on metallic surfaces. An application on 3D-shaped bodies is possible. By using direct dispensing, stepped index fibres can be produced. Therefore, the cladding material will be dispensed in the first step, afterwards the core material will be applied on top and finally the second cladding will cover the structure. In every step between the polymer is cured by UV-light. The integration of electro-optical components within directly dispensed structures is much easier. The preparation of end cross section surfaces is not necessary. Cutting down the production steps of cutting and polishing, the whole process can be economized [1]. Further investigations are related to a different production process. Within the knowledge of direct dispensing the topic on future investigations is the development of processes for printing polymer optical fibres. At this stage the potential of printing wave guides within gravure printing is shown.

2. Motivation

A main motivation for this project is the integration of light guiding structures directly in and on a 3D-shaped metallic surface. These structures should provide a high and secure data transfer, resistance to electromagnetic radiation and potential separation. The direct integration of electro-optical components is one of the goals of this project. Because of liquid processing the integration in direct dispensed structures is much easier, because there is no further preparation like cutting and polishing the end cross section necessary. In the next step alternative production processes should be carried out for low cost applications. Therefore, the idea of printing optical wave guides is the topic of future investigations. The knowledge of processing direct dispensed POFs will be used.

3. Required material characteristics

For direct dispensing of POFs the material characteristics can be divided in three different parts. Regarding light guiding characteristics the main optical aspect the refraction index has to be considered. With a refraction index difference of 4 %, for example a refraction index of 1.44 for cladding material and 1.50 for core material, the numerical aperture is 0.42. Different fibre parameters are dependent on the numerical aperture, for example injected light power and sensitivity for bended structures [2]. The transmittance should be 100 % for the guided wavelength. To secure resistance against estimated stress mechanical characteristics like temperature resistance have to be in a range of -40°C to 120°C. Different properties of POFs are important for their dispensing behaviour. Particularly the viscosity [3] and cure behaviour belonging to the most important facts [4]. These one component resins needs a low viscosity for the cladding and for core materials applied in 2D-trench structures. To apply core material on the surface it is necessary to use highly viscous polymers.

4. Technical implementation

High quality requirements regarding the dispensed structures of optical polymers refer to high demands in processing. Unevenness of the surface occurs in the range of up to 50 µm. As a result of a variation of dispensing distance during processing different effects can be seen. Long distances will cause separations of structures, which look like a row of pearls. If the distance is to short the dispensed structure will be destroyed by the needle. Differences of about 50 µm will produce structures with bad qualities. Both effects can be seen in **figure 1.**

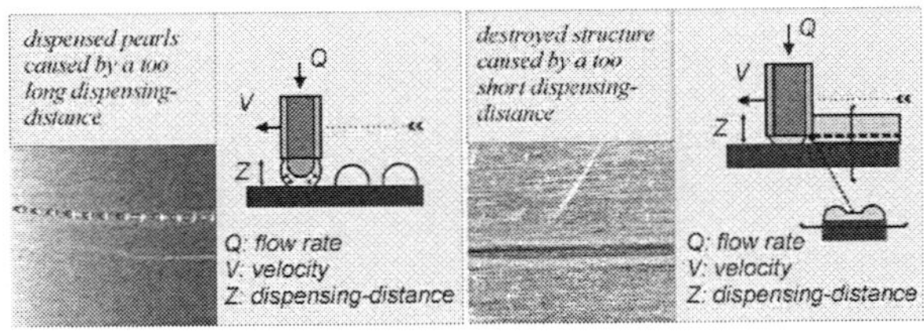

Figure 1: Effects of wrong dispensing distance

The implementation of a Z-resolution of 5 µm enables a control of the dispensing distance. For processing a surface scan is needed. The measured Z-profile is combined with the original CNC-file for producing light guiding structures. Within a cost-efficient-solution a XYZ-dispensing robot interprets these CNC-commands and corrects them with the measured Z-profile from the surface scan. **figure 2** shows the dispensing robot.

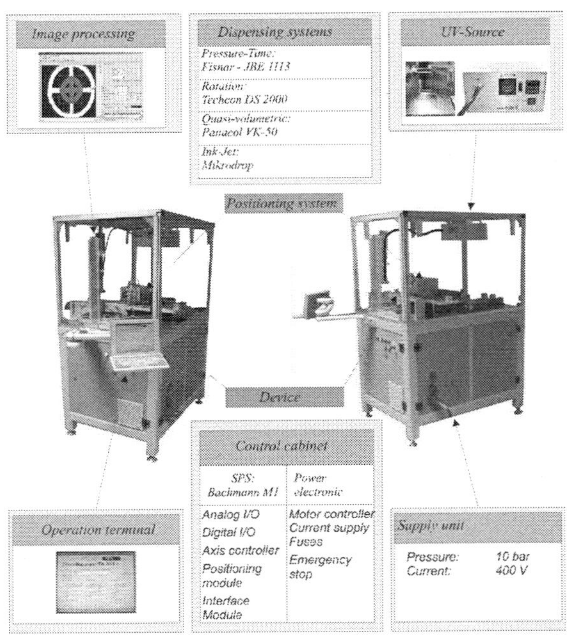

Figure 2: Dispensing robot

If the dispensing needle for 3D-applications has an orthogonal position to the surface, two additional rotation axes for the robot will be needed, which means an increase in production costs and handling efforts. Instead of using the orthogonal positioned needle the surface tension and contact angle (angle at which a liquid/vapor interface meets the solid surface) is used to allow self alignment. This contact angle is unique for every material combination which depends on the used polymer and the device material. Therefore, the acceptable pitch of the device surface has to be determined. For a flexible production process it is definitely a disadvantage that every material combination has another contact angle. In order to avoid different core-geometries by using different material combinations the cladding can be applied by a coating process. In this case the material combination is reproducible and the contact angle is independent of the device material.

With direct dispensing light guiding structures can be produced in different ways. They can be integrated in a device, can be applied on a device surface, or can be produced as dam structures. Integrated wave guides and their production steps are shown schematically in **figure 3**.

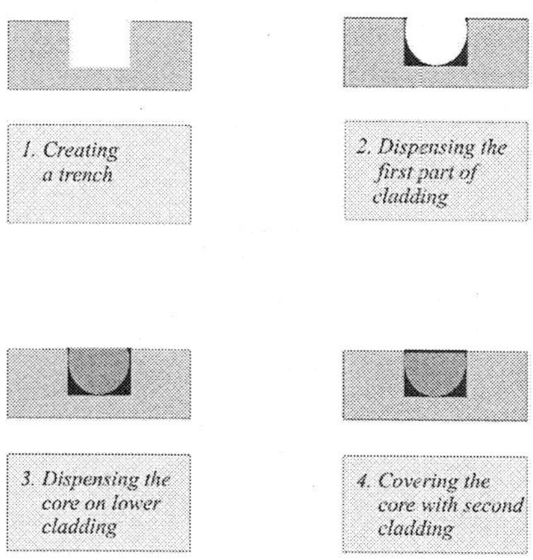

Figure 3: Production steps for surface integrated light guiding structures

After creating the trench structure in step one the first cladding will be dispensed. Due to low material viscosity and surface tension the adhesive is allocated regularly in the trench. In a third step, the core material is filled in the cured lower cladding. By covering the core with the second part of the cladding the structure is closed. After each step the applied polymers have to be cured with UV-radiation. **Figure 4** shows the four necessary production steps to create a surface applied light guiding structure.

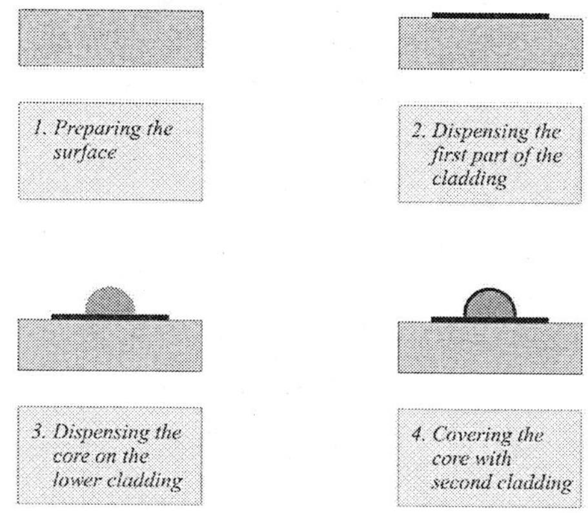

Figure 4: Production steps for surface applied light guiding structures

After preparing the device surface in step one, the cladding is dispensed directly on the surface area. Small irregularities can be compensated by surface tension of the lower cladding material. Before adding the core material on top the lower cladding has to be cured. Because of high requirements regarding the stability of profile shape a high viscous material has to be used. Processing parameters like dispensing speed and flow rate must be constant. After the core is cured by UV-light the second cladding to cover the core will be dispensed, finished by a final curing step. A different method to apply light guiding structures on surfaces is the dam structure, shown schematically in **figure 5**.

Parallel structures, to separate transmitter and receiver channel can be produced as dam structures. After cleaning the surface from different dust and grease deposits the cladding (high viscosity of material necessary) is created on the surface in shape of two side by side lying polymeric crawlers. As a result of surface tension there will be a curved shape created between these parallel structures. With this procedure it is possible to compensate scratches in the surface. The space between both structures will carry the core material. Requirements regarding the viscosity of core materials are not high. It is possible to use high and low viscous materials. After applying the core, the structure is covered by another cladding layer. The advantage of dam structures is the comparatively lower demand of position accuracy. Of course the distance of these parallel dam structures has to be constant.

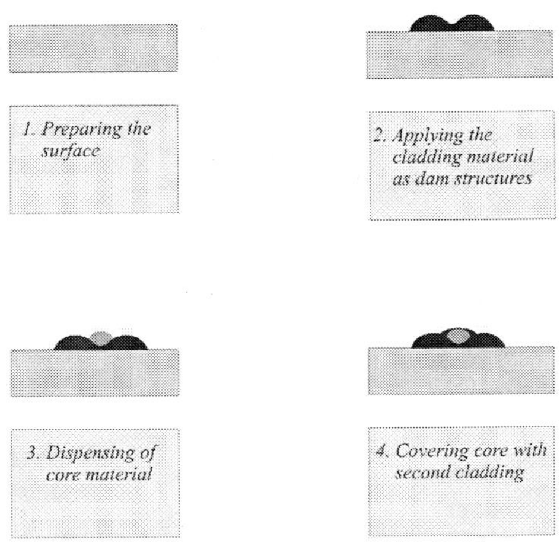

Figure 5: Production steps for surface applied dam structures

For surface applied light guiding structures dimensions of approximately 400 µm widths and 250 µm heights can be produced. This method is limited to a maximum radius for bended structures. To fulfil the requirements of total reflexion the maximum radius is described by the critical angle alpha total [5]. In case of exceeding alpha total an optical leak will be the result.

5. Structural properties

To ensure that the light is led through the fibre due to total reflection, the influence of the cladding material in two different straight structures is examined. One structure is produced without and the other structure with a cladding. The test shows that the radiation output for uncoated wave guides, measured in mV, has a much lower value and only a small peak area which can be seen in **figure 6**. The measured output of uncoated light guiding structures has a voltage of 350 mV. The compared structure with cladding has a higher output in generally from 800 mV and a bigger peak area. Furthermore there was carried out an investigation about curved structures with a radius up to 35 mm.

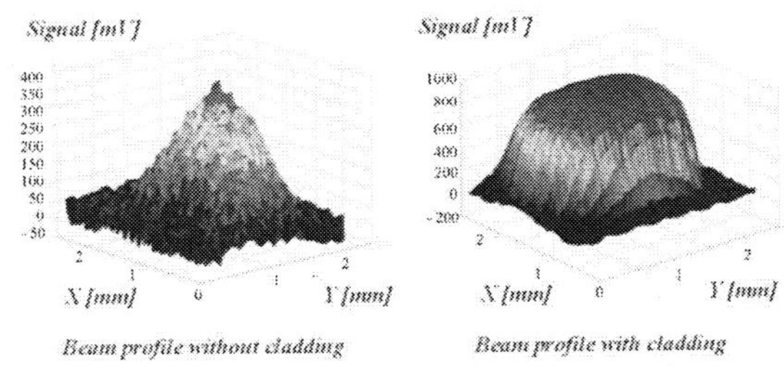

Figure 6: Radiation output of light guiding structures with and without cladding

The cross section reached dimensions from 2.00 mm to 0.70 mm. Over a distance from 150 mm the optical signal can be guided inside the structure. Compared to straight light guiding structures there is a loss of 40 % in transmission by using curved structures with a radius of 35 mm. Attenuation is measured by the cut-back-method, which means cutting the pattern in slices and comparing the outcome of every slice [6]. The investigated values are between 0.06 db/cm and 0.16 db/cm [7].

6. Printing technology

Differentiation between the main four printing processes is shown schematically in **figure 7.** The transfer of the printed media (black highlighted) depends on the difference of printing parts and nonprinting parts within the printing plate.

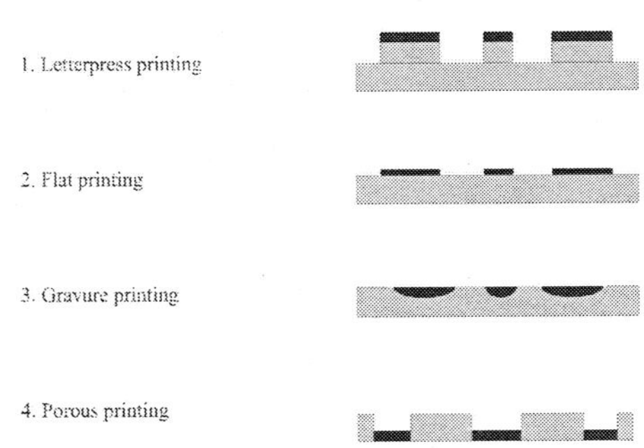

Figure 7: Main printing processes

These construction items result in the medial film thickness that can be transferred. The biggest film thickness can be achieved by using gravure printing and porous printing. By using stencil printing, which belongs to the group of porous printing processes wave guides were produced. The processing steps are quite similar to the processing of dispensed structures on surfaces. In the first step the lower cladding is applied to an even surface by a spray coating process. Afterwards the core material can be applied by stencil printing on the lower cladding. Therefore the printing plate is adjusted on the device and the optical polymer is applied. To guarantee the needs of a stable shape a high viscous polymer is used. Finally, the second cladding applied by spray coating covers the core material. Between every step the materials have to be cured by UV-light. An advantage of this technology is the application for low cost production. The printed wave guides are reproducible in the same shape and quality.

7. Summary

The presented technology shows the feasibility to create light guiding structures directly on and in a 3D-surface by using dispensing systems. Also it is possible to integrate this technology in a fully automated process. Also the feasibility of printing wave guides by using stencil printing was shown. For future applications, the light guiding structures can be integrated in automotive application. It is possible to create optical sensors by using the described technology.

References

[1] Fahlbusch, T.: Dispensieren polymerer Lichtwellenleiter, Berichte aus dem ITA, Band 01/2007

[2] Daum, W.: POF – Polymer Optical Fibers for Data Communication, Springer Verl., Heidelberg, Deutschland, 2002.

[3] Bachmann, A. G.: Advances in Light Curing Adhesives International Symposium on Optical Science and Technology, San Diego, CA, 1. August 2001, Paper 4444-20.

[4] Orthmann, K.: Kleben in der Elektronikanwendung, Aufbau, Applikation, Qualität, Zuverlässigkeit. expert Verl.; Renningen Malmsheim, Deutschland, 1995.

[5] Ziemann, O., Krauser, J., Zamzow, P. E., Daum, W.:POF Handbuch Optische Kurzstrecken-Übertragungssysteme, Springer Verlag 2001

[6] Eberlein, D.: Lichtwellenleiter-Technik, Expert Verlag 2003

[7] Fahlbusch, T., Franke, S.: Direct dispensing of 3D light guiding structures, 15th International Conference on Plastic Optical Fiber, Seoul, Korea, 2006

Innovative Production Machines and Systems
D.T. Pham, E.E. Eldukhri and A.J. Soroka (Eds.)

A Procedure for the Sensitization of α'- Martensite Evolution

B.-A. Behrens, K. Weilandt

Institute of Metal Forming and Metal-Forming Machines (IFUM)
Leibniz Universität Hannover
An der Universität 2
30823 Garbsen, Germany
weilandt@ifum.uni-hannover.de

Abstract

Timely identification of damage is crucial for the secure use of sheet metal components in lightweight constructions. This work offers a new non-destructive method that allows for damage identification considerably earlier than achieved by conventional inspection. The method is based on a new concept of part-inherent sensing capability. These sensors are small areas on a metastable austenitic steel sheet, which have been plastically deformed by means of metal forming. Due to the plastic deformation, these areas are sensitive to martensite evolution under applied elastic strains, and this sensitivity facilitates the correlation between the martensite content of the sensor and the strains it has undergone.
A mathematical description for the martensite evolution was analysed, based on the material model of Springub. This investigation was performed to clarify the influence of chemical fluctuations caused by different batches on the martensite evolution of the metastabile austenitic steel X5CrNi18-10 (EN 1.4301). In addition, a mathematical relation between the minor/major strain and the martensite content was derived.

Keywords: Martensite Evolution, Metal Forming, Stamping

1. Introduction

The industry expresses an increasing interest in information on manufactured components during their service life.

At the Institute of Metal Forming and Metal-Forming Machines (IFUM) investigations of metastable austenitic sheet metal components concerning the use of their base materials as sensors are done. Measuring the martensite evolution at sensitized reference points (in the following called sensors) of these components makes it possible to draw conclusions about the strain to which the components are subjected to. The local sensitizations are realized by imposing defined elongations and residual stresses onto the material using a metal forming process. This results in a component-inherent sensor acting as a damage-control element.

The sheet metal components occasionally exposed to high stresses are not equipped with additional sensors, but the material of the components itself stores all the information on past and present strains during the entire service life [1]. The important advantage of this approach is that strains and damage of the component can be diagnosed before they would be detected by a visual inspection. To achieve this, a pre-defined basic martensite content must be set at first. Its subsequent evolution by elastic and plastic strains is part of the investigations described in this paper, see also [2, 3] for further details.

2. Martensite Evolution

Because of their high corrosion resistance and deformation characteristics, stainless steels are of high importance for industrial applications. These materials are subjected to a phase transformation. The transformation of austenite to martensite occurs during deep-drawing processes. The martensite evolution causes a rising flow stress and higher strength of the material (Fig. 1).

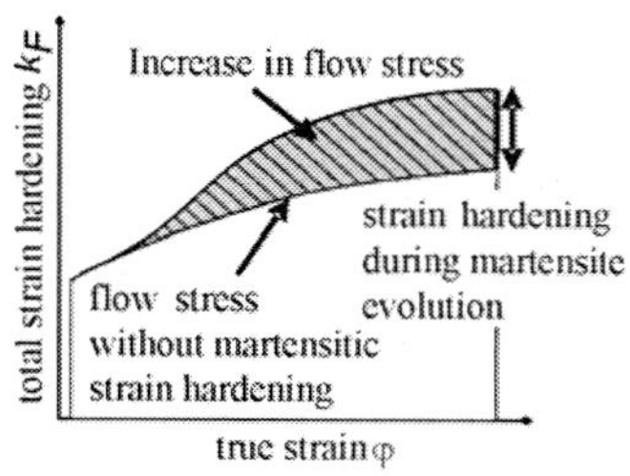

Fig. 1. Strain hardening caused by martensite evolution

The martensite microstructure can exist in two different lattices, shown in Fig. 2. The ε-martensite is characterized by a hexagonal packing lattice (hcp). Tetragonally distorted martensite shows a body-centered cubic lattice (bcc), which is called α'-martensite and shows ferromagnetic properties similar to α-ferrite ε-martensite does not have the same strength and thermodynamic stability as α'-martensite [4]. Due to these differences in material behavior, only the formation of α'-martensite is favored.

Stainless steel varies in its chemical composition. In this paper, the influence of different batches of the material 1.4301 is investigated.

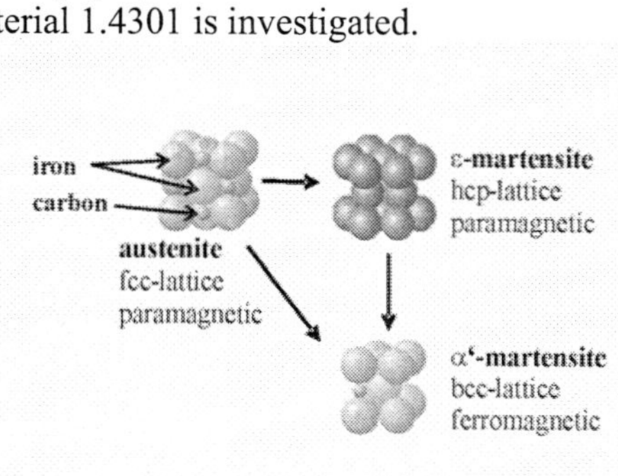

Fig. 2. Mechanism of martensite evolution

3. Experimental Investigations and Mathematical Formulation

Based on numerical and experimental studies involving the metastable austenitic steel X5CrNi18-10, a material model was developed and qualified at the IFUM [5]. The transformation depends on alloying constituents, transformation temperature, stresses, and strains, and thus, all of these factors had to be considered in the design of deep-drawing processes of stainless steels.

Below, a mathematical model for the calculation of the martensite evolution depending on these parameters is presented. This model is based on Tsuta [6] and considers the material's forming history [7].

In detail it includes the formation of α'-martensite depending on the temperature *T*, the true strain φ, the stress tensor $\underline{\underline{\Sigma}}$, and the stress condition in terms of the ratio between the principal stresses σ_1/σ_2. The calculation is done incrementally, but mostly independent of the step size in time. It is based on the following equations (Fig. 3), with the material constants *A, B, C, D, E, F, G, H (A-D* for the uniaxial, *A-H* for the biaxial stress condition*)*. The martensite evolution for the uniaxial stress condition as well as for the biaxial one can be described by this model.

The equation for the biaxial stress consists of two summands. The first one is the uniaxial ratio. The second is a product of the functions of the relevant influencing factors for the biaxial stress. The considered factors are the microstructure arrangement, the temperature *T*, the deformation degree φ, and the state of stress σ_1/σ_2.

$$\left(\frac{\partial f^{\alpha'}}{\partial \varphi}\left(f^{\alpha'},T,\varphi\right)\right)_{uniaxial} = \left(1-f^{\alpha'}\right)\cdot\frac{C\cdot\varphi^{-1}}{1-A\cdot e^{D+\frac{B}{T[K]}}\cdot\varphi^{-C}} \tag{1}$$

$$\left(\frac{\partial f^{\alpha'}}{\partial \varphi}\left(f^{\alpha'},T,\varphi,\underline{\underline{\Sigma}}\right)\right)_{biaxial} = f\left(f^{\alpha'},T,\varphi\right)_{uniaxial} + E\cdot\left(1+f^{\alpha' F}\right)\cdot\frac{e^{G-T[K]}}{1+e^{G-T[K]}}\cdot\sqrt{\varphi}\cdot\left(1+H\cdot\mathrm{signum}\left(\frac{\sigma_2}{\sigma_1}\right)\right) \tag{2}$$

In the following passage the experimental investigations and the method to gain the material parameters are described. The main aim of this investigation is to quantify the differences of behavior concerning martensite evolution caused by fluctuations between the batches, and thus the transferability for all material batches has to be verified.

The flow curve and the martensite fraction of five

different batches of the sheet metal material X5CrNi18-10 were derived by uniaxial tensile tests on the testing machine DYNA/Mess S100/ZD available at the IFUM. Two sheet thicknesses ($s_0 = 1.0$ mm and $s_0 = 0.8$ mm) were examined. The measurement of martensite fraction was based on the magnet-inductive measuring method with the aid of the Feritscope® MP3C. The results of the measurements and the approximation are shown in diagram martensite content over deformation degree (Fig. 3.) The curves for the five batches show differences. Particularly batch 1 has a remarkably lower martensite content. Batch 2 and 5 are the samples with the material thickness of $s_0 = 0.8$ mm.

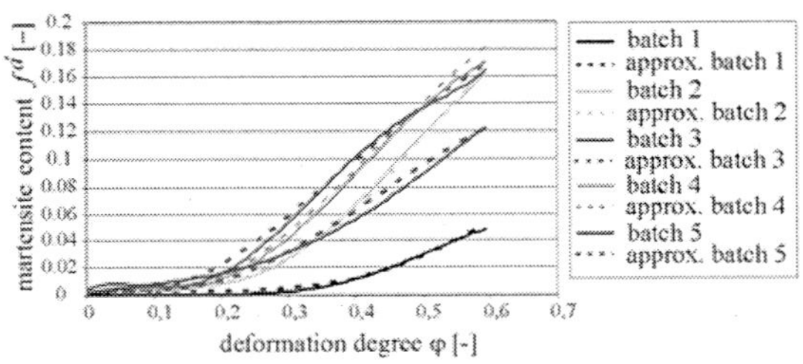

Fig. 3. Real and approximated content of martensite for the analysed batches

The incremental description of the martensite evolution with the constants *A, B, C,* and *D* (Fig. 3.) was in order to enable description of *t* martensite evolution in the tensile test. The material parameters of the developed model were determined according to the minimum square deviation method by means of the program Mathematica®. A good description of the martensite evolution as a function of the temperature and the plastic strain was obtained using the material parameters in Table 1.

Table 1. Uniaxial material parameters *A-D* of the five investigated batches

material	batch	thickness [mm]	A	B	C	D
1.4301	1	1.0	-0.043	-4988	6.24	18.26
1.4301	2	0.8	-6.53E-05	-5000	4.39	24.64
1.4301	3	1.0	-1.09E-05	-5325	3.63	24.85
1.4301	4	1.0	-1.09E-05	-5337	4.44	24.18
1.4301	5	0.8	-1.31E-05	-4998	3.89	24.27

In order to quantify the stress-dependence of the martensite evolution, deep-drawn rotationally symmetric cups were analyzed. During the deep drawing process, different stress and strain conditions occured (tensile stresses in radial direction and compression stresses in tangential direction in the flange, plane strain in the wall and tensile stresses in the bottom of the component), whose influence on the martensite evolution was examined.

The tests using the material of batch 1 with the initial sheet thickness $s_0 = 1.0$ mm were deep-drawn by a tool with a stamp diameter of $d_0 = 200$ mm, a blank holder force $F_N = 200$ kN and a temperature of 20°C. The cups of the material of batch 2, 3, 4, and 5 were deep-drawn on a rotationally symmetric tool system with the stamp diameter $d_0 = 50$ mm. The influence of the tensile stresses occurring in the bottom of the cup on the martensite fraction can be examined in this geometry. The martensite fraction in the cups was measured magnet-inductively at reference points in the rolling direction. Since the examined stainless steel shows small anisotropy, the martensite fraction is assumed to be corresponding in the whole cup [8]. The results of the experimental investigations and the mathematical approximated curves are shown in Fig. 4.

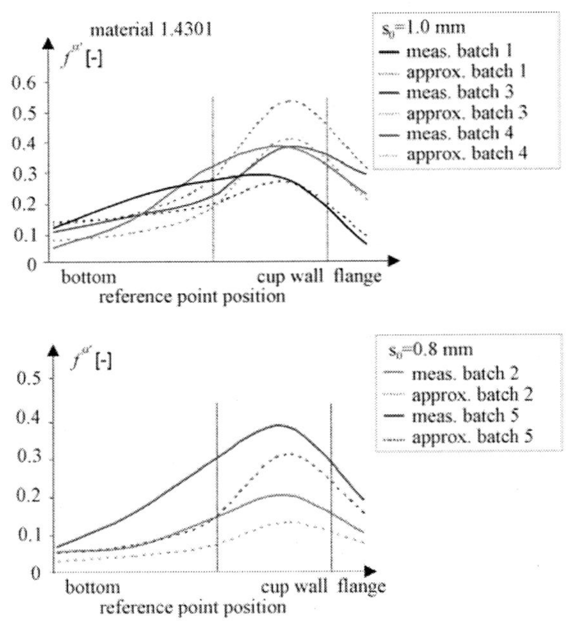

Fig. 4. Comparison of approximated and measured martensite content over cup contour for both thicknesses

The investigations show that the martensite content equation for the biaxial stress state (1) and (2) fits with the same material parameters *E, F, G, H* for the batches 1 to 5, shown in Table 2. This phenomenon occurs although the material behavior in the uniaxial tensile test shows significant differences.

It leads to the assumption that the influence of the biaxial stress for martensite evolution compensates the differences of the uniaxial one. Thereby the biaxial material parameters *E-H* of the five investigated batches are identical.

Table 2. Biaxial material parameters *E-H* of the five investigated batches

material	batch	thickness [mm]	E	F	G	H
1.4301	1	1.0	1.45	0.33	317.15	0.35
1.4301	2	0.8	1.45	0.33	317.15	0.35
1.4301	3	1.0	1.45	0.33	317.15	0.35
1.4301	4	1.0	1.45	0.33	317.15	0.35
1.4301	5	0.8	1.45	0.33	317.15	0.35

Basic experiments were done with the material X5CrNi18-10 to get information about the relations between the minor/major strain and the martensite content. The Marciniak test, used for forming limit diagrams, was extended by the martensite content. The strain and the martensite content are evaluated in the middle area of the specimen, which is a non-friction area due to the tool geometry. The results are presented in the following figures. The martensite content is depicted over minor and major strains in the 3D diagram of Fig. 5. To find an adequate plane that maps the state of strain more data points are needed, that is part of actual investigations.

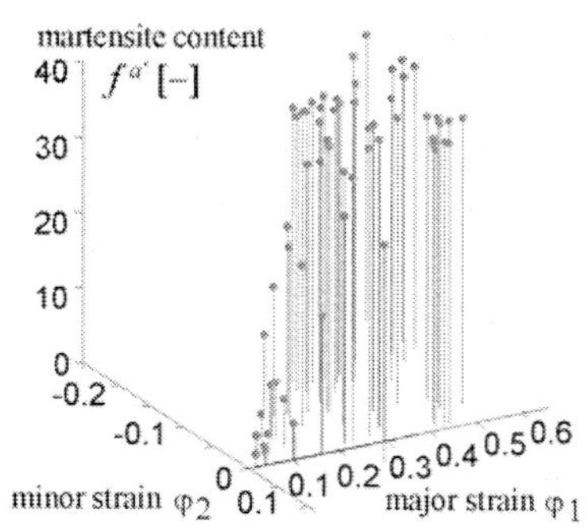

Fig. 5. Martensite content as a function of minor and major strain

Fig. 6. a 2D illustration depicts the martensite content as a functional of the effective strain. The mathematical relationship between minor, major and effective strain is:

$$\varphi_V = \sqrt{\frac{2}{3}(\varphi_1^2 + \varphi_2^2)} \tag{3}$$

Following, an equation and its curve were derived to approximate the gained data points [3]. For our investigation with the parameters a = 36.91, b = 11.89, c = 58.91, d = 3.42 best correlation is gained. The equation has to be investigated concerning the batch stability.

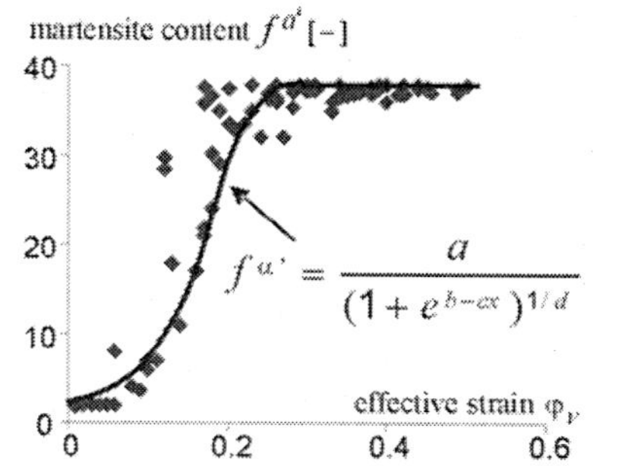

Fig. 6. Martensite content over effective strain

4. Detection of Damage in Sheet Metal Components

The service life of a component equipped with the part-inherent sensing capability is shown in Fig. 7. In the course of developing a new safety concept, a defined martensite content was brought into the metastabile austenitic steel component by stamping. During the components use, the martensite content changes due to the cyclic load, as a result this it is possible to draw conclusions from the martensite content to the load history. The martensite content is measured in regular intervals to test if the component has to be replaced or can continue its service life. In addition, the gained information can be used for designing the next generation of components.

In order to find out the forming process best suited to sensitize the material with regard to the martensite evolution, simulation models were set up in the explicit FEA software Abaqus / Explicit. The forming processes for deep drawing, stretch-forming, and stamping were investigated, and the results were evaluated. Fig. 8 illustrates the stress ratios of the three different forming processes. As a precondition for the adjustment of a locally defined initial martensite content, the stresses must be applied systematically to the component during the respective forming process.

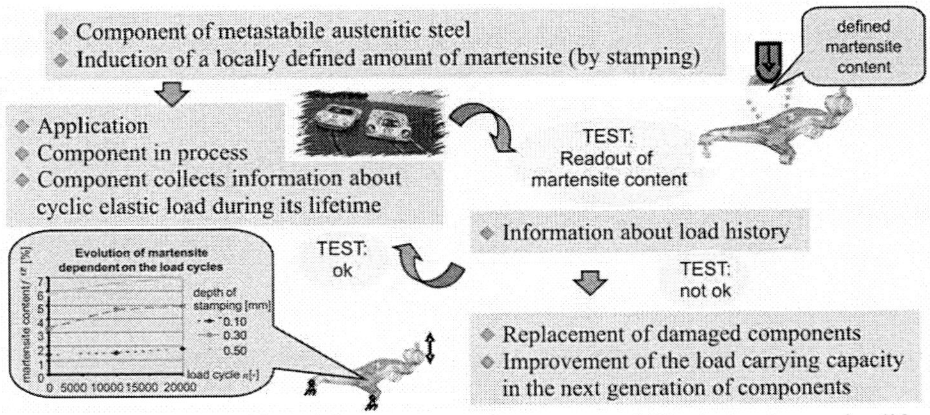

Fig. 7. Component with an inherent martensite senor capabilityduring its service life

The simulations were aimed at localizing the compressive and tensile stresses inside the specimen, thereby knowledge about the resulting α′- and ε-martensite areas was obtained. In order to attain a sensitization of the material, a high evolution of ε-martensite is desirable during the pre-elongation process. The intention is to have stable compressive stresses in the stamping area and locally restricted maximum stresses at the stamping position.

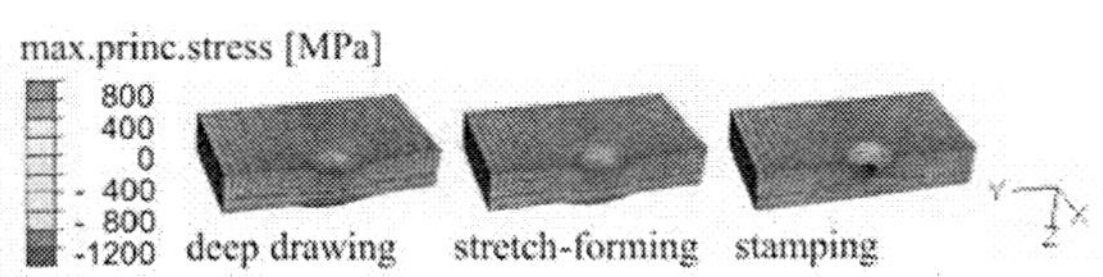

Fig. 8. Tensile and compressive stresses of the simulated metal forming processes

Stamping guarantees stable compressive stresses in the stamping area. Furthermore, the maximal stresses are locally restricted to the stamping position. The results of the simulation clearly indicate that neither the rear side of the stamping position nor the surrounding areas are affected by the stamping [1].

Based on the numerical results obtained, a new stamping tool has been constructed. It allows for the predefinition of specific martensite content in certain areas of the sheets. The martensite content is measured after stamping and elastic and plastic loading on the stamped sheet, for more details see [2,3]. As a result, the material-inherent sensorfunctions show a higher martensite content than the rest of the sheet (see Fig. 9).

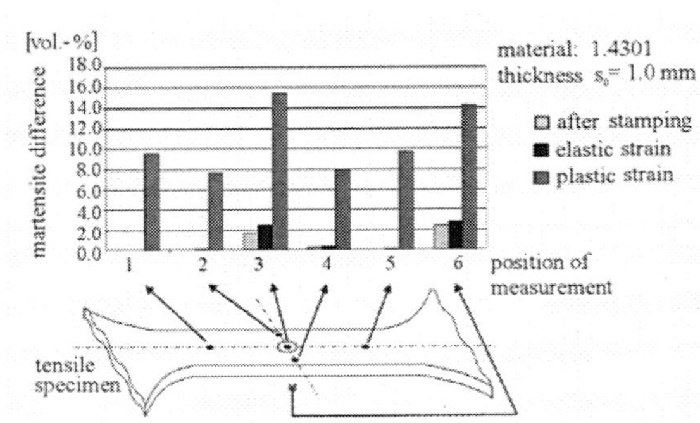

Fig. 9. Measured martensite contents on defined positions on the tensile specimen after different loadings

5. Conclusion

Martensite evolution is an effect that influences the strain hardening of stainless steel. For the simulation of forming processes in particular, the transformation-induced martensite and the effect of increasing hardening have to be considered, and thus experimental investigations were evaluated. A mathematical description of the martensite evolution was analysed, based on the material model of Springub [5]. This investigation was performed to clarify the influence of chemical fluctuations caused by different batches on the martensite evolution of the metastabile austenitic steel X5CrNi18-10 (EN 1.4301). The results show that the martensite content equation for the biaxial stress state (2) holds with the same material parameters *E, F, G, H* for all batches investigated, since the material behavior in the uniaxial tensile test shows significant differences.

Investigations were done to get information about the relation between the minor/major strain and the martensite content.

The obtained numerical results were used to extend an existing material model developed at the IFUM. Furthermore, a new stamping tool was designed based on this results. It allows the predefinition of a specific martensite content in certain areas of the sheet by stamping. The measurements after several subsequent elastic load cycles show a higher martensite content in the sensors than in the rest of the sheet. The aim for future work is to develop a mathematical model which allows for drawing conclusions about the load history of the sheet from the local increase of martensite content in the sensors. This way, a critical load limit could be detected, and damage of the component could be predicted.

Acknowledgements

We would like to thank the German Research Foundation DFG for funding this research work within the collaborative research centre 653.

References

[1] Behrens, B.-A.; Kamp. M.; Weilandt, K.: Umformtechnisches Verfahren zur Sensibilisierung der Martensitbildung unter elastischer Belastung, Metall, Heft 3, Giesel Verlag, 2007.

[2] Behrens, B.-A.; Weilandt, K.; Kamp. M.: Umformtechnische Herstellung mechanischer Belastungssensoren durch Einbringen lokaler Dehnungen – Einfluss hochgenauer lokaler Prägungen auf die Martensitbildung in metastabilem austenitischen Stahl, Tagungsband SFB 653 Kolloquium 2007, S.25-30, Hannover, April 2007.

[3] Behrens, B.-A.; Weilandt, K.; Voges-Schwieger, K.: Description of an adapted material model for the numerical determination of martensite evolution during deep drawing processes, Complas 2007, Computational Plasticity IX- Fundamentals and Applications, E. Onate, D.R.J. Owen and Suarez B. (Ed.), pp. 612-615, September 2007.

[4] Goodschild, D.; Roberts, W. T. and Wilson, D. V.: Mechanical behaviour of textured austenitic stainless steel, Journal of the Iron and Steel Institute, pp. 283-290, 1971.

[5] Springub, B.: Semi-analytische Betrachtung des Tiefziehens rotationssymmetrischer Bauteile unter Berücksichtigung der Martensitevolution, Ph.D. Thesis, Universität Hannover, Germany, 2005.

[6] Tsuta, T.; Cortes R., J.A.: Flow Stress and Phase Transformation Analyses in Austenitic Stainless Steel Under Cold Working, Part 2, JSME International Journal, Serie A, 36. no, Vol 1, pp. 63-72., 1993.

[7] Hänsel, A. H. C.; Hora, P.; Reissner, J.: Model for the kinetics of strain-induced martensitic phase transformation at non isothermal conditions for the simulation of sheet metal forming processes with metastable austenitic steels, Simulation of Materials Processing: Theory, Methods and Applications, Rotterdam, 1998.

[8] Behrens, B.-A.; Doege, E.; Springub, B.: Transformation induced martensite evolution in metal forming processes of stainless steels; steel research int. 75, 7/2004.

Innovative Production Machines and Systems
D.T. Pham, E.E. Eldukhri and A.J. Soroka (Eds.)

Automated Design of Gentelligent® Fixtures

B. Denkena[a], H.-C. Möhring[a], K. M. Litwinski[a], D. Heinisch[a]

[a] *Institute of Production Engineering and Machine Tools, Leibniz Universität Hannover, Germany*

Abstract

With the aim of improving the availability and performance of machine tools gentelligent® components of machine tools are developed at the Institute of Production Engineering and Machine Tools (IFW), starting with a gentelligent fixturing system. Being the interface between the machine tool and the workpiece, these systems allow a first evaluation of the potentials of a gentelligent production system.
This paper presents an approach for the design of the fixturing as a component of a sensing machine tool on the basis of finite elements simulations. The fixture design is adapted to the workpiece geometry and properties as well as to the machining process and the loads resulting from the machining process. This is achieved by minimizing the deformation of the workpiece and, thus, increasing the accuracy. The method developed comprises five stages: the analysis of fixture elements, the preprocessing of input (G-code), reduction of input data for the sake of computation time, computation of subconfigurations, and assessment and comparison of all subconfigurations.

Keywords: Fixture design, finite elements simulation, design for manufacturing

1. Introduction

With the aim of improving the availability and performance of machine tools gentelligent® components of machine tools are developed at the Institute of Production Engineering and Machine Tools (IFW) at the Leibniz Universität Hannover for the monitoring of the machine and machining processes. Such systems collect information over their life cycles and bequeath them onto later generations for purposes of optimization. Primarily a gentelligent fixturing system is developed. Being the interface between the machine tool and the workpiece these systems allow a first evaluation of the potentials of a gentelligent production system.

In a gentelligent® production system workpieces contain information from the beginning (primary shaping) of its life cycle [1]. This allows a unique identification of the workpiece during the whole production and lifecycle and makes accompanying label, as they are currently used, obsolete. The workpiece finds its way through the production [2]. The workpiece is processed by sensing machine tools, which adapt to the workpiece based on the experiences with former workpieces [3, 4, 5, 6]. The newly collected experiences obtained by the processing of the workpiece are collected in the machine tool for improving the processing of further workpieces and stored within the workpiece in order that they have a picture of their "quality". Particularly when problems during the further lifecycle of the workpiece occur, a correlation between "quality" and occurring problems might be realized.

This paper presents an approach of the adaption of the fixturing as a component of a sensing machine tool to the workpiece on the basis of experiences. Using this approach fixturing systems are designed depending on the loads exerted on the workpiece during the machining process. Thus, deformation of the workpiece is minimized and, accuracy is increased.

2. Fixture Design

Fixtures for workpieces are important components for machining processes. They define the quality of workpieces as they are located in the force flow between the machine tool and the workpiece [7]. Two different types of fixturing systems are differentiated: fixtures for rotational symmetric workpieces and fixtures for prismatic workpieces. The latter are subject of the research described in this paper. Fixtures may be divided into four groups [8]:

- Simple clamping elements and positioning parts (e.g. clamp straps, tensioning brackets, locating and supporting elements)
- Standard fixtures (e.g. vices, jaw chucks, swiveling and tilting tables)
- Undedicated/modular fixturing systems
- Dedicated fixtures.

The development of undedicated/modular fixturing systems represented an important step towards flexible manufacturing systems [9], which is a requirement for the adaption of machine tools in gentelligent production systems. According to Tuffentsammer, fixtures have three tasks to fulfill: locating, supporting and clamping [10].

The term *locating* describes the definition of position and orienttation of workpiece in respect to the tool. The fixture *supports* the workpiece in order to minimize deformations of the workpiece resulting from process and clamping forces. The *clamping* forces hold the workpiece in place and inhibit workpiece movements and vibrations [10].

In this research it was distinguished between two different types of clamping elements to fulfill these tasks of locating/supporting pins and clamps. The tasks of locating and supporting cannot be separated. As the location of a workpiece is not to be changed due to resulting process loads, locating elements have to be very stiff. In contrast to this, clamps fulfill the task of exerting a force on the workpiece, securing the contact of the workpiece to the locating elements. This force is supposed to remain constant even as the workpiece is deflected due to process or clamping forces or thermal effects. To keep the force constant, a low stiffness is mandatory. These opposing requirements on the fixturing elements lead to the fact that most of the forces exerted on the workpiece are absorbed and transmitted by the stiff locating/supporting elements.

In the past different approaches were pursued for the automated computer aided design of fixturing systems. These were mostly pure geometrical approaches [11] including force and moment equilibria and concepts of friction [12, 13]. Negele showed that workpiece deformation have a strong influence on the machinability and should be considered during the design of the fixtures [14].

3. Automated Clamping System Design

As explained above, the adaptation of machine tools to the individual demands of the processing of a workpiece and the ability of learning are fundamental steps in the development of gentelligent production. In the following a newly developed algorithm is described, which allows the automated design of clamping systems and demonstrates a possibility to feed back information collected during the machining of a workpiece for the development of further generations. Primarily a criterion is explained to assess a certain fixture configuration. Subsequently the procedure of determining optimal clamping configurations is explained and finally all stages are detailed.

3.1. Criterion for the Assessment of Fixture Configurations

In the past mainly geometrical characteristics and rigid body mechanics were used to optimize clamping configurations. In the method described here static workpiece deformation as a measure for rigidity against deflections during the processing and, thus, workpiece quality are introduced. The workpiece deflections might be calculated using the finite elements method (FEM). As for an automated fixture configuration method a criterion is developed for the interpretation of the simulation results. In equation (1) this criterion is given.

$$C = \sum_{j=1}^{n_p} \mathbf{u}_j^2 = \sum_{j=1}^{n_p} \left(u_{x,j}^2 + u_{y,j}^2 + u_{z,j}^2 \right) \tag{1}$$

where $\mathbf{u}_j$ describes the vector of displacement of the workpiece at the jth point of the n_p times discretized tool path. This criterion corresponds to the 2nd norm of the displacement along the tool path. The 2nd norm provides the advantage of averaging the values of the displacements neglecting their sign on the one hand and emphasizing peaks especially by the quadrature on the other hand. Thus, no point will exist, which extremely deviates from the nominal value.

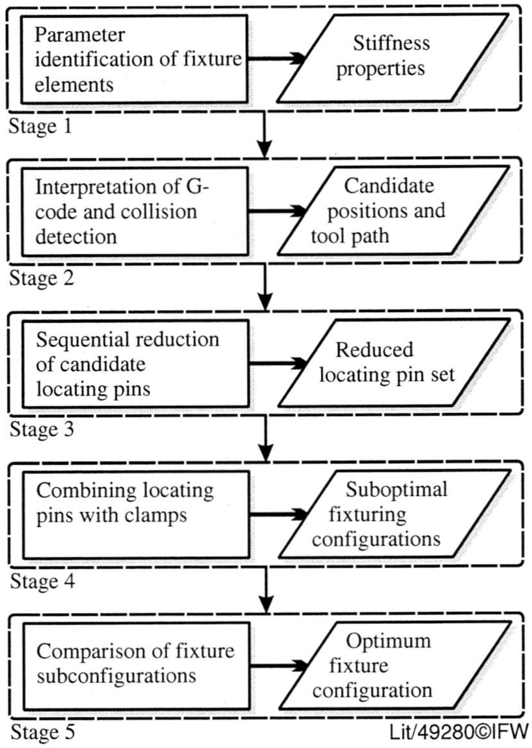

Fig. 1. five steps of automated fixture design

3.2. Procedure

The developed method is comprised of five stages (compare Fig. 1). In the first stage the stiffness matrices of the fixture elements are determined. In combination with the workpiece geometry and material properties these are important for the computation of deflections. In the second stage the G-code for the manufacturing process is interpreted and the tool path (relative movement between tool and workpiece) is extracted. It is used for a collision detection and avoidance by excluding candidate fixture positions, which collide with the tool. Furthermore, it is discretized into n_p steps to generate load cases. In the third stage the candidate positions for locating pins are analyzed. Because of the high influence of the stiffness primarily only the locating elements are considered. Since the complexity of the further algorithms depends strongly (by $O(n^3)$) on the number of candidate positions of locating elements their number is reduced. In the fourth stage sets of locating pins are generated and clamps are added to each set, which optimally ensure contact to the locating pins. In the final stage all previously computed fixture subconfigurations are compared and the optimal configuration is determined.

3.3. Parameterization of Fixture Elements

The parameter identification of the fixture elements is the aim of the first stage. Including the fixture elements in the FEM analysis would increase the computation time significantly without improving the accuracy notably. Therefore, the elements are substituted by stiffness matrices corresponding to the respective elements. As explained previously, two types of fixture elements are considered: locating/supporting pins and clamps. The determination of their stiffness is essential for calculating optimum fixture configurations. This can be achieved either by experimental investigations or finite elements analysis. The stiffness matrices are used for the further simulation steps.

3.4. NC-Interpreter

The algorithm has to be prepared for the calculation of clamping configurations. Apart from the workpiece geometry, the machining parameters and resulting loads have to be determined. An interpreter analyzes the G-code and calculates the tool path. The tool path is compared with the candidate positions of fixture elements. Collisions are avoided by ensuring a minimum distance between the tool path and the fixture elements. The tool path is further discretized into n_p points and combined with process forces obtained either by simulations [15] or force monitoring in previous machining cycles. This results in load cases representing the machining operation, which are input to the simulation.

3.5. Reducing Candidate Positions for Locating Pins

In general the vertical positioning of prismatic workpieces is achieved with three locating pins. Having a large number of initial candidate positions for these elements $n_{initial}$ the variety of possible configurations might be excessive and not feasible be handled as it increases cubically with the number of candidate positions. Therefore, in the third stage the number of candidate positions for locating pins is reduced by excluding positions, which have only a small influence on the rigidity of the workpiece. This is achieved by applying locating pins on all candidate positions and applying the process loads subsequently in finite elements simulations. The locating pin, which is least loaded in these simulations, is excluded. This procedure is repeated until only a small number of candidate positions for locating pins remains. In the investigations a residuum of eight candidate positions turned out as a good compromise between reduction of computation time and error from simplification. Thus, $3 \times n_p \times (n_{initial}-8)$ simulations are needed in this step. These are three simulations – one for each direction – at each point of the (discretized) tool path in order to apply the processing forces at each point independently.

3.6. Optimization of Fixture Configurations

From the previously determined set of candidate positions $\binom{8}{3}=56$ subsets are determined, each consisting of three locating pins. Primarily, they are tested for stability as it was proposed by Nee et. al. [13]. Each combination of three candidate positions of locating pins is tested, for whether the center of gravity of the workpiece lies within the triangle spanned by the three locating pins. If it doesn't, the configuration is discarded. Otherwise, the next step results in the calculation of optimum clamping elements. This might be reduced to a linear programming problem.

The clamping force exerted by the clamp results in a change of the contact forces between the workpiece and the locating pins on the one hand and in a deflection of the workpiece, decreasing the machining accuracy on the other hand. The deflection caused by a clamping force is calculated in finite elements analyses. Equation (1) is used as objective function. The square of the lengths of the deflection vectors caused by a unity clamping force at equidistant points along the tool path are calculated according to equation (2)

$$c_i = \sum_{j=1}^{n_p} \left(u_{x,j}^2 + u_{y,j}^2 + u_{z,j}^2 \right), \tag{2}$$

where i represents an index for the candidate position of a clamp, n_p is the number of discretization steps of the tool path, and $[u_{x,j}\ u_{y,j}\ u_{z,j}]$ is the displacement vector at the jth position of the tool path. Equation (2) is calculated for each of the n_{cl} candidate positions of a clamp and summarized in the vector

$$\mathbf{c} = \begin{bmatrix} c_1 & c_2 & \cdots & c_{n_{cl}} \end{bmatrix}. \tag{3}$$

With the simplification of the linear superposition, equation (4) is derived as quality function to be minimized using the simplex algorithm

$$Q(\mathbf{f}_{cl}) = \mathbf{c} \cdot \mathbf{f}_{cl} = Min! \tag{4}$$

Furthermore, the linear relationship between the clamping forces $\mathbf{f}_{cl}$ and the contact force between the three locating pins and the workpiece $\mathbf{f}_l$ can be stated in the form

$$\mathbf{f}_l = \mathbf{L}\mathbf{f}_{cl}, \tag{5}$$

where $\mathbf{L}$ is a $3 \times n_{cl}$ matrix mainly corresponding to the geometric properties (i.e. lever arms) of the workpiece. Equation (5) contains the constraints for the optimization algorithm; as the contact force on each locating pin has to exceed a threshold in order to ensure the ability of transmitting shear loads.

The minimization problem stated by the equations (4) and (5) is solved by using the simplex algorithm. Thus, up to three clamps and corresponding clamping forces are calculated for each subset of three locating pins. These subconfigurations cause a low deflection of the workpiece during machining, while ensuring high contact forces between the locating pins. For this step a maximum of $56 \times n_p$ simulations are required.

3.7. Comparison of Fixturing Subconfigurations

In the final stage of the method all fixturing subconfigurations are compared regarding the deflections caused by the process and the clamping forces using the load data corresponding to the machining process. From equation (1) equation (6) is derived to assess a certain fixture configuration.

$$U_{total} = \frac{1}{n_p} \sum_{i=1}^{n_p} \mathbf{u}_{k,i}^2 + \frac{1}{n_p} \sum_{i=1}^{n_p} \mathbf{u}_{cl,i}^2 \tag{6}$$

$\mathbf{u}_{k,i}$ is the displacement vector at the position k of the discretized tool path caused by the process force applied on the point i of the tool path and $\mathbf{u}_{cl,i}$ denotes the displacement vector at the point i of the discretized tool path caused by the clamping forces. The two components, clamping force induced displacement and displacement caused by the process loads, are not vectorially added. Only the squares of the resulting deflections are considered. This procedure accommodates the fact that the process load in contrast to clamping forces is not static but rather dynamic, particularly in milling processes. Thus, in total $(3 \times n_p)+1$ simulations are necessary for each of the subconfigurations. Three simulations – one for each force direction – at each point of the tool path and one for the deflections due to clamping forces. The maximum total amount of simulations therefore is $3n_p(n_{initial}-8)+56(4n_p+1)$.

4. Application for an Example Workpiece

The previously explained method is applied to the calculation of an optimal fixture configuration. The workpiece, which will be analyzed for fixturing, is part of a gearbox (see Fig. 2 a). Two types of fixture elements are used to fix the workpiece: a locating pin and a swing clamp (see Fig. 2 b).

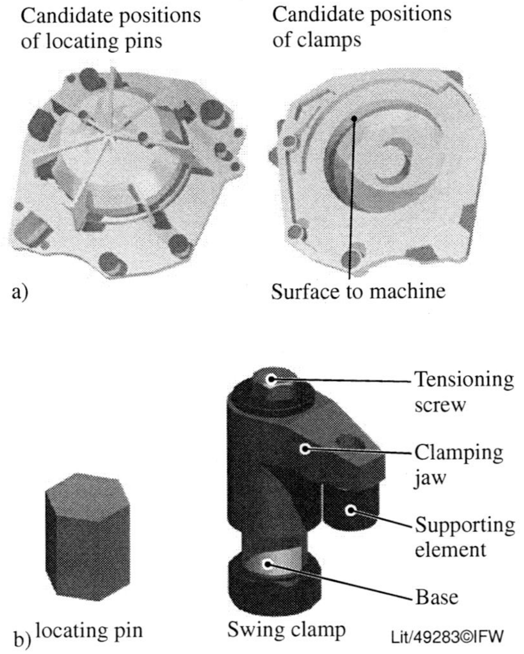

Fig. 2. Geometry of a) workpiece, b) fixture elements

4.1. Parameterization of Fixture Elements

The first step is the simulation of the fixture elements in order to parameterize the simulations. In Fig. 3 a) the simulation results of the locating pin are depicted. In Fig. 3 b) the results from prestressing the swing clamp with a force of $F_{prestress}$ = 1 kN are shown. The prestressing is necessary for the swing clamp because nonlinear contact has a strong influence on the stiffness of the clamp. Thus, the stiffness matrix is linearized around the working point by prestressing the system. The resulting stiffness matrices **K** defining the relationship of the forces acting on the fixture elements $\mathbf{f}_{cl}$ and the deflection of the contact point of the fixture elements $\mathbf{u}_{cl}$ are calculated according to

$$\Delta\mathbf{f}_{cl} = \mathbf{K} \cdot \Delta\mathbf{u}_{cl}$$

$$\mathbf{K} = \left[\frac{\partial \mathbf{f}_{cl}}{\partial u_{cl,x}} \quad \frac{\partial \mathbf{f}_{cl}}{\partial u_{cl,y}} \quad \frac{\partial \mathbf{f}_{cl}}{\partial u_{cl,z}} \right]\Bigg|_{\mathbf{f}_{cl}=\mathbf{F}_{prestress}} \quad . \tag{7}$$

4.2. Reducing Candidate Positions for Locating Pins

As explained previously, the computation time strongly depends on the number of candidate positions for locating pins. As can be seen in Fig. 2 a) initially 14 positions existed. The reduction of these positions led to eight candidate positions displayed in Fig. 4.

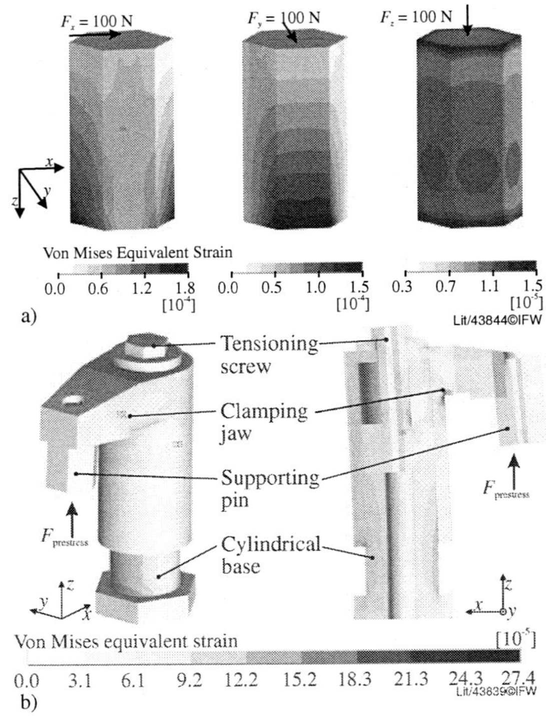

Fig. 3. Simulation results of a) locating pin, b) clamp

4.3. Optimization of Clamping Subconfigurations

Using the reduced set of candidate positions for locating pins, eighteen stable subconfigurations of three locating pins can be derived. For each of these configurations the optimal clamping positions are calculated using the simplex algorithm.

4.4. Comparison of Clamping Subconfigurations

In Fig. 5 the contact areas used for optimal fixture of the workpiece are displayed. The three locating pins span a wide almost equilateral triangle. The positions of the clamps are located closely to the locating pins causing very short force flows through the workpiece, and thus, causing only few deformations.

5. Summary

In this paper a method is explained for the automated fixture design based on FEM simulations. The fixture design is adapted to the workpiece geometry and properties as well as to the machining process and the loads resulting from the processing. The method uses finite elements simulations to determine the deformation of the workpiece caused by clamping forces and the process load.

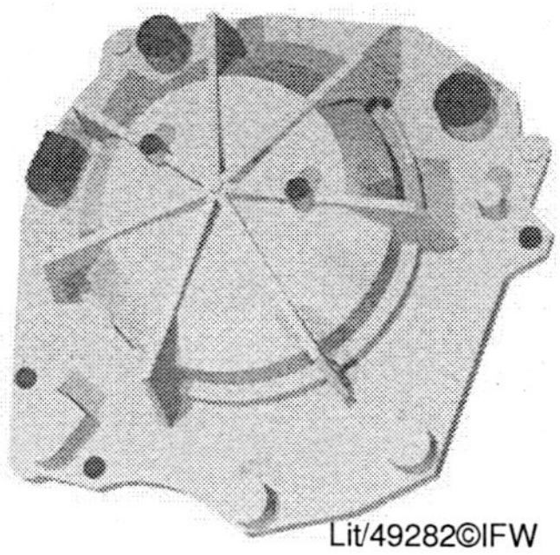

Fig. 4. Results of the reduction of the candidate positions for locating pins

In the context of the Collaborative Research Center 653 the abilities of learning and adaption are focused. In the design phase of workpieces information about the machining process are available only from simulations. In this phase the explained method of fixture design might be implemented on the basis of this knowledge. During the processing of the first generation of workpieces gentelligent feeling machine tools gather experience, particularly on the process forces. These data can be directly used to adapt the fixture to the process loads. But the greatest profit can be achieved in the next generation of workpieces, as these can be designed including the knowledge gained during the machining of the previous generation. The feed back of process loads in combination with the explained method allows the optimal design for manufacturing for the workpieces.

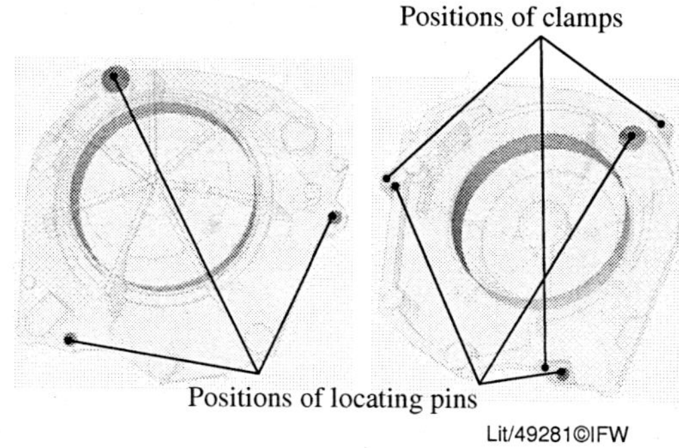

Fig. 5. Resulting fixture configurations

Acknowledgements

The authors want to thank the German Research Foundation (DFG) for funding this research within the Collaborative Research Center 653.

References

[1] Behrens, B.-A.; Gastan, E.; Lange, F. Experimental und numerical investigation on the marking of sintered parts by the targeted application of particles. Materials Science and Engineering Technology, 38, (2007), pp. 816 – 820, WILEY-VCH Verlag, Weinheim

[2] Denkena, B.; Schmidt, C.; Lorenzen, L.-E.: Virtuelle Nullserienfertigung in der Planung, wt-online – Ausgabe 3, Springer-VDI-Verlage (2008), pp. 149 – 155

[3] Denkena, B., Litwinski, K. M. Gentelligente Komponenten für Werkzeugmaschinen. 1st Kolloquium of CRC 653. (2007), Hannover, Germany

[4] Litwinski, K. M.; Denkena, B. & Möhring, H. Optimum Design of Multi Sensor Systems. 8th Heinz-Nixdorf Symposium, (2008), Paderborn, Germany

[5] Denkena, B.; Möhring, H. & Litwinski, K. M. Design of Multi Sensor Systems for Gentelligent Process Monitoring Systems. 2nd International Conference on Experiments/Process/System Modelling/Simulation & Optimization, (2007), Athens, Greece

[6] Denkena, B.; Möhring, H. & Litwinski, K. M. Simulationsbasierte Multisensorsystemauslegung. Autonome Mobile Systeme, (2007), Kaiserslautern, Germany, pp. 114 – 117

[7] Lünemann, M. Eigenschaftsermittlung von Konstruktionselementen in Werkstückspannsystemen. Dissertation, Leibniz-Universität Hannover, 2005

[8] Weck, M; Brecher, C. Werkzeugmaschinen – Maschinenarten und Anwendungsbereiche. (6. neu bearb. Auflage). Springer Verlag, Berlin Heidelberg, 2005

[9] Koch, D. H. Undedicated Fixturing International Manufacturing Technology Conference, 1990

[10]Tuffentsammer, K. Automatic Loading of Machining Systems and Automatic Clamping of Workpieces. Annals of the CIRP, Vol. 30/2, 1981

[11]Trappey, A. J. C., Matrubhutam, S. Fixture Configuration Using Projective Geometry. Journal of Manufacturing Systems 12 (1993), 486 – 495

[12]De Meter, E. C. Restraint Analysis of Machining Fixtures Transactions of NAMRI/SME 20 (1992), pp. 339 – 345

[13]Nee, A. Y. C.; Tao, Z. J.; Senthil Kumar, A. An Advanced Treatise on Fixture Design and Planning. Vol.1, World Scientific Publishing Co. Pte. Ltd., Singapore, 2004

[14]Negele, T. Simulationsmethoden zur optimierten Gestaltung von Spannmitteln. Praxistagung Werkstückspanntechnik – Sonderspannmittel und Sonderspannlösungen, Wolfsburg, Germany, 2007

[15]Denkena, B.; Schmidt, C., Lorenzen, L.-E. Zukunftsweisend Fertigen – Die Simulationsgestützte Planung und Überwachung spanender Fertigungsprozesse innerhalb der gentelligenten® Produktion. ZWF 10/07 (2007) pp. 626 – 629

Innovative Production Machines and Systems
D.T. Pham, E.E. Eldukhri and A.J. Soroka (Eds.)

Development of a System for Data Storage on Machine Components

K.-H. Wu, H.H. Gatzen

Institute for Microtechnology (imt), Center for Production Technology, Leibniz Universitaet Hannover, An der Universitaet 2, 30823 Garbsen, GERMANY

Abstract

A new approach for data recording through the entire manufacturing process is to apply the magnetic recording technology for the storage of information directly on the surface of the components themselves. Such a system capable of magnetic recording consists of a device for data recording, one for data retrieval, and a magnetic storage medium. For recording the data, the basic approach of inductive writing is used, using a recording head creating a magnetic flux, which performs the data storage. The storage medium of interest is "magnetic magnesium", i.e. magnesium filled with hard magnetic particles. This is the material the component consists of and makes any additional storage medium unnecessary. To ensure the application in a machine environment, a remote readout from a distance is desirable. For this purpose, the Magneto-optical Kerr Effect (MOKE), which senses the direction of the magnetization of a reflecting surface, is used. To allow a magneto-optical readout based on MOKE of a storage medium containing magnetic particle, the bulk storage material is covered with a soft magnetic layer. This "keeper layer" is necessary to generate a homogeneous magnetic and reflective surface, which are features not provided by the magnetic magnesium. This paper describes the developing and testing the magnetic recording system. The tests were done on a preliminary material, which has the same magnetic properties like magnetic magnesium.

Keywords: Data Storage, Recording Head, Magneto-optical Kerr Effect

1. Introduction

In any production process, the marking or serializing of components and hence the storage of information is a very important aspect of any production. Conventional systems use bar codes or Radio Frequency Identification (RFID) systems for tracking the components during the manufacturing process. These tags are usually adhesive labels or stickers attached to the machine component, which may be detached accidentally during the manufacturing process, e.g. in a cleaning process. Recording the data using an inductive write head is an approach well established due to the widespread of magnetic storage applications like Hard Disc Drives (HDD) and Video Tape Recorders (VTR). Magnetic recording systems are very reliable and cost-effective, and subject to continuous improvements [1, 2]. Such a system also seems advantageously for storing data directly on components, as long as they are hard magnetic.

Within the Collaborative Research Center 653 "Gentelligent Components" (genetically intelligent) [3], a new concept is developed to store information directly on the surface of the component, which means that can act as its own information carrier. To do so, the component material has to fulfill both structural and storage requirements. To fulfill this requirement, a new type of material was conceived: magnetic magnesium [4, 5]. It is magnesium filled with hard magnetic particles. Similarly to magnetic recording systems, this method allows for a dynamic storage of data and immediate changes of stored information (writing and overwriting of information). This way, for instance production data of gentelligent products may be updated at any time.

2. Data Storage Concept

The data storage system presented in this paper consists of three parts: a devise for data recording, a storage medium, and a device for data retrieval for the readout. For the magnetic writing, an approach using an inductive write head is chosen. This write head features an air gap, which creates a fringing field required for the magnetization of the storage medium.

As discussed previously, the storage medium to be used is magnetic magnesium. Figure 1 shows a milling head made out of this material.

Fig. 1. Milling head

For the readout of the stored information, a concept was chosen, which can detect data from a distance. In most cases, quasi contact or contact readout systems (like the one used in HDD or VTR, respectively) would not lend themselves for a use in a real machine component environment. Instead, a remote readout is more desirable, particularly since a reading of the data is expected more often than writing.

The chosen remote readout system takes advantage of the Magneto-optical Kerr Effect (MOKE). However, in order to apply a MOKE readout, the surface of the storage medium has to fulfil two requirements: it has to be magnetic as well as reflective. Being a particulate medium, magnetic magnesium does not meet either requirement. To resolve this issue, the magnetic magnesium surface is covered by a soft magnetic and reflective "keeper layer", which mirrors the magnetisation of the particulate material.

2.1. Data Writing

An inductive write head consists of a soft magnetic ring core with an air gap, as well as a coil for inducing a field. The purpose of the air gap is to create a fringing field, which magnetizes the storage medium. To do so, the recording head passes over the surface of the storage medium and magnetizes a data track. The data are encoded in flux reversals. Figure 2 shows the schematics of the write head positioned above the particulate storage medium. All the hard magnetic particles exposed to a fringing field greater than their coercivity are magnetized.

Based on simulation results, the geometry of the recording head was defined. The recording head consists of two symmetric C-cores, which were fabricated by a combination of profiling, dicing, and bonding. Each core carries a hand-wound coil. The write head has a footprint of 3.77 mm x 2.95 mm. The air gap length is 100 μm [6].

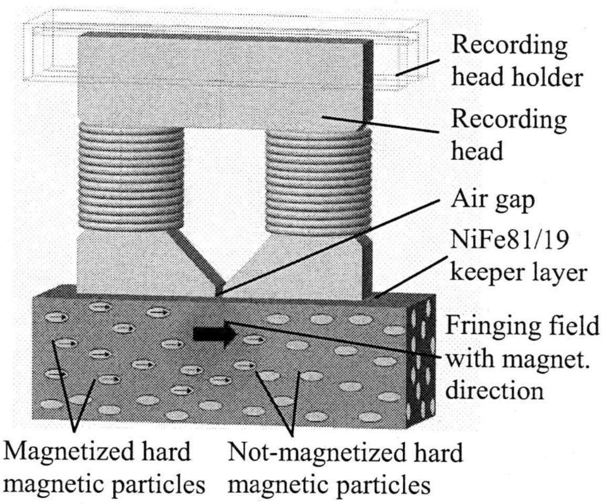

Fig. 2. Principle of magnetic recording

2.2. Storage Medium

As mentioned before, the particular storage medium used for the component level storage system is a magnetic magnesium, which is a mixture of the non-magnetic Mg bulk material and acicular hard magnetic γ-Fe_2O_3 particles. The coercivity H_c of the γ-Fe_2O_3 particles is 23 kA/m and the particle size is between 20 μm and 30 μm.

A keeper layer coats the surface of the magnetic magnesium to allow a MOKE readout. The keeper layer chosen is the permalloy NiFe81/19, which is not only soft magnetic, but also reflective and corrosion resistant. The saturation flux density B_s of NiFe81/19 is 1.0 T, while the relative permeability μ_r of this layer depends on its thickness and decreases with growing layer thickness [7].

For the writing of tracks, the keeper layer has to be saturated in order to penetrate and magnetize the material underneath. Otherwise, the fringing field would be shielded by this keeper layer and the system would not be able to store information. For the optimal head design, the interaction of keeper layer and recording head was considered in simulations [8].

For the initial tests, the magnetic magnesium was not yet available. Therefore, alternative storage media were used. They will be described below.

2.3. Data Readout

The non-contact readout MOKE system converts a magnetization into a Kerr rotation of the polarized detection beam reflected on the magnetic material (Fig. 3). The Kerr angle θ_k depends on the magnitude of the magnetization, but also on the magneto-optical properties of the reflecting magnetic material.

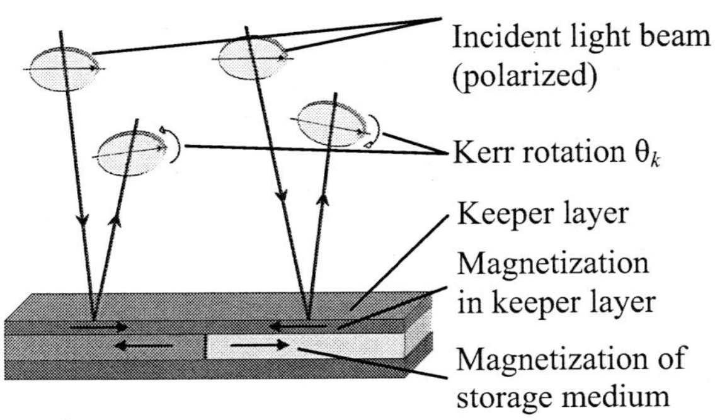

Fig. 3. MOKE readout principle

In order to use MOKE for data retrieval in magnetic recording, enable an optimal readout, two requirements have to be fulfilled. First, the keeper layer has to replicate the magnetic pattern of the magnetized storage medium underneath. The second requirement is the ability of the surface to reflect the incident light. The reflectivity of the surface is important for the precise determination of the Kerr angle.

The prototype of the readout system is a Kerr microscope. It is a combination of a microscope, a device for creating a polarized light beam, and an analyzer detecting the Kerr rotation of the reflected beam.

3. Experimental

3.1. Recording Set-up

For the writing test, a recording head was electrically connected and attached to a positioning stand, which moves the head along the storage media in a contact mode. Figure 4 shows the test set-up. The fine-pitch thread spindle of this positioning stand allows an accurate displacement of the recording head within a range of 5 µm. A parallelogram mechanism accomplishes a travel of the recording head parallel to the storage medium. This is accomplished by using elastic joints as shown in Fig. 4.

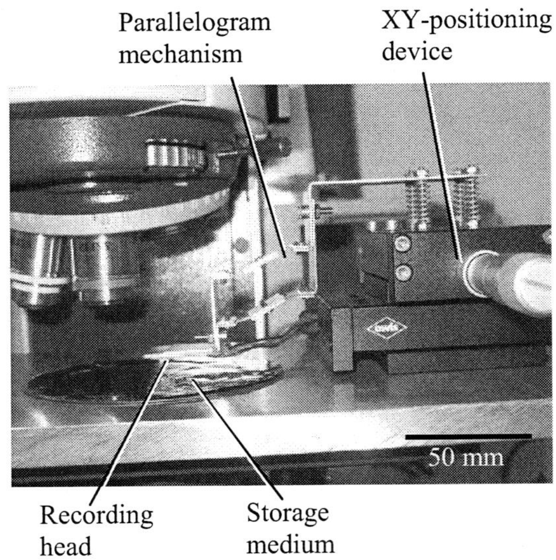

Fig. 4. Experimental Set-up

For the experiments, tracks were written on the surface of storage media. Among other, the tests aimed at investigating if the particulate storage medium can be saturated despite the presence of a keeper layer. Obviously, an analysis of the write data requires data retrieval capabilities.

3.2. Experimental Storage Medium

Since the magnetic magnesium was not available at the time when the initial tests were conducted, an alternative material with embedded particles was used. As long as the matrix material is non-magnetic, such a substitution is acceptable.

For the fabrication of these preliminary storage media with embedded particles, Crystalbond™, which is delivered as bulk material, was milled into a powder. Then, this powder was mixed with hard magnetic γ-Fe_2O_3 particles, with a volume ratio of 50:50. The result of this mixing was a nearly homogenous mixture of binder and storage paricles. After that, the mixture was heated up, until the melting point of Crystalbond™ was achieved at 140°C and stirred. Due to the stirring, the hard magnetic particles were distributed randomly in the Crystalbond™ matrix similar to the one expected in the magnetic magnesium. Then, a heated wafer was coated with the liquid, which allowed to coating the wafer surface. After solidifying, a homogenous layer was generated on the wafer. Next, the surface was planarized by grinding and polished with a soft polishing pad and DI water to achieve a flat and smooth surface. Next, a soft magnetic NiFe81/19 keeper layer was sputter deposited. The thickness of this layer is 200 nm.

A simple alternative was the application of commercial data storage material used in peripheral data storage. The materials used were commercial γ-Fe_2O_3 and CrO_2 media used in Flexible Disc Drives (FDD) and VTR. The difference between γ-Fe_2O_3 and CrO_2 is the coercivity H_c. CrO_2 has a higher coercivity than γ-Fe_2O_3. Such media provide only qualitative results since they do not feature a keeper layer.

3.3. Reading Set-up using MOKE

After the magnetization of the storage medium, the tracks were detected and analyzed. In initial tests, only a marginal contrast of the bitcells was observed in the Kerr microscope, clearly indicating that a further research on the detection method intended for use in the system is necessary.

To judge the data recording results, alternative indicator methods were applied. Two indirect methods for the detection of the written magnetization were used, a magnetic lens and an indicator layer. Both are able to detect the magnetization pattern on the surface. The magnetic lens is filled with a ferrofluid, which contains hard magnetic particles. By putting the lens on top of the storage medium, the particles are arranged along the direction of the magnetic field of the magnetized surface.

The second approach taken is the use of MOKE with an indicator layer. This layer is an auxiliary layer amplifying the Kerr angle. This method may be used in conjunction with a keeper layer. Figure 5 shows the schematics of such a layer.

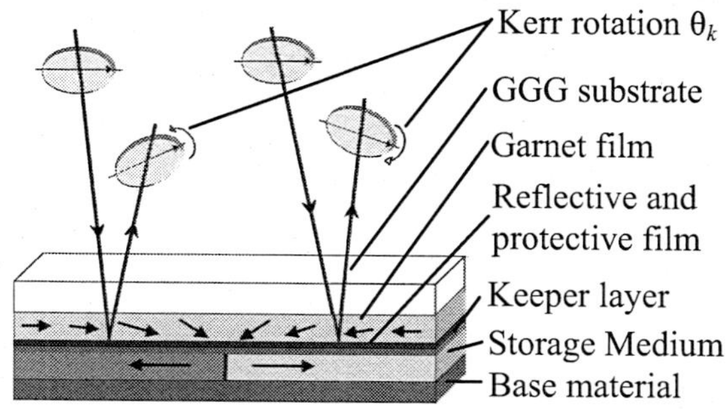

Fig. 5. Composition and application of an indicator layer

The indicator layer is composed as a stacked structure consisting of a GGG (gadolinium-gallium garnet) substrate, a Bi (bismuth) magnetic garnet film, a reflective film, and a protective film on top [9, 10]. The indicator layer is placed on top of the previously magnetized storage medium.

The magnetic field penetrates the reflective film and interacts with the garnet film. Polarized incident light penetrates the GGG substrate and the garnet film and is rotated by reflection at the reflective surface. This rotation can be detected clearly allowing to reproducing the magnetization of the surface (readout process).

4. Results

4.1. Writing and Reading with a Magnetic Lens

The readout tests were performed by putting the magnetic lens on the surface of the magnetized storage medium without a keeper layer. Figure 6 depicts the bits written on a CrO_2 medium, visible on the magnetic lens. At the boundary line between two bits magnetized oppositely, the magnetic field lines are closing perpendicular to the surface. These are depicted as white ovals.

Each bit has a length of 100 µm, which is shown as an ellipse surrounded by a bright field. Because of the agglomeration of the particles along the magnetic field lines, the bits appear as an ellipse, which is not the real bit geometry. Due to the distance between the storage medium and magnetic fluid, the measured magnetic signal seen on the top is lower. Therefore, the magnetic lens test evaluated the magnetization of the track, but not the geometry of the tracks and bits.

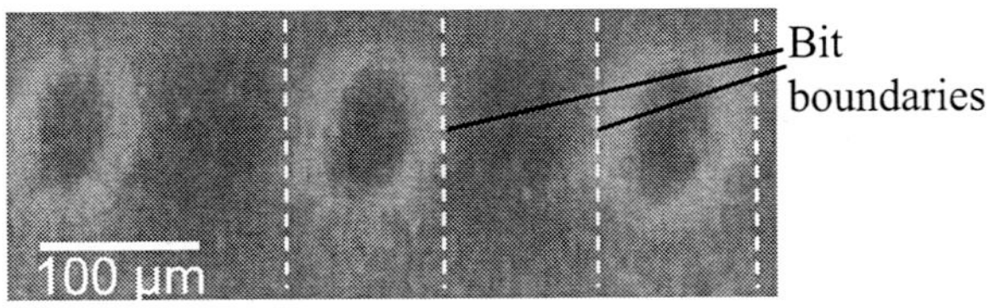

Fig. 6. Detection with magnetic lens

4.2. Writing and Reading with Kerr microscope and Indicator Layer

Similar to the magnetic lens, an indicator layer was put onto the surface of the storage medium. After this, the surface of the indicator layer was analyzed by a Kerr microscope. The results are shown in Fig. 7 and Fig. 8. Compared to the bits visible with the magnetic lens (Fig. 6), this method shows rectangular bits on the media. Figure 7 shows a magnetization track on a γ-Fe_2O_3 medium. The bits featuring a width of 500 µm and a length of 100 µm are clearly visible.

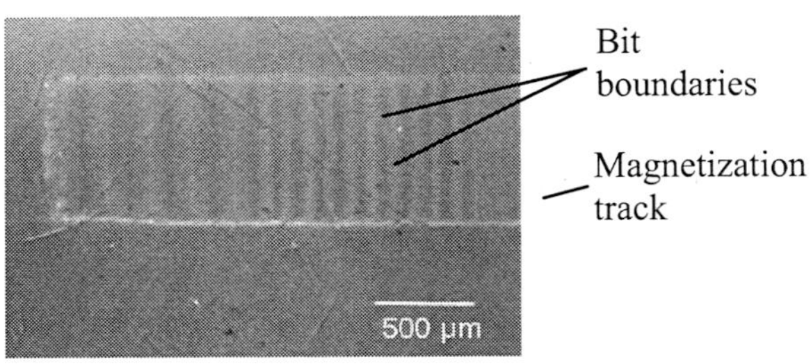

Fig. 7. Magnetization track of γ-Fe_2O_3 medium

Figure 8 depicts two tracks on a CrO_2 medium.

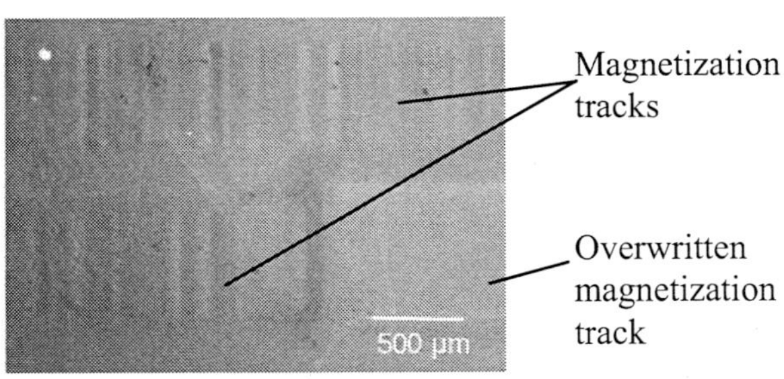

Fig. 8. Magnetization track of CrO_2 medium

The lower track is generated by the recording head switching the direction of magnetization. Finally, the hard magnetic particle mixture with keeper layer was magnetized and analyzed. Figure 9 shows the detected track.

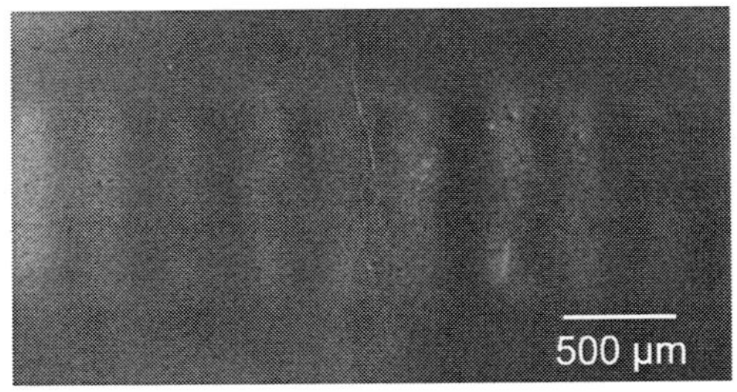

Fig. 9. Magnetization of γ-Fe_2O3 in a Crystalbond^{TM} matrix.

The bits are clearly visible verifying that the developed magnetic head generates a field strong enough to penetrate the keeper layer and to magnetize the preliminary storage medium.

5. Conclusion

This work demonstrates the feasibility of magnetic data storage on components made out of a hard magnetic material and a remote data readout. Due to the unavailability of the magnetic magnesium featuring hard magnetic particles in a Mg matrix, this storage medium was substituted for initial tests by using γ-Fe_2O_3 particles in a Crystalbond^{TM} matrix.

For the magneto-optical readout system, the machine part has to be coated with a keeper layer. It could be shown that the recording head can saturate the keeper layer during recording, which is a precondition to be able to magnetize the hard magnetic particles embedded in the matrix underneath the keeper layer. Because of the low intensity of the Kerr effect, the magnetization on the surface could not be detected by a direct method yet. However, by using an auxiliary indicator layer, these magnetization tracks could be detected. The feasibility of the concept was demonstrated.

Acknowledgements

This research is sponsored in part by the DFG (German Research Foundation) within the SFB (Collaborative Research Center) 653 "Gentelligent Components in their Lifecycle".

The authors would like to thank Dr. R. Schaefer, Institute for Metallic Materials, Leibniz Institute for Solid State and Materials Research, Dresden, Germany, for carrying out the magneto-optical investigations.

The authors would like to thank Dipl.-Phys. S.

Dreyer, Institute for Material Physics, Georg-August University, Goettingen, Germany, for performing the magneto-optical measurements and for providing the indicator layer required for the magneto-optical measurements.

References

[1] D. Z. Bai, et. al. Write Pole Tip Remanence in Perpendicular Recording. IEEE Trans. on Magnetics, Vol. 42, No. 3, 2006

[2] K. Takano, et. al. Relationship Between Head Design, Pole-Tip Magnetization, Head Field, and Medium Magnetization in Longitudinal Recording. IEEE Trans. on Magnetics, Vol. 42, No. 2, 2006

[3] B. Denkena, A. Battino, C. Liedtke. Optimization of Process Parameters with Gentelligent Components. Proc XVII Workshop on Superv. a. Diagnos. of Machin. Syst.. pp. 15-23, 2006

[4] Fr.-W. Bach, et. al. Casting of Magnesium alloys with magnetic properties, Magnesium-broad horizons. Giessereiindustrie, N2, ISSN 0024-449X, pp. 16-19, 2006

[5] Fr.-W. Bach, et. al. Magnesium Alloys with Magnetic Properties. TMS2006 135th Ann. Meet. & Exh., pp. 275, 2006

[6] K.-H. Wu, H.H. Gatzen. Fabrication and Assembly of a Magnetic Recording Head. Euspen08, Zürich, 2008

[7] M.C. Wurz, D. Dinulovic, H.H. Gatzen. Investigation of the Permeability on Electroplated and Sputtered Permalloy. Proc. 8th Int. Symp. on Mag. Mater., Proc. and Dev., pp. 525-536, 2004

[8] H.H. Gatzen, K.-H. Wu, O. Traisigkhachol. Development of a Recording Head for Magnetic Storage of Information on Machine Components. 210th Meet. o. Electroch. Soc. 2006, Vol. 3, Issue 25, pp. 191-202, 2007

[9] T. Terada, et. al. A Novel Flying Magnetic Recording Head Based on Magneto-optic Transfer. IEEE Trans. on Magnetics, Vol. 34, No. 4, 1998

[10] X.R. Zhao, et. al. Magneto-Optical Imaging for High-Throughput Characterization of Combinatorial Magnetic Thin Films. App. Surface Science 223, pp. 73-77, 2004

Innovative Production Machines and Systems
D.T. Pham, E.E. Eldukhri and A.J. Soroka (Eds.)

Gentelligent Factory Structures and Assembly Control

P. Nyhuis, F. Fisser, M. Schmidt

Institute of Production Systems and Logistics, Leibniz University Hanover, Germany

Abstract

This paper shows the basic concept of Gentelligence, which is based on information directly saved on a component itself without having further external storage devices. The advantages for the logistic and the impact on future factory structures are highlighted. These factories have to fulfil certain criteria which are demanded by the new components. Furthermore the components enable possibilities like a component driven assembly control. This new concept is described in its first steps and the expected impacts on the logistical performance are shown.

Keywords: Factory Planning, Production Planning and Control, Decentralized Information System

1. Introduction

The challenge for companies in these days is, to cope with internal and external material and information flows. These are getting more complex due to ongoing outsourcing processes and highly developed company networks. The coordination of the flows is unthinkable without the support of modern information technology and information systems [1, 2]. They form the basis for a successful supply chain management and therefore a high logistic efficiency. The information are used to adapt production planning and control as well as the factory or the structure of a factory respectively.

Commonly the information is stored on external systems. Gentelligent components have the ability to store information on themselves. These new components are currently developed in the Collaborative Research Centre 653 "Gentelligent Components in the Lifecycle"; which is sponsored by the Deutsche Forschungsgemeinschaft (DFG) and resides at the Leibniz Universitaet of Hanover.

The possibility of storing information on a part, without having further information carrier like RFID-chips, creates the demand for factory structures, which take advantage of the capabilities of these parts. Besides storing information on the components the resources in a factory should be able to communicate with the components. A totally new information structure has to be set up for future factories. This has the effect that new criteria for structuring a factory have to be developed besides the commonly know, like the two extremes of function orientation or the process orientation. Possible criteria are the type of information technology, the complexity of information or the need for Gentelligence itself. These new possibilities will be demonstrated in the following paragraphs.

2. Structural Factory Planning

The definition of "structure" is used differently in the habitual language and in the scientific community. In the organization theory dimensions of a structure are used for describing the characteristics of an organization [3, 4]. Henn and Kühnle use the word "structure" to define various material flows and building shapes. In the information technology, the combination of computers is defined as a network structure [5]. A universal definition is given by Ropohl:. a structure consists of elements which can be combined by relationships [6]. These elements may be defined and may be seen for example as atoms or biological cells but also as business units or production cells. The relationship between these elements can be expressed by mathematical equitations or stochastical laws. This depends on the definition of the elements and may take on different manifestations.

In the field of factory planning the definition for a structure is also different. Harms defined it as follows: A factory structure is formed by elements and their relationships. The smallest element in a factory structure is a manufacturing resource or a manual workplace. Together with the employees it creates a structure unit [7].

The factory structure planning is a significant phase in the planning process. It follows the preparation phase and delivers important information for the next step: the layout planning. There are different methods for the support of the structure planning. For example if the structure is defined by the principles of production, resources, employees and work piece are placed next to each other. Units are set up which have their own targets. Business units, Centre organization and the production segmentation are the most important appendages. The last one is characterized by product-orientated units which comprise more steps of the logistical chain of one product or product family. Other structuring methods use modules or autonomous production units to fulfil the demand for transformability.

All the planning philosophies have one thing in common: they use information about systems and technologies, but the information is not applied as a criterion for defining a factory structure. Information systems and technologies support the structuring but the structure itself is developed with other criteria. Because factory planning is by no means a task which has to be done only once, the structure of a factory has to be controlled continuously.

The agent theory has been used in the past to change structures of a factory actively based on negotiations of these agents. They were charged with economic and value adding tasks. Some prototype versions of such agent systems have been developed, but the problem was that the decisions of the agents were just based on production orders without looking at the factory structures. A first concept which uses agents to control a factory structure was introduced by Harms. But this concept was not implemented into information systems.

Other concepts for structure controlling which are often based on a control loop use the variance analysis for predefined target values. They examined these values but do not analyze the structures.

It can be stated recapitulatory that information technology and systems have mainly a supporting part in the structure planning. Often these systems are still under development. Existing information technologies do not allow storing all the information which is used for the factory structure controlling on the component.

2.1. The Concept of Gentelligence

"Gentelligence" is a neologism which consists of the parts "genetic" and "intelligent". It describes a component that has the ability to sense the environment and to store information. Key is that also lifecycle information of a component will be saved. In terms of the genetic part it is intended to information inheriting next component generation. These genetic components will be created within the collaborative research centre (SFB) 653 "Gentelligent Parts in the Lifecycle" (which is being conducted at the Leibniz Universitaet Hannover and sponsored by the Deusche Forschungsgemeinschaft – DFG).

The primary objective of this research centre is to integrate the component and the information carrier in one part. The information itself becomes an integral part of the component. This implies that this new component is able to read and write information. External Devices, like scanner, have to be created which fulfil this task. Also an internal information structure has to be set up to establish a communication with other Gentelligent parts or the read/write devices This is still part of the research program as well as the sensors which are located on the components to sense temperatures or stresses and strains. This lifecycle information will also be stored on the component.

2.2. Methods

The main methods which are involved in this project are the segmentation of the production and the simulation. With the segmentation a production is separated into different independent units. The units of a structure are often characterized by the integration of planning and indirect functions and often act as a cost or profit centre. Generally the segmentation is a method to develop a structure of a production based on criteria like customers, units, lot sizes, technologies or quality. Mainly two extremes are widely used to build a structure of a production: the function orientated and the process orientated approach. But often occurs a mixture of these two structures, for example a factory is divided into different levels: from the station / workplace to the plant. While the whole plant can have a process orientated approach with different lines for the products, the lines itself can be organized in a functional way. With the product orientation the main objective is to reduce the time and effort for the coordination compared to the production of multiple products. A requirement is a heterogeneous production program with high volumes. Advantages in the specialization of the segments will be achieved. Interferences between different segments will be avoided.

The second important method for the use of Gentelligent components for the segmentation in a production is the simulation. It is used to simulate the new control concept based on the Gentelligent parts. Factory structures are analyzed in terms of their logistical performance. It is planned to compare different control concepts: the concepts of today and the Gentelligent concepts which use the component inherent data.

2.3. Adapted Factory Structures

New factory structures will be developed which are based on the concept of Gentelligence. The use of the component inherent abilities and also the use of resources which use the Gentelligence create new criteria for the segmentation of factories. They go beyond the traditionally used criteria like function or process orientation. Potential criteria are for example the information technology, the amount of required information or the presence of Gentelligence itself. This means for a factory structure that it may have a basic segment in which the components are produced in a traditional manner with no variants. The other segment could be a "Gentelligent segment" which is characterized by variants. By having all the necessary information for the production on the component, the variants steer their own way through the production by the given information. This segment would not only use the Gentelligent information for the logistic. It also will include new equipment for using the Gentelligent information. This equipment will be necessary for the exchange of information. This segment will also lead to a cost saving compared to equipping the whole factory with Gentelligent information equipment.

3. Part Driven Assembly Control

Enterprises today are confronted with a constantly changing world of business and competition. The orientation necessary on increasing customer demands, on innovation and high quality products leads to growing complexity and variety in a company's linecard. The assembly as the last step in the value adding chain thus continually takes on greater significance. Due to the extent of the value adding at this state, logistic problems and poor quality during the assembly have a particularly strong impact. An enterprise's competitive edge in the areas of cost, time and quality is thus determined to an increasing degree by the assembly [10]. The logistic control of the assembly is therefore the focus of attention, since it decisively influences the assembly area's logistic efficiency.

3.1. Assembly Processes

Assembly means that a number of parts "n" is put together to one product. The premise that assembly processes can take place is that the required number of parts is allocated. The correct allocation of parts by an earlier production stage which is of highly importance is considered in this section. The delay of a single part leads to the delay of the associated customer order. Regarding a make-to-order process the maximum date deviation of allocated parts determines the extent of delay of the customer order if the pace of the assembly process itself can not be forced. However, all other parts need to be buffered, even if they are made available too late in relation to the planned demand date. The resulting situation can be summarized as follows: Despite high stock levels and resulting high capital tie up the customer order cannot be started punctually. The customer order is likely to be tardy.

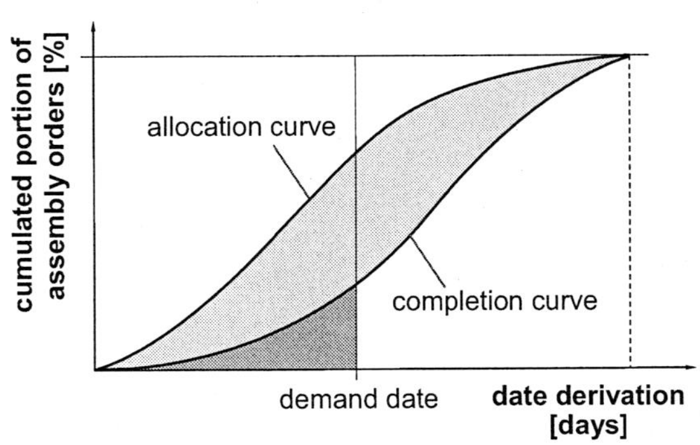

Fig. 1. Allocation Diagramm.

The customer order can only be fulfilled if all required parts manufactured in an earlier production stage as well as those delivered by a supplier are available. The effects (work in process - WIP buffers, date deviations of customer orders) of date deviations in earlier production stages as well as in sourcing can be monitored. To analyse the logistic coordination the entirety of customer orders must be observed. This can be done with the help of the Allocation Diagram, which is visualized in fig. 1. The Allocation Diagramm was developed at the Institute of Production Systems and Logistics to support the analysis of the allocation quality of several customer orders [11], [12]. In order to enable the logistic modelling of the allocation situation of multiple customer orders the planned demand date of each order is first normalized by being set to zero. To generate the Allocation Diagram the allocation time of each order will then be transferred into a diagram considering the normalized date required. In a first step the customer orders are sorted regarding the date of the last allocation. The connection of those dates defines the so called completion curve. In a second step the assembly orders are sorted regarding the date of the first allocation.

With the help of the resulting Allocation Diagram it is possible to make statements about the quality of the allocation for customer orders. The area between the completion curve and the allocation curve (light-grey area) corresponds to the stock of customer orders waiting for the allocation of parts.

The area defined by the normalized demand date and the completion curve describes the stock of assembly orders allocated with all required parts before the planned demand date (dark grey area). The value of the completion curve in the case of a date deviation of zero divides the analyzed customer orders into orders without delay and orders with delay caused by an inadequate allocation. Both areas described and the portion of customer orders without a delay in allocation are key performance indicators of logistic coordination of customer orders and thus of allocation quality.

3.2. A Part Driven Approach for Controlling the Assembly

The function of production control is to implement the given production plan even when there are disruptions, which are frequently unavoidable [13]. The assembly control now has the function of ensuring the precise synchronization of the procurement storage and manufacturing orders' timing both with regards to the scheduling of the assembly orders as well as the logistic costs. In particular, there is the question of how the enterprise's upstream internal/external processes can be controlled in order to improve their logistic coordination and to maintain the planned schedule.

A real-time capable method, which utilizes the Gentelligent Parts' special properties for controlling assembly processes, is illustrated in fig. 2 for a simple assembly process with three parts assembled to one

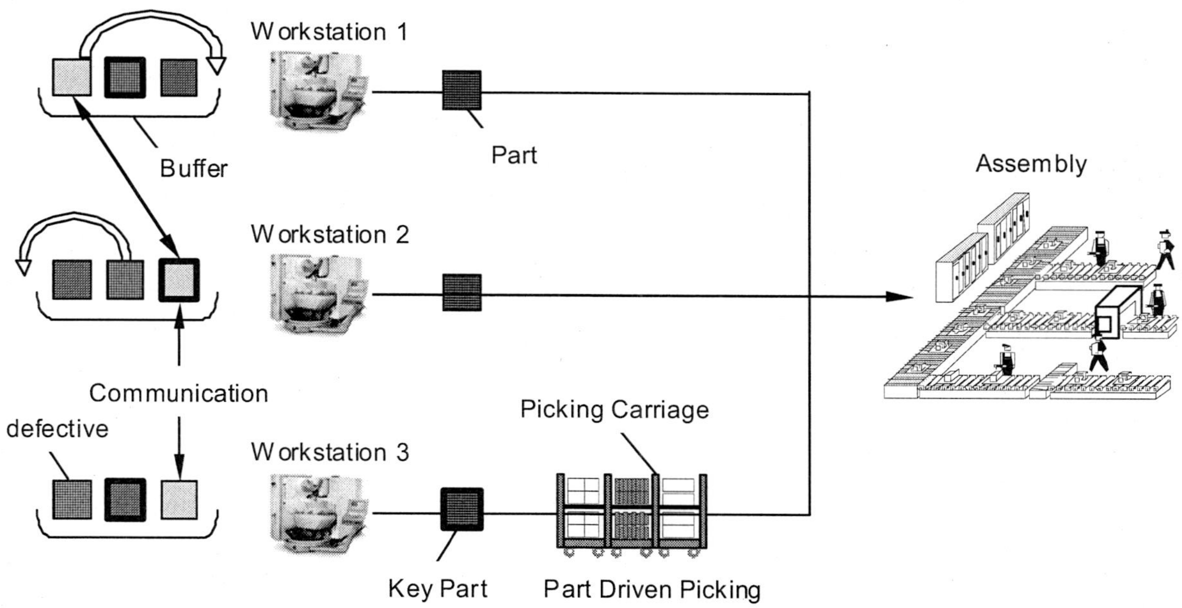

Fig. 2. Part Driven Production Planning and Control.

product. First of all, sequencing and capacity control functions are met. Regarding the requirement of assembly processes outlined in section 3.1 the method has to ensure the allocated parts reach the assembly area with no delay and not much too early. Furthermore, parts of one customer order have to be allocated at the same time.

For each of the products that are to be assembled, a key part is determined, which can communicate with the remaining product parts. The key part knows from the beginning of production, what the target completion date is for the module as well as the planned assembly start date viz. the demand date of the parts to be allocated.

Based on the inter-operation time which remains to the target completion date, the components will decide amongst themselves the sequencing of the processing on the workstations. The part with the least remaining inter-operation time is prioritized. Thus if a part falls behind, and others have enough time to wait, a decentralized order sequencing change will be carried out. If under no circumstances, one of the module's parts to be assembled can meet the scheduled start date (i.e. defective part), then all of the other parts in this module correspondingly reset their priority. This ensures that no unnecessary WIP accumulates on an assembly system with a high degree of added value, thus avoiding the uncalled for handling effort.

The goal of the algorithm that this is based on is to maintain the given targeted completion date and ensure that all parts in a group arrive at an assembly workstation simultaneously. The algorithm is based on the idea to assign a priority number to each Gentelligent Part to be assembled. Gentelligent Parts with a high priority possess a low priority number. The communication of parts is assured by reverting to the priority numbers of other Gentelligent Parts of the same product. Thus disturbances are considered and the requirement of simultaneous arrival of all parts can be achieved.

As soon as all of the parts to be manufactured and assembled have passed through the production, the key part will trigger the picking of the remaining parts from the stores based on the backed up part list for the assembly order. To avoid a shortage of allocated parts in a store the stock levels of parts allocated by a store have to be calculated appropriately [14].

The parts must meet requirement of timeliness, synchrony and punctuality. Considering the Allocation Diagram visualized in fig. 3 the allocation curve and the completion curve will rank closer and be precipitous in the area proximate left of the demand date. Thus all demand dates should be met and no parts should have to wait one other parts of one product to be assembled.

In order to avoid nervousness in the system a frozen zone can be defined for each buffer. In this

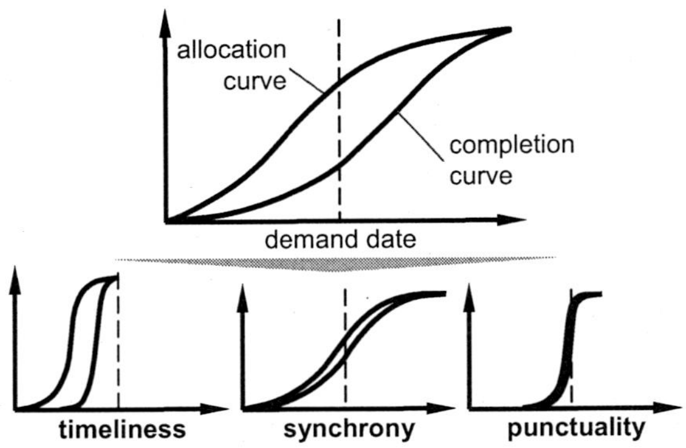

Fig. 3. Allocation Diagramm.

context defining the frozen zone means that no more sequence shifts are made by the above outlined algorithm. The dimension of the frozen zone can be defined by a number of orders, by an amount of work content or by the remaining inter-operation time at the regarded work station. The method introduced here should increase the scheduling reliability and reduce both the WIP and the handling effort required during the assembly and the allocation.

4. Results

As pointed out, the logistic efficiency of the production demanded by the customers is largely dependant on the availability of materials for the assembly. This is in turn determined through the logistic coordination of the assembly with material supply process. At the interface between the assembly and the upstream manufacturing, procurement and storage processes, a number of supply orders generally converge into one assembly order. This type of material flow is referred to as converging or synthetic. The complete availability of all the materials necessary for the assembly order at the planned time is required in order for an enterprise to achieve a superior logistic performance. This will be achieved by the use of the special properties of the Gentelligent Parts.

The central challenge of assembly control is coordinating the key dates for the material delivery process with the assembly orders. One the one hand, prematurely delivered procurement orders or completed individual manufacturing orders create inventories as long as the remaining material for the assembly order is not available. On the other hand, material which is available too early has to be buffered before the assembly, resulting in costs for processing and tied-up capital.

The new real-time capable control method, combined with new factory structures optimized for Gentelligent components will lead to new material flow concepts which offer from todays point of view cost savings in the production.

Acknowledgements

The research results presented here were acquired within the context of the Collaborative Research Centre 653, sponsored by the German Research Foundation.

References

[1] Baumgarten, H.; Thoms, J.: Trends und Strategien in der Logistik: Supply Chains im Wandel. Bereich Logistik TU Berlin, Berlin 2002.

[2] Nienhaus, J; Schnetzler, M.; Sennheiser, J.; Weidemann, M.; Glaubitt, K.; Pierpapoli, F.; Heinzel, H.: Trends im Supply Chain Management: Ergebnisse einer Studie mit mehr als 200 Unternehmen. Zentrum für Betriebswissenschaften der ETH Zürich, Zürich 2002.

[3] Frese, E.: Grundlagen der Organisation: Konzept – Prinzipien – Struktu-ren. 6. überarb. Aufl., Gabler Verlag, Wiesbaden 1995.

[4] Kieser, A., Kubicek, H.: Organisation. 2. überarb. Aufl., De Gruyter-Verlag, Berlin 1983.

[5] Henn, G.; Kühnle, H.: Strukturplanung. In: Schuh, G. und Eversheim, W. (Hrsg.): Betriebshütte Produktion und Management, 7. Aufl., Springer-Verlag, Berlin 1996.

[6] Ropohl, G.: Allgemeine Technologie – Eine Systemtheorie der Technik. 2. Aufl., Carl Hanser Verlag, München Wien 1999.

[7] Harms, Th: Agentenbasierte Fabrikstrukturplanung, In: Berichte aus dem IFA, Verlag: Produktionstechnisches Zentrum GmbH, Band 02-2004, Diss. Universität Hannover.

[8] Fahlbusch, T.; Franke, S.; Overmeyer, L.: Direct dispensing of 3D light guiding structures; POF & MOC 2006 the joint international conference on plastic optical fiber & microoptics 2006, Seoul, Korea, September 2006.

[9] Hansen, S.; Rissing, L.; Gatzen H.-H.: A Concept For a Toolbox For Modular Magnetic Microsensors- "210th Meeting of the Electrochemical Society", October, Cancune, Quintana Roo, Mexico.

[10] K. Feldmann, S. Slama, "Highly flexible Assembly – Scope and Justification", Annals of the CIRP, vol. 50, no. 2, pp. 489-498, 2001.

[11] H. Kettner, H. „Neue Wege der Bestandsanalyse im Fertigungsbereich", DGfB, Berlin, 1976.

[12] P. Nyhuis, M. Schmidt, F. S. Wriggers, "Reduction of Capital Tie Up for Assembly Processes", Annals of the CIRP, vol. 57 no. 1, 2008.

[13] H.-H. Wiendahl, G. von Cieminski, H.-P. Wiendahl, "Stumbling blocks of PPC: Towards the holistic configuration of PPC systems", Production Planning and Control, vol. 16, no. 7, Taylor and Francis, pp. 634-675, 2005.

[14] S. Lutz, H. Lödding, H.-P. Wiendahl, „Logistics-oriented inventory analysis", International Journal of Production Economics, vol. 85, pp. 217-231, 2003.

Innovative Production Machines and Systems
D.T. Pham, E.E. Eldukhri and A.J. Soroka (Eds.)

Inherent Mechanical Data Storage on Micro-Structured Surfaces

R. Dragon[a], I. Steinbrenner[b], J. Ostermann[a], B. Denkena[b]

[a] *Institut für Informationsverarbeitung, Leibniz Universität Hannover, Appelstr. 9a, 30167 Hannover, Germany*
[b] *Institute of Production Engineering and Machine Tools, Leibniz Universität Hannover, An der Universität 2, 30823 Garbsen, Germany*

Abstract

The Collaborative Research Centre "Gentelligent Components in their Lifecycle" deals with the development of techniques for the creation and usage of innovative mechanical components. Developing manufacturing methods which enable data storage on the component itself is one major aspect. Here an approach using micro structures on component surfaces to carry the coded information is presented. These micro structures are created by a fast-tool-servo during a longitudinal turning process. Data, that has been stored in this way, can be read out by a newly developed optical method using directed illumination. The complete data storage system is described with an emphasis on micro-structuring and the readout process.

Keywords: Production Process, Surface Structuring, Mechanical Information Storage, Surface Reconstruction

1. Introduction

In the context of the Collaborative Research Center (CRC) 653 "Gentelligent Components in their Lifecycle", components with novel properties as well as concepts, methods and techniques for their manufacturing and utilization in production engineering are developed [1]. The interdisciplinary CRC's long-term objective is the integration of components with their corresponding information for reproduction as well as stress information from their lifecycle. The term "gentelligent component" was introduced for these components in order to refer to the genetic and intelligent character of these innovative components.

A gentelligent component inherently stores all information necessary for its distinct identification, reproduction as well as for its manufacturing process. Unlike exherent data storage like on centralized servers or attached RFID tags, information on a gentelligent component is stored during the whole component lifetime. Thus, the information cannot be mixed up or get lost. Furthermore, inherent data storage provides an excellent basis to secure from plagiarism. By using adequate materials and sensors, gentelligent components are enabled to detect, process, store and interpret specific information collected during their use. Examples of these pieces of information are applied loads, accelerations and temperatures. Having both the initial state as well as information about the component's lifetime, it will be possible to replace the component on the eve of its failure and not after a certain age. Thus, storing and gathering information is the main focus of the CRC.

In this paper, a system for inherent data storage

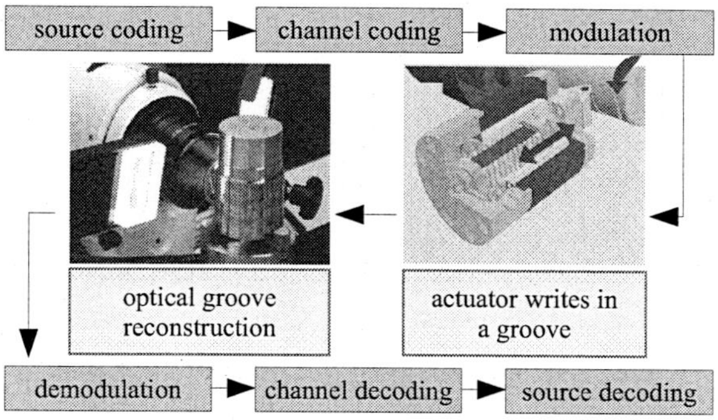

Fig. 1: Procedure for digital data storage.

is presented. The information is saved inside the micro-structured surface of a component. Typically, surface structuring in mechanical engineering is used for functional reasons such as reduced flow resistance [2], improved tribological characteristics [3, 4] or optimized mechanical properties. The structures mentioned in this paper are generated during a turning process by a fast-tool-servo. This tool technology is characterized by high tool acceleration and is usually used in machining precisely contoured and structured optical surfaces [5, 6, 7]. By variation of the excitation of the tool in phase and amplitude, different surface micro structures are produced. In this way a macroscopically non-visible groove is inserted into the surface which can be read out by optical means later. It is the same principle as gramophone records or the first video data media like the capacitive electronic disc [8] are using. As the storage principle is designed for simple and fast writing during the production and insensitivity to mechanical load, data density is quite low but sufficient for gentelligent components.

2. System Overview

For the whole storage system, the standard model for digital data transmission over a channel is used (Fig. 1). Only the channel, where data may be disturbed or corrupted, differs from ordinary data transmission. Here it consists of the actuator's writing, the transmitted signal on the component's surface and of the optical readout, which represents the reconstruction of that signal. The advantage of using this standard model is that source coding, channel coding and the modulation method may be re-used from existing systems.

Source coding is the efficient digital representation of the information being transmitted. For portability, this could be a string list representation in the form *keyN=valueN* to assign the *N*th key a value. If hierarchical data is used, xml [9] would be a flexible standard solution for source coding. The data size can be reduced by an additional compression using e.g., the bzip2 algorithm [10].

Following, redundancy is added during channel coding to protect the data against corruption during the transmission. Here, *Turbo-Codes* [11] are used which make sense for high data rates in combination with low signal-to-noise ratios.

Modulation is the step from the digital to the analog domain. Digital symbols are converted to analog signal runs according to the characteristic of the data transmission. The analog signal is used to control the piezo actuator cutting in a groove with varying depth.

Now, the component's surface is characterized by a groove with varying depth containing the stored data. To reconstruct it during the readout process, optical methods were examined using a microscope with small magnification.

Then, the signal read out is demodulated. This is not inverse to the modulation as the actuator adds (mostly linear) distortion to the signal which needs to be compensated. Knowing the actuator's characteristic i.e. by also writing in an impulse or a chirp, the distortion can be removed. The demodulation then recovers digital symbols as well as an estimation of the bit error rate. As there is redundancy, the channel decoder may use this to find the most probable decoding of the symbols. Finally, source decoding recovers the original information.

3. Micro-Structuring - Writing Data

3.1. Fast-Tool-Servo

In many occasions fast-tool-servos are used to finish free-formed optical surfaces but another application in turning operations is active vibration control as described by Denkena in [12]. Thus, they are designed to withstand relatively high cutting forces in comparison to the needs of a pure structuring tool. So the first natural frequency in advance direction is low, which does not allow high data density storage. In order to realize a significantly higher frequency range, a new tool called *Structool* has been developed with a first natural frequency of 6 kHz. This was mainly achieved by a guiding unit out of titanium, which contributes a significant reduction of the moving masses. The link between the static part and the moving part of the titanium frame is of flexure hinge design, thus reducing additional membranes and screw connections (Fig. 2).

The tool design allows the use of different configurations of piezo-electric elements with up to

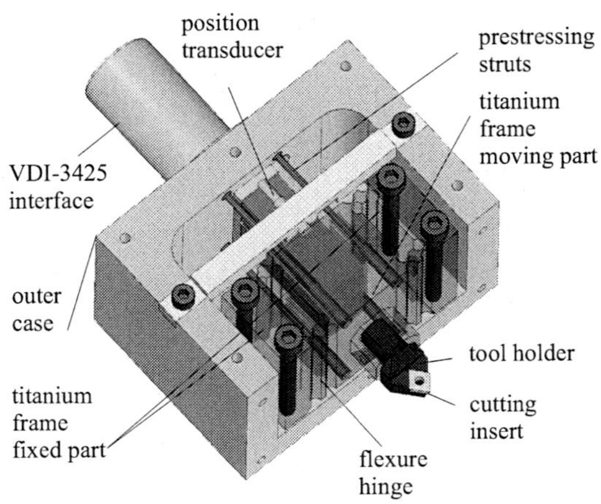

Fig. 2: The Structool.

four elements used at a time. Each element has a dimension of 14 x 14 x 20 mm and a stiffness of 500 N/µm. The maximum force generation per element is 11 kN, the resonance frequency is 32 kHz. The tool can be used with only one piezo-electric element, with two elements mechanically connected in series or in parallel as well as with a combination of these. The first structuring results were achieved with only one piezo-electric element and may thus be improved by amplitude or bandwidth.

3.2. Modulation Method

For best data transmission, the coded data have to be modulated according to the channel's characteristic, which is nearly the characteristic of the actuator in this case as the readout hardly adds systematic distortion. It was found out that the actuator behaves as an linear time-invariant system that can be described by its impulse response or by its transfer function.

In Fig. 3, the actuator's transfer function is displayed. The resonance at the first natural frequency of approximately 6.5 kHz can be seen clearly. Also important is another smaller resonance at around 5 kHz. Until that frequency, the phase difference lowers linearly from -90° near 0 Hz. The reason for the 90° phase offset is that the actuator's input signal is a control current

$$I = C\frac{dU}{dT}$$

and the measured excitation is proportional to the impressed voltage *U*. So without any mechanical restrictions, an input cosine will become an output sine wave.

For digital data transmission, there are two constraints that have to be fulfilled. First, the

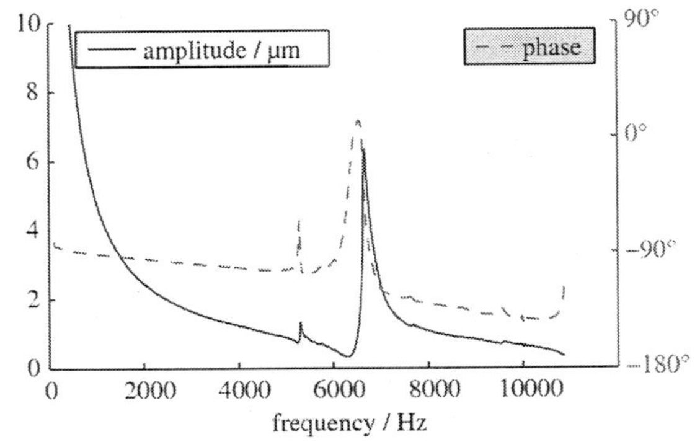

Fig. 3: Transfer function of the acutator.

amplitude of the transmitted signal should be clearly distinguishable from distortions to achieve a good signal-to-noise ratio. Until 5 kHz the baseband would be sufficient with the lowest amplitude at about 1 µm. The amplitude could be further improved by using two piezo-electric elements and pre-emphasis on higher frequencies.

The second constraint is that there should not be any phase distortion that jams digital data transmission. Phase distortion is generated by a non-constant group delay, which is the negative deviation of the phase response. Because of the 90° phase offset around 0 Hz baseband signals would disturb the transmission. This means that the modulated signal run has to be mean-free. The same occurs with signals over the resonance frequency of around 5 kHz. In the residual band with approximately 5 kHz bandwidth, 5 kBit/s may be stored with a binary modulation method. That bandwidth could be extended using two piezo-electric elements in parallel. Further bandwidth extensions could be achieved using multi-carrier modulation methods which are able to occupy a second frequency band e.g., from 7 to 9.5 kHz, for data transmission.

3.3. An Example for Written Data

In the following, an example for writing data is made. The source-coded information is the uncompressed ASCII string "SFB 653" (means CRC 653) of 56 bit data size. There was no channel-coding used for better visibility. As modulation method, a simple mean-free translation from the digital symbols to analog signal runs was chosen. Every logical one physically is assigned a full sine wave leading to a positive sine wave movement of the actuator. A logical zero means no movement of the tool. The run of the actuator is shown in Fig. 4.

The green line shows the position of the tool. The red crosses show the target position in terms of

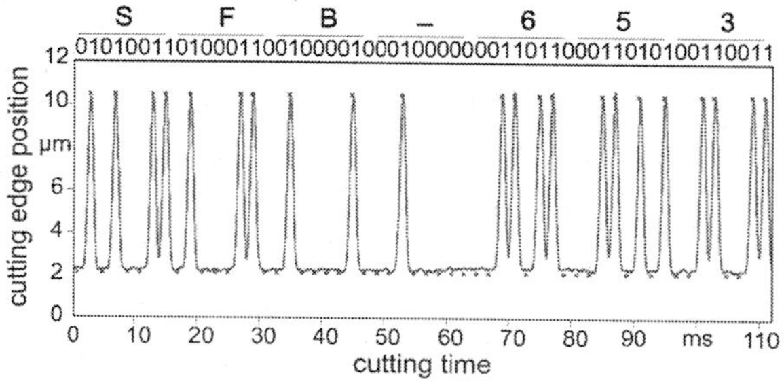

Fig. 4: Example for a signal written as a groove onto the surface.

displacement, which represents either a zero with a position value of 2 µm or a one with a position value of 10 µm. The demodulation in this case is simply distinguishing these two states. It can be seen, that their positions clearly differ at the chosen sampling frequency of 500 Hz and thus the demodulation will succeed. At higher frequencies the maximum movement of the cutting edge decreases. According to the transfer function, a double-amplitude movement of 2 µm can be obtained at a sampling frequency of 5 kHz, so that the resultant structures can be recognized by optical readout methods.

4. Reading Data

4.1. Readout Method

The readout here is the reconstruction of the groove's run. This could be achieved using expensive surface reconstruction instruments. But the readout should be possible in a simple way so that a mechanical component could e.g., be inspected at a service station. Thus, shape reconstruction methods with a customary microscope with low magnification were evaluated, which later can be adapted to cheap hand-held units. As multiple views from the surface are needed for the readout, the setup here includes an apparatus for rotating and translating the mechanical component. A special channel coding could later be used so that the whole information may be retrieved from one view.

The developed readout method uses directed illumination to recover the groove's depth. The surface of the mechanical component here is

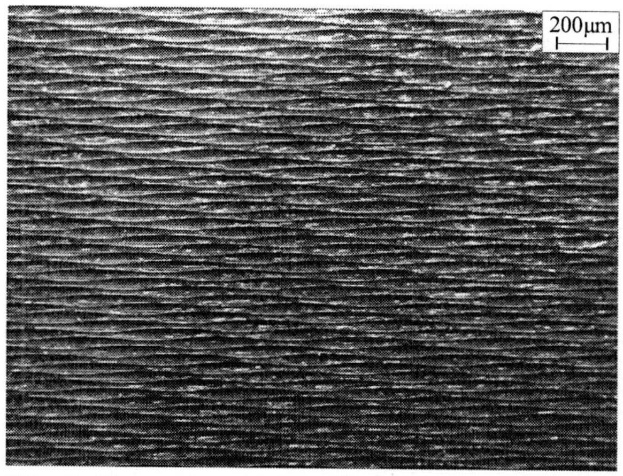

Fig. 5: The surface of the component under directed illumination from the top. The grooves run horizontal. The topmost groove is magnified in Fig. 6.

illuminated with parallel light from the top. As it is highly reflective, it only appears light at positions with a specific surface normal. Here, it is assumed the groove consists of two flanks which meet in the middle of the groove and have a constant surface normal. The angle of the incident light is thus chosen that way, the lower flank reflects into the direction of the camera. The depth of the groove can then be determined from the flank's width which is approximately proportional to its depth. It should be noted that this is just a simple model for the groove. Variations will be corrected during the demodulation as they are included in the system's transfer function. In practice, there are many disturbances which also appear light, but the grooves (Fig. 5) as well as their varying widths (Fig. 6) are clearly recognizable.

4.2. Groove Reconstruction

Under the assumption that the grooves of the components run horizontal in the microscope image, a one-dimensional frequency analysis in vertical image direction is performed. As the feed of the turning process is constant, the vertical offset between the grooves is constant, so a vertical frequency analysis should find a peak at that distance.

Thus, the first step is to partition a microscope image $I(x, y)$ such that each part contains only one

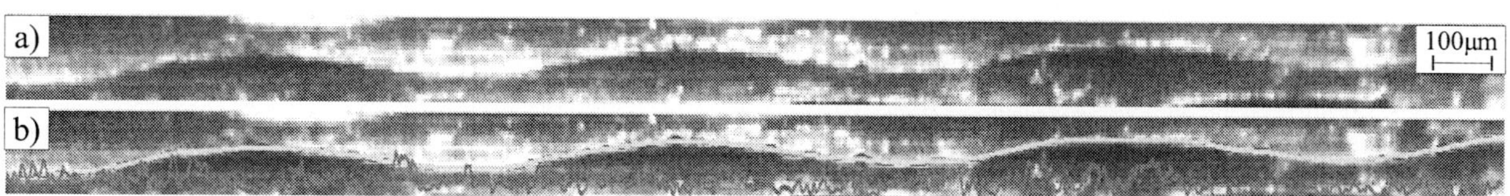

Fig. 6: a) Original image of the groove and b) its reconstructed run (red, smoothed in yellow). Green is the probability density of the Markov model. The images were stretched vertically by two for better visibility.

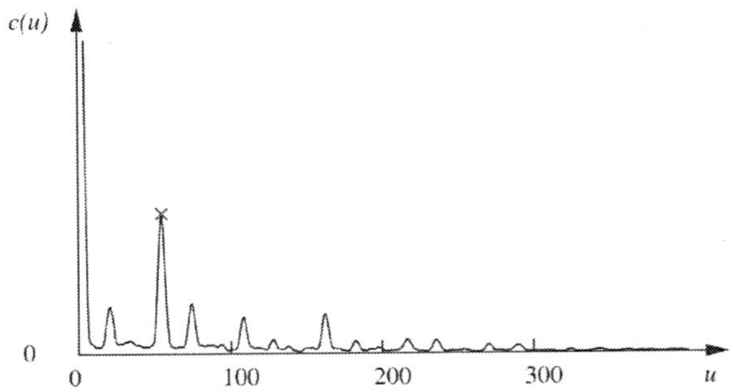

Fig. 7: Amplitude c of frequency components u of Fig. 5 in vertical direction. The peak at $u=54$ corresponds to a groove distance of 14.2 px.

groove. First, several windowed one-dimensional discrete Fourier transformations are averaged in an appropriate manner to make the algorithm robust with respect to image noise and visible cutting artifacts. The highest amplitude peak $c(u_0)$ besides the mean value at frequency $u=0$ is searched for. A typical example for a frequency distribution is displayed in Fig. 7 where the peak is visible at $u=54$. From the frequency u_0 at that peak, the groove distance may be determined. Finally with the help of the phase offset $\varphi(u_0)$, the exact position of the border between the grooves is computed. A microscope image containing several grooves may now be partitioned into images containing only one groove like displayed in Fig. 6.

It has to be considered that only the lower flank is illuminated and that the flank's width is proportional to the groove's depth. As the middle of the groove is known from the previous step, the lower groove edge has to be tracked to reconstruct its depth. A heuristic p for the edge being at a specific position (x_i, y_i) is the negative discrete deviation of the image in *y*-direction

$$p(x_i, y_i) = -\left.\frac{\partial I(x_i, y)}{\partial y}\right|_{y=y_i}.$$

To enhance the continuity of the found edge, a Markov model with a regular topology is introduced. Each state S_{x_i, y_i} stands for the edge passing the pixel at position (x_i, y_i). Like shown in Fig. 8, state transitions are only allowed from x_i to x_i+1. Their transition probability is assigned a normal distribution with a mean value of y_i. A Viterbi algorithm is then able to find the most probable run of the edge from the left to the right side of the image. It can be seen from Fig. 6 that even in the case of heavy perturbation, the found edge is correct.

Finally, all groove pieces from multiple microscope views are put together. To determine the offset between two images, normalized cross-corre-

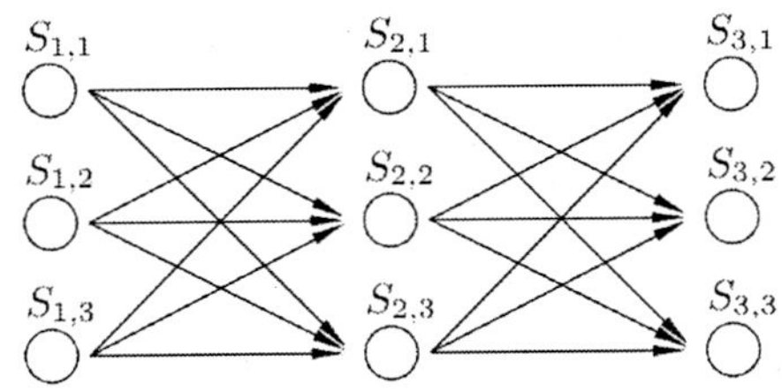

Fig. 8: The Markov model used with a regular topology. To model the run of the groove, transitions are allowed from all states S_{x_i, y_i} to $S_{x_i+1, y_{i+1}}$.

lation coefficients are minimized. When the data is rearranged according to the groove- and image offsets, the written signal is reconstructed. Finally, the demodulation- and decoding algorithms will recover the stored data. Redundancy is chosen so that disturbed or destroyed surface regions can be recovered until the readout method itself fails. Tests have shown that oxidation has no influence on the data integrity. So the here-presented readout method is practicable.

5. Conclusion

A complete system for data storage on the surface of mechanical components was presented. The standard model for digital data transmission was applied. The channel of the communication system here consists of a fast-tool servo writing in data during a turning process and of the optical reconstruction of the written groove. The other parts of the communication system, which are source coding, channel coding and modulation, thus do not have to be reinvented.

Only the modulation part which has to take care of the characteristic of the channel has to be adapted. It was shown that for the current configuration a frequency band with a size of 5 kHz may be used for data transmission.

To read the data written in the groove of varying depth, a three-step method utilizing directed illumination is applied. First, the image is partitioned using frequency analysis in vertical direction. Then, the groove is tracked, greatly enhanced by the application of a Markov model with a regular topology. To put the groove parts exactly together to the one-dimensional signal written in by the actuator, overlapping regions of several images are taken.

Thus it was successfully demonstrated that inherent data storage on surfaces using simple means is possible. To determine optimal writing and reading parameters as well as the best-suited modulation method, further investigations are carried out.

Acknowledgment

Research was undertaken within the Collaborative Research Center 653 (SFB 653) which is founded by the German Research Foundation (DFG).

References

[1] Denkena, B., Hasenfuß, K., Liedtke, C., 2005, Gentelligente Bauteile - Genetik und Intelligenz in der Produktionstechnik, ZWF (10), pp. 569-572

[2] Denkena, B., Reichstein, M., Wang, B., 2006, Manufacturing of Micro-Functional Structures by Grinding, Annals of the German Academic Society for Production Engineering (WGP), Vol. XIII/1, pp. 31-34

[3] Lierse, T., Friemuth, F., 1998, Schleifen keramischer Funktionsflächen. IDR (32) 3, pp. 228-235

[4] Tönshoff, H.K., Denkena, B., Friemuth, T., Reichstein, M., 2003, Precision Grinding of Components for Aerostatic Micro Guidance, Precision Engineering 27, pp. 185-188

[5] Miller, A.C.; Cuttino, F., 1997, Performance Evaluation and Optimization of a Fast Tool Servo for Single-Point Diamond Turning Machines, Proc. of SPIE, Optical Manufacturing and Testing II, Vol. 3134, pp. 318-328

[6] Lu, X.; Trumper, D.L., 2005, Ultra Fast Tool Servos for Diamond Turning, CIRP Annals Vol. 54/1, pp. 383-388

[7] Lu, X.; Trumper, D.L., 2004, High Bandwidth Fast Tool Servo Control, Proc. of American Control Conference, pp. 734-739

[8] Isailovic, J., 1985, Videodisc and Optical Memory Systems, Prentice-Hall, chapter 2.4: Capacitive Videodiscs

[9] W3C, 1997, Extensible Markup Language (XML), http://www.w3.org/XML/Core/

[10] Seward, J., 1996, A program and library for data compression, http://www.bzip.org

[11] Berrou, C., Glavieux, A., Thitimajshima, P., 1993, Near Shannon Limit Error Correcting Coding and Decoding: Turbo Codes, International Conference on Communication, Geneva, pp. 1064-1071

[12] Denkena, B., Harms, A., Lhermet, N., 2004, Tool Adaptor for Active Vibration Control in Turning Operations, 9th International Conference on New Actuators, "Actuator 2004", Bremen, 14-16 June 2004

Innovative Production Machines and Systems
D.T. Pham, E.E. Eldukhri and A.J. Soroka (Eds.)

Integration of a Temperature Sensor into a Modular, Multifunctional Micro Sensor Family

T. Griesbach, L. Rissing, H.H. Gatzen

Institute for Microtechnology, Center for Production Technology, Leibniz Universitaet Hannover, Garbsen, Germany

Abstract

A new approach for the development and fabrication of micro sensors is applying a modular sensor concept. By using a toolbox of standardized basic sensor elements, a multitude of sensor variations can be realized. So far, a family of modular magnetic micro sensors, covering both strain measurements and magnetic material analysis, was developed. With this work, the sensor family is augmented by thermal measurement capabilities. To do so, a positive temperature coefficient (PTC) Pt temperature micro sensor was developed and integrated into the magnetic sensor family. The work aims at supporting "genetically intelligent" (gentelligent) machine tools.

Keywords: Temperature sensor, magnetic micro sensors, modular sensor concept

1. Introduction

Modern production technology requires the acquisition of a multitude of process data. On one hand, this information serves to ensure the process reliability during the production and on the other hand, it help to verify the quality of a product at the end of the production.

A novel approach is purposed within the DFG´s (German Research Foundation) Collaborative Research Center 653. To come up with "genetically intelligent" (gentelligent) machine tools or components, various sensors will be developed to gather process data not only during the production but also in excess of the entire life cycle [1]. In order to achieve this goal, the sensors will be integrated into machine tools or components to acquire process data at the device directly under testing conditions. Thus, the sensors are intended for controlling material conditions during the whole life cycle of a component. In contrast to existing systems, these sensors will be fabricated completely by applying Micro Electro-mechanical Systems (MEMS) technology. As a result, the small dimensions of such micro sensors enable the collection of information at places so far not accessible.

To easily combine various sensor types, a modular design is of advantage. A main benefit of such a modular design is that it often allows to fabricating a number of different sensors with similar fabrication sequences. A challenge is to avoid process compatibility issues when fabricating sensors requiring rather different process sequences.

To realize a variety of different sensor functions, the concept of a modular multifunctional micro sensor family was developed [2]. The basis for this micro sensor family is the development, fabrication, and characterization of functional layers, which represent the basic sensor elements of the modular toolbox. This paper describes the integration of a temperature sensor into a family of inductive micro sensors. Depending on their layout, the latter ones either serve as strain gauge or *B-H* loop sensor.

2. Modular Micro Sensor Concept

Figure 1 presents the main elements of the modular component toolbox. The combination of these elements allows to realizing a multitude of sensor variations. So far, investigations focused on magnetoelastic, magnetoresistive, insulating, and temperature sensitive layers.

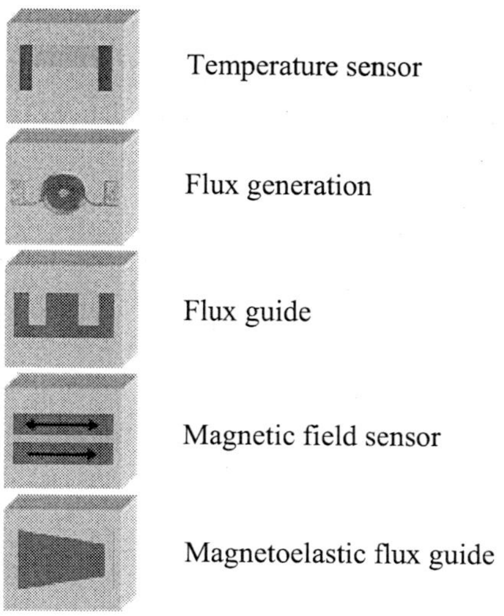

Fig. 1. Toolbox of sensor elements for modularly designed micro sensors

Magnetoelastic materials can be used in sensor systems to measuring force and strain. Magnetoresistive multilayer systems with layer thicknesses in the nanometer range can be applied to measure small magnetic fields. Conductors and insulators serve as elements for the fabrication of coil systems. Temperature sensitive layers permit the measurement of thermal effects.

To demonstrate the use of the tool box, two different micro sensor combinations were created [3]. By combining the appropriate sensor elements of the modular tool box, a strain gauge sensor for force measurement and a *B-H* loop sensor for acquisition of magnetic material properties were designed (Fig. 2). For both sensors, an E-shaped flux guide was used. The core geometry allows to focusing the flux directly into the magnetoelastic sensing film (strain gauge sensor), which is forming one leg of a transformer, or into the probe material (*B-H* loop sensor). In both cases, a single turn excitation coil was used, which wraps around a center pole of an E-shaped core. A great cross section allows to using a high excitation current. For signal detection of both the strain gauge and the *B-H* loop sensor, a multi turn spiral coil is used. Like the excitation coil, it wraps around the center pole as well.

The first sensor realized by applying the modular micro sensor concept was a strain gauge sensor shown in Fig. 3 [3]. It serves as a basis for a new modular sensor family, which not only measures strain or magnetic film properties, respectively, but also measures temperature. Such an additional measurement capability allows to augmenting mechanical measurement capabilities by thermal ones. The temperature measurement provided may be either used to "merely" monitor the temperature at the micro sensor location. However, the data may also be used to compensate for the temperature drift of the inductive micro sensors. A temperature compensation for the magnetic micro sensors is necessary if a high accuracy is required. The reason for the temperature drift in these sensors is the temperature dependence of the magnetic material's relative permeability μ_{rel} and the coil resistance.

(a) Strain gauge sensor

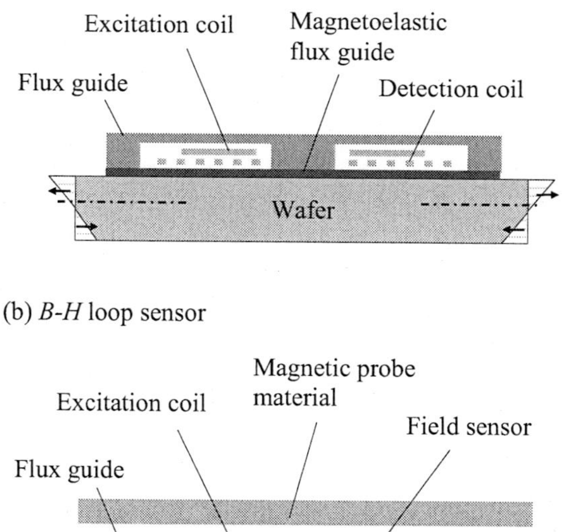

(b) *B-H* loop sensor

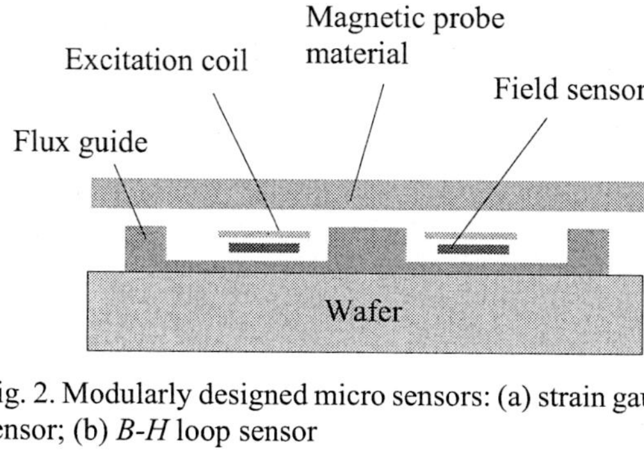

Fig. 2. Modularly designed micro sensors: (a) strain gauge sensor; (b) *B-H* loop sensor

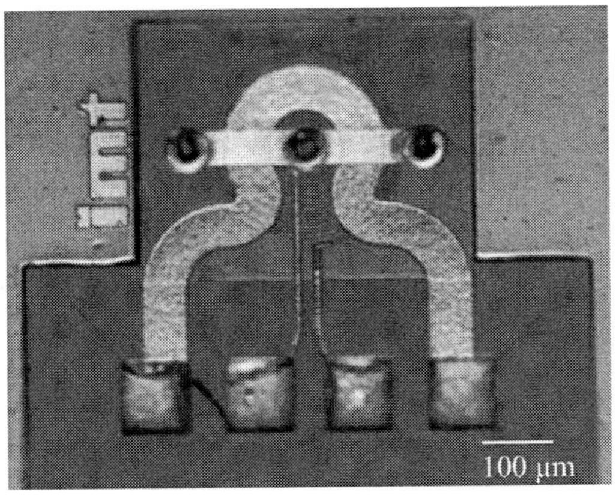

Fig. 3. Micrograph of a completed strain gauge sensor

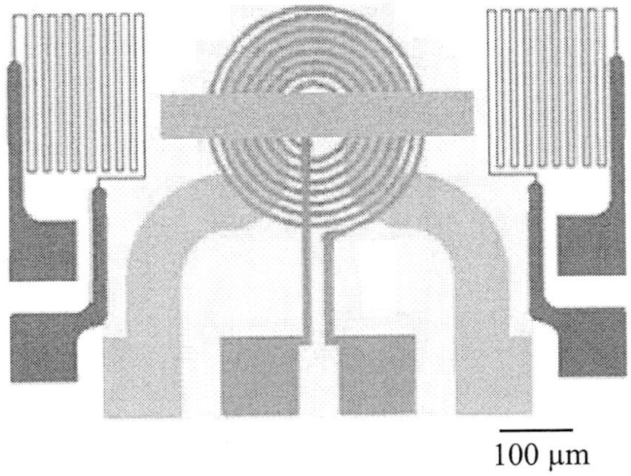

Fig. 4. Schematic representation of the integrated strain gauge sensor with temperature compensation elements

3. Design and Fabrication of the Temperature Sensor

The measurement principle of a temperature sensor is based on a change of the electrical resistance of a conductor as a function of the temperature. For metals, the resistance rises with an increase of the temperature, a behavior resulting in a positive temperature coefficient (PTC). As a resistor material, platinum (Pt) was chosen, which features an excellent resistance against corrosion and can be used for the measurement of temperatures from -50°C to +600°C [4, 5].

As temperature sensor geometry, a meander was chosen. Figure 4 presents a schematic representation of the strain gauge sensor design with two meander structures. The accumulated length of the meander amounts to 3,910 µm. The width of the leads is 5 µm, and the insulation distance between the leads also amounts to 5 µm.

Initial investigations concentrated on only fabricating the temperature sensor itself. For depositing the Pt leads, electron beam vapor deposition was chosen. For developing the process technology, investigations were carried out determining process parameters for the vapor deposition of Pt. As the optimal rate, a deposition of 3 Å/s was identified. A further optimization was required to achieve a sufficient adhesion of the Pt. Therefore, as an adhesion layer, Cr was chosen. Various adhesion layers of thicknesses and substrate temperatures were evaluated. As part of the investigations, Cr adhesion layers with 50 nm and 100 nm were deposited under various substrate temperatures (20°C, 100°C, and 150°C). An optimal adhesion for the Pt layers could be reached by choosing a 100 nm thick Cr layer and depositing it at a substrate temperature of 150°C.

Figure 5 shows the fabrication sequence for the temperature sensor. As a first step, the Cr adhesion layer was sputter deposited, followed by an Au layer serving as a conductive seed layer for the following electroplating process of the contact pads. To pattern the micromolds for electroplating, a positive photoresist was used. Next, the Cu contact pads were electroplated (Fig. 5a). Subsequently, the photoresist was stripped. A removal of the seed layer was also required to inhibit short circuits (Fig. 5b). It was executed by Ion Beam Etching (IBE). Afterwards, a Cr adhesion layer was deposited, followed by a 300 nm thick Pt layer for the temperature sensing.

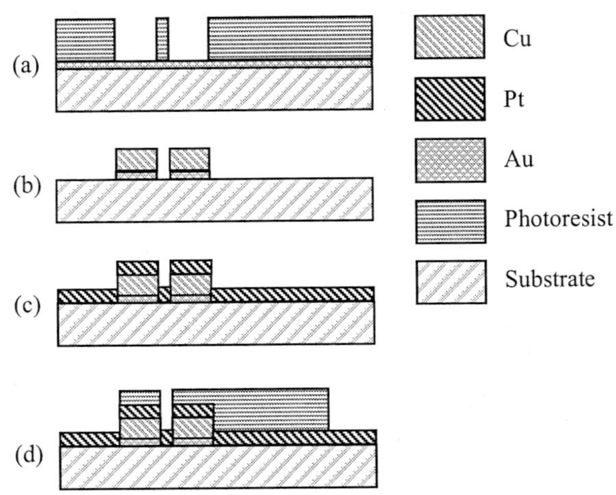

Fig. 5. Fabrication sequence of the temperature sensor

Both layers were deposited by electron beam evaporation (Fig. 5c). The etch mask for patterning the meander was created by photolithography, using an image reversal photoresist. The patterning itself was executed by an IBE process (Fig. 5d). As the next step, the photoresist was removed. Figure 6 depicts micrographs of the photoresist mask for the IBE process (Fig. 6a) and the finished meander temperature micro sensor (Fig. 6b).

(a)

(b)

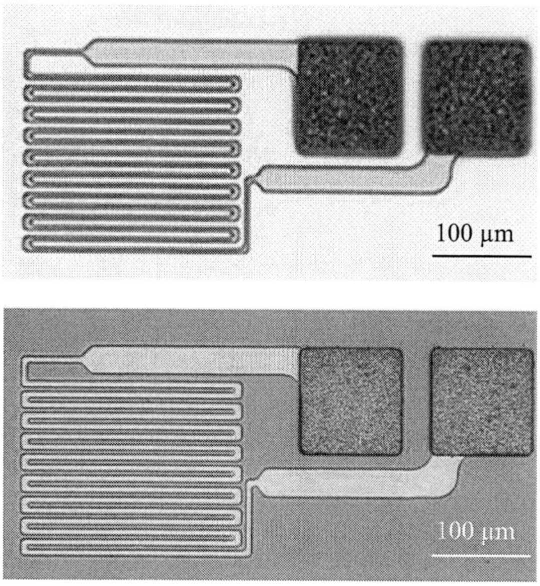

Fig. 6. Micrographs of the temperature sensor: (a) photoresist mask for IBE process; (b) completed meander structure

4. Measurements and Comparison to Analytical Calculations

To evaluate the fabricated temperature sensors, electrical resistance measurements were conducted. The aim of these investigations was the characterization of the temperature dependence of the electrical resistance R and the determination of the specific electrical resistivity ρ. The results were compared to theoretical values.

The resistance measurements were conducted by heating up the temperature sensors on a hot plate and conducting successive measurements of the ohmic resistance with an ohmmeter. The temperature range investigated was 20°C to 150°C. Figure 7 depicts the electrical resistance variation of a fabricated Pt meander structure as a function of the temperature.

The average resistance R at room temperature (20°C) was 831 Ω. A temperature rise to 100°C resulted in a nearly linear increase of the resistance to 896.7 Ω. For temperatures higher than 100°C, a non-linear increase of the resistance occurred.

The specific electrical resistance ρ was calculated

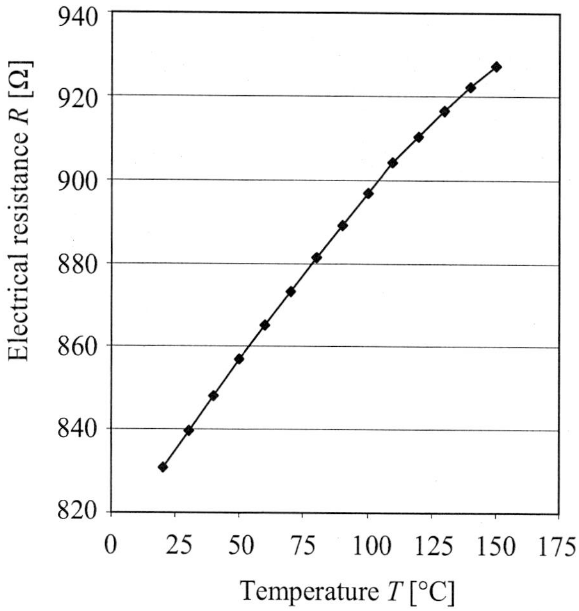

Fig. 7. Resistance variation of a Pt temperature sensor

using Eq. 1, where R is the electrical resistance, A the cross sectional area, and l the length of a uniform lead.

$$\rho = \frac{R \cdot A}{l} \qquad (1)$$

The calculated specific electrical resistivity ρ at room temperature was 31.9 µΩ·cm for the electron beam deposited Pt layers. In comparison, the specific resistivity ρ of the Pt bulk material is 11 µΩ·cm [4]. This result agrees with sheet resistance measurements on bulk and thin-film Pt which shows variations of up to 300 % [6, 7]. These deviations between bulk and thin-film material were caused by pinholes whose occurrence is typical for electron beam deposited metals.

5. Conclusion and Outlook

A temperature sensor applying a meander structure for compensation thermal effects in a strain gauge sensor during force measurements was successfully developed and fabricated. Investigations demonstrate that the electron beam vapor deposition of Pt for temperature sensitive layers is a suitable approach for such a temperature sensor. Experimental investigations proved the feasibility of the sensor principle. With such an addition, mechanical and material data measurements of the modular sensor family are augmented by thermal data acquisition capabilities.

The sensors are intended to be ultimately integrated into gentelligent machine tools or components to gather process data during the entire life cycle.

Acknowledgements

This research was sponsored in part part by the DFG (German Research Foundation) within the Collaborative Research Center (SFB) 653 "Gentelligent Components in their Life Cycle".

References

[1] B. Denkena, K. Hasenfuß, C. Liedtke, Zeitschrift für wirtschaftlichen Fabrikbetrieb, Spur, G. (Hrsg.), Vol .100, No. 10, 569, München, 2005.

[2] S. Hansen, L. Rissing, H.H. Gatzen: A Concept for a Toolbox of Components for Modular Magnetic Microsensors. Symposium on Magnetic Materials, Processes and Devices, 210th Meet. of The Electrochemical Society 2006, Cancun, Mexico, 2006, ecs transactions, Vol. 3, Issue 25, pp. 235-246, 2007.

[3] S. Hansen, L. Rissing, H.H. Gatzen, Modularly Designed Magnetoelastic Micro Strain Gauge, IEEE Trans. on Magnetics, Vol. 43, No. 6, 2385, 2007.

[4] C. M. Soriano, Platinum, IEEE, Potentials Magazine, Vol. 16, Issue 5, 29, 1998.

[5] X. Li, G.C.M Meijer, A low-cost, high-precision sensor interface for platinum temperature sensors, ASIC, Proceedings, 5th International Conference on Vol. 1, Issue 21-24, 2003.

[6] S. Smith, A.J. Walton, S. Bond, A.W.S. Ross, J.T.M. Stevenson, A.M. Gundlach, Electrical Characterization of Platinum Deposited by Focused Ion Beam, IEEE Trans. on Semiconductor Manufacturing, Vol. 16, No. 2, 2003.

[7] S. Smith, A.J. Walton, S. Bond, A.W.S. Ross, J.T.M. Stevenson, A.M. Gundlach, Test Structures for the Electrical Characterisation of Platinum Deposited by Focused Ion Beam, Proc. IEEE, Int. Conference on Microelectronic Test structures, Vol.15, 2002.

Innovative Production Machines and Systems
D.T. Pham, E.E. Eldukhri and A.J. Soroka (Eds.)

Investigation of magnetic magnesium alloys

Fr.-W. Bach, M. Schaper, D. Bormann, W. Reimche, G. Mroz, M. Rodman,

Leibniz Universität Hannover, Institute of Materials Science, An der Universität 2, 30823Garbsen, Germany

Abstract

Here are the test results of alloy produced by alloying the magnesium with magnetic particles. There is an influence of cooling rate during casting on the distribution of the magnetic precipitation. The resistance and the corrosion properties of the alloy were collected too.

Keywords: magnesium, magnetic properties, hardness, corrosion

Introduction

Within the framework of the special research area SFB 653 "Gentelligent components' life cycle - Use of inheritable, inherent component information in manufacturing", components with new types of properties are to be developed together with concepts, processes and technologies for their manufacture and use.

At the Institute of Materials Science of the Leibniz University Hanover, magnesium alloys were developed which exhibit magnetic properties [1].

Various manufacturing parameters were investigated which have an influence on the magnetic properties of finished magnesium alloys. The results of these investigations are presented in this article.

Magnesium, with a density of 1.74 g/cm^3, has outstanding weight specific material properties and is the lightest structural material. For this reason it is mainly used in light-weight structures and for highly accelerated components. These outstanding properties and the fields of application are broadened by augmenting with magnetic properties. Prospective components from magnetic magnesium alloys will thus have the possibility to store data in the component itself or to be used as a sensor. By using these sensoring properties, it is possible to determine a component's location and velocity or, by exploiting magnetostriction (also known as the Villari-effect), the component's extension. Furthermore, an eddy-current method [2] can be used whereby it is possible to determine a component's loading.

The aim of these investigations is to characterize the parameters which have an influence on the magnetic properties of the alloys. These are the alloy's magnetic material content as well as the distribution of magnetic phases.

The magnesium alloys developed by the Institute of Materials Science [1] are further investigated with respect to their hardness values and corrosion resistant properties. Moreover, those manufacturing parameters are investigated which have an influence on the formation and properties of the magnetic phases. Among these are the alloy's magnetic powder content, the cooling rate, the corrosion resistance and the hardness values.

A high-speed milling cutter, which is to be produced using die-casting, was selected for subsequent use as a component made from magnetic magnesium alloys.

Initial materials and testing method

The magnetic magnesium alloys, containing from 10 to 40 wt % SmCo [1], developed at the Institute of Materials Science are here the subject of further investigations.

The influence of cooling rate was investigated by casting the alloys using different casting processes. Die casting, centrifugal casting and melt spinning were employed. The cooling rates for die casting are $<10\ Ks^{-1}$, for centrifugal casting they lie between 10 and $10^2\ Ks^{-1}$ and for melt spinning between 10^3 und $10^4\ Ks^{-1}$. Figure 1 shows the equipment employed as well as a casting, which was produced by centrifugal casting process.

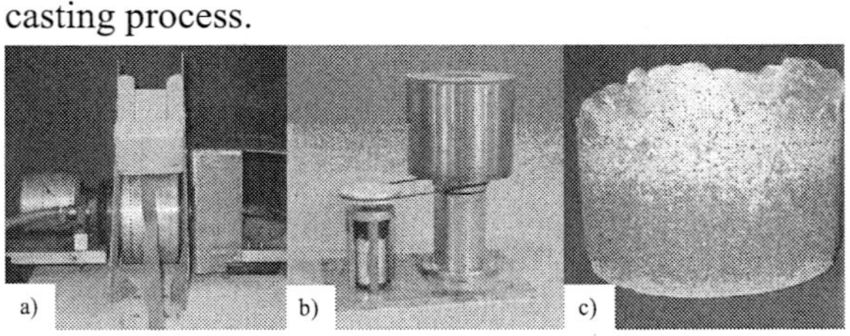

Figure 1: a) Melt spinning equipment b) Centrifugal casting mould c) Centrifugal casting, wall thickness 5-6mm

The specimens were examined using metallographic sections, electron beam microanalysis EBMA, (JXA-8900R, JEOL) and computer tomography, CT, (MicroCT80, ScancoMedical).

The examination of the corrosion resistance was carried out in the coolant KSM560. This corrosive medium was selected to emphasis the subsequent alloys' use as HS milling cutters.

Results of the present investigations

Magnesium was alloyed with SmCo-powder thereby varying the alloy's content between 10 and 40 wt. %. Other magnetic powders based on rare earths were not investigated since satisfactory results regarding the magnetic properties could only be obtained by using SmCo. The manufactured specimens were magnetised by means of two magnetic fields offset by 90° with strengths of $540\ kAm^{-1}$. Following this and after cessation of the external magnetic field, the remaining field strength in the component was determined using a Hall-probe. The magnetic field strength can be seen as a function of the alloy's SmCo content in figure 2.

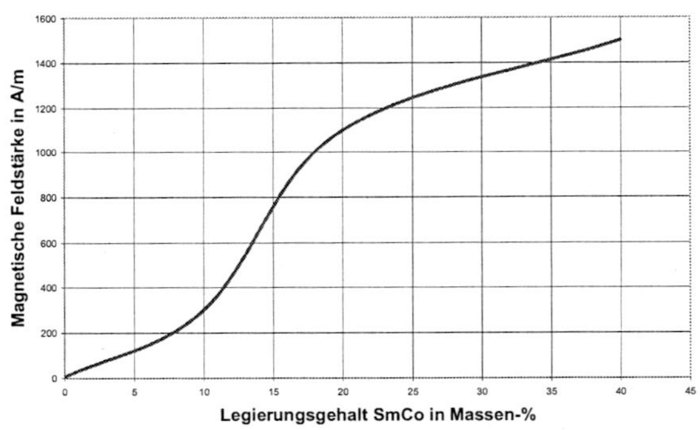

Figure 2: Magnetic field strength as a function of the alloy's content

It can be clearly seen that the magnetisability, that is, the magnetic field strength in the component, does not increase proportionally with the alloy's content. The largest increase is obtained for SmCo contents up to 20 wt. %; thereafter the increase is significantly lower. The maximum obtained field strength is $1500\ Am^{-1}$ for a SmCo content of 40 wt. %.

For investigating the cooling rate's influence on the microstructure and the distribution of the precipitates containing SmCo in the magnesium matrix, specimens were produced using various casting processes and subsequently examined by means of metallography, CT and EBMA. The results of the CT-examinations of the castings, which were manufactured by melt spinning, centrifugal and die casting, can be seen in figure 3.

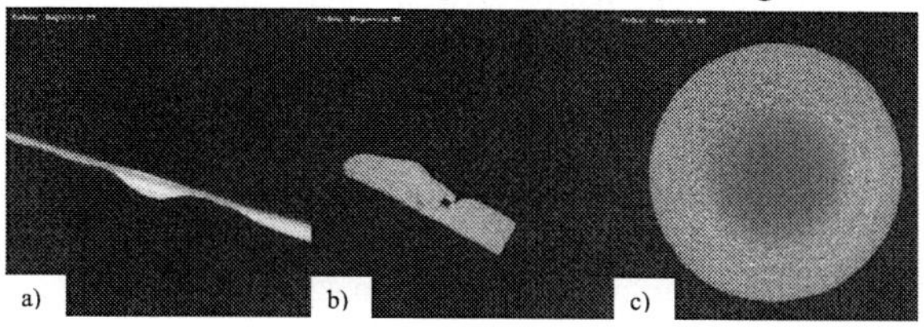

Figure 3: Results of the CT-examinations of Mg + 20 wt. %: a) Melt spinning b) Centrifugal casting c) Die casting

It can be seen that at higher cooling rates, a far more homogenous material and better casting quality results. The inhomogeneities, which occur at relatively low cooling rates, can be clearly seen in figure 3 c).

Metallographic sections were prepared and EBM analyses (line-scan) were carried out on the castings which were manufactured by melt spinning and centrifugal casting. The results of these examinations are depicted in figure 4. The weight percentage of SmCo is 20 %.

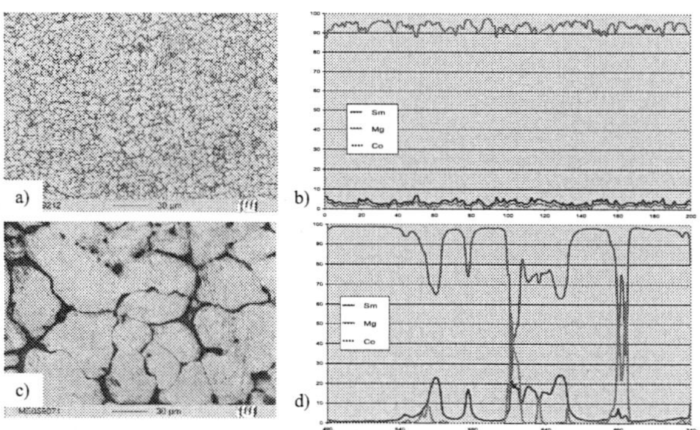

Figure 4: Metallographic section: a) Melt spinning c) Centrifugal casting; EBMA line-scans: b) Melt spinning d) Centrifugal casting

As can be seen here, a higher cooling rate produces a more homogenous distribution. This can be explained by the magnetic phase precipitating at the grain boundaries and the higher cooling rate leading to grain refining, thereby also increasing the volume at the grain boundaries.

In order to investigate the corrosion resistance, disc shaped specimens of 27mm diameter and 5 mm thick were produced from magnesium, as a reference material, and magnesium with various amounts of SmCo. These specimens were weighed and then subjected to the coolant KSM 560 for 504 hours and weighed again. The corrosion specimens are shown in figure 5; in their initial state (fig. 5a) and with their respective SmCo contents after 504 hours.

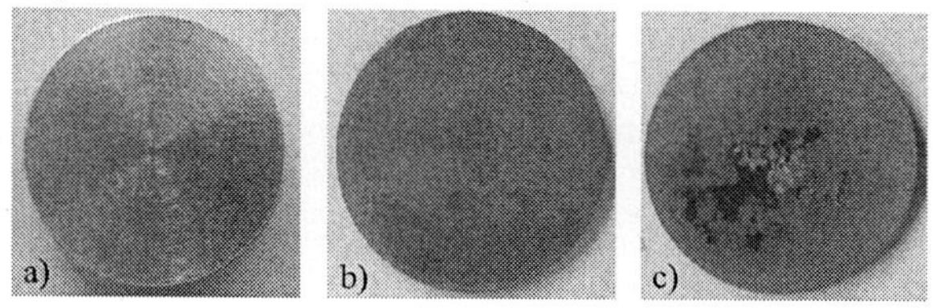

Figure 5: Corrosion specimens: a) Initial state b) Mg + 20 wt. % SmCo (after 504h) c) Mg+ 40 wt.% SmCo (after 504h)

The corrosion specimens' weight-loss depicted in figure 6 clearly shows that magnesium containing up to 20% SmCo exhibits no significant difference to pure magnesium regarding its corrosion behaviour in KSM560. However, when the content of SmCo exceeds 20%, the corrosion resistance drops considerably. This is expressed in figure 6 by the high weight-loss which initiates after approx. 250h. The weight-loss is up to 13% for magnesium with 40 wt.% SmCo.

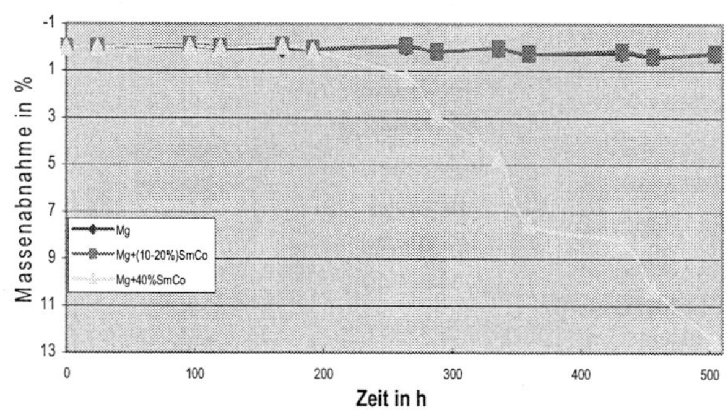

Figure 6: Weight-losses of the individual corrosion specimens

Furthermore, pitting occurs on the specimens which can extend through the entire specimen thickness (see figure 5 c)). Measurements of the corrosion solution's pH-value have shown here that this value increases from 8.8, at the beginning, to 10 at the end of the test duration. This and a chemical analysis of the corrosion products indicate the presence of magnesium hydroxide in these products. This presence is also substantiated by black deposits in the corrosion solution.

In order to investigate the alloy's SmCo content on the mechanical properties of the developed magnesium alloys, the hardness value was initially measured. For this purpose, specimens of pure magnesium, as a reference material, and magnesium with SmCo contents between 10 and 20 wt. % were manufactured. The hardness (HV10) of individual specimens were measured prior to and after a heat-treatment (350°C for 8 hours). The results of the hardness measurements are depicted in figure 7. The hardness values also increase with rising SmCo content. In comparison to pure magnesium, this amounts to an increase by a factor of 1.7 for magnesium with 20 wt. % SmCo. This is comparable to rolled magnesium products of the alloy AZ31. Furthermore, it is clear that the heat-treatment has a considerable influence on the hardness. The hardness values drop by a factor of approx. 1.3 following heat-treatment.

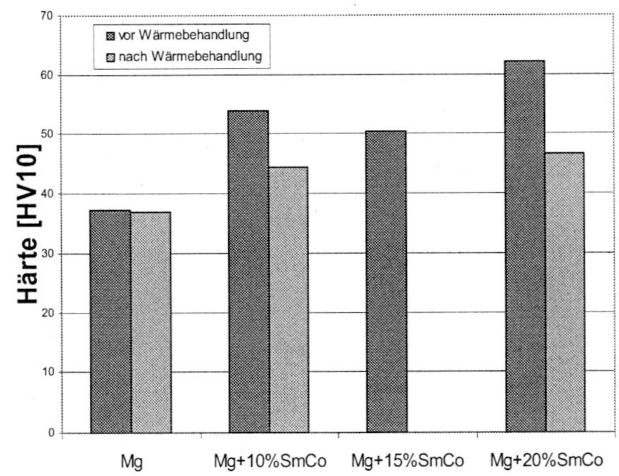

Figure 7: Hardness prior to and after heat-treatment

Summary

Within the framework of this article, the influence of the cooling rate on the magnetic phase formation was firstly investigated. Secondly, various parameters were examined which exert an influence on the magnetic properties, the corrosion behaviour and the mechanical properties of the developed magnesium alloys.

Investigations into the influence of the alloy's composition on the magnetic properties have shown that increasing the SmCo content enhances the magnetic field strength subsequent to magnetisation. Nevertheless, this enhancement does not increase proportional to the alloy's content. The largest increase takes place at a SmCo weight content of 20%.

Investigations into the influence of the cooling rate have shown that by increasing the rate a better casting quality is obtained as well as an increase of the magnetic phase's homogeneity, which can be attributed to a grain refinement.

Furthermore, the corrosion resistance of magnetic magnesium alloys with SmCo was investigated in coolant KSM 560. According to a test period of 504 hours, it appears that magnetic magnesium alloys with up to 20 wt. % SmCo have the same corrosion resistance as pure magnesium. At a higher SmCo weight content, the corrosion resistance drops significantly.

The hardness investigations of magnetic magnesium alloys with SmCo have shown that these alloys exhibit good properties, which are comparable with those of formed AZ31 alloys. The hardness values increase together with the SmCo weight content. Subsequent heat-treatment again decreases the alloy's hardness. Heat-treatment thus has a decisive influence on hardness values.

References

[1] Bach, Fr.-W.; Schaper, M.; Bosse M.; Gershteyn, G.; Nowak, M.: *Casting of Magnesium alloys with magnetic properties - "Magnesium-broadhorizons"*, Moskau, 29.11-1.12. 2005, Giessereiindustrie, Moskau, **2006**, N2, S. 16-19, ISSN 0024-449X.

[2] Heutling, B.: Zerstörungsfreie Online-Materialcharakterisierung von Stahlfeinblechen mittels Harmonischen Analyse von Wirbelstromsignalen, PZH-Verlag, Hannover 2004

Innovative Production Machines and Systems
D.T. Pham, E.E. Eldukhri and A.J. Soroka (Eds.)

Production of PM components with integrated information

B.-A. Behrens, F. Lange, E. Gastan

Institute of Metal Forming and Metal-Forming Machines (IFUM), Leibniz Universität Hannover (LUH), An der Universität 2, 30823 Garbsen, Germany

Abstract

The industry is interested in optimizing their products as well as the production process. An example of product optimization is the use of light alloy materials to reduce the component's weight. An example of an efficient process is the Powder metallurgical (PM) manufacturing, due to its high utilization of material and the near-net-shape production. This paper describes a modified PM process that allows the storage of information inside sintered light alloy components. Examples for the stored information are the workpiece identification number and the batch number. The process is based on the insertion of foreign particles within the powder at predefined positions, which can be translated into numerical data.
Both experimental and numerical analyzes are carried out. The experimental analysis proved the feasibility of the process by the production of specimen and the readout of the stored information. The focus of the numerical analysis was on the distribution of the porosity and the safety against crack initialization.

Keywords: Powder metallurgy (PM), light metal, information storage

1. Introduction

Light weight materials, e.g. magnesium or aluminum, provide a high potential for the weight reduction of components. This is of particular interest for automotive and aircraft industries, for reduced craft weight decreases the fuel consumption and thus also reduces toxic emissions.

Powder metallurgy (PM) offers itself to the processing of these materials, as it combines a large number of advantages. Notable economic advantages are the high utilization of material, which is particularly important for mass production and cost-intensive materials, and the near-net-shape production, which to some extent makes further processing unnecessary. Also, the energy requirement is lower compared to casting because the material does not have to be heated to fluidity. An important technical advantage is the possibility of generating alloys which cannot be produced by means of melting metallurgy [1].
This paper describes a procedure which allows for the storage of information inside sintered light alloy components. By this, component information can be stored decentrally, and at the same time the components are marked for plagiarism prevention.

The first part of the paper covers the manufacturing principle as well as the encoding of the data and their readout. After that, the numerical model is described, followed by the results for predicting the local distribution of the relative density within the components and their cracking safety as well. The relative density, the ratio of particle volume V_{particle} and total volume V_{total}, describes the porosity of a component.

2. Insertion of information and storage principle

2.1. Pressing process with insertion of information

The information is inserted within the component by means of foreign particles which are placed at predefined positions. Figure 1 shows the pressing process.

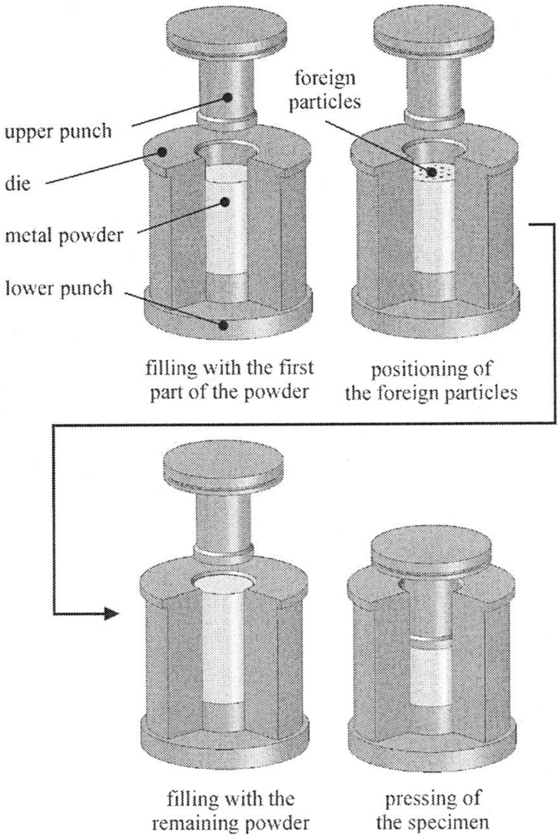

Fig. 1. Pressing process

The die is filled with a part of the metal powder, followed by the positioning of the foreign particles with a template. The powder used for the experiments was the magnesium wrought alloy AZ61 (6 % aluminium, 1 % zinc) from the ZFW GmbH. It was produced by injection under protective gas. The average grit size of the powder is 73 µm (Fig. 2, left). The spherical particles consist of polished steel with a diameter of 1 mm (Fig. 2, right).

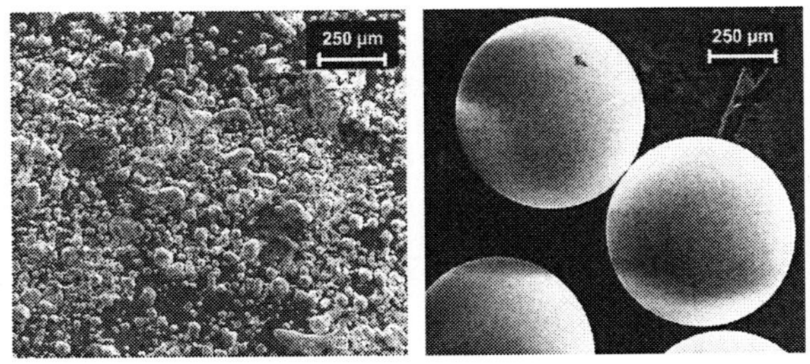

Fig. 2. Basic powder and foreign particles

For an automated insertion of the particles a positioning device is currently being set up. After the filling with the remaining powder, a compact – the so-called green compact – is pressed from the powder and the foreign particles, and the component is sintered under vacuum thereafter. For the experiments the velocity of the upper punch was constant at 10 mm s^{-1}. The maximum pressure was 450 N mm^{-2} at the end of the compaction process.

Due to the sintering process, the powder and the foreign particles are connected inseparably. This ensures that the stored information cannot be manipulated subsequently. Thus, the procedure offers not only a possibility to insert information, but an effective protection against plagiarism as well.

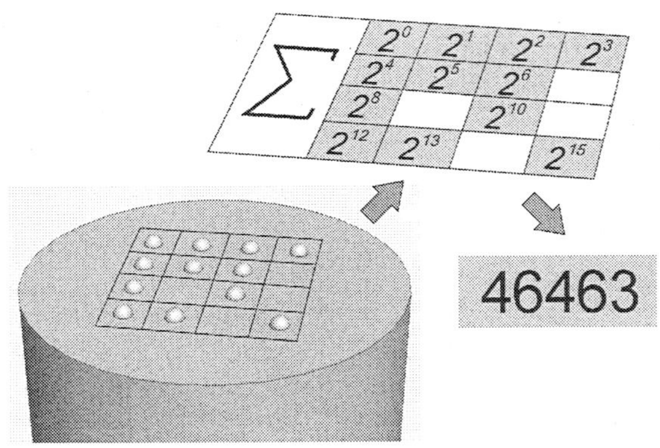

Fig. 3. Storage principle

The foreign particles are inserted as a matrix. A 4x4 matrix is already capable of representing numerical values from 1 to 65535. Figure 3 shows the storage principle. Each position in the matrix represents a power of two. The stored number results from the summation of all positions marked by a particle. Depending on the component, it may be necessary to predefine the matrix orientation. This, for example, can be done by inserting a larger foreign particle. By means of increasing the number of columns and rows, the possible amount of data inserted in the component can be enlarged accordingly. A 6x6 matrix yields the possibility of storing more than 68.7 million different number combinations. Another option to increase the storage capacity is the multi-plane arrangement of the particles.

2.2. Sintering process

The sintering process increases the components strength and toughness. It is a heat treatment below the melting point of the material. Diffusion processes are activated by the heating and material flows into voids, causing a decrease in overall volume. This leads to a decrease of the porosity as well. The ideal sintering temperature for metallic materials is approx. 75-85 % of the melting temperature (in degrees Celsius). In this case, the alloy AZ61 was sintered for 30 minutes at 420°C under vacuum of $1 \cdot 10^{-4}$ mbar. The sintering furnace used, Xretort1200 made by Xerion, is a three-zone chamber oven with two vacuum pumps and active cooling.

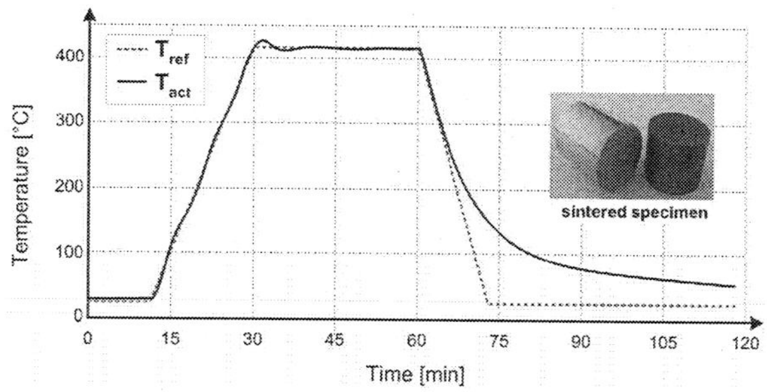

Fig. 4. Temperature during the sintering process

Figure 4 shows the temperature of the oven during the sintering process of the specimen. After sintering powder and foreign particles are joined.

2.3. Readout of the information

The information readout is based on the different physical properties of the basic material and the foreign particles. Depending on the material combination, computer tomography (CT) as well as radiotechnology or ultrasonics can be employed due to, for example, differences in density or varying degrees of magnetizability.

In this project, aluminum and magnesium as the powder material and a steel alloy as the foreign particle material are examined. Figure 5 presents X-ray images of a cylindrical Mg specimen produced at the Institute of Materials Science, Leibniz Universität Hannover. The top view is shown on the left, the side view on the right. It can be seen that the foreign particles are clearly detectable regarding shape and position. Planar particle arrangements can be detected by a standard X-ray device. For a 3D detection, a CT can be used.

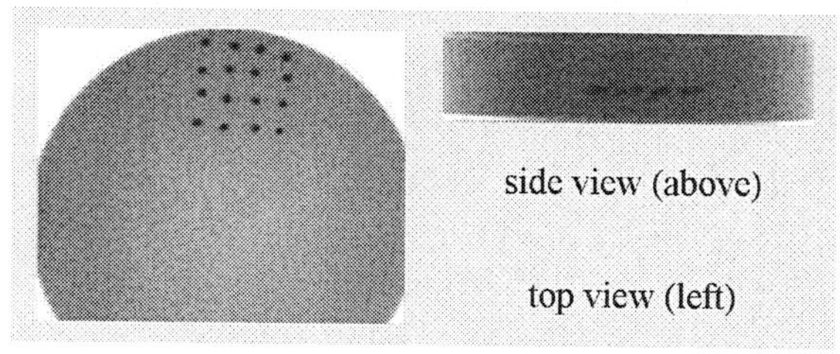

Fig. 5. X-ray images of a cyclic Mg specimen

3. Numerical investigation

3.1. Computational model

For the manufacturing of sintering components, it is important to aim at a density distribution as homogeneous as possible in order to avoid subsequent distortion due to sintering later on. The inserted foreign particles, however, cause a disturbance of the homogeneous powder distribution during compaction. By means of the finite element analysis (FEA), the densification of the powder can be described. The aim of the numerical investigation is the prediction of the density distribution, especially in the surrounding area of a foreign particle, and a statement about crack safety. Below, the model for the simulation-based layout is described. The FE system Abaqus is used for the analysis.

For the material modeling, the Drucker-Prager cap model is used. This model describes the material as an elastic-plastic, compressible continuum. Figure 6 shows the material model plotted in the plane of the hydrostatic pressure p and the von Mises equivalent stress q [3].

The left bound of the elastic region represents the failure straight. It is described by the shear strength d and the inner friction angle β. When reached, softening within the material occurs, thus causing the risk of crack initiation. The elastic region's right bound is an elliptically-shaped cap. When the cap is reached, densification and thus hardening of the material occurs [3].

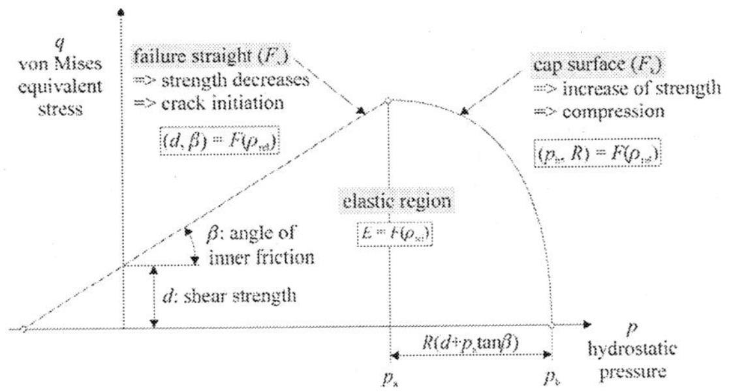

Fig. 6. Drucker-Prager cap model

The standard formulation of the Drucker-Prager cap model implemented in ABAQUS does not describe the model-relevant parameters as functions of the porosity, which is characterized by the relative density. These dependencies are extended by means of user-defined subroutines to increase the accuracy of the simulation. The data used are taken from [4] and [5]. Likewise, the powder's elastic modulus is formulated against the relative density.

The inserted steel particles are modeled as elastic bodies with an elastic modulus of $E = 210{,}000$ N/mm^2 and a Poisson's ratio of $\nu = 0.3$. For a realistic modeling of the friction, the friction model according to Coulomb, extended by the dependency on the relative density, is applied [6].

3.2. Density distribution results

In the first part of the simulation, a cylindrical aluminum powder specimen (diameter: 20 mm, height after pressing: 25 mm) is examined. For basic investigations concerning the influence of the foreign particles (diameter 1 mm), specimens with and without inclusions are compared. The underlying models are set up 2d-axisymmetrically. The computational results are shown in Figure 7.

Due to the wall friction, a density gradient symmetric to the horizontal midplane results. The friction interferes the powder movement at the die wall. This phenomenon can be observed for both specimens. The foreign particle influences the relative density distribution only locally. Its sphere of influence is equivalent to approximately twice the particle diameter. At the particle surface the powder movement, as at the die wall, is hindered. For this reason, the powder is densified the least in the horizontal direction.

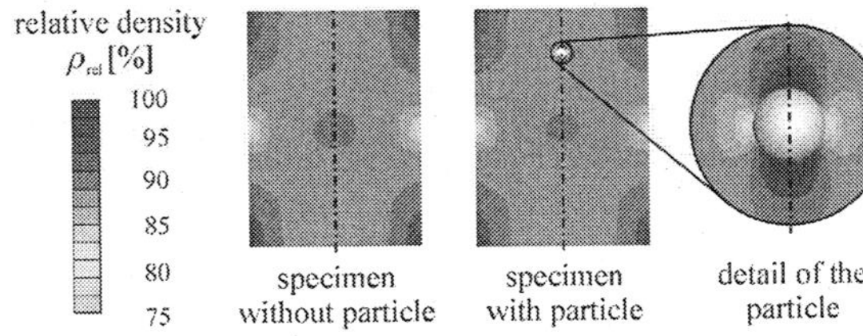

Fig. 7. Comparison of specimens with and without particles

In order to predict the interaction of several foreign particles a three dimensional model was created. The particles are positioned as a 4x4 matrix as shown in the storage principle (Fig. 3). In Figure 8 the results of the relative density for two perpendicular sections are presented.

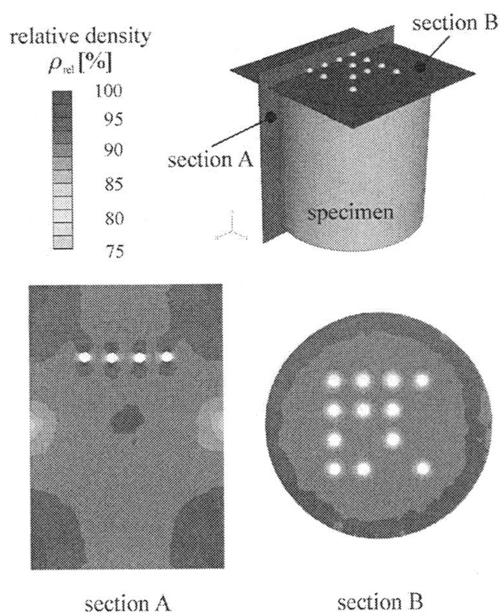

Fig. 8. Density distribution of a cylindrical specimen with a 4x4 information matrix

According to the density distribution the interaction between the particles is negligible. The distance between the particles is about 1.5 times of their diameter.

Figure 9 represents the density distribution for the 3d model of a complex sintering component. It shows a toothed belt disk as used in an automobile, for instance. In this model, four extrusion rams are driven independently.

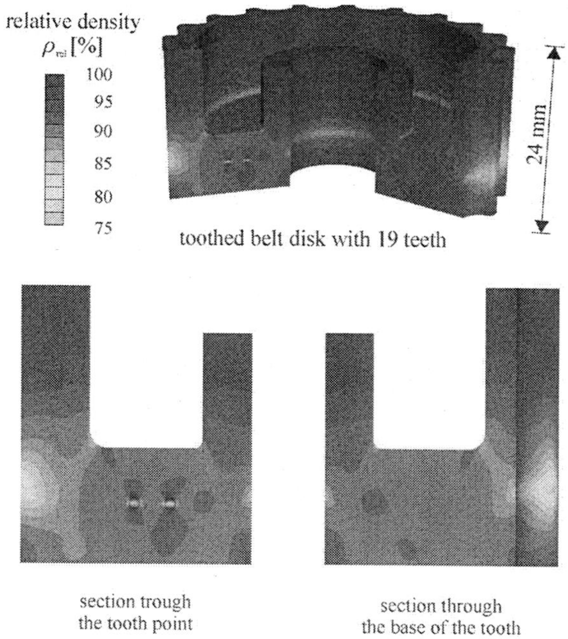

Fig. 9. Density distribution of a toothed belt disk

It turns out that, with adequate distance in the radial direction, the foreign particles do not influence each other as shown in the simple model before. Nevertheless the distance between the foreign particles as well as their distance to the component's surface have to be chosen such that disadvantageous distributions of the relative density are not additionally amplified by overlap.

Further results concerning the influence of the friction and the particle shape on the density distribution can be found in [7].

3.3. Cracking safety results

For the evaluation of a possible crack initiation, a cracking criterion according to Hanini [6] is used. It describes the position of a stress condition in the elastic region of the Drucker-Prager cap model (Figure 10).

The crack criterium is defined as:

$$R_s = B_i \cdot (A_i + B_i)^{-1} \cdot 100\%$$

Here, A_i represents the perpendicular distance from the axis of the hydrostatic pressure, while B_i represents the perpendicular distance from the failure straight.

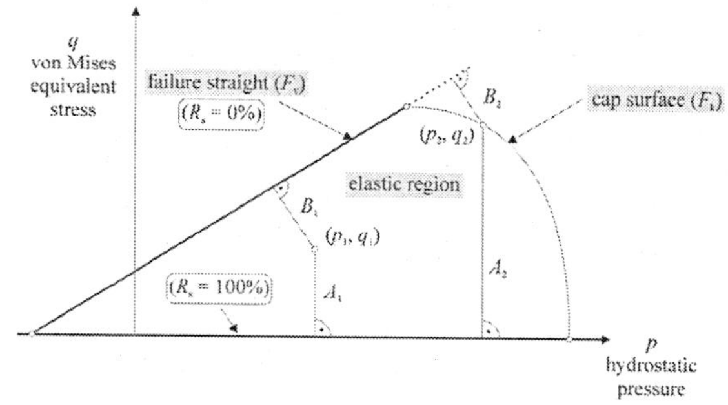

Fig. 10. Crack criterium

For stress conditions on the failure straight, the first term of the definition equation becomes zero. The cracking criterion thus has the value 0%, which shows that there is a chance of crack initiation. The other border case of a stress condition on the axis of the hydrostatic pressure causes the product of the first two terms of definition equation to become one, yielding a value of 100% for R_s. There is thus no risk of cracking in this case.

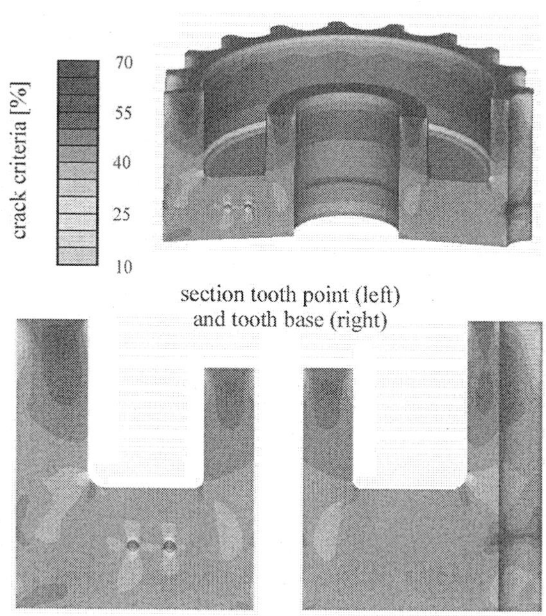

Fig. 11. Crack criterium of a toothed belt disk

In Figure 11, the cracking criterion is shown for the toothed belt disk. As for the density distribution as well, the effect of the foreign particles is locally. It can be clearly seen that there is an average value of approximately 40% for the cracking criterion. The distribution of the values over the component section is nearly homogeneous. Local minima of the cracking criterion are located between the component's toothing and the recess in the center as well as above the inserted foreign particles (here hidden in the graphical representation for the sake of clarity). Due to the fact that the cracking criterion does not reach a value of 0 % at any point in the component (lowest value is 10%), no crack initiation can be expected.

4. Conclusion

Within this paper, a procedure allowing for the insertion of information in a powder metallurgical manufactured component was presented. The stored information is secure against subsequent manipulation and thus contributes to the protection against plagiarism.

By producing sample specimen and then reading out the stored information, the procedure's feasibility was experimentally verified. The application to an industrially relevant component is the subject matter of future research.

A numerical simulation based on the finite element analysis was used to supply information about the density distribution in the press-formed component and its safety against cracking. The results show that the particles' influence on the density distribution and the cracking safety is localized.

5. Acknowledgements

The presented results were generated in the project E3 “Sintering gentelligent parts from aluminum and magnesium powder” at the Institute of Metal Forming and Metal-Forming Machines (IFUM) of the Leibniz Universität Hannover (LUH). It is a subproject of the collaborative research center 653 “Gentelligent Components in their Lifecycle - Utilization of Inheritable Component Information in Production Engineering”. The X-ray images were produced at the Institute of Materials Science (IW).

The authors would like to thank the German Research Foundation (DFG) for their financial support.

6. References

[1] Schatt, W.; Wieters, K. P.: Pulvermetallurgie - Technologien und Werkstoffe, VDI Verlag, Düsseldorf 1994

[2] SFB653: http://www.sfb653.uni-hannover.de/ as from 30th march 2008

[3] Kraft, T.: Optimizing press tool shapes by numerical simulation of compaction and sintering—application to a hard metal cutting insert, Modelling Simul. Mater. Sci. Eng. 11 (2003) pp. 381–400, 2004

[4] Behrens, B.-A.; Bouguecha, A.: Ermittlung der Versagenslinie von Aluminiumpulver für das Drucker-Prager-Kappenmodell. wt Werkstattstechnik online 10, 2004

[5] Behrens, B.-A.; Bouguecha, A.: Beschreibung des Verfestigungsverhaltens von Aluminiumpulver im Drucker-Prager-Kappenmodell, UTF science 2/2004, 2004

[6] Hanini, K.: Simulationsgestützte Algorithmen für die automatisierte Auslegung von Pulverformgebungsprozessen, doctorial thesis, Universität Hannover, 2006

[7] Behrens, B.-A.; Gastan, E.; Lange, F.: Experimental und numerical investigation on the marking of sintered parts by the targeted application of particles, Materials Science and Engineering Technology, issue 38, 10/2007, S. 816-820, WILEY-VCH Verlag, Weinheim, October 2007

Innovative Production Machines and Systems
D.T. Pham, E.E. Eldukhri and A.J. Soroka (Eds.)

RF communication into metallic components

E. B. Kaldjob[a], S. Franke[b], B. Geck[a], L. Overmeyer[b]

[a] *Institute of Radiofrequency and Microwave Engineering, Leibniz Universität Hannover, Appelstr. 9a, Hanover, Germany*
[b] *Institute of Transport and Automation Technology, Leibniz Universität Hannover, An der Universität 2, Garbsen, Germany*

Abstract

This article describes a technology enabling metallic components to communicate by radiofrequency waves, which is a key feature for gentelligent components. Gentelligence is a concept to enable the components of a product to feel, react and talk to their surrounding. The realization, integration and packaging of an electronic circuit and a waveguide in the metallic component are presented. Solutions for the filling of the waveguide are illustrated. The wireless transmission of power to the circuit is explained.

Keywords: Energy transmission, RF communication, Electronic packaging

1. Introduction

For quick and efficient production detailed information about the processes and the product are essential. Systems like Siemens TIA [1] focus on monitoring the production machinery. Each machine sends data to a central data base. The more assembly steps are taken and the longer the assembly line the more complex the automation system gets.

Storing and handling the data locally would permit to limit this complexity. All data about an individual gentelligent component would be stored on the component, each machine adds process data and the component itself could sense process parameters by the means of integrated sensors. This is, what gentelligent components are going to do. The project described in this article focuses on the realization of a communication into gentelligent components.

Data can be transmitted wireless to an arbitrary component using radiofrequency waves. It can be stored there and on demand, be changed or updated. This technology is the basis of Radio Frequency IDentification (RFID). Transponders, tags or smart labels are usually placed on or little below the surface of a component. To ensure a safe coupling of component and information and for the protection of the circuitry, system integration deep within the component is required. Metallic components are a special challenge since they prevent penetration of electromagnetic waves because of the skin effect.

2. Transponder for waveguide

Unlike common transponders consisting of a chip connected to a printed dipole antenna, the integration of transponders into metallic objects needs new considerations as shown in [2]. The transponder has to be integrated in a hole made in the metallic component's surface (Fig. 1). This cavity works as waveguide for electromagnetic waves between the air interface and the integrated transponder.

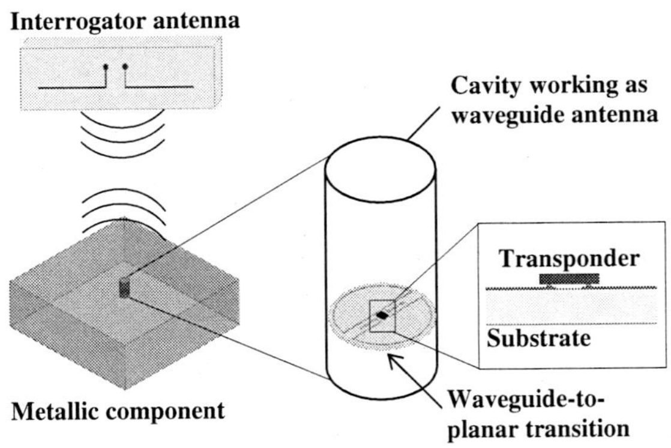

Fig. 1. Wireless identification of metallic components in factories.

However, it should be regarded that the smallest cavity diameter is inverse proportional to the lowest signal frequency of propagable waves. Once the frequency and the cavity geometry are adjusted to each other, a planar waveguide-to-planar transition as presented in [3] can be used to convert the waveguide mode to a mode which suits to the chip input. Thus, the waveguide including this transition is the antenna feeding the transponder with the received power. Fig. 2 shows the final system block diagram.

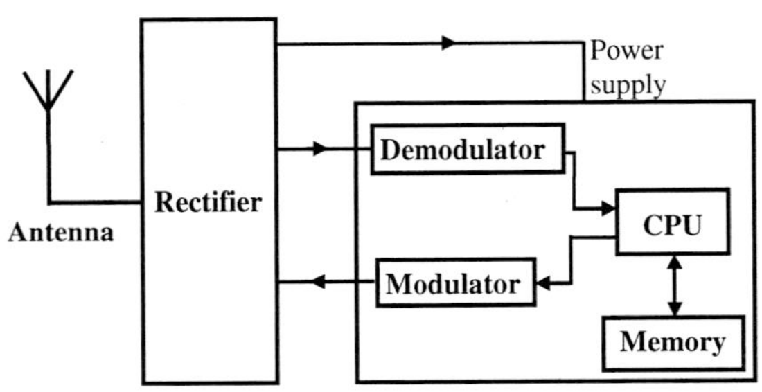

Fig. 2. Transponder block diagram.

The transponder consists of a rectifier, a modulator, a demodulator, a CPU and a memory. The rectifier including two Schottky barrier diodes and two capacitors converts the received radiofrequency power to DC voltage, which supplies the rest of the circuit (Fig. 3). To enable long read/write ranges, two main criteria should be regarded. First, the rectifier conversion performances should be optimized by choosing low-loss Schottky barrier diodes with very low forward voltage. Secondly, the system power consumption should be minimized by carefully choosing the microcontroller. In the framework of this work, the microcontroller PIC12F510 from the manufacturer Microchip has been used because of its low power consumption (< 300 μW) even at clock frequencies up to several megahertz.

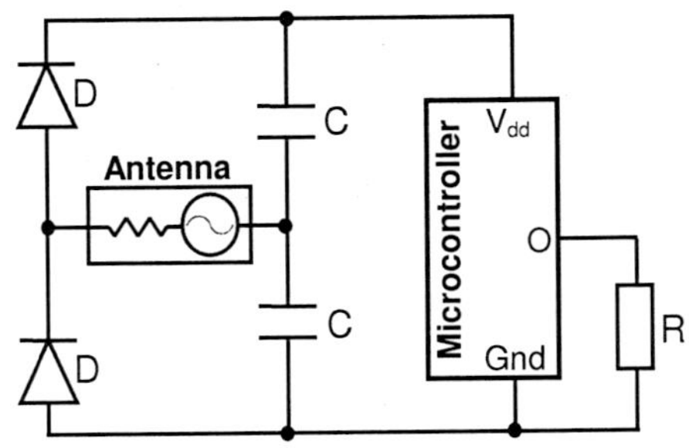

Fig. 3. Transponder circuit schematic.

In this work, the interrogator sends information using amplitude shift keying modulation (ASK), so the information is contained in the DC voltage and can be processed by the CPU.

To send information to the interrogator, the transponder CPU achieves a backscatter modulation by controlling the rectifier input reflection coefficient. For this purpose, the CPU drives the resistor R with a pulse sequence identical to the bit sequence to be sent. Integrated transponders in metallic components operating in the 2.45 GHz and 5.8 GHz ISM Bands have been realized (Fig. 4) and the transmission system described in Fig. 1 has been tested.

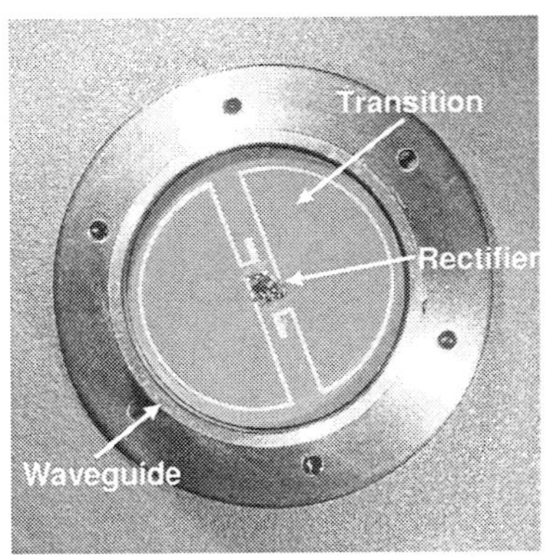

Fig. 4. Transponder integrated in cylindrical waveguide.

The cavities diameters have been chosen to 85 mm and 36 mm for the 2.45 GHz and 5.8 GHz bands respectively. Here the transponder modulator has generated a bit sequence to modulate the backscattered signal. Fig. 5 shows the measured modulator pulses as well as the envelope of the backscattered signal received by the interrogator.

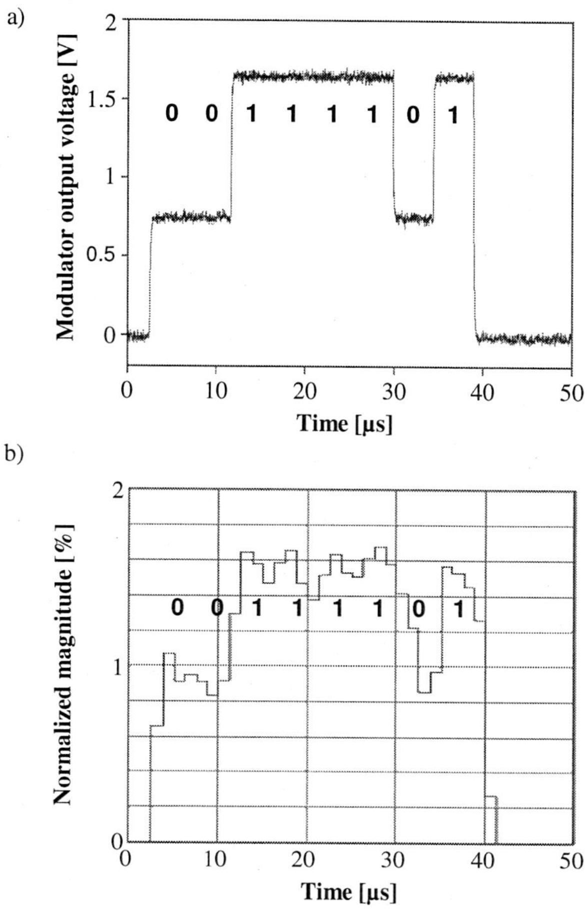

Fig. 5. a) Modulator output signal; b) Envelope of demodulated backscattered signal.

It should be mentioned that the power attenuation over the air interface is proportional to the square of the signal frequency. Hence, the smaller the cavity diameter, the higher the required frequency and thus the lower the received RF power is. With respect to the regulations about the interrogator maximum radiated power in Germany, transponder power supply has been achieved over a distance of 80 cm and 51 cm for the 2.45 GHz and 5.8 GHz system respectively. Data transmission rates up to 2 Mb/s have been achieved. The use of a programmable microcontroller allows simple system enhancements, such as controlling of future added integrated sensor applications.

However, the cavity dimensions mentioned above are not technically interesting for most situations. The next steps in investigation are the communication at same frequency bands by reduced cavity diameters. This is achievable by filling the cavity with dielectric materials with permittivity higher than 1.

3. Power supply

Power is needed in the gentelligent component for signal transmission, signal processing and possible sensor operations. Depending on the application a permanent power source is needed (e.g. for sensor operations) or power should be supplied only during the signal transmission. Fig. 6 shows the ways power can be supplied.

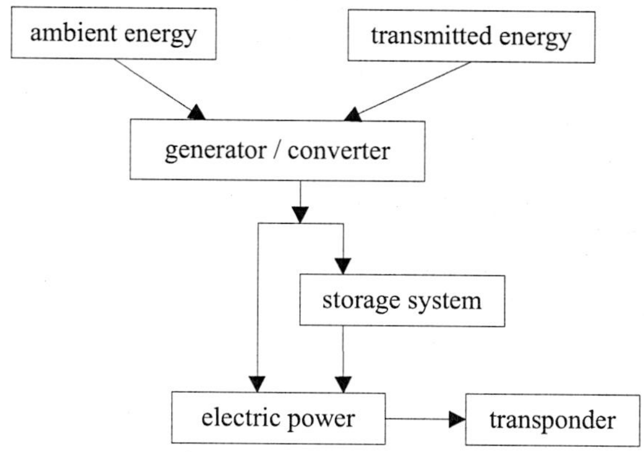

Fig. 6. Power supply concept for gentelligent components.

For a permanent supply a storage system is required. This storage system has to store all energy needed for the whole life time of the component or it has to be recharged on a regular base. This could be done using ambient energy by energy scavenging.

If the circuit has to work, only, while it is in range of a reader, the power can be transmitted from the reader. In most cases, this is sufficient for the electronics in the gentelligent component. Commonly power is transmitted to a transponder by using the electromagnetic field, as mentioned before.

Since the maximum power for a transmitter of radio frequency waves limits the maximum energy to be received, another form of energy has to be used for transmission, if the read range or the power demand increase.

There are several reasons to use light. In contrast to sound, it has a fairly high energy density. The intensity of sun light can reach 1 mW/mm², so the maximum power of light that may be emitted from a reader unit is high. Converters from light to electricity like solar cells and photo diodes are available in a big variety. Efficiency factors in the order of magnitude of 10% are realizable. The maximum power received by a photodiode is not defined by its volume, but by its surface. Hence, a sufficient power supply does not require much space within the component.

Photodiodes are integrated in the waveguide to protect them from harsh environment and to reduce constrains to the components construction. In the first step they are placed on the waveguide-to-planar transitions level, because the transitions substrate is opaque Arlon. In the next step transparent substrate will be used and diodes will be placed on the bottom of the waveguide. To reduce influence on the functionality of the waveguide, bare diodes are being used. The volume of bare diodes is half the one of housed photo diodes.

For experimental results Ashine 043PD photo diodes were used. They have a size of 1.1 mm x 1.1 mm and a height of 0.3 mm. Five of them were connected in series and were illuminated by a 20 W halogen lamp and a 150 W cold light from different distances. The results are shown in Fig. 7.

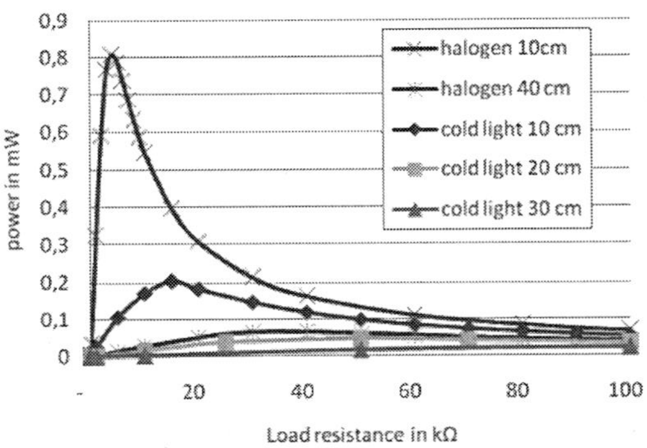

Fig. 7. Power over load resistance for five photo diodes

Although the illuminace of the cold light, in a distance of 10 cm is 130000 lx vs. 90000 lx for the halogen lamp, the halogen lamp leads to more power, since its infrared light contributes to the power generation in the silicon semiconductor.

The maximum received power was 0.8 mW at a voltage of 1.8 V, which is enough to supply the PIC12F510 directly. Still the efficiency of this power transmission is low. The received power decreases quadratically with increasing distance and most of the light does not reach the active area of the diodes. Using lasers promises a higher efficiency, but it demands an exact alignment of the component to the reader. Another measure to improve efficiency will be applying an infrared emitter. Since the wavelength of infrared light is closer to the band gap of silicon semiconductors.

Future research will be about focusing light within the waveguide on the photo diodes.

4. Assembly

Integration of the transponder in metal is simple. The waveguide is a round blind hole. It can be drilled or milled in any metallic material. In the blind hole a package is integrated. It consists of the dielectric filling the waveguide-to-planar transistion and the power supply. Fig. 8 shows the individual parts of the prototype of a screw as gentelligent component.

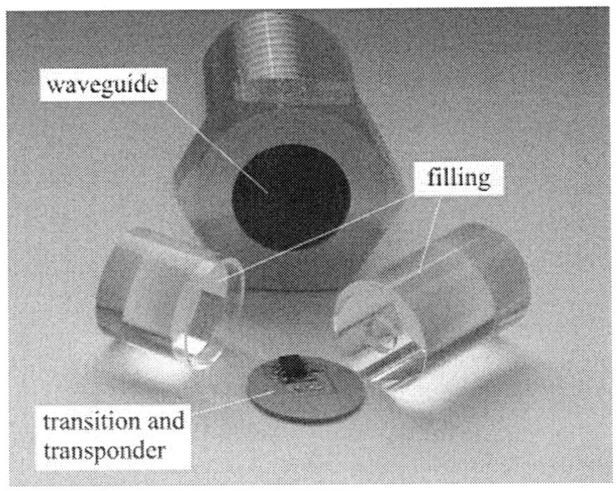

Fig. 8. Individual parts of a gentelligent component

The package has to be transparent to allow light to pass trough to the photodiodes. There are several transparent plastics available. Polymethyl methacrylate (PMMA), Polymethylpentene (PMP) and Polycarbonate (PC) were investigated and tested.

Table 1 summarizes the relevant properties of these possible fillings. The permittivity of the materials was measured by using TE11p modes in circular cylindrical cavities [4].

PMMA, also known as Plexiglas, is a transparent thermoplastic at visible light. Though, it is barely transparent for wavelengths of 300-350 nm. PMMA is easy to process. It can be milled and polished.

PMP, also known as TPX, is a transparent thermoplastic, too. PMP can not be milled as easy and its price is higher than that of PMMA and PC. On the other hand, the low permittivity leads to higher waveguide diameters. It may be interesting for future application in thinner waveguides at higher frequencies.

PC is more expensive and less transparent than

Table 1
Properties of plastics for the filling

	PMMA	PMP	PC
permittivity at 5.5 GHz	2.66	2.11	2.84
cavity diameter range for 5.8 GHz band (mm)	18.9-23.9	21.2-26.8	18.3-23.1
transparency for visible light	excellent	excellent	moderate
transparency for wavelength 300-350 nm (ultraviolet)	poor	good [5]	poor
processability	good	poor	good
temperature stability	poor	good [5]	moderate
cost	low	high	moderate

PMMA. Its resistance to heat is higher.

PMMA was chosen to be the best material for the package by now, because the diameter needed for the waveguide is small and the transparency for visible light is excellent.

The plastic has to be placed below and above the waveguide-to-planar transition. All parts have to be fixed in the waveguide. Form closure is not possible, as the form of the hole is firmly defined by its functionality as a waveguide. Therefore the filling has to be bonded to the metal using adhesives. As the photodiodes are molded by the adhesive, transparent adhesives are used. They need to have a low viscosity to reduce air bubbles.

To cure the adhesive there are four possibilities. Two-component adhesives harden when the two components are mixed. Adhesives with one component are activated by humidity from the air, heat or light.

Using two-component adhesives is not the best option because they are hard to handle. Humidity activated adhesives start curing when they get contact to air. Viscosity of both two-component and humidity activated adhesives increases during the assembly already. This is a problem, because the package is inserted in the wave guide after the adhesive and the adhesive has to be squeezed out of the blind hole to coat the walls.

Because PMMA is transparent, but not heat resistant hardening using light, in contrast to heat, is a good option. The wavelength of light needed to cure the adhesive is an important parameter. Standard wavelengths for UV curing of 350 nm are blocked by PMMA. Some adhesives can be activated at 450 nm. Using Delo PB 4442 led to good results, stable connections, little air bubbles and high transparency. Its refraction index is very close to that of PMMA, so almost no refraction and reflection occurs. Adhesives with lower refraction indexes are investigated to achieve total reflexion inside the waveguide [6].

5. Conclusion and future work

The methods outlined in this paper allow the realization of communication into gentelligent components. Wireless communication and power supply can be achieved. The integrated controller can store and compute data.

For greater read ranges special optics within the waveguide are required to focus light on the photodiodes.

Since the diameter of 20 mm is not small enough for most components a further reduction by increasing the frequency is the main focus in future work.

Acknowledgements

This research was supported by the German Research Foundation (DFG).

References

[1] Wucherer, K. Totally Integrated Automation: A concept for manufacturing and process industries, ARC's Automation Strategies Forum in Boston, 24.6.1997

[2] Kaldjob E.B., Gamal El Din M. Wellenleiteranordnung zur Übertragung elektromagnetischer Wellen mit einem Hohlleiter und einem im Hohlleiter angeordneten planaren Leiter", German patent 10 2006 053 389, 10. Nov. 2006.

[3] Kaldjob E.B., Gamal El Din M., Geck B. A novel planar cylindrical waveguide-to-planar transition for transceivers at 2.45/5.8 GHz band, Int. Symp. Antennas and Propagation, Niigata, Japan, August 2007.

[4] Zinal, S.; Boeck, G. Complex permittivity measurements using TE/sub 11p/ modes in circular cylindrical cavities, IEEE Transactions on Microwave Theroy and Techniques, vol. 53, no. 6, part 1, June 2005, pp. 1870-1874.

[5] Mitsui Chemicals. TPX Data Sheet, http://www.mitsui.de/pdf/de/aktiv/kunststoffe/tpx.pdf March 2008

[6] Fahlbusch, T.; Franke, S.; Overmeyer, L. Direct dispensing of 3D light guiding structures. In: POF & MOC 2006 the joint international conference on plastic optical fiber & microoptics, Seoul, Korea, Sept. 2006

Innovative Production Machines and Systems
D.T. Pham, E.E. Eldukhri and A.J. Soroka (Eds.)

Virtual pilot production in process planning

B. Denkena, L.-E. Lorenzen, C. Schmidt

Institute of Production Engineering and Machine Tools, Leibniz University Hannover, An der Universität 2, 30823 Garbsen, Germany

Abstract

Human abilities like intelligence and learning aptitude are increasingly adapted to production planning systems. The Institute of Production Engineering and Machine Tools (IFW) is developing a new system for simulation-based planning and controlling of cutting processes. Therefore, a novel system based on process simulation integrated in a process planning system will be set up. This system allows an optimized cutting process right from the beginning.

Keywords: process planning, process simulation, process monitoring

1. Introduction

For European manufacturers it becomes increasingly important to be able to react to an increasing variability and complexity of the manufactured products as well as to varying lot sizes. All company departments involved in the product development have to meet this demand and ensure the required degree of flexibility and reactivity. The Collaborative Research Centre 653 (CRC 653) "Gentelligent Components in their Lifecycle" in Hannover addresses this demand among other things. Thus the area for process planning treated in this publication plays a special role. At this early stage of the product development the respective processing steps, their order, as well as the required and available production resources are determined based on the construction information.

A process planning system must fulfil three essential prerequisites in order to achieve the highest degree of flexibility and reactivity possible while constantly considering the overall efficiency and the deadlines for delivery. Firstly it has to be possible to show all executable process sequences that are necessary due to the respective processing steps and the available production resources based on the information deposited in the system. This results in a complete overview of all possible alternatives of the process sequence, which can be adapted to the current capacity utilization during the actual production. Secondly, it has to be possible to evaluate the planning output under realistic conditions.

This way, the determined process sequences and parameters can be checked on the basis of a simulation carried out parallel to the production planning and adjusted if necessary. At the same time, the process understanding gained in the simulation is used to parameterize the process monitoring systems used in the production. Thirdly, a process planning system has to be, at any time and without great effort, able to change the basis for decision-making used for the planning as well as to adapt it to the conditions in the enterprise and the real process courses.

In order to ensure the highest flexibility and reactivity possible in a versatile production environment with variable lot sizes, a novel process planning method is required.

2. The Gentelligent Production

A novel approach to the "simulation based planning and control of cutting processes" is developed within the framework of CRC 653. Constructional specifications and up-to-date information from the production are used in connection with the results of a machining simulation aimed at predicting the process course and the process variables, for a semi-automated process planning. This novel approach to the process modeling and simulation-based process planning as well as to the associated detailed planning of the production process is called "virtual planner" [1]. The focus is on the using intrinsic information of gentelligent components in order to plan and control the production processes required for their machining as soon as possible. A major quality of the arising GI-technology is that the data required for the production of the workpiece are inherently and inseparably stored in it. In the following, the qualities of the "virtual planner" are to be introduced (see. section 2.1).

2.1. Planning and Control of Cutting Processes on the Basis of Intrinsic Information

The planning and controlling of the cutting processes is carried out by means of the "virtual planner". This is the combination of a process planning system with a process simulation system. The concept of the interconnected systems processes the manufacturing of gentelligent components in two steps. First, the virtual - and therefore computer internal - planning and execution of the production process are carried out based on the intrinsic information. This step allows to plan the process and to verify it by means of the simulation (Fig. 1, left).

Then follows the real production process based on the process parameters that have been planned and verified by the simulation in the previous step (Fig. 1, right).

Part of the "virtual planner" is the semi-automated process planning system, which ends the classical functional separation between process planning and NC-programming and functions as a communication interface between gentelligent component, utilization schedule, machine control and simulation system. In the first step of the process planning, possible process sequences are determined by means of geometry information of the component as well as technological and economic planning parameters of the available tools and machines (1). That means so-called nonlinear process chains are developed, which represent alternative processing steps and orders and to which different production resources are assigned in addition. At first, the capacity utilization and the availability of particular resources are not taken into consideration. As a result, the following planning stages possess more freedom increasing the planning reliability as well as flexibility and reactivity. The compiled information of alternative process sequences is then stored as an operation sequence sheet n the structure of the gentelligent component.

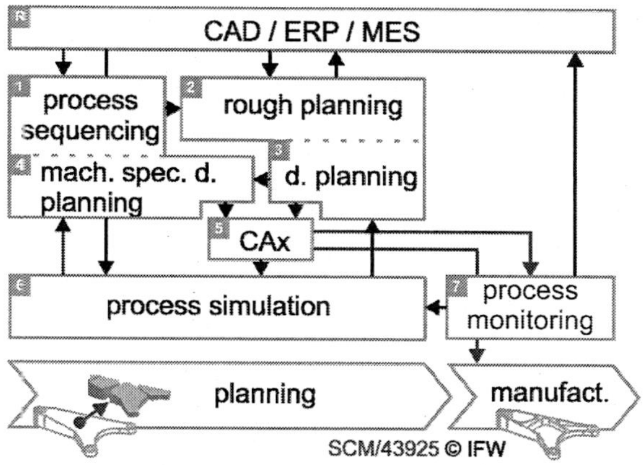

Fig. 1. Planning and Controlling of a production process with gentelligent components.

After that a rough, fine and machine-specific detailed planning of the following processing steps are carried out on the basis of the compiled nonlinear process chains (2, 3, 4). The adjustment of the level of detail of the processing steps complies with the concept of the "adaptive process planning" and, is therefore carried out as late as possible and parallel to production [2]. In this step it is necessary to identify the process chain among the possible alternatives that complies with the existing production data, such as future capacity utilization or current availability of particular production resources and to integrate the found process chain into the current production flow. In this way, each predefined processing step is assigned a concrete production resource while at the same time the machine-specific process parameters spindle speed, cutting speed, cutting depth and contact width are determined for each production resource.

When the process parameter specifications are known, the production flow is drafted step by step in (5). In addition to the data required for the production, like required tools, clamping plans and the processing order, the NC-code for the control of the production resource is generated in this step. The virtual planer verifies the selected parameters before the data are transmitted to the production resource (6). In a virtual pilot run (see section 4.1), the production process is simulated in detail. This results in a virtual component generated according to the technological limitations. The following evaluation of the process is the basis for the decision if the preset parameters are valid on the basis of simulated process state variables and the virtual component. If this is not the case, a new production flow based on the result of the evaluation is drafted using the adjusted parameters. Degree and direction and kind of the parameter adjustment are fixed by the planner.

Provided that a valid parameter setting is found, the planner transmits the process planning data to the production system. Apart from the production specific information listed above, this information also includes data needed for the parameter setting the process control system (7) (see section 4.2).

After the whole production process is completed, the selected process chain, the adjusted processing parameters, as well as a description of the processing result also become an inherent part of the gentelligent component. In this way, it is later possible to evaluate the processes as well as to adjust the planning basis.

3. Generation, Evaluation and Optimization of Alternative Process Sequences

First task of the "virtual planner" is the generation and evaluation of alternative process sequences, which are then stored in form of nonlinear process chains in the structure of the gentelligent component. Subsequent to the completion of the entire production process, the thus determined target specifications are compared with the eventually selected process sequence, the corresponding resources, the adjusted process parameters, as well as with the recorded process sequences (see section 4.3).

An essential basis of semi-automated process planning systems is the database used for the planning and organization as well as the available enterprise-specific information on the capacities of the production resources at hand. In order to develop an alternative process sequence, the process planner needs a multitude of production specific information.

The procedure of the semi-automated generation and evaluation of alternative process sequences can be subdivided into two main sections (see Fig. 2). At first, an application-specific planning basis must be established that is then used in the second section for the generation and evaluation of alternative process sequences. In order to enable the semi-automated decision-making, a formalized representation and processing of the production-specific knowledge and information structure in an enterprise is required. The concept of merging ontologies is used for the information modeling and processing. Ontologies are information models that define the concepts corresponding to a context, their relations to each other and valid rules by means of semantic connections.

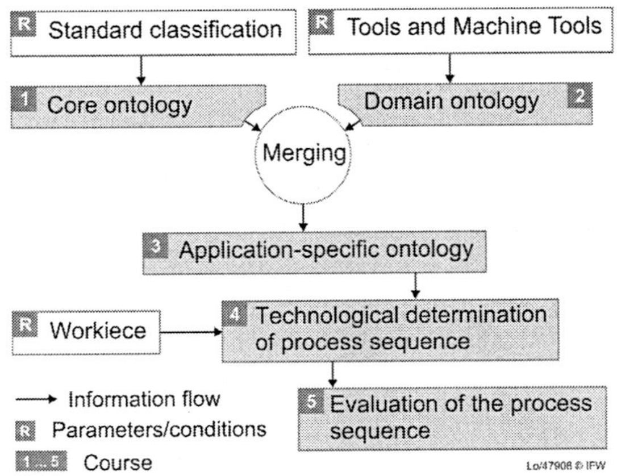

Fig. 2. Determination of the process sequence.

That means that within the core ontology (1) the relevant concepts of the process planning and production environment are integrated into a universal concept hierarchy based on standard classifications. This concept hierarchy, in turn, is extended by descriptions, attributes, relations and rules. The standards refer to nationally and internationally standardized classifications of processing features, manufacturing methods, tools and machine tools. Thus it is possible to list all generally possible manufacturing methods as well as applicable tools and machine tools based on a processing feature. At the same time, enterprise-specific production structural data, that is a formalized overview of tools and machine tools available in the enterprise, are gathered in a domain ontology (2). Then, the both original ontologies are merged into a third application-specific ontology (3). Based on this third ontology in combination with the workpiece-specific processing features, it is possible to determine all technologically possible process sequences that can be produced according to qualitative criteria (4). Finally, all process sequences are evaluated regarding their compliance with appointed completion dates/deadlines as well as with planned target costs and are then listed in a weighted order (5). An essential evaluation factor is the already existing empirical knowledge about individual process steps. If particular processing features have already been produced with the existing resources, the gathered information is used as the basis for the identification of the knowledge factor. This describes the degree of the confidence in the applicability of the planned process parameters, based on the collected experiences. An increasing knowledge factor determines the degree of the optimization of the chosen process sequence and of the adjusted process parameters and, thus, represents the confidence in them. Thus, the alternative with the best overall result in the evaluation is the first choice for the process sequence that is selected later in the production and is subsequently evaluated by the simulation of the production process.

4. Virtualization of the Pilot Production

In order to meet the demands on the production of gentelligent components, a high process stability already has to be ensured at the production of the first component generation. Therefore, the run-in of the milling process, which cannot be avoided, has to be done before the actual manufacturing process. For this purpose, the identified process sequence which promises to achieve the best result compared to all alternatives is run in by the "virtual planner" in a virtual manufacturing process. The simulation of the physical production process [4] is carried out under consideration of the adjusted process parameters in order to calculate process state variables relevant for a process evaluation. In order to ensure that the calculations are sufficiently exact, a selection of process-relevant phenomena whose effects have to be represented in the model is needed. The "virtual planner" selects these phenomena under consideration of the technological basic conditions (see Fig. 2). Depending on the process results it might be necessary to readjust the parameters. Like a real run-in procedure, this process results in a parameter combination that ensures a failure-free production process.

Since not all eventualities can be reproduced by a process simulation or its process model, the real process sequence is monitored additionally to ensure safety.

4.1. Simulation-based Process Monitoring

A variety of monitoring strategies are available. Which of these strategies is preferred in the end, depends - among other things - on the desired automation degree of the production facility [5]. Apart from the technical requirements, the process security and reliability, as well as the quality requirement made on the component production play an important role. The tasks of the process monitoring are essentially to avoid tool damages caused by overloads or wear as well as to recognize collisions in the workspace. Apart from damage prevention, it also supports the recognition of process instabilities that would not inevitably damage components or machines, but could massively disturb the production result [6].

A differentiated investigation of failure states is therefore indispensable due to the spectrum of possible failures. For this purpose the failure states that are to be recognized have to be systematically characterized using their individual features and converted into a fact database. This does not only help to specifically identify occurring failures but also to influence their occurrence during planning. Using IF-AND-THEN relations, the monitoring system infers from the available facts on a failure cause. Thus, feature extraction is one of the central operations in a process monitoring system.

One of the processes of time-domain feature extraction is the method of enveloping curves. Here, the signal is continuously compared with suitable parameterized threshold values that cannot be overrun or under-run.

4.2. Calculation and Parameterization of Threshold Envelopes

In this case, the term "threshold envelope" means an envelope which encloses a given time signal at a specified interval 1/2 a (Fig. 3a).

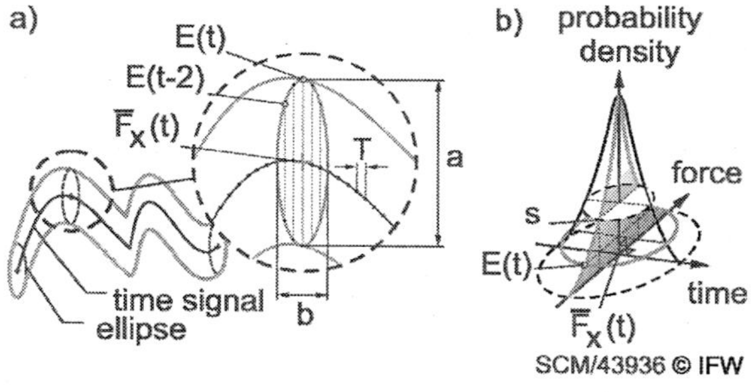

Fig. 3. Dimensioning of the elliptical filter.

This signal envelope is formed by means of an ellipse moving along the discreet time signal $F_x(t)$. The ellipse forms extreme values E(t) both above and below the force signal at constant intervals T. For every time step these values are stored in a data vector. Due to the ellipse width b which is many times bigger than T, there are always more than two entries in the data vector. The smallest and the biggest entries determine the lowest and highest threshold values respectively for the time t. They are added to the traverse of the signal envelope and result in the desired envelope curve if the procedure is carried out along the entire signal length.

The envelope has the same point density as the signal it encloses. Depending on the simulation step size and the length of the process to be simulated, the number of points can become very large. Regarding data storage, that means that the memory requirements are increased by increasing process time. Since the signal dynamics of an enveloping curve is low compared to that of the enclosed force signal (Fig. 4a), a local point reduction would not have any perceptible effect on the information content of the envelope, but lead to a considerable reduction of the memory requirements. An already established method for the reduction of the curve complexity is known from geoinformatics. Douglas and Peucker [7] developed this method for almost loss-free point thinning and curve smoothing. This method was used in the example described below (Fig. 4a), and allowed to reduce the number of points by 99,9575%. Thus, it successfully contributes to the reduction of the memory requirements.

The process of milling is characterized by constantly changing process conditions, which become apparent in the form of time-variant material removal rates, lead to cyclic as well as acyclic loads on the cutter and whose direction of influence changes permanently. This has to be taken into consideration when the parameter of the envelopes is dimensioned.

Figure 4a shows the simulated run of the cutting force. The figure shows that the signal $F_x(t)$ consists of two parts: the almost static or low-frequency part, indicated here by the arithmetic mean (black), and a dynamic, high-frequency part (blue). The degree of the dynamics is important for the parameterization of the enveloping curve, or more precisely of the tolerance band width since process stages with high variations in amplitude, that is with a high signal dynamics, have to be monitored with a broader tolerance range. A measure for this dynamics is the standard deviation s of the signal $F_x(t)$. In order to determine this, the signal is divided into sections of defined length. Within a section the arithmetic mean is almost constant. The individual data points of the signal section group around this mean $F_{x,a}(t)$ correspond to a probability density function (Fig. 3b). Thus, the standard deviation describes the dynamics of the signal within the considered section and can therefore be used for the automatic parameter setting of the width of the tolerance band (Fig. 3a).

4.3. Multi-area Process Monitoring and Feature Extraction

In principle, a trouble-free process is characterized by the fact that the measurement signal stays within the predefined tolerances and, thus, within the envelope. Force amplitudes transgressing the threshold values are, therefore, an evidence for a failure in the production process. This assumption, however, cannot be generalized. A reason for it lies in the precision of the simulation model, which, by definition, represents a simplification of reality. In view of it, transgressions of threshold values have to be differentiated making the generation of several features necessary. The appearance of the measurement signal over several zones can be observed by the use of multi-area-envelopes. Moreover, it seems reasonable to take into account not only the current condition but also the history of the signal for the evaluation of the process state.

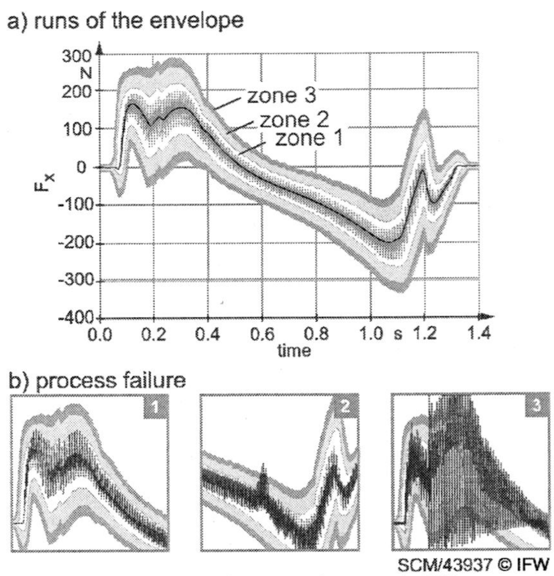

Fig. 4. Course of an envelope and different process failure.

The example in Figure 4 illustrates this procedure. Figure 4a shows a three-area envelope derived from a simulation signal. Due to the changing signal dynamics, the width of the tolerance band varies in the process sequence (see section 4.2). Figure 4b shows three process states that would lead to the signaling of a process failure if a simple envelope were used. Nevertheless, the signal course represented in b1 does not result from a failure but only from an imprecise reproduction of the simulation model causing the deviation of the power signal and the envelope. Thus, the process state would be misinterpreted. In order to avoid this, the interpretation of the measurement signal is carried out on the basis of several features in different envelope zones so that state 1 is evaluated as a failure-free process while states 2 and 3 are evaluated as showing failures. Furthermore, the distribution of the measuring data over the areas provides different results for states 2 and 3, so that the material inhomogeneity, which led in state 2 to the signal outbreak, is to be clearly distinguished from the cutting edge breakage in state 3.

The failures found in this way have to be added to the process information, which in turn is fed back to the process planning as a gain in insight and, thus, improves the degree of optimization with regard to process sequences and parameters to be chosen (knowledge factor). Moreover, this information serves the generation of a process quality seal, which is transmitted to the component.

5. Summary

Within the framework of the collaborative research center 653 "Gentelligent Components in their Lifecycle", components with novel properties as well as production concepts, methods and technologies for their manufacturing and use are developed. Within the phases of the component production, the concept introduced here concentrates on a new method for the simulation-based process planning that guarantees a high degree of flexibility and reactivity in a versatile production environment with variable lot sizes. The construction of an application-specific planning basis by means of which all alternative process sequences can be determined and evaluated semi-automatically is a part of this procedure. The basis for decision-making is formed by an information model that reproduces both universal and enterprise-specific properties of the production and its environment via the combination of specific ontologies. Moreover, the model depth of reproduced process phenomena within the simulation model contributes to the sufficiently accurate representation of machining operations within a process sequence. The process knowledge gained in this way constitutes the output value for an optimization of process parameters and is also used in the parameter setting of monitoring systems of the production resource. The process-accompanying monitoring and the feedback of process-specific knowledge in the form of identified failure conditions enable the information circulation from the planning via the production to quality control and back.

Acknowledgements

The results presented in this paper were obtained within the framework of the collaborative research center 653 "Gentelligent Components in their Lifecycle". The authors would like to thank the German Research Foundation for its financial and organizational support of this project. The IFW is a member of the EU-Network of Excellence I*Proms.

References

[1] Denkena B. and Battino A. and Lorenzen L.-E. and Schmidt C. Simulation based Process Planning for Gentelligent Production. In: 40th CIRP International Seminar on Manufacturing Systems (ISMS´07), Liverpool, UK, pp. PS5 1-6, 2007

[2] Denkena B. and Tracht K. and Battino A. Dynamic Analysis of Process Chains as an Enabler for the Adaptive Process Planning. In: 2nd International Conference on Changeable, Agile, Reconfigurable and Virtual Production (CARV´07), Toronto, Kanada, pp. 677-686, 2007

[3] Gruber TR. A Translation Approach to Portable Ontology Specifications. Knowledge Acquisition. Vol. 5, Nr. 2, pp. 199-220, Academic Press Ltd., London 1993.

[4] Denkena B. and Schmidt C. Experimental Investigations and Simulation of Machining Thin-Walled Workpieces, Production Engineering WGP XIV/4, 2007

[5] Kaever M. Steuerungsintegrierte Fertigunsprozess-überwachung bei spanender Bearbeitung, Books on Demand GmbH, Norderstedt, Aachen, 2004

[6] Sick B. Signalinterpretation mit Neuronalen Netzen unter Nutzung von modellbasiertem Nebenwissen am Beispiel der Verschleißüberwachung von Werkzeugen in CNC-Drehmaschinen, Passau, 2000

[7] Wu S. and Gonzales Marquez M. A non-self-intersection douglas-peucker algorithm, XVI Brazilian Symposium on Computer Graphics and Image Processing (SIBGRAPI'03), 2003

Innovative Production Machines and Systems
D.T. Pham, E.E. Eldukhri and A.J. Soroka (Eds.)

A comparison between two incremental forming processes for the achievement of a component with concave-convex shape

A. Formisano, M. Durante, A. Langella, F. Capece Minutolo

Dept. of Materials and Production Engineering, University of Naples Federico II, P.le Tecchio 80, 80125 Naples, Italy

Abstract

Incremental forming is a sheet metal forming process that offers flexibility, given that it requires no high capacity press or set of dies with specified shape for the product, and therefore meets the ever increasing demand for small batch production and rapid prototyping. The simplest type of this process, known as single point incremental forming, has shown itself to be unsuitable for manufacturing components with concave-convex shape.

The aim of this study is to research two versions of the process under examination: the multistep process and the process that uses a pattern to manufacture a component with both concave and convex features, starting with aluminum alloy sheets.

FE analyses were carried out in order to provide for the correct implementation of the process. A number of observations regarding the two process typologies and geometrical analyses, used to compare the experimental profiles with their corresponding design profiles, allowed us to highlight their advantages and disadvantages.

Keywords: Incremental forming, FE analysis, Geometrical accuracy

1. Introduction

Incremental forming is a sheet metal forming process that can be summed up in few words: a simple shaped tool, controlled by a general purpose CNC machine, progressively produces a local plastic deformation in a metal sheet, in order to create the designed shape of a component. This process has the advantage of high flexibility; it is, in fact, possible to create components with different shapes, simply by operating on the tool path [1]. For this reason, the process under examination is recommended for small batch production and rapid prototyping. On the other hand, it cannot be used in its traditional typology for manufacturing components with both concave and convex features.

The purpose of this work is to research two versions of the process under examination: the multistep process and the process that uses a pattern to manufacture a component with both concave and convex features, starting with aluminum alloy sheets.

Following FE analyses, useful in this work for providing information about the proper implementation of the process in terms of working parameters, an experimental campaign was carried out, in order to manufacture a component with both concave and convex features, by using the process in two steps, as well as the one with a pattern. An evaluation of the geometrical features related to the components manufactured, as well as a number of considerations regarding the two typologies, mean that we have been able to draw some interesting conclusions on the same process.

2. Component to be manufactured

The component, manufactured in order to carry out the analyses under examination, has an axialsymmetric shape with both concave and convex features. Fig. 1 shows the profile, the rotation of which around the z axis generates the designed shape of the component. It consists of three parts: the first one is a segment with a slope angle of 60°; after this segment another one follows, perpendicular to the rotation axis; the third part consists of a segment with a slope of 60° and an arc of circumference, extending for 60° as far as the rotation axis of the geometry. From the rotation of these three segments around the z axis, convex, flat and concave parts are generated respectively.

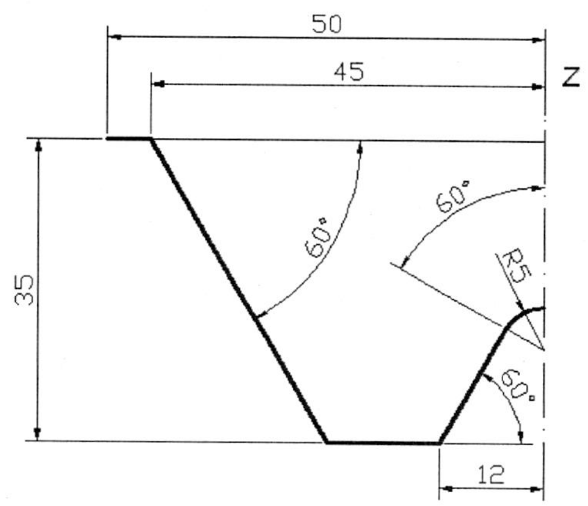

Fig. 1. Profile of the component under examination.

3. Choice of the process parameters and description of the forming typologies

The material used in the experiments is an AA 7075-T0; sheet properties are shown in Table 1. Mechanical characterization was carried out by means of tensile tests with an MTS Alliance RT/50 materials testing machine.

Table 1
AA 7075-T0 sheet properties.

t (mm)	E (GPa)	Y (MPa)	UTS (MPa)	εr (%)
1.2	75	100	200	22

The tests, with controlled feed, were performed on specimens with two different orientations: that is, with an angle of 0° and 90°, as regards the rolling direction. No differences were found between the stress-strain curves for the two different orientations; therefore the material displays plane-anisotropy. The characteristic parameters of the plastic behavior were evaluated, with the result that it can be described using the following power law (Eq. 1):

$$\sigma=330\varepsilon^{0.19} \quad (1)$$

Sheet formability, which, for aluminum, can be quantified using a scalar number equal to the sum of the maximum and minimum strain (ε_{max},ε_{min}) [2], was, in this specific case, determined in [3] and can be expressed (in percentage terms) by (Eq. 2):

$$\varepsilon_{max}+\varepsilon_{min}=115 \quad (2)$$

The sheet to be formed is clamped along its periphery by a blocking system, with a forming area of 100mm·100mm; the component has, therefore, a flange which extends up this square and does not come into contact with the tool, a steel cylindrical stylus with a hemispherical head of 5 mm in diameter.

The component was manufactured using two incremental forming versions. In both cases, a general purpose CNC machine was employed, and a tool path was set, alternating between circular coils and the vertical feed rate p=0.5 mm, with a feed of 1000 mm/min. The tool-sheet interface was lubricated using oil, and the tool was set in rotation at 600 rpm to reduce the friction and, consequently, the risk of cracks in the sheet. A description of the two typologies is presented below.

3.1. Multistep incremental forming

This typology of incremental forming provides different forming steps for the different geometrical features to be created. Between one forming step and another, a blocking system or sheet handling operation is required. For the case in question, two forming steps are required, in order to create first the internal, and then the external part of the component (Fig. 2), along with a sheet handling operation.

3.2. Incremental forming with pattern

Incremental forming with a pattern allows the manufacture of components with both concave and convex features in only one forming step (Fig. 3). Obviously, unlike the above mentioned typology, it requires an additional component, the CAD model of which, for this specific case, is shown in Fig. 4.

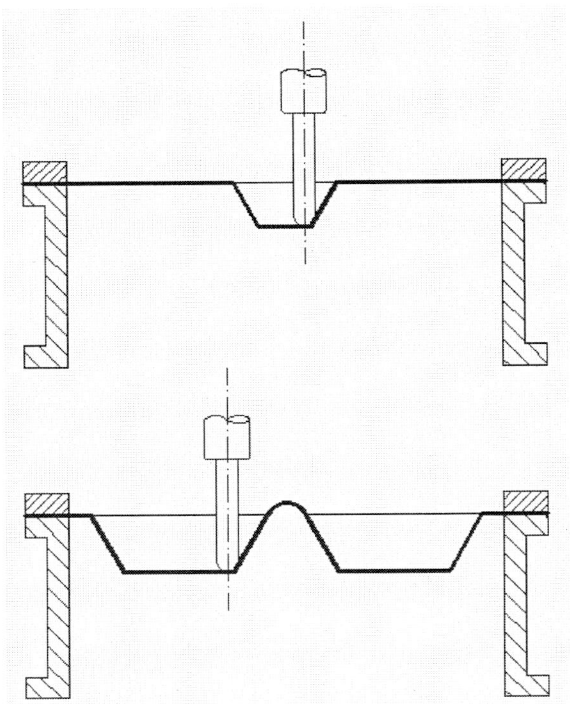

Fig. 2. Multistep incremental forming.

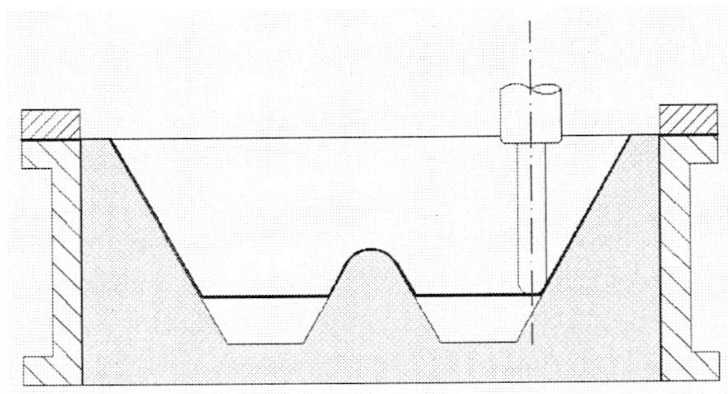

Fig. 3. Incremental forming with pattern.

Fig. 4. CAD model of the pattern.

4. FE analyses

FE analyses were carried out in order to provide for the correct implementation of the process under examination, in terms of suitability of the chosen working parameters and, for the typology with the pattern, to ensure the quality of the tool path and to avoid interference between the tool and the pattern. Moreover, the analyses enabled preliminary information to be provided about the geometrical quality of the manufactured components.

The analyses were carried out with LS DYNA, an explicit finite element code widely used to study automotive crashes. For this reason, it was necessary to choose input parameters correctly, in order to optimize metal forming simulations [3-8]. The run time, otherwise high in explicit forming simulations, was reduced by setting a mass scaling time step size parameter. All the process parts were modeled using shell elements: the default type, with two thickness integration points for the punch and the surfaces of the pattern that come into contact with the sheet, and a fully integrated shell element type, with five thickness integration points for the sheet. Various aspects associated with numerical choices, as well as materials and process parameters, play a vital role [9-10]. For the tool and the pattern, a rigid material model was defined, as in sheet metal forming problems, the tooling can properly and accurately be treated as rigid, whereas an isotropic plasticity model, which uses a power law hardening rule, was assigned for sheet elements. A fix constraint, applied along the sheet periphery, replaced the effect of the blocking system; specified displacement-time curves were assigned in order to prescribe the translations of the tool to the letter. A penalty-based contact interface was chosen, as it is much more accurate in applications such as metal forming; a friction coefficient value of 0.06, measured in experimental tests [11], was assigned to its control card.

The multistep process simulation during the two forming steps, a section of the equipment at the beginning and the component at the end of the typology with the pattern, are shown respectively in Figs. 5, 6 and 7. These simulations enabled us to affirm that the process in its two typologies can be carried out correctly. In fact, it was checked that the tool path had been appropriately set, given that no interference was brought to light between the tool and the pattern. Furthermore, the sheet does not incur failures, since the points of the forming limit diagrams are contained within the experimental curve, expressed by Eq. 2, as shown in Fig. 8.

5. The manufacture of the components

The experimental phase followed the numerical analyses. The process, in both its versions, was carried out using a general purpose CNC machine. Figs. 9 and 10 reproduce the sheet after both the forming steps related to the multistep forming typology. Fig. 11 shows the manufacture of the component using the pattern (Fig. 12), which was created starting with hard wood (but it would have been possible to create it using other materials, without the necessity for excessive care, such as materials for rapid prototyping) by the same CNC machine. The simulation parameters were set for the manufacture of the components.

The components were manufactured without incurring fractures, as predicted by the simulations, which also provide a realistic image of them. Fig. 13 shows the comparison between experimental and designed profiles, and brings to light closer correspondence for the profile created by the version with the pattern. It is evident, in fact, that the pattern allows a significant reduction in the main geometrical errors typical of an incremental forming process, that is rounded edges and springback.

Other observations can be summarized as follows:

- Using the pattern, it is possible to create a concave-convex component in only one forming step;
- The version with the pattern allows deformation of the sheet on one face;
- The forming process with the pattern requires an additional component compared with the multistep forming, which preserves the flexibility typical of the incremental forming processes, since it requires no more components for the equipment;
- The forming process with the pattern is faster and more accurate, since it does not require the sheet handling operation.

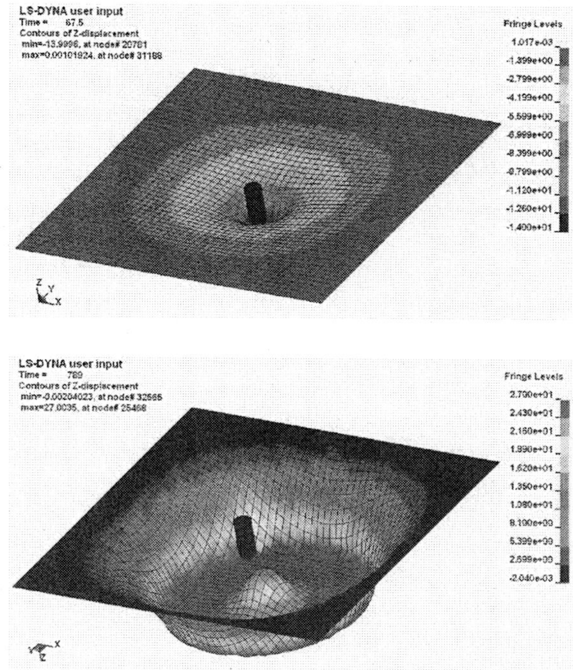

Fig. 5. Simulation of the multistep process.

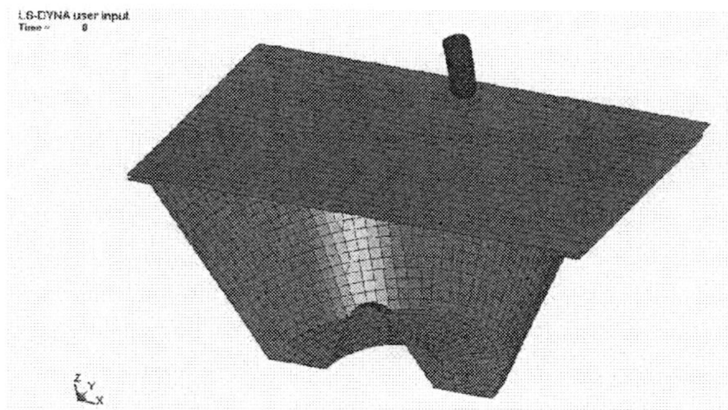

Fig. 6. A section of FE equipment model of the process with pattern.

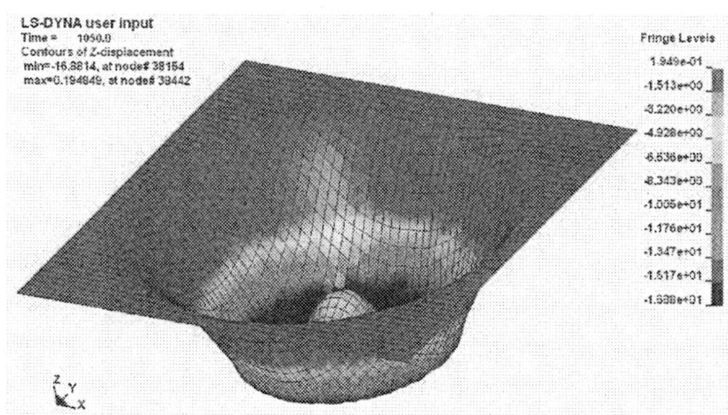

Fig. 7. Simulation of the component at the end of the process with pattern.

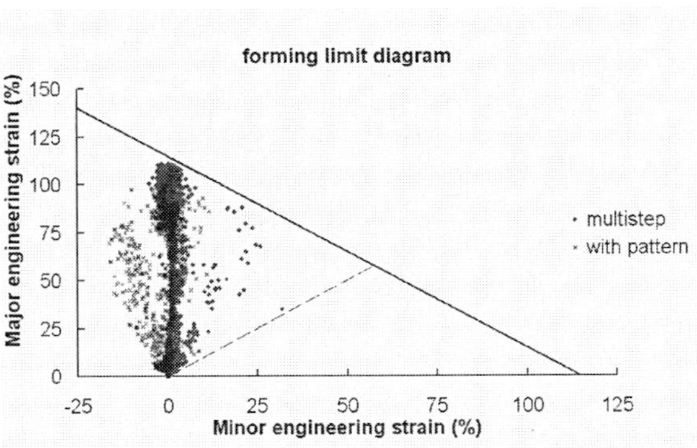

Fig. 8. FLD's for the two forming typologies.

Fig. 11. The manufacture of the component with the pattern.

Fig. 9. The sheet after the first forming step.

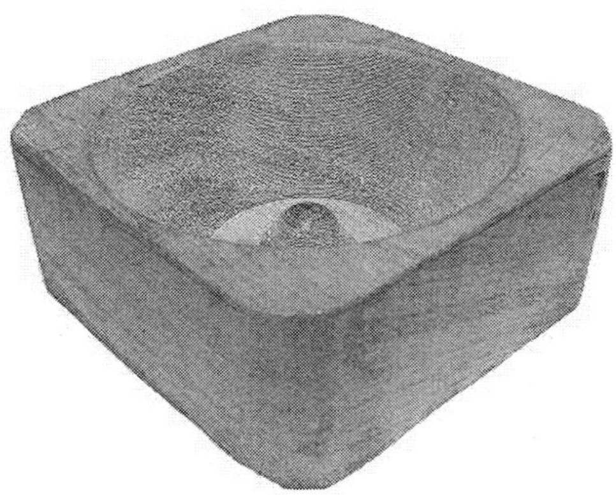

Fig. 12. The pattern used in this study.

Fig. 10. The sheet after the second forming step.

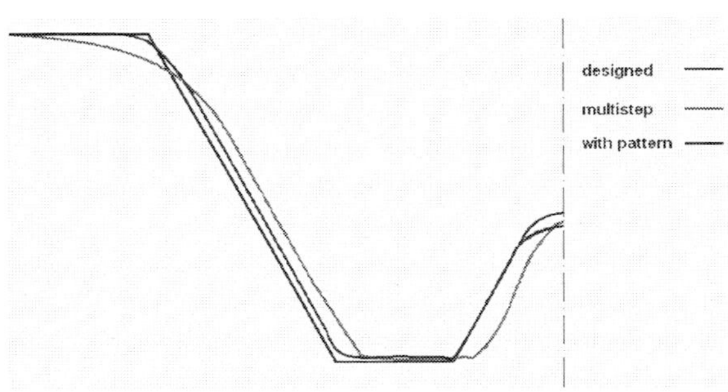

Fig. 13. Comparison between experimental and designed profiles.

6. Conclusions

Research was carried out into two incremental forming versions, i.e. the multistep process and the process that uses a pattern. The analyzed components, with concave-convex geometry, were created starting with aluminum alloy sheets.

FE numerical analyses, performed by means of a program validated in previous papers, have been an effective support to this study, particularly in providing for the correct implementation of the process.

The experimental campaign has highlighted the limits, in terms of geometrical accuracy, associated with this process, particularly for the multistep version, which maintains higher flexibility, but proves to be less accurate than the typology with the use of a pattern.

References

[1] Shim MS and Park JJ. The formability of aluminum sheet in incremental forming. Journal of Materials Processing Technology 113 (2001) 654-658.

[2] Iseki H and Kumon H. Forming limit of incremental sheet metal stretch forming using spherical rollers. Journal of the Japan Society for Technology of Plasticity 35 (1994) 1336-1341.

[3] Capece Minutolo F, Durante M, Formisano A and Langella A. Evaluation of the maximum slope angle of simple geometries carried out by incremental forming process. Journal of Materials Processing Technology 194 (2007) 145-150.

[4] Capece Minutolo F, Durante M, Formisano A and Langella A. Forces analysis in sheet incremental forming and comparison of experimental and simulation results. Proceedings of the 2nd IPROMS Virtual International Conference (2006) 229-234.

[5] Nielsen KB, Jensen MR and Danckert J. Optimization of sheet metal forming processes by a systematic application of finite element simulations. Proceedings of the 2nd European LS-DYNA User Conference (1999) 3-16.

[6] Formisano A, Durante M, Langella A and Capece Minutolo F. A complex geometry carried out from steel sheets by incremental forming. Proceedings of the 3rd IPROMS Virtual International Conference (2007) 619-623.

[7] Yamashita M, Gotoh M and Atsumi S. Numerical simulation of incremental forming of sheet metal. Journal of Materials Processing Technology 199 (2008) 163-172.

[8] Capece Minutolo F, Durante M, Formisano A and Langella A. Optimization of a hydroforming process to realize asymmetrical aeronautical components by FE analysis. Proceedings of the 2nd IPROMS Virtual International Conference (2006) 241-246.

[9] He S, Van Bael A, Van Houtte P, Tunckol Y, Duflou J, Henrard C, Bouffioux C and Habraken AM. Effect of FEM choices in the modelling of incremental forming of aluminium sheets. Proceedings of the 8th ESAFORM Conference on Material Forming (2005) 711-714.

[10] Bambach M and Hirt G. Error analysis in explicit finite element analysis of incremental sheet forming. AIP Conference Proceedings Volume 908 (2007) 859-864.

[11] Formisano A, Durante M, Capece Minutolo F and Langella A. The influence of tool rotation in an incremental forming process. Proceedings of the 8th AITeM Conference (2007) Poster session.

Innovative Production Machines and Systems
D.T. Pham, E.E. Eldukhri and A.J. Soroka (Eds.)

A simulation model of an industrial network

L. Iandoli[a], C. Ponsiglione[b], G. Zollo[c]

[a, b, c] *Department of Business and Managerial Engineering, University of Naples Federico II, Naples, Italy*

Abstract

This paper presents a theoretical and computational model to analyze the dynamics of inter-organizational relationships among firms of a complex network, such as an industrial district. The aim of the proposed model is to verify, through generative experiments, the effects of some determinants of coordination mechanisms on firms and on the network's performances. The conceptual model was implemented on Agentsheets platform and some experimental sets were developed in order to analyze the evolution of the network in different contexts.

Keywords: Agent-based simulation, Collective memory, Firms' network.

1. Introduction

The aim of this paper is to study the dynamics of inter-firm relationships within firms' networks, such as Industrial Districts (IDs), in order to analyze and describe their role in influencing ID's competitive and innovative performances.

An ID can be defined as a cluster of small and medium sized firms, belonging to the same geographical area and characterized by high degrees of specialized production [1]. Traditionally, industrial districts' competitiveness was based on some specific factors as acknowledged by literature:

- the existence of external economies for the firms, but internal to the district, that balance the disadvantage of scale economy lacks [2, 1];
- the flexible specialization, that creates competitive advantages in terms of production efficiency [3];
- the intensive exchange activities between small and medium enterprises (SMEs) of the ID, that create a network able to develop incremental innovation [4].

The rising complexity of the competitive scenario implied the search of new way of development. In the current unstable scenario, IDs firms are more and more requested to activate learning and change mechanisms in order to combine tacit and explicit knowledge, past experiences and business diversification.

In section 2 we provide a short background on agent-based simulation in social sciences. In section 3 and 4 we describe an agent-based model of an ID. In section 5 we outline the main results obtained by the model simulation. Finally, in section 6 we discuss research implications and conclusions.

2. Background on agent-based simulation

By "computational approach" we mean a computer-based way of formalizing evolutionary interacting agent-based models [5]. Such models assume heterogeneity, localization and interdependence of agents and decentralized mechanisms of interaction [6, 7].

Agent-based simulation enables to analyse self-organising systems, learning systems and systems with high complexity.

The Industrial District can be defined as an evolving network of heterogeneous, localized and functionally integrated interacting firms. Thus agent-based modelling seems particularly suitable to model and to analyze IDs. To build up our simulation model we used Agentsheets platform, which is simple to use, although not very flexible. An agent-based model is developed in two steps: the definition of a conceptual model of the system/phenomenon to be studied (meta-model), the implementation of this conceptual model on an agent-based software platform (computational model).

3.The meta - model

According to literature on agent-based models of firms networks [5, 8, 9], we structure the meta-model into five principal building blocks (Fig1). In the First Building Block, *Classes of Agents*, we define two classes of firms (agents): final and subcontracting firms. To define the Second Building Block, *Agents inner models,* agents firms are described by a set of variables, evaluation and decision rules, all concurring in defining agent social behavior. The third building block represents *the context of interaction*: the inter-firm scaffolding structures. These structures are necessary to co-ordinate the whole production process. To translate the Fourth Building Block, *the environment structure*, we assume that the market carries all the needed information to develop attractive products. At some point in the simulation, product quality requirements increase, driving firms to improve their competencies, in order to achieve the new quality standards. In the proposed model for the fifth building block, *Structure of the information flow,* we assume that final firms have higher strategic capabilities than subcontracting ones; so final firms convert market inputs (competences target levels) into internal information [5]. Then, subcontracting firms receive product requirements interacting with final firm. The ambiguity and vagueness of information flow can create gaps in the agents' routines. To reduce the uncertainty of the environment and the ambiguity of decisional situations we introduced scripts and routines contained in the *collective memory* of the district. For this reason, we added a new building block to the general design of the agent-based model: the specific context of the industrial district (the shared memory of the district) (Fig. 1). In this way, the structure of ID collective memory is translated into agents firms characteristics and environment and institutions laws and frameworks.

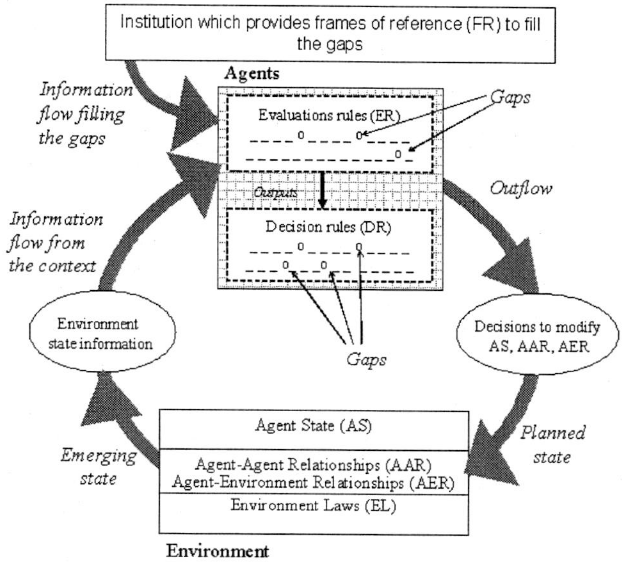

Fig.1 The proposed model

4.The computational model

4.1 Agents

Agents are all the elements of the simulation model. We classify them into two typologies: *principal and second order agents.*

4.1.1 Principal agents

According to theories of flexible specialization [3], we modeled ID as a system of small and medium firms, which interact to build up production chains. We define three classes of principal agents: final firms, subcontracting firms and production chains. Final firms principal objective (*fin*) is to meet market requirements; they manage it building up efficient production chains. Production chain efficiency is measured in terms of achieved quality during the production phase. Subcontracting firms (*sub*) are phase firms; they provide materials, components and semi-finished products to final firms. Their main objective is to find an adequate final firm in order to build-up a production chain. Thus, the interaction structure is based on district production chains (*Pch*): one final and one subcontracting specialized firm compose every production chain.

4.1.2 Second order agents

They have been created to translate the fourth and the sixth building blocks of the meta-model: the structure of environment and the specific context of ID. They are designed with a top-down perspective [7]. This means that we already know how they vary during simulations. In fact, we are interested in analyzing the phenomena, which emerge from the interactions among firms (principal agents) and between firms and environment (second order agents).

ID super - agent

To translate the specific context of the industrial district, we introduced a super - agent called *ground* in the model. Through its laws and framework of references (EL and FR in Fig.1) the super-agent diffuses and stores information (e.g. the market demand, target competencies levels, firms agents competencies levels and so on) in the district allowing firms agents to interact among them and take decisions.

Market Segments

To translate, in computational terms, the role of environment structure we define three market segments: S_1, S_2, S_3. These segments represent the potential markets for district firms. Different levels and types of competencies characterize each market segments: $S_i = (M_i, T_i, P_i)$. Where M_i, T_i, P_i represent the highest level of competencies required from the market at iteration *i*. We assume that at the beginning of a simulation exists only the S_1 segment. At certain iteration of the simulation, first S_2 and then S_3 appear on the district.

4.2 Agents inner models

Agents are modeled, in computational terms, as state variables, competencies, and evaluation and decision-making rules. Each agent is characterized by an internal state, which is an instantiation of its state variables and competencies levels. This state progressively changes in each iteration (called 'cycle') of the simulation, accordingly to firm choices and interactions with the other agents of the model. Thus, for each cycle *i*, firm internal state (IS(S_i))is function of: IS(S_i)= f (m_i, t_i, p_i, opp_i, $risk_i$, bdg_i) Where:

- m_i represents the level of market competencies. Market competencies group all the skills, knowledge and capabilities of the firm to sell its product to the market.
- t_i represents the level of technical competencies. These are strictly related to product quality. Firms with higher level of t_i are able to manufacture better products.
- p_i represents the level of relational competencies. In this work, with relational competencies we mean organizational competencies within production phases.
- opp_i Degree of Opportunism. This variable affects final and subcontracting firms' aptitude in building up a production chain; while, for production chain this variable influences its aptitude in breaking up the chain.
- $risk_i$ Risk Propensity. This variable indicates agent inclination to carry out risky investments.
- bdg_i The budget function. It computes the amount of economic resources of the firm. For each cycle, the function output (budget value) increases or decreases accordingly to firms choices; as follows: $bdg_i = bdg_{i-1} - C_{exp} - C_{inv} + PFT_i$, where: bdg_{i-1} is the budget value in the cycle *i*-1. C_{exp} represents firm's *costs of exploration*. An agent spends economic resources for every movement he takes in the environment space. This cost is constant during all the simulation. C_{inv} is firm's *costs of investment*. Firms have to invest some economic resources for every competencies improvement they perform. This cost is function of the level of competencies and decreases as competencies levels increase: $C_{inv} = f(C_{inv\_MAX} / m_i, t_i, p_i)$. PFT_i indicates firm *profit* at cycle *i*. Profit is function of product quality. Firms that sell higher quality products will receive higher profit.

When a simulation starts we assume that it is populated only by final and subcontracting firms, whose initial values of competencies levels and economic resources (m_0, t_0, p_0, bdg_0) are settled externally. Instead, for production chains the initial values are determined when the chain is built up.

4.3 Simulation Steps

At this point, we have to translate the information and production flows of the meta-model. This means explicating the events that occur in each iteration of the simulation. We summed up these events into five principal steps:

1. *Internal state check:* at the beginning of each cycle, firms check their internal state: IS(S_i)= *f*(m_i, t_i,

p_i, opp_i, $risk_i$, bdg_i). To limit complexity we assume that some of the state variables (opp_i, $risk_i$) do not change their values during simulation. Instead, the levels of economic resources and competencies vary and derive from the previous cycles.

2. *Evaluations:* firms evaluate their levels of competencies comparing them to the target levels given by the market.

3. *Improvements strategies:* at this step, firms choose their improvement strategies, according to evaluation outputs, objectives, state variables values and environment information. Firms have to decide which competencies to improve, among those indicated in evaluation output.

4. *Production chains generation or break up:* a fundamental objective for final and subcontracting firms is to become part of a production chain. Partner selection is based on competencies levels comparisons: each final firm looks for a subcontracting firm with competency levels greater or equal than its levels. Subcontracting firms are not able to select for final firm. In this step, production chain might decide to break up the chain, basing this decision on their opportunism degree.

5. *Product sale:* every production chain assembles and sells its product to the market and receives a profit when the market buys the output of its production process. The gained profit is function of market segment profitability and of product quality; the latter is function of the competencies gap of the firm. When market refuses the product, the chain does not receive any profit and it cannot increase its economic resources.

The hypotheses that we are going to verify through simulation can be synthesized as follow:

- When the degree of cohesion of ID collective memory is too high or too low, the industrial district risks to close on itself and experiences its inability to adapt to market evolution.
- When the collective memory of the district keeps a certain level of diversity among agents inner models and guarantees, at the same time, the integrity of the district identity, the system evolve, and is able to learn and innovate.

5.The experimental sets

We assume that at the beginning of each simulation run, final and subcontracting firms are characterized by the same level of competencies and of economic resources, but they show different values for risk propensity (risk) and opportunism degree (opp). Thus, each agent assumes values in one of the four quadrants of the following matrix (Table 1).

Table 1. Agents values

Risk	**H/L**	**H/H**
0.5	**L/L**	**L/H**
	0.5	*Opp*

We define the different starting population changing the agents' distributions in the four quadrants of the above matrix. The higher is the presence of the agents in one of the four quadrants, the higher will be the degree of cohesion of the collective memory of the population. According to these issue, we name the starting population distribution as *closed*, when the 70% of the starting population takes values in one of the four quadrants of the matrix above. We say that a starting population is *open*, when at least one of the two variables opp and risk is equally distributed in the quadrants. And finally, we say that a starting population is *intermediate* when there is a sloght prevalence of the agents in one of the quadrants. In this way, we built up one *open*, two *intermediate* and four *closed* starting population distributions,. Each population has been tested within two environments, which differ for the type of market evolution. The two environments are:

- Stable Market: this means slow changes in market requirements.
- Turbulent Market: this situation indicates quick changes in market requirements.

The different market evolutions are translated in terms of temporal distance (number of cycles) that elapses between two changes of market conditions. In the stable case this distance is constant during simulation (20 cycle), while in turbulent case it is variable.

6. Results, analysis and discussion

We performed 50 run per experimental set, in order to reduce the influence of agents starting position on the results. Each run was composed by 40 firms: 20 final firms and 20 subcontractors and lasts for 180 cycles.

The indicators used to analyze and compare different simulations were:

- **P**: this represents the overall economic performance of the district. It is computed as the sum of the budget of the firms (final firms, subcontracting firms and production chains) at each simulation cycle.
- **N**: this stands for the number of firms (final firms, subcontracting firms and production chains) at each simulation cycle.

To make our analysis and comparisons we used the final value of these variables, which is the value computed at the 180th cycle. For each multiple–simulations we obtained 50 pairs of values for P and N. At first, the 50 pairs have been ordered according increasing P. Secondly, we drew the 50 pairs on orthogonal axis. Then for each of the experimental sets we performed the analysis of relative frequencies, we computed the average for the two variable P and N (Table 2) and, finally, we compared the results.

The second and the fourth experimental set are characterized by high opp levels and low risk levels. We name these last two set as "Cooperative Districts (CD)". On the contrary, we called the third and fifth sets, which are characterized by high levels of risk and low opportunity, "Non Cooperative Districts (NCD)". Through the analysis of table 3, we compare the behaviours of these two types of districts. Some considerations according the analysis of the table 3:

- Cooperative Districts: from closed to the intermediate population distribution, the higher diversity among firms rewards more in the turbulent case.
- Non Cooperative Districts: from closed to the intermediate population distribution, diversity rewards only the turbulent case.
- Moving from closed to intermediate population distributions, which means increasing diversity among firms population, the highest performance increment is computed for CD in the turbulent case.
- In turbulent cases an increase in diversity is always rewarded with an increasing in performance, and this effect is higher in areas like CD.

Table 2: experimental sets

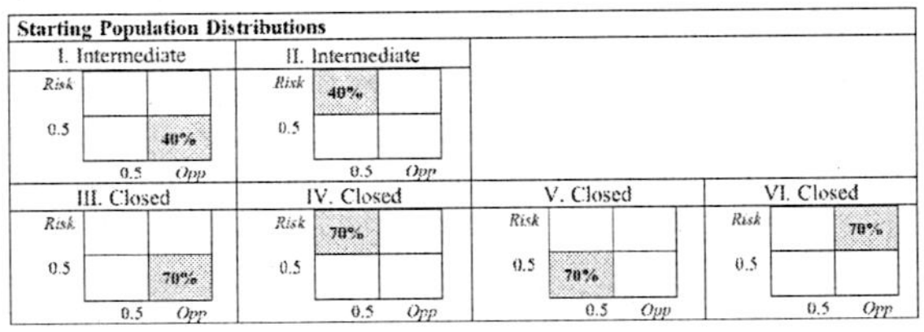

Table 3: Simulations results

Environment \ Population		Intermediate N	Intermediate P		Closed N	Closed P
Stable	LCD	11.96	79.25	LCD	11.34	66.86
	HCD	11.66	79.78	HCD	10.94	88.57
Turbulent	LCD	7.40	31.03	LCD	6.42	19.10
	HCD	8.20	40.42	HCD	6.98	33.63

References

[1] Becattini G., 1989. Riflessioni sul distretto industriale marshalliano come concetto socio economico *Stato e Mercato*, n. 25, pp. 111-128.

[2] Marshall A., 1919. Industry and Trade, London, MacMillan.

[3] Piore M.J., Sabel C.F., 1984. The Second Industrial Divide: Possibilities for Prosperity, New York, Basic Books.

[4] Becattini G., 1979. Dal settore industriale al distretto industriale. Alcune considerazioni sull'unità di indagine dell'economia industriale. L'Industria, n. 1.

[5] Boero R., Squazzoni F., 2001. A computational prototype of industrial district. Workshop on "Complexity and Industrial Districts", Università di Modena e Reggio Emilia.

[6] Arthur W.B., Durlauf S.N., Lane D.A. (eds), 1997. The Economy as an Evolving Complex System II, Redwood (CA), Addison-Wesley Longman.

[7] Conte, R., (1999). "Social Intelligence Among Autonomous Agents". Journal of Computational & Mathematical Organization Theory 5(3), 203-228.

[8] Péli G., Nooteboom B., 1997. Simulation of Learning in Supply Partnership, Computational & Mathematical Organization Theory, vol. 3(1), pp. 43-66.

[9] Strader T. J., Lin F., Shaw M. J., 1998. Simulation of Order Fulfillment in Divergent Assembly Supply Chain, JASSS - Journal of Artificial Societies and Social Simulation, vol. 1(2).

Innovative Production Machines and Systems
D.T. Pham, E.E. Eldukhri and A.J. Soroka (Eds.)

FE Analysis of a Splitting Rod Rolling Pass

F. Capece Minutolo, M. Durante, F. Lambiase and A. Langella

Department of Materials and Production Engineering
University of Naples Federico II
Piazzale Tecchio 80, 80125 Naples, Italy

Abstract

The chief rolling designer's goal is the reduction of the number of passes accounting on the process technological limits. Therefore, many efforts are spending to develop new types of passes with higher elongation ratios, just like split sequences.
The present paper has analyzed a two-step split sequence used by a company which produces rebars for reinforced concrete. The experimental data were used to validate a FE model; then, as the experimental rolled bars presented material overflowing (the material flows outside the rolls groove), the FE model was used to change the rolls groove geometries.

Keywords: roll pass, FE models, elongation coefficient, sequence optimization.

1 Introduction

The rolling process must be subdivided into several passes, given the high cross- sectional reduction to be achieved (usually ranging between 30 and 400). To improve the process efficiency in terms of manufacturing time and cost, designers must reduce the number of sequence steps by adopting new passes with high elongation ratios (ER). This goal has, of late, been achieved through the adoption of split passes, which provides an extra virtual elongation ratio, by splitting the incoming bar into two parts.
Generally, the design of rolling passes involves several structural (e.g. loads on rolls, rolls axial flexion, motor sizing, etc.), processing (e.g. rolls speed, workpiece temperature, etc.) and geometrical issues.
Material overflowing (the material which flows outside the rolls groove) and rolls groove voids can do prejudice the process outcome (often causing bar entrance problems), and they must, therefore, be accurately predicted and prevented.
So far, many approaches have been attempted to design rolling stands methodically (e.g. Holzweier [1], for Diamond-Square-Diamond, Square-Flat Oval-Square and Square-Oval-Round sequences, and Mercader [1] for Square-Oval-Square sequences).
These models used a set of empirical rules to guide designers without providing a real/physical understanding of process parameters; thus, changing and generalizing these formulas for other passes was impossible.
To overcome this limit, many studies focused on geometrical aspects: Kemp [2], for the evaluation of the bar's cross section strain, Moses [3], Shinokura et al. [4] and Lee et al. [5] who proposed a model for predicting the radius of curvature in round-oval passes. Although analytical models provide a great physical insight into the influence of factors on the process result, none can be adapted to split sequences.
For forecasting the rolled bar shape after a split sequence, and finding out whether geometrical defects occur, FE simulations represent, perhaps, the best solution; indeed, they allow a reduction in testing, allowing us to change process parameters easily and inexpensively, and to test different

alternatives with minimum effort [6-10].
In this work, an industrial split sequence, consisting of two passes, was analyzed, corrected and optimized by the application of FE techniques. After validating the FE model through a ready comparison with the actual process, FEM was employed to adapt the rolls gap and the incoming bar dimensions.
Finally, a new roll geometry was proposed and compared with the original one, in order to simplify the original rolls groove profile while, at the same time, reducing the cost of production and maintenance.

2 Experimental data

Experimental tests were performed with IMISUD Inc., which produces rebars for reinforced concrete. The split is used in the pre-finishing steps that come after a flat-oval pass (32((20), as shown in Figs. 1. The workpiece material is a low carbon steel B450C; the process temperature before the first split pass is about 1050°C (the change of the bar's temperature between going into the first pass and coming out of the second, measured by means of an optical pyrometer, was not appreciable) This can be ascribed to heat generation, produced by distortion work that equilibrates heat dispersion and high bar velocity (6-8 m/s).
A photo of the rolled specimen after the first pass is shown in Fig. 2; it can be seen that the material flows outside the rolls; this phenomenon was accentuated as the bar passed between the second split pass rolls. This mistake was ascribed to an erroneous pairing of incoming workpiece dimensions with split rolls dimensions.

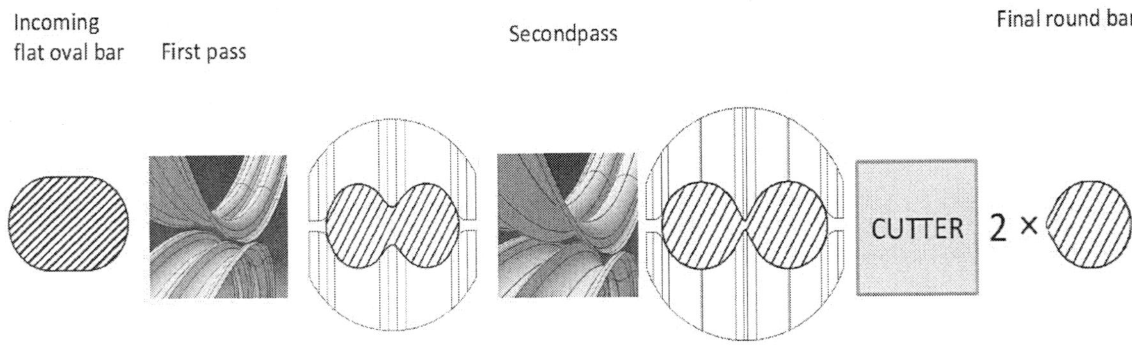

Fig. 1. Representation of the entire split sequence

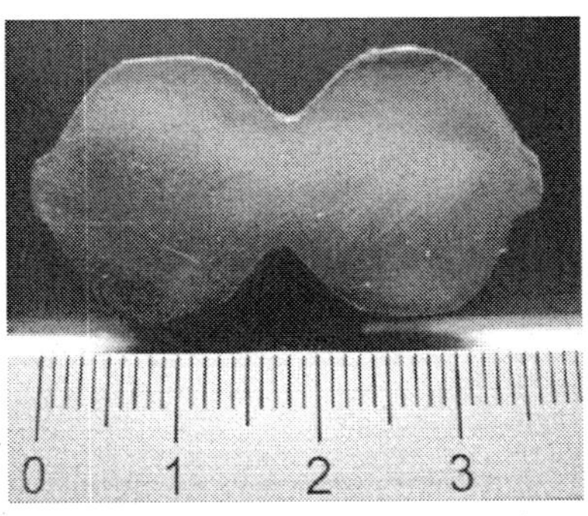

Fig. 2. Cross section of the experimental specimens after the first split pass.

3 Numerical models

Numerical models were used to analyze these split passes. To validate the FE results, a simulation of the first pass was conducted using the LS-DYNA program. The bar cross sections shapes (particularly area and spread) simulated by FEM were subsequently measured and compared with experimental ones.

A quantitative comparison was available only for the first pass, because no samples coming from the second were obtainable; this is due to the fact that, in the actual process, the bar passed immediately through the cutter.

Two planes of symmetry were adopted in the FE model to save computation time. Rolls motion was attained by imposing a constant prescribed motion speed, while a pushing plane was used to bring the bar and rolls into contact. The friction between bar and rolls was modeled with a Coulomb coefficient of 0.30. This value was calculated using the formula (1) [1]:

$$F = (1.05 - 0.0005 \cdot T) \cdot K_1 \qquad (1)$$

where T represents the temperature in degrees centigrade, and K_1 is a coefficient depending on the process speed (table 1 reports the values of K_1 versus this factor).

An elastic-plastic law was used to describe the material behaviour of the bar. Young's modulus values versus temperature for low carbon steel are reported in [11].

The yield stress was calculated according to Eq. 2 [12]:

$$\sigma = \sigma_f \cdot f\left(\frac{\dot{\varepsilon}}{10}\right)^m \qquad (2)$$

where $\dot{\varepsilon}$ is the strain rate, σ_f , f and m are factors depending on the temperature and carbon content [12]. On the other hand, the rolls and the pushing plane were simulated with rigid elements.

The bar's temperature was fixed at 1050°C; heat transfer was, besides, ignored, because the temperature changes were insignificant.

To keep the conditions of the process constant, the simulated bar was 60 mm long (the arc of contact is about 20 mm long). The accuracy of the FE model was verified by comparing the width of the FE bars with that of the experimental one. The FE model error in computing spread is less than 3% (38 and 36.9 mm experimental and simulated width respectively, as shown in Fig. 5 and Fig. 6).

Both simulated and experimental bars showed, therefore, the same edge defects: material overflowing on the sides and partial voids in the upper groove region, as shown in Fig. 6.

Table 1

Values of the parameter K_1 versus the process speed.

V (m/s)	Up to 2	3	4	5	6	7	8	9	10	12	14	16	18	20 to 22
K1	1	0.9	0.8	0.72	0.66	0.6	0.57	0.55	0.52	0.47	0.45	0.43	0.42	0.41

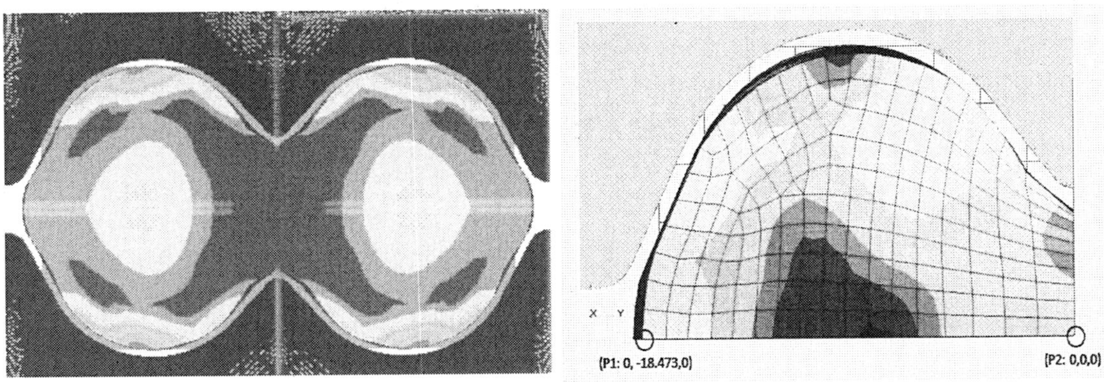

(a) b)

Fig.3. a) Cross-section of the rolled bar simulated by FEM at the first pass. b) Co-ordinates of the two nodes used to evaluate the bar's spread.

4 Correction of the original rolls groove by the implementation of FEM.

Once the FE model had been validated, it was employed to correct the geometrical defects (especially the overflowing) which arose in both passes. The rolls gap increase and the incoming bar width reduction were tested, in particular, since both provide a solution for the above matters, without changing the rolls profile, and, at the same time, allow savings in both time and cost.

4.1 Change of the rolls gap

Some simulations were performed by changing the rolls gap to avoid material overflowing. Fig. 7 shows the cross-sectional spread calculated by FE simulations versus rolls gap increase. The rolls gap must be increased up to the point at which the bar's width becomes smaller than that of the rolls groove. Therefore, in order to avoid material overflowing completely, it was necessary to increase the rolls gap by 2 mm and 4 mm for the two rolls stands respectively.

4.2 Change of incoming bar dimensions

Even though material overflow was eliminated, the more the rolls gap increased, the greater was the outgoing bar cross section area. Thus, to attain the final bar geometry through these two steps, the cross sectional area of the incoming bar had to be reduced (by adopting another step before the first split pass).

To find the correct incoming workpiece dimensions, the width was incrementally reduced by 2 mm for every simulation, until no overflowing occurred throughout the two steps.

Incoming bar width reduced to 30 mm would avoid material overflowing in the first step but not in the second. On the other hand, a further bar width decrease to 28 mm meant that material overflow could also be avoided in the second pass. The FE models results are summarized in Fig. 8.

The reduction of the incoming bar width improves the rolls groove filling behavior on the contrary of the increasing of rolls gap .

5 Conclusions

An industrial split sequence made up of two passes has been analyzed. Initially, as the original design of rolls grooves was incorrect (causing material overflow laterally and from the voids in the upper area), initial efforts were concentrated on removing this defect. Indeed, although lateral flash at the first pass was relatively low (less than 2 mm), it increased critically after the second pass, often causing entrance problems in the following passes of the industrial process.

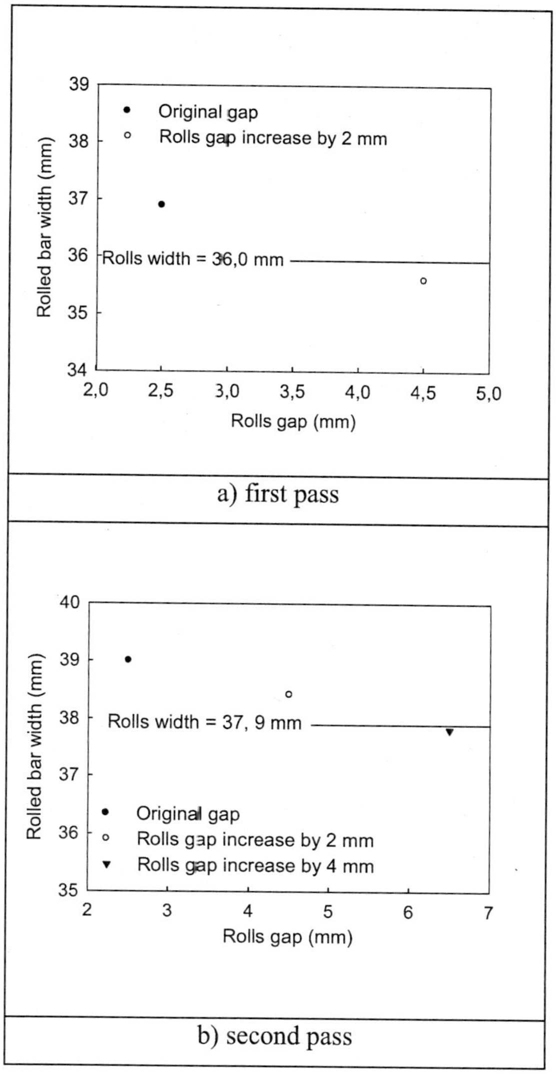

Fig.5. Values of the width rolled bar versus rolls gap: after the first pass a) and after the second pass b).

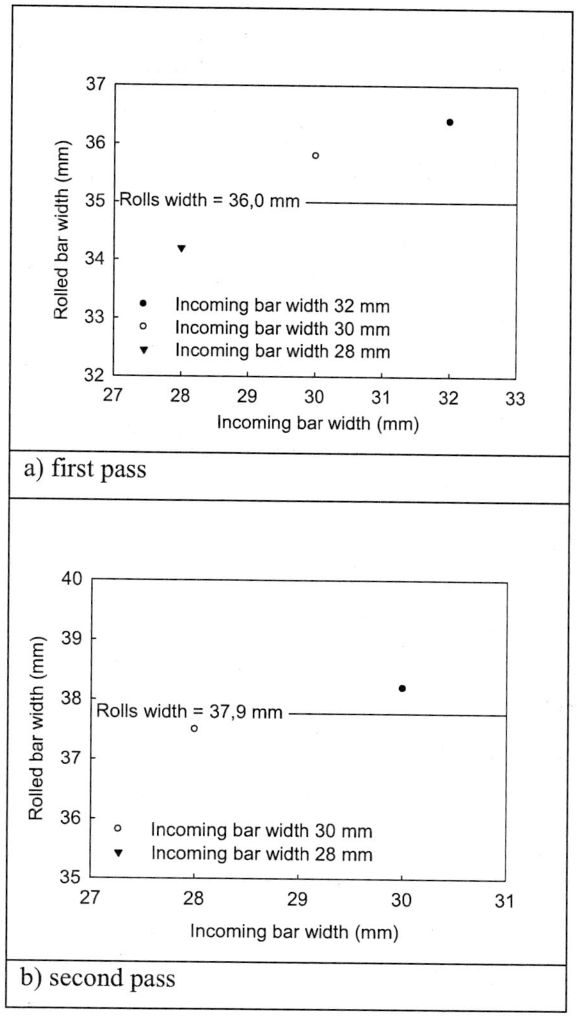

Fig.6. Values of the width of the rolled bar versus incoming bar width: after the first pass a) and after the second pass b).

This also occurred in the finite element model of the process, which was validated using only the first pass cross-sectional profile, due to the fact that no samples of the outgoing bar coming from the second pass were available (the bar having passed immediately through the cutter).
It was evident that there was a flaw in the rolling pass design, which was intended for an excessive elongation ratio. Thus, in order to remove this flaw quickly (avoiding long production downtime), it was necessary to redesign those roll passes without changing the rolls geometries.
Two courses of action were taken to achieve this purpose: increasing the rolls gap and reducing the dimensions of the incoming bar. The first action is more simple, but generates a decrease in area reduction of the bar for the split passes; the second action is more complex but produce a correct and complete employment of the split passes. However, the FE analysis let to calculate the numerical values of the factors that must be changed to remove the defects. The employment of FE models allowed a reduction in what-if analysis time, and the relative cost, meaning that these changes could be optimized.

Acknowledgments

The authors wish to thank IMISUD inc. and, in particular, Eng. R. Habetswallner, for their invaluable providing of experimental data and specimens.

References

[1]Z. Wusatowski, Fundamental of Rolling, Pergamon Press, London-New York, (1969)

[2]I.P. Kemp, Model of Deformation and Heat Transfer in Hot Rolling of Bar Sections, Journal of Iron Making and Steelmaking 17 (1990) 139-143

[3]L. Moses, Mathematical Aids to Determine Spread and Reduction Rates in Rolling, Iron and Steel Engineer (1959) 65-77.

[4]T. Shinokura and K.A. Takai, A New Method for Calculating Spread in Rod Rolling, Proceedings of Experimental Verification of Process Models, ASM, Cincinnati, (1983) 175-188.

[5]Y. Lee, S. Choi, Y.H. Kim, Mathematical Model and Experimental Validation of Surface Profile of a Workpiece in Round-Oval-Round Pass Sequence, Journal of Materials Processing Technology 108 (2000) 87-96.

[6]S.Y. Kim, Y.T. Im, Three-Dimensional Finite Element Analysis of Non-Isothermal Shape Rolling, Journal of Materials Processing Technology 127 (2002) 57–63.

[7]N.S. Kim, S. Kobayashi, T. Altan, Three-Dimensional Analysis and Computer Simulation of Shape Rolling by the Finite and Slab Element Method, International Journal of Machine Tools and Manufacturing 31 (1991) 553–563.

[8]H.J. Kim, T.H. Kim, S.M., Hwang, A New Free Surface Scheme for Analysis of Plastic Deformation in Shape Rolling, Journal of Materials Processing Technology 104 (2000) 81-93.

[9]Y.Lee, H.J. Kim, S.M. Hwang Analytical Model for the Prediction of Mean Effective Strain in Rod Rolling Process, Journal of Materials Processing Technology 114 (2001) 129-138.

[10] J.H. Min, H.C. Kwon, Y. Lee, J.S. Woo, Y.T. Im, Analytical Model for Prediction of Deformed Shape in Three-Roll Rolling Process, Journal of Materials Processing Technology 140 (2003) 471-477.

[11] K. Lange, Handbook of Metal Forming, McGraw-Hill Book Company, ISBN 0-07-036285-8, New York (1985).

[12] M. Pietrzyk, J. G. Lenard, Thermal-Mechanical Modeling of the Flat Rolling Process, Spring-Verlag, ISBN 3-540-53316-8, Berlin (1991).

Innovative Production Machines and Systems
D.T. Pham, E.E. Eldukhri and A.J. Soroka (Eds.)

Innovation in global competition: a simulation model of an industrial network

L. Iandoli[a], C. Ponsiglione[b], G. Zollo[c]

[a, b, c] *Department of Business and Managerial Engineering, University of Naples Federico II, Naples, Italy*

Abstract

This paper presents a theoretical and computational model to analyze the dynamic of inter-organizational relationships among firms of a complex network, such as an industrial district. The aim of the proposed model is to verify, through generative experiments, the effects of some determinants of coordination mechanisms on firms and on the network's performances. The conceptual model was implemented on Agentsheets platform and some experimental sets was developed in order to analyze the evolution of the network in different contexts.

Keywords: Agent-based simulation, Collective memory, Firms' network.

1. Introduction

The aim of this paper is to study the dynamics of inter-firm relationships within firms' networks, such as Industrial Districts (IDs), in order to analyze and describe their role in influencing ID's competitive and innovative performances.

An ID can be defined as a cluster of small and medium sized firms, belonging to the same geographical area and characterized by high degrees of specialized production [1]. Traditionally, industrial districts' competitiveness was based on some specific factors as acknowledged by literature:

- the existence of external economies for the firms, but internal to the district, that balance the disadvantage of scale economy lacks [2, 1];
- the flexible specialization, that creates competitive advantages in terms of production efficiency [3];
- the intensive exchange activities between small and medium enterprises (SMEs) of the ID, that create a network able to develop incremental innovation [4].

The rising complexity of the competitive scenario implied the search of new way of development. In the current unstable scenario, IDs firms are more and more requested to activate learning and change mechanisms in order to combine tacit and explicit knowledge, past experiences and business diversification.

2. Background on agent-based simulation

For "computational approach" we mean a computer-based way of formalizing evolutionary interacting agent-based models [5]. Such models assume heterogeneity, localization and interdependence of agents and decentralized mechanisms of interaction [6, 7].

Agent-based simulation enables to analyse self-organising systems, learning systems and systems with high complexity.

The Industrial District can be defined as an evolving network of heterogeneous, localized and functionally integrated interacting firms. Thus agent-based modelling seems particularly suitable to model and to analyze IDs. To build up our simulation model we used Agentsheets platform, which is simple to use, although not very flexible. An agent-based model is developed in two steps: the definition of a conceptual model of the system/phenomenon to be studied (meta-model), the implementation of this conceptual model on an agent-based software platform (computational model).

3.The meta - model

According to literature on agent-based models of firms networks [5, 8, 9], we structure the meta-model into five principal building blocks (Fig1). In the First Building Block, *Classes of Agents*, we define two classes of firms (agents): final and subcontracting firms. To define the Second Building Block, *Agents inner models,* agents firms are described by a set of variables, evaluation and decision rules, all concurring in defining agent social behavior. The third building block represents *the context of interaction*: the inter-firm scaffolding structures. These structures are necessary to co-ordinate the whole production process. To translate the Fourth Building Block, *the environment structure*, we assume that the market carries all the needed information to develop attractive products. At some point in the simulation, product quality requirements increase, driving firms to improve their competencies, in order to achieve the new quality standards. In the proposed model for the fifth building block, *Structure of the information flow,* we assume that final firms have higher strategic capabilities than subcontracting ones; so final firms convert market inputs (competences target levels) into internal information [5]. Then, subcontracting firms receive product requirements interacting with final firm. The ambiguity and vagueness of information flow can create gaps in the agents' routines. To reduce the uncertainty of the environment and the ambiguity of decisional situations we introduced scripts and routines contained in the *collective memory* of the district. For this reason, we added a new building block to the general design of the agent-based model: the specific context of the industrial district (the shared memory of the district) (Fig. 1). In this way, the structure of ID collective memory is translated into agents firms characteristics and environment and institutions laws and frameworks.

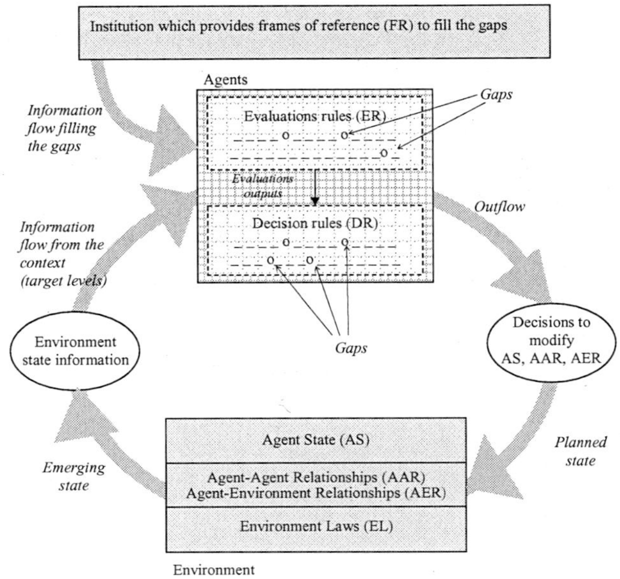

Fig.1 The proposed model

4.The computational model

4.1 Agents

Agents are all the elements of the simulation model. We classify them into two typologies: *principal and second order agents.*

4.1.1 Principal agents

According to theories of flexible specialization [3], we modeled ID as a system of small and medium firms, which interact to build up production chains. We define three classes of principal agents: final firms, subcontracting firms and production chains. Final firms principal objective (*fin*) is to meet market requirements; they manage it building up efficient production chains. Production chain efficiency is measured in terms of achieved quality during the production phase. Subcontracting firms (*sub*) are phase firms; they provide materials, components and semi-finished products to final firms. Their main objective is to find an adequate final firm in order to build-up a production chain. Thus, the interaction structure is based on district production chains (*Pch*): one final and one subcontracting specialized firm compose every production chain.

4.1.2 Second order agents

They have been created to translate the fourth and the sixth building blocks of the meta-model: the structure of environment and the specific context of ID. They are designed with a top-down perspective [7]. This means that we already know how they vary during simulations. In fact, we are interested in analyzing the phenomena, which emerge from the interactions among firms (principal agents) and between firms and environment (second order agents).

ID super - agent

To translate the specific context of the industrial district, we introduced a super - agent called *ground* in the model. Through its laws and framework of references (EL and FR in Fig.1) the super-agent diffuses and stores information (e.g. the market demand, target competencies levels, firms agents competencies levels and so on) in the district allowing firms agents to interact among them and take decisions.

Market Segments

To translate, in computational terms, the role of environment structure we define three market segments: S_1, S_2, S_3. These segments represent the potential markets for district firms. Different levels and types of competencies characterize each market segments: $S_i = (M_i, T_i, P_i)$. Where M_i, T_i, P_i represent the highest level of competencies required from the market at iteration *i*. We assume that at the beginning of a simulation exists only the S_1 segment. At certain iteration of the simulation, first S_2 and then S_3 appear on the district.

4.2 Agents inner models

Agents are modeled, in computational terms, as state variables, competencies, and evaluation and decision-making rules. Each agent is characterized by an internal state, which is an instantiation of its state variables and competencies levels. This state progressively changes in each iteration (called 'cycle') of the simulation, accordingly to firm choices and interactions with the other agents of the model. Thus, for each cycle *i*, firm internal state (IS(S_i))is function of: IS(S_i)= f (m_i, t_i, p_i, opp_i, $risk_i$, bdg_i) Where:

- m_i represents the level of market competencies. Market competencies group all the skills, knowledge and capabilities of the firm to sell its product to the market.
- t_i represents the level of technical competencies. These are strictly related to product quality. Firms with higher level of t_i are able to manufacture better products.
- p_i represents the level of relational competencies. In this work, with relational competencies we mean organizational competencies within production phases.
- opp_i Degree of Opportunism. This variable affects final and subcontracting firms' aptitude in building up a production chain; while, for production chain this variable influences its aptitude in breaking up the chain.
- $risk_i$ Risk Propensity. This variable indicates agent inclination to carry out risky investments.
- bdg_i The budget function. It computes the amount of economic resources of the firm. For each cycle, the function output (budget value) increases or decreases accordingly to firms choices; as follows: $bdg_i = bdg_{i-1} - C_{exp} - C_{inv} + PFT_i$, where: bdg_{i-1} is the budget value in the cycle *i*-1. C_{exp} represents firm's *costs of exploration*. An agent spends economic resources for every movement he takes in the environment space. This cost is constant during all the simulation. C_{inv} is firm's *costs of investment*. Firms have to invest some economic resources for every competencies improvement they perform. This cost is function of the level of competencies and decreases as competencies levels increase: $C_{inv} = f(C_{inv\_MAX} / m_i, t_i, p_i)$. PFT_i indicates firm *profit* at cycle *i*. Profit is function of product quality. Firms that sell higher quality products will receive higher profit.

When a simulation starts we assume that it is populated only by final and subcontracting firms, whose initial values of competencies levels and economic resources (m_0, t_0, p_0, bdg_0) are settled externally. Instead, for production chains the initial values are determined when the chain is built up.

4.3 Simulation Steps

At this point, we have to translate the information and production flows of the meta-model. This means explicating the events that occur in each iteration of the simulation. We summed up these events into five principal steps:

1. *Internal state check:* at the beginning of each cycle, firms check their internal state: $IS(S_i)=f(m_i, t_i, p_i, opp_i, risk_i, bdg_i)$. To limit complexity we assume that some of the state variables (opp_i, $risk_i$) do not change their values during simulation. Instead, the levels of economic resources and competencies vary and derive from the previous cycles.

2. *Evaluations:* firms evaluate their levels of competencies comparing them to the target levels given by the market.

3. *Improvements strategies:* at this step, firms choose their improvement strategies, according to evaluation outputs, objectives, state variables values and environment information. Firms have to decide which competencies to improve, among those indicated in evaluation output.

4. *Production chains generation or break up:* a fundamental objective for final and subcontracting firms is to become part of a production chain. Partner selection is based on competencies levels comparisons: each final firm looks for a subcontracting firm with competency levels greater or equal than its levels. Subcontracting firms are not able to select for final firm. In this step, production chain might decide to break up the chain, basing this decision on their opportunism degree.

5. *Product sale:* every production chain assembles and sells its product to the market and receives a profit when the market buys the output of its production process. The gained profit is function of market segment profitability and of product quality; the latter is function of the competencies gap of the firm. When market refuses the product, the chain does not receive any profit and it cannot increase its economic resources.

The hypotheses that we are going to verify through simulation can be synthesized as follow:

- When the degree of cohesion of ID collective memory is too high or too low, the industrial district risks to close on itself and experiences its inability to adapt to market evolution.
- When the collective memory of the district keeps a certain level of diversity among agents inner models and guarantees, at the same time, the integrity of the district identity, the system evolve, and is able to learn and innovate.

5.The experimental sets

We assume that at the beginning of each simulation run, final and subcontracting firms are characterized by the same level of competencies and of economic resources, but they show different values for risk propensity (risk) and opportunism degree (opp). Thus, each agent assumes values in one of the four quadrants of the following matrix (Table 1).

Table 1. Agents values

Risk	**H/L**	**H/H**
0.5	**L/L**	**L/H**
	0.5	*Opp*

We define the different starting population changing the agents' distributions in the four quadrants of the above matrix. The higher is the presence of the agents in one of the four quadrants, the higher will be the degree of cohesion of the collective memory of the population. According to these issue, we name the starting population distribution as *closed*, when the 70% of the starting population takes values in one of the four quadrants of the matrix above. We say that a starting population is *open*, when at least one of the two variables opp and risk is equally distributed in the quadrants. And finally, we say that a starting population is *intermediate* when there is a sloght prevalence of the agents in one of the quadrants. In this way, we built up one *open*, two *intermediate* and four *closed* starting population distributions,. Each population has been tested within two environments, which differ for the type of market evolution. The two environments are:

- Stable Market: this means slow changes in market requirements.
- Turbulent Market: this situation indicates quick changes in market requirements.

The different market evolutions are translated in terms of temporal distance (number of cycles) that elapses between two changes of market conditions. In the stable case this distance is constant during simulation (20 cycle), while in turbulent case it is variable.

6. Results, analysis and discussion

We performed 50 run per experimental set, in order to reduce the influence of agents starting position on the results. Each run was composed by 40 firms: 20 final firms and 20 subcontractors and lasts for 180 cycles.

The indicators used to analyze and compare different simulations were:

- **P**: this represents the overall economic performance of the district. It is computed as the sum of the budget of the firms (final firms, subcontracting firms and production chains) at each simulation cycle.
- **N**: this stands for the number of firms (final firms, subcontracting firms and production chains) at each simulation cycle.

To make our analysis and comparisons we used the final value of these variables, which is the value computed at the 180th cycle. For each multiple–simulations we obtained 50 pairs of values for P and N. At first, the 50 pairs have been ordered according increasing P. Secondly, we drew the 50 pairs on orthogonal axis. Then for each of the experimental sets we performed the analysis of relative frequencies, we computed the average for the two variable P and N (Table 2) and, finally, we compared the results.

The second and the fourth experimental set are characterized by high opp levels and low risk levels. We name these last two set as "Cooperative Districts (CD)". On the contrary, we called the third and fifth sets, which are characterized by high levels of risk and low opportunity, "Non Cooperative Districts (NCD)". Through the analysis of table 3, we compare the behaviours of these two types of districts. Some considerations according the analysis of the table 3:

- Cooperative Districts: from closed to the intermediate population distribution, the higher diversity among firms rewards more in the turbulent case.
- Non Cooperative Districts: from closed to the intermediate population distribution, diversity rewards only the turbulent case.
- Moving from closed to intermediate population distributions, which means increasing diversity among firms population, the highest performance increment is computed for CD in the turbulent case.
- In turbulent cases an increase in diversity is always rewarded with an increasing in performance, and this effect is higher in areas like CD.

Environment	Stable Case		Turbulent Case	
Experimental Set	N	P	N	P
I – Open	11.80	78.43	7.20	30.55
II – Intermediate	11.96	79.25	7.40	31.03
III – Intermediate	11.66	79.78	8.20	40.42
IV – Closed	11.34	66.86	6.42	19.10
V– Closed	10.94	88.57	6.98	33.63

Table 2: experimental sets

Environment		Population: Intermediate			Closed	
		N	P		N	P
Stable	***CD***	11.96	79.25	***CD***	11.34	66.86
	NCD	11.66	79.78	***NCD***	10.94	88.57
Turbulent	***CD***	7.40	31.03	***CD***	6.42	19.10
	NCD	8.20	40.42	***NCD***	6.98	33.63

Table 3. Comparison between CD and NCD results

References

[1] Becattini G., 1989. Riflessioni sul distretto industriale marshalliano come concetto socio economico *Stato e Mercato*, n. 25, pp. 111-128.

[2] Marshall A., 1919. Industry and Trade, London, MacMillan.

[3] Piore M.J., Sabel C.F., 1984. The Second Industrial Divide: Possibilities for Prosperity, New York, Basic Books.

[4] Becattini G., 1979. Dal settore industriale al distretto industriale. Alcune considerazioni sull'unità di indagine dell'economia industriale. L'Industria, n. 1.

[5] Boero R., Squazzoni F., 2001. A computational prototype of industrial district. Workshop on "Complexity and Industrial Districts", Università di Modena e Reggio Emilia.

[6] Arthur W.B., Durlauf S.N., Lane D.A. (eds), 1997. The Economy as an Evolving Complex System II, Redwood (CA), Addison-Wesley Longman.

[7] Conte, R., (1999). "Social Intelligence Among Autonomous Agents". Journal of Computational & Mathematical Organization Theory 5(3), 203-228.

[8] Péli G., Nooteboom B., 1997. Simulation of Learning in Supply Partnership, Computational & Mathematical Organization Theory, vol. 3(1), pp. 43-66.
[9] Strader T. J., Lin F., Shaw M. J., 1998. Simulation of Order Fulfillment in Divergent Assembly Supply Chain, JASSS - Journal of Artificial Societies and Social Simulation, vol. 1(2).

Innovative Production Machines and Systems
D.T. Pham, E.E. Eldukhri and A.J. Soroka (Eds.)

Intelligent evaluation of ultimate strength prediction of composites and fabrics' extensibility

D. D'Addona[a], R. Teti[a], P.N. Koustoumpardis[b], N.A. Aspragathos[b]

[a] *Department of Materials and Production Engineering, University of Naples Federico II, Naples, Italy*
[b] *Department of Mechanical & Aeronautics Engineering, University of Patras, Patras, Greece*

Abstract

An approach to intelligent evaluation of residual flexural strength of glass fiber reinforced plastic composites and to tensile test for the estimation of the fabrics extensibility is proposed.

The glass fiber reinforced plastic composites were subjected to hydrothermal aging cycles in water at 60 °C. Sensor fusion of acoustic emission and load data is carried out through neural networks to obtain a prediction of ultimate flexural strength as early as possible in the loading history of glass fiber reinforced plastic composites.

A robotized system is used to perform the fabrics tensile test and the extensibility of the samples is estimated using a feed-forward neural network while trying to imitate the human expert valuation.

The results show that neural network processing provides an effective method for the monitoring of composite material strength behavior based on acoustic emission detection and analysis and for the estimation of the extensibility of new fabrics.

Keywords: Residual strength, Composite materials, Acoustic emission, Fabric testing, Extensibility, Neural networks

1. Introduction

1.1. Residual flexural strength of glass fiber reinforced plastic composites

In most engineering usage of fiber reinforced plastic composites, the structure of the composite is chosen such that stiffness and strength are controlled by fiber modulus and strength, respectively. Thus, some matrix softening can be accomodated in such applications without serious consequences. If considerable matrix softening occurs, however, the ability of the resin to support the fibers is severely reduced, along with the ability to transfer load through the matrix to the fibers. The result is a change in failure mode from filament dominated to matrix dominated.

A classic example of failure mode change is the unidirectional flexural test. This test is commonly used for quality control and 0° flexural strength is considered to be a filament dominated property. For state-of-the-art high performance polymers, 0° dry flexural strength is relatively insensitive to temperatures below 150 °C. But with increasing moisture absorption, a considerable strength reduction can occur for much lower temperatures [1]. Such strength degradation is accompanied by a change from filament dominated to matrix dominated failure mode.

Nondestructive evaluation (NDE) methods for composite materials are generally capable of providing information on defects generated during fabrication or service [2, 3]. Most of these methods can determine the type and location of damage in the material: it is then necessary to correlate the detected damage with residual mechanical properties. But, despite the efforts of many researchers [4, 5], completely satisfying analytical methods for strength prediction after material degradation are not yet available.

A totally different NDE approach is based on acoustic emission (AE) testing techniques. The detection of AE, i. e. transient stress waves released in materials undergoing permanent deformation and fracture, presents the unique advantage of sensing failure processes as they occur in the material under loading [3, 6]. This allows for on-line decisions on damage development and real-time corrective actions.

Caprino and Teti [7] showed that the residual strength of quasi-isotropic glass fiber reinforced plastic (GFRP) composites with center holes or impact damage could be evaluated through the analysis of AE detected during tensile testing. In [7], a reliable prediction of the residual strength for these materials was achieved through neural network (NN) processing of AE data.

In this paper, AE monitoring of the strength behavior of GFRP material subjected to accelerated hydrothermal aging in water at 60 °C for 6 months is used for flexural strength prediction at an early stage of the AE response based on a NN approach.

1.2. Fabrics tensile test

In the clothing industry, the experts traditionally estimate the fabric properties by hand actions, using their tactile feelings, which is a subjective measurement of fabric properties. Their evaluation is expressed through qualitative linguistic variables such as hardness, roughness, stiffness etc. In the research community as well as in the clothing industry, there is a wide interest for objective measurements of the fabric properties. Therefore, a variant of testing methods and instruments have been developed.

The objective measurement of the fabric properties has been attracted the interest of researchers from the beginning of the previous century. Kawabata [8] introduced the widespread Kawabata Evaluation System-KES available from Kato Tekko [9], which is the first integrated system for low-stress fabric testing. The KES system can be considered as a real progress towards the objective measurements of fabric properties. Four parameters or tensile properties are extracted from the two curves representing the extension and the recovery fabric behaviour, namely the linearity (LT), the tensile energy (WT), the tensile resilience (RT) and the extensibility (EMT). The first three of these parameters are used to estimate the primary hand values of the fabric and afterwards the total hand value of the fabric. The KES system is the most common low-stress testing system but it requires a considerable time to measure the fabric properties and it is composed of very high cost instruments.

Recently, artificial intelligent based techniques have been introduced for the evaluation of the fabric properties. Gong and Chen [10] used neural networks to predict the performance of fabrics based on fabric mechanical properties measured on the KES-FB system. Park et. al. [11, 12] proved that the neural networks are more efficient, than the KES evaluation approach, for determining the total hand value of fabrics. Their approach is focused on the data processing after the test rather than on the test process.

The majority of the industries producing or using fabrics have a separate laboratory for testing textile products. A testing laboratory is equipped with a variant of fabric testing instruments and operated by trained staff [13], which adds a noticeable cost to the product cost. The information acquired from these tests is used for improving the fabric's quality but is not used or can not be directly used, for the handling process of fabrics in clothing industry. This fabric property expresses the ability of a fabric to be extended under low-stresses. The tensile properties (LT, WT, RT and EMT) derived from the tensile test using the above described instruments have to be interpreted by an expert in order to be used.

In this paper, a new approach is introduced for the estimation of the fabrics extensibility while the fabric tensile test is performed in a robotic cell that could be the same with that used for the sewing process. The acquired force-strain data from the tensile test are directly fed into a neural network, throughout the test. The neural network estimates the fabric extensibility according to its training which is based on experts' estimations for the extensibility of a representative set of fabric samples.

2. Experimental

2.1 Bending tests

Three-point bending tests were carried out according to the UNI EN 63 standard on 162 mm long, 12.7 mm wide, and 3.2 mm thick unidirectional GFRP samples fabricated by milling pulltruded bars.

A total number of 13 valid bending tests were performed on an Instron machine using a test span of 102 mm and a displacement speed of 1.5 mm/min. Six virgin samples and seven samples aged at 60 °C for six months were tested to failure (Tab. 1).

AE was detected during bending tests using a 150 kHz resonant sensor; amplification was 40 dB, threshold level 1 V, and high-pass filter cut-off frequency 100 kHz. AE event counts, Nt, was the AE count-based parameter considered for GFRP composite material strength behavior prediction. An AE Nt versus time curve was available for each test. The Instron machine provided the load vs. displacement curve. By dividing load by the transverse section and displacement by the displacement speed (1.5 mm/min), the stress vs. time curve was obtained. By combining the Nt vs. time and the stress vs. time curves, the Nt vs. stress curve was built (Fig. 1). By plotting the AE event counts detected during flexural tests of GFRP samples versus applied stress, typical N_t - σ curves were obtained [3, 14] (Fig. 1). All Nt - σ curves display a similar trend: at first the AE activity increases slowly with applied stress, then an 'elbow' is evidenced in the curve and the AE activity increases rapidly up to the final failure. However, curve characteristics depend on material condition: for aged samples, ultimate strength is significantly lower and the elbow in the Nt - σ curve occurs much earlier (Fig. 1).

A notable result illustrated in Fig. 1 is that aged and virgin samples form two distinct groups of N_t - σ curves, with aged samples displaying a higher AE N_t for a given stress level.

Tab. 1.
Summary of experimental results.

Sample id.	Material condition	σ_{max} [N/mm^2]	N_{tmax} [counts]
A1	Aged	920	2649
A2	Aged	877	2680
A3	Aged	858	1144
A4	Aged	823	274
A5	Aged	861	417
A6	Aged	811	519
A7	Aged	909	161
V8	Virgin	1078	993
V9	Virgin	1209	1526
V10	Virgin	1120	602
V11	Virgin	1128	1626
V12	Virgin	1092	1651
V13	Virgin	1268	1940

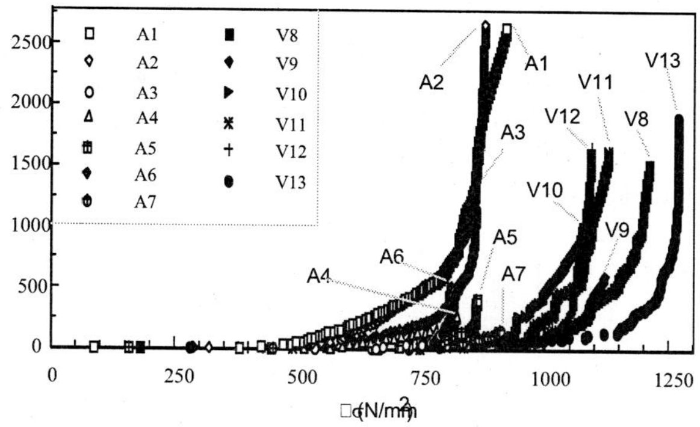

Fig. 1. N_t - σ curves for aged (A1, ..., A7) and virgin (V8, ..., V13) GFRP samples.

Moreover, AE activity trend appears to be dependent on utimate strength as determined by the aging cycle: N_t - σ curve elbowing occurs for distinct σ ranges in the case of aged and virgin samples

2.2. Tensile test

Fourteen fabric samples spanning a wide range of fabrics' extensibility are chosen to form the training set for the neural network. The extensibility of each fabric sample is estimated by the team of experts.

The experts are asked to characterize the samples using the linguistic values: "very low", "low", "medium", "high" and "very high" extensibility which represented by the following numbers: 0.1, 0.3, 0.5, 0.7 and 0.9 respectively in order to facilitate the neural network training.

The curves of the 14-fabrics cover the Force-Strain area adequately, starting with a fabric with a very low extensibility (No. 1) and come to a fabric with a very high extensibility (No. 14) while in a few cases two fabric curves cover nearby areas. Therefore, the choice of the 14-fabrics seams to be adequate for training a neural network that could estimate the fabrics extensibility into this area.

The force/strain curve for each of the samples are illustrated in Fig. 2. The measured force and strain for the first fourteen samples are fed into the neural network for training. It is evident that the curves shown in Figure 2 can be grouped in three categories A, B and C according to the type of fabrics.

Figure 2 shows that two (or more) curves could intersect in one (or more) points, which might cause problems to the performance of the neural network.

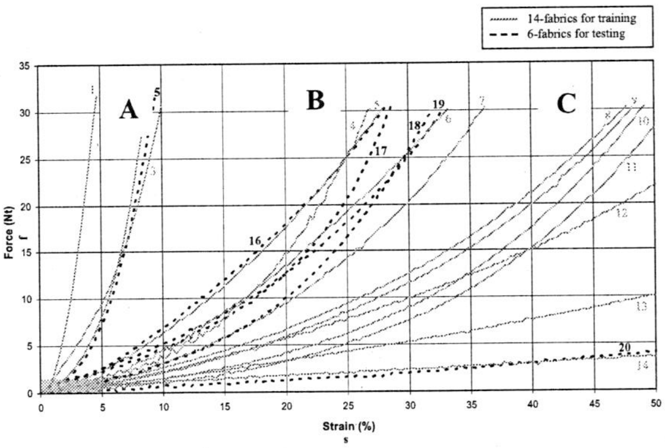

Fig. 2. Fabrics' force/strain curves.

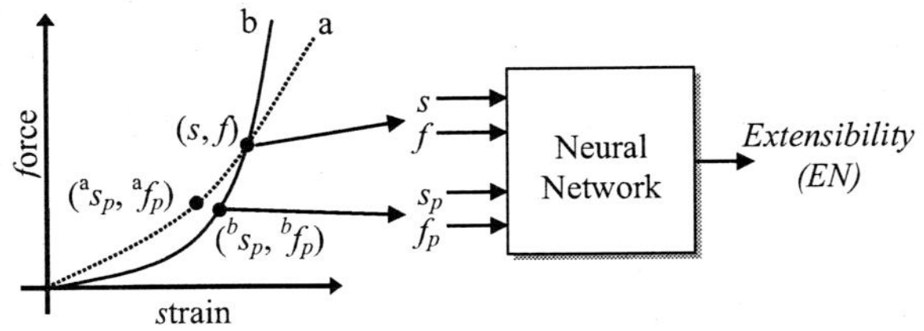

Fig. 3 Improving the network performance by feeding with current and previous measurements.

The common point of the two intersecting curves a and b is (s, f) where f is the measured force and s is the strain of the fabric, as it is shown in Figure 3. Since, the point (s, f) is one of the input pair for training the network and this input set corresponds to two desired outputs, the extensibility corresponding to each curve, the training of the network can be obstructed. This problem is faced by feeding the network with the local gradient of the curves, but to avoid differentiation, in each instance the neural network is fed with the current point (s, f) and the previous one (s_p, f_p).

3. Neural network processing

3.1 AE data

NN processing was applied to predict GFRP residual strength at an early stage of the loading history. A pattern vector representing the N_t - σ curve was fed to the input of a NN to obtain the predicted value of ultimate strength from the NN output.

If the evaluation of ultimate strength is carried out after the entire N_t - σ curve is available, the GFRP material would be already failed and the information would be useless. If a correct evaluation is performed at a stress level lower than the ultimate stress, the prediction of ultimate strength would allow for composite material repair or substitution.

A total number of 13 valid flexural tests on aged and virgin GFRP samples were considered for NN processing (Tab. 1). For each flexural test, an N_t - σ experimental curve, consisting in a sequence of data points each identified by an AE event count and its corresponding stress value, was available. AE event counts increase from 1 to the total number of events at failure by increments of 1 event. Thus, the N_t - σ curve can be represented by a vector, the components of which are the stress values for each AE event; the position of the stress value in the curve vector corresponds to its associated AE event count.

The last component of the curve vector is the aged or virgin GFRP ultimate strength, σ_{max}, and the length of the vector represents the total number of AE events at failure, N_{tmax}. Curve vectors have different length as both σ_{max} and N_{tmax} vary significantly with material conditions (Fig. 1).

Four feed-forward back-propagation 3-layer NN configurations were used to produce a mapping from input vectors to output values [15]: the curve data points were the input and the aged or virgin GFRP sample ultimate strength was the output.

The number of hidden nodes was chosen according to a "cascade learning" procedure [16]: hidden units are added one at a time until an acceptable training speed is achieved. Weights and thresholds were randomly initialized between -1 and +1. Learning coefficients were: learning rate between input and hidden layer η = 0.3, learning rate between hidden and output layers η = 0.15, momentum α = 0.4. The learning rule was the Normal Cumulative Delta Rule and the transfer function applied to the nodes was the sigmoid function $f(x) = 1/(1+e^{-X})$ [16]. The number of learning steps for a complete training was between 10000 and 30000, according to the time to convergence. Epoch size, i.e. the number of training presentations between weight updates, was 6. NN training was carried out by the "leave-k-out" method, which is particularly useful when dealing with small training sets [15]. One pattern vector represented by one experimental curve (k = 1) was held back in turn for the recall phase, and the other pattern vectors were used for learning: 13 different learning and recalling procedures were carried out for each NN configuration.

In the first NN configuration, single data points from each experimental curve were utilized for NN training and testing, according to the scheme illustrated in Fig. 4a. Each data point, consisting of a stress value and its associated AE event count, was sequentially presented to the NN input layer and the corresponding ultimate strength was fed to the output layer for NN training. The NN structure was 2-5-1: the input layer had 2 nodes for applied stress value and AE event count of each data point, the hidden layer had 5 nodes, and the output layer 1 node for ultimate strength prediction.

The second and third NN configurations were trained by presenting to the NN input layer a group of either three or ten consecutive data points from each experimental curve, according to the training scheme illustrated in Fig. 4b.

The second NN configuration had structure 6-8-1: the input layer had 6 nodes, one for each of the three stress values and three associated AE event counts, the hidden layer had 8 nodes, and the output layer 1 node for ultimate strength prediction.

The third NN configuration had structure 20-10-1: the input layer had 20 input nodes, one for each of the ten stress values and ten associated AE event counts, the hidden layer had 10 nodes, and the output layer 1 node for ultimate strength prediction.

The fourth NN configuration utilized a different NN training approach to simulate a "memory" effect in the NN learning. Groups of ten sequential data points from each experimental curve were utilized as input vectors but each next group was shifted by one data point from the previous one, according to the training scheme in Fig. 4c.

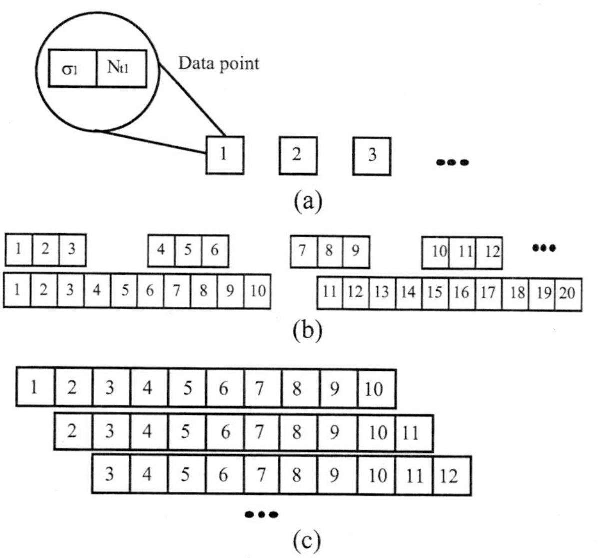

Fig. 4. NN training schemes: (a) training by single data points (NN configuration # 1),
(b) training by groups of three or ten consecutive data points (NN configurations # 2 and # 3),
(c) "memory" effect training (NN configuration # 4).

The predicted strength, σ_{pred}, obtained as output from the learned NN after sequentially inputting the single or the group of data points of the held back N_t - σ curve were normalized ($\sigma_{pred}/\sigma_{max}$) and plotted vs. normalized AE event counts (N_t/N_{tmax}).

In Fig. 5, the normalized predicted strength $\sigma_{pred}/\sigma_{max}$ is reported versus the normalized AE counts N_t/N_{tmax} for typical N_t - σ curves obtained from an aged sample and a virgin sample after NN processing with the 4 NN configurations.

In the case of aged material (Fig. 5a), predicted residual strength is affected by a large error in the first part of the NN response up to $N_t = 0.19\ N_{tmax}$, corresponding to an applied stress $\sigma = 0.86\ \sigma_{max}$. Then, the error tends to decrease with increasing applied stress. However, for all NN configurations, a prediction of residual strength with precision $\geq$ 92% can be obtained for $N_t \leq 0.50\ N_{tmax}$, corresponding to $\sigma \leq 0.95\ \sigma_{max}$. In the case of virgin material (Fig. 5b), predicted strength is affected by an even larger error in the very first part of the NN response. Accuracy, however, increases rapidly to 92% for $N_t = 0.06\ N_{tmax}$, corresponding to $\sigma = 0.83\ \sigma_{max}$. Then, strength prediction accuracy grows with oscillations as applied stress increases up to $N_t = 0.35\ N_{tmax}$, corresponding to $\sigma = 0.95\ \sigma_{max}$. Finally, accuracy stays constant or grows slightly till ultimate failure. For all NN configurations, a prediction of residual strength with precision $\geq$ 96% can be obtained for $N_t \leq 0.35\ N_{tmax}$, corresponding to $\sigma \leq 0.95\ \sigma_{max}$.

The aged and virgin material behavior described above is verified for all 4 NN configurations. However, as the number of simultaneous input data points increases by using NN configurations # 1, # 2 or # 3, the global NN response accuracy is gradually improved (Fig. 5). Moreover, NN configuration # 4 always yielded the best global performance (Figs. 5a and 5b), emphasizing the effectiveness of the application of the "memory" effect approach.

The higher initial incertitude in the NN strength prediction for virgin material is due to its higher σ_{max} mean value (1149 N/mm^2) than for aged material (866 N/mm^2) (Tab. 2), which involves a higher difference between input σ value(s) and desired output σ_{max} value during NN processing. The higher prediction accuracy for higher applied stress values verified for virgin material can be attributed to the reduced dispersion of N_{tmax} values for virgin material (31%) in comparison with aged material (98%), also considering that the dispersions of σ_{max} values for aged (4%) and virgin (6%) materials are practically coincident (Tab. 2).

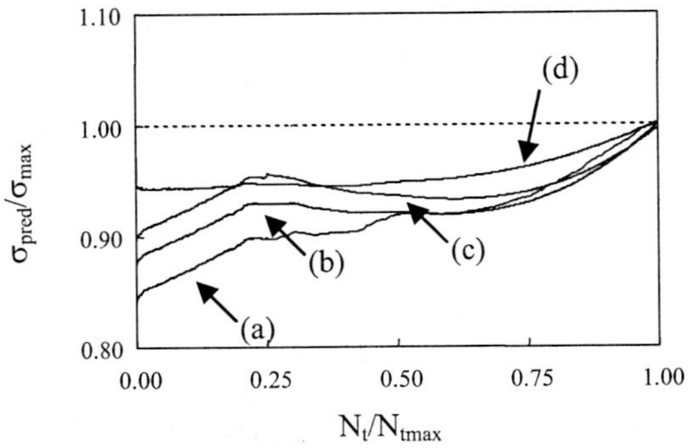

(a) Aged sample A1. NN configuration (a) # 1, (b) #2, (c) #3, (d) #4

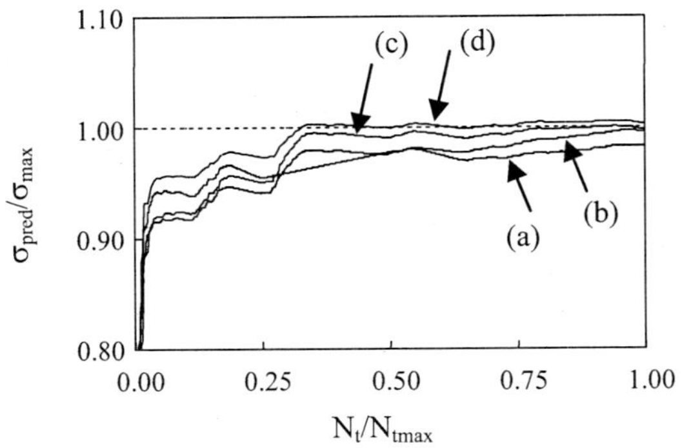

(b) Virgin sample V11. NN configuration (a) # 1, (b) #2, (c) #3, (d) #4

Fig. 5. Normalized predicted strength $\sigma_{pred}/\sigma_{max}$ versus normalized AE event counts N_t/N_{tmax} for typical N_t - σ curves after data processing with the 4 NN configurations: (a) aged material, (b) virgin material.

Tab. 2.
Result statistics for aged and virgin samples.

Material	σ_{max}			N_t/N_{tmax}		
	Mean [N/mm^2]	STD	Disp. [%]	Mean [N/mm^2]	STD	Disp. [%]
Aged	866	37	4	1121	1100	98
Virgin	1149	74	6	1606	495	31

3.2 Tensile test data

The extensibility estimation is a classification problem, since the desired outputs (estimations of extensibility) for each input (data of tensile test) as well as the number of classes are known from the available experts' knowledge. The use of neural networks in classification requires supervised training in order to assign instances to predefined classes, which is followed in the proposed approach.

The tensile test data (pairs of strain and force) are fed in a feed-forward neural network. Each input set is composed by four elements:

$$({}^{j}s_i, {}^{j}f_i, {}^{j}s_{i-1}, {}^{j}f_{i-1}) \qquad (1)$$

where, j is the number of the curve corresponding to the jth fabric sample, $j=1-14$, ${}^{j}s_i, {}^{j}s_{i-1}$ are the ith measured strain and the previous one, ${}^{j}f_i, {}^{j}f_{i-1}$ are the ith measured force and the previous one.

The range of i varies for each curve j since the duration of the tensile test and hence the number of the measurements varies from fabric to fabric. For a fabric with very low extensibility a low number of measurements are recorded while for a fabric with very high extensibility a large number of measurements are recorded.

For each input set a network output (EN) is calculated which is compared with the desired output or target set (ET) and produces the error in order to train the network.

The target ${}^{j}ET$, is the experts' estimation about the extensibility for the jth fabric sample.

The error used for the supervised training of the network is:

$$error = {}^{j}ET - {}^{j}EN_i \qquad (3)$$

where, ${}^{j}EN_i$ is the output of the network for the ith pair of measurements of the jth fabric sample.

Many high performance algorithms have been proposed [17] with much higher convergence rate than the standard back-propagation algorithm. The Levenberg-Marquardt algorithm [18] is the fastest algorithm for moderately sized networks and function approximation problems. However, as the number of weights increases the algorithm's advantages decrease.

All the training algorithms provided by the neural network Toolbox of MatLab have been tested in order to find the most appropriate. Finally, the Levenberg-Marquardt algorithm is used for the training. The starting value of the μ coefficient [18] is set equal to 0.001. The neural network is trained for 1000 epochs, which means that the input set (all the measurements for all fabrics) is fed 1000 times in the network.

The Mean Square Error-MSE (sum of squares of errors over all inputs) is about 0.00041 at the end of the training. For our case, this MSE is very satisfactory since the trained network can estimate the extensibility for each of the fourteen samples with accuracy about

±0.001. The trained neural network is used to estimate the extensibility of the six fabric samples (No. 15-20), which were not used for training.

Tensile test is performed for each sample and the measured forces and strains are fed on-line into the trained network. The output of the network is the estimation of the extensibility for the tested sample. The response of the network for each of the six fabrics is shown in Fig. 6.

For each of the samples the horizontal solid line is the extensibility estimated by the experts, and the dashed line is the output of the network. The network is capable to make a rough estimation from the very beginning of the tensile test, and its estimation is improved as the test is progressed.

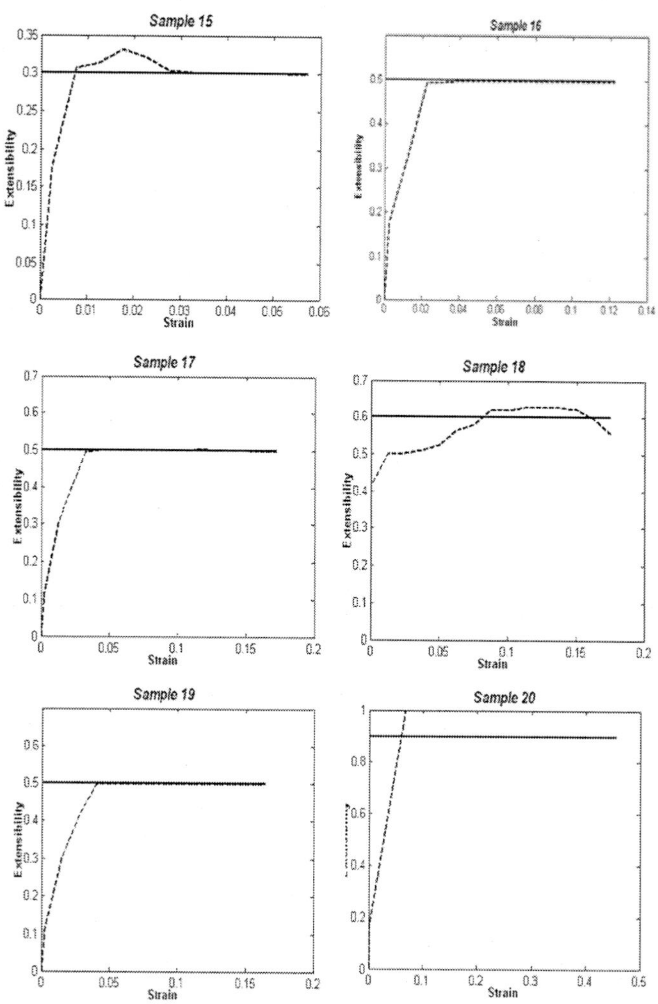

Fig. 6. The neural network estimations of the extensibility for samples 15-20.

It is important to remind that the training of the network was performed with data greater than 0.05 for the strain and greater than 1.5 Nt for the force. Nevertheless, the network makes satisfactory estimations even for this ignored data range, while quite far before the end of the test (approximately in the 20-30% of the total test time) the desired estimation is closely approximated.

Each pair of the measured force and strain is fed immediately into the trained neural network, during the tensile test; consequently it is not necessary to complete the tensile test in order to estimate the fabric extensibility.

As it is illustrated in Fig. 6, the responses for the samples 15 and 20 are not as satisfactory as the other ones since their force-strain curves are close to the limits of the training set. It is well known that when the fed sample is close to the limits of the training set the neural network performance is deteriorated.

Moreover, the samples 15 and 20 are fabrics with a very low and a very high extensibility respectively, where a small number of training samples was used compared to the other regions. This problem could be faced if new samples of fabrics with tensile curves outside these limits are used to train the network.

4. Conclusions

Experimental curves of AE event counts vs. stress obtained from unidirectional flexural tests on hydrothermally aged and virgin GFRP materials were utilized as input patterns to different NN configurations. The capability of NN processing to effectively predict material residual strength at an early stage in the AE response evolution and composite loading history was verified and critically assessed.

The proposed robotic system performs the tensile test of fabrics and estimates their extensibility using a simple feed-forward neural network. This approach is close to the experts' way of testing and evaluating the fabric extensibility. The specifications of the tensile test derived by the observation of the respective tests carried out by experts using their hands, senses and brain. The intelligent information processing, through neural networks obtains results close to the experts estimations.

The results obtained emphasize the usefulness of NN computation in materials technology problems where analytic solutions are not available.

Acknowledgements

This research work was carried out with partial support from the EC FP6 Networks of Excellence on "Innovative Productions Machines and Systems (I*PROMS)" and "Multi-Material Micro Manufacture (4M)".

References

[1] Wilson, D.W., Wetherhold, R.C., Cadot, H.M., McCullough, R.L., Byron Pipes, R. (Eds.) (1984). Composite Design Encyclopedia, University of Delaware, Newark, DE, USA.

[2] Teti, R. (1990). Ultrasonic Identification and Measurement of Defects in Composite Material Laminates, Annals of the CIRP, Vol. 39/1, pp. 527-530.

[3] Harris, B. and Phillips, M.G. (1983). Nondestructive Evaluation of the Quality and Integrity of Reinforced Plastics, In: Development in GRP Technology, (B. Harris, Ed.), Applied Science Publishers, London, pp. 191-247.

[4] Caprino, G. (1984). Residual Strength Prediction of Impacted CFRP Laminates, J. Composite Materials, Vol. 18, pp. 508-518.

[5] Cantwell, W.J. and Morton, J. (1990). An Assessment of the Residual Strength of an Impact-Damaged Carbon Fiber Reinforced Epo¬xy, Composite Structures, Vol. 14, pp. 303-317.

[6] Matthews, J.R. (Ed.), (1983). Acoustic Emission, Gordon and Breach Science Publishers, New York.

[7] Caprino, G. and Teti, R. (1994). Quantitative Acoustic Emission for Fracture Behavior Prediction of Center-Hole GFRP Laminates, J. Composite Materials, Vol. 28, No. 13, pp. 1237-1249.

[8] Kawabata, S. (1980), The Standardization and Analysis of Hand Evaluation, 2nd ed., The Textile Machinery Society of Japan, Osaka, Japan.

[9] Manuals for the Kawabata Evaluation System (1980), Kato Tekko Ltd, Kyoto, Japan.

[10] Gong, R. H. and Chen, Y. (1999), "Predicting the Performance of Fabrics in Garment Manufacturing with Artificial Neural Networks", Textile Research Journal, Vol. 69, No. 7, pp. 477-482.

[11] Park, S.W., Hwang, Y.G., Kang, B.C. and Yeo, S.W. (2000), "Applying fuzzy logic and neural networks to total hand evaluation of knitted fabrics", Textile Research Journal, Vol. 70, No. 8, pp. 675-681.

[12] Park, S.W., Hwang, Y.G., Kang, B.C. and Yeo, S.W. (2001), "Total handle evaluation from selected mechanical properties of knitted fabrics using neural network", International Journal of Clothing Science and Technology, Vol. 13, No. 2, pp. 106-114.

[13] Saville, B.P. (1999), Physical testing of textiles, Woodhead Publishing Ltd, ISBN 1 85573 367 6, Cambridge, England.

[14] Teti, R. and Caprino, G. (1995). Evaluation of Impacted Laminate Residual Strength through Neural Networks, Proc. 10th Int. Conf. on Appl. of Artif. Intell. in Engineering (AIENG '95), Udine, 4-6 July, pp. 411-418.

[15] Masters, T. (1993). Practical Neural Network Recipes in C++, Academic Press, San Diego.

[16] Fahlman, S.E. and Lebiere, C. (1990). An Empirical Study of Learning Speed in Back Propagation Networks, CMU Technical Report, CMU-CS-88-162.

[17] Haykin, S. (1999), Neural Networks: A Comprehensive Foundation, Prentice Hall, ISBN: 0-13-273350-1.

[18] Hagan, M. T. and Menhaj, M. (1994), "Training feedforward networks with the Marquardt algorithm", IEEE Transactions on Neural Networks, Vol. 5, No. 6, pp. 989-993.

Innovative Production Machines and Systems
D.T. Pham, E.E. Eldukhri and A.J. Soroka (Eds.)

Neuro fuzzy network for composites manufacturing time estimation

A. De Simone, L. Nele

Department of Materials and Production Engineering – University of Naples "Federico II" – P.le Tecchio 80, 80125 Naples-Italy – e-mail: nele@unina.it

Abstract

Composite materials offer great promise to all future air vehicles by allowing engineers to design and build aircrafts that weigh less, have better performance and lower lifecycle cost.
However even though these materials have been used by the industry for over a decade, there still does not exist an accurate means to estimate their cost. The primary aim of this study is to develop a cost estimation system in manufacture of advanced composite materials for aeronautic use.
The result of this study is the development of a predictive model that relates cost in direct labor hours. An artificial intelligence system has been successfully developed with the ANFIS Editor GUI of the Matlab Fuzzy Logic Toolbox.

Keywords: Neuro-Fuzzy systems, ANFIS, manufacturing time

1. Introduction

Cost estimation plays an important role in the product development cycle. Typically, the product cost is derived from the summation of the various cost components such as material, machine hours, direct labor, administration and engineering costs. During the early stage of product engineering it is always difficult to obtain these cost elements accurately. Moreover, to allow the designer to make a good trade-off decision, product cost must be identified early, during the designing stage where a large amount of the manufacturing cost is determined. The work reported in this paper is a result of an industrial project with an aeronautic industry. A new cost estimation method for advanced composite materials has been established to replace the company's current method which is based on the experience of the operators.

This paper discusses how the Neuro-Fuzzy systems can be applied to product cost estimation.

A Neuro-Fuzzy system is based on a fuzzy system which is trained by a learning algorithm derived from neural network theory. A possible architecture of a Neuro-Fuzzy hybrid system is shown in Figure 1.

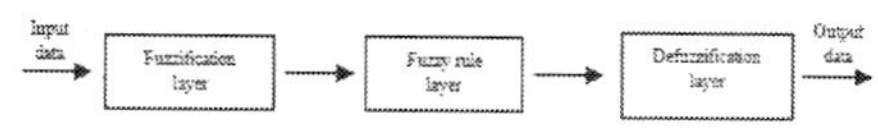

Fig. 1. Schematic diagram of a neuro-fuzzy system

The system contains the following three different layers:

- fuzzification layer
- fuzzy rule layer
- defuzzification layer

In a *fuzzification layer* each neuron represents an input membership function of the antecedent of a fuzzy rule.

In a *fuzzy inference layer* fuzzy rules are fired and the value at the end of each rule represents the initial weight of the rule, and will be adjusted to its appropriate level at the end of training. In the *defuzzification layer* each neuron represents a consequent proposition and its membership function can be implemented by combining one or two sigmoid functions and linear functions. The weight of each output link here represents the centre of gravity of each output membership function of the consequent and is trainable. After getting the corresponding output the adjustment is made in the connection weights and the membership functions in order to compensate the error and produce a new control signal.

One can create neuro-fuzzy systems with adaptive techniques like ANFIS (Adaptive Neuro-Fuzzy Inference Systems), which is available in Fuzzy Logic Toolbox. Using a given input/output data set, the toolbox function ANFIS constructs a fuzzy inference system (FIS) whose membership function parameters are tuned using either only a backpropagation algorithm, or a backpropagation algorithm coupled with a least squares type of method. This allows the fuzzy systems to learn from the data they are modeling [1].

The ANFIS architecture has five layers, each of which performs fuzzy rule-based system operations. Let us consider a single node from each layer. Let us denote f as the transfer function (which is a weighted sum in a normal neural network), a as the activation function, u^k_p as the inputs to the node, where $k=1...5$, p and i are the total number of the input and output variables respectively. The process of each node can be described as follows:

Layer 1: consists of the input nodes and directly transmits the input linguistic variables to the next layer.

The link weight, w^1_i, at the layer one is unity

$$f = u_i^1 \quad a = f \tag{1}$$

Layer 2: is the input term layer where each node acts as a membership function and represents the terms of the respective linguistic variables. Instead of calculating a weighted sum as in a normal neural network model, it calculates, for example, in terms of a bell-shaped fuzzy membership function,

$$f = M^j_{x_i}(m_{ij}, \sigma_{ij}) = -\frac{(u_i^2 - m_{ij})}{\sigma_{ij}^2} \quad a = e^f \tag{2}$$

where, m_{ij} and s ij are the centres (or mean) and the width (or variance) of the bell-shaped function of the jth term of the ith input linguistic variable xi. The link's weight at the second layer (w^2_{ij}) can be interpreted as m_{ij}. Also a set of nodes can represent a membership function in which case the nodes act as a standard neuron model as in the above equation.

Layer 3: this layer is a fuzzy rule layer where each node represents one fuzzy rule. The links in this layer are used to perform antecedent matching fuzzy logic rules and also the fuzzy *AND* operation. The link weight, w^3_i, at the third layer is also equal to unity.

$$f = \min(u_1^3, u_2^3, \ldots, u_p^3) \quad a = f \tag{3}$$

Layer 4: the links to this layer define consequences of the rule nodes. The links in the third and fourth layer, in association, function as a fuzzy inference engine. The nodes in this layer have two operation modes: the *down-up* transmissions and the *up-down* transmission mode. In the *down-up* mode, the links in the layer should perform the fuzzy *OR* operation to integrate the fired rules which have the same consequence.

$$f = \sum_{i=1}^{p} u_i^4 \quad a = \min(1, f) \tag{4}$$

The link weight, w^4_i, at the fourth layer is also equal to unity. In the *up-down* mode, the nodes in the layer 4 and the layer 5 function exactly the same way of the nodes in the layer 2 except that only a single node is used to perform a membership function for the output linguistic variables.

Layer 5: is an output linguistic layer, which has two nodes for each input variable. The first type of node is used to feed the networks for training it again.

$$a = f \quad f = u_i^5 \tag{5}$$

The second type of node is used for deriving a real output or a decision signal.

$$a = \frac{f}{\sum \sigma_{ij} u_i^5} \quad f = \sum w_{ij}^5 u_i^5 = \sum (m_{ij} \sigma_{ij}) u_i^5 \tag{6}$$

However, ANFIS only supports Sugeno-type systems, and these must have the following properties:
- Be first or zeroth order Sugeno-type systems.
- Have a single output, obtained using weighted average defuzzification. All output membership functions must be the same type and either be linear or constant.
- Have no rule sharing. Different rules cannot share the same output membership function, namely the number of output membership functions must be equal to the number of rules.
- Have unity weight for each rule.

2.Identification of a Takagi-Sugeno model

For a given data set, different TS models can be constructed using different identification methods. In this work a subtractive clustering method has been used to identify the antecedent membership functions.

2.1 Subtractive clustering

SC was introduced by Chiu [3]. For this method, data points have to be rescaled to [0,1] in each dimension. Each data point $z_j=(x_j,y_j)$ is assigned a potential P_j, according to its location with respect to all other data points:

$$P_j = \sum_{k=1}^{N} e^{-(4/r_a^2)\|z_j - z_k\|^2}, \tag{7}$$

with N the number of data points and r_a a positive constant called *cluster radius*. The potential of a data point to be a cluster center is higher when more data points are closer. The data point with the highest potential, denoted by P^*_1 is considered as the first cluster center $c_1=(d_1,e_1)$. The potential is then recalculated for all other points excluding the influence of the first cluster center according to:

$$P_j^{new} = P_j^{old} - P_1^* e^{-(4/r_b^2)\|z_j - c_1\|^2} \tag{8}$$

with $r_b=\eta r_a$ the radius defining the neighborhood that will have measurable reductions in potential, and η a positive constant, called the *squash factor*. Again, the data point with the highest potential P^*_2 is considered to be the next cluster center c_2, if

$$P_2^* > \bar{\varepsilon} P_1^* \tag{9}$$

with $\bar{\varepsilon}$ the *accept ratio*. If this is not the case but the following condition holds:

$$\frac{d_{min}}{r_a} + \frac{P_2^*}{P_1^*} \geq 1 \tag{10}$$

with d_{min} the minimal distance between c_2 and all previously found cluster centers, the data point is still accepted as the next cluster center c_2. Further iterations can then be performed to obtain new cluster centers c_l. If a possible cluster center does not fulfill the above described conditions, it is rejected as a cluster center and its potential is set to 0. The data point with the next highest potential P^*_k is selected as the new possible cluster center and re-tested.

The clustering ends if the following condition is fulfilled:

$$P_k^* < \underline{\varepsilon} P_1^* \tag{11}$$

with $\underline{\varepsilon}$ the *reject ratio*. Each cluster center is considered as a fuzzy rule that describes the system behavior.

2.2 Computing the consequent parameters

A network-type structure similar to that of a neural network, which maps inputs through input membership functions and associated parameters, and then through output membership functions and associated parameters to outputs, can be used to interpret the input/output map.

The parameters associated with the membership functions will change through the learning process. The computation of these parameters (or their adjustment) is facilitated by a gradient vector, which provides a measure of how well the fuzzy inference system is modeling the input/output data for a given set of parameters. Once the gradient vector is obtained, any of several optimization routines could be applied in order to adjust the parameters so as to reduce some error measure (usually defined by the sum of the squared difference between actual and desired outputs). ANFIS uses either backpropagation or a combination of least squares estimation and backpropagation for membership function parameter estimation.

3.Training Set

3.1 Input data definition

Input/output data in a form that will be usable by ANFIS for training have been collected. The manufacturing cycle of the firm has been studied. For every activity those characteristics that mostly affect manufacturing time have been chosen. Particularly, the cycle analysis has shown that the network inputs are:

Material;
Presence or absence of honeycomb;
Number of plies;
Geometric difficulty;
Number of holes;
Presence or absence of painting.

The material has been encoded in the following way:

Fiberglass = 0,35
Kevlar + Fiberglass = 0,5
Carbonfiber + Fiberglass = 0,6
Kevlar = 0,7
Carbonfiber + Kevlar = 0,8
Carbonfiber = 1.

Every material has a different difficulty in his lay-up sequence. The fiberglass is easier to lay, so it has been encoded with the lowest value. The carbonfiber is stronger, so it has been encoded with the highest value.

The presence or absence of honeycomb has been encoded with a boolean variable (0/1).

The number of plies has been normalized from 0 to 1.

Every material is positioned manually in the open mold. The process difficulty is that, when draping a woven cloth over a curved shape, some wrinkles can result. A degree of difficulty, based on amount of curvatures on the molds, has been considered. Particularly, a method to encode angles has been elaborated, both for concave angles and for convex angles:

A1/B1 for angles in range [180°,120°];
A2/B2 for angles in range]120°,60°];
A3/B3 for angles in range]60°,0°[.

Figure 2 shows some examples:

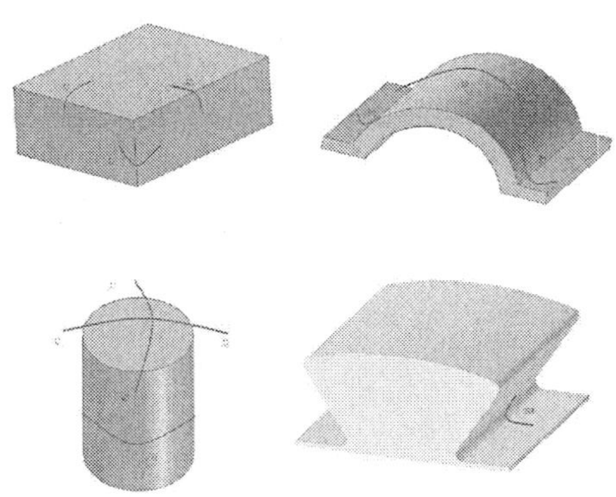

Fig. 2. Models of difficulty

A degree of difficulty equal to 1 corresponds to every A1/B1.

A degree of difficulty equal to 2 corresponds to every A2/B2.

A degree of difficulty equal to3 corresponds to every A3/B3.

The total degree of difficulty results from the sum of all angles on the molds.

The number of holes has been normalized from 0 to 1.

The presence or absence of painting has been encoded with a boolean variable (0/1).

3.2 Output data definition

The network output is the time (standard time) required for completion of the work by a qualified worker when working at a defined level of performance.

First a preliminary observation of the work has been undertaken to identify suitable elements which can be clearly recognized on subsequent occasions and are convenient, in terms of their length, for measurement. Then, each occurrence of each element has been timed using a stopwatch while at the same time an assessment of the worker's rate of working has been made. This assessment has been based on the factors involved in the work, such as effort, dexterity, speed of movement and consistency. The assessment has been made on a rating scale, of which there are three or four in common usage. Thus on the 0-100 scale a judgement of the worker's rate of working as a percentage of the standard rate of working (100) has been made. The rating has been then used to convert the observed time into the basic time using the simple formula:

Basic time = observed time * observed rating/ standard rating

Once a basic time for each element has been determined, allowances are added to derive a standard time.

4. Network parameters

The initial model parameters have been formed using subtractive clustering. In order to find the optimal model, the parameters of the subtractive clustering have been varied between 0.2 and 0.5 for the range of influence and 1 and 1.5 for the squash factor. The accept ratio and reject ratio have been inizialized to 0.5 and 0.15.

Two configurations have been used for And, Or and Defuzzification method: the first configuration with prod, probor and wtaver (weight-average), the second one with min, max and wtaver.

The Error Tolerance and the number of epochs, used to create a training stopping criterion, have been equaled to zero and to 100.

For each combination of these parameters a model has been built and trained.

However, in some cases data can be collected using noisy measurements and the training data cannot be representative of all features of the data that will be presented to the model. So, a checking data set has been used to check the generalization capability of the resulting fuzzy inference system and the potential for the model overfitting the data. The checking data has the same format as that of the training data and its elements are distinct from those of the training data. When checking data is presented to ANFIS as well as training data, the FIS model is selected to have parameters associated with the minimum checking data model error [8].

5.Results

For each trial the network error on the checking data has been calculated using the following formula:

$$e = \frac{|T(cr) - T(calc)|}{T(cr)} \tag{12}$$

where T(cr) is the checking data output value and T(calc) is the output of the fuzzy inference system. Particularly, the mean and max error has been calculated.

Figure 3 and 4 show mean and max error value in each configuration. The range of influence value is on the abscissa while the mean and max error percentage is on the ordinate. The squash factor value is on the histograms.

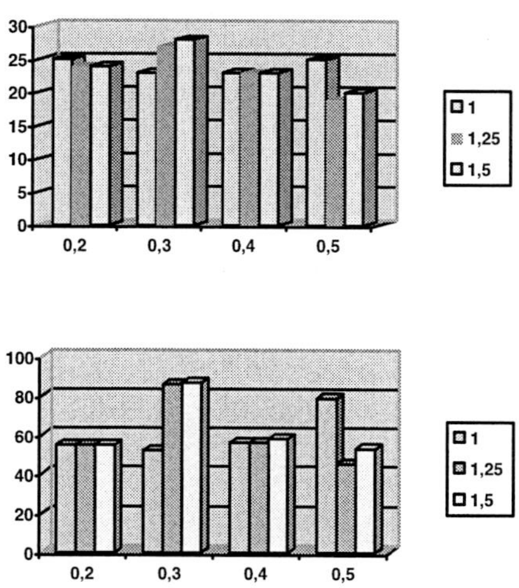

Fig.3. Mean and max error in the first configuration (prod, probor, wtaver)

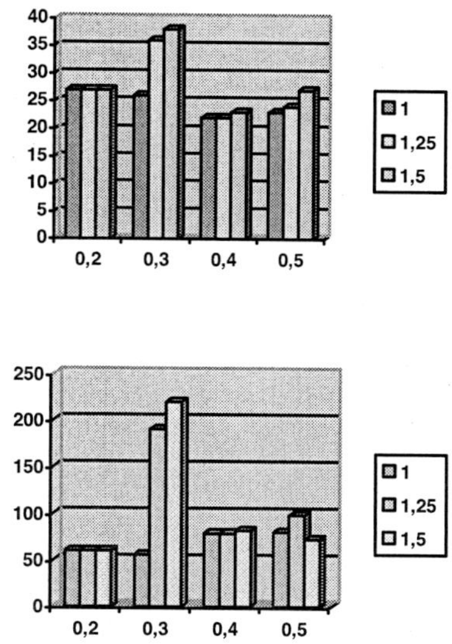

Fig.4. Mean and max error in the second configuration (min, max, wtaver)

Finally, the network with lower mean and max error has been chosen. The optimal configuration is the first one with the following parameters:

Range of influence = 0.5
Squash factor = 1.25
Accept ratio = 0.5
Reject ratio = 0.15

For this configuration the mean and max error are equal to 19% and 45%.

Figure 5 shows the predicted time associated to the training and checking data.

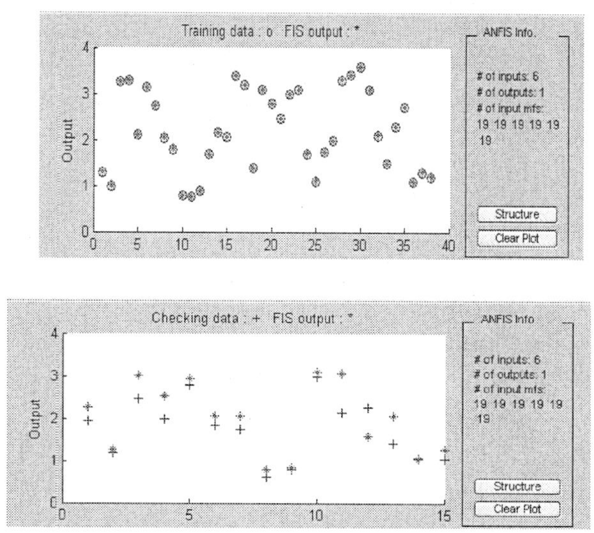

Fig. 5. Training and checking data

6.Conclusions

A novel neuro-fuzzy model, the adaptive network based fuzzy inference system (ANFIS), has been used for prediction in composite manufacturing time. By employing the hybrid learning algorithm, ANFIS can obtain the optimal distributed membership functions to describe the mapping relation in the input process parameters and output process parameters.

The trained network is reasonably accurate in comparison with the actual values and the use of ANFIS network in lieu of the numerical models reduces and simplifies the computational time.

References

[1] Jang J.-S. R., ANFIS – Adaptive-Network-Based Fuzzy Inference Systems, IEEE Trans. Man & Cybernetics, (1993), vol. 23, pp. 665-685

[2] T. Takagi and M. Sugeno, Fuzzy identification of systems and its applications to modeling and control, IEEE Trans. Syst., Man, Cybern., vol. 15, pp. 116132, (1985).

[3] Chiu, S. L. (1994), Fuzzy model identification based on cluster estimation. Journal of Intelligent and Fuzzy Systems, 2.

[4] S.L. Chiu, Selecting input variables for fuzzy models, J. Intell. Fuzzy Systems 4 (1996) pp. 243–256.

[5] D. Nauck, R. Kruse, Neuro-fuzzy systems for function approximation, Fuzzy Sets and Systems 101 (1999), pp. 261–271.

[6] L.A. Zadeh, Outline of a new approach to the analysis of complex systems and decision processes, IEEE Trans. Systems, Man Cybernet. 3 (1973), pp. 28–44.

[7] R. Pytelkova, Modelling and control with neuro-fuzzy system. Helsinki Summer School, (2001).

[8] Fuzzy Logic ToolBox, User's Guide

[9] Hertz, J., Krogh, A. and Palmer, R.G., Introduction to the Theory of Neural Computation, Addison-Wesley, New York, 1991.

[10] H. J. Zimmerman, Fuzzy Set Theory and its Application. Kluwer, (1990).

Innovative Production Machines and Systems
D.T. Pham, E.E. Eldukhri and A.J. Soroka (Eds.)

On mechanical characterization of Tungsten Inert Gas and Friction Stir Welding butt joints of 6061 T6 aluminium alloy

A. Squillace[a], G. Giorleo[a], U. Prisco[a], L. Nele[a], E. Sarno[a]

[a] *Department of Materials and Production Engineering, University of Naples Federico II, Piazzale Tecchio 80, 80125 Naples, Italy*

Abstract

The results of an experimental investigation on butt weld joints of AA 6061 T6, realized using two different welding techniques: Tungsten Inert Gas (TIG) and Friction Stir Welding (FSW), are reported. A comparison is proposed between performances and behaviour shown by the two kinds of joints in terms of microstructure, modifications induced with respect to parent alloy and of mechanical properties. Microstructure has been investigated by mean of optical microscopy and micro-hardness measurements. Mechanical properties have been evaluated by means of tensile tests. Bending tests have been carried out to further characterise the residual ductility of joints. Results show the influent of the initial heat treatment of the alloy, especially in terms of modifications induced to the microstructure of different zones of the joints and in terms of loss of mechanical properties. FSW joints, however, show better results than TIG ones; the influence of FSW tool dimensions has been put in evidence too.

Keywords: aluminum alloy, heat treatment, Tungsten Inert Gas, Friction Stir Welding

1. Introduction

The wider and wider use of aluminium and, above all, of its alloys is strictly related to the achievement of both good mechanical and electrochemical properties. Strength improvement is a major objective in the design of aluminium alloys because the low strength of pure aluminium limits its industrial effectiveness. The alloy under investigation in the present paper belongs to so-called heat-treatable or precipitation-hardening alloy. Such treatment consists in dissolving alloying elements into solid solution and precipitating them as coherent sub-microscopic particles [1]. In case of AA 6061, the solution heat-treatment and artificially aging, designed as T6 condition, results in a remarkable improvement of ultimate tensile strength and, above all, yield strength [2]. On the other hand, since temperatures involved in described treatments are quite low, these alloys can be used for temperature up to 150°C. When temperatures overcome 200°C, in fact, the precipitation reactions continue into softening regime: this represents a great problem in the welding of these alloys. Anyway, this is just one of the great number of problems related to the welding of aluminium and its alloys: the main ones being the oxide layer, hydrogen solubility, electrical conductivity and thermal characteristics; they all result in crack sensitivity (both solidification and liquation cracking), porosity, heat affected zone (HAZ) degradation, and so on.

The welding of aluminium alloys has been widely investigated in literature. The specific alloy used in this research, has been made object of several researches involving different welding processes, and, as consequence, lots of results are available, both in terms of mechanical and electrochemical properties, and in terms of microstructure analysis; notwithstanding, due to its short life and, above all, thanks to the chance it gives to realise sound joints, FSW, in last years, appears most attractive.

Liu et al. [3] have widely described the microstructure of 6061 T6 FSW butt joints, showing an evident reduction in grain size of weld bead with respect to parent alloy, and remarkable lower values of micro hardness in the bead and in the HAZ; they also put in evidence the complex evolution experienced by hardening precipitates. Heinz et al. [4], studying FSW joints of an aluminium alloy very similar to that under investigation here, found one more time a fine equiaxed grain structure in weld bead with dissolution of precipitates, while temperatures experienced by the HAZ lead to over aging and transformations of parent precipitates, anyway resulting in a general decay of mechanical properties. Reynolds et al. [5], carrying out an experimental characterisation of FSW joints on a variety of aluminium alloys, including AA 6061 T6, show a great number of numerical results, in good agreement with those proposed in present paper. Liu et al. [6], on the same alloy, have proposed an interesting correlation between mechanical properties of FSW butt joints and some process parameters. Shigematsu et al. [7] have investigated also joints of dissimilar alloys, while Krishnan [8] has conducted an interesting research on AA 6061, starting in annealed condition, but evaluating the influence of different post welding heat treatments on microstructure and some mechanical properties of joints. Cavaliere et al. [9] studied the effect of FSW speed (with advancing speed in the range 40–460mm/min) on themechanical and microstructural properties of AA6082. They observed a strong variation in the nugget mean grain size by increasing the advancing speed from 40 to 165mm/min up to a plateau corresponding to no further variations by increasing the speed up to 460 mm/min. The yield strength was recorded to increase strongly from the lower speeds to 115mm/min and after starts to decrease by increasing the advancing speed, the ductility of the material followed the same behaviour but restarted to increase after 165 mm/min. The material welded with the advancing speed of 115mm/min exhibited the best fatigue properties and the higher fatigue limit, while a very narrow similar behaviour in the low cycle regime, differing strongly by decreasing the stress amplitude up to the fatigue limit, was observed in all the configurations. In [10] Monetta et al. assessed the correlation between FSW parameters and corrosion behaviour for friction stir welded 6056 aluminium alloy. For the welding parameters studied, the data strongly indicate that sensitivity to corrosion is linked to the welding parameters employed. They found substantial differences for the higher and lower speed welding parameters, with the lower speed parameters apparently resulting in less corrosion resistance. The most widespread form of corrosion by Monetta et al. observed was pitting; however, intergranular corrosion (IGC) was also seen. While the T78 thermal treatment appears to decrease the corrosion resistance of the parent alloy, the weld region tends to improve, with the nugget being cathodic.

The aim of this study, instead, consists in a comparison of mechanical properties provided by butt weld joints, realised using two different welding processes (the former, TIG, being a conventional fusion welding process, the latter, FSW, being a solid state joining process), both in terms of numerical results and in terms of macroscopic behaviour, related to microstructures and modifications induced by investigated joining processes.

The paper is organized as follows: Section 2 surveys materials used, experimental and measuremental procedure. Welding process is accurately described as well the mechanical and metallurgical tests. Section 3 deals with the experimental data. Here the experimental results are exploited to better understand the physical phenomena accompanying the Friction Stir Welding process and its effects on the studied alloy. Finally, Section 4 summarizes the most important issues of this experimental investigation.

2. Materials and Methods

2 mm thick, rolled plates of AA 6061, in T6 conditions, made by Alcoa, have been used in present research; all weld joints have been realised along rolling direction.

Full penetration TIG butt joints (with a main bead on upper side and a further bead on the reverse one) have been manually made using a STEL 200 H welding machine in AC mode; a water-cooled torch has been adopted; argon as shielding gas and no filler material have been used.

Two different single pass FSW butt joints have been realised, using two different steel tools: they are similar in geometric features, but they differ in main dimensions. The tools are graphically depicted in Fig. 1, while their main dimensions are collected in Table 1.

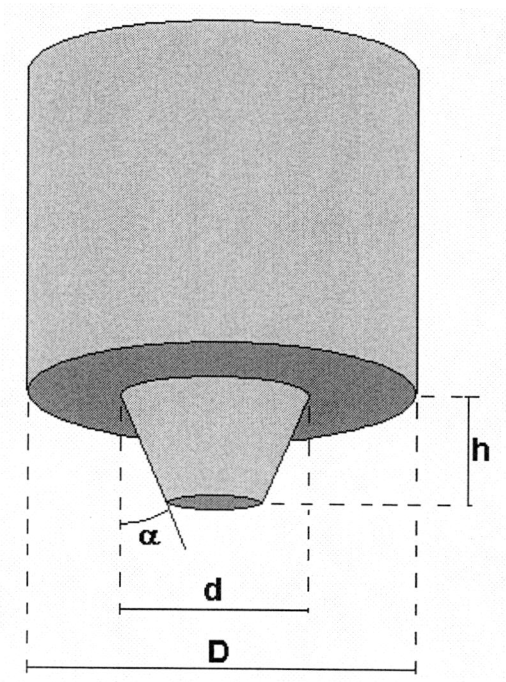

Fig. 1. Geometry of FSW tools.

Table 1
FSW tools main dimensions.

Tool designation	D (mm)	d (mm)	h (mm)	α (°)
T1	8.0	3.0	1.9	10
T2	11.2	4.0	1.9	10

Process parameters adopted in welding execution are: a tool welding speed of 78 mm/min, a rotational speed of 900 rpm and a tilt angle of 2°.

Experimental characterisation of joints has been carried out by mean of macrographs; Vickers micro-hardness profiles; tensile tests (with weld bead perpendicularly disposed with respect to loading direction); bending tests up to 90°.

Macrographs of cross section of the welded plates have been realised using a Leitz MM6 optical microscopy coupled to a Nikon Coolpix 5000 digital camera. Vickers micro-hardness profiles have been performed with a Leitz Miniload 2. For the TIG welded joint, a single profile has been measured, at a depth of 1 mm; for FSW welded joint, instead, in order to keep in account the truncated cone shape of the pin, two different profiles have been measured: the former at 0.5 mm of depth and the latter at 1.5 mm, with respect to upper surface of welded plates; in all cases, the distance between two consecutive indentations has been set equal to 0.5 mm.

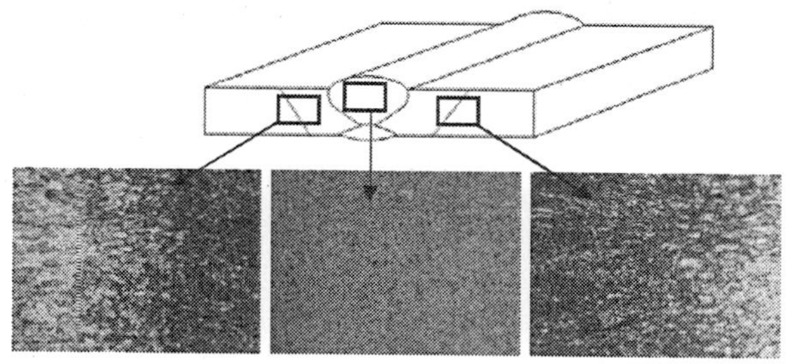

Fig. 2. Through the thickness images of TIG joint.

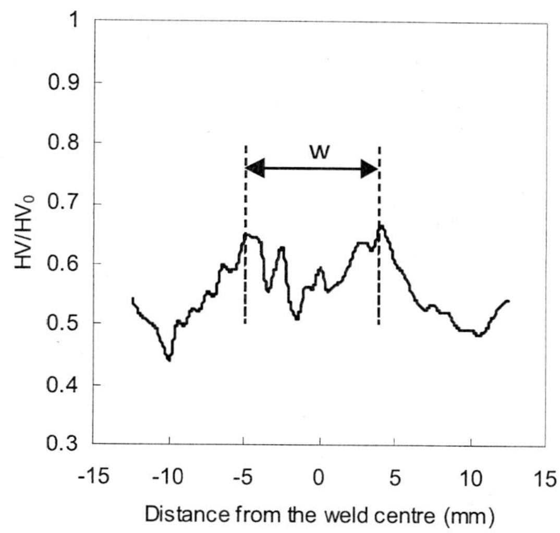

Fig. 3. Micro-hardness profile of TIG joint; "w" indicates the width of weld bead.

Tensile tests have been carried out on rectangular specimens; their main dimensions are: 100 x 12 x 2 mm. The same dimensions have been set for specimens tested in bending way. For a full comparison, further tests have been performed on parent material, with specimens characterised by same dimensions.

3. Results and Discussions

Fig. 2 shows three zones of the TIG joint cross section and a sketch indicating their location. It is possible to appreciate the evident modifications induced by fusion welding, especially in transition zones (from parent alloy to HAZ); in these zones a grain refinement is appreciable.

The core of the weld bead shows a fine grain structure, even if a great number of defects (mostly

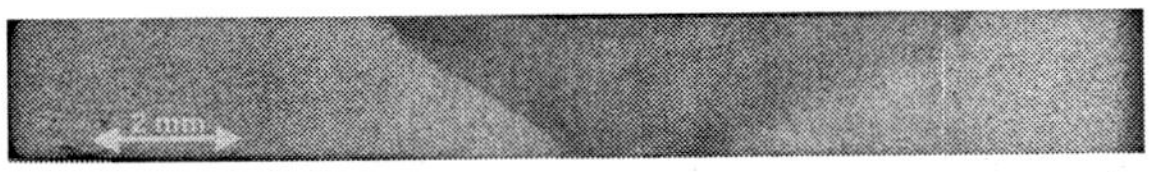

Fig. 4. Cross section of FSW joint realized using T1 tool.

Fig. 5. Cross section of FSW joint realized using T2 tool.

porosities and inclusions) are evident. According to alloy under investigation and, above all, to its heat treating conditions, the heat provided to realize the welding and to melt material causes the total loss of benefits related to initial heat treatment.

This is corroborated by Fig. 3, in which Vickers micro-hardness profile, measured on TIG joint, is shown (micro-hardness values are reported as a ratio with respect to value of parent alloy).

In the Fig. 3 it's possible to appreciate the evident loss of hardness in both weld bead and HAZ with respect to parent alloy; though the measures have been performed in a wide range (25 mm, centred on weld bead), no recovery of original hardness can be appreciate, resulting in a very wide HAZ associated to actual welding process. Micro-hardness values of HAZ, likewise to results available in literature [1, p. 532], are lower than those shown by weld bead.

Figures 4 and 5 show cross section images of FSW joints realised using respectively T1 and T2 tool. It's evident the wider zone interested by modifications when a greater tool is adopted.

In Fig. 6 and 7, for the two FSW joints (for both the one realised using T1 tool and the one realised using T2 tool), micro-hardness profiles are reported. From comparison of these figures, a confirmation of narrower extension of modified zone, when littler tool is used, is evident. Another evident aspect, emerging from both figures, is the value attained by micro-hardness in the core of weld bead: even if in the FSW process the maximum temperature experienced by material is lower than melting one, it is high enough to bring to the almost complete loss of properties provided by parent heat treatment. Anyway, the lowest values attained by FSW joints are greater than those attained by TIG joint, and this is due to the severe plastic deformation induced by tool motion, resulting in the recrystallization of a very fine grain structure [3, 4]. The two figures show also how the width of range in which micro-hardness attain their minimum value is very close to tool shoulder diameter; such range, however, decreases as depth of measure increases.

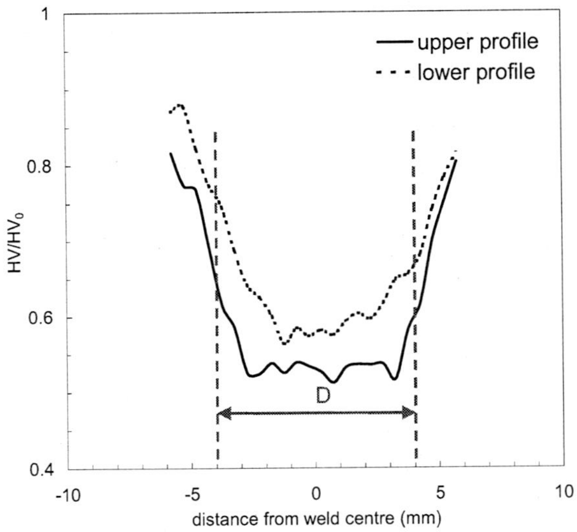

Fig. 6. Micro-hardness profiles of FSW T1 joints.

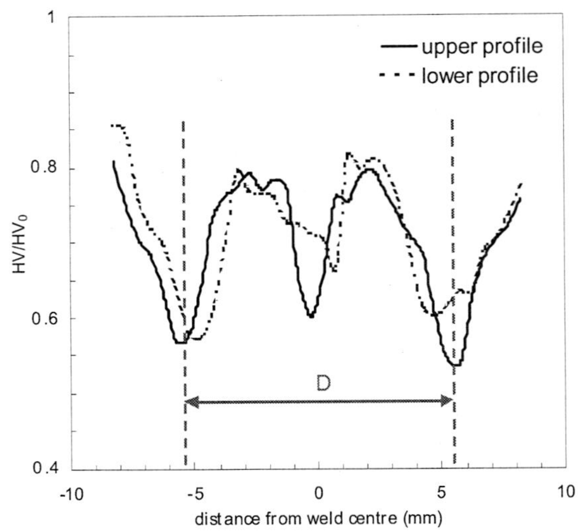

Fig. 7. Micro-hardness profiles of FSW T2 joints.

In Fig. 7, a partial recovery of micro-hardness value can be appreciated in the bead: such situation has been often observed in literature [3, 4], but it is not evident in Fig. 6.

Tensile tests (numerical results are reported in Table 2) allow many interesting considerations. First of all, tests conducted on parent alloy provide results very close to those available in literature [2, pp. 102-103]. Both welding techniques under investigation induce modifications resulting in an evident loss of mechanical properties: performances provided by weld joints are far from results attained by parent alloy both in terms of yield strength and UTS and elongation.

TIG weld joints show a mean value of residual yield strength of about 27% and a mean value of residual UTS of 42%, while the loss in terms of elongation attains 72%; what's more, a large scatter is evident on numerical values of both yield strength and UTS: probably this has been caused by manual execution of welds; notwithstanding, the fracture occurs, for all specimens, in the HAZ, quite far from the centre of weld bead; it takes place as 45° shear fracture and is accompanied with negligible necking phenomenon.

Also FSW joints show, with respect to parent alloy, a generally decay of mechanical properties, but in this case it's important, as a comparison with TIG joints, to put in evidence two fundamental aspects: the former is that the residual strengths (both in terms of yield and UTS) are generally greater than those provided by TIG joints; the latter is that the scatter shown by tests is significantly narrower than that shown by TIG joints, and this is evident, independently by tool used. What's more, it's interesting to underline the best results provided by FSW joints realised using T2 tool.

Liu et al. [6] have shown the influence of FSW process parameters on residual strength of welded joints: in this study, keeping constant just process parameters, the influence of tool dimension emerges. In terms of elongation, results provided by FSW joints are similar to those shown by TIG joints and to those observed in [4] and in [6], and they all testimony the dramatic loss of ductility of weld joints with respect to parent alloy; Shigematsu et al [7], instead, for the same alloy, found a remarkable value of residual elongation of about 11%.

Focusing attention exclusively on FSW joints, however, an aspect appears interesting: the place in which the fracture occurred. The great part of cited authors have found that fracture always occurs in retreating side, at a distance from weld centre depending on process parameters.

Liu et al. [6], for example, found that fracture occurred always in retreating side (similarly to behaviour observed on another heat treatable aluminium alloy investigated by same authors [11]), at a distance from weld centre that increases as revolutionary pitch decreases, anyway inside the zone interested by tool shoulder, analogously to what observed in present study.

In this study, instead, in joints realised with T1 tool fracture has always occurred in the centre of weld bead, while, in joints realised with T2 tool, fracture has randomly occurred both in the centre of bead and in retreating side and in advancing side. Such a situation could be encouraged by evident local reduction of thickness that has been appreciated in specimens under investigation, and that is strictly related to the actual joining process. All FSW specimens have shown, however, a 45° shear fracture coupled to necking.

Table 2
Tensile tests results.

Test	Joint	Tool	Yield strength (MPa)		UTS (MPa)		Fracture place	Elongation (%)	
			actual	mean	actual	mean		actual	mean
01	Parent alloy	---	270	277	339	342	---	17	18
02			290		348			18	
03			270		339			20	
04	TIG	---	60	74	120	143	HAZ	4	5
05			76		147			5	
06			86		161			6	
07	FSW	T1	108	112	174	178	Weld bead	3	3
08			115		182			3	
09			112		179			3	
10		T2	110	113	195	202	Weld bead	4	5
11			118		210		Advancing side	5	
12			112		202		Retreating side	5	

Table 3
Comparison of tensile test results of both machined and un-machined FSW joints.

Test	Tool	Machining	UTS (MPa)		Fracture place
			actual	mean	
07	T1	No	174	178	Weld bead
08			182		
09			179		
13		Yes	186	203	
14			221		
10	T2	No	195	202	Weld bead
11			210		Advancing side
12			202		Retreating side
15		Yes	235	223	Retreating side
16			210		

To better evaluate the influence of above-mentioned step, a further series of tensile test has been carried out on FSW joints. This time all specimens have, previously, been machined on the upper side, to have a smooth surface and to avoid the presence of that little step produced by tool shoulder. Unfortunately, the lack of material, has forced the execution of only two tests for each tool. Most significant results are briefly reported in Table 3.

It's evident that performances provided by machined specimens are better than those shown by un-machined joints, and this is an evident indication of the influence of the above-mentioned step; on the other side, joints realised with T1 tool show, one more time, the occurrence of fracture just in the centre of weld bead, while both machined specimens, realised using

T2 tool, show fracture on retreating side.

To conclude, to better characterise the ductility of different joints, three points bending tests have been carried out. The results can briefly reported by assessing that all joints have been easily bended till a 90° angle, and any crack has been observed on specimens.

4. Conclusions

On the basis of the experimental characterisation conducted on weld joints of AA 6061 T6, the following conclusions can be drawn:

- Irrespective of welding technique, all weld joints have shown an evident loss of strength and, above all, of ductility: this can be ascribed to the complete loss of benefits related to initial heat treatment. It's important to underline that, for the alloy under investigation, the achievement of a temperature of about 200°C, for a short period too, involves the above-mentioned total loss.
- FSW joints allow the achievement of better performances than TIG ones; the scatter appreciated in results shown by the former technique appears narrower than that shown by the latter.
- Performances provided by FSW joints, seem to be affected not only by loss of parent heat-treating conditions, but also by evident microstructure modifications occurring, at different levels, in nugget and flow arm zones with respect to both TMAZ and HAZ: this is corroborated by the frequent fracture occurrence in the centre of weld bead.
- Performances achievable by FSW joints seem to be influenced by tool main dimensions: in present research, joints realized using greater tool have shown best results.
- Even if the inadequate number of specimens doesn't allow a proper comparison, tensile strength of FSW welded joints seems to be affected by step produced on the top surface of bead by tool shoulder; specimens subjected to tensile tests, which have been previously machined in order to eliminate such step, have shown an appreciable improvement in UTS, higher than 10%.

References

[1] ASM Metals Handbook, *Welding, Brazing and Soldering*, vol. 6, 10th edition, 1993, p. 529.

[2] ASM Metals Handbook, *Properties and selection: nonferrous alloys and special-purpose materials*, vol. 2, 10th edition, 1990, p. 102.

[3] G. Liu et al, *Scripta Materialia*, Elsevier science ltd., 37 (3), 1997, pp. 355-361.

[4] B. Heinz et al., *Material Science Forum*, Trans Tech Publications, 331-337, 2000, pp. 1757-1762.

[5] A. P. Reynolds et al., *Material Science Forum*, Trans Tech Publications, 331-337, 2000, pp. 1719-1724.

[6] H. Liu et al., *Journal of Materials Science Letters*, Kluwer Academic Publisher, 22 (15), 2003, pp. 1061-1063.

[7] I. Shigematsu et al., *Journal of Materials Science Letters*, Kluwer Academic Publisher, 22 (5), 2003, pp. 353-356.

[8] K. N. Krishnan, *Journal of Materials Science*, Kluwer Academic Publisher, 37 (3), 2002, pp. 473-480.

[9] P. Cavaliere et al., *Journal of Materials Processing Technology*, 200 (2008), pp. 364–372.T.

[10] C. Monetta et al., *Advanced Materials Research*, 38 (2008), pp. 285-297.

[11] L.Nele et al., Proceedings of 6th AITEM International Conference, Gaeta, Italy, September 8th to 10th 2003.

Innovative Production Machines and Systems
D.T. Pham, E.E. Eldukhri and A.J. Soroka (Eds.)

Quick Response Manufacturing Approach for Lead Time Reduction in PCB Fabrication

D. D'Addona, R. Teti, R. Carbone

Dept. of Materials and Production Engineering, University of Naples Federico II, Naples, Italy

Abstract

In recent years, manufacturing companies have been affected by the economic slump due to the challenge of competition by low-wage countries in the globalized market. This has pushed them to look inward, struggling to find ways to reduce response time, improve quality and cut costs. In this framework, the development of the concept of Quick Response Manufacturing showed that, to compete on speed, industries must change the management mind-set and reduce lead times throughout the organization. The key issue is that a factory's lead times, work-in-process, and actual capacity are the result of complex dynamics and interactions in the manufacturing system. Thus, to predict whether a decision will improve lead times, it is necessary to be able to forecast both process and queue times, which means that any lead time reduction tool must be able to model these dynamics and interactions. An interesting approach based on established techniques in queuing theory, called Rapid Modelling Technology (RMT), was developed for complex shop floor operations modelling with mathematical equations and fast lead time calculations on a PC; a user-friendly program based on RMT is the MPX© software package. In this paper, the evaluation of human and economic resources of a printed circuit board manufacturing company is analysed through RMT using MPX©.

Keywords: PCB Fabrication, Lead Time Reduction, Quick Response Manufacturing, Rapid Modelling Technology

1. Introduction

Recently, many manufacturing companies affected by the economic slump due to the challenge of competition by low-wage countries in the globalised market have looked inward, struggling to find ways to reduce response time, improve quality and cut costs. As a feedback, the concept of Quick Response Manufacturing (QRM) was developed as a companywide strategy addressing two key factors [1-4]: 1) the external aspect of responding quickly to customers by rapidly estimating, designing and manufacturing customized or new products; and 2) the internal aspect, focusing on lead time reduction for all tasks throughout the organization. Thus, the application of QRM aims at simultaneously reducing response times, improving quality and lowering costs.

Manufacturing is the natural place to look for lead time reduction. In fact, manufacturing has been studied more than any other area of an industrial business. Producing more products, in greater variety, with lower cost, in less time requires manufacturers to rethink and change in fundamental ways the critical policies of capacity planning, lot sizing, and factory organization.

To show the need for these changes, there are dedicated software packages showing the impact of alternative policies on a factory's lead times. The latter are, however, the result of complex dynamics.

Previously, analytical queuing models could only analyze simple manufacturing systems, so that the dynamics for complex factories could be modelled only through sophisticated simulation software codes. The use of such codes, however, requires time-consuming training and long computing times.

2. Rapid modelling technology

An interesting approach based on techniques in queuing network modelling, called Rapid Modelling Technology (RMT), was presented in [5, 6] for complex shop floor operations modelling with pre-built mathematical equations and fast lead time calculations on a PC. An user-friendly program based on RMT is the MPX© software package [7, 8], a tool to assist in achieving quick response in manufacturing and in analyzing production operations to find opportunities for improvements related to capacity, work-in-process, labour allocation, and new products introduction.

In this paper, the evaluation and optimization of human and economic resources of a printed circuit board (PCB) manufacturing company is analysed through RMT using MPX©. In particular, the lead time in PCB fabrication for a multimeter instrumentation was studied by accounting for all production phases, labour resources, machines and equipments utilised in the PCB manufacturing process, and solutions to the changes needed to achieve lead time reduction are presented and evaluated.

The MPX© is used to conduct 'what if' scenarios which impact a variety of manufacturing parameters, including parts routing, labour, equipment, equipment failure/repair, set-up, run time and lot size. The software model assists in evaluating the effects of alternative management decisions, and is of particular importance when analysing the operation of cells. It helps in obtaining an insight into the factors that influence the lead time performance of cells. MPX© can be used to establish whether the business will benefit from manufacturing in cells, and if so, what would be the ideal cell configuration.

Lead time depends on both processing time and queue time (this is the time parts waiting to get their turn at machines, when the machines are busy). While processing time may be known based on the machining parameters, queue time depends on many 'dynamic' factors such as, which other parts are already in queue to use the machine, whether the machine has broken down, whether an operator is available, and so on.

In order to predict whether a decision will improve lead times, it is necessary to be able to predict these queue times, which means any lead time reduction tool must model these dynamics and interactions. The RMT technology in MPX, models these complex dynamics of the manufacturing facility in terms of mathematical equations.

3. Printed circuit board manufacturing

Multimeter (MTM) PCBs are fabricated through a sequence of main production phases: (1) purchase of components; (2) incoming component inspection; (3) kitting; (4) PCB assembly; (5) cleaning; (6) visual inspection; (7) rework; (8) automatic testing of single components; (9) PCB functional testing; (10) issue of final testing certificate.

Phase 1. Purchasing of components is carried out through an MRP procedure on the basis of customer orders that include the list of components for a single PCB and the number of PCBs to be fabricated. The MRP output reports include proposed purchase order releases and production plan proposals to be validated by the responsible staff.

Phase 2. All incoming components are checked against the delivery documents to coincide with the ordered components and are then tested for conformity.

Phase 3. Kitting follows the shop order setup that lists the number of PCBs to be fabricated and the number of components for each single component code to be retrieved from the warehouse to make up the kit (i.e. the set of components necessary to achieve the fabrication of a complete PCB batch). The components are manually stowed in different containers according to their codes and PCB insertion positions.

Phase 4. PCB component assembly is carried out according to either traditional PTH (plated-through-hole) or combined PTH - SMT (surface-mount technology) assembly technology [9]. The PTH assembly method utilises components with pins or leads that are preformed and manually inserted into the PCB through-holes from the top side; then, a mechanised wave soldering process is applied. The combined assembly method utilises both traditional PTH components and SMT components mounted on the PCB bottom side by adhesive bonding and reflow soldering.

Phase 5. After soldering, contaminants such as flux, oil, grease, salts or dirt are removed from the PCB by hand cleaning and vapour degreasing operations.

Phases 6, 7. Visual inspection is used to detect for board substrate damage, missing or damaged components, soldering faults, and similar quality defects. If inspection indicates that one or more components on the board are defective, or that certain solder joints are faulty, the assembly is repaired by manual reworking operations that include touchup of solder faults, replacement of defective or missing components, pin cutting to size, and repair of copper film that has lifted from the substrate surface.

Phases 8, 9, 10. Automatic testing of single components as well as PCB functional testing is carried out through dedicated equipment. A testing certificate is then issued reporting the PCB specifications and the tests that were performed and passed.

4. Mpx© modelling of the PCB manufacturing system

The MTM PCB manufacturing process analyzed through MPX© represents a model as close as possible to the actual production.

The main difference is the fact that the MTM PCBs are not manufactured on dedicated lines with complete batch transfer from one workstation to the next, as required by MPX©, but the actual production is rather organized in mixed mode, allowing for incomplete batch transfer.

The main objectives of the MPX© utilization are: (1) to verify if the analysed PCB manufacturing system can meet with the predefined lead time reduction goals; (2) to find solutions to optimize the performances of the PCB manufacturing system.

4.1 Input Data

To examine the throughput process of the MTM PCBs, each manufacturing task was analysed by considering the related workstations, machines, equipments, and human resources. The data acquisition for each single PCB of the MTM was the first step in the analysis. The input data necessary for MPX© modelling are reported below in MPX© table form for the categories of Labour, Equipment, Product and Operation data.

4.1.1 Labour Group Data

Figure 1 reports the data for all Labour Groups involved in fabricating the MTM PBCs.

Labour Name: name of the Labour Group.

No. in Group: number of workers that are present at one time.

Overtime %: time the Labour Group works beyond the standard time scheduled.

% Time Unavailable: time the Labour Group is unavailable, e.g. for meal breaks.

Comments: allows the user to add comments about the Labour Group.

4.1.2 Equipment Data

Figure 2 shows the information on equipments used in fabricating the MTM PCBs.

Equipment Name: name of the Equipment Group (an "equipment group" is a number of machines that can be considered "identical" for modelling purposes).

Equipment Type: it defines the operational characteristics of the equipment; for the developed model, all equipment have the "standard" characteristics of requiring a setup first, and then processing one piece at a time.

No. in Group: number of individual pieces of equipment in the group.

MTTF: Mean Time to Failure, i.e. the average time that a given piece of equipment operates before it breaks down (for manual operations, MTTF is formally set = 1).

MTTR: Mean Time to Repair, i.e. the average time it takes to repair one of the broken down equipment.

Overtime %: time the Equipment Group works beyond the standard time scheduled.

Assigned Labour: Labour Group assigned to work on this Equipment Group.

Comments: allows the user to add comments about the Equipment Group.

Labor Data Table Format

ADVANCED DATA | Show Form

LABOR NAME	No. in Group	Overtime %	% Time Unavailable	Comment
Assemblers	1	0	8	
Inspectors	6	0	8	
Kitters	2	0	8	
Packagers	1	0	8	
Prefomers	2	0	8	
PTH operators	3	0	8	
SMT operators	3	0	8	
Solderers	2	0	8	
Warehousemen	2	0	0	

Fig. 1. Table of Labour Data.

4.1.3 Product Data

The Product data are reported in Figure 3 with reference to the following items:

Product Name: name of the product (also called "part").

End Demand: end use demand (quantity shipped) for each product (the demand for components of each product will be calculated by MPX©).

Average Lot Size: average lot (or batch) size for each product, i.e. average number of pieces completed for each operation before the equipment is setup for a new lot.

Comments: allows the user to add comments about the product.

4.1.4 Product Operation Data

In Figure 4, the Product Operation data to fabricate the MTM PCBs are reported.

Operation Name: name of the operation.

Operation Number: sequence number of the operation.

Equipment Name: Equipment Group where this operation will be performed.

% Assigned: percent of time this operation will be done on this Equipment Group.

Equipment Setup Time: average time to setup the equipment for a product lot.

Equipment Run Time: average equipment time to complete one piece of this product (thus, total run time = lot size multiplied by the number in this data field).

Labour Setup Time: average labour time to setup the equipment for a product lot.

Labour Run time: average labour time to complete one piece of this product.

Equipment Table Format

ADVANCED DATA | Show Form | MTTF and MTTR Time units are MIN(s)

EQUIPMENT NAME	Equip Type	No. in Group	Overtime %	Mean Time to Failure	Mean Time to Repair	Labor Assigned	Comment
Packaging bench	Standard	1	0	1	0	Packagers	
Preforming station	Standard	2	0	1	0	Prefomers	
PTH insertion station	Standard	3	0	1	0	PTH operators	
SMT placement station	Standard	5	0	2400	60	SMT operators	
Soldering station	Standard	2	0	2400	60	Solderers	
Testing and rework station	Standard	6	0	1	0	Inspectors	
Unpacking	Standard	2	0	1	0	Warehousemen	

Fig. 2. Table of Equipment Data.

Production Data Table Format

ADVANCED DATA | Show Form | Show Operations | Show IBOM

Copy Product And Associated Data | Production period is 1 YEAR

PRODUCT NAME	End Demand	Lot Size	Comment
32AL-AA90028TF01	0	1800	
33AL-AA90027TF01	0	600	
34AL-AA10186NN01	0	600	
Power unit	7200	600	
Bus	7200	600	
Display	7200	600	
Microcontroller	7200	600	
Relais	7200	600	

Fig. 3. Table of Product Data (for brevity reasons, only the first 3 of the 96 component codes are reported).

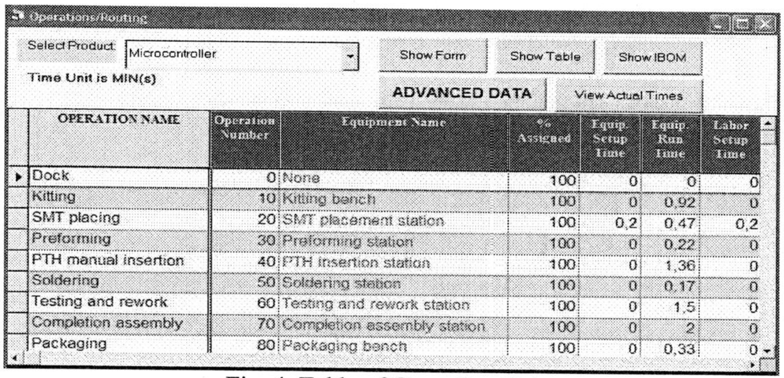

Operations/Routing

Select Product: Microcontroller

Time Unit is MIN(s)

Show Form | Show Table | Show IBOM | ADVANCED DATA | View Actual Times

OPERATION NAME	Operation Number	Equipment Name	% Assigned	Equip. Setup Time	Equip. Run Time	Labor Setup Time
Dock	0	None	100	0	0	0
Kitting	10	Kitting bench	100	0	0,92	0
SMT placing	20	SMT placement station	100	0,2	0,47	0,2
Preforming	30	Preforming station	100	0	0,22	0
PTH manual insertion	40	PTH insertion station	100	0	1,36	0
Soldering	50	Soldering station	100	0	0,17	0
Testing and rework	60	Testing and rework station	100	0	1,5	0
Completion assembly	70	Completion assembly station	100	0	2	0
Packaging	80	Packaging bench	100	0	0,33	0

Fig. 4. Table of Operation Data.

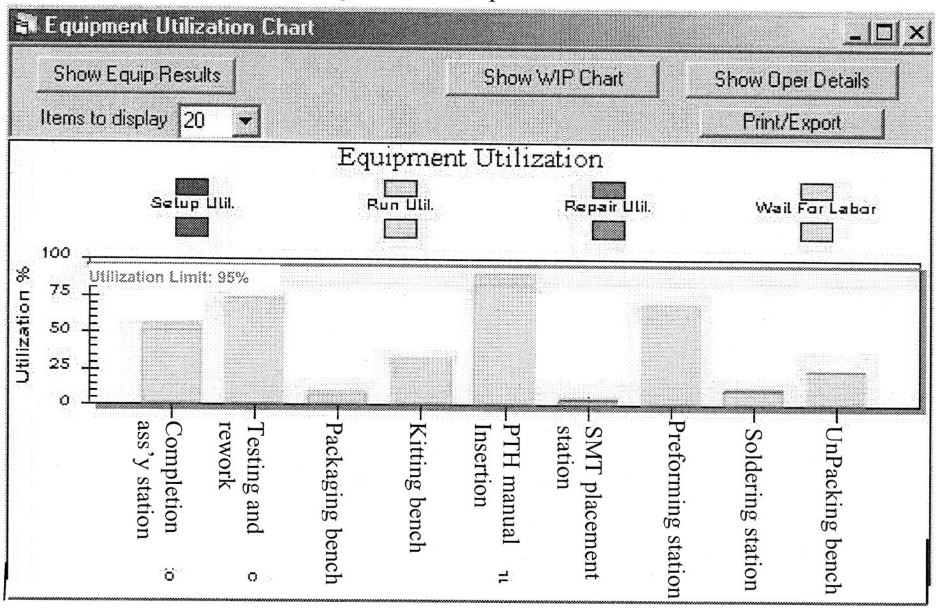

Fig. 5. Equipment Utilization Chart.

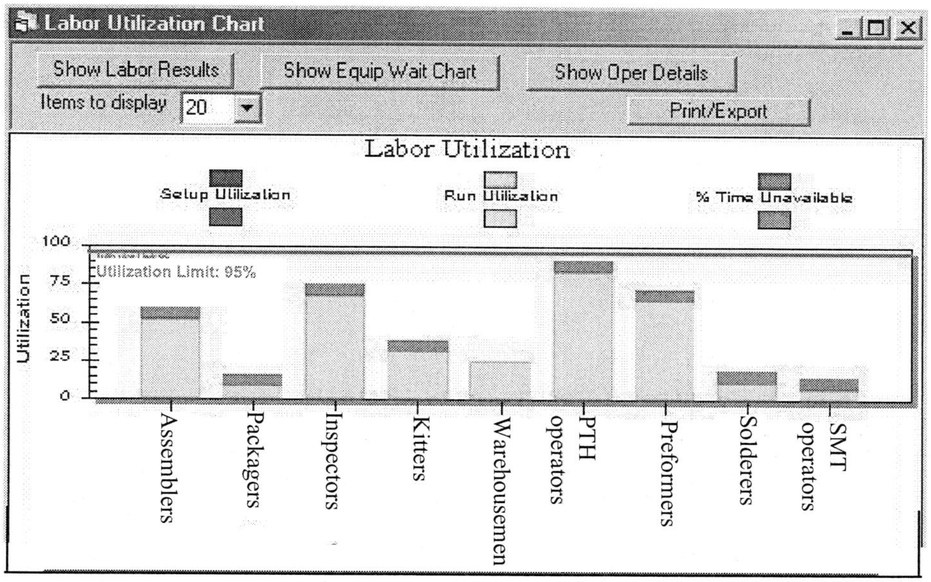

Fig. 6. Labour Utilization Chart.

4.2 Running the MPX© Model

When the description of the input data is complete, MPX© is able to evaluate the system performances and also obtain insights about performance improvements.

The main MPX© output are the Equipment Utilization Chart and the Labour Utilization Chart (Figures 5 and 6).

Figure 5 shows the total % utilization of each Equipment Group in the manufacturing system, broken out into the categories of Setup Utilisation, Run Utilisation, Repair Utilisation, and Wait for Labour. The latter is the result of dynamic interactions between Labour Group and Equipment Group and is evidenced by horizontal yellow stripes on top of the utilisation bars.

Figure 6 shows the total % utilisation of each Labour Group, broken out into the categories of Setup Utilisation, Run Utilisation, and % Time Unavailable.

The horizontal red line labelled "Utilization Limit: 95%" shows the maximum allowable utilization of any Equipment or Labour Group, i.e. the sum of all utilization categories. If this sum is > 95% for any Equipment or Labour Group, MPX© warns that production cannot be achieved in due time.

Figure 5 evidences the main Wait for Labour problem for the largely over-utilised PTH Insertion Stations (90%). In contrast, SMT Placement Stations are under-utilised (15%).

Figure 6 shows that PTH Operators are over-utilised (91%) and they have over 9,1% of idle time, whereas SMT Operators have the lowest utilisation (5%).

To reach the second objective, the impact of alternative decisions on the performance of the system will be evaluated in the future using an important feature of MPX: a pull down menu dedicated to performing a "what-if" analysis.

5. Conclusions

The RMT technology that models the complex dynamics of manufacturing facilities in terms of queuing theory mathematical equations was applied, through the use of the MPX© software code, for the evaluation and optimisation of the human and economic resources of a PCB manufacturing company.

The MPX© utilisation allowed for the analysis and verification of the manufacturing system capability to meet predefined lead time reduction goals and the finding of opportunities for performance improvements in the production system.

Acknowledgements

This research work was carried out with partial support from the EC FP6 Networks of Excellence on "Innovative Productions Machines and Systems (I*PROMS)" and "Multi-Material Micro Manufacture (4M)".

This paper draws on the thesis work of Gianni Milano whose contribution is acknowledged.

Special thanks are due to Prof. Rajan Suri of the University of Wisconsin-Madison for making available the MPX© software package and for his constant encouragement, precious advice and kind support both during his sabbatical at the University of Naples and from his Laboratory in the USA.

References

[1] Suri R., Wemmerlov U., Rath F., Gadh R., Veeramani R., 1996. Practical Issues in Implementing Quick Response Manufacturing: Insights from Fourteen Projects with Industry, MSOMS Conf., Dartmouth, NH, pp. 56-59.

[2] Suri R., 1998. Quick Response Manufacturing: A Companywide Approach to Reducing Lead Times, Productivity Press, Portland, OR, USA.

[3] Harding L., 2002. QRM: An Enabler on the Road to Agility? Quick Response Manufacturing – Part 1, Control-Journal of the Institute of Operations Management, pp. 20-22.

[4] Harding L., 2002. QRM: An Enabler on the Road to Agility? Quick Response Manufacturing – Part 2, Control-Journal of the Institute of Operations Management, pp. 15-18.

[5] Rehman A., Diehl M.B., 1993. Rapid Modeling Helps Focus Setup Reduction at Ingersole, Industrial Engineering, vol. 25/11, pp. 52-55.

[6] Suri R., 1996. Using Queueing Models to Support Quick Response Manufacturing, International IIE Conference, Minneapolis, MN, pp. 110-115.

[7] MPX© Users Manual, 1999, Network Dynamics Inc., MA, 01803, USA.

[8] Suri R., Diehl G.W.W., de Treville S., 1995. From CAN-Q to MPX©: Evolution of Queueing Software for Manufacturing, Interfaces, vol. 25/5, pp. 128-150

[9] Groover M.P., 2007, Fundamentals of Modern Manufacturing, J.Wiley & Sons.

Innovative Production Machines and Systems
D.T. Pham, E.E. Eldukhri and A.J. Soroka (Eds.)

Sensor Monitoring Based Optimisation during Turning of Titanium Alloys

R. Teti[a], T. Segreto[a], R. Neugebauer[b], C. Harzbecker[b]

[a] *Dept. of Materials and Production Engineering, University of Naples Federico II, Naples, Italy*
[b] *Fraunhofer Institute for Machine Tools and Forming Technology (IWU), Chemnitz, Germany*

Abstract

This paper focuses on the development and application of sensor monitoring of machining processes carried out on difficult-to-machine materials with the aim of process optimisation. In particular, a turning process of a Ti base alloy is considered. The research activities have been carried out by two Laboratories: Machining Technology Laboratory at the Fraunhofer Institute for Machine Tools and Forming Technology (IWU), Chemnitz, Germany, and Laboratory for Advanced Production Technology (LAPT) at the Dept. of Materials and Production Engineering, University of Naples Federico II, Italy. The joint research work included: (a) detection and conditioning of acceleration sensor signals during turning of Ti alloys with variable machining parameters yielding diverse process conditions, and (b) machining process optimisation based on advanced signal processing, feature extraction, and pattern recognition, with the scope to classify Acceptable / Not Acceptable process conditions.

Keywords: Difficult-to-machine materials, turning, acceleration, sensor monitoring, neural networks

1. Introduction

In the aircraft industry, Ti alloys have been widely used because of their low weight, high strength or high temperature stability.

Ti alloys have also seen an growth in demand due to the increased efficiency and higher operating temperatures of aero-gas turbine engines. Dating back to the 1960s, the use of Ti alloys has increased while the use of steel and aluminum has decreased.

With the growing application of Ti alloys, the ability to produce parts with high productivity and good quality becomes challenging. However, these materials are known to be difficult-to-machine [1].

During machining of Ti alloys, tool wear develops in a very rapid way reducing tool life down to few minutes, chip form is continuous and therefore difficult to remove from the machining area, and, in order to fulfil the quality requirements of the product, process parameter values must be kept low against productivity [2]. The optimization of these machining processes is decidedly important in industrial environments but, since reliable machining models are scarce for these materials, process parameters are selected in an empirical way on the basis of skilled operators experience.

The utilisation of sensor monitoring systems for process conditions during machining represents an effective solution to the problem of their optimization and production cost reduction [3 - 8].

In this work, two Laboratories are involved in a collaborative research activity: the Machining Technology Laboratory at the Fraunhofer Institute for Machine Tools and Forming Technology (IWU),

Chemnitz, Germany, and the Laboratory for Advanced Production Technology (LAPT) at the Department of Materials and Production Engineering, University of Naples Federico II, Italy [9].

The main activities of the joint research are:

(i) turning of Ti alloys with detection, and conditioning of acceleration sensor signals;

(ii) cutting process optimisation based on advanced signal processing, feature extraction, and pattern recognition.

The IWU Lab provided the acceleration signals from turning tests of Ti alloys under variable cutting conditions, using commercial instrumentation for sensor signal detection and storage.

The LAPT Lab applied advanced signal analysis procedures including acceleration signal specimen conditioning, feature extraction, and pattern recognition to achieve correlations between sensor data and process conditions.

The principal objective and scope of this work is the reliable and robust classification of Acceptable / Not Acceptable process conditions [10].

2. Experimental procedures

Turning tests were performed at the IWU Lab on a CNC lathe (ZMM CU500) during dry cutting of a difficult-to-machine Ti alloy (TiAl6V4) under the following process conditions (see Table 1):

- cutting speed: v_c = 85, 110, 145, 185 m/min
- feed rate: f = 0.1, 0.15, 0.20 mm/rev
- depth of cut: d = 0.4 mm

The cutting tool had these characteristics:

- coated carbide insert: Kendex TPGN160308
- coating: KC5010 (PVD TiAlN)
- rake angle: $\gamma = 0°$
- clearance angle: $\alpha = 11°$
- no chip breaker

During the cutting tests, the following measurements were carried out:

- three acceleration components a_x, a_y, and a_z using a 3D vibration sensor (ICP 604A61 SN 183) bolted onto the tool holder near the cutting insert.

3. Signal processing methodology

Sensor signal analysis procedures were carried out for the identification of the correlations between sensor data and process conditions during turning of Ti alloys.

Table 1
Test programme: v_c = cutting speed, f = feed rate, d = depth-of-cut.

Test ID	v_c (m/min)	f (mm/rev)	d (mm)
Ti-85-10	85	0.10	0.40
Ti-85-15	85	0.15	0.40
Ti-85-20	85	0.20	0.40
Ti-110-10	110	0.10	0.40
Ti-110-15	110	0.15	0.40
Ti-110-20	110	0.20	0.40
Ti-145-10	145	0.10	0.40
Ti-145-15	145	0.15	0.40
Ti-145-20	145	0.20	0.40
Ti-185-10	185	0.10	0.40
Ti-185-15	185	0.15	0.40
Ti-185-20	185	0.20	0.40

The procedures adopted for the characterization of sensor signals detected during the experimental tests are based on frequency domain analysis.

The sensor signal spectral estimation was carried out through a parametric method allowing for feature extraction from the signal frequency content [11].

Decision making on process conditions acceptability was performed using a supervised expert knowledge based neural network approach [12].

3.1. Signal conditioning

The raw acceleration sensor signals contain signal portions not related to the cutting test under regime conditions. In particular, their initial and final portions, connected with transient conditions at the beginning and the end of the cutting test, need to be removed as they may easily contain spurious information. The spurious signal portion removal was carried out by:

- examination of the raw signal to identify the signal spurious portions;
- selection of the initial and final time windows for spurious signal portion removal;
- visualisation of the conditioned signal for final verification and signal acceptance.

After spurious signal portion removal, the signals presented different lengths expressed in terms of number of samples.

The shortest length, equal to 62500 samples, was selected as the common length of all the conditioned signals. Thus, only the first 62500 samples of all filtered signals were considered for further analysis.

3.2. Signal analysis

The conditioned acceleration component signals are sub-divided into 5 equal parts of 62500/5 = 12500 samples (acceleration signal specimens) for further signal analysis procedures.

The analysis of the acceleration signal specimens was carried out by achieving signal spectral estimation through a parametric method [11].

In this procedure, the signal spectrum is assumed to take on a specific functional form, the parameters of which are unknown.

The spectral estimation problem, therefore, becomes one of estimating these unknown parameters of the spectrum model rather than evaluating the spectrum itself.

From the signal specimen (measurement vector), p features or predictor coefficients $\{a_1, \ldots, a_p\}$ (feature vector), characteristic of the spectrum model, are obtained through linear predictive analysis (LPA).

Feature extraction was implemented through the application of Durbin's algorithm with p = 4, 8, 16. Accordingly, the feature vector extracted from each sensor signal specimen through this procedure is a 4, 8, 16-component vector.

4. Neural network sensor data processing

Decision making on process conditions is performed using supervised neural network approaches applied to:

- single acceleration component signal specimens;
- sensor fusion of the 3 acceleration component signal data.

The quality parameter selected for process condition characterization was the vibration level and the chip type ranked between 1 (good) and 5 (bad) on the basis of expert turning operator knowledge.

The rule-of-the-thumb to classify process conditions as Acceptable or Not Acceptable is the following: whenever a 4 or 5 ranking is verified either for Vibration Level or Chip Type, process conditions are deemed Not Acceptable (see Table 2).

Neural network (NN) sensor data analysis under supervised training allows for pattern recognition in high dimensions feature spaces using the 4-, 8-, and 16-component feature vectors extracted from the acceleration signal specimens.

Three-layer feed-forward back-propagation NN were built with the following configurations (Figure 1).

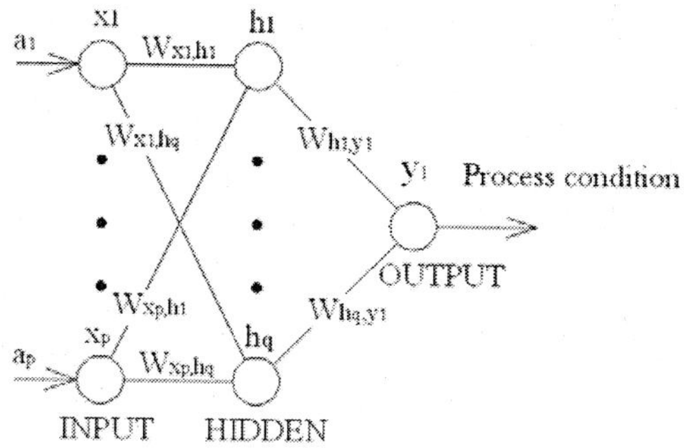

Fig. 1. Neural network configuration.

Table 2
Process condition classification and NN coding. Vibration Level and Chip Type are ranked between 1 (good) and 5 (bad).

Test ID	Vibration Level	Chip Type	Process Condition	NN Coded Value
Ti-85-10	1	2	Acceptable	0
Ti-85-15	2	2	Acceptable	0
Ti-85-20	2	3	Acceptable	0
Ti-110-10	1	3	Acceptable	0
Ti-110-15	2	3	Acceptable	0
Ti-110-20	5	4	Not Acceptable	1
Ti-145-10	1	5	Not Acceptable	1
Ti-145-15	2	3	Acceptable	0
Ti-145-20	4	3	Not Acceptable	1
Ti-185-10	1	3	Acceptable	0
Ti-185-15	1	3	Acceptable	0
Ti-185-20	5	2	Not Acceptable	1

For single acceleration components (a_x, a_y, a_z):

- the input layer has a number of processing elements (nodes) equal to 4, 8 or 16, according to the number of input feature vector components;
- the number of nodes at the hidden layer is 4, 8, 16, 32 or 64, depending on the number of input nodes;
- the output layer contains only one node, yielding a coded value related to the process condition: 0 = Acceptable; 1 = Not Acceptable (see Table 2).

For sensor fusion of acceleration (ax + ay + az):

- the input layer has a number of processing elements (nodes) equal to 4 x 3 = 12 or 8 x 3 = 24, according to the number of input feature vector components;
- the number of nodes at the hidden layer is 12, 24, 32, 48 or 64, depending on the number of

input nodes;

- the output layer contains only one node, yielding a coded value related to the process condition 0 = Acceptable; 1 = Not Acceptable (see Table 2).

Training and testing of the NN was performed using training sets made of the 4-, 8-, 16-component feature vectors, in the case of single component signal data, and the 12-, 24-component feature vectors, in the case of sensor fusion of signal data. The leave-k-out method with k = 1 was applied: one homogeneous group of k patterns, extracted from the training set, was held back in turn for testing and the rest of the patterns was used for training [11].

5. Results and discussion

Signal processing for feature extraction and NN based pattern recognition analysis was performed with the aim to classify Acceptable and Not Acceptable process conditions on the basis of vibration sensor monitoring and measurements.

Experimental sensor signal data were subdivided into 12500 points signal specimens to construct full-bodied training sets comprising a total of 60 training cases including 40 Acceptable and 20 Not Acceptable process conditions cases (see Table 3).

Process conditions classification was carried out through NN sensor data processing of:

- single acceleration components (a_x, a_y, a_z);
- sensor fusion of the three acceleration components ($a_x + a_y + a_z$).

The NN output is correct if the actual output, O_a, is equal to the desired output, O_d, ± 0.50 % of the difference between the numerical codes for Acceptable and Not Acceptable process conditions, which was 1. By setting error $E = (O_a - O_d)$, the process conditions identification is correct if $-0.5 \leq E \leq +0.5$; otherwise, a misclassification case occurs. The ratio of correct classifications over total training cases yields the NN success rate (SR).

In Figure 2, the 4-8-1 NN classification error, E, is reported vs. the number of training cases for:

(a) single ax acceleration component;
(b) single ay acceleration component;
(c) single a_z acceleration component.

In Figure 3, the NN classification error, E, is reported vs. the number of training cases for:

(a) sensor fusion of the 3 acceleration components ($a_x + a_y + a_z$) as obtained through NN configuration 12-12-1;

Table 3
NN training for acceleration signal specimens.

Process condition	Coded value	Number of training cases
Acceptable	0	40
Not Acceptable	1	20

Table 4
NN Success Rate (SR) for single acceleration components (a_x, a_y, a_z).

NN Configuration	Process Condition	a_x	a_y	a_z
4-4-1	Acceptable	97	90	92
	Not Acceptable	75	60	50
	Global	90	80	78
4-8-1	Acceptable	97	92	95
	Not Acceptable	75	65	60
	Global	90	83	83
4-16-1	Acceptable	100	92	92
	Not Acceptable	65	50	60
	Global	88	78	82
8-8-1	Acceptable	95	85	85
	Not Acceptable	70	85	65
	Global	87	85	78
8-16-1	Acceptable	80	90	87
	Not Acceptable	60	70	70
	Global	73	83	82
8-32-1	Acceptable	87	87	85
	Not Acceptable	80	70	65
	Global	85	82	78
16-16-1	Acceptable	82	90	82
	Not Acceptable	75	70	70
	Global	80	83	78
16-32-1	Acceptable	87	82	90
	Not Acceptable	70	80	55
	Global	82	82	78
16-64-1	Acceptable	90	82	80
	Not Acceptable	70	80	50
	Global	83	82	70

(b) sensor fusion of the 3 acceleration components ($a_x + a_y + a_z$) as obtained through NN configuration 24-32-1.

The corresponding SR values are reported in the captions of Figures 2 and 3.

In Tables 4 and 5, the NN SR for Acceptable, Not Acceptable, and Global (= all cases) process conditions identification is reported for:

- single acceleration components (a_x, a_y, a_z) (see Table 4);
- sensor fusion the 3 acceleration components ($a_x + a_y + a_z$) (see Table 5).

Table 5
NN Success Rate (SR) for sensor fusion of the three acceleration components ($a_x + a_y + a_z$).

NN Configuration	Process Condition	$a_x + a_y + a_z$
12-12-1	Acceptable	100
	Not Acceptable	75
	Global	92
12-24-1	Acceptable	97
	Not Acceptable	65
	Global	87
12-32-1	Acceptable	100
	Not Acceptable	65
	Global	88
12-48-1	Acceptable	92
	Not Acceptable	65
	Global	83
24-24-1	Acceptable	92
	Not Acceptable	85
	Global	90
24-32-1	Acceptable	100
	Not Acceptable	100
	Global	100
24-48-1	Acceptable	92
	Not Acceptable	90
	Global	91
24-64-1	Acceptable	95
	Not Acceptable	90
	Global	93

In Figure 4, the best NN SR values are reported for each input feature vector dimension: 4, 8 and 16 input vector sizes for single signal components; 12 and 24 input vector sizes for sensor fusion of the three signal components.

From Tables 4 and 5 and Figure 4, the following considerations can be done.

The NN SR values for Acceptable process conditions identification are always significantly higher than for the Not Acceptable process conditions.

The NN SR values for Global process conditions classification are typically intermediate between Acceptable and Not Acceptable process condition.

The NN SR for Global and Acceptable process conditions identification are always higher than 70% - 80%, respectively, representing very interesting classification performance values (see Tables. 4 and 5).

The NN SR for Not Acceptable process conditions identification can be, in the worst monitoring cases, as low as 50% representing undesirable classification performance values (see Tables 4 and 5).

In the best monitoring cases (see Figure 4), these minimum NN SR values are still rather low in the range 60%, representing barely sufficient classification performance values.

This is due to the fact that the number of training cases for Not Acceptable process conditions is only 50% of the number of training cases for Acceptable process conditions (see Figures 2 and 3): the low number of Not Acceptable cases in the NN training sets is responsible for insufficient NN learning and consequent higher numbers of Not Acceptable misclassifications cases.

A way to improve this inconvenient is represented by the generation of more balanced training sets that should include an approximately equal number of cases for all types of process states.

This could be achieved by extending the experimental testing programme.

It is interesting to note that, by resorting to sensor fusion of the 3 acceleration components, the NN SR values tend to increase also for Not Acceptable process conditions identification.

This emphasizes the NN capability to efficiently realize the concept of sensor fusion [13 - 14] as well to deal with incomplete or noisy data sets, yielding satisfactory results also under adverse situations by synergically combining the knowledge extracted from multiple sources of information [15].

6. Conclusions

Turning tests were carried out on a difficult-to-machine Ti alloy (TiAl6V4) under diverse cutting conditions.

Sensor monitoring of process conditions was performed through acceleration sensor signal detection and analysis.

Decision making on process conditions acceptability was obtained through supervised expert knowledge NN pattern recognition procedures.

The obtained results showed that the NN SR for Acceptable process conditions identification are always significantly higher than for Not Acceptable process conditions.

By resorting to sensor fusion of the 3 cutting force components (ax + ay + az) instead of using single cutting force components (ax, ay, az), the NN SR tend to increase also for Not Acceptable process conditions identification.

This emphasizes the NN capability to realize the concept of sensor fusion.

Acknowledgements

This research work was carried out with support from the CRUI-DAAD Italy/Germany "VIGONI" Project on Forming and Cutting of Special Alloys for Highly Precise Hollow Shapes, the EC FP6 NoE on Innovative Productions Machines and Systems "I*PROMS" and the EC FP6 NoE on Multi-Materials Micro Manufacture "4M".

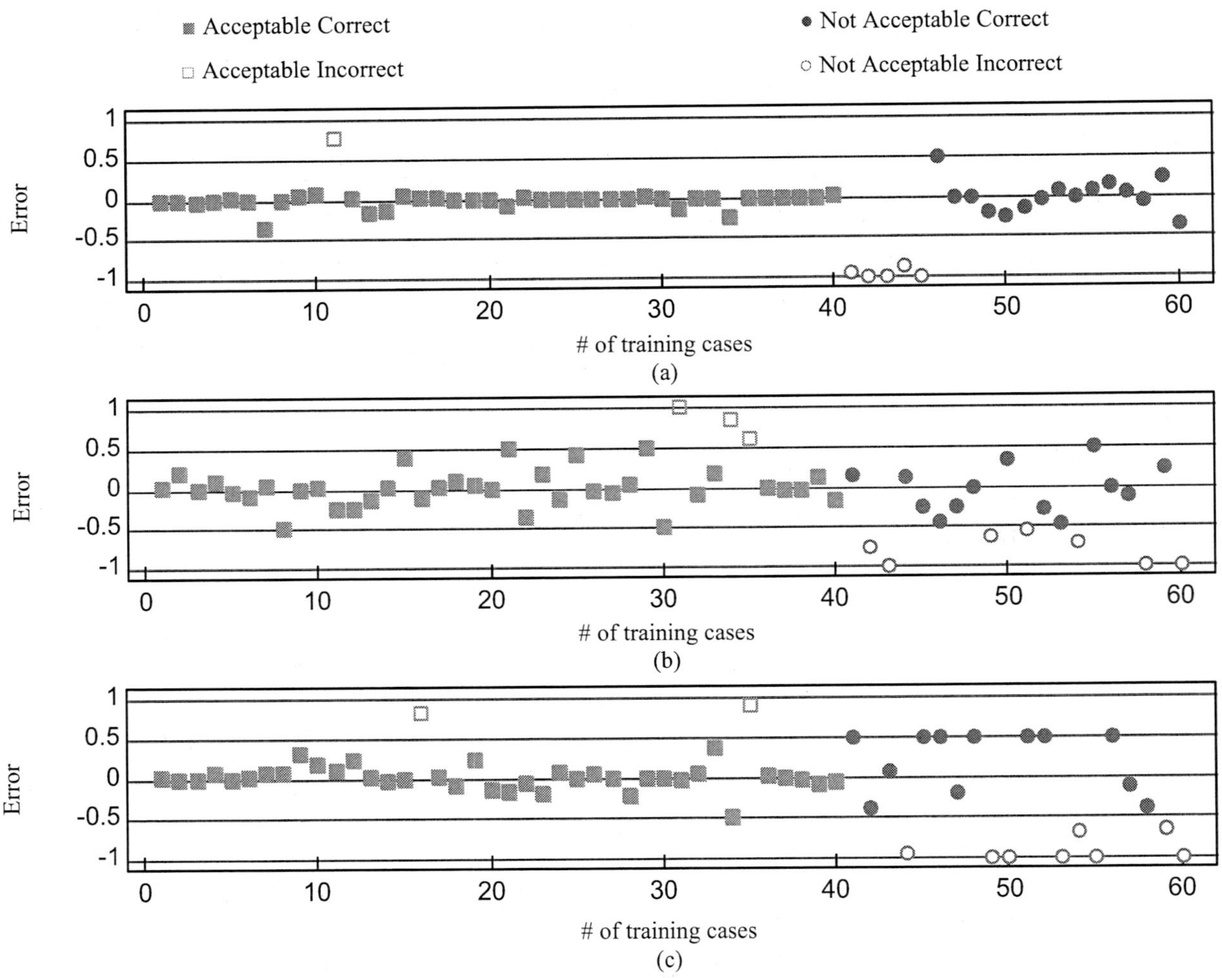

Fig. 2. NN classification error versus number of training cases for:
(a) single a_x acceleration component (SR = 90 %, NN configuration 4-8-1);
(b) single a_y acceleration component (SR = 83 %, NN configuration 4-8-1);
(c) single a_z acceleration component (SR = 83 %, NN configuration 4-8-1).

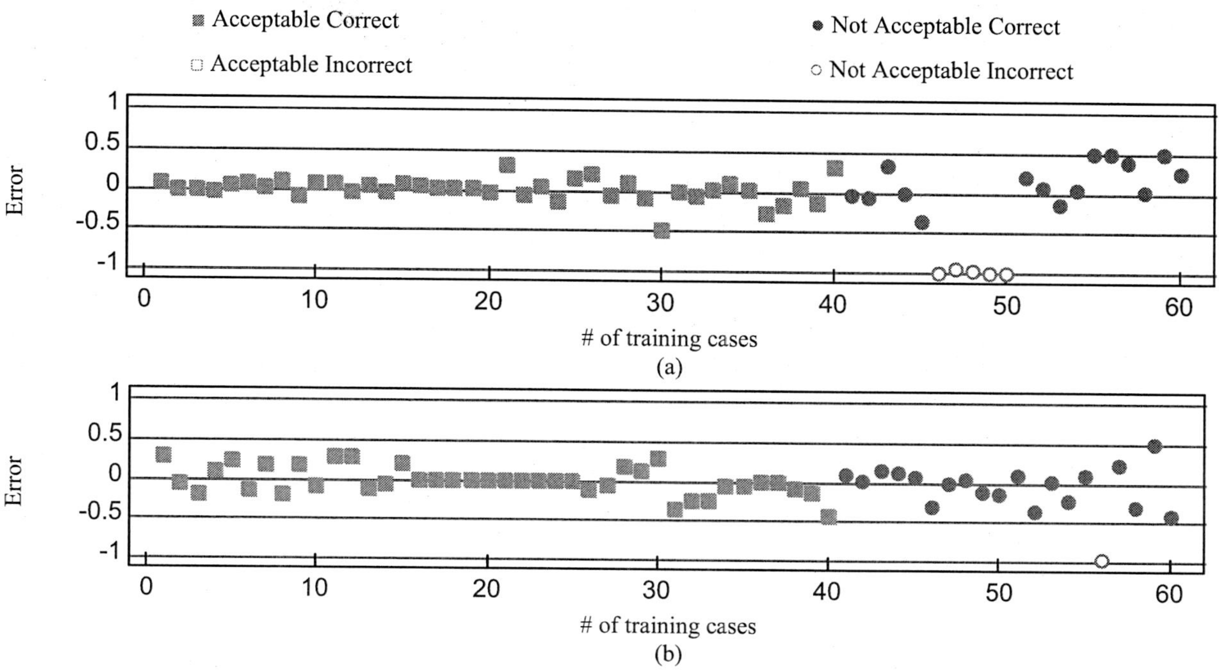

Fig. 3. NN classification error versus number of training cases for:
(a) sensor fusion of the 3 acceleration components ($a_x + a_y + a_z$): SR = 92 %, NN configuration 12-12-1;
(b) sensor fusion of the 3 acceleration components ($a_x + a_y + a_z$): SR = 100 %, NN configuration 24-32-1.

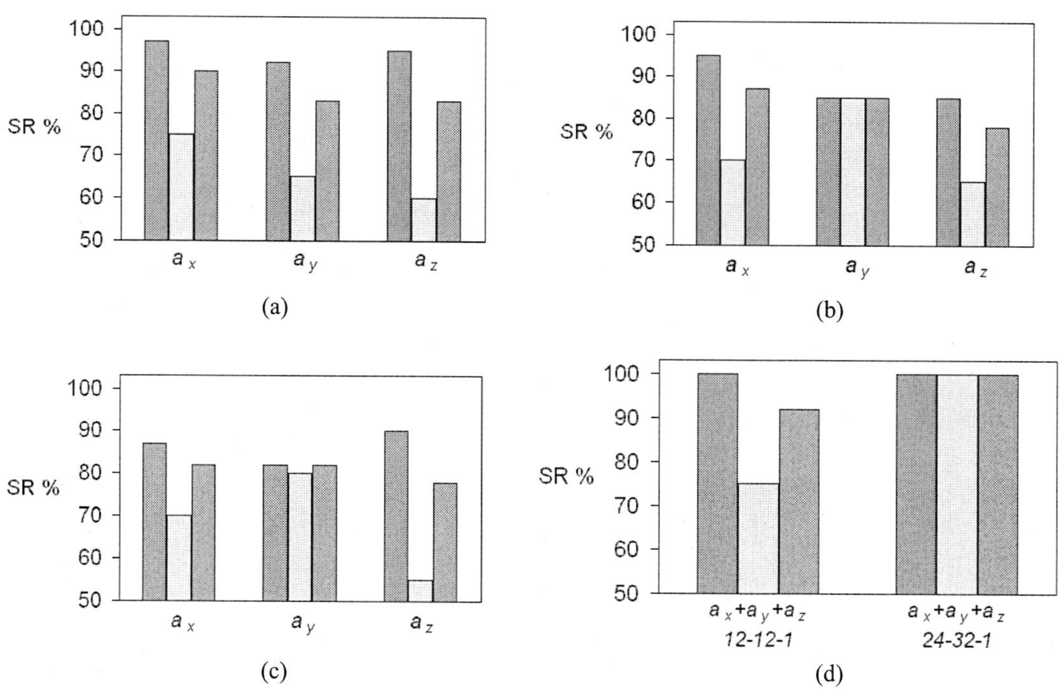

Fig. 4. The best NN SR values are reported for: (a) NN configuration 4-8-1; (b) NN configuration 8-8-1;
(c) NN configuration 16-32-1 for single acceleration components (a_x, a_y, a_z);
(d) NN configuration 12-12-1 and 24-32-1 for the 3 acceleration components (a_x+a_y+a_z).
Acceptable, Not Acceptable, Global

References

[1] Rahman, M., Wang Z.G., Wong, Y.S., 2006, A Review of High-Speed Machining of Titanium Alloys, JSME Int. J. Ser. C: Mech. Syst. Mach. Elem. Manuf., Vol. 49: 11-20.

[2] Teti, R., 2007, Difficult to Machine Materials, 2nd Mftg. Eng. Soc. Int. Conf. – CISIF/MESIC '07, 9-11 July, Madrid, CD ROM.

[3] Segreto, T., Teti, R., 2007, Applications of Intelligent Sensor Monitoring for Machining Processes, 8th Int. Conf. on Lamdamap, 25-28 June, Cardiff: 388-399.

[4] Teti, R., Jawahir, I.S., Jemielniak, K., Segreto, T., Chen, S., Kossakowska, J., 2006, Chip Form Monitoring through Advanced Processing of Cutting Force Sensor Signals, Annals of CIRP, 55/1: 75-80.

[5] Rubio, E.M., Teti, R., Baciu, I.L., 2006, Advanced Signal Processing in Acoustic Emission Monitoring Systems for Machining Technology, 2nd Int. Virtual Conf. on IPROMS, 3-14 July: 189-192.

[6] Teti, R., Baciu, I.L., Rubio, E.M., 2005, Sensor Monitoring of Machining based on Audible Sound Detection and Analysis, 18th Int. Conf. on ICPR, Salerno, 31 Jul. - 4 Ago, CD ROM.

[7] Teti, R., Baciu, I.L., 2004, Neural Network Processing of Audible Sound Signal Parameters for Sensor Monitoring of Tool Conditions, 4th CIRP ICME Int. Sem., Sorrento, 30 June - 2 July: 385-390.

[8] Teti, R., Buonadonna, P., Nele, L., Dentale, F., 1999, Monitoring of Grinding Machines through Vibration Sensors, Int. Conf. on Quality Mftg., Stellenbosch, 13-15 Jan.: 109-117.

[9] Teti, R., Segreto T., Neugebauer, R., Harzbecker, C., 2008, Process Acceptability in Turning of Titanium Alloys Based on Cutting Force Sensor Monitoring, 3rd CIRP International Conference on High Performance Cutting, Vol. 1: 241-250.

[10] Ezugwu, E.Q., Wang, Z.M., 1997, Titanium Alloys and Their Machinability", Int. J. Mater. Process. Technol., Vol. 68: 262-274.

[11] Teti, R., Buonadonna, P., 1999, Round Robin on Acoustic Emission Monitoring of Machining, Annals of CIRP, Vol. 48/3: 47-69.

[12] Segreto, T., Andreasen, J. L., De Chiffre, L., Teti, R., 2005, Chip Form Monitoring in Turning Based on Neural Network Processing of Cutting Force Sensor Data, 1st I*PROMS Virtual International Conference on Intelligent Production Machines and Systems, IPROMS 2005, 4-15 July: 609-614.

[13] Haykin, S., 1998,Neural Networks: A Comprehensive Foundation, Pearson Books.

[14] Hall, D.L. and Llinas, J., 2001, Handbook of Multisensor Data Fusion, CRC Press LLC, USA.

[15] Rafiq, M. Y., Bugmann, G., Easterbrook D.J., 2001, Neural network design for engineering applications, Computers & Structures, Vol. 79/17: 1541-1552.

Innovative Production Machines and Systems
D.T. Pham, E.E. Eldukhri and A.J. Soroka (Eds.)

Spectrum estimation and processing of cutting force sensor signals for chip form monitoring and classification

D. D'Addona[a], A. Keshari[b], R. Teti[a], A. De Maio[a]

[a] *Department of Materials and Production Engineering, University of Naples Federico II, Italy*
[b] *Department of Manufacturing Engineering, NIFFT, Ranchi, India*

Abstract

Dealing with a sensor controlled longitudinal turning operation of carbon steel, yielding different chip forms; authors have presented advanced sensor signal analysis for chip form classification and its identification during turning operation. The cutting force sensor signals have been obtained through the spectrum estimation. For all cutting force signal and associated chip form, a set of features, corresponding to the characteristic parameters of the spectrum model, are obtained by linear predictive analysis. To classify the chip form, decision making and analysis of sensor signal features are performed using an unsupervised neural network methodology based on Kohonen maps.

Keywords: Chip form classification, Sensor monitoring, Neural networks

1. Introduction

Achieving the higher material removal rates with high degrees of automation in truly untended manufacturing systems has always been an important issue of research. To design the high degrees of automation, manufacturing units requires the development of very reliable machining processes. One of the aspects notably affecting the efficiency of machining processes is the monitoring and control of the chip form. The normal variations of process conditions, due to the inherent variation of work material properties, thermal expansion, tool wear development, etc., can produce changes of the chip form or shape during a machining operation. Problems in surface finish, work-piece accuracy and tool life can be caused by even minor changes in chip forms. Moreover, unacceptable chip shapes can cause injuries to operators and damage to cutting tools, work-pieces and machines.

All these negative effects can lead to additional costs due to scrap parts, lost machining time, and delays in part delivery.

Efficient chip form monitoring and control is therefore needed to allow for the formation of favourable chip shapes that can be easily and reliably evacuated from the working zone. This would contribute to the improvement of machining process reliability, the production of high quality machined surfaces, the increase of productivity, and the enhancement of operation safety (including operator safety) and tools and machines protection [1].

Several systems for sensor based chip form monitoring have been proposed. The main physical principles used are: cutting force analysis, chip-tool contact conditions determination, chip flow direction assessment, hot chip infrared emission, and acoustic emission [2- 13].

In this work, sensor monitoring of chip form during turning of carbon steel was carried out using cutting force sensor signals and advanced analysis for signal spectrum characterization and feature extraction.

The analysis of sensor signal features is performed by graphical examination through pair-wise plotting in a 2-D feature space and neural network processing in a higher dimension feature space in order to identify the chip form through a cutting force sensor monitoring approach.

Description of experimental setup and detail about the material and turning operation are described in section 2. However, section 3 provides a more detailed description about the utilization to sensor signal and data extraction. Section 4 describes the 2-D graphical representations and its inferences.

The chip form identification carried out through an unsupervised neural network (NN) paradigm based on Kohonen or self-organizing maps (SOM) has been illustrated in section 5; and finally section 6 concludes the work by pointing out the specific contribution of the proposed methodology [14].

2. Materials and experimental procedures

The Cutting force Sensor Signals and the spectrum have been collected from the activities of the STC-C CIRP Co-Operative Work on "Round Robin on Sensor Monitoring of Chip Form during Machining" [15]. In particular, sensor data were gathered during turning tests generating different chip forms, classified according to ISO 3685 standard annex G as: snarled, short, and short spiral (Table 1) [16].

The three components of the cutting force (Fc, Ff, Fp) were measured using a 3-channel piezoelectric dynamometer. The cutting force component signals were digitized at 2500 Hz for 3 s providing a data sequence of 7500 points. The cutting parameters were:

- Cutting speed = 150, 250 m/min
- Feed = 0.10, 0.20, 0.30, 0.35 mm/rev
- Depth of cut = 1.0, 1.2, 1.3, 1.4 mm.

3. Signal analysis of cutting force signal specimens

The analysis of cutting force signal (CFS) specimens is carried out by achieving spectral estimation through a parametric method [17].

In this procedure, the signal spectrum is assumed to take on a specific functional form, the parameters of which are unknown. The spectral estimation problem, therefore, becomes one of estimating these unknown parameters of the spectrum model rather than the spectrum itself. From the (measurement vector), p features or predictor coefficients (feature vector), {a1, ..., ap}, characteristic of the spectrum model, are obtained through linear predictive analysis (LPA). Feature extraction is executed through the application of Durbin algorithm and the p value is chosen by examining the plot of the normalized RMS prediction error vs. the order of the model [17].

4. Chip form identification through graphical examination

The 4-component feature vectors extracted from the available signal specimens were utilized in pattern recognition procedures aiming at recognizing the chip form through cutting force sensor signal analysis based on graphical data examination.

No single feature of the 4-component pattern vector allows for chip form identification. Figure 1 shows that overlapping of data points occurs in all cases for at least two of the four groups identifying the different chip forms. Thus, pattern recognition was attempted in a higher dimension feature space. Graphical representation of data in a 4-dimension feature space is not feasible. Therefore, pairwise plotting of feature vector components was used to present the data in a 2-D feature space. Figure 2 reports the pairwise plots for longitudinal turning of carbon steel work material yielding different chip forms.

Table 1.
ISO 3685 chip form classification.

Cutting		Favourable	Unfavourable	
Straight	1 Ribbon Chips	1.2 Short		1.1 Long/1.3 Snarled
Mainly up curling	2 Tubular Chips	2.2 Short	2.1 Long	2.3 Snarled
	3 Spiral Chips		3.1 Flat/3.2 Conical	
Mainly Side curling	4 Washer type Chips	4.2 Short	4.1 Long	4.3 Snarled
Up and Side Curling	5 Conical Helical Chips	5.2 Short	5.1 Long	5.3 Snarled
	6 Arc Chips	6.2 Loose/6.1 Conn		
	7-8 Natural Broken Chips	7 Elemental		8 Needle

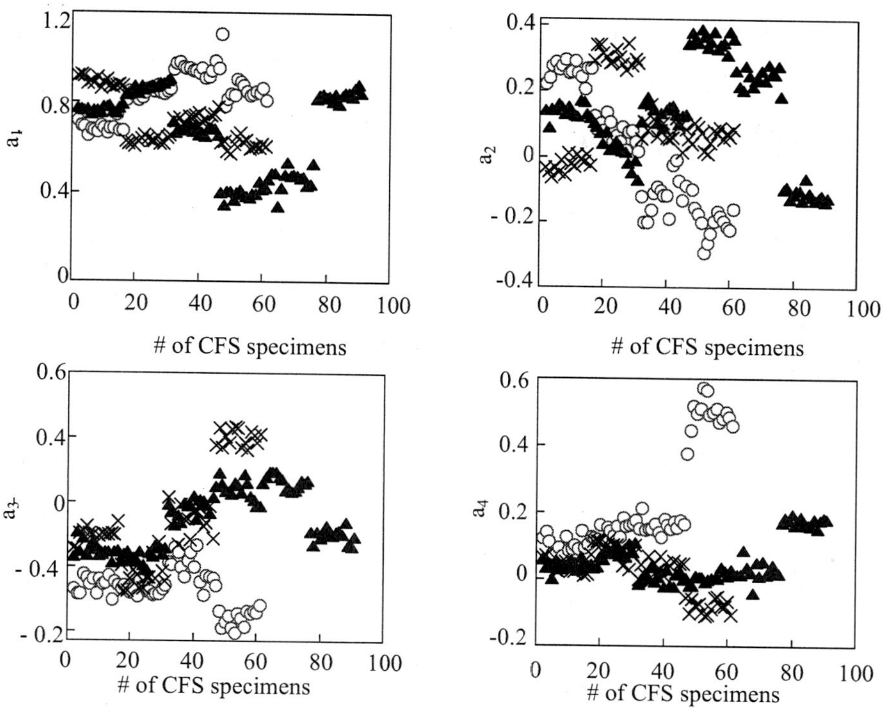

Fig. 1. Plots of single predictor coefficients a1, ..., a4 for all chip forms. O = Snarled , ▲ = Short , ×= Short spiral

5. SOM NN data processing

Exploratory data analysis through graphical "detective" work is an excellent tool to elucidate as much as possible the information contained within the data. However, numerical, statistical or intelligent computing methods of data classification are probably more attractive for in-process, real time monitoring of manufacturing operations. It is also worth noting that, although the available feature vectors contain 4 components, graphical data analysis could only be carried out in a 2-dimensional feature space. SOM NN sensor data analysis was utilized to carry out a clustering procedure using the cutting force component, 4-element feature vectors {a1, ..., a4} [17, 18]. The software package used was the SOM MatLab Toolbox.

The SOM architecture comprises the input layer and a two-dimensional map, called competitive layer. Since input layer and competitive layer are fully connected, each unit of the competitive layer represents a vector with the same dimension as the input layer.

A SOM can be considered as a two-dimensional scene in which the observations are classified so that those which share related characteristics are located in the same zone of the map. A SOM map is formed by a grid of neurons, also called nodes or units, which the stimuli are presented to.

A stimulus is a vector of dimension d which describes the case to be classified. In this work, a stimulus is identified by the cutting force component 4-element feature vectors {a1, ..., a4}.Each stimulus, or input vector, is associated with its corresponding encoded chip form (label): short, short spiral or snarled. Each unit of the grid is linked to the input vector (stimulus) by means of d synapses of weight w. Thus, each unit is associated with a vector of dimension d which contains the weights w.

The 4-element input vectors {a1,..., a4} for cutting force Fc components are utilized for training and testing of the SOM NN with optimal configuration. The sequence of 210 rows of 4-element input vectors made up a set of 210 stimuli to be applied to the SOM NN [19, 20]. The input data is in an ASCII file, few lines of which are shown in Table 2. The first line contains the names of the variables. Each of the following lines gives one data sample, beginning with numerical variables and followed by labels (chip form).

After training, it is important to verify whether the SOM NN has properly adapted itself to the training set data. Thus, the SOM NN is verified using the training set. At the end of the verifying phase, the number of input vectors assigned to each node is obtained where each node are leveled associated with its corresponding encoded chip form (figures 3-4). An example of the test procedure is presented in what follows.

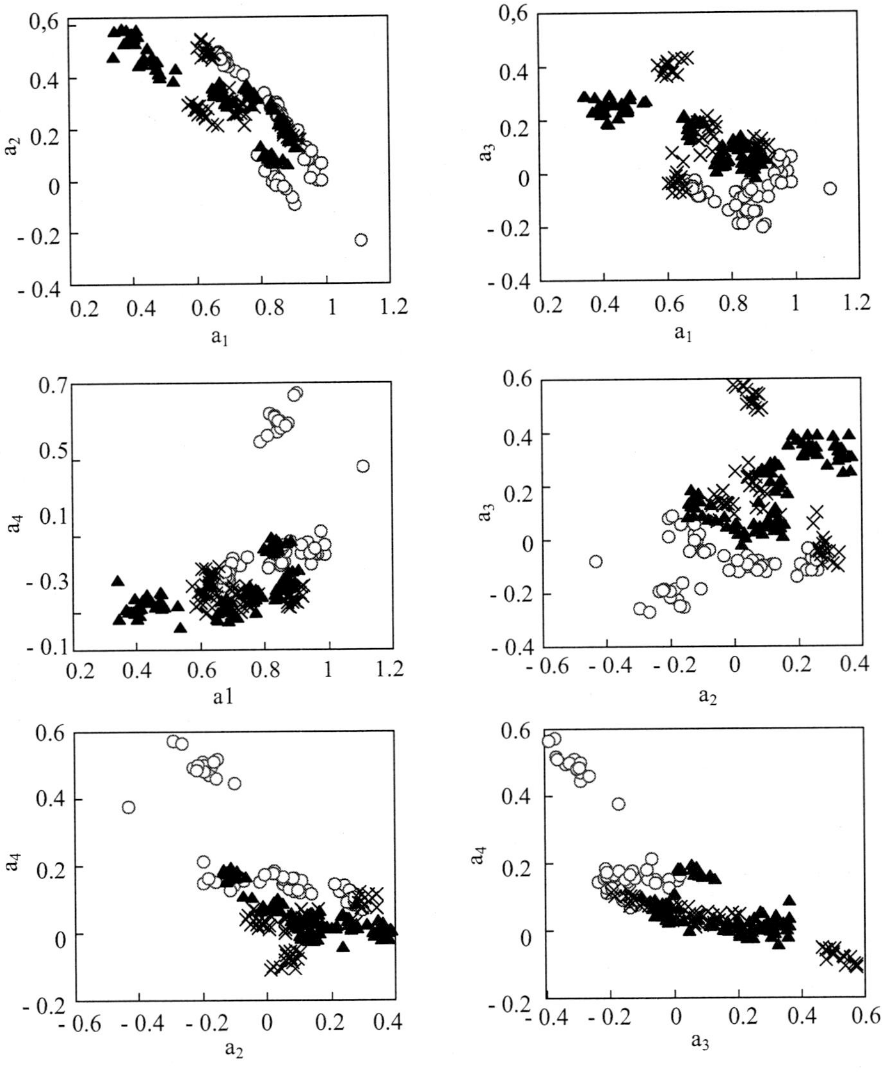

Fig. 2. Plots of ai - aj for all chip forms (O = Snarled , ▲ = Short , ×= Short spiral).

Table 2.
4-element input vectors and corresponding labels
(Sna = Snarled; Sho = Short; Sh.Sp = Short Spiral)

#n	a1	a2	a3	a4	Label
1	0.721	0.222	-0.069	0.124	Sna
2	0.699	0.241	-0.084	0.143	Sho
...	...	...	...	...	...
210	0.765	0.053	0.024	0.032	Sh.Sp

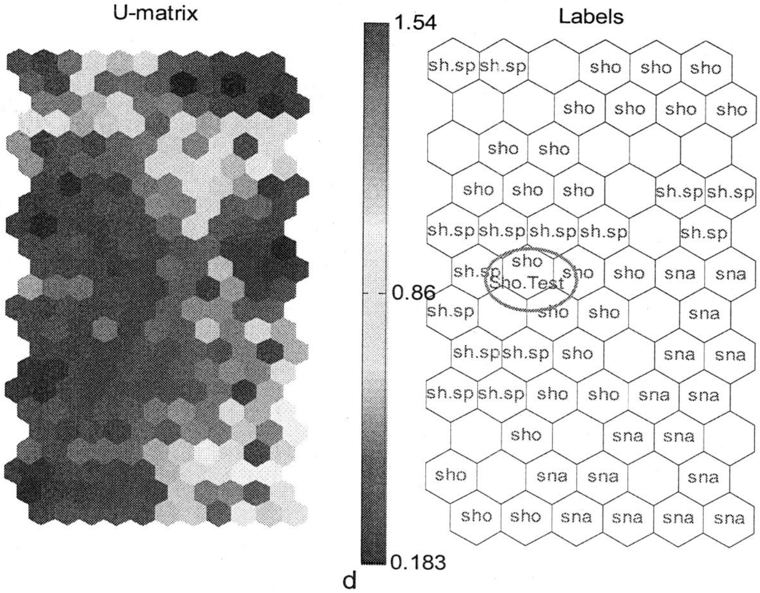

Fig. 3. U-matrix and trained labeled SOM. Cutting force component: Fc.
Labels: Sho = Short; Sna = Snarled; Sh.Sp = Short Spiral. Chip form test : "Sho_Test".

The SOM NN has been trained with all input vectors in the training set except for the test vector which was used to test the classification performance.

Figure 3 is the visualization of SOM for the combined effect of the all signal parameter (a1, a2, a3, a4) over the chip formation (Snarled, Short and Short Spiral chip-form) . Right side of the figure 3, levels are representing as Map level units with its associated levels. However, U-matrix reveals the distinguishing presence of the map units with help of variation of its colour. Colour variation is measure on the scale of Euclidian-distance.

The test results for input test vector (Sho.Test) are pointed out in the figure3. In the figure, "Sho.Test" is correctly lying in right Map unit. Same tests have been performed for several different input test vectors.

For all test instances, the testing vectors are correctly responded onto the relevant map unit corresponding to the related chip form, providing for a 100% success rate in the SOM NN classification of chip form.

The chip form distributions are well separated in the map and therefore this SOM can be the basis for the classification of further unknown chip forms.

6. Conclusions

Sensor monitoring of chip form was performed through cutting force sensor signal detection and analysis. Classification of chip form was obtained on the basis of SOM NN processing of cutting force sensor data.

The obtained results showed that the unsupervised NN approach allows for excellent data group separability and chip form identification.

On this basis, an on-line real time chip form classification, making use of cutting force sensor signal features, could be developed and implemented for industrial applications.

Acknowledgements

The research work at the University of Naples Federico II was carried out with support from the FP6 EC NoE on "I*PROMS" and the MIUR PRIN 2005 "ASMIM" Project. Prof. K. Jemielniak, Warsaw University of Technology, Poland, is gratefully acknowledged for providing cutting force sensor signals within the STC-O CIRP Co-Operative Work "Round Robin on Sensor Monitoring of Chip Form".

References

[1] Byrne, G., Dornfeld, D., Inasaki, I., Ketteler, G., Koening, W., Teti, R., 1995, Tool Condition Monitoring: The Status of Research and Industrial Application, Annals of the CIRP, 44/2, pp. 541-567.

[2] Spaans, C., 1970, A Systematic Approach to Three Dimensional Chip Curl, Chip Breaking and Chip Control, SME Paper, MR 70-241.

[3] Spur, G., Druminski, R. and Mainusch, M., 1979, Automatisches Erkennen von gebrochenen und ungebrochenen Spanformen beim Drehen, Industrieanzeiger, HGF 1321 (59).

[4] Koenig, W., Kluft, W. and Wicklaus, R., 1980, Sensor zur automatischen Spanformerkennung bei der Derhbearbeitung, Industrieanzeiger, HGF 1321, Bericht 80/2 (1).

[5] Nakayama, K, 1984, Chip Control in Metal Cutting, Bulletin JSPE, 18 (2), pp. 97-103.

[6] Dornfeld, D.A. and Pan, C.S., 1985, A Study of Continous/Discontinus Chip Formation Using Acoustic Emission, J. Applied Metalwork, 4 (1), pp. 18-29.

[7] Krause, W., et al., 1985, Infrarotsensor zur Spanbrucherkennung, Feingerätetechnik, Berlin, 3, pp. 99-100.

[8] Sukvittayong, S. and Inasaki, I., 1991, Identification of Chip Form in Turning Process, ISME Series III, 34 (4), pp. 553-560.

[9] Koenig, W., Kutzner, K. and Schehl, U., 1991, Spanformerkennung beim Drehen - Überwachungssystem nutz Korperschallsignale beim Spanbruch, VID-Z, 133 (nr. 6), pp. 121-124.

[10] Moriwaki, T. and Mori, Y., 1992, Sensor Fusion for In-process Identification od Cutting Process Based on Neural Network Approach, Proc. IMACS/SICE Int. Symp. on Robotics, Mechatronics and Manufacturing Systems, Kobe, Japan, pp. 245-250.

[11] Andreasen, J.L., De Chiffre, L., 1993, Automatic Chip-Breaking Detection in Turning by Frequency Analysis of Cutting Force, Annals of CIRP, Vol. 42/1, pp. 45-48.

[12] Teti, R., Buonadonna, P., D'Addona, D., 2001, Chip Form Monitoring through Cutting Force Sensor Signal Processing, V AITEM Conf., Bari, 18-20 Sept., pp. 21-30.

[13] Segreto, T., Teti, R., Applications of Intelligent Sensor Monitoring for Machining Processes, 8th Int. Conf. and Exhibition on Laser Metrology, Machine Tool, CMM & Robotic Performance, 25-28 June 2007, Cardiff, Wales, pp. 388-399.

[14] Kohonen, T. (1999), Self-Organizing Maps, Springer, Berlin, Heidelberg.

[15] Minutes of the STC-O CIRP Co-Operative Work Meeting, CIRP January Meeting, Paris, 24-26 Jan. 2001.

[16] ISO 3685, 1993, Tool-Life Testing with Single-Point Turning Tools, Annex G: 41.

[17] Teti, R., Buonadonna, P., 1999, Round Robin on Acoustic Emission Monitoring of Machining, Annals of CIRP, Part 3, Int. Doc. & Rep., pp. 47-69.

[18] Teti, R., Segreto, T., D'Addona, D., 2004, Unsupervised Neural Network Monitoring of Tool Wear in the Machining of Composite Materials, 4th CIRP Int. Sem. on Intelligent Computation in Manufacturing Engineering – CIRP ICME '04, Sorrento, 30 June - 2 July, pp. 362-368.

[19] Merkl and Rauber, (1997). Alternative ways for cluster visualization in self-organizing maps, Proc. of WSOM'97, Workshop on Self-Organizing Maps, Espoo, Finland, June 4-6, pp. 106-111.

[20] Masters, T., 1993, Practical Neural Network Recipes in C++, Academic Press, San Diego, CA.

Innovative Production Machines and Systems
D.T. Pham, E.E. Eldukhri and A.J. Soroka (Eds.)

A two-phase approach for configuration and location of mobile robots in a hospital setting

H. Baalbaki[a], M. Lamiri[a], X. Xie[a,b]

[a] Department of Healthcare system operations
Ecole Nationale Supérieure des Mines de Saint Etienne, Saint Etienne 42023, FR
[b] *INRIA-Lorraine, Ile du Saulcy, 57045 Metz, FR*

Abstract

In order to equally distribute the workload among a set of reconfigurable mobile robots, this paper addresses the problem of the location and configuration of robots assisting the medical staff in a hospital setting. The robotic system is consisted of a set of identical base robots that can be configured with different functional modules. The system is designed to perform a set of nursing and logistic activities, each requiring a robot with specific modules. The problem consists in determining the configuration of each base robot and assigning each robot to some given home stations from a set of possible location across the hospital floor. In this paper, we propose a two-phase approach that first solves the configuration problem taking into consideration the distribution of the workload and then in a second phase the location problem of robots among a set of potential sites. A linear programming formulation is proposed to find a solution in each phase.

Keywords: Location problems, configuration, robots, linear programming.

1. Introduction

According to the estimation of International Federation of Robotics, the number of operational industrial robots is over one million and will surpass the 1.2 million by 2010. And the number of service robots sold will reach 5.6 millions units. The IWARD project funded by EC explores the use of mobile service robots to attend health personal in their daily activities. IWARD is an on-going three-year project that intends to develop an intelligent robot swarm approach for attendance, recognition, cleaning and delivery for health care systems. The goal of the IWARD project is to develop a system of mid-size mobile robots that can be equipped with additional modules, connected through a failsafe communication network, coordinated by some distributed/hybrid control strategies, and capable of performing different types of missions/tasks in the hospital context and in order to optimize some key performance indicators. Plug and play technologies will be developed such that modules can be easily inserted and removed from base robots by nurses without the need to reprogram the robot or the modules. Figure 1 shows the first IWARD robot prototype.

Based on a preliminary study of hospital requirements and specification, IWARD project choses to target five types of missions/tasks including (i) delivery of medicines, notes, X-rays and personal items, (ii) regular and unplanned cleaning of patient rooms and spillage, (iii) guidance of patient/visitors inside a hospital, (iv) patient monitoring and patient-doctor teleconferencing, (v) hospital surveillance.

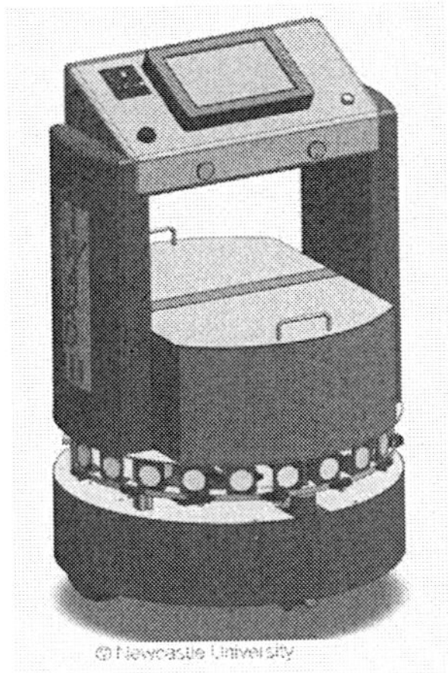

Fig. 1. IWARD robot .

This paper is part of our effort in developing operation strategies of the mobile robot platform for efficient assistance of the day-to-day basic nursing and logistic activities. Decisions raised in the operation of such a robotic system includes (i) configuration and location of different robots according to the needs of the hospital, (ii) power management, (iii) assignment of scheduled and unscheduled missions/tasks to different robots, and (iv) real time control.

This paper focuses on the configuration and location of robots. Such decision usually rises at a shift basis. As a matter of fact, the requirements are not the same during the day and during the night. For example, there are more delivery missions during the day than in the night while there are more patient monitoring missions during the night when there are less nurses in the hospital. As a result, robots should be configured according to the requirements of each shift. During the day, more robots with delivery modules are needed and during the night, more robots with monitoring modules are needed. Further, robots should be located appropriately across the hospital in order to be present at the place of mission quickly when a request for robot is issued. For example, we would locate the delivery robots close to the pharmacy if most delivery missions concern the delivery of medicines.

To the best of our knowledge, the problem of configuration and location of heterogeneous robots has never been addressed in the literature. The most relevant work in the literature includes facility location and ambulance location problems. The facility location problem has been extensively addressed in the literature; typical location models include p-median model, capacitated facility location model, network design model and more recently inventory-location models. Extensive literature surveys can be found in [1, 2, 3, 4, 5]. The application of facility location research to health care systems including ambulance location has led to different location models (see [6] for a survey). The Location Set Covering Model (LSCM) developed in [7] is used to determine the number of emergency service provider needed to cover all demand area. The Maximal Covering Location Problem (MCLP) proposed in [8] involves the notion of coverage that is usually used in the ambulance location problem, where each ambulance can only cover a certain area respecting the maximum time delay needed by the ambulance from its base (home) station to arrive to the demand zone. In this model we tend to put the service provider near the zone that generates a lot of demands. Several formulations emerged fro the initial MCLP like the backup problem coverage BACOP1 and BACOP2 introduced in [9] and the DSM introduced in [10] to solve the location problem using tabu search method. The TEAM model and the FLEET model of [11] are extensions of the MCLP model.

The rest of the paper is organized as follows. Section II describes the problem of configuration of mobile robots and locating their home stations in hospitals. Section III proposes a linear programming model for the configuration problem. Section IV presents a model for the location problem. Section V presents some numerical results. Section VI concludes the paper and discusses possible extensions of this work.

2. Configuration and location of robots

The robotic system considered in this paper is composed of a set N of identical base robots. Each base robot can be configured with additional modules. There is a set M of different types of modules and N_m copies of module m.

Let C be the set of all possible configurations of a base robot. Each configuration $c \in C$ is defined by the set M_c of modules mounted on the base robot. Here we assume implicitly that at most one module of each type is mounted on a base robot. This assumption trivially holds in our application scenarios.

Let B be the set of home stations in which robots can be located. When a task/mission is assigned to a robot, the robot first moves from its home station to the starting point of the task, performs the task and then

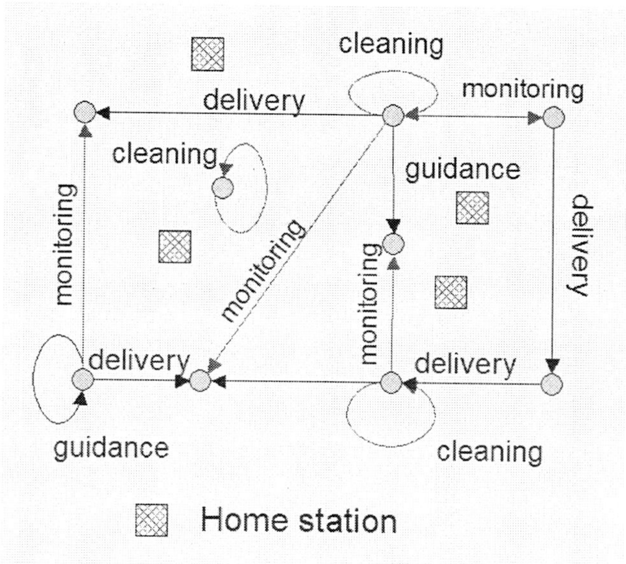

Fig. 2 Task representation.

returns to its home station.

As a result, the problem considered in this paper consists of selecting for each base robot a configuration in the set *C* and then a home location in the set *B*. Constraints and criteria will be described.

The robotic system should be configured to perform a set T of tasks for the upcoming shift. Each task $t \in T$ is characterized by the following data:

f_t: the frequency of task t

T_t: the duration of task t

s_t: the starting position of task t

e_t: the ending position of task t

C_t: the set of configurations having the capability to perform task t.

Figure 2 illustrates a set of tasks, and it clearly shows the difference between the nature and the characteristics of tasks where some tasks start and end at the same point like spillage cleaning, whereas other tasks start and end at different points. The cleaning tasks are typically of the first type while the delivery and guidance tasks belong to the second type.

We further consider the following two notations:

- A_{bt}: time needed for a robot located at home station b to reach the starting point of task t,
- R_{bt}: time needed for a robot to return to its home station b after the completion of task t.

These time durations can be derived from the position of home station b, the starting and the ending positions of task *t* , i.e. s_t and e_t.

As we can see from Figure 3, the execution of a task/mission can be decomposed into 3 periods: the reaction time A_{bt}, the execution time T_t and the returning time R_{bt}. We notice the choice of the home station doest not have any influence on the execution time that is the same for all the possible home stations. And the choice of home station only influences on the

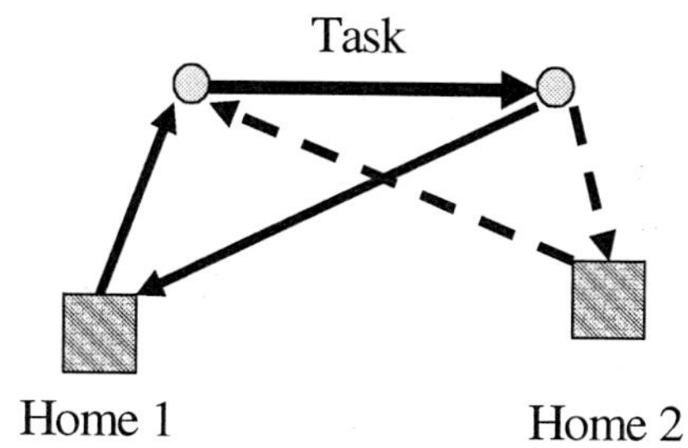

Fig. 3. Task segmentation.

reaction and the returning times. As the robot can sometime directly starts another mission without returning to the home station, the reaction time is the period that interests us the most. And if we simplify the problem by discarding the execution time and the returning back trip we get an extension of the ambulance location problem.

If a task t is assigned to a robot located at home station b, then the total workload of the robot due to task *t* is $f_t \times (A_{bt} + T_t + R_{bt})$. We assume that the maximum allowable workload of a robot is T_{max}.

In this paper, the following assumptions are made.

- All missions should be assigned to the available robots.
- There is no cost for the configuration of robots.
- There is no cost of robot location and relocation.
- Each task is executed by a single robot.
- Each task requires at most one module of each type.
- Each home station b can holds at most H_b robots.

In this paper, the goal is to minimize the total workload of all robots which is equivalent to minimize the total time of empty trips for reaching the starting points of different tasks and for returning to the home stations. Other relevant criteria include (i) minimization of the reaction time, (ii) minimization of the return trips, (iii) workload balancing of different robots, and (iv) minimization of the reaction times of backup robots.

Hereafter we propose a two-phase approach. Phase 1 focuses on the configuration problem without taking into account the location of robots. The goal is to best balance workload of the robots without taking into account the reaction times and returning times. Phase 2 addresses the location of robots with configuration determined in phase 1.

3. Mathematical Model of the Configuration problem

The goal of this section is to provide a formal mathematical formulation of the configuration problem of the robotic system under consideration in order to equally distribute the tasks among the robots.

Two sets of decision variables need to be considered:

$$y_{nc} = \begin{cases} 1, & \text{if robot } n \text{ selects configuration } c \\ 0, & \text{otherwise.} \end{cases}$$

$$z_{nt} = \begin{cases} 1, & \text{if task } t \text{ is assigned to robot } n \\ 0, & \text{otherwise.} \end{cases}$$

The following notation is used to simplify the presentation

$$a_{mc} = \begin{cases} 1, & \text{if a module } m \text{ is needed in configuration } c \\ 0, & \text{otherwise.} \end{cases}$$

Clearly a_{mc} can be easily derived from the set M_c.

The configuration problem of robotic system can be formulated as follows.

$$Minimize\, W \quad (1)$$

subject to:

$$\sum_{c\in C} y_{nc} = 1, \quad \forall n \in N \quad (2)$$

$$\sum_{n\in N} z_{nt} = 1, \quad \forall t \in T \quad (3)$$

$$\sum_{c\in C}\sum_{n\in N} a_{mc} y_{nc} \le N_m, \ \forall m \in M \quad (4)$$

$$\sum_{t\in T} f_t T_t z_{nt} \le W, \quad \forall n \in N \quad (5)$$

$$\sum_{c\in C_t} y_{nc} \ge z_{nt}, \forall t \in T, \forall n \in N \quad (6)$$

$$y_{nc}, z_{nt} \in \{0,1\}, \forall b \in B, \forall t \in T, \forall n \in N \text{ and } W > 0 \quad (7)$$

The criterion (1) is the minimization of the maximum workload assigned to a robot. Constraint (2) ensures that each base robot is assigned to a configuration. Constraint (3) ensures that each task is assigned to one robot. Constraint (4) is due to the availability of modules. Constraint (5) determines the maximal workload W. Constraint (6) ensures that each task is assigned to a robot with appropriate configuration. Constraint (7) is the integrity constraints of the decision variables.

The special case of single configuration is a parallel machine problem and any parallel-machine problem can be transformed into such a special case. As the parallel-machine problem is NP-hard, the configuration of the robotic system NP-hard.

4. Mathematical Model of the Location problem

After mounting the functional modules chosen in phase 1 on base robots, the goal of this phase is to allocated each robot to a certain home station. This section provides a formal mathematical formulation of the location problem of robots.

Two sets of decision variables need to be considered:

$$x_{nb} = \begin{cases} 1, & \text{if robot } n \text{ is located at home station } b \\ 0, & \text{otherwise.} \end{cases}$$

$$z_{nt} = \begin{cases} 1, & \text{if task } t \text{ is assigned to robot } n \\ 0, & \text{otherwise.} \end{cases}$$

Note that by considering z_{nt} as decision variables of phase 2, we choose to reassign tasks to robots by taking into account locations.

In order to obtain a LP formulation, we also need an auxilliary variable $r_{nbt} = z_{nt}x_{nb}$ which is equal to 1 if the task t is assigned to the robot n located at the home station b. Let y_{nc} be the optimal configuration of phase 1.

The location of robotic system can be formulated as follows.

$$Minimize \sum_{n\in N}\sum_{b\in B}\sum_{t\in T} f_t (A_{bt} + T_t + R_{bt}) r_{nbt} \quad (8)$$

subject to:

$$\sum_{b\in B} x_{nb} = 1, \quad \forall n \in N \tag{9}$$

$$\sum_{n\in N} z_{nt} = 1, \quad \forall t \in T \tag{10}$$

$$\sum_{n\in N} x_{nb} \le H_b, \forall b \in B \tag{11}$$

$$\sum_{b\in B}\sum_{t\in T} f_t T_t r_{nbt} \le T_{\max}, \forall n \in N \tag{12}$$

$$r_{nbt} \ge (x_{nb} + z_{nt}) - 1, \forall t \in T, \forall n \in N \tag{13}$$

$$\sum_{c\in C_t} y_{nc} \ge z_{nt}, \forall t \in T, \forall n \in N \tag{14}$$

$$x_{nb}, z_{nt}, r_{nbt} \in \{0,1\}, \forall b \in B, \forall t \in T, \forall n \in N \tag{15}$$

The criterion (8) is the minimization of the total workload of all robots. Constraint (9) ensures that each base robot is assigned a location (a home station). Constraint (10) ensures that each task is assigned to one and only one robot. Constraint (11) guaranties the maximum allowable number of robots at each home station is not exceeded. Constraint (12) guaranties that the maximal allowable workload of a robot is not exceeded where $T_{max} = \alpha W$ with W the maximum workload of phase 1 and $\alpha \ge 1$ is a control parameter. Constraint (13) ensures that if r_{nbt} is equal to 1 then task t is assigned to the robot n located at the home station b. Constraint (14) ensures that each task is assigned to a robot with appropriate configuration. Constraint (15) is the integrity constraints of the decision variables.

This complexity of the location problem of robots can be established through the transformation to ambulance location problem. As the ambulance location problem is strongly NP-hard, the joint location and configuration of the robotic system is strongly NP-hard.

5. Numerical Results

The section presents an illustrative case to highlight the important steps in the configuration phase. The numerical experiments presented in this section were carried out on a 1.7 GHz Pentium 4 PC with the memory of 256 MB., and running Windows XP. The algorithm was implemented in MS Visual C++ .NET 2005 and linked with the COIN CLP solver and COIN CBC library.

The problem instance, illustrated by Figure 4, consists of: 3 possible home stations {H1, H2, H3}, 2 base robots {R1, R2}, 2 transportation modules $N_{transp} = 2$, one cleaning module $N_{clean} = 1$, and 7 cleaning or transportation tasks to be performed. Characteristics of tasks are given in TABLE I. For each task, we provide its type, cleaning {*clean*} or transportation {*transp*} task, and the time duration needed to perform the task (execution time).

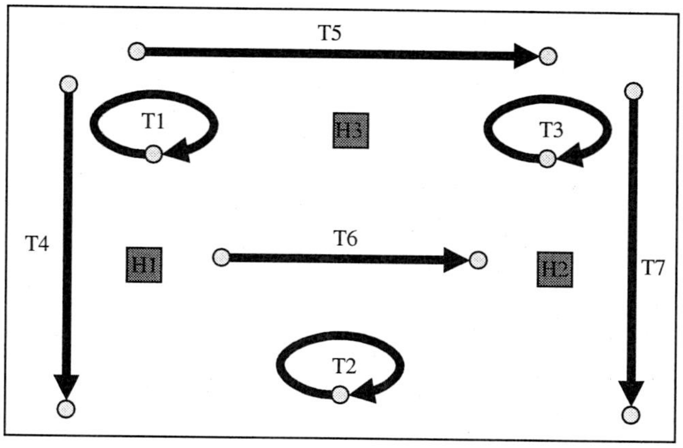

Fig. 4 An illustrative example.

The set of possible configurations of a base robot is C={{*clean*}, {*transp*}, {*clean, transp*}}, i.e., each base robot can be equipped with just one module {*clean*} or {*transp*}, or with two modules {*clean, transp*}.

Table 1
Characteristics of the case study

Task	Type	Duration(seconds)
T1	*clean*	100
T2	*clean*	100
T3	*clean*	100
T4	*transp*	200
T5	*transp*	250
T6	*transp*	150
T7	*transp*	200

The solution of the configuration problem (1)-(7) implemented using the COIN CBC library as MIP solver is presented in table 2.

Table 2
Solution of the case study

Robot	Configuration	Tasks	Workload(seconds)
R1	*Clean, transp*	T1,T2,T3,T5	550
R2	*transp*	T4,T6,T7	550

In this example, the objective of equally distributing the workload among the set of robots was reached and the appropriate configuration was associated to each robot.

6. Further work and conclusion

In this paper, we have considered the problem of locating and the configuring a set of modular and mobile robots in a hospital. The problem consists on determining the configuration, the location, and the set of tasks that would be performed by each robot in order to achieve an efficient assistance with daily logistic and nursing activities.

In this work, the problem has been divided into 2 part where in a first time the problem of finding the appreciate configuration of each robot is addressed, and after reconfiguration of the robot the problem of assigning to each robot a home station on hospital floor is addressed. These 2 problems were modeled as mathematical linear problems and were solved using a mixed integer solver.

The proposed model and the solution approach can be extended to other problems related to modular and mobile service providers providing different types of service to distributed customers.

In a previous article the problem of joint location and configuration model was studied in detail and in the current study the problem was addressed differently and it was modeled as 2 different problems solved sequentially. A issue for future research is to study the direct impact the those 2 strategies on the scheduling and dispatching problems, also the impact of on power management strategies that is a important avenue of our short-term research activities.

Acknowledgements

This work was supported in part by the European Commission under the project IWARD Intelligent Robot Swarm for Attendance, Recognition, Cleaning and Delivery (Contract 045245, 01.01.2007-31.12.2009).

We would also like to thanks our partners from the university of Newcastle for providing the sketch of the IWARD base robot.

References

[1] M. Daskin, C. Coullard and Z. Shen. "An Inventory-Location Model: Formulation, Solution Algorithms and Computational results." Annals of Operations Research, pp. 83-106, 2001.

[2] A. Geoffrion qnd R. Powers, Twenty years of strategic distribution system design: an evolutionary perspective, Interfaces, Vol. 25, pp. 105-128, 1995.

[3] A. Klose and A. Drexl (2005). Facility location models for distribution system design, European Journal of Operational Research, Vol. 162, pp. 4-29, 2005.

[4] C.S. ReVelle and H. A. Eiselt. "Location analysis: A synthesis and survey," European Journal of Operational Research, Vol. 165, pp. 1-19, 2005.

[5] G. A. Tanonkou, L. Benyoucef and X. Xie. "Distribution network design with random demand and supply lead-time: A Lagrangian relaxation approach. " IEEE Transaction on Automation Science and Engineering. 2007, to appear.

[6] Daskin, M. S. and L. K. Dean, "Location of Health Care Facilities," chapter 3 in the Handbook of OR/MS in Health Care: A Handbook of Methods and Applications, F. Sainfort, M. Brandeau and W. Pierskalla, editors, Kluwer, pp. 43-76, 2004.

[7] Toregas, C.S.R, C. ReVelle and L. Bergman. "The location of emergency service facilities," Operations Research, 19, 1363-1373, 1991.

[8] Church, R.L., ReVelle, C.S.. "The maximal covering location problem," Papers of the Regional Science Association, 32, 101-118, 1974.

[9] Hogan, K., ReVelle, C.S, "Concepts and applications of backup coverage," Management Science, 34, 1434-1444, 1986.

[10] Gendreau , M., Laporte, G., Semet, F, "Solving an ambulance location model by Tabu search," Location Science, 5, 75-88, 1997.

[11] Schilling, D.A., Elzinga, D.J., Cohon, J., Church,R.L., Revelle, C.S., "The TEAM/FLEET models for simultaneous facility and equipment sitting," Transportation Science, 13, 192-200, 1979.

Innovative Production Machines and Systems
D.T. Pham, E.E. Eldukhri and A.J. Soroka (Eds.)

Development of a surveillance module for hospital robots

K. A. Mamun, S. Rayhan, R. Sherry, T. Szecsi

School of Mechanical and Manufacturing Engineering
Dublin City University, Dublin 9, Ireland

Abstract

This paper specifies the requirements of sensors for a surveillance module for hospital robots of the IWARD project, identifying the responsibilities for each of these requirements and their application. The IWARD swarm robot is based on an autonomous multi-purpose platform. A modular design philosophy will be adopted to ensure compliance with requirements established within the project, and to allow for ease of modification and/or adaptation, as the project progresses. This includes necessity of different sensors and their specification which will be used for the surveillance modules. The actions to be carried out in the project will ensure the development and establishment of a cost effective solution and a safe and dependable, robust design. Different sensors from each category are compared and selected for the project. The sensors are to be selected based on the limitations and requirements such as data communication, selected operating system and software (Linux, player and orca 2).

Keywords: Requirements, Surveillance module, Sensors and Application

1. Overview of the IWARD project

IWARD stands for 'Intelligent Robot Swarm for Attendance, Recognition, Cleaning and Delivery'. According to the IWARD project specification all sensors should be compatible with the Player and Orca2 software. The IWARD project has two main goals:

1. Development of a robust mobile robotic platform within Service Robotics.
2. Application of the developed platform to form a swarm of robots, in order to execute the functions defined in the scenario specifications.

Selection of sensors is an important part of this project to accomplish the above goals. The IWARD swarm robot is to be based on an autonomous multi-purpose platform. This is to include all sensors and their specification. To summarize, the actions to be carried out in the project will ensure the development and establishment of: a cost effective solution, a safe and dependable, robust design. Finally the current research compares and selects different sensors from each category according to availability, compatibility, feasibility, cost effectiveness and performance for the surveillance module. There are few limitations, such as cost, interface with Linux platform and others.

2. Sensors

A critical point here is the proper selection of sensors for the surveillance module, as it will greatly influence if it can properly accomplish the tasks it must perform. Additionally, selection of sensors will be influenced by the functional requirements of the robot modules (a typical scenario of this surveillance module is detailed in the following section).

If we classify sensing according to its purpose, we can say that sensors can serve for 'proprioception (estimation of the internal state of the robot, such as configuration, temperature, current, speed of axis and so on) or 'exteroception' (estimation of the state of the environment with respect to robot). On the other side, sensors are also divided according to measurement principle, and so they can be: 'passive' (use ambient energy to perform the measurement) or 'active' (transmit energy into environment to allow measurements). Sensor fusion is one of the main challenges in this current research.

2.1. Scenario: surveillance (Routine monitoring of the environment) modules

In this scenario the moving robot equipped with a surveillance module performs routine monitoring of the parameters of the environment: ambient temperature, humidity, smoke/fire detection. The module cyclically reads the sensors and based on the current location of the robot it evaluates the sensor data using the database at the nurses' terminal. The database contains the allowable ranges of the environmental parameters for each location. In case of detected smoke an alarm signal is generated. If the temperature and/or humidity level is out of range the module informs the nurses' station about it. Another task of the surveillance module is intruder monitoring. In case of motion detection by a PIR (Passive infrared) sensor it triggers the camera and after analysing the image of the source of motion the system could take a decision whether the moving person is an intruder or a legal person in the ward. If an intruder is detected the system informs the nurses' station and stores the event in an event log. A distance measurement sensor gives another parameter (distance of the person) which helps to take self (Robot itself) protection (not to be damaged by an intruder). This mobile platform based (i.e. Robot base) sensors will provide dynamic data in the event of condition and surveillance monitoring task compare to approach of using stationary sensors (i.e. Traditional approach).

3. Sensors requirement for each individual Module

Based on the scenarios, the following sensors and hardware are necessary to accomplish the specific tasks of surveillance module:
•Camera (Including PTZ mechanism)
•Distance Measurement sensor (i.e. Laser Scanner)
•Motion detector (PIR sensor)
•Temperature and Humidity sensors
•Fire and Smoke detectors
•DAQ board
•Unique color LED/ Self reflective surface (Identifying and Image Analysis purpose)

3.1 List of different sensors for each category and their specification

For each type of sensor an extensive literature survey was carried out during this research. In relation to this project, the most appropriate sensors of each category are listed in this report. During sensor selection few factors were accountable which are mentioned below:

- Type of interface/communication (i.e. USB, Serial, Parallel, Wifi/Bluetooth and so on)
- Software compatibility (i.e. Player, Orca2 and Linux Ubuntu OS supported)
- Requirement of Power supply (Preferred 'low')
- Output/Input signal (i.e. voltage, current)
- Cost effective (Lower price)
- Capacity or Range of each sensor (Minimum and Maximum, accuracy, resolution and so on)

4. Surveillance module

General tasks of the Surveillance module are as follows:
(A) Detect a motion within the range (a PIR sensor is able to do this task).
(B) Identify the source and distance of the motion using a laser scanner.
(C) Verify that the source of motion is either an allowed person or an intruder (using a camera and image analysis software; where an allowed person will use a badge made of unique coloured reflective surface of a unique coloured LED). In the event of intruder, the camera will take either a still image or stream a video clip and store it for later analysis (by human interaction).
(D) For condition monitoring (i.e. Fire/smoke, temperature and humidity control) the following sensors are able to do the required task; a fire/smoke detector, a temperature and humidity sensor.

In order to detect that a person / intruder is entering the hospital environment, numerous technical solutions based on a large variety of physical phenomena are available and are still being refined. Apart from the widespread passive infrared (PIR) detectors, glass-break sensors and vibration shock sensors, active systems using infrared light, ultrasound or microwaves are used. Also, video cameras are widely applied in security systems. Various access control systems including electronic keys (transponders, touch memory (iButton®) and systems for biometrical identification (fingerprint recognition, handwriting recognition, etc.) are developed. For the user a maximum sensitivity and reliability is most important since missing alarms question the purpose and false alarms are costly and reduce the system's credibility. With a multisensor approach the false-alarm probability can be reduced significantly. Other important issues are the detection range of the devices, their ability to distinguish objects due to their size, form or distance to the sensor and, last but not least, their cost [1].

The following section includes detailed information of all sensors which are essential to achieve the goals of this project (Surveillance Module). For each types of sensor different options are reviewed and from those ranges of selection the best one was suggested for this project; considering that it will be able to satisfy the primary objective. Some changes have occurred during purchasing of these sensors due to budgetary reasons. Sensors of each category which will be used in the current research are listed in the following section.

4.1. Motion detector (PIR sensor)

KC7786 is a PIR sensor module [2] which is very small in size (27 square mm) with built in amplifier and logic circuit. The fresnel lens is included for the detection of human body movement at a distance of 5 meters. This is suitable for automatic illumination systems and intruder alarm systems.

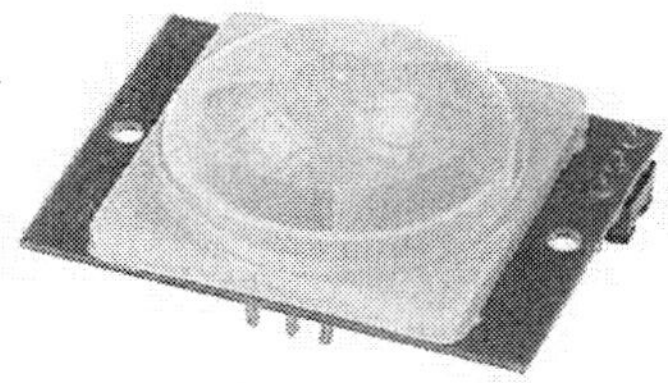

Figure 1: PIR sensor module (KC7786) [2]

Features and electrical specifications

The main features of the selected PIR sensor are single voltage operation (12 VDC), open collector output, 3 pin terminal, high sensitivity with special noise cancellation design, high immunity to RFI, delay to get signal after 25 sec when power up, signal output 0.5 sec minimum, and low operational power consumption.

4.2. Distance measurement sensor

With the efector pmd 3d [3], ifm electronic is entering the next dimension of optoelectronics. Efector pmd 3d (Fig. 2) is the first industrial 3D sensor that can detect objects in three dimensions at a glance. Efector pmd 3d is based on the technology of time of flight measurement. The measurement and the evaluation of the time of flight are integrated on one chip. The dimensions of the extraordinarily compact sensor housing are just 60 x 42 x 35 mm. The resolution of the sensor is 64 x 48 pixels. Each pixel of this matrix evaluates its distance to the object. This results in 3072 distance values at the same time. The image of the object on the matrix and the respective distance values correspond to a 3D image. The geometrical characteristics of the object result from these values.

Figure 2: PMD 3D sensor [3]

Main characteristics:

Resolution: 64 x 48 pixels.
Range: ~100...2000 (max.7500) mm.
FOV: 30° side-to-side; 40° top-to-bottom.
Input: trigger for object
Output: binary and analogue (4...20 mA / 0...10 V)
Setting: Ethernet / display menu
Protection R.: IP65

4.3. Smoke/Fire detector

An ionization type smoke detector [4] is cheaper than an optical detector. However, it is sometimes rejected for environmental reasons. It can detect particles of smoke that are too small to be visible. It includes a tiny mass of radioactive americium-241, which is a source of alpha radiation. The radiation passes through an ionization chamber, which is an air-filled space between two electrodes, and permits a small, constant current between the electrodes. Any smoke that enters the chamber absorbs the alpha particles, which reduces the ionization and interrupts this current; activate the alarm (Fig. 3).

Figure 3: Inside a basic ionization smoke detector [4]

4.4. Temperature sensor

The LM35 [5] series of temperature sensors (Fig. 4), selected for the surveillance module are precision integrated-circuit temperature sensors, whose output voltage is linearly proportional to the Celsius (Centigrade) temperature. The LM35 thus has an advantage over linear temperature sensors calibrated in ° Kelvin, as the user is not required to subtract a large constant voltage from its output to obtain convenient Centigrade scaling.

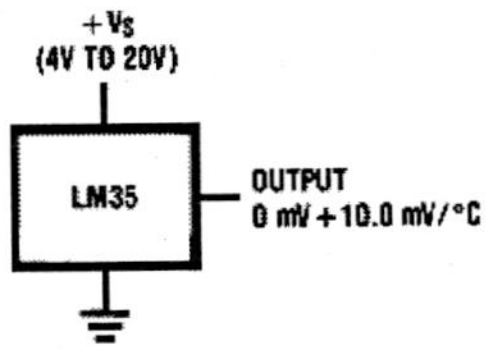

Figure 4: Basic centigrade sensor [5]

4.5. Humidity transmitter

The 808H5V5 module humidity transmitter [6] is designed based on a capacitive humidity sensor (Fig. 5).

Figure 5: Humidity sensor (808H5V5) [6]

Specifications:

Measuring range: 0~100%RH
Signal output: 0.8~3.9V @25
Accuracy: ≤ ± 4%RH （at 25°, 30~80%RH, when the power supply is 5VDC）
Electrical current (Ic): Typical current: 0.38 mA ; max current <0.5mA
Power supply: 5V DC ±5%
Operating environment: -40°~+85°
Responding time: <15s
Interface: 3 pin 2.54mm between, SIP

4.6. Color LED / Badge made of Color reflective material

For identifying a person in the hospital, infrared coloured LED and reflective material surface will be used as an identifying badge.

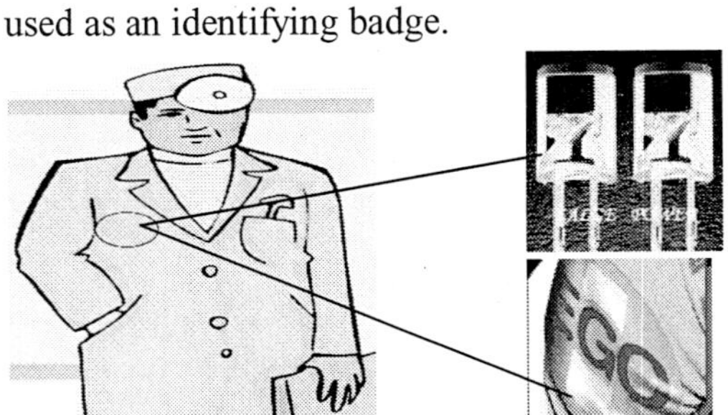

Figure 6: Identifying badge used made by either reflective surface or Infrared coloured LED

4.7. Camera (PTZ mechanism included)

Canon's latest Pan/Tilt/Zoom (P/T/Z) camera, the VC-C4 [7] combines great looks, compact design, and ease-of-use. It can be controlled through a serial port interface.

Figure 7: Cannon VC-C4Camera [7]

Features:

Video Standard: NTSC / PAL
Total Number of Pixels: 410,000 (380,000 effective pixels)
Horizontal Field of View: 3 to 47.5° (65° with Wide angle lens adapter)
White Balance: Auto
Exposure: Auto/Manual
Fosus: Auto/Manual and Zoom: 16x
Focus Length: 4 to 64mm, f/1.4 to 2.8
Pan Angle: +/- 100° (+/- 170° VC-C4R)
Pan Speed: 1 to 90°/sec (+10°, -90° VC-C4R)
Tilt Angle: +90/-30°
Tilt Speed: 1 to 70°/sec
Pan/Tilt Control: RS232 Serial, RTS/CTS has to be active (up to 19.2 kbps)
Power: 13V, 12W

4.7. DAQ Board

There are few DAQ cards available which are supported by the Linux OS. National instrument (NI), Iotech and Lab Jack are such boards. This research uses Lab Jack U3 which also works on the Linux platform. An NI USB-6008 board was also tried in this project; however, it was only used for prototyping purposes as their software doesn't support Linux (Ubuntu) version yet.

Lab Jack

The Lab jack U3 [8] USB DAQ device is with up to 16 12-bit analog inputs, up to 20 digital I/O, 2 10-bit analog outputs, up to 2 counters, and up to 2 timers.

Figure 8: Lab Jack (U3) **[8]**

Features:

- 16 Flexible I/O (Digital Input, Digital Output, or Analog Input)
- Up to 2 Timers (Pulse Counting/Timing, PWM Output, Quadrature Input)
- Up to 2 Counters (32-Bits Each)
- 4 Additional Digital I/O
- 12-bit Analog Inputs (0-2.4 V or 0-3.6 V, SE or Diff.)
- 1 Dedicated Analog Output (8-Bit, 0-5 volts)
- 2nd Analog Output Available in Certain Configurations
- Supports SPI, I2C, and Asynchronous Serial Protocols (Master Only)
- Maximum Input Stream Rate of 2.5-50 kHz (Depending on Resolution)
- Capable of Command/Response Times Less Than 1 Millisecond
- Built-In Screw Terminals for Some Signals
- Powered by USB Cable
- Drivers Available for Windows, Linux, Mac and Pocket PC
- Rated for Industrial Temperature Range (-40 to +85 Degrees C)

5. Research and application

The aim of this research is to develop a surveillance module with two distinct functions: intruder detection using image analyzing technique and hospital condition monitoring. Figure 9 shows the pioneer robot platform onto which the surveillance module will be attached.

Figure 9: Pioneer robot platform used in the IWARD project

The hospital condition monitoring system is equipped with three different sensors (temperature, humidity and smoke detector). Figure 10 shows this process; when the sensory data is received the application programme analyses those signals according to the pre set range for each sensor and create an output signal to take necessary action. Using the basic algorithm this research successfully tested this system under the Windows platform whereas developing the programme under the Linux platform is ongoing.

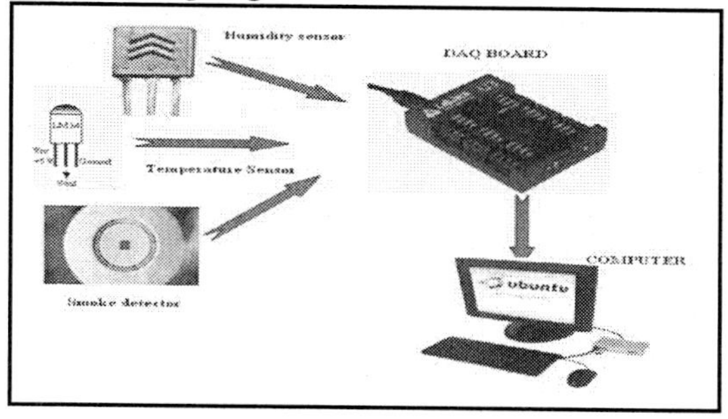

Figure 10: Graphical presentation of the Hospital condition monitoring system.

The intruder monitoring system is equipped with two primary sensors: a PIR sensor and a PMD 3D distance measurement sensor to detect human motion and trace the source and distance of motion. Figure 11 and 12 show the intruder detection system.

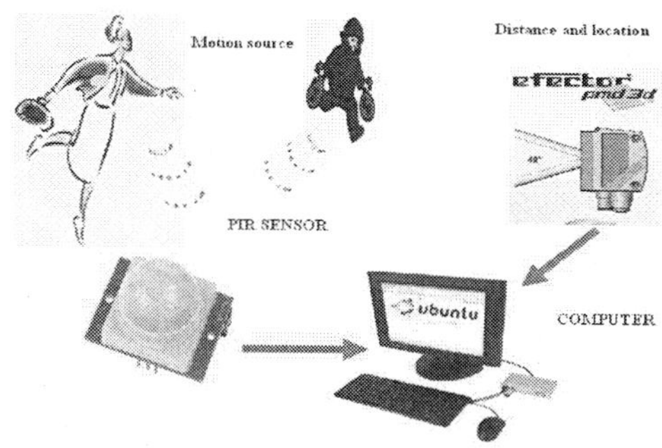

Figure 11: Motion detection and distance measurement

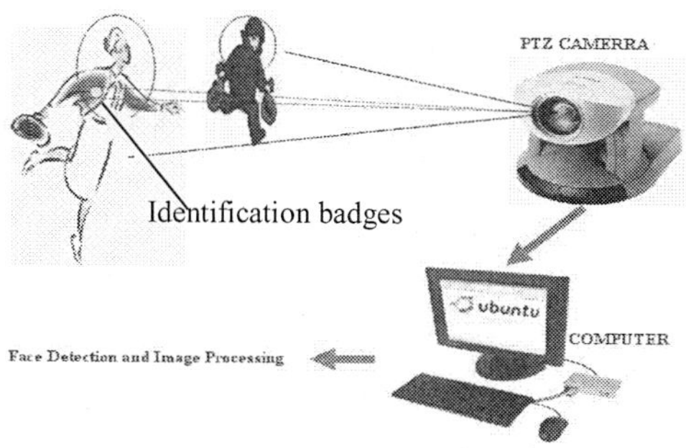

Figure 12: Face detection and Image analysis techniques

When a motion is detected by the PIR sensor the output signal triggers the camera and the PMD 3d sensors. The PMD analyses the data and trace the location and distance of the motion source. Then it sends the signal through an Ethernet communication port. Based on this data the PTZ camera take the image (zoomed according to face detection) of the motion source. Then it analyses the image based on a colour identifying process using Open CV [9] (this research uses Infrared LED and reflective surfaces for identification badges to define if a person is legal). Monitoring nighttime events for real surveillance systems is an important application. For this purpose, a colour restoration algorithm [10, 11] has been proposed based on color instances. In the event of an intruder the PTZ camera follows that person and the programme saves the recorded stream (preferred option in this research) or still images for security information and also creates a log file of the event. A video sequence is a much richer source of visual information than a still image [12]. This research tests all the sensory data and performs experiment for multisensor fusion. The surveillance module is still under development.

6. Conclusion

A prototype of a surveillance module for hospital robots is designed and the development process is ongoing. The success of this research will create a great impact on hospital environment in terms of surveillance and condition monitoring.

Acknowledgment

Dublin City University is a partner of the EU-funded FP6 project called Intelligent Robot Swarm for Attendance, Recognition, Cleaning and Delivery (iWARD) (www.iward.eu). Dublin City University is a partner of the EU-funded FP6 Innovative production Machines and Systems (I*PROMS) Network of Excellence (www.iproms.org).

References

[1] Presence detection and person identification in Smart Homes, [1]Neubiberg, University of Bundeswehr Munich, [2]Passau, FORWISS, University Passau.

[2] Electronic123, Online, "PIR sensors", http://www.electronics123.com/s.nl/it.A/id.62/.f, Accessed March 2008.

[3] Ifm electronics, online, "PMD 3D", http://www.ifmelectronic.com/ifmgb/news/news__6nzcj3.htm, Accessed March 2008.

[4] Online, "Smoke detector", http://en.wikipedia.org/wiki/Smoke_detector, Accessed March 2008.

[5] Online, "Temperature sensor", http://www.chipcatalog.com/Datasheet/FEE0878DF555FDFD1F46C41A7F735068.htm, Accessed March 2008.

[6] Sencera Co., Ltd., Online, "Humidity sensor", http://www.sensorelement.com/humidity/808H5V5.pdf, Accessed march 2008.

[7] Cannon, Online, "Cammera", http://www.canondv.com/consumer/controller?act=ModelInfoAct&tabact=DownloadDetailTabAct&fcategoryid=362&modelid=7402#ModelDetailAct, Accessed March 2008.

[8] Lab Jack, Online, "DAQ Card", http://www.labjack.com/files/LabJack_U3_Users_Guide.pdf, Accessed March 2008.

[9] Open CV, Online, http://en.wikipedia.org/wiki/Open_CV, Accessed March 2008.

[10] Y. Hatakeyama, K. Kawamoto, H. Nobuhara, S. Yoshida, K. Hirota, Color restoration algorithm for dynamic images under multiple luminance conditions using correction vectors, Pattern Recogn. Lett. 26 (2005) 1304–1315.

[11] Y. Hatakeyama, K. Kawamoto, H. Nobuhara, S. Yoshida, K. Hirota, Color instance-based reasoning and its application to dynamic image restoration under low luminace conditions, J. Adv. Computat. Intell. Intell. Inform. 8 (6) (2004) 639–648.

[12]Handbook of image and video processing, 2nd edition, 2005, pp-253-274.

Innovative Production Machines and Systems
D.T. Pham, E.E. Eldukhri and A.J. Soroka (Eds.)

Development of an automated cleaning system for hospitals.

C. Griffin, T. Szecsi

School of Mechanical & Manufacturing Engineering, Dublin City University, Dublin 9, Ireland

Abstract

In recent years so-called hospital superbugs; bacteria which are resistant to antibiotics, have become a major issue in medical facilities across the country. This bacteria can cause serious illness, and in some cases death in patients. In 2007 alone 432 cases of the MRSA bug were officially reported between the months of January and September, showing the extent of the problem. The Health Service Executive (HSE) has pledged to spend close to €171million this year on cleaning in Irish hospitals following an independent survey published late last year which found that not one hospital of the 51 surveyed nationally had a rating of "very good" hygiene.

As hygiene is becoming an ever more important issue there is a need for a cheap, simple automated cleaning system that could be installed in hospitals to tackle the growing problem. To achieve this goal an existing cleaning robot, the iRobot Roomba, was used as the platform to build upon. This paper documents the work carried out thus far on the cleaning system which involves testing the capabilities of the Roomba vacuum & development of software in both a Windows and Linux environment.

Keywords: Hygiene, Hospital, Roomba

1 . Introduction

Over the past three years cleaning costs in Irish hospitals has risen sharply from €99.7m to €157.2m in the battle against hospital acquired infections, yet this dramatic increase in expenditure has had little effect on the problem. An audit carried out during July & August of 2005 consisted of spot checks of 51 hospitals across the country focussing on eight different areas at each site. Examples of these categories tested are hand hygiene, safe handling/disposal of sharps & handling/disposal of linen. This audit was carried out by an English consultancy firm on behalf of the Health Service Executive to determine the standards of cleanliness of Irish hospitals. The results were not encouraging. The overall scores of all hospitals were: 9% received a good rating, 43% received a fair rating & 48% received a poor rating [1].

This serious problem highlights the need for a new system to be developed to try and combat the problem and help to reduce the risk to patients in a hospital and prevent the spread of diseases. An automated floor cleaning system was conceptualised where routine cleaning could be carried out in wards, corridors and hallways. The second requirement was that certain spots could be cleaned should a spillage occur in a certain area. An iRobot Roomba XE [2] was chosen to carry out the routine cleaning duties & was selected based on a number of criteria including cost, size, simplicity of design, ease of maintenance & ability to be integrated into the project.

Due to the low computational power of the Roomba mobile vacuum unit it had very limited decision making capabilities and did not allow the user to integrate software onto the Roomba's computer to install external sensors. Since the vacuum does not possess on-board navigational software a Pioneer 3DX [3] mobile robot platform would be used to guide the cleaning robot to its required destination since it is already intended to be used within the iWARD project. The iWARD project intends to develop a number of these base robots, each one capable of fulfilling different functions but all possessing the ability to safely navigate through the hospital environment. Restrictions meant that the Pioneer base robot could not carry the vacuum unit which is why the Roomba must have the capability to follow a lead robot.

This paper will discuss the early stages of development of the cleaning system and will briefly introduce areas in which future development will take place and the intended direction for further research. Work carried out to date focuses on testing the capability of the Roomba vacuum unit to ensure that it is sufficiently capable of carrying out the desired task before too much time is invested in the robot. In the event that this concept proves unsuccessful a contingency plan is being researched at a high level so that a cleaning system can be developed by using more than one concept if required.

2. Scenario

The following is a brief description of the scenario that was decided upon that the automated cleaning system should carry out for scheduled cleaning once the prototype is in operation.

While not in use the Roomba is located at the charging base station which is at a fixed position in the hospital. The base robot arrives and instructs the Roomba to follow, the base robot then departs with the vacuum robot following closely behind; the intended method of following is detailed below in section 3. When the base robot reaches the intended destination it issues a command to the vacuum robot to begin a normal cleaning cycle of the area. Once a predetermined amount of time has elapsed the base robot returns to retrieve the Roomba and returns it to the docking station once more.

3. Roomba following pioneer

The Roomba carries a distance measurement device (DMD), a camera and a small Linux based computer (possibly a Gumstix [4]). The base robot is required to carry a distinctly recognisable light emitter on the rear of the platform. The camera can isolate the light and orientate the Roomba based on the vision analysis. The DMD ensures a constant distance is kept between the lead and following robots. The Linux computer holds the operating programs for both the camera & DMD, and carries out the decision making process based on the inputs received after which commands are sent to the Roomba's onboard computer in order to drive and steer etc.

3.1 Cricket device

The principle on which the cricket device works is as follows: One device is assigned as a listener & a second is assigned as a beacon. The beacon periodically transmits an ultrasonic & RF signals at the same time. The listener detects both the ultrasonic & RF signals; based on the difference of arrival times of the two signals it can calculate the distance from the listener to the beacon.

When two or more identical devices are used in conjunction with each other the linear distance between two crickets can be measured. By means of an algorithm developed in MIT [5] the exact position of a robot can be determined in an indoor environment using a larger number of cricket devices. However for the purposes of this project the linear distance is all that will be required. A number of tests & experiments are planned to investigate the accuracy and repeatability of the cricket devices and to determine how effective this system would be under real world conditions.

3.2 RFid

Radio Frequency identification works on the principle of a tag transmitting a unique RF signal that is capable of being read by a reader at a considerable distance. This system can often be used to track multiple objects throughout a designated area. Previous research [6] has proven that the system can track tagged objects as long as the object in question is within range of at least one reader, however those projects were not concerned with exact distance from a reader, merely presence or absence of a tag in an area.

Extensive testing is required to determine whether or not RFid can prove accurate enough to be used for this cleaning system. The software provided with the RFid kit provides different types of data, however the one of particular interest in the forth-coming testing is the RSSI value. The relative signal strength indicator is a value returned by the system based on the quality of the RF signal received from the tag. The goal is to examine whether there is any correlation between the RSSI value and the distance from the reader to the tag. This testing will include repeatability testing to check if the RFid system can reliably output a distance measurement between the Roomba and the Pioneer base robot.

3.3 Vision system

In order to develop a low cost cleaning system a sophisticated vision component is not possible. As a result the abilities of this section of the robot will be somewhat restricted. The aim is to attach a relatively cheap, generic camera to the top of the vacuum in a forward pointing position. No pan, tilt or zoom functions will be present as all orientation will be carried out by the Roomba itself. The aim is to use filter(s) over the lens of the camera to remove as much as possible from the image without removing the LED on the rear of the base robot. For this purpose infrared LED's may suffice.

Once a software programme is developed the camera will distinguish the LED on the Pioneer robot from its surroundings. The image on view will be sectioned into three zones as illustrated in figure 1 below. (Please note this is merely a representation for explanation purposes). If the LED is shown to be in Zone 1 (far left) then the robot must rotate counter-clockwise until the LED on the base robot is shown to be in the Zone 2 (centre section). Should the LED appear in Zone 3 (far right) as is in the illustration then the robot must turn clockwise. This simple algorithm will ensure that the Roomba roughly follows the base robot in a straight line.

The outputs from the vision system algorithm will be used by the onboard computer to correct the course of the vacuum robot during travel from one destination to another.

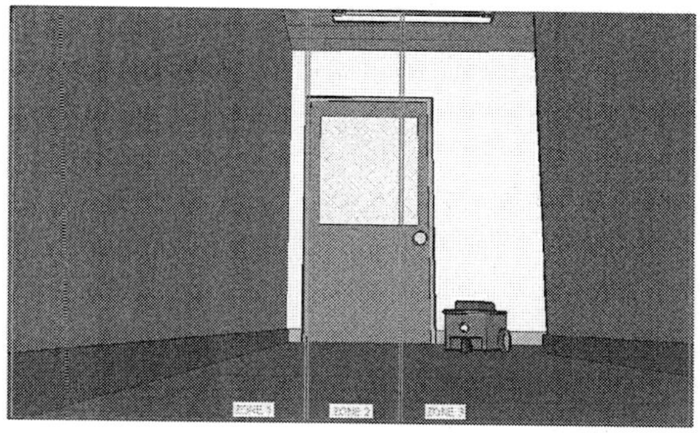

Fig 1. Robot vision system

4. Testing of Roomba $^{\underline{XE}}$ vacuum

A variety of software testing was carried out in increasing stages of difficulty, each one getting closer to the end result of full control of the Roomba unit through the Ubuntu Linux environment. Detailed below are the chronological investigations carried out to date.

4.1 SCI testing

Initial testing of the Roomba unit was carried out through the Serial Command Interface (SCI) that was available for free download [7] in the Windows environment. Communication was carried out using a Bluetooth dongle connected to both the desktop PC and the vacuum itself. This testing ensured that the user had full control of the unit through the wireless link and investigated the full capabilities of the Roomba prior to any further development taking place.

4.2 Labview testing

The purpose of work being carried out using the National Instruments Labview program was to determine the ease with which custom applications could be developed by the user and also to obtain assurance that full control of the Roomba could be maintained when the free graphical user interface (GUI) was no longer being utilised. Research showed that Labview enabled the user to have a much more versatile control of the vacuum unit.

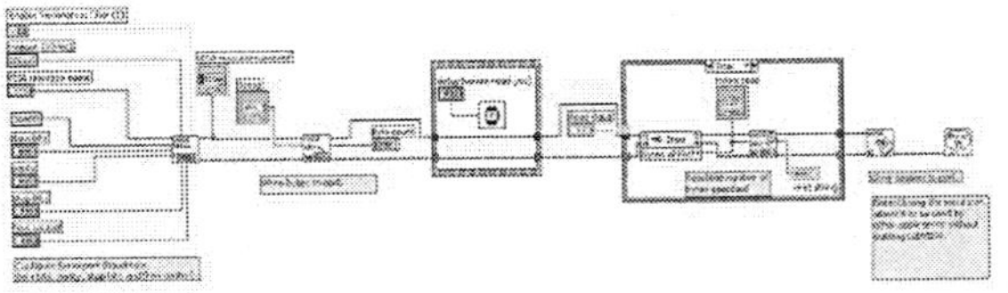

Fig 2. Labview block diagram

Figure 2 shows a reduced sized image of the block diagram of the program used to send and receive data to and from the Roomba. Here is a brief description of the block diagram (left to right): the set up for the serial port communication is carried out using baud rate, stop bits, parity, time out and flow control. Once the connection is established a string is written to the serial port, in this case a hexadecimal command obtained from the Roomba user manual. A 500ms delay is included before the "read" action takes place, assuming the toggle switch has been set to activate the read blocks. Data returned from the vacuum unit is displayed on screen as a string of characters, after which the serial communication session is closed. A larger version of Fig. 2 is shown in Appendix A.

5. Cable construction

Bluetooth was the original means by which communication was carried out between the Roomba and the PC. A Bluetooth dongle purchased with the Roomba and a USB-Bluetooth dongle plugged into the PC allowed for reliable and mobile testing without restrictions on movement during testing. However future plans intended that a smaller Linux based computer would be mounted on the vacuum to provide more computational power resulting in a greater amount of independence for the Roomba. For this situation a physical rather than wireless connection was more favourable. A voltage regulator circuit was therefore built to allow a connection between the RS232 port on a desktop PC and the 14.4VDC connection on the Roomba. The circuit had to enable two way communications between the PC and vacuum to send the commands and receive the data from the sensors during various stages of testing. Labview had already proven to be a reliable testing environment therefore this was chosen as the means to ensure the voltage regulator circuit was operating correctly.

6. Linux Ubuntu testing

The aim of the project was to have all computer programming, all equipment drivers and all communication occurring through the medium of C++ in a Linux operating system. This was to ensure that all parties involved in the development of the iWARD project were able to aid the other partners should help be required and also to eliminate any problems with compatibility between various components.

The terminal in Ubuntu served as the platform for testing the C++ code and the Roomba unit. For the duration of the Linux testing the physical cable was used to communicate with the vacuum robot. This ensured that no problems could arise from complications with Bluetooth signals or issues arising from using the USB-Bluetooth dongle that had been previously used in the Windows environment. The format of the C++ code established a connection through the serial port using the relevant settings after which a small number of characters were sent in hexadecimal format to activate the Roomba and change the mode to instruct the vacuum to be prepared to accept user inputs from an external source, in this case a desktop PC.

As had been done before each command was sent individually to ascertain what level of control was possible through this method. The results were not as positive as had been received using the Windows environment. Testing, however, is not complete at the time of writing this report and the investigation is set to continue to develop a more robust system of control over the Roomba robot.

7. Contingency design

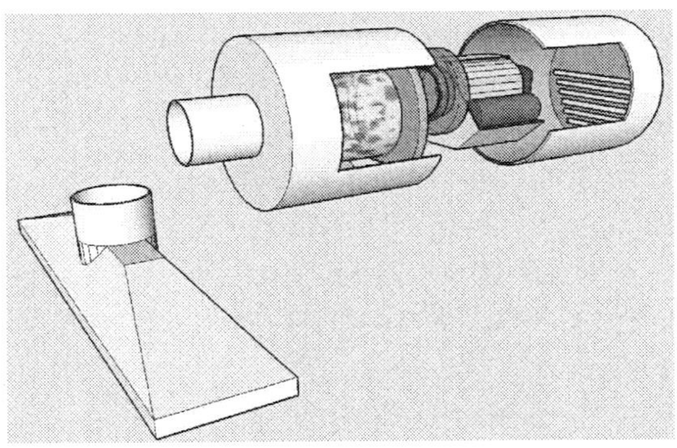

Fig 3. Alternative vacuum system

To ensure success in the project a contingency plan was put in place in the event that the Roomba cleaning system was unsuccessful by the deadline. This second cleaning system involved mounting a custom designed vacuum unit directly to the Pioneer base robot and eliminating the need for a second, separate cleaning robot. Figure 3 is an assembly drawing of the proposed vacuum unit with cut-away areas to show the interior parts. Left to right: the wide nozzle placed close to the ground to collect dust (flexible hose not included), dust collection chamber with air filter included inside, a motor with fan & batteries attached to provide suction & on the far right the exterior cover for the motor, shown with slots for air to escape.

The design is based on parts from a "Hoover" hand-held car vacuum cleaner. The long life rechargeable batteries, light weight construction and compact design make it an ideal donor for such a cleaning system. The main body of the unit is a cylinder that measures approximately 10cm diameter and 20cm in length meaning it could be mounted vertically on the rear of the Pioneer robot without much difficulty.

Shown in figure 4 is an initial design for the vacuum attachment to be mounted on the front of the Pioneer robot. The numbered parts of the design are as follows: (1) points to the vacuum head surrounded by a transparent plastic protective bumper that was attached to the front of the Pioneer robot to protect the sensitive electronics from impact. (2) is a pull-action linear solenoid designed to lift the vacuum head off the ground while not in use. (3) is another solenoid with a spring return motion designed to hold the vacuum head in the UP position after being lifted by the vertical solenoid. Retracting this solenoid would allow the vacuum head to drop to the ground again. (4) is the arm that retains the vacuum head in the UP position. (5) shows the sheet metal that is cut, then bent to shape that holds all other components in place. (6) is the mounting point for the vacuum head assembly onto the front of the Pioneer robot. This method of attachment has not yet been decided upon.

This design does however have a number of minor drawbacks associated with it, but also some advantages too. The base robot is larger in height than the Roomba and as a result it is unable to clean underneath any beds or a low table which is a significant flaw with this system.

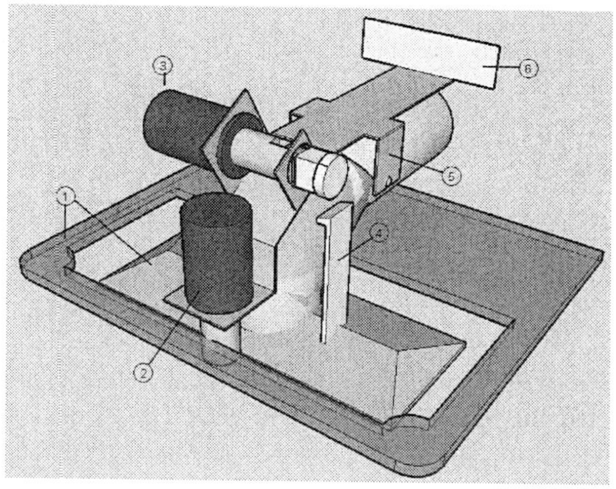

Fig 4. Vacuum attachment

The Roomba is sold with a cleaning algorithm already included as standard, with 3 separate modes available for selection.

To utilise this contingency plan an algorithm and required C++ code would have to be written to enable effective cleaning to occur. This would involve reading sensory information from the navigational array (such as a compass, vision system and sonar) and enabling the software to react accordingly to provide adequate cleaning coverage of the room. The drawbacks mentioned are off-set against the advantages of eliminating a rather difficult set of problems which dealt with leading the Roomba from one destination to the next.

8. Conclusion

This paper began with the problem of hygiene currently in Irish hospitals and the large amounts of funding being spent to address the problem. In an attempt to aid in the solution to this problem an automated cleaning system was proposed. Thus far the concept is in the early development stages and much work is required before an active prototype is possible. For this reason the future aims and developments were outlined but no experimental results or concrete conclusions were documented.

The three systems to allow the vacuum robot to successfully follow the Pioneer base robot are the three key areas to be investigated in the near future. These components will provide the system with the ability to navigate through the hospital without human interference thus creating an automated cleaning system. The scheduling, navigation and obstacle avoidance will all be present on the base robot which is in development by a partner in the iWARD project.

Care will need to be taken to ensure that the ultrasonic and RF signals being transmitted to and from the vacuum robot will not interfere with any of the existing systems on the base robot or any equipment currently in use in a hospital environment. As this is only a prototype certain assumptions are being made; since the vacuum can only clean dry areas it is assumed no wet areas are brought to attention for cleaning by the Roomba. A separate, very similar system will be required for wet cleaning, however once proof of concept is shown a second system could be constructed.

Since the spread of infections and bacteria is a large reason for the conception of the project sterilisation must be taken into account. If not the Roomba and base robot could become carriers of infections themselves. A simple solution could be to add a small bottle of disinfectant that would periodically release a small amount of fluid to clean the wheels of the robot.

Ultimately the goal of this automated cleaning system is to help reduce the problem of infections being spread in hospitals.

Acknowledgements

Dublin City University is a partner of the EU-funded FP6 project called Intelligent Robot Swarm for Attendance, Recognition, Cleaning and Delivery (iWARD) (www.iward.eu).

Dublin City University is a partner of the EU-funded FP6 Innovative production Machines and Systems (I*PROMS) Network of Excellence (www.iproms.org).

I would like to extend my thanks to Mohammed Anwarul Islam & Kamrul Hassan who are colleagues of mine who are also working on aspects of this project.

Appendix A. Labview block diagram

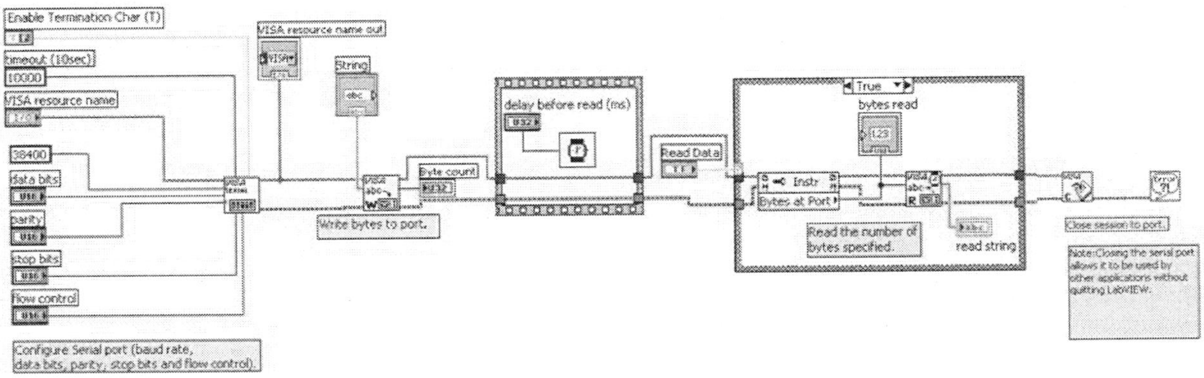

References

[1] Report on a National Acute Hospitals Hygiene Audit undertaken on behalf of the National Hospitals Office, Health Service Executive. Desford Consultancy Limited. Oct 2005

[2] http://www.irobot.com/

[3] http://www.activrobots.com/ROBOTS/p2dx.html

[4] http://www.gumstix.com/

[5] N.B. Priyantha The Cricket Indoor Location System. Massachusetts Institute of Technology. June 2005

[6] J. Mortensen/T. Pedersen. Possible use of RFID Technology in Support of Construction Logistics. Agder University College. Master Thesis. June 2004

[7]http://www.irobot.com/images/consumer/hacker/Roomba_SCI_Spec_Manual.pdf

Innovative Production Machines and Systems
D.T. Pham, E.E. Eldukhri and A.J. Soroka (Eds.)

Development of a cleaning module for hospital robots

A. Islam and T. Szecsi

School of Mechanical and Manufacturing Engineering
Dublin City University, Dublin 9, Ireland

Abstract

This paper describes a cleaning module which is a part of IWARD Project (Intelligent Robot Swarm for Attendance, Recognition, Cleaning and Delivery) where a stand-alone robot will perform a scheduler or regular cleaning and spot cleaning in a hospital ward. For this instant the roomba needs to follow the robot swarm to reach the specific place to clean as roomba doesn't has any navigation system. To accomplish this task it is required to orient roomba towards the robot when the roomba is following the base robot, establish distance measurement between roomba and the robot, and communicate between roomba and the robot. To achieve these different types of sensors is going to be attached to the roomba and the base robot and the working procedures are described in this paper.
Keywords: Roomba, stand-alone robot, hospital cleaning module

1. Introduction

There are significant problems with recruiting cleaning staff in hospitals and if this pilot with the robot cleaner works out might become a valuable member of a cleaning team [1]. Transitions research corporation has been working on autonomous mobile robots since 1985, focusing particularly on floor cleaning [2]. Intelligent robot swarm for attendance, recognition, cleaning and delivery (IWARD) targets mainly hospitals and healthcare centers to overcome the shortages of healthcare staff - a major issue in european healthcare. Our aging society and economic pressure increase the patients-to-medics' ratio, having an adverse effect on healthcare quality and performance. Not being able to attend all patients at the right time and not keeping the hospitals clean enough (e.g. MRSA Transmission) also increases recovery time and cost. To improve the quality of healthcare, one of these focal issues emerges: effective cleaning in hospitals. To achieve this, IWARD presents a robot swarm delivering support to oversee activities in healthcare environments, providing a multipurpose, cost-effective and scalable solution to enhance quality of healthcare. Four major tasks are: attendance, recognition, communication and support (assisting /cleaning). Current research is focused on developing the cleaning module. The robots can be equipped with different adaptable hardware components for floor cleaning which will shorten the reaction time, reducing human error and increasing efficiency to deliver better patient care in hospital environment.

2. Scenario

In this scenario, the cleaning module (roomba) performs regular, scheduled cleaning of one room. The normal (parking) position of the roomba is near its docking (charging station). The docking station has a fixed location. The robot picks up the roomba from this location. The roomba follows the robot using an 'electronic rubber band' consisting of an orientation and a distance measurement device (DMD). The distance measurement device has an emitter on the robot and a receiver on roomba. It measures the distance between them. If the distance is too large roomba instructs the robot to slow down. If the distance is too small the roomba stops. The orientation device consists of a distinct light-emitter on the robot and a vision system (camera) on roomba. The roomba orients itself towards the light emitter using image analysis from the camera. When the robot and the roomba reach the location the robot leaves the roomba to perform regular cleaning while the robot can do some other task. When the cleaning cycle is ready the robot returns to the same room and guides the roomba back to its docking station.

3. Roomba® cleaning robot [3]

The cleaning module is a standalone small cleaning robot called roomba (see Fig. 1), commercially available from iRobot®. Versions of iRobot® roomba® vacuuming robot manufactured after October, 2005 contain an electronic and software interface that allows users to control or modify roomba's behavior and remotely monitor its sensors. This interface is called the iRobot roomba serial command interface or roomba SCI. Roomba SCI is a serial protocol that allows users to control a roomba through its external serial port (Mini-DIN connector).

The SCI includes commands to control all of roomba's actuators (motors, LEDs, and speaker) and also to request sensor data from all of roomba's sensors. Using the SCI, users can add functionality to the normal roomba behavior or they can create completely new operating instructions for roomba.

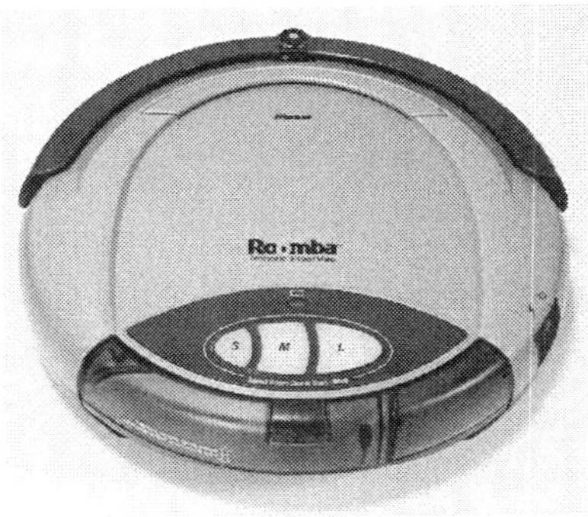

Fig. 1. iRobot roomba

Since roomba does not have navigation capabilities, it can only start its cleaning activities relative to a start position. In order roomba to find this start position it has to be guided by the base robot (for simplicity we will refer to it as the robot). A robot can navigate to any position in the hospital as long as the position is a node on the location graph. Roomba has to follow the robot until the cleaning start position is reached. In order to accomplish this, three major tasks have to be solved: orienting roomba towards the robot, distance measurement between roomba and the robot, and communication between roomba and the robot.

3.1 Orienting roomba towards the base robot

In order roomba to be able to follow the robot it is necessary to continuously orient roomba towards it. Then roomba can be issued a command for a straight linear forward motion in this direction. There are several ways how this can be achieved. They all use still images from a camera installed on roomba:

3.1.1. Image analysis of the robot

Using still images captured by the camera, a vision analysis software identifies the shape of the robot in the frame. Since the image of the robot depends on many factors (orientation of the robot relative to roomba, magnification, lighting condition), and also the background image might interfere with the image of the robot, this task can be quite difficult to achieve. Also, finding the middle of the robot from a varying image makes orientation of roomba more unreliable. As an alternative, the vision system may attempt to find some characteristic parts on the robot (like the long pole that holds the compass).

3.1.2. Recognisable pattern

A pattern that is relatively easy to recognise is placed at the rear of the robot. An unusual pattern (see Fig. 2) that is unlikely to exist in the background image may be recognised through image analysis. Since the exact image of the pattern depends on the orientation of the robot, several patterns might be needed. This, however, makes finding the middle of the robot more difficult.

Fig. 2. Pattern for Image recognition

3.1.3. Visible light signal

A visible light source (bulb) with relatively high intensity is placed in the middle of the rear of the robot. The intensity and the colour of the light make it easy to apply filtering effect to the image and eliminate the background images. Since the light source appears as a characteristic spot in the frame (see Fig. 3), it makes orientation of roomba easy. If the camera on roomba is positioned symmetrically forward, an off-centre light spot defines the direction of the necessary rotation of roomba.

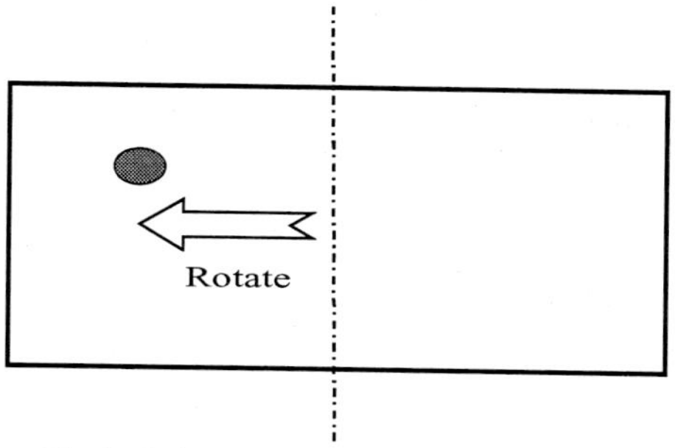

Fig. 3. Orienting roomba using visible light

3.1.4. Infrared light signal

This method is similar to the previous one but the light source is infrared. This makes filtering easier since the background image is in the visible range and an infrared-pass filter would filter them out. However, this requires a camera with an extended wavelength range. Many cameras, however, in order to reduce infrared contamination artifacts, are equipped with special internal infrared cut filters. Also, reflections from artificial lighting (like incandescent (tungsten) lighting) and from the sun might interfere with the filtered image.

3.2 The pololu IR beacon [4]

The pololu IR beacon is a small device that allows pairs of autonomous robots to detect each other. The beacon operates by transmitting and detecting modulated infrared light. The beacon provides four outputs, which correspond to the four cardinal directions. When another beacon is detected, the output for the appropriate direction is activated. This simple interface allows even the most basic robot controller to connect to the beacon.

The pololu beacon (see Fig. 4) is ideal for robot contests and other applications in which small autonomous robots interact. The devices are transceivers meant to be used in pairs to give robots a simple means for detecting each other.

Fig. 4. Pololu IR beacon

One can use the beacons to build pairs of robots that interact or chase one another or to make a robot that can identify and return to a home base.

The IR beacons feature round PCBs with six wide-angle emitters to provide uniform broadcasting in all directions. The operating voltage range has increased to 6 - 16 V, and on-board voltage monitoring enables more consistent output brightness over the voltage range. Most components are pre-assembled, so all is needed is to solder in the capacitor, connectors, and sensors.

The beacons work by transmitting and detecting infrared light, much like a television remote control. Each beacon has four IR emitters and four IR detectors (see Fig. 5). The beacons alternate between transmitting and receiving so that they never get confused by reflections of their own transmissions.

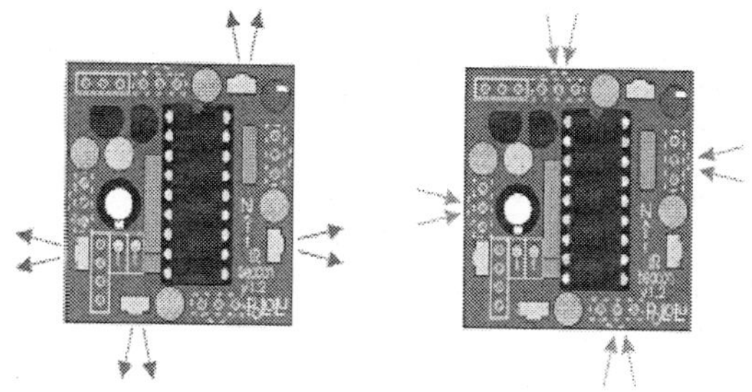

Fig. 5. IR emitters and detectors on polulu

The transmit and detect cycle is carried out more than one thousand times per second, and a small microcontroller monitors all four detectors to decide the direction to the other beacon. The beacons have four red LEDs that indicate the direction to the other beacon; if one takes two beacons and rotates them, the LEDs will always keep lighting up in the direction of the other beacon (see Fig. 6).

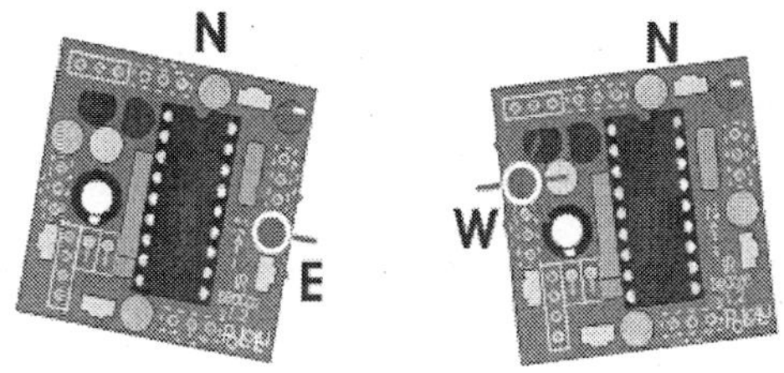

Fig. 6. Orienting the Polulu

Interfacing to the beacon is simple — it has four digital outputs that indicate which of its four sides detects the other beacon the strongest. One can establish the direction to another beacon to within a few degrees by rotating the beacon back and forth and noting the point where the output switches from one side to another. An enable input lets one select between active mode and a low-power mode. It is important to note that the IR beacon achieves its brightness by pulsing the IR LEDs with very high currents; turning on the IR LEDs continuously will burn them out. When testing the IR beacon, a digital camera can be useful in determining whether the IR LEDs are on. However, some cameras (usually more expensive ones) have better filters that block IR.

This IR beacon can be attached to the base robot and the roomba. The beacon operates by transmitting and detecting modulated infrared light. The beacon attached with the base robot emits the signal and the beacon with the roomba detects the signals and follows the base robot. The same IR beacon can be used in guidance module. Since the Polulu beacons can only detect one of four cardinal directions, its accuracy is more limited than that of a camera-based vision system.

3.3 Distance measurement between roomba and the base robot

In order to maintain a constant following distance between the robot and roomba, it is necessary to continuously measure the distance between the two. Based on the distance measurement, Roomba can instruct to speed up or slow down the robot.

3.3.1 Wavetrend active rfid [5]

The Wavetrend L-RX201 readers detect and decode RF transmitted signals from the wavetrend family of tags. The reader incorporates a BNC connector to allow for attachment of an external antenna to extend the reader's range.

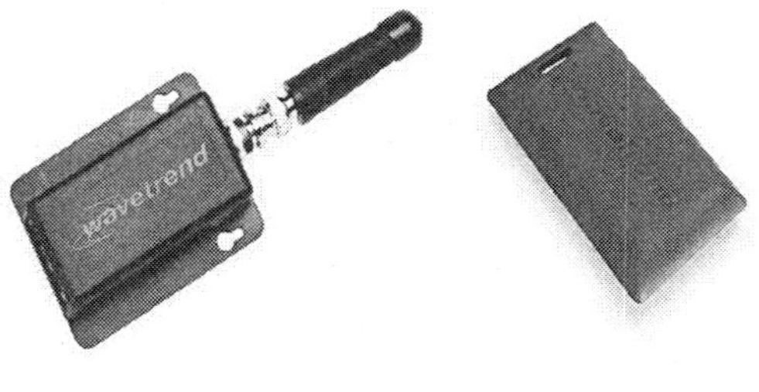

Fig. 7. L-RX201 Active reader with attached AN200 Stub Antenna (left), TG501 Personnel Tag (right).

The L-AN200 433 MHz Wavetrend stub antenna shown in (see Fig. 7) has been designed for optimal RF receiving performance with the wavetrend reader range. This antenna has an omni-directional reception pattern around the vertical axis with null points being at the extreme ends of the stub. The RX201 type readers can be used singly, or networked together to form a reader network. The reader connects to a PC or intelligent hardware controller via RS232, RS485 or TTL.

Wavetrend tag TG501

This tag transmits a radio frequency (RF) signal at a pre-set time-interval. The tag life is estimated at 5 years at a transmission time interval of approximately 1.5 seconds.

The transmitted tag data could include optional customer site code (CSC), tag ID, tag age counter value, movement alarm and tamper alarm status, tamper logged status, etc. The tag has an anti-tamper status that is set whenever the tag is tampered with. This status is not re-settable. The tag has an auto wake function. When it is secured with the anti-tamper kit it will automatically 'wake-up' after 5 minutes and start transmitting its pre-configured data.

3.3.2 Cricket device:

Cricket is intended for use indoors or in urban areas where outdoor systems like the global positioning system (GPS) don't work well. It can provide distance ranging and positioning precision of between 1 and 3 cm, so applications that benefit from better accuracy that the cellular E-911 services and GPS will also find cricket useful. Cricket is designed for low-power operation and can be used as a location-aware sensor computing node (running TinyOS), to which a variety of sensors can be attached. [6]

It can provide location information such as distance from a known point, a space name and coordinates. The component for this system is just one device called cricket this device can be configured to act as a beacon or listener. Fig. 8 shows the cricket device. [7]

Fig. 8. A cricket hardware unit; this unit can function as either a beacon or a listener under software control, and can also be used in a more symmetric way as both listener and beacon. [6]

Beacon

Beacon is an active device. Its function is to send radio frequency signal (RF) and ultrasound signal (US) at the same time. The RF signal range is around 20 meter, while the US signal range is around 10 meters.

Listener

Listener is passive device which receives the RF and US signal from the beacon, and use these signals to calculate the distance between itself and the source beacon which sent these signals.

Distance calculation is done by the listener. When the beacon send RF and US signal at the same time, due to the different in the speed travel the RF will arrive to the listener before the US. Once the listener receives the RF signal it starts a counter to count the time difference between the arrival of the RF and the US. When the listener receives the US, it stops the counter and uses that time to do the distance calculation. Figure 9 shows the RF and US signals timing for both the beacon and listener. [7]

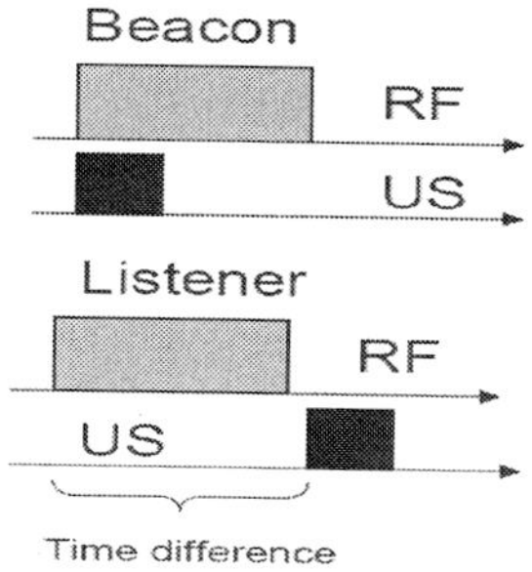

Fig. 9. The RF and US signals timing for both the beacon and listener.

Examples of location-aware applications that can be developed using cricket including resource discovery, human/robot navigation, physical/virtual computer games, location-aware sensing, hospital/medical applications (e.g., equipment and patient tracking/monitoring), stream migration and pose-aware applications like the software flashlight/marker, etc.

Its radio runs at a frequency of 433 MHz, with the default transmit power level and antennas providing a range of about 30 meters indoors when there is no obstacle. The maximum ultrasound range is 10.5 meters when the listener and the beacon are facing each other and there are no obstacles between them. Our measurements show that the distance accuracy of the cricket hardware is on the order of 1 centimeter at a distance of up to 3.5 meters, and 2 centimeters in the rest of the 10.5-meter range. [6]

3.3.3 Distance measurement by camera:

As an alternative, the camera system used for orienting roomba towards the robot could be used (see Fig. 10).

Since the vertical viewing angle α of the camera is constant, the distance between the camera and the robot L can be calculated based on the position of the light spot in the picture frame (see Eq. 1).

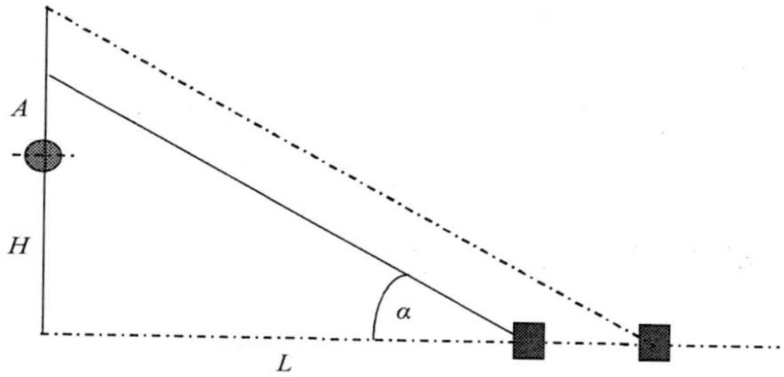

Fig. 10. Distance Measurement with a Camera

$$L = H \cdot (n+1) \cdot tg\alpha \qquad (1)$$

Where H is the height of the light bulb relative to the camera and n is the ratio between H and A in the picture frame (see Eq. 2).

$$n = \frac{A}{H} \qquad (2)$$

However, the accuracy of this method decreases at larger distances and is also affected by the accuracy of finding the centre of the light spot in the picture frame.

4. Communication between roomba and the robot

Roomba communication with the base robot can be accomplished by attaching a gumstick with the roomba. Gumstick will be used to store all of the programming on linux based platform so that roomba can fulfil the cleaning scenario according to the base robot. There are 3 options described on the gumstix [8] website:

- Pre-made computer (Full assembled computer pack with casing)
- Cost saving (Pre-selected bundle "Gumstix Pack" consist of separate parts to assemble)
- Individual components (Purchase components separately)

The Individual components type is the best option due to the fact that we want to choose the type of expansion board based on what we want the embedded computer to perform.

Motherboard selection is one of the most vital factor where the central processing unit (CPU) is based as well as on board RAM memory and flash memory. As verdex (XL6P) produced by gumstix has three pin connectors along with 128mb SDRam memory which will be more appropriate option for our project that we can use based on the expansion boards [9].

There are many expansion boards available on the Gumstix website. These expansion boards are design to fit onto the motherboard. Based on what was chosen so far before the 60 pins expansion board analysis. The capabilities that have been satisfied are:

- CPU
- On board Ram memory
- On board flash memory
- Expansion memory
- Wi-Fi/Ethernet

The serial port on roomba is 5V TTL which only the robostix matched that 5V TTL. Therefore the robostix is preferred over the breakout-vx and console-vx. The remaining problem is to avail of a USB port. The verdex motherboard actually support USB [10] host signal. Even though there is no direct physical output port available.

As for the final criteria of communicating with a PC to transfer program codes etc, we can simply use the tweener that sits between robostix and the verdex motherboard as it has both the female and male connector. Hence every aspect of the required criteria has been fulfilled.

5. Spot cleaning

This scenario is similar to that of regular cleaning. However, instead of cleaning a whole room in a planned manner, the roomba is instructed to clean a localised spillage. In order to identify the location of the spillage, a nurse or auxiliary marks the area with a special cone with a light-emitting bulb on its top. The cone is placed at the far side of the spillage relative to the location from which roomba is to arrive (see Fig. 11.). Once the robot guides roomba into the room, roomba orients itself towards the light emitter the same way as towards the robot itself during following it.

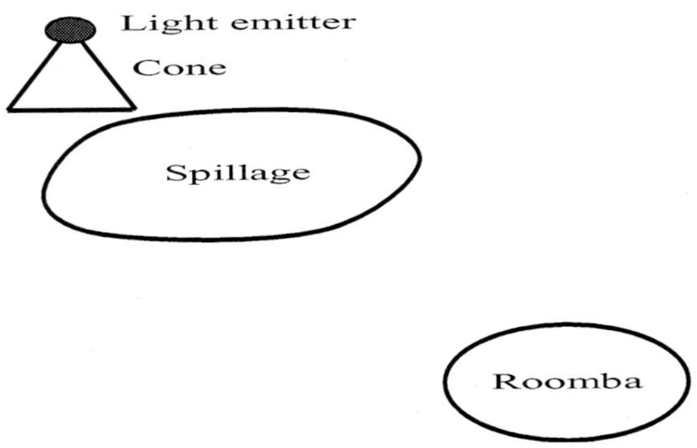

Fig. 11. Spot cleaning with roomba

For avoiding ambiguity, the light source on the cone is with a different colour from the one at the rear of the robot. However, the sensor and camera system for this task is the same as for following the robot. In order to avoid placing distance measurement sensors on each cone, measuring the distance between roomba and the cone (in order to position roomba near the spillage) can be done using the camera system (as described above).

6. Conclusion

A hospital cleaning module design has been described and the development process is ongoing. The hospital cleaning procedure will be more organized by fulfilling this project and that can meet the basic requirement (cleaning staff) of the hospital ward cleaning procedure.

Acknowledgment

Dublin City University is a partner of the EU-funded FP6 Innovative Production Machines and Systems (I*PROMS) network of excellence (www.iproms.org). This is a EU-funded FP6 project called Intelligent Robot Swarm for Attendance, Recognition, Cleaning and Delivery (iWARD) (www.iward.eu).

References

[1] J Clarke. IrishHealth.com. Hospital hires robot cleaner, http://www.irishhealth.com , accessed 15th April 2008.

[2] J.M.Evans, An autonomous mobile robot courier for hospitals, Vol 3 (12-161994). pp.1695 – 1700.1994

[3] iRobot Corporation [online], http://www.irobot.com/, Accessed 10th December 2007.

[4] Pololu, USA, [online], http://www.pololu.com, Accessed 8th December.

[5] [Wavetrend] Wavetrend, UK, [online], http://www.wavetrend.co.uk/ (Accessed 4th December 2007)

[6] MIT, USA, [online], http://cricket.csail.mit.edu/v2man.pdf, Accessed 4th December 2007.

[7] Z. Al Sabbagh, PhD," Real-time Tracking for Assisted Living", Distinction Project Thesis, Ohio State University, Ohio, USA.

[8] Gumstix, http://www.gumstix.com, accessed 9th December 2007.

[9] Gumstix Motherboard Platform, http://www.gumstix.com/platforms.html, accessed 9th December 2007.

[10]Gumstix Support Wiki [USB], http://docwiki.gumstix.org/USB_24_Pin, accessed 10th December 2007.

Innovative Production Machines and Systems
D.T. Pham, E.E. Eldukhri and A.J. Soroka (Eds.)

Developing a patient guidance module for hospital robots

M. K. Hasan and T. Szecsi

*School of Mechanical and Manufacturing Engineering,
Dublin City University, Dublin 9, Ireland*

Abstract

In this paper the development of a patient guidance module for an Intelligent Robot Swarm for Attendance, Recognition, Cleaning and Delivery (IWARD) hospital robot is described. The proposed prototype module features an Active RFID reader and active Tags. While a patient is guided to the X-ray room, or a visitor to a patient's ward, the speed of the robot can be adjusted automatically according to the pace of the person using the system. The reader is to be installed on a Pioneer 3DX base robot and active tags can be worn by the follower. Furthermore, a device driver will be developed for the reader in the Player work environment, which is a language and platform independent network server for robot control [1].

Keywords: Guide Robot, Active RFID, Hospital Robot

1. Introduction

The Guidance module in the project, called IWARD will guide patients and visitors in the hospital. They can follow the moving robot to their destination. For example, a patient can be guided to the X-ray room, or a visitor to a patient's ward. The speed of the robot can be adjusted automatically according to the pace of the person [2].

The remainder of the paper is organised as follows. Section 2 clarifies the reasons for speed adjustment in the guidance task. Section 3 describes the detailed scenario of the patient guidance task in a hospital environment. Section 4 consists of some common distance measurement theory. Section 5 outlines the background related to the theme of the paper. Section 6 includes a brief note on the Wavetrend Active RFID reader and Tags and describes how the guidance task will be performed using them. Section 7 describes the software architecture. Finally, Section 8 gives the conclusions.

2. Reasons for speed adjustment

In a hospital environment patients with very different physical abilities may require guiding assistance from the robot. Different patients can follow a moving robot with different speed. Setting one robot speed for all patients, therefore, is not the optimum solution. Also, a patient may be temporarily hindered from following the robot due to obstacles and other people and robots in the hospital. In order to ensure that a patient is able to follow the robot, speed control of the robot is necessary.

The speed control is based on continuously measuring the distance between the moving robot and patient. If the patient is lagging far behind the robot the latter may slow down or stop waiting for the patient. This simulates a virtual electronic rubber-band between the robot and the patient. The robot should be able to measure distance regardless of the orientation of the patient relative to the robot.

3. Scenario: robot guides patient to specified location

An authorized staff member (nurse, doctor, receptionist) requests a patient guidance robot from the Patient program on the nurses' terminal, by specifying the start and end location. The robot management system selects a suitable robot and sends it to the start point. An emitter tag is placed on the person to be guided. The robot identifies the signal from the emitter tag .The patient's ID is linked to this tag and the robot only guides a patient with this tag. The robot moves to the final destination; the patient follows the robot with a 'virtual elastic band'. The Patient Guidance Module (PGM) measures the distance between the robot and the patient. The speed of the robot is adjusted according to the following distance.

- 0-3 meters: the robot keeps the human walking speed
- 3-5 meters: the robot slows down and waits for the tag to come within the 3 meter range
- 5+ meters: the robot stops and waits for the tag to come in its range.

For the patient guidance module sensors are required for the following tasks: identification of the patient and speed control of the robot.

4. Common distance measurement techniques

PGM has to continuously measures the distance between the robot and the patient so that the speed of the robot can be adjusted according to the required distance. There are three common techniques for measuring the distance to an object from a given reference point.

4.1 Time-of-flight

This method determines the time t taken for a given signal to negotiate a linear course between two points (stated as a designated reference marker and the target object). Denoting the signal speed to be v, the distance d is calculated as $d = v \times t$. To propose a practical exemplar, GPS systems use the flight time of radio frequency (RF) signals to approximate the distance between GPS satellites and the GPS receiver [3]. Time of arrival technique require very fast processors with extremely accurate clocks. GPS works in similar fashion to this but only because the signals are travelling quite a large distance from satellite to receiver. But for PGM scenario TOA is really impractical at these short ranges. It takes 300 nanoseconds for a signal to traverse 100 meters (approx.) and there are no processors or clocks in any embedded controllers that can run fast enough to time the signal including all associated instruction clock cycles for taking a measurement as well as intervening operating system clock cycles nor any methodology that calibrates the origination time of the transmission per tag without having substantial impact on the anti-collision properties of multi-RFID. The reason GPS works is the signal takes up to 30 microseconds to arrive from satellites, well within the clocking capabilities of high powered embedded controllers which would account for the high cost of GPS units .[4]

4.2 Time-Difference-of-Arrival (TDOA)

By using two signals of disparate speed negotiating identical paths between two designated points, TDOA [5] based methods allow measurement to be calculated determining the distance between two points. Consider two signals A and B with speeds vA and vB sent simultaneously by a transmitter. If $vA < vB$, then signal B lags behind signal A as they propagate. Let t denote this temporal delay at a receiver located at a distance d from the transmitter, Then,

$$d = \frac{t}{1/vA - 1/vB} \quad (1)$$

One example is the Cricket device developed by MIT which is a transceiver and can be configured to act as a beacon or listener [6]. Cricket uses time-difference-of-arrival between RF and ultrasound to obtain distance estimate as shown in Fig. 1.

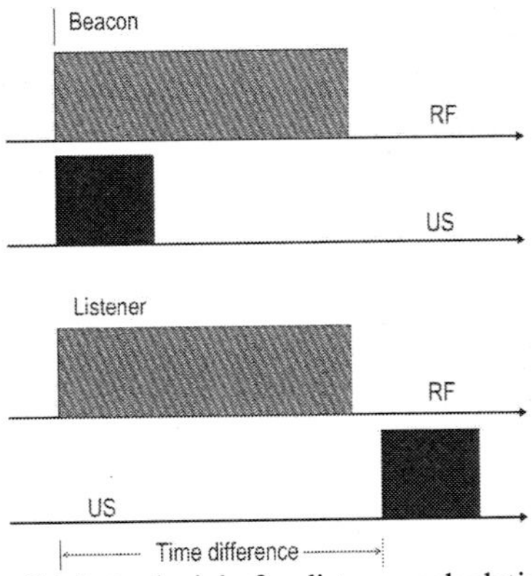

Fig. 1. Cricket principle for distance calculation [10]

However, in such systems the optical line of sight is needed for transmitting the ultrasonic pulse [9,10]. This might not be possible for the IWARD guidance scenario as the patient or the robot may both turn around. Also, it may be the case that the cricket's ultrasonic pulse interferes with other ultrasonic sensors of the Pioneer 3DX robot which are used for the robot navigation in order to avoid obstacles.

4.3 Image processing.

Firstly, a Charge Coupled Device (CCD) sensor

array is placed behind a motor-driven lens whose operation is regulated through the use of a micro processing unit. Once image focus has been established for a particular lens, the calculated expanse between lens 1 and the CCD shall represent the distance d to the target object; the law of geometric optics being used to determine the distance d from l and the F quantity of the lens. This method is extensively used in passive auto-focus cameras [8].

5. Related work

Within the last few years various RF-based techniques have been developed to solve indoor localization problems. Most methods use the known locations of reference stations, where the transponder position is reported by nearby stations [9].

Kartoun [10] developed a system where a robot follows a human Using Parco UWB RFID Tags where both active tags are worn by a man and the robot. He has achieved tracking based on dynamic calculations between the x and y coordinates of a tag located on the human and x and y coordinates of a tag placed on the robot. But in his experiments, the speed of the robot was kept constant. To locate a specific tag requires triangulation from multiple readers that are placed in very specific patterns. A large number of readers are required to provide locations in a large area [11].

This approach had the problem of cost and space with the number of stations and the interval between the stations affected the accuracy [9].

For the PGM only the distance between the robot and the follower is sufficient rather than getting the dynamic location of them. Moreover, all the sensors and devices required for the IWARD project should only be on to robot. It is not possible to install several readers in the various location within the hospital due to project limitations.

Kim and Chong [9] have developed a prototype system that guide an RFID based mobile robot to a stationary target. Their system consists of a 315MHz RFID reader that reads the tag data simultaneously and picks up the direction of the tag using the received signal strength pattern. Based on the angle of signal arrival they have developed a guidance strategy that enables a robot to find its way to the tag position.

In the IWARD project, both the robot and the patient wearing tag move around. Moreover, considering that the RSSI value for distance measurement is not reliable, as it can vary and be distorted significantly depending on several factors [11].

To deal with the problem of getting dynamic distance between the robot and the follower we should consider the system's ease of use and range of applicability. Hence the proposed system employs only active RFID tags and a reader without additional external sensors and/or reference stations.

6. Distance Calculation technique for PGM

6.1 Active RFID kit

Active tags and Readers (Receivers) can give an RSSI (Received Signal Strength Indication) value that can be indicative of the spacing between the patient and the robot. However, it is necessary to investigate how to adjust the power levels and receive antenna gain to control the signal strength within the span of range (2 - 5 meters). By the patient just turning around it may affect the RSSI value since his position and orientation of the tag he is carrying will affect relative signal strength. The application software on the robot's platform will have to meter the gradient of change in signal strength and adjust its speed algorithm accordingly.

Wavetrend 433 MHz, L-RX201 Readers were selected for the PGM which detect and decode RF transmitted signals from the Wavetrend family of tags. The reader incorporates a BNC connector to allow for attachment of an external antenna to extend the reader's range. The L-AN200 433 MHz Wavetrend stub antenna shown in Fig.2

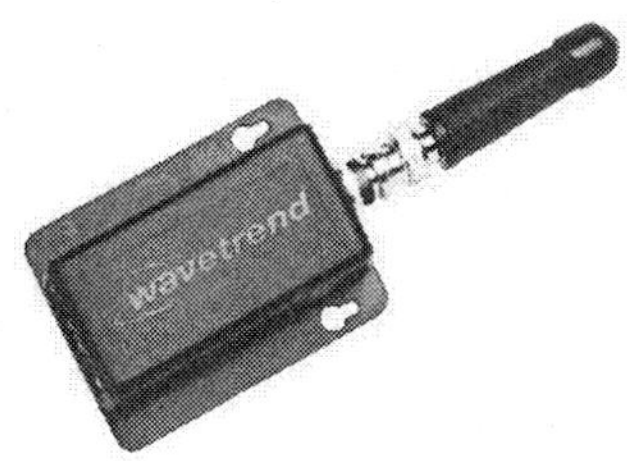

Fig. 2. L-RX201 Active reader with attached AN200 Stub Antenna [12]

Typical read ranges of Wavetrend Tags with various antennas are shown in Table 1 [12]. This is suitable for the IWARD project as combining these two will give an approximately 30m range and would be sufficient for the IWARD PGM scenario.

The RSSI will vary significantly depending on the spatial configuration of line of sight impediments from tag to reader. The RSSI will vary significantly depending on the spatial configuration of line of sight impediments from tag to reader. It also varies based on the CW configuration or modulation technique utilized by the manufacturer. Additionally, objects to which the tag is mounted and the orientation of the tag to the reader depending on antenna configurations will dramatically impact the RSSI output [4].

Table 1
Typical read ranges of Wavetrend Tags with various antennas [a]

Antenna	Min. Range (meter)	Max. Range (meter)
None	0.2	7
L-AN200	1	30
L-AN100	3	35
L-AN300	3	100

[a] Distances may vary depending on the ambient RF environment. The following tests were conducted in free air.

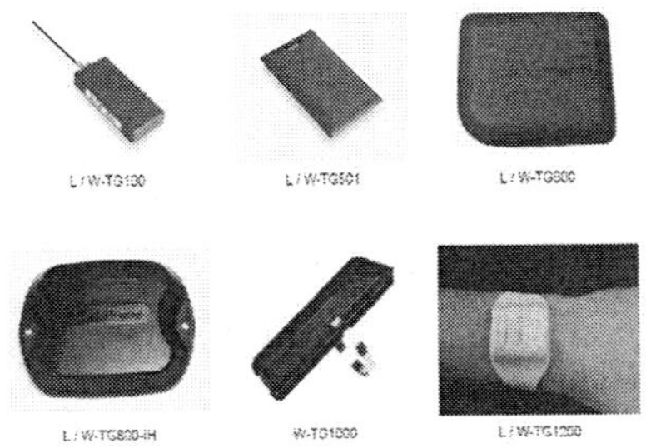

Fig. 3. Wavetrend active tags family [12]

6.1.1 Calculate distance using RSSI value

Tests were carried out in an effort to relate the RSSI to distance. Extensive tests, under different circumstances, were carried out in which the following section will outline.

6.1.1.1 Test 1

The constraint to this test was that the antenna of the reader was facing towards the tags, while the distance of the tags varied. The following graph shows the change in RSSI values with respect to a change in the distance, from reader to tag, from 1 meter to 5 meters.

The ideal plotted line for each tag would be a linear decreasing line (from the top left to the bottom right), but as can be seen from graph 1, this is not the case. It can be seen from the graph that the tags do tend to take this curve, having similar values for 1m and 2m then taking a slump to 3m and levelling out then. While the majority of the tags RSSI values decrease over the distance, tag 475689 was only detected at 1 and 2 meters, the reader could not detect them at any further distance. The highest value for this experiment was tag 611631 with an RSSI value of 107.

6.1.1.2Test 2

The constraint to this test was that the antenna of the reader was facing away from the tags, while the distance of the tags varied.

This graphs values took a more desired route than test 1. Tags 113630, 611631 and 556086 took the desired route for which distance could be related to the RSSI values. Tag 508433 almost had no change from 1m to 5m except for one high reading. The reason for this was unknown, as there were no known interferences that could have done this. Like before tag 508408 decrease to 3m and then increase as the distance was extended. Tag 475689 again appeared to be unreliable as this time at 4m it could not be detected, while at 5m it could. The highest value for this experiment was tag 113630 with an RSSI value of 113.

6.1.1.3Test 3

Like test 1, test 3 was the same, except this time an extra constraint was added, this being that a male of average height and build was blocking the signal.As can be seen from the graph not one of the tags followed the desired route. Two of the tags 508433 and 508408 had higher RSSI values at 5m then they did at 1m. Like the two tests before, tag 475689 again was not detected at 4m but at all other values between 1m and 5m.

6.1.1.4Test 4

Like test 2, test 4 was the same, except this time an extra constraint was added, this being that a male of average height and build stood in the path of the signal.

It can be seen from the graph that the similar tags have similar values at 1 m, but then take on their own characteristic curves as the distance increases. Again like the three tests before, tag 475689 was not detected at 4m but at all other values between 1m and 5m.

6.1.2 Alternative approcah

Active RFID systems seem applicable only to detect whether the tags are in its range and are not reliable to calculate the exact distance using RSSI value. There are several types of active tags shown in Fig. 3. Each type has a certain range. The idea is based on having two different range tags added together and worn by the follower. For an example, Tag 1 which has approximately a range of 2m and

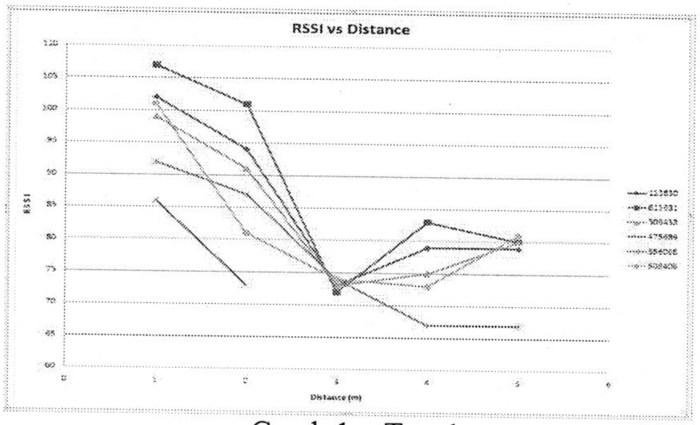

Graph 1 – Test 1

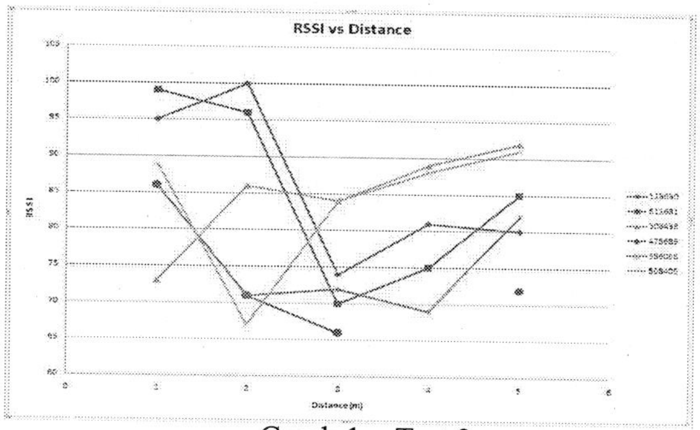

Graph 1 – Test 3

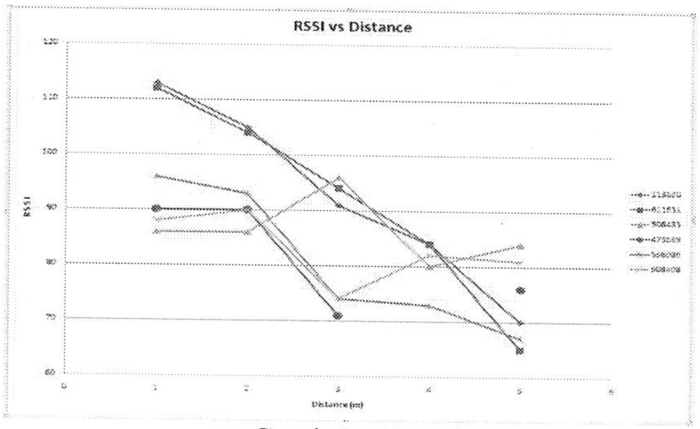

Graph 2 – Test 2

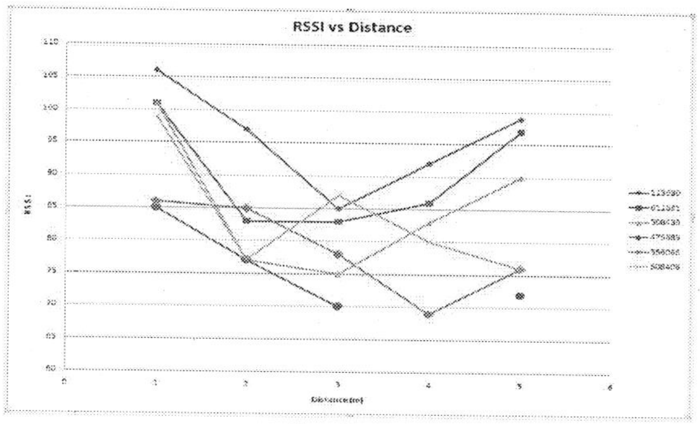

Graph 1 – Test 4

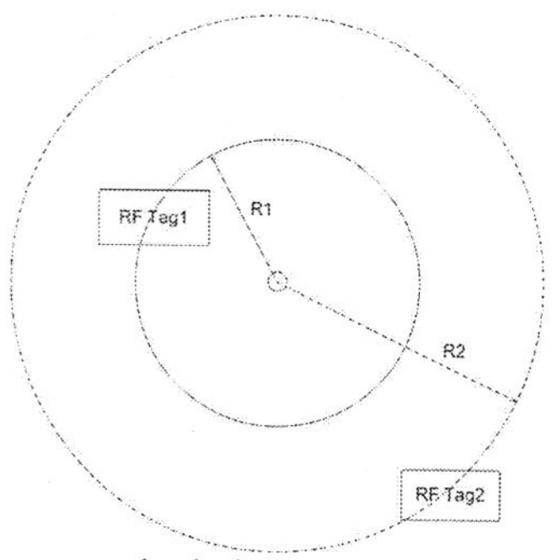

Fig. 4. Distance calculation technique using two tags

Tag 2 which has approximately a range of 5m are shown in Fig. 4. If the system detects both Tagl & Tag 2 then the PGM will consider that the patient is within 2 m range. If it detects Tag2 but not Tag 1 then it is in between 2-5m. Finally if neither tags is detected then the patient is out of range and eventually the PGM request the robot to stop and wait for the patient to come within its range. The proposed system is shown in Fig. 5. Wavetrend tags transmit a Radio Frequency (RF) signal at a pre-set time-interval. For the prototype purpose experiments will be carried out with TG100 and TG801 series Tags.

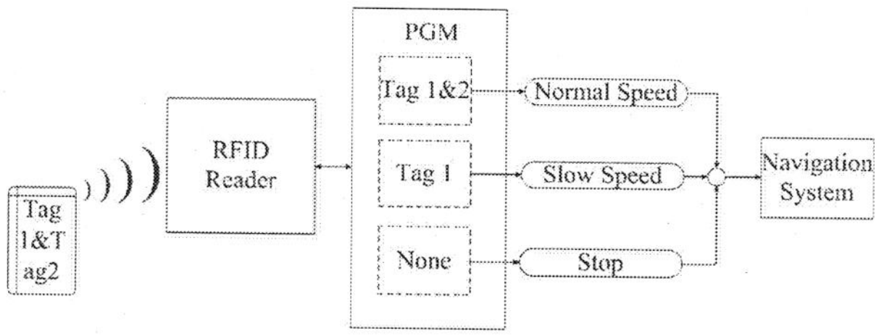

Fig. 5. Proposed Guidance flow chart

7. Software architecture

Eventually, the entire PGM assembly would be installed upon a Pioneer 3DX base robot with onboard PC. The IWARD software architecture is based on two main frameworks, Player and Orca2. It was proposed to select the hardware considering its compatibility with these frameworks. Player is a socket-based device server that allows control of a wide variety of robotic sensors and Actuators [1]. Each IWARD robot will contain a Player server, which is responsible for "serving" through TCP/IP the data acquired by the Player drivers (connected to the robot's actuators and sensors) to all the clients.

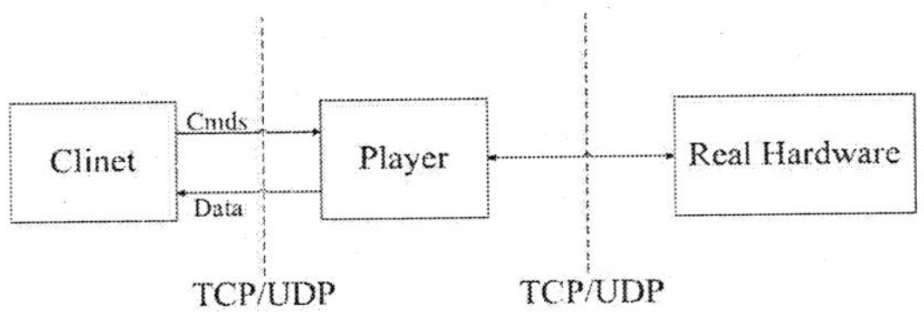

Fig . 6. Integration between Player and real hardware

The client software will simply access the hardware of the robot through Player as shown in Fig. 6.

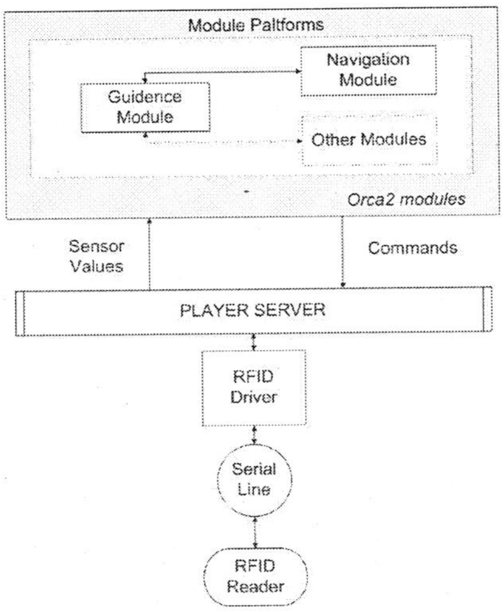

Fig. 7. Software architecture for guidance module

Currently there is no player compatible driver for Wavetrend active readers and all software is windows based.

13.56 MHz inside Contactless M300RFID and SkyeTek M1/M1-mini RFID reader's player drivers are available. Only the ISO 15693 tags are supported by them, but their maximum read distance is only 1-1.5 meters [1]. Hence A Player driver would be developed for the Wavetrend Active reader.

As the client software accesses the hardware of the robot through Player and Player's external interface is simply a TCP socket, client programs can be written in any programming language that provides socket support. A client Programme will also be developed for the PGM using C/C++.

In order to complete a guidance task the guidance system needs to interact with different systems such as Navigation, Human Robot Interface (developed by IWARD partners) etc. As a middleware, Orca2 [13] will be used to interact between each other, as shown in Fig. 7. Orca2 is an open-source framework for developing component-based robotic systems. Different IWARD system developers can compose a system from any set of components.

8. Conclusions

For the guidance operation, Wavetrend Active RFID reader and tags are selected for the prototype experiment. The player driver and client programs are under developing. In an iterative implementation process the sensor selection and software design will be evaluated and improved continuously.

Acknowledgements

Dublin City University is a partner of the EU-funded FP6 project called Intelligent Robot Swarm for Attendance, Recognition, Cleaning and Delivery (IWARD) www.iward.eu & Innovative production Machines and Systems (I*PROMS) Network of Excellence www.iproms.org

References

[1] Player Project. Sourceforge.net. http: / / playerstage . sourceforge. net (Accessed 9th August 2007).

[2] Szecsi T. Robonurse leads the way, DCU Times, Summer: 0507,pp 49-50.

[3] I. Getting. The Global Positioning System. IEEE Spectrum, 30(12):36–47, December 1993.

[4] RFIDTalk. http://www.rfidtalk.com/showthread.php?

[5] Priyantha N. The Cricket Indoor Location System. Ph.D. Thesis.MIT. 2005.

[6] Cricket v2 User Manual.MIT.2004

[7] Sabbagh Z .Real-time Tracking for Assisted Living. Distinction Project, Ohio State University. USA.

[8] E. Hecht. Optics. Addison-Wesley, Reading, MA, fourth edition, 2002.

[9] Kim M and Chong N .RFID-based mobile robot guidance to a stationary target. Mechatronics . Vol. 17(4-5). pp. 217-229. ,2007.

[10] Kartoun U. Human-Robot Following Using Parco UWB RFID Tags. Ben Gurion University of the Negev,Israel.2005

[11] Smith R. RFID: A Brief Technology Analysis CTOnet.org. 2007

[12] Active RFID Manual. Wavetrend. USA

[13] Orca-robotics project, Sourceforge.net. http://orca-robotics.sourceforge.net/ (Accessed 9th August 2007).

Innovative Production Machines and Systems
D.T. Pham, E.E. Eldukhri and A.J. Soroka (Eds.)

Ultrasonic distance scanning techniques for mobile robots

D.T. Pham, Z. Ji and A. Soroka

Manufacturing Engineering Centre, Cardiff University, Cardiff, CF24 3AA, UK

Abstract

This paper discusses the use of ultrasound in mobile robot navigation. Two methods employing an array of transducers are described, namely the Phased-Array (PA) method and the Maximum-Likelihood-Estimation (MLE) method. These two methods are compared with the conventional Time-of-Flight (ToF) method. The paper shows that the PA method has advantages such as giving focused beams and allowing efficient beam steering. Limitations of the method are discussed, including the need to assume a far field and the unavailability of suitably-sized transmitters. The MLE method has the advantage of only involving one transmitter. Experimental results show that the locations of the reflectors employed with this method can be represented in a probability map and reliability can be improved through the use of multiple receivers.

Keywords: ultrasound, phased array, robot navigation

1. Introduction

Ultrasound has been widely applied to distance measurement for mobile robots and in many other applications. Its advantages, including low cost, reasonably good accuracy and robustness, over other techniques are significant. Despite these advantages and its popularity, its importance has declined, as alternative techniques, such as laser rangers and vision techniques, have started playing a major role in mobile robot navigation tasks.

In recent years, laser range-finders have been considered as standard navigation sensors that are equipped on many mobile robots for accurate sensing. However, the technology is limited to non-transparent objects, such as walls, because materials such as glass will allow light to pass through. The performance will also be influenced in environments where the laser range-finder is exposed to strong direct light. Vision-based techniques have been actively researched in the past few decades and applied to mobile robots, either entirely relying on mono-camera or stereo-cameras, or working together with other methods, such as sonar or infrared. However, due to their immaturity and the complexity of the operating environments, there is still a long way before they are usable in the real world.

From a practical point of view, ultrasound is not replaceable by any other of the available technologies at present. There are a number of reasons why it has not been utilised as extensively as other techniques in the past. One main reason is that to implement an advanced signal processing algorithm, there is an associated high cost and complexity of hardware which has become a major barrier. This kind of hardware will normally require a high sampling rate and multiple channel data acquisition system with a high speed computation unit. Nowadays, the decreasing cost and fast growth of such devices has made it feasible to improve the performance of ultrasound sensing with more advanced techniques.

This paper will first review the Time-of-Flight (ToF) method and its improvement, and then will discuss two approaches using ultrasound transducer arrays. The first one is an active phased array, which is similar to the techniques that have been widely used in medical human body scanning. The other one is a novel passive technique based on the statistical Maximum-Likelihood-Estimation (MLE) method.

2. Time of Flight (ToF) and its improvement

The most commonly used method of ultrasound range-finding for robot navigation is the ToF method, which simply emits a short period of pulses and receives the first reflected signal by a threshold detection criterion. As the time of emission is known and the time of arrival can be measured by taking the first threshold crossing, the distance can be then calculated by the following equation:

$$l = c(t_r - t_0)/2 \quad (1)$$

where l is the distance, c is the sound speed in air, t_r and t_0 represent the times of arrival and emission respectively.

Some researchers have attempted to improve the performance of the ToF method. By observing the actual sonar waveforms, a parametric model of such waveforms can be estimated. Barshan improved the ToF method by employing a parabolic fitting method, which used a pre-defined wave-front model to numerically fit the received signal [1]. Barshan has also compared four different ToF methods [2], and concluded that a more sophisticated method of applying cross-correlation to localise the wave arrival outperformed the other three methods.

Instead of only measuring the distance, many researchers have also tried to use the ToF method to identify the geometrical features of objects, such as walls, edges, and corners. These capabilities will be very useful for navigation algorithms that are based on geometric features. Binaural structures with one transmitter and two receivers have been studied for this purpose [1]. In [3], a wide beamwidth ultrasound transducer arrangement was used to measure the geometric position of plane reflectors. Similarly, tri-aural systems have also been of interest to distinguish corners or walls from ambiguous measurements [4].

Moreover, some work has been undertaken on the recognition of object geometric features by only using the conventional ToF method [5]. Instead, to identify the features, multiple measurements must be performed from various positions and angles. The shape of an object will be identified with a spatial consistency consideration.

Although the ToF method has received a great deal of interest from researchers due to its simplicity and reasonably good accuracy, there are obviously some deficiencies, including vulnerability to the effects of external noise, interference when multiple channels are used and low consistency of timing threshold crossing.

3. Active phased array

3.1. Theory

One main property of an ultrasound transmitter is its radiation directivity, which can be considered as a cone shape at its corresponding angle. For navigation applications, a narrower beam of radiation will be better in determining the location of a detected obstacle. In contrast, a wider beam will cover more area and will be able to detect more objects, but with the disadvantage of a lower resolution. For a low-cost ultrasonic transmitter, the bearing angle is usually wide, and consequently it will be difficult to provide a high resolution of object detection. This will make navigation difficult for a robot. An ultrasound phased array can be used to overcome this problem by narrowing the beam width to some extent. On the other hand, the phased array will allow the radiation beam to be steered electronically. This is an important feature that will help a robot scan its surrounding environment efficiently.

The principle of the ultrasound phased array is similar to the radar theory, and it has been used in some medical and industrial inspection applications, such as Non-Destructive Testing (NDT). In-air applications have not been investigated and researched as extensively as have those for liquid or solid mediums. In [6], a phased array has been introduced for the mobile robot navigation application.

Fig. 1 depicts the structure of a phased array. The ultrasound transmitters in the array are placed at an equal aperture d in a line, and will be driven in sequence with specific time-delays.

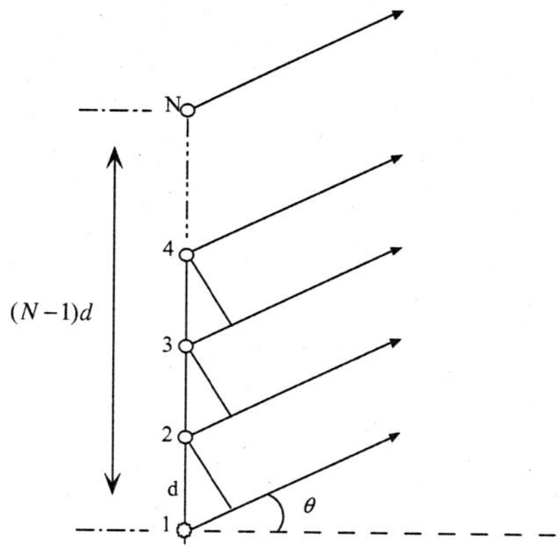

Fig. 1. Linear array of equally spaced elements

For an excitation to all transmitters with an initial phase and unit amplitude, the transmitted signals in the far-field area can considered as a planar wave, and can be formulated as [6][7]:

$$E(\theta)=\sum_{i=1}^{N} e^{j(i-1)(kd\sin\theta)} \tag{2}$$

where θ is the angle of the observing point, N is the number of elements, and k is the wave number. The intensity pattern can be derived from the above equation and simplified into the form:

$$|E(\theta)|=\left|\frac{\sin[Nkd\sin\theta/2]}{\sin(kd\sin\theta/2)}\right| \tag{3}$$

In this equation, each element is assumed to be isotropic. When the direction radiation profile $f(\varphi)$ and the normalisation are taken into consideration, the intensity pattern can be expressed as:

$$|E_n(\theta)|=\frac{1}{N}\left|f(\varphi)\frac{\sin[Nkd\sin\theta/2]}{\sin(kd\sin\theta/2)}\right| \tag{4}$$

To steer the main beam electronically, the array elements can be driven at varying times, which can be determined by the phase difference. For a steering angle θ_0, the phase difference between two adjacent elements will be $kd\sin(\theta_0)$. Then, the normalised radiation pattern can be expressed as:

$$E(\theta,\theta_0)=\frac{1}{N}\left|f(\varphi)\frac{\sin[(Nkd/2)(\sin\theta-\sin\theta_0)]}{\sin[(kd/2)(\sin\theta-\sin\theta_0)]}\right| \tag{5}$$

In order to obtain the distance of an object, the conventional ToF method will not be suitable anymore. Instead, it is suggested that one receiver be mounted at the centre of the array and the peak value of the received signal used to estimate the distance [6]. The relationship between the peak time t_p and the time of threshold crossing t_{TOF} is derived, as:

$$t_p = t_{TOF} + 2/a_1 \tag{6}$$

where a_1 is a parameter associated with the assumed model of the received signals. Details of these derivations can be found from [6], where a simple experiment has also been introduced.

3.2. Discussion

To verify the feasibility of using the ultrasound phased array in air, the quality of the steered radiation pattern is essential. It can be seen, from Eq. 5, that the pattern is related to the selection of the following two factors: the number of elements *N* and the aperture between two adjacent elements *d*. In addition, it should be noted that the phased array can be only held with the far-field assumption, which also relates to the value of *Nd*.

Eliminating the effect of the non-isotropic properties of the transmitters, the radiation pattern can be calculated. Fig. 2 shows an example of such an array of 8 elements and 16 elements respectively. The pattern is symmetric because it is only dependent on the value of $\sin\theta$. It can be seen that the main beam is concentrating at the angle 0. More elements will narrow the beam width.

The aperture d between the two elements will also affect the result. Fig. 3 shows two radiation patterns with a steering angle of 20°, where d is set to one wavelength and a half wavelength respectively. It can be seen that the first one has a slightly narrower beam, but with a very strong side lobe, which will cause uncertainties in measurement. By contrast, the second one has a distinct main lobe at the desired angle.

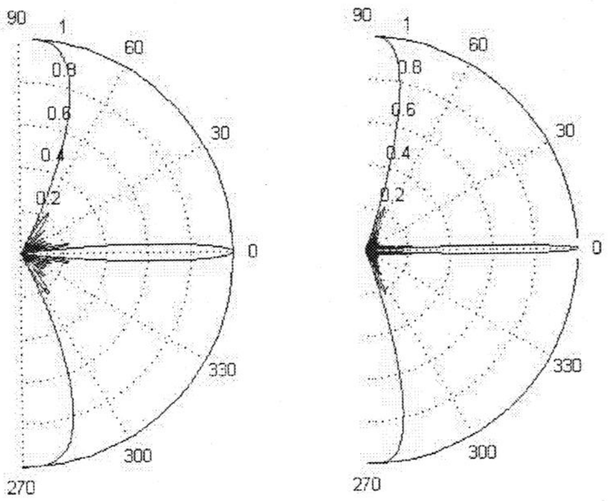

Fig. 2. Normalised intensity patterns of 8 and 16 element phased arrays

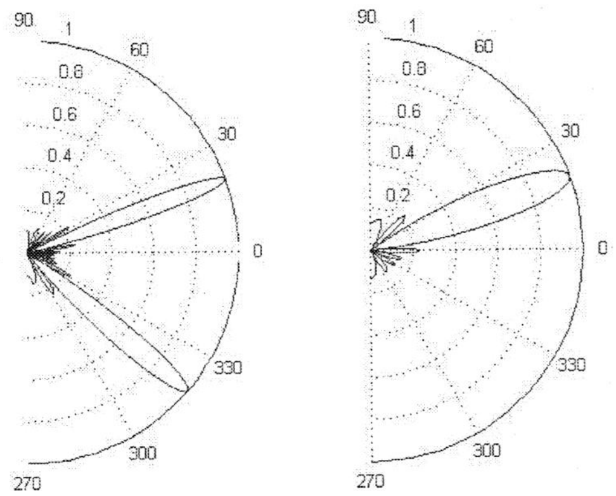

Fig. 3. Radiation patterns of two 8 element phased arrays, steering at the angle of 20°.

However, the ultrasound transducers in this range are not produced in as small a size as required in the above theory. Therefore, the most miniature ones that are commonly available in the market will be evaluated. The diameter of the transmitter chosen for evaluation is *9mm*, which is similar to the wavelength of its resonant frequency at *40KHz*. To optimise the configuration, all eight elements are placed with no gap in a line, so that *d* will be in effect *9mm*.

In this case, strong side lobes will be unavoidable. However, taking into account the directivity properties of the transmitters, careful engineering selection of the scanning area will help minimise the uncertainty. This will need to be further verified with experiments.

In addition, as this is based on the far-field assumption, it is necessary to evaluate the configuration and validate the effective area that satisfies the far-field requirement. Therefore, the size of the whole array aperture *Nd* needs to be much smaller than the effective distance. Rigorous derivations can been found in several places in the literature.

In this configuration, the array size *Nd* is $8 \times 9 = 72mm$. For example, if the restriction is set to 8 times more than *Nd*, then the minimum reliable distance would be $576mm$. In this case, it is slightly bigger than that used in the conventional ToF method, which normally allows closer measurement of at least $150mm$. However, it is still acceptable in most indoor or outdoor environments.

4. A passive array technique based on MLE

4.1. Theory

As mentioned before, the cross-correlation technique has been proved to be an accurate means of determining the time of arrival. The peak value of the correlation coefficients indicates the time lag of the main reflector, and other reflectors can be represented with some local corresponding maximums in the correlation coefficients.

However, performance of the ultrasound method can be prone to specular reflections, weak diffused reflections, and external noise. Moreover, multiple objects will result in multiple echoes in the received signals. All these will cause uncertainties in determining the locations of reflectors.

To reduce the ambiguities, the Maximum-Likelihood-Estimation (MLE) method can be applied to enhance the contrast between the real reflections and ambiguities. One very popular method of localising sound sources is using an array of sensors, such as the beam forming method or other similar schemes. However, due to the unsuitability of the high-end hardware requirement of acquiring signals from a large number of sensors at a high sampling rate, here the number of sensors is reduced to four.

In this configuration, four receivers with one transmitter are used, as shown in Fig. 4. When a series of pulses is emitted, the four sensors will receive signals reflected by objects from different directions. In the following experiments, a high speed data acquisition system is used to capture the signals at a sampling rate of *500 kHz*.

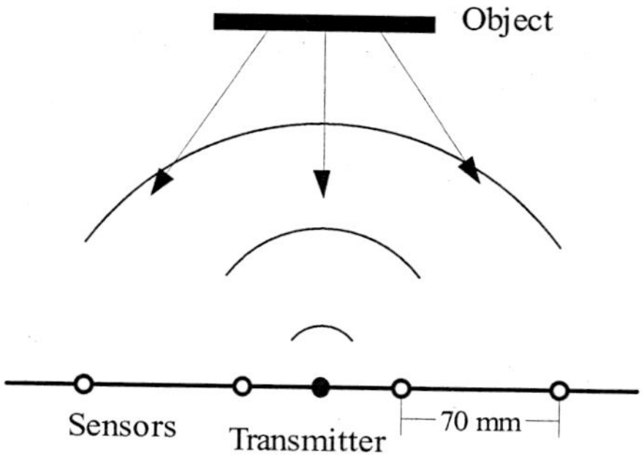

Fig. 4. Array of four receivers and one transmitter.

The likelihood distribution can be described with the Bayesian rule that the posterior probability of a source at location L is expressed as [8]:

$$P(L,s \mid x_1,x_2,x_3,x_4)=\frac{P(x_1,x_2,x_3,x_4 \mid L,s)P(L,s)}{P(x_1,x_2,x_3,x_4)} \tag{7}$$

In the simplified condition, the signal model at the i th sensor is assumed as:

$$x_i(t)=a_i s(t-\tau_i)+n_i(t) \tag{8}$$

where a_i denotes the attenuation factor during the wave propagation and $s(t-\tau_i)$ is the source signal. $n_i(t)$ and τ_i represent the external noise and the time delay respectively. Hence, the probability can be expressed with the Gaussian distribution as:

$$\begin{aligned} P(x_1,x_2,x_3,x_4 \mid L,s) &= \prod_{i=1}^{4} P(x_i \mid L,s) \\ &= \prod_{i=1}^{4} e^{-\int_{0-W}^{0+W}\frac{[x_i(t+\tau_i)-a_i s(t)]^2}{2\sigma_i^2}dt} \end{aligned} \tag{9}$$

where

$$\begin{aligned} &(x_i(t+\tau_i)-a_i s(t))^2 \\ &= x_i^2(t+\tau_i)+a_i^2 s^2(t)-2a_i x_i(t+\tau_i)s(t) \end{aligned} \tag{10}$$

The second term $s(t)$ is a constant and independent from the estimated location L. Similarly, $x(t+\tau_i)$ is the received signal with a shift of τ_i, the integral of it over the whole acquired length will have little effect to the result. Therefore, these two terms can be ignored. The logarithmic likelihood function can thus be yielded as:

$$\ln P=\frac{1}{2\sigma^2}\sum_{i=1}^{4}\int_{0-W}^{0+W}2a_i x_i(t+\tau_i)s(t)dt+A \tag{11}$$

where A represents the value of the ignored terms, and σ_i is assumed the same for all four channels and is represented by σ here. To further simplify the equation, a_i is also assumed equal. Then, the likelihood will totally rely on the summation of the correlation coefficients between the source signal and the received signals from all channels $\sum_{i=1}^{4} R_{sx_i}(\tau_i)$.

4.2. Experiments and discussion

In order to evaluate the performance the proposed method, a good model of the source signal will be necessary. Here, a real received signal is used instead of directly employing an estimated model. Fig. 5 shows a typical source model and Fig. 6 shows an example of a received signal with multiple echoes, where the first object (a piece of paper of A4 size) was placed *1000mm* away in front of the array.

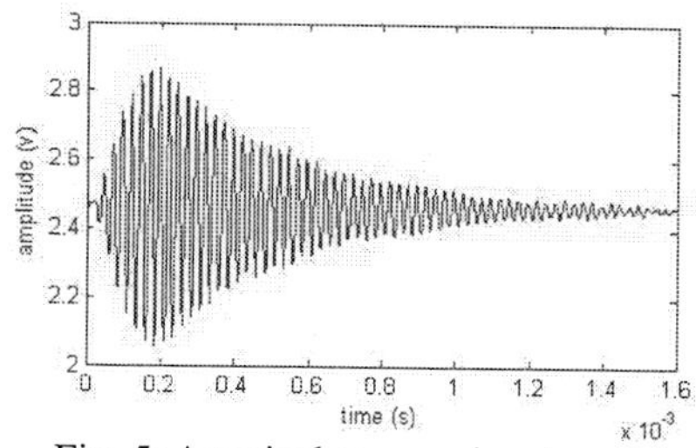

Fig. 5: A typical source signal model.

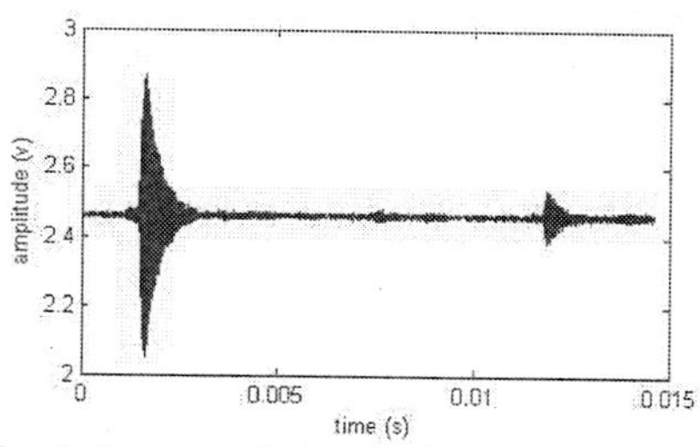

Fig. 6. A received signal with multiple echoes.

Fig. 7 illustrates the result from the MLE calculation. Higher amplitude indicates a higher probability of the target presence.

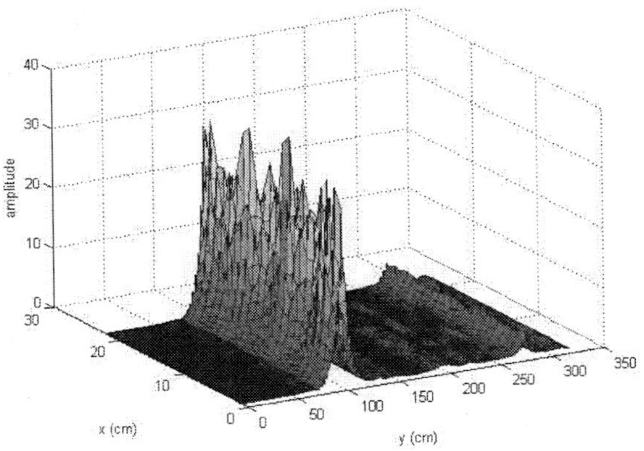

Fig. 7. Result of the MLE method.

This experiment has shown a higher probability at the distance of *1000mm*, although ambiguities have also been introduced at some other locations. They will need to be filtered out by using some additional processes, such as choosing proper values for a_i. A second lobe can be seen at the distance of about *2700mm*, corresponding to the second echo in the received signal (Fig. 6). Moreover, this experiment can actually be considered as a multiple objects problem, because the reflections to different sensors were induced at different positions on the planar object. A smaller object will produce a higher intensity of the MLE at the corresponding position. With regard to the resolution, the widths of the lobes in the MLE highly depend on the length of the transmitted pulses. Another issue associated with resolution is the sampling rate and the signal frequency. However, it is not practical to use a very fast data acquisition system on a mobile robot. Despite these problems, the robustness of detection has been improved considerably.

5. Conclusion

This paper has proposed and discussed two novel approaches using ultrasound for robot navigation. The feasibility of applying the phased array for in-air applications has been discussed. It has been shown that the number of transmitters and the transmitter diameter need to be carefully selected. The MLE method has been discussed with both theoretical and experimental analyses. The reliability and robustness can be enhanced considerably with this method. The potential of identifying multiple objects has also been discussed.

Acknowledgements

This work was financed by the European FP6 IST Project "Intelligent Robot Swarm for Attendance, Recognition, Cleaning and Delivery (IWARD)". The support of the European Commission is gratefully acknowledged. The MEC is the coordinator of the I*PROMS Network of Excellence, being funded by the EC under its FP6 programme.

References

[1] Barshan B and Kuc R. A bat-like sonar system for obstacle localization. IEEE Trans. Systems, Man and Cybernetics.vol. 22(4) 1992.
[2] Barshan B and Ayrulu B. Performance comparison of four time of flight estimation methods for sonar signals. Electronics letters, vol. 34(16) 1998.
[3] Yang M, Hill S L, and Gray J O. Localization of plane reflectors using a wide-beamwidth ultrasound transducer arrangement. IEEE Trans. Instrumentation and measurement. Vol. 46 (3) 1997.
[4] Peremans H and Van Campenhout J. Tri-aural perception on a mobile robot. IEEE international conference on robotics and automation. Vol. 1, 1993.
[5] Choi J, Ahn S, and Chung WK. Robust sonar feature detection for the SLAM of mobile robot. IEEE/RSJ international conference on intelligent robots and systems, 2005.
[6] Pham DT, Yang M and Wang Z. Airborne ultrasound linear phased array for localising solid objects. 7th International conference on signal processing (ICSP). Vol. 3, 2003.
[7] Mahafza B R. Radar systems analysis and design using Matlab. Chapman & Hall/CRC, USA. 2000.
[8] Birchfield S T and Gillmor D K. Fast Bayesian acoustic localization, IEEE International Conference on Acoustic, Speech, and Signal Processing, 2002.

Author Index